Modern Pharmaceutics

Third Edition, Revised and Expanded

DRUGS AND THE PHARMACEUTICAL SCIENCES

A Series of Textbooks and Monographs

edited by

James Swarbrick
AAI, Inc.
Wilmington, North Carolina

1. Pharmacokinetics, *Milo Gibaldi and Donald Perrier*
2. Good Manufacturing Practices for Pharmaceuticals: A Plan for Total Quality Control, *Sidney H. Willig, Murray M. Tuckerman, and William S. Hitchings IV*
3. Microencapsulation, *edited by J. R. Nixon*
4. Drug Metabolism: Chemical and Biochemical Aspects, *Bernard Testa and Peter Jenner*
5. New Drugs: Discovery and Development, *edited by Alan A. Rubin*
6. Sustained and Controlled Release Drug Delivery Systems, *edited by Joseph R. Robinson*
7. Modern Pharmaceutics, *edited by Gilbert S. Banker and Christopher T. Rhodes*
8. Prescription Drugs in Short Supply: Case Histories, *Michael A. Schwartz*
9. Activated Charcoal: Antidotal and Other Medical Uses, *David O. Cooney*
10. Concepts in Drug Metabolism (in two parts), *edited by Peter Jenner and Bernard Testa*
11. Pharmaceutical Analysis: Modern Methods (in two parts), *edited by James W. Munson*
12. Techniques of Solubilization of Drugs, *edited by Samuel H. Yalkowsky*
13. Orphan Drugs, *edited by Fred E. Karch*
14. Novel Drug Delivery Systems: Fundamentals, Developmental Concepts, Biomedical Assessments, *Yie W. Chien*
15. Pharmacokinetics: Second Edition, Revised and Expanded, *Milo Gibaldi and Donald Perrier*
16. Good Manufacturing Practices for Pharmaceuticals: A Plan for Total Quality Control, Second Edition, Revised and Expanded, *Sidney H. Willig, Murray M. Tuckerman, and William S. Hitchings IV*
17. Formulation of Veterinary Dosage Forms, *edited by Jack Blodinger*
18. Dermatological Formulations: Percutaneous Absorption, *Brian W. Barry*
19. The Clinical Research Process in the Pharmaceutical Industry, *edited by Gary M. Matoren*
20. Microencapsulation and Related Drug Processes, *Patrick B. Deasy*
21. Drugs and Nutrients: The Interactive Effects, *edited by Daphne A. Roe and T. Colin Campbell*
22. Biotechnology of Industrial Antibiotics, *Erick J. Vandamme*
23. Pharmaceutical Process Validation, *edited by Bernard T. Loftus and Robert A. Nash*
24. Anticancer and Interferon Agents: Synthesis and Properties, *edited by Raphael M. Ottenbrite and George B. Butler*
25. Pharmaceutical Statistics: Practical and Clinical Applications, *Sanford Bolton*

26. Drug Dynamics for Analytical, Clinical, and Biological Chemists, *Benjamin J. Gudzinowicz, Burrows T. Younkin, Jr., and Michael J. Gudzinowicz*
27. Modern Analysis of Antibiotics, *edited by Adjoran Aszalos*
28. Solubility and Related Properties, *Kenneth C. James*
29. Controlled Drug Delivery: Fundamentals and Applications, Second Edition, Revised and Expanded, *edited by Joseph R. Robinson and Vincent H. Lee*
30. New Drug Approval Process: Clinical and Regulatory Management, *edited by Richard A. Guarino*
31. Transdermal Controlled Systemic Medications, *edited by Yie W. Chien*
32. Drug Delivery Devices: Fundamentals and Applications, *edited by Praveen Tyle*
33. Pharmacokinetics: Regulatory · Industrial · Academic Perspectives, *edited by Peter G. Welling and Francis L. S. Tse*
34. Clinical Drug Trials and Tribulations, *edited by Allen E. Cato*
35. Transdermal Drug Delivery: Developmental Issues and Research Initiatives, *edited by Jonathan Hadgraft and Richard H. Guy*
36. Aqueous Polymeric Coatings for Pharmaceutical Dosage Forms, *edited by James W. McGinity*
37. Pharmaceutical Pelletization Technology, *edited by Isaac Ghebre-Sellassie*
38. Good Laboratory Practice Regulations, *edited by Allen F. Hirsch*
39. Nasal Systemic Drug Delivery, *Yie W. Chien, Kenneth S. E. Su, and Shyi-Feu Chang*
40. Modern Pharmaceutics: Second Edition, Revised and Expanded, *edited by Gilbert S. Banker and Christopher T. Rhodes*
41. Specialized Drug Delivery Systems: Manufacturing and Production Technology, *edited by Praveen Tyle*
42. Topical Drug Delivery Formulations, *edited by David W. Osborne and Anton H. Amann*
43. Drug Stability: Principles and Practices, *Jens T. Carstensen*
44. Pharmaceutical Statistics: Practical and Clinical Applications, Second Edition, Revised and Expanded, *Sanford Bolton*
45. Biodegradable Polymers as Drug Delivery Systems, *edited by Mark Chasin and Robert Langer*
46. Preclinical Drug Disposition: A Laboratory Handbook, *Francis L. S. Tse and James J. Jaffe*
47. HPLC in the Pharmaceutical Industry, *edited by Godwin W. Fong and Stanley K. Lam*
48. Pharmaceutical Bioequivalence, *edited by Peter G. Welling, Francis L. S. Tse, and Shrikant V. Dinghe*
49. Pharmaceutical Dissolution Testing, *Umesh V. Banakar*
50. Novel Drug Delivery Systems: Second Edition, Revised and Expanded, *Yie W. Chien*
51. Managing the Clinical Drug Development Process, *David M. Cocchetto and Ronald V. Nardi*
52. Good Manufacturing Practices for Pharmaceuticals: A Plan for Total Quality Control, Third Edition, *edited by Sidney H. Willig and James R. Stoker*
53. Prodrugs: Topical and Ocular Drug Delivery, *edited by Kenneth B. Sloan*
54. Pharmaceutical Inhalation Aerosol Technology, *edited by Anthony J. Hickey*
55. Radiopharmaceuticals: Chemistry and Pharmacology, *edited by Adrian D. Nunn*
56. New Drug Approval Process: Second Edition, Revised and Expanded, *edited by Richard A. Guarino*
57. Pharmaceutical Process Validation: Second Edition, Revised and Expanded, *edited by Ira R. Berry and Robert A. Nash*

58. Ophthalmic Drug Delivery Systems, *edited by Ashim K. Mitra*
59. Pharmaceutical Skin Penetration Enhancement, *edited by Kenneth A. Walters and Jonathan Hadgraft*
60. Colonic Drug Absorption and Metabolism, *edited by Peter R. Bieck*
61. Pharmaceutical Particulate Carriers: Therapeutic Applications, *edited by Alain Rolland*
62. Drug Permeation Enhancement: Theory and Applications, edited by *Dean S. Hsieh*
63. Glycopeptide Antibiotics, *edited by Ramakrishnan Nagarajan*
64. Achieving Sterility in Medical and Pharmaceutical Products, *Nigel A. Halls*
65. Multiparticulate Oral Drug Delivery, *edited by Isaac Ghebre-Sellassie*
66. Colloidal Drug Delivery Systems, *edited by Jörg Kreuter*
67. Pharmacokinetics: Regulatory · Industrial · Academic Perspectives, Second Edition, *edited by Peter G. Welling and Francis L. S. Tse*
68. Drug Stability: Principles and Practices, Second Edition, Revised and Expanded, *Jens T. Carstensen*
69. Good Laboratory Practice Regulations: Second Edition, Revised and Expanded, *edited by Sandy Weinberg*
70. Physical Characterization of Pharmaceutical Solids, *edited by Harry G. Brittain*
71. Pharmaceutical Powder Compaction Technology, *edited by Göran Alderborn and Christer Nyström*
72. Modern Pharmaceutics: Third Edition, Revised and Expanded, *edited by Gilbert S. Banker and Christopher T. Rhodes*

ADDITIONAL VOLUMES IN PREPARATION

Microencapsulation: Methods and Industrial Applications, *edited by Simon Benita*

Modern Pharmaceutics

Third Edition, Revised and Expanded

edited by

Gilbert S. Banker
University of Iowa
Iowa City, Iowa

Christopher T. Rhodes
University of Rhode Island
Kingston, Rhode Island

MARCEL DEKKER, INC.   NEW YORK · BASEL

Library of Congress Cataloging-in-Publication Data

Modern pharmaceutics / edited by Gilbert S. Banker, Christopher T.
  Rhodes.—3rd ed., rev. and expanded.
      p.   cm.—(Drugs and the pharmaceutical sciences ; v. 72)
    Includes bibliographical references and index.
    ISBN 0-8247-9371-4 (alk. paper)
    1. Drugs—Dosage forms.   2. Biopharmaceutics.
  3. Pharmacokinetics.   4. Pharmaceutical industry—Quality control.
  I. Banker, Gilbert S.   II. Rhodes, Christopher T.   III. Series.
  RS200.M63   1995
  615′.1—dc20   95-33238
                                                            CIP

Marcel Dekker, Inc.
270 Madison Avenue, New York, New York 10016

Current printing (last digit)
10   9   8   7   6

PRINTED IN THE UNITED STATES OF AMERICA

Preface

The primary purpose of this third edition of *Modern Pharmaceutics* is to provide a basic background in the design and evaluation of modern pharmaceutical dosage forms, with emphasis on both the manner in which the quality of the dosage forms may be assessed and the relationship between quality features and actual drug product performance.

The core of the work is devoted to the various pharmaceutical dosage forms and drug product classes, which are organized as far as possible according to routes of administration. A major objective of the book is to demonstrate the profound effect the twin disciplines of biopharmaceutics and pharmacokinetics have had on the design of pharmaceutical products and the development of new dosage forms. Also stressed is the importance of physical pharmacy in the design of drug products that will be simultaneously safe, effective, and reliable.

The book begins with background chapters on the principles of drug absorption, pharmacokinetics, the factors influencing drug absorption and drug availability, the effect of route of administration and distribution on drug action, and the effect of chemical kinetics on drug stability. Each background chapter highlights fundamental principles that impact on drug properties and drug performance. A chapter on preformulation then describes the manner in which drugs are characterized by their physical, chemical, and pharmacokinetic properties, as a basis for pursuing rational drug product design to meet a wide range of product design objectives.

Other texts in the pharmaceutics field do not provide the connections between drug and drug product features and subsequent drug performance, or they do not connect pharmacodynamic and physicochemical drug properties to drug product design and performance, as does this book. *Modern Pharmaceutics* couples the disciplines of biopharmaceutics, pharmacokinetics, and physical pharmacy with pharmaceutical technology to provide, in a fully cross-referenced text, the information necessary for a thorough understanding of modern pharmaceutical dosage forms.

Chapter 1 reviews the role of drugs and drug products in the treatment of disease, summarizes major quality features of drug products, and examines their current and future status as drug delivery systems. Its historical perspective provides an appreciation of the rapid advances that have been made in chemotherapy over the last 50 years, advances that have resulted in its becoming the preeminent method of treating disease today.

The following chapters complete the treatment of modern pharmaceutical dosage form design and evaluation. These chapters include what is thought to be the most comprehensive chapter in print on the packaging of various pharmaceutical dosage forms; a chapter on optimization techniques in pharmaceutical formulation and processing, whereby complex drug products may be designed and manufactured with the aid of mathematical models; a chapter that reviews food and drug laws that affect drug design, manufacture, and distribution; and a concluding chapter that offers a view to the future and examines the impact of biotechnology on the new revolution in drug discovery.

Contributors to *Modern Pharmaceutics* have been recruited from both university and corporate settings. The most knowledgeable experts in their fields have been brought together in our team. In planning and preparing for the third edition, the editors not only encouraged substantial modification and updating of chapters that were included in earlier editions, but also invited new authors to cover areas of drug use and development that are of contemporary importance.

Because of the growing role of the European Union as a world center for pharmaceutical production, and because of the importance of international harmonization of drug standards and the international character of major drug companies today, a new chapter on regulatory aspects of the industry in the EU has been added. We believe that many readers will find this new chapter of value.

During the past decade, increasing attention has been given in North America, the European Union, and elsewhere to the pharmaceutical needs of special populations. Therefore, a chapter on pediatric and geriatric aspects of pharmaceutics has been added. We are convinced that the development of specialized dosage forms catering to the unique needs of different subcohorts of the general population will become increasingly common.

Protein drugs are of growing importance. The therapeutic modalities made possible by protein drugs are impressive. The Biotechnology Revolution will greatly impact science and technology as we move into the 21st century. Nowhere is this impact expected to be greater than in the treatment and prevention of disease. We expect the new chapter on biotechnology-based pharmaceuticals to be of great value to our readers. Drug delivery is often a challenge with protein drugs. Advances in targeted drug delivery are expected to be an outcome of biotechnology research. A new chapter on target-oriented drug delivery systems has been added, which focuses on biological processes and events involved in drug targeting, pharmacodynamic considerations, and the many delivery systems being employed and under development.

In view of the role of pharmacotherapy in maintaining the health of both small companion animals and farm animals, a new chapter on veterinary pharmaceuticals has been added. This chapter describes some of the special challenges of designing drug products for animals and provides basic information for pharmacists on the use of drug products in animal care.

This book thus provides an integrated, sequenced series of topics that span the field of pharmaceutics. This approach provides a way for the student, the pharmacy practitioner, and the pharmaceutical scientist to develop a thorough understanding of the pharmacodynamic and physicochemical factors influencing drug action according to various routes of administration.

As with our first two editions, we are grateful to all those who have helped with this edition, and we thank those readers who sent us their comments on earlier editions. It is a source of great satisfaction to us that our book has established itself internationally as both a reference and teaching text. We are committed to a continuing process of updating and improving this volume.

Gilbert S. Banker
Christopher T. Rhodes

Contents

Preface   *iii*
Contributors   *vii*

1. Drug Products: Their Role in the Treatment of Disease, Their Quality,
   and Their Status as Drug Delivery Systems   1
   Gilbert S. Banker

2. Principles of Drug Absorption   21
   Michael Mayersohn

3. Pharmacokinetics   75
   David W. A. Bourne and Lewis W. Dittert

4. Factors Influencing Drug Absorption and Drug Availability   121
   Betty-ann Hoener and Leslie Z. Benet

5. The Effect of Route of Administration and Distribution on Drug Action   155
   Svein Øie and Leslie Z. Benet

6. Chemical Kinetics and Drug Stability   179
   J. Keith Guillory and Rolland I. Poust

7. Preformulation   213
   Jens T. Carstensen

8. Cutaneous and Transdermal Delivery: Processes and Systems of Delivery   239
   Gordon L. Flynn

9. Disperse Systems 299
   S. Esmail Tabibi and Christopher T. Rhodes

10. Tablet Dosage Forms 333
    Edward M. Rudnic and Mary Kathryn Kottke

11. Hard and Soft Shell Capsules 395
    Larry L. Augsburger

12. Parenteral Products 441
    James C. Boylan, Alan L. Fites, and Steven L. Nail

13. Design and Evaluation of Ophthalmic Pharmaceutical Products 489
    Gerald Hecht, Robert E. Roehrs, John C. Lang,
    Denise P. Rodeheaver, and Masood A. Chowhan

14. Pharmaceutical Aerosols 547
    John J. Sciarra

15. Sustained- and Controlled-Release Drug Delivery Systems 575
    Gwen M. Jantzen and Joseph R. Robinson

16. Target-Oriented Drug Delivery Systems 611
    Vijay Kumar and Gilbert S. Banker

17. Packaging of Pharmaceutical Dosage Forms 681
    Donald C. Liebe

18. Optimization Techniques in Pharmaceutical Formulation and Processing 727
    Joseph B. Schwartz and Robert E. O'Connor

19. Food and Drug Laws that Affect Drug Product Design,
    Manufacture, and Distribution 753
    Garnet E. Peck

20. European Aspects of the Regulation of Drug Products 773
    Brian R. Matthews

21. Pediatric and Geriatric Aspects of Pharmaceutics 809
    Michele Danish and Mary Kathryn Kottke

22. Biotechnology-Based Pharmaceuticals 843
    S. Kathy Edmond Rouan

23. Veterinary Pharmaceutical Dosage Forms: An Overview 875
    J. Patrick McDonnell

24. A View to the Future 887
    Gilbert S. Banker and Christopher T. Rhodes

Index 907

Contributors

Larry L. Augsburger, Ph.D. Professor, Department of Pharmaceutical Sciences, University of Maryland School of Pharmacy, Baltimore, Maryland

Gilbert S. Banker, Ph.D. Dean and John Lach Distinguished Professor, College of Pharmacy, University of Iowa, Iowa City, Iowa

Leslie Z. Benet, Ph.D. Professor and Chairman, Department of Pharmacy and Pharmaceutical Chemistry, School of Pharmacy, University of California, San Francisco, California

David W. A. Bourne, Ph.D. Associate Professor of Pharmacy, Department of Medicinal Chemistry and Pharmaceutics, University of Oklahoma, Oklahoma City, Oklahoma

James C. Boylan, Ph.D. Director, Pharmaceutical Technology, Hospital Products Division, Abbott Laboratories, Abbott Park, Illinois

Jens T. Carstensen, Ph.D. Professor of Pharmacy, School of Pharmacy, University of Wisconsin, Madison, Wisconsin

Masood A. Chowhan, Ph.D. Associate Director, Department of Formulation Development, Alcon Laboratories, Inc., Fort Worth, Texas

Michele Danish, Pharm.D. Clinical Coordinator, Department of Pharmacy, St. Joseph Hospital, Providence, Rhode Island

Lewis W. Dittert, Ph.D. Professor of Pharmacy, College of Pharmacy, University of Kentucky, Lexington, Kentucky

Alan L. Fites, Ph.D. Fites Consulting, Greenwood, Indiana

Gordon L. Flynn, Ph.D.   Professor of Pharmaceutics, College of Pharmacy, The University of Michigan, Ann Arbor, Michigan

J. Keith Guillory, Ph.D.   Professor Emeritus, College of Pharmacy, University of Iowa, Iowa City, Iowa

Gerald Hecht, Ph.D.   Senior Director, Department of Ophthalmology, Research and Development, Alcon Laboratories, Inc., Fort Worth, Texas

Betty-ann Hoener, Ph.D.   School of Pharmacy, University of California, San Francisco, California

Gwen M. Jantzen   School of Pharmacy, University of Wisconsin, Madison, Wisconsin

Mary Kathryn Kottke, Ph.D.   Department of Formulation Development, AutoImmune, Inc., Lexington, Massachusetts

Vijay Kumar, Ph.D.   Clinical Assistant Professor, Division of Pharmaceutics, College of Pharmacy, University of Iowa, Iowa City, Iowa

John C. Lang, Ph.D.   Assistant Technical Director, Department of Drug Delivery and Formulation Research, Alcon Laboratories, Inc., Fort Worth, Texas

Donald C. Liebe, Ph.D.   Manager, Pharmaceutical Package Development, Schering-Plough Research Institute, Kenilworth, New Jersey

Brian R. Matthews, Ph.D.   Director, EC Registration, Alcon Laboratories, Croydon, England

Michael Mayersohn, Ph.D.   Professor, Department of Pharmaceutical Sciences, College of Pharmacy, University of Arizona, Tucson, Arizona

J. Patrick McDonnell   Senior Auditor, Department of Quality Assurance and Quality Control, Solvay Animal Health, Inc., Charles City, Iowa

Steven L. Nail, Ph.D.   Associate Professor, Department of Industrial and Physical Pharmacy, Purdue University, West Lafayette, Indiana

Robert E. O'Connor, Ph.D.*   R. W. Johnson Pharmaceutical Research Institute, Raritan, New Jersey

Svein Øie, Ph.D.   Professor, Departments of Pharmacy and Pharmaceutical Chemistry, School of Pharmacy, University of California, San Francisco, California

Garnet E. Peck, Ph.D.   Professor and Director of Industrial Pharmacy Laboratory, Purdue University, West Lafayette, Indiana

---

Current affiliation: Whitehall-Robins Healthcare, Hammonton, New Jersey

Rolland I. Poust, Ph.D. Professor, Division of Pharmaceutics, College of Pharmacy, University of Iowa, Iowa City, Iowa

Christopher T. Rhodes, Ph.D. Professor, Department of Applied Pharmaceutical Sciences, University of Rhode Island, Kingston, Rhode Island

Joseph R. Robinson, Ph.D. Division Chair and Professor of Pharmacy, School of Pharmacy, University of Wisconsin, Madison, Wisconsin

Denise P. Rodeheaver, Ph.D. Manager, Department of Toxicology, Alcon Laboratories, Inc., Fort Worth, Texas

Robert E. Roehrs, Ph.D. Vice President, Drug Products, Department of Regulatory Affairs, Alcon Laboratories, Inc., Forth Worth, Texas

S. Kathy Edmond Rouan, Ph.D. Project Director, Department of Project Management, SmithKline Beecham, King of Prussia, Pennsylvania

Edward M. Rudnic, Ph.D. Vice President, Department of Pharmaceutical Research and Development, Pharmavene, Inc., Rockville, Maryland

Joseph B. Schwartz, Ph.D. Philadelphia College of Pharmacy and Science, Philadelphia, Pennsylvania

John J. Sciarra, Ph.D. President and Director of Research and Development, Sciarra Laboratories, Inc., Hicksville, New York

S. Esmail Tabibi, Ph.D.* Associate Professor, Department of Pharmaceutics, University of Rhode Island, Kingston, Rhode Island

---

Current affiliation: Pharmaceutical Resources Branch, National Cancer Institute, National Institutes of Health, Bethesda, Maryland

Modern Pharmaceutics

Third Edition, Revised and Expanded

Drug Products: Their Role in the Treatment of Disease, Their Quality, and Their Status as Drug Delivery Systems

Gilbert S. Banker
University of Iowa, Iowa City, Iowa

I. ROLE OF DRUGS AND DRUG PRODUCTS IN THE TREATMENT AND PREVENTION OF DISEASE

The current methods of treating illness and disease include the use of the following techniques or forms of therapy: (a) surgery, including organ transplantation; (b) psychotherapy; (c) physical therapy; (d) radiation; and (e) chemotherapy. Of these various methods, chemotherapy (treatment with drugs) is the most frequently used technique for treating disease, has the broadest range of application over the greatest variety of disease states, and is frequently the preferred treatment method. Although surgery is the preferred method of treating some ailments or disease states, when alternative methods are available, these methods (usually chemotherapy) will be employed first or as the method of choice, if at all feasible, in the initial attempt to secure satisfactory relief or control of the condition or a complete cure. As chemotherapy continues to improve, it is replacing other forms of treatment as the preferred method of therapy. Chemotherapy is, for example, increasingly becoming the treatment of choice in treating various forms of cancer, including breast cancer, replacing the use of radical surgery. Pharmacotherapy is now an effective option to surgery in the treatment of some forms of prostate disease. When cure rates or reliability of disease control by chemotherapy can match surgical treatment (e.g., prostate surgery or radical mastectomy), most patients will strongly prefer the chemotherapeutic approach, or the use of chemotherapy combined with less radical surgical approaches.

In some surgical procedures, such as organ transplantation, the success of that procedure will be only as great as the course of chemotherapy that follows. Organ transplant recipients are required to continue drug therapy for the balance of their lives for control of their immune systems and to prevent organ rejection.

Chemotherapy is also very important in the prevention of disease, since vaccines and other immunizing agents are drug products. Some diseases that previously killed or crippled tens of millions of people worldwide, often reaching epidemic proportions, are now virtually unknown in most of the world. Table 1 shows the average number of deaths in the United States per

Table 1 Causes of Death in the United States per Million Population in Nonepidemic Years from Infectious and Other Identifiable Diseases

Disease	Year						
	1900	1920	1940	1960	1975	1985	1990
Influenza and pneumonia	2030	2080	700	310	370	250	313
Tuberculosis (all forms)	2020	1150	460	220	50	8	7
Diarrheas and intestinal	1330	540	100	50	40	1	0
Kidney diseases	890	890	820	210	110	85	83
Bronchitis	460	130	30	20	30	16	1
Diphtheria	430	260	10	3	0	0	0
Typhoid and paratyphoid	360	80	10	0	0	0	0
Syphilis	120	160	140	50	20	0	0
Measles	120	90	5	3	0	0	0
Whooping cough	120	120	20	7	0	0	0
Appendicitis	100	130	100	20	0	2	1
Scarlet fever	100	50	5	0	0	0	0
Malaria	80	40	10	0	0	0	0
Smallpox	20	6	0	0	0	0	0

million people as a result of various diseases in a nonepidemic situation, from 1900 through 1990. Diseases that have been obliterated or nearly obliterated in the United States through drug immunology include poliomyelitis, measles, whooping cough, typhoid, diphtheria, and smallpox.

Other diseases in Table 1 have now been brought largely under control or eradicated by the discovery and effective use of anti-infective drugs than can combat bacterial infections. Examples are pneumonia, tuberculosis, certain diarrheal and intestinal disorders, and bronchitis.

Yet other diseases have been largely controlled by improved sanitation and public health procedures, alone or in combination with chemotherapy. Bubonic plague (the ''black death''), malaria, and typhus are in this category. Largely through chemotherapy, including the use of immunological agents, epidemics of life-threatening diseases are nearly a thing of the past in all but the least developed regions of the world. An exception is acquired immune deficiency syndrome (AIDS), a virally transmitted disease of fairly recent origin, for which the development of an effective vaccine has proved to be difficult, given the retrograde character of the virus, and for which no drug or other cure currently exists. Although AIDS is or has been epidemic or nearly so in some segments of society and regions of the world, it is now sufficiently well understood that it can be avoided or exposure greatly reduced by the majority of the population in this country. Nevertheless, it is predicted that AIDS will be one of the leading three or four causes of death in the United States by the years 2000–2010, since it is estimated that several million Americans are now carrying the virus.

It is difficult for us to appreciate the terror and sense of helplessness that individuals and families must have felt as little as a few generations ago, when epidemics swept entire countries and continents. In one 8-year period in the fourteenth century when bubonic plague was epidemic throughout Europe, two-thirds of the population were infected, half of whom died, totaling 25 million deaths. In 1918–1919 an influenza epidemic swept most of the world, causing 20 million deaths, with more than a half-million deaths occurring in the United States.

Another impact of chemotherapy not generally recognized has been its effect on life expectancy and health in the newborn, infants, and children. Some of the earliest valid historical statistics on the death rates of children are found in parish records of London, England. One parish, St. Botolph's, "which was bordered by the city wall and eastern gate, the Tower of London and the Thames," has detailed records from 1558 to 1626, which have survived the years intact [1]. The population of this region probably enjoyed better medical care than did their rural neighbors. The stillborn death rate in the parish fluctuated between 40.8 and 133.4, averaging 71.6:1000 christenings. The current overall average stillborn death rate in the United States is about 20:1000 births. Infants dying in the first month in Shakespeare's day in London were known as chrisoms. The average chrisom death rate from 1584 to 1598 was more than 162:1000 (about four times the average stillborn rate). Thus, about one child out of four was stillborn or failed to survive the first month. By contrast, in the United States today, fewer than 10 deaths per 1000 live births occur during the first year of life. Survival rates of infants and older children were equally grim in Shakespeare's England. Of every 100 children born in the late sixteenth century, only about 70 survived to the 1st birthday, about 48 to their 5th, and only 27–30 survived to their 15th birthday.

The death rate statistics of newborns, infants, and older children have greatly improved from the sixteenth, or even the nineteenth century, to the last decade of the twentieth century. An interesting exercise when next you visit an old cemetery would be to read tombstones that predate 1900, or even 1940. You will find that one grave marker out of every two or three is for a newborn, infant, or a young child. This was largely related to the inability of the medical and pharmaceutical professions of that day to effectively combat infectious and children's diseases. Many of these diseases numbered children almost exclusively among their fatality victims (see Table 1), including measles, scarlet fever, polio, and whooping cough. Added to the infectious "children's diseases" problem was that these diseases often left their young survivors with permanent physiological damage, such as scarred heart valves, brain damage, poorly developed limbs or paralysis, and other defects that remained for the balance of life. The dramatic improvement in the infant mortality rate in the United States over the past 50 years is shown in Fig. 1 [2]. Although dramatic advances have been made, a number of countries in Western Europe have rates that are substantially better than those of the United States.

The increasing life expectancy and the growing number of the elderly in our citizenry over the last century, or even in recent decades, are well known (Figs. 2 and 3) [2]. A person born in 1920 could expect to live only 54.1 years. Today we can expect to live over 75 years (about 73 years for men and nearly 80 years for women). The life expectancy for women helps explain the fact that octogenarians (persons over 80) are the fastest-growing segment of any age group in our population. Figure 3 shows the very rapid growth in the over-65 population as a percentage of the U.S. population since 1960 and projected to the year 2000. The exact contribution of modern pharmaceutics to our increased longevity can be only estimated and weighed in comparison with improved diet, sanitation, housing, and generally improved public health. However, advances in chemotherapy have certainly been the major factor in extending our life expectancy. Similarly, the contribution of chemotherapy to an improved quality of life in recent decades can be only estimated—but it has been a very major factor. It is clearly a leading factor in the well-being of the elderly, allowing them to remain active and essentially healthy through more years and over a greater fraction of their total life span. The role of chemotherapy in improving the quality of life of the mentally ill is also clearly evident. Hundreds of thousands of mentally ill patients in the United States alone are currently being treated as outpatients and can remain with their families through the use of chemotherapy. Many of the patients would require institutionalization or at least short- to midterm hospitalization without the availability

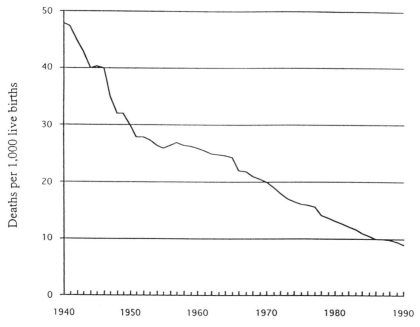

Fig. 1 Infant (under 1 year) mortality rates: United States, 1940–1990.

of effective psychotherapeutic drug agents. Other diseases that formerly required long-term hospitalization or complete isolation include tuberculosis and the dreaded leprosy. Only a generation or two ago, for patients to be told that they had such diseases was equivalent to receiving a death verdict, or worse. These diseases are totally curable today by means of chemotherapy, and the patient no longer needs to be isolated in a sanitarium. Other diseases, such as rheumatoid arthritis, frequently drove patients to suicide. Today, even though we still lack cures for some of these diseases, nevertheless, we can contain and control them, permitting patients to lead essentially normal lives.

Great strides have been made in chemotherapy since World War II and in the decades following the war. Antibiotics and other anti-infective drugs, steroids, psychotherapeutic agents, many new immunizing agents, important cardiovascular agents, antineoplastic agents, and numerous other drug classes and agents have appeared in the last four to five decades. Given the rapid advances in biotechnology, new drug innovation is entering another period of revolutionary growth. Nevertheless, pharmaceutical scientists have no cause for complacency. We cannot yet cure the debilitating diseases of cystic fibrosis or muscular dystrophy. Many forms of cancer are treatable with only low to moderate success if detected early; that battle is far from won. Over one-half million Americans are dying each year from cancer, the number 2 cause of death in this country. Although the mortality rates are consistently continuing to decline (Fig. 4), we must continue the battle to effectively combat many degenerative diseases affecting our growing elderly population, notably heart disease, the number 1 killer, which kills three-quarters of a million Americans a year. The current third leading cause of death, also heavily affecting the elderly, is cerebrovascular disease, which kills about 150,000 Americans a year. Accidents is the fourth leading cause of death at 90,000–95,000/year. Chemotherapy will provide major answers to most, if not all of these and many other disease challenges in the years to come.

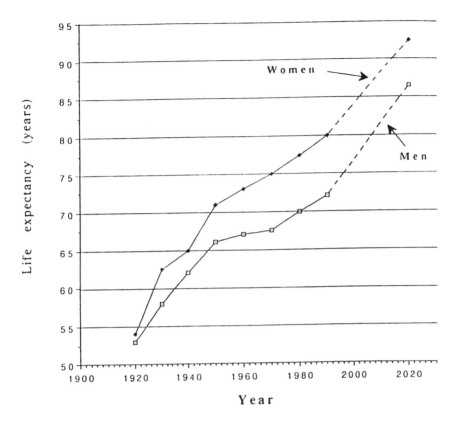

Fig. 2   Expectation of life in years at birth in the United States.

II.   DRUG AND DRUG PRODUCT QUALITY AND ITS EVALUATION

A.   Reasons for the Drug Product Quality Question

The quality of drugs, which used to be an important discussion topic only for pharmaceutical manufacturers and experts in education, compendial standards, and regulatory enforcement, has now been placed in the spotlight of public attention. The reasons for the broad-based interest in drug product quality are based on the following factors, at least in part: (a) a clear realization that drug products are different from other consumer products; (b) rapidly increasing health care costs over the last several decades (Fig. 5); (c) increasing public advertising of drug prescription prices, with identification of price differentials among chemically equivalent generic products; (d) increasing payment of health care costs by third parties (Fig. 6); (e) promulgation of the federal ''maximum allowable cost'' (MAC) regulations; and (f) effects of health care reform.

The greatest difference between drug products and other consumer products is that the principle of *caveat emptor* (''let the buyer beware'') cannot operate in the usual way when the layperson acquires prescription drugs. On the one hand, laypeople lack the skills and sophistication to evaluate the quality of prescription drug products, whereas they do have some ability

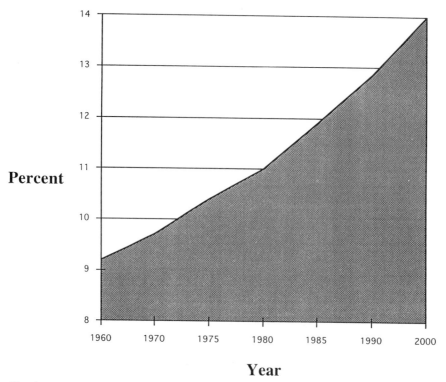

Fig. 3 Percentage of the total U.S. population older than 65 years, from 1960 and projected to the year 2000. (From Ref. 2.)

to evaluate most other consumer products. On the other hand, consumers do not select prescription products as they do nearly every other product they purchase or use. A consumer is often compelled to buy a specific drug product, whereas he or she has more freedom of choice about when or if they buy other products. In the process, the layperson must usually trust the physician who prescribes the medication and the pharmacist who selects it (if a generic drug), and dispenses it. Other differences are that drug products are typically more critical to the consumer's well-being than are other products. A poor-quality drug product can have more serious consequences than a poor-quality consumer product of nearly any other category. Drug products are also more complex than nearly any other class of consumer product. These last two factors are why drug products are subjected to many more tests and controls than are other types of products.

The increasing cost of national health care expenditures from 1929 to 1995, from all types of health-related transactions, including dental, medical, hospital, prescription, and over-the-counter (OTC) drugs, is illustrated in Fig. 5 in dollars and as a percentage of the gross national product (GNP). As shown in the figure, total health expenditures reached 1 trillion dollars and over 14% of the GNP in 1995.

There are many reasons for the cost increases in health care, in addition to inflation, including the following: (a) there are more and more older people, requiring more care; (b) new and additional types of care and treatment are available; and (c) the quality of treatment has been improving. Aggregate expenditures in the health sector have been rising 10–15% or more a year over the last 10–20 years. It is estimated that less than one-half of the increase in

Drug Products

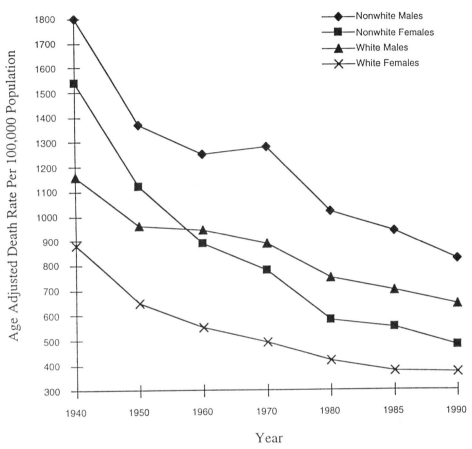

Fig. 4 Age-adjusted death rates in the United States from all causes by sex and by race (white and all others), 1940–1990. (Courtesy of National Center for Health Statistics.)

expenditures since 1965 is due to price increases. Health care expenditure are now approaching 15% of the GNP (see Fig. 5), and continue to climb. Over the last 45 years, since 1950, health care expenditures have increased about 300-fold, or by 30,000%. The distribution of personal health care expenditures by source of funds is shown in Fig. 6 and explains the reason behind recent attempts of federal and state governments to control health care costs more effectively. In 1929 over 85% of health care expenditures were paid directly by the individual consumer. Direct patient payments now account for only about one-fourth of all health care expenditures, even when the growing shift of health care plan costs to the employees of the country is factored in. State and local expenditures as a percentage of the total have remained nearly constant since 1940, but the expenditures on a dollar basis have increased 40-fold.

The two major federal health care benefits programs are Medicaid and Medicare. Currently Medicaid provides a prescription drug benefit; Medicare does not. Medicaid has grown from a 21.6-billion–dollar program (that included 1.3 billion for prescription drugs) in 1980, to a 96-billion–dollar program (including about 7 billion for prescription drugs) in 1995. This

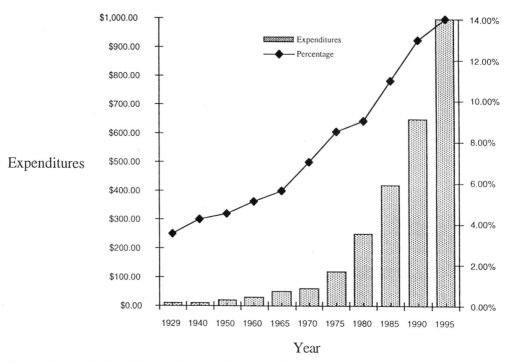

Fig. 5 National total health expenditures in billions of dollars and as a percentage of the gross national product, 1929–1995. (From Ref. 3.)

makes the drug component of the Medicaid program about 7.0% of the total. Medicaid accounts for 7% of the 1.5-trillion–dollar federal budget. The 177 billion expended on Medicare will constitute 11% of the 1995 federal budget. These two rapidly growing federal health care programs are now approaching one-fifth of the total federal budget. Social Security accounts for 21% of the fiscal year 1995 budget, meaning that health care and elderly benefits are now approaching 40% of the federal budget.

As discussed in the last chapter, ''A View to the Future,'' future legislation is expected to add a drug benefit to the federal Medicare program. This will very dramatically increase the dollars committed, probably both at the federal and state levels, to drug benefits for the American people.

B. The Pharmacist's Responsibility and Role in Drug Product Selection

The pharmacist bears a heavy responsibility for quality of the drug products that he or she dispenses. The Code of Ethics of the American Pharmaceutical Association states (in Section 2): ''The pharmacist should never condone the dispensing of drugs and medications which are not of good quality or which do not meet standards required by law.'' Pharmacists are frequently obliged to make judgments concerning the quality of individual drug products and the various dosage forms and possible presentations available for individual drugs. This occurs as pharmacists serve on formulary committees, therapeutics committees, and in other ''official'' roles, as well as on a day-to-day basis, in evaluating manufacturers' and other data to select

$3.2 $3.4 $10.4 $22.7 $33.5 $60.1 $117.1 $219.7 $471.4 $827.6   Billions of Dollars

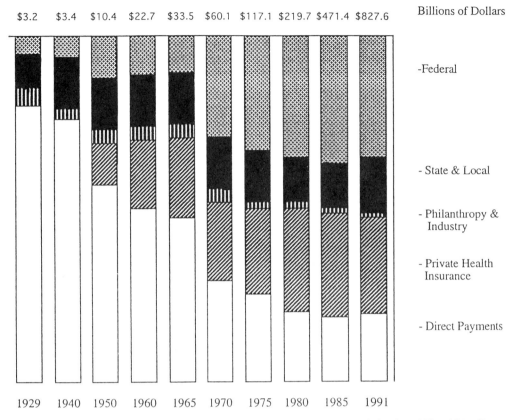

-Federal

- State & Local

- Philanthropy & Industry

- Private Health Insurance

- Direct Payments

1929   1940   1950   1960   1965   1970   1975   1980   1985   1991

Fig. 6 Distribution of personal health care expenditures by source of funds, 1929–1991. (From Ref. 3.)

drug products for their patients that are not only cost-competitive, but are also safe and effective. The pharmacist's role in drug and drug product selection has increased dramatically in recent decades with the growth of third-party payments for prescription drugs. Fortunately, the pharmacist is the most knowledgeable expert on available drugs and drug products and criteria affecting drug product quality. Pharmacists are also better qualified and more knowledgeable than any other health care professional when it comes to drug products, including various product forms available, relative merits of different dosage forms, even within a given route of administration, and possible or most likely side effects within a given population group such as the elderly. Furthermore, pharmacists are well acquainted with the storage requirements of various drugs and drug products, and with the physical signs by which deterioration may be detected. The pharmacist is the health care professional in the best position today to know the various drugs a given patient is taking and thus, to be able to avert adverse therapeutic effects that can arise from many sources. This is because many patients see more than one physician, and they are increasingly taking a broader range of OTC products.

The questions of drug quality, drug cost, drug selection, and proper drug utilization have become a matter of widespread public interest. It is thus very important for pharmacists to be highly knowledgeable concerning all aspects of drug product quality and optimal drug utilization on an individual patient basis.

C. History and Evolution of Drugs and Drug Products

Every pharmacist and pharmacy student should read one of the many available books on the
history of drugs and the drug industry [4–9]. A pictorial history such as *Drugs* [4] in the Life
Science Library Series or one of the several illustrated books on the patent remedy era [5,6]
is entertaining as well as educational reading. When the history of drugs and drug products is
expressed as a time continuum, as shown in Table 2, it is very apparent that most drug and
drug product development has occurred in the twentieth century. Before the early twentieth
century, there were no purified organic chemical substances used in chemotherapy, other than
aspirin, quinine, and morphine. For the preceding 4000 years, drugs changed relatively little.
Early Egyptian physicians recorded over 70 "drugs" in 800 remedies administered in 14
different forms from pills to ointments and salves and poultices. The drugs were all from
natural sources, ranging from spider webs, to animal excretions, to packs of mud, to poppy
seeds. Between the eighth and thirteenth centuries, Arabic alchemists greatly advanced the
pharmaceutical art by introducing extraction and distillation to concentrate and purify natural
products. However, Arabic drugs did not reach Christian Europe until the late Middle Ages
(thirteenth to fifteenth centuries). Pharmaceutical art in colonial America some 200 years ago
was little better than that of the Arabic alchemists of 1000 years earlier. The patent remedy

Table 2 Periods and Notable Events in the History of the Development of Drugs and Drug Products

2000 B.C.		First drug records
Ancient times to Middle Ages	Ancient to medieval pharmacy and medicine	Witch doctors Religious healers
1700		Natural products development
1800		First compendium (U.S.)
1850	Patent remedy era	First U.S. drug companies
1900		Development of analytical standards
1906		Wiley Act—first food and drug law
1938		Second major FDA legislation
1945–1965	Golden age of discovery	Development of sulfa anti-infectives (antibiotics, steroids, etc.)
1962	New regulations	Kefauver-Harris Amendment
		Accelerating development of cancer chemotherapy
		New drug delivery systems
1977		Additional drug regulatory legislation—Phase IV testing, etc.
1980s	Pharmaceutics advances	Full implementation of bioavailability standards
		Additional new drug delivery capabilities
		Mathematical optimization of drug product safety, effectiveness, and reliability

1980s–2000s	Biotechnology era beginning	First recombinant DNA products
		  Human insulin
		  Human growth hormone
		  Interferons, etc.
		Monoclonal antibodies
		Nucleotide blockage
		Gene therapy

era that flourished in the second half of the nineteenth century was a colorful period in pharmaceutical history when "cure-alls" were marketed by "pitchmen" from the backs of wagons, often as a follow-up to some free entertainment used to gather the "locals." As depicted in the advertised indications from the label of one popular patent remedy of the day (Fig. 7), the only limit on claims appeared to be the imagination of the label writers [4]. As noted on the advertising copy of Fig. 7, these remedies were expensive in the days when an average wage earner made 15–20 dollars a week. Grandmother Pinkham used pictures of her grandchildren on her advertising and label copy; others, such as the medicinal syrups millionaire G. G. Green, placed on their advertising and labeling illustrations of the mansions that their high-profit cure-alls brought them. Most patient remedies contained common plant extracts, such as taraxicum (common dandelion weed), at least 15% alcohol, and occasionally opium or other narcotic or addictive substances to help assure repeat sales. It is interesting to speculate on the number of persons in the temperance era who kept all those around them "dry," and then enjoyed their evening toddy of Lydia Pinkhams or stronger products before retiring. The Food, Drug, and Cosmetic Act of 1906 was intended to combat the abuses of the patent remedy era, at least in part by requiring labeling of the active ingredients content of all pharmaceuticals, and by broadly limiting fraudulent practices. The 1938 act went further, requiring that at least some principles of rational therapeutics be applicable to all products and their claims. The 1962 act required many more proofs of safety and effectiveness. (See Chapter 21 for a more detailed description of the evolution of drug laws.)

One interesting way of examining drug and drug product quality is to analyze the changes in drugs and drug products over the centuries, especially the very rapid changes in the last half century (see Table 2). Anyone who imagines that current drug products are optimum as to quality features, or that we have reached the ultimate in chemotherapeutic capabilities, is assuming that the history of drugs is now standing still, as well as ignoring the science of pharmaceutics as we currently know it. Although the Middle Ages ended in about 1450, historically speaking, drugs and drug products did not progress substantially above the quality and knowledge level of the medieval period until the late nineteenth and early twentieth centuries (see Table 2). Drug and drug product advancement during the two decades of the golden age of drug discovery (from about 1945 to about 1965) surpassed the total advancement of the entire previous 4000-year history of drugs.

D. Criteria of Drug and Drug Product Quality

Compendial standards and government regulations require that all drug products, whether ethical prescription or OTC products, meet strict standards of identity, potency, and purity. From about 1900 (see Table 2) until recent decades, standards of identity (the product is what it is actually labeled to be), potency (the active ingredient is present in the labeled amount), and purity [basically limiting nondrug materials as well as describing the amount of active ingredient(s) in natural substances] were thought to define drug quality adequately and were enforced under evolving law and Food and Drug Administration (FDA) regulation. The addition of a few physical tests, such as weight variation and disintegration time to compendial products, such as tablets and capsules, was thought to accurately define the quality of these products. We now know that drug products require very careful evaluation to accurately reflect their quality and performance in clinical roles, and that earlier concepts of evaluation required expansion. The designation "quality," applied to a drug product, according to a modern definition, requires that the product:

> Contain the quantity of each active ingredient claimed on its label, within the applicable limits of its specifications

Fig. 7 Example of the advertised indications from the labeling of a widely used patent remedy, before enactment of the Food, Drug and Cosmetic Act of 1906. (From Ref. 4.)

Contain the same quantity of active ingredient from one dosage unit to the next

Be free from extraneous substances

Maintain its potency, therapeutic availability, and appearance until used

Upon administration, release the active ingredient for full biological availability

In the contemporary definition of quality [7] we see that the concepts of identity, potency, and purity are retained in the first three criteria, but that potency maintenance (including maintenance of pharmaceutical elegance and therapeutic availability or full biological availability) are added. The definition recognizes that drug products may undergo changes with time that result in a loss of biological and therapeutic activity, even though the product complies completely with the original potency and purity standards and no significant drug decomposition has occurred. Such losses in therapeutic activity, without any chemical potency change, may occur as a result of a variety of causes, including

1. Physical changes in the dosage form (moisture loss or gain, crystal changes in excipients, tablet hardening, loss of disintegration/deaggregation properties, etc.)
2. Physical changes in the drug (conversion of a more stable, less readily soluble polymorph, etc.)
3. Chemical changes or interactions involving excipients (esterification of coatings, rendering them less polar and less soluble)

As noted in several chapters that follow, in vitro tests may not themselves be adequate to assure that a product possesses adequate or full bioavailability and therapeutic activity. Whether or not in vitro tests are adequate depends on their sensitivity to pick-up aging or to environmental exposure effects that produce the type of changes noted in the preceding paragraphs and that are of consequence in therapy. The effects on drug potency of environmental stress conditions (e.g., increased temperature, increased humidity, or combinations of the two; temperature cycling) are known for most drug products. The effects of such stress conditions on bioavailability or therapeutic activity are much less well known currently for many commercial products.

A drug product can be of no higher quality than the quality of the drug(s) and excipients (nondrug additives) from which the product is made. It is possible, however, for a product to be of much lower quality than its components, since quality and clinical performance are also related to

1. The rationale of the dosage form design (e.g., if a drug is rapidly destroyed in solution at a pH of 2 or below, the design of an oral product must enable the drug to get through the stomach without substantial dissolution)
2. The method(s) of product manufacture
3. In-process and final quality control procedures to assure that the quality designed and manufactured into the product is actually there
4. Reasonable convenience and ease of product use to assure patient compliance with prescribed dosages

Some of the factors and considerations in the design of high-quality drug products are shown in Table 3. Various physicochemical and pharmacokinetic properties of drugs that affect dosage form design, and their influence on design of high-quality products, are described in Chapter 7 on preformulation. The criteria and properties defining the quality of various dosage forms are discussed in the relevant chapters on dosage forms. The features of an optimized drug product and the concepts of true drug delivery systems are described in the next section.

Table 3 Factors in the Design and Production of High-Quality Drug Products

Input factors	Output factors
A sound development/design manufacturing base	Effectiveness
The preformulation research database	Safety
Physiochemical properties of the drug	
Pharmacokinetic characterization of the drug	Reliability
A rational dosage form design	Stability
Formulation of a stable, reliable system	Physical
	Chemical
Objective preclinical and clinical testing	Microbiological
	Bioavailability
A precise, reproducible manufacturing scheme	
Well-controlled manufacturing steps	Pharmaceutical
Coordinated manufacturing sequences	Appearance
Efficient, sanitary operation	Organoleptic properties
Modern plant and equipment	
	Convenience
A sensitive product control system	East of use
Raw material control	Dosing frequency
Processing controls	Consumer acceptance
Final product controls	
Chemical control standards	
Physical properties standards	
Biological and microbiological standards	
Informed, qualified, and responsible personnel	
Management	
Research and development	
Quality control	
Production	
Services	

III. THE DRUG PRODUCT AS A DELIVERY SYSTEM

A. Drug Products, Drug Delivery Systems, and Therapeutic Systems

Drug substances in their purified state usually exist as crystalline or amorphous powders or as viscous liquids. The majority of drug substances exist as white or light-colored crystalline powders. Although drugs were dispensed as such in powder papers as recently as several decades ago, this practice is virtually unknown in pharmacy practice today. With the possible exception of the anesthetic gases, all drugs in legitimate commerce are now presented to the patient as drug products. It is now well recognized that the therapeutic efficacy and the therapeutic index [ratio of LD_{50} (lethal dose in 50% of the subjects) to ED_{50} (effective dose in 50% of the subjects)] of a drug product is not totally defined by the chemical constitution of the drug and its inherent pharmacokinetic profile. The actual performance of many drugs in clinical practice is now known to be greatly affected by the method of presentation of the drug to the patient. Factors affecting the presentation include

The portal of drug entry in the body
The physical form of the drug product
The design and formulation of the product
The method of manufacture of the drug product
Various physicochemical properties of the drug and excipients
Physicochemical properties of the drug product
Control and maintenance of the location of the drug product at the absorption site(s)
Control of the release rate of the drug from the drug product

In the late 1940s and early 1950s, sustained-release products appeared as a major new class of pharmaceutical product in which product design was intended to modify and improve drug performance by increasing the duration of drug action and reducing the required frequency of dosing. In the mid- to late-1960s, the term "controlled drug delivery" came into being to describe new concepts of dosage form design, which also usually involved controlling and retarding drug dissolution from the dosage form, but with additional or alternative objectives to sustained drug action. These new objectives included safety, enhancing bioavailability, improving drug efficiency and effectiveness, reliability of performance, facilitating patient use and compliance, or other beneficial effects. In the 1970s, yet another term and concept of drug product design and administration appeared: the therapeutic system. The objective of the therapeutic system is to optimize drug therapy by design of a product that incorporates an advanced engineering systems control approach. Three types of therapeutic systems have been proposed, the first of which is already in use: (a) the "passive preprogrammed" therapeutic system—one containing a controlling "logic element," such as a membrane or series of plastic laminates, which preprograms at the time of fabrication or assembly a predetermined delivery pattern (usually constant zero-order release) that is ideally independent of all in vivo physical, chemical, and biological processes; (b) the "active, externally programmed or controlled" therapeutic system—wherein the logic element is capable of receiving and converting a signal (such as an electromagnetic signal) sent from a source external to the body to control and properly modulate drug release from the device within the body; and (c) the "active, self-programmed" therapeutic system—containing a sensing element that responds to the biological environment (such as blood sugar concentration in diabetes) to modulate drug delivery in response to that information. Before the sustained-release concepts of the 1940s and 1950s, which also included depot forms of parenteral products, no significant new oral drug delivery concepts had occurred in the preceding 75 years (since the enteric-coating concept).

B. Concept of the Optimized Drug Product

The optimized drug product may be viewed as a drug delivery system for the one or more drugs that it contains. The goal of this drug delivery system is to release the drug(s) to produce the maximum simultaneous safety, effectiveness, and reliability, as depicted in Fig. 8. Various physicochemical product properties that influence the quality features of safety, effectiveness, and reliability are shown in Table 4. Some physicochemical properties can affect two or all three quality features of Fig. 8. For example, consider chemical stability. As a drug decomposes, if the decomposition products are inactive, this is equivalent to a reduction of the drug dose remaining in the product—or, in other words, a reduction in product reliability and eventually effectiveness. If the decomposition products are toxic or irritating to the body, product safety is also reduced as the product degrades. In examining Table 4, the manner in which each physicochemical, physiological, or therapeutic property affects the various quality features will generally be apparent. You will think of and read about many other physicochemical properties that influence the quality features of drugs and drug products as you read various chapters of

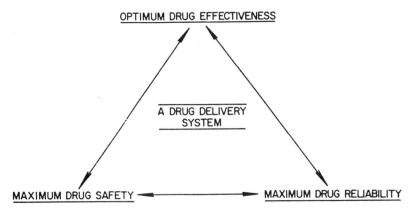

Fig. 8 Features of the optimized drug product.

Table 4 Factors Affecting Drug Product Safety, Effectiveness, and Reliability

Safety	Effectiveness	Reliability
Acute safety quantification Therapeutic index = LD_{50}/ED_{50}	Clinical effectiveness Generic effectiveness	Chemical stability Physical stability

Long-range safety considerations
    Onset of side effects
    Accumulation
	Blood levels
	Urinary elimination
	Pharmacological
	    response(s)
		Microbiological stability
		
		Unit-dose precision

Nature of side effects
    Severity
    Reversibility
	Bioavailability	Patient acceptance
		    Convenience
		    Pharmaceutical elegance

Frequency of side effects		Bioavailability

Untoward and other reactions
    Idiosyncratic responses
    Anaphylaxis
    Tolerance
    Addiction
		High percentage
		    Uniformity
		    Stability

Drug interactions
    Number of drugs involved
    Probability of interaction in therapy
    Severity of the interactions
    Frequency of interaction		

Stability considerations
    Chemical stability
    Physical stability
    Microbiological stability
    Bioavailability stability		

this book. It should also be noted that the three basic quality features of Fig. 8 are connected by double-headed arrows. Thus, as the pharmaceutical formulator modifies the design of a drug product or its method of manufacture to improve one quality feature or one physicochemical property related primarily to one quality feature, the other properties or quality features may be, and usually are, altered. As an example, it may be our goal to increase the hardness of a tablet by formulation (adding more binder) and processing (compressing harder) to improve tablet gloss and appearance or to reduce tablet friability (powdering and chipping in the bottle). This is a worthy objective, but may also reduce the rate and extent of drug dissolution from the tablet. This, in turn, could reduce the reliability of drug absorption and drug performance from patient to patient, or influence transit rate and drug dissolution along the gastrointestinal tract within a given patient, or even reduce effectiveness if the dissolution rate now limits or reduces bioavailability. In the example just cited, maximizing tablet hardness and appearance is a "competing objective" to maximizing drug dissolution and bioavailability.

In Fig. 8 we see the definition of the optimized drug product as the drug delivery system that balances all these factors against each other to produce the maximum possible effectiveness as the primary objective, while producing the best possible simultaneous safety and reliability as secondary objectives, with mathematical certainty. An alternative optimization approach would be to produce the maximum possible (optimized) product safety as the primary objective, while producing the best effectiveness and reliability as the secondary objectives. Yet a third approach would be to optimize safety and effectiveness as equally weighted primary objectives, while maximizing reliability as the secondary objective. Chapter 20 is devoted to the topic of optimization and treats the manner in which experiments must be designed to establish the necessary factors and relationships between factors (independent and controllable processing, formulation, and other variables) as these influence one another and the product quality features (dependent or response variables). Optimization methods then treat this database to design and manufacture the best possible product from an overall standpoint, considering quality features, which may be competing (i.e., as you improve one feature, another degrades), taking into account primary versus secondary features and numerous possible trade-off decisions. Although it is true that the vast majority of drug products that are on the market today are reasonably safe and effective, it is also true that very few products have been designed as optimized systems. Indeed, until about a decade ago, formal optimization methods were unknown in the pharmaceutical and most other industries. The significance of drug products not being optimum systems varies with drug product class. For drugs and drug product classes with a high therapeutic index (ratio of LD_{50} to ED_{50}), maximizing safety is of less concern, and if the drug is well absorbed, a good, stable, pure, and potent drug product that is reasonably reliable may be nearly optimum. For drugs that have less of a safety margin, it may be argued that the conventional, rapidly releasing, effective, stable, and typical reliable product that is currently marketed is not optimum.

Figure 9 illustrates the types of blood level or tissue concentration profiles that are produced with different doses of a rapidly releasing product versus a controlled-release product. This figure is representative of the oral route of administration, but may extrapolated to other routes of administration, with the tissue-level time frame simply being shifted to reflect changes in absorption (and possible distribution) patterns. For the rapidly releasing product, whether given in a single 100-unit dose (curve A) or three divided doses of 33 units (curve B), the inherent ability of the individual to absorb the drug determines the rate of absorption and the peak blood level obtained. (At the peak, the rates of absorption and elimination are equal.) The conventional rapidly releasing drug product is not controlling the blood or tissue level versus time profiles; such a product is simply an uncontrolled "dump system," dumping the drug in the stomach for rapid dissolution and uncontrolled absorption. The body's inherent ability to

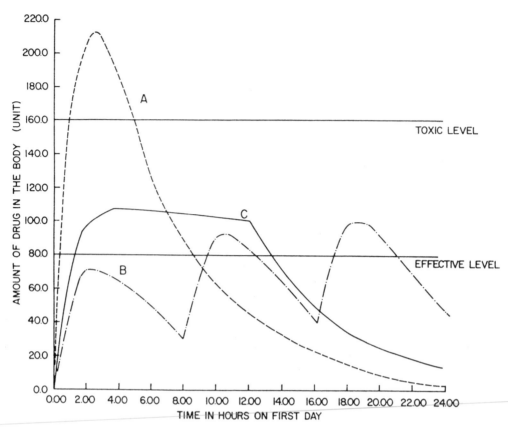

Fig. 9 Blood or tissue level versus time profile simulations following: (A) a single dose representing 100 units of a low therapeutic index drug from a rapid-releasing system; (B) three divided doses of 33 units each from the same rapidly releasing system; and (C) a single 100-unit dose from an optimized controlled-release system. The units of drug in the body may be total units absorbed or units/volume (concentration). A hypothetical effective tissue level (80 units) and toxic level (160 units) is depicted.

absorb the drug under the patient's physiological state at a particular time point and the drug's pharmacokinetics dictate the shape of the blood or tissue level versus time profile at any particular dosing level or dosing frequency. For the optimized controlled-release form (curve C in Fig. 9), the drug product, which is now a drug delivery system, is controlling the rate of release of drug in solution for absorption, and the release rate has been optimized to match drug inactivation and elimination, so that nearly constant tissue levels may be maintained while the drug is in an absorption region in the gut. It has been clearly demonstrated that controlled drug delivery can also substantially increase the therapeutic index and safety margins of certain drugs, while retaining full therapeutic effectiveness. This is because many controlled-release systems produce more rounded blood level versus time profiles, without sharp peaks (compare curves A and C in Fig. 9), so that much larger doses of the controlled-release product are required to reach toxic tissue levels. In one study [11], the effectiveness and duration of action of an antihistamine was followed in animals using a histamine vapor challenge test. The drug as a rapidly soluble dispersion or solution had a duration of action of 3.8 hr, whereas the controlled-release form provided protection to histamine vapor of 8.7 hr. Many antihistamines are depressant drugs and are dangerous on overdosage or in combination with other depressant drugs. In the same study, the drug in conventional form was lethal to 85% of the rats dosed within 30 min at a dose of 200 mg/kg. The same dose of drug when administered as the active controlled-release form killed none of the rats dosed (LD_0 at 24 hr) [11]. Similar studies have demonstrated the ability to improve the safety of barbiturate drugs. In one study, conventional phenobarbital had an LD_{50} in rats at a dose of 200 mg/kg. Twice the dose from an active controlled-release form was only an LD_{40} [12]. Although the cost of developing, testing, and gaining approval through the FDA of depressant or hazardous drugs as new products, optimized from a safety as well as effectiveness standpoint, is apparently currently too high to warrant much activity in the area; I predict that such products will be the rule, rather than the exception, at some point in the future. Not only would such products save lives as a result of the consequences of purposeful or accidental overdosing of potent, depressant, and low therapeutic index drugs, but they might also reduce the severity of some drug interaction effects (especially when alcohol is one of the drugs). Other controlled-delivery forms have the potential of reducing some drug abuse problems, since the slow drug dissolution from the forms makes it difficult to produce extemporaneous illegal and hazardous injectable solution forms.

Drug delivery systems have been designed that are held in the stomach or are bioadhesive at other absorption sites for prolonged and controlled time periods (some 6–10 hr) while releasing drug at controlled rates. Such concepts are bringing us closer to being able to achieve the ideal attributes of drug delivery systems (Table 5), including control of such factors as variable and uncontrolled gastric emptying and transit along the gut, which currently cause the oral route of administration to be the least reliable route of drug administration, even though it is the most popular method of achieving systemic drug effects. Future drug delivery systems that have controlled, prolonged retention in the stomach (or accessible body cavities), with simultaneously controlled delivery, will no doubt improve drug effectiveness for some agents by enhancing absorption reliability and efficiency. They also offer new possibilities of truly optimizing drug safety by prolonging the time over which the drug can be recovered and retrieved, not only on over-dosing, but in acute drug interaction episodes. Bioadhesive systems that can retain drugs on mucosal surfaces for prolonged periods, while promoting drug absorption and delivery, will play a growing role in developing future delivery systems and drug products for some proteins, peptides, and other biotechnology-generated drugs. During their working lifetimes, the pharmacy students of today are certain to see many new drug delivery

Table 5   Ideal Attributes of a Drug Delivery System

1. Capable of controlled-delivery rates to accommodate the pharmacokinetics of various drugs (flexible programming)
2. Capable of precise control of a constant-delivery rate (precise programming)
3. Not highly sensitive to physiological variables, such as
   Gastric motility and emptying, pH, fluid volume, and content of the gut
   Presence/absence or concentration of enzymes
   State of fasting and type of food present
   Physical position and activity of subject
   Individual variability
   Disease state
4. Predicated on physicochemical principles (not pharmaceutical art)
5. Capable of a high order of drug dispersion (the ultimate is molecular in scale)
6. Drug stability is maintained or enhanced
7. The controlling mechanism adds little mass to the dosage form
8. Applicable to a wide range and variety of drugs

concepts reach the marketplace which permit true optimization of drug action and the attainment of the ideal attributes of drug delivery.

REFERENCES

1. R. Forbes, Life and death in Shakespeare's London, *Am. Sci.*, 58, 511–520 (1970).
2. *Statistical Abstract*, U.S. Dept. of Commerce, Washington, DC, (1994).
3. *Health Care Financing Rev.*, Vol. 15 (1993). Division of National Cost Estimates, Health Care Financing Administration, Washington, DC.
4. W. Model and A. Lansing, *Drugs*, Life Science Library Series, Time-Life Books, New York, 1969.
5. A. Hechtlinger, *The Great Patent Medicine Era*, Galahad Books, New York, 1970.
6. G. Carson, *One for a Man, Two for a Horse*, Bramhall House, New York, 1971.
7. W. Screiber and F. K. Mathys, *Infectious Diseases in the History of Medicine*, Kreis & Co., Basel, 1987.
8. R. Carlisle, *A Century of Caring—The Upjohn Story*, Benjamin Co., Elmsford, NY, 1987.
9. P. Boussel, H. Bonnemain, and F. Bove, *History of Pharmacy and the Pharmaceutical Industry*, Asklepious Press, Paris, 1983.
10. The Academy of Pharmaceutical Sciences, The drug product quality statement, J. Am. Pharm. Assoc., 10, 107 (1970).
11. H. Goodman and G. Banker, Molecular-scale drug entrapment as a precise method of controlled drug release I: Entrapment of cationic drugs by polymeric flocculation, J. Pharm. Sci. 59, 1131–1137 (1970).
12. J. Boylan and G. Banker, Molecular-scale drug entrapment as a precise method of controlled drug release IV: Entrapment of anionic drugs by polymeric gelation, J. Pharm. Sci., 62, 1177–1184 (1973).

2

Principles of Drug Absorption

Michael Mayersohn
College of Pharmacy, University of Arizona, Tucson, Arizona

I. INTRODUCTION

Drugs are most often introduced into the body by the oral route of administration. In fact, the vast majority of drug dosage forms are designed for oral ingestion, primarily for ease of administration. It should be recognized, however, that this route may result in inefficient and erratic drug therapy. Whenever a drug is ingested orally (or by any nonvascular route), one would like to have rapid and complete absorption into the bloodstream for the following reasons:

1. If we assume that there is some relationship between drug concentration in the body and the magnitude of the therapeutic response (which is often the case), the greater the concentration achieved, the greater the response.
2. In addition to desiring therapeutic concentrations, one would like to obtain these concentrations rapidly. The more rapidly the drug is absorbed, the sooner the pharmacological response is achieved.
3. In general, one finds that the more rapid and complete the absorption, the more uniform and reproducible the pharmacological response becomes.
4. The more rapidly the drug is absorbed, the less chance there is of drug degradation or interactions with other materials present in the gastrointestinal tract.

In a broad sense, one can divide the primary factors that influence oral drug absorption and, thereby, govern the efficacy of drug therapy into the following categories: (a) physicochemical variables, (b) physiological variables, and (c) dosage form variables. For the most part, these variables will determine the clinical response to any drug administered by an extravascular route. Although often the total response to a drug given orally is a complex function of the aforementioned variables interacting together, the present discussion is limited to primarily the first two categories involving physicochemical and physiological factors. Dosage form variables influencing the response to a drug and the effect of route of administration are discussed in Chapters 4 and 5.

Almost all drugs in current use and those under development are relatively simple organic molecules obtained from either natural sources or by synthetic methods. This statement is especially true of those drugs administered orally, the route emphasized in this chapter. However, I would be derelict in not noting the virtual revolution in development of new therapeutic entities; those based on the incredible advances being made in the application of molecular biology and biotechnology. These new drugs, especially peptides and proteins, are not the small organic molecules stressed in this chapter. Indeed, those compounds have unique physicochemical properties that are quite different from those of small organic molecules, and they offer remarkable challenges for drug delivery. As a result, new and more complex physical delivery systems are being designed in conjunction with an examination of other, less traditional, routes of administration (e.g., nasal, pulmonary, transdermal, or other). Because of issues of instability in the gastrointestinal tract and poor intrinsic membrane permeability, it now appears unlikely that these new biotechnology-derived drugs will employ the oral route for administration to any appreciable extent. Numerous strategies, however, are being explored, and there is evidence that some measure of gastrointestinal absorption can be achieved for some peptides [1].

II. ANATOMICAL AND PHYSIOLOGICAL CONSIDERATIONS OF THE GASTROINTESTINAL TRACT

The gastrointestinal tract (GIT) is a highly specialized region of the body, the primary functions of which involve the processes of secretion, digestion, and absorption. Since all nutrients needed by the body, with the exception of oxygen, must first be ingested orally, processed by the GIT, and then made available for absorption into the bloodstream, the GIT represents an important barrier and interface with the environment. The primary defense mechanisms employed by the gut to rid it of noxious or irritating materials are vomiting and diarrhea. In fact, emesis is often a first approach to the treatment of oral poisoning. Diarrheal conditions, initiated by either a pathological state or a physiological mechanism, will result in the flushing away of toxins or bacteria or will represent the response to a stressful condition. Indeed, the GIT is often the first site of the body's response to stress, a fact readily appreciated by students taking a final examination. The nearly instinctive gut response to stress may be particularly pertinent to patients needing oral drug therapy. Since stress is a fact of our daily lives, and since any illness requiring drug therapy may, in some degree, be considered stressful, the implications of the body's response to stress and the resulting influence on drug absorption from the gut may be extremely important.

Figure 1 illustrates the gross functional regions of the GIT. The liver, gallbladder, and pancreas, although not part of the gut, have been included, since these organs secrete materials vital to the digestive and certain absorptive functions of the gut. The lengths of various regions of the GIT are presented in Table 1. The small intestine, comprising the duodenum, jejunum, and ileum, represents greater than 60% of the length of the GIT, which is consistent with its primary digestive and absorptive functions. In addition to daily food and fluid intake (about 1–2 liters), the GIT and associated organs secrete about 8 liters of fluid per day. Of this total, between 100 and 200 ml of stool water is lost per day, indicating efficient absorption of water throughout the tract.

A. Stomach

After oral ingestion, materials are presented to the stomach, the primary functions of which are storage, mixing, and reducing all components to a slurry with the aid of gastric secretions;

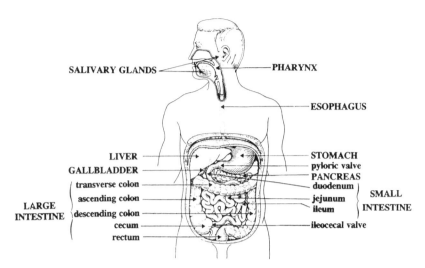

SALIVARY GLANDS — **PHARYNX**

ESOPHAGUS

LIVER
GALLBLADDER
transverse colon
ascending colon
LARGE descending colon
INTESTINE cecum
rectum

STOMACH
pyloric valve
PANCREAS
duodenum
jejunum **SMALL**
ileum **INTESTINE**
ileocecal valve

Fig. 1 Diagrammatic sketch of the gastrointestinal tract (and subdivisions of the small and large intestines) along with associated organs. (Modified from Ref. 2.)

and then emptying these contents in a controlled manner into the upper small intestine (duodenum). All of these functions are accomplished by complex neural, muscular, and hormonal processes. Anatomically, the stomach has classically been divided into three parts: fundus, body, and antrum (or pyloric part), as illustrated in Fig. 2. Although there are no sharp distinctions among these regions, the proximal stomach, made up of the fundus and body, serves as a reservoir for ingested material, and the distal region (antrum) is the major site of mixing motions and acts as a pump to accomplish gastric emptying. The fundus and body regions of the stomach have relatively little tone in their muscular wall, as a result these regions can distend outward to accommodate a meal of up to 1 liter.

A common anatomical feature of the entire GIT is its four concentric layers. Beginning with the luminal surface, these are the mucosa, submucosa, muscularis mucosa, and serosa. The three outer layers are similar throughout most of the tract; however, the mucosa has distinctive structural and functional characteristics. The mucosal surface of the stomach is lined by an epithelial layer of columnar cells, the surface mucous cells. Along this surface are many tubular invaginations, referred to as gastric pits, at the bottom of which are found specialized secretory cells. These secretory cells form part of an extensive network of gastric glands that produce and secrete about 2 liters of gastric fluid daily. The epithelial cells of the gastric mucosa represent one of the most rapidly proliferating epithelial tissues, being shed by the normal stomach at the rate of about a half million cells per minute. As a result, the surface epithelial

Table 1 Approximate Lengths of Various Regions of the Human Gastrointestinal Tract

Region	Length (m)
Duodenum	0.3
Jejunum	2.4
Ileum	3.6
Large Intestine	0.9–1.5

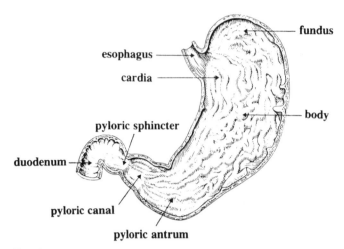

esophagus
cardia
pyloric sphincter
duodenum
pyloric canal
pyloric antrum
fundus
body

Fig. 2 Diagrammatic sketch of the stomach and anatomical regions. (Modified from Ref. 3.)

layer is renewed every 1–3 days. Covering the epithelial cell surface is a layer of mucus 1.0–1.5 mm thick. This material, made up primarily of mucopolysaccharides, provides a protective lubricating coat for the cell lining.

The next region, the muscularis mucosa, consists of an inner circular and an outer longitudinal layer of smooth muscle. This area is responsible for the muscular contractions of the stomach wall, which are needed to accommodate a meal by stretching, and for the mixing and propulsive movements of gastric contents. An area known as the lamina propria lies below the muscularis mucosa and contains a variety of tissue types, including connective and smooth muscles, nerve fibers, and the blood and lymph vessels. It is the blood flow to this region and to the muscularis mucosa that delivers nutrients to the gastric mucosa. The major vessels providing a vascular supply to the GIT are the celiac and the inferior and superior mesenteric arteries. Venous return from the GIT is through the splenic and the inferior and superior mesenteric veins. The outermost region of the stomach wall provides structural support for the organ.

B. Small Intestine

The small intestine has the shape of a convoluted tube and represents the major length of the GIT. The small intestine, comprising the duodenum, jejunum, and ileum, has a unique surface structure, making it ideally suited for its primary role of digestion and absorption. The most important structural aspect of the small intestine is the means by which it greatly increases its effective luminal surface area. The initial increase in surface area, compared with the area of a smooth cylinder, is due to the projection within the lumen of folds of mucosa, referred to as the folds of Kerckring. Lining the entire epithelial surface are fingerlike projections, the villi, extending into the lumen. These villi range in length from 0.5 to 1.5 mm, and it has been estimated that there are about 10–40 villi per square millimeter of mucosal surface. Projecting from the villi surface are fine structures, the microvilli (average length, 1 mm), which represent the final large increase in the surface area of the small intestine. There are approximately 600 microvilli protruding from each absorptive cell lining the villi. Relative to the surface of a smooth cylinder, the folds, villi, and microvilli increase the effective surface area by factors

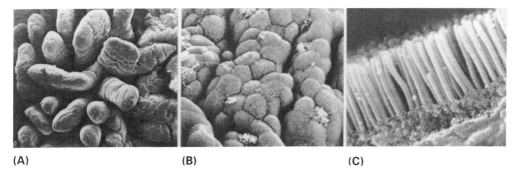

(A)   **(B)**   **(C)**

Fig. 3 (A) Photomicrograph of the human duodenal surface illustrating the projection of villi into the lumen (magnification ×75). The goblet cells appear as white dots on the villus surface. (B) Photomicrograph of a single human duodenal villus illustrating surface coverage by microvilli and the presence of goblet cells (white areas; magnification ×2400). (C) Photomicrograph illustrating the microvilli of the small intestine of the dog (magnification ×33,000). (From Ref. 4.)

of 3, 30, and 600, respectively. These structural features are clearly indicated in the photomicrographs shown in Fig. 3. A diagrammatic sketch of the villus is shown in Fig. 4.

The mucosa of the small intestine can be divided into three distinct layers. The muscularis mucosa, the deepest layer, consists of a thin sheet of smooth muscle three to ten cells thick and separates the mucosa from the submucosa. The lamina propria, the section between the muscularis mucosa and the intestinal epithelia, represents the subepithelial connective tissue space and, together with the surface epithelium, forms the villi structure. The lamina propria

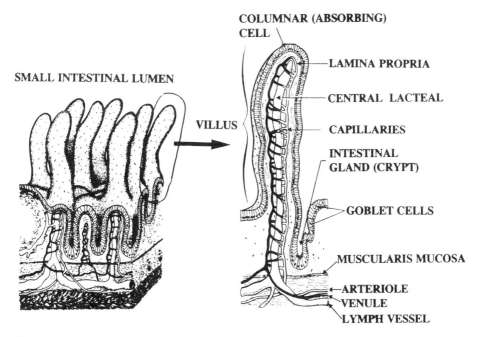

Fig. 4 Diagrammatic sketch of the small intestine illustrating the projection of the villi into the lumen (left) and anatomic features of a single villus (right). (Modified from Ref. 5.)

contains a variety of cell types, including blood and lymph vessels and nerve fibers. Molecules to be absorbed must penetrate this region to gain access to the bloodstream.

The third mucosal layer is that lining the entire length of the small intestine and represents a continuous sheet of epithelial cells. These epithelial cells are columnar, and the luminal cell membrane, upon which the microvilli reside, is called the apical cell membrane. Opposite this membrane is the basal plasma membrane, which is separated from the lamina propria by a basement membrane. A sketch of this cell is shown in Fig. 5. The primary function of the villi is absorption.

The microvilli region has also been referred to as the striated border. It is in this region that the process of absorption is initiated. In close contact with the microvilli is a coating of fine filaments composed of weakly acidic, sulfated mucopolysaccharides. It has been suggested that this region may serve as a relatively impermeable barrier to substances within the gut, such as bacteria and other foreign materials. In addition to increasing the effective luminal surface area, the microvilli region appears to be an area of important biochemical activity.

The surface epithelial cells of the small intestine are renewed rapidly and regularly. It takes about 2 days for the cells of the duodenum to be renewed completely. As a result of its rapid renewal rate, the intestinal epithelium is susceptible to various factors that may influence proliferation. Exposure of the intestine to ionizing radiation and cytotoxic drugs (such as folic acid antagonists and colchicine) reduce the cell renewal rate.

C. Large Intestine

The large intestine, often referred to as the colon, has two primary functions: the absorption of water and electrolytes, and the storage and elimination of fecal material. The large intestine, which has a greater diameter than the small intestine (ca., 6 cm), is connected to the latter at the ileocecal junction. The wall of the ileum at this point has a thickened muscular coat, called the ileocecal sphincter, which forms the ileocecal valve, the principal function of which is to prevent backflow of fecal material from the colon into the small intestine. From a functional point of view the large intestine may be divided into two parts. The proximal half, concerned

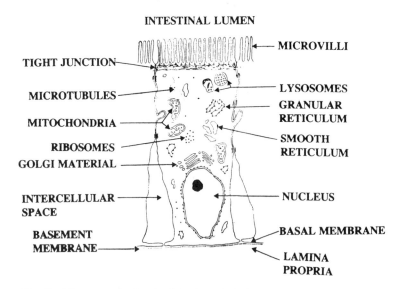

Fig. 5 Diagrammatic sketch of the intestinal absorptive cell. (Modified from Ref. 6.)

primarily with absorption, includes the cecum, ascending colon, and portions of the transverse colon. The distal half, concerned with storage and mass movement of fecal matter, includes part of the transverse and descending colon, the rectum, and anal regions, terminating at the internal anal sphincter (see Fig. 1).

In humans, the large intestine usually receives about 500 ml of fluidlike food material (chyme) per day. As this material moves distally through the large intestine, water is absorbed, producing a viscous and, finally, a solid mass of matter. Of the 500 ml normally reaching the large intestine, approximately 80 ml are eliminated from the gut as fecal material, indicating efficient water absorption.

Structurally, the large intestine is similar to the small intestine, although the luminal surface epithelium of the former lacks villi. The muscularis mucosa, as in the small intestine, consists of inner circular and outer longitudinal layers. Figure 6 illustrates a photomicrograph and diagrammatic sketches of this region.

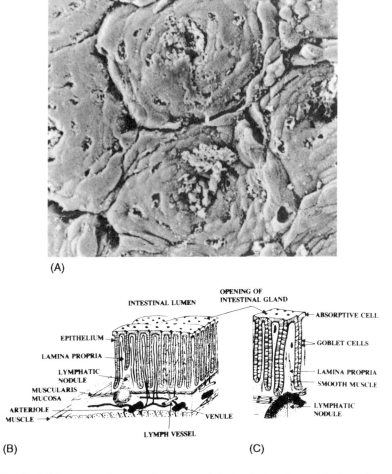

Fig. 6   (A) Scanning electron micrograph of the luminal surface of the large intestine (transverse colon; magnification ×60). (From Ref. 7.) (B) Schematic diagram showing a longitudinal cross section of the large intestine. (C) Enlargement of cross section shown in (B). (B and C modified from Ref. 8.)

D. Pathways of Drug Absorption

Once a drug molecule is in solution, it has the potential to be absorbed. Whether or not it is in a form available for absorption depends on the physicochemical characteristics of the drug (i.e., its inherent absorbability) and the characteristics of its immediate environment (e.g., pH, the presence of interacting materials, and the local properties of the absorbing membrane). If there are no interfering substances present to impede absorption, the drug molecule must come in contact with the absorbing membrane. To accomplish this, the drug molecule must diffuse from the gastrointestinal fluids to the membrane surface. The most appropriate definition of *drug absorption* is the penetration of the drug across the intestinal ''membrane'' and its appearance, unchanged in the blood draining the GIT. There are two important points to this definition: (a) It is often assumed that drug disappearance from the GI fluids represents absorption. This is true only if disappearance from the gut represents appearance in the bloodstream. This is often not the situation; for example, if the drug degrades in GI fluids or if it is metabolized within the intestinal cells. (b) The term *intestinal membrane* is rather misleading, since this membrane is not a unicellular structure, but really a number of unicellular membranes parallel to one another. In fact, relative to the molecular size of most drug molecules, the compound must diffuse a considerable distance. Thus, for a drug molecule to reach the blood, it must penetrate the mucous layer and brush border covering the GI lumen, the apical cell surface, the fluids within this cell, the basal membrane, the basement membrane, the tissue region of the lamina propria, the external capillary membrane, the cytoplasma of the capillary cell, and finally, the inner capillary membrane. Therefore, when the expression intestinal membrane is used, we are discussing a barrier to absorption consisting of several distinct unicellular membranes and fluid regions bounded by these membranes. Throughout this chapter, intestinal membrane will be used in that sense.

For a drug molecule to be absorbed from the GIT and gain access to the body (i.e., the systemic circulation) it must effectively penetrate all the regions of the intestine just cited. There are primarily three factors governing this absorption process once a drug is in solution: the physicochemical characteristics of the molecule, the properties and components of the GI fluids, and the nature of the absorbing membrane. Although penetration of the intestinal membrane is obviously the first part of absorption, we discuss the factors controlling penetration extensively in the following section. At this point, assume that the drug molecule has penetrated most of the barriers in the intestine and has reached the lamina propria region. Once in this region, the drug may either diffuse through the blood capillary membrane and be carried away in the bloodstream, or penetrate the central lacteal and reach the lymph. These functional units of the villi are illustrated in Fig. 4. Most drugs, if not all, reach the systemic circulation by the bloodstream of the capillary network in the villi. The primary reason for this route being dominant over lymphatic penetration is that the villi are highly and rapidly perfused by the bloodstream. Blood flow rate to the GIT in humans is approximately 500–1000 times greater than lymph flow. Thus, although the lymphatic system is a potential route for drug absorption from the intestine, under normal circumstances, it will account for only a small fraction of the total amount absorbed. The major exception to this rule will be drugs (and environmental toxicants, such as insecticides) that have extremely large oil/water partition coefficients (on the order of 10,000). By increasing lymph flow or, alternatively, by reducing blood flow, drug absorption by the lymphatic system may become more important. The capillary and lymphatic vessels are rather permeable to most low-molecular-weight and lipid-soluble compounds. The capillary membrane represents a more substantial barrier than the central lacteal to the penetration of very large molecules or combinations of molecules, as a result of frequent separations of cells along the lacteal surface. This route of movement is important for the absorption of triglycerides, in the form of chylomicrons, which are rather large (about 0.5 μm in the diameter).

III. PHYSICOCHEMICAL FACTORS GOVERNING DRUG ABSORPTION

A.   Oil/Water Partition Coefficient and Chemical Structure

As a result of extensive experimentation, it has been found that the primary physicochemical properties of a drug influencing its passive absorption into and across biological membranes are its oil/water partition coefficient ($K_{O/W}$), extent of ionization in biological fluids, determined by its pK_a value and pH of the fluid in which it is dissolved, and its molecular weight or volume. That these variables govern drug absorption is a direct reflection of the nature of biological membranes. The cell surface of biological membranes (including those lining the entire GIT) is lipid; as a result, one may view penetration into the intestine as a competition for drug molecules between the aqueous environment on one hand, and the lipidlike materials of the membrane, on the other. To a large extent, then, the principles of solution chemistry and the molecular attractive forces to which the drug molecules are exposed will govern movement from an aqueous phase to the lipidlike phase of the membrane.

At the turn of this century, Overton examined the osmotic behavior of frog sartorius muscle soaked in a buffer solution containing various dissolved organic compounds. He reasoned that, if the solute entered the tissue, the weight of the muscle would remain essentially unchanged; whereas, loss of weight would indicate an osmotic withdrawal of fluid and, hence, impermeability to the solute. He noted that, in general, the tissue was most readily penetrated by lipid-soluble compounds and poorly penetrated by lipid-insoluble substances. Overton was one of the first investigators to illustrate that compounds penetrate cells in the same relative order as their oil/water partition coefficients, suggesting the lipid nature of cell membranes. With animal or plant cells, other workers provided data in support of Overton's observations. The only exception to this general rule was the observation that very small molecules penetrate cell membranes faster than would be expected based on their $K_{O/W}$ values. To explain the rapid penetration of these small molecules (e.g., urea, methanol, formamide), it was suggested that cell membranes, although lipid, were not continuous, but were interrupted by small water-filled channels or ''pores''; such membranes are best described as being lipid-sieve membranes. As a result, one could imagine lipid-soluble molecules readily penetrating the lipid regions of the membrane while small water-soluble molecules pass through the aqueous pores. Fordtran et al. [9] estimated the effective pore radius to be 7–8.5 and 3–3.8 Å in human jejunum and ileum, respectively. There may be a continuous distribution of pore sizes; a smaller fraction of larger ones and a greater fraction of smaller pores.

Our knowledge of biological membrane ultrastructure has increased considerably over the years as a result of rapid advances in instrumentation. Although there is still controversy over the most correct biological membrane model, the concept of membrane structure presented by Davson and Danielli, as a lipid bilayer is perhaps the one best accepted [10,11]. The most current version of that basic model is illustrated in Fig. 7 and is referred to as the *fluid mosaic* model of membrane structure. That model is consistent with what we have learned about the existence of specific ion channels and receptors within and along surface membranes.

Table 2 summarizes some literature data supporting the general dependence of the rate of intestinal absorption on $K_{O/W}$, as measured in the rat [13,14]. As with other examples that are available, as $K_{O/W}$ increases, the rate of absorption increases. One very extensive study [15–

CARBOHYDRATES BOUND
TO LIPIDS AND TO PROTEINS

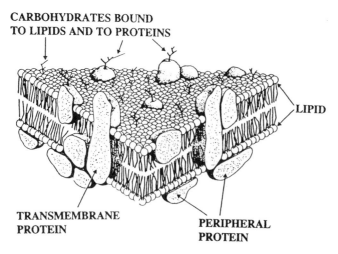

LIPID

TRANSMEMBRANE
PROTEIN

PERIPHERAL
PROTEIN

Fig. 7 Diagrammatic representation of the fluid mosaic model of the cell membrane. The basic structure of the membrane is that of a lipid bilayer in which the lipid portion (long tails) points inward and the polar portion (round head) points outward. The membrane is penetrated by transmembrane (or integral) proteins. Attached to the surface of the membrane are peripheral proteins (inner surface) and carbohydrates that bind to lipid and protein molecules (outer surface). (Modified from Ref. 12.)

17] has examined in depth the physicochemical factors governing nonelectrolyte permeability for several hundred compounds. This study employed an in vitro rabbit gallbladder preparation, the mucosal surface of which is lined by epithelial cells. The method used to assess solute permeability is based on measurement of differences in electrical potential (streaming potentials) across the membrane. The more permeable the compound, the smaller the osmotic pres-

Table 2 Influence of Oil/Water Partition Coefficient ($K_{O/W}$) on Absorption from the Rat Intestine

Compound	$K_{O/W}$	Percentage absorbed
Olive oil/water		
Valeramide	0.023	85
Lactamide	0.00058	67
Malonamide	0.00008	27
Chloroform/water		
Hexethal	> 100	44
Secobarbital	50.7	40
Pentobarbital	28.0	30
Cyclobarbital	13.9	24
Butethal	11.7	24
Allybarbituric Acid	10.5	23
Phenobarbital	4.8	20
Aprobarbital	4.9	17
Barbital	0.7	12

sure it exerts, and the smaller the osmotic fluid flow it produces in the opposite direction; this results in a small potential difference. If the compound is impermeable, it produces a large osmotic pressure and osmotic fluid flow, resulting in a large potential difference. Experimentally, one exposes the mucosal membrane surface to a buffer solution containing a reference compound to which the membrane is completely impermeable and measures the resulting potential difference. This is followed by exposing the same membrane to a solution of a test compound and again measuring the resulting potential difference. The ratio of the potential difference of the test compound to that of the reference compound is referred to as the *reflection coefficient* (σ). The reflection coefficient is a measure of the permeability of the test compound relative to a reference solute with the particular membrane being used. The less permeable the test compound, the closer the reflection coefficient approaches 1; the more permeable the test compound, the closer the coefficient approaches zero.

By using this method, Wright and Diamond were able to reach a number of important conclusions concerning patterns of nonelectrolyte permeability. In general, membrane permeability of a solute increases with $K_{O/W}$, supporting previous findings mentioned earlier. The two classes of exceptions to this pattern are (a) highly branched compounds, which penetrate the membrane more slowly than would be expected based on their $K_{O/W}$; and (b) smaller polar molecules, which penetrate the membrane more readily than would be expected based on their $K_{O/W}$. The latter observation has been noted by other workers and, as mentioned earlier, it has resulted in the development of the lipid-sieve membrane concept, whereby one envisions aqueous pores in the membrane surface. The authors postulate that these small, polar, relatively lipid-insoluble compounds penetrate the membrane by following a route lined by the polar groupings of membrane constituents (i.e., localized polar regions). This concept is an attractive structural explanation of what have been referred to as pores. The accessibility of this route would be limited primarily by the molecular size of the compound as a result of steric hindrance. In fact, it is the first one or two members of a homologous series of compounds that are readily permeable, but beyond these members, it is primarily $K_{O/W}$ that dictates permeability. Table 3 illustrates this effect for several members of various homologous series. Recall that

Table 3 Influence of Chain Length on Membrane Permeability Within Several Homologous Series[a]

Compound	Reflection coefficient, σ	
Urea	0.29	↑
Methyl urea	0.54	│
Ethyl urea	0.92	│
Propyl urea	0.93	—
Butyl urea	0.70	↓
Malononitrile	0.09	↑
Succinonitrile	0.30	—
Glutaronitrile	0.21	↓
Methylformamide	0.28	↑
Methylacetamide	0.51	—
Methylproprionamide	0.22	↓

[a]The reflection coefficient σ is defined in the text. The direction of the arrows indicates an increase in permeability from the least permeable member of the series.

the smaller the σ, the more permeable the compound. In each instance, permeability decreases after the first member, reaches a minimum, and then increases again.

The other anomalous behavior was the smaller-than-expected permeability of highly branched compounds. This deviation has been explained on the basis that membrane lipids are subject to a more highly constrained orientation (probably a parallel configuration of hydrocarbon chains of fatty acids) than are those in a bulk lipid solvent. As a result, branched compounds must disrupt this local lipid structure of the membrane and will encounter greater steric hindrance than will a straight-chain molecule. This effect with branched compounds is not adequately reflected in simple aqueous–lipid partitioning studies (i.e., in the $K_{O/W}$ value).

With the exception of rather small polar molecules, most compounds, including drugs, appear to penetrate biological membranes by a lipid route. As a result, the membrane permeability of most compounds is dependent on $K_{O/W}$. The physicochemical interpretation of this general relationship is based on the atomic and molecular forces to which the solute molecules are exposed in the aqueous and lipid phases. Thus, the ability of a compound to partition from an aqueous to a lipid phase of a membrane involves the balance between solute–water and solute–membrane intermolecular forces. If the attractive forces of the solute–water interaction are greater than those of the solute–membrane interaction, membrane permeability will be relatively poor and vice versa. In examining the permeability of a homologous series of compounds and, therefore, the influence of substitution or chain length on permeability, one must recognize the influence of the substituted group on the intermolecular forces in aqueous and membrane phases (e.g., dipole–dipole, dipole-induced dipole, or van der Waals forces). The membrane permeabilities of the nonelectrolytes studied appear to be largely determined by the number and strength of the hydrogen bonds the solute can form with water. Thus, nonelectrolyte permeation is largely a question of physical organic chemistry in aqueous solution. Table 4 summarizes some of the interesting findings of Diamond and Wright relative to the influence of substituent groups on membrane permeation. These data have been interpreted based on the solutes' ability to form hydrogen bonds with water.

Within a homologous series of compounds, the first few small members are readily permeable, owing to the polar route of membrane penetration. Permeability decreases for the next several members (i.e., σ increases), and then increases as the carbon chain length increases. The regular influence of chain length on permeability is a result not of increased solubility in the lipid phase of the membrane, but of the unique interaction of hydrocarbon chains with water. The nonpolar hydrocarbon molecules are surrounded by a local region of water that has a more highly ordered structure than bulk water. This "iceberg" structure of water results in increased $K_{O/W}$ and membrane permeability as the carbon chain length is increased owing to the compound being "pushed out" of the aqueous phase by the resulting gain in entropy.

There have been several, albeit limited, attempts to develop quantitative, structure–activity relationship in drug absorption [18,19]. Such relationships could prove extremely useful to produce optimum absorption characteristics in the early stages of drug design.

B. The pK_a and pH

Most drug molecules are either weak acids or bases that will be ionized to an extent determined by the compound's pK_a and the pH of the biological fluid in which it is dissolved. The importance of ionization in drug absorption is based on the observation that the nonionized form of the drug has a greater $K_{O/W}$ than the ionized form, and since $K_{O/W}$ is a prime determinant of membrane penetration, ionization would be expected to influence absorption. The observation that pH influences the membrane penetration of ionizable drugs is not a recent finding. At the turn of the century, Overton was able to relate pH to the rate of penetration of various alkaloids

Table 4 Influence of Chemical Substitution on the Membrane Permeability of Several Series of Nonelectrolytes

Substituent group	Influence on membrane permeability	Compound	Example	σ^a
Alcoholic hydroxyl group (—OH)	a. At any given chain length, permeability decreases as the number of —OH groups increases	n-Propanol 1,2-Propanediol Glycerol	$CH_3CH_2CH_2OH$ $CH_3CHOHCH_2OH$ $CH_2OHCHOHCH_2OH$	0.02 ↑ 0.84 0.95
	b. Intramolecular H-bonds formed between adjacent —OH groups result in greater permeability, compared with the same compound with nonadjacent —OH groups owing to decreased H-bond formation with water	2,3-Butanediol 1,3-Butanediol 1,4-Butanediol	$CH_3CHOHCHOHCH_3$ $CH_3CHOHCH_2CH_2OH$ $CH_2OHCH_2CH_2CH_2OH$	0.74 ↑ 0.77 0.86
Ether group (—O—)	Has less of an influence than an —OH group in decreasing permeability	n-Propanol Ethyleneglycol-methyl ether 1,2-Propanediol	$CH_3CH_2CH_2OH$ $CH_3—O—CH_2CH_2OH$ $CH_3CHOHCH_2OH$	0.02 ↑ 0.15 0.84
Carbonyl group Ketone (—C=O) Aldehyde (—HC=O)	Has less of an influence than an —OH group in decreasing permeability; difficulty in measuring permeability of these compounds per se, as many are unstable in solution forming diols and enolic tautomers	Acetone 2-Propanol 2,5-Hexanedione 2,5-Hexanediol	$CH_3\overset{O}{\overset{\|}{C}}CH_3$ $CH_3CHOHCH_3$ $CH_3\overset{O}{\overset{\|}{C}}CH_2\ CH_2\ \overset{O}{\overset{\|}{C}}CH_3$ $CH_3CHOHCH_2CH_2CHOHCH_3$	0.01 ↑ 0.10 0.00 ↑ 0.59
Ester group $\overset{O}{\overset{\|}{(—C—O—)}}$	Has less of an influence than an —OH group in decreasing permeability	1,2-Propanediol-1-acetate 1,5-Pentanediol	$CH_3\overset{O}{\overset{\|}{C}}—O—CH_2CHOHCH_3$ $CH_2OH(CH_2)_3CH_2OH$	0.31 ↑ 0.71
Amide group $\overset{O}{\overset{\|}{—C—NH_2}}$	Causes a greater decrease in permeability than any of the above groups	n-Propanol Acetone Ethyleneglycol-methyl ether Proprionamide	$CH_3CH_2CH_2OH$ $CH_3\overset{O}{\overset{\|}{C}}CH_3$ $CH_3—O—CH_2CH_2OH$ $CH_3CH_2\overset{O}{\overset{\|}{C}}NH_2$	0.02 ↑ 0.08 0.15 0.66
Urea derivatives $\overset{O}{\overset{\|}{R—NH—C—NH_2}}$	Have lower permeability than amides with the same number of carbons and are about as impermeable as the corresponding dihydroxyl alcohols	n-Butanol n-Butryamide 1,4-Butanediol n-propyl urea	$CH_3CH_2CH_2CH_2OH$ $CH_3CH_2CH_2\overset{O}{\overset{\|}{C}}—NH_2$ $CH_2OHCH_2CH_2CH_2OH$ $CH_3CH_2CH_2NH\overset{O}{\overset{\|}{C}}NH_2$	0.01 ↑ 0.42 0.86 0.89

Table 4 Continued

Substituent group	Influence on membrane permeability	Compound	Example	σ^a
α-Amino acids R—CHCOOH | NH$_2$	Have the lowest $K_{o/w}$ values of all organic molecules and are essentially impermeable owing to large dipole–dipole interactions with water	Proprionamide 1-Amino-2- propanol 1,3-Propanediol	$\overset{\displaystyle O}{\overset{\displaystyle \|}{CH_3CH_2CNH_2}}$ CH$_3$CHOHCH$_2$NH$_2$ CH$_2$OHCH$_2$CH$_2$OH	0.66 0.89 0.92
		Alanine	$\overset{\displaystyle H_2N\ \ O}{\overset{\displaystyle \|\ \ \ \|}{CH_3CHC-OH}}$	1.06
Sulfur Functional Groups				
Sulfur replacement of oxygen	a. Have greater $K_{o/w}$ values and permeate membranes more readily than the corresponding oxygen compound; a result of poor H-bond formation between sulfur and water compared with the oxygen analog	1-Thioglycerol Glycerol Thiodiglycol Diethylene glycol	CH$_2$OHCHOHCH$_2$SH CH$_2$OHCHOHCH$_2$OH (OHCH$_2$CH$_2$)$_2$S (OHCH$_2$CH$_2$)$_2$O	0.69 0.95 0.71 0.92
	b. Sulfoxides (R$_2$S=O) are less permeable than the corresponding ketone (R$_2$C=O), owing to stronger H-bond formation with water	Acetone Dimethyl sulfoxide	$\overset{\displaystyle O}{\overset{\displaystyle \|}{CH_3CCH_3}}$ $\overset{\displaystyle O}{\overset{\displaystyle \|}{CH_3SCH_3}}$	0.01 0.92

Source: Ref. 15–17.

[a]The reflection coefficient σ is defined in the text. The direction of the arrows indicates an increase in permeability.

into cells, and he noted the resulting influence on toxicity. Other investigators have made similar observations relative to the influence of pH on the penetration of alkaloids through the conjunctival and mammalian skin [20,21]. The rate of penetration of these weak bases is enhanced by alkalinization owing to a greater fraction of the nonionized species being present. Travell [22] examined the influence of pH on the absorption of several alkaloids from the stomach and intestine of the cat. After ligation of the proximal and distal ends of the stomach of an anesthetized cat, a 5.0-mg/kg solution of strychnine at pH 8.5 produced death within 24 min; however, the same dose at pH 1.2 produced no toxic response. Identical results were found with nicotine, atropine, and cocaine. The same trend was also seen when the drug solution was instilled into ligated intestinal segments and after oral administration (by stomach tube) to ambulatory animals. These results indicate that alkaloids, which are weak bases, will be more rapidly absorbed in the nonionized form (high pH) compared with the ionized form (low pH). This fundamental observation is sometimes overlooked in oral acute drug toxicity studies.

In 1940, Jacobs [23] made use of the Henderson–Hasselbalch equation to relate pH and pK_a to membrane transport of ionizable compounds. Extensive experimentation by a group of investigators in the early 1950s [14,23–28] quantitated many of the aforementioned observations concerning the influence of pH and pK_a on drug absorption from the GIT. These studies have resulted in the so-called pH-partition hypothesis. In essence, this hypothesis states that ionizable compounds penetrate biological membranes primarily in the nonionized form (i.e., nonionic diffusion). As a result, acidic drugs should best be absorbed from acidic solutions for which pH < pK_a, whereas basic compounds would best be absorbed from alkaline solutions for which pH > pK_a. The data in Table 5 illustrate this principle.

These investigators noted some inconsistencies in their data, however, as some compounds (e.g., salicylic acid) that were essentially completely ionized in the buffer solution, nevertheless, were rapidly absorbed. To explain these exceptions, it was suggested that there was a "virtual membrane pH" (about pH 5.3), different from the bulk pH of the buffer solution, which was the actual pH determining the fraction of drug nonionized and, hence, dictating the absorption pattern. Although there may indeed be an effective pH at the immediate surface of the intestinal membrane, different from the pH of solutions bathing the lumen, there is overwhelming experimental evidence indicating that many drugs in the ionic form may be well absorbed. Over the years, there has been an unqualified acceptance of the pH–partition hypothesis and, as a result, many texts and considerable literature on drug absorption indicate that acidic drugs are best absorbed from the acidic gastric fluids of the stomach, and basic drugs best absorbed from the relatively more alkaline intestinal fluids. If all other conditions were the same, the nonionized form of the solution would be more rapidly absorbed than the ionized form. However, conditions along the GIT are not uniform and, hence, most drugs, whether ionized or nonionized (i.e., regardless of pH), are best absorbed from the small intestine as a result of the large absorbing surface area of this region. A good example to illustrate this point is presented in Table 6. There are three important comparisons that should be made in examining these data:

1. By comparing gastric absorption at pH 3 and pH 6 when surface area and factors other than pH are constant, one sees that the general principle is supported; acid drugs are more rapidly absorbed from acidic solution, whereas basic drugs are more rapidly absorbed from relatively alkaline solution.
2. At the same pH (i.e., pH 6) acidic and basic drugs are more rapidly absorbed from the intestine compared with the stomach, by virtue of the larger intestinal surface area.

Table 5 Influence of pH on Drug Absorption from the Small Intestine of the Rat[a]

Drug	pK_a	Percentage absorbed			
		pH 4	pH 5	pH 7	pH 8
Acids					
5-Nitrosalicylic acid	2.3	40	27	<2	<2
Salicylic acid	3.0	64	35	30	10
Acetylsalicylic acid	3.5	41	27	—	—
Benzoic acid	4.2	62	36	35	5
Bases					
Aniline	4.6	40	48	58	61
Aminopyrine	5.0	21	35	48	52
p-Toluidine	5.3	30	42	65	64
Quinine	8.4	9	11	41	54

[a]Drug buffer solutions were perfused through the in situ rat intestine for 30 min and percentage of drug absorbed was determined from four subsequent 10-min samples of the buffer solution.

Table 6 Influence of pH on Drug Absorption from the Stomach and Intestine of the Rat[a]

| | Apparent first-order absorption rate constant (min^{-1}) | | |
| | Stomach | | Intestine |
Drug	pH 3	pH 6	pH 6
Acids			
Salicylic acid	0.015	0.0053	0.085
Barbital	0.0029	0.0026	0.037
Sulfaethidole	0.004	0.0023	0.022
Bases			
Prochlorperazine	<0.002	0.0062	0.030
Haloperidol	0.0028	0.0041	0.028
Aminopyrine	<0.002	0.0046	0.022

[a]Drug buffer solutions were placed into the GIT of an in situ rat preparation. The apparent first-order absorption rate constants are based on drug disappearance from the buffer solution.

3. Acidic drugs are more rapidly absorbed from the intestine (pH 6), although there is substantial ionization, compared with the rate of gastric absorption, even at a pH at which the drug is in a far more acidic solution (pH 3). Again, this is primarily a result of surface area differences.

Interestingly, in an analysis of the original data used in developing the pH–partition hypothesis, Benet [30] has shown that these data support the findings in point 3. The pH–partition hypothesis provides a useful guide in predicting general trends in drug absorption, and it remains an extremely useful concept. There are numerous examples illustrating the general relation among pH, pK_a, and drug absorption developed in that hypothesis [31–33]. The primary limitation of this concept is the assumption that only nonionized drug is absorbed, when, in fact, the ionized species of some compounds can be absorbed, albeit at a slower rate. There is also the presence of unstirred water layers at the epithelial membrane surface that can alter the rate of drug diffusion [34–37]. Furthermore, the hypothesis is based on data obtained from drug in solution. In a practical sense, there are other considerations that are more likely to govern the pattern of drug absorption, and these include dissolution rate from solid dosage forms, the large intestinal surface area, and the relative residence times of the drug in different parts of the GIT. These factors are discussed in the following section. Numerous authors have reviewed the inconsistencies in the pH–partition hypothesis, and they place the issue in its proper perspective [30,38,39]. In general, then, drug absorption in humans takes place primarily from the small intestine, regardless of whether the drug is a weak acid or a base.

C. Mechanisms of Drug Absorption

A thorough discussion of the mechanisms of absorption is provided in Chapter 4. Water-soluble vitamins (B_2, B_{12}, and C) and other nutrients (e.g., monosaccharides and amino acids) are absorbed by specialized mechanisms. With the exception of various antimetabolites used in cancer chemotherapy, L-dopa, and certain antibiotics (e.g., aminopenicillins, aminocephalosporins), virtually all drugs are absorbed in humans by a passive diffusion mechanism. Passive diffusion indicates that the transfer of a compound from an aqueous phase through a membrane may be described by physicochemical laws and by the properties of the membrane. The membrane itself is passive, in that it does not partake in the transfer process, but acts as a simple barrier to diffusion. The driving force for diffusion across the membrane is the concentration gradient (more correctly, the activity gradient) of the compound across that membrane. This mechanism of membrane penetration may be described mathematically by Fick's first law of diffusion, which has been simplified by Riggs [40] and discussed by Benet [30].

$$\left(\frac{dQ_b}{dt}\right)_{g \to b} = D_m A_m R_{m/aq}\left[\frac{C_g - C_b}{\Delta X_m}\right] \tag{1}$$

The derivative on the left side of the equation represents the rate of appearance of drug in the blood (amount/time) when the drug diffuses from the gut fluids g to the blood b. The expression reads, the rate of change of the quantity Q entering the bloodstream. The other symbols have the following meanings (and units): D_m, the diffusion coefficient of the drug through the membrane (area/time); A_m, the surface area of the absorbing membrane available for drug diffusion (area); $R_{m/aq}$, the partition coefficient of the drug between the membrane and aqueous gut fluids (unitless); $C_g - C_b$, the concentration gradient across the membrane, representing the difference in the effective drug concentration (i.e., activity) in the gut fluids C_g at the site of absorption and the drug concentration in the blood C_b at the site of absorption (amount/volume); and ΔX_m, the thickness of the membrane (length). This equation nicely explains several of the observations discussed previously. Thus, rate of drug absorption is directly dependent on the membrane area available for diffusion, indicating that one would expect more rapid absorption from the small intestine, compared with the stomach. Furthermore, the greater the membrane aqueous fluid partition coefficient ($R_{m/aq}$), the more rapid the rate of absorption, supporting the previous discussion indicating the dependence of absorption rate on $K_{O/W}$. We know that pH will produce a net effect on absorption rate by altering several of the parameters in Eq. (1). As the pH for a given drug will determine the fraction nonionized, the value of $R_{m/aq}$ will change with pH, generally increasing as the nonionized fraction increases. Depending on the relative ability of the membrane to permit the diffusion of the nonionized and ionized forms, C_g will be altered appropriately. Finally, the value of D_m may be different for the ionized and nonionized forms of the compound. For a given drug and membrane and under specified conditions, Eq. (1) is made up of several constants that may be incorporated into a large constant (K) referred to as the permeability coefficient:

$$\left(\frac{dQ_b}{dt}\right)_{g \to b} = K(C_g - C_b) \tag{2}$$

where K incorporates D_m, A_m, $R_{m/aq}$, and ΔX_m and has units of volume/time, which is analogous to a flow or clearance term. Since the blood volume is rather large compared with the gut fluid volume, and since the rapid circulation of blood through the GIT continually moves absorbed drug away from the site of absorption, C_g is much greater than C_b. This is often referred to as a *sink condition* for drug absorption, indicating a relatively small drug concentration in the bloodstream at the absorption site. As a result, Eq. (2) may be simplified:

$$\left(\frac{dQ_b}{dt}\right)_{g \to b} = KC_g \tag{3}$$

Equation (3) is in the form of a differential equation describing a first-order kinetic process and, as a result, drug absorption generally adheres to first-order kinetics. The rate of absorption should increase directly with an increase in drug concentration in the GI fluids.

Figure 8 illustrates the linear dependence of absorption rate on concentration for several compounds placed into the *in situ* rat intestine. The slopes of these lines represent the rate constant *K* for absorption in Eq. (3). Alternatively, one may express these data as the percentage absorbed per unit of time as a function of concentration or amount. Several examples illustrating such a treatment are listed in Table 7. As can be seen, for the compounds investigated, the percentage absorbed in any given period is independent of concentration, indicating that these compounds are absorbed by a passive diffusion, or first-order kinetic, process over the concentration ranges studied. Similar studies by other investigators employing an *in situ* rat intestine preparation indicate that several other drugs (see those listed in Table 6) are absorbed in a first-order kinetic fashion.

It is far more difficult to establish the mechanism(s) of drug absorption in humans. Most investigators analyze drug absorption data in humans (from blood or urine data) by assuming first-order absorption kinetics. For the most part, this assumption seems quite valid, and the results of such analyses are consistent with that assumption. As discussed in Chapter 3, one method used to assess the mechanism of drug absorption in humans is based on a pharmacokinetic treatment of blood or urine data and the preparation of log percentage unabsorbed *versus* time plots as developed by Wagner and Nelson [41]. If a straight-line relationship is found, this is indicative of an apparent first-order absorption process, for which the slope of that line represents the apparent first-order absorption rate constant. Some cautions must be taken in the application of this method, as pointed out by several investigators [42,43]. Although the overall absorption process in humans, for many drugs, appears consistent with the characteristics of a first-order kinetic process, there are some questions about which of the sequential steps in the absorption process is rate-limiting. As discussed in a thorough review of mass transport phe-

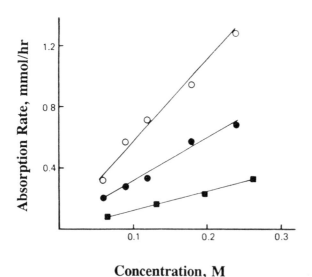

Concentration, M

Fig. 8 Influence of concentration on the rate of absorption from the in situ rat intestine. The linear dependence of absorption rate on concentration suggests an apparent first-order absorption process over the range studied. Absorption rates have been calculated from the data in Ref. 13 and the straight lines are from linear regression of the data: (open circle) erythritol; (solid circle) urea; (solid square) malonamide.

Table 7 Influence of Concentration on the Absorption of Various Solutes from the In Situ Rat Intestine

Compound	Concentration (mM)	Percentage absorbed	Compound	Concentration (mM)	Percentage absorbed
Urea[a]	60	20.9	Salicylic Acid[b]	1	12
	90	19.0		2	12
	120	17.0		10	13
	180	20.0			
	240	17.8	Aniline[b]	1	44
Erythritol[a]	60	54.1			
	90	65.0	Benzoic acid[b]	1	12
	120	62.2		2	12
	180	54.4		10	13
	240	55.5			
			Quinine[b]	1	20
Malonamide[a]	66	16.9		10	20
	132	16.8		0.1	58
	198	16.5	Aniline[c]		
	264	18.4		1	54
				10	59
				20	54

Source: [a]Ref. 13; [b]Ref. 14; [c]Ref. 27.

nomena [44], the oil/water partition coefficient of a solute ($K_{O/W}$) will govern its movement across a lipidlike membrane as long as the membrane is the predominant barrier to diffusion. However, for such membranes, when the $K_{O/W}$ becomes very large, the barrier controlling diffusion may no longer be the membrane, but rather, an aqueous diffusion layer surrounding the membrane. Thus, for some molecules, depending on their physicochemical characteristics, the rate-limiting step in membrane transport will be movement through or out of the membrane, whereas for other compounds the rate-limiting step will be diffusion through an aqueous layer. Several investigators have discussed such behavior for a variety of compounds [45–48]. Wagner and Sedman [38] have provided an extensive analysis of much of the previous literature on drug absorption, and based on their mathematical models, they suggest that absorption may be rate-limiting by drug transfer out of the membrane. Other investigators have formulated different models to help conceptualize the transport process [49,50]. Our incomplete understanding of drug transport across biological membranes is not that surprising, given the complexity of the system and the experimental requirements needed to make unequivocal statements about this process on a molecular level. More definitive data are needed to completely characterize and better understand the complex process of drug absorption.

The analysis of absorption data in humans in recent years has moved away from the more traditional modeling and data-fitting techniques [51]. Absorption processes are now more often characterized by a mean absorption (or input) time (i.e., the average amount of time that the drug molecules spend at the absorption site) or by a process called deconvolution. The former analysis results in a single value (similar to absorption half-life), and the latter results in a profile of the absorption process as a function of time (e.g., absorption rate vs. time). These approaches offer additional ways of interpreting the absorption process.

Equation (1) suggests that diluting the GI fluids will decrease drug concentration in these fluids (C_g), lower the concentration gradient ($C_g - C_b$), and thus reduce the rate of absorption.

In fact, oral dilution is often suggested as an emergency first aid approach in treating cases of oral overdose. There are data from experiments in rats that question the usefulness of this procedure. Henderson et al. [52] found that pentobarbital and quinine were absorbed more rapidly and to a greater extent from larger than from smaller volumes of water, whereas volume had no influence on aspirin absorption. These authors suggest that reduction in the concentration gradient is offset by a greater interfacial area between solution and membrane, as a result of the larger fluid volume. This effectively increases the surface area term (A_m) in Eq. (1). Although other factors may be involved, one has to question oral dilution procedures in treating oral overdose.

Most drugs appear to be absorbed in humans by passive diffusion. Since many essential nutrients (e.g., monosaccharides, amino acids, and vitamins) are water-soluble, they have low oil/water partition coefficients, which would suggest poor absorption from the GIT. However, to ensure adequate uptake of these materials from food, the intestine has developed specialized absorption mechanisms that depend on membrane participation and require that the compound have a specific chemical structure. Since these processes are discussed in Chapter 4, we will not dwell on them here. Absorption by a specialized mechanism (from the rat intestine) has been shown for several agents used in cancer chemotherapy (5-fluorouracil and 5-bromouracil) [53,54], which may be considered "false" nutrients, in that their chemical structures are very similar to essential nutrients for which the intestine has a specialized transport mechanism. It would be instructive to examine some studies concerned with riboflavin and ascorbic acid absorption in humans, as these illustrate how one may treat urine data to explore the mechanism of absorption. If a compound is absorbed by a passive mechanism, a plot of amount absorbed (or amount recovered in the urine) versus dose ingested will provide a straight-line relationship. In contrast, a plot of percentage of dose absorbed (or percentage of dose recovered in the urine) versus dose ingested will provide a line of slope zero (i.e., a constant fraction of the dose is absorbed at all doses). If the absorption process requires membrane involvement, the absorption process may be saturated as the oral dose increases, making the process less efficient at larger doses. As a result, a plot of amount absorbed versus dose ingested will be linear at low doses, curvilinear at larger doses, and approach an asymptotic value at even larger doses. One sees this type of relationship for riboflavin and ascorbic acid in Figs. 9A and C, suggesting non-passive absorption mechanisms in humans [55,56]. This nonlinear relationship is reminiscent of Michaelis–Menten saturable enzyme kinetics from which one may estimate the kinetic parameters (K_m and V_{max}) associated with the absorption of these vitamins. Figures 9B and D illustrate an alternative plot; percentage absorbed versus dose ingested. For a nonpassive absorption process, the percentage dose absorbed will decrease as the dose increases, as a result of saturation of the transport mechanism and of there being a reduction in absorption efficiency. It has been suggested [56] that one means of overcoming the decrease in absorption efficiency is to administer small divided doses, rather than large single doses, as illustrated later for ascorbic acid.

Several investigators suggest that L-dopa (L-dihydroxyphenylalanine) absorption may be impaired if the drug is ingested with meals containing proteins [57,58]. Amino acids formed from the digestion of protein meals, which are absorbed by a specialized mechanism, may competitively inhibit L-dopa absorption if the drug is also transported by the same mechanism. There is evidence (in animals) indicating a specialized absorption mechanism for phenylalanine and L-dopa, and there are data illustrating L-dopa inhibition of phenylalanine and tyrosine absorption in humans [59–61]. L-Dopa appears to be absorbed by the same specialized transport mechanism responsible for the absorption of other amino acids [62]. In a later section, we discuss several of the complicating factors in L-dopa absorption that influence therapy with this drug.

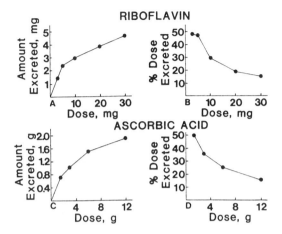

Fig. 9 Urinary excretion of riboflavin (A, B) and ascorbic acid (C, D) in humans as a function of oral dose. A and C illustrate the nonlinear dependence of absorption on dose, which is suggestive of a saturable specialized absorption process. B and D represent an alternative graph of the same data and illustrate the reduced absorption efficiency as the dose increases. (A and B based on the data in Ref. 55 and C and D based on the data in Ref. 56.)

In addition to some anticancer agents being absorbed by a specialized process in humans (e.g., methotrexate) [63], there is recent evidence to suggest that a similar mechanism exists for the absorption of aminopenicillins (e.g., amoxicillin) [64] and aminocephalosporins (e.g., cefixime) [65]. Absorption of these compounds appears to be linked to cellular amino acid or peptide transporters. Other compounds that have the requisite structural properties may also benefit from those transporting systems (e.g., gabapentin) [66]. This behavior may be an important observation for the new generation of drugs being developed through the application of biotechnology (e.g., peptides), assuming such compounds are sufficiently stable in the GIT. Calcium channel blockers, such as nifedipine, increase the absorption of amoxicillin and cefixime [64,65]. This may result from the role of calcium in the transport process, the inhibition of which (i.e., calcium channel blockers) enhances absorption. Another pertinent observation that may explain why some compounds are poorly absorbed is the discovery of a P-glycoprotein-mediated efflux system that resides near the surface of the intestinal epithelium [67]. The latter has been associated with multidrug resistance in tumor cells. Interestingly, this system, which "pumps" drug out of the cell, may be inhibited by the same calcium channel blockers just noted.

IV. PHYSIOLOGICAL FACTORS GOVERNING DRUG ABSORPTION

A. Components and Properties of Gastrointestinal Fluids

The characteristics of aqueous GI fluids to which a drug product is exposed will exert an important influence on what happens to that dosage form in the tract and on the pattern of drug absorption. To appreciate clearly how physiological factors influence drug absorption, one must consider the influence of these variables on the dosage form *per se*; that is, how these variables influence drug dissolution in the aqueous GI fluids, and finally, what influence these variables exert on absorption once the drug is in solution.

One important property of GI fluids is pH, which varies considerably along the length of the tract. The gastric fluids are highly acidic, usually ranging from pH 1 to 3.5. There appears to be a diurnal cycle of gastric acidity, the fluids being more acidic at night and fluctuating during the day, primarily in response to food ingestion. Gastric fluid pH generally increases when food is ingested and, then, slowly decreases over the next several hours, fluctuating from pH 1 to about 5 [68]. There is, however, considerable intersubject variation in GI fluid pH, depending on the general health of the subject, the presence of local disease conditions along the tract, types of food ingested, and drug therapy. Upper GI pH appears to be independent of gender.

An abrupt change in pH is encountered when moving from the stomach to the small intestine. Pancreatic secretions (200–800 ml/day) have a high concentration of bicarbonate, which neutralizes gastric fluid entering the duodenum and, thus, helps regulate the pH of fluids in the upper intestinal region. Neutralization of acidic gastric fluids in the duodenum is important to avoid damage to the intestinal epithelium, prevent inactivation of pancreatic enzymes, and prevent precipitation of bile acids, which are poorly soluble at acidic pH. The pH of intestinal fluids gradually increases when moving in the distal direction, ranging from approximately 5.7 in the pylorus to 7.7 in the proximal jejunum. The fluids in the large intestine are generally considered to have a pH of between 7 and 8.

Gastrointestinal fluid pH may influence drug absorption in a variety of ways. Because most drugs are weak acids or bases, and because the aqueous solubility of such compounds is influenced by pH, the rate of dissolution from a dosage form, particularly tablets and capsules, is dependent on pH. This is a result of the direct dependence of dissolution rate on solubility, as discussed in Chapter 6. Acidic drugs dissolve most readily in alkaline media and, therefore, will have a greater rate of dissolution in intestinal fluids than in gastric fluids. Basic drugs will dissolve most readily in acidic solutions and, thus, the dissolution rate will be greater in gastric fluids than in intestinal fluids. Since dissolution is a prerequisite step to absorption and is often the slowest process, especially for poorly water-soluble drugs, pH will exert a major influence on the overall absorption process. Furthermore, since the major site of drug absorption is the small intestine, it would seem that poorly soluble basic drugs must first dissolve in the acidic gastric fluids to be well-absorbed from the intestine, as the dissolution rate in intestinal fluids will be low (e.g., dipyridamole, ketaconazole, and diazepam). In addition, the disintegration of some dosage forms, depending on their formulation, will be influenced by pH if they contain certain components (e.g, binding agents or disintegrants) the solubility of which is pH-sensitive. Several studies [e.g., 69–71] have indicated that if the specific products being examined were not first exposed to an acidic solution, the dosage form would not disintegrate and, consequently, dissolution could not proceed.

A complication here, however, is noted with those drugs that exhibit a limited chemical stability in either acidic or alkaline fluids. Since the rate and extent of degradation is directly dependent on the concentration of drug in solution, an attempt is often made to retard dissolution in the fluid in which degradation is seen. There are preparations of various salts or esters of drugs (e.g., erythromycin) that do not dissolve in gastric fluid and, thus, are not degraded there, but that dissolve in intestinal fluid before absorption. A wide variety of chemical derivatives are used for such purposes.

As mentioned previously, pH will also influence the absorption of an ionizable drug once it is in solution, as outlined in the pH–partition hypothesis. Most drugs, however, are best absorbed from the small intestine, regardless of pK_a and pH. In some instances, especially lower down the GIT, there is the possibility of insoluble hydroxide formation of a drug or insoluble film formation with components of a dosage form, which reduces the extent of absorption of, for example, the pamoate salt of benzphetamine [72], aluminum aspirin (in chewable tablets) [73,74], and iron [75]. The coadministration of acidic or alkaline fluids with certain drugs may exert an effect on the overall drug absorption process for any of the foregoing reasons.

Moreover, in addition to pH considerations, the GI fluids contain various materials that influence absorption, particularly bile salts, enzymes, and mucin. Bile salts, which are highly surface-active, may enhance the rate or extent of absorption of poorly water-soluble drugs by increasing the rate of dissolution in the GI fluids. This effect has been noted in *in vitro* experiments and has also been seen with other natural surface-active agents (e.g, lysolecithin). Increased absorption of the poorly water-soluble drug griseofulvin after a fatty meal [76,77] may reflect bile secretion into the gut in response to the presence of fats, and the bile salts that are secreted increase the dissolution rate and absorption of the drug. The contrast agent, iopanoic acid, used in visualizing the gallbladder, dissolves more rapidly and is better absorbed from the dog intestine in the presence of bile salts. Studies in rats have indicated enhanced intestinal drug absorption from bile salt solutions; however, the implications of these findings for humans are uncertain. Bile salts may also reduce drug absorption (e.g., neomycin and kanamycin) through the formation of water-insoluble, nonabsorbable complexes.

Since intestinal fluids contain large concentrations of various enzymes needed for digestion of food, it is reasonable to expect certain of these enzymes to act on a number of drugs. Pancreatic enzymes hydrolyze chloramphenicol palmitate. Pancreatin and trypsin are able to deacetylate *N*-acetylated drugs, and mucosal esterases appear to attack various esters of penicillin. It has been suggested that the preparation of various fatty acid esters of a drug that can be hydrolyzed by the GI enzymes may provide a method for controlled drug release and absorption from the GIT. Some caution must be applied in the use of such a dosage form, the performance of which depends on specific physiological conditions, as there is likely to be considerable variation within the population relative to enzyme concentration and activity.

Mucin, a viscous mucopolysaccharide that lines and protects the intestinal epithelium, has been thought to bind certain drugs nonspecifically (e.g., quarternary ammonium compounds) and, thereby, prevent or reduce absorption. This behavior may partially account for the erratic and incomplete absorption of such charged compounds. Mucin may also represent a barrier to drug diffusion before reaching the intestinal membrane.

B.   Gastric Emptying

For many years, physiologists have been interested in factors that influence gastric emptying and the regulatory mechanisms controlling this process. Our interest in gastric emptying is because most drugs are best absorbed from the small intestine, any factor that delays movement of drug from the stomach to the small intestine will influence the rate (and possibly the extent) of absorption and, therefore, the time needed to achieve maximal plasma concentrations and pharmacological response. As a result, and in addition to rate of dissolution or inherent absorbability, gastric emptying may represent a limiting factor in drug absorption. Only in those rare instances when a drug is absorbed by a specialized process in the intestine will the amount of drug leaving the stomach exceed the capacity of the gut to absorb it.

Gastric emptying has been quantitated with a variety of techniques that use liquid or solid meals or other markers. Gastric emptying is quantitated by one of several measurements, including emptying time, emptying half-time ($t_{1/2}$), and emptying rate. Emptying time is the time needed for the stomach to empty the total initial stomach contents. Emptying half-time is the time it takes for the stomach to empty one-half of its initial contents. Emptying rate is a measure of the speed of emptying. Note that the last two measures are inversely related (i.e., the greater the rate, the smaller the value for emptying half-time).

Gastric emptying and factors that affect that process need to be understood because of the implications for drug absorption and in relation to optimal dosage form design [78]. Gastric-emptying patterns are distinctly different, depending on the absence or presence of food. In the absence of food, the empty stomach and the intestinal tract undergo a sequence of repetitious events referred to as the interdigestive migrating motor (or myoelectric) complex [79]. This complex results in the generation of contractions, beginning with the proximal stomach and ending with the ileum. The first of four stages is one of minimal activity, which lasts for about 1 hr. Stage 2, which lasts 30–45 min, is characterized by irregular contractions that gradually increase in strength, leading to the next phase. The third phase, although lasting only 5–15 min, consists of intense peristaltic waves that result in the emptying of gastric contents into the pylorus. The latter is sometimes referred to as the "housekeeper" wave. The fourth stage represents a transition of decreasing activity, leading to the beginning of the next cycle (i.e., phase 1). The entire cycle lasts for about 2 hr. Thus, a solid dosage form ingested on an empty stomach will remain in the stomach for a time period that depends on the time of dosing relative to the occurrence of the housekeeper. The gastric residence time of a solid dosage form will vary from perhaps 5 to 15 min (if ingested at the beginning of the housekeeper) to about 2 hr or longer (if ingested at the end of the housekeeper). It would not be surprising, however, for gastric residence time to range up to 8–10 hr among some subjects. The latter points undoubtedly explain some of the intersubject variation in rate of absorption, and it raises some question concerning the term, "ingested on an empty stomach." Although it is quite common in clinical research studies for a panel of subjects to ingest a solid test dosage form following an overnight fast and, therefore, on an "empty stomach," it is unlikely that all subjects will be in the same phase of the migrating motor complex. It appears to be the latter point, rather than an empty stomach *per se*, that will determine when emptying occurs and, consequently, when drug absorption is initiated. The foregoing considerations will not apply to liquid dosage forms, however, which are generally able to empty during all phases of the migrating motor complex.

Various techniques have been used to visualize the gastric emptying of dosage forms. Radiopaque tablets undergo relatively mild agitation in the stomach; a point that needs to be considered in the design and interpretation of disintegration and dissolution tests [80]. Although single, large, solid dosage forms (e.g., tablets and capsules) rely on the housekeeper wave for entry into the small intestine, some controversy remains about the influence of particle (or pellet) size (diameter and volume), shape, and density on gastric emptying. There has been a great deal of recent interest in this issue, which has been investigated primarily with use of gamma scintigraphy (a gamma-emitting material is ingested and externally monitored with a gamma camera). These studies are generally performed with the use of nondisintegrating pellets, so that movement throughout the tract may be estimated. It is generally claimed that particles must be smaller than about 1–2 mm to empty from the stomach; larger particles requiring additional digestion. This no longer appears to be correct, as particles as large as 5–7 mm may leave the stomach. Therefore, it is likely that there is a range of particle sizes that will empty from the stomach, rather than there being an abrupt cutoff value. The range of values among individuals will be affected by the size of the pylorus diameter and the relative force of propulsive contractions generated by the stomach. The interest in this issue stems from the desire to develop sustained-release dosage forms that would have sufficient residence time in the GIT to provide constant drug release over a long time. Experimental dosage forms that have been investigated include floating tablets, bioadhesives (to attach to the gastric mucosa), dense pellets, and large dimension forms.

Eating interrupts the interdigestive migrating motor complex. Gastric emptying in the presence of solid or liquid food is controlled by a complex variety of mechanical, hormonal, and neural mechanisms. Receptors lining the stomach, duodenum, and jejunum that assist in controlling gastric emptying include mechanical receptors in the stomach that respond to distension; acid receptors in the stomach and duodenum; osmotic receptors in the duodenum that respond to electrolytes, carbohydrates, and amino acids; fat receptors in the jejunum; and L-tryptophan receptors. Neural control appears to be through the inhibitory vagal system (the exact neurotransmitter is unknown, but may be dopamine and enkephalin). Hormones involved in controlling emptying include cholecystokinin and gastrin, among others.

As food enters the stomach, the fundus and body regions relax to accommodate the meal. After reaching the stomach, food tends to form layers that are stratified in the order in which the food was swallowed, and this material is mixed with gastric secretions in the antrum. Nonviscous fluid moves into the antrum, passing around any solid mass. Gastric emptying will begin once a considerable portion of the gastric contents become liquid enough to pass the pylorus. Peristaltic waves begin in the fundus region, travel to the prepyloric area, and become more intense in the pylorus. The antrum and pyloric sphincter contract, and the proximal duodenum relaxes. A moment later the antrum relaxes, and the duodenum regains its tone. The pyloric sphincter will remain contracted momentarily to prevent regurgitation, and the contents in the duodenum are then propelled forward. Emptying is accomplished by the antral and pyloric waves, and the rate of emptying is regulated by factors controlling the strength of antral contraction. Gastric emptying is influenced primarily by meal volume, the presence of acids, certain nutrients, and osmotic pressure. Distension of the stomach is the only natural stimulus known to increase the emptying rate. Fat in any form, in the presence of bile and pancreatic juice, produces the greatest inhibition of gastric emptying. This strong inhibitory influence of fats permits time for their digestion, as they are the slowest of all foods to be digested. Meals containing substantial amounts of fat can delay gastric emptying for 3–6 hr or more. These various factors appear to alter gastric emptying by interacting with the receptors noted earlier.

Other than meal volume *per se*, all of the other factors noted in the foregoing result in a slowing of gastric emptying (e.g., nutrients, osmotic pressure, and acidity). It is important to recognize that there are a host of other factors that are known to influence emptying rate. Thus, a variety of drugs can alter absorption of other drugs by their effect on emptying. For example, anticholinergics and narcotic analgesics reduce gastric emptying rate; whereas, metoclopramide increases that rate. A reduced rate of drug absorption is expected in the former instance and an increased rate in the latter. Among other factors that should be recognized: body position (reduced rate lying on left side); viscosity (rate decreases with increased viscosity); emotional state (reduced rate during depression, increased rate during stress). As an illustration, one recent report indicates that absorption rate (and potentially completeness of absorption) may be altered when comparing posture; lying on the left or right side [81]. Acetaminophen and nifedipine absorption rates were faster when the subjects were lying on the right, compared with the left side, suggesting more rapid gastric emptying. For nifedipine, the extent of absorption was greater when the subjects were lying on the right side, which may be due to transient saturation of a presystemic metabolic process (this is discussed in a later section). Miscellaneous factors, for which the exact effect on emptying may vary, include gut disease, exercise, obesity, gastric surgery, and bulimia. Emptying appears not to be influenced by gender, but there are age-related differences (discussed later).

Many investigators have suggested that gastric emptying takes place by an exponential (i.e., first-order kinetic) process. As a result, plots of log volume remaining in the stomach versus time will provide a straight-line relationship. The slope of this line will represent a rate constant associated with emptying. This relationship is not strictly linear, however, especially at early and later times, but the approximation is useful in that one can express a half-time for emptying ($t_{1/2}$). Hopkins [82] has suggested a linear relationship between the square root of the volume remaining in the stomach versus time. There may be a physical basis for this relationship, since the radius of a cylinder varies with the square root of the volume, and the circumferential tension is proportional to the radius. Methods for analyzing gastric-emptying data have been reviewed [83].

Gastric-emptying rate is influenced by a large number of factors, as noted earlier. Many of these factors account for the large variation in emptying among different individuals and variation within an individual on different occasions. Undoubtedly, much of this variation in emptying is reflected in variable drug absorption. Although gastric emptying probably has little major influence on drug absorption from solution, emptying of solid dosage forms does exert an important influence on drug dissolution and absorption. A prime example are enteric-coated tablets, which are designed to prevent drug release in the stomach. Any delay in the gastric emptying of these forms will delay dissolution, absorption, and the onset time for producing a response. Since these dosage forms must empty as discrete units, the drug is either in the stomach or the intestine. The performance of this dosage form can be seriously hampered if it is taken with or after a meal, as emptying is considerably delayed. Furthermore, if the drug is to be taken in a multiple-dosing fashion, there is a possibility that the first dose will not leave the stomach until the next dose is taken, resulting in twice the desired dose getting into the intestine at one time. Blythe et al. [84] administered several enteric-coated aspirin tablets containing $BaSO_4$ and radiologically examined emptying of these tablets. The tablets emptied in these subjects anywhere from 0.5 to 7 hr after ingestion. Tablets will empty more rapidly when given before a meal compared with administration after a meal. One potential way of improving the emptying and release pattern of enteric-coated products is to use capsules containing enteric-coated microgranules. The median time for 50 and 90% emptying of such a dosage form has been shown to be 1 and 3–3.5 hr, respectively [85].

Several publications have reviewed the effects of food on drug absorption in humans [86–89]. Although not a thorough compilation, the influence of food and several drugs that affect gastric emptying on drug absorption is summarized in Table 8. Food will exert an influence on drug absorption by its effect on gastric emptying and residence time in the GIT, but there are also interactions with various food components. One must also consider the type of food ingested (e.g., carbohydrate, protein, fat, fiber, and so on) and the time of food ingestion relative to drug administration, as these may have different effects on absorption. There is, in addition, the observation that food ingestion may alter hepatic drug extraction subsequent to absorption (i.e., the hepatic first-pass effect) [141]; this point will be discussed later. Foods include liquid nutrients as well as solids, and apple juice has been found to slow gastric emptying [142]. As a general rule, drugs should be ingested on an empty stomach with a glass of water to provide optimal conditions for dissolution and absorption. This rule is particularly important for those compounds unstable in gastric fluids (e.g., penicillin and erythromycin), enteric-coated dosage forms, and those compounds best absorbed in the lower portion of the intestine (e.g., vitamin B_{12}).

As with all general rules and as exemplified in Table 8, there are exceptions. These exceptions include compounds that are irritating to the tract (e.g., phenylbutazone or nitrofurantoin), those compounds absorbed high in the tract by a specialized mechanism (e.g., riboflavin), and those compounds for which the presence of certain food constituents are known to enhance absorption (e.g., griseofulvin). For those compounds that irritate the tract, perhaps the best recommendation is to ingest the drug with or after a light meal that does not contain fatty foods or constituents known to interact with the drug. Nitrofurantoin absorption is improved in the presence of food [121,122]. Riboflavin and ascorbic acid, which are absorbed by a specialized process high in the small intestine, are best absorbed when gastric emptying is delayed by the presence of food [55,93]. As the residence time of the vitamins in the upper

Table 8 Influence of Food and Drugs Affecting Gastric Emptying on Drug Absorption in Humans

Drug	Influence on drug absorption	Ref.
Food		
Acetaminophen	Reduced rate, but not extent	90,91
	Rapid emptying produces greater maximum plasma concentrations and shortens time to achieve maximum concentration; amount absorbed is greater with rapid emptying	92
Ascorbic acid	Increased extent	93
Aspirin	Reduced rate, but not extent	94,95
Bretylium tosylate	Reduced rate and extent	96
Captopril	Reduced rate and extent	97,98
Capuride	Reduced rate, but not extent	99
Cefurozime	Increased extent	100
Cephradine	Reduced rate, but not extent	101,102
Chlorophenoxylisobutyric acid	Increased extent	103
Clindamycin	Reduced rate, but not extent	104
Clobazam	Reduced rate, but not extent	105
Digoxin	Reduced rate, but not extent	106–108
L-Dopa	Any factor reducing emptying rate will reduce rate and extent of absorption	57,109
Ethanol	Reduced rate and extent	110
Fenoprofen	Reduced rate	111
Fenretinide	Increased extent, but not rate	112
Indomethacin	Reduced rate	113
Isoniazid	Reduced rate and extent	114
Isotretinoin	Increased extent	115
Ketoconazole	Reduced extent, but not rate	116,117
Lincomycin	Reduced rate and extent	118
β-Methyl digoxin	Reduced rate, but not extent	119
Methyldopa	No effect	120
Nitrofurantoin	Reduced rate, but increased extent	121,122
Phenylbutazone	No effect	123
Propylthiouracil	No effect	124
Pravastatin	Reduced extent and rate	125
Riboflavin	Increased extent	55
Rifampicin	Reduced rate and extent	126,127
Sulfaisodimidine	No effect	128
Zidovudine (AZT)	Reduced rate and extent	129,130
Drug		
Acetaminophen	Reduced rate, but not extent (propantheline)	131
	Increased rate, but not extent (metoclopramide)	131,132
Digoxin	Increased rate and extent for a slowly dissolving tablet; no effect for a rapidly dissolving tablet (propantheline)	133–135
	Reduced rate and extent (metoclopramide)	134
L-Dopa	Increased rate and extent (metoclopramide)	136
Pivampicillin	Increased rate (metoclopramide)	137
	Reduced rate (atropine)	137
Ranitidine	Increased extent (propantheline)	138
Riboflavin	Reduced rate but increased extent (propantheline)	139
Sulfamethoxazole	Reduced rate (propantheline)	140
Tetracycline	Increased rate (metoclopramide)	137
	Reduced rate (atropine)	137

portion of the intestine is prolonged, contact with absorption sites is increased and absorption becomes more efficient. The influence of food on the absorption of those vitamins is illustrated in Fig. 10, along with improved efficacy of ascorbic acid absorption achieved by administering divided doses. The absorption of griseofulvin, which is a very poorly water-soluble drug, is enhanced when it is coadministered with a fatty meal, as discussed previously [76,77]. The importance of gastric emptying can probably be most readily appreciated by those investigators who have examined drug absorption in patients after a partial or total gastrectomy. Muehlberger [143] notes that, following a near-total gastrectomy, patients often complain of a "sensitivity" to alcohol. This is probably best explained by ethanol moving rapidly from the poorly absorbing surface of the stomach to the small intestine, where absorption will be rapid. Gastric emptying is important in oral L-dopa therapy, and it has been noted [144] that patients with a partial gastrectomy or gastrojejunostomy exhibit a prompt response with less than average doses of the drug. This observation is consistent with rapid absorption from the small intestine in such patients and is essentially equivalent to introduction of the drug into the duodenum. Similar conclusions have been reached for aspirin and warfarin absorption [145,146].

For many drugs, as has been shown for acetaminophen, there will be a direct relation between gastric-emptying rate and maximal plasma concentration, and an inverse relation between gastric-emptying rate and the time required to attain maximal plasma concentrations. Those relations are illustrated in Fig. 11A and B. Also shown in that figure is the influence of a narcotic (heroin) on the gastric emptying and absorption of acetaminophen (see Fig. 11C and D). In attempting to predict such relations, however, it is essential that one consider the physicochemical characteristics of the drug. Whereas an increased gastric-emptying rate will probably increase the rate (and possibly the extent) of absorption for drugs best absorbed from the small intestine from rapidly dissolving dosage forms, the converse may be true in other circumstances. For example, if the dosage form must first be exposed to the acidic gastric fluids to initiate disintegration or dissolution, rapid emptying may reduce the rate and extent of absorp-

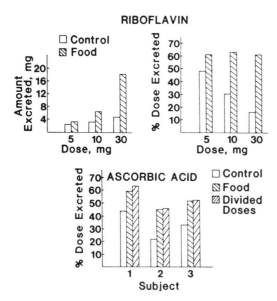

Fig. 10 (Top) Influence of food on the absorption of different doses of riboflavin; (Bottom) influence of food and divided doses on ascorbic acid absorption in three subjects. (Data from Refs. 55 and 93.)

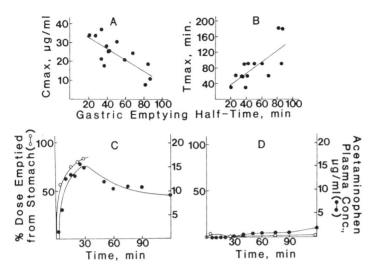

Fig. 11 (A, B) Maximum acetaminophen plasma concentration (C_{max}) and time to achieve that concentration (T_{max}) as a function of gastric emptying half-time. (C) Percentage of an acetaminophen dose emptied from the stomach (open circles) and acetaminophen plasma concentrations (solid circles) as a function of time in one subject. (D) The same plot and for the same subject as in (C) after a 10-mg intramuscular dose of heroin. (A, from Ref. 92; C and D from Ref. 147.)

tion. Similarly, if the drug dissolves slowly from the dosage form, a shortened residence time in the gut may reduce the extent of dissolution and absorption. Obviously, one needs a good deal of fundamental understanding of the chemistry of the drug, its dosage form, and the absorption mechanism before being able to anticipate or rationalize the influence of these various factors on the efficacy of absorption.

The gastric-emptying process can sometimes be observed in plots of log percentage of dose unabsorbed versus time, in which at early times a smaller slope is seen than at later times. The lag time for absorption can also reflect disintegration or dissolution processes, as well as slower absorption in the stomach before emptying.

A final point that should be mentioned here, although it has received relatively little attention, is that of esophageal transit. Delay in movement down the esophagus will delay absorption and, in addition, for certain drugs, may also cause local mucosal damage. Capsule disintegration has been observed to occur in the esophagus within 3–5 min. Esophageal transit is delayed when solid dosage forms are swallowed with little fluid, or when the subject is supine [148,149]. Antipyrine absorption from capsules [150] and acetaminophen absorption from tablets [151] was delayed when esophageal transit was prolonged. To avoid this delay, the dosage form should be swallowed with water or other fluids, and the subject should be in a standing or sitting position.

C. Intestinal Transit

Once a dosage form empties from the stomach and enters the small intestine, it will be exposed to an environment totally different from that in the stomach, as discussed previously. Since the small intestine is the primary site of drug absorption, the longer the residence time in this region, the greater the potential for complete absorption, assuming that the drug is stable in the intestinal fluids and will not form water-insoluble derivatives.

There are primarily two types of intestinal movements: propulsive and mixing. Propulsive movements, generally synonymous with peristalsis, will determine intestinal transit rate and, therefore, the residence time of a drug in the intestine. This time of residence is important, since it will dictate the amount of time the dosage form has in which to release the drug, permit dissolution, and allow for absorption. Obviously, the greater the intestinal motility, the shorter the residence time, and the less time there is for those processes to proceed. Intestinal motility will be most important for those dosage forms that release drug slowly (e.g., sustained-release products) or require time to initiate release (e.g., enteric-coated products), as well as those drugs that dissolve slowly or for which absorption is maximal only in certain regions of the intestine. Peristaltic waves propel intestinal contents down the tract at about 1–2 cm/sec. Peristaltic activity in increased after a meal as a result of the gastroenteric reflex initiated by distension of the stomach and results in increased motility and secretion.

Mixing movements of the small intestine are a result of contractions dividing a given region of the intestine into segments, producing an appearance similar to a chain of sausages. These contractions result in mixing of the intestinal contents with secretions several times a minute. This mixing brings the gut contents into optimal contact with the surface epithelium and, thereby, provides a larger effective area for absorption. In addition, the muscularis mucosa produces folds in the surface epithelium, resulting in an increased surface area and rate of absorption. The villi contract during this process, which results in a ''milking'' action, so that lymph flows from the central lacteal into the lymphatic system.

These mixing motions will tend to improve drug absorption for two reasons: (a) Any factor that increases rate of dissolution will generally increase the rate (and possibly the extent) of absorption, especially for poorly water-soluble drugs. Since rate of dissolution depends on agitation intensity, mixing movements will tend to increase dissolution rate and, thereby, influence absorption. (b) As rate of absorption depends directly on membrane surface area, and since mixing increases the contact area between drug and membrane, these motions will tend to increase rate of absorption.

As with gastric emptying, there are a variety of factors that will influence intestinal motility and, thereby, may influence drug absorption. Food, although it may be a bulk stimulant for intestinal transit, results in active mixing movements in the intestine. Although these movements may increase the rate of drug dissolution, the general recommendation made in the preceding section still applies; it is best to ingest a drug as much time before a meal as possible. Not only will this avoid problems associated with gastric emptying, but it will reduce potential drug interactions with food components in the small intestine, where these materials are in intimate and prolonged contact. In addition, the presence of food, which tends to provide a viscous environment in the gut, will reduce drug diffusion to the absorbing membrane. Exceptions to this rule were noted earlier. Sustained-release products may represent a general exception, since these dosage forms depend on longer residence time in the tract to completely release the drug. Since the performance of these products may be influenced by the presence of food, the Food and Drug Administration requires that such products be examined in fed and fasted subjects. This requirement was deemed necessary following the observation of ''dose-dumping.'' The latter occurred when the entire dose was released as a result of failure of the release mechanism in response to food.

As noted in Table 8 metoclopramide will increase the rate of gastric emptying, which often, but not always, will increase the rate of drug absorption. However, metoclopramide will also increase the rate of intestinal transit and thus reduce the residence time in the intestine. These two effects may have an opposing influence on absorption. The net effect on absorption depends on the characteristics of the drug and its dosage form as well as the mechanism of absorption. Metoclopramide or similar-acting drugs will probably have little if any effect on absorption of a drug given orally in solution, unless the drug (e.g., riboflavin) is absorbed by a specialized process high in the small intestine, in which circumstance there is likely to be a reduction in the amount absorbed. Metoclopramide will probably increase the rate of absorption of a drug from a solid dosage form because of its effect on gastric emptying if the drug is rapidly released and readily dissolved. On the other hand, if the drug dissolves slowly from the dosage form, the extent of absorption may be reduced as a result of shortened residence time in the intestine, even though gastric emptying rate is increased. Similar reasoning may be applied to the influence on drug absorption of various anticholinergics (e.g., atropine and propantheline) and narcotic analgesics that reduce gastric-emptying and intestinal transit rates. Although there will be a reduction in gastric-emptying rate and thus a delay in absorption, these compounds will increase intestinal transit time and possibly increase the extent of absorption, particularly for slowly dissolving drugs or dosage forms that release drug slowly.

Transit through the small intestine appears to be quite different in a variety of ways from movement through the stomach. Once emptied from the stomach, material (such as pellets and tablets) will move along the small intestine and reach the ileocecal valve in about 3 hr. Although this value may range from about 1 to 6 hr, intestinal residence time appears to be relatively consistent among normal subjects [152]. Values similar to this have been found for food and water movement along the small intestine. Transit appears to be less dependent on the physical nature of the material (liquid *vs.* solid and size of solids) than with the response of the stomach. Furthermore, food appears not to influence intestinal transit, as it does gastric emptying.

Three to four hours in the small intestine is a relatively short time for a dosage form to completely dissolve or release drug and then be absorbed. This time would be even more critical to the performance of poorly water-soluble drugs, slowly dissolving, coated dosage forms (enteric or polymer coated), and sustained-release forms. If one assumes minimal absorption from the colon (discussed later), gastric residence time may prove a critical issue to the performance of certain drugs and drug dosage forms (especially those in the latter categories) as a result of the relatively short intestinal residence time.

There is less information available concerning the factors that may influence intestinal transit time compared with what we know about gastric residence time. Although based on small populations, there appear to be no gender-related differences in intestinal transit time [153], and vegetarians appear to have longer intestinal transit times than nonvegetarians [154]. The latter point may have implications for drug therapy in the third world where the diet is primarily vegetarian. Other factors that result in an increased transit time include reduced digestive fluid secretion, reduced thyroxine secretion, and pregnancy [155–157].

The distal portion of the GIT, the colon (see Fig. 1), has as its primary function water and electrolyte absorption and the storage of fecal matter before it is expelled. The proximal half of the colon is concerned with absorption and the distal half with storage. Although there are mixing and propulsive movements in the colon, they tend to be rather sluggish. Large circular constrictions occur in the colon that are similar to the segmenting contractions seen in the small intestine. The longitudinal muscles lining the colon also contract, producing a bulging, similar in appearance to sacs, and referred to as haustrations. These movements increase the surface area of the colon and result in efficient water absorption.

Contents within the colon are propelled down the tract, not by peristaltic waves, but by a ''mass movement'' that occurs only several times a day, being most abundant the first hour after breakfast as a result of a duodenocolonic reflex. The greatest proportion of time moving down the GIT is spent by a meal moving through the colon. In the presence of a diarrheal condition, fluid absorption is incomplete, which results in a watery stool.

Colonic residence time is considerably longer than in other parts of the GIT, and it is also more variable. The transit time can be as short as several hours to as long as 50–60 hr. Transit along the colon is characterized by abrupt movement and long periods of stasis. In one study of 49 healthy subjects, the average colonic residence time was 35 hr, with the following times associated with different regions: 11 hr in the right (ascending) colon; 11 hr in the left (descending) colon; and 12 hr in the rectosigmoid colon [158]. The latter values do not appear to be influenced by particle size (i.e., pellets vs. tablet), but these times are highly variable and are shortened in response to ingestion of a laxative (average time for a 5-mm tablet in the ascending colon of 8.7 vs. 13.7 hr) [159,160]. Furthermore, the ingestion of food, which is known to increase colonic activity, does not appear to have a dramatic effect on the movement of dosage forms from the ileum into the colon, nor on the movement within the colon [161]. Any differences in colonic transit times as a function of age and gender are not clear at this time, owing to conflicting reports and investigation in small populations of subjects.

The colonic mucosal pH varies along the length of the colon: right colon, pH 7.1; transverse colon, pH 7.4; left colon, pH 7.5; sigmoid colon, pH, 7.4; rectum, pH 7.2. These values were determined in a group of 21 subjects (mean age 54 years), and they are somewhat higher than previous estimates (ca., pH 6.7 in the right colon) [162]. Those values contrast with the proximal small intestine with a pH of about 6.6 and the terminal ileum with a pH of about 7.4. This near-neutral pH in conjunction with low enzymatic activity has made the colon an interesting potential site for drug absorption. Indeed, there is active interest in delivery of drug dosage forms to the colon for site-specific absorption, especially for peptides [e.g., 78,163]. Characteristics of the colon that are thought to provide a good environment for drug absorption include a mild pH, little enzymatic activity, and long residence time. The disadvantage of the colon, however, include several considerations that substantially limit this area for providing good absorption: small surface area, relatively viscous fluidlike environment (which varies along the length of the colon), and the large colonies of bacteria. The latter factors would limit dissolution and contact with the absorbing surface membrane and may result in presystemic drug metabolism.

The intention of colon-specific drug delivery is to prevent the drug from being released from the dosage form (by coating or other release-controlling mechanism) until it reaches the distal end of the large intestine (i.e., the ileocecal valve). Drug release needs to be delayed for about 5 hr, but clearly this delay time will vary from patient to patient and will depend on a host of factors that may affect gastric emptying and intestinal transit (e.g., food or drugs). The dosage form should then release drug over the next 10–15 hr while in the colon. The results of studies that have examined colonic absorption are not that encouraging, although they do indicate that absorption does occur, but to a variable extent (depending on the drug). The hormone, calcitonin, provided an absolute bioavailability from the colon of less than 1%; however, no comparison to oral dosing was made [164]. The relative bioavailability of ranitidine solution from the cecum was about 15% of that following gastric or jejunal administration [165]. Benazepril's relative bioavailability following a colonic infusion was about 23% that of an oral solution [166]. Figure 12 illustrates the plasma concentration–time profiles for those two drugs. The long-lasting analog of vasopressin, (desmopressin; desamino-8-D-arginine vasopressin; dDAVP), a nonapeptide, had a relative bioavailability of about 17 and 21%, compared with duodenal and gastric (and jejunal) solution administration, respectively. Rectal administration provided absorption comparable with that from the colon [167]. Sumatriptan solution was absorbed from the cecum to an extent of about 23%, compared with an oral (and jejunal) dose [168]. In all cases, the rate of absorption is substantially slower than from the upper regions of the GIT. Furthermore, in some instances, the metabolite/parent drug concentration ratios change depending on the site of administration, which may reflect a number of causes (e.g., different extent of presystemic metabolism, differences in metabolite absorption). The latter needs to be a consideration for those compounds for which the metabolites are either pharmacologically active or toxic.

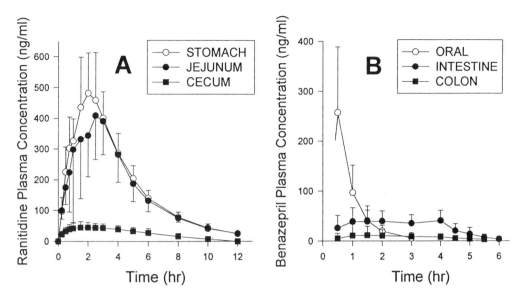

Fig. 12   (A) Ranitidine plasma concentrations as a function of time following administration of a solution into the stomach, jejunum, or cecum. Each value is the mean of eight subjects (the cross-hatched vertical bars are standard deviations). (B) Benazepril plasma concentrations as a function of time following a solution dose taken orally or administered as a 4-hr intestinal or 4-hr colonic infusion. Each value is the mean of 7–13 subjects. (Data from: A, Ref. 165; B, Ref. 166.)

A relevant consideration of absorption from the colonic area is rectal drug administration. Although this is not a frequently used route, it is employed to some extent, especially in infants, children, and those unable to swallow medication. Absorption from the rectum is generally considered to be relatively poor, at least in comparison with absorption from regions of the upper GIT. The reasons for this are essentially those outlined earlier for the colon; small absorbing surface area, little fluid content, and poor mixing movements. There are, in addition, two other considerations. First, the presence of fecal material may provide a site for adsorption that can effectively compete for absorption. Second, the extent of absorption will be dependent on the retention time of the dosage form in the rectum. This may be a critical issue for infants, who often have irregular bowel movements. The readers are referred to a review of this topic [169].

D.   Blood Flow

The entire GIT is highly vascularized and, therefore, well perfused by the bloodstream. The splanchnic circulation that perfuses the GIT receives about 28% of cardiac output, and this flow drains into the portal vein, and then, goes to the liver before reaching the systemic circulation. An absorbed drug will first go to the liver, which is the primary site of drug metabolism in the body; the drug may be metabolized extensively before systemic distribution. This has been referred to as the *first-pass* effect or presystemic elimination, and it has important implications in bioavailability and drug therapy.

The fact that the GIT is so well perfused by the bloodstream permits efficient delivery of absorbed materials to the body. As a result of this rapid blood perfusion, the blood at the site of absorption represents a virtual "sink" for absorbed material. Under normal conditions, then, there is never a buildup in drug concentration in the blood at the site of absorption. Therefore, the concentration gradient will favor further unidirectional transfer of drug from the gut to the blood. Usually, then, blood flow is not an important consideration in drug absorption. Generally, the properties of the dosage form (especially dissolution rate) or the compound's inherent absorbability will be the limiting factors in absorption.

There are circumstances, however, when blood flow to the GIT may influence drug absorption. Those compounds absorbed by active or specialized mechanisms require membrane participation in transport which, in turn, depends on the expenditure of metabolic energy by intestinal cells. If blood flow and, therefore, oxygen delivery is reduced, there may be a reduction in absorption of those compounds. This is the case in rats for the active absorption of phenylalanine [170].

The rate-limiting step in the absorption of those compounds that readily penetrate the intestinal membrane (i.e., have a large permeability coefficient) may be the rate at which blood perfuses the intestine. However, absorption will be independent of blood flow for those compounds that are poorly permeable. Extensive studies have illustrated this concept in rats [171,172]. The absorption rate of tritiated water, which is rapidly absorbed from the intestine, is dependent on intestinal blood flow; but a poorly absorbed compound, such as ribitol, penetrates the intestine at a rate independent of blood flow. In between these two extremes are a variety of intermediate compounds the absorption rate of which is dependent on blood flow at low-flow rates, but is independent of blood flow at higher flow rates. By altering blood flow to the intestine of the dog, as blood flow decreased, the rate of sulfaethidole absorption also decreased [173]. These relationships are illustrated in Fig. 13.

An interesting clinical example of the influence of blood flow on drug absorption is that provided by Rowland et al. [174]. After oral ingestion of aspirin, one subject fainted while a blood sample was being taken. Absorption ceased at that time, but continued when the subject recovered. Interestingly, there was no reduction in the total amount of aspirin absorbed, compared with another occasion when the subject did not faint. Another investigator observed a 3-hr delay in the absorption of sulfamethoxypyridazine in a patient who fainted [175]. The most reasonable explanation of these observations is that in a fainting episode blood is preferentially shunted away from the extremities and other body organs, including the GIT, thereby reducing blood perfusion of the tract and resulting in a decreased rate of absorption. It is

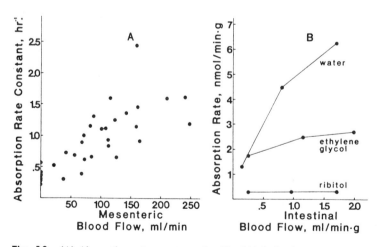

Fig. 13 (A) Absorption rate constant of sulfaethidole in dogs as a function of mesenteric blood flow. (B) Absorption rate of several compounds in rats as a function of intestinal blood flow. (Data from: A, Ref. 173; B, Ref. 172.)

possible that generalized hypotensive conditions may be associated with altered drug absorption. Therefore, consideration needs to be given to the influence of congestive heart failure and other disease conditions that will alter gut blood flow as well as the presence of other drugs that may alter flow. For example, it has been suggested that digoxin absorption is impaired in congestive heart failure, but improves after compensation [176]. The influence of such conditions on absorption has been reviewed elsewhere [177], but there is relatively little information available.

Blood flow to the GIT increases shortly after a meal and may last for several hours. Digestive processes, in general, seem to enhance blood flow to the tract. For the reasons discussed previously, however, coadministration of a drug with a meal would normally not be expected to improve drug absorption. Strenuous physical exercise appears to reduce blood flow to the tract and may reduce absorption.

V. COMPLICATING FACTORS IN DRUG ABSORPTION

A. Drug–Food and Drug–Drug Interactions

There are a variety of factors that may affect the rate or extent of absorption. Interactions in absorption are mediated by physicochemical or physiological factors. Physicochemical considerations include the characteristics of the dosage form and altered solubility, dissolution, and chemical stability within the GIT. Physiological factors include residence time in the tract (i.e., gastric emptying and intestinal transit rates) and blood flow to and the characteristics of the absorbing membrane. Drug–food and drug–drug interactions will alter absorption by one or more of the foregoing mechanisms.

As discussed previously, drug absorption is generally less efficient when food is present in the GIT, although there are several exceptions (see Table 8). Food will reduce the rate or extent of absorption by virtue of reduced gastric-emptying rate, which is particularly important for compounds unstable in gastric fluids and for dosage forms designed to release drug slowly. In addition, food provides a rather viscous environment that will reduce the rate of drug dissolution and drug diffusion to the absorbing membrane. Drugs may also bind to food particles or react with gastrointestinal fluids secreted in response to the presence of food. An interesting example of the influence of food and gastric emptying on the fate of a drug in the body is that of p-aminobenzoic acid. This compound is metabolized in the body (acetylation) by a saturable process. The rate of presentation (i.e., the rate of absorption) of the drug to its site of metabolism will influence the extent of acetylation. The more rapid the absorption, the less the extent of acetylation, since the capacity to metabolize is exceeded by the rate of presentation to the enzymatic system. When absorption rate is reduced by slowing gastric emptying (e.g., with fat or glucose), a greater fraction of the dose is metabolized [178]. Similar observations have been made for salicylamide's absorption rate and metabolism [179].

There are problems as well in the absorption of certain drugs in the presence of specific food components. L-Dopa absorption may be inhibited in the presence of amino acids formed from the digestion of proteins [61]. The absorption of tetracycline is reduced by calcium salts present in dairy foods and by several other cations, including magnesium and aluminum [180–182], which are often present in antacid preparations. In addition, iron and zinc reduce tetracycline absorption [183]. Figure 14 illustrates several of these interactions. It is thought that these materials react with tetracycline to form a water-insoluble and nonabsorbable complex. Obviously, these offending materials should not be coadministered with tetracycline antibiotics.

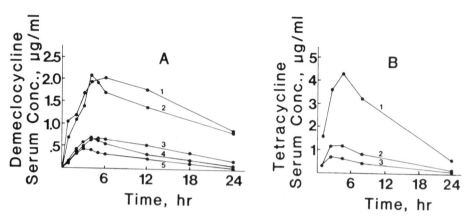

Fig. 14 (A) Demeclocycline serum concentrations as a function of time in four to six subjects after oral ingestion of demeclocycline in the absence or presence of dairy meals. Key: 1, meal (no dairy products); 2, water; 3, 110 g cottage cheese; 4, 240-ml buttermilk; 5, 240-ml whole milk. (B) Tetracycline serum concentrations as a function of time in six subjects after oral ingestion of tetracycline in the absence or presence of iron salts (equivalent to 40-mg elemental iron). Key: 1, control; 2, ferrous gluconate; 3, ferrous sulfate. (Data from: A, Ref. 182; B, Ref. 183.)

The tetracycline example just cited is one type of physicochemical interaction that may alter absorption. The relative influence of complexation on drug absorption will depend on the water-solubility of the drug, the water-solubility of the complex, and the magnitude of the interaction (i.e., the complexation stability constant). If the drug itself is poorly water-soluble, the absorption pattern will be governed by rate of dissolution. Often, such compounds are incompletely and erratically absorbed. As a result, complexation will probably exert more of an influence on the absorption of such a compound than on one that is normally well-absorbed, although this will depend on the nature of the complex. If the complex is water-insoluble, as with tetracycline interactions with various metal cations, the fraction complexed will be unavailable for absorption. Although most complexation interactions are reversible, the greater the stability constant of the complex, the greater the relative influence on absorption. Generally, however, because the interaction is reversible, complexation is more likely to influence the rate than the extent of absorption.

Drug complexation is sometimes used in preparing pharmaceutical dosage forms to improve stability or solubility, or to prolong drug release. There are several examples, however, for which a drug complex results in reduced absorption. Amphetamine interacts with sodium carboxymethyl cellulose to form a poorly water-soluble derivative and a decrease in absorption is seen [184]. Phenobarbital absorption is reduced as a result of interaction with polyethylene glycol 4000 [185]. These large macromolecules have the potential to bind many drugs.

Surface-active agents, because they are able to form micelles above the critical micelle concentration (CMC), may bind drugs either by inclusion within the micelle (solubilization) or by attachment to its surface. Below the critical micelle concentration, surfactant monomers have a membrane-disrupting effect that can enhance drug penetration across a membrane. The latter influence has been seen in drug absorption studies in animals. The influence of surface-active agents on drug absorption will depend on the surfactant concentration and the physicochemical characteristics of the drug. If the drug is capable of partitioning from the aqueous to the micellar phase, and if the micelle is not absorbed, the usual situation, there may be a reduction in rate of absorption. Micellar concentrations of sodium lauryl sulfate or polysorbate

80 (Tween 80) increase the rectal absorption rate of potassium iodide in the rat, but reduce the absorption rate of iodoform and triiodophenol [186,187]. Since potassium iodide is not solubilized by the micelle, the enhanced rate of absorption is attributed to the influence of the surfactant on the mucosal membrane. The other compounds, which partition into the micelle, exhibit a reduced rate of absorption, since there is a decrease in their effective concentration. Similar observations, from pharmacological response data in goldfish, have been made for several barbiturates in the presence of varying surfactant concentrations.

In addition to the aforementioned effects of surfactants, one must consider their influence on drug dissolution from pharmaceutical dosage forms. If the drug is poorly water-soluble, enhanced dissolution rate in the presence of a surface-active agent, even if part of the drug is solubilized, will result in increased drug absorption. The absorption rate of sulfisoxazole suspensions given rectally to rats increased with increasing polysorbate 80 concentration. At surfactant concentrations in excess of that needed to solubilize the drug completely, there was a reduced rate of absorption; however, the rate was greater than that from the control suspension (i.e., without surfactant) [188].

Another important type of physicochemical interaction that may alter absorption is that of drug binding or adsorption onto the surface of another material. As with complexation and micellarization, adsorption will reduce the effective concentration gradient between gut fluids and the bloodstream, which is the driving force for passive absorption. Although absorption frequently reduces the rate of absorption, the interaction is often readily reversible and will not affect the extent of absorption. A major exception is adsorption onto charcoal, which in many cases appears to be irreversible, at least during the time of residence within the GIT. As a result, charcoal often reduces the extent of drug absorption. Indeed, this, along with the innocuous nature of charcoal, is what makes it an ideal antidote for oral drug overdose. The effectiveness of that form of therapy will depend on the amount of charcoal administered and the time delay between overdose and charcoal dosing. Another interesting aspect of charcoal dosing is its influence on shortening the elimination half-life of several drugs. This is a particularly attractive noninvasive means of enhancing drug elimination from the body.

In addition to charcoal, adsorption is often seen with pharmaceutical preparations that contain large quantities of relatively water-insoluble components. A good example is antidiarrheal products and perhaps antacids. The importance of the strength of binding as it influences absorption has been illustrated by Sorby [189], who showed that both attapulgite and charcoal reduce the rate of drug absorption, but only charcoal reduced the extent of absorption. Lincomycin is an example of a drug for which absorption is impaired by an antidiarrheal preparation [190]. Another type of compound that has altered drug absorption by binding is the anion-exchange resins, cholestyramine and colestipol. The foregoing physicochemical interactions that may alter drug therapy may be minimized by not coadministering the interacting compounds at the same time, but separating their ingestion by several hours.

There are other drug–drug interactions in absorption that are mediated by alterations of gut physiology. The mechanism that has received the greatest attention is that associated with changes in gastric emptying and intestinal transit, as discussed previously and illustrated in Table 8. A review of interactions in absorption is available [191].

We noted previously, when discussing mechanisms of drug absorption, that certain calcium channel blockers enhance the gastrointestinal absorption of several aminopenicillin and aminocephalosporin derivatives [64,65]. This type of interaction, although perhaps not of practical clinical importance, is very intriguing in terms of promoting absorption efficiency, and it illustrates the useful (rather than deleterious) aspect of a drug–drug interaction. Another interesting and recently observed drug–food interaction is that between grapefruit juice and certain drugs, especially those with a high hepatic clearance that undergo substantial first-pass hepatic metabolism (discussed later). Grapefruit juice coadministration with felodipine results in a large increase in the bioavailability of the latter (ca., threefold) [192,193]. This effect is believed to be the result of the inhibition of the hepatic metabolism of the drug by components of the grapefruit juice (e.g., bioflavonoids). One such bioflavonoid, naringin, was directly tested by coadministration with felodipine, but it produced an effect smaller than that with grapefruit juice, suggesting that other factors contribute to the interaction [193]. Grapefruit juice also increases the absorption of similar compounds; nifedipine and nitrendipine [194,195]. Clearly, we need to be more aware of potential food interactions with drugs.

B. Metabolism

Drug metabolism may occur at various sites along the GIT, including within gut fluids, within the gut wall, and by microorganisms present in the low end of the tract. Several examples of enzymatic alteration of certain drugs in gut fluids have been noted previously. Gut fluids contain appreciable quantities of a variety of enzymes that are needed to accomplish digestion of food. An additional consideration, although not involving enzymatic action, is that of acid- or base-mediated drug breakdown. Numerous drugs are unstable in acidic media (e.g., erythromycin and penicillin) and, therefore, will degrade and provide lower effective doses, depending on the pH of the gastric fluid, solubility of the drug, and the residence time of the dosage form in the stomach. Chemical modification of the drug by, for example, salt or ester formation may provide a more stable derivative, the absorption of which will be influenced to a smaller degree by the aforenoted factors. Clorazepate is an interesting example of a prodrug that must first be acid-hydrolyzed to produce the active chemical form; hydrolysis in the gut fluids produces the active form, N-desmethyldiazepam. In this instance, unlike the examples cited earlier, acid hydrolysis is a prerequisite for absorption of the pharmacologically active form. As a result, pH of gastric fluids and gastric emptying time and variables that influence those factors are expected to affect the absorption profile of clorazapate. Greater concentrations of N-desmethyldiazepam are achieved at the lower gastric pH, which is consistent with the more rapid acid hydrolysis at acidic pH [196].

The mucosal cells lining the gut wall represent a major potential site of drug metabolism. The metabolic activity of this region has been studied by a variety of techniques, ranging from subcellular fractions, to tissue homogenates, to methods involving the whole living animal. Metabolic reactions include both phase I and II processes. It appears that the small intestine (duodenum and jejunum) has the greatest enzymatic activity, although most regions of the GIT can partake in metabolism. It is not a simple matter, especially in the whole animal, to distinguish between the sites of metabolism responsible for so-called presystemic elimination (or the first-pass effect). The latter refers to all processes of metabolism before the drug reaches the systemic circulation, which take place primarily in the gut and liver. It is this presystemic elimination that contributes to differences in drug effects as a function of route of administration and that may seriously compromise the clinical efficiency of certain drugs given orally. A thorough discussion of this topic is beyond the scope of this chapter and readers are referred to a review of gut wall metabolism of drugs [197]. Drugs that undergo, or that are suspected to undergo, metabolism in the gut wall include aspirin, acetaminophen, salicylamide, p-aminobenzoic acid, morphine, pentazocine, isoproterenol, L-dopa, lidocaine, and certain steroids. L-Dopa appears to be metabolized by decarboxylase enzymes present in the gastric mucosa, which, as discussed previously, suggests the importance of rapid gastric emptying to achieve maximal absorption of the unchanged compound [198]. Salicylamide and p-aminobenzoic acid are interesting examples because they illustrate another aspect of gut metabolism, that of saturation. In addition, factors affecting absorption rate will influence the fraction of the dose that reaches the systemic circulation in the form of intact drug. Figure 15 illustrates the relationship between the salicylamide plasma concentration–time curve (AUC) as a function of the oral dose of sodium salicylamide. Normally, that relationship is expected to be linear, and the line should go through the origin. The curvilinearity, especially at low doses, suggests some form of presystemic elimination that becomes saturated above a certain dose. Evidence in animals suggests that the drug is metabolized in the gut wall to sulfate and glucuronide conjugates, although metabolism in the liver also occurs on the first pass. Extrapolation of the straight-line segment to the x-axis in Fig. 15 gives an intercept that has been referred to as the *break-through* dose, which approximates the dose needed to saturate the enzyme system (about 1–1.5 g for salicylamide). Doses less than that value produce only small plasma concentrations of the unchanged drug. Another interesting aspect of this phenomenon is that the rate of drug presentation to the enzymatic system will influence the fraction of the dose reaching the systemic circulation unchanged and will alter the metabolic pattern. The latter factors, therefore, will be influenced by the dosage form characteristics and the rate of gastric emptying. The more rapid the rate of absorption, the more likely the enzymatic system will become saturated. This will result in greater plasma concentrations of unchanged drug and a metabolic pattern with a lower percentage of the drug recovered as the saturable metabolite. At any given dose, the more rapidly soluble sodium salicylamide produces greater plasma concentrations of unchanged drug than salicylamide. The more rapidly soluble dosage forms of salicylamide (solution and suspension) produce smaller fractions of the dose in the form of the saturable metabolite (sulfate conjugate) compared with a slowly dissolving tablet [200]. Observations similar to the latter point have been made for p-aminobenzoic acid. In that instance, however, a delay in gastric emptying produced a greater percentage of the dose in the form of the saturable metabolite (acetyl derivative), which is consistent with a reduced rate of absorption [201].

An interesting example of gut metabolism and gender-dependent differences is that of ethanol. Females appear to have higher blood ethanol concentrations following an oral dose than do males given the same dose. The latter is true even if the data are corrected for weight and lean body mass differences. Ethanol is metabolized by alcohol dehydrogenase present in the

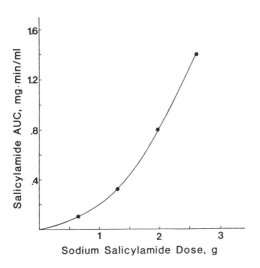

Fig. 15 Area under the salicylamide plasma concentration–time curve (AUC) as a function of the oral dose of sodium salicylamide. Each point is the average of five subjects. (Data from Ref. 199.)

gastric mucosa, and it appears that this enzyme is present in smaller quantities in females. This results in a greater fraction of the dose not being metabolized compared with males and a subsequent higher blood ethanol concentration. Females absorb about 91% of the dose as ethanol, compared with about 61% in males [202]. More recent evidence suggests that alcohol dehydrogenase activity is lower in young women (younger than about 50 years of age), elderly males, and in alcoholics [203]. A recent review has examined first-pass metabolism in oral absorption and factors that affect that process [204].

The gastrointestinal microflora is another site of drug metabolism within the GIT, and it has received some attention. In normal subjects, the stomach and proximal small intestine contain small numbers of microorganisms. Concentrations of these organisms increase toward the distal end of the intestine. A wide variety of aerobic and anaerobic organisms are present in the gut. The microflora, derived primarily from the environment, tend to adhere to the luminal surface of the intestine. Within an individual, the microflora tend to remain rather stable over long periods. The primary factors governing the numbers and kinds of microorganisms present in the tract include (a) the activity of gastric and bile secretions, which tend to limit the growth of these organisms in the stomach and upper part of the GIT; and (b) the propulsive motility of the intestine, which is responsible for continually cleansing the tract, thereby limiting the proliferation of microorganisms. Gastric atrophy permits increased numbers of microorganisms to pass into the small intestine, and reduced intestinal motility results in overgrowth.

A review of drug metabolism by intestinal microorganisms indicates that most studies have dealt with animals other than humans [205]. These studies indicate a wide range of primarily phase I metabolic pathways. Various drugs that are glucuronidated in the body are secreted into the intestine by the bile, and these are subject to cleavage by bacterial glucuronidase enzymes. The cleavage product may then be in a form available for absorption. Various drug conjugates may be similarly deconjugated by other bacterial enzymes (e.g., the glycine conjugate of isonicotinic acid). Although some drugs may be rendered inactive, bacterial metabolism of other drugs may give rise to more active or toxic products. The formation of the toxic compound cyclohexylamine from cyclamate is an example [206]. Salicylazosulfapyridine (sulfasalazine), which is used in treating ulcerative colitis, provides an interesting example of a drug for which its metabolites represent the active phamacological species. The parent drug is metabolized to 5-aminosalicylate and sulfapyridine. In conventional rats, both metabolites and their conjugates appear in urine and feces. In germ-free rats, however, the metabolites are not excreted. This suggests that the intestinal flora play a role in reducing the parent compound and formation of the two metabolites. If this is true, factors influencing the population and types of intestinal microorganisms may, in turn, influence the absorption and effectiveness of the drug. For example, concomitant antibiotic therapy, by reducing the population of microorganisms, may prevent the parent drug from being metabolized.

C. Disease States

Gastrointestinal disorders and disease states are likely to influence drug absorption. Although this important area has not been explored thoroughly, numerous studies have addressed this issue. One major concern in this area is that many of these studies have not been correctly designed. This has resulted in conflicting reports and in our inability to reach generally valid conclusions. The majority of these studies are conducted after administration of an oral dose, and the area under the plasma concentration–time curve (AUC) is measured. The latter parameter is frequently used in assessing bioavailability. The resulting AUC is compared with that from a control group of different subjects, or within the same subject, during the time the disorder is present, and compared with the value before or after the disorder is resolved. The problem here is that a value for AUC depends as much on the body's ability to clear or eliminate the drug as it does on absorption. Differences in the former parameter are likely to be present between subjects as well as within a subject from time to time (especially in the presence of a disease). Therefore, AUC values after oral dosing may lead to incorrect conclusions. To use such a value properly, one must be certain that drug clearance is not different between, or within, the subjects. In the ideal situation, an intravenous dose would be given to establish the correctness of that assumption. This is an approach, unfortunately, that is not generally used. A recent review of the influence of gastrointestinal disease on drug absorption has been published [207].

Elevated gastric pH is seen in subjects with achlorhydria as a result of reduced acid secretion. Aspirin appeared to be better absorbed in achlorhydric subjects than in normal subjects [208]. In contrast, the absorption of tetracycline, which is most soluble at acidic pH, appears to be unaffected by achlorhydria or after surgery in which the acid-secreting portion of the stomach was removed [209,210]. The absorption of clorazepate would be expected to be reduced in achlorhydria (for the reasons discussed earlier), but as yet, the data are not conclusive. The clinical significance of altered gut pH for drug absorption is not clearly established. Alterations in drug absorption caused by changes in gut pH will most likely be mediated by its influence on dissolution rate.

Changes in gastric emptying are expected to influence the rate and, possibly, the extent of absorption, for the reasons discussed previously. Emptying may be severely hampered and absorption altered soon after gastric surgery, or as a result of pyloric stenosis, or in the presence of various disease states. Riboflavin absorption is increased in hypothyroidism and reduced in hyperthyroidism, conditions that alter gastric emptying and intestinal transit rates [211]. There is the indication that absorption is impaired during a migraine attack, possibly as a result of reduced gastric-emptying rate, since metoclopramide administration increases the rate of drug absorption [212].

Diarrheal conditions may decrease drug absorption as a result of reduced intestinal residence time. The absorption of several drugs was decreased in response to lactose- and saline-induced diarrhea [213]. Digoxin absorption from tablets was impaired in one subject who developed chronic diarrhea as a result of x-ray treatment [214]. Abdominal radiation or the underlying disease reduces digoxin and clorazepate absorption [215]. A dosage form that provides rapid drug dissolution (e.g., solution) may partially resolve this problem.

There are various malabsorption syndromes known to influence the absorption of certain nutrients. Although not thoroughly investigated, such syndromes may exert an influence on the efficacy of drug absorption. Heizer et al. [216] have noted reduced absorption of digoxin in patients with sprue, with malabsorption syndrome, or with pancreatic insufficiency. The dosage form of digoxin, especially dissolution rate from tablets, will partially determine the influence of malabsorption states on absorption, the problem being compounded by poorly dissolving tablets. Phenoxymethyl penicillin absorption is reduced in patients with steatorrhea [217], and ampicillin and nalidixic acid absorption appears to be impaired in children with shigellosis [218].

There are a variety of other disease states for which influence on drug absorption has been reported, including cystic fibrosis, villous atrophy, celiac disease, diverticulosis, and Crohn's disease. The results of these studies are frequently divergent; therefore, general statements cannot be made. A thorough discussion of these findings is beyond the scope of this chapter, and the interested reader is referred to a recent review [207].

As most drugs are best absorbed from the small intestine, any surgical procedure that removes a substantial portion of the small intestine is likely to influence absorption; however, and as discussed previously, the characteristics of the dosage form may affect the findings. Although the procedure has fallen out of favor, intestinal bypass surgery has been used in treatment of the morbidly obese. A number of studies have been conducted to examine absorption before and after surgery. As noted in the introduction to this section, care must be exercised in study design and evaluation of data, as large weight loss may alter drug elimination from the body compared with the presurgery condition. Further complications include the time that the study is conducted relative to the time of surgery, and the length and sections of the intestine removed.

One excellent study [219] employed intravenous and oral dosing at each of several times after surgery (1–2 weeks, 6 and 12 months). This design permits valid conclusions about the absorption process. There was a significant reduction after surgery in ampicillin absorption, but no change in propylthiouracil absorption. Other studies suggest reduced absorption of hydrochlorothiazide and phenytoin [220,221], but these findings must be qualified by the aforementioned concerns.

In those instances during which a patient's response to a drug is less than expected and there is reason to believe that this is a result of impaired absorption owing to any of the pathological conditions or disease states cited earlier, a first attempt in seeking to improve drug therapy is to optimize absorption from the GIT. To do this, a practical approach might well be to administer the drug in a form readily available for absorption. Usually, if such a form is marketed or easily prepared, administration of a drug in solution will represent the best way to achieve maximal absorption, as this will eliminate the time for drug dissolution in the gut needed by solid oral dosage forms. When absorption cannot be sufficiently improved by use of a drug solution, alternative routes of administration must be considered (e.g., intramuscular).

D. Age

Most of the information discussed to this point and most of the literature concerned with drug absorption involve studies performed in young, healthy (usually male) adults. In contrast, there is considerably less information concerning absorption in subjects at either end of the age spectrum (i.e., pediatric and geriatric populations). For a variety of reasons, one would expect the absorption process in the latter groups to be different from that in young adults; unfortunately, there is as yet little information to present valid general statements.

The pediatric population (neonates, infants, and children) presents a particularly difficult group in which to conduct clinical experimentation because of ethical considerations. A further complication is the rapid development of organ function, which is likely to influence results, even over a relatively short experimental period (e.g, 2–4 weeks), especially in neonates and infants. An additional consideration in the latter groups is whether the neonate is premature or full-term. Most often, plasma AUC data are obtained after an oral dose for the purpose of estimating elimination half-life or to provide a basis for the development of a multiple-dosing regimen. Such data provide very limited information about rate or extent of absorption. Indeed, most reviews of drug disposition in the pediatric population indicate the lack of rigorous information on this topic.

Gastric fluid is less acidic in the newborn than in adults, since acid secretion is related to the development of the gastric mucosa. This condition appears to last for some time, as pH values similar to the adult are not reached until after about 2 years. The higher gastric fluid pH, along with a smaller gut fluid volume, may influence dissolution rate and the stability of acid-unstable drugs. The gastric-emptying rate appears to be slow, approaching adult values after about 6 months. An interesting example in support of that suggestion is a study that examined riboflavin absorption in a 5-day-old neonate and a 10-month-old infant [222]. The maximum urinary excretion rate was considerably greater in the infant, whereas excretion rate in the neonate was constant and prolonged. This suggests more rapid absorption in the infant, whereas absorption in the neonate proceeds for a longer time. For the reasons discussed previously, these data suggest slower emptying or intestinal transit rates in the neonate (recall that riboflavin is absorbed by a specialized process high in the small intestine). Intestinal transit tends to be irregular and may be modified by the type of food ingested and the feeding pattern.

Intestinal surface area and total blood flow to the GIT are smaller than in adults and may influence the efficiency of absorption. Relative to the use of rectal suppositories, one needs to remember that the completeness of absorption will be a function of retention time in the rectum. Since bowel movements in the young are likely to be irregular, the retention time may limit the efficiency of absorption by that route. In light of the little information available about absorption in the young, it would seem reasonable to attempt to optimize absorption by using solution, rather than solid, dosage forms.

Only in recent years has there been any substantial progress made in better understanding drug disposition in the elderly. Active research programs in gerontology have begun to provide more information about rational drug dosing in this population. There are several important and unique characteristics of the elderly that make a compelling argument for the need of such information (e.g., they ingest more drugs per capita, their percentage of the population is increasing, and they suffer from more disease and physical impairments). As noted for the pediatric population, there are a variety of complex issues associated with the conduct of research in the elderly. Careful consideration must be given to experimental design and data analysis. Some considerations include the appropriate definition of age, cross-sectional *versus* longitudinal study design, and health status of the subject [223].

There have been numerous statements in the literature to the effect that GI absorption in the elderly is impaired and less efficient than in young adults. Although there have been few data to support the suggestion, one basis for that statement has been the results obtained from the application of the so-called xylose tolerance test, which is often used in assessing malabsorption. This conclusion of impaired absorption in the elderly presents a good example of the need for careful study design and appropriate pharmacokinetic analysis of data. Figure 16 illustrates the results of several studies that have examined xylose absorption [224]. Most

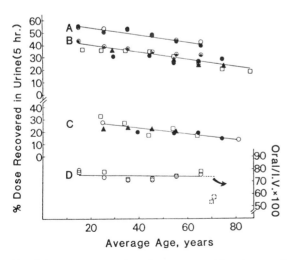

Fig. 16   Percentage xylose dose recovered in urine as a function of age after (A) a 5-g intravenous dose; (B) 5-g oral dose; and (C) 25-g oral dose. Line D is the ratio of urinary recoveries (oral to intravenous) after 5-g doses (*y*-axis on right). Symbols represent data obtained from different studies. (From Ref. 224.)

studies indicate an inverse relation between urinary xylose recovery and age after an oral dose (lines B and C). It is this observation that has suggested reduced absorption with age. However, the same inverse relation is found after an intravenous dose (line A), an observation that cannot be explained by impaired absorption, but rather, by reduced renal clearance of xylose. Line D shows the ratio of urinary recovery (oral to intravenous), which suggests that absorption is not altered with age. A more recent study in which each subject (age range, 32–85 years) received both an oral and an intravenous dose indicates no relation between xylose bioavailability and age [225].

There are substantial changes in a variety of physiological functions in the elderly that may influence drug absorption [223], including a greater incidence of achlorhydria, altered gastric emptying, reduced gut blood flow, and smaller intestinal surface area. One recent example indicates that gastric pH may be an important determinant of drug absorption in the elderly [226]. Dipyridamole is a poorly water-soluble weak base, the dissolution of which would be optimal in an acidic environment. Elevated gastric pH caused by achlorhydria (a condition that is more prevalent in the elderly than in the young) results in impaired absorption of dipyridamole. The ingestion of glutamic acid by achlorhydric subjects improves absorption. There are, in addition, other factors that may influence absorption, such as a greater incidence of GI disease, altered nutritional intake and eating habits, and ingestion of drugs that may affect the absorption of other drugs. Although data are still somewhat limited, the general impression is that the rate of absorption is frequently reduced, whereas there is little if any change in the extent of absorption [223]. This is a tentative statement that needs to be qualified for the specific drug and for the health status of the subject. For example, the absorption of drugs that undergo hepatic first-pass metabolism may be improved in the elderly (e.g., propranolol) as a consequence of reduced hepatic clearance with age.

REFERENCES

1. J. D. Verhoff, H. E. Bodde, A. G. deBoer, J. A. Bouwstra, H. E. Junginger, F. W. H. M. Merkus, and D. D. Breimer, Transport of peptide and protein drugs across biological membranes, Eur. J. Drug Metab. Pharmacokin. 15, 83–93 (1990).
2. J. W. Hole, *Human Anatomy and Physiology*, 2nd Ed., Wm. C. Brown Co., Dubuque, IA, 1981, p. 414.
3. H. Leonhardt, *Color Atlas/Text of Human Anatomy*, Vol. 2, 4th Ed., Thieme Medical Publishers, New York, 1993, p. 213.
4. T. Fujita, K. Tanaka, and J. Tokunaga, *SEM Atlas of Cells and Tissues*, Igaku-Shoin, New York, 1981, pp. 122, 123, 129.
5. J. W. Hole, *Human Anatomy and Physiology*, 2nd Ed., Wm. C. Brown Co., Dubuque, IA, 1981, p. 439.
6. J. S. Trier, Morphology of the epithelium of the small intestine, in *Handbook of Physiology*, Section 6, *Alimentary Canal*, Vol. 3, *Intestinal Absorption* (C. F. Code, ed.), American Physiological Society, Washington, DC, 1968, p. 1133.
7. T. Fujita, K. Tanaka, and J. Tokunaga, *SEM Atlas of Cells and Tissues*, Igaku-Shoin, New York, 1981, p. 135.
8. G. J. Tortora and N. P. Anagnostakos, *Principles of Anatomy and Physiology*, 6th Ed., Harper & Row, New York, 1990, p. 770.
9. J. S. Fordtran, F. C. Rector, Jr., M. F. Ewton, N. Soter, and J. Kinney, Permeability characteristics of the human small intestine, J. Clin. Invest., 44 1935–1944 (1965).
10. J. F. Danielli and H. Davson, A contribution to the theory of permeability of thin films, J. Cell. Comp. Physiol., 5, 495–508 (1935).
11. H. Davson and J. F. Danielli, *The Permeability of Natural Membranes*, 2nd Ed., Cambridge University Press, New York, 1952.

12. L. C. Junqueira, J. Carneiro, and R. O. Kelley, *Basic Histology*, 7th Ed., Appleton & Lange, East Norwalk, CT, 1992, p. 29.

13. R. Hober and J. Hober, Experiments on the absorption of organic solutes in the small intestine of rats, J. Cell. Comp. Physiol., 10, 401–422 (1937).

14. L. S. Schanker, Absorption of drugs from the rat colon, J. Pharmacol. Exp. Ther., 126, 283–290 (1959).

15. E. M. Wright and J. M. Diamond, Patterns of non-electrolyte permeability, Proc. Royal Soc. B, 172, 203–225 (1969).

16. E. M. Wright and J. M. Diamond, Patterns of non-electrolyte permeability, Proc. Royal Soc. B, 172, 227–271 (1969).

17. J. M. Diamond and E. M. Wright, Molecular forces governing nonelectrolyte permeation through cell membranes, Proc. Royal Soc. B, 172, 273–316 (1969).

18. E. J. Lien, Structure–activity relationships and drug disposition, Annu. Rev. Toxicol., 21, 31–61 (1981).

19. Y. Yoshimura and N. Kakeya, Structure–gastrointestinal absorption relationship of penicillins, Int. J. Pharm., 17, 47–57 (1983).

20. T. Sollmann, The comparative efficiency of local anesthetics, JAMA, 70, 216–219 (1918).

21. J. M. Faulkner, Nicotine poisoning by absorption through the skin, JAMA, 100, 1664–1665 (1933).

22. J. Travell, The influence of the hydrogen ion concentration on the absorption of alkaloids from the stomach, J. Pharmacol. Exp. Ther., 69, 21–33 (1940).

23. M. H. Jacobs, Some aspects of cell permeability to weak electrolytes, Cold Spring Harbor Symp. Quant. Biol., 8, 30–39 (1940).

24. P. A. Shore, B. B. Brodie, and C. A. M. Hogben, The gastric secretion of drugs: A pH partition hypothesis, J. Pharmacol. Exp. Ther., 119, 361–369 (1957).

25. L. S. Schanker, P. A. Shore, B. B. Brodie, and C. A. M. Hogben, Absorption of drugs from the stomach. I. The rat, J. Pharmacol. Exp. Ther., 120, 528–539 (1957).

26. C. A. M. Hogben, L. A. Schanker, D. J. Tocco, and B. B. Brodie, Absorption of drugs from the stomach. II. The human, J. Pharmacol. Exp. Ther., 120, 540–545 (1957).

27. L. S. Schanker, D. J. Tocco, B. B. Brodie, and C. A. M. Hogben, Absorption of drugs from the rat small intestine, J. Pharmacol. Exp. Ther., 123, 81–87 (1958).

28. C. A. M. Hogben, D. J. Tocco, B. B. Brodie, and L. S. Schanker, On the mechanism of intestinal absorption of drugs, J. Pharmacol. Exp. Ther., 125, 275–282 (1959).

29. J. T. Doluisio, N. F. Billups, L. W. Dittert, E. T. Sugita, and J. V. Swintosky, Drug absorption. I. An in situ rat gut technique yielding realistic absorption rates, J. Pharm. Sci., 58, 1196–1200 (1969).

30. L. Z. Benet, Biopharmaceutics as a basis for the design of drug products, in *Drug Design*, Vol. 4 (E. J. Ariens, ed.), Academic Press, New York, 1973, pp. 26–28.

31. W. G. Crouthamel, G. H. Tan, L. W. Dittert, and J. T. Doluisio, Drug absorption. IV. Influence of pH on absorption kinetics of weakly acidic drugs, J. Pharm. Sci., 60, 1160–1163 (1971).

32. A. J. Aguiar and R. J. Fifelski, Effect of pH on the in vitro absorption of flufenamic acid, J. Pharm. Sci., 55, 1387–1391 (1966).

33. K. Kakemi, T. Arita, R. Hori, and R. Konishi, Absorption and excretion of drugs. XXX. Absorption of barbituric acid derivatives from rat stomach, Chem. Pharm. Bull., 15, 1534–1539 (1967).

34. D. Winne, Shift of pH-absorption curves, J. Pharmacokinet. Biopharm., 5, 53–94 (1977).

35. M. J. Jackson and J. A. Dudek, Epithelial transport of weak electrolytes, Fed. Proc., 38, 2043–2047 (1979).

36. C.-Y. Tai and M. J. Jackson, Transport of weak bases across rat gastric mucosa in vivo and in vitro, J. Pharmacol. Exp. Ther., 222, 372–378 (1982).

37. M. L. Hogerle and D. Winne, Drug absorption by the rat jejunum perfused in situ: dissociation from the pH-partition theory and role of microclimate–pH and unstirred layer, Arch. Pharmacol., 322, 249–255 (1983).

38. J. G. Wagner and A. J. Sedman, Quantitation of rate of gastrointestinal and buccal absorption of acidic and basic drugs based on extraction theory, J. Pharmacokinet. Biopharm., 1, 23–50 (1973).

39. M. Gibaldi, Limitations of classical theories of drug absorption, in *Drug Absorption: Proceedings of the Edinburgh International Conference* (L. F. Prescott and W. S. Nimmo, eds.), Adis Press, New York, 1981, pp. 1–5.

40. D. S. Riggs, *The Mathematical Approach to Physiological Problems*, Williams & Wilkins, Baltimore, 1963, pp. 181–185.

41. J. G. Wagner and E. Nelson, Per cent absorbed time plots derived from blood level and/or urinary excretion data, J. Pharm. Sci., 52, 610–611 (1963).

42. J. C. K. Loo and S. Riegelman, New method for calculating the intrinsic absorption rate of drugs, J. Pharm. Sci., 57, 918–928 (1968).

43. D. Perrier and M. Gibaldi, Calculation of absorption rate constants for drugs with incomplete availability, J. Pharm. Sci., 62, 225–228 (1973).

44. G. L. Flynn, S. H. Yalkowsky, and T. J. Roseman, Mass Transport phenomena and models: theoretical concepts, J. Pharm. Sci., 63, 479–510 (1974).

45. G. L. Flynn and S. H. Yalkowsky, Correlation and prediction of mass transport across membranes. I. Influence of alkyl chain length on flux-determining properties of barrier and diffusant, J. Pharm. Sci., 61, 838–852 (1972).

46. N. F. H. Ho, W. I. Higuchi, and J. Turi, Theoretical model studies of drug absorption and transport in the GI tract. III, J. Pharm. Sci., 61, 192–197 (1972).

47. A. Suzuki, W. I. Higuchi, and N. F. H. Ho, Theoretical model studies of drug absorption and transport in the gastrointestinal tract. I, J. Pharm. Sci., 59, 644–651 (1970).

48. A. Suzuki, W. I. Higuchi, and N. F. H. Ho, Theoretical model studies of drug absorption and transport in the gastrointestinal tract. II, J. Pharm. Sci., 59, 651–659 (1970).

49. W. I. Higuchi, N. F. H. Ho, J. Y. Park, and I. Komiya, Rate-limiting steps and factors in drug absorption, in *Drug Absorption: Proceeding of the Edinburgh International Conference* (L. F. Prescott and W. S. Nimmo, eds.), Adis Press, New York, 1981, pp. 35–60.

50. G. L. Amidon, J. Kou, R. L. Elliott, and E. N. Lightfoot, Analysis of models for determining intestinal wall permeabilities, J. Pharm. Sci., 69, 1369–1373 (1980).

51. M. Mayersohn, Drug absorption, J. Clin. Pharmacol., 27, 634–638 (1987).

52. M. L. Henderson, A. L. Picchioni, and L. Chin, Evaluation of oral dilution as a first aid measure in poisoning, J. Pharm. Sci., 55, 1311–1313 (1966).

53. L. S. Schanker and J. J. Jeffrey, Active transport of foreign pyrimidines across the intestinal epithelium, Nature, 190, 727–728 (1961).

54. D. F. Evered and H. G. Randall, The absorption of amino acid derivatives of nitrogen mustard from rat intestine in vitro, Biochem. Pharmacol., 11, 371–376 (1962).

55. G. Levy and W. J. Jusko, Factors affecting the absorption of riboflavin in man, J. Pharm. Sci., 55, 285–289 (1966).

56. M. Mayersohn, Ascorbic acid absorption in man—pharmacokinetic implications, Eur. J. Pharmacol., 19, 140–142 (1972).

57. J. R. Bianchine, L. R. Calimlim, J. P. Morgan, C. A. Dujovne, and L. Lasagna, Metabolism and absorption of L-3,4-dihydroxyphenylalanine in patients with Parkinson's disease, Ann. N.Y. Acad. Sci., 179, 126–140 (1971).

58. N. G. Gillespie, I. Mena, G. C. Cotzias, and M. A. Bell, Diets affecting treatment of parkinsonism with levodopa, J. Am. Diet. Assoc., 62, 525–528 (1973).

59. G. Wiseman, *Absorption from the Intestine*, Academic Press, New York, 1964, p. 60.

60. D. N. Wade, P. T. Mearrick, and J. L. Morris, Active transport of L-dopa in the intestine, Nature, 242, 463–465 (1973).

61. M. H. Van Woert, Phenylalanine and tyrosine metabolism in Parkinson's disease treated with levodopa, Clin. Pharmacol. Ther., 12, 368–375 (1971).

62. H. Lennernas, D. Nilsson, S.-M. Aquilonius, O. Ahrenstedt, L. Knutson, and L. K. Paalzow, The effect of L-leucine on the absorption of levodopa, studied by regional jejunal perfusion in man, Br. J. Clin. Pharmacol., 35, 243–250 (1993).

63. J. Zimmerman, Methotrexate transport in the human intestine, Biochem. Pharmacol., 43, 2377–2383 (1992).

64. J.-F. Westphal, J.-H. Trouvin, A. Deslandes, and C. Carbon, Nifedipine enhances amoxicillin absorption kinetics and bioavailability in humans, J. Pharmacol. Exp. Ther., 255, 312–317 (1990).

65. C. Duverne, A. Bouten, A. Deslandes, J.-F. Westphal, J.-H. Trouvin, R. Farinotti, and C. Carbon, Modification of cifixime bioavailability by nifedipine in humans: involvement of the dipeptide carrier system, Antimicrob. Agents Chemother., 36, 2462–2467 (1992).

66. B. H. Stewart, A. R. Kugler, P. R. Thompson, and H. N. Bockbrader, A saturable transport mechanism in the intestinal absorption of gabapentin is the underlying cause of the lack of proportionality between increasing dose and drug levels in plasma, Pharmaceut. Res., 10, 276–281 (1993).

67. J. Hunter, B. H. Hirst, and N. L. Simmons, Drug absorption limited by P-glycoprotein-mediated secretory drug transport in human intestinal epithelial Caco-2 cell layers, Pharm. Res., 10, 743–749 (1992).

68. J. B. Dressman, R. R. Berardi, L. C. Dermentzoglau, T. L. Russell, S. P. Schmeltz, J. L. Barnett, and K. M. Jarvenpao, Upper gastrointestinal (GI) pH in young, healthy men and women, Pharm. Res., 7, 756–761 (1990).

69. R. A. O'Reilly, E. Nelson, and G. Levy, Physicochemical and physiologic factors affecting the absorption of warfarin in man, J. Pharm. Sci., 55, 435–437 (1966).

70. C. B. Tuttle, M. Mayersohn, and G. C. Walker, Biological availability and urinary excretion kinetics of oral tolbutamide formulations in man, Can. J. Pharm. Sci., 8, 31–36 (1973).

71. T. R. Bates, J. M. Young, C. M. Wu, and H. A. Rosenberg, pH-dependent dissolution rate of nitrofurantoin from commercial suspensions, tablets and capsules, J. Pharm. Sci., 63, 643–645 (1974).

72. W. I. Higuchl and W. E. Hamlin, Release of drug from a self-coating surface–benzphetamine pamoate pellet, J. Pharm. Sci., 52, 575–579 (1963).

73. G. Levy and B. A. Sahli, Comparison of the gastrointestinal absorption of aluminum acetylsalicylate and acetylsalicylic acid in man, J. Pharm. Sci., 51, 58–62 (1962).

74. G. Levy and J. A. Procknal, Unusual dissolution behavior due to film formation, J. Pharm. Sci., 51, 294 (1962).

75. C. E. Blezek, J. L. Lach, and J. K. Guillory, Some dissolution aspects of ferrous sulfate tablets, Am. J. Hosp. Pharm., 27, 533–539 (1970).

76. R. G. Crounse, Human pharmacology of griseofulvin. The effect of fat intake on gastrointestinal absorption, J. Invest. Dermatol., 37, 529–533 (1961).

77. M. Kraml, J. Dubuc, and D. Beall, Gastrointestinal absorption of griseofulvin. I. Effect of particle size, addition of surfactants and corn oil on the level of griseofulvin in the serum of rats, Can. J. Biochem. Physiol., 40, 1449–1451 (1962).

78. A. J. Moes, Gastroretentive dosage forms, Crit. Rev. Ther. Drug Carrier Syst., 10, 143–195 (1993).

79. H. Minami and R. W. McCallum, The physiology and pathophysiology of gastric emptying in humans, Gastroenterology, 86, 1592–1610 (1984).

80. G. Levy, Effect of certain tablet formulation factors on dissolution rate of the active ingredient. I, J. Pharm. Sci., 52, 1039–1046 (1963).

81. A. G. Renwick, C. H. Ahsan, V. F. Challenor, R. Daniels, B. S. MacKlin, D. G. Waller, and C. F. George, The influence of posture on the pharmacokinetics of orally administered nifedipine, Br. J. Clin. Pharmacol., 34, 332–336 (1992).

82. A. Hopkins, The pattern of gastric emptying: A new view of old results, J. Physiol., 182, 144–149 (1966).

83. J. D. Elashoff, T. J. Reedy, and J. H. Meyer, Analysis of gastric emptying data, Gastroenterology, 83, 1306–1312 (1982).

84. R. H. Blythe, G. M. Grass, and D. R. MacDonnell, The formulation and evaluation of enteric coated aspirin tablets, Am. J. Pharm., 131, 206–216 (1959).

85. M. Alpsten, C. Bogentoft, G. Ekenved, and L. Solveil, Gastric emptying and absorption of acetylsalicylic acid administered as enteric-coated micro-granules, Eur. J. Clin. Pharmacol., 22, 57–61 (1982).

86. P. G. Welling, Influence of food and diet on gastrointestinal drug absorption: A review, J. Pharmacokinet. Biopharm., 5, 291–334 (1977).

87. A. Melander, Influence of food on the bioavailability of drugs, Clin. Pharmacokinet., 3, 337–351 (1978).

88. R. D. Toothaker and P. G. Welling, The effect of food on drug bioavailability, Annu. Rev. Pharmacol. Toxicol., 20, 173–199 (1980).

89. L. Williams, J. A. Davis, and D. T. Lowenthal, The influence of food on the absorption and metabolism of drugs, Med. Clin. North Am., 77, 815–829 (1993).

90. G. L. Mattok and I. J. McGilveray, The effect of food intake and sleep on the absorption of acetaminophen, Rev. Can. Biol., 32 (Suppl.), 77–84 (1973).

91. J. M. Jaffe, J. L. Colaizzi, and H. Barry, Effect of dietary components on GI absorption of acetaminophen tablets in man, J. Pharm. Sci., 60, 1646–1650 (1971).

92. R. C. Heading, J. Nimmo, L. F. Prescott, and P. Tothill, The dependence of paracetamol absorption on the rate of gastric emptying, Br. J. Pharmacol., 47, 415–421 (1973).

93. S. Yung, M. Mayersohn, and J. B. Robinson, Ascorbic acid absorption in man: Influence of divided dose and food, Life Sci., 28, 2505–2511 (1981).

94. J. H. Wood, Effect of food on aspirin absorption, Lancet, 2, 212 (1967).

95. G. N. Volans, Effects of food and exercise on the absorption of effervescent aspirin, Br. J. Clin. Pharmacol., 1, 137–141 (1974).

96. C. T. Dollery, D. Emslie-Smith, and J. McMichael, Bretylium tosylate in the treatment of hypertension, Lancet, 1, 296–299 (1960).

97. R. Mantyla, P. T. Mannisto, A. Vuorela, S. Sundberg, and P. Ottoila, Impairment of captopril bioavailability by concomitant food and antacid intake, Int. J. Clin. Pharmacol. Ther. Toxicol., 22, 626–629 (1984).

98. S. M. Singhvi, D. N. McKinstry, J. M. Shaw, D. A. Willard, and B. H. Migdalef, Effect of food on the bioavailability of captopril in healthy subjects, J. Clin. Pharmacol., 22, 135–140 (1982).

99. P. C. Johnson, G. A. Braun, and W. A. Cressman, Nonfasting state and the absorption of a hypnotic, Arch. Intern. Med., 131, 199–201 (1973).

100. P. E. O. Williams and S. M. Harding, The absolute bioavailability of oral cefuroxime axetil in male and female volunteers after fasting and after food, J. Antimicrob. Chemother., 13, 191–196 (1984).

101. C. Harvengt, P. Deschepper, F. Lamy, and J. Hansen, Cephradine absorption and excretion in fasting and nonfasting volunteers, J. Clin. Pharmacol., 13, 36–40 (1973).

102. T. W. Mischler, A. A. Sugerman, D. A. Willard, L. J. Brannick, and E. S. Neiss, Influence of probenecid and food on the bioavailability of cephradine in normal male subjects, J. Clin. Pharmacol., 14, 604–611 (1974).

103. G. Houin, R. A. Shastri, J. Barre, B. Pinchon, and J. P. Tillement, Influence of food on the absorption of the p-chlorophenolic ester of chlorophenoxyisobutyric acid in man, Br. J. Clin. Pharmacol., 17, 341–345 (1984).

104. R. M. DeHann, W. D. Vanden Bosch, and C. M. Metzler, Clindamycin serum concentrations after administration of clindamycin palmitate with food, J. Clin. Pharmacol. J. New Drugs, 12, 205–211 (1972).

105. M. Divoll, D. J. Greenblatt, D. A. Ciraulo, S. K. Puri, I. Ho, and R. I. Shader, Clobazam kinetics: Intrasubject variability and effect of food on absorption, J. Clin. Pharmacol., 22, 69–73 (1982).

106. R. J. White, D. A. Chamberlain, M. Howard, and T. W. Smith, Plasma concentrations of digoxin after oral administration in the fasting and post prandial state, Br. Med. J., 1, 380–381 (1971).

107. N. Sanchez, L. B. Sheiner, H. Halkin, and K. L. Melmon, Pharmacokinetics of digoxin: interpreting bioavailability, Br. Med. J., 4, 132–134 (1973).

108. D. J. Greenblatt, D. W. Duhme, J. Koch-Weser, and T. W. Smith, Bioavailability of digoxin tablets and elixir in the fasting and postprandial states, Clin. Pharmacol. Ther., 16, 444–448 (1974).

109. J. R. Bianchine and L. Sunyapridakul, Interactions between levodopa and other drugs: Significance in the treatment of Parkinson's disease, Drugs, 6, 364–388 (1973).

110. Y.-J. Lin, D. J. Weidler, D. C. Garg, and J. G. Wagner, Effects of solid food on blood levels of alcohol in man, Res. Commun. Chem. Pathol. Pharmacol., 13, 713–722 (1976).

111. S. M. Chernish, A. Rubin, B. E. Rodda, A. S. Ridolfa, and C. M. Gruber, Jr., The physiological disposition of fenoprofen in man. IV. The effects of position of subject, food ingestion and antacid ingestion on plasma levels of orally administered fenoprofen, J. Med., 3, 249–257 (1972).

112. D. R. Doose, F. L. Minn, S. Stellar, and R. K. Nayak. Effect of meals and meal composition on the bioavailability of fenretinide, J. Clin. Pharmacol., 32, 1089–1095 (1992).

113. H. W. Emori, H. Paulus, R. Bluestone, G. D. Champion, and C. Pearson, Indomethacin serum concentrations in man; effect of dosage, food, and antacid, Ann. Rheum. Dis., 35, 333–338 (1976).

114. A. Melander, K. Danielson, A. Hanson, L. Jansson, C. Rerup, B. Schersten, T. Thulin, and E. Wahlin, Reduction of isoniazid bioavailability in normal men by concomitant intake of food, Acta Med. Scand., 200, 93–97 (1976).

115. W. A. Colburn, D. M. Gibson, R. E. Wiens, and J. J. Hanigan, Food increases the bioavailability of isotretinoin, J. Clin. Pharmacol., 23, 534–539 (1983).

116. P. T. Mannisto, R. Mantyla, S. Nykanen, U. Lamminsivu, and P. Ottoila, Impairing effect of food on ketoconazole absorption, Antimicrob. Agents Chemother., 21, 730–733 (1982).

117. T. K. Daneshmend, D. W. Warnock, M. D. Ene, E. M. Johnson, M. R. Potten, M. D. Richardson, and P. J. Williamson, Influence of food on the pharmacokinetics of ketoconazole, Antimicrob. Agents Chemother., 25, 1–3 (1984).

118. C. E. McCall, N. H. Steigbigel, and M. Finland, Lincomycin: activity in vitro and absorption and excretion in normal young men, Am. J. Med. Sci., 254, 144–155 (1967).

119. K. Tsutsumi, H. Nakashima, T. Kotegawa, and S. Nakeno, Influence of food on the absorption of beta-methyldigoxin, J. Clin. Pharmacol., 32, 157–162 (1992).

120. O. Stenbaek, E. Myhre, H. E. Rugstad, E. Arnold, and T. Hansen, The absorption and excretion of methyldopa ingested concomitantly with amino acids or food rich in protein, Acta Pharmacol. Toxicol., 50, 225–229 (1982).

121. T. R. Bates, J. A. Sequeira, and A. V. Tembo, Effect of food on nitrofurantoin absorption, Clin. Pharmacol. Ther., 16, 63–68 (1974).

122. H. A. Rosenberg and T. R. Bates, The influence of food on nitrofurantoin bioavailability, Clin. Pharmacol. Ther., 20, 227–232 (1976).

123. J. C. K. Loo, I. J. McGilveray, K. Midha, and R. Brien, The effect of an antacid and milk on the oral absorption of phenylbutazone tablets, Can. J. Pharm. Sci., 12, 10–11 (1977).

124. A. Melander, E. Wahlin, K. Danielson, and A. Hanson, Bioavailability of propylthiouracil: Interindividual variation and influence of food intake, Acta Med. Scand., 201, 41–44 (1977).

125. H. Y. Pan, A. R. De Vault, D. Brescia, D. A. Willard, M. E. McGovern, D. B. Whigan, and E. Ivashkiv, Effect of food on pravastatin pharmacokinetics and pharmacodynamics, Int. J. Clin. Pharmacol. Ther. Toxicol., 31, 291–294 (1993).

126. D. I. Siegler, D. M. Burley, M. Bryant, K. M. Citron, and S. M. Standen, Effect of meals on rifampicin absorption, Lancet, 2, 197–198 (1974).

127. K. Polasa and K. Krishnaswamy, Effect of food on bioavailability of rifampicin, J. Clin. Pharmacol., 23, 433–437 (1983).

128. A. Melander, E. Wahlin, K. Danielson, and C. Rerup, On the influence of concomitant food intake on sulfonamide bioavailability, Acta Med. Scand., 200, 497–500 (1976).

129. J. D. Unadkat, A. C. Collier, S. S. Crosby, D. Cummings, K. E. Opheim, and L. Corey, Pharmacokinetics of oral zidovudine (azidothymidine) in patients with AIDS when administered with and without a high-fat meal, AIDS, 4, 229–232 (1990).

130. E. Lotterer, M. Ruhnke, M. Trautman, R. Beyer, and F. E. Baver, Decreased and variable systemic availability of zidovudine in patients with AIDs if administered with a meal, Eur. J. Clin. Pharmacol., 40, 305–308 (1991).

131. J. Nimmo, R. C. Heading, P. Tothill, and L. F. Prescott, Pharmacological modifications of gastric emptying: effects of propantheline and metoclopramide on paracetamol absorption, Br. Med. J., 1, 587–589 (1973).

132. J. Nimmo, The influence of metoclopramide on drug absorption, Postgrad. Med. J., 49 (Suppl.), 25–28 (1973).

133. V. Manninen, A. Apajalahti, H. Simonen, and P. Reisell, Effect of propantheline and metoclopramide on absorption of digoxin, Lancet, 1, 1118–1119 (1973).

134. V. Manninen, A. Apajalahti, J. Melin, and M. Karesoja, Altered absorption of digoxin in patients given propatheline and metoclopramide, Lancet, 1, 398–399 (1973).

135. S. Medin and L. Nyberg, Effect of propantheline and metoclopramide on absorption of digoxin, Lancet, 1, 1393 (1973).

136. P. T. Mearrick, D. N. Wade, D. J. Birkett, and J. Morris, Metoclopramide, gastric emptying and L-dopa absorption, Aust. N. Z. J. Med., 4, 144–148 (1974).

137. G. Gothoni, P. Pentikainen, H. I. Vapaatalo, R. Hackman, and K. A. F. Bjorksten, Absorption of antibiotics: Influence of metoclopramide and atropine on serum levels of pivampicillin and tetracycline, Ann. Clin. Res., 4, 228–232 (1972).

138. K. H. Donn, F. N. Eshelman, J. R. Plachetka, L. Fabre, and J. R. Powell, The effects of antacid and propantheline on the absorption of oral ranitidine, Pharmacotherapy, 4, 89–92 (1984).

139. G. Levy, M. Gibaldi, and J. A. Procknal, Effect of an anticholinergic agent on riboflavin absorption in man, J. Pharm. Sci., 61, 798–799 (1972).

140. J. A. Antonioli, J. L. Schelling, E. Steininger, and G. A. Borel, Effect of gastrectomy and of an anticholinergic drug on the gastrointestinal absorption of a sulfonamide in man, Int. J. Clin. Pharmacol., 5, 212–215 (1971).

141. A. Melander and A. McLean, Influence of food intake on presystemic clearance of drugs, Clin. Pharmacokinet., 8, 286–296 (1983).

142. J. B. Dressman, R. R. Berardi, G. H. Elta, T. M. Gray, P. A. Montgomery, H. S. Lau, K. L. Pelekousas, G. J. Szpunar, and J. G. Wagner, Absorption of flurbiprofen in the fed and fasted states, Pharmaceut. Res., 9, 901–907 (1992).

143. C. W. Meuhlberger, The physiological action of alcohol, JAMA, 167, 1842–1845 (1958).

144. J. Fermaglich and S. O'Doherty, Effect of gastric motility on levodopa, Dis. Nerv. Syst., 33, 624–625 (1972).

145. M. Siurala, O. Mustala, and J. Jussila, Absorption of acetylsalicylic acid by a normal and atrophic gastric mucosa, Scand. J. Gastroenterol., 4, 269–273 (1969).

146. M. Kekki, K. Pyorola, O. Justala, H. Salmi, J. Jussila, and M. Siurala, Multicompartment analysis of the absorption kinetics of warfarin from the stomach and small intestine, Int. J. Clin. Pharmacol., 5, 209–214 (1971).

147. L. F. Prescott, W. S. Nimmo, and R. C. Heading, Drug absorption interactions, in *Drug Interactions* (D. G. Grahame-Smith, ed.), Macmillan, Baltimore, 1977, p. 45.

148. K. S. Channer and J. Virjee, Effect of posture and drink volume on the swallowing of capsules, Br. Med. J., 285, 1702 (1982).

149. H. Hey, F. Jorgenson, K. Sorensen, H. Hasselbalch, and T. Wamberg, Oesophageal transit of six commonly used tablets and capsules, Br. Med. J., 285, 171–179 (1982).

150. K. S. Channer and C. J. C. Roberts, Antipyrine absorption after delayed oesophageal capsule transit, Br. J. Clin. Pharmacol., 18, 250–253 (1984).

151. K. S. Channer and C. J. C. Roberts, Effect of delayed esophageal transit on acetaminophen absorption, Clin. Pharmacol. Ther., 37, 72–76 (1985).

152. S. S. Davis, J. G. Hardy, and J. W. Fara, Transit of pharmaceutical dosage forms through the small intestine, Gut, 27, 886–892 (1986).

153. J. L. Madsen, Effects of gender, age, and body mass index on gastrointestinal transit times, Dig. Dis. Sci., 37, 1548–1553 (1992).

154. J. M. C. Price, S. S. Davis, and I. R. Wilding, The effect of fibre on gastrointestinal transit times in vegetarians and omnivores, Int. J. Pharm., 76, 123–141 (1991).

155. F. Pirk, Changes in the motility of the small intestine in digestive disorders, Gut, 8, 486–490 (1967).

156. A. C. Guyton, *Basic Human Physiology: Normal Functions and Mechanisms of Disease*, W. B. Saunders, Philadelphia, 1971, p. 428.

157. E. Parry, R. Shields, and A. C. Turnbull, Transit time in the small intestine in pregnancy, J. Obstet. Gynaecol., 77, 900–901 (1970).

158. A. M. Metcalf, S. F. Phillips, A. R. Zinsmeister, R. L. MacCarty, R. W. Beart, and B. G. Wolff, Simplified assessment of segmental colonic transit, Gastroenterology, 92, 40–47 (1987).

159. P. J. Watts, L. Barrow, K. P. Steed, C. G. Wilson, R. C. Spiller, C. D. Melia, and M. C. Davies, The transit rate of different sized model dosage forms through the human colon and the effects of a lactulose-induced catharsis, Int. J. Pharm., 87, 215–221 (1992).

160. D. A. Akin, S. S. Davis, R. A. Sparrow, and I. R. Wilding, Colonic transit of different sized tablets in healthy subjects, J. Controlled Release, 23, 147–156 (1993).

161. J. M. C. Price, S. S. Davis, R. A. Sparrow, and I. R. Wilding, The effect of meal composition on the gastrocolonic response: implications for drug delivery to the colon, Pharm. Res., 10, 722–726 (1993).

162. C. J. McDougall, R. Wong, P. Scudera, M. Lesser, and J. J. De Cosse, Colonic mucosal pH in humans, Dig. Dis. Sci., 38, 542–545 (1993).

163. D. R. Friend, Colon-specific drug delivery, Adv. Drug Deliv. Rev., 7, 149–199 (1991).

164. K. H. Antonin, V. Saano, P. Bleck, J. Hastewell, R. Fox, P. Low, and M. MacKay, Colonic absorption of human calcitonin in man, Clin. Sci., 83, 627–631 (1992).

165. M. F. Williams, G. F. Dukes, W. Heizer, Y.-H. Han, D. J. Hermann, T. Lampkin, and L. J. Hak, Influence of gastrointestinal site of drug delivery on the absorption characteristics of ranitidine, Pharm. Res., 9, 1190–1194 (1992).

166. K. K. H. Chan, A. Buch, R. D. Glazer, V. A. John, and W. H. Barr, Site-differential gastrointestinal absorption of benazepril hydrochloride in healthy volunteers, Pharm. Res., 11, 432–437 (1994).

167. L. d'Agay-Abensour, A. Fjellestad-Paulsen, P. Hoglund, Y. Ngo, O. Paulsen, and J. C. Rambaud, Absolute bioavailability of an aqueous solution of 1-deamino-8-D-arginine vasopressin from different regions of the gastrointestinal tract in man, Eur. J. Clin. Pharmacol., 44, 473–476 (1993).

168. P. E. Warner, K. L. R. Brouwer, E. K. Hussey, G. E. Dukes, W. D. Heizer, K. H. Donn, I. M. Davis, and J. R. Powell, Sumatriptan absorption from different regions of the human gastrointestinal tract, Pharm. Res., 12, 138–143 (1995).

169. A. G. De Boer, F. Moolenaar, L. G. J. de Leede, and D. D. Breimer, Rectal drug administration: clinical pharmacokinetic considerations, Clin. Pharmacokinet., 7, 285–311 (1982).

170. D. Winne, The influence of blood flow on the absorption of L- and D-phenylalanine from the jejunum of the rat, Arch. Pharmacol., 277, 113–138 (1973).

171. D. Winne, Formal kinetics of water and solute absorption with regard to intestinal blood flow, J. Theor. Biol., 27, 1–18 (1970).

172. D. Winne and J. Remischovsky, Intestinal blood flow and absorption of nondissociable substances, J. Pharm. Pharmacol., 22, 640–641 (1970).

173. W. G. Crouthamel, L. Diamond, L. W. Dittert, and J. T. Doluisio, Drug absorption. VII. Influence of mesenteric blood flow on intestinal drug absorption in dogs, J. Pharm. Sci., 64, 664–671 (1975).

174. M. Rowland, S. Riegelman, P. A. Harris, and S. D. Sholkoff, Absorption kinetics of aspirin in man following oral administration of an aqueous solution, J. Pharm. Sci., 61, 379–385 (1972).

175. E. Kruger-Thiemer, Pharmacokinetics and dose–concentration relationships, in Physico-Chemical Aspects of Drug Action (E. J. Ariens, ed.), Pergamon Press, Elmsford, NY, 1968, pp. 63–113.

176. G. C. Oliver, R. Tazman, and R. Frederickson, Influence of congestive heart failure on digoxin level, in Symposium on Digitalis (O. Storstein, ed.), Gyldendal Norsk Forlag, Oslo, 1973, pp. 336–347.

177. N. L. Benowitz, Effects of cardiac disease on pharmacokinetics: Pathophysiologic considerations, in Pharmacokinetic Basis for Drug Treatment (L. Z. Benet, N. Massoud, and J. G. Gainbertoglio, eds.), Raven Press, New York, 1984, pp. 89–103.

178. M. M. Drucker, S. J. Blondheim, and L. Wislicki, Factors affecting acetylation in vivo of para-aminobenzoic acid by human subjects, Clin. Sci., 27, 133–141 (1964).

179. G. Levy and T. Matsuzawa, Pharmacokinetics of salicylamide elimination in man, J. Pharmacol. Exp. Ther., 156, 285–293 (1967).

180. K. E. Price, Z. Zolli, Jr., J. C. Atkinson, and H. G. Luther, Antibiotic inhibitors. I. The effect of certain milk constituents, Antibiot. Chemother., 7, 672–688 (1957).

181. K. E. Price, Z. Zolli, Jr., J. C. Atkinson, and H. G. Luther, Antibiotic inhibitors. II. Studies on the inhibitory action of selected divalent cations for oxytetracyctine, Antibiot. Chemother., 7, 689–701 (1957).

182. J. Scheiner and W. A. Altemeier, Experimental study of factors inhibiting absorption and effective therapeutic levels of declomycin, Surg. Gynecol. Obstet., 114, 9–14 (1962).

183. P. J. Neuvonen and H. Turakka, Inhibitory effect of various iron salts on the absorption of tetracycline in man, Eur. J. Clin. Pharmacol., 7, 357–360 (1974).

184. J. G. Wagner, Biopharmaceutics: Absorption aspects, J. Pharm. Sci., 50, 359–387 (1961).

185. P. Singh, J. K. Guillory, T. D. Sokoloski, L. Z. Benet, and V. N. Bhatia, Effect of inert tablet ingredients on drug absorption. I. Effect of polyethylene glycol 4000 on the intestinal absorption of four barbiturates, J. Pharm. Sci., 55, 63–68 (1966).

186. S. Riegelman and W. J. Crowell, The kinetics of rectal absorption. II. The absorption of anions, J. Am. Pharm. Assoc. Sci. Ed., 47, 123–127 (1958).

187. S. Riegelman and W. J. Crowell, The kinetics of rectal absorption. III. The absorption of undissociated molecules, J. Am. Pharm. Assoc. Sci. Ed., 47, 127–133 (1958).

188. K. Kakemi, T. Arita, and S. Muranishi, Absorption and excretion of drugs. XXVII. Effect of nonionic surface-active agents on rectal absorption of sulfonamides, Chem. Pharm. Bull., 13, 976–985 (1965).

189. D. L. Sorby, Effect of adsorbents on drug absorption. I. Modification of promazine absorption by activated attapulgite and activated charcoal, J. Pharm. Sci., 54, 677–683 (1965).

190. J. G. Wagner, Design and data analysis of biopharmaceutical studies in man, Can. J. Pharm. Sci., 1, 55–68 (1966).

191. P. G. Welling, Interactions affecting drug absorption, Clin. Pharmacokinet., 9, 404–434 (1984).

192. B. Edgar, D. Bailey, R. Bergstrand, G. Johnsson, and C. G. Regardh, Acute effects of drinking grapefruit juice on the pharmacokinetics and dynamics on felodipine and its potential clinical relevance, Eur. J. Clin. Pharmacol., 42, 313–317 (1992).

193. D. G. Bailey, J. M. O. Arnold, C. Munoz, and J. D. Spence, Grapefruit juice–felodipine interaction: Mechanism, predictability, and effect of naringen, Clin. Pharmacol. Ther., 53, 637–642 (1993).

194. D. G. Bailey, J. D. Spence, C. Munoz, and J. M. O. Arnold, Interaction of citrus juices with felodipine and nifedipine, Lancet, 337, 268–269 (1991).

195. P. A. Soons, B. A. P. M. Vogels, M. C. M. Roosemalen, H. C. Schoemaker, E. Uchida, B. Edgar, J. Lundahl, A. F. Cohen, and D. D. Breimer, Grapefruit juice and cimetidine inhibit stereo selective metabolism of nitrendipine in humans, Clin. Pharmacol. Ther., 50, 394–403 (1991).

196. C. W. Abruzzo, T. Macasieb, R. Weinfeld, J. A. Rider, and S. A. Kaplan, Changes in the oral absorption characteristics in man of dipotassium clorazepate at normal and elevated gastric pH, J. Pharmacokinet. Biopharm., 5, 377–390 (1977).

197. K. F. Ilett and D. S. Davies, In vivo studies of gut wall metabolism, in Presystemic Drug Elimination (C. F. George and D. G. Shand, eds.), Butterworth, Woburn, MA, 1982, pp. 43–65.

198. L. Rivera-Calimlim, C. A. Dujovne, J. P. Morgan, L. Lasagna, and J. R. Bianchine, Absorption and metabolism of L-dopa by the human stomach, Eur. J. Clin. Invest., 1, 313–320 (1971).

199. L. Fleckenstein, G. R. Mundy, R. A. Horovitz, and J. M. Mazzullo, Sodium salicylamide: relative bioavailability and subjective effects, Clin. Pharmacol. Ther., 19, 451–458 (1976).

200. G. Levy and T. Matsuzawa, Pharmacokinetics of salicylamide in man, J. Pharmacol. Ther., 156, 285–293 (1967).

201. M. M. Drucker, S. J. Blondheim, and L. Wislicki, Factors affecting acetylation in vivo of para-aminobenzoic acid by human subjects, Clin. Sci., 27, 133–141 (1964).

202. M. Frezza, C. Di Padova, G. Pozzato, M. Terpin, E. Baraona, and C. S. Lieber, High blood alcohol levels in women: The role of decreased gastric alcohol dehydrogenase and first-pass metabolism, N. Engl. J. Med., 322, 95–99 (1990).

203. H. K. Seitz, G. Egerer, U. A. Simanowski, R. Waldherr, R. Eckey, D. P. Agarwal, H. W. Goedde, and J.-P. von Wartourg, Human gastric alcohol dehydrogenase activity: Effect of age, sex and alcoholism, Gut, 34, 1433–1437 (1993).

204. Y. K. Tam, Individual variation in first-pass metabolism, Clin. Pharmacokinet., 25, 300–328 (1993).

205. A. G. Renwick, First-pass metabolism within the lumen of the gastrointestinal tract, in *Presystemic Drug Elimination* (C. F. George and D. G. Shand, eds.), Butterworth, Woburn, MA, 1982, pp. 3–28.

206. R. L. Smith, The role of the gut flora in the conversion of inactive compounds to active metabolites, in *A Symposium on Mechanisms of Toxicity* (W. N Aldridge, ed.), Macmillan, New York, 1971, pp. 228–244.

207. W. A. Ritschel and D. D. Denson, Influence of disease on bioavailability, in *Pharmaceutical Bioequivalence* (P. G. Welling, F. L. S. Tse, and S. V. Dighe, eds.), Marcel Dekker, New York, 1991, pp. 67–115.

208. A. Pottage, J. Nimmo, and L. F. Pescott, The absorption of aspirin and paracetamol in patients with achlorhydria, J. Pharm. Pharmacol., 26, 144–145 (1974).

209. P. A. Kramer, D. J. Chapron, J. Benson, and S. A. Merik, Tetracycline absorption in elderly patients with achlorhydria, Clin. Pharmacol. Ther., 23, 467–472 (1978).

210. H. R. Ochs, D. J. Greenblatt, and H. J. Dengler, Absorption of oral tetracycline in patients with Billroth-II gastrectomy, J. Pharmacokinet. Biopharm., 6, 295–303 (1978).

211. G. Levy, M. H. MacGillivray, and J. A. Procknal, Riboflavin absorption in children with thyroid disorders, Pediatrics, 50, 896–900 (1972).

212. G. N. Volans, Migraine and drug absorption, Clin. Pharmacokinet., 3, 313–318 (1978).

213. L. F. Prescott, Gastrointestinal absorption of drugs, Med. Clin. North Am., 58, 907–916 (1974).

214. W. J. Jusko, D. R. Conti, A. Molson, P. Kuritzky, J. Giller, and R. Schultz, Digoxin absorption from tablets and elixir: The effect of radiation-induced malabsorption, JAMA, 230, 1554–1555 (1974).

215. G. H. Sokol, D. J. Greenblatt, B. L. Lloyd, A. Georgotas, M. D. Allen, J. S. Harmatz, T. W. Smith, and R. I. Shader, Effect of abdominal radiation therapy on drug absorption in humans, J. Clin. Pharmacol., 18, 388–396 (1978).

216. W. D. Heizer, T. W. Smith, and S. E. Goldfinger, Absorption of digoxin in patients with malabsorption syndromes, N. Engl. J. Med., 285, 257–259 (1971).

217. A. E. Davis and R. C. Pirola, Absorption of phenoxymethyl penicillin in patients with steatorrhea, Aust. Ann. Med., 17, 63–65 (1968).

218. J. D. Nelson, S. Shelton, H. T. Kusmiesz, and K. C. Haltalin, Absorption of ampicillin and nalidixic acid in infants and children with acute shigellosis, Clin. Pharmacol. Ther., 13, 879–886 (1972).

219. J. P. Kampmann, H. Klein, B. Lumholtz, and J. E. M. Hansen, Ampicillin and propylthiouracil pharmacokinetics in intestinal bypass patients followed up to a year after operation, Clin. Pharmacokinet., 9, 168–176 (1984).

220. L. Backman, B. Beerman, M. Groschinsky-Grind, and D. Hallberg, Malabsorption of hydrochlorthiazide following intestinal shunt surgery, Clin. Pharmacokinet., 4, 63–68 (1979).

221. M. C. Kennedy and D. N. Wade, Phenytoin absorption in patients with ileojejunal bypass, Br. J. Clin. Pharmacol., 7, 515–518 (1979).

222. W. J. Jusko, N. Khanna, G. Levy, L. Stern, and S. J. Yaffe, Riboflavin absorption and excretion in the neonate, Pediatrics, 45, 945–949 (1970).

223. M. Mayersohn, Special pharmacokinetic considerations in the elderly, in *Applied Pharmacokinetics: Principles of Therapeutic Drug Monitoring*, 3rd Ed. (W. E. Evans, J. J. Schentag, and W. J. Jusko, eds.), Applied Therapeutics, Vancouver, WA, 1992, pp. 9-1–9-43.

224. M. Mayersohn, The ''xylose test'' to assess gastrointestinal absorption in the elderly: A pharmacokinetic evaluation of the literature, J. Gerontol., 37, 300–305 (1982).

225. S. L. Johnson, M. Mayersohn, and K. A. Conrad, Gastrointestinal absorption as a function of age: Xylose absorption in healthy adult subjects, Clin. Pharmacol. Ther., 38, 331–335 (1985).

226. T. L. Russell, R. R. Berardi, J. L. Barnett, T. L. O'Sullivan, J. G. Wagner, and J. B. Dressman, pH-related changes in the absorption of dipyridamole in the elderly, Pharm. Res., 11, 136–143 (1994).

3

Pharmacokinetics

David W. A. Bourne
University of Oklahoma, Oklahoma City, Oklahoma

Lewis W. Dittert
College of Pharmacy, University of Kentucky, Lexington, Kentucky

I. INTRODUCTION

Drug therapy is a dynamic process. When a drug product is administered, absorption usually proceeds over a finite time interval; and distribution, metabolism, and excretion (ADME) of the drug and its metabolites proceed continuously at various rates. The relative rates of these "ADME processes" determine the time course of the drug in the body, most importantly at the receptor sites that are responsible for the pharmacological action of the drug.

The usual aim of drug therapy is to achieve and maintain effective concentrations of drug at the receptor site. However, the body is constantly trying to eliminate the drug and, therefore, it is necessary to balance absorption against elimination to maintain the desired concentration. Often the receptor sites are tucked away in a specific organ or tissue of the body, such as the central nervous system, and it is necessary to depend on the blood supply to distribute the drug from the site of administration, such as the gastrointestinal tract, to the site of action.

Since the body may be viewed as a very complex system of compartments, at first, it might appear to be hopeless to try to describe the time course of the drug at the receptor sites in any mathematically rigorous way. The picture is further complicated because, for many drugs, the locations of the receptor sites are unknown. Fortunately, body compartments are connected by the blood system, and distribution of drugs among the compartments usually occurs much more rapidly than absorption or elimination of the drug. The net result is that the body behaves as a single homogeneous compartment relative to many drugs, and the concentration of the drug in the blood directly reflects or is proportional to the concentration of the drug in all organs and tissues. Thus, it may never be possible to isolate a receptor site and determine the concentration of drug around it, but the concentration at the receptor site usually can be controlled if the blood concentration can be controlled.

The objective of pharmacokinetics is to describe the time course of drug concentrations in blood in mathematical terms so that (a) the performance of pharmaceutical dosage forms can be evaluated in terms of the rate and amount of drug they deliver to the blood, and (b) the dosage regimen of a drug can be adjusted to produce and maintain therapeutically effective blood concentrations with little or no toxicity. The primary objective of this chapter will be to describe the mathematical tools needed to accomplish these aims when the body behaves as a single homogeneous compartment and when all pharmacokinetic processes obey first-order kinetics

On some occasions, the body does not behave as a single homogenous compartment and multicompartment pharmacokinetics are required to describe the time course of drug concentrations. In other instances, certain pharmacokinetic processes may not obey first-order kinetics and saturable or nonlinear models may be required. Readers interested in such advanced topics are referred to several texts that describe these more complex pharmacokinetic models in detail [1–5].

II. PRINCIPLES OF FIRST-ORDER KINETICS

A. Definition and Characteristics of First-Order Processes

The science of kinetics deals with the mathematical description of the rate of the appearance or disappearance of a substance. One of the most common types of rate processes observed in nature is the first-order process in which the rate is dependent on the concentration or amount of only one component. An example of such a process is radioactive decay, in which the rate of decay (i.e., the number of radioactive decompositions per minute) is directly proportional to the amount of undecayed substance remaining. This may be written mathematically as follows:

$$\text{Rate of radioactivity decay} \propto [\text{undecayed substance}] \tag{1}$$

or

$$\text{Rate of radioactive decay} = k(\text{undecayed substance}) \tag{2}$$

where k is a proportionality constant called the first-order rate constant.

Chemical reactions usually occur through collision of at least two molecules, very often in a solution, and the rate of the chemical reaction is proportional to the concentrations of all reacting molecules. For example, the rate of hydrolysis of an ester in an alkaline-buffered solution depends on the concentration of both the ester and hydroxide ion:

$$\text{Ester} + \text{OH}^- \rightarrow \text{acid}^- + \text{alcohol} \tag{3}$$

The rate of hydrolysis may be expressed as follows:

$$\text{Rate of hydrolysis} \propto [\text{ester}][\text{OH}^-] \tag{4}$$

or

$$\text{Rate of hydrolysis} = k[\text{ester}][\text{OH}^-] \tag{5}$$

where k is the proportionality constant called the second-order rate constant.

But, in a buffered system, $[\text{OH}^-]$ is constant. Therefore, at a given pH, the rate of hydrolysis is dependent only on the concentration of the ester and may be written:

$$\text{Rate of hydrolysis}_{(\text{pH})} = k^*[\text{ester}] \tag{6}$$

where k^* is the pseudo-first-order rate constant at the pH in question. (The pseudo-first-order rate constant k^*, is the product of the second-order rate constant and the hydroxide ion concentration: $k^* = k[\text{OH}^-]$).

Fortunately, most ADME processes behave as pseudo-first-order processes—not because they are so simple, but because everything except the drug concentration is constant. For example, the elimination of a drug from the body may be written as follows:

$$
\begin{bmatrix} \text{Drug} \\ \text{in} \\ \text{body} \end{bmatrix} + \begin{bmatrix} \text{enzymes;} \\ \text{membranes;} \\ \text{pH; protein} \\ \text{binding; etc.} \end{bmatrix} \rightarrow \begin{array}{l} \text{metabolized or} \\ \text{excreted drug} \end{array} \tag{7}
$$

If everything except the concentration of drug in the body is constant, the elimination of the drug will be a pseudo-first-order process. This may seem to be a drastic oversimplification, but most in vivo drug processes, in fact, behave as pseudo-first-order processes.

B. Differential Rate Expressions

In the previous discussion of radioactive decay it was noted that the rate of decay is directly proportional to the *amount* of undecayed substance remaining. In a solution of a radioactive substance, a similar relationship would hold for the *concentration* of undecayed substance remaining. If a solution of a radioactive substance were allowed to decay and a plot were constructed of the concentration remaining versus time, the plot would be a curve such as that shown in Fig. 1.

In this system, the rate of decay might be expressed as a change in concentration per unit time, $\Delta C/\Delta t$, which corresponds to the slope of the line. But the line in Fig. 1 is curved, which means that the rate is constantly changing and, therefore, cannot be expressed in terms of a finite time interval. By resorting to differential calculus, it is possible to express the rate of

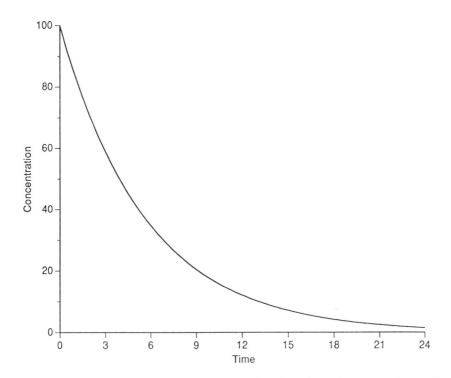

Fig. 1 Plot of concentration remaining versus time for a first-order process (e.g., radioactive decay).

decay in terms of an infinitesimally small change in concentration (dC) over an infinitesimally small time interval (dt). The resulting function, dC/dt, is the slope of the line, and it is this function that is proportional to concentration in a first-order process.

Thus,

$$\text{Rate} = \frac{dC}{dt} = -kC \tag{8}$$

The negative sign is introduced because the concentration is falling as time progresses.

Equation (8) is the differential rate expression for a first-order reaction. The value of the rate constant k could be calculated by determining the slope of the concentration versus time curve at any point and dividing by the concentration at that point. However, the slope of a curved line is difficult to measure accurately, and k can be determined much more easily using integrated rate expressions.

C. Integrated Rate Expressions and Working Equations

Equation (8) can be rearranged and integrated as follows:

$$\frac{dC}{C} = -k \, dt \tag{9}$$

$$\int \frac{dC}{C} = -k \int dt$$

$$\ln C = -kt + \text{constant} \tag{10}$$

where $\ln C$ is the natural logarithm (base e) of the concentration.

The constant in Eq. (10) can be evaluated at zero time when $kt = 0$ and $C = C_0$, the initial concentration. Thus,

$$\ln C_0 = \text{constant}$$

and since $\ln x = 2.30 \log x$, Eq. (10) can be converted to common logarithms (base 10) as follows:

$$2.30 \log C = -kt + 2.30 \log C_0 \tag{11}$$

$$\log C = \frac{-kt}{2.30} + \log C_0$$

Equation (11) is the integrated rate expression for a first-order process and can serve as a working equation for solving problems. It is also in the form of the equation of a straight line:

$$y = mx + b$$

Therefore, if $\log C$ is plotted against t, as shown in Fig. 2, the plot will be a straight line with an intercept (at $t = 0$) of $\log C_0$, and the slope of the line (m) will be $-k/2.30$. Such plots are commonly used to determine the order of a reaction; that is, if a plot of $\log C$ versus time is a straight line, the reaction is assumed to be a first-order or pseudo-first-order process.

The slope of the line and the corresponding value of k for a plot such as that shown in Fig. 2 may be calculated using the following equation:

$$\text{Slope } (m) = \frac{\log C_1 - \log C_2}{t_1 - t_2} = -\frac{k}{2.30} \tag{12}$$

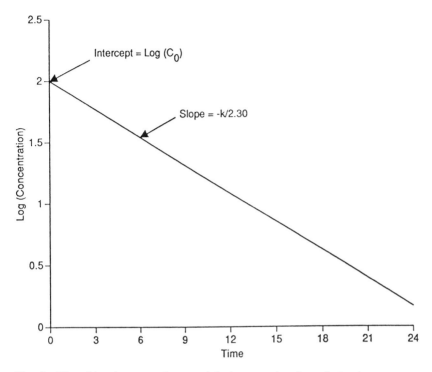

Fig. 2 Plot of log (concentration remaining) versus time for a first-order process.

EXAMPLE. A solution of ethyl acetate in pH 10.0 buffer (25°C) 1 hr after preparation was found to contain 3 mg/ml. Two hours after preparation, the solution contained 2 mg/ml. Calculate the pseudo-first-order rate constant for hydrolysis of ethyl acetate at pH 10.0 (25°C).

$$\text{Slope } (m) = \frac{\log 3 - \log 2}{(1 - 2) \text{ hr}} = -\frac{k}{2.30}$$

$$= \frac{0.477 - 0.301}{(1 - 2) \text{ hr}}$$

$$= -0.176 \text{ hr}^{-1} = -\frac{k}{2.30}$$

$$k = 0.176 \times 2.30 \text{ hr}^{-1}$$

$$= 0.405 \text{ hr}^{-1}$$

Note that since $\log C$ is dimensionless, the rate constant k has the dimensions of reciprocal time (i.e., day^{-1}, hr^{-1}, min^{-1}, sec^{-1}).

Another useful working equation can be obtained by rearranging Eq. (11) as follows:

$$\log C - \log C_0 = -\frac{kt}{2.30}$$

$$\log C_0 - \log C = \frac{kt}{2.30}$$

$$\log \frac{C_0}{C} = \frac{kt}{2.30} \tag{13}$$

Equation (13) shows that since $k/2.30$ is a constant for a given process, the ratio C_0/C is determined solely by the value of t. For example, C_0/C will be equal to 2 after the same length of time, no matter what was the value of the initial concentration (C_0).

EXAMPLE. For the foregoing ethyl acetate hydrolysis ($k = 0.405 \text{ hr}^{-1}$), if $C_0 = 3$ mg/ml, when would $C = 1.5$ mg/ml?

$$\log \frac{C_0}{C} = \frac{kt}{2.30}$$

$$\log \frac{3}{1.5} = \frac{0.405 \times t}{2.30}$$

$$\log 2 = \frac{0.405 \times t}{2.30}$$

$$0.301 = 0.176 \times t$$

$$t = 1.71 \text{ hr}$$

If $C_0 = 1.5$ mg/ml, when would $C = 0.75$ mg/ml?

$$\log \frac{C_0}{C} = \frac{kt}{2.30}$$

$$\log \frac{1.5}{0.75} = \frac{0.405 \times t}{2.30}$$

$$\log 2 = 0.176 \times t$$

$$t = 1.71 \text{ hr}$$

The time required for the concentration to fall to $C_0/2$ is called the half-life, and the foregoing example shows that the half-line for a first-order or pseudo-first-order process is a constant throughout the process; it also demonstrates that a first-order process theoretically never reaches completion, since even the lowest concentration would fall to only half its value in one half-life.

For most practical purposes, a first-order process may be deemed "complete" if it is 95% or more complete. Table 1 shows that five half-lives must elapse to reach this point. Thus, the elimination of a drug from the body may be considered to be complete after five half-lives have elapsed (i.e., 97% completion). This principle becomes important, for example, in cross-over bioavailability studies in which the subjects must be rested for sufficient time between each drug administration to ensure that "washout" is complete.

The half-life of a first-order process is very important. Since it is often desirable to convert a half-life to a rate constant, and vice versa, a simple relationship between the two is very useful. The relationship may be derived as follows:

$$\log \frac{C_0}{C} = \frac{kt}{2.30}$$

When $C_0/C = 2$ and $t = t_{1/2}$. Thus,

$$\log 2 = \frac{kt_{1/2}}{2.30}$$

Table 1 Approach to Completeness with Increasing Half-Lives

Number of half-lives elapsed	Initial concentration remaining (%)	"Completeness" of process (%)
0	100.0	0.0
1	50.0	50.0
2	25.0	75.0
3	12.5	87.5
4	6.25	93.75
5	3.13	96.87
6	1.56	98.44
7	0.78	99.22

$$0.301 = \frac{kt_{1/2}}{2.30}$$

$$kt_{1/2} = 0.693$$

$$k = \frac{0.693}{t_{1/2}} \tag{14}$$

$$t_{1/2} = \frac{0.693}{k} \tag{15}$$

D. Examples of Calculations

Equations (13), (14), and (15) can be used to solve three types of problems involving first-order processes. These types of problems are illustrated in the following examples:

Type 1

Given the rate constant or half-life and the initial concentration, calculate the concentration at some time in the future.

EXAMPLE. A penicillin solution containing 500 units/ml has a half-life of 10 days. What will the concentration be in 7 days?

$$k = \frac{0.693}{t_{1/2}} = \frac{0.693}{10 \text{ day}} = 0.069 \text{ day}^{-1}$$

$$\log \frac{C_0}{C} = \frac{kt}{2.30}$$

$$\log \frac{500 \text{ units/ml}}{C} = \frac{0.069 \text{ day}^{-1} \times 7 \text{ days}}{2.30} = 0.210$$

$$\frac{500 \text{ units/ml}}{C} = \text{antilog} (0.210) = 1.62$$

$$C = 308 \text{ units/ml}$$

Type 2

Given the half-life or rate constant and the initial concentration, calculate the time required to reach a specified lower concentration.

EXAMPLE. A penicillin solution has a half-life of 21 days. How long will it take for the potency to drop to 90% of the initial potency?

$$k = \frac{0.693}{21 \text{ days}} = 0.033 \text{ day}^{-1}$$

$$\log \frac{C_0}{C} = \frac{kt}{2.30}$$

$$\log \frac{100\%}{90\%} = \frac{0.033 \times t}{2.30}$$

$$t = 3.2 \text{ days}$$

Type 3

Given an initial concentration and the concentration after a specified elapsed time, calculate the rate constant or half-life.

EXAMPLE. A penicillin solution has an initial potency of 125 mg/5 ml. After 1 month in a refrigerator, the potency is found to be 100 mg/5 ml. What is the half-life of the penicillin solution under these storage conditions?

$$\log \frac{C_0}{C} = \frac{kt}{2.30}$$

$$\log \frac{125 \text{ mg/5 ml}}{100 \text{ mg/5 ml}} = \frac{k \times 30 \text{ day}}{2.30}$$

$$k = 0.0074 \text{ day}^{-1}$$

$$t_{1/2} = \frac{0.693}{0.0074 \text{ day}^{-1}} = 94 \text{ days}$$

For each type of problem the following assumptions are made: (a) The process follows first-order kinetics, at least over the time interval and concentration range involved in the calculations; and (b) all time and concentration values are accurate.

The latter assumption is particularly critical in solving problems such as type 3, for which a rate constant is being calculated. It would be unwise to rely on only two assay results at two time points to calculate such an important value. Normally, duplicate or triplicate assays would be performed at six or more time points throughout as much of the reaction as possible. The resulting mean assay values and standard deviation values would be plotted on semilogarithmic graph paper, and a straight line would be carefully fitted to the data points. The half-life could then be determined using Eq. (14).

EXAMPLE. A solution of ethyl acetate in pH 9.5 buffer (25°C) was assayed in triplicate several times over a 20-hr period. The data obtained are presented in Table 2. The results were plotted on semilogarithmic graph paper as shown in Fig. 3. Calculate the pseudo-first-order rate constant for the hydrolysis of ethyl acetate at pH 9.5 (25°C).

By fitting a straight line through the data points in Fig. 3 (this can be done by eye using a transparent straight edge) and extrapolating to $t = 0$, the intercept C_0 is found to be 3.13 mg/ml. The half-life is the time at which the concentration equals 1.57 mg/ml, and this is found by interpolation to be 2.4 hr. The value of k is then given by

$$k = \frac{0.693}{t_{1/2}} = \frac{0.693}{2.4 \text{ hr}} = 0.289 \text{ hr}^{-1}$$

Table 2   Assay of Ethyl Acetate

Time (hr)	Concentration (mg/ml) ± SD
2	1.83 ± 0.15
4	1.01 ± 0.09
6	0.58 ± 0.07
8	0.33 ± 0.06
10	0.18 ± 0.04
12	0.10 ± 0.02
16	0.031 ± 0.006
20	0.012 ± 0.002

Semilogarithmic graph paper is readily available from many graph paper manufacturers. It consists of a logarithmic scale on the y axis and a cartesian scale on the x axis (see Fig. 3). On the log scale, the spatial distribution of lines is such that the position of each line is proportional to the log of the value represented by the mark. For example, plotting a concentration of 1.83 mg/ml on semilog paper is equivalent to looking up the log of 1.83 and plotting it on a cartesian scale. This type of graph paper is extremely useful for kinetic calculations because raw concentration data can be plotted directly without converting to logs, and concentration values can be extrapolated and interpolated from the plot without converting logs to numbers.

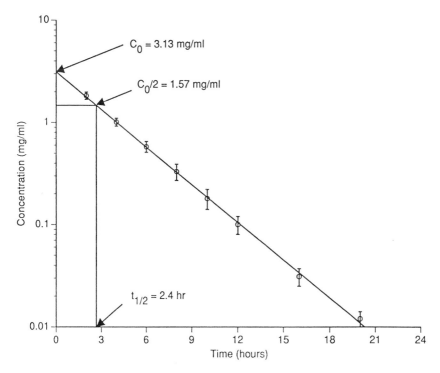

Fig. 3   Semilogarithmic plot of concentration versus time for the hydrolysis of ethyl acetate. (Data shown in Table 2; one standard deviation is indicated by error bars.)

For example, to determine the half-life in the preceding example, the C_0 value and the time at which $C = C_0/2$ were both read directly from the graph. If Fig. 3 had been a plot of log C (on a Cartesian scale) versus time, it would have been necessary to read log C_0 from the graph, convert it to C_0, divide by 2, convert back to log $(C_0/2)$, then read the half-life off the graph. If the rate constant is determined for this example using Eq. (12), the slope must be calculated. To calculate the slope of the line it is necessary first to read C_1 and C_2 from the graph and then take the logarithm of each concentration as described in Eq. (12).

III. FIRST-ORDER PHARMACOKINETICS: DRUG ELIMINATION FOLLOWING RAPID INTRAVENOUS INJECTION

It was mentioned previously that drug elimination from the body most often displays the characteristics of a first-order process. Thus, if a drug is administered by rapid intravenous (IV) injection, after mixing with the body fluids, its rate of elimination from the body is proportional to the amount remaining in the body.

Normally, the plasma concentration is used as a measure of the amount of drug in the body, and a plot of plasma concentration versus time has the same characteristics as the plot in Fig. 1. A semilogarithmic plot of plasma concentration versus time is a straight line, with a slope equal to $k_{el}/2.30$, where k_{el} is the overall elimination rate constant. The intercept at $t = 0$ is C_p^0, the hypothetical plasma concentration after the drug is completely mixed with body fluids, but before any elimination has occurred.

A typical semilog plasma concentration versus time plot is shown in Fig. 4. This figure shows that pharmacokinetic data can also be expressed in terms of a half life, called the *biological half-life*, which bears the same relation to k_{el} as that shown in Eqs. (14) and (15).

Since all the kinetic characteristics of the disappearance of a drug from plasma are the same as those for the pseudo-first-order disappearance of a substance from a solution by hydrolysis, the same working equations [Eqs. (11) and (13)] and the same approach to solving problems can be used.

EXAMPLE. A 250-mg dose of tetracycline was administered to a patient by rapid IV injection. The initial plasma concentration (C_p^0) was 2.50 µg/ml. After 4 hr the plasma concentration was 1.89 µg/ml. What is the biological half-life $(t_{1/2})$ of tetracycline in this patient?

$$\log \frac{C_p^0}{C_p} = \frac{k_{el}t}{2.30}$$

$$\log \frac{2.50}{1.89} = \frac{k_{el} \times 4}{2.30}$$

$$k_{el} = \frac{2.30 \times 0.121}{4}$$

$$= 0.0698 \text{ hr}^{-1}$$

$$t_{1/2} = \frac{0.693}{0.0698 \text{ hr}^-} = 9.93 \text{ hr}$$

Note that this approach involves the following assumptions: (a) the drug was eliminated by a pseudo-first-order process, and (b) the drug was rapidly distributed so that an "initial plasma concentration" could be measured before any drug began to leave the body. The latter assumption implies that the body behaves as a single homogeneous compartment throughout which the drug distributes instantaneously following IV injection. In pharmacokinetic terms,

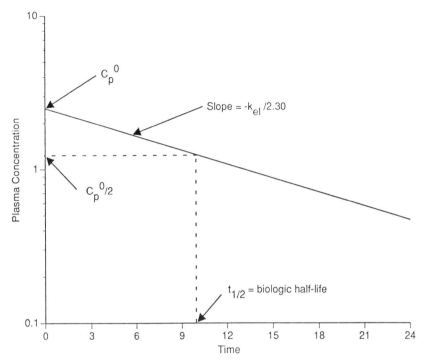

Fig. 4 Semilogarithmic plot of plasma concentration versus time for a drug administered by rapid intravenous injection.

this is referred to as the *one-compartment model*. Although most drugs do not, in fact, distribute instantaneously, they do distribute very rapidly, and the one-compartment model can be used for many clinically important pharmacokinetic calculations.

An important parameter of the one-compartment model is the apparent volume of the body compartment, because it directly determines the relation between the plasma concentration and the amount of drug in the body. This volume is called the *apparent volume of distribution, V_d*, and it may be calculated using the relationship:

$$\text{Volume} = \frac{\text{amount}}{\text{concentration}}$$

The easiest way to calculate V_d is to use C_p^0, the plasma concentration when distribution is complete (assumed to be instantaneous for a one-compartment model) and the entire dose is still in the body. Thus,

$$V_d = \frac{\text{dose}}{C_p^0}$$

EXAMPLE.   Calculate V_d for the patient in the previous example;

$$V_d = \frac{250 \text{ mg}}{2.50 \text{ }\mu\text{g/ml}}$$

$$V_d = 100 \text{ liters}$$

Note: Since 1 µg/ml = 1 mg/liter, dividing the dose in milligrams by the plasma concentration in micrograms per milliliter will give V_d in liters.

The apparent volume of distribution of a drug very rarely corresponds to any physiological volume and, even in cases where it does, it must never be construed as showing that the drug enters or does not enter various body spaces. For example, the 100-liter volume calculated in the foregoing example is much greater than either plasma volume (about 3 liters) or whole blood volume (about 6 liters) in a standard (70-kg) man; it is even greater than the extracellular fluid volume (19 liters) and total body water (42 liters) in the same average man. Based on the calculated value of V it cannot be said that tetracycline is restricted to the plasma, or that it enters or does not enter red blood cells, or that it enters or does not enter any or all extracellular fluids.

A discussion of all the reasons for this phenomenon is beyond the scope of this chapter, but a simple example will illustrate the concept. Highly lipid soluble drugs, such as pentobarbital, are preferentially distributed into adipose tissue. The result is that plasma concentrations are extremely low after distribution is complete. When the apparent volumes of distribution are calculated, they are frequently found to exceed total body volume, occasionally by a factor of two or more. This would be impossible if the concentration in the entire body compartment were equal to the plasma concentration. Thus, V_d is an empirically fabricated number relating the concentration of drug in plasma (or blood) with the amount of drug in the body. For drugs such as pentobarbital, the ratio of the concentration in adipose tissue to the concentration in plasm in much greater than unity, resulting in a large value for V_d. In calculating V_d from Eq. (16), the assumption is made that the drug concentration in the entire body equals that in plasma.

IV.   PHARMACOKINETIC ANALYSIS OF URINE DATA

Occasionally, it is inconvenient or impossible to assay the drug in plasma, but it may be possible to follow the appearance of the drug in urine. If the drug is not metabolized to any appreciable degree, the pharmacokinetic model may be written as shown in Scheme 1.

$$D_B \xrightarrow{k_{el}} D_U$$

Scheme 1

A plot of cumulative amount of drug appearing in urine (D_U) versus time will be the mirror image of a plot of amount of drug remaining in the body (D_B) versus time. This is illustrated in Fig. 5, which shows that the total amount of drug recovered in urine throughout the entire study (D_U^∞) is equal to the dose (D_B^0) and, at any time, the sum of drug in the body (D_B) plus drug in urine (D_U) equals the dose (D_B^0).

A kinetic equation describing urine data can be developed as follows.

If

$$\frac{dD_B}{dt} = -k_{el}D_B \qquad \text{then,} \qquad \frac{dD_U}{dt} = +k_{el}D_B$$

But,

$$D_B + D_U = D_U^\infty$$

$$= \text{amount recovered in urine}$$

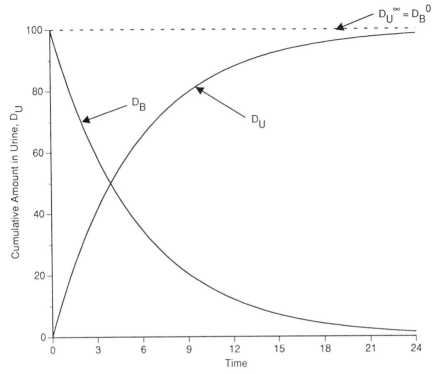

Fig. 5 Plot of cumulative amount of drug in urine, D_U (solid line), and amount of drug in body, D_B (second solid line), versus time according to Scheme 1.

Then,

$$D_B = D_U^\infty - D_U$$

Therefore,

$$\frac{dD_U}{dt} = k_{el}(D_U^\infty - D_U) \qquad \text{or} \qquad \frac{dD_U}{D_U^\infty - D_U} = k_{el}\,dt$$

Integration gives

$$\int \frac{dD_U}{D_U^\infty - D_U} = k_{el} \int dt$$

$$-\ln(D_U^\infty - D_U) + \ln(D_U^\infty - D_U^0) = k_{el}t$$

Since $\ln(x) = 2.30 \log(x)$ and $D_U^0 = 0$ (there is no drug in urine when $t = 0$),

$$\log(D_U^\infty - D_U) - \log D_U^\infty = -\frac{k_{el}t}{2.30}$$

$$\log(D_U^\infty - D_U) = -\frac{k_{el}t}{2.30} + \log D_U^\infty \qquad (17)$$

Equation (17) is in the form of the equation for a straight line ($y = mx + b$), where t is one variable (x), $-k_{el}/2.30$ is the slope (m), $\log D_U^\infty$ is the constant (b), and $\log(D_U^\infty - D_U)$ the other variable (y). Thus a plot of log ($D_U^\infty - D_U$) versus time is a straight line with a slope equal to $-k_{el}/2.30$ and an intercept of log D_U^∞. Since D_U^∞ is the total amount excreted and D_U is the amount excreted up to time t, $D_U^\infty - D_U$ is the amount remaining to be excreted (ARE). A typical ARE plot is shown in Fig. 6.

EXAMPLE.   The plot in Fig. 6 was constructed using the data shown in Table 3. Note that the concentration of the drug in each urine specimen is not the information analyzed. The total amount excreted over each time interval and throughout the entire study must be determined. As a result, the experimental details of a urinary excretion study must be very carefully chosen, and strict adherence to the protocol is required. Loss of a single urine specimen, or even an unknown part of a urine specimen, makes construction of an ARE plot impossible.

V.   CLEARANCE RATE AS AN EXPRESSION OF DRUG ELIMINATION RATE

A *clearance rate* is defined as the volume of blood or plasma completely cleared of drug per unit time. It is a useful way to describe drug elimination because it is related to blood or plasma perfusion of various organs of elimination, and it can be directly related to the physiological function of these organs. For example, the renal clearance rate (RCR) of a drug can be calculated using the following equation:

$$RCR = \frac{\text{amount excreted in urine per unit time}}{\text{plasma concentration}} \tag{18}$$

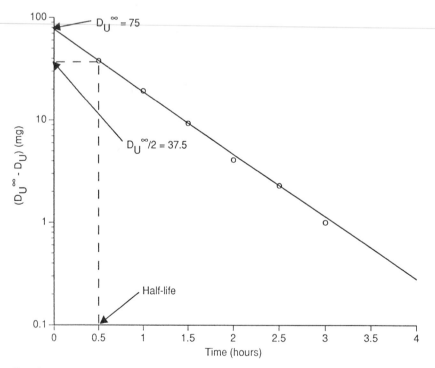

Fig. 6 Semilogarithmic plot of amount of drug remaining to be excreted (ARE) into urine, $D_U^\infty - D_U$, versus time.

Table 3 Drug Excreted into the Urine Versus Time

Time interval (hr)	Amount excreted (mg)	Cumulative amount excreted, D_U^a (mg)	$D_U^\infty - D_U$ (mg)
0.0–0.5	37.5	37.5	37.8
0.5–1.0	18.5	56.0	19.3
1.0–1.5	10.0	66.0	9.3
1.5–2.0	5.2	71.2	4.1
2.0–2.5	1.8	73.0	2.3
2.5–3.0	1.3	74.3	1.0
3.0–6.0	1.0	75.3	0.0
6.0–12.0	0.0	75.3	0.0

$^a D_U^\infty = 75.3$ mg.

EXAMPLE. In the example plotted in Fig. 6, the amount of drug excreted over the 0- to 0.5-hr interval was 37.5 mg. If the plasma concentration at 0.25 hr (the middle of the interval) was 10 µg/ml, what was the renal clearance rate? From Eq. (18),

$$RCR = \frac{37.5 \text{ mg/0.5 hr}}{10 \text{ µg/ml}}$$

$$= 7.5 \text{ liters/hr}$$

$$= 125 \text{ ml/min}$$

The glomerular filtration rate (GFR) in normal males is estimated to be 125 ml/min, and the results of the example calculation suggest that the drug is cleared by GFR. If the RCR had been less than 125 ml/min, tubular reabsorption of the drug would have been suspected. If it had been greater than 125 ml/min, tubular secretion would have been involved in the drug elimination.

Drugs can be cleared from the body by metabolism as well as renal excretion, and when this occurs, it is not possible to measure directly the amount cleared by metabolism. However, the total clearance rate (TCR), or total body clearance, of the drug can be calculated from its pharmacokinetic parameters using the following equation:

$$TCR = k_{el}V_d \tag{19}$$

EXAMPLE. The biological half-life of procaine in a patient was 35 min, and its volume of distribution was estimated to be 58 liters. Calculate the TCR of procaine.

$$k_{el} = \frac{0.693}{35 \text{ min}} = 0.0198 \text{ min}^{-1}$$

$$TCR = k_{el}V_d$$

$$= 0.0198 \text{ min}^{-1} \times 58 \text{ liters}$$

$$= 1.15 \text{ liters/min}$$

When a drug is eliminated by both metabolism and urinary excretion, it is possible to calculate the metabolic clearance rate (MCR) by the difference between TCR and RCR:

$$MCR = TCR - RCR \tag{20}$$

The RCR can be determined from urine and plasma data using Eq. (18), and the TCR can be determined from the pharmacokinetic parameters using Eq. (19). Alternatively, the RCR can be calculated by multiplying the TCR by the fraction of the dose excreted unchanged into urine, f_e:

$$RCR = f_e\ TCR \tag{21}$$

If it is assumed that the fraction of the dose not appearing as unchanged drug in urine has been metabolized, the MCR can be calculated as follows:

$$MCR = (1 - f_e)\ TCR \tag{22}$$

EXAMPLE.   Sulfadiazine in a normal volunteer had a biological half-life of 16 hr and a volume of distribution of 20 liters. Sixty percent of the dose was recovered as unchanged drug in urine. Calculate TCR, RCR, and MCR for sulfadiazine in this person.

$$k_{el} = \frac{0.693}{16\ hr} = 0.0433\ hr^{-1}$$

$$\begin{aligned}
TCR &= k_{el}V_d \\
&= 0.0433\ min^{-1} \times 20\ liters \\
&= 0.866\ liter/hr \\
&= 14.4\ ml/min
\end{aligned}$$

$$\begin{aligned}
RCR &= f_e\ TCR \\
&= 0.6 \times 14.4\ ml/min \\
&= 8.64\ ml/min
\end{aligned}$$

$$\begin{aligned}
MCR &= (1 - f_e)\ TCR \\
&= (1 - 0.6) \times 14.4\ ml/min \\
&= 5.76\ ml/min
\end{aligned}$$

It should be emphasized that the assumption that any drug not appearing as unchanged drug in urine has been metabolized may introduce a great amount of error into the values of the clearance rates estimated using Eqs. (21) and (22). By this assumption, unchanged drug eliminated in the feces would be included with metabolized drug, as would any orally administered drug that was unabsorbed.

VI. PHARMACOKINETICS OF DRUG ELIMINATED BY SIMULTANEOUS METABOLISM AND EXCRETION

Although some drugs are excreted unchanged in urine, most are partially eliminated by metabolism. Usually both the urinary excretion of unchanged drug and the metabolism are first-order processes, with the rate of excretion and metabolism dependent on the amount of unchanged drug in the body. This results in a "branch" in the kinetic chain representing exit of drug in the body as depicted in the accompanying pharmacokinetic model (Scheme 2).

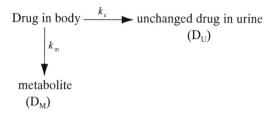

Scheme 2

In Scheme 2, the rate of loss of drug from the body is determined by both k_e and k_m, and this can be written in differential form as follows:

$$\frac{dD_B}{dt} = -k_e D_B - k_m D_B$$

$$= -(k_e + k_m)D_B \tag{23}$$

Thus, the overall elimination rate constant (k_{el}) here is the sum of the urinary excretion rate constant (k_e) and the metabolism rate constant (k_m):

$$k_{el} = k_e + k_m \tag{24}$$

For drugs that are both metabolized and excreted unchanged, semilogarithmic plots of plasma concentrations versus time have slopes equal to $-k_{el}/2.3$.

Urine data are required to determine the individual values of k_e and k_m. The required equations are derived next.

Derivation

From Scheme 2, the differential equation describing overall rate of disappearance of drug from the body may be written:

$$\frac{dD_B}{dt} = -k_{el}D_B$$

and the following integrated equation can be written [see also Eq. (10)]:

$$\ln D_B = \ln D_B^0 - k_{el}t$$

Taking antilogs yields

$$D_B = D_B^0 \exp(-k_{el}t) \tag{25}$$

It should be noted that Eq. (25) is another form of an integrated rate equation. This form makes use of an exp $(-x)$ term and may be referred to as an exponential rate expression. These expressions are useful for visualizing the characteristics of a first-order process. For example, when $t = 0$, $\exp(-k_{el}t) = 1$, and $D_B = D_B^0$. When $t = t_{1/2}$, $\exp(-k_{el}t_{1/2}) = 0.5$, and $D_B = 0.5 \times D_B^0$. When $t = \infty$, $\exp(-k_{el}t) = 0$, and $D_B = 0$. Thus, the value of $\exp(-k_{el}t)$ varies from 1 to 0 as time varies from 0 to ∞. At any time between 0 and ∞, the fraction of the dose remaining in the body is equal to $\exp(-k_{el}t)$.

Exponential rate expressions are also useful in deriving kinetic equations because they can be substituted into differential equations that can then be integrated. For example, from Scheme

2 the differential equation describing the rate of appearance of unchanged drug in urine may be written:

$$\frac{dD_U}{dt} = + k_e D_B \tag{25a}$$

Substituting Eq. (25) into Eq. (25a) gives:

$$\frac{dD_U}{dt} = + k_e[D_B^0 \exp(-k_{el}t)]$$

$$dD_U = + k_e[D_B^0 \exp(-k_{el}t)] \, dt$$

Integration yields:

$$D_U = -\frac{k_e}{k_{el}} D_B^0 \exp(-k_{el}t) + \text{constant}$$

at $t = 0$, $D_U = 0$, and $\exp(-k_{el}t) = 1$; therefore, the constant equals $(k_e/k_{el}) D_B^0$, and

$$D_U = \frac{k_e}{k_{el}} D_B^0[1 - \exp(-k_{el}t)] \tag{26}$$

At $t = \infty$ after elimination is complete, the total amount of drug excreted unchanged in urine (D_U^∞) can be calculated using Eq. (26) as follows:

$$D_U^\infty = \frac{k_e}{k_{el}} D_B^0(1 - 0)$$

$$\frac{D_U^\infty}{D_B^0} = \frac{k_e}{k_{el}} = f_e \tag{27}$$

Equation (27) shows that the function of the dose appearing as unchanged drug in urine (f_e) is equal to the fraction of k_{el} attributable to k_e. [An equation analogous to Eq. (27) for D_M^∞ and k_m could be derived in much the same way.]

Substituting Eq. (27) into Eq. (26) and rearranging gives

$$D_U^\infty - D_U = D_U^\infty \exp(-k_{el}t)$$

Taking logs yields

$$\log(D_U^\infty - D_U) = \frac{-k_{el}t}{2.30} + \log D_U^\infty \tag{28}$$

Equation (28) is identical with Eq. (17), for the case in which all eliminated drug was excreted unchanged in urine. $(D_U^\infty - D_U)$ is the amount remaining to be excreted (ARE), and Eq. (28) shows that an ARE plot of unchanged drug in urine versus time will be a straight line, with a slope equal to $-k_{el}/2.30$, even when the drug is partially eliminated by metabolism (see Figs. 6 and 7). With Eq. (27) and the total amount of unchanged drug excreted in urine (D_U^∞), it is possible to calculate k_e. Also, k_m can be calculated from Eq. (24). Thus, all the rate constants in Scheme 2 can be calculated solely on the basis of urinary excretion of unchanged drug.

EXAMPLE.   Five hundred milligrams of a drug was administered IV to a normal healthy volunteer, and various amounts of unchanged drug were recovered from the urine over the 24-hr postdrug period (Table 4). Calculate k_{el}, k_e, and k_m for this drug. A plot of $(D_U^\infty - D_U)$ on a log

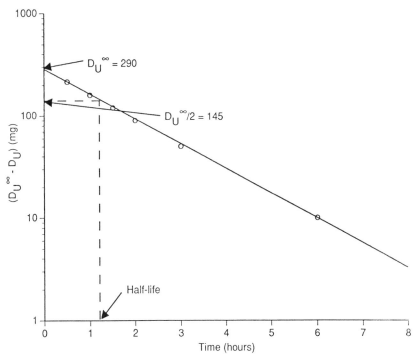

Fig. 7 Semilogarithmic plot of amount of unchanged drug remaining to be excreted into urine, $D_U^\infty - D_U$, versus time according to Scheme 2.

scale versus time is shown in Fig. 7. A half-life of 1.2 hr can be estimated from the line in Fig. 7.

$$k_{el} = \frac{0.693}{1.2 \text{ hr}} = 0.578 \text{ hr}^{-1}$$

Table 4 Drug Recovered in Urine

Time interval (hr)	Amount excreted unchanged (mg)	Cumulative amount excreted unchanged, D_U^a (mg)	$D_U^\infty - D_U$ (mg)
0.0–0.5	75	75	215
0.5–1.0	55	130	160
1.0–1.5	40	170	120
1.5–2.0	30	200	90
2.0–3.0	40	240	50
3.0–6.0	40	280	10
6.0–12.0	10	290	0
12.0–24.0	0	290	0

$^a D_U^\infty = 290$ mg.

From Eq. (27):

$$\frac{D_U^\infty}{D_B^0} = \frac{k_e}{k_{el}}$$

$$\frac{290 \text{ mg}}{500 \text{ mg}} = \frac{k_e}{0.578 \text{ hr}^{-1}}$$

$$k_e = 0.335 \text{ hr}^{-1}$$

From Eq. (24),

$$k_{el} = k_e + k_m$$

$$k_m = 0.578 - 0.335 = 0.243 \text{ hr}^{-1}$$

It is important to reemphasize the following assumptions inherent in this type of calculation:

1. It must be assumed that urine collections were accurately timed and that complete urine specimens were obtained at each collection time. It is also assumed that the assay procedure is accurate and reproducible.
2. It is assumed that all processes of elimination obey first-order kinetics.
3. It is assumed that any drug not appearing unchanged in urine has been metabolized. Furthermore, if the drug is not administered by IV injection, it must also be assumed that the dose is completely absorbed. (The IV route was chosen for the preceding example specifically to avoid the need to introduce this assumption.)

A. Significance of k_e and k_m in Patients with Kidney or Liver Disease

In the foregoing example, the drug was administered to a healthy subject who had normal kidney and liver function. The estimated biological half-life in this person was 1.2 hr. If the same drug were administered to a person with no kidney function, but with a normal liver, it would be impossible for this individual to excrete unchanged drug. They would, however, be able to metabolize the drug at the same rate as a normal individual. The net result would be that the overall k_{el} would be reduced to the value of k_m, and the biological half-life would increase to

$$\frac{0.693}{k_m} = \frac{0.693}{0.243 \text{ hr}^{-1}} = 2.85 \text{ hr}$$

Similarly, if the patient had no liver function, but normal kidney function, the half-life would increase to

$$\frac{0.693}{k_e} = \frac{0.693}{0.335 \text{ hr}^{-1}} = 2.07 \text{ hr}$$

Thus, the biological half-life of a drug can increase dramatically when the organs of elimination are diseased or nonfunctional; it may increase to varying degrees if these organs are partially impaired.

Currently, no simple relationship exists between clinical measurements of liver function and the value of k_m. Fortunately, kidney function can be measured quantitatively using standard clinical tests, and it is directly related to k_e for a number of drugs. Great success has been achieved in using kidney clearance measurements to predict the biological half-lives of several drugs. This is best illustrated with a drug that is eliminated exclusively by urinary excretion.

EXAMPLE. Kanamycin is a member of the aminoglycoside class of antibiotics, all of which are eliminated exclusively by glomerular filtration. Creatinine is a natural body substance that is cleared almost exclusively by glomerular filtration, and creatinine clearance rate is frequently used as a diagnostic tool to determine glomerular filtration rate. The relation between creatinine clearance rate and kanamycin clearance rate is shown in Fig. 8. Creatinine clearance rate can be determined as a standard clinical procedure, and the corresponding kanamycin clearance rate can be determined by interpolation on the plot in Fig. 8. Since clearance rate = $k_{el}V_d$ [see Eq. (19)], the kanamycin clearance rate can be converted to kanamycin elimination rate constant by dividing by the V_d value for kanamycin, estimated to be about 27% of the patient's body weight.

Although determination of creatinine clearance rate is a standard clinical procedure, it is difficult to carry out, mainly because accurate collection of total urine output over a 24-hr period is required. It is never certain that the patient (or the nurse) has met this requirement. Since creatine is produced continuously in muscle and is cleared by the kidney, renal failure is characterized by elevated serum creatinine levels. The degree of elevation is directly related to the degree of renal failure—if it is assumed that the production of creatinine in the muscle mass is constant and that renal function is stable. When these assumptions are valid, there is a direct relation between serum creatinine level and kanamycin half-life as shown in Fig. 9. The equation of the line in Fig. 9 is

Kanamycin half-life (hr) = 3 × serum creatinine concentration (mg/100 ml)

Thus, kanamycin half-lives (hr) can be predicted in patients with varying degrees of (stable) renal failure by multiplying the serum creatinine level (in mg/100 ml) by 3.

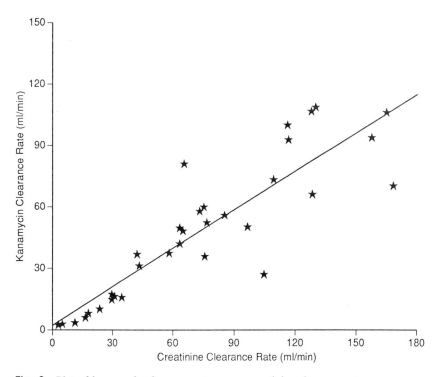

Fig. 8 Plot of kanamycin clearance rate versus creatinine clearance rate.

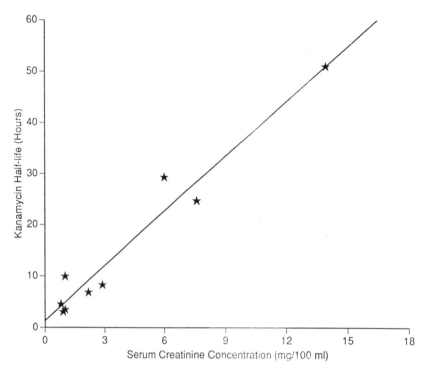

Fig. 9 Plot of kanamycin elimination half-life versus serum creatinine concentration in patients with varying degrees of (stable) renal failure.

VII. KINETICS OF DRUG ABSORPTION

For all commonly used routes of administration, except intravenous, the drug must dissolve in body fluids and diffuse through one or more membranes to enter the plasma. Thus, all routes except intravenous are classed as extravascular routes, and absorption is defined as appearance of the drug in plasma.

The most common extravascular route is oral. When a solution or a rapidly dissolving solid dosage form is given orally, the absorption process often obeys first-order kinetics. In these cases, absorption can be characterized by evaluating the absorption rate constant k_a, from plasma concentration versus time data.

A. The Method of "Residuals" ("Feathering" the Curve)

When absorption is first-order, the kinetic model may be written as shown in Scheme 3:

$$D_G \xrightarrow{k_a} D_B \xrightarrow{k_{el}} D_E$$
   Scheme 3

where

   D_G = drug at the absorption site (gut)
   D_B = drug in the body
   D_E = eliminated drug

k_a = first-order absorption rate constant
k_{el} = overall elimination rate constant

The differential equations describing the rates of change of the three components of Scheme 3 are

$$\frac{d\mathrm{D_G}}{dt} = -k_a\, \mathrm{D_G} \tag{29}$$

$$\frac{d\mathrm{D_B}}{dt} = k_a\, \mathrm{D_G} - k_{el}\, \mathrm{D_B} \tag{30}$$

$$\frac{d\mathrm{D_E}}{dt} = + k_{el}\, \mathrm{D_B} \tag{31}$$

To determine k_a from plasma concentration versus time data, it is necessary to integrate Eq. (30). This is best achieved through exponential expressions. First, integration of Eq. (29) gives

$$\mathrm{D_G} = \mathrm{D_G^0}\, \exp(-k_a t) \tag{32}$$

where $\mathrm{D_G^0}$ is the initial amount of drug presented to the absorbing region of the gut. ($\mathrm{D_G^0} =$ dose, if absorption is complete.)

Substituting Eq. (32) into Eq. (30) gives

$$\frac{d\mathrm{D_B}}{dt} = +k_a\, \mathrm{D_G^0}\, \exp(-k_a t) - k_{el}\, \mathrm{D_B} \tag{33}$$

Integration of Eq. (33) may be accomplished with Laplace transforms.* The result is

$$\mathrm{D_B} = \frac{\mathrm{D_G^0}\, k_a}{k_a - k_{el}} \left[\exp(-k_{el}t) - \exp(-k_a t) \right] \tag{34}$$

Thus, the amount of drug in the body following administration of an extravascular dose is a constant $[(\mathrm{D_G^0}\, k_a)/(k_a - k_{el})]$ multiplied by the difference between two exponential terms—one representing elimination $[\exp(-k_{el}t)]$ and the other representing absorption $[\exp(-k_a t)]$.

Dividing both sides of Eq. (34) by V_d yields an equation for plasma concentration versus time:

$$C_p = \frac{\mathrm{D_G^0}\, k_a}{V_d(k_a - k_{el})} \left[\exp(-k_{el}t) - \exp(-k_a t) \right] \tag{35}$$

Equation (35) describes the line in Fig. 10, which is a semilog plot of C_p versus time for an orally administered drug absorbed by a first-order process. The plot begins as a rising curve and becomes a straight line with a negative slope after 6 hr. This behavior is the result of the biexponential nature of Eq. (35). Up to 6 hr, both the absorption process $[\exp(-k_a t)]$ and the elimination process $[\exp(-k_{el}t)]$ influence the plasma concentration. After 6 hr, only the elimination process influences the plasma concentration.

This separation of the processes of absorption and elimination is the result of the difference in the values of k_a and k_{el}. If k_a is much larger than k_{el} (a good rule is that it must be at least five times larger), the second exponential term in Eq. (35) will approach zero much more

---

*Full details of this integration may be found in Mayersohn and Gibaldi, Am. J. Pharm. Ed., 34, 608 (1970), Eq. (27).

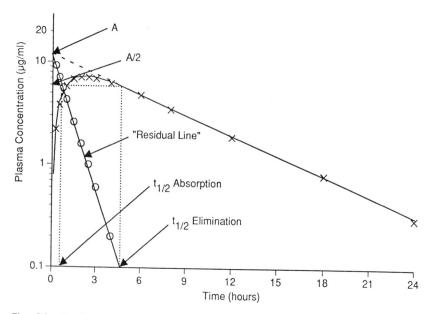

Fig. 10 Semilogarithmic plot of observed plasma concentrations (crosses) and "residuals" (circles), versus time for an orally administered drug absorbed by a first-order process.

rapidly than the first exponential term. And at large values of t, Eq. (35) will reduce to

$$C_p = \frac{D_G^0 \, k_a}{V_d(k_a - k_{el})} \, [\exp(-k_{el}t)] \tag{36}$$

or

$$C_p = A \exp(-k_{el}t)$$

where A is a constant term.

Converting to common logs we obtain

$$\log C_p = \frac{-k_{el}t}{2.30} + \log A \tag{37}$$

Thus, after 6 hr the semilog plot of C_p versus time shown in Fig. 10 becomes a straight line, with a slope of $-k_{el}/2.30$. Therefore, the overall elimination rate constant for a drug may be accurately determined from the "tail" of a semilog plot of plasma concentration versus time following extravascular administration, if k_a is at least five times larger than k_{el}.

The value of k_a can also be determined from plots like Fig. 10 using the following logic: In Fig. 10 the curved line up to 6 hr is given by

$$C_{p1} = A \exp(-k_{el}t) - A \exp(-k_a t)$$

The straight line after 6 hr and the extrapolated (dashed) line before 6 hr is given by

$$C_{p2} = A \exp(-k_{el}t)$$

The difference ("residual") between the curved line and the extrapolated (dashed) line up to 6 hr is given by

$$\text{Residual} = C_{p2} - C_{p1}$$

$$= A \exp(-k_a t)$$

Converting to common logs:

$$\log (\text{residual}) = \frac{-k_a t}{2.30} + \log A \tag{38}$$

As shown in Fig. 10, a semilog plot of residuals versus time is a straight line, with a slope of $-k_a/2.30$.

The intercepts (A) for both the extrapolated (dashed) line [Eq. (37)] and the residuals line [Eq. (38)] are the same and are equal to the constant in Eq. (35):

$$A = \frac{D_G^0 \, k_a}{V_d (k_a - k_{el})} \tag{39}$$

A is a function of the two rate constants (k_a and k_{el}), the apparent volume of distribution (V_d), and the amount of drug absorbed (D_G^0). After k_a and k_{el} have been evaluated and A has been determined by extrapolation, a value for V_d can be calculated if it is assumed that D_G^0 is equal to the dose administered (i.e., absorption is 100% complete).

EXAMPLE.   Fig. 10 is a plot of the data shown in Table 5. The extrapolated value of A is 11.8 μg/ml.

The $t_{1/2}$ (elimination) is the time at which the elimination line crosses $A/2 = 4.5$ hr:

$$k_{el} = \frac{0.693}{t_{1/2} \text{ (elimination)}} = 0.154 \text{ hr}^{-1}$$

Table 5 Plasma Concentrations and "Residuals" Versus Time

Time (hr)	Observed C_p (μg/ml)	Extrapolated C_p (μg/ml)	Residuals (μg/ml)
0.0	0.0	11.8	11.8
0.25	2.2	11.4	9.2
0.5	3.8	10.9	7.1
0.75	5.0	10.6	5.6
1.0	5.8	10.1	4.3
1.5	6.8	9.4	2.6
2.0	7.1	8.7	1.6
2.5	7.1	8.1	1.0
3.0	6.9	7.5	0.6
4.0	6.2	6.4	0.2
6.0	4.8	4.8	
8.0	3.5	3.5	
12.0	1.9	1.9	
18.0	0.8	0.8	
24.0	0.3	0.3	

The $t_{1/2}$ (absorption) is the time at which the residuals line crosses $A/2 = 0.7$ hr:

$$k_a = \frac{0.693}{t_{1/2} \text{ (absorption)}} = 0.990 \text{ hr}^{-1}$$

Assuming that the 100-mg dose of drug was completely absorbed, the V_d can be calculated from Eq. (39):

$$A = 11.8 \ \mu g/ml = \frac{100 \text{ mg} \times 0.990 \text{ hr}^{-1}}{V_d(0.990 - 0.154) \text{ hr}^{-1}}$$

$V_d = 10.0$ liters

This method of calculation is often referred to as the *method of residuals* or *feathering the curve*. It is important to remember that the following assumptions were made:

1. It is assumed that k_a is at least five times larger than k_{el}; if not, neither constant can be determined accurately.
2. It is assumed that the absorption and elimination processes are both strictly first-order; if not, the residuals line and, perhaps, the elimination line will not be straight.
3. It is assumed that absorption is complete; if not, the estimate of V_d will be erroneously high.

B. The Wagner-Nelson Method*

A major shortcoming of the method of residuals for determining the absorption rate constant from plasma concentration versus time data following administration of oral solid dosage forms is the necessity to assume that the absorption process obeys first-order kinetics. Although this assumption is often valid for solutions and rapidly dissolving dosage forms for which the absorption process itself is rate-determining, if release of drug from the dosage form is rate-determining, the kinetics are often zero-order, mixed zero- and first-order, or even more complex processes.

The Wagner-Nelson method of calculation does not require a model assumption concerning the absorption process. It does require the assumption that (a) the body behaves as a single homogeneous compartment, and (b) drug elimination obeys first-order kinetics. The working equations for this calculation are developed next.

Derivation

For any extravascular drug administration, the mass balance can be written as amount absorbed (A) equals amount in body (W) plus amount eliminated (E), or

$$A = W + E$$

Taking the derivative relative to time yields

$$\frac{dA}{dt} = \frac{dW}{dt} + \frac{dE}{dt}$$

But

$$W = V_d C_p \quad \text{or} \quad \frac{dW}{dt} = V_d \frac{dC_p}{dt}$$

---

*See Wagner and Nelson, J. Pharm. Sci. 53, 1392 (1964).

and

$$\frac{dE}{dt} = k_{el}W$$

$$= k_{el}V_dC_p$$

Therefore,

$$\frac{dA}{dt} = V_d\frac{dC_p}{dt} + k_{el}V_dC_p$$

$$dA = V_d dC_p + k_{el}V_dC_p\, dt$$

Integrating from $t = 0$ to $t = $ t

$$\int_0^t dA = V_d\int_0^t dC_p + k_{el}V_d\int_0^t C_p\, dt$$

$$A_t = V_dC_p^t + k_{el}V_d\int_0^t C_p\, dt$$

Rearranging we have

$$\frac{A_t}{V_d} = C_p^t + k_{el}\int_0^t C_p\, dt$$

where A_t/V_d is the amount of drug absorbed up to time t divided by the volume of distribution, C_p^t is plasma (serum or blood) concentration at time t, and $\int_0^t C_p\, dt$ is the area under the plasma (serum or blood) concentration versus time curve up to time t (see Sec. VIII.A). An equation similar to Eq. (40) can be derived by integration from $t = 0$ to $t = \infty$. Since $C_p = 0$ at $t = \infty$, the equation becomes

$$\frac{A_{max}}{V_d} = k_{el}\int_0^\infty C_p\, dt \tag{41}$$

where A_{max} is the total amount of drug absorbed from the dosage form divided by the volume of distribution; and $\int_0^\infty C_p\, dt$ is the area under the entire plasma (serum or blood) concentration versus time curve (Sec. VIII.A).

Equation (41) is useful for comparing the bioavailabilities of two dosage forms of the same drug administered to the same group of subjects. If it is assumed that k_{el} and V_d are the same for both administrations, it can be seen that the relative availabilities of the dosage forms is given by the ratio of the areas under the plasma concentration versus time curves:

$$\frac{A_{max_1}}{A_{max_2}} = \frac{\displaystyle\int_0^\infty (C_p\, dt)_1}{\displaystyle\int_0^\infty (C_p\, dt)_2} \tag{42}$$

Other methods of comparing bioavailabilities will be discussed in a later section.

A great deal can be learned about the absorption process by applying Eqs. (40) and (41) to plasma concentration versus time data. Since there is no model assumption relative to the

Table 6   Data Illustrating the Wagner-Nelson Calculation

Time (hr)	C_p (μg/ml)	$\int_0^t C_p \, dt$	$k_{el} \int_0^t C_p \, dt$	$\dfrac{A_t}{V_d}$	$\dfrac{A_{max}}{V_d} - \dfrac{A_t}{V_d}$
0.25	0.6	0.1	0.0	0.6	9.4
0.50	1.2	0.3	0.1	1.3	8.7
0.75	1.8	0.7	0.1	1.9	8.1
1.0	2.3	1.2	0.2	2.5	7.5
1.5	3.4	2.6	0.4	3.8	6.2
2.0	4.3	4.5	0.7	5.0	5.0
3.0	6.0	9.7	1.5	7.5	2.5
6.0	5.6	27.1	4.1	9.7	0.3
12.0	2.3	50.8	7.6	9.9	0.1
18.0	0.9	60.4	9.1	10.0	
24.0	0.4	64.3	9.6	10.0	

Note: $A_{max}/V_d = 10.0$.

absorption process, the calculated values of A_t/V_d can often be manipulated to determine the kinetic mechanism that controls absorption. This is best illustrated by an example.

EXAMPLE.   A tablet containing 100 mg of a drug was administered to a healthy volunteer and the plasma concentration (C_p) versus time data shown in Table 6 were obtained. Figure 11 shows a semilog plot of these C_p versus time data. The half-life for elimination of the drug can be estimated from the straight line "tail" of the plot to be 4.7 hr. The overall elimination

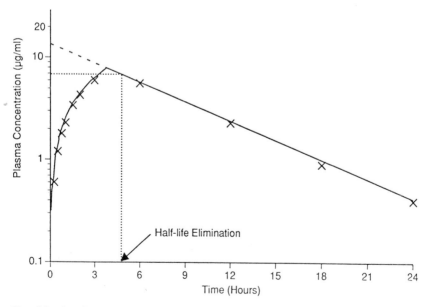

Fig. 11   Semilogarithmic plot of observed plasma concentrations (crosses) versus time for an orally administered drug absorbed by a zero-order process. (Data shown in Table 6).

rate constant is then

$$k_{el} = \frac{0.693}{4.7 \text{ hr}} = 0.147 \text{ hr}^{-1}$$

Table 6 illustrates the steps involved in carrying out the Wagner-Nelson calculation. The third column ($\int_0^t C_p \, dt$) shows the area under the C_p versus time curve calculated sequentially from $t = 0$ to each of the time points using the trapezoidal rule (see Sec. VIII.A). The fourth column ($k_{el} \int_0^t C_p \, dt$) shows each of the preceding areas multiplied by k_{el} (as estimated from the "tail") constituting the second term of the Wagner-Nelson equation [see Eq. (40)]. The fifth column (A_t/V_d) shows the sums of the values indicated in the second and fourth columns according to Eq. (40). A_{max}/V_d is the maximum value in fifth column (i.e., 10.0), and the sixth column shows the residual between A_{max}/V_d and each sequential value of A_t/V_d in the fifth column.

If the absorption process obeyed first-order kinetics, a semilog plot of the residuals in the sixth column would be a straight line with a slope of $-k_a/2.3$. However, the regular cartesian plot of the residuals shown in Fig. 12 is a straight line showing the absorption process obeys zero-order kinetics; that is, the process proceeds at a constant rate (25 mg/hr), stopping abruptly when the dose has been completely absorbed.

This example illustrates the usefulness of the Wagner-Nelson calculation for studying the mechanism of release of drugs from dosage forms in vivo. Whereas the absorption process itself usually obeys first-order kinetics, dissolution of capsules, tablets, and especially, sustained-release dosage forms often must be described by more complex kinetic mechanisms. Although pure zero-order absorption, such as that just illustrated, is almost never observed in

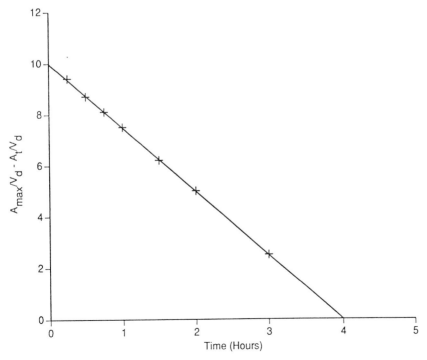

Fig. 12 Plot of the residuals (crosses) between A_{max}/V_d and A_t/V_d versus time (last column of Table 6).

practice, many sustained-release dosage forms are designed to produce as close to zero-order release as possible, since constant absorption produces constant plasma levels.

C. The Method of "Inspection"

Often, it is unnecessary to calculate an exact value for an absorption rate constant. For example, when several oral tablets containing the same drug substance are all completely absorbed, it may be sufficient to merely determine if the absorption rates are similar to conclude that the products would be therapeutically equivalent. In another instance, it would be possible to choose between an elixir and a sustained-release tablet for a specific therapeutic need without assigning accurate numbers to the absorption rate constant for the two dosage forms.

In these instances the *time of the peak* in the plasma concentration versus time curve provides a convenient measure of the absorption rate. For example, if three tablets of the same drug are found to be completely absorbed and all give plasma peaks at 1 hr, it can be safely concluded that all three tablets are absorbed at essentially the same rate. (In fact, if all tablets are completely absorbed and all peak at the same time, it would be expected that all three plasma concentration versus time curves would be identical, within experimental error.)

The time of the peak can also be used to roughly estimate the absorption rate constant. If it is assumed that k_a is at least $5 \times k_{el}$, then it can be assumed that absorption is at least 95% complete at the peak time; that is, the peak time represents approximately five absorption half-lives (see Table 1). The absorption half-life can then be calculated by dividing the time of the peak by 5, and the absorption rate constant can be calculated by dividing the absorption half-life into 0.693.

EXAMPLE.   Inspection of Fig. 10 gives a peak time of about 2.5 hr. The absorption half-life can be estimated to be 0.5 hr and the absorption rate constant, to be 1.4 hr^{-1}.

VIII.   BIOAVAILABILITY (EXTENT OF ABSORPTION)

If a drug is administered by an extravascular route and acts systemically, its potency will be directly related to the amount of drug the dosage form delivers to the blood. Also, if the pharmacologic effects of the drug are related directly and instantaneously to its plasma concentration, the rate of absorption will be important because the rate will influence the height of the plasma concentration peak and the time at which the peak occurs. Thus, the *bioavailability of a drug product is defined in terms of the amount of active drug delivered to the blood and the rate at which it is delivered.*

Whenever a drug is administered by an extravascular route, there is a danger that part of the dose may not reach the blood (i.e., absorption may not be complete). When the intravenous route is used, the drug is placed directly in the blood; therefore an IV injection is, by definition, 100% absorbed. The absolute bioavailability of an extravascular dosage form is defined relative to an IV injection. If IV data are not available, the relative bioavailability may be defined relative to a standard dosage form. For example, the bioavailability of a tablet may be defined relative to an oral solution of the drug.

In Sec. VII we dealt with methods of determining the rate (and mechanism) of absorption. In this section we will deal with methods of determining the extent of absorption. In every example, the calculation will involve a comparison between two studies carried out in the same group of volunteers on different occasions. Usually, it will be necessary to assume that the volunteers behaved identically on both occasions, especially relative to their pharmacokinetic parameters.

A. Area Under the Plasma Concentration Versus Time Curve

In the development of equations for the Wagner-Nelson method of calculation, the following equation was derived [see Eq. (42)]:

$$\frac{A_{\max_1}}{A_{\max_2}} = \frac{\displaystyle\int_0^\infty (C_p \, dt)_1}{\displaystyle\int_0^\infty (C_p \, dt)_2}$$

This equation shows that the amounts of drug absorbed from two drug products (i.e., the relative bioavailability of product 1 compared with product 2) can be calculated as the ratio of the areas under the plasma concentration versus time curves (AUCs), assuming k_{el} and V_d were the same in both studies. This assumption is probably valid when the studies are run with the same group of volunteers and within a few weeks of one another.

If dosage form 2 (Eq. (42)) is an intravenous dosage form, the *absolute bioavailability* of the extravascular dosage form (dosage form 1) is given by:

$$\begin{matrix} \text{Absolute bioavailability} \\ \text{(extravascular dosage form)} \end{matrix} = \frac{\text{AUC}_{\text{extravascular}}}{\text{AUC}_{\text{IV}}} \tag{43}$$

The AUC for a plasma concentration versus time curve can be determined by using the trapezoidal rule. For this calculation, the curve is divided into vertical segments, as shown in Fig. 13. The top line of each segment is assumed to be straight, rather than slightly curved, and the area of the segment is calculated as though it were a trapezoid; for example, the area of segment 10 is

$$\text{Area}_{10} = \frac{C_{p9} + C_{p10}}{2} \times (t_{10} - t_9) \tag{44}$$

The total AUC is then obtained by summing the areas of the individual segments. [Equation (44) can be programmed into a microcomputer that will calculate the areas and sum them as rapidly as the C_p values can be entered.]

It should be readily apparent that the trapezoidal rule does not measure AUC exactly. However, it is accurate enough for most bioavailability calculations, and the segments are chosen on the basis of the time intervals at which plasma was collected.

EXAMPLE.   The AUC for Fig. 10 can be calculated from the data given in Table 7.

Assuming that the AUC for a 100-mg IV dose given to the same group of volunteers was 86.7 hr · μg/ml, the absolute bioavailability of the extravascular dosage form is

$$\text{Absolute bioavailability} = \frac{\text{AUC}_\infty}{\text{AUC}_{\text{IV}}} \times 100 = \frac{67.2}{86.7} \times 100 = 77.5\%$$

It is not necessary to apply the trapezoidal rule to the entire plasma concentration versus time curve to calculate the total AUC. After the semilog plot becomes a straight line, the remaining area out to $t = \infty$ can be calculated from the following equation:

$$\text{AUC}_{(t \text{ to } \infty)} = \frac{C_p^t}{k_{el}} \tag{45}$$

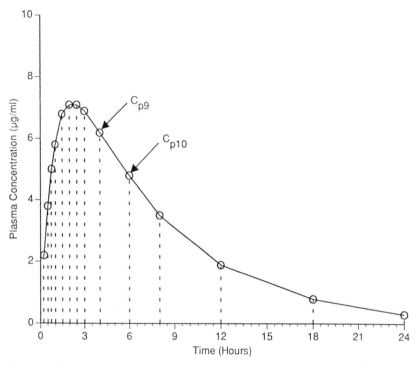

Fig. 13 Plot of plasma concentrations (circles) versus time with the curve divided into vertical segments. (Data shown in Table 7).

Table 7 Calculation of Area Under the Plasma Concentration Versus Time Curve (AUC) Using the Trapezoidal Rule

Time (hr)	C_p (μg/ml)	Area of segment (hr μg/ml)	Cumulative area up to time = t (hr μg/ml)
0.0	0.0	0.275	0.275
0.25	2.2	0.75	1.025
0.5	3.8	1.1	2.125
0.75	5.0	1.35	3.475
1.0	5.8	3.15	6.625
1.5	6.8	3.475	10.10
2.0	7.1	3.55	13.65
2.5	7.1	3.50	17.15
3.0	6.9	6.55	23.70
4.0	6.2	11.0	34.70
6.0	4.8	8.3	43.0
8.0	3.5	10.8	53.8
12.0	1.9	8.1	61.9
18.0	0.8	3.3	65.2
24.0	0.3	2.0	67.2[a]
∞	—	—	

[a]$AUC_\infty = 67.2$ hr $\cdot$ μg/ml.

Once a semilog plasma concentration versus time plot begins to follow simple first-order elimination kinetics, the remaining AUC can be calculated in one step from Eq. (45).

EXAMPLE. In the previous problem, the AUC from 24 hr to infinity is given by

$$\text{AUC}_{(24 \text{ hr to } \infty)} = \frac{0.3 \ \mu\text{g/ml}}{0.15 \ \text{hr}^{-1}} = 2.0 \ \text{hr} \cdot \mu\text{g/ml}$$

It follows that if the entire semilog plot were straight, as it would be for a one-compartment drug following IV administration, the total AUC would be given by

$$\text{AUC}_{\text{IV}} = \frac{C_{\text{p}}^{0}}{k_{\text{el}}} \tag{45a}$$

EXAMPLE. For IV administration in the foregoing problem, the AUC was calculated as follows:

$$\text{AUC}_{\text{IV}} = \frac{13.0 \ \mu\text{g/ml}}{0.15 \ \text{hr}^{-1}} = 86.7 \ \text{hr} \cdot \mu\text{g/ml}$$

B. Cumulative Urinary Excretion

In the development of equations for calculating urine data when the drug is partially metabolized and partially excreted unchanged in urine, the following equation was derived [see Eq. (27)]:

$$\frac{D_{\text{U}}^{\infty}}{D_{\text{B}}^{0}} = f_{e}$$

where D_{U}^{∞} is the amount of drug recovered from urine, D_{B}^{0} is the amount of drug absorbed, and f_{e} is the fraction of the absorbed amount recovered as unchanged drug in urine. Equation (27) may be rearranged and written for two dosage forms as follows:

$$D_{\text{U}_1}^{\infty} = D_{\text{B}_1}^{0} \times f_{e_1} \qquad \text{and} \qquad D_{\text{U}_2}^{\infty} = D_{\text{B}_2}^{0} \times f_{e_2}$$

Dividing the first equation by the second gives:

$$\frac{D_{\text{U}_1}^{\infty}}{D_{\text{U}_2}^{\infty}} = \frac{D_{\text{B}_1}^{0} \times f_{e_1}}{D_{\text{B}_2}^{0} \times f_{e_2}}$$

Assuming that $f_{e_1} = f_{e_2}$, we have

$$\frac{D_{\text{U}_1}^{\infty}}{D_{\text{U}_2}^{\infty}} = \frac{D_{\text{B}_1}^{0}}{D_{\text{B}_2}^{0}} = \text{relative bioavailability} \tag{46}$$

Similarly,

$$\frac{D_{\text{U(extravascular)}}^{\infty}}{D_{\text{U(IV)}}^{\infty}} = \text{absolute bioavailability} \tag{47}$$

Thus, if it is assumed that the same fraction of absorbed drug always reaches the urine unchanged, the bioavailability can be calculated as the ratio of total amounts of unchanged drug recovered in urine.

EXAMPLE. When potassium penicillin G was administered IV to a group of volunteers, 80% of the 500-mg dose was recovered unchanged in urine. When the same drug was administered orally to the same volunteers, 280 mg was recovered unchanged in urine. What is the absolute bioavailability of potassium penicillin G following oral administration? From Eq. (47),

$$\text{Absolute bioavailability} = \frac{280}{400} \times 100$$

$$= 70\%$$

For the calculation, it is unnecessary to assume that V_d or k_{el} or both, are the same for the two studies. It is necessary that only f_e be the same in both studies. This is usually a valid assumption unless the drug undergoes a significant amount of "first-pass" metabolism in the gut wall or liver following oral administration or a significant amount of decomposition at an intramuscular (IM) injection site. When this occurs, the availability of the extravascular dosage form may appear to be low, but the fault will not lie with the formulation. The bioavailability will be a true reflection of the therapeutic efficacy of the drug product, and reformulation may not increase bioavailability.

C. The Method of "Inspection"

Bioavailability studies are frequently carried out for the sole purpose of comparing one drug product with another, with the full expectation that the two products will have identical bioavailabilities; that is, their rates and extents of absorption will be identical. Such studies are called *bioequivalence studies* and are often employed when a manufacturer wishes to market a "generic equivalent" of a product already on the market. To take advantage of the safety and efficacy data the product's originator has filed with the FDA, the second manufacturer must show that his product gives an *identical plasma concentration versus time curve*.

In these cases, it is not necessary to determine the absolute bioavailability or the absorption rate constant for the product under study. It is necessary only to prove that the plasma concentration versus time curve is not significantly different from the reference product's curve. This is done by comparing the means and standard deviations of the plasma concentrations for the two products at each sampling time using an appropriate statistical test.

A discussion of the statistical methods used in analyzing the data from the bioequivalence studies is beyond the scope of this chapter. For a discussion of these considerations, the reader is referred to a description by Westlake [6].

IX.   MULTIPLE-DOSING REGIMENS (REPETITIVE DOSING)

Drugs are infrequently used in single doses to produce an acute effect, the way aspirin is used to relieve a headache. More often, drugs are administered in successive doses to produce a repeated or prolonged effect, the way aspirin is used to relieve the pain and inflammation of arthritis. A properly designed multiple-dosing regimen will maintain therapeutically effective plasma concentrations of the drug while avoiding toxic concentrations. Such regimens are easily designed if the pharmacokinetic parameters of the drug are known.

When drugs are administered on a multiple-dosing regimen, each dose (after the first) is administered before the preceding doses are completely eliminated. This results in a phenomenon known as *accumulation*, during which the amount of drug in the body (represented by plasma concentration) builds up as successive doses are administered. The phenomenon of accumulation for a drug administered IV is shown in Fig. 14.

Figure 14 shows that the plasma concentrations do not continue to build forever, but reach a plateau where the same maximum (C_{max}) and minimum (C_{min}) concentrations are reproduced over and over. The objectives of designing a dosing regimen are to keep C_{min} above the *min-*

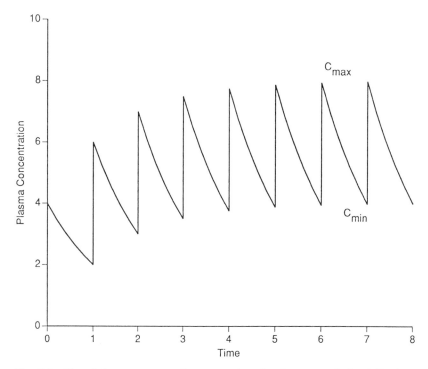

Fig. 14 Plot of plasma concentration versus time showing accumulation following multiple intravenous injections.

imum effective concentration (MEC) and to keep C_{max} below the *minimum toxic concentration* (MTC).

A. Repetitive Intravenous Dosing

The plasma concentrations in Fig. 14 can be calculated as follows: From Eq. (25) the plasma concentration at the *end of the first dosing interval* (T) is given by:

$$C_{p_1}^T = C_{p_1}^0 \exp(-k_{el}T) \tag{48}$$

Immediately after the second dose is given, the plasma concentration will be,

$$C_{p_2}^0 = C_{p_1}^T + C_{p_1}^0 = C_{p_1}^0 \exp(-k_{el}T) + C_{p_1}^0 \tag{49}$$

and so on.

It is now helpful to define the parameter R as the *fraction of the initial plasma concentration that remains at the end of any dosing interval*; R is given by the following equation:

$$R = \exp(-k_{el}T) = 10^{-k_{el}T/2.30} \tag{50}$$

As was pointed out in Sec. VI, when $T = t_{1/2}$, $R = 0.5$. The plot in Fig. 14 was constructed by using these conditions; therefore, the plasma concentration at the end of each dosing interval is half the concentration at the beginning of the dosing interval.

Equations (48) and (49) can be simplified to

$$C_{p_1}^T = C_{p_1}^0 R$$

for the *plasma concentration at the end of the first dosing interval,* and

$$C_{p_2}^0 = C_{p_1}^0 R + C_{p_1}^0$$

for the *plasma concentration at the beginning of the second dosing interval.*
The series can be carried further for more doses:

$$C_{p_2}^T = (C_{p_1}^0 R + C_{p_1}^0)R$$

$$C_{p_3}^0 = (C_{p_1}^0 R + C_{p_1}^0)R + C_{p_1}^0$$

$$C_{p_3}^T = [(C_{p_1}^0 R + C_{p_1}^0)R + C_{p_1}^0]R, \ldots, \text{etc.}$$

The plasma concentrations at the beginning and end of the *n*th dosing interval are given by the following power series:

$$\text{Beginning} = C_{p_1}^0 + C_{p_1}^0 R + C_{p_1}^0 R^2 + \cdots + C_{p_1}^0 R^{n-1} \tag{51}$$

$$\text{End} = C_{p_1}^0 R + C_{p_1}^0 R^2 + C_{p_1}^0 R^3 + \cdots + C_{p_1}^0 R^n \tag{52}$$

Since R is always smaller than 1, R^n becomes smaller as n increases. For example, if $R = 0.5$, $R^{10} = 0.001$. Therefore, the high power terms in Eq. (51) and (52) become negligible as n increases, and additional doses do not change the value of $C_{p_n}^0$ or $C_{p_n}^T$ significantly. This explains why the plasma concentrations reach a plateau instead of continuing to rise as more doses are given.

Hence, C_{max} and C_{min} (see Fig. 14) are defined as the plasma concentrations at the beginning and end, respectively, of the *n*th dosing interval after the *plateau* has been reached (i.e., $n = \infty$). When $n = \infty$, Eqs. (51) and (52) become

$$C_{max} = \frac{C_{p_1}^0}{1 - R} \tag{53}$$

$$C_{min} = C_{max}R = \frac{C_{p_1}^0 R}{1 - R} \tag{54}$$

Thus, the maximum and minimum plasma concentrations on the plateau of a repetitive IV dosing regimen can be calculated if the dosing interval (T), the overall elimination rate constant (k_{el}), and the initial plasma concentration (C_p^0) are known.

EXAMPLE. A drug has a biological half-life of 4 hr. Following an IV injection of 100 mg, C_p^0 is found to be 10 µg/ml. Calculate C_{max} and C_{min} if the 100-mg IV dose is repeated very 6 hr until a plasma concentration plateau is reached.

$$k_{el} = \frac{0.693}{4 \text{ hr}} = 0.173 \text{ hr}^{-1}$$

$$R = 10^{-k_{el}T/2.30}$$
$$= 10^{-0.173 \times 6/2.30} = 10^{-0.451}$$
$$= 0.354$$

$$C_{max} = \frac{10 \text{ µg/ml}}{1 - 0.354} = 15.5 \text{ µg/ml}$$

$$C_{min} = 15.5 \text{ µg/ml} \times 0.354 = 5.49 \text{ µg/ml}$$

EXAMPLE.   As indicated earlier, when $T = t_{1/2}$, $R = 0.5$. As a result, on the plateau in Fig. 14,

$$C_{max} = \frac{C_p^0}{1 - 0.5} = 2 \times C_p^0$$

$$C_{min} = \frac{C_p^0 \times 0.5}{1 - 0.5} = C_p^0$$

Thus, when a dose is administered every half-life, C_{max} will be twice C_p^0 and C_{min} will be half C_{max} or equal to C_p^0.

The second example illustrates a very simple and often-used dosage regimen; that is, administration of a *maintenance dose every half-life*. The calculations indicate that on this regimen, C_{min} will be $C_{max}/2$ and C_{max} will be $2 \times C_p^0$. Figure 14 indicates that approximately *five half-lives will be required to reach the plasma concentration plateau*. If the drug has a relatively long half-life, many hours, perhaps days, may be required for the plasma concentrations to reach the ideal range. If the patient's condition is serious, the physician may not want to wait for this to happen. It is under these circumstances that a *loading dose* is indicated. The loading dose immediately puts the plasma concentrations in the plateau range, and the maintenance dose maintains that condition.

For IV administration, the easiest way to determine the loading dose is in terms of C_p^0 and C_{max}. For example, if the desired C_{max} is 20 µg/ml and a dose of 100 mg gives a C_p^0 of 10 µg/ml, a loading dose of 200 mg should give a C_p^0 of 20 µg/ml, which is the desired C_{max}. If this loading dose is followed by maintenance doses of 100 mg every half-life, the plasma concentrations can be maintained at the plateau from the very beginning and throughout the entire dosing regimen. Thus, *for the maintenance dose every half-life regimen, the ideal loading dose is twice the maintenance dose*.

EXAMPLE.   *Kanamycin* is an aminoglycoside antibiotic that exerts a toxic effect on the hearing. If the plasma concentrations are allowed to remain above 35 µg/ml (MTC) for very long, permanent hearing loss may result. The minimum effective concentration (MEC) of kanamycin in plasma is estimated to be about 10 µg/ml for most organisms against which it is used. Thus, kanamycin is a classic example of a drug with a *narrow therapeutic index* for which a very precise dosing regimen is an absolute necessity. (In fact, this is true of all aminoglycosides.) When kanamycin is administered IV in a dose of 7.5 mg/kg to adults, it yields a C_p^0 of about 25 µg/ml and a half-life of about 3 hr. What would be a good dosing regimen for kanamycin?

Since 25 µg/ml is well above the MEC but below the MTC, a loading dose of 7.5 mg/kg might be given initially. After one half-life (3 hr), the plasma concentration should be 12.5 µg/ml. Since this is just above the MEC and corresponds to half the initial 25 µg/ml, a maintenance dose of 3.75 mg/kg could be administered. With repeated 3.75 mg/kg maintenance doses every 3 hr, C_{max} should be 25 µg/ml and C_{min} should be 12.5 µg/ml, which would allow some margin for error on either side.

EXAMPLE.   The kanamycin problem could be solved more aggressively as follows: Let $C_{max} = 35$ µg/ml and $C_{min} = 10$ µg/ml. From Eqs. (53) and (54), the value of R on the plateau may be calculated as follows:

$$C_{p_1}^0 = C_{max}(1 - R) = (35 \text{ µg/ml})(1 - R)$$

$$C_{p_1}^0 = \frac{C_{min}(1 - R)}{R} = \frac{(10 \text{ µg/ml})(1 - R)}{R}$$

$$35 \ \mu g/ml(1 - R) = \frac{(10 \ \mu g/ml)(1 - R)}{R}$$

$$R = \frac{10}{35} = 0.286$$

$$= 10^{-k_{el}T/2.30} \qquad \left(k_{el} = \frac{0.693}{3 \ hr} = 0.231 \ hr^{-1}\right)$$

$$0.286 = 10^{-0.231T/2.30}$$

$$= 10^{-0.100T}$$

$$-0.544 = 0.100T$$

$$T = 5.44 \ hr \ (dosing \ interval)$$

A loading dose that produces a $C_{p_1}^0$ of 35 $\mu g/ml$ is desired, and this can be calculated as follows [from Eq. (16)]:

$$\frac{Dose_1}{C_{p_1}^0} = \frac{Dose_2}{C_{p_2}^0}$$

$$\frac{7.5 \ mg/kg}{25 \ \mu g/ml} = \frac{x \ mg/kg}{35 \ \mu g/ml}$$

Loading dose = 10.5 mg/kg

The *amount of drug remaining in the body* at the end of the first dosing interval can be calculated in a similar way from the known C_{min}:

$$\frac{7.5 \ mg/kg}{25 \ \mu g/ml} = \frac{x \ mg/kg}{10 \ \mu g/ml}$$

Amount remaining = 3 mg/kg

The maintenance dose needed to replace the amount lost over the dosing interval is the difference between the loading dose and the amount remaining at the end of the interval:

Maintenance dose = (10.5 − 3) mg/kg = 7.5 mg/kg

Thus, the regimen would be a loading dose of 10.5 mg/kg, followed by maintenance doses of 7.5 mg/kg every 5.44 hr. This regimen is not only impractical but, were it carried out, it would produce C_{max} and C_{min} concentrations too close to the limiting values to allow for any errors. A better approach would be to define clinically relevant C_{max} and C_{min} values and use the approach in the previous section to develop a useful dosing regimen.

B. Repetitive Extravascular Dosing

Although the equations become considerably more complex than for the IV example, C_{max} and C_{min} can be calculated when the drug is administered by an extravascular route. The required equations may be developed as follows: The equation describing the plasma concentration versus time curve following one extravascular administration was discussed previously. Equation (35) may be written as follows:

$$C_p = \frac{FD}{V_d} \times \frac{k_a}{k_a - k_{el}} \left[\exp(-k_{el}t) - \exp(-k_a t)\right] \qquad (55)$$

where D is the dose administered and F is the fraction of the administered dose absorbed [$FD = D_G^0$ in Eq. (35)].

If n doses of the drug are administered at fixed time intervals (T), the plasma concentrations following the nth dose are given by

$$C_p = \frac{FD}{V_d} \times \frac{k_a}{k_a - k_{el}} \left[\frac{1 - \exp(-nk_{el}T)}{1 - \exp(-k_{el}T)} \exp(-k_{el}t') - \frac{1 - \exp(-nk_aT)}{1 - \exp(-k_aT)} \exp(-k_at') \right] \quad (56)$$

where t' is the time elapsed after the nth dose. When n is large (i.e., when the plasma concentrations reach a plateau), the terms $\exp(-nk_{el}T)$ and $\exp(-nk_aT)$ become negligibly small, and Eq. (56) simplifies to

$$C_p = \frac{FD}{V_d} \times \frac{k_a}{k_a - k_{el}} \left[\frac{\exp(-k_{el}t')}{1 - \exp(-k_{el}T)} - \frac{\exp(-k_at')}{1 - \exp(-k_aT)} \right] \quad (57)$$

Equation (57) can be used to calculate the C_{max} and C_{min} values on the plasma concentration plateau by substituting values for t' that correspond to the "peaks" and "valleys" in the C_p versus t curve. Thus, if $t' = t_{max}$ (the time of the peak), Eq. (57) gives C_{max}:

$$C_{max} = \frac{FD}{V_d} \times \frac{k_a}{k_a - k_{el}} \left[\frac{\exp(-k_{el}t_{max})}{1 - \exp(-k_{el}T)} - \frac{\exp(-k_at_{max})}{1 - \exp(-k_aT)} \right] \quad (58)$$

If $t' = 0$ (the time at which another dose is to be given), Eq. (57) gives C_{min}:

$$C_{min} = \frac{FD}{V_d} \times \frac{k_a}{k_a - k_{el}} \left[\frac{1}{1 - \exp(-k_{el}T)} - \frac{1}{1 - \exp(-k_aT)} \right] \quad (59)$$

EXAMPLE. The results of a single IM dose of kanamycin show that the dose is completely absorbed ($F = 1.0$), $V_d = 20$ liter, $k_{el} = 0.3$ hr^{-1}, and the time of the peak is about 1 hr ($k_a = 3.47$ hr^{-1}). If 800-mg doses of kanamycin are administered IM every 6 hr, what will C_{max} and C_{min} be when the plasma concentration plateau is reached?

$$C_{max} = \frac{1.0 \times 800 \text{ mg}}{20 \text{ liters}} \times \frac{3.47 \text{ hr}^{-1}}{(3.47 - 0.3) \text{ hr}^{-1}} \left[\frac{\exp(-0.3 \times 1)}{1 - \exp(-0.3 \times 6)} \right.$$
$$\left. - \frac{\exp(-3.47 \times 1)}{1 - \exp(-3.47 \times 6)} \right] = 37.5 \text{ }\mu\text{g/ml}$$

From Eq. (59),

$$C_{min} = \frac{1.0 \times 800 \text{ mg}}{20 \text{ liters}} \times \frac{3.47 \text{ hr}^{-1}}{(3.47 - 0.3) \text{ hr}^{-1}} \left[\frac{1}{1 - \exp(-0.3 \times 6)} \right.$$
$$\left. - \frac{1}{1 - \exp(-3.47 \times 6)} \right] = 8.67 \text{ }\mu\text{g/ml}$$

Note: The value of $\exp(-x)$ can be calculated directly with a scientific calculator or as the inverse or antilog(base e).

The foregoing example shows the calculation of the two most important features of a repetitive-dosing regimen, the maximum and minimum plasma concentrations on the plateau. But if Eq. (56) had been used, it would have been possible to calculate the plasma concentration at any time throughout an entire dosing regimen. Although these calculations are complex and laborious when done by hand, relatively inexpensive programmable calculators and now microcomputers can solve Eq. (56) in seconds. As a result, a plasma concentration versus time plot can be generated for an entire dosing regimen in a very short time. The next example illustrates such a calculation.

EXAMPLE.   The following microcomputer program, written in BASIC, will solve Eq. (56) with an added loading dose for any time during a repetitive dosing regimen.

Program Listing 1

```
10   TEXT : HOME
20   VTAB 10: PRINT "MULTIPLE DOSE - PLASMA CONCENTRATION"
100  VTAB 15: PRINT "NEW PATIENT"
110  PRINT : PRINT "ENTER ELIMINATION HALF-LIFE IN HOURS": INPUT T
120  KEL = 0.69 / T
130  PRINT : PRINT "ENTER ABSORPTION HALF-LIFE IN HOURS": INPUT T
140  KA = 0.69 / T
150  PRINT : PRINT "ENTER VOLUME OF DISTRIBUTION AS FRACTION OF
     BODY WEIGHT": INPUT T
160  PRINT : PRINT "ENTER PATIENT WEIGHT IN KG": INPUT TT
170  VD = T * TT
200  PRINT : PRINT "NEW DOSE OR REGIMEN"
210  PRINT : PRINT "ENTER FRACTION OF DOSE ABSORBED": INPUT T
220  PRINT : PRINT "ENTER LOADING DOSE IN MG": INPUT TT
230  LD = T * TT
240  PRINT : PRINT "ENTER MAINTENANCE DOSE IN MG": INPUT TT
250  MD = T * TT
260  PRINT : PRINT "ENTER DOSING INTERVAL": INPUT T
270  TAU = T
280  PRINT : PRINT "ENTER TIME INTERVAL BETWEEN POINTS": INPUT T
290  STP = T
300  PRINT : PRINT "NEW TIME POINT"
310  PRINT : PRINT "ENTER NUMBER OF MAINTENANCE DOSES GIVEN":
     INPUT T
320 N = T
400  PRINT : PRINT "PLASMA CONCENTRATION AFTER"
410  PRINT N;" MAINTENANCE DOSES"
420  PRINT : PRINT "TIME (HR)","CONCENTRATION (MG/L)": PRINT
430  FOR TIME = 0 TO TAU STEP STP
440  E1 = 0: IF (1 - EXP (KA * TAU)) = 0 THEN  GOTO 460
450  E1 = (LD *  EXP ( - N * KA * TAU) + MD * (1 - EXP ( - N * KA
     * TAU)) / (1 - EXP ( - KA * TAU))) *  EXP ( - KA * TIME)
460  E2 = 0: IF (1 - EXP (KEL * TAU)) = 0 THEN  GOTO 480
470  E2 = (LD *  EXP ( - N * KEL * TAU) + MD * (1 - EXP ( - N *
     KEL * TAU)) / (1 - EXP ( - KEL * TAU))) *  EXP ( - KEL *
     TIME)
```

```
480 CP = KA * (E1 - E2) / (VD * (KEL - KA))

490   PRINT TIME,CP: NEXT TIME

500   PRINT : PRINT "*** MENU ***": PRINT

510   PRINT "1) NEW PATIENT"

520   PRINT "2) NEW DOSING REGIMEN"

530   PRINT "3) NEW DOSE INTERVAL"

540   PRINT "4) QUIT PROGRAM"

550   PRINT : INPUT "ENTER CHOICE ";T

560   IF T < 1 OR T > 4 THEN   GOTO 500

570   ON T GOTO 10,200,300,600

600   END
```

Comments: This program has been written to be as flexible as possible for use in clinical situations; therefore, many parameters are entered in a form different from their form in Eq. (56). The following comments should clarify the relationships:

1. The program accepts half-lives for the elimination and absorption processes and converts them into rate constants, but both half-lives must be in the same time units. (If the half-life for absorption is unknown, the time of the peak divided by 5 gives a reasonable estimate; see Sec. VII.C.)
2. The fraction of the dose absorbed must be estimated in a separate study or from past experience.
3. The volume of distribution is expressed as a fraction of the body weight, and the program calculates V_d in liters using the next entry, subject weight, in kilograms.
4. Since loading doses are often used clinically, the program is written to accept both a loading dose and a maintenance dose. If no loading dose is given, these two entries are the same value.
5. The time units for the dosing interval (T) and the step interval (*int*) must be the same as those for the half-lives of elimination and absorption.
6. Once the number of maintenance doses (n) is entered the program calculates C_p for the nth-dosing interval.
7. This program can be used in the following ways:
   a. The entire plasma concentration versus time profile for a repetitive-dosing regimen can be calculated by starting with $n = 0$ and calculating C_p for the first-dosing interval. Then $n = 1$ is entered, and the second interval C_p values calculated. The process can be repeated for as many dosing intervals as desired.
   b. The plasma concentration versus time profile for a dosing interval on the plateau can be calculated by entering $n = 50$. The values of C_{min} and C_{max} can then be determined by noting the maximum and minimum values calculated. [*Note*: When $n = 50$, or some other large number, Eq. (56) becomes Eq. (57). Thus, calculations involving Eqs. (57), (58), and (59) can also be performed with this program if a large number is entered for n.]
   c. The entire plasma concentration versus time profile for a single-dose administration can be calculated if the dose is entered as the loading dose (D_0), 50 (or any large number) is entered for T, and 0 is entered for d_0 and n. [Equation (56) becomes Eq. (55).]

EXAMPLE. Calculate C_p at quarter-hourly intervals for the first 24 hr of the kanamycin dosing regimen described in the first example in this subsection. The entry data are as follows:

Half-life for elimination	= 2.3 hr
Half-life for absorption	= 0.2 hr
Fraction of body weight equal to volume of distribution	= 0.27
Subject weight	= 74 kg
Fraction of the dose absorbed	= 1.0
Loading dose (D_0)	= 800 mg
Maintenance dose (d_0)	= 800 mg
Dosing interval (T)	= 6 hr
Time interval	= 0.25 hr
Number of maintenance doses (n)	= 0–4

The results of the calculation are shown in Fig. 15. Note that C_{max} and C_{min} are identical with the values calculated in the aforementioned example.

C. Dosage Regimen Adjustment in Renal Failure

The influence of impaired renal or liver function on the rate of elimination of a drug was mentioned previously, and a method for predicting the half-life of kanamycin in patients with varying degrees of renal impairment was described earlier (see Sec. VI). If patients with im-

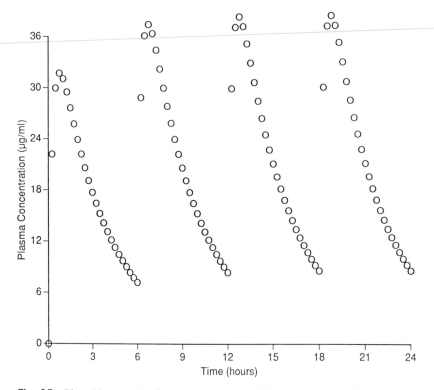

Fig. 15 Plot of kanamycin plasma concentrations (circles) versus time following multiple IM injections, calculated with Eq. (56).

paired renal function are given a normal-dosing regimen of kanamycin, they will soon build up toxic plasma concentrations of the drug. However, they can be dosed safely and effectively by adjusting the dosing regimen in accordance with the predicted elimination half-life. *Dosing regimen adjustment involves changing the dose or the dosing interval.*

Alteration of the Dosing Interval

The kanamycin package insert recommends that a dose of 7.5 mg/kg be administered every three half-lives, and that for patients with impaired renal function, the half-life in hours can be estimated by multiplying the serum creatinine level (mg/100 ml) by 3 (see also Sec. VI). Thus the dosing interval for a renal patient should be nine times serum creatinine level (mg/100 ml).

EXAMPLE.   A patient (74 kg) has a serum creatinine level of 6 mg/100 ml (elimination half-life ≈ 18 hr); therefore, the dosing regimen would be 7.5 mg/kg IM every 54 hr. Figure 16 shows the results of this dosing regimen in terms of the plasma concentrations of kanamycin produced in the renal patient (solid line) and a normal patient (dashed line). In both cases, the C_{max} values are below the MTC of 35 μg/ml, but the C_{min} values fall below the MEC of 10 μg/ml (Sec. IX.A).

For the normal patient, the time during which the plasma concentrations are below 10 μg/ml is only about 4 hr for each dosing interval; but for the renal patient, this time extends to about 18 hr. This gives the renal patient inadequate therapy for long time periods and can foster the development of strains of bacteria that are resistant to kanamycin.

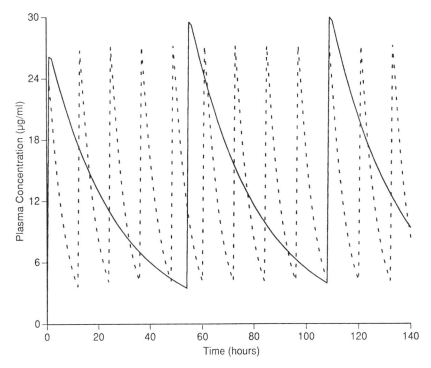

Fig. 16 Plot of kanamycin plasma concentration versus time following multiple IM injections in a normal patient (dashed line) and a renal patient (solid line), calculated with Eq. (56) and $T = 9 \times$ serum creatinine level (mg/100 ml) (*altered-dosing interval* approach).

Alteration of the Dose

In a previous example the "maintenance dose every half-life" regimen was proposed for kana-mycin because it maintained plasma concentrations between the MTC and MEC. The same regimen could be employed for IM injections of kanamycin in renal patients, but the logistical problems that arise in the clinic when the maintenance doses must be given at odd time intervals (e.g., 18 hr) make this regimen somewhat impractical. It would be much better if the maintenance doses could be administered at the same time other medication is given (i.e., every 4, 8, or 12 hr).

The same loading dose (in milligrams per kilogram) can be given to all patients regardless of their renal function because the loading dose is determined by the volume of distribution (or body weight) and not by the rate of elimination. However, if a normal-dosing interval is used, the amount of drug eliminated over one interval (T) will be much less for a patient with renal failure than for a patient with normal renal function. As a result, the maintenance doses must be reduced to replace only that amount of drug lost during the preceding dosing interval.

EXAMPLE.   The dosing regimen recommended for normal adults in the kanamycin package insert is 7.5 mg/kg every 12 hr. The insert also states that the half-life of kanamycin in a normal adult is about 4 hr. What should the 12-hr maintenance dose be for the individual in the previous example?

For a person whose half-life is 18 hr ($k_{el} = 0.0385$ hr^{-1}), the amount remaining at the end of a 12-hr–dosing interval is

$$\log \frac{A_0}{A} = \frac{k_{el}t}{2.30}$$

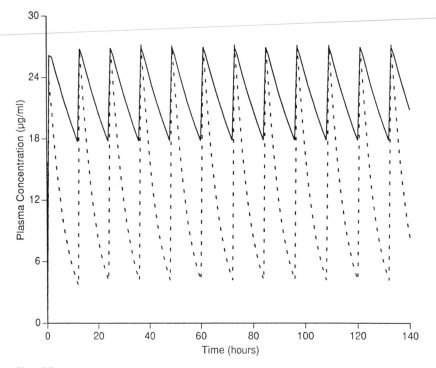

Fig. 17   Plot of kanamycin plasma concentration versus time following multiple IM injections in a normal patient (dashed line) and a patient with renal impairment (solid line) calculated with Eq. (56) and maintenance dose = amount lost over each 12-hr–dosing interval (*altered dose* approach).

$$\log \frac{7.5}{A} = \frac{0.0385 \times 12}{2.30}$$

$A = 4.72$ mg/kg

The amount lost over the 12-hr interval is

7.5 mg/kg $-$ 4.72 mg/kg = 2.78 mg/kg

Therefore, the maintenance dose = 2.78 mg/kg every 12 hr. The loading dose (same as normal) = 7.5 mg/kg.

The results of this dosing regimen in terms of plasma concentrations are shown as a solid line in Fig. 17. The dashed line in Fig. 17 shows the plasma concentrations that would be produced in a normal patient on a normal-dosing regimen (7.5 mg/kg every 12 hr). Figure 17 shows that the administration of 2.8 mg/kg maintenance doses every 12 hr to the aforementioned renal patient is a convenient dosing regimen that produces C_{min} plasma levels above the MEC and C_{max} plasma levels below the MTC.

REFERENCES

1. R. E. Notari, *Biopharmaceutics and Clinical Pharmacokinetics: An Introduction*, 4th Ed., Marcel Dekker, New York, 1987.
2. M. Gibaldi and D. Perrier, *Pharmacokinetics*, 2nd Ed., Marcel Dekker, New York, 1982.
3. J. G. Wagner, *Fundamentals of Clinical Pharmacokinetics*, Drug Intelligence Publications, Hamilton, IL, 1975.
4. J. G. Wagner, *Biopharmaceutics and Relevant Pharmacokinetics*, Drug Intelligence Publications, Hamilton, IL, 1971.
5. G. A. Portmann, Pharmacokinetics, in *Current Concepts in the Pharmaceutical Sciences: Biopharmaceuticals* (J. Swarbrick, ed.), Lea & Febiger, Philadelphia, 1970.
6. W. J. Westlake, The design and analysis of comparative blood-level trials, in *Current Concepts in the Pharmaceutical Sciences: Dosage Form Design and Bioavailability* (J. Swarbrick, ed.), Lea & Febiger, Philadelphia, 1973.

4

Factors Influencing Drug Absorption and Drug Availability

Betty-ann Hoener and Leslie Z. Benet
University of California, San Francisco, California

I. INTRODUCTION

In this chapter we examine dosage forms as drug delivery systems. The scope of the examination will, however, be limited to those oral dosage forms in which the drug is not in solution when taken by the patient. Usually, the absorption of drugs from aqueous solutions can be defined by the principles discussed in Chapter 2. However, occasionally, drugs in solution may precipitate in the gastrointestinal (GI) fluids and effectively be considered as suspensions. Nevertheless, the dosage forms considered in this chapter are tablets, capsules, and suspensions. In addition, the discussion is limited to drugs absorbed by passive diffusion that thus appear to obey first-order linear absorption kinetics.

The successful transposition of a drug from an oral dosage form into the general circulation can be described as a four-step process: first, delivery of the drug to its absorption site; second, getting the drug into solution; third, movement of the dissolved drug through the membranes of the gastrointestinal tract (GIT); and, finally, movement of the drug away from the site of absorption into the general circulation. Each of these four steps is considered in turn. However, the order of the first two steps is not absolute. That is, the drug may dissolve either before or after reaching its absorption site. It is imperative, however, as discussed in Chapter 2, that the drug be in solution before it can be absorbed. The slowest of the four steps will determine the rate of availability of the drug from an oral dosage form. The rate and extent of availability of a drug from its dosage form will be influenced by many factors in all four of these steps. Those factors related to the physicochemical properties of the drug and the design and production of the dosage form will be called pharmaceutical variables. Those variables resulting from the anatomical, physiological, and pathological characteristics of the patient are called patient variables. The pharmacist, who is the one member of the health care team knowledgeable about both the patient and the dosage form, is uniquely qualified to evaluate the influence of both the pharmaceutical and patient variables on rational drug therapy.

II. GETTING THE DRUG TO ITS SITE OF ABSORPTION

When an oral dosage form is swallowed by a patient it will travel through the gastrointestinal tract (Fig. 1). During this passage the dosage form will encounter great anatomical and physiological variations. These patient variables have been discussed in Chapter 2. Two of these variables are most important in effecting the delivery of the drug from its dosage form. The first variable, hydrogen ion concentration, exhibits a 10^7-fold difference between the mucosal fluids of the stomach and the intestine. The second variable, available surface area of the absorbing membranes, changes dramatically between different regions of the GIT (see Chapter 2).

A. Gastric Emptying

Since various factors, principally increased membrane surface area and decreased thickness of the membrane, favor the small intestine versus the stomach as the primary site for drug absorption, the rate at which the drug gets to the small intestine can significantly affect its rate of absorption. Hence, gastric-emptying rate may well be the rate-determining step in the absorption of a drug. As has been discussed in Chapter 2, light physical activity will stimulate stomach emptying, but strenuous exercise will delay emptying [2,3]. If a patient is lying on the left side, the stomach contents will have to move uphill to get into the intestine [2,3]. The emotional state of the patient can either reduce or speed up the stomach-emptying rate [2,3]. In addition, numerous pathological conditions may alter gastric-emptying rate [4]. Other drugs may affect the stomach-emptying rate by influencing GI motility. It should be noted that any substance in the stomach delays emptying and, therefore, can delay the absorption of a drug. The stomach serves to protect the intestine from extreme conditions. Thus, if the stomach contents differ appreciably in pH, temperature, osmolarity, or viscosity from those conditions normally expected in the intestine, the stomach will delay emptying until those conditions

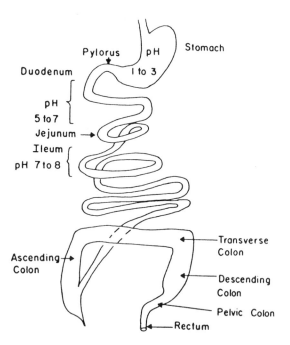

Fig. 1   Diagram of the gastrointestinal tract showing the variations of pH. (From Ref. 1.)

approach normal [2,3]. Food, by altering the foregoing factors, can affect the rate of stomach emptying. Moreover, the volume, composition, and caloric content of a meal can alter the stomach-emptying rate [2,3].

Thus, the timing of meals relative to the timing of the oral dosing of a drug can influence the rate, and possibly the extent, of drug availability. It can be anticipated that taking a drug shortly before, after, or with a meal may delay the rate of drug availability as a function of decreased-emptying rate. However, the effect of food on the extent of availability cannot be so readily predicted. Figure 2 illustrates the mean serum concentration versus time curves obtained following administration of three 2.5-mg tablets of methotrexate to normal healthy volunteers [5]. The tablets were taken on an empty stomach (x's) or with a standard, high-fat content breakfast (solid circles). Food significantly decreased the rate of availability of methotrexate from these tablets, as indicated by the shift to the right of the peak time. However, there was no significant change in the area under the curve and, therefore, no change in the extent of availability of methotrexate from these tablets.

In contrast with these results, Fig. 3 illustrates the decreased extent of availability of erythromycin when two 250-mg tablets of erythromycin were taken on a fasting stomach with 20 or 250 ml of water or with 250 ml of water immediately after high-fat, high-protein, and high-carbohydrate meals [6]. The rate of availability of this poorly absorbed water-soluble, acid-labile antibiotic was not affected by food (i.e., peak time approximately the same for all dosings), indicating that stomach emptying is not the rate-determining step in the absorption of erythromycin from this drug delivery system. The extent of availability was, however, markedly decreased. This decrease might be due to complexation between drug and food, or more likely, to degradation of the antibiotic when it is retained in the acid environment of the stomach for longer periods. (The effect of the volume of water is considered later in the chapter.)

A third possibility, an increase in the extent of availability of a drug taken with food, is illustrated in Fig. 4. Here the cumulative amount of unmetabolized nitrofurantoin excreted in

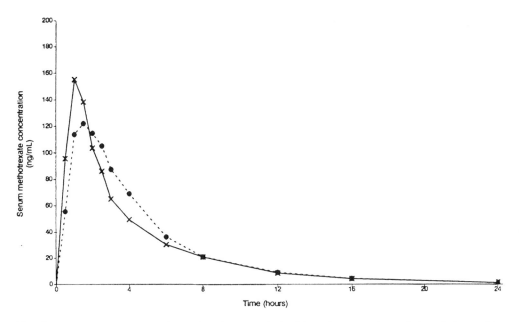

Fig. 2 Mean serum concentrations of methotrexate in healthy volunteers who took three 2.5-mg tablets: x, on a fasting stomach; •, with breakfast. (From Ref. 5.)

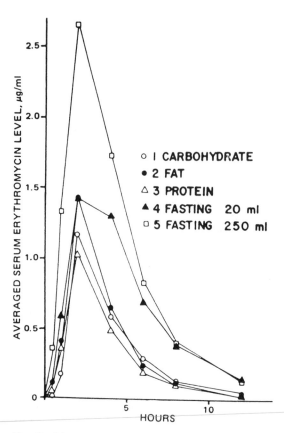

Fig. 3 Mean serum erythromycin levels in healthy volunteers given 500 mg of erythromycin stearate with 20 ml of water (▲), 250 ml of water (□), or 250 ml of water immediately after a high-carbohydrate meal (○), 250 ml of water immediately after a high-fat meal (●), and 250 ml of water immediately after a high-protein meal (△). (From Ref. 6.)

the urine is plotted against time. Healthy male volunteers took nitrofurantoin as a capsule containing 100 mg of the drug in a *macro*crystalline form (circles) or as a tablet containing 100 mg of the drug in the *micro*crystalline form (squares). (The significance of the macro- and microcrystalline forms are discussed later in this chapter.) The dosage forms were taken with 240 ml of water, either on an empty stomach (open symbols) or immediately after a standard breakfast (solid symbols). It can be seen that food delays the absorption of nitrofurantoin from both the tablet and the capsule dosage forms, as indicated by the initial phase of each plot. However, food enhanced the extent of availability of the drug from both dosage forms, as indicated by the cumulative amount of drug excreted at 24 hr. It appears that delaying the rate of transit of nitrofurantoin through the GIT gives this poorly soluble drug more time to dissolve in the GI fluids. Thus, more nitrofurantoin gets into solution and, therefore, more is absorbed.

In summary, food will delay stomach emptying and, as a result, may decrease the rate of availability of a drug from its oral dosage form. The extent of availability of that drug, however, may be increased, decreased, or unaffected by meals. Thus, it is important for the pharmacist to counsel patients on the importance of timing the taking of their medications relative to their mealtimes.

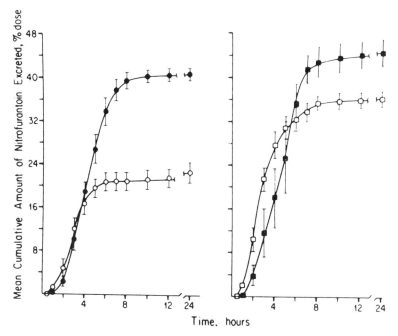

Fig. 4 Mean cumulative urinary excretion of nitrofurantoin after oral administration of a 100-mg macrocrystalline capsule to fasting (○) and nonfasting (●) subjects and a 100-mg microcrystalline tablet to fasting (□) and nonfasting (■) subjects. Vertical bars represent standard errors of the mean. (From Ref. 7.)

Large objects, such as nondisintegrating dosage forms, are handled by the stomach quite differently from liquids and particles smaller than about 2 mm in diameter [8]. These larger particles will remain in the stomach pouch until they are swept out by a series of contractions called a migrating motor complex (MMC). This MMC occurs about every 3 hr in the fasted stomach [9]. However, when a patient has eaten a meal, the MMC does not occur until after most of the food has left the stomach [8]. Thus, there could be a considerable delay in the absorption of drugs from solid dosage forms that do not disintegrate or release their contents by some other mechanism while in the stomach. This effect of size on the rate of delivery of the drug from its dosage form may or may not also affect the extent of availability.

B. Gastrointestinal Motility

Other patient variables may affect GI motility and, thereby, the extent or rate of availability of a drug from a delivery system. As illustrated in Chapter 2, the degree of physical activity, age, disease state, and emotional condition of a patient may increase or decrease GI motility. Other drugs taken concurrently may affect GI motility and, thus, indirectly affect the rate and extent of availability of a particular drug. This concurrent therapy may increase or decrease GI motility, thereby increasing or decreasing the rate at which a drug reaches its absorption site. Such changes in GI motility may increase, decrease, or have no effect on the extent of availability of a drug from an oral dosage form.

Average serum digoxin levels for an elderly female patient taking 0.375 mg of digoxin daily in a tablet dosage form are shown in Fig. 5A. When the patient was also given three daily 10-mg oral doses of metoclopramide, a drug that increases GI motility, the serum digoxin levels

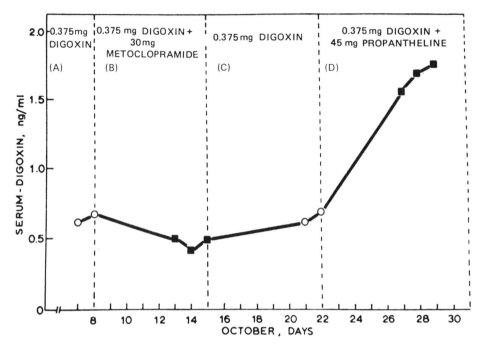

Fig. 5 Variations in serum digoxin concentration in an elderly female patient during treatment with digoxin and metoclopramide or propantheline. (From Ref. 10.)

dropped (see Fig. 5B), probably indicating a decrease in the extent of digoxin availability. It is possible that this decreased extent of availability occurred because there was insufficient time for the digoxin to be released from its dosage form, or to dissolve in the fluids of the stomach or small intestine, before it was moved completely through the GI tract.

An increase in GI motility will not, however, always decrease the extent of availability of a drug. Figure 6 illustrates the plasma level versus time curves obtained after oral administration of a 100-mg tablet of atenolol to six healthy male volunteers. When the atenolol was given 1 hr after a 25-mg dose of metoclopramide, a slight increase in the rate of availability was observed (i.e., shorter peak time), but the extent of availability of the atenolol was not significantly different from when atenolol was administered alone. Therefore it appears that, under these study conditions, stomach emptying is not the rate-determining step in absorption, nor does an increase in gastrointestinal motility significantly affect the amount of drug available to the general circulation.

Concurrent drug therapy may also decrease the motility of the GIT. Figure 5C shows again average digoxin serum levels in an elderly female patient receiving 0.375 mg of digoxin daily in tablet form. After receiving three daily doses of a 15-mg tablet of propantheline, a drug that slows GI motility, the patient's serum digoxin level increased significantly. Since the digoxin tablet was moving more slowly through its principal site of absorption, the small intestine, it is probable that there was more time for the tablet to disintegrate or for the digoxin to dissolve, thereby increasing the extent of availability.

Returning to Fig. 6, it can be seen that the oral administration of two 15-mg tablets of propantheline 1.5 hr before atenolol delayed the rate of availability of this β-blocker, while increasing its extent of availability [11]. This increased extent might be due to more complete dissolution of the drug, resulting from its increased time in the gastrointestinal tract.

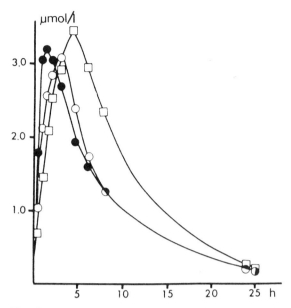

Fig. 6 Mean plasma levels of atenolol in six male volunteers after a 100-mg tablet alone (○), with 25 mg of metoclopramide (●), or with 30 mg of propantheline (□). (From Ref. 11.)

Blood levels versus the time curves obtained after oral administration of two 0.5-g tablets of sulfamethoxazole to male patients are pictured in Fig. 7. (Each subject took the drug with 300 ml of lightly sugared tea.) When a 15-mg tablet of propantheline was taken 30 min before the sulfamethoxazole was given, and a second propantheline tablet taken with it, both the rate and extent of availability of the sulfa drug from the tablet dosage form decreased (see dotted line in Fig. 7). Here, it is possible that slowing the passage of a drug through the GIT increased the opportunity for drug degradation to take place.

Thus, drugs that alter GI motility may increase or decrease the rate of availability of another drug. In addition, concurrent therapy may increase, decrease, or not affect the extent of availability of a drug. It should be noted that a knowledge of the effect of a drug on GI motility usually can lead to an accurate prediction of its potential effect on the rate of availability of a second drug, if, and only if, this is the rate-determining step in absorption, but this knowledge usually does not allow the pharmacist to predict how changes in motility may affect the extent of availability for the second drug. To make such a prediction, additional information, such as degradation and binding mechanisms along the GI tract, must be known.

Since gastric emptying can significantly affect the availability of a drug, it might be suspected that patients who have undergone partial or total gastrectomies will exhibit abnormal rates or extents of availability. Figure 7 illustrates the increased rate of availability of sulfamethoxazole in a patient who had previously undergone a partial gastrectomy. Since the patient had most of his stomach removed, the drug could reach its principal site of absorption, the small intestine, without delay. Some pathological conditions are also accompanied by altered gastrointestinal motility, which may affect availability [4].

The pharmacist, as a drug therapist, should advise both the physician and the patient of the potential problems involved in either initiating or discontinuing concurrent therapy of drugs known to alter GI motility. The pharmacist can also advise the physician on rational drug therapy in patients in whom the gastrointestinal tract has been altered surgically or by disease.

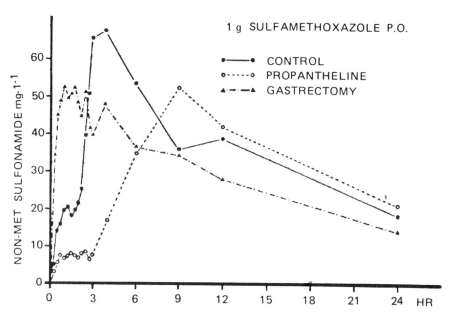

Fig. 7 Blood levels of nonmetabolized sulfamethoxazole following oral administration of 1.0 g, as two 0.5-g tablets, under three different conditions. (From Ref. 12.)

C. Degradation and Metabolism in the Gastrointestinal Tract

Absorption is not the only process that can occur along the GIT. A drug may be degraded or metabolized before it can be absorbed. Chemical degradation, especially pH-dependent reactions, can occur in the solubilizing fluids of the GIT. Drugs that structurally resemble nutrients, such as polypeptides, nucleotides, or fatty acids, may be especially susceptible to enzymatic degradation [2,3]. Digestion of these molecules would result as a consequence of the normal functions of the GIT. In addition to enzymatic metabolism in the GI fluids, drugs may be metabolized in the GI wall [13–15]. Moreover, the GIT is rich in microflora. These microorganisms could metabolize drug molecules before absorption. The variety of degradations attributable to bacteria has been reviewed [16]. In addition, both the fluids and the microflora of the GIT may be altered in disease states. As a result, drug therapy may need to be modified if the rate and extent of availability are affected by these changes [4].

It would seem that metabolism and degradation of a drug within the GIT would serve primarily to reduce its extent of availability. In some instances, however, these processes may enhance availability. For example, clindamycin palmitate is less soluble than clindamycin HCl. A suspension of the palmitate ester is a more stable, better-tasting dosage form than a solution of clindamycin HCl. The palmitate ester is rapidly hydrolyzed in the GIT to free the more soluble active parent antibiotic drug, which becomes rapidly available to the systemic circulation [13]. Molecules such as clindamycin palmitate are chemical derivatives of drugs. These derivatives, called prodrugs, are made to enhance the pharmaceutical properties of the parent molecule. These prodrugs may depend on degradation in the GIT to release the active parent molecule. Consequently, although metabolism or degradation in the GIT decreases the extent of availability for most drugs, with some prodrugs the degradative processes may be essential for complete bioavailability.

The pharmacist should certainly know which drug dosage forms have been designed as delivery systems for prodrugs and also understand the mechanism whereby the active form is made available within the body. Only then can the pharmacist counsel the physician about why a certain product may be inappropriate for such individuals as an achlorhydric patient, a patient with a gastrectomy, or even possibly a patient with a hepatoportal bypass, if liver metabolism is responsible for conversion of the prodrug into the active form.

III. GETTING THE DRUG INTO SOLUTION: FACTORS AFFECTING THE RATE OF DISSOLUTION

Once the dosage form reaches the absorption site, it must break down and release its therapeutic agent. Figure 8 depicts the disintegration and dissolution processes involved in the gastrointestinal absorption of a drug administered in a tablet dosage form. Figure 9 indicates more comprehensively the various solubility problems that may be encountered after the administration of a drug in an oral dosage form. Arrows drawn with heavy dark lines indicate primary pathways that most drugs administered in a particular dosage form undergo. Arrows drawn with dashed lines indicate that the drug is administered in this state in the dosage form. Arrows drawn as thin, continuous lines and labeled "precipitation" indicate situations in which a drug is already in solution, but then precipitates out as fine particles, usually owing to a change in pH of the aqueous environment that causes a change in drug solubility. Other thin arrows indicate secondary pathways that are usually inconsequential in achieving therapeutic efficacy. The dissolution process is primarily dependent on pharmaceutical variables, with the possible exception of a pH dependency, which may be a patient variable. The relative importance of the various processes in Fig. 9 may be explained in terms of the equation developed by Nernst and Brunner [18] whereby dissolution is described by a diffusion layer model:

$$\frac{dQ}{dt} = \frac{D}{h} S(C_s - C_g) \tag{1}$$

where

Q = amount of drug involved
t = time
D = diffusion coefficient of the drug in the solubilizing fluids of the GIT
S = effective surface area of the drug particles
h = thickness of a stationary layer of solvent around the drug particle
C_s = saturation solubility of the drug in the stationary layer, h
C_g = concentration of drug in the bulk fluids of the GIT

In deriving this equation, it was assumed that the drug dissolved uniformly from all surfaces of the particles. The particles were assumed to be spherical and all of the same size. In addition, h was assumed to be constant, and both h and C_s were assumed to be independent of particle size. Figure 10 schematically illustrates the relationship of the terms in Eq. (1) to the dissolution process. It should be emphasized that this crude model does not completely describe the dissolution process. The model has, however, helped to explain many experimental results. The following subsections will consider the pharmaceutical implications of the terms given in Eq. (1).

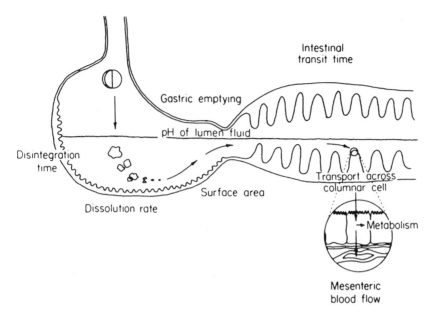

Fig. 8 Factors affecting the rate of absorption of drug from the gastrointestinal tract. (From Ref. 17.)

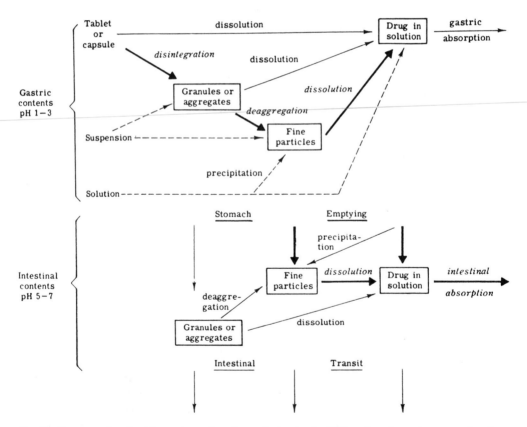

Fig. 9 Processes involved in getting a drug into solution in the GIT so that absorption may take place. Heavy arrows indicate primary pathways that the majority of drugs administered in a particular dosage form undergo. Dashed arrows indicate that the drug is administered in this state in the dosage form. Thin continuous arrows indicate secondary pathways, which are usually inconsequential in achieving therapeutic efficacy. (From Ref. 17.)

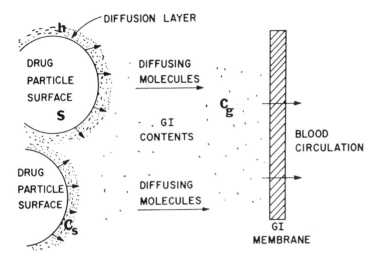

Fig. 10 Schematic diagram of the dissolution process. (From Ref. 1.)

A. Effective Surface Area of the Drug, S

Particle Size

The smaller the drug particles, the greater the surface area for a given amount of drug (e.g., 1 g). Thus Eq. (1) predicts that dissolution rate will increase as particle size decreases. Figure 11 confirms this expectation. The smaller granules of phenacetin dissolve more quickly than the larger granules, and there is a graded response; that is, for the five size ranges an increasing amount of drug dissolves over a specified time as the particle size decreases and surface area

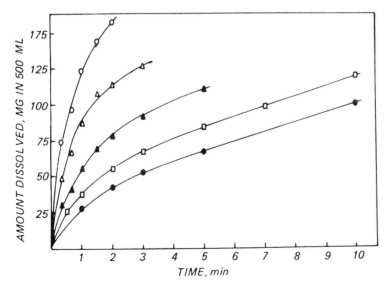

Fig. 11 Effect of particle size of phenacetin on dissolution of drug from granules containing starch and gelatin. ○, Particle size 0.11–0.15 mm; △, particle size 0.15–0.21 mm; ▲, particle size 0.21–0.30 mm; □, particle size 0.30–0.50 mm; ●, particle size 0.50–0.71 mm. (From Ref. 19.)

increases. However, in this example the hydrophobic drug phenacetin has been manipulated by a pharmaceutical manufacturing process called granulation (see Chapter 9 for further discussion) whereby the hydrophilic diluent gelatin has been incorporated into the particle. An opposite effect is seen when phenacetin particles themselves are dissolved in 0.1 N HCl (see the solid lines in the lower portion of Fig. 12). In this dissolution study of the hydrophobic drug, the dissolution rate increases with increasing particle size and decreasing particle surface area, in direct contradiction of Eq. (1). However, when a surface-active agent, Tween 80, is added to the 0.1 N HCl medium, the dissolution rate of phenacetin particles increases as the particle size decreases (see the dashed lines in the upper portion of Fig. 12). It is probable that decreasing the particle size of a hydrophobic drug actually decreases its *effective* surface area (i.e., the portion of the surface actually in contact with the dissolving fluids). In fact, the smaller phenacetin particles had more air adsorbed on their surfaces and actually floated on the dissolution medium. When a surface-active agent was added to the dissolution medium, the smaller particles were more readily wetted. Thus, their absolute surface area became their effective surface area. In Fig. 11 the granulation process has incorporated the hydrophobic phenacetin particles into the hydrophilic granules where the surface area of the granule approximated the effective surface area.

Figure 13 compares the dissolution rate of pure phenacetin particles with the granulated particles of the same size and with tablets prepared from these granules. These comparisons were made by using dilute gastric juice as the dissolution medium. Gastric juice has a relatively low surface tension, 42.7 dyn/cm, compared with water, which has a surface tension of approximately 70 dyn/cm. The low surface tension of the gastric juice aids in the wetting of both the hydrophobic particles and the hydrophilic granules. When these hydrophilic granules were compressed into tablets, the dissolution rate of phenacetin decreased, but remained greater than the rate of dissolution for the hydrophobic powder. It is possible that the large tablets do not rapidly disintegrate into the smaller granules, as indicated by the apparent lag time before

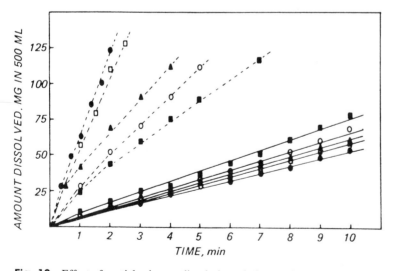

Fig. 12 Effect of particle size on dissolution of phenacetin. ●, Particle size 0.11–0.15 mm; □, particle size 0.15–0.21 mm; ▲, particle size 0.21–0.30 mm; ○, particle size 0.30–0.50 mm; ■, particle size 0.50–0.71 mm; ——, dissolution medium, 0.1 N HCl; - - - -, dissolution medium, 0.1 N HCl containing 0.2% Tween 80. (From Ref. 19.)

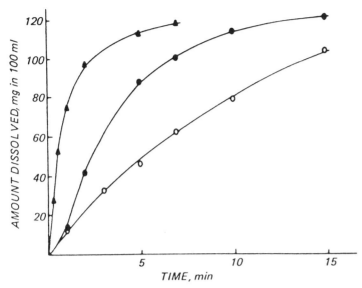

Fig. 13 Rate of dissolution of phenacetin from powder, granules, and tablets in diluted gastric juice (surface tension, 42.7 dyn/cm; pH 1.85). ○, Phenacetin powder; ▲, phenacetin granules; ●, phenacetin tablets. (From Ref. 19.)

dissolution begins. However, when the tablets do disintegrate, the hydrophilicity of the granules enables them to be more readily wetted than the powdered form.

In addition to these in vitro demonstrations of the importance of the effective surface area of drug particles on dissolution rate, many in vivo studies are available. Phenacetin plasma levels versus time are plotted for three different particle sizes of phenacetin in Fig. 14. Healthy adult volunteers received 1.5-g doses of phenacetin as an aqueous suspension on an empty stomach. The results show that both the rate and extent of availability of the phenacetin increase as particle size decreases. Since the large particles dissolve very slowly, the dosage form may pass through the GIT before dissolution is complete. When Tween 80 is added to the dosage form, the rate and extent of availability of the phenacetin increases even more, perhaps as a result of increasing the wettability of the particles, as shown in Fig. 12 for an in vitro study.

The effect of particle size reduction on the bioavailability of nitrofurantoin was shown in Fig. 4. The microcrystalline form (< 10 μm) is more rapidly and completely absorbed from the tablet dosage form than is the macrocrystalline form (74–177 μm) from the capsule dosage form. This is not a completely satisfactory illustration of the effect of particle size on the rate and extent of availabiilty, since other manufacturing variables have not been held constant. Nevertheless, it does suggest some correlation between particle size, dissolution rate, and rate of availability.

In summary, it is the effective surface area of a drug particle that determines its dissolution rate. The effective surface area may be increased by physically reducing the particle size, by adding hydrophilic diluents to the final dosage form, or by adding surface-active agents to the dissolution medium or to the dosage form.

Disintegration and Deaggregation

The rate of disintegration of the dosage form and the size of the resulting aggregates can be the rate-limiting step in the dissolution process. Disintegration is a particularly important step

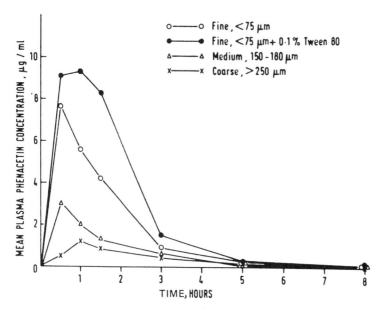

Fig. 14 Mean plasma phenacetin concentrations in six adult volunteers following administration of 1.5-g doses (in aqueous suspensions containing 200 mg of phenacetin per milliliter). (From Ref. 20.)

in the dissolution of drugs from coated dosage forms. Tablets or pellets of drugs contained within tablets or capsules are coated for several reasons, as will be discussed in Chapter 10. Rationales for the use of coating include the need to protect a drug during storage or from the very low pH of the stomach. Dosage forms may be coated so that they release their active ingredients slowly for prolonged action. Some drugs are coated to protect the patient's GIT from local irritation. Whatever the reason, these coats are made from materials having various degrees of hydrophilicity that may or may not break down and allow their active ingredients to dissolve. Some poorly formulated coated tablets do not break down at all, and these dosage forms can be recovered intact in the feces. Pharmacists must be aware of such pharmaceutical failures. They must assure themselves that the particular dosage forms they are dispensing to their patients meet all bioavailability standards, as discussed in Chapter 3.

As illustrated in Fig. 9, after a dosage form disintegrates into large particles, these large particles must deaggregate to yield fine particles. Hence, deaggregation may be a rate-limiting step in the dissolution process. Chloramphenicol plasma level versus time curves obtained after the oral administration to healthy volunteers of two 250-mg capsules of each of four different brands of chloramphenicol are plotted in Fig. 15. The in vitro deaggregation rate of the same four brands of chloramphenicol increased in the order D < B ≃ C < A. Thus, the in vivo rate and extent of availability of chloramphenicol from these capsule dosage forms correlates well with their in vivo deaggregation rate.

Table 1 summarizes in vitro studies on the effect of pH, a potential patient variable, on the dissolution rate of phenytoin from two different brands of capsules containing 100 mg of sodium phenytoin. Preparation A retained its compact capsule shape no matter what the initial pH of the dissolution medium. Moreover, the dissolution rate of phenytoin from preparation A was pH-independent. Preparation B, however, broke down into fine particles. When the initial pH was neutral, these particles rapidly dissolved. As the initial pH was lowered, these fine particles precipitated. Since these precipitated particles then dissolved very slowly, it would

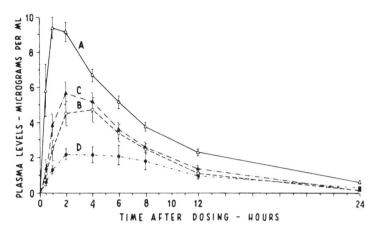

Fig. 15 Mean plasma levels for groups of ten human subjects receiving single 0.5-g doses (as two 250-mg capsules) of chloramphenicol: preparations A, B, C, or D. Vertical lines represent one standard error on either side of the mean. (From Ref. 21.)

appear that the precipitate was not the freely soluble sodium phenytoin, but the poorly soluble acid form of phenytoin. Since capsule A did not disintegrate or deaggregate, most of the sodium phenytoin was not exposed to the lower pH media and, as a result, was not converted to its very slowly dissolving acid form.

Effect of Manufacturing Processes

Various manufacturing processes can affect dissolution by altering the effective surface area of drug particles. Each of the individual processes mentioned here is discussed in more detail in Chapter 9. The effect of adding hydrophilic granulating agents to a dosage form has been discussed earlier (see Sec. III.A).

Lubricating agents are often added to capsule or tablet dosage forms so that the powder mass or the finished dosage forms will not stick to the processing machinery. When the hydrophilic lubricating agent sodium lauryl sulfate was added, 325-mg salicylic acid tablets dissolved in 0.1 N HCl more rapidly than did control tablets containing no lubricant, as shown

Table 1 In Vitro Dissolution Characteristics[a] of Two Commercial Phenytoin Sodium Capsule Preparations

Solvent system	Initial pH	Time for 50% dissolution (min); average of five capsules (SD)	
		Preparation A	Preparation B
175 ml 0.1 N HCl and 175 ml 0.1 N NaOH	~7	42.4 (5.6)	12.3 (3.2)
175 ml 0.1 N HCl for 30 min, then addition of 175 ml 0.1 N NaOH	~1	34.9[b] (5.6)	80.0[b] (26.2)
175 ml 0.01 N HCl for 30 min, then addition of 175 ml 0.01 N NaOH	~2	32.6[b] (10.4)	25.2[b] (8.6)

[a] Modified Levy-Hayes method, 55 rpm, 17°C.
[b] Time after addition of NaOH; less than 5% dissolved in HCl solution.
Source: Ref. 22.

in Fig. 16. If, however, the hydrophobic lubricant magnesium stearate was added, the dissolution rate decreased. Most lubricants, and all the effective lubricants, are very hydrophobic,
and they act by particle coating. Thus they must be properly formulated to avoid reducing
dissolution rate and bioavailability. Once again, increasing hydrophilicity of a dosage form
enhanced its dissolution rate in an aqueous medium, but increasing its hydrophobicity decreased
its dissolution rate.

The possible effects on dissolution rate of the force used in compressing the drug–diluent
mixture into a tablet dosage form have been summarized in Fig. 18. As compression force is
increased, the particles may be more tightly bound to one another. (Part I of Fig. 18 best
represents this possibility.) On the ther hand, it is also possible that higher pressures may
fracture the particles so that they break into yet smaller particles (see part III of Fig. 18).
Depending on which of these two extremes is dominant for a given formulation, any of the
combinations illustrated in Fig. 18 is possible. Furthermore, a sum of any of them may result.
Thus, the effect of the compression force on the dissolution rate of a tablet dosage form would
appear to be unpredictable.

The packing density, as illustrated in Fig. 19, may affect the rate of release of a drug (not
identified in these studies) from a capsule dosage form [25]. For the rapidly dissolving formulation (see Fig. 19a), packing density had no effect on release rate. For the slowly dissolving
formulation (see Fig. 19b), however, increasing packing density decreased dissolution rate. It
is probable that further studies of the effect of packing density on dissolution rate for different

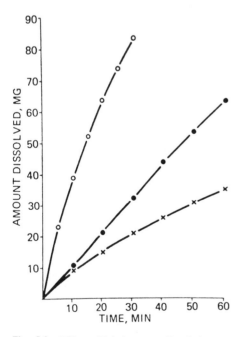

Fig. 16   Effect of lubricant on dissolution rate of salicylic acid contained in compressed tablets. ×, 3%
magnesium stearate; •, no lubricant; ○, 3% sodium lauryl sulfate. (From Ref. 23.)

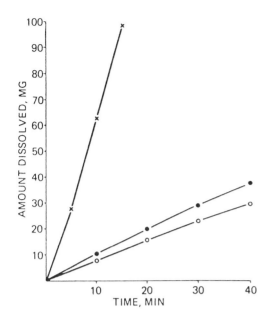

Fig. 17   Effect of starch content of granules on dissolution rate of salicylic acid contained in compressed tablets. ○, 5%; ●, 10%; ×, 20% starch in granules. (From Ref. 24.)

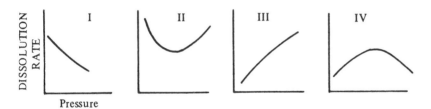

Fig. 18   Effect of compression pressure on dissolution rate. See the text for an explanation. (From Ref. 19.)

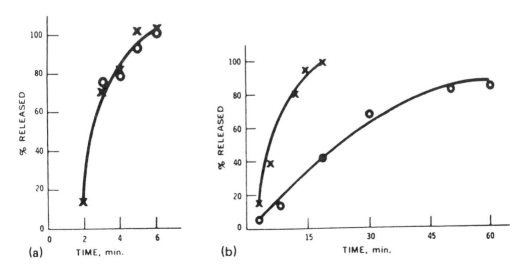

Fig. 19   Influence of packing density on dissolution of drug from capsules in simulated gastric fluid. ×, Regular packing: 355 mg/No. 2 capsule; ○, dense packing: 400 mg/No. 2 capsule. (a) Drug, CaHPO$_4$; (b) drug, CaHPO$_4$, 5% magnesium stearate. (From Ref. 25.)

drugs from different formulations would result in both increasing and decreasing dissolution rates.

From the very few studies discussed here, of the many that have been published, it would appear that manufacturing processes can determine the dissolution rate of a drug from its final dosage form. Whether changes in these manufacturing variables are beneficial or detrimental to the ultimate bioavailability of a drug depends on the physicochemical properties of the drug and its dosage form. Pharmacists should be aware of the possible effects that "inert" ingredients and manufacturing methods, which are usually carefully guarded trade secrets, may have on the bioavailability of the drug products they select to dispense to their patients.

B. Saturation Solubility of the Drug, C_s

The next term in Eq. (1) that can be manipulated is C_s, the saturation solubility of the drug. This variable can be influenced by both patient and pharmaceutical variables. The patient variables include the changes in pH as well as the amounts and types of secretions along the GIT. Additionally, both the physical and chemical properties of a drug molecule can be modified to increase or decrease its saturation solubility.

Salt Form of the Drug

Since the salt form of a drug is more soluble in an aqueous medium, the dissolution rate for the salt form of a drug should be greater than the dissolution rate of the nonionized form of the drug. However, the solubility of the salt depends on the counterion; generally, the smaller the counterion, the more soluble the salt. If dissolution is the rate-determining step in absorption, then, for a series of salts and the nonionized form of a drug, it can be anticipated that the rate of availability of the drug will increase as solubility increases. The weakly acidic drug p-aminosalicylic acid (PAS) is available as the potassium, calcium, or sodium salt. The solubility of nonionized PAS is 1 g/600 ml of water; of KPAS, 1 g/10 ml; of CaPAS, 1 g/7 ml; and of NaPAS, 1 g/2 ml. Healthy adult volunteers took, on a fasting stomach with 250 ml of water, tablets containing either 4 g of PAS or tablets containing 4 g of one of the salts (containing 2.6–2.8 g of PAS) [26]. Mean plasma concentrations (corrected to a 4-g dose of PAS) versus time curves are shown in Fig. 20. Although the rates of availability of the salts are not significantly different, their rank order does correlate with solubility. All the salt forms were more rapidly available than the nonionized PAS to a significant extent. Furthermore, the extent of availability of the acid form is only about 77% of that of the salt forms, indicating that the drug may not have completely dissolved before the dosage form moved through the GIT. Since commercial tablets were used, it is probable that other dosage form variables (as discussed in Sec. III.A) were not held constant. Thus, the decrease in rate and extent of availability of the PAS may not be exclusively due to differences in solubility.

pH Effect

As indicated, the ionized form of a drug will be more soluble than the nonionized form in the aqueous fluids of the GIT. The classic studies on the beneficial effects of changing nonionized drugs into salt forms were reported by Nelson for tetracycline [27], and Nelson et al. for tolbutamide [28]. Table 2 combines portions of the data from each study. Urinary excretion of the drug or its metabolite was taken as the in vivo measure of the relative absorption rate for the salt and the nonionized form of each drug. No comparison can be made between the two drugs, and they are combined here only to illustrate that the same principles hold for both positively and negatively charged drug ions. Note that the salt forms of the drug dissolve much faster than the nonionized forms (or zwitterion for tetracycline) in all media and that more of the salt forms of the drug are absorbed and subsequently excreted in each period. For a dis-

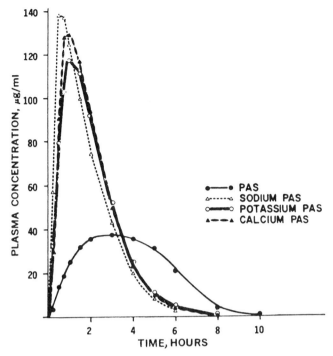

Fig. 20 Mean plasma concentrations of unchanged drug from 12 subjects following administration of four different preparations of *p*-aminosalicylic acid (PAS). Data were corrected to 70 kg of body weight and to a dose equivalent of 4 g of free acid. (From Ref. 26.)

Table 2 Correlation of Dissolution Rates with Biological Measurements for Tolbutamide and Tetracycline Absorption in Humans

Drug as nondisintegrating pellet	In vitro dissolution rate[a] (mg cm^{-2} hr^{-1}) 0.1 N HCl[b] or simulated gastric fluid[a]	pH 7.2 buffer[b] or simulated neutral intestinal fluid[c]	Average amount excreted (mg) to time indicated[d] 1 hr	2 hr	3 hr	Lowering of blood sugar level (mg/100 ml) after 1 hr
Tolbutamide	0.21 (N → N)	3.1 (N → I)	5	7	12	5.2
Sodium tolbutamide	1069 (I → N)	868 (I → I)	21	65	117	9.1
Tetracycline	2.6 (N → I)	0.001 (N → N)	0.2	1.5	3.3	
Tetracycline HCl	4.1 (I → I)	7.8 (I → N)	3.0	12.0	20.4	

[a] The N → N, etc. designations for the dissolution data are explained in the text.
[b] Tolbutamide study.
[c] Tetracycline study.
[d] Tolbutamide excretion measured as the carboxytolbutamine metabolite.
Source: Refs. 27 and 28.

sociable drug, either an acid or a base, such as tolbutamide or tetracycline, the pH of the GI fluids will determine whether the drug is ionized or nonionized. The dependence of the dissolution rate of a drug on pH is not evident in the Nernst-Brunner equation as written in Eq. (1). That equation may be rewritten as

$$\frac{dQ}{dt} = \frac{DS}{h}(C_h - C_g) \tag{2}$$

where Q, t, D, S, h, and C_g are defined as before, and C_h = saturation solubility of the drug in the boundary layer, h, at any particular pH; $C_h = C_s$ when the drug is dissolving in an aqueous solution in which it is totally nonionized. Then, for a weakly acidic drug:

$$AH \underset{}{\overset{K_a}{\rightleftharpoons}} A^- + H^+; \qquad K_a = \frac{[A^-][H^+]}{[AH]}$$

$$C_h = [AH] + [A^-]$$

$$C_h = [AH]\left(1 + \frac{K_a}{[H^+]}\right) \qquad \text{but,} \qquad C_s = [AH]$$

and, therefore,

$$C_h = C_s\left(1 + \frac{K_a}{[H^+]}\right)$$

so

$$\frac{dQ}{dt} = \frac{DS}{h}\left\{C_s\left(1 + \frac{K_a}{[H^+]}\right) - C_g\right\} \tag{3}$$

Therefore, as pH increases the dissolution rate of a weak *acid* increases. Similarly, for a weak base:

$$BH^+ \overset{K_a}{\rightleftharpoons} B + H^+; \qquad K_a = \frac{[B][H^+]}{[BH^+]}$$

$$C_h = [BH^+] + [B]$$

$$C_h = [B]\left(1 + \frac{[H^+]}{K_a}\right) \qquad \text{but,} \qquad C_s = [B]$$

and, therefore

$$C_h = C_s\left(1 + \frac{[H^+]}{K_a}\right)$$

so

$$\frac{dQ}{dt} = \frac{DS}{h}\left\{C_s\left(1 + \frac{[H^+]}{K_a}\right) - C_g\right\} \tag{4}$$

Thus, as the pH increases, the dissolution rate of a weak *base* decreases. Referring to Table 2, we can see that, for the weak acid tolbutamide, the dissolution rate increases as pH is increased, as predicted by Eq. (3). Additionally, for the weak base tetracycline, as predicted by Eq. (4), the dissolution rate decreases as pH is increased. Thus far, the more rapid dissolution of the salt forms of these two drugs and the direction of change of the dissolution rate with pH have been accounted for with Eqs. (1) to (4). However, there are six possible dissolution rate comparisons that can be made for each set of nonionized drug and its salt in the two buffers, as discussed by Benet [17]. The letters in Table 2 indicate similar measurements in the two studies. Thus N → N indicates a nonionized solid drug dissolving in a medium in which the dissolved solute in the bulk solution will be nonionized; N → I indicates a nonionized solid drug dissolving in a medium in which the dissolved solute in the bulk solution will be ionized; I → I indicates an ionized solid drug dissolving in a medium in which the dissolved solute will be ionized; and I → N indicates an ionized solid drug becoming nonionized solute after dissolution.

Process I → I should be faster than process N → N, since the solubility of the dissolving substance will be much greater for the salt than for the nonionized molecule. A similar explanation can be used for (I → N) > (N → N) and (I → I) > (N → N) and (I → I) > (N → I), but it must be remembered that according to Eqs. (1), (3), and (4), C_s is the saturation solubility of the drug in the diffusion layer, not in the bulk solution. It is believed that the dissolving solid acts as its own buffer and changes the pH of the liquid environment immediately surrounding the solid particle; thus, the dissolution rate should be governed by the solubility of the drug in the buffered diffusion layer. The I → N and N → N comparison is especially significant in the oral administration of weakly acidic drugs and their salts, since the acidic region of the stomach is the first solvent medium encountered following normal oral dosing. Frequently, administering the sodium or potassium salt of an acidic drug actually speeds up absorption by increasing the effective surface area of the solid drug according to the following hypothesized process. The salt acts as its own buffer in the diffusion layer and goes into solution in this layer. However, when the salt molecules diffuse out of the layer and encounter the bulk solution, they precipitate out as very fine nonionized prewetted particles. The large surface area thus precipitated favors rapid dissolution when additional fluid becomes available for one of the following reasons: (a) dissolved particles are absorbed; (b) more fluid accumulates in the stomach; or (c) the fine particles are emptied into the intestine. The classic study of Lee et al. [29] comparing serum levels of penicillin V following administration of the salt and free acid to dogs is explained by this phenomenon also. However, in at least three cases—aluminum acetylsalicylate [30], sodium warfarin [31], and the pamoate salt of benzphetamine [32]—administration of the salt slowed dissolution of the drug and subsequent absorption as compared with the nonionized form. This decrease appears to be due to precipitation of an insoluble particle or film on the surface of the tablet, rather than in the bulk solution. Precipitation of an insoluble particle or film onto the surface of the tablet decreases the effective surface area by preventing deaggregation of the particles.

The comparison of I → N and N → I may also be explained by the buffered pH in the diffusion layer and leads to an interesting comparison between a process under kinetic control versus one under thermodynamic control. Because the bulk solution in process N → I favors formation of the ionized species, a much larger quantity of drug could be dissolved in the N → I solvent if the dissolution process were allowed to reach equilibrium. However, the dissolution rate will be controlled by the solubility in the diffusion layer; accordingly, a faster dissolution of the salt in the buffered diffusion layer (process I → N) would be expected. In comparing N → I and N → N, or I → N and I → I, the pH of the diffusion layer is identical in each set, and the differences in dissolution rate must be explained either by the size of the diffusion layer, or by the concentration gradient of drug between the diffusion and the bulk solution. It is probably safe to assume that a diffusion layer at a different pH than that of the bulk solution is thinner than a diffusion layer at the same pH because of the acid–base interaction at the interface. In addition, when the bulk solution is at a different pH than that of the diffusion layer, the bulk solution will act as a sink and C_g can be eliminated from Eqs. (1),

(3), and (4). Both a decrease in the h and C_g terms in Eqs. (1), (3), and (4) favor faster dissolution in processes N → I and I → N as opposed to N → N and I → I, respectively.

Although the explanation for (N → I) > (N → N) and (I → N) > (I → I) is self-consistent for a nonionized drug and its salt form, and reflects the experimentally observed values in Table 2, Nelson [33] studied a series of weak organic acids and found (I → I) > (I → N) for the sodium salt of four of these compounds. For example, the dissolution rate of sodium benzoate in pH 6.83 buffer was 1770 mg/100 min · cm^2 versus 980 mg/100 mm · cm^2 in a pH 1.5 solution. Corresponding values of I → I and I → N were 820 versus 200 for sodium phenobarbital, 2500 versus 1870 for sodium salicylate, and 810 versus 550 for sodium sulfathiazole. The acid forms of these drugs all showed the expected (N → I) > (N → N) relationship, and we cannot yet explain the salt data [33]. As stated earlier, the preceding explanation of the data in Table 2 is presented with reference to a specific theory of dissolution. Although this theory may not be acceptable to some, it does provide a basis for understanding the general principles that dictate the (I → N) ≈ (I → I) > (N → I) > (N → N) relationship observed in dissolution rate measurements for a nonionized drug and its salt form.

The pH of a solution affects not only the active ingredients of a dosage form, but also the inert ingredients. Returning to Table 1, the pH dependence of two different commercial capsules of sodium phenytoin were studied [22]. For one capsule, dissolution rate was independent of the initial pH of the dissolution media. For the second capsule, dissolution rate was very much dependent, as has been discussed, on the initial pH. Similar studies have been made using three commercial dosage forms of nitrofurantoin (Table 3). Nitrofurantoin is a weak acid with $pK_a = 7.2$. Equation (3) predicts that as the pH of the dissolution medium increases, the dissolution half-life of nitrofurantoin should decrease. For the suspension of microcrystalline nitrofurantoin and for the capsule, containing 100 mg of nitrofurantoin in the macrocrystalline form, the experimental results follow this prediction. For the tablet, containing 100 mg of nitrofurantoin in the microcrystalline form, the dissolution half-life increases with increasing pH. However, the tablet did not disintegrate at pH 7.20. Thus, for the tablet, it would appear that the physicochemical properties of the dosage form, rather than physicochemical properties of the drug, determined the rate of release of the drug from its dosage form. Interestingly, both the capsule and tablet, which must both disintegrate, are less rapidly available than the suspension. (See the dissolution scheme shown in Fig. 9.)

It has been mentioned that the pH of the bulk fluids may not reflect the pH of the stationary diffusion layer. As the active or inert ingredients of a dosage form dissolve, they may alter the pH of the microenvironment of the stagnant diffusion layer, without significantly changing the

Table 3 Effect of pH on the Dissolution Half-Life of Nitrofurantoin from Commercial Dosage Forms at 37°C

	Mean dissolution half-life[a] (min)	
Commercial dosage form	pH 1.12	pH 7.20
Aqueous suspension[b]	12.5 (1.2)	2.64 (0.34)
Compressed tablet [a]	77.9 (19.0)	167.0 (35)
Gelatin capsule[d]	212.0 (44)	160.0 (24)

[a] Determined from log-normal probability plots of individual dissolution rate data. Mean of five determinations (SD).
[b] Furadantin suspension containing microcrystalline (<10 μm) drug.
[c] Furadantin tablets containing microcrystalline (<10 μm) drug.
[d] Macrodantin capsules containing macrocrystalline (74–177 μm) drug.
Source: Ref. 34.

pH of the bulk fluids. Moreover, it is possible to intentionally add buffering agents to a capsule of tablet dosage forms. The small amount of buffer may not alter the pH of the bulk fluids of the GIT, but can buffer the pH in the diffusion layer to a pH favoring rapid dissolution of the active ingredient. Such an effect was hypothesized when dissolution rates were compared for buffered versus plain (no alkaline additives) tablets of aspirin [35]. Later, more extensive studies on the effects of a variety of buffers on the dissolution of aspirin tablets were carried out [36]. Hydrophilic buffers and buffers that released carbon dioxide increased the dissolution rate; however, hydrophobic buffers, decreased the dissolution rate, possibly by effectively waterproofing the tablets.

In summary, the effect of pH on the dissolution rate of a drug from an oral dosage form depends on (a) the pH of the GI fluids, a patient variable; (b) the acid or base strength of the drug, a pharmaceutical variable; as well as (c) the physicochemical properties of the dosage form, another pharmaceutical variable. Furthermore, by intentionally designing the dosage form such that it buffers the diffusion layer, we can control a patient variable by a pharmaceutical variable.

Pharmacists are probably the only health professionals who, by training, understand pH effects on drug solubility and transport. They must be aware of why different drugs are prepared as different salt forms and how changes in the acid–base environment of the GIT will influence availability. Usually, drugs in a salt form will be more quickly available and, often, available to a greater extent. However, there are instances when the drug may not be formulated in the salt form, even when a proper, stable formulation of the salt may be produced. Such an example was shown in Table 2. Sodium tolbutamide dissolves faster than tolbutamide when both in vitro dissolution data and in vivo urinary excretion measurements are compared. As would be expected, the pharmacological effect (i.e., lowering of the blood sugar level, also is more pronounced after 1 hr for the salt form). However, the product is actually formulated containing the acidic nonionized tolbutamide, because the manufacturer does not want to induce such a rapid decrease in blood sugar, which might lead to diabetic coma.

Solvate Formation

Another variable that influences the saturation solubility of a drug molecule is its degree of solvation. Since the anhydrous, hydrated, and alcoholated forms of a drug have slightly different solubilities, they may well have different dissolution rates and, therefore, different rates of absorption. However, these differences may not be clinically significant [37].

Polymorphism

Yet another property of a drug that may affect its saturation solubility and, hence, its dissolution rate is its crystalline state. Many drugs exhibit polymorphism [38]; that is, they are available in an amorphous or several different crystalline states. Chloramphenicol palmitate is available in at least two different forms, A and B. The B form is the more soluble. Figure 21 shows the mean serum level of chloramphenicol versus time curves obtained after oral administration to male volunteers of chloramphenicol palmitate suspensions containing the equivalent of 1.5 g of chloramphenicol [39]. The fraction of the B form of chloramphenicol palmitate ranges from 0% in suspension M, to 100% in suspension L. Since the suspensions were identical except for the crystalline form, it would appear that the increase in rate and extent of availability with an increasing percentage of the B form is due to the increasing rate of dissolution. A warning, however, is necessary. In general, the more soluble polymorph is the least stable thermodynamically. Thus, older dosage forms may contain more of the more stable, but less soluble, polymorph. Aging may significantly affect the bioavailability of drugs exhibiting polymorphism. Pharmacists must, for this and many other reasons, be aware of the storage requirements

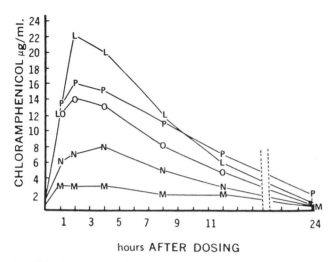

hours AFTER DOSING

Fig. 21 Comparison of mean blood serum levels obtained with chloramphenicol palmitate suspensions containing varying ratios of A and B polymorphs, following a single oral dose equivalent to 1.5 g of chloramphenicol. Percentage of polymorph B in the suspension: M, 0%; N, 25%; O, 50%; P, 75%; L,100%. (From Ref. 39.)

of all drugs dispensed. They must also counsel their patients on the proper storage of their medicines and the imortance of discarding out-of-date drugs.

Complexation

Several studies have examined the effects of complex formation on the rate and extent of drug availability. A drug may complex with both absorbable and nonabsorbable excipients in a dosage form. This complexation may occur within the dosage form or in the solubilizing fluids, and the resulting complex may be more or less soluble than the drug itself. The rate and extent of availability of promazine from a suspension of the drug adsorbed on either activated charcoal or the clay attapulgite were compared with the availability of the drug from an aqueous solution [40]. Healthy adults took, on an empty stomach, 45 ml of either the solution or one of the suspensions with 4 oz (0.12 liters) of water. The rate of availability of promazine from the attapulgite suspension was less than from the solution, but the extent of absorption was the same. When compared with the solution, the promazine–activated charcoal suspension showed both a decreased rate and a decreased extent of availability. However, the association constant for the promazine–attapulgite complex was greater than for the promazine–activated charcoal complex. Thus, it appears that it is not the magnitude of the association constant of the complex, but the rate at which the complex dissociates, that determines whether absorption of the drug is as rapid or as complete as in the absence of complex formation.

In addition to possible complexation with other ingredients in the dosage form, a drug may complex with the natural components of the GIT. The mucin, enzymes, bile, and physiological surfactants found in the mucosal fluids may interact with a given drug. The influence of these interactions on the availability of the drug will depend on the physicochemical properties of the drug and those of the endogenous compound, as discussed in Chapter 2. It seems probable that both increased and decreased drug solubilities will be encountered. Thus, for drugs for which the rate of availability is dissolution rate-limited, such interactions may be beneficial or detrimental. Once again, patient variables may control the availability of the drug from its dosage form.

Solid–Solid Interactions

Preparing a solid solution of a drug is one additional way of controlling its dissolution rate. For this chapter, the term *solid solution* will be used to describe any solid system in which one component is dispersed at the molecular level within another [41]. In deriving Eq. (1) it was assumed that C_s is independent of particle size. However, for extremely small particles, such as those found in solid solutions, the saturation solubility (C_s) of the drug may increase as particle size decreases. This increase in solubility can be described by the Kelvin equation [42]:

$$C_s^{micro} = C_s \exp \left(\frac{2\gamma M}{r\rho RT} \right) \tag{5}$$

where

C_s^{micro} = saturation solubility of the microscopic particle
C_s = saturation solubility of drug (macroparticles)
γ = interfacial tension between drug particles and the solubilizing fluids
M = molecular weight of the drug
r = radius of the microscopic drug particle
R = ideal gas constant
ρ = density of the microscopic drug particle
T = absolute temperature

Figure 22 depicts the dissolution rates of griseofulvin, an insoluble antifungal agent, and griseofulvin–succinic acid samples as described by Goldberg et al. [43]. Several conclusions can be drawn from these experiments. The first is that physically reducing the particle size of the griseofulvin, by micronization, increased the rate of dissolution in accordance with Eq. (1), but did not increase the concentration of drug above its equilibrium solubility. Second, physical mixtures of griseofulvin and succinic acid, at ratios corresponding to those in the solid solution and eutectic mixture, had dissolution rates similar to griseofulvin alone. Therefore, succinic acid does not itself increase the solubility of griseofulvin (by a complexation mechanism, for example). Finally, the solid solution and the eutectic mixture were very rapidly soluble. Both gave supersaturated solutions, as predicted by Eq. (5), because of the extremely small—approaching molecular—size of the griseofulvin particles in these states. Solid solutions and eutectic mixtures have been used, at least in research laboratories, to increase the rate of dissolution of drugs, presumably by decreasing the particle size of the drug molecules [44]. However, since the drug is in an unstable thermodynamic state, it is critical that the stability of the formulation be maintained. Otherwise, the preparation may revert to a state of having even less favorable dissolution properties than those of a micronized formulation, as shown by the dissolution characteristics of the physical mixtures in Fig. 22. Pharmacists should be aware of the potential problems of such formulations and may, once again, take an active role in advising patients of proper storage conditions.

C. Concentration of the Dissolved Drug in Bulk Solution, C_g

In Eq. (1), the saturation solubility of the drug in the diffusion layer does not of itself determine dissolution rate. The determinant is, rather, the difference between C_s and C_g, the concentration of dissolved drug in the bulk fluids of the GIT. Thus, the driving force for dissolution is the concentration gradient, $C_s - C_g$. It is usually assumed that C_g is much smaller than C_s, meaning that dissolution occurs under sink conditions. If a drug is absorbed through the GI membrane very slowly, drug concentration in the GIT fluids may build up. This buildup would, by decreasing the gradient, decrease the dissolution rate. Additionally, the bulk fluids may not be

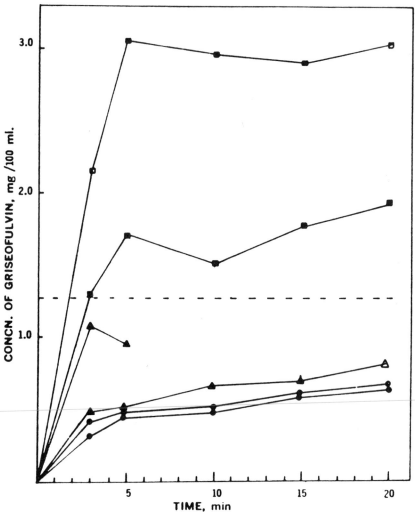

Fig. 22 Dissolution rates of various griseofulvin and griseofulvin-succinic acid samples as determined by the oscillating bottle method. ●, Griseofulvin, crystalline; ▲, griseofulvin, micronized; ■, eutectic mixture; ○, physical mixture at eutectic composition; □, solid solution; △, physical mixture at solid solution composition. The dashed line indicates the equilibrium solubility of griseofulvin in water. (From Ref. 43.)

identical with the solubilizing fluids of the diffusion layer. A drug may be more soluble in the boundary layer and then precipitate in the bulk fluids, especially if the pH differs between these two sites. However, these precipitated particles should be quite small, and thus may rapidly redissolve (see the dissolution scheme presented in Fig. 9). On the other hand, the drug may be more soluble in the bulk fluids than in the boundary layer because of a difference in pH or by complexation with other components. Here, the gradient may decrease and dissolution may slow down or even stop. However, the volume of the bulk fluids is much larger than the volume of the boundary layer. Thus, a large absolute amount of drug in the bulk fluids may still give a small value of C_g.

Figure 3 illustrates a situation in which this may not be true. When 250 ml of water was taken with the erythromycin tablets, the extent of absorption was much greater than when the tablets were taken with only 20 ml of water. In the latter case, dissolution probably did not occur under sink conditions. Hence, the dissolution rate decreased, and it appears that not all of the erythromycin had a chance to dissolve in the GIT. Note that the dissolution was not, however, the rate-determining step in absorption, since the time to reach the peak concentration was the same in all situations.

Usually, an increase in C_g that would affect the dissolution rate would occur only when another process, such as membrane transport or stomach emptying, becomes the rate-limiting step in drug absorption. As a general rule pharmacists should advise patients to take their oral medications with a full glass of water to ensure that dissolution occurs under optimal conditions.

D. Diffusion Coefficient Divided by the Thickness of the Stationary Diffusion Layer, D/h

Although it is possible to control the dissolution rate of a drug by controlling its particle size and solubility, the pharmaceutical manufacturer has very little, if any, control over the D/h term in the Nernst-Brunner equation, Eq. (1). In deriving the equation it was assumed that h, the thickness of the stationary diffusion layer, was independent of particle size. In fact, this is not necessarily true. The diffusion layer probably increases as particle size increases. Furthermore, h decreases as the "stirring rate" increases. In vivo, as GI motility increases or decreases, h would be expected to decrease or increase. In deriving the Nernst-Brunner equation, it was also assumed that all the particles were spherical and of the same size. In fact, particles in pharmaceutical systems are neither spherical nor uniform. Furthermore, pharmaceutical systems are usually mixtures of different compounds, the particle sizes of which are not identical. Thus pharmaceutical systems are polydisperse and multiparticulate [45]. Their size distribution in terms of number of particles tends to be skewed toward the smaller particles. Furthermore, as dissolution proceeds, the particles become smaller. Hence h can vary considerably initially and throughout the dissolution process.

The other virtually uncontrollable term in Eq. (1) is D, the diffusion coefficient of the drug. For a spherical, ideal drug molecule in solution,

$$D = \frac{kT}{6\pi\eta r}$$

where

k = Boltzmann's constant
T = absolute temperature
r = radius of molecule in solution
η = viscosity of the solution

Both k and 6π are constants. The radius r is a property of the drug molecule not subject to manipulation. The viscosity η of the GI fluids can vary. Increasing the viscosity will decrease dissolution. In addition, increasing the viscosity of the stomach contents will slow gastric emptying, thereby delaying delivery of the drug to the absorption site. Increasing the temperature of the GI fluids tends to increase diffusion. Patients might be advised to take oral dosage forms with warm liquids. However, extremely hot liquids will delay stomach emptying. In summary, D and h are largely uncontrollable factors, although they may influence drug availability.

IV. GASTROINTESTINAL MEMBRANE TRANSPORT

Once a drug is in solution at the absorption site, it must move through the GIT membrane and then into the general circulation. In Chapter 2 Fick's law was used to describe this transport:

$$\left(\frac{dQ_b}{dt}\right)_{g \to b} = D_m A_m R_{m/aq} \left[\frac{(C_g - C_b)}{\Delta X_m}\right] \tag{6}$$

Paradoxically, the larger the value for C_g, the more quickly the drug will move through the membrane. This C_g is the same concentration of drug in the bulk fluids that was minimized in Eq. (1) to increase the rate of dissolution. A second paradox is the relation between $R_{m/aq}$ and C_g. The bulk fluids of the GIT are aqueous. Thus C_g will increase as the water solubility of the drug increases; $R_{m/aq}$ will, however, decrease. Thus, to be absorbed, a drug cannot be so lipid-soluble that it will not dissolve in the aqueous fluids of the GIT, nor so water-soluble that it will not penetrate the lipid GI membrane. Although diffusion across the GI membrane is one of the major rate-limiting steps in defining the rate and extent of drug availability for oral dosage forms, this topic has been extensively covered in Chapter 2 and need not receive further treatment here.

V. MOVING THE DRUG AWAY FROM THE SITE OF ABSORPTION

The fourth basic process that may influence drug absorption is moving the drug away from the site of absorption. Drugs that have crossed the GI membrane are primarily removed as a function of blood flow [46]. It can be seen in Eq. (6) that if there were no blood flow, the concentration of drug in the blood, C_b, would quickly approach C_g, and net transfer of drug across the GIT would cease. Thus, a decreased blood flow might decrease the rate of removal of passively absorbed drugs [47,48]. Decreased flow could possibly also interfere with active transport systems owing to the reduction of the supply of oxygen to the tissues. Winne and Ochsenfahrt [48] and Winne [49] have developed models and derived equations for GI absorption considering blood flow and countercurrent exchanges, respectively. For the following theoretical discussion, a simplified equation is presented as a modification of Eq. (6) [50].

$$\left(\frac{dQ_b}{dt}\right)_{g \to b} = \frac{C_g - C_b}{1/P_m A_m + 1/\alpha BF} \tag{7}$$

where

$P_m = D_m R_{m/aq}/\Delta X_m$
  α = fraction of blood flowing through the capillaries near the GI membrane
  BF = GI blood flow

The other terms are defined in Chapter 2. The denominator of Eq. (7) can be interpreted as the resistance of the region between the GI lumen and the blood pool. Winne and Remischovsky [50] have divided this resistance into two parts (first and second terms of the denominator): (a) the resistance to transport of the region between the gastrointestinal lumen and the capillary blood (mainly the resistance of the membrane), and (b) the resistance to drainage by blood. Figure 23 indicates the influence of blood flow on the rate of intestinal absorption for eight substances from the jejunum of the rat. The absorption of highly permeable materials [those with a small first term in the denominator of Eq. (7)], such as very lipid-soluble or pore-diffusable substances should be blood flow-limited. Conversely, the absorption rate of drugs characterized by low membrane permeability [those with a large first term in the denominator

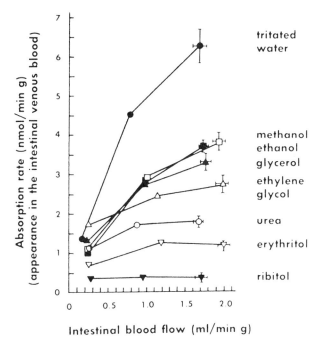

Fig. 23 Dependence of intestinal absorption on blood flow as reported by Winne and Remischovsky. All data are corrected to a concentration of 50 nmol/ml in the solution perfusing jejunal loops of rat intestine. Bracketed points indicate the 50% confidence intervals. (From Ref. 50.)

of Eq. (7)] may be independent of blood flow. From Fig. 23 it may be seen that the absorption of freely permeable tritiated water is very sensitive to blood flow, but that ribitol, a sugar that penetrates the GI membrane with great difficulty, is essentially unaffected by changes in the intestinal blood flow in the range studied. As would be expected from Eq. (7), the absorption rate of intermediate substances, such as urea, appear to be flow-limited at low blood flow rates, but then become insensitive to blood flow at higher rates. Winne and co-workers [48–50] have reported a blood flow dependence for several relatively small drug molecules. Crouthamel et al. [51] also noted a decrease in the absorption rate of sulfaethidole and haloperidol as a function of decreased mesenteric blood flow rates, using an in situ canine intestinal preparation. Haas and coworkers [52], using a guinea pig model, found a strong correlation between spontaneously varying portal blood flow and the amount of digitoxin and digoxin absorbed following intraduodenal infusion of the drug. As can be seen in Fig. 24, digitoxin, the most lipophilic of the three cardiac glycosides, showed the most pronounced effects as a function of the blood flow. Digoxin also showed increasing absorption rates as a function of blood flow. However, ouabain, the most hydrophilic of the three, showed no dependence on blood flow. These results are consistent with the predictions of Eq. (7) for the effects of blood flow on the absorption rate of drugs.

These animal studies should indicate to the pharmacist that blood flow can, under certain circumstances, be an important patient variable that may affect the absorption of drugs. Patients in heart failure would generally be expected to have a decreased cardiac output and, therefore, a decreased splanchnic blood flow. This could lead to a decreased rate of absorption for drugs when the blood flow rates in Eq. (7) become rate-limiting. In addition, redistribution of cardiac output during cardiac failure may lead to splanchnic vasoconstriction in patients [53]. Other

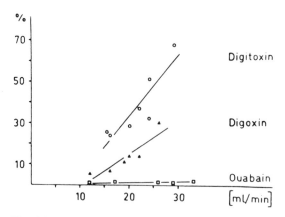

Fig. 24 Total amount of radioactivity absorbed during 1 hr, plotted versus portal blood flow. Ordinate: absorption as a percentage of the amount infused into the duodenum. Abscissa: mean portal blood flow in milliliters per minute. Each point represents one guinea pig. (From Ref. 52.)

disease states and physical activity can also decrease blood flow to the GIT [2–4]. Thus, the pharmacist must be aware of the possible importance of blood flow rate, especially alterations in that rate, on the availability of drugs.

VI. SUMMARY

Drug availability following oral dosing may be thought of as the resultant of four basic steps: (a) getting the drug to its absorption site, (b) getting the drug into solution, (c) moving the dissolved drug through the membranes of the GIT, and (d) moving the drug away from the site of absorption into the general circulation. Although steps a, c, and d were discussed briefly, these topics are also found in Chapters 2 and 5. Step b, the combination of factors influencing the dissolution rate of a drug from its dosage form, served as the major topic of this chapter. Dissolution rate was discussed in terms of the parameters found in the Nernst-Brunner equation. Although this equation [Eq. (1)] is derived in terms of a specific model for dissolution, which, in fact, may not accurately describe the physical process, we believe that the treatment gives pharmacists a point of reference from which they can predict and interpret availability data that are rate-limited by the dissolution step. Two terms in the Nerst-Brunner equation are of major importance and are susceptible to manipulation by the pharmaceutical manufacturer in preparing the dosage form, by the patient in the manner in which he or she takes the drug and stores it between drug dosing, and by the biological system, specifically the GIT of the patient, through interactions with the dosage form and the drug. These two variables are the surface area of the drug particles and the saturation solubility of the particular chemical form of the drug. We believe that pharmacists must be aware of these variables and how they can change in each of the three situations listed previously before they can make rational judgments about which of many pharmaceutical alternatives should be dispensed under a specific set of conditions. Knowledge of these variables and their potential for change also allows pharmacists to make a rational guess about possibly drug-related factors that may be responsible for inefficacious drug treatment. Under these conditions, pharmacists may be able to recommend an alternative formulation that will prove to be efficacious.

REFERENCES

1. D. E. Cadwallader, *Biopharmaceutics and Drug Interactions*, 3rd Ed., Raven Press, New York, 1983.
2. W. H. Bachrach, Physiology and pathologic physiology of the stomach, Ciba Clin. Symp., 11, 1–28 (1959).
3. H. W. Davenport, Gastric digestion and emptying: Absorption, in *Physiology of the Digestive Tract*, 3rd Ed., Year Book Medical Publishers, Chicago, 1971, pp. 163–171.
4. L. Z. Benet and B. Hoener, Pathological limitations in the application of rate control systems, in *Proceedings of the 2nd International Conference of Drug Absorption Rate Control in Drug Therapy*, (L. Prescott, ed.), Edinburgh, 1983, pp. 155–165.
5. G. D. Kozloski, J. M. De Vito, J. C. Kisicki, and J. B. Johnson, The effect of food on the absorption of methotrexate sodium tablets in healthy volunteers, Arthritis Rheum., 35, 761–764 (1992).
6. P. G. Welling, H. Huang, P. F. Hewitt, and L. L. Lyons, Bioavailability of erythromycin stearate: Influence of food and fluid volume, J. Pharm. Sci., 67, 764–766 (1978).
7. T. R. Bates, J. A. Sequeria, and A. V. Tembo, Effect of food on nitrofurantoin absorption, Clin. Pharmacol. Ther., 16, 63–68 (1974).
8. J. W. Fara, Physiological limitations: Gastric emptying and transit of dosage forms, in *Proceedings of the 2nd International Conference on Drug Absorption, Rate Control in Drug Therapy* (L. Prescott, ed.), Edinburgh, 1983, pp. 144–150.
9. C. P. Dooley, C. Di Lorenzo, and J. E. Valenzuela, Variability of migrating motor complex in humans, Dig. Dis. Sci., 37, 723–728 (1992).
10. V. Manninen, J. Melin, A. Apajalahti, and M. Karesoja, Altered absorption of digoxin in patients given propantheline and metoclopramide, Lancet, 1, 398–400 (1973).
11. C. G. Regardh, P. Lundborg, and B. A. Persson, The effect of antacid, metoclopramide and propantheline on the bioavailability of metoprolol and atenolol. Biopharm. Drug Dispos., 2, 79–87 (1981).
12. J. A. Antonioli, J. L. Schelling, E. Steininger, and G. A. Borel, Effect of gastrectomy and of an anticholinergic drug on the gastrointestinal absorption of a sulfonamide in man, Int. J. Clin. Pharmacol., 5, 212–215 (1971).
13. W. H. Barr and S. Riegelman, Intestinal drug absorption and metabolism. I. Comparison of methods and models to study physiological factors in vitro and in vivo intestinal absorption, J. Pharm. Sci., 59, 154–163 (1970).
14. W. H. Barr and S. Riegelman, Intestinal drug absorption and metabolism. II. Kinetic aspects of intestinal glucuronide conjugation, J. Pharm. Sci., 59, 164–168 (1970).
15. J. C. Kolars, P. Schmiedlin-Ren, J. D. Schuetz, C. Fang, and P. B. Watkins, Identification of rifampin-inducible P450IIIA4 (CYP3A4) in human small bowel enterocytes, J. Clin. Invest., 90, 1871–1878 (1992).
16. R. R. Scheline, Metabolism of foreign compounds by gastrointestinal microorganisms, Pharmacol. Rev., 25, 451–523 (1973).
17. L. Z. Benet, Biopharmaceutics as a basis for the design of drug products, in *Drug Design*, Vol. 4 (E. J. Ariens, ed.), Academic Press, New York, 1973, pp. 1–35.
18. W. Nernst and E. Brunner, Z. Phys. Chem., 47, 52–102 (1904).
19. P. Finholt, Influence of formulation on dissolution rate, in *Dissolution Technology* (L. J. Leeson and J. T. Carstensen, eds.), Academy of Pharmaceutical Sciences, American Pharmaceutical Association, Washington, DC, 1974, pp. 106–146.
20. L. F. Prescott, R. F. Steel, and W. R. Ferrier, The effects of particle size on the absorption of phenacetin in man, Pharmacol. Ther., 11, 496–504 (1970).
21. A. J. Glazko, A. W. Kinkel, W. C. Alegnani, and E. L. Holmes, An evaluation of the absorption characteristics of different chloramphenicol preparations in normal human subject, Clin. Pharmacol. Ther., 9, 472–483 (1968).
22. K. Arnold, N. Gerber, and G. Levy, Absorption and dissolution studies on sodium diphenlyhydantoin capsules, Can. J. Pharm. Sci., 5, 89–92 (1970).
23. G. Levy and R. H. Gumtow, Effect of certain tablet formulation factors on dissolution rate of the active ingredient. III. Tablet lubricants, J. Pharm. Sci., 52, 1139–1141 (1963).

24. G. Levy, J. M. Antkowiak, J. A. Procknal, and D. C. White, Effect of certain tablet formulation factors on dissolution rate of the active ingredients. II. Granule size, starch concentration, and compression pressure, J. Pharm. Sci., 52, 1047–1051 (1963).

25. J. C. Samyn and W. Y. Jung, In vitro dissolution from several experimental capsule formulations, J. Pharm. Sci., 59, 169–175 (1970).

26. S. H. Wan, P. J. Pentikainen, and D. L. Azarnoff, Bioavailability of aminosalicylic acid and its various salts in humans. III. Absorption from tablets, J. Pharm. Sci., 63, 708–711 (1974).

27. E. Nelson, Influence of dissolution rate and surface on tetracycline absorption, J. Am. Pharm. Assoc. Sci. Ed., 48, 96–103 (1959).

28. E. Nelson, E. L. Knoechel, W. E. Hamlin, and J. G. Wagner, Influence of the absorption rate of tolbutamide on the rate of decline of blood sugar levels in normal humans, J. Pharm. Sci., 51, 509–514 (1962).

29. C. C. Lee, R. O. Froman, R. C. Anderson, and K. C. Chen, Gastric and intestinal absorption of potassium penicillin V and the free acid, Antibiot. Chemother., 8, 354–360 (1958).

30. G. Levy and B. A. Sanai, Comparison of the gastrointestinal absorption of aluminum acetylsalicylic and acetylsalicylic acid in man, J. Pharm. Sci., 51, 58–62 (1962).

31. R. A. O'Reilly, E. Nelson, and G. Levy, Physicochemical and physiologic factors affecting the absorption of warfarin in man, J. Pharm. Sci., 55, 435–437 (1966).

32. W. I. Higuchi and W. E. Hamlin, Release of drug from a self-coating surface: Benzphetamine pamoate pellet, J. Pharm. Sci., 52, 575–579 (1963).

33. E. Nelson, Comparative dissolution rates of weak acids and their sodium salts, J. Am. Pharm. Assoc. Sci. Ed., 47, 297–299 (1958).

34. T. R. Bates, J. M. Young, C. M. Wu, and H. A. Rosenberg, pH-dependent dissolution rate of nitrofurantoin from commercial suspensions, tablets and capsules, J. Pharm. Sci., 63, 643–645 (1974).

35. G. Levy, J. R. Leonards, and J. A. Procknal, Development of in vitro dissolution tests which correlate quantitatively with dissolution rate-limited drug absorption in man, J. Pharm. Sci., 54, 1719–1722 (1966).

36. K. A. Javaid and D. E. Cadwallader, Dissolution of aspirin from tablets containing various buffering agents, J. Pharm. Sci., 61, 1370–1373 (1972).

37. J. W. Poole, G. Owen, J. Silverio, J. N. Freyhot, and S. B. Roseman, Physicochemical factors influencing the absorption of the anhydrous and trihydrate forms of ampicillin, Curr. Ther. Res., 10, 292–303 (1968).

38. J. Hableblian and W. McCone, Pharmaceutical applications of polymorphism, J. Pharm. Sci., 58, 911–929 (1969).

39. A. J. Aguiar, J. Krc, A. W. Kinkel, and J. C. Samyn, Effect of polymorphism on the absorption of chloramphenicol palmitate, J. Pharm. Sci., 56, 847–853 (1967).

40. D. L. Sorby, Effect of adsorbents on drug absorption. I. Modification of promazine absorption by activated attapulgite and activated charcoal, J. Pharm. Sci., 54, 677–683 (1965).

41. K. C. Kwan and D. J. Allen, Determination of the degree of crystallinity in solid–solid equilibria, J. Pharm. Sci., 58, 1190–1193 (1969).

42. E. N. Hiestand, W. I. Higuchi, and N. F. H. Ho, Theories of dispersion techniques, in *Theory and Practice of Industrial Pharmacy*, 2nd Ed. (L. Lachman, H. A. Lieberman, and J. L. Kanig, eds.), Lea & Febiger, Philadelphia, 1976, p. 159.

43. A. H. Goldberg, M. Gibaldi, and J. L. Kanig, Inreasing dissolution rates and gastrointestinal absorption of drugs via solid solutions and eutectic mixtures. III. Experimental evaluation of griseofulvin-succinic acid solid solution, J. Pharm. Sci., 55, 487–492 (1966).

44. A. Goldberg, Methods of increasing dissolution rates, in *Dissolution Technology* (I. J. Leeson and J. T. Carstensen, eds.), Academy of Pharmaceutical Sciences, American Pharmaceutical Association, Washington, DC, 1974, pp. 147–162.

45. L. Z. Benet, Theories of dissolution: Multi-particulate systems, in *Dissolution Technology* (I. J. Leeson and J. T. Carstensen, eds.), Academy of Pharmaceutical Sciences, American Pharmaceutical Association, Washington, DC, 1974, pp. 29–57.

46. L. Z. Benet, A. Greither, and W. Meister, Gastrointestinal absorption of drugs in patients with cardiac failure, in *The Effect of Disease States on Drug Pharmacokinetics* (L. Z. Benet, ed.), Academy of Pharmaceutical Association, Washington, D.C., 1976, pp. 33–50.

47. L. Ther and D. Winne, Drug absorption, Annu. Rev. Pharmacol., 11, 57–70 (1971).

48. D. Winne and H. Ochsenfahrt, Die formale Kinetic der Resorption unter Berücksichtigung der Darmdurchblutung, J. Theor. Biol., 14, 293–315 (1967).

49. D. Winne, The influence of villous countercurrent exchange on intestinal absorption, J. Theor. Biol., 53, 145–176 (1976).

50. D. Winne and J. Remischovsky, Intestinal blood flow and absorption of nondissociable substances, J. Pharm. Pharmacol., 22, 640–641 (1970).

51. W. G. Crouthamel, L. Diamond, L. W. Dittert, and J. T. Doluisio, Drug absorption. VII. Influence of mesenteric blood flow on intestinal drug absorption in dogs, J. Pharm. Sci., 64, 661–671 (1975).

52. A. Haas, H. Lullman, and T. Peters, Absorption rates of some cardiac glycosides and portal blood flow, Eur. J. Pharmacol., 19, 366–370 (1972).

53. J. Ferrer, S. E. Bradley, H. O. Wheeler, Y. Enson, R. Presig, and R. M. Harvey, The effect of digoxin on the splanchnic circulation in ventricular failure, Circulation, 32, 524–537 (1965).

5

The Effect of Route of Administration and Distribution on Drug Action

Svein Øie and Leslie Z. Benet
University of California, San Francisco, California

I. THE DOSE-EFFICACY SCHEME

When a health practitioner administers (or "inputs") a dose of drug to a patient, usually the ultimate goal is solely directed to the usefulness of the drug under abnormal conditions. That is, the drug must be efficacious and must be delivered to its site of action in an individual experiencing a particular physiological anomaly or pathological state. Pharmaceutical scientists, on the other hand, concentrate their attention to solving problems inherent in drug delivery to deliver the optimal dose to the site(s) of action.

The general pathway a drug takes from residence in a dosage form until its clinical utility is depicted in Fig. 1. Ideally, the drug should be placed directly at the site of action, as illustrated by the stippled arrow in Fig. 1, to maximize the effect and minimize side effects relating to unwanted responses at sites other than the target tissue. However, delivery directly to the site of action is more often than not, impractical or not possible. Instead, we have to settle for the most convenient routes of delivery. This is illustrated by the solid arrows in Fig. 1. That is, the drug is placed directly in the vascular systems or in close proximity to some biological membrane through which the drug can traverse to reach body fluids or the vascular system. The delivery system is generally designed to release the drug in a manner that is conducive to this passage through the membrane. Previous chapters have discussed how drug delivery systems may be optimized in terms of dissolution in the fluids surrounding the membrane to allow the desired rate of passage through the membrane. Subsequent chapters will deal with specific drug delivery systems and their optimization. Once the drug has passed through the membrane and into the blood stream adjacent to the site of absorption, a general distribution of the drug will take place throughout the biological system. As pointed out in Chapter 3, the degree of dilution (referred to as the apparent volume of distribution) will dictate the initial concentration of drug in the general circulation, as sampled from a peripheral vein. Usually, the dose of the drug administered to the patient was chosen to give sufficiently high

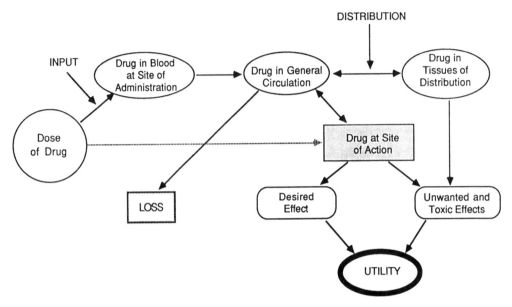

Fig. 1 A schematic representation of the dose–efficacy relationship for a drug.

blood levels so that an adequate quantity of the drug would reach the site of action. The rate of input needed to achieve adequate levels of the drug at the site of action is influenced not only by the distribution and general elimination in the body, but may also be modified by the loss processes that are unique to a specific route of administration. This chapter will deal primarily with the distribution and loss processes that result uniquely from the physiological parameters inherent in the use of a particular route of administration.

Unfortunately, no drug is yet so specific that it interacts with only the target site in the target tissue, and will not give rise to hyperclinical activity. Too much drug at the wrong place or too high a concentration at the right place may result in unwanted or toxic effects. Thus the practitioner must determine the usefulness of any dose of a drug from a particular drug delivery system by balancing the efficacy achieved from the clinical effect against the toxic reactions observed.

Most drug delivery systems achieve the required drug levels at the site of action as a result of attaining adequate blood levels in the general circulation (see Fig. 1, solid arrows). This process is followed because of the ease with which present drug delivery systems can ''input'' drugs into the general circulation and the inherent difficulties in delivering the drug selectively to a relatively inaccessible site (e.g., pituitary gland). In addition, for many compounds, the exact site of action is still unknown. However, when the site of drug action is sufficiently defined, Fig. 1 illustrates the advantage of delivering the drug directly to the site of action. By direct administration to the active site, a lower dose could be used to achieve the clinical effect because the drug no longer is diluted or eliminated en route. As a result, drug concentrations at unwanted sites of action could be kept to a minimum; in addition, clinically effective levels at the site of action might be attained much more rapidly, since the process of distribution throughout the entire body could be avoided. One should not forget that, in addition to the obvious clinical advantage of direct administration, there is also an economical one. By delivering the drug to the site of action, the amount of drug needed is much smaller than by more traditional delivery methods. This is particularly important for many of the newer recombinant compounds that can be very expensive. Much work is currently being carried out in an attempt to achieve such a selectiveness as that described in Fig. 1.

II. PHYSIOLOGICAL CONSIDERATIONS FOR THE VARIOUS ROUTES AND PATHWAYS OF DRUG INPUT

A. Drug Input at or Close to the Site of Action

Figure 2 illustrates a number of sites where drug delivery systems have historically been used to input drug directly to its site of action [1,2]. Various classic dosage forms were developed to take advantage of these input sites: eye, ear, and nose drops; inhalation, oral, topical, and vaginal aerosols; topical solutions, creams, and ointments; and rectal solutions, enemas, and suppositories. Each of the sites for local drug administration requires specific formulation to allow the drug to remain at the site of application for a sufficient length of time to allow the drug to penetrate through the particular membrane(s) so that it can reach the actual site of action adjacent to the site of application. For example, some ophthalmic preparations may be given to elicit a superficial anti-infective effect, such as treatment of an inflammation of the conjuctiva. Thus, only topical effects are desired, and there is no need for the drug to penetrate into the eyeball. Formulation of such products would be quite different than formulation of a drug delivery system for which the drug must be absorbed into the interior of the eye to produce

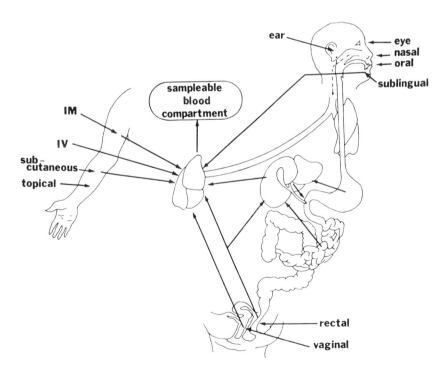

Fig. 2 Various routes and pathways by which a drug may be "input" into the body. The position of one lung is distorted to emphasize that the lungs are in an excellent position for cleansing the blood. The diagram is especially useful in explaining the first-pass effect following oral dosing, for which drug absorbed from the small intestine or stomach must first pass through the liver and, therefore, is subject to metabolism or biliary excretion before reaching the sampleable blood. (From Ref. 2.)

a response, such as miotics, mydriatics, anti-inflammatory drugs that act in the anterior segment of the eye, and, occasionally, drugs for treatment of infections. A detailed description of the factors involved in the development of such ophthalmic preparations, as well as the particular physiological characteristics of the eye, will be presented in Chapter 15. Similar types of design problems arise for many of the other sites that are traditionally treated by direct local application. One of the most difficult problems facing the formulator is that the behavior of the diseased tissue may be different from that for healthy individuals, and it may also change over the course of treatment. For example, diseased skin is often more permeable than healthy skin; therefore, the drug may disappear faster from the site of administration than desired, and the effect will be less than expected. Should the formulation be designed to accommodate this phenomenon, one must be mindful of the fact that as the pathological condition improves, the absorption may also change.

Although the classic dosage forms mentioned earlier can be used to put drug directly into the site of action, many of them have a degree of ''messiness'' that prevents good patient acceptance and adherence. Not only is there an initial psychological barrier that must be overcome, but the general public has an aversion to taking drugs by routes other than oral. There is, in addition, a general dislike for sticky creams, drippy drops, greasy ointments, and the like. Over the last two decades much work has been directed toward developing more acceptable delivery systems than the traditional ones. Emphasis has been placed on long-acting drug delivery systems that may be more convenient, since they would only require self-administration once a week or possibly at even longer intervals.

A large number of new devices have been developed, and new ones are constantly being investigated. Plastic disks for placement in the eye (similar to a contact lens) that slowly release drug into the humoral fluid; drug-impregnated plastic rings or loops that when placed in the uterus will release controlled amounts of contraceptive agents; bioadhesive tablets or disks that can be placed buccally, nasally, or vaginally for local release; hydrogels for slow release in the eye are examples of such new delivery systems that input drug directly to the site of action.

In a more ambitious move, many groups have also embarked on site-specific delivery to less accessible sites than those given in Fig. 2. Although numerous experimental systems have been designed, few have reached the clinical stage. The simplest and most direct method when a specific target organ can be located is cannulation (direct access port). A catheter is placed in an appropriate artery or vein. If a vein is used, the catheter has to reach the organ, or otherwise the drug will be flowing away from the target tissue, be diluted with blood from the rest of the body, and be not different than a systematic intravenous administration. Catheters can also be placed in the peritoneum, the bladder, and in the cerebrospinal fluid. A drug can now be administered directly into the desired tissues at a rate that can be well controlled. Although catheter delivery is a direct method, it is limited in that it is essentially restricted to inpatient use. Use of implants in the desired tissue, or a drug carrier (e.g., liposomes, nanoparticles, and such) that will either home in on the desired tissue by specific receptors, or release their content at the desired site by an external stimulus (e.g., magnetic fields, light, current), are drug systems currently being explored for target-specific delivery.

Although the method of direct delivery is a very attractive one, it also has its regulatory problems. Benet [1] has noted that assessing the bioavailability of this system can create difficulties because the manufacturer may not be able to devise a control procedure that can measure drug concentration at the site of action. For example, the extent and rate of availability of an orally administered drug can easily be assessed by measuring blood levels, whereas for a drug input into a site of action, significant blood levels would indicate distribution away from that site. Frequently, significant blood levels of a drug that is administered at a site of action (such as a topical preparation, an eye drop, a nasal insufflation, or an antibiotic that acts on intestinal flora) indicate either a poor drug delivery system or substantial overdosing. For this class of drug delivery systems, clinical efficacy necessarily has to serve as the best measurement of drug availability and dosage form efficacy.

B. Drug Input into the Systemic Circulation

The overwhelming majority of existing drugs are, however, given by general routes; that is, by routes that do not deliver the drug directly to the site of action. These modes of drug input rely on a passive delivery of drug through distribution by the vascular system. The most commonly accepted method is oral administration. As will be discussed later, oral administration is not ideal, as one needs to be concerned about whether the drug can be destroyed in the stomach, in the gastrointestinal fluid, in its passage through the gut wall, through the liver, or simply not be absorbed in time before it is expelled from the gastrointestinal tract. Several alternative routes of delivery are being used or are being developed to diminish these potential losses. The advantages and problems inherent in the individual routes of administration will now be discussed.

Parenteral Administration: Intravascular

Of the routes of input depicted in Fig. 2, intravenous (IV) administration yields one of the fastest and most complete drug availabilities. However, intra-arterial injections might be employed when an even faster and more complete input of drug to a particular organ is desired. By administering the drug through an artery, the total drug delivered will enter the organ or tissue to which the artery flows. Intravenously administered drug will first be diluted in the venous system as the venous blood is pooled in the superior and inferior vena cava. It then enters the heart, and is subsequently pumped to the lung before it can enter the arterial system and reach the target organ(s). In addition, the fraction of the drug reaching a desired site is dependent on the fraction of the arterial blood flow reaching that site. Additional drug can reach the target tissue only by being recirculated from the other organs. In comparison with intra-arterial administration, IV administration reaches the target slower, and initially at a lower concentration. Although intra-arterial injections appear superior, they are infrequently used because they are considered much more dangerous than IV administration. Intra-arterial administration has been associated with patient discomfort, bleeding, and thrombosis.

In addition to the dilution factor resulting from mixing with larger volumes of blood after intravenous administration, one also needs to consider the possibility of temporary or permanent loss of drug during its passage through the lung. The position of the lungs in Fig. 2 has been distorted to emphasize the point that the lungs are in an excellent strategic position for cleansing the blood, since all of the blood passes through the lungs several times a minute. Apart from their respiratory function and the removal of carbon dioxide from the pulmonary circulation, the lungs serve other important cleansing mechanisms, such as filtering emboli and circulating leukocytes, as well as excretion of volatile substances. The lungs also have metabolic capacity [3] and may serve as a metabolic site for certain drugs [4] or as an excretory route for compounds with a high vapor pressure. The lungs can also act as a good temporary storage site for a number of drugs, especially basic compounds, by partitioning of the drug into lipid tissues, as well as serving a filtering function for particulate matter that may be given by IV injection. Accumulation of lipophilic compounds and filtering of any compounds in solid form can be viewed as a temporary clearing or dilution of the drug, as it will eventually leach back into the vascular system. Thus the lung serves as a dampening or clearing device, that is not present following intra-arterial injection. Drugs given by the IV route may, therefore, not necessarily be completely available to the site of action, since a certain fraction of the drug could be eliminated by the lung before entering into the general circulation [5]. This might be called a "lung first-pass effect."

The foregoing concepts may be visualized by referring to Fig. 3. In this figure one can readily see the difference between intra-arterial and intravenous administration of drugs. Let us assume that compartment n is the target tissue. Administration into any vein (i.e., into any of the efferent arrows on the left-hand side of the figure) would lead the drug to the heart and, from there, to the lung. Drug that enters the lung can leave by only one of two routes, as illustrated in Fig. 4: by the blood that leaves the lung, or by being eliminated. The result is that there is a competition between the two routes for the drug, and the greater the ability of the lung to eliminate the drug in comparison with the pulmonary blood flow, the more drug will be extracted. If we assume that the pulmonary blood flow is Q_P and that the intrinsic elimination clearance of the organ is CL_{Int}, and no plasma protein binding occurs (Cu = C_{out}), then the extraction ratio can be expressed as

$$E = \frac{C_{out}\, CL_{Int}}{(C_{out}\, CL_{Int}) + (C_{out}\, Q_P)} = \frac{CL_{Int}}{CL_{Int} + Q_P} \tag{1}$$

In perfusion models, as depicted in Fig. 3, it is assumed that distribution into and out of the organ is perfusion rate-limited such that drug in the organ is in equilibrium with drug concentration in the emergent blood [6]. The intrinsic clearance of an organ is different from the value we normally think of as the clearance of the organ. The *clearance of the organ* is defined as the rate of loss in relation to the incoming concentration, whereas the *intrinsic clearance* is defined as the rate of loss in relation to the organ concentration (or exiting concentration). In addition, it is also clear that, of the drug that escapes elimination in the lung, only a small fraction goes to compartment n, the rest is distributed to other organs. Drugs that enter these organs will be exposed to elimination in these organs and must necessarily recirculate through the heart and lungs before they again have the opportunity to reach compartment n.

Parenteral Administration: Depot

The other parenteral routes depicted in Fig. 2, intramuscular (IM) and subcutaneous (SC) injections, may also be considered in terms of Fig. 3. Drug absorbed from the IM and SC sites into the venous blood will return to the heart and pass through the lungs before being distributed to the rest of the body. However, there will be an initial lag between the time when the drug is injected and when it enters the circulation. Thus, the kinetics for drugs administered by these parenteral routes would be expected to show a decreased *rate* of availability and may also show a decreased *extent* of availability in comparison with intravenous administration, if loss processes take place at the site of injection. For example, we could consider that drug is now injected directly into compartment m in Fig. 3 and that this compartment is the muscle. The rate at which the drug leaves the muscle will depend primarily on blood flow in relation to the size (apparent volume of distribution) of the organ.

Evans and co-workers [7] measured resting human muscle blood flow through the gluteus maximus, vastus lateralis, and deltoid muscles. Deltoid muscle blood flow was significantly greater than gluteus muscle blood flow, with vastus being intermediate. Because the two sites most commonly used for IM injections are the deltoid and the gluteus muscles, we might expect to see differences in rates of drug absorption following injection to these sites. Lidocaine is one drug that has been investigated for its effect in response to the site of injection [8,9]. Deltoid injection gave higher peak levels than lateral thigh injection which, in turn, gave higher levels than gluteal injection. Schwartz et al. [9] demonstrated that therapeutic plasma levels for a particular lidocaine dose were reached only when the deltoid injection site was used. Evans et al. [7] concluded, "This demonstrates that the site of injection can influence the

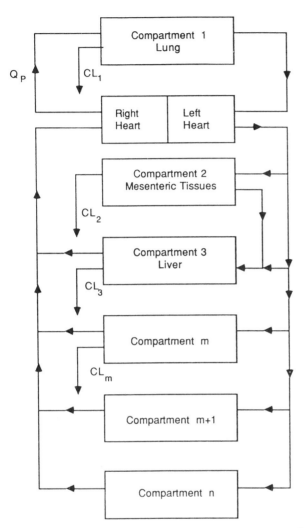

Fig. 3 The body depicted as a physiological perfusion model. Compartment m must be considered as a summation of the individual tissues that metabolize the drug and compartment m + 1 through n as noneliminating tissues.

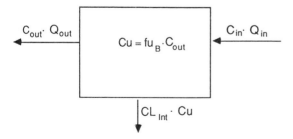

Fig. 4 Flow model of an eliminating organ. The drug enters the body by the organ blood flow, CL_{in} and is immediately mixed in the organ. The drug leaves the organ by either being eliminated ($CL_{Int}Cu$) or by the exiting blood flow.

plasma level achieved and that the deltoid muscle should be used to achieve therapeutic blood levels as rapidly as possible.'' Likewise, if a sustained or a prolonged release is desired, this would more readily be achieved by injection into a lower blood flow muscle, such as the gluteus.

Loss processes may also account for a decrease in the extent of availability following an IM injection. This can be visualized by assuming that the dose is injected into compartment m, which, as depicted in Fig. 3, is capable of eliminating the drug. As shown in Fig. 4, the drug can leave the tissue only by one of two routes, either by the blood leaving the organ, or by being eliminated by metabolism in the muscle. In addition, the drug that leaves the site of administration will also be subject to the additional distribution and elimination in the lung, similar to intravenous administration. In other words, drug given by intramuscular administration may be not only further delayed in its distribution to the target organ, but may also show a decreased extent of distribution to the organ, in comparison with the intravenous dose. For example, degradation can take place in the muscle, as shown by Doluisio et al. [10] for ampicillin. These workers found that only 77–78% of an IM dose of ampicillin sodium solution was absorbed, as compared with the IV solution. The most likely explanation is that the drug may have been decomposed chemically or enzymatically at the injection site. In addition, temporary losses may also occur. For example, intramuscular doses of phenytoin result in a marked decreased rate and extent of absorption in comparison with IV or oral doses. Wilensky and Lowden [11] demonstrated that this could be due to precipitation of the drug as crystals in the muscle. Although these crystals eventually dissolve, the drug is essentially lost during a normal dosing interval.

Oral Administration

First-Pass Effect. Metabolism in the Gastrointestinal Fluids and Membranes. When a dosage form is administered by the oral route, drug particles come in contact with varying pH solutions, different enzymes, mucus, gut flora, and bile, all of which may contribute to decreasing the extent of availability by degradation, binding, or sequestering mechanisms. These factors, as well as the possiblity of drug metabolism in the intestinal membrane itself, have been well covered in Chapter 2 and will not be discussed here.

Hepatic Metabolism: Linear Systems. As depicted in both Figs. 2 and 3, drug that is absorbed from the gastrointestinal tract must pass through the liver before reaching the sampleable circulation and the rest of the body. Thus, if a drug is metabolized in the liver or excreted into the bile, some of the active drug absorbed from the gastrointestinal tract will be inactivated by hepatic processes before the drug can reach the general circulation and be distributed to its site(s) of action. An exception would be if the liver itself were the target organ, as we then would have to contend with only losses in the gastrointestinal tract and in the gut wall before reaching the site of action.

For many drugs, the fraction of the dose eliminated on the first pass through the liver is substantial. The fraction eliminated is often referred to as the hepatic extraction ratio, designated herein as E_H. Many drugs are known or suspected to have a high hepatic extraction ratio. A short list of some of the better-known compounds is given in Table 1. The hepatic first-pass phenomenon is not restricted to any particular pharmacological or chemical group of drug substances and the foregoing list includes acids, bases, and neutral compounds.

The available fraction (F) of an oral dose appearing in the sampleable blood circulation will, therefore, be governed by not only the extent of drug absorbed from the gastrointestinal tract, as discussed in Chapter 4, but also by the fraction metabolized in the gut membranes (E_G) and the fraction metabolized or excreted into the bile following passage through the liver (E_H), where E_G and E_H are the extraction ratios for the gut and liver, respectively. If it is

Table 1 Drugs Suspected to Have High Hepatic Extraction Ratios

Acetylsalicylic acid	Alprenolol	Aldosterone
Cocaine	Desipramine	Doxorubicin
Fluorouracil	Isoproteranol	Imipramine
Lidocaine	Lorcainide	Morphine
Nitroglycerin	Prazepam	Propranolol

assumed that a drug is completely absorbed from the gastrointestinal tract and not degraded during passage through the gut membranes, this would be equivalent to an injection of the drug into the hepatic portal vein. Under these conditions the unmetabolized fraction (F_H) of an oral dose appearing in the sampleable blood circulation would equal $1 - E_H$. If the drug elimination in the liver follows first-order kinetics, Rowland [12] has shown that F_H, the fraction of the oral dose that is available following the liver first pass, may be related to liver blood flow (Q_H) and the hepatic clearance for an IV dose of the drug ($CL_{H,iv}$).

$$F_H = 1 - E_H = 1 - \frac{CL_{H,iv}}{Q_H} \tag{2}$$

This hepatic first-pass effect, as discussed by Boyes et al. [13], is well illustrated in Fig. 5. Identical doses of lidocaine hydrochloride were injected into dogs (beagles) by an exponential infusion process. The upper curve describes the mean levels found when the drug was infused into a peripheral vein, whereas the lower curve describes the lidocaine levels following hepatic portal vein infusion (thereby eliminating any effects caused by gastrointestinal degradation either in the fluids of the tract or in the intestinal membrane). The area under the curve (AUC)

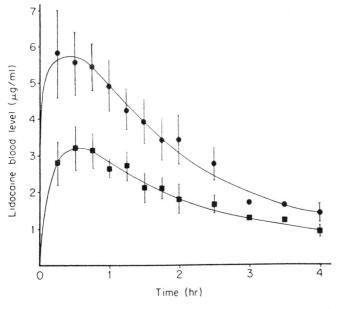

Fig. 5 Mean blood levels of lidocaine hydrochloride in five beagles after exponential IV infusion through a peripheral vein (●) and in the portal vein (■). Vertical bars represent standard errors of the mean. (From Ref. 13.)

measurements for the two curves show that the extent of bioavailability following portal vein infusion is only 60% of that found following infusion into a peripheral vein (i.e., $F_H = 0.6$). This "oral" availability could be predicted for a drug such as lidocaine for which drug elimination occurs predominantly in the liver when the hepatic clearance—calculated by dividing the dose by the area under the peripheral vein infusion curve—and the hepatic blood flow are substituted into Eq. (2). An even greater first-pass effect is found for oral administration of lidocaine to humans, $F_H = 0.25–0.48$ [14].

Thus, if the hepatic clearance for a drug is largely relative to the hepatic blood flow, the extent of availability for this drug will be low when it is given by a route that yields first-pass effects. The decrease in availability is a function of only the anatomical site from which absorption takes place, and no amount of dosage form redesign can improve the availability. Of course, therapeutic blood levels can be reached by this route of administration if larger doses are given, but the health practitioner and pharmacist must be aware that levels of the drug metabolite may increase significantly over that seen following IV administration.

Lidocaine is not normally administered by the oral route, but many of the drugs listed earlier are routinely given orally. For these drugs, analysis of Eq. (2) leads to the conclusion that small variations in plasma or blood clearance of a drug throughout a population may yield significant differences in availability when the drug is given by a route subject to significant first-pass effects. For example, data from a study of Shand et al. [15] for oral and IV administration of propranolol in five men are shown in Table 2. Although the clearance following an IV dose of the drug (column 6) varies by only 67% from the smallest to the largest, the oral availability (column 5) varies by 275%. As would be expected from Eq. (2), the oral availability decreases as the IV clearance increases.

Hepatic Metabolism: Nonlinear Systems. The discussion of the first-pass effect has thus far assumed linear first-order kinetics. Under this condition, the hepatic extraction will be independent of the rate of drug availability. That is, no matter when a drug molecule is absorbed from the gastrointestinal tract and at whatever dose administered, the hepatic extraction and the extent of availability [see Eq. (2)] for that drug will remain constant. This would not be true for a drug for which saturation of the hepatic enzymes is a possibility. Under such a condition, the extraction ratio would vary depending on the concentration of drug in the hepatic portal vein. If the concentration were high, the hepatic enzymes would become partially saturated, and a large amount of drug would pass through the liver without being metabolized (i.e., E_H would be low). Saturable metabolism during the first passage through the liver is not

Table 2 Peak Plasma Levels and Areas Under Plasma Concentration Time Curves Following Oral and Intravenous Administration to Men

Subject	Propranolol, 80 mg fasting orally		Propranolol, 10 mg IV	$\dfrac{AUC_{Oral}}{AUC_{IV}} \cdot \dfrac{10}{80} \cdot 100$	Clearance, IV (ml/min)
	Peak (ng/ml)	Area (ng/ml·hr)	Area (ng/ml·hr)		
OF	212	1400	292	60	570
DS	100	480	220	30	756
GY	94	510	200	32	833
JC	45	290	183	20	909
JF	36	220	175	16	950

Source: After Ref. 15.

an unusual occurrence even for a drug that, after intravenous administration, fails to show saturable metabolism. The main reason is that drug absorbed from the gastrointestinal tract is diluted to only a small degree before it enters the liver (in the portal blood only), and the concentrations can be relatively high. On the other hand, intravenously administered drug is generally quickly distributed to individual tissues, and the concentration reaching the liver during the elimination process is sufficiently low that no saturation of the metabolism is achieved. However, if the concentration in the hepatic portal vein were low, either because of administration of a low dose or because of very slow absorption of the drug, then the enzymes would not be saturated, and most of the drug could be metabolized on passage through the liver (i.e., E_H would approach 1). In a similar manner, metabolism in the gastrointestinal membranes could also be saturated or not, depending on the dose and the rate of absorption (i.e., E_G would vary from 0 to 1).

Salicylamide appears to be a drug for which the first-pass extraction may be dose-dependent. When a 300-mg oral dose of salicylamide is administered as a solution, the area under the plasma concentration time curve is less than 1% of the area seen following a similar IV dose [16]. Even following a 1-g oral dose, most of the drug is found in the systemic circulation as the inactive glucuronide and sulfate conjugates. However, when a 2-g dose was given to a particular subject (Fig. 6), the area under the plasma concentration time curve for unmetabolized salicylamide increased dramatically (> 200 times increase over that seen for a 1-g dose). This very large increase is probably related to saturation of enzymes in the gastrointestinal mucosa as well as in the liver. Saturation of the hepatic enzymes not only increases the extent of oral availability, but a sufficient amount of unchanged drug is allowed to enter the systemic circulation, thereby decreasing the rate of drug metabolism during the postabsorptive phase as well. This phenomenon is due to enzyme saturation. Thus, the extraction ratio [see Eq. (1)] and, thereby, the drug clearance, change continuously during both the absorption and elimination processes. Under these conditions AUC measurements can no longer be used to deter-

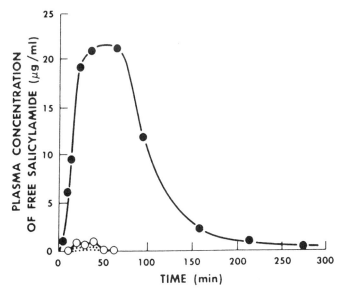

Fig. 6 Comparison of plasma concentrations of intact salicylamide when given as a 1-g (○) and a 2-g (●) dose in solution to the same subject. Dotted lines show the plasma concentrations of a 0.5-g dose and 0.3-g dose. (From Ref. 16.)

mine the extent of drug availability (as was previously described in Chapter 3 for linear systems.)

The effects of stomach emptying on drug availability are discussed in Chapters 2 and 4, relative to processes within the gastrointestinal system. In addition, Benet [1] has interpreted data for p-aminobenzoic acid (PABA) as reflecting an example for which the rate of drug absorption modified by changes in stomach emptying causes changes in the extent of drug availability owing to saturation of first-pass metabolism. Table 3 represents a situation in which the drug PABA is completely absorbed from the gastrointestinal tract, yet there is a decrease in the extent of drug available to the systemic circulation as a function of food [17].

The data for the oral solution, single-dose studies suggest that all of the drug is being absorbed, since the total amount of PABA found in the urine (both as unchanged drug and acetyl metabolite) equals the dose administered. It appears that the decreasing fraction of the acetyl metabolite found in the urine with increasing doses would indicate that a saturable metabolic process was operable. For the second series of studies listed in Table 3, the oral solution dose remains constant at 1 g, while increasing amounts of sweet cream are added to the solution. With increasing amounts of fat, there is a concomitant increase in the percentage of dose excreted as the acetyl metabolite. This may be explained by assuming that the fat decreases the rate of stomach emptying, causing the drug to be emptied from the stomach more slowy and to be absorbed over a longer time period. If plasma concentration were maintained at lower levels by slowing absorption, the metabolic site would not reach the same degree of saturation and a greater fraction of the metabolite should appear in the urine. Even if gastric absorption of PABA did occur, owing to retention of the drug in the stomach, the absorption rate should be significantly lower than that seen for the oral solutions without fat, thereby maintaining plasma concentrations of PABA at lower levels. The prolonged administration of the smaller oral and IV doses in the third part of Table 3 yields high levels of metabolite in the urine, which is consistent with the saturable enzyme hypothesis (i.e., when plasma concentrations are maintained at a low level, extensive metabolism will occur).

Table 3 Extent of Urinary Excretion of p-Aminobenzoic Acid (PABA) and Its Acetyl Metabolite as a Function of Route of Administration and Ingestion of Fat

Route	Dose Na-PABA to 61-kg man (g)	Total PABA in urine as percentage of dose in 24 hr	Acetyl-PABA in urine as percentage of total PABA excreted in 24 hr
Oral solution			
Single dose	1	103	51
	2	103	47
	4	102	36
	8	102	30
Fat added to oral dose			
60 g sweet cream	1	95	76
90 g sweet cream	1	104	83
120 g sweet cream	1	99	90
Prolonged administration			
10 oral doses given every 30 min	0.365	95	97
Intravenous infusion, 270 min	0.365	90	95
Intravenous bolus	1	102	51

Source: Ref. 17.

Biliary Excretion. The effects of significant hepatic extraction as a result of biliary secretion, with or without metabolism, would be expected to follow the same principles just outlined for hepatic metabolism. In fact, a whole class of compounds that serve as biliary contrast agents for radiological examination depend on significant first-pass biliary secretion to be effective.

Several studies in rats have shown that certain acidic and basic compounds can be actively secreted into the bile. Thus, one might expect to see saturation of the biliary excretion process, although data in humans describing this phenomenon have not, as yet, been reported for orally dosed drugs.

Oral Dosing Without a First-Pass Effect. If a drug is not metabolized in the gut wall or the liver and if the drug is not subject to biliary excretion, there will be no first-pass effect following oral dosing. In addition, with drugs for which the hepatic clearance is significantly less than hepatic blood flow [see Eq. (2)], the hepatic extraction will be negligible (i.e., $F_H \approx$ 1). Most orally administered drugs used today fall into this latter category. Examination of Fig. 2 indicates that another form of oral administration, sublingual dosing, may avoid first-pass metabolism as well as the degradation process that may occur in the gastrointestinal fluids. This route of administration has been used predominantly in dosing organic nitrates in patients experiencing angina.

Rectal Administration

As can be seen in Fig. 2, the first-pass effect can be partially avoided by rectal administration. The capillaries in the lower and middle sections of the rectum drain into the inferior vena cava, thus bypassing the liver. However, suppositories tend to move upward in the rectum into a region where veins (such as the superior hemorrhoidal vein) drain predominantly into the portal circulation [18]. In addition, there are extensive anastomoses between the middle and superior hemorrhoidal veins. Thus, Schwarz [19] has suggested that only about 50% of a rectal dose can be assumed to bypass the liver and its first-pass extraction. Again referring to Fig. 2, we can see that absorption of drugs from the vagina would bypass the first-pass effect, although almost no drugs where systemic levels are desired have been formulated using this route of administration.

Other Routes of Administration

It is clear from the foregoing discussion that none of the methods traditionally used for systemic administration are ideal. Intravenous administration requires hospital staff and, therefore, is rarely used in an outpatient setting. Intramuscular and subcutaneous doses may be metabolized at the site of administration, may be significantly delayed, and may run the risk of infections and meet with public reluctance for self-administration, although subcutaneous administration has been met with great success in insulin-dependent diabetics. Oral administration must run through a series of elimination sites before the drug can enter the general circulation. Rectal administration can also be exposed to hepatic first-pass and must also contend with a public reluctance in its usage. Because of these factors, several other routes for systemic administration of drugs have been explored over the years. Significant acceleration in these area has occurred as a response to the increased interest in the use of peptide and protein drugs made by recombinant techniques. Peptides and proteins are notoriously prone to degradation in the gastrointestinal tract, and the only modes of administration have been parenteral administrations. This limits the use by outpatients, and great efforts are being made to develop alternative routes of delivery.

Nasal Administration. The nasal mucosa is relative permeable to small molecular weight compounds. The most notorious example is cocaine. Cocaine that is snorted is both rapidly and extensively absorbed. Small peptides have also been successfully administered nasally, although the bioavailability is low. However, where the availability is not critical, nasal administration of peptides has been successful. The best example is calcitonin, which shows an activity after nasal administration, similar to that seen after intravenous administration, although the plasma concentrations achieved after nasal delivery were lower [20]. This is a compound that has a relatively large therapeutic index for which a variable bioavailability is not critical.

During the absorption process through the nasal lining, the drug has to cross not only lipophilic barriers, but it must also pass through the nasal epithelia, which have a significant capability to metabolize drugs. Nasally administered drug can, in addition to this loss by metabolism, also be removed by mucous flow and ciliary movement and be swallowed. Nasal administration can, under circumstances during which significant amounts are swallowed, be thought of as being similar to administering the drug in part as an oral dose. To increase the absorption by the nasal epithelia, the use of bioadhesive drug delivery forms and many exciting absorption enhancer techniques are being explored. Chemicals that disrupt the lipophilic membranes as well as opening the tight junction between the cells have resulted in a dramatic increase in the availability. The opening of the tight junctions is particularly interesting, as the compounds pass between the cells, thereby avoiding exposure to the metabolizing enzymes in the nasal epithelia. Whether these absorption enhancement techniques are the ways of the future as a means to reduce first-pass elimination, or if, in the end, they will be judged to invoke too much tissue damage for continuous use, still needs to be evaluated.

Transdermal Absorption. If transdermal delivery of drugs to the systemic circulation is to be successful, it must mimic subcutaneous injection in terms of yielding minimal first-pass skin elimination. The main barrier to absorption is the thick lipophilic keratin layer of the skin. Few nonlipophilic compounds penetrate the skin to a sufficient degree that this mode of delivery can be used for systematic absorption without modification. Highly lipophilic drugs, on the other hand, penetrate the skin with relative ease, although the absorption usually takes a long time. Several drugs are successfully given by dermal administration (e.g., nitroglycerin, scopolamine, nicotine, progesterone). To increase the absorption of hydrophilic compounds, including peptides and proteins, strategies similar to those described previously for nasal absorption have been studied (i.e., use of hydration and chemical enhancers). In addition, use of nonchemical enhancement methods (i.e., iontophoresis) is also being explored. This technique uses a low electric charge to force fluid and solutes to cross the skin. Although it was introduced more than 200 years ago [21], it is only recently that the method appears to have reached a practical stage [22]. These methods not only allow the drug to bypass the gastrointestinal tract and the liver, but they also may be employed to achieve very attractive long-term sustained delivery.

Pulmonary Inhalation. Although aerosol preparations now serve primarily as a convenient drug delivery system that can input drug directly to its site of action, new interest has recently been generated in this delivery system as a potential route for systemic administration of drug. The lung has a relatively large surface area and is relatively permeable to lipophilic compounds and, to some degree, even to protein [23]. Several barriers to absorption by the lungs do exist. The barrier to absorption is greatest in the upper bronchi and decreases in the alveolie. In the upper bronchi the mucus is relatively thick, the surface area small, and ciliar movement tends to move impacted particles up the bronchi and into the esophagus. The particle size of therapeutic inhalation aerosols determines the site of deposition in the lungs and, thus, the clinical effectiveness of a particular formulation. Particles that are too large will impact in the upper bronchi, and those too small will not readily impact on the wall of the lung and, therefore, will simply be exhaled. Particle sizes in the order of $0.3-1$ μm are usually considered to be most effective. Sciarra [24] has suggested that almost all drugs given by IV injection can be reformulated into a suitable aerosol, provided that the drug is capable of being deposited in the respiratory tract and is nonirritating. However, the total availability of the pulmonary route will, to a large degree, depend on how much drug is deposited in the lung and, again, how much metabolized during first pass. It is expected that the first-pass bioavailability will be lower than that seen after intravenous administration because the drug has to pass the epithelial cell layer before it can reach the general circulation. From intravenous administration, only the amount of drug that is actually taken up by the lung tissue will be exposed to metabolism and exhalation.

III. DRUG DISTRIBUTION

Figure 1 indicates that distribution will take place as the drug reaches the general circulation. This will dilute the drug and influence the levels at the site of action. Thus, an understanding of drug distribution is critical in designing appropriate drug dosage regimens. This has led to the determination of ''apparent'' volumes of distribution (as discussed in Chapter 3), which can be used to relate the amount of drug in the body (or in a hypothetical compartment) to a measured plasma or blood concentration. The volume of distribution is a function of four major factors: (a) the size of the organs into which the drug distributes; (b) the partition coefficient of drug between the organ and the circulating blood; (c) the blood flow to the distributing organs; and (d) the extent of protein binding of the drug both in the plasma and in various tissues.

A. Organ Size, Blood Flow, and Partition Coefficient

A particular organ in the body may act as a site of distribution, or as both a site of distribution and elimination. The relative importance of the various organs as storage or elimination sites depends on how fast the drug gets to each organ and how much space or volume is available to hold the drug. Table 4 presents a compilation of the volumes and blood flows of the different regions of the human body for a standard man, as compiled by Dedrick and Bischoff [25] using the mean estimates of Mapleson [26].

The various regions of the body are listed in decreasing order relative to blood flow per unit volume of tissue (adrenals highest and bone cortex lowest). This value essentially describes how fast a drug can be delivered to a body region per unit volume of tissue, and reflects the relative rates in which tissues may be expected to come to equilibrium with the blood. How much drug can be stored or distributed into a tissue will depend on the size of the tissue (volume) and the ability of the drug to concentrate in the tissue (i.e., the partition coefficient between the organ and blood, $K_{O/B}$). For example, the blood flow per unit volume of thyroid gland (see region C in Table 4) is one of the highest in the body, whereas the gland itself is quite small. Thus, if partition of the drug between the thyroid and blood were approximately 1, we would expect to see that the drug in the tissue would rapidly come into equilibrium with that in the blood, but that relatively little drug would be found in the thyroid. However, for certain drugs containing iodine moieties, $K_{O/B}$ is enormous, and a significant amount of drug will distribute into this small gland relatively rapidly. In addition, Table 4 lists the volume of blood contained within each tissue and believed to be in equilibrium with the tissue. Thus, the volume of the thyroid in Table 4 is considered to be 20 ml of tissue and 49 ml of blood. Note that total volume of all tissues in column 3 is 70 liters, including the 5.4 liters of blood volume. This blood volume is broken down into 1.4 liters of arterial blood (which is listed in the last column as being in equilibrium with the air in the lungs) and 4 liters of venous blood (which

Table 4 Volumes and Blood Supplies of Different Body Regions for a Standard Man[a]

Tissue	Reference letters	Volume (liters)	Blood flow (ml/min)	Blood flow (ml/100 ml)	Volume of blood in equilibrium with tissue (ml)
Adrenals	A	0.02	100	500	62
Kidneys	B	0.3	1240	410	765
Thyroid	C	0.02	80	400	49
Gray matter	D	0.75	600	80	371
Heart	E	0.3	240	80	148
Other small glands and organs	F	0.16	80	50	50
Liver plus portal system	G	3.9	1580	41	979
White matter	H	0.75	160	21	100
Red marrow	I	1.4	120	9	74
Muscle	J	30	300/600/1500	1/2/5	185/370/925
Skin				1/2/5	18/37/92
Nutritive	K	3	30/60/150		
Shunt	L		1620/1290/300	54/43/10	
Nonfat subcutaneous	M	4.8	70	1.5	43
Fatty marrow	N	2.2	60	2.7	37
Fat	O	10.0	200	2.0	123
Bone cortex	P	6.4	≈ 0	≈ 0	≈ 0
Arterial blood	Q	1.4			
Venous blood	R	4.0			
Lung parenchymal tissue	S	0.6			
Air in lungs	T	2.5 + half			1400[b]
		Tidal volume			999/795/185[c]
Total		70.0[d]	6480		5400

[a]Standard man = 70-kg body weight, 1.83-m^2 surface area, 30–39 years old.
[b]Arterial blood.
[c]Skin-shunt venous blood.
[d]Excluding the air in the lung.

is in equilibrium with tissues A through O). Since different muscle masses throughout the body receive different blood flows (as discussed in Sec. II.B, with reference to drug input following IM injection), Mapleson [26] lists only a range for this tissue as well as for the skin. Dedrick and Bischoff [25] suggest an average blood flow value of 3.25 ml/(100 ml tissue × min) for tissues J and K, corresponding to average total flows of 980 and 98 ml/min for muscle and skin, respectively. Note that total blood flow in column 4 corresponds to the cardiac output, 6.48 liters/min.

When discussing drug distribution, it is often convenient to lump various tissue regions into general categories. For example, following an IV bolus injection, the heart, brain, liver, and kidneys achieve the highest and earliest drug concentrations, with equilibrium between these tissues and blood being rapidly achieved. Thus, Dedrick and Bischoff [25] have combined these tissues and other well-perfused regions (see A through H in Table 4) into a well-perfused compartment that they designate as viscera. Similarly, regions I, J, K, and M are lumped into a less well-perfused compartment, the lean tissues, whereas poorly perfused regions N and O are designated as the adipose compartment. Blood flows, volumes, and such, for these lumped compartments can be calculated by summing the appropriate terms in Table 4. By using a perfusion model containing these three lumped compartments and a blood compartment, Bischoff and Dedrick [27] were able to describe thiopental concentrations in various tissues as shown in Fig. 7.

Levels in the dog liver, representative of the visceral tissues, are already at a maximum by the time the first sample is taken, since very rapid equilibrium is achieved between these tissues and the blood. Drug uptake into the less well-perfused skeletal muscle, representative of the lean tissue, is slower—peaking at about 20 min, but still achieving apparent distribution equilibrium between 1 and 2 hr. Uptake into the poorly perfused adipose tissue is even slower. In fact, peak levels in this tissue have not even been reached by the time the last samples are taken.

Since the site of action for the barbiturates is the brain, we might expect the pharmacological action to correspond to the time course of thiopental concentrations in the viscera which, in turn, would be reflected by blood levels, since a rapid equilibrium is attained between viscera and blood. Although the pharmacological action may terminate quickly (within an hour) owing to decreased blood levels, traces of the drug may be found in the urine for prolonged periods (days) owing to accumulation in the fatty tissues. Note that the partition coefficient for the drug between tissues and blood is greater than 1 for all three tissue groups (i.e., at distribution

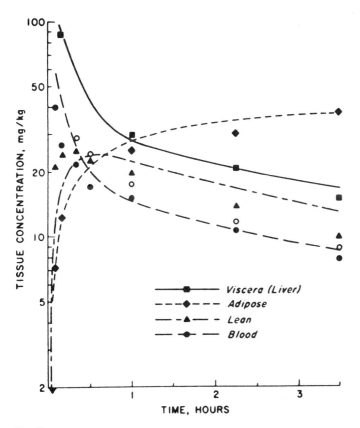

Fig. 7 Thiopental concentrations in various tissues following 25-mg/kg IV bolus doses. Solid symbols indicate data in dogs; the open circles are from data in humans. Lines correspond to predicted values in various tissues using a perfusion model containing compartments corresponding to the blood, viscera, lean, and adipose tissues. (From Ref. 26.)

equilibrium tissue concentrations greater than blood concentrations) but that the $K_{O/B}$ for adipose tissue is very large. This can be deducted from the fact that drug continues to distribute into adipose tissue, even when the concentration in the fat is significantly greater than the blood concentration. At $3^{1}/_{2}$ hr, most of the lipid-soluble thiopental left in the body is in the fat, and at later times, this percentage may even increase before distribution equilibrium is reached. At these later times the removal rate of the barbiturate from the body will be controlled by the slow movement of drug out of the fatty tissue, as a result of the high partition into the fat and the low blood flow to this region.

B. Protein Binding

One of the major determinants of drug distribution is the extent of protein binding. Many drugs bind to plasma proteins, mainly to albumin and α_1-acid glycoprotein (orosomucoid), but sometimes to lipoproteins, various globulins, and specific binding proteins as well. Up to the present time, most studies of drug protein binding have examined the interaction between drugs in the plasma and plasma albumin and plasma α_1-acid glycoprotein. Total plasma albumin for a 70-kg man is about 120 g. Total interstitial albumin is approximately 156 g [28]. Thus almost 60% of total albumin in the body is found outside the plasma. The total extracellular α_1-acid glycoprotein is approximately 3 g, of which approximately 45% is located extravascularly [29]. α_1-Acid glycoprotein has also been found intracellularly and exists as a membrane-bound form [30], but its total cellular amount is unknown. The extracellular amount of α_1-acid glycoprotein can increase significantly in many diseases (inflammation, infections, cancers) as well during trauma. The increase is variable and usually averages a twofold change, although increases up to fivefold have been reported. Binding to tissue components is also assumed to be important for the overall distribution, although direct evidence is difficult to obtain. However, albumin and α_1-acid glycoprotein represent only approximately 1% of the "dry tissue" in the body, of which a large fraction is proteins. One might, therefore, expect that the binding of drugs to tissue would affect drug distribution significantly and frequently much more than the binding of drugs to plasma proteins. The binding of drugs to plasma proteins has been studied extensively, primarily because the experiments can be easily carried out. Tissue-binding studies do not have this advantage, and thus knowledge of the qualitative and quantitative aspects of the binding of drugs to tissue components is poorly understood. The partition coefficients between body organs and blood, $K_{O/B}$, discussed in the previous section will be considerably influenced by binding. It is probably for this reason that attempts to correlate drug distribution, drug action, and membrane transport with oil/water partition coefficients have succeeded only infrequently, necessitating the use of different organic solvents in different correlations to obtain at least an approximate rank order correlation.

Effect on Distribution

The extent of distribution to an organ is usually expressed as the apparent volume of distribution of the organ, V_i. Because the blood concentration is used as a reference, the apparent volume of an organ, V_i, can usually be expressed in the following way:

$$V_i = V_T K_{O/B} = V_T \left(\frac{fu}{fu_T}\right) \qquad (3)$$

where V_T is the physical volume of the tissue, fu is the unbound fraction of drug in the blood and fu_T is the unbound fraction of drug in the tissue. This relationship indicates that the partition coefficient $K_{O/B}$ essentially reflects the strength of binding to the tissues in comparison with the binding to blood proteins, which will be particularly true for hydrophilic compounds.

Decreased binding to proteins in a tissue will increase the value of fu_T, as the fraction of drug in the tissue that is not bound is now increased, and the apparent volume of the tissue decreases. Similarly, if the binding to blood proteins is decreased, the relative affinity for the tissue increases, and the apparent volume of distribution is increased. Often, volumes of distribution and protein binding are defined in terms of plasma measurements, rather than in blood, as used here. Such an approach is correct as long as one defines explicitly the reference fluid of measurement. We prefer to use blood measurements throughout, because, as described in the next section, the relationship of clearance to organ elimination capacity must be defined in terms of the total flow to each organ.

Effect on Clearance

The *clearance* of an organ can be viewed as the volume of blood completely cleared of drug per unit time. Given this definition, the blood clearance of an organ will be equal to the blood flow through the organ multiplied by the extraction ratio (equal to the fraction of the blood completely cleared of drug). From Eq. (1) we, therefore, obtain

$$CL = Q\,E = Q\left(\frac{CL_{Int}}{Q + CL_{Int}}\right) = \frac{Q\,CL_{Int}}{Q + CL_{Int}} \tag{4}$$

Equation (1) was derived assuming no protein binding, whereas, in fact, CL_{Int} should be defined in terms of the maximum ability of the organ to remove *unbound* drug because only unbound drug can bind to enzymes and excretory molecules. By including the unbound fraction of drug in Eq. (4) one obtains

$$CL_{organ} = \frac{Q\,fu\,CL_{Int}}{Q + fu \cdot CL_{Int}} \tag{5}$$

where fu is the fraction of drug unbound in the blood. [Note that all the discussions in this chapter have related clearances and flows to blood concentrations (i.e., drug concentrations in the actual transporting fluid). Blood concentrations must be used when any physiological interpretation is placed on the clearance process. However, it is easier to carry out studies measuring plasma concentrations and plasma protein binding. These measured parameters may be converted to blood values if one knows the blood/plasma concentration ratio.]

When the intrinsic clearance of an organ is very high compared with blood flow, such that $fu\,CL_{Int} \gg Q$, the extraction ratio approaches 1, and $CL_{organ} \to Q$. Thus, according to Eq. (5), if a drug is cleared exclusively in the liver with an extraction ratio approaching unity, organ clearance will be very sensitive to changes in liver blood flow and essentially independent of blood binding [31]. Likewise, for drugs eliminated in the kidney by active-transport processes, elimination appears to be independent of the extent of blood binding. Under these conditions, kidney blood flow becomes the rate-limiting step.

However, for many drugs, blood flow is significantly greater than the intrinsic organ clearance ($Q \gg fu\,CL_{Int}$) and then Eq. (5) reduces to

$$CL_{organ} = fu\,CL_{Int} \tag{6}$$

indicating that organ clearance is dependent on binding in blood and is proportional to the unbound fraction of drug in blood. This phenomenon is valid for both hepatic and renal elimination. For example, the glomerulus in the kidneys will filter approximately 120 ml of plasma per minute. High molecular weight compounds (i.e., most of the plasma proteins) will not be filtered. Likewise, drugs bound to these proteins will be retained. In other words, one could view the filtration process as being filtration of plasma water and of the small molecular weight compounds dissolved therein. Plasma water contains only unbound drug. Increases in binding will lower the fraction of the drug in plasma in unbound form and, therefore, decrease the filtration clearance. For hepatic elimination, data for several compounds have been published over the years that show a correlation between the clearance and the unbound fraction in plasma [32]. We expect similar relationships between the organ clearance and the unbound fraction to be also valid for other organs.

Effect on Availability

Should changes in protein binding have a significant effect on first-pass availability? Combining Eqs. (2) and (5) we obtain:

$$F = 1 - E = 1 - \frac{CL}{Q} = 1 - \left[\frac{fu\ CL_{Int}}{Q + fu \cdot CL_{Int}} \right] = \frac{Q}{Q + fu \cdot CL_{Int}} \tag{7}$$

This indicates that the first-pass availability is a function of organ flow, protein binding, and intrinsic clearance of the organ. When $fu\ CL_{Int} \gg Q$ (i.e., when we have relatively large extraction ratios), the first-pass bioavailability is equal to

$$F = \frac{Q}{fu\ CL_{Int}} \tag{8}$$

Under this circumstance, the first-pass bioavailability is inversely proportional to the unbound fraction, and changes in the binding are expected to have a significant effect. It is also clear that changes in both the blood flow and the intrinsic clearance of the first-pass organ may have a significant effect when the extraction ratio is high ($fu\ CL_{Int} \gg Q$). On the other hand, if $Q \gg fu\ CL_{Int}$ then Eq. (7) simply says that the first-pass bioavailability is approximately 1 (i.e., little or no drug is eliminated in a first pass), and changes in binding, blood flow, and intrinsic clearance are not expected to have any effect on F.

Combined Kinetic Effect of Binding Alterations

Changes in binding can occur in many situations. In some diseases, the concentrations of binding proteins may be altered, and accumulation of endogenous inhibitors of binding may occur. For example, in renal failure the albumin level may decrease, resulting in decreased binding. In renal failure, accumulation of waste products also occurs, and this, in turn, may further suppress the binding. In other diseases (e.g., inflammations), the level of α_1-acid glycoprotein increases, which may lead to increased drug binding. Concomitant administration of compounds that compete for the same binding sites may also decrease the binding and increase the unbound fraction. These changes will lead to changes in the apparent volume of distribution, in clearance, and in the first-pass availability, as discussed earlier, but will they lead to changes in the time required for the drug to reach the target organ and in the activity of the drug? The answer is not simple, as it depends on the actual values of clearance and volume of distribution, as well as on whether the target tissue is among the highly or poorly perfused tissues.

To begin answering this question we must realize that the effect of drugs in the body is related to the unbound and not the total concentration in the body [33]. Only non–protein-bound drug can interact with receptors and, therefore, elicit an effect. Therefore, it is important to look at how the *unbound* concentration changes. Let us assume that a situation of decreased binding exists in blood, and that no changes in the tissue binding have occurred. If a dose is given, we expect no changes in the first-pass bioavailability, if the extraction ratio in the first-pass tissue is low. If the extraction ratio is high, on the other hand, we expect a significant decrease in the first-pass bioavailability [see Eq. (8)] both of an IV administration (if first-pass in the lung is significant) or an oral administration (first-pass effect both in the liver and lung). When it comes to other routes of administration, we need to be more cautious. For example, let us look at dermal administration. If the major first-pass elimination occurs before the drug reaches the blood (i.e., in its passage through the epithelial cells), we do not expect the blood-binding changes to be important. On the other hand, if the major first-pass elimination relates to cell downstream from where the drug enters the blood, changes in the binding will cause changes predictable from Eq. (8).

Will the drug be delayed on its way to the target tissue in the foregoing example? A decreased blood binding is equivalent to increasing the $K_{O/B}$ value. Distribution into various tissues, including the target tissue, therefore, is expected to be more extensive and swifter. However, because the drug usually has to pass the lungs before it can reach the target tissue, high $K_{O/B}$ values in the lung will significantly reduce the amount available to other tissues when the $K_{O/B}$ value increases for the lung. Consequently, a delay of distribution here can occur. On the other hand, if the $K_{O/B}$ value in the lung is low, the amount sequestered by the lung will be too small to delay the distribution. In this situation, no delay is expected in distribution of drug to the target tissue, if the target tissue is highly perfused. However, if the target tissue is a poorly perfused tissue, a delay may still occur. In this situation, a higher $K_{O/B}$ value will mean that the highly perfused tissues can take up more of the initial drug presented to them. In turn, the diffusion out of the highly perfused tissues will be slower, and may take longer before the redistribution to poorly perfused tissues is completed. Therefore, if the effect is in the poorly perfused tissues, the effect will lag.

On continuous administration, the drug's effect is dependent on the average unbound steady-state concentration. The unbound concentration is dependent on the average rate of dosing, bioavailability, clearance, and degree of binding in plasma. Let us, for the moment, assume that the bioavailability and dosing rate remain constant. The change in the steady-state unbound concentration now depends on whether the extraction ratio is low or high (or whether $Q \gg fu\,CL_{Int}$ or $fu\,CL_{Int} \gg Q$) in addition to the alteration of the binding. For a low-extraction ratio compound the average unbound concentration, Cu_{ss}, is

$$Cu_{ss} = \frac{F\,(\text{rate of dosing})}{CL_{Int}} \qquad (9)$$

For a high-extraction ratio, the value is

$$Cu_{ss} = \frac{F\,(\text{rate of dosing})}{Q/fu} \qquad (10)$$

Under the assumptions of constant bioavailability and dosing rate, a high-extraction ratio compound is expected to have an increased unbound steady-state concentration when the plasma binding is decreased, and a low-extraction ratio compound is not affected by binding changes. Now, however, we also need to evaluate whether the first-pass availability is affected by a change in the binding and adjust our expectation accordingly. Under special circumstances for which the first-pass organ is the major metabolizing organ, some simplified concepts can be established. For a low-extraction ratio compound in this situation, neither the first-pass bioavailability, nor the clearance relative to unbound drug will be affected, and the result is that there is no overall effect on the average unbound steady-state concentration and the activity of the drug. For a high-extraction ratio compound both the clearance relative to unbound drug [Eq. (5)] and the first-pass bioavailability [see Eq. (7)] will increase proportional to the increase in the unbound fraction in plasma, and the result is again that the average unbound concentration is not affected by a change in the plasma binding. If the first-pass organ is different from the major eliminating organ, the extraction ratio in these organs may be different, and their dependence on binding may differ (i.e., the unbound clearance and bioavailability may be affected to a different extent). If saturation of the first-pass takes place, there is also likely to be a different effect of binding on the first-pass bioavailability and unbound clearance, even if the first-pass organ is the major eliminating organ.

Changes in tissue binding have effects on only the tissue distribution and not on clearance. Tissue-binding changes, therefore, are not expected to affect the first-pass bioavailability, but are expected to alter the distribution [see Eq. (3)] and the $K_{O/B}$ value. If the value of $K_{O/B}$ is reduced, we expect a smaller sequestering of drug in its passage through the body, higher initial concentrations will reach the target tissue, and the effect will occur more swiftly and, initially, more potently. But because the apparent volume of distribution is smaller, the half-life will also be smaller:

$$t_{1/2} = \frac{0.693\ V}{\text{CL}} \tag{11}$$

and the concentration and the effect will fall off faster. Documentation for such changes is difficult to obtain, because we cannot measure the tissue binding directly, and we can make only inferences from overall changes in the kinetic parameters and changes in plasma-binding values [34].

A third possibility, that the binding in plasma and tissue changes to the same degree, is not expected to affect the apparent volume of distribution to the individual organs [see Eq. (3)], and the value of $K_{O/B}$ will not be changed. The clearance, on the other hand, is affected only by changes in binding in blood and will change as described in Sec. III.B, the section describing the effect of protein-binding changes on clearance. Assuming that the first-pass bioavailability is not affected, this will mean that sequestering of drug will not increase, and the total concentration reaching the target tissue should not be significantly affected. However, because the unbound fraction in plasma is increased (decreased), the unbound concentration will be increased (decreased), the unbound concentration initially reaching the target tissue will be higher (lower), and the initial pharmacological activity will be higher (lower). The duration of this change will depend on the half-life of the drug, which will be altered according to Eq. (11).

IV. SUMMARY

The extent and time course of drug action can be markedly affected by the route of drug administration into the patient as well as the pattern of drug distribution within the patient. Drug efficacy can be improved, and drug toxicity probably decreased, if the drug can be administered directly to its site of action. However, several factors prevent direct application of drugs to the site of action, including incomplete knowledge about the action site and also poor patient adherence owing to the inconvenience of using direct application formulations. Because of these factors, most drug products have been formulated as oral, solid dosage forms.

Drugs that are rapidly cleared by hepatic processes will show a decreased extent of availability following oral administration owing to metabolism of the drug on its first pass through the liver. The magnitude of this first-pass effect will depend on the blood flow to the liver and the intrinsic clearing ability of the liver (i.e., the ability of the organ to eliminate the drug independently of the rate at which drug is brought to the organ). This first-pass elimination by metabolic or biliary excretion processes can be excluded if the drug is absorbed from a sublingual site. The rectal administration of drugs eliminates approximately one-half the first-pass metabolism. Absorption through nasal, dermal, and other sites may also give rise to lower first-pass elimination and higher bioavailability than oral administration if the metabolic capacity of these sites for the drug in question is small.

Drug distribution in the patient will depend on the blood blow to various sites in the body, as well as on the partition coefficient of the drug between the blood and distributive organs. Protein binding, both in the blood and in the tissues, will markedly affect this distribution. However, free drug concentrations are generally believed to be the effective determinant in drug therapy. Often a redistribution owing to changes in protein binding will have little effect on the therapeutic efficacy, since, although total drug distribution changes, the average unbound concentrations at steady state in blood remain essentially similar. An understanding of the effects of the route of administration as well as the distribution of the drug within the body is critical to the pharmacist in planning appropriate drug dosage regimens.

REFERENCES

1. L. Z. Benet, Biopharmaceutics as a basis for the design of drug products, in *Drug Design*, Vol. 4, (E. Ariens, Ed.), Academic Press, New York, 1973, pp. 1–35.
2. L. Z. Benet, Input factors as determinants of drug activity: route, dose, dosage regimen, and the drug delivery system, in *Principles and Techniques of Human Research and Therapeutics*, Vol. 3 (F. G. McMahon, Ed.), Futura, New York, 1974, pp. 9–23.
3. T. E. Gram, The metabolism of xenobiotics by the mammalian lung, in *Extrahepatic Metabolism of Drugs and Other Foreign Compounds* (T. E. Gram, Ed.), S.P. Medical and Scientific Books, New York, 1980, pp. 159–209
4. J. R. Vane, The role of the lungs in the metabolism of vasoactive substances, in *Pharmacology and Pharmacokinetics* (T. Teorell, R. L. Dedrick, and P. G. Condliffe, Eds.), Plenum Press, New York, 1974, pp. 195–207.
5. W. L. Chiou, Potential pitfalls in the conventional pharmacokinetic studies: Effects of the initial mixing of drug in blood and the pulmonary first-pass elimination, J. Pharmacokinet. Biopharm., 7, 527–536 (1979).
6. M. Rowland, L. Z. Benet, and G. G. Graham, Clearance concepts in pharmacokinetics, J. Pharmacokinet. Biopharm., 1, 123–136 (1973).
7. E. F. Evans, J. D. Proctor, M. J. Fratkin, J. Velandia, and A. J. Wasserman, Blood flow in muscle groups and drug absorption, Clin. Pharmacol. Ther., 17, 44–47 (1975).
8. L. S. Cohen, J. E. Rosenthal, D. W. Horner, Jr., J. M. Atkins, O. A. Matthews, and S. F. Sarnoff, Plasma levels of lidocaine after intramuscular administration, Am. J. Cardiol., 29, 520–523 (1972).
9. M. L. Schwartz, M. B. Meyer, B. G. Covino, R. M. Narange, W. Sethi, A. J. Schwartz, and P. Kemp, Antiarrhythmic effectiveness of intramuscular lidocaine: Influence of different injection sites, J. Clin. Pharmacol., 14, 77–83 (1974).
10. J. T. Doluisio, J. C. LaPiana, and L. W. Dittert, Pharmacokinetics of ampicillin trihydrate, sodium ampicillin, and sodium dicloxacillin following intramuscular injection, J. Pharm. Sci., 60, 715–719 (1971).
11. A. J. Wilensky and J. A. Lowden, Inadequate serum levels after intramuscular administration of diphenylhydantoin, Neurology, 23, 318–324 (1973).
12. M. Rowland, Influence of route of administration on drug availability, J. Pharm. Sci., 61, 70–74 (1972).
13. R. N. Boyes, J. H. Adams, and B. R. Duce, Oral absorption and disposition kinetics of lidocaine hydrochloride in dogs, J. Pharmacol. Exp. Ther. 174, 1–8 (1970).
14. M. Rowland, Effect of some physiologic factors on bioavailability of oral dosage forms, in *Dosage Form Design and Bioavailability* (J. Swarbrick, Ed.), Lea & Febiger, Philadelphia, 1973, pp. 181–222.
15. D. G. Shand, E. M. Nuckolls, and J. A. Oates, Plasma propranolol levels in adults with observations in four children, Clin. Pharmacol. Ther., 11, 112–120 (1970).
16. W. H. Barr, Factors involved in the assessment of systemic or biologic availability of drug products, Drug Inf. Bull., 3, 27–45 (1969).

17. M. Drucker, S. H. Blondheim, and L. Wislicki, Factors affecting acetylation in vivo of *para*-aminobenzoic acid by human subjects, Clin. Sci., 27, 133–141 (1964).

18. A. G. De Boer, D. D. Breimer, H. Mattie, J. Pronk, and J. M. Gubbens-Stibbe, Rectal bioavailability of lidocaine in man: Partial avoidance of "first-pass" metabolism, Clin. Pharmacol. Ther., 26, 701–709 (1979).

19. T. W. Schwarz, in *American Pharmacy*, 6th ed., (J. B. Sprowls, Jr. and H. M. Beal, Eds.), J. B. Lippincott, Philadelphia, 1966, pp. 311–331.

20. A. E. Pontiroli, M. Alberetto, and G. Pozza, Intranasal calcitonin and plasma calcium concentrations in normal subjects, Br. Med. J., 290, 1390–1391 (1985).

21. Y. W. Chien and K. Banga, Iontophoretic (transdermal) delivery of drugs: Overview of historic development, J. Pharm. Sci., 78, 353–354 (1989).

22. D. Parasrampuria and J. Parasrampuria, Percutaneous delivery of proteins and peptides using iontophoretic techniques, J. Clin. Pharm. Ther. 16, 7–17 (1991)

23. D. T. O'Hagan and L. Illum, Absorption of peptides and proteins from the respiratory tract and the potential for development of locally administered vaccine, Crit. Rev. Ther. Drug Carrier Syst., 7, 35–97 (1990).

24. J. J. Sciarra, Aerosols, in *Prescription Pharmacy*, 2nd ed. (J. B. Sprowls, Jr., Ed.), J. B. Lippincott, Philadelphia, 1970, pp. 280–328.

25. R. L. Dedrick and K. B. Bischoff, Pharmacokinetics in applications of the artificial kidney, Chem. Eng. Progr. Symp. Ser., 64, 32–44 (1968).

26. W. W. Mapleson, An electric analogue for uptake and exchange of inert gases and other agents, J. Appl. Physiol., 18, 197–204 (1963).

27. K. B. Bischoff and R. L. Dedrick, Thiopental pharmacokinetics, J. Pharm. Sci., 57, 1347–1357 (1968).

28. J. G. Wagner, *Fundamentals of Clinical Pharmacokinetics*. Drug Intelligence Publishers, Hamilton, IL, 1975, pp. 24–26.

29. F. Bree, G. Houin, J. Barre, J. L. Moretti, V. Wirquin, and J.-P. Tillement, Pharmacokinetics of intravenously administered ^{125}I-labelled human alpha-1-acid glycoprotein, Clin. Pharmacokinet, 11, 336–342 (1986).

30. C. G. Gahmberg and L. C. Anderson, Leucocyte surface origin of human alpha-1-acid glycoprotein (orosomucoid), J. Exp. Med., 148, 507–521 (1978).

31. T. W. Guenthert and S. Øie, Effect of plasma protein binding on quinidine kinetics in the rabbit, J. Pharmacol. Exp. Ther., 215, 165–171 (1980).

32. T. F. Blaschke, Protein binding and kinetics of drugs in liver disease, Clin. Pharmacokinet, 2, 32–44 (1977).

33. S. Øie and J.-D. Huang, Binding, should free drug levels be measured? in *Topics in Pharmaceutical Sciences 1983* (D. D. Breimer and P. Speiser, Eds.), Elsevier, Amsterdam, 1983, pp. 51–62.

34. B. Fichtl, Tissue binding of drugs: Methods of determination and pharmacokinetic consequences, in *Plasma Binding of Drugs and Its Consequences* (F. Belpaire, M. Bogaert, J. P. Tillement, and R. Verbeeck, Eds.), Academia Press, Ghent, 1991, pp. 149–158.

6

Chemical Kinetics and Drug Stability

J. Keith Guillory and Rolland I. Poust
College of Pharmacy, University of Iowa, Iowa City, Iowa

I. INTRODUCTION

In the rational design and evaluation of dosage forms for drugs, the stability of the active components must be a major criterion in determining their suitability. Several forms of instability can lead to the rejection of a drug product. First, there may be chemical degradation of the active drug, leading to a substantial lowering of the quantity of the therapeutic agent in the dosage form. Many drugs (e.g., digoxin and theophylline) have narrow therapeutic indices, and they need to be carefully titrated in individual patients so that serum levels are neither so high that they are potentially toxic, nor so low that they are ineffective. For these drugs, it is of paramount importance that the dosage form reproducibly deliver the same amount of drug.

Second, although chemical degradation of the active drug may not be extensive, a toxic product may be formed in the decomposition process. Dearborn [1] described several examples in which the products of degradation are significantly more toxic than the original therapeutic agent. Thus, the conversions of tetracycline to epianhydrotetracycline, arsphenamine to oxophenarsine, and *p*-aminosalicylic acid to *m*-aminophenol in dosage forms give rise to potentially toxic agents that, when ingested, can cause undesirable effects. Recently, Nord et al. [2] reported that the antimalarial chloroquine can produce toxic reactions that are attributable to the photochemical degradation of the substance. Phototoxicity has also been reported to occur following administration of chlordiazepoxide and nitrazepam [3]. Another example of an adverse reaction caused by a degradation product was provided by Neftel et al. [4], who showed that infusion of degraded penicillin G led to sensitization of lymphocytes and formation of antipenicilloyl antibodies.

Third, instability of a drug product can lead to a decrease in its bioavailability, rather than to loss of drug or to formation of toxic degradation products. This reduction in bioavailability can result in a substantial lowering in the therapeutic efficacy of the dosage form. This phenomenon can be caused by physical or chemical changes in the excipients in the dosage form, independent of whatever changes the active drug may have undergone. A more detailed discussion of this subject is given in Sec. II.B.

Fourth, there may be substantial changes in the physical appearance of the dosage form. Examples of these physical changes include mottling of tablets, creaming of emulsions, and caking of suspensions. Although the therapeutic efficacy of the dosage form may be unaffected by these changes, the patient will most likely lose confidence in the drug product, which then has to be rejected.

A drug product, therefore, must satisfy stability criteria chemically, toxicologically, therapeutically, and physically. Basic principles in pharmaceutical kinetics can often be applied to anticipate and quantify the undesirable changes so that they can be circumvented by stabilization techniques. Some chemical compounds, called prodrugs [5,6], are designed to undergo chemical or enzymatic conversion in vivo to pharmacologically active drugs. Prodrugs are employed to solve one or several problems presented by active drugs (e.g., short biological half-life, poor dissolution, bitter taste, inability to penetrate through the blood–brain barrier, and others). They are pharmacologically inactive as such, but are converted back in vivo to their parent (active) compounds. Naturally, the rate and extent of this conversion (which are governed by the same laws of kinetics that will be described in this chapter) are the primary determinants of the therapeutic efficacy of these agents.

In the present chapter, stability problems and chemical kinetics are introduced and surveyed. The sequence employed is as follows: first, an overview of the potential routes of degradation that drug molecules can undergo; then, a discussion of the mathematics used to quantify drug degradation; a delineation of the factors that can affect degradation rates, with an emphasis on stabilization techniques; and, finally, a description of stability-testing protocols employed in the pharmaceutical industry. It is not the intent of this chapter to document stability data of various individual drugs. Readers are referred to the compilations of stability data [7] and to literature on specific drugs [e.g., Ref. 8 and earlier volumes] for this kind of information.

II. ROUTES BY WHICH PHARMACEUTICALS DEGRADE

Since most drugs are organic molecules, it is important to recognize that many pharmaceutical degradation pathways are, in principle, similar to reactions described for organic compounds in standard organic chemistry textbooks. On the other hand, it is also important to realize that different emphases are placed on the types of reactions that are commonly encountered in the drug product stability area, as opposed to those seen in classic organic chemistry. In the latter, reactions are generally described as tools for use by the synthetic chemist; thus, the conditions under which they are carried out are likely to be somewhat drastic. Reactive agents (e.g., thionyl chloride or lithium aluminum hydride) are employed in relatively high concentrations (often > 10%) and are treated using exaggerated conditions, such as refluxing or heating in a pressure bomb. Reactions are effected in relatively short time periods (hours or days). In contrast, reactions occurring in pharmaceuticals often involve the active drug components in relatively low concentrations. For example, dexamethasone sodium phosphate, a synthetic adrenocorticoid steroid salt, is present only to the extent of about 0.4% in its injection, 0.1% in its topical cream or ophthalmic solution, and 0.05% in its ophthalmic ointment. The decomposition of a drug is likely to be mediated not by reaction with another active ingredient, but by reaction with water, oxygen, or light. Reaction conditions of interest are usually ambient or subambient. Reactions in pharmaceuticals ordinarily occur over months or years, as opposed to the hours or days required for completion of reactions in synthetic organic chemistry.

Reactions such as the Diels-Alder reaction and aldol condensations, which are important in synthetic and mechanistic organic chemistry, are of only minor importance when drug degradation is being considered. Students need to refocus their attention on reactions such as hydrolysis, oxidation, photolysis, racemization, and decarboxylation, the routes by which most pharmaceuticals degrade.

A cognizance of reactions of particular functional groups is important if one is to gain a broad view of drug degradation. It is a difficult task to recall degradative pathways of all commonly used drugs. Yet, through the application of functional group chemistry, it is possible to anticipate the potential mode(s) of degradation that drug molecules will likely undergo. In the following discussion, therefore, degradative routes are demonstrated by calling attention to the reactive functional groups present in drug molecules. The degradative routes are described, through the use of selected examples, as *chemical* when new chemical entities are formed as a result of drug decomposition, and as *physical* when drug loss does not produce distinctly different chemical products.

A. Chemical Degradative Routes

Solvolysis

In this type of reaction, the active drug undergoes decomposition following reaction with the solvent present. Usually, the solvent is water; but sometimes the reaction may involve pharmaceutical cosolvents, such as ethyl alcohol or polyethylene glycol. These solvents can act as nucleophiles, attacking the electropositive centers in drug molecules. The most common solvolysis reactions encountered in pharmaceuticals are those involving "labile" carbonyl compounds, such as esters, lactones, and lactams (Table 1).

Although all the functional groups cited are, in principle, subject to solvolysis, the rates at which they undergo this reaction may be vastly different. For example, the rate of hydrolysis of a β-lactam ring (a cyclized amide) is much greater than that of its linear analog. The half-life (the time needed for half the drug to decompose) of the β-lactam in potassium phenethicillin at 35°C and pH 1.5 is about 1 hr. The corresponding half-life for penicillin G is about 4 min [9]. In contrast, the half-life for hydrolysis of the simple amide propionamide in 0.18 molal H_2SO_4 at 25°C is about 58 hr [10]. It has been suggested that the antibacterial activity of β-lactam antibiotics arises from a combination of their chemical reactivity and their molecular recognition by target enzymes. One aspect of their chemical reactivity is their acylating power and, although penicillins are not very good acylating agents, they are more reactive than simple, unsubstituted amides [11]. Unactivated or "normal" amides undergo nonenzymatic hydrolysis slowly, except under the most extreme conditions of pH and temperature, because the N—C(O) linkage is inherently stable, yet when the amine function is a good leaving group (and particularly if it has a pK_a greater than 4.5), amides can be susceptible to hydrolysis at ordinary temperatures. [For a recent review on this subject see Ref. 12.] Acyl-transfer reactions in peptides, including the transfer to water (hydrolysis), are of fundamental importance in biological systems in which the reactions proceed at normal temperatures, and enzymes serve as catalysts.

The most frequently encountered hydrolysis reaction in drug instability is that of the ester, but certain esters can be stable for many years when properly formulated. Substituents can have a dramatic effect on reaction rates. For example, the *tert*-butyl ester of acetic acid is about 120 times more stable than the methyl ester, which, in turn, is approximately 60 times more stable than the vinyl analog [13]. Structure–reactivity relationships are dealt with in the discipline of physical organic chemistry. Substituent groups may exert electronic (inductive and resonance), steric, or hydrogen-bonding effects that can drastically affect the stability of compounds. Interested students are referred to a recent review by Hansch and Taft [14], and to the classic reference text written by Hammett [15].

Table 1 Some Functional Groups Subject to Hydrolysis

Drug type		Examples
Esters	RCOOR' $ROPO_3 M_x$ $ROSO_3 M_x$ $RONO_2$	Aspirin, alkaloids Dexamethasone sodium phosphate Estrone sulfate Nitroglycerin
Lactones		Pilocarpine Spironolactone
Amides	$RCONR'_2$	Thiacinamide Chloramphenicol
Lactams		Penicillins Cephalosporins
Oximes	$R_2C = NOR$	Steroid oximes
Imides		Glutethimide Ethosuximide
Malonic ureas		Barbiturates
Nitrogen mustards		Melphalan

A dramatic decrease in ester stability can be brought about by intramolecular catalysis. This type of facilitation is affected mostly by neighboring groups capable of exhibiting acid–base properties (e.g., $-NH_2$, $-OH$, $-COOH$, and COO—). If neighboring-group participation leads to an enhanced reaction rate, the group is said to provide anchimeric assistance [16]. For example, the ethyl salicylate anion undergoes hydrolysis in alkaline solution at a rate that is 10^6 times larger than the experimental value for the uncatalyzed cleavage of ethyl p-hydroxy-benzoate. The rate advantage is attributed to intramolecular general base catalysis by the phenolate anion [17].

Oxidation

Oxidation reactions are important pathways of drug decomposition. In pharmaceutical dosage forms, oxidation is usually mediated through reaction with atmospheric oxygen under ambient conditions, a process commonly referred to as autoxidation. Oxygen is, itself, a diradical, and most autoxidations are free-radical reactions. A free radical is a molecule or atom with one or more unpaired electrons. Of considerable importance to pharmaceutical scientists is a reliable method for determining and controlling oxygen concentration in aqueous solutions [18]. A thorough review of autoxidation and of antioxidants has been published [19].

The mechanisms of oxidation reactions are usually complex, involving multiple pathways for the initiation, propagation, branching, and termination steps. Many autoxidation reactions are initiated by trace amounts of impurities, such as metal ions or hydroperoxides. Thus, ferric ion catalyzes the degradation reaction and decreases the induction period for the oxidation of the compound procaterol [20]. As little as 0.0002 M copper ion will increase the rate of vitamin C oxidation by a factor of 10^5 [21]. Hydroperoxides contained in polyethylene glycol suppository bases have been implicated in the oxidation of codeine to codeine-N-oxide [22]. Peroxides apparently are responsible for the accelerated degradation of benzocaine hydrochloride in aqueous cetomacrogol solution [23] and of a corticosteroid in polyethylene glycol 300 [24,25]. Many oxidation reactions are catalyzed by acids and bases [26].

A list of some functional groups that are subject to autoxidation is shown in Table 2. The products of oxidation are usually electronically more conjugated; thus, the appearance of, or a change in, color in a dosage form is suggestive of the occurrence of oxidative degradation.

Photolysis

Normal sunlight or room light may cause substantial degradation of drug molecules. The energy from light radiation must be absorbed by the molecules to cause a photolytic reaction. If that

Table 2 Some Functional Groups Subject to Autoxidation

Functional group		Examples
Phenols		Phenols in steroids
Catechols		Catecholamines (dopamine, isoproterenol)
Ethers	$R-O-R'$	Diethylether
Thiols	RCH_2SH	Dimercaprol (BAL)
Thioethers	$R-S-R'$	Phenothiazines (chlorpromazine)
Carboxylic acids	$RCOOH$	Fatty acids
Nitrites	RNO_2	Amyl nitrite
Aldehydes	$RCHO$	Paraldehyde

energy is sufficient to achieve activation, degradation of the molecule is possible. Saturated molecules do not interact with visible or near-ultraviolet light, but molecules that contain π-electrons usually do absorb light throughout this wavelength range. Consequently, compounds such as aromatic hydrocarbons, their heterocyclic analogues, aldehydes, and ketones, are most susceptible to photolysis. In general, drugs that absorb light at wavelengths below 280 nm have the potential to undergo decomposition in sunlight, and drugs with absorption maxima greater than 400 nm have the potential for degradation both in sunlight and room light.

A dramatic example of photolysis is the photodegradation of sodium nitroprusside in aqueous solution. Sodium nitroprusside, $Na_2Fe(CN)_5NO\cdot2H_2O$, is administered by intravenous infusion for the management of acute hypertension. If the solution is protected from light, it is stable for at least 1 year; if exposed to normal room light, it has a shelf life of only 4 hr [27].

Photolysis reactions are often associated with oxidation because the latter category of reactions can frequently be initiated by light. But, photolysis reactions are not restricted to oxidation. For sodium nitroprusside, it is believed that degradation results from loss of the nitro-ligand from the molecule, followed by electronic rearrangement and hydration. Photo-induced reactions are common in steroids [28]; an example is the formation of 2-benzoylcholestan-3-one following irradiation of cholest-2-en-3-ol benzoate. Photoadditions of water and of alcohols to the electronically excited state of steroids have also been observed [29].

Dehydration

The preferred route of degradation for prostaglandin E_2 and tetracycline is the elimination of a water molecule from their structures. The driving force for this type of covalent dehydration is the formation of a double bond that can then participate in electronic resonance with neighboring functional groups. In physical dehydration processes, such as those occurring in theophylline hydrate and ampicillin trihydrate [30], water removal does not create new bonds, but often changes the crystalline structure of the drug. Since it is possible that anhydrous compounds may have different dissolution rates compared with their hydrates [31,32], dehydration reactions involving water of crystallization may potentially affect the absorption rates of the dosage form.

Racemization

The racemization of pharmacologically active agents is of interest because enantiomers often have significantly different absorption, distribution, metabolism, and excretion, in addition to differing pharmacological actions [33]. The best-known racemization reactions of drugs are those that involve epinephrine, pilocarpine, ergotamine, and tetracycline. In these drugs, the reaction mechanism appears to involve an intermediate carbonium ion or carbanion that is stabilized electronically by the neighboring substituent group. For example, in the racemization of pilocarpine [34], a carbanion is produced and stabilized by delocalization to the enolate. In addition to the racemization reaction, pilocarpine is also degraded through hydrolysis of the lactone ring.

Most racemization reactions are catalyzed by an acid or by a base. A notable exception is the "spontaneous" racemization of the diuretic and antihypertensive agent, chlorthalidone, which undergoes facile S_N1 solvolysis of its tertiary hydroxyl group to form a planar carbonium ion. Chiral configuration is then restored by nucleophilic attack (S_N2) of a molecule of water on the carbonium ion, with subsequent elimination of a proton [35].

Incompatibilities

Chemical interactions between two or more drug components in the same dosage form, or between active ingredient and a pharmaceutical adjuvant, frequently occur. An example of drug–drug incompatibility is the inactivation of cationic aminoglycoside antibiotics, such as kanamycin and gentamicin, by anionic penicillins in IV admixtures. The formation of an inactive complex between these two classes of antibiotics occurs not only in vitro, but apparently also in vivo in patients with severe renal failure [36]. Thus, when gentamicin sulfate was given alone to patients on long-term hemodialysis, the biological half-life of gentamicin was greater than 60 hr. But, when carbenicillin disodium (CD) was given with gentamicin sulfate (GS) in the dose ratio CD/GS = 80:1, the gentamicin half-life was reduced to about 24 hr.

Many pharmaceutical incompatibilities are the result of reactions involving the amine functional group. A summary of the potential interactions that can occur between various functional groups is given in Table 3.

Other Chemical Degradation Reactions

Other chemical reactions, such as hydration, decarboxylation, or pyrolysis, also are potential routes for drug degradation. Thus, cyanocobalamin may absorb about 12% of water when exposed to air, and p-aminosalicylic acid decomposes with evolution of carbon dioxide to form m-aminophenol when subjected to temperatures above 40°C. The temperature at which pyrolytic decomposition of terfenadine occurs has been used as a criterion for determining which of several tablet excipients will be perferable for long-term stability of the drug substance [37].

B. Physical Degradative Routes

Polymorphs are different crystal forms of the same compound [38]. They are usually prepared by crystallization of the drug from different solvents under diverse conditions. Steroids, sulfonamides, and barbiturates are notorious for their propensity to form polymorphs [39]. Yang and Guillory [40] attempted to correlate the occurrence frequency of polymorphism in sulfonamides with certain aspects of chemical structure. They found that sulfonamides that did not exhibit polymorphism have somewhat higher melting points and heats of fusion than those that were polymorphic. The absence of polymorphism in sulfacetamide was attributed to the stronger hydrogen bonds formed by the amide hydrogen in this molecule. These stronger hydrogen bonds were not readily stretched or broken to form alternate crystalline structures.

Since polymorphs differ from one another in their crystal energies, the more energetic ones will seek to revert to the most stable (and the least energetic) crystal form. When several polymorphs and solvates (substances that incorporate solvent in a stoichiometric fashion into the crystal lattice) are present, the conditions under which they may interconvert can become quite complex, as is true of fluprednisolone [41].

Polymorphs may exhibit significant differences in important physicochemical parameters, such as solubility, dissolution rate, and melting point [42]. Thus, the conversion from one polymorph to another in a pharmaceutical dosage form may lead to a drastic change in the physical characteristics of the drug. A well-known example of this phenomenon is the conversion of a more soluble crystal form (form II) of cortisone acetate to a less soluble form (form V) when the drug is formulated into an aqueous suspension [43]. This phase change leads to caking of the cortisone acetate suspension.

Another physical property that can affect the appearance, bioavailability, and chemical stability of pharmaceuticals is the degree of crystallinity. It has been reported that crystalline insulin [44] and crystalline cyclophosphamide [45] are much more stable than their amorphous counterparts.

Vaporization

Some drugs and pharmaceutical adjuvants possess sufficiently high vapor pressures at room temperature that their volatilization through the container constitutes a major route of drug

Table 3 Some Potential Drug Incompatibilities

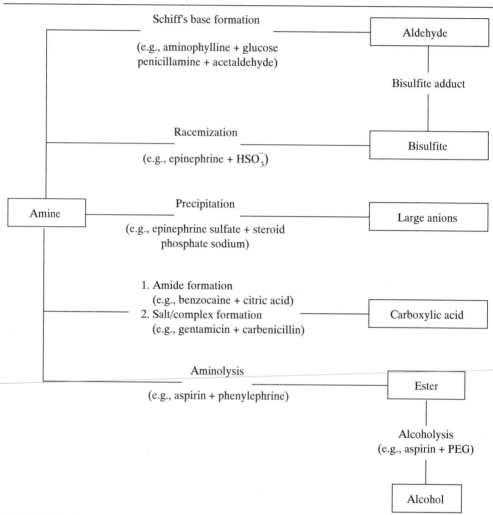

loss. Flavors, whose constituents are mainly ketones, aldehydes and esters, and cosolvents (low molecular weight alcohols) may be lost from the formulation in this manner. The most frequently cited example of a pharmaceutical that "degrades" by this route is nitroglycerin, which has a vapor pressure of 0.00026 mm at 20°C and 0.31 mm at 93°C [46]. Significant drug loss to the environment can occur during patient storage and use. In 1972, the Food and Drug Administration (FDA) issued special regulations governing the types of containers that may be used for dispensing sublingual nitroglycerin tablets [47].

Reduction of vapor pressure, and thereby of volatility, of drugs such as nitroglycerin can be achieved through dispersion of the volatile drug in macromolecules that can provide physicochemical interactions. The addition of macromolecules, such as polyethylene glycol, polyvinylpyrrolidone, and microcrystalline cellulose, allows preparation of "stabilized" nitroglycerin sublingual tablets [48,49]. A β-cyclodextrin–nitroglycerin tablet is currently being marketed in Japan to achieve the same purpose.

Another aspect of nitroglycerin instability has been observed by Fusari [49]. When conventional (unstabilized) nitroglycerin sublingual tablets are stored in enclosed glass containers, the high volatility of the drug gives rise to redistribution of nitroglycerin among the stored tablets. Interestingly, this redistribution leads to an increase in the standard deviation of the drug contents of the tablets, rather than the reverse. This migration phenomenon results in a deterioration in the uniformity of the tablets on storage.

Aging

The most interesting, and perhaps the least-reported, area of concern about the physical instability of pharmaceutical dosage forms is generally termed *aging*. This is a process through which changes in the disintegration or dissolution characteristics of the dosage form are caused by subtle, and sometimes unexplained, alterations in the physicochemical properties of the inert ingredients or the active drug in the dosage form [50]. Since the disintegration and dissolution steps may be the rate-determining steps in the absorption of a drug, changes in these processes, as a function of the "age" of the dosage form, may result in corresponding changes in the bioavailability of the drug product.

An example of this phenomenon was provided by deBlaey and Rutten-Kingma [51], who showed that the melting time of aminophylline suppositories, prepared from various bases, increased from about 20 min to over an hour after 24 weeks of storage at 22°C. Like the dissolution time for solid dosage forms, the melting time for suppositories can be viewed as an in vitro index of drug release. Thus, an increase in melting time can perceivably lead to a decrease in bioavailability. The mechanism responsible for this change appeared to involve an interaction between the ethylenediamine in aminophylline and the free fatty acids present in the suppository bases. Interestingly, no increase in melting time was detected when the suppositories were stored at 4°C, even up to 15 months.

Aging of solid dosage forms can cause a decrease in their in vitro rate of dissolution [52], but a corresponding decrease in in vivo absorption cannot be assumed automatically. For example, Chemburkar et al. [53] showed that, when a methaqualone tablet was stored at 80% relative humidity for 7–8 months, the dissolution rate, as measured by in vivo absorption, was not affected. This lack of in vitro dissolution–in vivo absorption correlation for the aged product was observed even through the particular dissolution method (that of the resin flask) was shown by the same workers to be capable of discriminating the absorption of several trial dosage forms of the same drug.

Adsorption

Drug–plastic interaction is increasingly being recognized as a major potential problem when intravenous solutions are stored in bags, or when they are infused through administration sets that are made from polyvinyl chloride (PVC). For example, up to 50% drug loss can occur after nitroglycerin is stored in PVC infusion bags for 7 days at room temperature [54]. This loss can be attributed to adsorption, rather than to chemical degradation, because the drug can be recovered from the inner surface of the container by rinsing with a less polar solvent (methanol here). A diverse array of drugs, including diazepam [55], insulin [56], isosorbide dinitrate [57], and others [58], have shown substantial adsorption to PVC. The propensity for significant adsorption is related to the oil/water partition coefficient of the drug, since this process depends on the relative affinity of the drug for the hydrophobic PVC (dielectric constant of about 3) and the hydrophilic aqueous infusion medium.

Physical Instability in Heterogeneous Systems

The stability of suspensions, emulsions, creams, and ointments is dealt with in other chapters. The unique characteristics of solid-state decomposition processes have been described in reviews by Monkhouse [59,60] and in the more recently published monograph on drug stability by Carstensen [61].

III. QUANTITATION OF RATE OF DEGRADATION

Before undertaking a discussion of the mathematics involved in the determination of reaction rates is undertaken, it is necessary to point out the importance of proper data acquisition in stability testing. Applications of rate equations and predictions are meaningful only if the data used in such processes are collected using valid statistical and analytical procedures. It is beyond the scope of this chapter to discuss the proper statistical treatments and analytical techniques that should be used in a stability study. But, some perspectives in these areas can be obtained by reading the comprehensive review by Meites [62] and from the section on statistical considerations in the stability guidelines published by the FDA in 1987 [63].

A. Kinetic Equations

Consider the reaction

$$a\text{A} + b\text{B} \rightarrow m\text{M} + n\text{N} \tag{1}$$

where A and B are the reactants; M and N, the products; and a, b, m, and n, the stoichiometric coefficients describing the reaction. The rate of change of the concentration C of any of the species can be expressed by the differential notations $-dC_A/dt$, $-dC_B/dt$, dC_M/dt, and dC_N/dt. Note that the rates of change for the reactants are preceded by a negative sign, denoting a decrease in concentration relative to time (rate of disappearance). In contrast, the differential terms for the products are positive in sign, indicating an increase in concentration of these species as time increases (rate of appearance). The rates of disappearance of A and B and the rates of appearance of M and N are interrelated by equations that take into account the stoichiometry of the reaction:

$$-\frac{1}{a}\frac{dC_A}{dt} = -\frac{1}{b}\frac{dC_B}{dt} = \frac{1}{m}\frac{dC_M}{dt} = \frac{1}{n}\frac{dC_N}{dt} \tag{2}$$

The Rate Expression

The rate expression is a mathematical description of the rate of the reaction at any time t in terms of the concentration(s) of the molecular species present at that time. By using the hypothetical reaction $a\text{A} + b\text{B} \rightarrow$ products, the rate expression can be written as

$$-\frac{dC_A}{dt} = -\frac{dC_B}{dt} \propto C_{A(t)}^a \, C_{B(t)}^b \tag{3}$$

Equation (3) in essence states that the rate of change of the concentration of A at time t is equal to that of B, and that each of these rate changes at time t is proportional to the product of the concentrations of the reactants raised to the respective powers. Note that $C_{A(t)}$ and $C_{B(t)}$ are time-dependent variables. As the reaction proceeds, both $C_{A(t)}$ and $C_{B(t)}$ will decrease in magnitude. For simplicity, these concentrations can be denoted simply by C_A and C_B, respectively.

$$-\frac{dC_A}{dt} = -\frac{dC_B}{dt} = kC_A^a\, C_B^b \qquad (4)$$

where k is a proportionality constant, commonly referred to as the reaction rate constant or the specific rate constant. The format for rate expressions generally involves concentration terms of only the reactants and very rarely those of the products. The latter occurs only when the products participate in the reaction once it has been initiated.

The order of the reaction, n, can be defined as $n = a + b$. Extended to the general case, the order of a reaction is the numerical sum of the exponents of the concentration terms in the rate expression. Thus if $a = b = 1$, the reaction just described is said to be second-order overall, first-order relative to A, and first-order relative to B. In principle, the numerical value of a or b can be integral or fractional.

Special attention is directed to those instances in which the rate of reaction is apparently independent of the concentration of one of the reactants, even though this reactant is consumed during the reaction. For example, in the reaction between an ester and water (hydrolysis) in a predominantly aqueous environment, the theoretical rate expression for the ester can be written in terms of the concentrations of the ester (C_E) and water (C_W):

$$-\frac{dC_E}{dt} = kC_E C_W \qquad (5)$$

If the initial concentration of the ester is 0.5 M or less, complete hydrolysis of the ester will bring about a corresponding decrease in the concentration of water of 0.5 M or less. Since the initial water concentration is 1000/18, which is about 55 M for an aqueous solution, the loss of water through reaction is insignificantly small and C_W can be considered a constant throughout the entire course of the reaction. Thus, in practice,

$$-\frac{dC_E}{dt} = k_\pi C_E \qquad (6)$$

where $k_\pi = kC_W$. The reaction is thus apparently first-order relative to ester and zero-order relative to water; the overall reaction is known as a pseudo-first-order reaction and k_π the pseudo-first-order constant.

This type of kinetics is observed whenever the concentration of one of the reactants is maintained constant, either by a vast excess initial concentration, or by rapid replenishment of one of the reactants. Thus, if one of the reactants is the hydrogen ion or the hydroxide ion, its concentration, though probably small when compared with that of the drug, can be kept constant throughout the reaction by using buffers in the solution. Similarly, the concentration of an unstable drug in solution can be maintained invariant by preparing a drug suspension, thus providing excess solid in equilibrium with the drug in solution.

Simple Reactions

It is obvious that to quantify the rate expression, the magnitude of the rate constant k needs to be determined. Proper assignment of the reaction order and accurate determination of the rate constant is important when reaction mechanisms are to be deduced from the kinetic data. The integrated form of the reaction equation is easier to use in handling kinetic data. The integrated kinetic relationships commonly used for zero-, first-, and second-order reactions are summarized in Table 4. [The reader is advised that basic kinetic theory is also extensively exploited in pharmacokinetics; for further information on this subject, see Chapter 3.] The

Table 4 Rate Expressions for Zero-, First-, and Second-Order Reactions

	Zero-order	First-order	Second-order $a = b = c_0$	Second-order $a \neq b$
Differential rate expression	$-\dfrac{dc}{dt} = k$	$-\dfrac{dc}{dt} = kc$	$-\dfrac{dc}{dt} = kc^2$	$-\dfrac{dc}{dt} = kc_a c_b$
Integrated rate expression	$k = \dfrac{c_0 - c}{t}$	$k = \dfrac{1}{t} \ln \dfrac{c_0}{c}$	$\dfrac{1}{c} - \dfrac{1}{c_0} = kt$	$k = \dfrac{1}{t(a-b)} \ln \dfrac{b(a-x)}{a(b-x)}$
$t_{1/2}$	$\dfrac{c_0}{2k}$	$\dfrac{0.693}{k}$	$\dfrac{1}{c_0 k}$	(i) When $x = 0.5a$ $\dfrac{1}{k(a-b)} \ln \dfrac{0.5ab}{a(b-0.5a)}$ (ii) When $x = 0.5b$ $\dfrac{1}{k(a-b)} \ln \dfrac{b(a-0.5b)}{0.5ab}$
$t_{90\%}$	$\dfrac{c_0}{10k}$	$\dfrac{0.105}{k}$	$\dfrac{0.11}{c_0 k}$	(i) When $x = 0.1a$ $\dfrac{1}{k(a-b)} \ln \dfrac{0.9ab}{a(b-0.1a)}$ (ii) When $x = 0.1b$ $\dfrac{1}{k(a-b)} \ln \dfrac{b(a-0.1b)}{0.9ab}$

concentration symbols in Table 4 are defined as follows: c is the concentration of the drug at any time t and c_0 is the initial concentration. In the last column describing a second-order reaction in which the reactants A and B do not have the same initial concentrations, these are designated as a and b, respectively; x is the concentration reacted at time t.

In a reaction of either zero-order or second-order, the time to reach a certain fraction of the initial concentration, for example, $t_{1/2}$ or $t_{90\%}$ [the time required for the drug concentration to decrease to 90% of its original value (i.e. 10% degradation)] is dependent on c_0. This is illustrated in Fig. 1, in which a zero-order reaction (see Fig. 1a) and a second-order reaction (see Fig. 1b) are plotted with two initial concentrations. It is readily seen that for a zero-order reaction, the $t_{1/2}$ increases with a higher initial concentration. Conversely, for a second-order reaction, $t_{1/2}$ decreases with increasing initial concentration. For a reaction obeying first-order kinetics, the $t_{1/2}$ or $t_{90\%}$ is independent of c_0.

Complex Reactions

Parallel First-Order Reactions. In many instances, the active drug may degrade through more than one pathway:

$$A \begin{cases} \xrightarrow{k_1} B \\ \xrightarrow{k_2} C \\ \xrightarrow{k_3} D \end{cases} \tag{7}$$

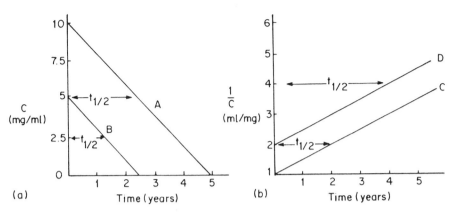

Fig. 1 Effect of initial concentration on the half-life of (a) a zero-order and (b) a second-order reaction. In (a), $k = 2$ mg/year-ml; curve A, initial concentration $c_0 = 10$ mg/ml, $t_{1/2} = 2.5$ years; curve B, $c_0 = 5$ mg/ml, $t_{1/2} = 1.25$ years. In (b), $k = 0.5$ ml/mg-year; curve C, $c_0 = 1$ mg/ml, $t_{1/2} = 2$ years; curve D, $c_0 = 0.5$ mg/ml, $t_{1/2} = 4$ years.

If the concentration of the active drug, A, can be monitored, the composite rate constant, $k' = k_1 + k_2 + k_3$, can easily be determined from the relationship $[A] = [A]_0 e^{-k't}$, where $[A]_0$ is the initial concentration and $[A]$ is the concentration at time t. If the concentrations of A cannot be determined because of assay difficulties, it is still possible to determine k' by monitoring one of the degradation products. For example, if the concentrations of B can be assayed as a function of time, and the concentration of B at time infinity, $[B]_\infty$, is also determined, the following relationships can be derived:

$$[B] = \frac{k_1}{k'} [A]_0(1 - e^{-k't}) \tag{8}$$

$$[B]_\infty = \frac{k_1}{k'} [A]_0 \tag{9}$$

$$\ln\left(1 - \frac{[B]}{[B]_\infty}\right) = -k't \tag{10}$$

Approach to Equilibrium Through First-Order Reactions. This type of reaction can be represented by Eq. (11):

$$A \underset{k_2}{\overset{k_1}{\rightleftharpoons}} B \tag{11}$$

The concentrations of A and B as a function of time can be derived:

$$[A] = \frac{k_2}{k_1 + k_2} [A]_0 + \frac{k_1}{k_1 + k_2} [A]_0(e^{-(k_1 + k_2)t}) \tag{12}$$

$$[B] = \frac{k_1}{k_1 + k_2} [A]_0(1 - e^{-(k_1 + k_2)t}) \tag{13}$$

The combined constants $(k_1 + k_2)$ can be obtained through Eq. (14), and the individual rate constants k_1 and k_2 can now be calculated through Eq. (15):

$$\ln([A] - [A]_\infty) = \ln[B]_\infty - (k_1 + k_2)t \tag{14}$$

$$k_1[A]_\infty = k_2[B]_\infty \tag{15}$$

Fractional Order. In the decomposition of pure solids, the kinetics of reactions can often be more complex than simple zero- or first-order processes. Carstensen [61] has reviewed the stability of solids and solid dosage forms as well as the equations that can be used in these cases. In addition to zero- and first-order kinetics, solid-state degradations are often described by fractional-order equations.

More complicated reactions schemes, including first-order reversible consecutive processes and competitive consecutive reactions, are considered in a textbook by Irwin [62]. Professor Irwin's textbook also includes computer programs written in the BASIC language. These programs can be used to fit data to the models described.

B. Energetics of Reactions

According to the transition state theory, the reaction between two molecules, A and B, to form products C and D proceeds through a transition state, X:

$$A + B \underset{}{\overset{K^{\ddagger}}{\rightleftharpoons}} X \longrightarrow C + D \tag{16}$$

Here $K^{\ddagger}$ is a thermodynamic equilibrium constant that can be expressed as a function of the activities [Eq. (17)] or of the activity coefficients γ_X, γ_A, and γ_B [Eq. (18)]:

$$K^{\ddagger} = \frac{a_X}{a_A a_B} \tag{17}$$

$$K^{\ddagger} = \frac{[X]}{[A][B]} \frac{\gamma_X}{\gamma_A \gamma_B} \tag{18}$$

The rate of the reaction $-d[A]/dt$ is proportional to the concentration of the transition state

$$-\frac{d[A]}{dt} = k'[X] \tag{19}$$

where k' is a proportionality constant. Combining Eqs. (18) and (19) yields

$$-\frac{d[A]}{dt} = k'K^{\ddagger}[A][B] \frac{\gamma_A \gamma_B}{\gamma_X} \tag{20}$$

If the activity coefficients are assumed to be unity, the specific rate constant k is then identical to $k'K^{\ddagger}$. It can be shown that

$$k' = \frac{k_B T}{h} \tag{21}$$

where k_B is Boltzmann's constant, h is Planck's constant, and T is the absolute temperature. Thus,

$$k = \frac{k_B T}{h} K^{\ddagger} \tag{22}$$

and

$$\ln k = \ln \frac{k_B}{h} + \ln T + \ln K^{\ddagger} \tag{23}$$

Differentiating relative to T, we obtain

$$\frac{d \ln K^{\ddagger}}{dT} = \frac{d \ln k}{dt} - \frac{1}{T} \tag{24}$$

Since

$$\frac{d \ln K^{\ddagger}}{dT} = \frac{\Delta H^{\ddagger}}{RT^2} \tag{25}$$

where $\Delta H^{\ddagger}$ is the enthalpy of activation, Eq. (26) can be obtained by combining Eqs. (24) and (25):

$$\frac{d \ln k}{dT} = \frac{\Delta H^{\ddagger} + RT}{RT^2} \tag{26}$$

The classic Arrhenius equation is given by Eq. (27), where E_a is the energy of activation:

$$\frac{d \ln k}{dT} = \frac{E_a}{RT^2} \tag{27}$$

On comparing Eqs. (26) and (27), it follows that

$$\Delta H^{\ddagger} = E_a - RT \tag{28}$$

The other thermodynamic parameters, $\Delta G^{\ddagger}$ and $\Delta S^{\ddagger}$, the free energy and entropy of activation, respectively, can also be obtained from the foregoing relationships:

$$\Delta G^{\ddagger} = -RT \ln K^{\ddagger} = -RT \ln \frac{kh}{k_B T} \tag{29}$$

and

$$\Delta S^{\ddagger} = -\frac{\Delta G^{\ddagger} + \Delta H^{\ddagger}}{T} = R \ln \frac{kh}{k_B T} + \frac{E_a - RT}{T} \tag{30}$$

The magnitudes of the thermodynamic parameters, $\Delta H^{\ddagger}$ and $\Delta S^{\ddagger}$, sometimes provide evidence supporting proposed mechanisms of drug decomposition. The enthalpy of activation is a measure of the energy barrier that must be overcome by the reacting molecules before a reaction can occur. As can be seen from Eq. (28), its numerical value is less than the Arrhenius energy of activation by the factor RT. At room temperature, RT is only about 0.6 kcal/mol. The entropy of activation can be related to the Arrhenius frequency factor (i.e., the fraction of molecules possessing the requisite energy that actually reacts). This parameter includes steric and orientation requirements of the reactants, the transition state, and the solvent molecules surrounding them. For unimolecular reactions, $\Delta S^{\ddagger}$ has a value of near zero or slightly positive. For bimolecular reactions, $\Delta S^{\ddagger}$ is more negative. For example, in the hydrolysis of esters and anhydrides, the entropy of activation is on the order of -20 to -50 entropy units, reflecting a transition state in which several solvent molecules are immobilized for solvation of the developing charges [66].

IV. THE ARRHENIUS EQUATION AND ACCELERATED STABILITY TESTING

The Arrhenius equation (27) may be integrated and rewritten as Eqs. (31) and (32):

$$k = Ae^{-E_a/RT} \tag{31}$$

$$\ln \frac{k_1}{k_2} = \frac{E_a}{R} \left(\frac{1}{T_2} - \frac{1}{T_1} \right) \tag{32}$$

where E_a is a constant and the subscripts 1 and 2 denote the two different temperature conditions. A plot of $\ln k$ as a function of $1/T$, referred to as the *Arrhenius plot*, is linear according to Eq. (31), if E_a is independent of temperature. Thus, it is possible to conduct kinetic experiments at elevated temperatures and obtain estimates of rate constants at lower temperatures by extrapolation of the Arrhenius plot. This procedure, commonly referred to as accelerated stability testing, is most useful when the reaction at ambient temperatures is too slow to be monitored conveniently and when E_a is relatively high. For example, for a reaction with an E_a of 25 kcal/mol, an increase from 25° to 45°C brings about a 14-fold increase in the reaction rate constant. In comparison, a rate increase of just threefold is obtained for the same elevation in temperature when E_a is 10 kcal/mol. The magnitude of E_a for a reaction can be obtained from the slope of its Arrhenius plot. Hydrolysis reactions typically have an E_a of 10–30 kcal/mol, whereas oxidation and photolysis reactions have smaller energies of activation [67].

An underlying assumption of the Arrhenius equation is that the reaction mechanism does not change as a function of temperature (i.e., E_a is independent of temperature). Since accelerated stability testing of pharmaceutical products normally employs a narrow range of temperature (typically, 35° to at most 70°C), it is often difficult to detect nonlinearity in the Arrhenius plot from experimental data, even though such nonlinearity is expected from the reaction mechanism [68]. Thus, even complex biological processes may show Arrhenius behavior within certain temperature ranges; Laidler [69] cited such phenomena as the frequency of flashing of fireflies and the rate of the terrapin's heartbeat as examples.

Non-Arrhenius behavior, has been observed in pharmaceutical systems [70]. This may be attributed to the possible evaporation of solvent, multiple reaction pathways, change in physical form of the formulation, and so on, [71] when the temperature of the reaction is changed. An interesting example of non-Arrhenius behavior is the increased rate of decomposition of ampicillin on freezing. Savello and Shangraw [72] showed that for a 1% sodium ampicillin solution in 5% dextrose, the percentage of degradation at 4 hr is approximately 14% at −20°C, compared with 6% at 0°C and 10% at 5°C. This decrease in stability in frozen solutions is most frequently observed when the reaction obeys second- or higher-order kinetics. For example, the formation of nitrosomorpholine from morpholine and nitrite obeys third-order kinetics [73], and the rate of nitrosation is drastically enhanced in frozen solutions (Fig. 2). A marked acceleration in the hydrolytic degradation of methyl, ethyl, and n-propyl 4-hydroxybenzoates in the frozen state has also been reported by Shiva et al. [74]. These authors found that although pseudo-first-order conditions found in the liquid state are also observed in the frozen state, the rate of reaction under frozen-state conditions showed very much less dependency on the initial hydroxide ion concentration.

The mechanism for rate enhancement in frozen solutions has been reviewed by Pincock [75]. In reactions following second- or higher-order kinetics, an increase in rate may be brought about by concentration of the reactants in the liquid phase, the solute molecules being excluded from the ice lattice when the solution freezes. Occasionally, an increase in rate may be due to a change in pH on freezing. Fan and Tannenbaum [73] reported that citrate-sodium hydroxide and citrate-potassium phosphate buffers do not change pH on freezing, but citrate-sodium

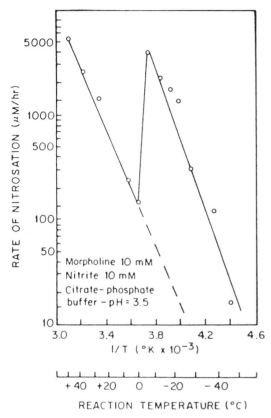

Fig. 2 Effect of temperature on the rate of nitrosation of morpholine with nitrite in citrate-sodium phosphate buffer for temperatures above and below freezing temperature. (From Ref. 73.)

phosphate buffer at pH 8 decreases to pH 3.5 and sodium hydrogen phosphate at pH 9 decreases to pH 5.5 on freezing. A possible explanation for this phenomenon is now available [76]. Monosodium phosphate forms supersaturated solutions on cooling that become amorphous, with no precipitation of the salt. The disodium and monopotassium salts, on the other hand, readily precipitated when the initial solution concentration was > 0.2 *M*. The possibility of a pH change and rate acceleration should be considered when evaluating the stability of freeze-dried products. Proteins are particularly sensitive to changes in pH, folding or unfolding to varying degrees in response to such changes. Proteins tend to be most stable at their isoelectric point owing to electrostatic interactions [77]; however, when a solution is adjusted to the optimum pH for stability at room temperature with buffers, that pH may not be maintained throughout the lyophilization cycle, and the protein may aggregate or undergo denaturation.

Considerable interest has been generated in the use of accelerated stability testing that is based on a single condition of elevated temperature and humidity. For Abbreviated New Drug Applications (ANDAs) the FDA stability guidelines [63] suggest that a tentative expiration date of 24 months may be granted for a drug product if satisfactory stability results can be documented under a stressed condition of 40°C *and* 75% relative humidity. The simplicity of such a guideline is naturally attractive because a substantial saving in time can be obtained in advancing a drug product to the marketplace. However, the reliability of prediction can be subjected to question under certain circumstances. An analysis of the use and limitation of this approach has been presented elsewhere, and interested readers may refer to it for further information [78].

V. ENVIRONMENTAL FACTORS THAT AFFECT REACTION RATE

A rational way to develop approaches that will increase the stability of fast-degrading drugs in pharmaceutical dosage forms is through a thorough study of the factors that can affect such stability. In this section, the factors that can affect decomposition rates are discussed; it will be seen that, under certain conditions of pH, solvent, presence of additives, and so on, the stability of a drug may be drastically affected. Equations that may allow prediction of these effects on reaction rates are discussed.

A. pH

The pH of the drug solution may have a very dramatic effect on its stability. Depending on the reaction mechanism, a change of more than tenfold in rate constant may result from a shift of just 1 pH unit. When drugs are formulated in solution, it is essential to construct a pH versus rate profile so that the optimum pH for stability can be located. Many pH versus rate profiles are documented in the literature, and they have a variety of shapes. The majority of these pH versus rate profiles can be rationalized by an approach in which the reaction of each molecular species of the drug with hydrogen ion, water, and hydroxide ion is analyzed as a function of pH. The discussion that follows is divided according to the ionization capability of a drug.

1. When the drug is nonionizable: In water, three hydrolytic pathways are available [Eq. (33)], it can degrade by specific acid catalysis represented by the first kinetic term in Eq. (33), water hydrolysis (second term), and specific base catalysis (third term):

$$-\frac{dc}{dt} = k_1[H^+]c + k_2c + k_3[OH^-]c \tag{33}$$

Equation (33) can be rearranged to give Eq. (34):

$$-\frac{dc}{c\,dt} = k_{obs} = k_1[H^+] + k_2 + k_3[OH^-] \tag{34}$$

Note that k_1 and k_3 are second-order rate constants, whereas k_2 is a pseudo-first-order constant. The pH versus rate profile can be constructed by considering, in turn, that one of the three kinetic terms is predominating, thus:

(a) When $k_1[H^+] >> k_2 + k_3[OH^-]$, $k_{obs} = k_1[H^+]$ and $\log k_{obs} = \log k_1 - pH$. (35)

(b) When $k_2 >> k_1[H^+] + k_3[OH^-]$, $k_{obs} = k_2$ and $\log k_{obs} = \log k_2$. (36)

(c) When $k_3[OH^-] >> k_1[H^+] + k_2$, $k_{obs} = k_3[OH^-]$ and $\log k_{obs} = \log k_3 + pH$. (37)

Equations (35) through (37) are plotted and shown in Fig. 3a. The lines are stipled to indicate that the relative positions of the lines are not fixed, but are dependent on the relative magnitudes of the rate constants. For example, when $k_3[OH^-] > k_2 >> k_1[H^+]$, a log k_{obs} versus pH profile such as the one depicted in Fig. 3b may result. On the other hand, if $k_1[H^+]$ and $k_3[OH^-]$ are both much greater than k_2, a log k_{obs} versus pH profile may resemble the curve shown in Fig. 3c. When $k_1[H^+] > k_2 >> k_3[OH^-]$, the log k_{obs} versus pH profile will be a mirror image of Fig. 3b; and when $k_2 >> k_1[H^+] + k_3[OH^-]$, the rate constant will be pH-independent.

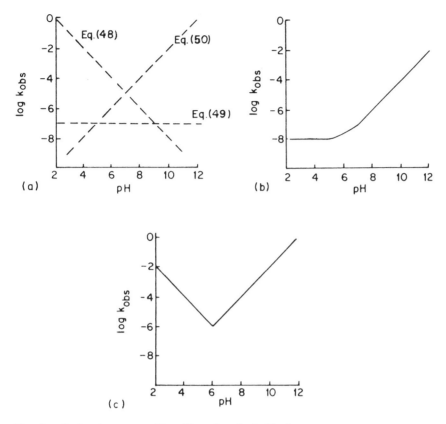

Fig. 3 The log k_{obs} versus pH profiles of nonionizable drugs.

2. When the drug is either monoacidic or monobasic, an equation similar to Eq. (34) can be written. Here, however, three kinetic terms are written for the acidic form of the drug, HA, and three terms for the basic form, A (electronic charges on HA and A are not designated here because either HA or A can be charged):

$$k_{obs} = k_1[H^+]f_{HA} + k_2 f_{HA} + k_3[OH^-]f_{HA} + k_4[H^+]f_A + k_5 f_A + k_6[OH^-]f_A \qquad (38)$$

where

$$f_{HA} = \frac{[HA]}{[HA] + [A]} = \frac{[H^+]}{[H^+] + K_a} \qquad (39)$$

and

$$f_A = \frac{[A]}{[HA] + [A]} = \frac{K_a}{[H^+] + K_a} \qquad (40)$$

Again, Eq. (38) can be analyzed by considering each individual term as a function of pH. Since the magnitudes of both f_{HA} and f_A are dependent on the relative magnitudes of K_a and H^+, the kinetic terms can be evaluated under three conditions: (a) when $[H^+] \gg K_a$, (b) when $[H^+] = K_a$, and (c) when $[H^+] \ll K_a$ (Table 5). The log k_{obs} versus pH profile for each kinetic term is shown in Fig. 4, using a hypothetical pK_a of 6 and the condition that $k_1 = 10^7 k_2 = k_3$

Table 5 Kinetic Expressions for Each Term in Eq. (38)

Logarithm of kinetic term	$\log k_{obs}$		
	When $[H^+] \gg K_a$	When $[H^+] = K_a$	When $K_a \gg [H^+]$
$\log k_1[H^+]f_{IIA}$	$\log k_1 - pH$	$\log \dfrac{k_1 K_a}{2}$	$\log \dfrac{k_1}{K_a} - 2\,pH$
$\log k_2 f_{IIA}$	$\log k_2$	$\log \dfrac{k_2}{2}$	$\log \dfrac{k_2}{K_a} - pH$
$\log k_3[OH^-]f_{IIA}$	$\log k_3 K_w + pH$	$\log \dfrac{k_3 K_w}{2K_a}$	$\log \dfrac{k_3 K_w}{K_a}$
$\log k_4[H^+]f_A$	$\log k_4 K_a$	$\log \dfrac{k_4 K_a}{2}$	$\log k_4 - pH$
$\log k_5 f_A$	$\log k_5 K_a + pH$	$\log \dfrac{k_5}{2}$	$\log k_5$
$\log k_6[OH^-]f_A$	$\log k_6 K_w K_a + 2\,pH$	$\log \dfrac{k_6 K_w}{2K_a}$	$\log k_6 K_w + pH$

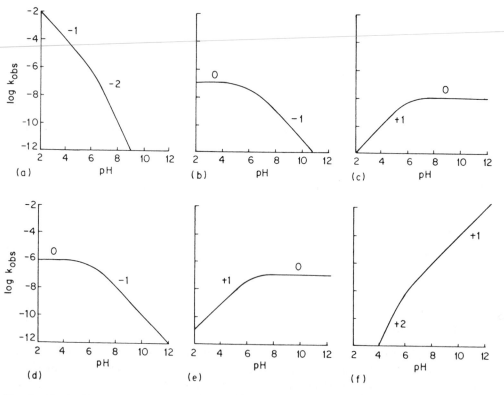

Fig. 4 The $\log k_{obs}$ versus pH profile for each kinetic term in Eq. (38): $k_1 = 10^7$, $k_2 = k_3 = k_4 = 10^7$, $k_5 = k_6 = 1$; $K_a = 10^{-6}$. Each number next to the curve indicates the slope of that portion of the curve.

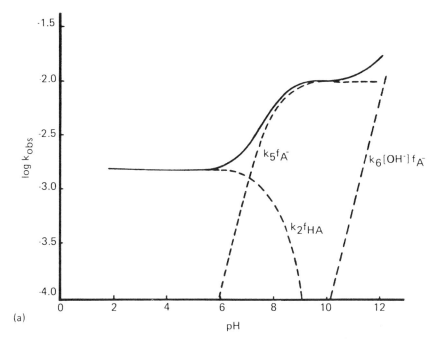

(a)

Fig. 5 (a) The pH versus log hydrolysis rate constant profile of idoxuridine at 60°C (solid line). The dashed lines indicate the individual contributions of each kinetic term. (b) pH versus log hydrolysis rate constant profile of acetylsalicylic acid (solid line). The dashed lines indicate the individual contribution of each kinetic term.

$= k_4 = 10^7 \, k_5 = k_6 = 1$. Compared with the curves shown in Fig. 3a, the profiles in Fig. 4 show one break each in the lines, with a change of slope of 1 unit at the breaks. It is also seen that term (b) is equivalent to term (d) and that term (c) is equivalent to term (e), as far as their dependency on pH is concerned (see Table 5 and Fig. 4). These terms, therefore, are kinetically equivalent and are indistinguishable from each other in a rate expression. Equation (38), then, can be reduced to a combination of only four terms. The shape of the overall log k_{obs}/pH profile of any drug is determined by the relative magnitudes of the four kinetic terms over the pH range considered. Each log k_{obs} versus pH profile of a monoacidic or monobasic drug can be adequately described by a combination of no more than four terms. Figure 5 illustrates this principle by showing the log k_{obs} versus pH profiles of idoxuridine [79] and acetylsalicylic acid [80]. The hydrolysis of idoxuridine (see Fig. 5a) as a function of pH can be rationalized by the equation $k_{obs} = k_2 f_{HA} + k_5 f_{A^-} + k_6[OH^-]f_{A^-}$ (three kinetic terms), whereas the hydrolysis of acetylsalicylic acid (see Fig. 5b) can be decided only by using all four kinetic terms ($k_{obs} = k_1[H^+]f_{HA} + k_4[H^+]f_{A^-} + k_5 f_{A^-} + k_6[OH^-]f_{A^-}$.

Some compounds exhibit pH behavior in which a bell-shaped curve is obtained with maximum instability at the peak [81]. The peak corresponds to the intersection of two sigmoidal curves that are mirror images. The two inflection points imply two acid and base dissociations responsible for the reaction. For a dibasic acid (H_2A) for which the monobasic species (HA^-) is most reactive, the rate will rise with pH as [HA^-] increases. The maximum rate occurs at pH = ($pK_1 + pK_2$)/2 (the mean of the two acid dissociation constants). Where an acid and base react, the two inflections arise from the two different molecules. The hydrolysis of penicillin G catalyzed by 3,6-bis(dimethylaminomethyl)catechol [82], is a typical example.

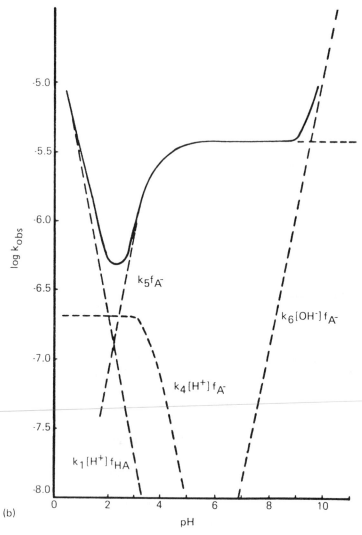

Fig. 5 Continued

B. Solvent

In many pharmaceutical dosage forms, it may be necessary to incorporate water-miscible solvents to solubilize the drug. These solvents are generally low molecular weight alcohols, such as ethanol, propylene glycol, and glycerin; or polymeric alcohols, such as the polyethylene glycols. Solvent effects can be quite complicated and difficult to predict. In addition to altering the activity coefficients of the reactant molecules and the transition state, changes of the solvent system may bring about concomitant changes in such physicochemical parameters as pK_a, surface tension, and viscosity, that indirectly affect the reaction rate. In some cases, an additional reaction pathway may be generated, or there may be a change in the product mix. The angiotensin-converting enzyme inhibitor, moexipril, undergoes hydrolysis as well as a cyclization reaction, leading to the formation of diketopiperazines. In mixed solvent (75–90% ethanol) systems the hydrolysis reaction is suppressed, but the rate of the cyclization reaction increases by 5.5-fold to 29-fold [83]. In the presence of increasing concentrations of ethanol in the solvent, aspirin degrades by an extra route, forming the ethyl ester of acetylsalicylic acid [84]. On the other hand, a solvent change may bring about stabilization of a compound. The hydrolysis of barbiturates occurs 6.7-fold faster in water than in 50% ethanol, and 2.6-fold faster in water than in 50% glycerol [85].

Many approaches have been used to correlate solvent effects. The approach used most often is based on the electrostatic theory, the theoretical development of which has been described in detail by Amis [86]. The reaction rate is correlated with some bulk parameter of the solvent, such as the dielectric constant or its various algebraic functions. The search for empirical parameters of solvent polarity and their application in multiparameter equations has recently been intensified, and this approach is described in the book by Christian Reichardt [87]. Although the solvent effect on reaction rate could, in principle, be large, the limited availability of nontoxic solvents suitable for pharmaceutical products has rendered this stabilization approach somewhat impractical in most circumstances.

C. Solubility

As mentioned earlier in this chapter, penicillins are very unstable in aqueous solution by virtue of hydrolysis of the β-lactam ring. A successful method of stabilizing penicillins in liquid dosage forms is to prepare their insoluble salts and formulate them in suspensions. The reduced solubility of the drug in a suspension decreases the amount of drug available for hydrolysis. An example of improved stability of a suspension over that of a solution is illustrated in Fig. 6, in which a hypothetical drug is formulated as a 10-mg/ml solution (curve A) and as a suspension containing the same total amount of drug, but with a saturated solubility of 1 mg/ml (curve B). It is seen that the drug in solution undergoes first-order degradation with a half-

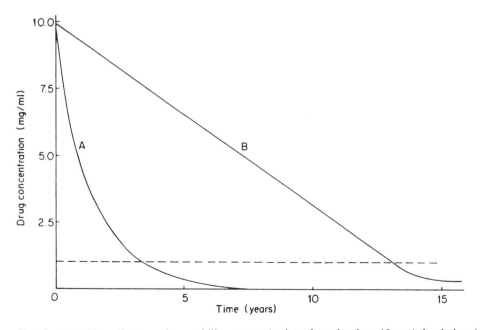

Fig. 6 Solubility effects on drug stability: curve A, drug formulated as 10 mg/ml solution ($t_{1/2} = 1$ year); curve B, drug formulated as a suspension with a saturated solubility of 1 mg/ml ($t_{1/2} = 7.3$ years).

life of 1 month. In the suspension, the drug degrades through zero-order kinetics until there is no more excess solid present, after which point first-order kinetics is operative.

D. Additives

Buffer Salts

In most drug solutions, it is necessary to use buffer salts to maintain the formulation at the optimum pH. These buffer salts can affect the rate of drug degradation in several ways. First, a primary salt effect results because of the effect salts have on the activity coefficient of the reactants. At relatively low ionic strengths, the rate constant, k_μ, is related to the ionic strength, μ, according to

$$\ln k_\mu = \ln k_0 + 1.02\, z_A z_B \sqrt{\mu} \tag{41}$$

where k_0 is the rate constant at $\mu = 0$; z_A and z_B are the ionic charges of reactants A and B, respectively, and the constant 1.02 is applicable to aqueous solutions at 25°C. According to Eq. (41), a plot of $\ln k_\mu$ as a function of $\sqrt{\mu}$ should yield a slope approximately equal to the product $z_A z_B$, which, in theory should always be an integer because both z_A and z_B are integral numbers. In practice, however, the slope is often fractional. The sign of the slope is sometimes informative in identifying the reactants that participate in the rate-limiting step in the reaction mechanism. But, one must avoid drawing such conclusions in instances where a choice is to be made between kinetically equivalent rate terms [88].

Buffer salts also can exert a secondary salt effect on drug stability. From Table 5 and Fig. 5, it is clear that the rate constant for an ionizable drug is dependent on its pK_a. Increasing salt concentrations, particularly from polyelectrolytes such as citrate and phosphate, can substantially affect the magnitude of the pK_a, causing a change in the rate constant. [For a review of salt effects, containing many examples from the pharmaceutical literature see Ref. 90].

Lastly, buffer salts can promote drug degradation through general acid or general base catalysis. In these cases, the rate expression will contain additional kinetic terms describing the applicable reactions between different molecular species of the drug and buffer components. The efficiency of general acid or base catalysis by the buffer components is often described by the Brønsted relationship:

$$k_A = G_A K_a^\alpha \quad \text{and} \quad k_B = G_B K_B^\alpha \tag{42}$$

where k_A and k_B are the catalytic constants for general acid and base catalysis, respectively; K_a and K_a^α are the acid and base ionization constants, respectively; and G_A, G_B, α, and β are constants characteristic of the reaction, the solvent, and the temperature [91,92].

Surfactants

Addition of surface-active agents may accelerate or decelerate drug degradation. Because micellar catalysis may provide a model for enzyme reactions, acceleration of rate owing to the presence of surfactants, is well documented [93]. By comparison, stabilization of drugs through the addition of surfactants, is less frequently reported. An example in which both effects are observed is especially rare. The hydrolysis of aspirin in the plateau region (pH 6–8) is inhibited by the presence of micelles of cetyltrimethylammonium bromide and cetylpyridinium chloride, whereas in the region where the normal base-catalyzed reaction occurs (pH greater than 9) the reaction is catalyzed by micelles of these same surfactants. The mechanism of hydrolysis in the plateau region involves intramolecular general base catalysis by the adjacent ionized carboxyl group, both in the presence and absence of micelles. This reaction is inhibited in the presence of micelles because the substrate molecules are solubilized into the micelle, and water is less available in this environment than in normal aqueous solutions [94].

Complexing Agents

Higuchi and Lachman [95] pioneered the work of improving drug stability by complexation. They showed that aromatic esters can be stabilized in aqueous solutions in the presence of xanthines such as caffeine. Thus, the half-lives of benzocaine, procaine hydrochloride, and tetracaine are increased by approximately two- to five-fold in the presence of 2.5% caffeine. This increase in stability is attributed to the formation of a less reactive complex between caffeine and the aromatic ester. Connors has written a comprehensive textbook that describes methods for the measurement of binding constants for complex formation in solution—along with discussions of pertinent thermodynamics, modeling statistics, and regression analysis [96]. The various experimental methods useful for measuring equilibrium constants are also discussed. A good deal of attention has recently been directed at the use of derivatives of cyclodextrin for the solubilization and stabilization of pharmaceuticals [97]. One cautionary note: complexation may adversely affect the dissolution or permeability characteristics of the drug, thereby possibly decreasing drug bioavailability.

Antioxidants and Chelating Agents

Antioxidants and chelating agents are used to protect drugs against autoxidation. Mechanistically, some antioxidants, such as ascorbic acid, ascorbyl palmitate, sodium bisulfite, sodium metabisulfite, sodium sulfite, acetone sodium bisulfite, sodium formaldehyde sulfoxylate, thioglycerol, and thioglycolic acid, act as reducing agents. They are easily oxidized, preferentially undergo autoxidation, thereby consuming oxygen and protecting the drug or excipient. They are often called oxygen scavengers because their autoxidation reaction consumes oxygen. They are particularly useful in closed systems in which the oxygen cannot be replaced once it is consumed [19]. Primary or true antioxidants act by providing electrons or labile H^+, which will be accepted by any free radical to terminate the chain reaction. In pharmaceuticals, the most commonly used primary antioxidants are butylated hydroxytoluene (BHT), butylated hydroxyanisole (BHA), the tocopherols (vitamin E), and propyl gallate. Chelating agents act by forming complexes with the heavy metal ions that are often required to initiate oxidation reactions. The chelating agents used most often are ethylenediaminetetraacetic acid (EDTA) derivatives and salts, citric acid, and tartaric acid.

E. Light and Humidity

The mathematical relationship between light intensity and drug degradation is much less developed than those describing pH and temperature effects. Part of the reason, perhaps, is that light effects on stability can be substantially avoided by using amber containers that shield off most of the ultraviolet light. Regulatory authorities usually require a statement on the photostability of products and the means of protection, if required. Often both daylight and artificial light sources are employed for tests on drug substances [98].

Humidity is a major determinant of drug product stability in solid dosage forms. Elevation of relative humidity usually decreases stability, particularly for those drugs highly sensitive to hydrolysis [99]. In addition, increased humidity can also accelerate the aging process [52,53] through interaction(s) with excipients. Humidity does not always affect drug stability adversely. Cyclophosphamide, in lyophilized cakes containing mannitol or sodium bicarbonate, undergoes rapid ($t_{90} \approx 15$ days) degradation in the solid state. The cyclophosphamide was in the amorphous state in these formulations. However, on exposure to high humidity, the cyclophosphamide was converted to the crystalline monohydrate form, which exhibited greatly improved stability [45]. Reviews dealing with the effects of moisture on the physical and chemical stability of drugs are available [100,101]. As peptides and proteins have become more important as therapeutic agents, the role residual moisture plays in their stabilization has attracted a good deal of attention [102,103].

VI. STABILITY TESTING IN THE PHARMACEUTICAL INDUSTRY

Stability testing of drug substances and drug products begins as part of the drug discovery and synthesis or development–preformulation effort and ends only with the demise of the compound or commercial product. Activities include testing of drug substance, of compatibility with excipients, of preclinical formulations, of Phase I formulations and modifications, of the final, NDA (commercial) formulation, and of postapproval formulation changes. The regulatory basis for the various aspects of stability testing is established in 21 CFR 211.137, 211.160, 211.170, 211.190, 314.50, 314.70, and 314.81 [104–110]. In addition, FDA guidelines for submitting stability data were published several years ago [63].

The final draft of the ICH Harmonized Tripartite Guideline "Stability Testing of New Drug Substances and Products" was issued by the International Conference on Harmonization (ICH) Expert Working Group of the ICH on technical requirements for the registration of pharmaceuticals for human use, in October 1993 [111]. These guidelines provide definitions of key terms and principles used in the stability testing of drug substance and drug product and cover a scope of issues similar to those appearing in the FDA guideline [63]. The draft has been recommended for adoption to the regulatory bodies of the participating parties.

Another valuable source of information on stability testing can be found in Carstensen's recent book [61].

A. Resources

Personnel

The number and types of personnel are dictated by the size of the program, the functions contained within the program, and the nature of the program. Some companies maintain separate development and commercial product programs; other integrate the two. It is important, both for operating efficiency and regulatory compliance, that all personnel, regardless of their function, receive adequate and well-documented training in both current Good Manufacturing Practices (cGMPs) and in the technical aspects of their jobs.

Education and Experience.   The program is generally headed by a professional with several years experience in the company. Experience in some aspect of formulation development or in the stability-testing program itself may be of equal importance to the educational level of the person heading the program. Successful programs have been led by people at the bachelors level with a degree in pharmacy, chemistry, or related science, as well as by people with masters or doctoral level degrees. If the size of the program warrants, intermediate-level scientists, usually at the bachelors level, may assume responsibility for specific functions within the program (e.g., chemical and physical testing, documentation, and so on). Technicians generally have a high-school education or equivalent with clerical or scientific experience or interest. Clerical and data-entry personnel have traditional training and experience in their respective areas.

Organization.   Functions are often divided into chemical and physical testing, documentation, and clerical–computer operations, if the information system is computerized. Successful manual or paper systems are possible without the aid of computers. However, custom-designed software or commercially available database programs can also be programmed to automate the program. The documentation function usually consists of one or more persons who prepare the stability sections of regulatory documents. Persons specifically trained in technical writing or scientists with an interest and talent in document preparation generally perform well in this capacity.

Facilities

Storage Chambers. Chambers capable of accurately maintaining freezer conditions ($-20°$ to $-10°C$), refrigerator ($2°-8°C$), and controlled room temperature ($15°-30°C$) are a necessity. Also, elevated temperature ($40°C$ and $50°C$) and humidity ($40°C/75\%$ RH) conditions should be available. Finally, a high-intensity light cabinet and a cycling chamber capable of cycling both temperature and humidity are needed. These chambers should be capable of controlling temperature within $\pm 2°C$ and humidity within $\pm 5\%$. They should be calibrated periodically according to a standard operating procedure and records of these calibrations maintained in a log book for each chamber.

The ICH Tripartite Guidelines have established that long-term stability testing should be done at $25°C/60\%$ RH. Stress testing should be done at $40°C/75\%$ RH for 6 months. If "significant change" occurs at these stress conditions, then the formulation should be tested at an intermediate condition (i.e., $30°C/75\%$ RH). *Significant change* is defined in the guidelines.

Storage chambers should be validated for their ability to maintain the desired conditions and, if so equipped, the ability to sound an alarm if a mechanical or electrical failure causes the temperature to deviate from preestablished limits. They should also be equipped with recording devices that will provide a continuous and permanent history of their operation. Log books should be maintained and frequent readings of mercury-in-glass, National Institute of Science and Technology-traceable thermometers recorded.

Bench Space. Adequate laboratory bench, desk, and file space are needed for physical, chemical, and microbiological testing, for documentation, and for storing records, respectively.

Equipment

Chemical Testing. Adequate instrumentation for a variety of different test methods should be available. Most stability-indicating chemical assays are performed by high-performance liquid chromatography (HPLC). Occasionally, gas chromatography, infrared spectrophotometry, or spectrofluorimetry are used. Test methods should be validated [112–114] and stability-indicating (i.e., able to distinguish the active ingredient from its degradation products) so that the active can be accurately measured. Also, methods are needed for identifying and quantitating degradation products that are present at levels of 0.1% or greater.

Biological Testing. A portion of the laboratory may be reserved for biological testing, or this work can be done by the company's microbiological laboratory. The ability to perform sterility, pyrogen, limulus amebocyte lysate (LAL), preservative challenge, and bioburden tests is needed to support the stability program. As in the chemical assays, test methods should be validated and operator familiarity should be documented.

Physical Testing. Equipment and trained personnel should be available for performing such tests as pH, tablet hardness, etc. One important and sometimes overlooked aspect of physical testing is the recording of product appearance. Carefully defined descriptions of appearance and standard descriptions of changes in appearance should be developed, especially when there is a high probability that the person who made the observation at the previous sampling time will not be the person making the observation at the next sampling time. Some companies maintain samples at a lower-than-label storage condition (e.g., refrigeration) to use as standards, assuming that minimal or no appearance change will occur at this condition. The same argument for standard nomenclature applies to other test parameters that are subjective in nature.

Computers. A certain number of personal computers are necessary for report generation and regulatory submission preparation. In addition, these may be useful for record keeping, depending on the type of stability information system that the company chooses to use. Alternatively, if the information system is intended to be accessible (read only) to many users, it may be more efficient to develop a mainframe system, rather than a local area network of minicomputers. The size of the database will help determine the nature of the software and hardware configuration used for this function.

B. Program

Scope and Goals

Activities encompassed by the stability program include sample storage of either development or production batches (or both); data collection and storage–retrieval; physical, chemical, and microbiological testing; document preparation of regulatory submissions; and package evaluation. In certain companies, some of these functions (e.g., regulatory document preparation), may be performed by personnel in separate departments. Nonetheless, the function is part of the company's overall stability program.

Protocols

The FDA stability guidelines [63] and ICH Harmonized Tripartite Guidelines [111] are rather detailed concerning sampling times, storage conditions, and specific test parameters for each dosage form. Generally, samples stored at the label storage condition, controlled room temperature for most products, are tested initially and after 3, 6, 9, 12, 18, and 24 months; and annually, thereafter. Accelerated testing is generally done more frequently and for a shorter duration (e.g., 1, 2, 3, and 6 months). Three batches should be tested to demonstrate batch-to-batch uniformity. The number three represents a compromise between a large number desired for statistical precision and the economics of maintaining a manageable program. Generally, real-time data obtained at the label storage conditions on the final formulation in the final packaging configuration(s) are needed for an NDA. Supportive data obtained from drug substance stability studies, preformulation studies, and investigational formulations tested during clinical trials and formulation development may be used to supplement primary stability data. Requirements for the IND are less defined, the only requirement being that there should be adequate data to support the clinical batch(es) for the duration of the trials.

There are instances, especially with solid, oral dosage forms, for which several package types and configurations are desired by marketing, and three or more strengths are needed for flexibility in dosing. In these situations, it may be feasible to apply the principles of bracketing or matrixing to reduce the amount of testing. Bracketing refers to reduced testing of either an intermediate dosage strength or package size when the formulation characteristics of all strengths are virtually identical, or when the same container–closure materials are used for all package sizes. Matrixing refers to reduced testing, regardless of strength or container, in situations for which there are similarities in formulation or container–closure. Bracketing and matrixing are acceptable only when the product is chemically and physically very stable and does not interact with the container–closure. Demonstration of this chemical and physical stability must be documented by preformulation, drug substance stability, and early formulation stability data. Although not as common, it may be possible to employ bracketing and matrixing with other types of dosage forms. It is imperative that discussions of such strategies with FDA should always occur before implementation.

Documentation

The need for adequate documentation of laboratory operations is established not only by good science, but also by regulatory requirements [107].

Documentation of all facets of the operation is necesary. This includes validation and periodic calibration of storage chambers, instrumentation, and computer programs. Log books for the storage chambers and instruments are also necessary. Standard operating procedures are needed for, among other things, the stability program itself, use of instrumentation, documentation of experiments and their results, determination of expiration dates, investigation of specification failures, and operation of a computerized record-keeping system.

Many companies have developed or purchased computer software for the purpose of storing stability data for a large number of studies. One example of a commercially available system is called "Stability System" [115]. This system can perform other functions as well, including work scheduling, preparation of summaries of selected or all studies in the system, tabulation of data for individual studies, label printing, statistical analysis and plotting, and search capabilities. Such systems should be validated to keep pace with current regulatory activity [116].

C. Regulatory Concerns

Current Good Manufacturing Practice Compliance

Current Good Manufacturing practices [105] establish the requirements for maintaining a stability program and require that most pharmaceutical dosage forms have an established expiration date, supported by test data [104]. There are few allowable exceptions.

Food and Drug Administration Stability Guidelines

The guidelines under which stability programs operate and corresponding documentation is prepared were issued in 1987 [63]. Revisions of these guidelines are currently in preparation. Although the agency emphasizes that these are guidelines, and not regulations, it is generally prudent to follow specific recommendations as indicated in the guidelines. Deviations or omissions should be addressed, and the reasons should be supported with data when applicable.

Regulatory Submissions

An easy-to-read stability summary document will go a long way toward rapid approval of any regulatory submission. Such a document should include a number of items. A clear statement of the objective(s) of the studies included in the submission and the approach that was taken to achieve the objective(s) is critical. This statement of objective(s) should accompany basic information including product and drug substance names, dosage forms and strengths, and type(s) of container–closure systems. Although the objective is usually stated in the summary letter accompanying the submission, a brief reminder to the reviewing chemist is helpful.

A discussion of each of the parameters that were tested in the course of the evaluation, including test methods and specifications for each, should then follow. These parameters should follow those recommended in the stability guidelines [63] for the specific dosage form. It is especially important to provide a rationale for those parameters not studied. Next, should come the study design itself, which should include a list of batch identification number, size, and date of manufacture, as well as packaging configuration, storage conditions, and sampling times for each batch. The strategy and rationale for any bracketing or matrixing should also be presented.

The actual data, including replicates, mean, and range, in tabular form should follow accompanied by a brief discussion of the data. It is important to explain any out-of-specification data. Statistical analyses for all parameters that lend themselves to such analyses, along with conclusions, should be incorporated into the document at this point. These statistical analyses should be accompanied by the results of experiments conducted to determine the "poolability" of batches, or commonality of slopes and intercepts of individual batches. Graphs of these data should be included as part of the documentation.

Protocols for these batches and a commitment to continue them along with a "tentative" expiry date should also be included. Approval of these protocols will allow extension of the expiry date without a special supplement as long as the data remain within specifications. These data will ultimately be reported to FDA as part of periodic reports following NDA approval. Protocols intended for use on commercial batches should also be submitted.

Finally, the three-part commitment to mount studies for the first three production batches and a statistically determined number (at least one) each year, to update current studies in annual reports, and to withdraw any lots not meeting specifications should appear in the submission. Statistical sampling of production batches is usually based on $\log N$, $\sqrt{N}$, $\cdots$, where N is the number of batches produced per year. These batches are generally spread over various package types and manufacturing campaigns. There should be a standard operating procedure to handle specification deviations, including confirmation of the results, cause-and-effect investigation, impact analysis, final report to management, and field alert or batch recall notice to FDA.

Annual Product Review

Once a product gains FDA approval for marketing, the sponsor should maintain a readily retrievable profile of commercial batches. This includes individual batch release data and stability data. These data should be compiled throughout the year and tabulated before the anniversary of NDA approval for submission in the annual product report to FDA. By maintaining an ongoing database that is reviewed as new information is added, changing trends in the data can be observed and management notified if any of these trends are unfavorable.

REFERENCES

1. E. H. Dearborn, in *The Dating of Pharmaceuticals* (J. J. Windheuser and W. L. Blockstein, eds.), University Extension, University of Wisconsin, Madison, WI, 1970, p. 29.
2. K. Nord, J. Karlsen, and H. H. Tønnesen, Int. J. Pharm., 72, 11 (1991).
3. P. J. G. Cornelissen, G. M. J. Beijersbergen van Henegouwen, and K. W. Gerritsma, Int. J. Pharm., 1, 173 (1978).
4. K. A. Neftel, M. Walti, H. Spengler, and A. L. deWeck, Lancet, 1, 986 (1982).
5. V. J. Stella, T. J. Mikkelson, and J. D. Pipkin, in *Drug Delivery Systems* (R. L. Juliano, ed.), Oxford University Press, New York, 1980.
6. A. A. Sinkula, in *Sustained and Controlled Release Drug Delivery Systems* (J. R. Robinson, ed.), Marcel Dekker, New York, 1978.
7. K. A. Connors, G. L. Amidon, and V. J. Stella, *Chemical Stability of Pharmaceuticals: A Handbook for Pharmacists*, 2nd ed., John Wiley & Sons, New York, 1986.
8. H. G. Brittain, ed., *Analytical Profiles of Drug Substances and Excipients*, Vol. 21, Academic Press, San Diego, 1992.
9. M. A. Schwartz, A. P. Granatek, and F. H. Buckwalter, J. Pharm. Sci., 51, 523 (1962).
10. V. K. Krieble and K. A. Holst, J. Am. Chem. Soc., 60, 2976 (1938).
11. L. A. Casey, R. Galt, and M. I. Page, J. Chem. Soc. Perkin Trans., 2, 23 (1993).
12. R. S. Brown, A. J. Bennet, and H. Slebocka-Tilk, Acc. Chem. Res., 25, 481 (1992).
13. H. B. Mark, Jr. and G. A. Rechnitz, in *Chemical Analysis*, Vol. 24 (P. J. Elving and I. M. Kolthoff, eds.), Wiley-Interscience, New York, 1970.
14. C. Hansch, A. Leo, and R. W. Taft, Chem. Rev., 91, 165 (1991).
15. L. P. Hammett, *Physical Organic Chemistry*, 2nd Ed., McGraw-Hill, New York, 1970.
16. B. Capon and S. P. McManus, *Neighboring Group Participation*, Plenum Press, New York, 1976.
17. M. N. Khan and S. K. Gambo, Int. J. Chem. Kinetics, 17, 419–428 (1985).
18. R. E. Lindstrom, S. N. Patel, and P. K. Wilkerson, J. Parenter. Drug Assoc., 34, 5 (1980).

19. D. M. Johnson and L. C. Gu, Autoxidation and antioxidants, in *Encyclopedia of Pharmaceutical Technology* (J. Swarbrick and J. C. Boylan, eds.), Marcel Dekker, New York, 1988, pp. 415–449.
20. T. M. Chen and L. Chafetz, J. Pharm. Sci., 76, 703 (1987).
21. P. Finholt, H. Kristiansen, L. Kyowezynski, and T. Higuchi, J. Pharm. Sci., 55, 1435 (1966).
22. J. Schulz and K.-H. Bauer, Acta Pharm. Technol., 32, 78 (1986).
23. R. Hamburger, E. Azaz, and M. Donbrow, Pharm. Acta Helv., 50, 10 (1975).
24. J. W. McGinity, J. A. Hill, and A. L. La Via, J. Pharm. Sci. 64, 356 (1975).
25. J. W. McGinity, T. R. Patel, A. H. Naqvi, and J. A. Hill, Drug Dev. Commun., 2, 505 (1976).
26. L. Gu, H.-S. Chiang, and D. M. Johnson, Int. J. Pharm., 41, 105 (1988).
27. M. J. Frank, J. B. Johnson, and S. H. Rubin, J. Pharm. Sci., 65, 44 (1976).
28. K. Thoma and R. Kerker, Pharm. Ind., 54, 551 (1992).
29. J. A. Waters, Y. Kondo, and B. Witkop, J. Pharm. Sci., 61, 321 (1972).
30. E. Shefter, H.-L. Fung, and O. Mok, J. Pharm. Sci., 62, 791 (1973).
31. E. Shefter and T. Higuchi, J. Pharm. Sci., 52, 781 (1963).
32. S. R. Byrn, *Solid-State Chemistry of Drugs*, Academic Press, New York, 1982.
33. F. Jamali, R. Mehon, F. M. Pasutto, J. Pharm. Sci., 78, 695 (1989).
34. M. A. Nunes and E. Brochmann-Hanssen, J. Pharm. Sci., 63, 716 (1974).
35. G. Severin, Chirality, 4, 111–116 (1992).
36. L. J. Riff and G. G. Jackson, Arch. Intern. Med., 130, 887 (1972).
37. M. D. Santos-Buelga, M. J. Sanchez-Martin, and M. Sanchez-Camazano, Thermochim. Acta, 210, 255 (1992).
38. J. Haleblian and W. McCrone, J. Pharm. Sci., 58, 911 (1969).
39. M. Kuhnert-Brandstätter, *Thermomicroscopy in the Analysis of Pharmaceuticals*, Pergamon Press, Oxford, 1971, pp. 37–42.
40. S. S. Yang and J. K. Guillory, J. Pharm. Sci., 61, 26 (1972).
41. J. Haleblian, R. T. Koda, and J. A. Biles, J. Pharm. Sci., 60, 1485 (1971).
42. D. J. W. Grant and T. Higuchi, *Solubility Behavior of Organic Compounds*, John Wiley & Sons, New York, 1990.
43. T. J. Macek, U.S. Patent 2,671,750, March 9, 1954.
44. J. Brange and L. Langkjaer, Acta Pharm. Nord., 4(3), 149–158 (1992).
45. T. R. Kovalcik and J. K. Guillory, J. Parenter. Sci. Technol., 42, 29 (1988).
46. S. Budavari, ed., *The Merck Index*, 11th Ed., Merck & Co., Rahway, NJ, 1989.
47. Fed. Regist., 37, 15959 (1972).
48. H.-L. Fung, S. K. Yap, and C. T. Rhodes, J. Pharm. Sci., 63, 1810 (1974).
49. S. A. Fusari, J. Pharm. Sci., 62, 2021 (1973).
50. Z. Chowhan, Pharm. Technol. 6(9), 47–65 (1982).
51. C. J. deBlaey and J. J. Rutten-Kingma, Pharm. Acta Helv., 51, 186 (1976).
52. S. T. Horhota, J. Burgio, L. Lonski, and C. T. Rhodes, J. Pharm. Sci., 65, 1746 (1976).
53. P. B. Chemburkar, R. D. Smyth, J. D. Buehler, P. B. Shah, R. S. Joslin, A. Polk, and N. H. Reavey-Cantwell, J. Pharm. Sci., 65, 529 (1976).
54. B. L. McNiff, E. F. McNiff, and H.-L. Fung, Am. J. Hosp. Pharm., 36, 173 (1979).
55. W. A. Parker, M. E. Morris, and C. A. Shearer, Am. J. Hosp. Pharm., 36, 505 (1979).
56. J. I. Hirsch, J. H. Wood, and R. B. Thomas, Am. J. Hosp. Pharm., 38, 995 (1981).
57. P. A. Cossum and M. S. Roberts, Eur. J. Clin. Pharmacol., 19, 181 (1981).
58. E. A. Kowaluk, M. S. Roberts, H. D. Blackburn, and A. E. Pollack, Am. J. Hosp. Pharm., 38, 1308 (1981).
59. D. C. Monkhouse and L. Van Campen, Drug Dev. Ind. Pharm., 10, 1175 (1984).
60. D. C. Monkhouse, Drug Dev. Ind. Pharm., 10, 1373 (1984).
61. J. T. Carstensen, *Drug Stability, Principles and Practices*, Marcel Dekker, New York, 1990.
62. L. Meites, CRC Crit. Rev. Anal. Chem., 8, 55 (1979).
63. Guidelines for Submitting Documentation for the Stability of Human Drugs and Biologics, February, 1987, Center for Drugs and Biologics, Food and Drug Administration, Rockville, MD.
64. J. T. Carstensen, J. Pharm. Sci., 63, 1 (1974).

65. W. J. Irwin, *Kinetics of Drug Decomposition, Basic Computer Solutions*, Elsevier Science, Amsterdam, 1990.
66. W. P. Jencks, *Catalysis in Chemistry and Enzymology*, McGraw-Hill, New York, 1969.
67. E. R. Garrett, Adv. Pharm. Sci., 2, 1 (1976).
68. H.-L. Fung and S.-Y. P. King, in *Pharm Tech Conference '83 Proceedings*, Aster Publishing, Springfield, OR, 1983.
69. K. J. Laidler, J. Chem. Educ., 49, 343 (1972).
70. M. J. Pikal, A. L. Lukes, and J. E. Lang, J. Pharm. Sci., 66, 1312 (1977).
71. A. J. Woolfe and H. E. C. Worthington, Drug Dev. Commun., 1, 185 (1974).
72. D. R. Savello and R. F. Shangraw, Am. J. Hosp. Pharm., 28, 754 (1971).
73. T.-Y. Fan and S. R. Tannenbaum, J. Agric. Food Chem., 21, 967 (1973).
74. R. Shija, V. B. Sunderland, and C. McDonald, Int. J. Pharm., 80, 203 (1992).
75. R. E. Pincock, Acc. Chem. Res., 2, 97 (1969).
76. N. Murase, P. Echlin, and F. Franks, Cryobiology, 28, 364–375 (1991).
77. T. Chen, Drug Dev. Ind. Pharm., 18, 1311 (1992).
78. H.-L. Fung and S.-Y. P. King, in *Pharm Tech Conference '83 Proceedings*, Aster Publishing, Springfield, OR, 1983.
79. L. J. Ravin, C. A. Simpson, A. F. Zappala, and J. J. Gulesich, J. Pharm. Sci., 53, 106 (1964).
80. E. R. Garrett, J. Am. Chem. Soc., 79, 3401 (1957).
81. J. I. Wells, *Pharmaceutical Preformulation*, Ellis Horwood, West Sussex, UK, 1988.
82. M. A. Schwartz, J. Pharm. Sci., 53, 1433 (1964).
83. L. Gu and R. G. Strickley, Int. J. Pharm., 60, 99 (1990).
84. E. R. Garrett, J. Org. Chem., 26, 3660 (1961).
85. K. Thoma and M. Stuve, Pharm. Ind., 47, 1078–1081 (1985).
86. F. S. Amis, *Solvent Effects on Reaction Rates and Mechanisms*, Academic Press, New York, 1966.
87. C. Reichardt, *Solvents and Solvent Effects in Organic Chemistry*, 2nd Ed., VCH Verlagsgesellschaft, Weinheim, 1988.
88. K. A. Connors, *Chemical Kinetics: The Study of Reaction Rates in Solution*, VCH Publishers, New York, 1990, p. 411.
89. J. W. Moore and R. G. Pearson, *Kinetics and Mechanism*, Wiley-Interscience, New York, 1981.
90. J. T. Carstensen, J. Pharm. Sci., 59, 1140 (1970).
91. B. G. Cox, A. J. Kresge, and P. E. Sørensen, Acta Chem. Scand., A24, 202 (1988).
92. A. J. Kresge, Chem. Soc. Rev., 2, 475 (1973).
93. C. A. Bunton and G. Savelli, in *Advances in Physical Organic Chemistry*, Vol. 22 (V. Gold and D. Bethell, eds.), Academic Press, Orlando, FL, 1986, pp. 213–309.
94. T. J. Broxton, Aust. J. Chem., 35, 1357 (1982).
95. T. Higuchi and L. Lachman, J. Am. Pharm. Assoc. (Sci. Ed.), 44, 521 (1955).
96. K. A. Connors, *Binding Constants: The Measurement of Molecular Complex Stability*, John Wiley & Sons, New York, 1987.
97. O. Bekers, E. V. Uijtendaal, J. H. Beijnen, A. Bult, and W. J. M. Underberg, Drug Dev. Ind. Pharm., 17, 1503 (1991).
98. N. H. Anderson, D. Johnson, M. A. McLelland, and P. Munden, J. Pharm. Biomed. Anal., 9, 443 (1991).
99. D. Genton and U. W. Kesselring, J. Pharm. Sci., 66, 676 (1977).
100. J. T. Carstensen, Drug Dev. Ind. Pharm., 14, 1927 (1988).
101. C. Ahlneck and G. Zografi, Int. J. Pharm., 62, 87 (1990).
102. T. Chen, Drug Dev. Ind. Pharm., 18, 1311 (1992).
103. M. J. Hageman, Drug. Dev. Ind. Pharm., 14, 2047 (1988).
104. *Code of Federal Regulations*, Title 21, *Food and Drugs*, Part 211, Current good manufacturing practice for finished pharmaceuticals, Subpart G, §211.137 Expiration Dating.
105. *Code of Federal Regulations*, Title 21, *Food and Drugs*, Part 211, Current good manufacturing practice for finished pharmaceuticals, Subpart I, §211.166 Stability Testing.

106. *Code of Federal Regulations*, Title 21, *Food and Drugs*, Part 211, Current good manufacturing practice for finished pharmaceuticals, Subpart I, §211.170 Reserve Samples.

107. *Code of Federal Regulations*, Title 21, *Food and Drugs*, Part 211, Current good manufacturing practice for finished pharmaceuticals, Subpart J, §211.194 Laboratory Records.

108. *Code of Federal Regulations*, Title 21, *Food and Drugs*, Part 314, Applications for FDA approval to market a new drug or antibiotic drug, Subpart B, §314.50 Content and format of an application.

109. *Code of Federal Regulations*, Title 21, *Food and Drugs*, Part 314, Applications for FDA approval to market a new drug or antibiotic drug, Subpart B, §314.70 Supplements and other changes to an approved application.

110. *Code of Federal Regulations*, Title 21, *Food and Drugs*, Part 314, Applications for FDA approval to market a new drug or antibiotic drug, Subpart B, §314.81 Other postmarketing reports.

111. ICH Expert Working Group, Stability testing of new drug substances and products, International Conference on Harmonisation of Technical Requirements for the Registration of Pharmaceuticals for Human Use, October, 1993.

112. Guidelines for submitting samples and analytical data for methods validation, February, 1987, Center for Drugs and Biologics, Food and Drug Administration, Rockville, MD.

113. E. Debesis, J. P. Boehlert, T. E. Givand, and J. C. Sheridan, Pharm. Technol. 6(9), 120–137 (1982).

114. U.S. Pharmacopeial Convention, *United States Pharmacopeia, XXII*, <1225> pp. 1710–12, 1990.

115. Stability System (1989), ScienTek Software, P.O. Box 323, Tustin, CA 92681.

116. R. F. Tetzlaff, Pharm. Technol., 16(5), 70 (1992).

7

Preformulation

Jens T. Carstensen
University of Wisconsin, Madison, Wisconsin

I. INTRODUCTION

Historically, preformulation evolved in the late 1950s and early 1960s as a result of a shift in emphasis in industrial pharmaceutical product development. Until the mid-1950s, the general emphasis in product development was to develop elegant dosage forms, and organoleptic considerations far outweighed such (as yet unheard of) considerations about whether a dye used in the preparation might interfere with stability or with bioavailability.

In fact, pharmacokinetics and biopharmaceutics were in their infancy and, although stability was a serious consideration, most analytical methodology was such that even gross decomposition often went undetected.

It was, in fact, improvement in analytical methods that spurred the first programs that might bear the name ''preformulation.'' Stability-indicating methods would reveal instabilities not previously known, and reformulation of a product would be necessary. When faced with the problem of attempting to sort out the component of incompatibility in a ten-component product, one might use many labor hours. In developing new products, therefore, it would be logical to check ahead of time which incompatibilities the drug exhibited (testing it against common excipients). This way the disaster could be prevented in advance.

A further cause for the birth of preformulation was the synthetic organic programs started in many companies in the 1950s and 1960s. Pharmacological screens would show compounds to be promising, and pharmacists were faced with the task of rapid formulation. Hence, they needed a fast screen (i.e., a preformulation program) to enable them to formulate intelligently. The latter adverb implies that some of the physical chemistry had to be known, and this necessitated determination of physicochemical properties, a fact that is also part of preformulation. The approach was so logical, indeed, that it eventually became part of official requirements for INDs and NDAs [1]:

> . . . New drug substances in Phase I submission. For the drug substance, the requirement includes a description of its physical, chemical, or biological characteristics. We in the reviewing divisions regard

stability as one of those characteristics. The requirement for NDA submissions . . . of the rewrite stability information is required for both the drug substance and drug product. A good time to start to accumulate information about the appropriate methodology and storage stations for use in dosage form stations for use in dosage form stability studies, therefore, is with the unformulated drug substance. . . . Stress storage conditions of light, heat and humidity are usually used for these early studies, so that the labile structures in the molecule can be quickly identified. . . . If degradation occurs, the chemical reaction kinetics of the degradation should be determined. . . . Physical changes such as changes from one polymorph to another polymorph should be examined. . . . With the drug substance stability profile thus completed, the information should be submitted in the IND submission.

II. TIMING AND GOALS OF PREFORMULATION

The goals of the program, therefore, are (a) to establish the necessary physicochemical parameters of a new drug substance, (b) to determine its kinetic rate profile, (c) to establish its physical characteristics, and (d) to establish its compatibility with common excipients.

To view these in their correct perspective, it is worthwhile to consider where (i.e., at what time) in an overall industrial program performulation takes place. The following events take place between the birth of a new drug substance and its eventual marketing (but, first of all, most investigational drug substances never make it to the marketplace for one reason or another):

1. The drug is synthesized and tested in a pharmacological screen.
2. The drug is found sufficiently interesting to warrant further study.
3. Sufficient quantity is synthesized to (a) perform initial toxicity studies, (b) do initial analytical work, and (c) do initial preformulation.
4. Once past initial toxicity, Phase I (clinical pharmacology) begins, and there is a need for actual formulations (although the dose level may not yet be determined).
5. Phase II and III clinical testing then begins, and during this phase (preferably phase II) an order of magnitude formula is finalized.
6. After completion of the foregoing, an NDA is submitted.
7. After approval of the NDA, production can start (product launch).

III. PHYSICOCHEMICAL PARAMETERS

Physicochemical studies are usually associated with great precision and accuracy and, for a new drug substance, would include (a) pK (if the drug substance is an acid or base), (b) solubility, (c) melting point and polymorphism, (d) vapor pressure (enthalpy of vaporization), (e) surface characteristics (surface area, particle shape, pore volume), and (f) hygroscopicity. Unlike usual physicochemical studies, an abundance of material is usually not at hand for the first preformulation studies: in fact, at the time this function starts, precious little material is supplied; therefore, the formulator will often settle for good estimates, rather than attempting to generate results with four significant figures.

There is another good reason not to aim "too high" in the physicochemical studies of the first sample of drug substance: Usually, the synthesis is only a first scheme; in later scale-up it will be refined and, in general, the first small samples contain some small amount of impurities, which may influence the precision of the determined constants. But it is neessary to roughly know important properties such as solubility, pK, and stability. These will now be dealt with in order.

A. pK_a and Solubility

One important goal of the preformulation effort is to devise a method for making solutions of the drug. Frequently, the drug is not sufficiently soluble in water itself to allow the desired concentrations: for example, for injection solutions. Solubilities are determined by exposing an excess of solid to the liquid in question and assaying after equilibrium has been established. This usually is in the range of 60–72 hr, and to establish that equilibrium indeed has been attained, sampling at earlier points is necessary. Unstable solutions pose a problem here; this subject is dealt with in more detail later. Solubilities cannot be determined by precipitative methods (e.g., by solubilizing an acid in alkali and then lowering the pH to the desired pH) because of the so-called metastable (solubility) zone [2]. In the discussion to follow, drug substances are subdivided into two categories: ionizable substances, and (virtually) nonionizable substances.

Ionizable Substances

For substances that are carboxylic acids (HA), it is advantageous to determine the pK_a, since this property is of importance in a series of considerations. For carboxylic acid, the species A^- usually absorbs in the ultraviolet (UV) region, and its concentration can be determined spectrophotometrically [3]; on the other hand, HA will absorb at a different wavelength.

The molar absorbances of the two species at a given wavelength are denoted ϵ_0 and ϵ_- (and it is assumed that at the wavelength chosen $\epsilon_0 < \epsilon_-$) and it can be shown that if the solution is m_0 molar in total A, then

$$\frac{A^-}{HA} = \frac{\epsilon - \epsilon_0 m_0}{m_0 \epsilon_- - \epsilon} \tag{1}$$

so that the ratio A^-/HA can be determined in a series of buffers of different pH. Hence, the pK can be found as the intercept by plotting pH as a function of $\log[(A^-)/(HA)]$ by Henderson-Hasselbach:

$$pH = pK + \log (A^-/(HA)) \tag{2}$$

If several buffer concentrations are used, extrapolation can be carried out to zero ionic strength, and the pK_a can be determined. For initial studies, however, a pK in the correct range (i.e., $\pm$ 0.2 unit) will suffice, so that the foregoing determination can be performed at one buffer concentration only.

The conventional approach is to do titrations (Fig. 1), and this will yield graphs of the fraction neutralized (x) as a function of pH. Usually, the water is titrated as well [4], and what is presented in Fig. 1 is the "difference." The pK is then the pH at half neutralization (which is also the inflection point). The pH–solubility curve can now be constructed simply by determining the solubility of HA (at low pH) and A^- (e.g., of NaA) at high pH (e.g., at pH 10). Note that, at a given pH, the amount in solution in a solubility experiment is

$$S = S_{HA} + C_{A^-} \tag{3}$$

where S denotes solubility. The last term can be determined from knowledge of the pH and use of Eq. (2).

For drugs that are amines, the free base is frequently poorly soluble and, in this event, the pK is often estimated by performing the titration in a solvent containing some organic solvent (e.g., ethanol). By doing this at different organic solvent concentrations (e.g., 5, 10, 15, and 20%), extrapolation can be carried out to 0% solvent concentration to estimate the aqueous pK.

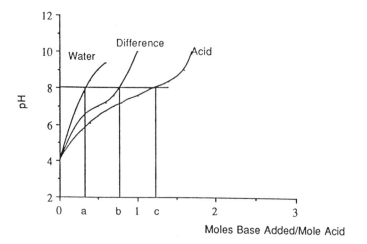

Fig. 1 Typical titration curves. The water curve indicates the amount of alkali needed to titrate the water, and the acid curve is a conventional titration curve. The difference curve is the horizontal difference between the acid and the water curve, and is the adjusted titration curve (e.g., the point *b* is *c* minus *a*). The p*K* is the point of inflection, which is also the point at which half of the acid is neutralized.

Nonionizable Substances

For hydrophobic, (virtually) nonionizable substances [i.e., those that show no ionic species of significance in the pH range 1–10 (e.g., diazepam)], solubility can usually be improved by addition of nonpolar solvents. Aside from solubility, stability is also affected by solvents, either in a favorable or in a nonfavorable direction [5]. Theoretical equations for solubility in water [6] and in binary solvents [7] have been reported in the literature, but, in general, the approach in preformulation is pseudoempirical. Most often the solubility changes as the concentration of nonpolar solvent, C_2, increases. For binary systems, it may simply be a monotonely increasing function [18], as shown in Fig. 2. The solubility is usually tied to the dielectric constant,

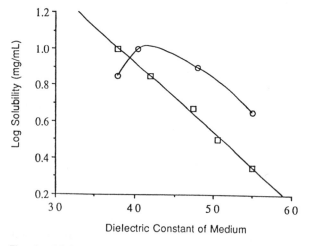

Fig. 2 (○) Solubility of 7-chloro-1,3-hydro-5-phenyl-2*H*-1,4-benzodiazepine-2-one-4-oxide in aqueous propylene glycol. (Data from Ref. 8.) (□) Solubility of another benzodiazepine. (Unpublished data.)

and in a case, such as that shown in curve A, the solubility is often log-linear when plotted as a function of inverse dielectric constant, E, that is,

$$\ln S = \frac{-e_1}{\epsilon + e_2} \tag{4}$$

where ϵ is the dielectric constant, and the e-terms are constants [3].

Frequently, however, the solubility curve has a maximum (as shown in curve B in Fig. 2) when plotted as both a function of C_2 and ϵ [9]. In either case, it is possible to optimize solubility by selection of a solvent system with a given value of ϵ; that is, once the curve has been established, the optimum water/solvent ratio for another solvent can be calculated from known dielectric constant relationships [10].

Ternary Systems and Optimization

Frequently, *ternary* solvent systems are resorted to. Examples are water–propylene glycol–benzyl alcohol or water–propylene glycol–ethanol. In such cases the solubility profile is usually presentable by a ternary diagram [11]. This type of diagram usually demands a fair amount of work; that is, the solubility of the drug substance in many solvent compositions must be determined. A priori, it would seem, therefore, that they would be out of place in a situation where only limited quantities of drug are available. However, their principle gives some validity to optimization procedures.

The diagram can be of one of two types, as shown in Figs. 3 and 4. In the first type, the solubility may be assumed to be of the type

$$S = a_{10} + a_{11}C_1 + a_{12}C_2 \tag{5}$$

where C denotes concentrations of nonaqueous solvents. An example of this is shown in Fig. 3 inset. Here the subscripts to C denote the two nonaqueous solvents. Hence, three solubility experiments would determine the relationship (with zero degrees of freedom). It is usual to do at least five, and determine possible curvature [i.e., inclusion of more terms in Eq. (5)].

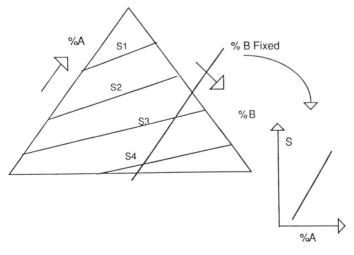

Fig. 3 Ternary diagram of solubility of a compound in a ternary mixture with linear solubility response. Inset: Concentration of drug in compositions with constant concentration of B. The composition of the solute is the constant concentration of B, the concentration of A in the abscissa, and the complement concentration of the third constituent. The drug solubility response is linear in the A-concentration here.

In the second case, in Fig. 4 component A, each tie line will give a parabolic-type curve as shown in the inset. Hence, at a given concentration of C_2 the solubility can be approximated by

$$S = b_{10} + b_{11}C_1 + b_{12}C_1^{\,2} \tag{6}$$

where, in the simplest case,

$$b_{10} = C_{20} + C_{21}C_2 \tag{7}$$

Hence, optimization can be achieved by five (or more) experiments, with zero (or in general $n - 5$) degrees of freedom.

Prediction of Solubility

It is advantageous, with a new drug substance, to be able to estimate what its solubility might be, before carrying out dissolution experiments. There are several systems of solubility prediction, notably those published by Amidon and Yalkowsky [12–14] in the 1970s. Their equation, for solubility of *p*-aminobenzoates in polar and mixed solvents, is a simplified two-dimensional analog of the Scatchard-Hildebrand equation and is based on the product of the interfacial tension and the molecular surface area of the hydrocarbon portion of a molecule.

More recently, Bodor and Huang [15] have developed a semiempirical solubility predictor based on 20 variables (S = molecular surface in Å^2, I_a = indicator variable for alkanes, D = calculated dipole moment in Debyes; Q_n = square root of sum of squared charges on oxygen atoms; Q_o = square root of sum of squared charges on non-oxygen atoms; V = molecular volume in Å^2; S_2 = square of molecular surface; C = constant; MW = molecular weight; $\{O\}$ = ovality of molecule; A_{bh} = sum of absolute values of atomic charges on hydrogen atoms; A_{bc} = sum of absolute values of atomic charges on carbon atoms, A_m = indicator variable for aliphatic amines; and N_h = number of N-H single bonds in the molecule.

The aqueous solubilities, W, of 331 compounds followed the following equation (with tolerances omitted):

$$
\begin{aligned}
\log W = {}& - 56.039 + 0.32235D - 0.59143I_a + 38.443Q_n^4 - 51.536Q_n^2 \\
& + 18.244Q_n + 34.569Q_o^4 - 31.835Q_o^2 + 15.061Q_o + 1.9882A_m \\
& + 0.15689N_h + 0.00014102S^2 + 0.40308S - 0.59335A_{bc} \\
& - 0.42352V + 1.3168A_{bh} + 108.80\{O\} - 61.272\{O\}^2
\end{aligned} \tag{8}
$$

Of the parameters listed, only the ovality and the indicator value for the alkanes I_a are unfamiliar entities that are obtained from the literature [15].

Dissolution

The importance of dissolution will be discussed in more detail later. A short note on the topic is, however, necessary for the further development at this point. According to Noyes-Whitney [16]

$$\frac{dm}{dt} = \frac{VdC}{dt} = -kA(S - C) \tag{9}$$

where

    m = mass not dissolved
    V = liquid volume
    t = time

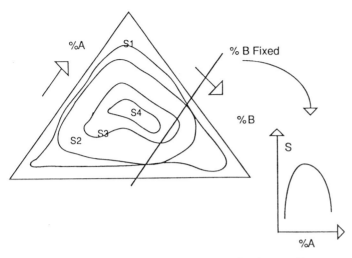

Fig. 4   Ternary diagram and tie-line concentration in a nonlinear system.

k = so-called intrinsic dissolution rate constant (cm/sec)
A = surface area of the dissolving solid

Many criticisms have been voiced against Eq. (9), but, in general, it is correct, and it will be assumed to be so in the following. If an experiment is carried out with constant surface (as e.g., using a Wood's apparatus [17], or with smaller amounts, making a small pellet and encasing it in wax, and exposing only one face to a dissolution medium, or if an excess of solid prevails throughout the dissolution experiment, then Eq. (9) may be integrated to give

$$\ln\left\{1 - \left(\frac{C}{S}\right)\right\} = \left(\frac{kA}{V}\right)t \tag{10}$$

$$C = S[1 - \exp(-\{kA/V\}\,t)] \tag{11}$$

A typical curve following Eq. (11) is shown in Fig. 5.

Solubility of Unstable Compounds

Quite often a compound is rather unstable in aqueous solution. Hence, the long exposure to liquid required for traditional solubility measurements will cause decomposition, and the resulting solubility results will be unreliable. In this particular instance, a method known as Nogami's method may be used. If a solution experiment is carried out as a dissolution experiment, with samples taken at equal time intervals, δ, it can be shown [18] that when the amount dissolved at time $t + \delta$ is plotted versus the amount dissolved at time t a straight line will ensue (Fig. 6).

Metastability is evident in dissolution rates as well. If dissolution of a metastable form is monitored at equal time intervals, δ, then, when the concentration at time $t + \delta$ is plotted versus the concentration the following relationship holds:

$$C(t + \delta) = S[1 - \exp(-k\delta)] + \exp(-k\delta)C_t \tag{12}$$

Hence, such as plot (as shown in Fig. 6) will give k from the slope, and inserting this in the intercept expression will give S. The advantage of the method is that it can be carried out in

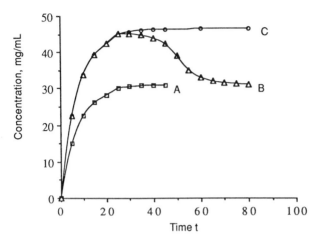

Fig. 5 Dissolution profiles obtained from the solubility determination of two polymorphic forms of the same drug substance. A is the stable form with solubility 31 mg/ml; C is the profile of the metastable form with solubility 46 mg/ml. This solubility (circles) is not achieved in many instances, and precipitation of the stable form occurs at a point beyond the solubility of A, and the trace becomes B.

a short time, and can reduce the effect of decomposition; the disadvantage is that it is not as precise as ordinary solubility determinations.

Solubility of Metastable Polymorphs

Polymorphism is an important aspect of the physical properties of drugs. One of the characteristics of a metastable polymorph (to be discussed in some detail at a later point) is that it is more soluble than its stable counterpart. The solubility profile of the drug will be as shown in Fig. 5; A is the stable form with solubility 31 mg/ml; B is the profile of the metastable form with solubility 46 mg/ml. This solubility (circle) is usually not achieved, and precipitation of the stable form occurs at a point beyond the solubility of A, and the trace becomes B.

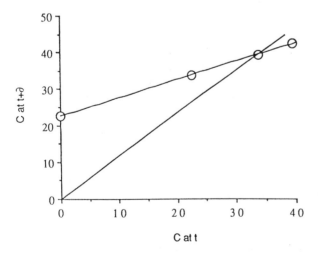

Fig. 6 The Nogami method applied to the data in Fig. 5.

In such cases, the Nogami method can be applied to the early points curve (see Fig. 6), and the solubility S', of the polymorph can be assessed. One of the important aspects of metastable polymorphs in pharmacy is exactly their higher solubility, since the dissolution rate will also be higher [see Eq. (9)]. Hence, the bioavailability will be increased where this is dissolution rate-limited [19].

Polymorphism

Solids exist as either amorphous compounds or as crystalline compounds. In the latter, the molecules are positioned in lattice sites. A lattice is a three-dimensional array, and there are eight systems known. Compounds often have the capability of existing in more than one crystalline form, and this phenomenon is referred to as *polymorphism.*

If a compound exhibits polymorphism, one of the forms will be more stable (physically) than the other forms; that is, of n existing forms, $n - 1$ forms will possess thermodynamic tendency to convert to the nth, stable form (which then has the lowest Gibbs energy; however, in the preformulations stage, it is not known whether or not the form on hand is the stable polymorph).

One manner in which different polymorphs are created is by way of recrystallizing them from different solvents, and at a time point when sufficiently material (and this need not be very much) is available, the preformulation scientist should undertake recrystallization from a series of solvents.

Knowledge of polymorphic forms is of importance in preformulation because suspension systems should never be made with a metastable form (i.e., a form other than the stable crystalline form). Conversely, a metastable form is more soluble than a stable modification, and this can be of advantage in dissolution [see Eq. (11)]. There are two types of polymorphism, a fact illustrated in the following discussion.

If the vapor pressure or solubility of a compound is plotted as a function of temperature, a plot, such as that shown in Fig. 7, will result. Here form I is the form that is stable at temperature T_1. If the compound exists in the two forms I and II, the phenomenon is referred to as an enantiotropic system, since, on heating to the temperature T_2, form I will transform into form II. Form II may exist below temperature T_2, but perturbations (e.g., presence of moisture) will convert it to form I, and the energy involved in the transformation will be

$$E = RT \ln \left[\frac{S_2}{S_1} \right] \tag{13}$$

If the compound is present as form I at room temperature and heated up fast it will melt at T', which is lower than the melting point of form II (T'').

A different situation exists if the compound exists as form I and form III. This is referred to as a monotropic system, and here, III is unstable relative to I over the whole solid range. In this case, however, the melting point of the "unstable" polymorph is lower than that of the stable one (T''' is lower than T').

If a metastable polymorph is kept dry, it may be stable for eons and, therefore, it is not referred to as an "unstable," but rather, as "metastable". An even more energetic state is represented by amorphous forms, which may be considered supercooled liquids. Today, polymorphism is checked for in two fashions. Thermal methods will give information about whether a polymorph is stable, enantiotropic, or monotropic. If the system possesses a transition point, G is zero at this point. In a fashion similar to the melting process, where G is also zero, the transition is associated with an enthalpy change, which is endothermic. Hence, if a sample of a compound is heated up and a change in enthalpy occurs below the melting point, it is an enantiotropic transformation if the point is reproducible and if it occurs again after the sample

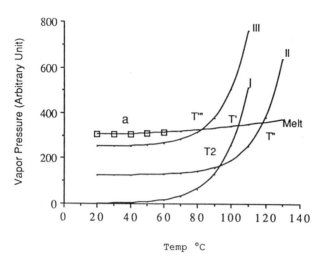

Fig. 7 Dependence of vapor pressure and solubility for an enantiotropic pair (I/II), a monotropic pair (I/III), and an amorphous compound (a).

has been cooled back down and shows the same enthalpy change on heating. Otherwise, it is monotropic and, in this case, as shown in Fig. 8, there will either be a (lower) melting point (T''') with a single endotherm, or there will be a melting followed by a recrystallization into the more stable forms that then melt [denoted III (Alternate)] in Fig. 8.

An amorphous compound has no melting point, and (above glass transition temperatures) its vapor pressure curve simply continues into that of the melt.

Second, x-ray diffraction will give spacings directly, in the crystal and reveal differences between samples. Finally, solubility curves can be carried out and, if a nick in the solubility curve is found (Fig. 9), this is a transition temperature. If no nick is found, there is no transition temperature, but if the dissolution curves are as shown in Fig. 5, there is polymorphism and, by indirect argumentation, it may be concluded that a monotropic system exists.

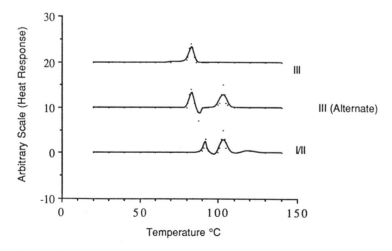

Fig. 8 DSC tracings of the polymorphs shown in Fig. 7.

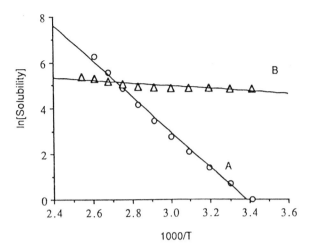

Fig. 9 Solubility curves of an enantiotropic pair.

C. Partition Coefficient

Partition coefficients between water and an alkanol (e.g., octanol) should be determined in preformulation programs [20]. The partition coefficient of a compound that exists as a monomer in two solvents is given by

$$K = \frac{C_1}{C_2} \tag{14}$$

If it exists as an n-mer in one of the phases, the equation becomes

$$K = \frac{(C_1)^n}{C_2} \tag{15}$$

or

$$\log k = n \log C_1 - \log C_2 \tag{16}$$

The easiest way to determine the partition coefficient is to extract V_1 ml of saturated aqueous solution with V_2 ml of solvent, and determine the concentration C_2 in the latter. The amount left in the aqueous phase is $(C_1 V_1 - C_2 V_2) = M$, so that the partition coefficient is given by

$$K = \frac{M}{V_1 C_2} \tag{17}$$

If it is assumed that the species is monomeric in both phases, the partition coefficient becomes the ratio of the solubilities, and it is simply sufficient to determine the solubility of the drug substance in the solvent (since it is assumed that the solubility is already known in water):

$$K = \frac{S_1}{S_2} \tag{18}$$

D. Vapor Pressure

In general, vapor pressures are not all that important in preformulation, but it should always be kept in mind that a substance may have a sufficiently low vapor pressure that it will (a) become lost to a large enough extent that apparent stability and content uniformity problems will result; and (b) it will exhibit a potential for interaction with other compounds and adsorption onto or sorption into package components [21].

Most drug substances are not substantially volatile. As an initial screen, it can be determined whether the drug is sufficiently volatile to cause concern by placing a weighed amount of it in a vacuum desiccator and weighing it daily for a time. It is better to have a high-vacuum system for this, and the use of a vacuum electrobalance is best for this purpose. A good estimate of the vapor pressure can be obtained [22] by using a pierced thermal analysis cell, placing it on a vacuum electrobalance, and monitoring the rate of weight loss. A substance with known vapor pressure can then be used for calibration, the loss rates being proportional to the vapor pressures.

E. Surface Characteristics

The surface characteristics of a batch of a drug substance may greatly influence its properties in processing (flow, dissolution). Crystals may form in different habits (plate, needle, cube) and these may not be due to morphology; that is, depending on crystallization circumstances, they could all be the same crystal form, but of different habit [23].

It is a good practice, both during development of a new drug and through to the NDA, to take photomicrographs of each new batch of drug substance delivered to the product development department. In this manner, there will be a permanent reference record, and when deviations from expected behavior occur during the product development sequence, the photomicrograph will be one record that may throw light on the problem. Aside from this, the specific surface area (A'', cm^2/g) of each batch of drug substance should be measured.

Shape and Fractal Dimension

Shape is of great interest and affects many properties; therefore, it is important to have a record of how a shape changes as the synthesis of the raw material undergoes changes during the developmental process. In the simplest form, microscopy of all batches used in product development should be carried out to determine the ratio of longest to shortest dimension (average of ten measurements). This is a type of shape factor.

However, there are more sophisticated methods that may be used to attain a good feel for the shape factor; namely, its fractal dimension. This is most conveniently carried out by use of imaging techniques.* The general principle of this is shown in Figs. 10 and 11.

As an example (and this is a hypothetical example only), a particle is shown in Fig. 10, such as it might appear on a microscope slide. This particle is gridded out in the form shown in the grid in the upper left-hand corner of Fig. 10. The number of squares in which parts of the trace of the particle are located is counted. This number is N, and the length of the grid size is g. The grid size is arbitrarily set equal to 1 in this example. The grid length is now halved, and the number of squares counted again; this is then repeated several times. The fractal equation is then given by

$$\ln[N] = -n \ln[g] + q \qquad (19)$$

---

*An example of an excellent system is that of Universal Imaging Concepts Image 1 Software 486/66 computer (Data Store, Inc); Hamamatsu C2400-77H B&W chip camera (CCD Camera Control); Sony 19″ PVM1943MD color monitor; UP5000 (Sony) color printer; Nikon model Labophot-2 40x ocular x 10 on-screen magnification; Nikon polarizing microscope, model POH-II.

14 Squares

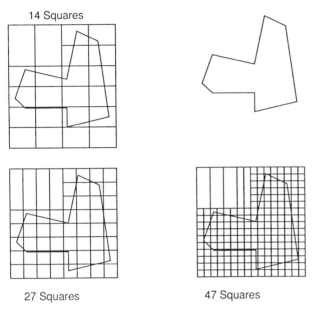

27 Squares   47 Squares

Fig. 10 Dividing the field containing a particle into more and more squares, by dividing the original grid length, a, by 2.0 in each step. The length would be $L = Na\sqrt{2}/g$ if it were a line going through corners of the squares, so that if it were a straight line, then $\ln[L]$ should equal $\ln[N] - \ln[g] + \ln[a\sqrt{2}]$, but the slope differs from one that is the ''natural'' dimension of a line.

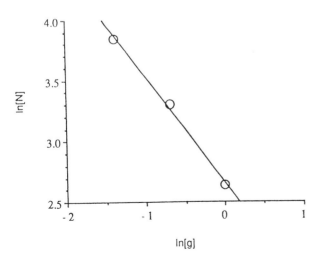

Fig. 11 Fractal graph of the data in Fig. 10. The least squares fit for the equation is $\ln[N] = 2.6575 - 0.8735 \ln[g]$.

where n and q are constants, and where $n + 1$, the slope of the line is the fractal dimension [24]. This dimension is characteristic for shapes, and its determination is worthwhile to keep record of how shape may change with synthesis changes.

The number of decreases in the so-called measuring stick (here, the grid) is of great importance, and it is the behavior at the smaller end of the measuring stick size that is important. In fact the small number of iterations leading to Eq. (19) may well be fallacious but at the lower end of the scale the relationship will still be correct (but the true fractal dimension may or will be different from the value found in Fig. 10).

Particle Size and Size Distributions

The "size" of the particles is of great importance, and particle size determinations should be carried out in preformulation as well in formulation functions. For small particle sizes, simple microscopy [25] may be used, but again, imaging techniques, particularly with motorized stages, are more representative and much easier to carry out [26].

When characterizing a powder, it is important to make certain what type of diameter is being described, since various techniques give different "types of diameter."

The volume-surface mean diameter, d_{sv}, can be determined by permeamitry (e.g., Fisher Sub-Sieve Sizer) for a fine powder [25]. It is given by

$$d_{sv} = \frac{6\,V}{A''} \tag{20}$$

where V is the solids volume of the sample and A'' is the geometric surface area. A'' can be obtained for a coarse powder by sieve analysis:

$$A'' = \{6/\rho\}\ \Sigma w_i/d_i \tag{21}$$

where ρ is the solids density (g/cm^3), d_i the mean diameter of the ith mesh cut, and w_i the weight retained by the ith screen. If the drug substance is not porous, the ratio

$$G = \frac{A}{A''} \tag{22}$$

is the rugosity and is a measure of the surface roughness.

Pore Size Distribution

With the advent of mercury intrusion porosimeters, it is advantageous to perform a pore size distribution of investigational batches of a drug [27]. The Washburn equation [28] states that the pressure P necessary to intrude a pore is given by

$$\ln P = -q\gamma\left(\frac{1}{r}\right)\cos[\theta] \tag{23}$$

where

q = a constant
γ = interfacial tension
θ = contact angle
r = radius of the pore being penetrated

Since mercury has a contact angle with most solids of about 140°, it follows that its cosine is negative (i.e., it takes applied pressure to introduce mercury into a pore). In a mercury porosimeter, a solid sample is evacuated in a cell, mercury is then intruded, the volume V is noted (it actually reads out), and the pressure P is then increased stepwise. In this fashion it is possible to deduce the pore volume of a particular radius [corresponding to P by Eq. (23)]. A pore size distribution will give the total internal pore area as well, which can be of importance in dissolution.

Hygroscopicity

Hygroscopicity is an important characteristic of a powder. It can be shown, roughly, for a fairly soluble compound that the hygroscopicity is related to its solubility [29,30], although the heat of solution plays an important part in what is conceived as ''hygroscopicity'' [31–33]. A hygroscopicity experiment is carried out most easily by exposing the drug substance to an atmosphere of a known relative humidity (e.g., storing it over saturated salt solutions in desiccators). Each solution will give a certain relative humidity (RH), and the test is simply to weigh the powder from time to time and determine the amount of moisture absorbed (weight gained). This does not work with drug substances that decompose (e.g., effervescent mixtures will start losing weight owing to carbon dioxide evolution [34]).

If the air space is agitated sufficiently to prevent vapor pressure gradients, the initial uptake rate (g, H_2O/g solid per hour) is related to relative humidity by

$$L = a_{21}[RH - RH_0] \tag{24}$$

where RH_0 is the vapor pressure of a saturated solution of the drug substance in water. The latter can be estimated by an ideality assumption; that is, if the solubility is expressed as a mole fraction X_s, the vapor pressure over a saturated solution will be P' given by

$$P' = (1 - X_s)P^* \tag{25}$$

where P^* is water's vapor pressure at that temperature.

The foregoing experiments are rather easy to carry out and should always be part of a preformulation program, since hygroscopicity can be so important that it will dictate whether or not a particular salt should be used. Flurazepam (Dalmane), for instance, is a monosulfate, and is used as such, since the disulfate, desirable in many other respects, is so hygroscopic that it will remove water from a hard-shell capsule and make it exceedingly brittle.

IV. COMPATIBILITY TESTS

It should again be emphasized that, at the onset of a new drug program, there are only small amounts of drug substance at hand. One of the first tasks for the preformulation scientist is to establish the framework within which the first clinical batches can be formulated. To this end, it is important to know with which common excipients the drug is compatible. The distinction will be made between solid and liquid dosage forms in the following sections.

A. Compatibility Test for Solid Dosage Forms

It is customary to make a small mix of drug substance with an excipient, place it in a vial, place a rubber stopper in the vial, and dip the stopper in molten carnauba wax (to render it hermetically sealed). The wax will harden and form a moisture barrier up to 70°C. A list of common excipients characteristic of this type of test is shown in Table 1. At times, it is possible to obtain quantitative relationships to excipient characteristics and interaction rates [35,36].

In addition to the test, as described, a similar set of samples are set up to which 5% moisture is added. A storage period of 2 weeks at 55°C (except for stearic acid and dicalcium phosphate, for which 40°C is used) is employed, after which time the sample is observed physically for (a) caking, (b) liquefaction, (c) discoloration, and (d) odor or gas formation. It is then assayed by thin-layer chromatography (or HPLC).

Table 1 Categories for Two-Component Systems

Additive		Identical	17–27 mo at 25°C	Worse 10 days at 55°C	Total 25°C	Score 55°C
Drug per se	Dry	15	4	1	38	31
	5% H_2O	9	8	3	49	38
+ Magnesium stearate	Dry	16	3	1	34	30
	5% H_2O	15	4	1	43	35
+ Calcium stearate	Dry	13	4	3	37	32
	5% H_2O	12	5	3	38	35
+ Stearic acid	Dry	15	5	0	42	31
	5% H_2O	7	11	2	60	38
+ Talc	Dry	14	5	1	38	30
	5% H_2O	10	8	2	45	34
+ Acid-washed talc	Dry	12	8	0	44	31
	5% H_2O	10	9	1	49	35
+ Lactose	Dry	12	5	3	38	32
	5% H_2O	9	7	4	65	56
+ $CaHPO_4$, anhydrous	Dry	12	6	2	46	36
	5% H_2O	9	8	3	66	53
+ Cornstarch	Dry	12	5	3	39	34
	5% H_2O	10	5	5	40	37
+ Mannitol	Dry	10	7	3	39	31
	5% H_2O	8	7	5	47	45
+ Terra alba	Dry	14	6	0	41	28
	5% H_2O	11	6	3	50	45
+ Sugar 4x	Dry	12	6	2	41	34
	5% H_2O	9	7	4	63	61

Source: Ref. 35.

Note that one of the samples set up is the drug by itself. This is done for several reasons, one of which is that it is now required by the FDA for IND submissions [1]. One more reason is that, at the onset of a program, the organic synthesis of the compound may lack the refinement it will later have, and it is not uncommon that there will be several weak spots (impurities) on a TLC chromatogram of a compound obtained by initial laboratory synthesis. Hence, in selecting the excipients with which the drug substance is deemed to be compatible, it is customary to use as criteria that (after accelerated exposure of a drug–excipient mix) no new spots have developed, and that the intensity of the spots in the drug that has been stored under similar conditions (2 weeks at 55°C) are the same as in the acceptable excipient. This type of program is used by many companies with good success (i.e., the formulas developed based on the findings from the compatibility program are stable).

Liquefaction occurs at times because of eutectic formation (e.g., often with caffeine combinations), and this may not necessarily be associated with decomposition. On the other hand, discoloration (e.g., amines and sugars) usually is.

Finally, the reason for not forcing dicalcium phosphate (a very valuable formulation aid in direct compression) beyond 40°C is that at higher temperatures (actually above 70°C [37]) it converts to the anhydrate, a conversion that is, curiously enough, catalyzed by water. In other words, the dihydrate will be autocatalytic at elevated temperatures, and it should not be ruled out based on high-temperature findings.

B. Kinetic pH Profiles

Frequently, a broad screen of stability is performed on the first small sample used for initial preformulation; this is frequently referred to as "forced decomposition studies" [38]. In this the drug is exposed to "acid degradation," "base degradation," "aqueous degradation," "drug powder degradation," and "light degradation." More refined studies are eventually needed.

For any compound marketed by a pharmaceutical concern, at one time during its development, there should be a concerted project to establish a very exact pH profile. To do this correctly is a time-consuming undertaking. However, the information that can be gleaned from it is very important for formulations; therefore, it is customary to carry out an approximate kinetic pH profile [39] early in the developmental stage. This will allow formulation of solutions for injections, and for oral products as well, at a pH and with buffers that will give the best stability. Without it, formulation is essentially guesswork.

Most drug decompositions are hydrolyses, during which the drug concentration C decreases with time through

$$\frac{dC}{dt} = -k_2 C_{H_2O} C \tag{26}$$

Since the water concentration hardly changes, this (bimolecular) reaction scheme is reduced to the pseudo-first-order expression

$$\frac{dC}{dt} = -KC \tag{27}$$

where K is the first-order rate constant. This integrates to the well-known form

$$\ln\left[\frac{C}{C_0}\right] = -Kt \tag{28}$$

Hence, semilog plotting of concentration versus time (see Fig. 12) will give a straight line, with a slope from which K is calculated. But, most reactions are catalyzed by buffers, by hydrogen ions, and by hydroxyl ions, so K will be of the form

$$K = k + k_+(H^+) + k_-(OH^-) + k_B(B) \tag{29}$$

where B denotes buffer concentration and k is a rate constant. A decomposition experiment is now carried out at, say, five pH values, each using two buffer concentrations. A graph is drawn of k (Fig. 13) versus B at the various pH values. A line is drawn through the two at low pH (at which hydroxyl ion concentration can be disregarded), and Eq. (29) becomes

$$K = k_+(H^+) + k_B(B) \tag{30}$$

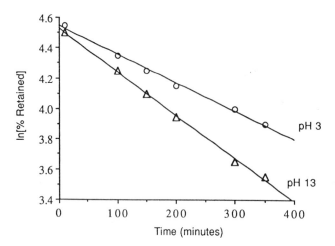

Fig. 12 Pseudo-first-order decompositions of carbuterol at 85°C at an ionic strength of 0.5. (Graphs plotted from data in Ref. 40.)

so that the plot in Fig. 13 will give

$$K = k_+(H^+) \tag{31}$$

as intercept and k_B as slope. The foregoing allows assessment of the effect of the buffer (which is an important point in the buffer selection). Taking the 10 log of Eq. (31) now gives

$$\log K = -pH + \log(k_+) \tag{32}$$

A similar argument will show that at high pH

$$\log K = pH - 14 + \log (k_-) \tag{33}$$

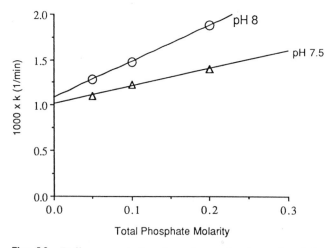

Fig. 13 Buffer concentration dependence of carbuterol at 85°C at an ionic strength of 0.5. (Graphs plotted from data in Ref. 40.)

This explains why the extremes of a pH profile (see Fig. 14) often have slopes of plus or minus unity [40]. The horizontal part is due to the uncatalyzed rate constant, k_+, in Eq. (29). A pH profile can be done at, for example, six pH values and, since there are two kinetic points (times) and two buffer concentrations at each, a total of 24 assays are needed, which is not insurmountable. This number may be minimized and optimized by careful selection of pH and buffer concentrations [39]. Later in the program the pH profile should be repeated, but with multiple points and several buffer concentrations, but this is beyond the point of preformulation.

C. Liquid Compatibilities

The pH profile is the most important part of liquid compatibilities. However, two-component systems are set up in aqueous (or other types of) solutions and treated as in Sec. IV.A. This is required in the 1987 stability guidelines [1], which state that "it is suggested that the following conditions . . . be evaluated in studies on solutions or suspensions of bulk drug substances: acidic and alkaline pH, high oxygen and nitrogen atmospheres, and the presence of added substances, such as chelating agents and stabilizers," and it is suggested "that stress testing conditions . . . include variable temperature (e.g., 5, 50, and 75°C)."

Aqueous Solution Capability

In general, such studies are carried out by placing the drug in a solution of the additive. This can be (and usually is) a heavy metal (with or without chelating agents present) or an anti-oxidant (in either oxygen or nitrogen atmosphere). Usually, both flint and amber vials are used and, in many cases, an autoclaved condition is included. This will answer questions about susceptibility to oxidation, to light exposure, and to heavy metals. These are important questions as far as injectable compatibilities are concerned. Exposure to various plugs is frequently included at this point so that early injectable preparations can be formulated.

For preparations for oral use, knowledge of the desired dosage form is important, but compatibility with ethanol, glycerin, sucrose, corn syrup, preservatives, and buffers is usually carried out. This type of study also gives an idea of the activation energy E of the predominant reaction in solution. The Arrhenius plots (Fig. 15) for compounds in solution are usually quite

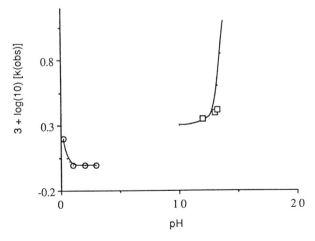

Fig. 14 Kinetic pH profile at low and high pH of carbuterol at 85°C. (Graphs plotted from data in Ref. 40.)

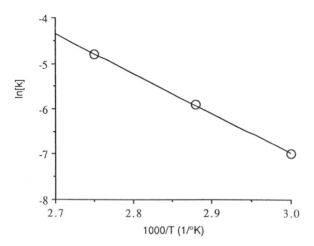

Fig. 15 Arrhenius plotting of carbuterol rate constants at pH 4 of carbuterol. (Graph plotted from data in Ref. 40.)

precise; that is, denoting the rate constant by k, absolute temperature by T, the gas constant by R, and a collision factor by Z:

$$k = Z^{-E/(RT)} \tag{34}$$

or its more useful logarithmic cousin,

$$\ln k = \ln Z - \left(\frac{E}{R}\right)\left(\frac{1}{T}\right) \tag{35}$$

Nonaqueous Liquid

With transdermal dosage forms being of great importance of late, it is advisable to test for compatibilities with "ointment" excipients and with polymers (e.g., ethylvinyl polymer, if that is the desired barrier). With transdermals, the dosage form is either directly placed in a stirred liquid, or it is placed in a cell with an appropriate membrane (e.g., cadaver skin) to estimate the release characteristics of the drug from the ointment [41].

  If the overall flux is J, then

$$\frac{1}{J} = \left[\frac{1}{J_{\text{ointment}}}\right] + \left[\frac{1}{J_{\text{membrane}}}\right] \tag{36}$$

where subscripts refer to the respective phase. J_{membrane} can be obtained from curves, such as that shown in Fig. 16, in the fashion that the overall flux is first obtained (with the membrane in place), giving the value of J. Then the release is obtained without the membrane in place, giving J_{ointment}, that is,

$$J = \left(\frac{1}{A}\right)\left(\frac{dm_1}{dt}\right) \tag{37}$$

and

$$J_{\text{ointment}} = \left(\frac{1}{A}\right)\left(\frac{dm_2}{dt}\right) \tag{38}$$

J_{membrane} is then obtained as the reciprocal of the difference.

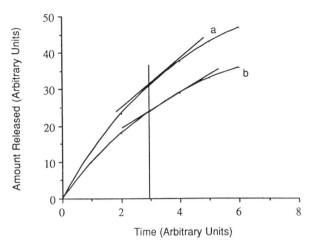

Fig. 16 Slope determination of flux of ointment release and release from ointment + membrane.

In vivo testing is usually carried out by applying the dosage form to hairless rats followed by subsequent sacrifice. Since the skin consists of several layers with differing hydrophilicity, the overall fate of the drug is of importance.

V. DISSOLUTION OF DRUG SUBSTANCE AND DOSAGE FORM

In the time path, solid dosage forms (tablets or capsules) must eventually be manufactured for the clinic (e.g., in Phase II). If possible, the drug substance per se is subjected to a dissolution test in a Wood's apparatus [17]. This test is useful, although quite dependent on hydrodynamic conditions. It consists of placing the powder in a type of tablet die, compressing the tablet, and exposing the flat, exposed side of the tablet (with surface area A) to a dissolution liquid (usually water or $N/10$ HCl) in which it has a solubility S. Under these conditions [42] the intrinsic dissolution rate constant (cm/sec) can be obtained by Eq. (9), which under sink conditions* (i.e., where C is less than 15% of S) becomes

$$C = \left(\frac{SkA}{V}\right)t \tag{39}$$

It has been suggested [43] that if k is obtained under sink conditions over a pH range of $1-8$ at $37°C$ in a USP vessel by way of Eq. (39) at 50 rpm, then if the dissolution rate constant (kA/V) is greater that 1 mg min^{-1} cm^{-2}, the drug is not prone to give dissolution–rate-limited absorption problems. On the other hand, if the value is less than 0.1, such problems can definitely be anticipated, and compounds with values of kA/V of from 0.1 to 1 mg min^{-1} cm^{-2} are in a gray area. For compound selectivity it is frequently useful to express dissolution findings in terms of k (i.e., in cm/sec).

For a small amount of powder, dissolution of the particulate material can often be assessed (and compared with that of other compounds) by placing the powder in a calorimeter [44] and

---

*Strictly speaking, sink conditions are when the amount dissolved plotted versus time yields a line that, within experimental error, is linear. When the surface area, A, is constant, then this corresponds to 15% dissolved. When the surface area changes (e.g., during particulate dissolution), then this number may be smaller.

measuring the heat evolved as a function of time. The surface area must be assessed microscopically (or by image analyzer), and the data must be plotted by a cube root equation [45]:

$$1 - \left[\frac{M}{M_0}\right]^{1/3} = \left(\frac{2kS}{\rho r}\right)t \tag{40}$$

where M is mass not dissolved; M_0, the initial amount subjected to dissolution, ρ is true density; S is solubility, and r is the mean "radius" of the particle. The method is simply comparative, not absolute, owing to the hydrodynamics being different in the calorimeter than it would be in a dissolution apparatus.

A. Biopharmaceutical Aspects

One important aspect of drug dosage form development is to obtain a dosage form that is absorbed in a desired fashion. This usually implies a rapidly and completely absorbed dosage form. This means that it is necessary to test the drug substance itself for in-vivo release characteristics. A good indication of whether a drug may give this type of problems is a comparison of LD_{50} values by the parenteral and by the oral route. If the former toxicity is much greater than the latter, there is often an absorption problem. In the following, it is assumed that the problems are dissolution dictated.

B. Partial-In Vivo Testing

The general goal is to submit an IND for the drug and to get it into testing in humans in the clinic. Frequently, biological absorption characteristics are checked by such procedures as the everted sac technique [46]. Here, a segment of the small intestine of a rat is everted. The ends are then tied off, and physiological fluid containing no drug (placebo liquid) is filled into the sac. It is then placed in a vessel containing a solution of the drug in a buffer solution. The setup is kept at 37°C, while oxygen is constantly being supplied to the solution. After a given interval, the contents of the sac are assayed for content of drug. This can then be repeated for other times. Collections of several samples from the same intestine segment is possible [47], and this greatly facilitates the procedure. Other methods, such as the method suggested by Dolusio et al. [48], exist and are used in preformulation efforts. Here, rats are anesthetized, and the ileal and duodenal ends of their intestines are then cannulized, allowing sampling and liquid introduction.

C. In Vivo Testing

For preformulation purposes, some animal testing is usually performed before Phase I. This could be in rats, dogs, or other species. The animals are being tested by being given a specific regimen (e.g., fasting, single-dose, or after each meal), and blood is then collected at various intervals. In this fashion a blood-level curve is obtained (Fig. 17). The ultimate value of this is not an absolute. To extrapolate from one species to another is a dubious undertaking, and the ramifications of this are beyond the scope of the chapter. General conclusions can be drawn, and methods are briefly described in the following.

It is customary to do one set of tests by the parenteral route and, at this stage (see Fig. 17), to assume or define that 100% of the dose is absorbed [49]. Depending on pathways, this may later be modified, but the correctness of the assumption is less important now, as will be seen shortly. Next, a solution (or if solubility is limited, a suspension) is administered, giving a different blood level (see curve II in Fig. 17). Finally, the drug is administered dry (e.g., in a capsule), giving rise to curve III in Fig. 17.

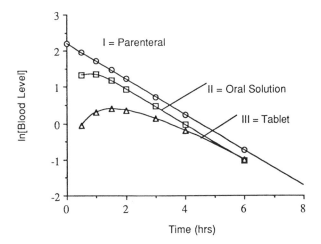

Fig. 17 Blood level curves of identical doses of a drug by parenteral route (I), by oral route as a solution (II), and orally as a tablet (III).

If the areas under the non-logarithmic curves are denoted by A, then AII/AI is the fraction absorbed by oral route. $AIII/AII$ is the fraction efficiency of the solid dosage form. The latter is because the solid dosage form has to dissolve before the drug contained in it is available for absorption. It is the latter ratio that is important to the investigating pharmaceutist; therefore, the outcome of the parenteral form is actually not a consideration from a formulations point of view. It is critical overall and, if it is low, it may, at the point of parenteral data acquisition, be advisable to stop the program and evaluate the possibility of derivatives that would give better availability.

There are large volumes of literature written on this subject, but the simplest manner of evaluating the formula efficiency by way of blood level data is the following simplified model: It is assumed that on ingestion the tablets or capsules (a) disintegrate into (N) particles, and (b) the drug then dissolves from these particles [50,51]. Each particle will release its content in T time units. The number of particles not dislodged from the dosage form at time t is

$$N_t = N_0 e^{-kt} \tag{41}$$

For this type of situation the blood levels B will be a function of time by an equation of the type

$$B = A[\exp(-k_1 t) - \exp(-k_2 t)] \tag{42}$$

where k_1 is an elimination rate constant, and k_2 is a function of dissolution and absorption rate constants. Figure 17 shows a case in point, where

$$BII = 7[\exp(-0-494t) - \exp(-2-813t)] \tag{43}$$

$$BIII = 4.48[\exp(-0.472t) - \exp(-0.85t)] \tag{44}$$

The k_1 values are about the same as they should be, and the difference, k'', between the k_2 values,

$$k'' = 2.81 - 0.85 = 2.0 \tag{45}$$

is a parameter indicative of the retardation-caused disintegration and dissolution. The areas under the non-logarithmic curves, Q, are

$$Q = \left(\frac{1}{k_1}\right) - \left(\frac{1}{k_2}\right) \tag{46}$$

That is, in Fig. 17, the areas would be 0.942 μg hr/cm^3 for the tablet and 1.67 for the solution. The formulation efficiency is 0.94/1.67 = 56% by this criterion, and the tablet formulation could stand some improvement.

REFERENCES

1. R. C. Schultz, Stability of dosage forms, FDA–Industry Interface Meeting, Washington, DC, Oct. 7, 1983; Stability guidelines, *Congressional Record*, May 7, 1984.
2. N. Rodriguez-Hornedo, Ph.D. Thesis, University of Wisconsin (1984).
3. W. J. W. Underberg and H. Lingeman, J. Pharm. Sci., 72, 553 (1983).
4. T. Parke and W. Davis, Anal. Chem., 25, 642 (1954).
5. S. K. Bakar and S. Niazi, J. Pharm. Sci., 72, 1024 (1983).
6. S. H. Yalkowsky and S. C. Valvani, J. Pharm. Sci., 72, 912 (1983).
7. W. E. Acree and J. H. Rytting, J. Pharm. Sci., 72, 293 (1983).
8. J. T. Carstensen, K. S. Su, P. Maddrell, and H. Newmark, Bull. Parenter. Drug Assoc., 25, 193 (1971).
9. A. N. Paruta and S. A. Irani, J. Pharm. Sci., 54, 1334 (1964).
10. G. Cave, F. Puisieux, and J. T. Carstensen, J. Pharm. Sci., 68, 424 (1979).
11. D. Scroby, R. Bitter, and J. Webb, J. Pharm. Sci., 52, 1149 (1963).
12. S. H. Yalkowsky, G. L. Flynn, and G. L. Amidon, J. Pharm. Sci., 61, 983 (1972).
13. G. L. Amidon, S. H. Yalkoskky, and S. Leung, J. Pharm. Sci., 63, 1858 (1974).
14. S. H. Yalkowsky, G. L. Amidon, G. Zografi, and G. L. Flynn, J. Pharm. Sci., 64, 48 (1975).
15. N. Bodor, Z. Gabanyi, and C.-K. Wong, J. Am. Chem. Soc., 111, 3783 (1989).
16. A. Noyes and W. Whitney, J. Am. Chem. Soc., 23, 689 (1897).
17. J. H. Wood, G. Catacalos, and S. Lieberman, J. Pharm. Sci., 52, 296 (1963).
18. H. Nogami, T. Nagai, and A. Suzuki, Chem. Pharm. Bull., 14, 329 (1966).
19. M. Shibata, H. Kokobu, K. Morimoto, K. Morisaka, T. Ishida, and M. Inoue, J. Pharm. Sci., 72, 1436 (1983).
20. S. H. Yalkowsky, S. C. Valvani, and T. J. Roseman, J. Pharm. Sci., 72, 866 (1983).
21. M. Pikal and A. L. Lukes, J. Pharm. Sci., 65, 1269 (1976).
22. J. T. Carstensen and R. Kothari, J. Pharm. Sci., 70, 1095 (1981).
23. J. T. Carstensen, *Pharmaceutics of Solids and Solid Dosage Forms*, Wiley-Interscience, New York, 1977, pp. 6, 41.
24. J. T. Carstensen and M. Franchini, Drug Dev. Ind. Pharm., 18, 85 (1992).
25. B. Kaye, in *The Fractal Approach to Heterogeneous Chemistry* (D. Avnir, Ed.), John Wiley & Sons, Chichester, UK, 1989, p. 62.
26. J. T. Carstensen, *Pharmaceutics of Solids and Solid Dosage Forms*, John Wiley & Sons, New York, 1977, p. 56, 226, 228.
27. J. T. Carstensen and X. P. Hou, J. Pharm. Sci., 74, 466 (1985).
28. E. H. Washburn, Phys. Rev., 17, 273 (1921).
29. J. T. Carstensen, *Pharmaceutics of Solids and Solid Dosage Forms*, Wiley-Interscience, New York, 1977, pp. 11–15.
30. L. Van Campen, G. Zografi, and J. T. Carstensen, Int. J. Pharm., 5, 1 (1980).
31. L. Van Campen, G. L. Amidon, and G. Zografi, J. Pharm. Sci., 72, 1381 (1983).
32. L. Van Campen, G. L. Amidon, and G. Zografi, J. Pharm. Sci., 72, 1388 (1983).
33. L. Van Campen, G. L. Amidon, and G. Zografi, J. Pharm. Sci., 72, 1394 (1983).
34. J. T. Carstensen and F. Usui, J. Pharm. Sci., 74, 1293 (1984).

35. J. T. Carstensen, J. B. Johnson, W. Valentine, and J. Vance, J. Pharm. Sci., 53, 1050 (1964).
36. P. R. Perrier and U. W. Kesselring, J. Pharm. Sci., 72, 1072 (1983).
37. A. D. F. Toy, Inorganic phosphorous chemistry, in *Comprehensive Inorganic Chemistry* (J. C. Bailar, Jr., H. J. Emelius, R. Nyholm, and A. F. Trotman-Dickenson, Eds.), A. Wheaton & Co., Exeter, UK, 19??, pp. 389–543.
38. J. E. Bodnar, J. R. Chen, W. H. Johns, E. P. Mariani, and E. C. Shinal, J. Pharm. Sci., 72, 535 (1983).
39. J. T. Carstensen, M. Franchini, and K. Ertel, J. Pharm. Sci., 81, 303 (1992).
40. J. T. Carstensen, *Drug Stability*, Marcel Dekker, New York, 1991, p. 60.
41. L. J. Ravin, E. S. Rattie, A. Peterson, and D. E. Guttman, J. Pharm. Sci., 67, 1528 (1978).
41. Y. W. Chien, P. R. Keshary, Y. C. Huang, and P. P. Sarpotdar, Drug Dev. Ind. Pharm., 72, 968 (1983).
42. J. T. Carstensen, in *Dissolution Technology* (L. Leeson and J. T. Carstensen, Eds.), The Academy of Pharmaceutical Sciences, American Pharmaceutical Association, Washington, DC, 1974, p. 5.
43. S. Riegelman, *Dissolution Testing in Drug Development and Quality Control*, The Academy of Pharmaceutical Sciences Task Force Committee, American Pharmaceutical Association, 1979, p. 31.
44. K. Iba, E. Arakawa, T. Morris, and J. T. Carstensen, Drug Dev. Ind. Pharm., 17, 77 (1991).
45. A. Hixson and J. Crowell, Ind. Eng. Chem., 23, 923 (1931).
46. T. H. Wilson and G. Wiseman, J. Physiol., 123, 116 (1954).
47. R. K. Crane and T. H. Wilson, J. Appl. Physiol., 12, 145 (1958).
48. J. T. Dolusio, N. F. Billups, L. W. Dittert, E. J. Sugita, and J. V. Swintosky, J. Pharm. Sci., 58, 1196 (1969).
49. J. T. Carstensen, *Pharmaceutics of Solids and Solid Dosage Forms*, Wiley-Interscience, New York, 1977, pp. 99, 101.
50. J. T. Carstensen, J. L. Wright, K. W. Blessel, and J. Sheridan, J. Pharm. Sci., 67, 48 (1978).
51. J. T. Carstensen, J. L. Wright, K. W. Blessel, and J. Sheridan, J. Pharm. Sci., 67, 982 (1978).

8

Cutaneous and Transdermal Delivery: Processes and Systems of Delivery

Gordon L. Flynn
College of Pharmacy, The University of Michigan, Ann Arbor, Michigan

I. INTRODUCTION

The skin forms the body's defensive perimeter against what is in reality a hostile external environment. As such, in the normal course of living, it suffers more physical and chemical insult than any other tissue of the body. It is inadvertently scraped, abraded, scratched, bruised, cut, nicked, and burned. Insects bite it, sting it and, occasionally, furrow through it. At times it is exposed to detergents, solvents, waterborne pollutants, and myriad other chemicals and residues. Bacteria, yeasts, molds, and fungi live on its surface and within its cracks and crevices. It is brushed, smeared, dusted, sprayed, and otherwise anointed with toiletries, cosmetics, and drugs. Any of these exposures can rile the skin or provoke allergy. If there is only minor damage associated with such insults, the skin repairs itself in short order, without a trace left of the injury. If the insult is severe, its reconstruction takes far longer and may occur with scarring. Such repair is essential, for humans cannot survive for long with an extensively damaged skin. In its intact state the skin is a formidable barrier, impenetrable to otherwise life-threatening microorganisms and resistant to chemicals and tissue-harmful ultraviolet rays. The skin also keeps us from losing life's essential chemicals and fluids to the external environment. To perform these necessary functions, the skin has to be tough and, at the same time, flexible, for it is stretched and flexed continually as we move around within it. In its healthy state, it is thus a remarkable fabric, strong and far more complex than any man-made material [1].

A myriad of medicated products are applied to the skin or readily accessible mucous membranes that in some way either augment or restore a fundamental function of the skin, or pharmacologically modulate an action in the underlying tissues. Such products are referred to as *topicals* or *dermatologicals.* A *topical delivery* system is one that is applied directly to any external body surface by inunction (spreading the formula with the fingers and rubbing it in), by spraying or dusting it on, or by instilling it (as with a dropper). Thus, the term topical is frequently used in contexts for which the application is to the surface of the eye (cornea and conjunctival membranes), the external ear, the nasal mucosa or the lining of the mouth (buccal mucosa), or even the rectum, the vagina, or the urethral lining. The term, *dermatological,* on the other hand, is limited to products that are applied to the skin or the scalp and an *external use only* label is used to denote such restricted use. The distinction between general topical use and external use is not trivial. Mucous membranes offer topically applied drugs ready access to the systemic circulation, whereas normal skin is relatively impenetrable. Therefore, many drugs and chemicals can be applied to the external skin surface safely that are unsafe to place in contact with mucosal barriers. The external use only label identifies these important differences and alerts a patient to restrict the use of a product to the true skin. This chapter deals mainly with dermatologicals, but the general concepts and rationale are applicable to other modes of topical therapy as well.

The distinctions pharmacists have to make concerning topical dosage forms and their suitability for use obviously go far deeper than merely appreciating the significance of external use only labels. Pharmacological, toxicological, and risk–benefit valuations have to be made for every drug product, but the industry together with the U.S. Food and Drug Administration (FDA) has these in hand long before the pharmacist ever sees a product. Today's dosage forms, the delivery systems for the drugs, are also industry creations subject to FDA sanction. The dispensing pharmacist's lack of input into drug development does not abrogate his or her responsibility for assuring that dispensed products conform to high standards, however. Consequently, the consummate professional takes every opportunity to evaluate, by literature or by observation, how the performances of various dosage forms measure up to absolute standards and stack up against one another. A pharmacist should thus be seeking answers to the following questions: Is the drug bioavailable as administered? Are the drug and the dosage form stable? Is the formulation free of contamination? Is it pharmaceutically elegant? The particulars relative to these attributes depend on the type of dosage form and the route of administration. Regardless, a *no* to any of the first three questions is reason to remove a product from distribution. Elegance may be sacrificed some for function, but only in degree. One of the goals here is to explicate the performance attributes of dermatological dosages forms to use as a basis for product evaluation.

II. THE STRUCTURE AND FUNCTION OF SKIN

To answer questions about the therapeutic and cosmetic uses of the myriad dermatological concoctions that are available, a pharmacist must be knowledgeable about the anatomical structure and physiological functions of the skin and the chemical compositions and physicochemical properties of its constituent phases. Furthermore, some understanding of how the latter attributes are affected by disease and damage is a must, as is knowledge of how the skin's physiology and function vary with age, environmental conditions, and other factors. Rational approaches to topical therapy rest on such insight.

A. Skin Functions

General functions of the skin are outlined in Table 1. These functions include containment of tissues and organs, multifaceted types of protection, environmental sensing, and body heat regulation. Some skin functions are inextricably entwined. For instance, containment and the barrier functions are to some extent inseparable. Active sweating is accompanied by increased peripheral blood flow which, in turn, is tied in with vascular nourishment of the cells of the skin as needed to promote their proliferation, differentiation, and specialization.

Table 1 Functions of the Skin

Containment of body fluids and tissues
Protection from harmful external stimuli (barrier functions)
    Microbial barrier
    Chemical barrier
    Radiation barrier
    Thermal barrier
    Electrical barrier
Reception of external stimuli
    Tactile (pressure)
    Pain
    Thermal
Regulation of body temperature
Synthesis and metabolism
Disposal of biochemical wastes (through secretions)
Intraspecies identification and or attraction (apocrine secretions)
Blood pressure regulation

Source: Refs. 2 and 3.

Let us consider how the skin is structured so that we might better understand how it performs some of its vital functions, beginning with the cross section of the skin sketched in Fig. 1. This illustration shows the readily distinguishable layers of the skin are, from the outside of the skin inward: (a) the ≈ 10-μm–thin, devitalized outer epidermis, called the stratum corneum; (b) the ≈ 100-μm–thin living epidermis; and (c) the ≈ 1000-μm–thin (1-mm-thin) dermis, all these stated thicknesses being representative only, for the actual thicknesses of these strata vary severalfold from place to place over the body. Dispersed throughout the skin, in various numbers and size, depending on body site, one finds several skin glands and appendages, namely (a) hair follicles and their associated sebaceous glands (pilosebaceous glands); (b) eccrine sweat glands; (c) apocrine sweat glands; and (d) nails (finger and toe). Each has unique population densities and distributions at disparate body locations. There are characteristic differences in appearances of the structures from place to place on the body as well.

A highly complex network of arteries, arterioles, and capillaries penetrates the dermis from below and extends up to the surface of, but not actually into, the epidermis. A matching venous system siphons the blood and returns it to the central circulation. The flow of blood through this vasculature is integrated with the production and movement of lymph through the dermal lymphatics. The dermis is laced with tactile, thermal, and pain sensors.

B. Stratum Corneum

The outermost layer of the skin appearing in the exploded epidermal sketch of Fig. 1 represents the stratum corneum or *horny layer* of the skin; the main element of the skin's permeation barrier, it is a multicellular, essentially metabolically inactive tissue comprised of acutely flattened, stacked, hexagonal cell building blocks formed from once-living cells. These cellular building blocks are layered 15–25 cells deep over most of the body [2]. Sometimes the cells appear stacked one on top of the next in neat columns, but on most occasions they are irregularly arranged. The stratum corneum exhibits regional differences in thickness, being as thick as several hundred micrometers on the friction surfaces of the body such as the palms and soles. However, over most of the body the thickness is approximately 10 μm, less than a fifth the thickness of an ordinary piece of paper [2,4]. It is a dense tissue, about 1.4 g/cm^3 in the

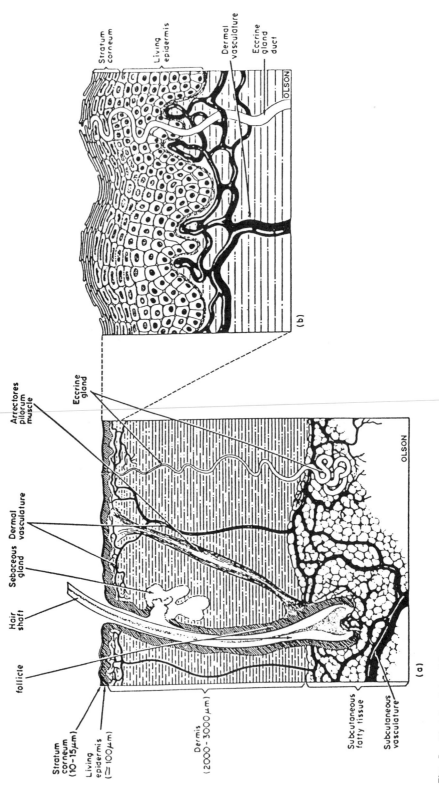

Fig. 1 Sketch of the skin.

dry state. Consequently, the stratum corneum is also occasionally referred to as the stratum compactum.

The stratum corneum is under continuous formation. Cells that are worn off the surface are replaced from beneath, one for one, with complete turnover of the layer occurring every 2 weeks in normal individuals [5]. In humans the cells that give rise to the stratum corneum originate exclusively in the basal (basement) layer of the epidermis. Thus, this layer is often referred to as the germinative (proliferative) layer. Mitosis begins an extraordinary process in which daughter cells are pushed outward, first to form the so-called spinous or prickle cell layer, and then serially, the granular, lucid, and horny layers. During their transit through the epidermal mass, the cells flatten acutely and internally synthesize the protein and lipid components that eventually characterize the fully differentiated horny layer. Individual strands of structural protein are formed, even while the cells reside in the basal layer. As the cells progress upward through the epidermis to take positions in the stratum corneum, their protein content expands, so much so that massed proteins of several kinds are distinguishable by the time the cells reach the granular layer. A basic protein here, which stains deeply to give the granular layer its characteristic histological appearance, is filagrin. Filagrin (*fil*ament *agg*regating pro-te*in*), which is released in one of the culminating events of formation of the stratum corneum, induces individual helical strands of prekeratin protein to twist together into multistranded fibers that themselves have helical geometry. These fibers, in turn, are spontaneously bundled and concentrated so that the intracellular space of the fully differentiated horny cell is literally packed full with this semicrystalline α-keratin and its amorphous keratin counterpart (β-keratin). Indeed, nothing else is seen there in the electron microscope. The intracellular space, therefore, is dense and, to a great degree, chemically impenetrable.

Lipid is also synthesized during a keratinocyte's epidermal transit and is collected in small vesicles visible in the granular layer. These were designated membrane-coating granules by microscopists long before the their content and function were known. As the granular cells transform, these vesicles migrate to the upper cell membrane, at which point their contents are passed exocytotically into the intercellular space. This lipid functionally becomes a mortar that seals the horny structure, making the stratum corneum, pound for pound, an incredibly efficient moisture barrier. Virtually all the lipid of the stratum corneum is in this interstitial space, much of it being there in liquid crystalline, bilayer arrays [6]. An exoskeleton (infrastructure) consisting of residual cell membranes, bound together by desmosomes and tonofibrils, separates these keratin and lipid domains. The lipid content of the horny layer represents about 20% of the stratum corneum's dry weight, whereas the endoskeleton amounts to roughly 5%, as indicated in Table 2 [2,5,7].

In its normal state at ordinary relative humidities, the stratum corneum takes up moisture to the extent of 15–20% of its dry weight [2]. However, on some areas of the body, this water content can increase to several multiples of the dry weight when the skin is waterlogged. The stratum corneum also becomes highly hydrated when natural evaporation of water from the skin's surface, so-called insensible perspiration, is blocked by an occlusive dressing. Regardless of how its water content is increased, this tissue becomes more pliable and molecules diffuse through it with greater facility. It is likely that some substances become more soluble in it as well. Conversely, as the stratum corneum dries out, it becomes brittle. Ultradry, inelastic horny tissue tends to split and fissure when stretched, giving rise to conditions we know as chapped lips, windburn, and dishpan hands.

The stratum corneum is thus a dense, polyphasic, epidermal sheathing made from dehydrated and internally filamented former cells held together by desmosomes, tonofibrils (intercellular anchors), and interstitial lipid. It has been estimated that the stratum corneum contains ten times the fibrous material of the living epidermis in roughly one-tenth the space [8]. At its

Table 2 Composition of the Stratum Corneum

Tissue component	Gross composition	Percentage of dry weight
Cell membrane	Lipid, protein	≈ 5
Intercellular space	Mostly lipid, some protein and polysaccharide	≈ 20
Intracellular space	Fibrous protein (≈ 65–70%), non-fibrous (soluble) protein (≈ 5–10%)	≈ 75
Overall protein	Water soluble (10%), keratin (≈ 65%), cell wall (≈ 5%)	70–80
Overall lipid		10–20
All other		Up to 10
Water (normal hydration)		15–20
Water (fully hydrated)		Upwards of 300

Source: Data from Refs. 2 and 4.

undersurface the stratum corneum is in contact with the living epidermal mass, and at its other surface with the environment. The underlying cells contain water with the high thermodynamic activity of the physiological milieu. As a rule, air at the surface of the skin is dry and of low water activity. Consequently, water diffuses outwardly (*down* this activity gradient), eventually escaping into the environment, a process known as *insensible perspiration*. Over normal skin about 5 ml of water is lost this way per square meter of body surface per hour (or 0.5 mg cm^{-2} hr^{-1}) [9]. Since an adult has upward of 2 m^2 of body surface area, whereas an infant has roughly 0.25 m^2 and a 2-year-old 0.75 m^2 or so, an adult loses about 250 ml (8 oz) of water per day by this mechanism. A small child loses less than half this amount. Such water loss can increase alarmingly and to as much or more than 100 ml hr^{-1} m^{-2} over skin ravaged by disease or damage. High evaporation rates such as this, over expansive areas, carry away sufficient body heat to lower the core temperature to dangerous levels.

C. Viable Epidermis

The animate cells of the epidermis have a readily definable upper interface with the lifeless stratum corneum and an even more clearly demarcated, deep interface with the dermis (see Fig. 1). In drug delivery considerations, the whole of this tissue is considered as a singular diffusional field, although, when viewed under microscope, it is clearly multilayered. The identifiable strata are, from bottom to top, the basal layer [stratum (s.) germinativum], a mono-layer of cubical or columnar cells, otherwise unremarkable in appearance; the spinous or prickle layer (s. spinosum), where cells exhibit sharp surface protuberances; the granular layer (s. granulosum), which takes up stain to yield a mottled appearance; and, in some histological displays, the lucid layer (s. lucidum), which appears translucent. These layers reflect a progressive differentiation in the cells, which eventuates in their death and placement within the horny structure. When physicochemically considered, the viable epidermis is nothing more than a wedge of tightly massed cells, literally small compartments of cytoplasm encapsulated within delicate cell membranes, held together by tonofibrils. Here, as elsewhere in the body, water has an activity equivalent to that of a highly dilute NaCl solution (0.9% NaCl). Relative to diffusion, the density and consistency of this cell composite are only a little greater than found with water.

The viable epidermis makes a flat interface with the stratum corneum. Its interface with the dermis is papillose (mounded). Myriad tiny bulges of the epidermis fit with exacting reciprocity over dermal depressions and ridges. It is these ridges that give the friction surfaces of the body their distinctive patterns (e.g., fingerprints). And, since hair follicles and eccrine glands have epidermal origins, epidermal cells actually extend through the dermis and into subcutaneous tissue by way of these tiny glands (see Fig. 1). This has a generally unappreciated significance in terms of the self-repair capabilities of the skin. As long as an injury does not extend to the base of the glands, islands of vital cells capable of regenerating a scar-free skin surface remain available for repair. Discounting these deep rootages, the epidermis is on the order of 100 μm thick [10].

The principal cells of the epidermis are its keratinocytes. One also finds Langerhans cells —cells of white blood cell progeny—at regular intervals within the bulk of the epidermal mass, and melanocytes strategically placed in the basal layer just above the epidermal–dermal junction (Table 3). Langerhans cells function as antigen-presenting cells in the skin's immunological responses. Under the influence of melanocyte-stimulating hormone (MHS), melanocytes synthesize the pigment that gives the races of humans their unique skin colorations. Melanocytes are also set into action by ultraviolet radiation, leading to suntanning. Other cells, occasionally seen in skin sections, are migrant macrophages and lymphocytes. These are particularly numerous when the skin is traumatized.

D. Dermis

The dermis, as depicted in Fig. 1, is a nondescript region lying between the epidermis and the subcutaneous, fatty region. In reality it is a complex structure, consisting mainly of a meshwork of structural fibers, collagen, reticulum, and elastin, filled with a mucopolysaccharidic gel, called the ground substance [2]. Approximate proportions of these phases are indicated in Table 4. The dermis ranges from 1 mm (1000 μm) to 5 mm in thickness [11]. The upper one-fifth or so of the tissue, the papillary layer by name, is finely structured and is the support for the delicate capillary plexus that nurtures the epidermis. The papillary dermis eventually merges with the far coarser fibrous matrix of the reticular dermis. This deeper layer is the main structural element of the dermis and, for that matter, of the skin. Of considerable importance, the microcirculation that subserves the skin is entirely located in the dermis. The dermis is also penetrated by a network of sensory nerves (pressure, temperature, and pain) and a rich lymphatic network. Numerous fibroblasts, cells that synthesize the structural fibers, are found here [2], and one also finds mast cells scattered about (see Table 3). The latter are thought to play

Table 3 Cells of the Skin

Cell type	Principal function
Cells of the epidermis	
Keratinocytes	Form keratinized structures
Langerhans cells	Antigen presentation
Melanocytes	Pigment synthesis
Macrophages, lymphocytes	Migrant cells, immune responses
Cells of the dermis	
Fibroblasts	Fiber synthesis
Mast cells	Make ground substance, histamine
Blood cells	
Endothelial cells	Form the blood vessels
Nerve cells and endings	Sensors

Source: Data from Refs. 2 and 11.

Table 4 Composition of the Dermis

Component	Approximate % composition
Collagen	75.0
Elastin	4.0
Reticulin	0.4
Ground substance	20.0

Source: Data from Refs. 2 and 11.

a role in synthesizing ground substance and are known to be a source of the histamine that is released when the skin is immunologically provoked.

E. Skin's Circulatory System

Arteries entering the skin arise from more substantial vessels located in the subcutaneous connective tissue. These offshoots form a plexus just beneath the dermis [11]. Branches from this subcutaneous network supply blood directly to the hair follicles, the glandular appendages, and the subcutaneous fat. Branches to the upper skin from this deep plexus divide again within the lower dermis, forming a deep subpapillary network. Arterioles reaching the upper dermis out of this more distal plexus are on the order of 50 μm in diameter and exhibit arteriovenous anastomoses, shuntlike connections that link the arterioles directly to corresponding venules. The dermal arterioles then branch to form a shallower, subpapillary plexus that supplies twigs to the dermis and fine-branched capillary loops to the papillae at the dermal–epidermal interface. The epidermis itself is avascular.

The veins of skin are organized along the same lines as the arteries in that there are both subpapillary and subdermal plexuses [11]. The main arteriole communication to these is the capillary bed. Copious blood is passed through capillaries when the core body is either feverish or overheated, far more than needed to sustain the life-force of the epidermis, and this rich perfusion lends a red coloration to skin. When there is the opposite physiological need, the capillary bed is short-circuited, and blood is passed directly into the venous drainage by way of the arteriovenous anastomoses. Fair skin blanches when this occurs. These mechanisms act, in part, to regulate body temperature and blood pressure.

The vascular surface available for exchange of substances between the blood and local tissue has been estimated to be of the same magnitude as that of the skin (i.e., 1–2 cm^2/cm^2 of skin). At room temperature, about 0.05 ml of blood flows into the skin per minute per gram of the tissue; the supply increases considerably when the skin is warmer [4,12]. Ordinarily, sufficient blood reaches to within 150 μm or so of the skin's surface to efficiently draw chemicals into the body that have percutaneously gained access to this depth [7]. Interestingly, this local circulation is turned down by vasoconstrictors (e.g., glucocorticoids) and up by vasodilators (e.g., nicotine), respectively. Such responses are so reliable that vasoconstriction (blanching) has become an FDA-sanctioned index of corticosteroid penetration of the skin [13,14]. The relation between blood flow and local clearance of percutaneously absorbed drugs, including aspects of vasoconstriction and vasodilation, is not well known.

Lymphatics of the skin extend up and into the papillary layers of the dermis. A dense, flat meshwork of lymphatic capillaries is found here [11]. These lymphatic vessels pass to a deeper network at the lower boundary of the dermis. Serum, macrophages, and lymphocytes easily pass through the interfaces of the skin's lymphatic and vascular networks.

F. Skin Appendages

Hair follicles and their associated sebaceous glands (pilosebaceous glands), eccrine glands, apocrine glands, and nail plates are referred to as the skin's appendages. Hair follicles are found within the skin everywhere except the soles, the palms, the red portion of the lips and the external genitalia. They are formed from epidermal cells in fetal life. From place to place, the follicles and the hair they produce differ markedly in prominence. Delicate primary hair is found on the fetus; secondary hair or *down* covers the adult forehead; terminal hair ordinarily blankets the scalp, and it covers the pubic region and underarms [2]. A hair (hair shaft) emerges from each *follicle*, as shown in Fig. 1. The follicle itself lies within the skin and consists of concentric layers of cellular and noncellular components positioned in the skin at a slight angle. Each follicle is anchored to the surrounding connective tissue by an individual strand of smooth muscle, the arrector pilorum, contraction of which causes the hair to stand upright, raising goose pimples on human skin. In animals like cats, hair stands on end as part of the flight-or-flight response.

The hair shaft is formed continuously by cell division, differentiation, and compaction within the bulb (base) of each active hair follicle, a process that is completed deep in the follicle. Hair, like stratum corneum, is thus a compact of fused, keratinized cells. Collectively, hair follicles occupy about 1/1000 of the skin's surface [10,15], a factor that sets a limit on the role they can play as a route of penetration. Each hair follicle possesses one or more flasklike sebaceous glands (see Fig. 1). These have ducts that vent into the open space surrounding the hair shaft just below the skin's surface. Just as with keratinocytes, the cells of sebaceous glands, *sebocytes*, are programmed to differentiate and die. Before they die and disintegrate, they pack themselves full of lipid-containing vesicles. The residue left behind at their death is mixed with other follicular debris deep within the follicular orifice to form the actual substance, sebum, expressed onto the skin. Sebum is then forced upward around the hair shaft and onto the skin's surface through outlets having diameters ranging from 200 to 2000 μm (2 mm), depending on body location [11]. Glands with the largest openings are found on the forehead, face, nose, and upper back. These contain only a tiny hair if they contain one at all.

Eccrine or salty sweat glands are found over the entire body, except the genitalia. Of fetal epidermal origin, they consist of tubes extending from the skin surface to the footings of the dermis. Here the tube coils into a ball roughly 100 μm in diameter (see Fig. 1) [11]. By anatomical count, there are between 150 and 600 glands per cubic centimeter of body surface, depending on the site [16]. They are particularly concentrated in the palms and soles, attaining a densities in these locations well in excess of 400 glands per centimeter. However, since many of the glands remain dormant, estimates of their numbers are appreciably lower if based on actual sweating. Each gland has a microscopic orifice within the surface of the skin of about 20 μm diameter from which its secretions are spilled. In total, these glandular openings represent approximately 1/10,000 of the skin's surface [10]. Eccrine sweat is a dilute (hypotonic), slightly acidic (pH $\approx$ 5.0 owing to traces of lactic acid) aqueous solution of salt. Its secretion is stimulated when the body becomes overheated through warm temperatures or exercise. Evaporation of the water of the sweat cools the body's surface and, thereby, the body. Since the gland is innervated by the autonomic nervous system, eccrine sweating is also stimulated emotionally (the clammy handshake).

Apocrine glands have highly regionalize locations and are found only in the axillae (armpits), in anogenital regions, and around the nipples. Along with other secondary sexual characteristics, the glands develop at puberty. We know that they are innervated emotionally and through concupiscence. In the mature female, they exhibit cyclic activities in harmony with the menstrual cycle. Similar to eccrine glands, they are coiled tubular structures, but they are roughly ten times larger. Therefore, they extend entirely through the dermis and well into the subcutaneous layer [2,11]. Each gland is paired up with a neighboring hair follicle, and its secretion is vented into the sebaceous duct of the follicle beneath the surface of the skin. Because the secretion of the apocrine gland is combined with sebum before reaching the skin's surface, its chemical makeup is a enigma. What is not a mystery is that bacterial decomposition of the secretion is responsible for human body odor.

III. SKIN FUNCTIONS

The skin's main physiological roles are outlined in Table 1. Of these, the chemical barrier function is central to the use of topical drugs because deposition of a topical drug into the deeper, living strata of skin is a prerequisite for achieving its pharmacological effect. Degeneration in some of the functions can be pathognomonic of disease. Even when specific functions do not relate materially to the skin's state of health, they are tied in with cosmetic practices and, thereby, are of interest to pharmacists.

A. Containment

The containment function relates specifically to the ability of the skin to confine underlying tissues and restrain their movements. The skin draws the strength it needs to perform this mechanical role from its tough, fibrous dermis [2]. Ordinarily, the skin is taut, even when under resting tension, yet it stretches easily and elastically when the body is in motion, quickly returning to normal contours when the stretching ceases. This extensibility of the skin is attributable to an alignment of collagen fibers, under tension and in the direction of a load, that are otherwise nonaligned in the ground tension state. Elastin fibers attached to individual collagen strands relax and, in doing so, restore the irregular order of the restive state. As one ages, the resilience of these dermal fibers decreases and the tensile strength of the tissue increases. Eventually, the skin becomes stretched beyond its ability to elastically restore its initial condition and it folds over itself or *wrinkles*. Lost elasticity is advanced through extended exposure to ultraviolet radiation (sunlight); thus wrinkling is often pronounced on dedicated sunbathers.

The behavior of the epidermis when distended is also of importance. Obviously, this layer should not be torn or broken when placed under mechanical stress, for an intact epidermis is the body's first line of defense against infection. It is the stratum corneum's role to fend against tearing [2]. Pound for pound, this tissue is actually stronger than the dermal fabric and, as a rule, it is sufficiently elastic to adjust to stretching. Its pliability, however, is conditional, and it fissures and cracks if stretched when excessively dry. Arid atmospheres alone can produce this condition (windburn). Detergents and solvents, which extract essential, water-sequestering lipids from the stratum corneum, and diseases, such as psoriasis, that are associated with a malformed horny structure, render the stratum corneum brittle and prone to fissuring.

Although much is still to be learned about the factors that contribute to the pliability of the stratum corneum, it is generally accepted that its elasticity is dependent on a proper balance of lipids, hygroscopic, water-soluble substances, and water, all in conjunction with its keratin proteins. Water is its principal plasticizer, or softening agent, and it takes roughly 15% moisture to maintain adequate pliability. The capacity of the stratum corneum to bind and hold onto water is greatly reduced by extracting it with lipid solvents such as ether and chloroform. Moreover, there is a further significant decrease in the water-binding capacity of callus, a thickened stratum corneum found on the palms and soles, when it is extracted with water after having first been treated with a lipid solvent. The latter observation seems to tell us that amino acids, hydroxy acids, urea, and inorganic ions, cosmetically referred to at the skin's *natural moistening factor*, and the stratum corneum's lipids both assist the stratum corneum in retaining moisture necessary to plasticize its mosaic, filamented matrix. In effect the water makes the tissue less crystalline through its interposition between polymer strands.

B. Microbial Barrier

Normal stratum corneum, taken in its entirety, is a dense, molecular continuum penetrable only by molecular diffusion. It is virtually an absolute barrier to microbes, preventing them from reaching the viable tissues and an environment suitable for their growth. The outermost stratum corneum is continuously being shed in the form of microscopic scales (natural desquamation) and, to a limited depth, is laced with tiny crevices. Many microorganisms—pathogens and harmless forms alike—are found in these rifts. Surgeons know well that superficial washing is insufficient to remove these surface microbes; therefore, the surgical scrub is an energetic and intense cleansing with a disinfectant soap. The microorganisms residing on and in the skin can and do initiate infections if seeded into living tissues as a result of abrasive or disease-induced stratum corneum damage. Consequently, antiseptics and antibiotics are widely used to chemically sanitize wounds.

Beyond physical barrier protection, several natural processes lead to skin surface conditions unfavorable to microbial growth. Both sebaceous and eccrine secretions are acidic, lowering the surface pH of the skin below that welcomed by most pathogens. This *acid mantle* (pH ≈ 5) [16] is moderately bacteriostatic. Sebum also contains a number of short-chain fungistatic and bacteriostatic fatty acids, including propanoic, butanoic, hexanoic, and hepatanoic acids [17]. That the skin's surface is dry also offers a level of protection. It comes as no surprise that fungal infections and other skin infections are more prevalent in the skin's folds during warm weather, as intensified sweating leaves the skin continually moist in these regions.

Glandular orifices provide possible entry points for microbes. The duct of the eccrine sweat gland is tiny and generally evacuated. Experience tells us that this is not an easy portal of entry, although localized infection is seen occasionally in infants suffering prickly heat. Pilosebaceous glands seem more susceptible to infection, particularly those on the forehead, face, and upper back, referred to as sebaceous follicles. Glands at these specific locations have an almost imperceptible hair surrounded by a massive sebaceous apparatus and are especially prone to occlusion and subsequent infection (acneform pimples and blackheads). Such sebaceous gland infections are usually localized. However, if the infected gland ruptures and spews it contents internally, deep infection is possible. The body defends against this by walling off the lesion (forming a sac or *cyst*) and then destroying the eliminating the infected tissue. The destruction caused by cystic acne is deep, so much so that facial scarring is associated with it. In hair follicles containing prominent hairs, the growing hair shaft acts as a sebum conveyer that unblocks the orifice. It may be strictly coincidental, but such follicles seem less prone to clogging and infection.

C. Chemical Barrier

The intact stratum corneum also acts as a barrier to chemicals brought into contact with it. Its diffusional resistance is orders of magnitude greater than found in other barrier membranes of the body. Externally contacted chemicals can, in principle, bypass the stratum corneum by diffusing through the ducts of the appendages. The ability of each chemical to breach the skin and the diffusional route or routes it takes are dependent on its own physicochemical properties and the interactions it has within the skin's various conduit regimes. Being central to the effectiveness of dermatological products, exposition of the skin's barrier properties is made in a following section.

D. Radiation Barrier

Ultraviolet wavelengths of 290–310 nm from the UV-B band of radiation constitute the principal tissue-damaging rays of the sun that are not fully atmospherically filtered. One hour of exposure to the summer sun and its damaging rays can produce a painful burn, with a characteristic erythema. The skin has natural mechanisms to prevent or minimize such sun-induced trauma, but it takes time to set these into place. On stimulation by ultraviolet rays, particularly longer, lower-energy rays above 320 nm, melanocytes at the epidermal–dermal junction produce the pigment, melanin. Melanin's synthesis begins in the corpus of the melanocyte, with pigment-forming granules migrating outward to the tips of the long protrusions of these starlike cells. Adjacent epidermal cells endocytotically engulf these projections. Through this cellular cooperation, melanin, which absorbs and diffracts harmful UV rays, becomes dispersed throughout the epidermis, and a person tans, with his or her capacity to sunburn declining accordingly. It should be realized that tanning takes time, several days in fact, and is incapable of protecting a person on first exposure. Damaging ultraviolet exposure also stimulates epidermal cell division and thickening of the epidermis (acanthosis). Such thickening, too, takes several days. When effected, it also lends protection to the underlying tissues.

Pharmacists should tell their sun-deprived, fair-skinned patrons not to spend more than 15–20 min in the midday sun (10:00 a.m. to 3:00 p.m.) on first exposure when traveling to vacation spots such as Florida [18]. This is ample, safe exposure to initiate the tanning response in those who are able to tan. Exposures can be increased incrementally 15 min/day until a 45-min tolerance is developed, which is generally an adequate level of sun protection in conjunction with the use of sunscreens. It should be obvious that dark-skinned people are already heavily pigmented and, thus, far less susceptible to burning. Other individuals do not tan at all and must apply sunscreens with high protection factors before sunbathing.

E. Electrical Barrier

Dry skin offers a high impedance to the flow of an electrical current [3]. Stripping the skin, by successively removing layers of the stratum corneum with an adhesive tape, reduces the electrical resistance about sixfold, which tells us that the horny layer is the skin's prime electrical insulator. Its high impedance complicates the measurement of body potentials, as is done in electroencephalograms and electrocardiograms. Consequently, electrodes having large contact areas are used to monitor the brain's and the heart's electrical rhythms. Granular salt suspensions or creams and pastes containing high percentages of electrolytes are placed between the electrode surface and the skin to assure the electrical conductance is adequate to make the measurements.

F. Thermal Barrier and Body Temperature Regulation

The body is basically an isothermal system fine-tuned to 37°C (98.6°F). The skin has a major responsibility in temperature maintenance. When the body is exposed to chilling temperatures that remove heat faster than the body's metabolic output can replace it, changes take place in the skin to conserve heat. Conversely, when the body becomes overheated, physiological processes come into play that lead to cooling.

The skin's mechanism of heat conservation involves its very complex circulatory system [2,3]. To conserve heat, blood is diverted away from the skin's periphery by way of the arteriovenous anastomoses. The blood's external-most circulation is effectively shut down, leading to a characteristic blanching of the skin in fair-skinned individuals. Less heat is irradiated and convectively passed into the atmosphere. Furry mammals have yet another mechanism to conserve body heat. Each tiny arrector pilorum stands its hair up straight, adding appreciable thickness to the insulating air layer entrapped in the fur, reducing heat loss.

When the body is faced with the need to cast out thermal energy, the circulatory processes are reversed, and blood is sent coursing through the skin's periphery, maximizing radiative and convective heat losses. This process produces a reddening in light skin, a phenomenon that is particularly noticeable following strenuous exercise. Exercise also leads to profuse eccrine sweating, a process that is even more efficient in heat removal. Watery sweat evaporates, with the heat attending this process (heat of vaporization) cooling the skin's surface. Factors that accelerate the evaporative process, such as the gentle flow of air produced by a fan, accelerate cooling. Low humidity favors evaporation, and one is more comfortable and sweating is less noticeable when the air's moisture content is low. Pharmacist's should be aware that eccrine sweating is a vital process not to be tampered with. Coverage of the body with a water-impermeable wrapping, as has occasionally been done in faddish weight control programs, may result in hyperthermia, particularly if there is concurrent exercise. In its extreme, hyperthermia can be fatal.

IV. RATIONALE FOR TOPICALS

One's grasp of topical dosage forms and their functioning can be nicely organized into several broad usage categories. For instance, many products exist to augment the skin barrier (Table 5). Sunscreens and anti-infectives obviously do this. The barrier is made pliable and restored in function by emollients. Pastes are sometimes used to directly block out sunlight and, at other times, to sequester irritating chemicals that would otherwise penetrate into the skin. Even insect repellants add function to the barrier.

A second general purpose of topical application involves the selective access drugs have to epidermal and dermal tissues when administered this way. Penetration of the skin can drench the local tissues with the drug before its systemic dissemination and dilution. As a result, the drug's systemic levels are kept low and pharmacologically inconsequential. In contrast, systemic treatment of local conditions bathes highly blood-perfused tissues with the drug first, with the drug's systemic effects or its side effects sometimes overpowering the actions sought for it in the skin.

In a few instances, drugs are applied to the skin to actually elicit their systemic effects. This is called transdermal therapy. Transdermal therapy is set apart from local treatment on several counts. It is only possible with potent drugs that are also highly skin permeable. To be used transdermally, compounds must be free of untoward cutaneous actions as well. When these

Table 5 Barrier Augmentation by Topical Products

Product type	Barrier effect
Sunscreens	Enhance radiation barrier
Topical anti-infectives	Augment microbial barrier
Emollients	Moisturize stratum corneum, restore barrier
Insect repellents	Add a chemical barrier to insects
Poison ivy products	Negate antigens, augment chemical barrier
Diaper rash products	Build up moisture and chemical barrier

demanding conditions are met, transdermal therapy offers an excellent means of sustaining the action of a drug. Transdermal delivery also skirts frequently encountered oral delivery problems, such as first-pass metabolic inactivation and gastrointestinal upset. Transdermal therapy is actually an old medical strategy, as compresses and poultices have been used for centuries, although never with certainty of effect. The current, effective use of small adhesive patches to treat systemic disease or its symptoms has revolutionized the practice.

A. Therapeutic Stratification of the Skin

How does a person best organize his or her thinking relative to these different rationales? One can start by asking what the topical drug is supposed to do. Is it to be applied to suppress inflammation?—eradicate infectious microorganisms?—provide protection from the sun?—stop glandular secretions?—provide extended relief from visceral pain? Regardless of which feat the drug is to perform, the answer to the question directs us to where and sometimes how the drug must act to be effective or to a *target* for the drug. Once knowing the locus of action, one can then consider its accessibility. Clearly, if the drug cannot adequately access its target, little or no therapeutic benefit will be realized.

Sundry drug targets exist on, within, or beneath the skin. These include (a) the skin surface itself (external target), (b) the stratum corneum, (c) any one of several levels of the live epidermis, (d) the avascular, upper dermis, (e) any one of several deeper regions of the dermis, (f) one or another of the anatomically distinct domains of the pilosebaceous glands, (g) eccrine glands, (h) apocrine glands, (i) the local vasculature, and, following systemic absorption, (j) any of numerous internal tissues. As these targets become increasingly remote, delivery to them becomes sparser as the result distributional dilution and, consequently, adequacy of delivery becomes less certain. Moreover, the specific properties of these targets and their negotiability are very much determined by the state of health of the skin. Disease and damage alter the barrier characteristics of the skin and, therefore, target accessibility itself.

Causes of skin damage or eruptions are diverse and may alternatively be traced to mechanical damage, irritant or allergic reactions, an underlying pathophysiological condition, or an infection. Depending on the problem, the entire skin, or only a small part of it, may be involved. Moreover, disease may be manifest in one part of a tissue as a consequence of a biochemical abnormality in another. For instance, the cardinal expression of psoriasis is its thickened, silvery, malformed stratum corneum (psoriatic scale), but the disease actually results from maverick proliferation of keratinocytes in the germinal layer of the epidermis. Mankind suffers many skin problems, such as this, each unique in expression to the well-trained eye. The names of some common afflictions are listed in Table 6, with indication of the tissue source of the problems. Table 7 adds to the lexicon pathophysiological terms used to describe the expressions of disease. Irrespective of their fundamental tissue origins, most diseases fan out and involve other tissue components. Inflammation and skin eruption are common sequelae. The nature and developing pattern of a skin eruption become the determinants of its diagnosis. There are subtleties, and it often takes a dermatologist to make a proper differential diagnosis.

The pharmacist will, from time to time, be called on to examine an eruption or condition and make recommendation for treatment. If, and only if, the condition is unmistakable in origin, delimited in area, and of modest intensity, should the pharmacist recommend an over-the-counter (OTC) remedy for its symptomatic relief. Physicians neither need nor want to see inconsequential cuts, abrasions, or mosquito bites, nor unremarkable cases of chapped skin, sunburn, or poison ivy eruption, and so on. However, if infection is present and at all deep-seated, or if expansive areas of the body are involved, otherwise minor problems can pose a

Table 6 Common Afflictions: Brief Outline of Common Dermatological Disorders and Other Common Skin Problems

Skin problems	Examples
I. General involvements	
A. Physical damage	
1. Blunt instrument	Contusion, bruise
2. Sharp instrument	Cut, nick, animal bite
3. Scraping, rubbing	Abrasion, blister
4. Heat	Burns (1°, 2°, 3°), blister
5. Ultraviolet radiation	Sunburn
6. Insects	Mosquito bite, bee sting, ticks, mites (chiggers), lice, crab lice
B. Chemical damage	
1. Contact dermatitis	Poison ivy, poison oak
2. Contact allergy	Cosmetic dermatitis
3. Solvent extraction	"Dishpan hands"
II. Abnormalities of the epidermis	
A. Stratum corneum	
1. Tardigrade sloughing and thickening	Ichthyosis
2. Hyperdryness	Chapping, windburn
3. Hyperproliferative thickening, abnormal structural organization	Psoriasis
B. Viable epidermis	
1. Cell damage and inflammation	Eczema, general dermatitis
2. Fluid collection	Blister
3. Abnormal cell growth (not division)	Keratosis
4. Thickening of granular layer	Lichen planus
5. Hyperproliferation, incomplete keratinization	Psoriasis
6. Malignancy	Epithelioma
III. Abnormalities of the dermis and dermal–epidermal interface	
A. Melanocyte abnormalities	
1. Hyperfunction	Tanning, chloasma, freckles
2. Hypofunction	Vitiligo
3. Abnormal growth	Mole
4. Malignancy	Melanoma
B. Dermal–epidermal interface	
1. Lifting of the epidermis	Dermatitis hypetiformis
2. Overgrowth of papillary layer	Warts
C. Dermis	
1. Vascular reactions	Urticaria, hives
2. Abnormal growth of fibrinocyte	Scar, keloid
3. Abnormal polymerization	Scleroderma, lupus erythematosus
IV. Abnormalities of the glands (appendages)	
A. Hair follicle	
1. Hyperactivity	Hirsutism
2. Hypoactivity	Alopecia, baldness
B. Sebaceous glands	
1. Hyperactivity	Seborrhea
2. Occlusion	Acne, pimples
C. Eccrine sweat gland	
1. Hyperactivity	Hyperhidrosis
2. Occlusion, inflammation	Miliaria (pricky heat, heat rash)
V. Infectious diseases	
A. Bacterial	Carbuncles (boils)
B. Fungal	Athlete's foot, ringworm
C. Viral	Chickenpox, herpes simplex (cold sores)
D. Protozoal	Topical amebiasis

Source: Refs. 16 and 19.

Table 7 Pathophysiological Terms: Brief Definitions of Select Pathophysiological Terms

Term	Definition
Acne	Inflammatory disease of the sebaceous glands characterized by papules, comedones, pustules, or a combination thereof
Alopecia	Deficiency of hair
Bulla	Large blister or vesicle filled with serous fluid
Chloasma	Cutaneous discoloration occurring in yellow-brown patches and spots
Comedo (pl. comedones)	Plug of dried sebum in the sebaceous duct; blackhead
Dermatitis	Inflammation of the skin
Dermatitis herpetiformis	Dermatitis marked by grouped erythematous, papular, vesicular, pustular, or bullous lesions occurring in varied combinations
Eczema	An inflammatory skin disease with vesiculation, infiltration, watery discharge, and the development of scales and crusts
Hirsutism	Abnormal, heavy hairiness
Ichthyosis	A disease characterized by dryness, roughness, and scaliness of the skin caused by hypertrophy of the stratum corneum
Infiltration	An accumulation in a tissue of a foreign substance
Keloid	Growth of the skin consisting of whitish ridges, nodules, and plates of dense tissue
Keratosis	Any horny growth
Lichen planus	Inflammatory skin disease with wide, flat papules occurring in circumscribed patches
Lupus erythematosus	A superficial inflammation of the skin marked by disklike patches; with raised reddish edges and depressed centers, covered with scales or crusts
Miliaria	An acute inflammation of the sweat glands, characterized by patches of small red papules and vesicles, brought on by excessive sweating
Nodule	Small node that is solid to the touch
Papilla	Small, nipple-shaped elevation
Psoriasis	A skin disease characterized by the formation of scaly red patches, particularly on the extensor surfaces of the body (elbows, knees)
Pustule	Small elevation of the skin filled with pus
Scleroderma	A disease of the skin in which thickened, hard, rigid, and pigmented patches occur with thickening of the dermal connective tissue layer
Seborrhea	A disease of the sebaceous glands marked by excessive discharge
Urticaria	Condition characterized by the appearance of smooth, slightly elevated patches, whiter than the surrounding skin
Vesicle	Small sac containing fluid; a small blister
Vitiligo	A skin disease characterized by the formation of light-colored (pigment-free) patches

serious threat, and physician referral is mandatory. Patients should also be directed to counsel with a physician whenever the origins of a skin problem are in question.

B. Surface Effects

Of the many possible aforementioned dermatological targets, the skin surface is clearly the easiest to access. Surface treatment begins at the fringe of cosmetic practice. Special cosmetics are available to hide unsightly blemishes and birthmarks. These lessen self-consciousness and are psychologically uplifting. Applying a protective layer over the skin is sometimes desirable. For example, zinc oxide pastes are used to create a barrier between an infant and its diaper which adsorbs irritants found in urine, ameliorating diaper rash. These same pastes literally block out the sun and, at the same time, hold in moisture, protecting the ski enthusiast from facial sun and wind burns on the high slopes. Transparent films containing ultraviolet light-absorbing chemicals are also used as sunscreens. Lip balms and like products lay down occlusive (water-impermeable) films over the skin, preventing dehydration of the underlying stratum corneum and, thereby, allaying dry skin and chapping. The actions of calamine lotion and other products of the kind are limited to the skin's surface. The suspended matter in these purportedly binds urushiol, the hapten (allergen) found in poison ivy and oak. However, these may best benefit the patient by drying up secretions, relieving itchiness. In all these instances for which the film itself is therapeutic, bioavailability has little meaning.

Bioavailability does matter with topical antiseptics and antibiotics, even though these also act mainly at the skin's surface. These anti-infectives are meant to stifle the growth of surface microflora; thus, formulations that penetrate into the cracks and fissures of the skin where the microorganisms reside are desirable. The extent to which the surface is sanitized then depends on uptake of the anti-infective by the microbes themselves. Slipshod formulation can result in a drug being entrapped in its film and inactivated. For instance, little to no activity is to be expected when a drug is placed in a vehicle in which it is highly insoluble. Ointment bases that contain salts of neomycin, polymyxin, and bacitracin are suspect here, in that hydrocarbon vehicles are extremely poor solvents for such drugs. Inunction (rubbing in) may release such drugs, but the pharmacist should seek evidence that such formulations are effective before recommending them.

Deodorants are also targeted to the skin surface to keep microbial growth in check. Here they slow or prevent rancidification of the secretions of apocrine glands found in and around the axillae (armpits) and the anogenital regions. Medicated soaps also belong in this family.

C. Stratum Corneum Effects

The stratum corneum is the most easily accessed part of the skin itself, and there are two actions targeted to this tissue: namely, emolliency, the *softening of the horny tissue*, which comes about through remoisturizing it; and keratolysis, the chemical digestion and removal of thickened or scaly horny tissue. Tissue needing such removal is found in calluses, corns, and psoriasis, and as dandruff. Common agents such as salicylic acid and, to a lesser extent, sulfur, cause lysis of the sulfhydral linkages holding the keratin of the horny structure together, leading to its disintegration and sloughing.

It has been mentioned that elasticity of the stratum corneum depends on its formation and on the presence of adequate natural lipids, hygroscopic substances, and moisture [19,20]. Remoisturization (emolliency) can be induced by simply occluding the surface and blocking insensible perspiration. However, it is best accomplished by lotions, creams, or waxy formulations, or combinations thereof (e.g., lip balms) that replenish lost lipid constituents of the stratum corneum. The fatty acids and fatty acid esters these contain partly fill the microscopic cracks and crevices in the horny layer, sealing it off, stabilizing its bilayer structures, and allowing it to retain moisture. Many emollient products also contain hygroscopic glycols and polyols to replenish and augment natural moisturizing factors of this kind, also assisting the stratum corneum in retaining moisture.

The introduction of moisturizing substances into the stratum corneum is ordinarily a straightforward process. Deposition of keratolytics, on the other hand, is not as easily achieved, as these agents must penetrate into the horny mass itself. Some salicylic acid-containing corn removers, therefore, are made up as concentrated nonaqueous solutions in volatile solvents. As these volatile solvents evaporate, drug is concentrated in the remaining vehicle and, thereby, thermodynamically driven into the tissue. These many examples illustrate that when the therapeutic target is at the skin's surface or is the stratum corneum, the therapeutic rationale behind the treatment usually involves enhancing or repairing or otherwise-modulating barrier functions (see Table 5).

D. Drug Actions on the Skin's Glands

A few products moderate operation of the skin's appendages. These include antiperspirants (as opposed to deodorants), which use the astringency of chemicals such as aluminum chloride to reversibly irritate and close the orifice of eccrine glands [21], impeding the flow of sweat. Astringents also decimate the population of surface microbes, explaining their presence in deodorants. The distinction between antiperspirants and deodorants is legally significant, as antiperspirants alter a body function and are regulated as drugs, whereas deodorants are classified as cosmetics. Thus, measurably reduced sweating has to be scientifically proved before it can be claimed for a product. Nevertheless, given the similarities in the compositions of deodorants and antiperspirants, they are likely functionally equivalent. Since eccrine glands are mediated by cholinergic nerves, sweating also can be shut off by anticholinergic drugs administered systematically [22] or topically. However, such drugs are too toxic for routine use as antiperspirants even when administered topically.

Acne is a common glandular problem arising from hyperproliferative closure of individual glands in the unique set of pilosebaceous glands located in and around the face and across the upper back. Irritation of cells lining the ducts of such glands initiates the formation of a lesions. Sheets of sloughed, sebum-soaked, keratinized cells that grow out from the walls surrounding the sebaceous duct are what clog the duct. Still-forming sebum is then trapped behind the obstruction, oftentimes bulging out the skin and giving rise to an observable lesion (papule). This may become infected and fill with a purulent exudate (pustule) or, after infection has set in, it may internally rupture and thereby begin the processes that lead to ulceration and scarring. Alternatively, the buildup of concentric sheets of sloughed cells may widen the glandular opening, with melanin in the widening plug darkening to the point of being black (blackhead).

Soap and water is considered a therapeutic treatment in acne when it is used to unblock the pores. Sebum is emulsified, and it and other debris is removed. Alcoholic solvents, often packaged as moistened pledgets, are used for the same purpose. With either treatment care must be exercised not to dry out and further irritate the skin. Both local and systemic antibiotics and antiseptics suppress the formation of lesions. It is believed these attenuate the population of anerobic microorganisms that are deep-seated in the gland, the metabolic by-products of which irritate the lining of the gland, setting off lesion formation. Mild cases of acne improve and clear under the influence of astringents, possibly for the same reason. Retinoids, oral and topical, reset the processes of epidermal proliferation and differentiation. Through such dramatic influences on cell growth patterns, they actually prevent the formation of lesions. However, because of concerns over toxicity, they tend to be used only in the most severe cases of acne and thus are prescribed for those patients whose acne lesions progress to cysts.

Hair is a product of the pilosebaceous apparatus and in this sense is glandular. It often grows out visibly in places where such display is unwanted. It may be shaved, but chemical hair removers (depilatories) along with other products are also used to remove it. In the main, the use of depilatories is cosmetic, rather than therapeutic. However, depilatories may be prescribed in hirsutism, when the existence of coarse, dark facial hair is psychologically distressing to the female patient for whom shaving is an anathema. Hair, like the stratum corneum is composed of layers of dead, keratinized cells. However, its keratin is more susceptible to the action of keratolytics because it is structured in ways that make it more chemically pervious. Thioglycolate-containing, highly alkaline creams generally dissolve hair in short order, without doing great harm to surrounding tissues. Facial skin is delicate, however, and depilatories must be used carefully here.

E. Effects in Deep Tissues

Local, Regional, and Systemic Delivery

When the target of therapy lies beneath the stratum corneum, topical drug delivery is more difficult and becomes more uncertain. Therefore, many potentially useful drugs find no place in topical therapy owing to their inability to adequately penetrate the skin. Nevertheless, a number of pathophysiological states can be controlled through local administration and subsequent percutaneous absorption. For example, most skin conditions are accompanied by inflammation of the skin; topical corticosteroids and nonsteroidal anti-inflammatory drugs alike are used to provide symptomatic relief in such instances. Corticosteroids are also used in psoriasis for which, in addition to suppressing inflammation, they somehow act on the basal epidermal layer to slow proliferation and restore the skin's normal turnover rhythm [23]. Pain originating in the skin can be arrested with locally applied anesthetics. Over-the-counter benzocaine and related prescription drugs are used for this purpose. Hydroquinone is applied to the skin to lighten excessively pigmented skin by oxidizing melanin deep within the surface. Another treatment that involves percutaneous absorption is the application of fluorouracil (5-FU) for the selective eradication of premalignant and basal cell carcinomas of the skin [22]. In all of these examples, the key to success is the ability we have to get therapeutic amounts of the drugs through the stratum corneum and into the viable tissues.

Systemic actions of some drugs can also be achieved by local application, in which case, their delivery is known as transdermal delivery. The application of warmed, soft masses of medicated bread meals and clays over wounds or aching parts of the body dates to antiquity. However, few such plasters and poultices (cataplasms) have survived into modern medicine. When analytical developments just past the middle of the twentieth century made it possible to measure the exceedingly low circulating levels of drugs that build up in the body during the course of therapy, research was begun on ways drugs might be delivered to lower their risks and extend their durations of action. Novel delivery systems involving nontraditional routes of administration were subsequently conceived, constructed, and put to test. Transdermal delivery with adhesive patches evolved as one of the innovations.

The possibilities for transdermal delivery might have been seen long ago in the systemic toxicities of certain topically contacted chemicals. As long as a century ago it was known that munitions workers who handle nitroglycerine suffered severe headaches and ringing in the ears (tinnitus). These same effects are experienced to a degree by those taking nitroglycerin to alleviate angina. The association between therapy and the inadvertent percutaneous absorption of nitroglycerin was finally made in the 1970s and a nitroglycerin ointment was introduced, producing peak blood levels comparable with those attained on sublingual administration of traditional tablet triturates, but levels that were also sustained. Since the permeability of human skin is highly variable and patient needs themselves vary, patients using nitroglycerin ointments were (and are) instructed to gradually lengthen the ribbon of ointment expressed from the tube and rubbed into the skin until tinnitus and headache are experienced, and then back off on the dosage. The lastingness of the effects of nitroglycerin administered in this fashion freed patients from a fear of wakening in the middle of the night with a heart attack (angina). Consequently, nitroglycerin ointments became the first commercially successful, therapeutically proved transdermal delivery systems. But ointments are greasy and, as with all semisolids, suffer variability in their dosing, even with dose titration, given that different patients apply semisolids more or less thinly and, therefore, over more or less area.

Since about 1980, sophisticated adhesive patches for transdermal delivery of scopolamine (motion sickness), nitroglycerin (anginal symptoms), clonidine (regulation of blood pressure), β-estradiol (menopausal symptoms), and fentanyl (cancer pain) have been introduced into medicine [24–26]. These patches are affixed to an appropriate body location and deliver drug continuously for periods ranging from a day (nitroglycerin) to a week (clonidine). To achieve such long periods of delivery, the patches contain reservoirs of their respective drugs. In one early type of system, nitroglycerin was formulated into a liquid-filled sponge held in direct contact with the skin by an adhesive band around the periphery of the patch. In more recent patch designs a membrane is placed over the delivery surface of the patch. Adhesive covering this membrane anchors the patch to the skin over the entire contact area of the patch. This interfacing membrane can be turned into a *rate-controlling membrane* to regulate delivery and prevent dose dumping, should the patch be inadvertently placed over a site of inordinately high permeability. Actually, this rate-controlling concept was used in the design of the first patch, the scopolamine transdermal system. The development of transdermal patches for yet other drugs is an active research area.

It is obvious from the foregoing that the skin is a formidable barrier, irrespective of whether therapy is to be local, regional, or systemic, and the first concern in topical delivery is sufficiency of delivery. With local therapy, the aim is to get enough drug into the living epidermis or its surroundings to effect a pharmacological action there without producing a systemically significant load of the drug. The latter is actually a rare occurrence, except when massive areas of application are involved. Regional therapy involves effects in musculature and joints deep beneath the site of application. To be successful, this requires a greater delivery rate, because an enormous fraction of the drug that passes through the epidermis is routed systemically by the local vasculature. Indeed, the levels of drug reached in deep local tissues have proved to be only a few multiples higher than those obtained after systemic administration of the drug [27]. Even more drug has to be delivered per unit area to transdermally effectuate a systemic action.

Factors Affecting Functioning of the Skin Barrier

A matter of considerable consequence in topical delivery is variability in skin permeability between patients, which may be as much as tenfold. The underlying sources of this high degree of variability are thought to be many and diverse. Humans differ in age, gender, race, and health, all of which are alleged to influence barrier function. Yet, insofar as can be told, a full-term baby is born with a barrier-competent skin and, barring damage or disease, the skin remains so through life. There is little convincing evidence that senile skin, which tends to be dry, irritable, and poorly vascularized, is actually barrier-compromised [28]. However, premature neonates have inordinately permeable skins. The incubators used to sustain such infants provide a humidified environment—which abates insensible perspiration—and a warm one, conditions that not only make the baby comfortable, but that also forestall potentially lethal dehydration and hypothermia [29].

Gender, too, affects the appearance of human skin. Nevertheless, there is little evidence that the skins of male and female differ much in permeability. However, there are established differences in the barrier properties of skin across the races of man. Although the horny layers of whites and blacks are of equal thickness, the latter has more cell layers and is measurably denser [30]. As a consequence, black skin tends to be severalfold less permeable [30,31].

Humidity and temperature also affect permeability. It has long been known that skin hydration—however brought about—increases skin permeability. Occlusive wrappings, therefore, are placed over applications on occasion to seal off water loss, hydrate the horny layer, and increase drug penetration. In the absence of such intervention, the state of dryness of the stratum corneum is determined by the prevailing humidity, explaining why the condition "dry skin" is exacerbated in the winter months in northern climates.

Temperature influences skin permeability in both physical and physiological ways. For instance, activation energies for diffusion of small nonelectrolytes across the stratum corneum lie between 8 and 15 kcal/mol [4,32]. Thus thermal activation alone can double the rate of skin permeability when there is 10° change in the surface temperature of the skin [33]. Additionally, blood perfusion through the skin, in terms of amount and closeness of approach to the skin's surface, is regulated by its temperature and also by an individual's need to maintain the body's 37°C isothermal state. Since clearance of percutaneously absorbed drug to the systemic circulation is sensitive to blood flow, a fluctuation in blood flow might be expected to alter the uptake of chemicals. No clear-cut evidence exists that this is so, however, which seems to teach us that even the reduced blood flow of chilled skin is adequate to efficiently clear compounds from the underside of the epidermis.

Above all else, the health of the skin establishes its physical and physiological condition, and, thus, its permeability. Consequences attributable to an unhealthy condition of skin can be subtle or exaggerated. Broken skin represents a high-permeability state, and polar solutes are several log orders more permeable when administered over abrasions and cuts. Irritation and mild trauma tend to increase the skin's permeability, even when the skin is not broken, but such augmentation is far less substantial. Sunburn can be used to illustrate many of the barrier-altering events that occur in traumatized skin. Vasodilation of the papillary vasculature, with marked reddening of the skin, is among the first signs that a solar exposure has been overdone. In its inflamed state, the skin becomes warm to the touch. After 1 or 2 days, epidermal repair begins in earnest, and the tissue is hyperproliferatively rebuilt in its entirety. It doubles in thickness, and a new stratum corneum is quickly laid down [34]. Because the newly formed stratum corneum's anchorage to existing tissues is faulty, the preexisting horny layer often eventually peels. Of more importance, hyperplastic repair leads to a poorly formed horny structure of increased permeability to water (as measured by transepidermal water loss) and presumably other substances. Given these events surrounding irritation, since many chemicals found in the workplace and home are mildly irritating—including the soaps we use to bathe and the detergents we use to clean house and clothes—is it really a wonder that the permeability of human skin is so demonstrably variable?

Some chemicals have prompt, destructive effects on the skin barrier. Saturated aqueous phenol, corrosive acids, and strong alkali instantly denature the stratum corneum and destroy its functionality, even as their corrosive actions stifle the living cells beneath. Although the stratum corneum may appear normal following such damage, the skin may be only marginally less permeable than denuded tissue [35]. Furthermore, permeability remains high during the full duration of wound repair and until a competent stratum corneum is laid down over the injured surface. Other chemicals are deliberately added to formulations to raise the permeability of skin and improve drug delivery. For obvious reasons, these are referred to as skin penetration enhancers. More will be said of these later.

Thermal burning produces comparably high states of permeability immediately following burning, providing that the surface temperature of the skin is raised above 80°C, a temperature on the lower side of temperatures able to denature keratin [36]. However, burning temperatures below 75°C, although fully capable of deep tissue destruction in seconds, leave the structure of the stratum corneum itself relatively unscathed. Burn wounds of this kind remain impermeable until tissue repair and restructuring processes get under way and the necrotic tissue with its horny capping is sloughed. The differences in permeability state of burned skin immediately following the trauma can be highly consequential in terms of drug delivery. In the instance of deep burns obtained at lower than keratin-denaturing temperatures, topical delivery of antibiotics and antiseptics into a wound, as may be necessary to control wound sepsis, remains difficult for as long as the stratum corneum over the wound stays in place, and aggressive use of antiseptics is warranted. In the other extreme, there is risk of toxic systemic accumulation of the antiseptics and antibiotics, particularly in major burns covering 20% or more of the body surface area. Conservative treatment is warranted. Since burns are rarely well characterized relative to their permeability dimension, attending physicians and pharmacists need to monitor antiseptic usages carefully to control wound sepsis without poisoning the patient. Finally, if not surgically debrided, the necrotic tissue is eventually walled off, enzymatically digested, loosened, and sloughed, producing a denuded, open, granulating surface. Small full-thickness wounds are closed and sealed off quickly by reconstruction of the epidermis from the edges of the wound. Large full-thickness injuries take too long to heal by this process and require grafting. All such wounds remain highly permeable until covered over again with a healthy, fully differentiated epidermis.

As with burns, physical disruption of the stratum corneum opens the skin in proportion to the extent of damage. Cuts and abrasions are associated with high permeability at and around such injuries. Eruption of the skin in disease has a similar effect, at least to the extent that the stratum corneum's integrity is lost. The skin over eczematous lesions should be considered highly permeable. Not all skin diseases raise permeability, however. The states of permeability of ichthyosiform, psoriatic, and lichenified skin have not been well characterized, but in all likelihood are low for most drugs. It has proved difficult to get potent corticosteroids through psoriatic plaque, for instance, and occlusive wrapping is often called for.

Percutaneous Absorption: The Process

The process of percutaneous absorption can be described as follows. When a drug system is applied topically, the drug diffuses passively out of its carrier or vehicle and, depending on where the molecules are placed down, it partitions into either the stratum corneum or the sebum-filled ducts of the pilosebaceous glands. Inward diffusive movement continues from these locations to the viable epidermal and dermal points of entry. In this way, a concentration gradient is established across the skin up to the outer reaches of the skin's microcirculation, where the drug is swept away by the capillary flow and rapidly distributed throughout the body. The volume of the epidermis and dermis beneath a 100-cm^2 area of application, roughly the size of the back of the hand, is approximately 2 cm^3. The total aqueous volume of a 75-kg ($\approx$165-lb) person is about 50,000 cm^3, yielding a systemic-to-local dilution factor well in excess of 10,000. Consequently, systemic drug levels are usually low and inconsequential. Thus, selectively high epidermal concentrations of some drugs can be obtained. However, if massive areas of the body ($\geq$20% of the body surface) are covered with a topical therapeutic, systemic accumulation can be appreciable. For instance, corticosteroids have produced serious systemic toxicities on occasion when they have been applied over large areas of the body [37].

Moreover, as has already been pointed out, if the stratum corneum is not intact, many chemicals can gain systemic entrance at alarming rates. Together these factors may place a patient at grave risk and should always be taken into account when topical drugs are put in use. The pharmacist, therefore, should carefully measure how topicals are to be applied and be on alert for untoward systemic responses when body coverages are unavoidably extensive.

The events governing percutaneous absorption following application of a drug in a thin, vehicle film are illustrated in Fig. 2. The important processes of dissolution and diffusion within the vehicle are cataloged. These will be discussed later. Two principal absorption routes are indicated in the sketch: (a) the transepidermal route, which involves diffusion directly across the stratum corneum; and (b) the transfollicular route, for which diffusion is through the follicular pore. Many words have been written concerning the relative importances of these two pathways. Claims that one or the other of the routes is the sole absorption pathway are ground-

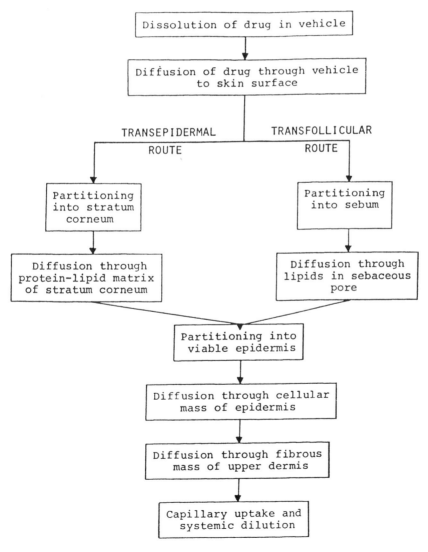

Fig. 2 Events governing percutaneous absorption.

less, since percutaneous absorption is a spontaneous, passive diffusional process that takes the path of least resistance. Therefore, depending on the drug in question and the condition of the skin, either or both routes can be important. There are also temporal dependencies to the relative importances of the routes. Corticosteroids breach the stratum corneum so slowly that clinical responses to them, which are prompt, are reasoned to be due to follicular diffusion [4].

One should not lose sight of the fact that the chemical barrier of the skin actually consists of all skin tissues between the surface and the systemic entry point. Although it is true that the stratum corneum is a source of high diffusional resistance to most compounds and, thus, the skin's foremost barrier layer, exceptional situations exist for which it is not the only or even the major resistance to be encountered. For example, extremely hydrophobic chemicals have as much or more trouble passing across the viable tissues lying immediately beneath the stratum corneum and above the circulatory bed, because such drugs have little capacity to partition into these tissues. Backing for the latter assertion comes from extensive clinical experience, as well as from physical modeling of percutaneous absorption. Consider that ointments can be used safely over open wounds, for their hydrocarbon constituents are not transported significantly across even denuded skin. Similarly, the skin is considerably more impermeable to octanol and higher alkanols than is the stratum corneum alone because of the presence of the viable tissue layer beneath.

Model of the Skin Barrier

The percutaneous absorption picture can be qualitatively clarified by considering Fig. 3, in which the schematic skin cross section is placed side by side with a simple model for percutaneous absorption patterned after an electrical circuit. In absorption across a membrane, the

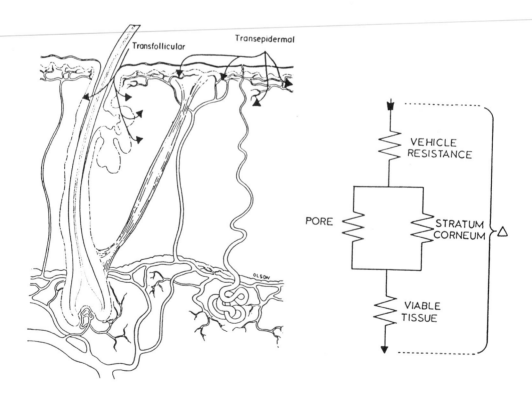

Fig. 3 Skin cross section beside a simple model.

current or flux is in terms of matter or molecules, rather than electrons, and the driving force is a concentration gradient (technically, a chemical potential gradient), rather than a voltage drop. Each layer of a membrane acts as a *diffusional resistor*. The resistance of a layer is proportional to its thickness (symbol = h); inversely proportional to the diffusive mobility of a substance within it, as reflected in a diffusion coefficient (D); inversely proportional to the capacity of the layer to solubilize the substance relative to all other layers, as expressed in a partition coefficient (K); and inversely proportional to the fractional area of the membrane occupied by the diffusion route (f) if there is more than one route in question [38]. In general, an individual resistance in a set may be represented by:

$$R_i = \frac{h_i}{f_i D_i K_i} = \frac{[\text{thickness}]}{[\text{fractional area}][\text{diffusion coefficient}][\text{partition coefficient}]} \tag{1}$$

The overall phenomenon of percutaneous absorption is describable after recognizing that the resistances of phases in series (phases encountered serially) are additive, and that diffusional currents (fluxes) through routes in parallel (differing routes through a given phase) are additive. Such considerations applied to skin allow one to explain, in semiquantitative terms, why percutaneous absorption through intact skin is slow for most chemicals and drugs, and why disruption of the horny covering of the skin profoundly increases the permeabilities of the ordinary run of solutes.

First, consider the transepidermal route. The fractional area of this route is virtually 1.0, meaning the route constitutes the bulk of the area available for transport. Molecules passing through this route encounter the stratum corneum and then the viable tissues located above the capillary bed. As a practical matter, the total stratum corneum is considered a singular diffusional resistance. Because the histologically definable layers of the viable tissues are also physicochemically indistinct, the set of strata represented by viable epidermis and dermis is handled comparably and treated as a second diffusional resistance in series. The estimated diffusion coefficients in the stratum corneum are up to 10,000 times smaller than found anywhere else in the skin, partly reflecting the considerable denseness of this tissue. Presuming diffusion to be through the intercellular lipid regime within the horny tissue, the estimates, which range from 1×10^{-9} cm^2/sec to a low of 1×10^{-13} cm^2/sec, have to be tempered with the knowledge that path tortuosity and excluded volume were not accounted for when estimating them. Regardless, small values such as this are associated with high resistance and low penetrability [4]. The breadth of the stratum corneum, fixed by nature, is an exceedingly thin 1×10^{-3} cm. The most variable parameter in the stratum corneum resistance equation is the partition coefficient, as this can take a value several log orders less than one when the permeant is a highly polar molecule, such as glucose, or a value several log orders greater than one when a hydrophobic molecule, such as β-estradiol, is involved. The wedge of living tissue between the stratum corneum and the capillaries is roughly 100 μm thick (1×10^{-2} cm). Permeation of this regime is facile and without great molecular selectivity, with diffusion coefficients being about one-tenth the magnitude of those found in a simple liquid such as water [39].

The follicular route can be analyzed similarly. The fractional area available for penetration by this route is on the order of 1/1000 [4], clearly a restricting factor. Here, partitioning is into sebum, and the distance that has to be traveled through the sebaceous medium filling the follicular duct can only be guessed at, but has to be greater than the thickness of the stratum corneum; 50 μm seems a reasonable estimate, the actual value almost certainly being within a factor of 2 of this. Diffusion coefficients in the quasiliquid sebum are from 100-fold to 1000-fold or more greater than in the stratum corneum [4,10]. Because sebum is lipoidal, partition coefficients for entering this route must also range widely. A thickness of viable tissue, perhaps comparable with that found along the transepidermal route, must also be diffusionally negotiated before drug reaches the microcirculation.

The net chemical penetration of the skin is simply the sum of the accumulations by each of the mentioned routes and by other routes, for instance eccrine glands, where these contribute. The latter tiny glands are ubiquitously distributed over the body, but are generally discounted in importance owing to the limited fractional area they occupy and their unfavorable physiological states, either empty or profusely sweating.

All the salient features described here can be incorporated into a quantitative framework that takes into account stratification within the tissues and the parallel pathways [38]. It is instructive to consider even the simplest such model based on these descriptions and embodying only transepidermal and transfollicular routes. We can assume each distinct tissue acts as a homogeneous phase, a gross distortion of reality, but an assumption that, nevertheless, leads to a useful conceptual description. The resistance by the transepidermal route would be:

$$R_{\text{Transepidermal}} = R_{\text{stratum corneum}} + R_{\text{viable tissues transepidermally}}$$

or

$$R_{\text{TE}} = R_{\text{sc}} + R_{\text{vt}-\text{TE}} \tag{2}$$

Similarly, the transfollicular resistance would be

$$R_{\text{TF}} = R_{\text{seb}} + R_{\text{vt}-\text{TF}} \tag{3}$$

Since these routes are in parallel, the total resistance on combining them is:

$$R_{\text{total}} = \cfrac{1}{\left(\cfrac{1}{R_{\text{sc}} + R_{\text{vt}-\text{TE}}}\right) + \left(\cfrac{1}{R_{\text{seb}} + R_{\text{vt}-\text{TF}}}\right)} \tag{4}$$

The mass transfer coefficient (permeability coefficient; P) of a route is the reciprocal of the resistance of that route and, thus, we can amend Eq. (4) to read:

$$R_{\text{total}} = \frac{1}{P_{\text{TE}} + P_{\text{TF}}} \tag{5}$$

Similarly, the overall permeability coefficient; taking both routes into account, is the reciprocal of the total resistance and thus P_{total} is:

$$P_{\text{total}} = P_{\text{TE}} + P_{\text{TF}} \tag{6}$$

Before proceeding farther, we should add in the fact that water is invariably the medium used to obtain permeability coefficients and, accordingly, water is assumed to be the vehicle under consideration. This choice of vehicle effectively sets the partition coefficients between the aqueous tissues and the vehicle to unity. It is important to realize that the model becomes general, even when based on a water vehicle, when saturated aqueous solutions, which operate at the thermodynamic activity of the solid drug, are brought into the analysis. Substituting fractional areas, thickness, and diffusion coefficients for all the phases, but partition coefficients for only the stratum corneum and sebum, leads to the following expression for the overall permeability coefficient:

$$P_{\text{total}} = f_{\text{TE}}\left(\frac{D_{\text{sc}}K_{\text{sc/w}}D_{\text{vt}}}{D_{\text{sc}}K_{\text{sc/w}}h_{\text{vt}-\text{TE}} + D_{\text{vt}}h_{\text{sc}}}\right) + f_{\text{TF}}\left(\frac{D_{\text{seb}}K_{\text{seb/w}}D_{\text{vt}}}{D_{\text{seb}}K_{\text{seb/w}}h_{\text{vt}-\text{TF}} + D_{\text{vt}}h_{\text{seb}}}\right) \tag{7}$$

Finally, a general expression describing the steady state flux across a membrane, $\partial M/\partial t$, can be written as:

$$\frac{\partial M}{\partial t} = AP_{total}\,\Delta C \tag{8}$$

This equation teaches us that the total steady-state flux (total rate of permeation across a membrane in the steady state of permeation), $\partial M/\partial t$, is proportional to the involved area (A) and the concentration differential expressed across the membrane, ΔC. In an experiment, flux is the experimentally measured parameter, whereas A and ΔC are fixed in value before starting the experiment. The value of the permeability coefficient, P_{total}, is what is calculated at the end of the experiment using Eq. (8). The permeability coefficient, besides having the other attributes already ascribed to it, can be looked at as merely being the number that is needed to make an equality out of the combined area and concentration proportionalities of flux. It has units of distance/time (cm/hr) and is effectively the average velocity of the molecules penetrating a membrane, irrespective of the complexity of the membrane. Its magnitude depends on the properties of the vehicle, membrane, and permeant. Moreover, when the membrane consists of several phases, the permeability coefficient is also dependent on the juxtaposition of the phases. We can now write for the skin:

$$\frac{\partial M}{\partial t} = A\left\{f_{TE}\left(\frac{D_{sc}K_{sc/w}D_{vt}}{D_{sc}K_{sc/w}h_{vt-TE} + D_{vt}h_{sc}}\right) + f_{TF}\left(\frac{D_{seb}K_{seb/w}D_{vt}}{D_{seb}K_{seb/w}h_{vt-TF} + D_{vt}h_{seb}}\right)\right\}\Delta C \tag{9}$$

In Eq. (9), A is the involved area of the skin and the term ΔC is the permeant's concentration differential across the skin. In clinical situations ΔC is usually well approximated by the actual concentration in the topical vehicle because dilution by way of systemic absorption of the permeant is so great.

Equation (9) defines the steady state flux in terms of physically meaningful parameters. In other words, it is an anatomically based, mathematical representation (model) of the skin barrier. The first collection of terms in the greater parentheses defines the role of the transepidermal route, and the second, the role of the transfollicular pathway. Representative values for some of the parameters of Eq. (7) needed to probe the model are given in Table 8. The listed fractional areas, diffusivities, and strata thicknesses are based on the best information available. They are approximate at best, some being only guesses. Regardless, the impression generated when they are substituted in the conceptual model is consistent with what is known about percutaneous absorption. For instance, if the higher values of the diffusion coefficients found in Table 8, that is, 10^{-9} cm²/sec for the stratum corneum, 10^{-7} cm²/sec for sebum, and 10^{-6} cm²/sec for the viable tissue, are converted to square centimeters per hour (cm²/hr) and then substituted into Eq. (7) along with the ordinary thicknesses of the respective tissues (cm) and the respective fractional areas of the routes, and if the partition coefficients are set to unity, the resulting magnitudes arrived at for the bracketed transepidermal and transfollicular contributions suggest that the steady-state flux through the transepidermal route is roughly 30 times greater than the steady-state flux through the follicular pores. We learn from this that high sebum diffusivity, by itself, is not enough to offset the small fractional area of the follicular route. Partitioning tendencies favoring either stratum corneum or sebum accumulation could magnify or shrink this ratio. Unfortunately we do not have the kind of information needed to pass judgment on how these distribution coefficients generally relate to one another.

Arguably, the stratum corneum harbors a minor polar pathway, an assertion that is admittedly still being debated. There is evidence the horny tissue supports disproportionally high fluxes of polar solutes such as methanol, ethanol, propylene glycol, glycerol, and even glucose. Ions

Table 8   Representative Parameters to Probe Model

	Diffusion coefficient, D $(cm^2/sec)^a$[4,8]
Stratum corneum	$10^{-9}-10^{-13}$
Water	$\simeq 10^{-9}$
n-Alkanols (hydrated tissue)	$\simeq 10^{-9}$
n-Alkanols (dry tissue)	$\simeq 10^{-10}$
Small nonelectrolytes	$10^{-9}-10^{-10}$
Progesterone	$\simeq 10^{-11}$
Cortisone	$\simeq 10^{-12}$
Hydrocortisone	$\simeq 10^{-13}$
Follicular pore (sebum)	$10^{-7}-10^{-9}$
Viable tissue	$\simeq 10^{-6}$

	Tissue thickness, h (μm) [2,8,11]
Stratum corneum	
Dry (normal state)	~10
Hydrated (as by occlusion) state	20–30
Pore diffusional length	Approximately two to five times greater than the stratum corneum thickness
Viable tissue stratum	150–2000[b] (200)

	Fractional area of the routes, F [4,8]
Transepidermal	~1
Transfollicular	$\sim 10^{-3}$
Transeccrine	$<10^{-4c}$

	Tissue/vehicle partition coefficient, K
Stratum corneum	From <1 to >>1[d]
Sebum	From <<1 to >>1[d]
Viable tissue	
Aqueous vehicle	~1
Nonaqueous vehicle	From <<1 to >>1[d]

[a]These diffusivities are estimates obtained by in vitro experiment (stratum corneum) or by comparison with small tissues in which diffusivities have been measured (all others). They do not account for regional variations across the body surface, so on both counts must be considered highly approximate.

[b]Highly approximate and variable, depending on blood flow patterns.

[c]This is sufficiently small to discount transeccrine diffusion contributions in the general treatment.

[d]All depend on the physiocochemical nature of the drug and vehicle as well as the physicochemical nature of the respective tissues.

diffuse through it, too. Sebum, on the other hand, although actually not well characterized in terms of its prevailing physicochemical state, is seemingly a nonpolar composite, a condition that would make it unsuitable for solubilization of such compounds. On these meager grounds, the transepidermal route should dominate for small, nonpolar nonelectrolytes, and it appears that it does. Analysis definitively identifies the stratum corneum as the main diffusional resistor in such instances, explaining why it has become known as the *barrier layer* of the skin.

As compounds of increasing hydrophobicity are brought under consideration, distribution coefficients of the substances into both stratum corneum and sebum are commensurately increased, as each medium is a lipoidal matrix to good first approximation. A useful feature of homologous compounds formed by extending alkyl chain length is that oil/water partition coefficients of series members grow exponentially, affording probing of the partitioning dependencies mass transport. The slope of a log-linear plot of partition coefficient against alkyl chain length yields the sensitivity of partitioning to a methylene group, the so-called π-value. Based on *n*-alkanol data, the π-value of normal human stratum corneum appears to be about 0.3 [4]. It is 0.5 for octanol/water partitioning, and over 0.6 for hexane/water partitioning. These differing values have been interpreted to mean that the stratum corneum's lipoidal phase is considerably more polar (less sensitive to a $-CH_2-$group) than either of the example organic solvents. The view held of sebum suggests its π-value might be higher than the stratum corneum's as well, but this is purely speculative assertion [40,41]. This does not necessarily cause a shift in the relative importances of the two identified routes, as judged through modeling, however, for when $K_{seb/w}$ reaches 1, the transfollicular route is already approaching viable tissue control of permeation. Consequently, further increases in this partition coefficient would not be experienced as increases in permeation through the pathway. On the other hand, according to the model, $K_{sc/w}$ must approach 50 before the viable tissue becomes the rate-controlling part of the barrier. This is the published value for octanol, which actually does appear to be at the threshold of viable tissue control of its permeation.

The story becomes quite different when minimum values for diffusivities are incorporated in the steady-state model (10^{-13} cm^2/sec for the stratum corneum and 10^{-9} cm^2/sec for the follicular shunt route). Keeping everything else constant and inserting these values produces a substantial upgrading of the importance of the transfollicular contribution. However, given laboratory data with steroids, the transepidermal route seems to retain its dominant position in the steady state even here.

We can now use the model to illustrate what happens to permeability when the stratum corneum is damaged. One only needs to use zero for the value of h_{sc}, the stratum corneum thickness, to assess the effect of extreme damage. This choice functionally denudes the skin and the transepidermal contribution takes the simple form, D_{vt}/h_{vt-TE}. Consequent increases in permeability can be as much or more that 1000-fold for polar compounds. However, for highly nonpolar compounds, $K_{sc/w}$ is so large that only small increases in permeability are projected. The implication here is that the increase in permeability that accompanies damage depends on how lipophilic a compound is. This is true, for as $K_{sc/w}$ is increased, the stratum corneum resistance is reduced proportionally, whereas the resistance in the living tissue, which sets the upper limit on permeability, is not. Thus, the more nonpolar a compound is, the less its permeation is affected by stratum corneum damage.

So far only steady-state conditions for permeation have been considered. There is strong evidence that the nonstationary-state period, that is, the time it takes to build up the gradient in the tissues, plays an important role in some drug delivery situations [42]. This is because the shunt route can be breached in a far shorter period than the transepidermal route, as molecular mobilities (D) are larger here. The time it takes to build up gradients is characterized in terms of the diffusional lag time, as illustrated in Fig. 4. The lag time can be seen to be the

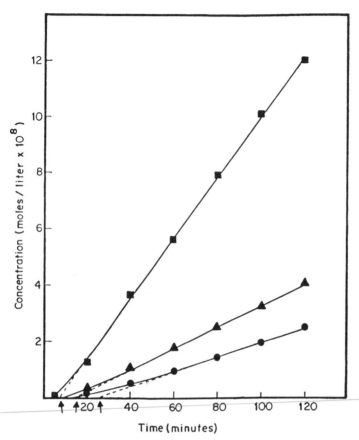

Fig. 4 Generalized permeation profile. From left to right the data are for n-butanol permeating hairless mouse skin at 20°, 25° and 30°C, respectively. Increasing temperature raises the flux (slope) and shortens the lag time.

intercept on the time axis found by extrapolating the steady-state line of a plot of cumulative amount of drug penetrated versus time. This lag time, t_L, is related to the diffusivity of a simple isotropic membrane by:

$$t_L = \frac{h^2}{6D} \tag{10}$$

where h is again used to signify the membrane's thickness and D is the diffusion coefficient. Lag times are different for parallel routes. Even though the diffusion coefficients we have for the skin are effective values embodying factors beyond molecular mobility, we can still estimate specific lag times for the two pathways under consideration through the skin using the diffusion coefficients tabulated in Table 8. Lag times for the transepidermal route would seem to range from minutes for small nonelectrolytes to multiple days for corticosteroids. On the other hand, transfollicular values seem to range from seconds to minutes. On this basis it appears likely that the first molecules to reach the viable epidermis come by the follicular path, although the amounts of drug reaching the lower epidermis early on may not be large owing to the limited area of the path. Nevertheless, it is this kind of thinking that has led scientists to the idea that some clinical responses are due to diffusion through the follicular shunt. When the lag time through the stratum corneum pathway itself is short, transient diffusion through the shunt is far less likely to be clinically significant. Moreover, since it is the stratum corneum that is responsible for the long lag times by the transepidermal route, the foregoing nonstationary-state considerations do not apply when it is impaired.

Equations (8) and (9) do underscore the point that percutaneous absorption is proportional to area. Transdermal patches of nitroglycerin and other drugs are provided in different sizes to take advantage of this simple fact to adjust dosages. It cannot be stressed too strongly that a pharmacist must be cognizant of this area dependency for another, more compelling reason. Even if a drug if not readily percutaneously absorbed, its systemic effects are magnified in proportion to the area over which it is applied. No other factor is so frequently associated with the untoward actions of topically applied drugs. Consider that there have been unfortunate poisonings of infants following the liberal use of talcum powders containing borates as lubricants, perhaps but not necessarily, in conjunction a diminished skin barrier property resulting from diaper irritation. Borates are now outlawed for this purpose. Babies have been poisoned by bathing them with a hexachlorophene-containing liquid soap; bathing involves total body coverage. Hexachlorophene is no longer used in the manner. The inflammation associated with diaper rash is extensive, and topical corticosteroids have also been used too liberally on the bodies of infants and small children to treat this and other conditions, to the point of inducing serious toxicity. These kinds of problems are not limited to small children. Systemic toxicities have accompanied the treatment of psoriatic lesions with the keratolytic, salicylic acid, under circumstances when the lesions have been massive. And there are now documented cases of abuse of transdermal patches, including one in which a patient, confused about the manner of use, succumbed to fentanyl by wearing four patches simultaneously. Topically, area is dose.

Phenomenological Considerations in Percutaneous Delivery

The model that has been presented is useful for ferreting *patterns* of permeability as determined by chemical structure and by the physiological state of the skin. It would be inappropriate to use it to calculate actual permeability coefficients, however, because so many iffy approximations are made. Therefore, we must take a different tack to gain a sense of the absolute limits of drug delivery by way of cutaneous absorption. The lower limit of flux of compounds across the skin would be none at all; surprisingly, a situation that rarely appears to be seen. Even proteins penetrate the intact skin to a minuscule extent. However and obviously, it is more instructive to know the upper limit of achievable flux. Some idea of this can be gained from examining the rates of delivery of nitroglycerin and similarly facile penetrants of skin.

Nitroglycerin, a liquid at room temperature, is a relatively lipophilic nonelectrolyte compound of only 227 molecular weight, all properties that tend to make it an ideal skin penetrant. It is formulated in transdermal systems virtually as a neat liquid and, thus, near its upper attainable thermodynamic activity. It diffuses through the skin at between 0.02 and 0.04 mg cm^{-2} hr^{-1} from the transdermal reservoirs in which it is placed, rates that equate to the delivery of 0.5–1.0 mg cm^{-2} day^{-1} [24]. Consequently, 20 cm^2 patches are used to provide a daily delivery of between 10 and 20 mg of the drug [24]. Nicotine, another low molecular weight, somewhat hydrophobic, liquid compound at 25°C, permeates at comparably high rates from its patch delivery systems. Selegiline is another facile skin penetrant, having a similar battery of physical properties and the same limiting rate. Thus, by this analysis, 1 mg cm^{-2} day^{-1} looks to be about the upper achievable limit of delivery of drugs through the skin. This claim is made even in the face of the fact that water diffuses through the skin much faster than any of these species. Water typically diffuses out of the skin into a dry atmosphere at a rate of 0.5 mg/cm^{-2} hr^{-1}, a flux over an order of magnitude greater than that seen with the compounds mentioned. Physiological water inside the skin has a thermodynamic activity equivalent to that of 0.9% NaCl, a highly dilute solution to say the least; thus, the driving force for water's escape from the body is not all that different than if cellular water were in a pure state. But water is a very unique molecule, both for its size and its interactions within the horny layer. Appearances are that its diffusion is not restricted to the intercellular domain of the stratum corneum, but rather, some water almost certainly works its way into and up through the keratin amassed in the intracellular space. It seems a major fraction of insensible perspiration exits the skin this way. Diffusion of nitroglycerin and the other drugs mentioned, on the other hand, is believed to be restricted to the lipoidal, intercellular domain of the stratum corneum.

If all nonpolar drugs were as skin-permeable as nitroglycerin, there would be many more topical and transdermal delivery systems. A good question is: Why are there so few? One reason is crystallinity for, in a general way, this limits the activity gradient that can be expressed across a membrane in a transport situation. We can superimpose a hypothetical crystallinity on nitroglycerin to illustrate the effect this might have on its activity and, thus, its delivery. By thermodynamic derivation the activity of the crystalline form of a compound relative to that of its liquid state (a *supercooled* liquid) is given by

$$\ln a_2 = -\frac{\Delta H_f(T_f - T)}{RT_f T} \tag{11}$$

where a_2 is the solid's activity, ΔH_f is its heat of fusion, T_f is its melting point (temperature of fusion), T is the experimental temperature (nominally 25°C), and R is the gas constant (1.987 in cal/mol per degree). By simple calculation, on assigning a melting point of 100°C and a plausible heat of fusion of 5000 cal/mol (about 20,000 J/mol) to our hypothetical crystalline form of nitroglycerin, its activity drops a little over fivefold. It would be possible to delivery only 200 µg/cm^{-2} day^{-1}, or only about 4 mg from a 20 cm^2 patch if this were true. If nitroglycerin melted at 200°C, all else equal, it would have only 4% of the activity it has as a liquid, and the delivery rate would now be only 40 µg/cm^{-2} day^{-1} (1 mg for a 20 cm^2 patch). Since most drugs are crystalline, some more so than in the illustration, delivery of 1 mg of drug per day from a 20 cm^2 area (roughly the area of a Ritz cracker) can be a true feat, to some extent making it clear why only very potent drugs are taken seriously for transdermal delivery. The underlying phenomenon here is that solubilities in all solvents are reduced by crystallinity over what they would otherwise be, including a drug's solubility in its delivery system matrix and, for that matter, in the skin's surface. Referring back to the model, the impact of this is directly on ΔC in Eq. (9), for the upper limit ΔC is the solubility of the compound. In other words, it is solubility that sets the upper limit on achievable delivery from the concentration gradient standpoint. These influences of crystallinity become particularly important when selecting compounds from the many that may be available to develop for topical or transdermal purposes. Everything else equal, low-melting compounds are far easier to deliver at therapeutically adequate rates.

Solubility in a particular solvent is also determined by the net interactions the solute has with the solvent in the solution phase. Generally, the activity of a dissolved solute is related to its concentration in solution through an activity coefficient; that is,

$$a_2 = X_2 \gamma_2 \tag{12}$$

Here, X_2 is the mole fraction of the solute in the saturated solution, this manner of expressing concentration being made on theoretical grounds, and γ_2 is its activity coefficient. The activity coefficient is nothing more nor less than the number needed to establish an equality between activity and concentration. Returning to the solubility expression we now have

$$\ln X_2 = \left[-\frac{\Delta H_f(T_f - T)}{RT_fT} \right] - \ln \gamma_2 \tag{13}$$

The right-hand side of Eq. 13 teaches us that solubility in a particular solvent depends on melting parameters, the same way activity does. However, the magnitude of the solubility also depends on the activity coefficient, as shown in the second right-hand term of Eq. (13). If the solute's solution phase interactions are strong (energetic), the activity coefficient is less than 1, and the solubility of the compound exceeds that of its ideal solution (where $\gamma_2 = 1.0$ and $a_2 = X_2$). Relatively weak interactions between the solute and solvent, on the other hand, are marked with activity coefficients greater than unity. Under such circumstance solubilities are lower than ideal. What one learns from all of this is that a high level of crystallinity is associated with low activity and vice versa, a factor set apart from the fact that compounds display as many solubilities as there are solvents to dissolve them in. The thermodynamic activity of a drug is the same in every one of its saturated solutions even though concentrations themselves are different from medium to medium owing to differing solute–solvent interactivity, the latter being couched in the activity coefficient term. One should not jump to the conclusion that drug delivery is the same from every saturated solution, however, as the solid to solution equilibria can rapidly kinetically break down on application of a dosage form, owing to drug dissolution in the vehicle not keeping up with partitioning of the drug into the skin. Additionally, when skin is the membrane, the enhancing attributes of different vehicles must also be considered.

The other two physical parameters of a drug that control its skin permeability are its size and its lipophilicity. It has only recently become clear that these, more than anything else, determine the magnitude of cutaneous permeability coefficients [43,44], as found in Eqs. (7) and (8). When extant human permeability coefficients (numbering over 90 compounds) were recently analyzed [44] by multiple linear regression within the following semiempirical equation:

$$\log_{10}(P) = \log_{10}\left(\frac{D^0}{h}\right) + \alpha \log_{10}(K_{oct/w}) - \beta(MW) \tag{14}$$

it was found that almost 70% of the variability in them is explicable by molecular size and partitioning attributes. For the record, D^0 in Eq. (14) is the hypothetical diffusivity of a molecule having zero molecular volume; h, as before, is the effective diffusion path length; α is a proportionality factor relating $K_{sc/w}$, as we have already defined it, to $K_{oct/w}$ or the respective octanol/water partition coefficient; β is a constant carried along from the dependency of diffusion coefficients on molecular size; and MW represents molecular weight (a surrogate for molecular volume). The computer-fit expression arising from a regression analysis involving all studied compounds is [43]

$$\log_{10}(P) = -6.3 + 0.71(K_{oct/w}) - 0.0061(MW) \tag{15}$$

Equation (15) provides its estimates of the permeability coefficients of compounds in units of centimeters per second (cm/sec). It can be used to make a ''guesstimate'' of the permeability coefficient of a compound through human skin from the compound's molecular weight and octanol/water partition coefficient alone. As with any parameter drawn out of statistically drawn relationships, such estimates must be used cautiously, for the absolute error of estimation for a single compound can be large.

V. UNIQUE PHYSICOCHEMICAL SYSTEMS USED TOPICALLY

As one scans the products at the drug counter, one finds an enormous variety of formulation types available for topical therapy or for cosmetic purposes. Solutions are commonly found. They come in packages that allow them to be rubbed on, sprayed on by aerosol and atomizers, painted on, rolled on, swabbed on by premoistened pledgets, and dabbed on from applicators. Assorted medicated soaps are available for a range of purposes. Emulsions for the skin are found in the form of shampoos and as medicated lotions. Powders to soothe and lubricate are placed in sprinkling cans, whereas others containing drugs are formulated into aerosols to be sprayed on the skin. There are numerous fluid suspensions to be used as makeup or for therapeutic purposes. Clear and opaque gels are also to be found in both cosmetic and therapeutic spheres, as are assorted semisolid creams, ointments, and pastes. The physical natures of these latter systems range from soft semisolids, intended to be squeezed out of tubes, to hardened systems, suitable for application in stick form. There are therapeutic and cosmetic oils for the bath. The list of products and formulation types is nearly endless.

Of all these formulations, it is the diverse semisolids that stand out as being uniquely topical. Semisolid systems fulfill a special topical need, as they cling to the surface of the skin to which they are applied, generally until being washed off or worn off. In contrast, fluid systems have poor substantivity and readily streak and run off the desired area. Similarly, powders have poor staying properties. Importantly, the fundamental physicochemical characteristics of solutions, liquid emulsions, and suspensions, and powders are independent of their route of application, and are discussed adequately elsewhere in this text and need not be reconsidered. This is not to say the compositions of such systems cannot be uniquely topical, for there are chemicals that can be safely applied to the skin, but that are unsafe to use systemically. There is need to elaborate the properties of semisolids.

A. General Behavior of Semisolids

The term *semisolid* infers a unique rheological character. Like solids, such systems retain their shape until acted on by an outside force, whereupon, unlike solids, they are easily deformed. Thus, a finger drawn through a semisolid mass leaves a track that does not fill up when the action is complete. Rather, the deformation made is, for all practical purposes, permanent, an outcome physically characterized by saying semisolids deform plastically. Their overall rheological properties allow them to be spread over the skin to form films that cling tenaciously.

To be semisolid, systems must have a three-dimensional structure that is sufficient to impart solidlike character to the undistributed system, but that is easily broken down and realigned under an applied force. The semisolid systems used pharmaceutically include ointments and solidified water-in-oil (w/o) emulsion variants thereof, pastes, oil-in-water (o/w) creams with solidified internal phases, o/w creams with fluid internal phases, gels, and rigid foams. The natures of the underlying structures differ remarkably across all these systems, but all share the property that their structures are easily broken down, rearranged, and re-formed. Only to the extent that one understands the structural sources of these systems does one understand them at all.

B. Ointments

Unless expressly stated otherwise, ointments are hydrocarbon-based semisolids containing dissolved or suspended drugs. They comprise fluid hydrocarbons, C_{16} to perhaps C_{30} straight-chain and branched, entrapped in a fine crystalline matrix of yet higher molecular weight hydrocarbons. The high molecular weight fraction precipitates out substantially above room temperature, forming interlocking crystallites [45]. The extent and specific nature of this structure determine the stiffness of the ointment. It follows directly from this that hydrocarbon-based ointments liquify on heating, for the crystallites melt. Moreover, when cooled very slowly, they assume a fluidity much greater than when rapidly cooled, because slow cooling leads to fewer and larger crystallites and, therefore, less total structure. Ordinary white and yellow petrolatum are examples of such systems.

Several alternative means of forming hydrocarbon ointments illustrate their structural properties. Ointments can be made by incorporating high-melting waxes into fluid mineral oil (liquid petrolatum) at high temperature. On cooling, interlocking wax crystallites form, and the system sets up. Polyethylene, too, can gel mineral oil if dissolved into this vehicle at high temperature and the solution is then forced cooled [46]. A network of polyethylene crystallites provides the requisite solidifying matrix. This polyethylene-gelled system is more fluid on the molecular level than are the semisolid petroleum distillates, while at the same time, macroscopically behaving as an ointment. Consequently, diffusion of drugs through this vehicle is more facile, and drug release is somewhat greater than found with petrolatum-based systems [47]. Plastibase (Squibb) is the commercially available base of polyethylene-gelled mineral oil. It is useful for the extemporaneous preparation of ointments by cold incorporation of drugs. Pharmacists should not melt down this base to incorporate drugs, for its gelled state cannot be restored without special processing equipment.

If a material other than a hydrocarbon is used as the base material of an ointmentlike system, the ointment bears the name of its principal ingredient. There are silicone ointments that contain polydimethylsiloxane oil in large proportion. These reportedly act as excellent water barriers and superior emollients. Some are actually used to protect skin from undesirable influences of long immersion in water.

Ointments of the specific kinds just mentioned are taken to be good vehicles to apply to dry lesions, but not to moist ones. All the foregoing are also greasy and stain clothes. The principal ingredients forming the systems, hydrocarbons and silicone oils, are generally poor solvents for most drugs, seemingly setting a low limit on the drug delivery capabilities of the systems. This solubility disadvantage can be offset somewhat if hydrocarbon-miscible solvents are blended into the systems to raise solvency. Alternatively, they can be made over into emulsions to raise their abilities to dissolve drugs. Along these lines, absorption bases are conventional ointments that contain water-in-oil (w/o) emulsifiers in appreciable quantity. A water-in-oil emulsion is formed when an aqueous medium, perhaps containing the drug in solution, is worked into the base. Such emulsions are still ointments, as structurally defined, for it is the external phase of the formed emulsion that imparts the structure, and this retains its ointment-like character. The term *absorption base* refers to a water incorporation capacity and infers nothing about bioavailability. This is not to say that it is not better to have water-soluble drugs emulsified than to have them as undissolved solids in such systems from the bioavailability standpoint. For optimum results, the internal, presumably aqueous phase, should be close to saturated. Diverse additives are used to emulsify water into these systems, including cholesterol, lanolin (which contains cholesterol, cholesterol esters, and other emulsifiers), semisynthetic lanolin derivatives, and assorted ionic and nonionic surfactants, singularly or in combination.

Polyethylene glycol ointment is a water-soluble system that contains fluid, short-chain polyoxyethylene polymers (polyethylene glycols) in a crystalline network of high-melting, long-chain polyoxyethylene polymers (Carbowaxes, Union Carbide). The structure formed is totally analogous to that of the standard ointment. In one variation, this system functions well as a suppository base. Liquid polyethylene glycols are fully miscible with water, and many drugs that are insoluble in petroleum vehicles readily dissolve in the polar matrix of this base. In fact, with some drugs, delivery (bioavailability) can be compromised by an excessive capacity of the base to dissolve substances, resulting in poor vehicle-into-skin partitioning. Since polyethylene glycols are highly water-soluble, bases formed from them literally dissolve off the skin when placed under a stream of running water.

C. Pastes

Pastes are basically ointments into which a high percentage of insoluble particulate solids have been added—as much or more than 50% by weight in some instances. This extraordinary amount of particulate matter stiffens the systems through direct interactions of the dispersed particulates and by adsorbing the liquid hydrocarbon fraction within the vehicle onto the particle surfaces. Insoluble ingredients, such as starch, zinc oxide, calcium carbonate, and talc, are used as the dispersed phase. Pastes make particularly good protective barriers when placed on the skin for, in addition to forming an unbroken film, the solids they contain can absorb, and thereby neutralize, certain noxious chemicals before they ever reach the skin. This explains why they are used to ameliorate diaper rash for, when spread over the baby's bottom, they absorb irritants (ammonia, others?) formed by bacterial action on urine. Like ointments, pastes form an unbroken, relatively water-impermeable film on the skin surface and, thus, are emollients; unlike ointments, the film is opaque and, therefore, an effective sun block. Accordingly, skiers apply pastes around the nose and lips to gain a dual protection. Pastes are actually less greasy than ointments because of the adsorption of the fluid hydrocarbon fraction to the particulates.

D. Creams

Creams are semisolid emulsion systems that have a creamy appearance as the result of reflection of light from their emulsified phases. This contrasts them with simple ointments, which are translucent. Little agreement exists among professionals concerning what constitutes a cream; therefore, the term has been applied both to absorption bases containing emulsified water (w/o emulsions) and to semisolid o/w systems, which are physicochemically totally different, strictly because of their similar creamy appearances. Logically, classification of these systems should be based on their physical natures, in which case, absorption bases would be ointments, and the term *cream* could be reserved exclusively for semisolid o/w systems, which in all instances derive their structures from their emulsifiers and internal phases.

The classic o/w cream is vanishing cream, which contains only 15% stearic acid or its equivalent as the internal phase. Vanishing cream and its variants are first prepared as ordinary liquid emulsions at high temperature; the structure that gives them their semisolid character forms as the emulsions cool. Both the aqueous and stearic acid phases are heated above the point at which the waxy components liquify and are then emulsified. Sufficient emulsifier is either formed in situ or added in to create a substantial micellar phase to exist in equilibrium with the liquified internal phase of the hot emulsion. In the instance of the classic vanishing cream, about 20% of the stearic acid it contains is neutralized with strong alkali to form the surfactant. Portions of the waxy alcohols or undissociated waxy acids are solubilized within such micelles. As these systems are then cooled, their emulsion droplets solidify and the micellar structures linking all together take on a liquid crystalline character [48]. The latter three-dimensional matrix has been referred to as *frozen micelles* and is what actually solidifies such creams. The compositions and amounts of both the internal phase and emulsifiers determine the extent and qualities of the structure. Creams such as this are more or less stiff depending on the level of micellar solubilization of and the melting properties of the internal waxy component [48]. Within the family of such creams, the internal phase ranges in composition from about 12 to 40% by weight.

Stiff o/w emulsions can also result from droplet interactions of the internal phase, but this requires emulsifying such a huge amount of internal phase that the droplets exceed close spherical packing. In this state, the emulsified particles are squashed together, losing their sphericities, producing large interfacial areas of contact at the sites where the droplets meet. A fragile structure is obtained somewhere between that of a highly viscid liquid and a true semisolid. This cream type is far less common than systems built around frozen micellar structures.

A semisolid cream of the o/w type containing a solidified, liquid crystalline internal phase is an elegant topical system, preferred by many for general purposes. Such o/w systems are readily diluted with water and, thus, easily rinsed off the skin and are generally nonstaining. After application, volatile components of the cream, which may constitute as much as 80% of the total system, evaporate, and the thin application shrinks down into an even thinner layer. Stearic acid creams are particularly interesting in this regard. The small amount of internal phase they contain causes them to evaporate down to near nothingness. The dry, nontacky, translucent nature of the stearic acid crystals left on the skin contributes to the sense of their indetectability. Most hand lotions and creams and foundation creams used to make face powders adherent to the face are variants of the vanishing cream formula.

Through evaporation, the drug in a spread is concentrated in its forming film, a process that can be orchestrated to program drug delivery. If no thought is given to the consequences attending drying out of the formulation, however, the drug is just as likely to precipitate out, in which instance drug delivery comes to an abrupt stop. Therefore, one must ensure that the formed film has some capacity to dissolve its drug. To this end, low volatility, water-miscible solvents, such as propylene glycol, are added to many cream formulations. When ingredients such as water and alcohol evaporate, the film left after applying these creams becomes a rich concentrate of drug, its internal phase, and its less volatile external phase components. One strives to add just enough cosolvent to keep the drug solubilized in the equilibrium film, but also near saturation. It should be kept in mind that, unless the internal phase liquifies at body temperature, the waxy constituents cannot act as a solvent for the drug and, accordingly, do not lend the film much capacity for delivery.

The typical cream, a soft, emulsified mass of solidified particle in an aqueous, micelle-rich medium, does not form a water-impermeable (occlusive) film on the skin. Nevertheless, creams contain lipids and other moisturizers that replace substances lost from the skin in the course of everyday living. Creams thus make good emollients because, by replenishing lipids and in some instances also polar, hygroscopic substances, they restore the skin's ability to hold onto its own moisture.

The oleaginous phases of creams differ compositionally from hydrocarbon ointments. Many creams are patterned after vanishing cream and contain considerable stearic acid, but not all. In lieu of some or all of the stearic acid, creams sometime contain long-chain waxy alcohols (cetyl, C_{16}; stearyl, C_{18}), long-chain esters (myristates, C_{14}; palmitates, C_{16}; stearates, C_{18}), other long-chain acids (palmatic acid), vegetable and animal oils, and assorted other waxes of both animal and mineral origin.

Properly designed o/w creams are elegant drug delivery systems, pleasing in both appearance and feel post application. They are nongreasy and are rinsable. They are good for most topical purposes and are considered particularly suited for application to oozing wounds.

E. Gels (Jellies)

Gels are semisolid systems in which a liquid phase is trapped within an interlocking, three-dimensional polymeric matrix of a natural or synthetic gum. A high degree of physical or chemical cross-linking of the polymer is involved. It only takes from 0.5 to 2.0% of the most commonly used gelants to set up the systems. Some of these systems are as transparent as water itself, an aesthetically pleasing state. Others are turbid, as the polymer is present in colloidal aggregates that disperse light. Clarity of the latter ranges from slightly hazy to a whitish translucence not unlike that observed with petrolatum.

Agarose gels admirably illustrate the properties and, to an extent, the structural characteristics of most gels. Agarose solutions are water-thin when warm, but solidify near room temperature to form systems that are soft to rubbery, depending on the source and concentration of the agarose. A structure is set up as the result of the entwining of the ends of different polymer strands into double helices. Kinks in the polymer mark the terminal points of these windings. Because individual polymer strands branch to form multiple endings, a three-dimensional array of physically cross-linked polymer strands is formed. The process of physical cross-linking is actually a crystallization phenomenon tying polymeric endings together, fixing the strands in place, yielding a stable, yet deformable, structure [49]. Less extensive structure than that found in agar growth media results in a spreadable semisolid suitable for medical application.

The structure should persist to temperatures exceeding body temperature for the gelled systems to be the most useful. Importantly, gellation is never a result of mere physical entanglement of polymer strands, or otherwise, the systems would be only highly viscid. The polymers used to prepare pharmaceutical gels include natural gums, such as tragacanth, pectin, carrageen, agar, and alginic acid, and synthetic and semisynthetic materials, such as methylcellulose, hydroxyethylcellulose, carboxymethylcellulose, and carboxypolymethylene (carboxy vinal polymers sold under the name Carbopol, B. F. Goodrich).

Gels or jellies are used pharmaceutically as lubricants and also as carriers for spermacidal agents to be used intravaginally with diaphragms as an adjunctive means of contraception. Since the fluid phase of a gel does not have to be strictly water, but can contain appreciable amounts of water-miscible organic solvents, gels hold considerable potential for an even wider range of uses, for, by blending solvents, it is possible to form gelatinous films with varying evaporation rates, solvencies, and other release attributes. A good example of what can be accomplished is found in Topsyn Gel (Syntex), which contains the anti-inflammatory steroid fluocinonide in a propylene glycol matrix gelled with Carbopol [50]. This product is used to treat inflammatory reactions of the scalp in lieu of creams and ointments, as the latter have proved too greasy for the purpose.

F. Rigid Foams

Foams are systems in which air or some other gas is emulsified in a liquid phase to the point of stiffening. As spreadable topical systems go, medicated foams tend toward the fluid side, but, like some shaving creams, they can be stiffer and approximate a true semisolid. Like the second type of o/w emulsion that only borders on semisolidity, these derive structure from an internal phase—bubbles of an entrapped gas—so voluminous that it exceeds close spherical packing. Consequently, the bubbles interact with their neighbors over areas, rather than at points of contact. The interactions are often sufficient to provide a resistance to deformation and something approaching semisolid character. Whipped cream is a common example of this type of system. Here air is literally beaten into the fluid cream until it becomes stiff. Aerosol shaving creams and certain medicated quick-breaking antiseptic foams are examples of the foams currently found in cosmetic and therapeutic practice. These are supplied in pressurized cans which have special valves that emulsify gas into the extruded preparations.

G. Common Constituents of Dermatological Preparations

So many materials are used as pharmaceutical necessities and as vehicles in topical systems that they defy thorough analysis. Nevertheless, the pharmacist should make some effort to learn of the more common constituents and their principal functions. This can be done by reading labels and studying the compositions of formulations, including those presented in Table 9.

Table 9 Prototype Formulations

I. *Ointment* (white ointment, USP)

White petrolatum	95% (w/v)
White wax	5%

Melt the white wax and add the petrolatum; continue heating until a liquid melt is formed. Congeal with stirring. Heating should be gentle to avoid charring (steam is preferred), and air incorporation by too vigorous stirring is to be avoided.

II. *Absorption ointment* (hydrophilic petrolatum, USP)

White petrolatum	86% (w/w)
Stearyl alcohol	3%
White wax	8%
Cholesterol	3%

Melt the stearyl alcohol, white wax, and cholesterol (steam bath). Add the petrolatum and continue heating until a liquid melt is formed. Cool with stirring until congealed.

III. *Water-washable ointment* (hydrophilic ointment, USP)

White petrolatum	25% (w/w)
Stearyl alcohol	25%
Propylene glycol	12%
Sodium lauryl sulfate	1%
Methylparaben	0.025%
Propylparaben	0.015%
Purified water	37%

Melt the stearyl alcohol and white petrolatum (steam bath) and warm to about 75°C. Heat the water to 75°C and add the sodium lauryl sulfate, propylene glycol, methylparaben, and propylparaben. Add the aqueous phase and stir until congealed.

IV. *Water-soluble ointment* (polyethylene glycol ointment, USP 14)

Polyethylene glycol 4000 (Carbowax 4000)	50%
Polyethylene glycol 400	50%

Melt the PG 4000 and add the liquid PG 400. Cool with stirring until congealed.

V. *Cream base, w/o* (rose water ointment, NF 14)

Oleaginous phase	
Spermaceti	12.5%
White wax	12.0%
Almond oil	55.58%
Aqueous phase	
Sodium borate	0.5%
Stronger rose water, NF	2.5%
Purified water, USP	16.5%
Aromatic	
Rose oil, NF	0.02%

Melt the spermaceti and white wax on a steam bath. Add the almond oil and continue heating to 70°C. Dissolve the sodium borate in the purified water and stronger rose water, warmed to 75°C. Gradually add the aqueous phase to the oil phase with stirring. Cool to 45°C with stirring and incorporate the aromatic (rose oil).

Note: This is a typical cold cream formulation. The cooling effect comes from the slow evaporation of water from the applied films. The aromatic is added at as low a temperature as possible to prevent its loss by volatilization during manufacture.

Table 9 Continued

VI. *Cream base, o/w* (general prototype)

Oleagenous phase		
Stearyl alcohol		15%
Beeswax		8%
Sorbitan monooleate		1.25%
Aqueous phase		
Sorbitol solution, 70% USP		7.5%
Polysorbate 80		3.75%
Methylparaben		0.025%
Propylparaben		0.015%
Purified water, q.s. ad		100%

Heat the oil phase and water phase to 70°C. Add the oil phase slowly to the aqueous phase with stirring to form a crude emulsion. Cool to about 55°C and homogenize. Cool with agitation until congealed.

VII. *Cream base, o/w* (vanishing cream)

Oleagenous phase		
Stearic acid		13%
Stearyl alcohol		1%
Cetyl alcohol		1%
Aqueous phase		
Glycerin		10%
Methylparaben		0.1%
Propylparaben		0.05%
Potassium hydroxide		0.9%
Purified water, q.s. ad		100%

Heat the oil phase and water phase to about 65°C. Add the oil phase slowly to the aqueous phase with stirring to form a crude emulsion. Cool to about 50°C and homogenize. Cool with agitation until congealed.

Note: In this classic preparation, the stearic acid reacts with the alkaline borate to form the emulsifying stearate soap.

VIII. *Paste* (zinc oxide paste, USP)

Zinc oxide	25%
Starch	25%
Calamine	5%
White petrolatum, q.s. ad	100%

Titrate the calamine with the zinc oxide and starch and incorporate uniformly in the petrolatum by levigation in a mortar or on a glass slab with a spatula. Mineral oil should *not* be used as a levigating agent, since it would soften the product. A portion of the petrolatum can be melted and used as a levigating agent is so desired.

IX. *Gel* (lubricating jelly)

Methocel 90 H.C. 4000	0.8%
Carbopol 934	0.24%
Propylene glycol	16.7%
Methylparaben	0.015%
Sodium hydroxide, q.s. ad	pH 7
Purified water, q.s. ad	100%

Disperse the Methocel in 40 ml of hot (80°–90°C) water. Chill overnight in a refrigerator to effect solution. Disperse the Carbopol 934 in 20 ml of water. Adjust the pH of the dispersion to 7.0 by adding sufficient 1% sodium hydroxide solution (about 12 ml is required per 100 ml) and bring the volume to 40 ml with purified water. Dissolve the methylparaben in the propylene glycol. Mix the Methocel, Carbopol 934, and propylene glycol fractions using caution to avoid the incorporation of air.

Because of the many materials that are used in topical preparations and the diverseness of their physical properties, the formulation of topicals tends to be something of an art, perfected through experience. Only by making myriad recipes does one eventually gain insight about the materials and their use in the design of new formulations. Such insight allows the experienced formulator to manipulate the properties of existing formulations to gain a desired characteristic. Often one finds good recipes to use as starting points for formulations in the trade literature. Two factors have to be kept in mind when borrowing the compositions of such trade formulations: (a) trade recipes (recipes supplied with advertising material touting specific components) are often inadequately tested in terms of their long-term stability and (b) the dominant features used in judging the merits of trade-promoted formulas tend to be the initial appearance and overall elegance. Little to no attention can be paid to the drug delivery attributes of the prototypical systems when they are first prepared in the suppliers laboratories, because the drug delivery attributes are so compound-specific. Thus, it is left up to the pharmacist (industrial research pharmacist) to make adjustments in the formulas that are consistent with good delivery of specific drugs. Each drug requires unique adjustments, in accord with its singular set of physicochemical properties.

H. General Methods of Preparation of Topical Systems

Irrespective of whether the scale of preparation is large or small, ointments, pastes, and creams tend to be produced by one or the other of two general methods. Either they are made at high temperatures by blending the liquid and heat-liquefiable components together and then dispersing other solids (often including the drug) within the oily melt or, in the instance of emulsions, within the aqueous phase of the emulsion or the freshly formed emulsion itself (fusion methods); or the drug is incorporated in the already solidified base (cold incorporation). As earlier pointed out, the first of these methods is commonly used to make o/w creams of the vanishing cream type. The fusion method is also used to prepare many ointments. Cold incorporation comes into play in large-scale manufacture when the systems in preparation contain heat-labile drugs, in which instance the drug is first crudely worked into an existing ointment or cream base using a serial dilution technique and is then distributed uniformly with the aid of a roller mill. Cold incorporation is also mandated when the base itself is destroyed by heat, as happens with Plastibase (Squibb).

In the fusion method for ointments, mineral oil, petrolatum, waxes, and such other ingredients as belong in the formulation are heated together to somewhere between 60° and 80°C, depending on the components, and mixed to a uniform composition while in the fluidized state. Cooling is then effected using some short of a heat exchanger. To prevent decomposition, drugs and certain delicate adjuvants are added sometime during the cooling process. If insoluble solids need to be dispersed, the system is put through a milling process (colloid mill, homogenizer, ultrasonic mixer, or other) to disperse them fully. A hand homogenizer works well at the prescription counter for small-volume, extemporaneously prepared systems. Systems in preparation are always cooled with mild stirring until they are close to solidification. The rate of cooling is important, for rapid cooling, as mentioned, imparts a finer, more rigid structure. Stirring should be set to minimize vortexing and, thereby, prevent air incorporation into the solidifying system. Representative formulations with more system-specific, detailed directions are given in Table 9 for ointments and the other semisolid systems of note.

The fusion method for preparing creams is a bit more complex. In this instance the aqueous and oil phases are heated separately to somewhere between 60° and 80°C. As a general rule the oil phase is heated to 5°C above the melting point of the highest-melting waxy ingredient, and the water phase is heated to 5°C above the temperature of the oil phase, the latter to prevent premature solidification during the emulsification process. Water-soluble ingredients are dissolved in the heated aqueous phase, and oil-soluble ingredients are dissolved in the oily melt, but only as long as they are heat-stable and not too volatile. If an o/w system is to be made, the emulsifiers are added to the aqueous phase and the emulsion is formed by slow addition of the oil phase. In the industry the crude emulsion is then passed through a high-shear mixer to form a finely divided emulsion state. Following this, the emulsion is cooled with gentle stirring until congealed, again taking care not to whip air into the formulation. Typically, the emulsions solidify between 40° and 50°C. If a w/o emulsion is to be made, the addition steps are usually reversed. Therefore and generally, the discontinuous phase is added to the continuous, external phase containing the emulsifier. However, methods vary here and, for a particular formula, the reverse order of addition may work best. Any means that reliably leads to a good emulsion is obviously acceptable.

A solid can be cold-incorporated directly into an already congealed system several ways. This is accomplished on a small scale by levigating the solid with a small portion of the total base it is to be suspended in to obtain a pastelike mass. The drug is worked into the base on a glass plate with the aid of a spatula, or is triturated in by using a mortar and pestle. After the initial mix is made smooth, a portion of the vehicle, roughly equal in bulk to that of the pasty mass, is added and blended in. This latter procedure is repeated several times more (geometric dilution) until the drug is uniformly dispersed throughout its total vehicle. In large-scale manufacture, solids are crudely dispersed into the base using a blender and then roller mills, in which a film of the formulation is passed from one roller to another and so on, in each passage with kneading and mixing, are used to obtain fine dispersion. As outlined when discussing absorption bases, the drug may also be dissolved in water to form a solution to be levigated into an ointment base or cream. Such addition softens creams even to the point of converting them to thick lotions. The chosen vehicle must have an inherent capacity to emulsify or otherwise take up the solution. Aromatic materials, such as essential oils, perfume oils, camphor, and methol, which volatilize if added when the base is hot, are incorporated into these semisolids while they are still being mixed, but near the temperature at which a particular system starts to congeal. Volatile materials are often introduced into the formulation as hydroalcoholic solutions.

The preparation of gels can also involve high-temperature processing. It is easier to disperse methylcellulose in hot than in cold water, for instance. The polymer then goes into solution and thickens or sets up as the temperature is lowered. Adding the hot methylcellulose dispersion to ice water gets one quickly to the final equilibrium state. Tragacanth gels, on the other hand, must be prepared at room temperature owing to the extreme heat lability of this natural gum. A little alcohol or propylene glycol can be mixed into this gum before adding water to it to facilate wetting and its dispersion. In contrast with these two materials, Carbopols are gelled by uniformly dispersing the polymer in an acidic medium and then neutralizing the medium with strong alkali. The alkali ionizes carboxyl groups on the polymer, instantaneously drawing the polymer into solution. Organic solvents can be gelled with Carbopol as well by selecting soluble amines for the neutralization.

Several prototype gel formulations are given in Table 9 to illustrate general compositional requirements and manufacturing methods. The design of specific systems tailored to meet predetermined, demanding performance criteria, particularly for bioavailability, generally requires modification of published formulations or a totally original approach.

VI. PERFORMANCE OF TOPICAL THERAPEUTIC SYSTEMS

Topical preparations, like all other dosage forms, must be formulated, manufactured, and packaged in a manner that assures that they meet general standards of bioavailability, physical

(physical system) stability, chemical (ingredient) stability, freedom from contamination, and elegance. Like all other pharmaceuticals, these factors must remain essentially invariant over the stated shelf life of the product, and they must be reproducible from batch to batch.

A. Bioavailability

Chemical Structure, Delivery, and Clinical Response

Much has already been said concerning the chemical structural dependencies of skin permeation. However, the goal of all treatment is successful therapy, not delivery per se, and consequently the intrinsic activities of the drugs must also be taken into account when selecting compounds for dermatological and transdermal development. The pharmacological response depends on delivering sufficient drug of a given activity to the target zone. Clearly, the more potent a compound is, the less of it that needs to be delivered. Since topical delivery is difficult at best, potency often dictates which compound from within a family of drugs should be developed, for the highly potent analog, reasonably formulated, offers the best chance of obtaining clinically sufficient delivery. Conversely, marginally potent analogs, even when expertly formulated, often fail because of inadequate delivery. An excellent example of this principle is found with the narcotic analgesics. Because of its extraordinary potency, fentanyl, with a daily palliative requirement of 1 mg, and not morphine, which requires between 60 and 120 mg to alleviate pain over the course of a day, is what has made its way into transdermal use. The fact that fentanyl is also physicochemically more suited to transdermal delivery than is morphine does not controvert the axiom concerning potency.

Unlike mass transport across membranes, which relates to chemical structure in predictable ways, the potencies of drugs as seen in pharmacological, pharmacodynamic, or other tests are highly structurally specific within a class of drugs and are without commonality across classes. A drug's activity involves a complex merging of these separate structural influences, with bioavailability always one of the concerns. Such concern is minimal when a truly superficial effect is involved. For example, the most potent antiseptic, as measured in the test tube, is also likely to have nearly the highest topical potency. The intrinsic activities of compounds may be poor indicators of relative topical potentials when deep skin penetration is required, however, because the structural features benefiting the biological response are often distinct from those that favor permeation. Thus, tissue permeability can be an important and sometimes a dominant factor in the clinical structure–activity profile.

We have seen that the determinants of skin permeation are the activity (concentration) of a drug in its vehicle, the drug's distribution coefficients between the vehicle and the skin and across all phases the skin, and the drug's diffusion coefficients within the skin strata. Congeners, if comparably sized, exhibit little variance in their diffusion coefficients. However, the structural differences seen within congeneric families profoundly affect the solubility, partitioning, and in-transit binding tendencies of the family members, in addition to determining their binding with receptors. Drug delivery and resulting clinical effectiveness are captive of the former phenomena [51]. For example, the 21-ester of hydrocortisone is more hydrophobic than its parent, its ether/water partition coefficient being about 18 times that of hydrocortisone [52]. Given the strong parallels in partitioning behaviors that exist across partitioning systems, it stands to reason that similar order-of-magnitude increases exist for the acetate's stratum corneum/water and sebum/water partition coefficients. At the same time, acetylating hydrocortisone at the 21-position increases the melting point 12°C. Consequently, not only does derivatization drop the aqueous solubility precipitously, but it depresses solubility in all other solvents as well [53]. Although the increase in partition coefficient raises the permeability coefficient relative to hydrocortisone, this effect is more than offset by reduced solubility, and far less of the acetate derivative can be delivered through the skin from respective saturated solutions [54]. However, as the aklyl chain length of the ester is methodically extended (C_3, C_4, . . . , C_7), the growing bulkiness of alkyl group increasingly interferes with crystalline packing. Consequently, melting points fall incrementally from the 224°C peak of the acetate ester to 111°C when a chain length of 7, the heptanoate ester, is reached [53]. An especially sharp drop of 69°C is experienced between chain lengths 5 and 6. Because of declining crystallinity beyond the chain length of 2, solubilities of the esters in organic solvents rise markedly. Moreover, aqueous solubilities, although methodically depressed by increasing hydrophobicity, remain many times higher than they otherwise would be. The net effect of these concerted forces is that the hexanoate and heptanoate esters of hydrocortisone are well over an order of magnitude more skin permeable than hydrocortisone when they are administered as saturated solutions [53,54].

Armed with this insight, we can examine the pharmacological ramifications of esterifying hydrocortisone. In Fig. 5 the ability of hydrocortisone esters to suppress inflammation induced by tetrahydrofurfural alcohol, which acts simultaneously as irritant and vehicle, is shown as a function of the alkyl chain length of the esters [55]. An optimum chain in effect is seen at an alkyl chain length of 6 (hexanoate), with substantially longer and shorter esters being measurably less effective. The behavior is exactly what would be predicted from partitioning and solubility considerations. That this is not an isolated behavioral pattern with corticosteroids can be seen in Fig. 6 in which vasoconstriction data of McKenzie and Atkinson for three betamethasone ester families, 21-esters, 17,21 ortho-esters, and 17-esters, are shown as functions of the ether/water partition coefficients of the compounds [56]. Vasoconstriction, blanching of the skin under the site of steroid application, is a proved index of a steroid's combined potency and ability to permeate through skin. Maxima are apparent in the data for the first two of these series and the indications are that the 17-ester series is also peaking. Both maxima lie between ether/water partition coefficients of 1,000 and 10,000 [52] as, interestingly and probably significantly, does the optimum ether/water partition coefficient of the hydrocortisone esters. The differing shapes and heights of the curves are not readily quantitatively explained, but reflect differences in intrinsic vasomotor activities of each ester type. The coincidence of the maxima on the partitioning scale, on the other hand, seemingly relates to an optimum lipophilicity for delivery.

The decline in activity at the longer chain lengths (see Fig. 6) also has a plausible explanation. Two factors are reasoned to be operative here, declining solubilities, coupled with changes in the absorption mechanism associated with stratification of the barrier. Aqueous solubilities of homologs exponentially decline as alkyl chain length is extended [57,58]. Whereas melting points are negatively affected and depressed early in the series, the crystalline structure eventually accommodates the alkyl chain, and further increases in chain length reverse the trend, reducing all solubilities [58]. Through all of this, o/w partition coefficients, which are unaffected by crystallinity, increase exponentially, which has the effect of exponentially increasing the ability of the stratum corneum and sebum phases to transport steroids relative to that of the viable tissue layer. In other words, the resistance of the stratum corneum drops precipitously without a commensurate drop in the resistance of the viable tissue layer. As a result, the latter takes control of the permeation process [10,37,59]. Once this change in mechanism is manifest, the permeation of homologs from their saturated solutions mirrors the downward trend in aqueous solubilities, with further increases in chain length (hydrophobicity) marked with exponential declines in steady-state fluxes. The homologs quickly become inactive [59]. In effect, the homologs become so insoluble in water that they are thermodynamically restrained from partitioning into the viable tissues on breaching the stratum corneum (or pore liquids). These features are in total agreement with expectations drawn from the earlier presented skin permeability model.

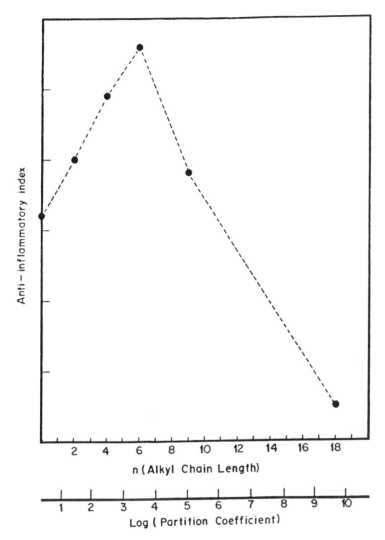

Fig. 5 Ability of hydrocortisone esters to suppress inflammation.

Clearly, the physicochemical properties of a drug are a decisive factor in its overall activity. When possible, molecular structures should be optimized to obtain the best clinical performance. Rarely does an oral drug have physicochemical features most suitable for topical or transdermal therapy, and it can take a great deal of systematic research to identify where the best balance of activity and permeability lies. Experience with the corticosteroids suggests that as much as 100-fold improvement in clinical activity may be attainable through molecular design, for today's most potent topical corticosteroids are more active than hydrocortisone by a factor at least this large.

Vehicle Properties and Percutaneous Absorption
The role solubility plays relative to maximal flux across membranes is clear from the preceding paragraphs. To kinetically reach the skin's surface, an appreciable fraction of a drug must also be in solution in the vehicle designed around it. Otherwise, diffusion of the drug through the

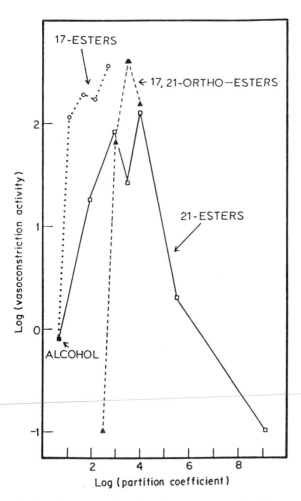

Fig. 6 Vasoconstriction data of betamethasone family.

vehicle to its interface with the skin may not completely compensate for drug lost through partitioning into the skin, kinetically dropping the drug's activity within this critical juncture of the formulation and the skin below saturation, lowering the drug's release into the surface tissues. Taken to an extreme, low vehicle solubility sets up a situation in which drug dissolution within, and diffusion through, the vehicle becomes delivery rate-controlling [60,61]. In instances for which a drug's pharmacological activity depends on getting all the drug that is possible into the tissues, this is a problem. The outcome is similar when the drug is formulated in a highly unsaturated state in the first place. Again, it will not partition into the skin to the fullest possible extent, resulting in less than maximal bioavailability. Assuming maximal delivery is the goal, the optimum between these extremes is achieved by adjusting the solvency of the vehicle so that all or most all the drug is in solution, but at the same time, the vehicle is saturated or close thereto. This has the effect of balancing the kinetic and thermodynamic factors. It is for this reason that solvents such as propylene glycol are added to topical formulations. Slowly evaporating propylene glycol provides a chemical environment in which drugs dissolve or remain dissolved, facilitating delivery. Therefore, one frequently finds propylene glycol (from 5 to 15%) in topical corticosteroid creams and other formulations.

These principles, which have been clinically validated, establish the critical role the vehicle plays in a drug's activity. Although pharmacists do not have the wherewithal to actually test products at the dispensing counter, nevertheless, they should be aware of these principles to select and dispense products from manufacturers who can demonstrate that such formulation factors have been given due consideration. These delivery dependencies also stand to caution the pharmacist. Extemporaneous mixing of commercial products, for example, one containing a steroid and another an antibiotic, and the diluting of products with homemade vehicle are suspect practices, because the compositional changes associated with such blending are likely to adversely affect the delivery attributes of otherwise carefully designed systems [62].

There is another way vehicles can influence percutaneous absorption, which is by altering the physicochemical properties of the stratum corneum. In the main, modification of the barrier results in increased skin permeability, but a buttressing effect is also achievable with substances having the capacity to solidify the horny structure. To repeat a point, simply hydrating the stratum corneum promotes absorption. This may be accomplished by covering the skin with a water-impermeable bandage or other wrapping (an occlusive dressing). The blockage of evaporative water loss leads to hydration of the stratum corneum, softening it and increasing the diffusive mobilities of chemicals through it. The occlusive covering also prevents evaporation of volatile vehicle components, compositionally stabilizing a spread film, maintaining its solvency for the drug. It is estimated that occlusive hydration increases percutaneous absorption from five- to tenfold, enlargements that are often clinically significant [63]. The technique has been used with corticosteroids in refractory dermatoses, such as psoriasis.

The following interesting phenomena associated with the occlusion of corticosteroids are enlightening. When applied under an occlusive dressing corticosteroids induce vasoconstriction at lower concentrations than when applied in the open. When the dressing is removed, vasoconstriction subsides in a few hours. However, as many as several days later blanching can sometimes be restored simply by rewrapping the area of application [64]. This suggests that steroid molecules somehow bottled up in the stratum corneum are released when occlusion is reestablished. It appears that, as the stratum corneum dehydrates and returns to its normal state, substances such as the corticosteroids that may be present are entrapped within one of its physical domains, freezing them in place until either the stratum corneum is sloughed or until occlusive hydration is reinstated. The phenomenon is referred to as the skin's *reservoir effect.* The application of drugs dissolved in volatile solvents, such as actone and ethanol, also creates reservoirs in the stratum corneum, for as the solvents evaporate, the concentrating drug is driven into the skin's surface. Certain solvents also momentarily increase the solvency of the stratum corneum [65].

A few water-miscible organic solvents are taken up by the stratum corneum in amounts that soften its liquid crystalline, lipoidal domain [65], particularly when applied in concentrated form. If used very liberally (under laboratory conditions), these so-called skin penetration enhancers even elute interstitital lipids and denature keratin [66,67]. Under these admittedly artificial conditions, the increases in percutaneous absorption resulting from their actions can be dramatic. Dimethyl sulfoxide (DMSO), dimethylacetamide (DMA), and diethyltoluamide (DEET) are key examples. Dimethyl sulfoxide has long been touted as a skin penetration enhancer. Experimental studies indicate that it reversibly denatures keratin, opening up the protein matrix, facilitating permeation [66,67]. It also extracts lipids from the skin when applied liberally. Even when used sparingly, neat DMSO is imbibed to a degree by the stratum corneum, increasing the ability of the tissue to dissolve substances of all kinds [68]. This favors the absorption of drugs by allowing more drug to dissolve in the tissue, steepening diffusion gradients that can be expressed across the tissue. Moreover, a crossflow of DMSO (inward) and water (outward) is set up when concentrated DMSO is placed over the skin, which delaminates the stratum corneum, apparently with a pooling of solvent between the separated layers. Of all the possible actions of DMSO, the latter two seem to be the most important because extraction of skin lipids and denaturation of keratin require far more DMSO than found in a topical application of ordinary thickness (20–30 μm). These collective factors, in the extreme, effectively chemically remove the stratum corneum as a contributing part of the skin barrier [33]. In reality, however, the limited amounts of enhancer actually applied limit enhancement. Moreover, much of the enhancement capacity is lost if the solvents are in any way diluted. Thus, the use of neat organic solvents as skin penetration enhancers is only a sometimes practice.

It has long been known that certain surfactants (e.g., sodium lauryl sulfate) are skin irritants, even when in relatively diluted states, in part because they impair the barrier function of the stratum corneum, facilitating their own absorption. Concern over irritation precludes serious consideration of agents such as these as enhancers. However, certain weaker amphiphilic substances (e.g., methyl oleate, glyceryl monolaurate, propylene glycol monolaurate), some of which have long been used as ingredients in cosmetics if not therapeutic systems, are showing they have an unrealized potential as enhancers. Of special importance, surface-active (amphiphilic) substances such as these are effective in the small amounts that can actually be applied to skin in spread films. Amphiphilic molecules penetrate into and blend with the stratum corneum's own lipids, which themselves are polar, amphiphilic substances. Thermoanalytical and spectroscopic evidence indicates that, in doing so, they relax the ordered structure of the stratum corneum's natural lipids, facilitating diffusion through existing channels and perhaps freeing up new channels [69–71]. Moreover, the reduction in liquid crystallinity invariably increases the capacity of the stratum corneum to dissolve substances, further magnifying the effect. It appears that relatively short alkyl chains (C_{10}, C_{12}) and relatively weak polar end groups favor enhancement.

These emerging structural requirements of enhancement have launched a quest for new, even more powerful enhancers. Several potent amphiphiles have surfaced from this pursuit which, at their worst, are only mildly tissue provocative, key examples being N-dodecylazacycloheptan-2-one (Azone) and methyl decyl sulfoxide. Each of these example compounds contains a short alkyl chain (Azone = C_{12}; decyl methyl sulfoxide = C_{10}) attached to a highly water-interactive but nonionic head group. Neither compound is at all water-soluble of itself, indicating that neither has the amphiphilic balance to form its own micelles. Of the two compounds, Azone has been the most scrutinized. It promotes the absorption of polar solutes at surprisingly low percentage concentrations. Its effects on animal skins have been especially profound; up to several hundred-fold improvements in the in vitro permeation rates of highly polar cyclic nucleosides through hairless mouse skin have been reported [72], for example. However, Azone does not appear to be comparably effective on human skin, but those actions it has are effected at low concentrations [73]. As with any agent of this kind, its actions are dependent on formulation and how this affects the thermodynamic activity of the enhancer in the delivery system. Concern over toxicity and the availability of alternative substances with established safety pedigrees have become impediments to the introduction of enhancers having new, totally unfamiliar chemical structures.

The effects of skin penetration enhancers on the stratum corneum may or may not be lasting, depending on the degree of chemical alteration of the stratum corneum that is experienced. Irreversibility is a perceived problem to the extent that the skin is left vulnerable to the absorption of other chemicals that come in contact with the conditioned area for as long as the area remains highly permeable. The fear is that such vulnerability will stay high until the greater part of the stratum corneum is mitotically renewed, which minimally takes several whole days. Moreover, the enhancing solvents are themselves absorbed to some degree, another source of toxicological concern. DMSO is known to increase intraocular pressure; DMA has been associated with liver damage; Azone may irritate, but its real liability is that its chemical structure is totally novel and without toxicological precedent. Although worry over toxicity may be out of proportion to the actual degrees of exposure attending the ordinary circumstances of clinical use of dermatological products, nevertheless, concern is warranted, given the occasional use of products over expansive areas. Consequently, compounds of proved safety, and their structural kind, are factoring out as the enhancers of the 1990s.

Transdermal Delivery: Attributes of Transdermal Delivery Systems

We are learning more and more that the conditions of use of topical delivery systems has profound influence on their performance. Transdermal systems, specifically the adhesive patches that are used to treat systemic disease, and dermatologicals, are subject to very different operating environments and conditions. Transdermal delivery is aimed at achieving systematically active levels of a drug. A level of percutaneous absorption that leads to appreciable systemic drug accumulation is absolutely essential. Ideally, one would like to avoid any buildup of a drug within the local tissues, but, nevertheless, buildup is unavoidable, for the drug is driven through a relatively small diffusional area of the skin defined by the contact area (absorption *window*) of the application. Consequently, high accumulations of drug in the viable tissues underlying the patch are preordained by the nature of the delivery process. Irritation and sensitization can be associated with such high levels; therefore, careful testing is done to rule out these complications before a transdermal delivery system gets far along in development.

Table 10 outlines general expectations associated with transdermal delivery. The water-impermeable backing materials of present, and we can presume, most future, transdermal systems cause them to operate occlusively. There is good reason for this. Foremost is again that occlusion facilitates drug delivery. In laboratories where these systems are designed, it has been learned that occlusion is often essential to achieving adequate rates of delivery. Furthermore, transdermal drugs, such as nitroglycerin and nicotine, are themselves relatively volatile compounds. Although they can be packaged in a fashion that prevents drug loss, a backing material that is substantially impermeable is also needed to prevent these compounds from evaporating

Table 10 Listing of the Norms of Operation of Transdermal Patches

Occluded applications
Composition relatively invariant in use
System size (area) predetermined
Specific site prescribed for application
Application technique highly reproducible
Delivery is sustained
Generally operate at unit drug activity, at least operate at steady activity
Delivery is zero-order
Serum levels related to product efficacy
Bioequivalency based on pharmacokinetic (blood level) endpoint
Unavoidable local tissue levels consequential only to system toxicity
Individual dose interruptable
Whole system removed when spent
Delivery efficiency is low (only a fraction of drug content is delivered)

off into space after placing patches containing them on the skin. Impermeable polymer or foil backings also block the diffusive transport of body water to the atmosphere by way of the patch. Insensible perspiration at the site of the patch is thus held in check, but not without creating a substantially moist environment at the interface the patch has with the skin. Consequently, if given enough time, organisms already in the skin can colonize within this interface.

Other than for possibly the insensible perspiration they absorb, transdermal patches tend to operate as thermodynamically static systems, meaning as compositionally fixed systems, from the moment they are applied until their removal. Marketed ethanol-driven estradiol and fentanyl patches are exceptions here, as these meter out ethanol and drive it into the stratum corneum to propel the absorption process. Compositional steadfastness is still the rule, however, and it is this feature that bestows the zero-order delivery attribute on the ordinary transdermal patch. Drug is present within the patches in reservoir amounts, irrespective of whether or not the reservoir compartment is easily distinguished, for enough drug has to be present to sustain delivery over the full course of patch wear, no matter if 1 day, 3 or 4 days, or 7 days is the time objective.

In some prototypes (e.g., the nitroglycerin transdermal systems), huge excesses of drug are placed in the patch to assure that the drug's activity remains essentially level during the patch's wear. Only a small fraction of the drug, well under 50% of the patch's total content, is actually delivered during the prescribed time the patch is to be worn. In situations in which the drug is prohibitively expensive or prone to abuse (e.g., fentanyl), efficiencies have to be raised to the maximum that are physically achievable, and the fractional delivery of formulated drug has been made to approach or exceed 50%. Part of the inherent stability of the delivery environment of patches results because their main materials of construction are polymers, fabricating laminates, and adhesives, all of which tend to be chemically robust. Solubilizing solvents (e.g., ethanol) and skin penetration enhancers (e.g., propylene glycol monolaurate) may also be present, and their absorption into the skin may change compositions, but even here, the processes are carefully orchestrated to gain a stable, long-term delivery environment.

A transdermal patch is a self-contained system that is applied as it is packaged, with its only manipulation being removal of the release liner to expose and ready its adhesive surface. The size of a patch, meaning its area of contact with the skin, is determined even before it is made. All of this area, or only an inner portion of it, may actually be involved in drug delivery, but, either way, the area is fixed. Since absorption is proportional to area, to meet the differing drug requirements of individual patients, patches of different sizes are generally made available. The application site is also a constant of therapy in that a specific site or sites are recommended for use (not always for scientifically supportable reasons, e.g., nitroglycerin patches are worn over the heart). Users tend to follow such dictates. Beyond this, the manner of application is also highly reproducible. Thus, there is as tight a control over absorption area and application site variables here as can be found in all of therapy. The only variability not customarily controlled for is that associated with the skin's permeability itself, but even here some systems have been made to operate with high-delivery precision by incorporating a rate-controlling membrane into them. Altogether the manner of use of the systems is highly reproduced from one application to the next.

Measures of function of transdermal systems distinguish them among the systems we use topically. Since systemic actions are sought, blood levels of the drug in question must reach and remain within therapeutic bounds. More often than not the requisite blood level is known from a drug's use by other routes of administration. Thus, a clear systemic target level usually exists, and an absolute rate of delivery commensurate with reaching this is a built-in feature of the patch. The requisite delivery rate can be estimated in several ways, even before any attempt is made to design a transdermal system. For instance, once an upper size limit is set for the patch, the total daily oral requirement of the drug can be used to calculate the minimal delivery rate in milligrams per square centimeter per hour. This is done after making an appropriate downward adjustment in the transdermal dose to account for oral drug losses attributable to first-pass metabolism. Alternatively, the rate can be estimated from an established blood level and a known rate of systemic clearance (as the product of these). Comparing performances of transdermal delivery systems is also a straightforward matter. Bioequivalency of different systems built around a specific drug are easily measured in terms of the blood levels they produce. And if therapy is not going well, one can bring delivery to a reasonably abrupt halt by simply removing a patch.

Topical Delivery: Attributes of Topical Delivery Systems

Topical delivery systems fill an important niche in therapy. Although not an efficient means of delivery in the sense that as little as 1% and usually no more that 15% of the drug in a dermatological application is systemically absorbed (systemically recoverable), nevertheless, topical delivery allows one to achieve total tissue levels of a drug far in excess of those achievable by the drug's systemic administration. At the same time, systemic toxicities of the drug are rarely encountered with topical administration, with the exceptions occurring when dermatological formulations are used liberally over extensive areas. Because only small amounts of a drug are ordinarily applied topically, in most instances the systemic levels achieved are so limited that one has trouble even measuring them. Thus, albeit imprecise, topical therapy actually represents a brute force form of drug targeting and has been discussed in this context. The principal drug delivery systems for this purpose are ointments, creams, and gels, with miscellaneous other powder, liquid, and semisolid vehicles sometimes being employed. The norms of topical delivery, which are in striking contrast with the norms of transdermal delivery, are outlined in Table 11.

We tend to think of them as being much the same, but the functioning of semisolid dermatologicals stands in stark contrast with that of transdermal delivery systems. To begin with, most topical applications are left open to the atmosphere. Amounts applied per unit area depend on the individual making the application. Of singular importance relative to system function, extraordinary physicochemical changes accompany the evaporative concentration of these formulations, possibly including the precipitation of the drug or other substances that were com-

Table 11 Listing of the Norms of Operation of Dermatological Formulations

Open application
Experience profound compositional shifts in use
May experience phases changes in use
Site is the disease's location
Operate at variable drug activity
Highly nonstationary state kinetics
Application technique and amount are highly individualized
Applications short-acting
Local tissue levels tied to efficacy
Used on diseased, damaged skin
No easy bioequivalency endpoint
Systemic absorption absolutely undesirable, but some unavoidable
Therapy interruptable by washing off application
System removal inadvertent—*wear and tear*
Delivery efficiency is low (only a small fraction of drug is delivered)

fortably in solution at the moment of their application. Evaporative concentration can also upset the oil-to-water balance of emulsions, destabilizing them, and at times, causing them to break or invert. In a matter of hours, if not just minutes, a surface film or dry residue having a totally different delivery faculty than the bulk formulation may be all that is left of the application. Such precipitous changes, if out of control, can bring drug delivery to an abrupt halt.

The amounts of ointments and creams people apply are highly individualized. So are the techniques of application. Some patients use a vigorous inunction, whereas others just work the application until it is more or less uniform over the desired site and stop there. Although pharmacokinetic assessments of a system's delivery attributes are ordinarily done using normal skin (in vitro) or on healthy volunteers (in vivo), the site of its clinical deployment is usually anything but normal. Rather, it is determined by the skin condition to be treated. Clearly, the manufacturer is without control over how a disease is expressed on a particular patient. For many diseases, disease manifestation can be anywhere on the body. Moreover, from individual to individual it varies in intensity and vastness. Thus, more area may be involved in one person than in another, and the barrier function of the skin may be more or less intact in any instance. The net result of this creates a set of imponderables relative to delivery, efficacy, and safety.

The removal of the dermatological applications is rarely deliberate. Rather, some substance is usually transferred to clothing and such; some is absorbed; some evaporates; and some is inadvertently removed by bathing or other means. Applications can be deliberately washed from the skin if one wishes to terminate therapy. Partly because of their temporal inhabitancy, local applications tend to be short-acting relative to transdermal delivery systems. Other factors here are the finite doses that are actually administered and the oftentimes rapid evaporative concentration of such films to compositions that cease supporting dissolution of the drug and its diffusion to the skin's surface. Consider that the application of a topical product to the skin in a representative, 20-μm–thick layer places only 20 μg of drug over each square centimeter of skin when the drug is formulated at the relatively high concentration of 1%. Roughly 10 mg of stratum corneum covers each square centimeter of skin, enough for 20 μg of drug to effectively get lost in. Consequently, only a fraction of the drug that does enter the skin actually reaches the live tissues. Such finite doses do not sustain delivery and, thus, delivery wanes after several hours irrespective of the wearability of the application and of processes attending its evaporative shrinkage. Since all these attending processes defy quantification, there is precious little existing information to guide one concerning the fitting regimen of application for most topical dosage forms. Rather, dosing regimens have evolved historically from collective clinical experience. All in all, topical therapy is an extraordinarily complex operation.

Compositional changes following the application of certain topical systems are unavoidable. Many o/w creams contain as much as 80–85% external phase, usually primarily water. Lotions and gels also contain volatile constituents in large proportion. All rapidly evaporate down after their application and, consequently, the drug delivery system is the formed, concentrated film that develops on the skin and not the medium as packaged in the tube, jar, or bottle. Ingredients should be chosen to assure that compositional changes, as invariably occur, interfere as little as possible with delivery and therapy. The rate at which the volatile components evaporate to form the equilibrium film can itself be a factor in bioavailability [74]. It has been reported, for instance, that a thinly applied corticosteroid preparation produced greater vasoconstriction than did thicker applications of the same material [75]. Although the total amounts of drug per unit area were greater with the thick films, responses were less in their case because either evaporative concentration of the steroid proceeded more slowly or dilution by insensible perspiration was more rapid. Even without knowing the mechanistic details, we can conclude from this that less steroid was driven into the skin from the thick applications in the course of the test. It has also be demonstrated that vasoconstriction is more pronounced at low concentrations when steroid is applied in volatile ethanol than when applied in propylene glycol [55]. Even though differences in solvency play their role here, it is also clear that the rapid evaporation of solvents such as ethanol drives drug into the skin. Such observations emphasize the importance of distinguishing between the system as packaged and the transitional system following application. Unfortunately, this distinction is not always made, and much topical delivery research aimed at assessing the relative abilities formulations have to deliver drug has been performed by placing extraordinarily thick layers of formulation over the skin. Such thick applications do not even remotely simulate the clinical release situation, especially when it comes to creams and gels. This area of drug delivery is in need of much research.

Knowledgeable formulators can use the tendency of creams, gels, and other systems to evaporatively concentrate to advantage. Solvents are chosen and blended so that the drug remains soluble in the formed film long after application is made. This can be accomplished by replacing a fraction of the water or other highly volatile solvent found in these systems with solvents of far lower volatility; and, as pointed out previously, propylene glycol is found in many topical corticosteroid creams and lotions just for this reason.

In summary, the way a topical drug is formulated has a great deal to do with its clinical effectiveness, a nonsurprising conclusion given what is known about the relations between bioavailability and formulation for other modes of administration. Yet, in the area of topical drug performance, antiquated concepts and approaches to system design linger on. In the days when topical bioavailability was little understood and, therefore, ignored, formulators concentrated on vehicle elegance and stability. Attempts were made to design vehicles compatible with all types of drugs, so-called universal vehicles. Universal vehicles are still discussed in many standard texts. Today's technology and science clearly indicate that the universal vehicle is akin to a unicorn: beautiful, but totally mythical. In the real world, each system must be designed around the drug it contains to optimize the clinical potential of the active ingredient. The duration of action will depend on how long the drugs remains appreciably in solution within its spread film.

B. Aspects of Physical and Chemical Stability

Concern for that physical and chemical integrity of topical systems is no different from that for other dosage forms. However, there are some unique and germane dimensions to stability associated with semisolid systems. A short list of some of the factors to be evaluated for semisolids is given in Table 12. All factors must be acceptable initially (within prescribed specifications), and all must remain so over the stated lifetime for the product (the product's *shelf life*).

The chemical integrities of drug, preservatives, and other key adjuvants must be assessed as a function of time to establish a product's useful shelf life from the chemical standpoint. Semisolid systems provide us with two special problems here. First, semisolids are chemically complex, to the point that just separating drug and adjuvants from all other components is an analyst's nightmare. Many components interfere with standard assays and, therefore, difficult separations are the rule before anything can be analyzed. Also, since semisolids undergo phase changes on heating, one cannot use high-temperature kinetics for stability prediction. Thus stability has to be evaluated at the storage temperature of the formulation, and this takes a long time. Under these circumstances, problematic stability may not be evident until studies have been in progress for a year or more. Be this as it may, stability details are worked out in the laboratories of industry, the pharmacist ordinarily accepting projected shelf lives as fact. Some qualitative indicators of chemical instability that the pharmacist might

Table 12 Factors for Evaluation of Semisolids

Stability of the active ingredient(s)
Stability of the adjuvants
Visual appearance
Color
Odor (development of pungent odor or loss of fragrance)
Viscosity, extrudability
Loss of water and other volatile vehicle components
Phase distribution (homogeneity or phase separation, bleeding)
Particle size distribution of dispersed phases
pH
Texture, feel upon application (stiffness, grittiness, greasiness, tackiness)
Particulate contamination
Microbial contamination and sterility (in the unopened container and under conditions of use)
Release and bioavailability

look for are the development of color (or a change in color or its intensity) and the development of an off odor. Often products yellow or brown with age as the result of oxidative reactions occurring in the base. Discolorations of this kind are commonly seen when natural fats and oils (e.g., lanolins) are used to build the vehicle. Extensive oxidation of natural fatty materials (rancidification) is accompanied by development of a disagreeable odor. One may also notice phase and texture changes in a suspect product. Pharmacists should take note of the appearances of the topical products they dispense, removing all those from circulation that exhibit color changes or become fetid. Changes in product pH also indicate chemical decompositions, most probably of a hydrolytic nature, and if somehow detected are reason to return a product.

Time-variable rheological behavior of a semisolid may also signal physical or chemical change. However, measures such as spreadability and feel on application are probably unreliable indicators of a changing rheology and more exacting measurements are necessary. A pharmacist does not ordinarily have the tools at hand to make accurate rheological assessments, but the equipment to do so is generally available and used within the development laboratories of the industry. One may find there exquisitely sensitive plate and cone research viscometers which, in principle, precisely quantify viscosity, or simply utilitarian rheometers. The latter include extrusion rheometers, which measure the force it takes to extrude a semisolid through a narrow orifice; penetrometers, which characterize viscosity in terms of the penetration of a weighted cone into a semisolid; and Brookfield viscometers, with spindle and helipath attachments, which measure the force it takes to drive a spindle helically through a semisolid. As used with semisolids, the utilitarian rheometers provide only relative, although quite useful, measures of viscosity. Increases (or decreases) in viscosity by any of these measuring tools indicate changes in the structural elements of the formulation. The gradual transformations in semisolid structure that take place are more often than not impermanent, in which event, the systems are restored to their initial condition simply by mixing them. Substantial irreversible rheological changes are a sign of poor physical stability.

Changes in the natures of individual phases of or phase separation within a formulation are reasons to discontinue use of a product. Phase separation may result from emulsion breakage, clearly a critical instability. More often, it appears more subtly as *bleeding*, the formation of visible droplets of an emulsion's internal phase in the continuum of the semisolid. This problem is the result of slow rearrangement and contraction of internal structure. Eventually, here and there, globules of what is often clear liquid internal phase are squeezed out of the matrix. Warm storage temperatures can induce or accelerate such structural crenulation; thus, storage of dermatologicals in a cool place is prudent. The main concern with a system that has undergone such separation is that a patient will not be applying a medium of uniform composition. Because of unequal distribution between phases (internal partitioning), one phase will invariably have a high concentration of the drug relative to the other. Therefore, since semisolid emulsions, unlike liquid emulsions, cannot be returned to an even distribution by shaking, formulations exhibiting separation are functionally suspect and should be removed from circulation.

Pharmacists should also take a dim view of changes in the particle size, size distribution, or particulate nature of semisolid suspensions. They are the consequence of crystal growth, changes in crystalline habit, or the reversion of the crystalline materials to a more stable polymorphic form. Any crystalline alteration can lead to a pronounced reduction in the drug delivery capabilities and therapeutic usefulness of a formulation. Thus, products exhibiting such changes are seriously physically unstable and unusable.

A more commonly encountered change in formulations is the evaporative loss of water or other volatile phases from a preparation while it is in storage. This can occur as the result of inappropriate packaging or a flaw made in packaging. Some plastic collapsible tubes allow diffusive loss of volatile substances through the container walls. One will find this phenomenon occasionally in cosmetics, which are hurried to the market place without adequate stability assessment, but rarely in ethical pharmaceuticals, which are time-tested. However, a bad seal may occur in any tube or jar, irrespective of its contents, with eventual loss of volatile ingredients around the cap or through the crimp. Such evaporative losses cause a formulation to stiffen and become puffy, and its application characteristics change noticeably. There is corresponding weight loss. Under this influence the contents of a formulation may shrink and pull away from the container wall. These phenomena are most likely to be seen in creams and gels owing the high fractions of volatile components that characterize them. Problems here are exacerbated when products are stored in warm locations.

Gross phase changes are detectable by eye on close inspection of products. The package may get in the way of such analysis, but if a product is truly suspect, it should be closely examined by opening and inspecting the full contents of the container. A jar can be opened and its contents probed with a spatula without wrecking the container. Close inspection of the contents of a tube requires destruction of the package, however. The easy way to do this is to scissor off the seal along the bottom of the tube and then make a perpendicular cut up the length of the tube to the edge of the platform to which the cap is anchored. Careful further trimming a quarter of the way around the platform in each direction creates left and right panels that can be pealed back with tweezers to expose the tube's contents. Textural changes such as graininess, bleeding, and other phase irregularities are easily seen on the unfolded, flat surface. Normally it takes a microscope to reveal changes in crystalline size, shape, or distribution, but palpable grit is a sure sign a problem of the kind exists. Weight loss of a product, which is easily checked at the prescription counter, clearly indicates the loss of volatile ingredients (the weight of a suspect tube can be directly compared with the weight of a fresh tube). On the rare occasions when deteriorations such as these are noted by the pharmacist in the course of handling products, or are reported to the pharmacist by knowledgeable patients, the suspect packages should be removed from circulation, and the manufacturer informed of the action. If a problem seems general, rather than isolated (i.e., to a single bad package) the FDA should be notified as well to best safeguard the public. This agency will determine if a product has gone bad and general recall is warranted.

C. Freedom from Contamination

Particulates

Numerous topical preparations contain finely dispersed solids. Pastes, for example, contain as much or more than 50% solids dispersed in an ointment medium. Powders themselves are used topically. Many dermatological liquids and semisolids contain suspended matter. However present, the particles should be impalpable (i.e., incapable of being individually perceived by touch) so that the formulations do not feel gritty. The palpability of a particle is a function of its hardness, shape, and size. The pharmacist can manipulate only the latter; thus, it is important to prepare or use finely subdivided solids when making topical dosage forms. Individual particles larger than 50 μm in their longest dimension can be individually perceived by touch. The surface of the eye is substantially more sensitive, and a 10-μm particle can be distinguished here. Clearly, the presence of hard, palpable particulates in semisolids makes them abrasive, particularly when applied to disease- or damage-sensitized skin. Severe eye irritation is possible if ophthalmic ointments contain them. One particularly troublesome source of particulate contamination is flashings (tiny metals slivers and shavings) left over from the production of tin and aluminum collapsible tubes. These often adhere electrostatically and tenaciously to tubing walls following cutting of the containers down to a particular size. Some escape removal in washing and rinsing done to cleanse the empty containers. Consequently, a jet of exceedingly high-velocity air is blown into the open end of tubes just before their filling to remove all particulates. If this precaution is not taken, tiny metal slivers may be packaged with the product, posing the threat that they will become dislodged and instilled into the eye while the product is in use. For reasons as this, the *United States Pharmacopeia/National Formulary (USP/NF)* has a particulate test for ophthalmic ointments. In this test the ointments are liquefied in a petri dish at high temperature, 85°C for 2 hr, and then solidified by cooling. Particles that have settled to the bottom of the shallow glass container are counted by microscopic scanning at 30-times magnification. The requirements are met if the total number of particles 50 μm or larger in any dimension does not exceed 50 in the ten tubes tested and if not more than 8 particles are found in a single tube. Products that are put into the distribution channels have to meet this test. Nevertheless, the pharmacist should be on the lookout for particulate problems associated with commercial products. The pharmacist must also take measures to ensure that extemporaneously compounded formulations are free of particulates. Particular attention must be paid to the cleaning of collapsible tubes and other package parts before their use.

Microbial Specifications and Sterility

As of the *USP XIX*, it is legally required that ophthalmic ointments be prepared and dispensed as sterile products (until opened for use). Presently, in the United States, nonophthalmic topical preparations do not need to be sterile, although they cannot contain pathogens and must have low microbial counts. The reasons ophthalmic sterility requirements were broadened to cover ointments are enlightening. In the mid-1960s there was an outbreak of extremely serious pseudomona eye infections in the Scandinavian block of countries, in some instances with loss of sight. The source of the contamination was traced to antibiotic-containing ophthalmic ointments made by a regional manufacturer known for its high standards of manufacturing and quality control [76]. Pathogenic pseudomonads were found in both the products and in the manufacturing facilities where the ointments were prepared. It was widely believed up until this time that pathogens could not and would not survive and grow in ointments and similar media. The presence of antibiotics in the preparations could only have added to the false sense of security this company had. This incident sent shock waves throughout the pharmaceutical world and spawned revisions in all world compendia. In the United States, ophthalmic ointments have to be sterile when dispensed. In Europe dermatological products that are to be used over broken skin also have to be sterile.

The foregoing incident has special meaning to the dispensing pharmacist. Unopened ophthalmic ointments should be dispensed for each condition and should be given very short datings. Patients should be advised to discard unused quantities of old preparations and to return for fresh supplies if and when chronic symptoms reappear. Similar advice and precautions are good practice with dermatologicals such as ointments that do not ordinarily contain microbial preservatives. Lotions, creams, and topical solutions that contain preservatives tend to remain pathogen-free after their packages have been opened, providing an extra measure of safety.

Preservatives have an important purpose in topical medications. Systems containing them tend to remain aseptic. Even if a few organisms subsist in the presence of the preservatives, these tend to be nonvegetative. Importantly, no pathogenic forms survive to cause problems. Preservatives are necessary for systems that have an aqueous phase, for water offers the most conducive environment for microbial growth. Therefore, all emulsions and aqueous solutions and suspensions should be preserved. However, choosing a preservative is no easy task, for the physical systems tend to be compositionally complex and polyphasic, affording many possible means for specific preservatives to be inactivated. In mass-produced products, the effectiveness of the preservation system of formulations is checked by the *USP* preservative challenge test.

D. Pharmaceutical Elegance

There are a number of attributes of the topical drug systems that may be classified as cosmetic, that make patients more or less willing to use their medications (compliant). These include the ease of application, the feel of the preparation once it is on the skin, and the appearance of the applied film. Ideally, the application should be undetectable to the eye and neither tacky nor greasy. Certain items, such as ointments and pastes, are intrinsically greasy, and suspensions of all types tend to leave an opaque, easily detectable film. Thus, the extent to which the cosmetic features can be idealized is dependent on the nature and purpose of the dosage form.

The ease of application and method of application of a formulation depend on the physicochemical attributes of the system involved. Solutions and other highly fluid systems may be swabbed on, sprayed on, or rolled on. A cotton pledget or other applicator is often necessary to obtain an even application. Soft semisolid systems, on the other hand, may be spread evenly and massaged into the skin with the fingers, a procedure technically referred to as inunction. The spreadability is a rheological quality related to the nature and degree of internal structure of the formulation. Formulations such as pastes that are very stiff tend to be hard to apply; their application over broken or irrigated skin can be disagreeable. The stiffness of a preparation can be up-regulated or down-regulated by manipulating the amounts of structure-building components of a vehicle and, in some instances, by adjusting the phase/volume ratio of semisolid emulsions. Thus, for ointments, increased spreadability can be obtained by decreasing the ratio of the waxy components (waxes and petrolatum) to fluid vehicle components (mineral oil, fixed oils). Greasiness of such preparations goes in the opposite direction. For o/w creams, decreasing the ratio of the internal phase to the external phase tends to make the systems more fluid. Substitution of more liquid oils for some of the high-melting waxy components of creams achieves the same end.

Tackiness and greasiness are determined by physicochemical properties of the vehicle constituents that compose the formed film on the skin. A sticky film is extremely uncomfortable and, generally, considerable effort is directed to minimizing this inelegant feature. When creams are concerned, waxy ingredients, such as stearic and cetyl alcohol, produce noticeably nontacky films. Stearic acid is the principal internal-phase component of vanishing creams, systems that are virtually undetectable visually or by touch after inunction. On the other hand, propylene glycol, which may be added to creams and gels to solubilize a drug, tends to make these systems tacky. The synthetic and natural gums used as thickening and suspending agents in gels and lotions tend to increase their tackiness and, therefore, these materials are used as sparingly as function allows.

Creams tend to be invisible on the skin. The same is true for ointments, although the oiliness of ointments causes them to glisten to an extent. Whatever opacity creams and ointments have is primarily due to the presence of insoluble solids. These often imbue applications with a powdery or even crusty appearance. Dispersed solids are usually functional, as in calamine lotion, zinc sulfide lotion, zinc oxide paste, and so on, and are an implacable feature of these preparations. However, sometimes insoluble solids are added as tints to match the color of the skin and to impart opacity. Since individual skins vary widely in hue (pigmentation) and texture, tinting to a single color and texture is generally unsuccessful.

Evaluation of the cosmetic elegance of topical preparations can be accomplished scientifically, but it is questionable whether physical experiments on system rheology and the like offer appreciable advantage over the subjective evaluations of the pharmacist, the formulator, or other experienced persons. Persons who use cosmetics are particularly adept and helpful as evaluators.

E. Skin Sensitivity: A Specific Toxicological Concern

One further problem of topical formulations associated with many ingredients, and of special concern with preservatives, is the development of skin sensitivities [77]. The skins of some individuals are particularly susceptible to an allergic conditioning to chemicals that is known as type IV contact hypersensitivity. Haptens (chemicals like urushiol found in poison ivy) are absorbed through the skin and, while in the local tissues, chemically react with local proteins. Langerhans cells, the local cells involved in immunological surveillance, identify these now denatured proteins as foreign (nonhost). The Langerhans cells then leave the dermis by way of the lymphatics and enter the draining lymph node, where they complete the sensitization process by passing the allergen message on to resident lymphocytes (antigen presentation). Once sensitized, subsequent contact with the offending chemical (hapten) leads to inflammation and skin eruption. Many of the preservatives used in pharmacy are phenols and comparably reactive substances, compounds that have a high propensity to sensitize susceptible individuals. The pharmacist should be alert to this possibility and prepared to recommend discontinuance of therapy and physician referral when allergic outbreak is evident or suspected. Moreover, the pharmacist should be ready to recommend alternative products that do not contain an allergically offending substance once it has been identified, assuming that a therapeutically suitable alternative exists.

Allergic incidents are widespread and, from an allergy standpoint, it is useful that the ingredients of dermatological medications are listed on the package or in the package insert. This allows the pharmacist to screen products for their suitability for individuals with known sensitivities. Over-the-counter medications and cosmetics also contain a qualitative listing of their ingredients. The pharmacist thus has access to critical information he or she needs to safeguard patients relative to their known hypersensitivities.

REFERENCES

1. A. P. Lemberger, in *Handbook of Non-Prescription Drugs*, American Pharmaceutical Association, Washington, DC, 1973, p. 161.

2. G. L. Wilkes, I. A. Brown, and R. H. Wildnauer, CRC Crit. Rev. Bioeng., Aug., 453 (1973). (*Note*: an excellent review).

3. R. F. Rushmer, K. J. K. Buettner, J. M. Short, and G. F. Odland, Science, 154, 343 (1966).

4. R. J. Scheuplein and I. H. Blank, Physiol. Rev., 51, 762 (1971) (*Note*: an excellent review).

5. W. Montagna, *The Structure and Function of Skin*, 3rd Ed., Academic Press, New York, 1974.

6. P. M. Elias, S. Grayson, M. A. Lampe, M. L. Williams, and B. E. Brown, in *Stratum Corneum* (R. Marks and G. Plewig, eds.), Springer-Verlag, New York, 1983, pp. 53–67.

7. J. E. Tingstad, D. E. Wurster, and T. Higuchi, J. Am. Pharm. Assoc., 47, 187 (1958).

8. R. J. Scheuplein, J. Invest. Dermatol., 47, 334 (1965).

9. H. Baker and A. M. Klingman, Arch. Dermatol., 96, 441 (1967).

10. R. J. Scheuplein, J. Invest. Dermatol., 48, 79 (1967).

11. R. T. Woodburne, *Essentials of Human Anatomy*, Oxford University Press, New York, 1965, p. 6.

12. S. Rothman, *Physiology and Biochemistry of the Skin*, University of Chicago Press, Chicago, 1954, p. 26.

13. A. E. McKenzie, Arch. Dermatol., 86, 611 (1962).

14. R. B. Stoughton, South. Med. J., 55, 1134 (1962).

15. G. Sazbo, Adv. Biol. Skin, 3, 1 (1962).

16. M. Katz and B. J. Poulsen, *Handbook of Experimental Pharmacology*, New Series, Vol. 28, Springer-Verlag, Berlin, 1971, p. 103 (*Note*: an excellent review).

17. S. M. Peck and W. R. Russ, Arch. Dermatol. Syphilol., 56, 601 (1947).

18. M. R. Liggins, Patient Care, July 1, 56, (1974).

19. B. Idson, J. Soc. Cosmet. Chem., 24, 1972 (1973).

20. K. Laden, Am. Perfum. Cosmet., 82, 77 (1967).

21. J. N. Robinson, in *Handbook of Non-Prescription Drugs*, American Pharmaceutical Association, Washington, DC, 1973, p. 209.

22. R. B. Stoughton, Clin. Pharmacol. Ther., 16, 869 (1974).

23. J. J. Voorhees, E. A. Duell, M. Stawiski, and E. R. Harrell, Clin. Pharmacol. Ther., 16, 919 (1973).

24. W. R. Good, Med. Device Diag. Ind., Feb., 35 (1986).

25. G. W. Cleary, in *Medical Applications of Controlled Release*, Vol. 1, (R. S. Langer and D. L. Wise, eds.), CRC Press, Boca Raton, FL, 1983, pp. 203–251.

26. Y. W. Chien, *Novel Drug Delivery Systems*, Marcel Dekker, New York, 1982, pp. 149–217.

27. J. P. Marty, R. H. Guy, and H. I. Maibach, *Percutaneous Absorption*, 2nd Ed., (R. L. Bronaugh and H. I. Maibach, eds.), Marcel Dekker, New York, 1989, pp. 511–529.

28. C. R. Behl, N. H. Bellantone, and G. L. Flynn, in *Percutaneous Absorption* (R. L. Bronaugh and H. I. Maibach, eds.), Marcel Dekker, New York, 1985, pp. 183–212.

29. L. B. Fisher, *Percutaneous Absorption*, 2nd Ed. (R. L. Bronaugh and H. I. Maibach, eds.), Marcel Dekker, New York, 1989, pp. 213–222.

30. D. A. Weingand, C. Haygood, J. R. Gaylor, and J. H. Anglin, Jr., in *Current Concepts in Cutaneous Toxicity* (V. A. Drill and P. Lazar, eds.), Academic Press, New York, 1980, pp. 221–235.

31. D. A. Weingand, C. Haygood, and J. R. Gaylor, J. Invest. Dermatol., 62, 563, (1974).

32. H. H. Durrheim, G. L. Flynn, W. I. Higuchi, and C. R. Behl, J. Pharm. Sci., 69, 781 (1980).

33. G. L. Flynn, E. E. Linn, T. Kurihara-Bergstrom, S. K. Govil, and S. Y. E. Hou, in *Transdermal Delivery of Drugs*, Vol. 2, CRC Press, Boca Raton, FL, 1987.

34. C. G. Toby Mathias, in *Dermatotoxicology*, 2nd Ed. (F. N. Marzulli and H. I. Maibach, eds.), Hemisphere Publishing, New York, 1983, pp. 167–183.

35. M. S. Roberts, A. A. Anderson, J. Swarbrik, and D. E. Moore, J. Pharm. Pharmacol., 30, 486 (1978).

36. H. P. Bader, L. A. Goldsmith, and L. Bonar, J. Invest. Dermatol., 60, 215 (1971).

37. J. A. Keipert, Med. J. Aust., 1, 1021 (1971).

38. G. L. Flynn, S. H. Yalkowsky, and T. J. Roseman, J. Pharm. Sci., 63, 479 (1974).

39. R. J. Scheuplein and R. L. Bronaugh, *Biochemistry and Physiology of the Skin*, Vol. 2 (L. A. Goldsmith, ed.), Oxford University Press, New York, 1983, pp. 1255–1295.

40. T. Higuchi and S. S. Davis, J. Pharm. Sci., 59, 1376 (1970).

41. S. S. Davis, T. Higuchi, and J. H. Rytting, Adv. Pharm. Sci., 4, 73 (1974).

42. R. J. Scheuplein, I. H. Blank, G. J. Brauner, and D. J. MacFarlane, J. Invest. Dermatol., 52, 63 (1969).

43. G. L. Flynn, *Principles of Route-to-Route Extrapolation for Risk Assessment* (T. R. Gerrity and C. J. Henry, eds.), Elsevier, New York, 1990, pp. 93–128.

44. R. O. Potts and R. H. Guy, *Pharm. Res*, 9:663–669, 1992.

45. N. Z. Erdi, M. M. Cruz, and O. A. Battista, J. Colloid Interface Sci., 28, 36 (1968).

46. P. Thau and C. Fox, J. Soc. Cosmet. Chem., 16, 359 (1965).

47. S. Foster, D. E. Wurster, T. Higuchi, and L. W. Busse, J. Am. Pharm. Assoc. (Sci. Ed.), 40, 123 (1951).

48. B. W. Barry, J. Pharm. Pharmacol., 21, 533 (1969).

49. S. Arnott, A. Fulmer, W. E. Scott, I. C. M. Dea, R. Moorhouse, and D. A. Rees, J. Mol. Biol., 90, 269 (1974).

50. *Physician's Desk Reference*, 28th ed., Medical Economics, Oradell, NJ, 1974, p. 1460.

51. G. L. Flynn and T. J. Roseman, J. Pharm. Sci., 60, 1778 (1971).

52. G. L. Flynn, J. Pharm. Sci., 60, 345 (1971).

53. T. A. Hagen, Physicochemical study of hydrocortisone and hydrocortisone *n*-alkyl-21-esters, Thesis, University of Michigan, 1979.

54. W. M. Smith, An inquiry into the mechanism of percutaneous absorption of hydrocortisone and its al-*n*-alkyl esters, Thesis, University of Michigan, 1982.

55. C. A. Schlagel, Adv. Biol. Skin, 12, 339 (1972).

56. A. W. McKenzie and R. M. Atkinson, Arch. Dermatol., 89, 741 (1964).

57. G. Saracco and E. Spaccamella-Marcheti, Ann. Chem., 48, 1357 (1958).

58. S. H. Yalkowsky, G. L. Flynn, and T. G. Slunick, J. Pharm. Sci., 61, 852 (1972).

59. G. L. Flynn and S. H. Yalkowsky, J. Pharm. Sci., 61, 838 (1972).

60. D. E. Wurster, Am. Perfum. Cosmet., 80, 21 (1965).

61. T. Higuchi, J. Soc. Cosmet. Chem., 11, 85 (1960).

62. K. H. Burdick, B. J. Poulsen, and V. A. Place, JAMA, 211, 462 (1970).

63. A. W. McKenzie and R. B. Stoughton, Arch. Dermatol., 86, 608 (1962).

64. C. F. H. Vickers, Adv. Biol. Skin, 12, 177 (1972).

65. R. B. Stoughton, Arch. Dermatol., 91, 657 (1965).

66. R. J. Scheuplein and L. Ross, J. Soc. Cosmet. Chem., 21, 853 (1970).

67. S. G. Elfbarum and K. Laden, J. Soc. Cosmet. Chem., 19, 163 (1968).

68. R. Jones, Excipient effects on topical drug delivery, paper presented at the Industrial Pharmaceutical Technology Section Symposium, 121st Annual Meeting of the American Pharmaceutical Association, Chicago, *Abstracts*, Vol. 4, p. 24.

69. M. Goodman and B. W. Barry, *Percutaneous Absorption*, 2nd Ed. (R. L. Bronaugh and H. I. Maibach, eds.), Marcel Dekker, New York, 1989, pp. 567–593.

70. W. Abraham and D. T. Downing, *Prediction of Percutaneous Absorption*, (R. C. Scott, R. H. Guy, and J. Hadgraft, eds.), IBC Technical Services, London, 1990, pp. 110–122.

71. B. D. Anderson, W. I. Higuchi, and P. Raykar, *Pharm. Res.*, 5, 566–573 (1988).

72. W. I. Higuchi, W. W. Shannon, J. L. Fox, G. L. Flynn, W. F. H. Ho, R. Vaidyanathan, and D. C. Baker, in *Recent Advances in Drug Delivery Systems* (J. M. Anderson and S. W. Kim, eds.), Plenum Press, New York, 1984, pp. 1–7.

73. S. Y. E. Hou and G. L. Flynn, Pharm. Res., 3(Suppl.), 525 (1986).

74. M. F. Goldman, B. J. Poulsen, and T. Higuchi, J. Pharm. Sci., 58, 1098 (1969).

75. B. J. Poulsen, E. Young, V. Coquilla, and M. Katz, J. Pharm. Sci., 57, 928 (1968).

76. L. O. Killings, O. Ringerts, and L. Silverstolpe, Acta Pharm. Suc., 3, 219 (1966).

77. A. A. Fisher, F. Pascher, and N. B. Kanot, Arch. Dermatol., 104, 286 (1971).

9

Disperse Systems

S. Esmail Tabibi* and Christopher T. Rhodes
University of Rhode Island, Kingston, Rhode Island

I. INTRODUCTION

Disperse systems encompass a variety of heterogeneous, multiphase systems in which one homogeneous phase (dispersed or discontinuous) is intimately distributed, in discrete units, within the second phase (dispersing or continuous). There is wide variety of combinations, depending on the state of the constituent phases, as listed in Table 1; but emulsions, suspensions, and liposomes are of considerable pharmaceutical interest; thus, we will give particular attention to these systems.

This chapter is, therefore, designed to provide a general understanding of the principles used in the formulation of pharmaceutical disperse systems. We suggest that the reader may find it useful to use this chapter in conjunction with Chapters 8 and 12, which deal with topical and parenteral products, since these topics include some of the most important applications of disperse systems as dosage forms. Even though liposomes, suspensions, and emulsions have been commonly used within the conventional formulations for topical, oral, and even parenteral applications, today their most important potential is directed toward the systemic circulation for either passive- or active-targeted delivery.

II. FUNDAMENTAL PROPERTIES

A. Classification

The size of the individual units of the dispersed phase can vary considerably. For example, particles of colloidal gold can be smaller than 10 nm [1 nanometer (nm) = 10^{-9} meter (m)] in diameter, whereas particles of some other pharmaceutical dispersions are larger than 10 μm [1 micrometer (μm) = 10^{-6} m]. The properties of disperse systems have been very extensively

---

**Current Affiliation*: National Cancer Institute, National Institutes of Health, Bethesda, Maryland.

Table 1 Types of Disperse Systems

Dispersed phase	Dispersing phase	Type
Gas	Gas	None exists
Liquid	Gas	Liquid aerosol[a]
Solid	Gas	Solid aerosol[a]
Gas	Liquid	Foam
Liquid	Liquid	Emulsion[a]
Solid	Liquid	Suspension[a]
Gas	Solid	Solid foam
Liquid	Solid	Solid emulsion
Solid	Solid	Solid suspensions

[a]Systems of particular pharmaceutical interest.

studied by many researchers, including various pharmaceutical scientists, for many years. We have now assembled an impressive body of theory that has been exploited by many pharmaceutical research scientists to elucidate the principles underlying the use of disperse systems as dosage forms. We will discuss some of these approaches in more detail in later parts of this chapter. However, in general, the sizes of particles, globules, or vesicles in a pharmaceutical disperse preparation are of supracolloidal dimensions. Thus, there is always some uncertainty in applying the findings of colloidal science to pharmaceutical systems. This point will also be discussed in more detail in a later section of this chapter.

The term *colloidal* is applied to disperse system in which the particle size of the dispersed phase is very fine—in the range of less than 1 μm. As is shown in Table 1, the dispersed and the dispersing phases can be solids, liquids, or gases, although it is impossible to have a disperse system in which both phases are gases.

It is common for colloid scientists to classify dispersions as being *lyophilic* (solvent loving) or *lyophobic* (solvent fearing) [1], with the lyophilic systems exhibiting much greater stability than do the lyophobic systems. Furthermore, dispersions can be classified as molecular or micellar. In *molecular dispersion*, such as protein and polymer solutions, units of the dispersion phase are composed of single macromolecules, whereas in *micellar* (association) *systems* the units of the dispersed phase comprise several molecules. Alternatively, one may classify disperse systems on the basis of the shape of the units in the dispersed phase (e.g., spherical or linear). Of the various types of disperse systems listed in Table 1, emulsions, vesicles, and suspensions are of particular pharmaceutical importance; accordingly, particular attention will be given to these systems here.

B. Particle Size

The mean and range of particle size (the particle size spectrum) in a pharmaceutical disperse system can often have a profound effect on the properties of such systems. Therefore, it is appropriate to consider some of the basic principles of micromeretics, the science and technology of fine particles, in this section. Furthermore, since determination of particle size and its distribution is of considerable importance in the evaluation of pharmaceutical disperse systems, special attention will be given to techniques presently available for the determination of particle size.

Several methods for determining particle size result in the accumulation of data in the form of ''equivalent spherical diameters.'' Obviously, this may be quite suitable for sizing and presenting the particles of emulsions and liposomes, but when classifying nonspherical particles, such as those in many other pharmaceutical suspensions, the use of the term *diameter* is somewhat artificial. For such particles there are at least four types of equivalent spherical diameters, which are presented in Table 2. Surface diameter, d_s, is the diameter of a sphere having the same surface area. Volume diameter, d_v, is the diameter of a sphere having the same volume. Projected diameter, d_p, is the diameter of a sphere having the same observed area as the particle when the particle is viewed in its most stable plane. Finally, Stokes' diameter, d_{st}, defines an equivalent sphere undergoing sedimentation at the same velocity as the asymmetrical particle.

There are numerous mean values that can be calculated from a particle size spectrum. The two values most used in pharmaceutics are the volume surface mean, which is inversely related to specific surface area, and the volume number mean, which is inversely related to the number of particles per gram of material. A fuller discussion of the calculation of mean particle size values is to be found in the text by Martin and associates [2].

Particle Size Determination

Traditionally, the pharmacist has made considerable use of sieves for particle size evaluation and control. A nest of five or six sieves is capable of providing a quick estimate of the particle size spectrum of a powdered drug; however, the method has disadvantages: friable particles may undergo size reduction during analysis; difficulty in evaluating irregular-shaped particles; the data will vary greatly between laboratories; and even with relatively new semimicrosieves, one cannot obtain data on particles smaller than about 5 μm (the more typical lower limit is 30–50 μm).

Scientists concerned with the formulation and evaluation of pharmaceutical disperse systems have made much more use of instruments employing optical or electrolyte displacement (the Coulter principle) methods. The most frequently used particle size determination method to characterize colloidal-sized particles (the size range from few nanometers to few microns) is the quasi-elastic light scattering (QELS) technique. Lack of space prevents a full appraisal of all the techniques used in the evaluation of pharmaceutical disperse systems; thus, attention will be concentrated on electrolyte displacement, optical, and QELS methods.

The electrolyte displacement method was first developed and patented by Coulter, and for many years the only commercial apparatus applying this principle was the Coulter counter. The principle behind the use of this equipment, sometimes termed *impeded conductance*, is very simple. The sample to be analyzed is suspended in an electrolyte solution in which dips a glass tube full of the same electrolyte solution. Two electrodes are immersed in the electrolyte, one inside and one outside the glass tube. The suspension is forced to flow through the glass tube; and as each particle passes through the orifice, the volume of electrolyte between the

Table 2 Particle Size Data: Cumulative Percentage Oversize

Cumulative percentage oversize (by weight)	Particle size fraction (μm)
0	300
5	250–300
17	200–250
41	150–200
58	100–150
85	50–100
100	0–50

electrodes is reduced. The reduction in volume causes a reduction in the resistance between the electrodes. This resistance change is converted to a voltage pulse, the amplitude of which is proportional to the particle volume. Thus, the Coulter counter measures d_v, the equivalent spherical volume diameter. Pulses are amplified and fed into an electronic counting unit that has a threshold setting so adjusted that all particles higher than the threshold value are counted. Even with a very dilute suspension or emulsion there is finite possibility of more than one particle passing through the orifice at the same time. Fortunately, mathematical techniques are available to correct for this problem, which is known as coincidence. The Coulter counter has been used by many pharmaceutical scientists in their investigations of disperse systems [3–5]. A range of different-sized orifice tubes are available and, with the smallest of these, one may be able to count particles smaller than 2 μm in diameter [6]. Polar, highly water-soluble materials may be difficult to analyze using the Coulter counter method, since nonsolvents in which the particle may be suspended do not provide adequate conductance.

Also of proved value in the examination of disperse systems are optical methods, which range in complexity and cost from a simple laboratory microscope, to rather refined pieces of automated equipment. Optical methods allow evaluation of d_p, the equivalent projected spherical diameter. When using a simple microscope, one must be sure to count a sufficiently large number of particles, or globules, so that a statistically valid conclusion may be drawn. One advantage of the optical method is that one can obtain direct and often valuable information about particle shape, presence of agglomeration, agglomerate character, and other properties. The scanning electron microscope may provide basic information on very fine particle and particle surface characteristics. Even when more sophisticated methods used to provide particle size characterization, microscopic visualization of the particles should not be overlooked [7].

The quasi-elastic light-scattering technique, also called photon correlation spectroscopy (PCS), is employed to determine the mean particle diameter and size distribution. This method is applicable for measuring the particle size ranging from 5 nm to approximately 3 μm. At this size range, the particles will exhibit significant random movement (brownian motion), owing to collisions with the molecules of surrounding liquid medium, this causes fluctuation in the intensity of scattered light. QELS consists of a laser light source, a temperature-controlled sample compartment, and a photo multiplier to detect the scattered light at a certain angle (usually 90°). The fluctuation in scattered light intensity is transferred to a digital correlator to calculate the correlation function. A microprocessor then calculates the diffusion coefficient of the particles from the correlation function. The diffusion coefficient is related to the particle diameter by means of Stokes-Einstein equation:

$$D = \frac{kT}{3\pi\eta d} \tag{1}$$

where D is diffusion coefficient; k is Boltzmann's constant; T absolute temperature; η is the viscosity of dispersing medium; and d is the particle diameter. The data so obtained is then used to determine the size distribution. Typically, a few drops of disperse system is diluted in about 20 ml of double-filtered diluent (usually made up of dispersing phase) to create a slightly turbid to clear sample. The sample is then placed in the sample compartment, and particle diameters are determined at a set time limit. The microprocessor will then take over to calculate the size and size distribution (polydispersity) [8]. This method can readily and accurately determines the size of the collection of uniform particles, but broadly distributed particles cannot easily be characterized by this system [9]. This problem has been overcome by Caldwell and Li [10], by combining sedimentation field-flow fractionation and QELS, which presents a detailed record of the particles' size at each size interval [10].

Presentation of Particle Size Data

There are several methods, both graphic and digital, by which particle size data may be presented. Perhaps the simplest is that sometimes employed when sieves are used as a relatively crude particle size control device. The statement: "The drug to be used in the suspension should be between 100 mesh and 200 mesh," provides only the general limits of a particle size range, and a possible estimate of the mean particle size if the sample is normally distributed (about 113 μm, the average of 100 mesh and 200 mesh screen apertures). Considerably more information is available if a number of percentage undersized (or percentage oversized) values are available for different particle size cuts, as obtained with a nest of sieves or by other methods. Such data can be tabulated (Tables 2 and 3), or presented as a graph or histogram.

Considerable care must be exercised in the interpretation of particle size data, such as are shown in Tables 2 and 3. Notice that in Table 2 the data are presented in terms of *weight* of particles, whereas in Table 3 the data refer to *number* of particles. Obviously, this makes a very considerable differences to the form of particle size spectrum—many very fine particles will make very little contribution in terms of weight. Conversely, a heavier weight of large particles will make a small contribution in terms of percentage number. Also, whereas in Table 2, the particle fractions are equal in size (50 μm), in Table 3 the fractions vary from 10 to 50 μm. Were the data in Table 3 plotted in the form of a histogram, the bars would be of different thickness, and this can be somewhat misleading to the untrained observer. One should always check to see what type of equipment has been used to obtain the data on particle size, since different methods yield different equivalent spherical diameters. Particle weight data may be rather more useful in sedimentation studies, whereas number data are of particular value in surface-related phenomena such as dissolution.

Although graphic methods for the presentation of particle size data can be very useful, there are occasions when it is more convenient to digitalize (i.e., convert into numbers) the data. The simplest case is when the particle size spectrum may be described by the normal distribution equation, for which the data can be fully specified by means of the mean and standard deviation. It is unfortunate that, for many pharmaceutical disperse systems, the normal equation is quite inappropriate. In these systems, the log-normal equation is more commonly used. However, it has been indicated that to fit some pharmaceutical particle size data to the log-

Table 3 Particle Size Data: Cumulative Percentage Undersize

Cumulative percentage undersize (by number)	Particle size fraction (μm)
100	300
95	250–300
73	210–250
58	170–210
38	140–170
25	110–140
10	90–110
5	70–90
1	60–70
0	50–60

normal equation, it may be necessary to neglect points outside the arbitrary limits of 20 and 80%. For some systems, therefore, there can be considerable advantage in using the Weibull equation:

$$y = \exp \left. -\left(\frac{X - G}{N - G}\right)^{B} \right. \tag{2}$$

where Y is the weight or number proportion percentage oversized; X is the equivalent spherical diameter; G, B, and N are the location, shape, and scale parameters, respectively. This equation can be readily solved by a digital computer, and Short et al. [9] have shown that it can define pharmaceutical particle size data with precision. For those readers interested in a more detailed discussion of the interpretation of pharmaceutical particle size data, the following references are recommended [12,13].

C. Surface Properties

One of the most obvious properties of a disperse system is the vast interfacial area that exists between dispersed and dispersing phases. Therefore, it is not surprising that surface properties of the disperse systems are of particular importance. Dispersed particles tend to become charged by adsorption of ions from solution, or by ionization of functional groups on the surface of the particles. Another factor that can play a major role in the development of charge on a particle is the difference in dielectric constant between the particles and the bulk of continuous phase. Consider a particle of no intrinsic charge, such as a rubber particle in water. The particle will probably adsorb OH^- in preference to H_3O^+ because of the asymmetric nature of the hydroxyl ion.

The charge on the particle can also result from ionization of surface functional groups or by lattice substitution. Ionizable groups include COOH and NH_2 on proteins. Example of materials exhibiting lattice substitution include the clays, in which Si^{4+} may be replaced by Al^{3+}, giving a net deficit of positive charge. Ionic surfactants can also be adsorbed onto the particle surface. In the atmosphere surrounding a negatively charged disperse particle there will be a preponderance of positive charges attracted to this particle. Some of the charges are very closely bound in the fixed (or Stern) layer. Peripheral to this area are further positive charges that are relatively mobile and, because of thermal energy, are in a constant state of motion into and from the main body of the continuous phase. Between the Stern layer and this partially mobile diffuse layer will be a shear plane at some (unknown) distance from the particle surface. The two layers of charge are known jointly as diffuse double layer.

The work required to bring unit charge from infinity to the edge of the fixed layer is defined as the *zeta (ζ) potential*. This value has a considerable influence on the physical stability of many pharmaceutical disperse systems. It can be measured by microelectrophoresis [14], in which a sample of disperse system is mounted on a special microscope slide across which a known potential is applied. The speed of movement of the particles across the field is a function of the ζ potential and is determined visually. The apparatus is standardized by use of particles of known ζ potential. Rabbit erythrocytes are commonly used for this purpose. Alternatively, more elaborate semiautomated and fully automated equipment are available. More comprehensive information on the surface properties of particles and their effect on the physicochemical [15,16] and physiological [17] properties of disperse systems has been presented.

Fricke and Huettenrauch [18] have recently studied the influence of water structure in the continuous phase of disperse systems on their ζ potential. With bentonite suspension as model, they showed a marked reduction in negative potentials of all solids on promoting the water structure [18]. On the other hand, the water structure of a model oil-in-water emulsion did not show any effect on the ζ potential of the system [19].

D. Rheological Properties

Rheological, or flow, properties of pharmaceutical disperse systems can be of particular importance, especially for topical products. Such systems often possess quite complicated and fascinating rheological properties, and pharmaceutical scientists [20–23] have made several fundamental investigations in this area. If we consider a fluid flowing through a pipe under conditions of streamline flow, we can readily appreciate that liquid adjacent to the walls of the pipe will be flowing slower than that in the center. The difference in velocity, dv, between two parallel planes for liquid separated by a distance, dr, is the velocity gradient, dv/dr, or *rate of shear*. The force per unit area, F/A, required to cause flow is called the *shearing stress*.

Sir Isaac Newton realized that the higher the velocity of a liquid, the greater the force per unit area (i.e., shearing stress) needed to produce a given rate of shear. Thus,

$$\frac{F}{A} = \eta \left(\frac{dv}{d_r}\right) \tag{3}$$

where η is the coefficient of viscosity. For simple newtonian fluids a plot of rate of shear against shearing stress gives a straight line; thus, η is a constant. However, pharmaceutical disperse systems rarely exhibit simple newtonian flow.

Plastic flow is shown in Fig. 1a. If we extrapolate the straight-line portion back to the shearing stress axis, we obtain what is known as the *yield value*. The rheological properties of such a system can thus be defined in terms of the plastic viscosity and the yield value. The plot of shearing stress versus rate of shear for such nonlinear systems is termed a *rheogram*. Plastic flow is often observed in flocculated suspensions (flocculation is discussed in some detail in Sec. IV.D, which is specifically concerned with suspensions). In such systems we may consider the yield value as a function of the attractive forces between the particles, which must be overcome before flow can start.

Pseudoplastic flow is shown by various substances that have been used as pharmaceutical thickening agents (e.g., polymer solutions). It is believed that, with linear polymers, the axes of such molecules become more and more oriented in the direction of flow as shearing stress is increased; thus, the resistance to flow decreases. Obviously, the rheological properties of such systems cannot be defined simply in terms of one value. For some pseudoplastic systems, certain equations have been developed to define their properties, but two or more terms are required. Similarly, dilatant flow (in which the system appears to become more structured and viscous with increase in shearing stress) requires more than one term to define its properties. Suspensions with high solid content sometimes show dilatant flow. Thixotropy, which may be seen in both plastic and pseudoplastic systems, is characterized by the fact that the rate of shear at any given shearing stress can vary depending on whether the rate of shear is increasing or decreasing. In other words, the up and down curves in Fig. 1d are not superimposable.

E. Measurement of Rheological Properties of Disperse Systems

Viscometers may be divided into two types: (a) those that operate at a single rate of shear, and (b) those that allow more than one rate of shear to be examined. The first type includes various capillary viscometers that have considerable use for newtonian systems. They are sometimes also used for quality control purposes in the formulation and evaluation of pharmaceutical disperse systems. However, it will be appreciated that, when dealing with plastic, pseudoplastic,

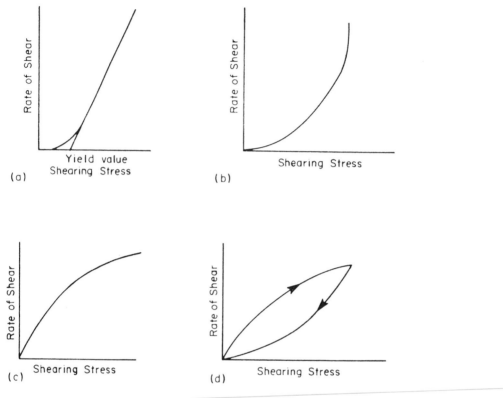

Fig. 1 Plots of rate of shear as a function of shearing stress for (a) plastic, (b) simple pseudoplastic, (c) dilatant, and (d) thixotropic flow.

or dilatant systems, the usefulness of this type of equipment is very limited, since they provide only a single point on the rheogram, and a thorough evaluation of the properties of such systems can only be effected by use of the second type of viscometer. Fortunately, a number of viscometers capable of operating at various rates of shear are available and some, such as the Farranti-Shirley, are found in many pharmaceutical laboratories. For some disperse systems, other rather empirical rheological data can also be useful. For example, "pourability" can be determined on a series of test formulations by measuring the volume of emulsion flowing from a bottle held at 45° to the horizontal in a standard time period. Fuller discussions of rheological aspects of pharmaceutical disperse systems have been presented [21,23,24].

III. SOLUBILIZED SYSTEMS

Solubilization is the process by which the apparent solubility of an otherwise sparingly soluble substance is increased by certain manipulations. These manipulations can be summarized as follows: (a) the presence of surfactant micelles, (b) the presence of cosolvents, (c) solubility alteration by means of complexation, (d) enhancement of solubility by solid-state changes, (e) the presence of hydrotropic agents, and (f) prodrug formation [25]. In this section we will concern with the solubilization in the presence of surfactant micellar systems (i.e., *micellar solubilization*). This should not be confused with microemulsion, which is a two-phase thermodynamically stable system and has been introduced by Schulman [26]. To discuss the process of solubilization and microemulsion, it is first necessary to discuss the fundamental properties of surface-active agents. This discussion also has relevance to the subsequent sections concerning liposomes, emulsions, and suspensions.

A. Surfactant Classification and Properties

The well-known colloid scientist Winsor defined *surfactants* as compounds "which possess in the same molecule distinct regions of lipophilic and hydrophilic character." For example, in the oleate ion, there is an alkyl chain that will interact, to only a very limited extent, with water and is thus basically hydrophobic (lipophilic) and COO^- headgroup that is hydrophilic by reason of ion–dipole and other interactions.

When the variation of any colligative property of a surfactant in aqueous solution is examined, two types of behavior are apparent. At low concentration, properties approximate those to be expected from ideal behavior. However, at a concentration value that is characteristic of a given surfactant system (the critical micelle concentration, or CMC), an abrupt departure from such behavior is observed. The literature is replete with descriptions of the methods available for the experimental evaluation of CMC values [27–29]. At concentration above the CMC, transient molecular aggregates, termed micelles, are formed. By increasing the concentration of the amphiphile (surfactant), depending on the chemical and physical nature of the surfactant molecule, structural changes to a more orderly state than micellar structure in the molecular assembly of amphiphiles occur. These are, generally, hexagonal, cubic, and lamellar liquid crystalline structures; and their occurrence relates mostly to the concentration of the surfactant in the system [30,31]. Liquid crystals have been known for more than a century and have been part of many creams and lotions, even without the knowledge of the formulator. Recently, owing to their particular effect and characteristics, they have been intentionally introduced into the cosmetic systems [32].

In aqueous solution of relatively low surfactant concentration it is now generally agreed that micelles are most likely to be spherical, with the narrow hydrocarbon "tails" of the molecule in the central nucleus and the wider hydrated hydrophilic head groups on the surface.

Surfactants are widely used in pharmacy, not only for solubilization, but also for wetting and the interfacial stabilization of the emulsions and suspensions. Surfactants may be classified chemically or physically.

Chemical classification is normally based on the nature of the hydrophilic headgroup, and surfactants can be classified as anionic, cationic, nonionic, or amphoteric. Sodium laurate is an example of an anionic surfactant; soaps such as sodium laurate are sometimes prepared in situ in a pharmaceutical preparation by reaction between a fatty acid and an alkali such as sodium hydroxide. Cationic surfactants such as cetyltrimethylammonium bromide (CTAB), are of special pharmaceutical interest, since they often possess antimicrobial activity. *N*-Alkyl polyoxyethylene surfactants, of general formula $CH_3(CH_2)_n(OCH_2CH_2)_mOH$, for which n is often between 10 and 18 and m may be between 6 and 60, are good examples of nonionic surfactants. In recent years, increasing pharmaceutical use has been made of nonionic surfactants, such as the Tweens and arlacels (Spans) as surfactants; Brij and glycerol monostearate as vesicle-forming amphiphiles [33]. Amphoteric surfactants include substances like *N*-dodecyl-*N*,*N*-dimethylglycine, $C_{12}H_{25}N^+(CH_3)_2COO^-$.

Particularly useful is the physical classification of surfactants derived from the concept of hydrophilic–lipophilic balance (HLB). Consider a homologous series of surfactants, $(CH_2)_nX^+Y^-$. As we increase the value of n there is an increase in the relative proportion of hydrophobic groups within the molecule, the value of the hydrophilic function X^+ remaining constant. Quantitatively we can conveniently express this balance between the hydrophilic and lipophilic parts of the surfactant molecule using a numerical scale, as in Table 4.

Reference has already been made to the fact that aqueous solutions of surfactants can increase the apparent solubility of substances of low water solubility; it is now appropriate to consider the mechanism whereby this change is effected and discuss the pharmaceutical implications of this phenomenon. One can consider the hydrocarbon nuclei of micelles as "island of low polarity in an aqueous sea of high polarity." They can provide a suitable environment for many hydrophobic drugs. There is evidence that some solubilized species may be incorporated ("dissolved") in the center of the micelles, whereas others may be adsorbed on the surface [34–36]. Thus, if we have an aqueous surfactant solution (above the CMC) in which a reasonably hydrophobic drug is dissolved, the following equilibrium will exist:

$$[Dr_{aq}] \overset{K_s}{\longleftrightarrow} [Dr_m] \tag{4}$$

where $[Dr_{aq}]$ is the concentration (activity) of free or non–micellar-bound drug, K_s is the solubilization equilibrium constant, and $[Dr_m]$ is the concentration of micellar-bound drug. Often, $[Dr_m]$ will increase linearly as surfactant concentration is increased, and, the greater the value of K_s, the greater the value of $[Dr_m]$. Solubilization, therefore, brings into existence a new micellar-bound species of drug that exists in addition to the normal aqueous species.

B. Properties of Pharmaceutical Solubilized Systems

Obviously, the opportunity that solubilization offers the pharmaceutical formulator is the ability to dissolve a drug in a much smaller volume than could be used if a purely aqueous solvent were used. Extensive studies of solubilization by pharmaceutical scientists [37,38] have shown that a wide variety of types of drugs can be efficiently solubilized by surfactants.

The formulator must keep in mind that the properties of micellar-bound drug, Dr_m, can sometimes vary significantly from those of the simple aqueous drug, Dr_{aq}. Since micellar-bound drug is located in an environment differing substantially from water (e.g., the dielectric constant is much lower), it is not surprising that several reports clearly show that a drug's resistance to hydrolysis or oxidation can be changed radically by micellar solubilization [36,39,40]. The extent of the change produced is likely to depend on the exact location of the solubilized drug within the micelle and to steric factors related to micellar architecture. Thus, if an ester is solubilized into the hydrocarbon core of a nonionic micelle, we would expect a greater degree of protection against hydrolysis than if it were located in the hydrated polyoxyethylene chains of the palisade layer or adsorbed on the micelle surface. Another interesting problem in the

Table 4 Relationship Between HLB Value and Use of Surfactants[a]

Range of HLB	Use of surfactant
3–6	Water-in-oil emulsifier
6–9	Wetting agent
8–13	Oil-in-water emulsifier
13–14	Detergent
15–18	Solubilizing agent

[a]Those readers interested in more information on the HLB concept may refer to the excellent technical brochures provided by Surfactant Division of ICI Americas, Inc.

pharmaceutical formulation development related to solubilization is the formulation of anti-neoplastic and anti-HIV (human immunodeficiency virus) agents. Most of these materials are water-insoluble and water-unstable, which can be solubilized and stabilized by micellar solution techniques [36]. It is conceivable that, if micellar solubilization results in a rather rigid orientation of drug at a micelle surface, steric factors might tend to promote rate of degradation.

Also of considerable importance is the effect that the presence of micelles can have on the action of preservatives in pharmaceutical systems. Many of the preservatives used to inhibit microbial growth in dosage forms (e.g., the parabens, benzoic acid, and chlorobutanol) have a considerable degree of hydrophobic character and, therefore, are quite likely to be micellar-bound to a greater or lesser extent. This complicates the preservation of solubilized and other disperse systems, since there is good reason to believe that the preservative action is primarily a function of the free (nonmicellar) preservative concentration. It might be thought that since such techniques as equilibrium dialysis, gel filtration, and potentiometry [41–43] can be used to evaluate the extent of the micellar-binding equilibrium shown in Eq. (3), it would be a relatively simple task to predict, from results obtained by use of such physicochemical methods, the amount of preservative needed in solubilized or other disperse systems. It has been suggested that the formulator could make use of the saturation ratio (R) concept as defined by Mitchell and Brown [44]:

$$R = \frac{C}{C_s} \tag{5}$$

where C is the drug concentration and C_s is the saturation solubility. A simple approach to the use of the saturation ratio technique for the calculation of the amount of preservative to be included in a solubilized system might be as follows. Suppose that a preservative P has an aqueous solubility of 1% w/v, and it gives adequate preservative protection in aqueous systems at a concentration of 0.02% w/v. Let us assume that, in a solubilized system, P has a solubility of 5% w/v. Thus, by using the saturation ratio concept we would estimate that a concentration of 0.1% w/v would provide adequate preservative protection since R would be constant. Implicit in the use of this approach is the assumption that the solubilization isotherm, a plot of $[Dr_m]$ as a function of $[Dr_{aq}]$ is linear. For many cases this assumption may be valid. However, studies in various undersaturated systems have shown that this is not always true [45]. In such a system the ratio of $[Dr_m]$ to $[Dr_{aq}]$ varies with the degree of saturation, and the aforementioned type of calculation would thus be invalid. Even in those systems in which it is known that $[Dr_m]$ and $[Dr_{aq}]$ are linearly related, direct testing (in terms of antimicrobial activity) of the foregoing approach has not been encouraging. There are probably several reasons for this finding. First, surfactants can directly affect the permeability of biological membranes. It is apparent that if the surfactant modifies the rate of transport across the membranes or interaction with a bacterial membrane, this factor will further complicate the situation. Furthermore, it will be appreciated that the equilibrium between micellar-bound and free preservative is dynamic; thus, if free preservative becomes bound to bacterial material, the micellar preservative can act as a reserve of preservative. Such a reserve is not available in a purely aqueous system. In other disperse systems, such as emulsions, suspensions, and in particular, liposomes, the situation is even more complex. Bean and his co-workers have made some most useful fundamental studies on the action of antimicrobial substances in disperse systems [46]. Thus, at best, the saturation ratio concept should probably be applied to give only a preliminary estimate of the range of concentrations that might usefully be tested in disperse systems. The situation in multicomponent solubilized systems becomes increasingly complex. For example, Crooks and Brown [47] have investigated the solubilization of several pairs of preservatives by a nonionic surfactant. The addition of a second preservative always altered the equilibrium solubility of the first.

Some idea of the problems that may be involved in the formulation of pharmaceutical solubilized systems can be obtained by considering the formulation of vitamin preparations for parenteral nutrition. Table 5 summarizes the recommendation for parenteral vitamin intake.

The difficulties in formulating all these vitamins into one preparation have caused the recommendation that the water-soluble and fat-soluble vitamins be placed in separate containers. To prepare a basically aqueous solution of acceptable volume of the fat soluble vitamins, it is necessary to make use of the solubilizing properties of surfactants. However, the formulation problems in this instance are indeed awe inspiring and do not yet seem to have been solved entirely satisfactorily. The various vitamins show different pH solubility and stability profiles. Obviously, for long-term intravenous use, it is essential that the surfactants be of known nontoxicity; but unfortunately, chronic toxicity data for surfactants are not always available. Table 6 presents a partial list of surfactants used in the formulation of solubilized drugs for parenteral administration, still there is a real need for further work in this area.

IV. SUSPENSIONS

A. Advantages and Disadvantages of Suspensions as a Dosage Form

Even those of us who have invested considerable time and labor in the formulation of suspensions must admit that the suspension has several disadvantages as a dosage form. First, uniformity and accuracy of dose, even when the preparation is nurse administered, is unlikely to compare favorably with that obtainable by the use of tablets, capsules, or solutions. Sedimentation and compaction of sediment cause problems that are by no means always easy to solve. Additionally, the product is liquid and relatively bulky; these properties are disadvantageous to both pharmacist and patient. Formulation of an effective and pharmaceutically elegant sus-

Table 5 Suggested Composition for Intravenous Multivitamin Preparation for Daily Maintenance of Adequate Vitamin Status

Vitamin	Units	Pediatric (<10 yr)	Adult
A	IU	2500	4000
D	IU	400	400
E	IU	7	15
K	mg	0.2	2.8
Thiamin	mg	1.2	2.8
Riboflavin	mg	1.4	3.6
Niacin	mg	17	40
B_6	mg	1	4
Pantothenic Acid	mg	6	15
Folacin	g	140	400 ?
Ascorbic Acid	mg	100	9 ?
B_{12}	g	0.7	6 ?
Biotin	g	20	60 ?

Source: AMA Nutrition Advisory Group, 1974.

Table 6 A Partial List of Approved Surfactants for Parenteral Administration of Pharmaceuticals

Drugs	Surfactant	Route
Chlordiazepoxode HCI	Polysorbate 80	IM
Cortisone acetate		IM
Dexamethasone acetate		IM
Etoposide		IV
Medroxyprogesterone acetate		IM
Methylprednisolone acetate		IM
Loxapine HCl		IM
Vitamin A		IM
Ampicillin	Polysorbate 40	IM and IV
Penicillin G benzathine		IM
Testolactone	Polysorbate 20	IM
Amphotericin B	Na desoxycholate	IV
Phytonadione	Polyoxyethylated fatty acid derivatives	IM, SC, and IV
Cyclosporine	Polyoxyethylated Castor Oil	IV
Penicillin G benzathine	Polyoxyethylated sorbitan	IM
Penicillin G procaine	Monopalmitate	IM

Source: PDR, 1992.

pension is usually much harder to achieve than that of a tablet or capsule of the same drug. However, suspensions do have some advantages that can, under certain circumstances, outweigh their disadvantages.

Many of the more recently developed drugs are basically hydrophobic and, therefore, have very low aqueous solubilities. Thus, solutions of these drugs, containing an appropriate dosage, would be of an unacceptably large volume. Suspensions allow the development of a liquid dosage form containing an appropriate quantity of drug in a rasonable volume. Moreover, resistance to hydrolysis and oxidation is generally good compared with that observed in aqueous solution. Suspensions can also be used to mask the taste of drugs. Also, there is a significant proportion of the population, especially very young children and elderly patients, who have difficulty in swallowing tablets or capsules. In recent years, increasing attention has been given to the use of suspension in intramuscular injection for depot therapy. For example, a recently FDA (1992) approved contraceptive is a suspension for deep intramuscular injection to provide 3 months of contraceptive protection [48]. The use of suspensions for intravenous injection has also been suggested, provided that the small and nonaggregating particles can be produced. Violante and co-workers [49–51], in a series of patents and in publications, have disclosed a method for preparation of suspensions containing such small, uniform, and nonaggregating particles from contrast agent or antineoplastic materials administered intravenously.

B. Physical Stability of Suspensions

Pharmaceutical suspensions are thermodynamically unstable systems. Aggregation of suspended particles and sedimentation (and possibly impaction of the sediment) present real problems to the pharmaceutical formulator. Much of the theory relevant to the formulation of acceptable pharmaceutical suspensions is derived from the findings of colloid scientists who have studied model systems. There are, however, several important differences between model colloidal systems and pharmaceutical suspensions, some of the more important of which are shown in Table 7.

Repulsive and Attractive Forces Between Particles

Much of the present-day theory concerning the charge on suspended particles results from the independent work of four scientists: Derjaguin and Landau in Russia, and Vervey and Overbeek from the Netherlands. The theory is thus often referred to as the DLVO theory. This theory allows us to develop insight into the factors responsible for controlling the rate at which particles in a suspension will come together, or aggregate, to form duplets (two particles), triplets (three particles), and so on. The process of aggregation will accelerate sedimentation and affect redispersibility and, therefore, is important to the pharmaceutical scientist formulating a suspension. The total energy of interaction, V_T, between two particles is defined as

$$V_T = V_R + V_A \qquad (6)$$

where V_R and V_A represent the repulsive and attractive forces, respectively. (It is possible to estimate V_R and V_A; see Matthews and Rhodes [52]). Figure 2 exemplifies some interaction energy curves, in which curve A applies when $V_R > V_A$, that is to indicate the high potential at the double layer owing to the positive sign of interaction energy. In such cases, a suspension would exhibit very good resistance to aggregation (i.e., flocculation or coagulation), provided that the particles are not sufficiently large that they sediment under gravity.

Curve B shows a high potential energy barrier, V_M, which must be surmounted if the particles are to approach one another sufficiently closely to enter the deep primary energy minimum at P. If the height of the energy barrier V_M greatly exceeds the mean thermal energy of the particles, they will not be able to enter P. The value V_M required to just prevent this is probably equivalent to a ζ potential of about 50 mV. Thus, in formulating a pharmaceutical suspension, it is often useful to aim at a system with a ζ potential of more than 50 mV. Aggregates that do form at P are likely to be very tightly bound together, since H, the interparticle distance, is small, and the energy well at P is often quite deep. Accordingly, in a pharmaceutical preparation, very vigorous shaking would be required to disperse the product. Also, there is a secondary energy minimum at S. If this trough is sufficiently deep, loose aggregates can form at this point; these will usually be easy to redisperse.

Table 7 Partial List of Amphiphiles Used in the Preparation of Various Lipid Structures and Their Designated Names

Lipid structure used	Designated name	Ref.
Phospholipids[a]	Liposome	119–120
Derivatives of ammonium		156
Amphiphiles (double-headed)	Niosomes	157
Oleic acid	Ufasomes	158
Sucrose fatty acids (double-tailed)		159
Cationic surfactants (double-tailed)		160
Cationic with amino acid moiety		161
Cationic and anionic surfactants (single-tailed)		144
Polyoxyethylene alkyl ethers	Novasomes	144
Sphingolipids and ceramides	Sphingosomes	162
Switerionic (double-chain)		161

[a]See text for details. Source of phospholipids may also vary, and includes egg, soybean, and synthetic sources, with a different degree of purity, depending on use. Various ingredients in this category have been used.

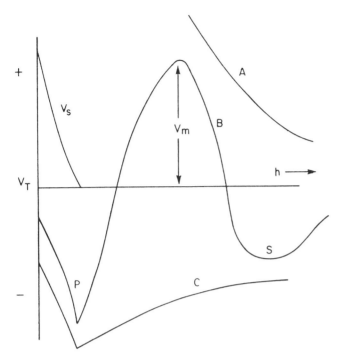

Fig. 2 Total energy of interaction curve between suspended particles (h is the interparticulate distance).

Curve C shows the situation that exists when attractive forces completely overwhelm repulsion forces (i.e., $V_A > V_R$). Under such conditions, very rapid aggregation will occur.

The V_s curve shows the stabilizing effect of surfactants adsorbed on the surface of the suspended particles; it shows a quite sharp cutoff point at $2d$, where d is the thickness of the absorbed surfactant layer. The strongly hydrated nature of the headgroups of surfactants impedes particle–particle contact, which would result in aggregation. Thus, even nonionic surfactants can be used to stabilize suspensions. However, in some systems, an excessive quantity of surfactant can have a significantly adverse affect on stability.

Aggregation Kinetics

The aggregation of particles in a suspension can be termed flocculation or coagulation. The term *coagulation* should be used when the forces involved are primarily physical owing to reduction in the repulsive forces at the double layer. The term *flocculation* is applied to those cases in which weak "bridging" occurs among the particles. However, since in many pharmaceutical systems the exact nature of the forces is somewhat obscure, we shall restrict ourselves here to use of the term *aggregation*. By using simple diffusion theory [53], Von Smoluchowski derived equations for both rapid aggregation (when all particle–particle collisions result in aggregation) and slow aggregation (in which only a fraction, α, of all particle–particle collisions results in the formation of aggregates). Pharmaceutical scientists are concerned with slow aggregation, since the aggregation in suspensions of drugs is mainly slow. The $t_{1/2}$ time for the initial number of single particles (singlets) in a suspension to decrease by 50% because of aggregation, is given by

$$t_{1/2} = \frac{1}{4D_1RN_0\alpha} \tag{7}$$

where D_1 is the diffusion coefficient of the singlet, R is the gas constant, N_0 is the initial number of single particles, and α is the collision efficiency. It is sometimes convenient to determine W, the stability ratio, which is the reciprocal of collision efficiency. This relatively simple approach to the kinetics of aggregation includes assumptions that are not always applicable to pharmaceutical systems [54]. In particular, a mixture of large and small particles, such as may be commonly found in a pharmaceutical suspension, will show a different rate [55] and pattern of aggregation from that of a similar system containing particles of only one size (homodisperse).

For a detailed information on the aggregation kinetics within the dispersed systems, the interested reader may refer to an excellent review chapter by Tadros and Vincent [56].

Wetting

Many drugs formulated in suspensions are basically hydrophobic; thus, in the absence of a suitable wetting agent, the drug particle would merely float on the aqueous phase. Surface-active agents, particularly anionic or nonionic, therefore, are included in the formulation of the pharmaceutical suspensions. A bridge is made by the adsorption of the hydrocarbon tail of the surfactant onto the particle, with the hydrated headgroup forming a link in the aqueous medium.

C. Brownian Motion and Sedimentation

If a suspended particle is sufficiently small, it will follow a random motion owing to molecular bombardment, which at times will be greater on one facet of the particle than on the others. The brownian motion of the particle (sometimes known as the ''drunken walk'') will be sufficient to prevent sedimentation. The distance moved or displacement, D_i, caused by brownian motion is given by

$$D_1^2 = \frac{RTt}{N3\pi\eta \, r} \qquad (8)$$

where R is the gas constant, T is absolute temperature, N is Avogadro's number, η is the viscosity, t is the time, and r is the particle's radius. Note that the larger the value of the r, the smaller the value of D_i.

The basic equation controlling sedimentation for spherical and monodisperse particles is defined by Stokes. It states that the velocity of sedimentation, v, is governed by the formula

$$v = \frac{2(\rho_1 - \rho_2)r^2g}{9\eta} \qquad (9)$$

where the ρ_1 and ρ_2 are the density of the particle and continuous phase, respectively, and g is the acceleration cause by gravity; the other symbols have the same meaning as in Eq. (8). Few if any pharmaceutical powders are spherical, and the equation is rigorously applicable only to systems in which the settling particles do not interfere with one another. Since, in many pharmaceutical suspensions, the solids content is high, hindered settling can be expected to occur. Thus, some workers in this field have applied modified Stokes' law and relationships, such as the Kozeny equation [2], without considering important physicochemical properties of the continuous medium, such as surface tension or electric conductivity. Alexander et al. [57] have recently shown that the dielectric constant of the continuous phase plays a major role on the velocity of sedimentation of concentrated aluminum and magnesium hydroxide suspensions. However, consideration of Stokes' law does allow an understanding of some of the techniques that the pharmaceutical formulator can use to produce an acceptable suspension. It can be seen from Stokes' law that both viscosity and density can affect the velocity of sedimentation. A wide variety of suspending agents are available for the use in the formulation. Some, such as cellulose derivatives, have a pronounced effect on viscosity, but scarcely any on density. Others, such as sorbitol, modify both density and viscosity. It can be seen from Eq. (9), that if the density difference between the particles and continuous medium can be eliminated, sedimentation can be prevented. However, it is seldom if ever possible to increase vehicle density above about 1.3 and, accordingly, it is not normally possible to match particle density, although it is possible to reduce the density difference.

Examination of Eqs. (8) and (9) reveals that, as the radius of a suspended particle is increased, brownian motion becomes less important and sedimentation more dominant. For any given system; we can define a "no sedimentation diameter" (NSD), below which value brownian motion will be sufficient to keep the particles from sedimentation. The value of NSD will obviously depend on the density and viscosity values of any given system.

The data shown in Table 8 were estimated by Matthews and Rhodes [58]. Calculations of the type made by these workers may be of really practical value in future formulation work. Thus, it can be seen from Table 8 that if one had a powdered drug smaller than 7 μm in diameter, a 70% glycerol medium would apparently be of sufficient viscosity and density to prevent sedimentation.

Crystal Growth (Ostwald Ripening)

The surface free energy for small particles is greater than for large particles. In some systems, therefore, small particles will be appreciably more soluble than large particles. For such systems small fluctuation in temperature will result in crystal growth as the small particles dissolve with a temperature increase; and then crystallize on the surface of existing particles, with a temperature drop. Thus, the larger particles will grow in size at the expense of the smaller ones. The suspension will become coarser as the mean of the particle size spectrum shifts to higher values. Many gums adsorb onto crystal surfaces and, thus, can be used to inhibit crystal growth. Freeze–thaw as well as more elevated temperature-cycling tests can provide a useful technique for evaluating crystal growth and crystal growth inhibitors.

Table 8 Partial List of Preparation Methods and Entrapped Active Ingredients

Preparation technique	Type	Entrapped agent	Ref.
Mechanical processing	MLV	Carboxyfluorescein	165
Hydration by injection	MLV	Pesticides	145
Solvent evaporation, sonication	MLV	Oxytetracycline	119–163
Dehydration–rehydration, sonication	NLV	Melphalin, vincristine	164
Freeze–thaw, sonication	MLV	Asparaginase	163
Freeze–thaw technique	MLV	Inulin	165
Solvent evaporation	LUV	Many drugs	119–120
Detergent removal technique	LUV	Cytochrome *C*	120
Reverse-phase evaporation	LUV	Insulin	166
Extrusion	LUV	Methotrexate	167
Sonication	SUV	Cytarabine (cytosine arabinoside)	167
Detergent removal	SUV	Carboxyfluorescein	120
French press	SUV	Carboxyfluorescein	120

D.   Methods of Evaluating Suspensions

Sedimentation Volume

The measurement of the volume of sediment produced by a given formulation of a suspension has been used by a number of workers in attempts to evaluate suspensions [59,60]. Most commonly the sedimentation volume V_s is defined as

$$V_S = \frac{100H}{H_0} \tag{10}$$

where H is the ultimate settled height and H_0 is the original height of the suspension before settling. For example, if 100 ml of a well-shaken test formulation is placed in a graduate cylinder and the ultimate height of the sediment is at the 20-ml line, then V_S is 20. As will be shown in a later discussion of controlled aggregation formulation (see Sec. IV.D), it is normally found that the greater the value of V_S for a given mass of drug, the more stable the product.

Ease of Redispersibility

Obviously, if a pharmaceutical suspension produces a sediment on storage, it is essential that it should be readily dispersable so that uniformity of dose is assured. The amount of shaking required to achieve this end should be minimal; the formulator should consider the problems of an 80-year-old, somewhat absent-minded arthritic. Stanko and DeKay [61] were probably the first workers to propose the use of redispersibility test for pharmaceutical suspensions. Various modifications of the basic principle have been described; for example, the suspension may be placed in a 100-ml graduate cylinder which, after storage and sedimentation, is rotated through 360° at 20 rpm. The endpoint is taken as being when the inside of the base of the graduate cylinder is clear of sediment. The ultimate test of redispersibility is the uniformity of suspended drug dosage delivered from a suspension product, from the first volumetric dose out of the bottle to the last, under one or more standard-shaking conditions as just described.

Particle Size Measurements

Particle size measurement will allow the aggregation or crystal growth to be evaluated. The particle size spectra of dispersed systems are often employed as fundamental quality control standards for pharmaceutical suspensions. The particle size of disperse systems also affects the biological availability of the parenterally administered drugs and in vitro evaluation. Rudt and Mueller [62] showed that the total particle uptake increased with an increase in particle size, confirming the similar observations by others.

Rheological Studies

As indicated in an earlier part of this chapter, rheological evaluations can be rapid and simple, or detailed and complex, depending on the exact nature of the product [63]. They are also an important quality control feature of all dispersed systems. The adequacy of hydration and quality control of the gums used as viscosity-imparting vehicles in suspensions is best confirmed by a rhelogicall check. All of the gums, whether natural (acasia, tragacant, plum, or other), modified natural (celluose, alginate derivatives, and such), or synthetic (carboxyl, vinyl, pyrolidone, or other) are polymers or contain polymer constituents that vary in molecular weight distribution from lot to lot. This, in turn, may have a material effect on vehicle viscosity. In addition, the degree of dispersion in disperse system, particle size, and particle size distribution influence viscosity, providing a usable method of quickly verifying dispersion consistency. Also formulation additives, such as sweeteners, surface active agents, and flocculating materials, influence the rheological properties of the dispersions, which play a major role in the formulation of suspensions [64–67].

Temperature and Gravitational Stress Tests

Some experts in the field of suspension formulation place great reliance on temperature and gravitational stress tests, whereas others have considerable reservations about their use, based on concerns of subjecting the systems to forces they will never encounter in practice. The latter view may have changed by now, for parenteral dispersions (suspensions, emulsions, or others) are subject to heat sterilization and autoclaving or freeze-drying [68]. If a disperse system can withstand the harsh environment of heat sterilization or freeze-drying, it can definitely tolerate temporary temperature extremes. Apte and Turco [69] studied the effect of various processing variables on the particle size and size distribution of a 40% v/v perfluorocarbon emulsion after autoclaving. They showed that such variables as time of emulsification, time of autoclaving, emulsifier concentration, and autoclaving temperature, all influence the formation of large particles and, thereby, the emulsion stability and clinical acceptability.

The use of storage at elevated temperatures is less popular than temperature cycling, and certainly a freeze–thaw type of test (e.g., $-5°C$ to $+40°C$ in 24 hr) can be of a real value in crystal or particle growth studies.

Centrifugation at a relatively low rate may also be efficacious in creating an in-use condition, thus predicting stability; however, there is a marked paucity of published data concerning this point. It is possible that testing suspensions at relatively low rates of centrifugation may have a sound theoretical basis [70]. However, there seems to be little predictive value in high centrifugal force stress studies for predicting suspension stability.

Zeta Potential Determinations

As discussed earlier (see Sec. II.C), determination of ζ potential can be of value in the development of pharmaceutical suspensions, particularly if the controlled aggregation approach, described immediately hereafter, is used.

E. Controlled Aggregation Formulation of Suspensions

The application of ζ potential measurements to the formulation of pharmaceutical suspensions is generally attributed to Haines and Martin [71]. They pointed out that if a suspension were prepared with a high ζ potential, the mutual repulsion between particles would be strong; thus, one would expect the rate of sedimentation to be slow. One's first reaction, therefore, might be to prepare suspensions with as high a ζ potential as possible. However, because the sedimentation rate is slow, the particles in the sediment will have plenty of time to pack tightly by falling over one another to form an impacted bed. The sedimentation volume of such a system is low, and the sediment is difficult to redisperse. If, however, the amount of electrolytes and, hence, the ζ potential, is reduced somewhat, the system may still have sufficient physical stability for the patient to obtain a uniform dose after shaking; but the sedimentation will be comparatively rapid, and the aggregates would not have time to impact at the base of the container. The sedimentation volume is high, and redispersion is relatively easy. Basically, therefore, the controlled aggregation theory produces pharmaceutically elegant products, as far as ease of redispersion is concerned, by control of ζ potential "not too little, not too much" [72]. Such systems may develop some region of clear supernatant solution above the loose sediment and, accordingly, may look less uniform on standing, even though they provide the greatest ease of redispersion and best dose uniformity.

Guidelines for Suspension Formulation Using the Controlled Aggregation Technique

1. Select a nontoxic surfactant for wetting the drug. Normally, an anionic material might be preferred, but nonionics may also be used.

2. If necessary, add a suspending agent (e.g., sodium carboxymethylcellulose). Care should be taken to ensure that there is no chemical or physicochemical interaction among surfactant, suspending agent, and any electrolyte that may be used.
3. Add just sufficient nontoxic electrolyte to produce aggregation.
4. Check that additional components, such as flavors, colors, humectants (to prevent the closure and the bottle from locking or sticking), or preservatives, do not substantially modify the properties of the system.
5. Evaluate the product after storage for, say 1 month by
    a. Redispersibility tests
    b. Sedimentation volume tests
    c. Particle size measurements before and after temperature stress tests
    d. Rheological tests
    e. Cyclic temperature stability tests
6. Check the suspension stability in the final package.

If the controlled aggregation approach is used, it will often be found that storage for long periods, say, a year or more, will make little if any difference in the results.

Optimization techniques (see Chapter 19) have been applied to suspension formulation as well [73].

F. Preparation of Suspensions

Once a satisfactory laboratory-scale formulation has been developed, the problem of pilot- and manufacturing-scale production must be solved. In the research and development (R&D) laboratory we will probably be producing quantities in the region of 200 mi (or less) up to 1 or 2 liters, whereas in industry much larger quantities will be produced (several decaliters up to kiloliters). At this scale-up point we may well meet problems of mixing and dispersion not met in small-scale production [74]. Thus, the new formulation is often taken to a pilot-scale plant, where larger quantities are produced and scale-up problems may become evident as an additional quantum jump in scale-up is met. In many pharmaceutical companies, therefore, it is normal for the R&D department to maintain control and responsibility until three or four full production batches have been produced.

Suspensions can be prepared by either dispersing finely divided powders in an appropriate vehicle, or by causing precipitation within the vehicle. The precipitation method is somewhat complex and may involve a controlled pH change of the solvent. Double decomposition has also been used (e.g., preparation of white lotion). The community or hospital pharmacist has an important role in preparing antibiotic suspensions for individual patients. This is a relatively simple operation that may involve just adding a premixed vehicle to the powdered drug and then dispersing by vigorous shaking. This technique helps to reduce stability problems. However, the pharmacist filling such a preparation has a special responsibility for ensuring that the product is used correctly. There are instances when a community or hospital pharmacist may be involved in preparing suspensions from existing solid dosage forms. In such extemporaneous suspensions, the drug stability may be compromised. Alexander et al. [75] studied acetazolamide in a suspension made from tablets. By applying an Arrhenius plot they predicted that a 25-mg/ml acetazolamide oral suspension is stable for 79 days at temperature range of 5–30°C, if the pH is maintained at 4–5, and the suspension is protected from light. Therefore, it is advisable to consider a smaller preparation to avoid a possible subpotent dose or unwanted side effects owing to the presence of degradation by-products. A study of earlier chapters in this book will reveal the importance of adhering to the appropriate dosage regimen, but also the patient should be advised about the necessity of shaking the bottle to obtain uniformity of dose. Furthermore, for some products, storage in a refrigerator is required. The practice of hoarding unused quantities of such preparations is to be discouraged because of their short shelf life. Time taken to explain the reason for this advice is always well spent.

On the industrial scale, production batches of 300 liters or more of suspension may be produced. If the suspension is prepared by the dispersion process, the milling of the solid is often effected by a micronization technique. This method can reduce particle sizes to values significantly less than 10 μm [76]. Basically, micronization is achieved by forcing a coarse powder into a turbulent air chamber where collision between particles results in fragmentation. If the suspension is to be prepared by controlled crystallization, a supersaturated solution is prepared and then quickly cooled by rapid stirring; by this method we obtain a large number of small crystals. Homogenization of the suspension will normally be required when the various other components are added. On the laboratory scale, an ultrasonic generator can be used; for industrial production, a conventional colloid mill may be suitable. Further details of industrial-scale production and equipment are provided elsewhere [77–79].

V. EMULSIONS

A. Uses of Emulsions in Pharmacy

The emulsion is a dosage form that has had considerable traditional use in pharmacy. Indeed, there are numerous advances in emulsion technology that are directly attributable to pharmaceutical scientists. Although emulsions can be designed for the oral (e.g., cod liver oil emulsion) or parenteral (e.g. fat emulsion for parenteral nutrition) route of administration, their major use in present-day pharmaceutical practices is topical preparations. They can also be used as drug delivery systems to carry drug molecules to the body, either for systemic or topical effect [80–83]. Recently, an alcohol-free emulsion formulation containing an antimicrobial agent for the oral cavity has been developed [84,85]. Radiopaque emulsions can be used as diagnostic agents for x-ray examinations. The technology of simple emulsions, that is oil-in-water (o/w) or water-in-oil (w/o), is reasonably clearly established. In recent years, however, increasing attention has been given to multiple emulsions such as water-in-oil-in-water (w/o/w), in which water droplets are dispersed in oil droplets which, in turn, are dispersed in water. There are considerable technical difficulties in the preparation of such systems; however, a series of patents covering such products have been issued, and it is known that several groups are considering the use of such systems for pharmaceutical purposes. It is possible that such multiple emulsion systems may be used for prolonged-action oral products or intramuscular depot therapy [33]. Release of a drug from the central aqueous phase can be controlled by several factors, including pH and the nature and thickness of the oil phase. Recently, Roy and coworkers [86] have used the multiple emulsion technology to develop drug-loaded microspheres from two different polymers.

Formulation of many cosmetic products is basically similar to formulation of topical pharmaceutical products, and thus the data presented in this chapter and Chapter 8 are of some relevance to cosmetic products. Cosmetic products and creams may contain various natural and synthetic waxes as the oil phase, in which case, the products are made at an elevated temperature that must exceed the melting point of the ingredient with the highest melting temperature.

B. Major Factors Affecting the Formulation of Emulsions

A good deal of the research discussed in earlier sections of this chapter is also of relevance to emulsions. In general, the globules in an emulsion have diameters ranging from 0.2 to 50 μm.

Thus, their properties can overlap the colloidal and supracolloidal areas. Because of their very large interfacial area and their heterogeneous nature, emulsions are basically unstable. The formulator, therefore, must use considerable skill in preparing an emulsion in which globule coalescence (permanent union of two or more globules to form one large globule) and other physical changes are minimized. To manufacture an o/w (or w/o) dispersion in a manner to produce a reasonably stable emulsion, a surfactant, or a mixture of surfactants, to reduce interfacial tension is normally essential. The surfactant molecules are adsorbed at the oil–water interface, with the hydrocarbon tails of the surfactant in the oil phase and the hydrated polar headgroups in the water. The amphiphilic surfactant can be considered as forming a link between the two phases of markedly different polarity. The type of emulsion formed (i.e., w/o or o/w) depends on various factors. Of particular importance is the phase volume ratio (the ratio of the quantities of oil and water in the system) and the hydrophilic–lipophilic balance (HLB) value of the surfactants. The HLB value generally reflects the relative solubility of surfactant in each phase; therefore, it provides a rational means by which one can select an emulsion-stabilizing surfactant for the emulsion type desired. For surfactants for which the HLB system does not apply, such as block copolymers, their solubility in the continuous phase plays a major role in selecting an emulsion stabilizer [78]. Surfactant molecules in an emulsion are packed "shoulder to shoulder" in a condensed film at the oil–water interface. Silvestri et al. [87] have presented a theoretical model in which a modified Langmuir adsorption isotherm is used to determine nonlinear adsorption of surfactant to an oil–water interface. They were able to predict the oil droplet radius from the bulk surfactant concentration. Thus, the mean droplet size in an emulsion can be affected by the amount of surfactant present. If only a limited amount of surfactant is present, it will not be sufficient to form a condensed film over a large number of small globules.

Parrott [88] gives the following example. The molecular weight of sodium oleate is 304.4 and the molecular cross-sectional area is 22×10^{-6} cm^2. Thus, the minimum amount of sodium oleate required to emulsify 100 cm^3 of oil to a globule with diameter of 1 μm can be calculated using simple spherical geometry. In this example, 1.37 g of surfactant is needed. Although this type of calculation is an oversimplification in that there is always a range of globule size in an emulsion and there will always be some surfactant, both micellar and nonmicellar, in the bulk phase, it does provide a useful approximation.

It is common in many emulsions to use a mixture of two or more surface-active agents, rather than a single surfactant. The HLB of a mixture of surfactants may be calculated from the HLB of the components and the relative amount of each present [88,89]. Consider a mixture of 12% polysorbate 80 and 88% sorbitan sesquioleate, the HLB values of which are, respectively, 15.0 and 3.7. The HLB value of the mixture is calculated thus:

HLB of mixture = $(15 \times 0.12) + (3.7 \times 0.88) = 5.05$

Therefore, this surfactant mixture would provide a good HLB value for an o/w emulsion. Many active formulators in the area of pharmaceutical emulsions make considerable use of the HLB concept in their work; however, some research, such as that published by Parkinson and Sherman [90], has indicated limitations of this concept.

In addition to forming a condensed film at the oil–water interface, emulsifying agents can assist in stabilizing emulsions by modifying the viscosity or density of the continuous phase. (The effect of Stokes' law and brownian motion on the physical stability of disperse systems has already been discussed in Sec. IV.B).

Selection of the Oil Phase

Selection of the substance that will be used for the oil phase is determined by several factors, including the use and desired physical properties of the product, the potential toxicity of the oil relative to the route of administration, the solubility of active product in the oil relative to phase volume ratio, the consistency required of the product, and any possible incompatibilities [91]. A variety of substances have been used for this purpose, including, but not limited to, fixed oils (corn, olive, soybean, peanut, safflower), aliphatic hydrocarbons, beeswax, spermaceti, and various long- and medium-chain glycerides, fatty acids, and alcohols. Many of the foregoing substances are prone to oxidation; thus, it is often necessary to include a suitable antioxidant in the formulation. The choice of the phase volume ratio depends on a number of factors, including the required consistency. However, if the stabilizing forces depend heavily on rheological characteristics of the product, it is generally most inadvisable to attempt to formulate emulsions containing less than about 25% of disperse phase. Such products are very susceptible to severe creaming or sedimentation problems. Conversely, products containing a high percentage of disperse phase (more than about 70%) are likely to exhibit phase inversion (i.e., the disperse phase becomes the continuous phase). With a combination of proper emulsifying agent and suitable processing technology, one will be able to prepare emulsions with as little as 10% oil phase without a considerable stability problems [81,92].

Emulsifying Agents

There are a very wide variety of emulsifying agents now available. For detailed consideration of the properties of the substances available, the reader is referred to other published texts [114–116]. Apart from the ability of emulsifying agents to assist in stabilizing the emulsion, the selection of these materials will be affected by other factors, such as cost, toxicity, and resistance to chemical or microbial attack. Often it can be advantageous to use combination of macromolecules, such as gelatin, with surfactants. These materials may not reduce the interfacial tension to the extent that surfactants can, but high interfacial viscosities at the oil–water interface can be attained [93,94]. In the past, pharmaceutical emulsions have often contained emulsifying agents of natural origin, such as gelatin, acacia, or tragacanth. These substances are nontoxic and relatively inexpensive; however, they do show considerable batch-to-batch variation and have the major disadvantage of readily supporting microbial growth [95]. For the large-scale manufacturer who is concerned with production of a uniform product, the batch-to-batch variation of natural products can present serious problems. The differences in property may be subtle and difficult to detect, and the result may become apparent only several months after production. Concern has also been expressed about the contamination of natural products by bacteria and yeasts. Although most such microbial contamination is nonpathogenic, there is every possibility that regulatory agencies such as the (FDA) may, in the near future, exercise further control over this situation. The FDA policies that require process validation also put pressure on manufacturers to reduce or eliminate their use of natural products, because of their batch-to-batch variability.

Rheological Properties

Rheological properties of emulsions can require special attention from the formulator. Basically, we require a certain consistency compatible with the properties demanded by the route of administration for which the product has been designed. Emulsions generally show considerable deviation from newtonian flow [96]. If we require a reasonably high consistency, one approach that can be used is to have a flocculated system that provides a quite rigid network. Such a system can be rigid at rest, but flow quite easily when agitated. This technique is most useful for emulsions designed for oral or parenteral use; it is obviously useful to have a high consistency during storage, but ready flow from a bottle or through a syringe when needed. Importantly, the shear-thickening type of consistency should be avoided for injectable emulsions, for they will block the needle owing to the cross-sectional differences between syringe barrel and the needle. Also, this type of emulsion cannot be sterile-filtered because of shear thickening within the filter pores [97].

Emulsions for external use generally require quite considerable consistency or "body," and creams are based on a semisolid character. A variety of thickening agents can be used for this purpose [98], in addition to using waxes as the oil phase, as described earlier. Rheological properties of the emulsions for topical use may affect the release of active ingredient into the site of action. In vitro permeation of an antiacne drug, through hairless abdominal skin, from high viscosity microemulsions has been compared with a gel formulation, by Gasco and co-workers [99]. Although there was a lag time in both formulations, the percentage of transported active drug through the skin from the microemulsion was considerably higher than that of the gel formulation.

Preservatives and Antioxidants

The preservation of pharmaceutical disperse systems against microbial attack is often difficult. Emulsions are especially susceptible to contamination by fungi and yeasts. Apart from the pathogenic hazard, this contamination can result in discoloration or cracking (separation into two bulk phases). The carbon dioxide produced in an emulsion by microbial growth has resulted in exploding bottles. Components, such as polypeptides, carbohydrates, sterols, and some surfactants, such as lecithin, provide an almost ideal culture medium for many types of microrganisms. Establishment of microbiological standards for raw materials and strict adherence to rules of current Good Manufacturing Practice (cGMP) during production can help reduce the severity of the problem. However, the addition of an antimicrobial preservative is often required. Substances that have been used as preservatives for disperse systems include quaternary ammonium compounds (benzalkonium chloride and bezothenium 15), benzoic acid, phenylmercuric nitrate, parabens, and others [100]. Their use is generally limited to products that are not intended for intravenous injection. Intravenous injectable products should be sterile and pyrogen-free. These products, therefore, should be prepared according to stringent sterile conditions from pyrogen-free raw materials, and normally, be terminally sterilized by either autoclaving or a filtration technique [97].

It is also often necessary to add antioxidants to emulsions. Antioxidants are added to pharmaceutical formulations as a redox system having higher redox potential than the drug or other substance that they are designed to protect; thus, they tend to "soak up" free radicals. Antioxidants that give protection primarily in the aqueous phase include sodium metabisulfate, ascorbic acid, thioglycerol, and cysteine hydrochloride. Oil-soluble antioxidants include lecithin, propyl gallate, ascorbyl palmitate, butylated hydroxytoluene, and others. Vitamin E has also been used, but its virtues as a "natural" antioxidant have been the subject of some controversy [101].

C. Method of Evaluating Emulsions

Before the final choice of the formula for an emulsion is made, it is often desirable to evaluate several possible test products. No emulsion can really be said to be truly stable; however, some products are very much more stable than others. Obviously, in selecting an emulsion from a group of test preparations, one would reject any showing phase inversion, cracking, or signs of microbiological attack. Minor problems, such as creaming, may not be of great importance for a pharmaceutical product, whereas, for a cosmetic preparation, appearance is of profound importance.

There are several relatively simple techniques available for examining coalescence and phase separation. Visual observation, before and after shaking, by the experienced worker can sometimes be remarkably useful. Photomicrography can also be a useful technique for testing emulsions for coalescence, and the particle size counters (either Coulter or light-scattering methods) can be used for emulsion particle size determination and particle size distribution analysis.

Tingstad [102] proposed the use of the following test to evaluate the coalescence tendency. An oil phase containing all the hydrophobic components of the emulsion is very carefully poured over the aqueous phase, which contains all the water-soluble substances. A drop of the oil is then transferred by means of a syringe into the aqueous phase and released at a specific distance below the phase boundary. It then floats up to the phase boundary. The time required for the droplet to coalesce with the bulk of the oil phase is recorded. The longer the coalescence time, the more stable the emulsion. Caldwell and co-workers [103] have combined photon correlation spectroscopy and sedimentation field-flow fractionation to demonstrate the effect of small quantities of some additives on the particle size and particle size distribution of a parenteral fat emulsion composed of 10 or 20% oil phase. They then related the results to the emulsion stability.

Both centrifugation and temperature stress tests have also been used for emulsion stability tests [104–107]. Unfortunately, there is now no general agreement on the design of such tests. Thus, in one laboratory a centrifugation test may involve a 5-min test using a high-speed centrifuge, whereas in another a 20-min test using a low-speed centrifuge may be used. The predictive value of such tests is by no means always clearly apparent. Similarly, temperature stress tests are, like those used for suspensions, variable. In one laboratory -5 to $+40°C$ on a 24-hr cycle is used for 24 cycles, whereas in another 5 to 35°C on a 12-hr cycle for 10 cycles is used.

Petrowski [108] has described the use of microwave irradiation to determine emulsion stability. This technique appears to merit further investigation. After treatment by microwave irradiation, the surface temperature of the emulsion tends to be highest and the temperature gradient between the surface and the bottom of the emulsion to be smallest for the more stable emulsions. Those interested in detailed studies of emulsion stability are referred to publication by groups such as Groves [109,110], Bennita [92], and Anderson et al. [103].

Phase inversion can usually be readily seen by eye. However, other tests available include conductiometry (if water is the continuous phase, the emulsion will conduct electricity). Clausse has an excellent review on this subject [113]. Alternatively, a few drops of water-soluble dye can be placed on the surface of the emulsion: if the emulsion is o/w, the dye will rapidly diffuse throughout the system; if the emulsion is w/o, the dye will not disperse. Microelectrophoretic measurements have also sometimes been used to aid in the evaluation of emulsions.

As with suspensions, it is useful to perform some of the tests for the emulsions in the final container. Also, as with suspensions, the type of rheological tests needed can vary with the nature of the product.

D. Preparation of Emulsions

It is rather rare for the present-day community or hospital pharmacist to have to prepare an emulsion, although compounded dermatological emulsions are still prescribed and the pharmacist still requires some knowledge of emulsion preparation. The pharmacist can use the English or the Continental method for the extemporaneous preparation of emulsions [114]. Large-scale production methods show considerable variation. The oil and water phases containing the hydrophobic and hydrophylic components, respectively, are often heated separately in large tanks. When waxes are present, both phases must be heated above the highest melting point of any component present. One phase is then pumped into the tank containing the second phase, constant agitation being provided throughout the time of addition. After cooling, the product is homogenized and then packaged. As with suspensions, scale-up problems quite possibly may arise between the laboratory and the large-scale production. Tabibi [78] has provided a complete review on emulsion preparation and related recent processing technologies.

VI. LIPOSOMES

A. Liposomes in Pharmacy

Although Bangham and associates were the first to make liposomes after developing a laboratory technique to study the function of the cell wall in 1965 he [117] observed that on proper hydration of dried thin films of phospholipids, spherules of concentric shells will spontaneously form and stated that "... phase structure appear to be that of a layer lattice giving rise to spherulites ... consisting of many concentric bimolecular layers of lipid each separated by an aqueous compartment." However, they were first prepared in 1911 by Lehmann and termed *kunstaliches zellen*, meaning artificial cells. Weissmann [118] coined the term *liposome* to define phospholipid spherules. The creation of the word liposome probably stems from the terminology of subcellular particles such as glycosome, lysosome, and ribosome. The term *liposome*, meaning lipid body, may generally be defined as, any fluid-filled closed spherical vesicle in which the amphiphilic molecules (mostly lipidic structures) form one or more bimolecular (bilayer or lamelar) structures, with significant barrier properties, surrounding the interior fluid. This structure is an overall hydrophilic membranelike assembly, in which apolar (i.e., lipophilic) tails of the amphiphilic molecules point inward and polar (i.e., hydrophilic) headgroups point outward within the lamellar structure. Because of these characteristics, they have been considered, from the time they were rediscovered by Bangham, as potential drug delivery systems to either reduce the side effect of drugs, such as doxorubicin (Adriamycin) and amphotericin B, or to increase their efficacy [119–121]. Their pharmaceutical uses are very widespread and cover parenteral administration of lipophilic drugs [122,123], ionophores [124], targeted delivery systems [125], potentiating activity of drugs by liposomal use of a toxic material [126], vaccine adjuvants [127], inhalation therapy [128], delivery of diagnostic agents and enhancement of their effect [129], topical delivery of drugs to enhance penetration [130–132], ophthalmic delivery [133], and finally, antibody-guided, targeted delivery of drugs, such as the antilaminin receptor monoclonal antibody, to cancer cells [134,135]. They have also been used as controlled drug-delivery systems, mostly for topical or intramuscular administration [136–139]. Liposomes have also been used in cosmetics [140–142] and in other fields, such as consumer products [143] and pesticides [144,145].

It is important to realize that many thousands of scientific articles containing the terms *liposome* or *vesicles*, along with other specific publications have been published. Therefore, the following section will not be, by any means, a comprehensive treatise on the subject; rather, it will concentrate on the basic understanding of this topic and will discuss the use of liposomes as drug delivery system. Interested readers should refer to the scientific and review publications on this subject [117–145]. These reviews indicate that liposomes display therapeutic advantage in such areas as antifungal, anticancer, antiviral, and antimicrobial therapy. As with other dispersed drug-delivery systems (e.g., emulsions), on intravenous administration, uptake of liposomes by the reticuloendothelial system (RES) limits their therapeutic applications. However, recently, the use of more hydrophilic surfactants makes them sterically hindered, thereby creating a system almost invisible to the RES. These liposomes show a prolonged circulation in blood, enhanced tissue distribution, and an improved therapeutic index owing to protection against RES uptake [146].

Liposomes have even been investigated for the topical delivery of drugs to enhance the penetration of the active ingredients deep into the stratum corneum and the cornea [147–149].

Although claims of penetration enhancement have been increased, it is not yet clear whether the difficulties in production of liposomes outweigh the simplicity of manufacturing emulsions.

Definitions and Classifications

Different amphiphilic materials have been tried in creating liposomal (vesicular) structures and, as a result, a variety of brand names have been designated to differentiate between them. Table 7 presents some of the amphiphiles used in preparation of these vesicular structures, along with the designated name or term. Generally, liposomes consist of bilayer-forming lipidic amphiphiles, cholesterol, and a charge-generating molecule. However, it should be emphasized that only the bilayer-forming lipid is the essential part of the lamellar structure and the other components are added to impart certain characteristics to the vesicles. For example, cholesterol or its derivatives are added to stabilize the bilayer structure, rendering it rigid and, as a result, reducing its permeability.

Phospholipids were among the first lipidic amphiphiles used to produce bilayer structures to mimic the cell membrane [117,118], and soon the potential of these structures as a drug delivery system was realized. The chemical structure of a typical phospholipid molecule (i.e., phosphatidylcholine), is presented.

Structure of phosphatidyl choline

There are various terms and methods used in the literature to classify the different morphological categories of vesicles or liposomes. The following is a classification that includes most of the definitions:

Multilamellar vesicles (MLV), in which a multiple "onionlike" bilayer structure surrounds a relatively small internal core, as defined and produced by Bangham [117].

Oligolamellar vesicles, (OLV), in which the large central aqueous compartment is surrounded by two to ten bilayer structures, also on some rare ocassions called paucilamellar vesicles (PLV) [145].

Unilamellar vesicles, (ULV), in which there is only a single bilayer structure surrounding the internal aqueous core. This particular category has several subcategories based on their size:

Small unilamellar vesicles (SUV), with the size range of 20–40 nm, with little use in drug delivery.

Medium unilamellar vesicles (MUV), with the size range of 40–80 nm.

Large unilamellar vesicles (LUV), with a large internal aqueous core having the size
   range of 10 ~ 1000 nm.
Giant unilamellar vesicles (GUV), for which the size is larger than 1000 nm, and
   probably most unstable from physicomechanical consideration
Multivesicular vesicles (MVV), in which a large vesicle contains smaller and, usually, unilamellar vesicles.

Although the methods for preparation of MLVs were described by Bangham [117], various
preparation techniques have been reviewed by many investigators in the field. Most of these
methods have been developed to produce multilamellar and unilamellar liposomes and are
partially listed in Table 8 [150]. Some of these methods will be considered in some detail later
in this chapter.

Characterization Methods

Because there are different categories of liposomes, the characterization of these lipidic vesicles
become extremely important if one desires reproducible results. Four major characterization
methods are briefly described here.

Size and Size Distribution. Most phospholipids adapt a bilayer structure spontaneously on
dispersing in water, which is not necessarily true for other amphiphiles. The resultant liposomes, irrespective of the lipid used in their preparation, may be large or small. Depending on
various factors, including the composition of bilayer-forming materials, their structure, and the
processing technique, they may differ in size and in size distribution. Therefore, it is important
to determine the size and size distribution of liposomal preparations and to determine their
stability after storage under various conditions for a certain time period. The size determination
and size stability are important, because the size and size distribution of a batch of liposome
may change during storage, thereby affecting their function. The laser light-scattering technique
is the most common technique used to determine size and size distribution of liposomal
preparations.

Lamellarity. Different types of liposomes have various degrees of lamellarity; thus, the
characterization of the lamellarity of liposomes seems very reasonable. A variety of experimental procedures have been employed in lamellarity determination, such as labeling and
binding studies. The use of labeling with ^{31}P and then employing nuclear magnetic resonance
(NMR) to determine the phosphorus signal intensity in phospholipid liposomes is one of the
best and most precise methods of characterizing the lamellarity of liposomes. It is obvious that
this labeling method will not be useful for nonphospholipid vesicles; therefore; the presence
of ^{14}C in the headgroup of the amphiphile may be required.

   Another method, but not as precise as labeling, for characterizing lamellarity, is scanning
electron microscopy.

Entrapped Volume. The entrapped volume is generally defined as the amount of entrapped
volume per mole of lipid. This parameter can vary from about 0.5 to 30 liters/mol of lipid,
which is normally expressed as microliters per micromole ($\mu l/\mu mol$). Also, the entrapped
volume is much smaller for MLV liposomes. Entrapped volume is generally determined by
entrapping an impermeable radiolabeled molecule, such as insulin, within the liposome; removing the external radioactivity by such techniques as dialysis, gel filtration, or centrifugation;
and then, determining the residual radioactivity. This method assumes that there is no binding
of insulin to the bilayer region(s) of the liposomes.

Solute Distribution. Partitioning of solute between the lipidic and aqueous phases is not
specific to liposomes, but differing degrees of distribution of the solute among the various
bilayers of the MLV system is a subject to consider. This varying degree of solute distribution

is probably due to the hydration sequence of dry lipid film on contact with aqueous solution. Also, various materials with different degrees of diffusivity permeate in different rates through the hydrating sequence of lipid bilayers in the MLV system. The solute distribution can be determined by NMR.

Although various factors affect the aforementioned liposomal characteristics, the pharmacokinetics and biodistribution of entrapped active ingredient generally depends on the route of administration. For example, owing to their particulate nature, liposomes, on intravenous injection, will be taken up by the reticuloendothelial system, rendering little sustained-release activity. On the other hand, if liposomes are injected intramuscularly or subcutaneously, they may work as sustained-release delivery system. It has been reported that one may dissolve a water-insoluble protein in a lipid-based solubilizer and then entrap the solubilized protein within a liposomal structure to provide controlled-release activity. This composition is then injected either intramuscularly or subcutaneously, providing sustained-release action for a period of about 1 month [151]. On the other hand, topical application of liposomes may cause penetration enhancement owing to possible fusion of liposomal lipids with cell membrane lipids of the stratum corneum or the cornea. Also reported is a sustained-release liposomal ophthalmic delivery system that will deliver an antimicrobial agent to the infected eye for 3 days, to treat infectious bovine keratoconjunctivitis [145–148]. The major advantages of liposomal drug delivery systems are their ability to either deliver drugs to a target organ [152,153], or to reduce drug delivery to a certain organ [154]. The reduction in side effects, for example, cardiotoxicity of doxorubicin in liposomal formulation [155], is probably due to a dramatic reduction in binding affinity to heart tissue of the liposome-encapsulated doxorubicin.

Preparation Methods

Solvent Evaporation Technique.   In this method, the bilayer-forming formulation, which is generally composed of phospholipids and cholesterol along with lipophilic active ingredient, is dissolved in a suitable organic solvent, such as chloroform, carbon tetrachloride, or ether. The solvent is then evaporated in a rotavapor under reduced pressure to create a thin layer of lipidic film deposited on the inner wall of the round-bottom flask. The produced film is then hydrated with either distilled water or a proper solution of drug, depending on the nature and concentration of the active ingredient, phospholipid/cholesterol ratio, nature and concentration of the lipidic ampliphiles, rate of hydration, and so on. The noncaptured or free drug is then removed by various methods, such as dialysis, gel filtration, or centrifugation. The advantages of this method of liposomal preparation are ease of production (scale-up possibility), relative storage stability, and suitability for encapsulation of large molecules. On the other hand, the liposomes suffer from size heterogeneity, low encapsulation capacity, and risk of material degradation by solvents or heat.

REFERENCES

1. B. Jirgensons and M. E. Straumanis, *A Short Textbook of Colloid Chemistry*, Pergamon Press, Elmsford, NY, 1969.
2. A Martin, J. Swarbrick, and A. Cammarata, *Physical Pharmacy, Physical Chemical Principles in the Pharmaceutical Sciences*, 3rd Ed., Lea & Febiger, Philadelphia, 1983, pp. 445–521.
3. B. A. Matthews and C. T. Rhodes, J. Pharm. Sci., 57, 557 (1968).
4. P. M. Short, E. T. Abbs, and C. T. Rhodes, Can. J. Pharm. Sci., 4, 8 (1969).
5. B. A. Matthews and C. T. Rhodes, J. Colloid Interface Sci., 32, 339 (1970).
6. Coulter Electeronics, Coulter Counter Operation Manual, Edison, New Jersey, (1991).
7. M. E. Houghton and G. E. Amidon, Pharm. Res., 9, 856 (1992).

8. B. B. Weiner and W. W. Tscharnuter, in *Particle Size Distribution, Assessment and Characterization* (T. Provder, Ed.), ACS Symposium Series 332, Washington, DC, 1987, p. 48.
9. A. A. El-Saged and A. J. Repta, Int. J. Pharm., 13, 303 (1983).
10. K. D. Caldwell and J. M. Li, J. Colloid Interface Sci., 132, 256 (1989).
11. P. M. Short, S. V. Lincoln, and C. T. Rhodes, Pharmazie, 6, 319 (1969).
12. C. Orr, in *Encyclopedia of Emulsion Technology,* Vol. 1, *Basic Theory* (P. Becher, ed.), Marcel Dekker, New York, 1983, p. 369.
13. I. C. Edmundson, *Advances in Pharmaceutical Sciences*, Vol. 2 (H. S. Bean, J. E. Carless, and A. H. Beckett, eds.), Academic Press, London, 1967, p. 95.
14. A. P. Black and A. L. Smith, J. Am. Water Works Assoc., 58, 445 (1966).
15. T. F. Tadros and B. Vincent, in *Encyclopedia of Emulsion Technology,* Vol. 1, *Basic Theory* (P. Becher, ed.), Marcel Dekker, New York, 1983, p. 129.
16. M. Clausse, in *Encyclopedia of Emulsion Technology,* Vol. 1, *Basic Theory* (P. Becher, ed.), Marcel Dekker, New York, 1983, p. 483.
17. R. H. Muller, *Colloidal Carriers for Controlled Drug Delivery and Targetting, Modification, Characterization and in vivo Distribution*, CRC Press, Boca Raton, FL, 1991.
18. S. Fricke and R. Huettenrauch, Eur. J. Pharm. Biopharm., 37, 55 (1991), cited in Chem. Abstr., 115:99079k.
19. S. Fricke and R. Huettenrauch, Eur. J. Pharm. Biopharm., 37, 60 (1991), cited in Chem. Abstr., 115:119908k.
20. S. S. Davis, Pharm. Acta Helv., 49, 161 (1974).
21. J. T. Carstensen, *Theory of Pharmaceutical Systems*, Vol. 2, Academic Press, New York, 1973.
22. A. Martin, G. S. Banker, and A. H. C. Chun, in *Advances in Pharmaceutical Sciences*, Vol. 1 (H. S. Bean, J. E. Carless, and A. H. Beckett, eds.), Academic Press, London, 1964, p. 1.
23. B. W. Barry, in *Advances in Pharmaceutical Sciences*, Vol. 4 (H. S. Bean, J. E. Carless, and A. H. Beckett, eds.), Academic Press, London, 1974, p. 1.
24. P. Sherman, in *Encyclopedia of Emulsion Technology*, Vol. 1, *Basic Theory* (P. Becher, ed.), Marcel Dekker, New York, 1983, p. 405.
25. S. H. Yalkowsky, ed., *Techniques of Solubilization of Drugs*, Marcel Dekker, New York, 1981.
26. S. Friberg, J. Soc. Cosmet. Chem., 41, 155 (1990).
27. M. E. L. McBain and E. Hutchinson, *Solubilization and Related Phenomena*, Academic Press, New York, 1955.
28. K. Shinoda, T. Nakagawa, B. Tamamushi, and T. Isemura, *Colloidal Surfactants*, Academic Press, New York, 1963.
29. K. Shinoda and S. Friberg, *Emulsions and Solubilization*, Wiley-Interscience, New York, 1986.
30. G. H. Brown and P. P. Crooker, Chem. Eng. News, 61(5), 24 (1983).
31. S. E. Tabibi, D. F. H. Wallach, and C. Yiournas, in *Interplex. USA, Proceedings of the 1991 Technical Program*, 1991, p. 61.
32. S. E. Friberg and M. A. El-Nokaly, in *Surfactants in Cosmetics* (M. M. Riger, ed.), Marcel Dekker, New York, 1985, p. 55.
33. T. Tice and S. E. Tabibi, in *Colloidal Drug Release Technologies* (A. Kydonieus, ed.), Marcel Dekker, New York, 1991, p. 315.
34. P. Molyneux and C. T. Rhodes, Kolloid Z. Z. Polym. 250, 886 (1972).
35. B. D. Anderson, R. A. Conradi, K. E. Knuth, and S. L. Nail, J. Pharm. Sci. 74, 75, (1985).
36. I. Oh, S. C. Chi, B. R. Vishnuvajjala, and B. D. Anderson, Int. J. Pharm., 73, 23, (1991).
37. B. A. Mulley, in *Advances in Pharmaceutical Sciences*, Vol. 1 (H. S. Bean, J. E. Carless, and A. H. Beckett, eds.), Academic Press, London, 1964, p. 86.
38. L. M. Prince, ed., *Microemulsions*, Academic Press, New York, 1977.
39. A. G. Mitchell, J. Pharm. Pharmacol., 15, 761 (1963).
40. J. Swarbrick and J. E. Carless, J. Pharm. Sci., 59, 1427 (1970).
41. K. J. Humphreys and C. T. Rhodes, J. Pharm. Sci., 57, 79 (1968).
42. M. Donbrow, E. Azaz, and R. Hamburger, J. Pharm. Sci., 59, 1427 (1970).
43. M. Donbrow and C. T. Rhodes, J. Chem. Soc., 2 (Suppl.), 6166 (1964).

44. A. G. Mitchell and K. F. Brown, J. Pharm. Pharmacol., 18, 115 (1966).
45. J. W. Bradshaw, C. T. Rhodes, and G. Richardson, J. Pharm. Sci., 61, 1163 (1972).
46. H. S. Bean, S. M. Heman-Ackah, and J. Thomas, J. Soc. Cosmet. Chem., 16, 15 (1965).
47. M. J. Crooks and K. F. Brown, J. Pharm. Pharmacol., 25, 281 (1973).
48. J. E. F. Reynolds, ed., *Martindale The Extrapharmacopoeia*, 23rd Ed., The Pharmaceutical Press, London, 1986, pp. 1392, 1402.
49. M. R. Violante, K. J. Parker, and H. W. Fischer, Invest. Radiol. 23, 294 (1988).
50. M. R. Violante, R. T. Steigbergel, U. S. Patent 4,783,484, November 8, 1988.
51. M. R. Violante and H. W. Fischer, U. S. Patent 4,826,689, May 2, 1989; U. S. Patent 4,997,454, March 5, 1991.
52. B. A. Matthews and C. T. Rhodes, J. Pharm. Sci., 59, 521 (1970).
53. H. R. Kruyt, *Colloid Science*, Elsevier, Amsterdam, 1953.
54. B. A. Matthews, J. Pharm. Sci., 62, 173 (1973).
55. B. A. Matthews and C. T. Rhodes, J. Colloid Interface Sci., 32, 332 (1970).
56. T. F. Tadros and B. Vincent, in *Encyclopedia of Emulsion Technology,* Vol. 1, *Basic Theory* (P. Becher, ed.), Marcel Dekker, New York, 1983, p. 129.
57. K. S. Alexander, D. Dollimore, S. S. Tata, and A. S. Savitri, J. Pharm. Sci., 81, 787, (1992).
58. B. A. Matthews and C. T. Rhodes, Pharm. Acta Helv., 45, 52 (1969).
59. J. G. Nairn, in *Remington's Pharmaceutical Sciences*, 18th Ed., (A. R. Genero, ed.), Mack Publishing, Easton, PA, 1990, p. 1519.
60. R. D. C. Jones, B. A. Matthews, and C. T. Rhodes, J. Pharm. Sci. 59, 518 (1970).
61. G. L. Stanko and H. G. Dekay J. Am. Pharm. Assoc. (Sci. Ed.), 47, 104 (1958).
62. S. Rudt and R. H. Mueller, J. Controlled Release, 22, 263–271 (1992).
63. P. C. Hiemenz, *Principles of Colloid and Surface Chemistry*, Marcel Dekker, New York, 1977.
64. R. S. Okar, Pharm. Res., 10, 220 (1992).
65. M. El-Khawas, Alexandria J. Pharm. Sci. 6, 233–237 (1992), cited in Chem. Abstr., 118: 45591q.
66. S. L. Hem, in *Current Concepts in Pharmaceutical Sciences, Dosage Form Design and Bioavailability* (J. Swarbrick, ed.), Lea & Febiger, Philadelphia, 1973, p. 77.
67. J. T. Carstensen, *Theory of Pharmaceutical Systems*, Vol. II: *Heterogeneous Systems*, Academic Press, New York, 1973, p. 2.
68. L. Labarquilla, S. E. Tabibi, and L. Shargel, Pharm. Res., 6, S68 (1989).
69. S. Apte and S. Turco, J. Parenter. Sci. Technol., 46, 12 (1992).
70. B. A. Matthews, J. Pharm. Sci., 62, 173 (1973).
71. B. A. Haines, Jr. and A. N. Martin, J. Pharm. Sci., 50, 228 (1961).
72. M. P. Short and C. T. Rhodes, Can. J. Pharm. Sci. 8, 46 (1973).
73. J. R. Buck, G. E. Peck, and G. S. Banker, Drug Dev. Commun., 1, 89 (1975).
74. S. Harder and G. van Buskirk, in *The Theory and Practice of Industrial Pharmacy*, 3rd Ed. (L. Lochman, H. A. Lieberman, and J. L. Kanig, eds.), Lea & Febiger, Philadelphia, 1986, p. 681.
75. K. D. Alexander, R. P. Haribhakti, and G. A. Parker, Am. J. Hosp. Pharm., 47, 1241 (1991).
76. B. A. Matthews and C. T. Rhodes, J. Pharm. Sci., 56, 838 (1967).
77. J. B. Boyett and C. W. Davis, in *Pharmaceutical Dosage Forms,* Vol. 2: *Disperse Systems* (H. A. Lieberman, M. M. Reiger, and G. S. Banker, eds.), Marcel Dekker, New York, 1989, p. 379.
78. S. E. Tabibi, in *Specialized Drug Delivery Systems: Manufacturing and Production Technology* (P. Tyle, ed.), Marcel Dekker, New York, 1989, p. 317.
79. R. R. Scott, in *Pharmaceutical Dosage Forms,* Vol. 2: *Disperse Systems* (H. A. Lieberman, M. M. Reiger, and G. S. Banker, eds.), Marcel Dekker, New York, 1989, p. 1.
80. S. Silvestri, L. L. Wu, and B. Bowser, J. Pharm. Sci., 81, 413, (1992).
81. K. Westesen and T. Wehler, J. Pharm. Sci., 81, 777 (1992).
82. T. T. Kararli, T. E. Needham, M. Griffin, G. Schoenhard, L. J. Ferro, and L. Alcorn, Pharm. Res., 9, 888 (1992).
83. A. Rubinstein, Y. V. Pathak, J. Kleinstern, A. Reches, and S. Benita, J. Pharm. Sci., 80, 643, (1991).
84. S. E. Tabibi and A. A. Siciliano, U. S. Patent 4,971,788, 1990.
85. S. E. Tabibi and A. A. Siciliano, U. S. Patent 5,130,122, 1992.

86. S. Roy, M. Pal, and B. K. Gupta, Pharm. Res., 9, 1132 (1992).
87. S. Silvestri, N. Ganguly, and E. Tabibi, Pharm. Res., 9, 1347 (1992).
88. E. L. Parrott, *Pharmaceutical Technology*, Burgess, Minneapolis, 1970.
89. ICI Americaas, *The HLB System, a Time-Saving Guide to Emulsifier Selection*, ICI Americas, Wilmington, DE, 1984.
90. J. B. C. Parkinson and P. Sherman, J. Colloid Interface Sci., 41, 666 (1972).
91. G. Zografi, in *Theory and Practice of Industrial Pharmacy* (L. Lochman, H. A. Lieberman, and J. L. Kanig, eds.), Lea & Febiger, Philadelphia, 1970, Chap. 16.
92. M. Y. Levy and S. Benita, J. Parenter. Sci. Technol., 45, 101 (1991).
93. E. Shotton, K. Wibberly, and A. Vaziri, in *Proceeding of the 4th International Congress on Surface Activity*, Vol. 3, 1965, p. 1.
94. E. Shatton and R. F. White, in *Rheology of Emulsions* (P. Sherman, ed.), Macmillan, New York, 1963.
95. K. H. Wallhausser, in *Surfactants in Cosmetics* (M. M. Riger, ed.), Marcel Dekker, New York, 1985, p. 211.
96. W. C. Griffin, in *Encyclopedia of Chemical Technology*, 2nd Ed., Vol. 3 (A. Standen, H. F. Mark, and D. F. Othmer, eds.), Wiley-Interscience, New York, 1965, p. 117.
97. D. B. Lidgate, T. Trattner, R. M. Schultz, and R. Maskiewicz, Pharm Res., 9, 860, (1992).
98. H. M. Fishman, Happi, April 28 (1992).
99. M. R. Grasco, M. Gallarate, and F. Pattarino, Int. J. Pharm., 69, 193, (1991).
100. H. Takruri and C. B. Anger, in *Pharmaceutical Dosage Forms: Disperse Systems*, Vol. 2 (H. A. Lieberman, M. M. Riger, and G. S. Banker, eds.), 1989, p. 73.
101. H. C. Carson, Happi, May, 53 (1992).
102. J. F. Tingstad, J. Pharm. Sci., 53, 935 (1964).
103. J. Li, K. D. Caldwell, and B. D. Anderson, Pharm. Res. 10, 535 (1993).
104. S. I. Rehfeld, J. Colloid Interface Sci., 46, 448 (1974).
105. K. Ridgway, Pharm. J., 212, 583 (1974).
106. B. Idson, Drug Cosmet. Ind., Jan./Feb. (1993).
107. G. Zografi, J. Soc. Cosmet. Chem., 33, 345 (1982).
108. M. J. Groves and D. C. Freshwater, J. Pharm. Sci., 57, 1273 (1968).
109. M. J. Groves, R. M. Mustafa, and J. E. Carless, J. Pharm. Pharmacol., 26, 264 (1974).
110. M. J. Groves and R. M. Mustafa, J. Pharm. Pharmacol., 26, 671 (1974).
111. M. M. Riger, in *The Theory and Practice of Industrial Pharmacy*, 3rd Ed. (L. Lochman, H. A. Lieberman, and J. L. Kanig, eds.), Lea & Febiger, Philadelphia, 1986, p. 502.
112. R. H. Muller, S. S. Davis, and W. Niemann, Acta Pharm. Technol., 34, 17S (1988).
113. M. Clausse, in *Encyclopedia of Emulsion Technology*, Vol. 1, *Basic Theory* (P. Becher, ed.), Marcel Dekker, New York, 1983, p. 481.
114. D. Myers, *Surfactant Science and Technology*, VCH Publishers, New York, 1988.
115. P. Becher, ed., *Encyclopedia of Emulsion Technology*, Vol. 1, *Basic Theory*, Marcel Dekker, New York, 1983.
116. M. M. Rieger, Ed., *Surfactants in Cosmetics*, Marcel Dekker, New York, 1985.
117. A. D. Banghm, M. M. Standish, and J. C. Watkins, J. Mol. Biol., 13, 238 (1965).
118. G. Sessa and G. Weissmann, J. Lipid Res., 9, 310 (1968).
119. M. J. Ostro, ed., *Liposomes, From Biophysics to Therapeutics*, Marcel Dekker, New York, 1987.
120. G. Gregoriadis, ed., *Liposome Technology*, Vols. 1–3, CRC Press, Boca Raton, FL, 1986.
121. J. Weinstein, Cancer Treat. Rep., 68, 127 (1990).
122. J. P. Sculier, A. Coune, C. Brassine, et al., J. Clin. Oncol., 4, 789 (1986).
123. H. Sasaki, T. Kakutani, M. Hashida, et al., J. Pharm. Pharmacol., 37, 461 (1985).
124. S. S. Daoud and R. L. Juliano, Cancer Res., 46, 5518 (1986).
125. K. Iga, Y. Ogawa, and H. Taguchi, Pharm. Res., 9, 658 (1992).
126. T. Griffin, M. E. Rybak, L. Recht, et al., JNCI, 85, 292 (1993).
127. D. Davis and G. Gregoriadis, Immunology, 61, 229 (1987).
128. M. Ausborn, B. V. Wichert, M. T. Carvajal, et al., Proc. Int. Symp. Control Rel. Bioact. Mater., 18, 371 (1991).

129. S. A. Baker, K. M. G. Taylor, and M. D. Shot, Proc. Int. Symp. Control Rel. Bioact. Mater., 18, 289 (1991).

130. V. Masini, F. Bonte, A. Meybeck, and J. Wepierre, J. Pharm. Sci., 82, 17 (1993).

131. M. Mezei and V. Gulasekharam, J. Pharm. Pharmacol. 34, 473 (1982).

132. M. Jacob, G. P. Martin, and C. Mariott, J. Pharm. Pharmacol., 40, 829 (1988).

133. E. Hirnle, P. Hirnle, and J. K. Wright, J. Microencapsulation, 8, 391 (1991).

134. A. Rahman, M. Panneerselvam, R. Gurgins, et al., JNCI, 81, 1794 (1989).

135. Y. Watanabe and T. Osawa, Chem. Pharm. Bull., 35, 740 (1987).

136. S. E. Tabibi, R. Mathur, and D. F. H. Wallach, 83rd Annual Meeting of AACR, San Diego, CA, 1992.

137. A. L. Weiner, S. S. Carpenter-Green, E. C. Soehngen, et al., J. Pharm. Sci., 74, 922 (1985).

138. G. Blume, G. Cevc, Bochim. Biophys. Acta, 1029, 91 (1990).

139. V. M. Knepp, R. S. Hinz, F. C. Szoka, and R. H. Guy, J. Controlled Release, 5, 211 (1988).

140. G. J. Brooks and R. C. McManus, INFORM, 1, 891 (1990).

141. H. Fishman, Happi, Aug., 20 (1992).

142. H. C. Carson, Happi, Mar., 53 (1993).

143. S. E. Tabibi, in *Second Workshop and Exhibition on Controlled Delivery in Consumer Products*, Secaucus, NJ, 1992.

144. S. E. Tabibi, R. Mathur, S. Henderson, et al., Proc. Int. Symp. Controlled Release Bioact. Mater., 18, 231 (1991).

145. S. E. Tabibi, J. D. Sakura, R. Mathur, et al., in *Pesticide Formulations and Application Systems*, 12th Vol., ASTM STP 1146 (B. N. Devisetty, D. G. Chasin, and P. D. Berger, eds.), American Society for Testing and Materials, Philadelphia, PA, 1993, p. 155.

146. A. Gabison and D. Paphadjopoulos, Proc. Natl. Acad. Sci., USA, 85, 6949 (1988).

147. M. G. Ganesan, N. D. Weiner, G. L. Flynn, and N. F. H. Ho, Int. J. Pharm., 20, 139 (1984).

148. G. Strauss, in *Annual Scientific Seminar, Society of Cosmetic Chemists*, Minneapolis, 1988.

149. J. Kreuter, in *Ophthalmic Drug Delivery. Biopharmaceutical, Technological and Clinical Aspects* (M. S. Saettone, G. Bucci, and P. Speiser, eds.), Liviana Press, Padova, 1987, pp. 101.

150. S. E. Tabibi, in *Formulating Delivery Systems for Cosmetics and OTC Dermatologicals*, Technomic Publishing., Princeton, NJ, 1992.

151. T. R. Tice and S. E. Tabibi, in *Treatise on Controlled Drug Delivery, Fundamentals, Optimization, Applications*) (A. Kydonieus, Ed.), Marcel Dekker, New York, 1992, pp. 315–339.

152. V. V. Ranade, J. Clin, Pharmacol., 29, 685 (1988).

153. G. Gergoriadis, J. Drug Targetting, 1, 3 (1993).

154. C. A. Raymond, JAMA, 257, 1143, (1987).

155. A. Rahman, A. Kessler, N. More, B. Sikic, G. Rowdan, P. Wooley, and P. S. Schein, Cancer Res., 40, 1532 (1980).

156. R. M. Handjani-vila, A. Ribier, and G. Vanlenberghe, in *Les Liposomes*, Technique et Documentation Lavoisier, Paris, 1985, pp. 297–312.

157. T. Kunitake and Y. Okahata, J. Am. Chem. Soc., 99, 3860 (1977).

158. J. M. Gebicki and M. Hicks, Nature, 243, 232 (1973).

159. Y. Ishigami and H. Machida, J. Am. Oil Chem. Soc., 66–599 (1989).

160. K. Kano, A. Romero, B. Djermouni, H. J. Ache, and J. F. Fendler, J. Am. Chem. Soc., 101, 4030 (1979).

161. Y. Murakami, A. Nakano, and H. Ikeda, J. Org. Chem., 47, 2137 (1982).

162. D. Brunke, SOFW, 116, 53 (1990).

163. T. Ohsawa, H. Miura, and K. Harada, Pharm. Bull., 33, 2916 (1985).

164. C. J. Kirby and G. Gregoriadis, Biotechnology, 2, 979 (1984).

165. L. D. Mayer, J. J. Hope, R. P. Cullis, and A. S. Janoff, Biochim. Biophys. Acta, 817, 193 (1985).

166. F. Szoka and D. Papahadjopoulos, Proc. Natl. Acad. Sci. USA, 75, 4194 (1978).

167. M. J. Hope, M. B. Bally, G. Webb, and R. P. Collis, Biochim. Biophys. Acta, 812, 55 (1985).

10

Tablet Dosage Forms

Edward M. Rudnic
Pharmavene, Inc., Rockville, Maryland

Mary Kathryn Kottke
AutoImmune, Inc., Lexington, Massachusetts

I. INTRODUCTION

During the past three and a half decades, the pharmaceutical industry has invested vast amounts of time and money in the study of tablet compaction. This expenditure is quite reasonable when one considers how valuable tablets, as a dosage form, are to the industry. As oral dosage forms can be self-administered by the patient, they are obviously more profitable to manufacture than parenteral dosage forms, which usually must be administered by trained personnel. This is reflected by the fact that well over 80% of the drugs in the United States that are formulated to produce systemic effects are marketed as oral dosage forms. Compared with other oral dosage forms, tablets are the manufacturers' dosage form of choice because of their relatively low cost of manufacturing, packaging, and shipping; increased stability, and virtual tamper resistance (i.e., most tampered tablets either become discolored or disintegrate).

II. DESIGN AND FORMULATION OF COMPRESSED TABLETS

A. General Considerations

The most common solid dosage forms in contemporary practice are *tablets*, which may be defined as unit forms of solid medicaments prepared by compaction. Most consist of a mixture of powders that are compacted in a die to produce a single, rigid body. The most common types of tablets are those intended to be swallowed whole and then disintegrate and release their medicaments in the gastrointestinal tract (GIT). A less common type of tablet is formulated to allow dissolution or dispersion in water before administration. Ideally, for this type of tablet, all ingredients should be soluble, but frequently, a fine suspension has to be accepted. Many tablets of this type are formulated to be effervescent, and their main advantages include rapid release of medicament and minimization of gastric irritation.

Some tablets are designed to be masticated (chewed). This type of tablet is often used when absorption from the buccal cavity is desired, or to enhance dispersion before swallowing.

Alternatively, a tablet may be intended to dissolve slowly in the mouth (e.g., lozenges) to provide local activity of the drug. A few tablets are designed to be placed under the tongue (sublingual) or between the teeth and gum (buccal) and rapidly release their medicament into the bloodstream. Buccal or sublingual absorption is often desirable for drugs subject to extensive hepatic metabolism by the first-pass effect (e.g., nitroglycerin, testosterone). Recently, a lozenge on a stick, or "lollipop," dosage form of fentanyl was developed for pediatric use.

There are now many types of tablet formulations that provide for the release of the medicament to be delayed or to control the rate of the drug's availability. Some of these preparations are highly sophisticated and are rightly referred to as complete "drug-delivery systems."

"Sustained-release" tablets can encompass a broad range of technologies. Since the concepts of prolonged drug delivery are the subjects of Chapter 15, the strategies of these systems will not be discussed here. However, solid dosage formulators must be aware of the various options available to them.

For example, some water-soluble drugs may need to be formulated so that their release and dissolution is controlled over a long period. For these, certain water-insoluble materials will have to be coformulated with the drug. If the dose of this drug is high, the drug will dictate the tableting properties of the formula. If the drug exhibits poor compactibility, hydrophobic agents, such as waxes, will surely make matters worse. To solve such a problem, the formulators would have to turn to other types of water-insoluble materials, such as polymers, to achieve drug release and tableting goals.

Some tablets combine sustained-release characteristics with a rapidly disintegrating tablet. Such products as K-Dur (Key Pharmaceuticals) combine coated potassium chloride (KCl) crystals in a rapidly releasing tablet. In this particular instance, the crystals are coated with ethylcellulose, a water-insoluble polymer and are then incorporated in a rapidly disintegrating microcrystalline cellulose matrix. The purpose of this tablet is to minimize GI ulceration, commonly seen with KCl therapy. This simple, but elegant, formulation is a masterpiece of solid dosage form strategy to achieve clinical goals.

Thus, the single greatest challenge to the tablet formulator is in the definition of the purpose of the formulation and the identification of suitable materials to achieve developmental objectives. To do this properly, the formulator must know the properties of the drug, the materials to be coformulated with the drug, and the important aspects of the granulation, tableting, and coating processes.

Pharmaceutical compressed tablets are prepared by placing an appropriate powder mix, or granulation, in a metal die on a tablet press. At the base of the die is a lower punch, and above the die is an upper punch. When the upper punch is forced down on the powder mix (single-punch press), or when the upper and lower punches squeeze together (rotary press), the powder is forced into a tablet. Despite that powder compaction has been observed for millennia, scientists still debate the exact mechanisms behind this phenomenon.

Perhaps the most significant factor in the tableting process arises from the need to produce tablets of uniform weight. This is achieved by feeding constant volumes of homogeneous material to the dies. Such an approach is necessary because direct weighing at rates commensurate with modern tablet press operation is impossible. This requirement immediately places demands on the physical characteristics of the feed and on the design of the tablet press itself. In the former, precompression treatment of the granulation is one of the most common ways of minimizing difficulties arising from this source.

The great paradox in pharmaceutical tableting is the need to manufacture a compact capable of reproducibly releasing the drug that is of sufficient mechanical strength to withstand the rigors of processing and packaging. Usually, the release of the drug is produced by the penetration of aqueous fluids into the fine residual pore structure of the tablet and the contact of these fluids with components that either swell or release gases.

The selected precompression treatment, if any, markedly affects the manufacture of tablets. In particular, one must determine whether a mixture of powdered ingredients is to be tableted directly, or if an intervening wet granulation step is to be introduced. This decision is influenced by many factors, including the stability of the medicament to heat and moisture, the flow properties of the mixed ingredients, and the tendency of the granulation to segregate. Currently, there are also two conflicting considerations that tend to play a major role in this choice. These are the reluctance to change the traditional methods employed by the company, versus the economic advantages of omitting complete stages in the production sequence. In wet granulation, the components of the formulations are mixed with a granulating liquid, such as water or ethanol, to produce granules that will readily compress to give tablets. Wet granulation methods predominate in the manufacture of existing products, whereas the trend for new products is to use direct compression procedures. Although many steps are eliminated when using direct compression, some formulators have found that wet granulated products are more robust and able to accommodate variability in raw materials. Thus, for some companies, the trend is reverting to the formulation of tablets by wet granulation.

B. Desirable Properties of Raw Materials

Most formulations will be composed of one or more medicaments plus a variety of excipients. Irrespective of the type of tablet, general criteria for these raw materials are necessary. To produce accurate, reproducible dosage forms it is essential that each component be uniformly dispersed within the mixture and that any tendency for component segregation be minimized. In addition, the processing operations demand that the mixture be both free-flowing and cohesive when compressed.

Particle Size

In general, the tendencies for a powder mix to segregate can be reduced by maintaining similar particle size distribution, shape and, theoretically, density of all the ingredients. Flow properties are enhanced by using regular-shaped, smooth particles with a narrow size distribution, together with an optimum proportion of "fines" (particles < 50 μm). If such conditions cannot be met, then some form of granulation should be considered.

Particle size distribution and, hence, surface area of the drug itself, is an important property that has received considerable attention in the literature. For many drugs, particularly those for which absorption is limited by the rate of dissolution, attainment of therapeutic levels may depend on achieving a small particle size [1]. In fact, it has been suggested that, for such drugs, standards for specific surface areas and the number of particles per unit weight should be developed. However, the difficulty in handling very fine powders, as well as the possibility of altering the material in other ways, has shifted the emphasis towards producing an optimum, rather than a minimum, particle size. For instance, several researchers have found that decreasing particles size produces tablets of increased strength, as well as reduced tendency for lamination [2–5]. This is probably due to the minimization of any adverse influences that a particular crystal structure may have on the bonding mechanism. On the other hand, samples of milled digoxin crystals prepared by various size-reduction techniques have been reported to elicit different equilibrium solubilities [1]. This suggests that the method of grinding may well affect the dissolution behavior of certain medicaments.

The effect of particle size on the compaction characteristics of two model sulfonamide drugs, one exhibiting brittle fracture and the other being compressed chiefly by plastic deformation, has been reported [3]. In particular, the tensile strength of tablets made from the brittle material were more sensitive to the drug's particle size than that of tablets made from the plastically deforming material. In addition, larger granules possess better flow, whereas small aggregates deform during compaction (e.g., spray-dried lactose) [6].

An alternative approach aimed at reducing the segregation tendencies of medicaments and excipients involves milling the former to a small particle size and, then, physically absorbing it uniformly onto the surface of the larger particles of an excipient substrate. By these means *ordered*, as opposed to *random*, mixing is realized, and dissolution is enhanced as a result of the fine dispersion [7].

Moisture Content

One of the most significant parameters contributing to the behavior of many tablet formulations is the level of moisture present during manufacture, as well as that residual in the product. In addition to its role as a granulation fluid and its potentially adverse effects on stability, water has some subtle effects that should not be overlooked. For example, there is increasing evidence to suggest that moisture levels may be very critical in minimizing certain faults, such as lamination, that can occur during compression. Moisture levels can also affect the mechanical strength of tablets and may act as an internal lubricant. For example, Fig. 1 illustrates the effect of moisture content on the compactibility of anhydrous lactose [8]. As the moisture content increases, it is absorbed by the lactose, thereby converting it from the anhydrous to the hydrous form. During this transformation, the β-form of lactose most probably changes to the α-form and, thus, produces changes in compactibility.

Accelerated aging and crystal transformation rates have also been traced to high residual moisture content. Ando et al. studied the effect of moisture content on the crystallization of anhydrous theophylline in tablets [9]. Their results also indicate that anhydrous materials convert to hydrates at high levels of relative humidity. In addition, if hygroscopic materials [e.g., polyethylene glycol (PEG) 6000] are also contained in the formulation, needlelike crystals form at the tablet surface and significantly reduce the release rate of the theophylline.

In many products, it seems highly probable that there is a narrow range of optimum moisture content that should be maintained. More specifically, the effect of moisture on microcrystalline

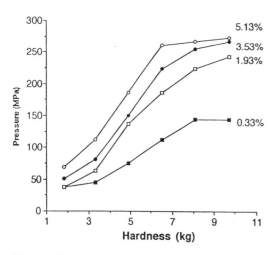

Fig. 1 The effect of moisture content on the compactibility of anhydrous β-lactose tablets. (From Ref. 8.)

cellulose (MCC)-containing tablets has been the subject of an investigation that demonstrates the sensitivity of this important excipient to moisture content [10]. Differences exist in both the cohesive nature and the moisture content of two commercial brands of MCC [10]. A very useful report on the equilibrium moisture content of some 30 excipients has been compiled by a collaborative group of workers from several pharmaceutical companies [11]. The information garnered from this study now appears in the *Handbook of Pharmaceutical Excipients* [12].

Crystalline Form

Selection of the most suitable chemical form of the active principle for a tablet, although not strictly within our terms of reference here, must be considered. For example, some chloramphenicol esters produce little clinical response [13]. There is also a significant difference in the bioavailability of anhydrous and hydrated forms of ampicillin [14]. Furthermore, different polymorphic forms, and even crystal habits, may have a pronounced influence on the bioavailability of some drugs, owing to the different dissolution rates they exhibit. Such changes can also give rise to manufacturing problems. Polymorphism is not restricted to active ingredients, as shown, for example, in a report on the tableting characteristics of five forms of sorbitol [15].

Many drugs have definite and stable crystal habits. Morphological changes rarely occur in such drugs as the formulation process is scaled up. However, some drugs exhibit polymorphism, or have different identifiable crystal habits. Chan and Doelker reviewed several drugs that undergo polymorphic transformation when triturated in a mortar and pestle [16]. Some of their conclusions are listed in Table 1 and illustrated in Fig. 2. In addition, several researchers have concluded that both polymorph and crystal habit influence the compactibility and mechanical strength of tablets prepared from polymorphic materials [16–21]. York compared the compressibility of naproxen crystals that had been spherically agglomerated with different solvents and found that significant differences existed between the various types of agglomerates (see Fig. 3) [21]. Other investigators have found that, in some instances, there is a correlation between the rate of reversion to the metastable form during dissolution and the crystal growth rate of the stable form [22]. These polymorphic changes may have a profound effect on tablet performance in terms of processing, *in vitro* dissolution, and *in vivo* absorption. In fact, a major clinical failure of generic carbamazepine tablets can be directly linked to

Table 1 Some Drugs That Undergo Polymorphic Transition When Triturated

Drug	Number of polymorphs before trituration	Number of polymorphs after trituration
Barbitone	2	1
Caffeine	2	1
Chlorpropamide	3	2
Clenbuterol HCl	2	3
Dipyridamole	2	1
Maprotiline HCl	3	1
Mebendazole	4	5
Nafoxidine HCl	4	3
Pentobarbitone	3	2
Phenobarbitone	2	1
Sulfabenzamide	2	1

Source: Ref. 16.

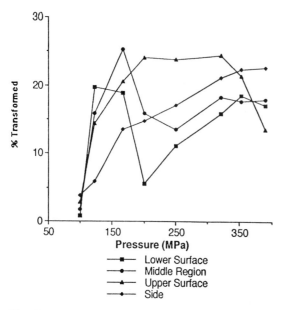

Fig. 2 Percentage of caffeine "form A" transformed versus applied pressure. (From Ref. 16.)

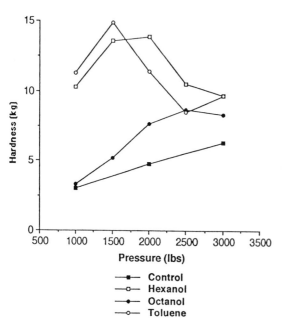

Fig. 3 Intrinsic compressibility of nonagglomerated naproxen (control) and of naproxen that has been spherically agglomerated with different solvents. (From Ref. 21.)

polymorphic changes (dihydrate formation) that led to altered dissolution of the tablets and, ultimately, disastrous clinical consequences. Thus, formulators of solid dosage forms must be aware of a subject compound's propensity for polymorphic transition so that a rational approach to formulation can be followed.

Hiestand Tableting Indices

Materials that do not compress will produce soft tablets; brittle crystalline materials will yield brittle tablets. Hiestand was the first pharmaceutical scientist to quantify rationally the compaction properties of pharmaceutical powders [23–28]. The results of this work are three indices known as the Hiestand Tableting Indices. The strain index (SI) is a measure of the internal entropy, or strain, associated with a given material when compacted. The bonding index (BI) is a measure of the material's ability to form bonds and undergo plastic transformation to produce a suitable tablet. The third index, the brittle fracture index (BFI), is a measure of the brittleness of the material and its compact. Table 2 lists these indices for several drugs and excipients. For most materials, the strength of the tablet is a result of competing processes. For example, erythromycin is a material known for its tendency to cap and laminate when tableted. On the basis of its BI value, one might expect relatively good bonding. However, the very high strain index associated with this drug appears to overcome its bonding abilities. Microcrystalline cellulose, on the other hand, has very high strain index, but its bonding index is exceptionally high and compensates for this effect.

Other investigators have evaluated the potential for these indices. In their studies, Williams and McGinnity have concluded that evaluation of single-material systems should precede binary or tertiary powder systems [29]. A full discussion of compaction mechanisms is given later in this chapter.

Variability

The effect of raw material variability of tablet production [2,30,31] and suggestions for improving tableting quality of starting materials [21] have been the subject of recent publications. Table 3, which lists the characteristics of different sources of magnesium stearate, clearly illustrates the variability of this material [32]. Phadke and Eichorst have also confirmed that significant differences can exist between different sources, and even different lots, of magnesium stearate [33]. Given that the effectiveness of magnesium stearate is primarily due to its large surface area, these variations should not be overlooked. In addition, studies assessing raw material variability emphasize the need for physical, as well as chemical, testing of raw materials to ensure uniformity of the final product.

Purity

Raw material purity, in general, must also be given careful attention. Apart from the obvious reasons for a high level of integrity, as recognized by the regulatory requirements, we should be aware of more subtle implications that are perhaps only just beginning to emerge. For instance, small proportions of the impurity acetylsalicylic anhydride reduces the dissolution rate of aspirin itself (Fig. 4) [34].

Another area of interest is that of microbiological contamination of solid dosage forms, which is thought to arise chiefly from raw materials, rather than the manufacturing process [35,36]. Ibrahim and Olurinola monitored the effects of production, environment, and method of production, as well as microbial quality of starting materials, on the microbial load during various stages of tablet production [35]. Although high levels of contamination were present during the wet granulation process, these levels were significantly reduced during the drying process. Thus, products derived from natural origins, such as gelatins and starch, are sometimes heavily contaminated.

Table 2 Hiestand Compaction Indices for Some Drugs and Excipients

Material	Bonding index	Brittle fracture index	Strain index
Aspirin	1.5	0.16	1.11
Caffeine	1.3	0.34	2.19
Croscarmellose sodium NF	2.7	0.02	3.79
Dicalcium phosphate	1.3	0.15	1.13
Erythromycin dihydrate	1.9	0.98	2.13
Hydroxypropyl cellulose	1.6	0.04	2.10
Ibuprofen			
A	1.9	0.05	0.98
B	1.8	0.57	1.51
C	2.7	0.45	1.21
Lactose USP			
Anhydrous	0.8	0.27	1.40
Hydrous Fast-Flo	0.4	0.19	1.70
Hydrous bolted	0.6	0.12	2.16
Hydrous spray process	0.6	0.45	2.12
Spray dried			
A	0.6	0.18	1.47
B	0.5	0.12	1.81
Mannitol			
A	0.8	0.19	2.18
B	0.5	0.15	2.26
Methenamine	1.6	0.98	0.84
Methyl cellulose	4.5	0.06	3.02
Microcrystalline cellulose NF			
Avicel PH 102 (coarse)	4.3	0.04	2.20
Avicel PH 101 (fine)	3.3	0.04	2.37
Povidone USP	1.7	0.42	3.70
Sorbitol NF	0.9	0.16	1.70
Starch NF			
Corn	0.4	0.26	2.48
Pregelatinized	1.8	0.14	2.02
Pregelatinized compressible	1.2	0.02	2.08
Modified (starch 1500)	1.5	0.27	2.30
Sucrose NF			
A	1.0	0.35	1.45
B	0.8	0.42	1.79
C	0.5	0.53	1.55

Source: Refs. 23–28.

Compatibility

One final area that should be considered when choosing the excipients to be used in the tablet formulation is that of drug–excipient interactions. There is still much debate about whether excipient compatibility testing should be conducted before formulation [37–39]. These tests most often involve the trituration of small amounts of the active ingredient with a variety of excipients. Critics of these small-scale studies argue that their predictive value has yet to be established and, indeed, they do not reflect actual processing conditions [37]. Instead, they

Table 3 Average Particle Data for Different Sources of Magnesium Stearate

Source	Size (μm)	Surface area (m²/g)	Pore radius (Å)
United States	1.5–3.2	13.4	50
Great Britain	2.1–5.2	12.2	68
Germany	4.1–6.9	7.4	61
Italy	5.5–9.1	4.6	36

Source: Ref. 32.

suggest a sound knowledge of the chemistry of the materials used in conjuncture with ''mini-formulation'' studies as a preferable method for investigation of drug–excipient interactions.

C. Tablet Components

Conventional solid dosage forms can be divided into two classes: those that disintegrate, and those that do not. Disintegrating dosage forms release their medicaments by breaking down the physical integrity of the dosage form, usually with the aid of solid disintegrating agents or gas-releasing effervescent agents. Nondisintegrating tablets are usually made of soluble drugs and excipients that will rapidly dissolve in the mouth or gastrointestinal tract (GIT) on ingestion.

In recent years, the arrival of new prolonged-release dosage forms has caused some pharmaceutical scientists to consider conventional disintegrating dosage forms as ''non–controlled-release.'' This term is a misnomer, since, with the aid of modern tablet disintegrants and other excipients, the disintegration of these dosage forms can be controlled, both quantitatively and qualitatively. Moreover, there are still many drugs in which rapid attainment of therapeutic levels, rather than controlled release, is required. Analgesics, antibiotics, and drugs for the immediate treatment of angina pectoris are prime examples. These tablets need to be designed so that the drug is liberated from the dosage form in such a manner that dissolution of the

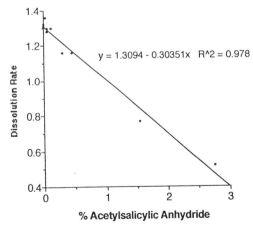

y = 1.3094 - 0.30351x   R^2 = 0.978

Fig. 4 Effect of acetylsalicylic anhydride impurity on the dissolution rate of aspirin tablets. (From Ref. 34.)

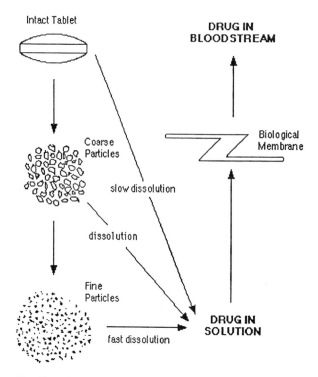

Fig. 5 Absorption of a drug from an intact tablet.

drug is maximized. Very often, this means that disintegration of the tablet must be followed by granular disintegration (Fig. 5) to promote rapid dissolution and, hence, absorption.

The ingredients, or excipients, used to make compressed tablets are numerous. They can be classified by their use, or function, as in Table 4. Keep in mind, however, that not all formulations need contain all the types of ingredients listed in this table. Certain excipients, such as antioxidants and wetting agents, are used only in situations for which they are expressly needed to assure the stability and solubility of the active ingredients. Other excipients, such as dissolution modifiers, are used primarily in controlled-release formulations. In fact, by reducing the number of ingredients in a formulation, one will generally reduce the number of problems that may arise in the manufacturing process. Hence, many formulators adhere to the motto "Keep it simple."

Table 4 Ingredients Used in Tablet Formulation

Active Ingredient (drug)	Antiadherants
Fillers	Wetting agents
Binders (dry and wet)	Antioxidants
Disintegrants	Preservatives
Dissolution retardants	Coloring agents
Lubricants	Flavoring agents
Glidants	

Because of the nature of modern pharmaceutical systems, formulators have made more complete investigations of the materials they use. This interest has identified several materials that may have more than one use in tableted systems. The type of effect that an excipient will produce is often dependent on the concentration in which it is used. For example, Table 5 lists some multiuse excipients and the corresponding concentration ranges required for their various applications.

Active Ingredients

The dose of the drug to be administered has a profound effect on the design and formulation of a dosage form. Content uniformity and drug stability become very important issues when the dose of the drug is very small (e.g., oral contraceptives). However, the effect of the drug's properties on the tablet, in this case, is minimal. In general, as the dosage increases, so does the effect of the drug's attributes on the tablet.

Sometimes processing can affect the particle morphology of the active ingredient. This may lead to adverse effects on mixing and tableting operations. In particular, micronization may cause crystals to change their shape, even though polymorphism is not evidenced.

Fillers

An increasing number of drugs are used in very low dosages. To produce tablets of a reasonable size (i.e., minimum diameter of 3 mm), it is necessary to dilute the drug with an inert material. Such diluents should meet important criteria, including low cost and good-tableting qualities.

Table 5 Some Multiple-Use Excipients for Tablet Formulation

Excipient/concentration in formula (%)	Use
Glyceryl behenate	
0–5	Lubricant
5–30	Controlled-release excipient
Hydroxypropylmethyl cellulose (HPMC), low viscosity	
0–5	Wet binder
5–20	Film former
5–26	Controlled-release excipient
Microcrystalline cellulose (MCC)	
0–8	Improve adhesion of film coat to core
5–15	Disintegrant
5–95	Binder/filler
Polyethylene glycol	
0–10	Lubricant
5–40	Controlled-release excipient
Polyvinylpyrrolidone (PVP)	
0–15	Wed binder
5–10	Coating excipient
5–30	Disintegrant
10–35	Controlled-release excipient
Starch	
0–5	Intragranular binder/disintegrant
5–10	Wet binder
5–20	Disintegrant

It may be possible, in some instances, to combine the role of diluent with a different property, such as a disintegrant or flavoring agent.

Commonly used fillers and binders and their comparative properties are listed in Table 6. As can be seen by this list, both organic and inorganic materials are used as fillers and binders. The organic materials used are primarily carbohydrates because of their general ability to enhance the product's mechanical strength as well as their freedom from toxicity, acceptable taste, and reasonable solubility profiles.

One of the most commonly used carbohydrates in compressed tablets is lactose. Work by Bolhuis and Lerk [40] and Shangraw et al. [6] has demonstrated that all lactoses are not alike, chemically, physicochemically, or functionally. Besides various size grades of normal hydrous lactose, one can purchase spray-dried lactose, which is an agglomerate of α-lactose monohydrate crystals, with up to 10% amorphous material. Spray-dried lactose has very good flow properties, but its poor compression characteristics require the addition of a binder, such as microcrystalline cellulose. However, one particular brand of spherical crystalline/amorphous agglomerate, Fast-Flo (NF hydrous), possesses superior compressibility and dissolution characteristics. The spherical nature of the crystals make them more compressible than spray-dried agglomerates of lactose [41]. One must also give attention to this component's stability, as aging may adversely affect these properties. Anhydrous lactose has also been used as a diluent, particularly in direct compression formulations for which low moisture content is desirable, since it has very good stability and a reduced tendency to color with aging. Another advantage in the use of anhydrous lactose is that its insensitivity to temperature changes allows it to be reworked with relative ease. Unfortunately, its flow properties are not particularly good, and its compressibility is inferior to other forms of lactose.

Some other sugars are now being produced in special grades to meet the needs of the pharmaceutical industry. Most of these products contain combinations of sucrose with invert sugar or modified dextrins and are of particular value in the formulation of chewable tablets.

Starch is often cited as a filler, but it is more commonly used in its dry state as a disintegrating agent. However, modified starches such as StaRx 1500 and National 1551 (partially

Table 6 Comparative Properties of Some Directly Compressible Fillers[a]

Filler	Compactibility	Flowability	Solubility	Disintegration	Hygroscopicity	Lubricity	Stability
Dextrose	3	2	4	2	1	2	3
Spray-dried lactose	3	5	4	3	1	2	4
Fast-Flo lactose	4	4	4	4	1	2	4
Anhydrous lactose	2	3	4	4	5	2	4
Emdex (dextrates)	5	4	5	3	1	2	3
Sucrose	4	3	5	4	4	1	4
Starch	2	1	0	4	3	3	3
Starch 1500	3	2	2	4	3	2	4
Dicalcium phosphate	3	4	1	2	1	2	5
Avicel (MCC)	5	1	0	2	2	4	5

[a]Graded on a scale from 5 (good/high) down to 1 (poor/low); 0 means none.

hydrolyzed starch) are marketed for direct compression and appear to offer the advantage of substantial mechanical strength and rapid drug release.

Certain inorganic salts are also used as fillers. Some common examples are listed in Table 7, together with their comparative properties. Among the most popular is dicalcium phosphate dihydrate, a comparatively low-cost, insoluble diluent, with good powder flow potential, but inherently poor compression characteristics [4]. Importantly, this material is slightly alkaline and, thus, must not be used where the active ingredient is sensitive to pH values of 7.3 or above. Special formulations in which unmilled dicalcium phosphate is the main ingredient are available under the trade name Emcompress and contain 5–20% of other components designed to improve compaction and disintegration performance [6,40,42]. In accelerated stability studies, Shah and Arambulo [43] found Emcompress to be unsuitable for use with ascorbic acid and thiamine hydrochloride owing to deteriorating hardness and disintegration characteristics evidenced, as well as the chemical degradation of ascorbic acid. In addition, calcium salts may adversely affect the absorption profile of certain drugs [44].

The influence of the actual manufacturing process can also affect the contribution of the diluent to the final characteristics of the product. For instance, recent work by Shah et al. [45] has shown that the release of drug from tablets formulated with soluble excipients may be more prompt than from those formulated with insoluble excipients (Fig. 6). However, the method of preparing the triturates was also very important. In this example, ball or muller milling gave the best overall results.

Few tablets intended for oral administration are totally soluble in aqueous media, but if such a product is needed, then soluble excipients are employed. These include dextrose, lactose, mannitol, and sodium chloride, the last of these sometimes acting as its own lubricant. Urea may also be used, but because of its known pharmacological effects, it is less desirable than the other soluble compounds cited.

Binders and Granulating Fluids

Most binders used in wet granulation tend to be polymeric. The binders most commonly used are from natural sources, such as starch or cellulose derivatives. Typically, these agents are dispersed or dissolved in water or a hydroalcoholic medium. The binders can be sprayed, poured, or admixed into the powders to be agglomerated. The methods of incorporating these materials can be classified into low-shear, heat-shear, and atomization methods. As illustrated in Fig. 7, one can see that the concentration of binder used and its method of addition (as a

Table 7 Comparative Properties of Some Inorganic Fillers[a]

Filler	Availability	Mechanical	Solubility	Absorbency	Acid/base	Abrasiveness	Lubricity
Calcium carbonate	2	4	0	3	Base	2	1
Dicalcium phosphate	2	2	0	4	Base	2	0
Calcium triphosphate	3	2	0	5	Base	2	0
Magnesium carbonate	2	2	0	4	Base	1	1
Sodium chloride	5	5	5	1	Neutral	3	2

[a]Graded on a scale from 5 (good/high) down to 1 (poor/low); 0 means none.

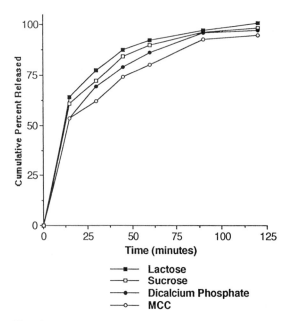

Fig. 6 Dissolution of digoxin tablets containing different fillers (in simulated gastric fluid at 37°C). (From Ref. 45.)

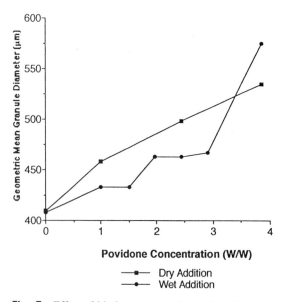

Fig. 7 Effect of binder concentration and method of addition on granule size. (From Ref. 46.)

dry powder or as a granulating fluid) can significantly affect granule size [46]. Moreover, some researchers have found that increasing the amount of granulating fluid used can have profound effects on a tablet's mechanical strength and disintegration time [47]. The equipment and processes used to incorporate these will be discussed later in this chapter; however, some commonly used binders are listed in Table 8.

Seager et al. [48] showed that a binder can be useful for a given process, but may not be universally useful. They studied gelatin in granulations made by roller–compaction, conventional wet granulation, and spray-drying. These researchers found spray-drying to be the preferred method of granulation for gelatin-granulated acetaminophen (paracetamol). They hypothesized that this was due to the improved distribution of the binder in this system. Iyer et al. [49] investigated the effects of rotogranulation on the performance of hydroxypropylmethyl cellulose (HPMC), gelatin and polyvinylpyrrolidone (PVP). In this process, all three binders produced similar results. However, HPMC was preferred owing to prolonged drug-release profiles, smaller particle size, and better content uniformity.

Chowhan showed that considerable variability can occur when scale-up of a granulation takes place. He demonstrated, quite convincingly, that a major factor in scale-up is the drug itself, which may require more or less (usually more) granulating fluid and binder to make a suitable tablet. He also found that this can impinge on drug release and, thereby, the bioavailability of the dosage form [50].

York reviewed the solid-state properties of solids and showed that both intrinsic and induced solid properties can have a profound effect on wettability and processing; this is an excellent review of the literature [51]. In addition, Lerk investigated the surface characteristics of several drugs and showed that the contact angles can vary greatly depending on the drug (Table 9). Of interest is the difference in wettability of different crystalline forms of the same drug [52,53]. These properties will have a profound effect on the ability of various binders to function, as well as change the processing parameters needed to effect proper granulation.

Disintegrants

For most tablets, it is necessary to overcome the cohesive strength introduced into the mass by compression. Therefore, it is usual practice to incorporate an excipient, called a disintegrant, which will induce this process. Several types, acting by different mechanisms, may be distin-

Table 8 Some Commonly Used Wet Binders and Granulating Fluids

Name	Strength (%)	Comments
Acacia mucilage[a]	10–20	Produces hard, friable granules
Cellulose derivatives[a]	5–10	HPMC[b] is most common
Ethanol		Must be applied to easily hydratable material
Gelatin solutions	10–20	Gels when cold, therefore use warm; strong adhesive; used in lozenges; less attractive in warm moist climates
Glucose syrups	25–50	Strong adhesive; tablets may soften in high humidity
Polyvinylpyrrolidone	5–20	Different MW grades give varying results
Starch mucilage	5–10	One of best general binders; better when used warm
Sucrose syrups[a]	65–85	Strongly adhesive; tablets may soften in high humidity
Tragacanth mucilage[a]	10–20	Produces hard, friable granules
Water		Must be applied to easily hydratable material

[a]May also be added as dry powder to the formulation, but this is less efficient than liquid preparation.
[b]Hydroxypropylmethylcellulose.

Table 9 Contact Angles for Some Powders

Material	Contact angle (ϕ)
Acetylsalicylic acid	74
Aminophylline	47
Ampicillin, anhydrous	35
Ampicillin, trihydrate	21
Calcium stearate	115
Chloramphenicol	59
Chloramphenicol palmitate (α-form)	122
Chloramphenicol palmitate (β-form)	108
Diazepam	83
Digoxin	49
Indomethacin	90
Lactose	30
Magnesium stearate	121
Phenylbutazone	109
Prednisolone	43
Prednisone	63
Stearic acid	98
Sulfacetamide	57
Theophylline	48
Tolbutamide	72

Source: Refs. 52 and 53.

guished: (a) those that enhance the action of capillary forces in producing a rapid uptake of aqueous liquids; (b) those that swell on contact with water; (c) those that release gases to disrupt the tablet structure; and (d) those that destroy the binder by enzymatic action.

The method of addition has also received attention. In particular, during wet granulation, the addition of the disintegrant before (intragranular) or after the granulation (extragranular) process has been investigated (Figs. 8 and 9) [54]. The extragranular portion ensures rapid disintegration, whereas the intragranular fraction leads to harder tablets and a finer size distribution on dispersion. The consensus from the published papers on this topic appears to be that there are advantages to be gained in dividing the disintegrant into both extra- and intragranular portions, with between 20 and 50% being external. For example, 2.5% intragranular and 12.5% extragranular disintegrant produced the best overall performance in tablets of calcium orthophosphate [55].

Types of Disintegrants

Disintegrants That Propagate Capillary Effects. There is evidence to suggest that water uptake caused by capillary forces is the crucial factor in the disintegration process of many formulations. In such systems the pore structure of the tablet is of prime importance, and any inherent hydophobicity of the tablet mass will adversely affect it. Therefore, disintegrants in this group must be able to maintain a porous structure in the compressed tablet and show a low interfacial tension toward aqueous fluids. Rapid penetration by water throughout the entire tablet matrix to facilitate its break up is thus achieved. Concentrations of disintegrant that ensure a continuous matrix of disintegrant are desirable, and levels of between 5 and 20% are common.

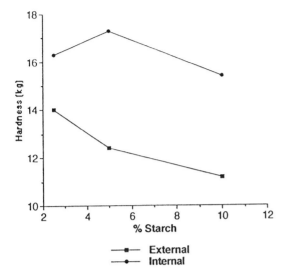

Fig. 8 Effect of starch concentration and location on tablet hardness. (From Ref. 54.)

Starch was the first disintegrant used in tablet manufacture and still enjoys wide use today [56]. Its mode of action is probably through the induction of water uptake into the tablet, rather than by the swelling action previously ascribed to it [57–63]. Other workers [64] still consider that the hydration of the hydroxyl groups causes them to move apart, yet it appears that starch swells little in water at body temperature. There is some evidence to suggest that the fat content of starch can also influence its performance as a disintegrant. In addition, since starch possesses poor binding characteristics, once the tablets containing it become thoroughly wetted, they

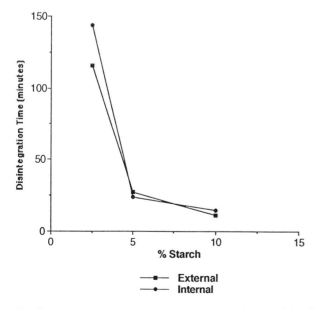

Fig. 9 Effect of starch concentration and location on tablet disintegration time. (From Ref. 54.)

break up easily [59,60,62]. Varieties of starch containing large grains are preferred for other reasons, but, in the present context, a large particle size may provide the optimum pore size distribution within the tablet and, thus, promote capillary action (e.g., potato starch).

Some forms of microcrystalline cellulose (MCC) are highly porous, with strong ''wicking'' tendencies, thereby making them good disintegrants. This is a fortuitous finding, since they also serve as excellent binders and are able to substantially improve the mechanical strength of some weak formulations. One disadvantage of using MCC, however, is that dissolution performance may be adversely affected at higher compression forces. Another disintegrant group, the insoluble cationic-exchange resins, typified by polyacrylin, exhibits better dissolution characteristics when subjected to higher pressures. Comparisons of disintegrant action, however, will be valid only if carried out under the same controlled conditions.

Some disintegrants propagate capillary effects, but also swell or dissolve to enhance disintegration behavior. Sodium starch glycolate and insoluble cationic-exchange resins are two examples that have been extensively studied by Khan and Rhodes [65,66], who demonstrated their superiority to sodium carboxymethyl cellulose and corn starch. Their results correlated well with the comparative release patterns of a dye from tablets containing these materials. In addition, they were able to show the long-term deterioration in hardness and disintegration time when the tablets containing the more effective disintegrants were subjected to high humidities. A comparative evaluation of sodium starch glycolate against cross-linked carboxymethylcellulose and sodium glycine carbonate has been carried out by Bavitz et al. [67], who determined that cross-linked carboxymethylcellulose compared very favorably with sodium starch glycolate in formulations of four different actives over long test periods. Colloidal silicon dioxide has been investigated as a disintegrant and, although capable of absorbing approximately nine times as much water as starch, the process is several times slower; the hindered action counteracts much of the advantage of its greater absorbency.

Disintegrants That Swell. One general problem with this group of disintegration is that on swelling, many disintegrants produce a sticky or gelatinous mass that resists breakup of the tablet, making it particularly important to optimize the concentration present. Although untreated starches do not swell sufficiently, certain modified forms, such as sodium starch glycolate, do swell in cold water and are better as disintegrants. Various cellulose derivatives, including methylcellulose and carboxymethylcellulose have been used in this role, but with limited success owing to the marked increase in viscosity they produce around the dispersing tablet mass.

Some powdered gums, such as agar, karaya, or tragacanth, swell considerably when wet, but their pronounced adhesiveness limits their value as disintegrants and restricts the maximum concentration at which they can be effectively used to approximately 5% of tablet weight. These substances also exhibit considerable lot-to-lot variabilities, are prone to microbiological contamination, and in recent years, have become very expensive. However, alginic acid and its sodium salt possess the best combinations of sufficient swelling, with minimum stickiness, and concentrations as low as 4 or 5% are often adequate. The small amounts used compensate for the somewhat high cost of these materials.

Gas-Producing Disintegrants. Gas-producing disintegrants are used when extrarapid disintegration or a readily soluble formulation is required [68]. They have also been of value when poor disintegration characteristics have resisted other methods of improvement. Their main drawback is the need for more stringent control over environmental conditions during the manufacture of tablets made with these materials. In particular, gas-producing disintegrants are quite sensitive to small changes in humidity levels. For this reason, these disintegrants are often incorporated immediately before compression, when the moisture content can be controlled more easily, or they can be added to two separate fractions of the formulation. Composition is based on the same principles as those used for effervescent tablets, the most common being mixtures of citric and tartaric acids plus carbonates or bicarbonates.

In many instances, lower concentrations of gas-producing disintegrants can be used than are required by other disintegrating agents. This is a distinct point in their favor. Certain peroxides that release oxygen have been used for this purpose, but they do not perform as well as those releasing carbon dioxide.

Enzymes. When tablets are not naturally very cohesive and have thus been manufactured by a wet granulation process involving one of the binders listed in Table 8, addition of small quantities of appropriate enzyme may be sufficient to produce rapid disintegration. It has also been proposed that disintegration action might result from expansion of the entrapped air owing to generation of ''heat of wetting'' when the tablet is placed in a fluid. This concept has received little attention.

Several published studies have described the use of various substances as tablet disintegrants. Reviews by Lowenthal and Kanig and Rudnic have been published on the various agents used to bring about tablet disintegration [68,69]. Shangraw et al. reviewed the modern so-called super disintegrants and compared their relative morphological properties [6]. Most published studies have attempted to explain mechanisms that relate to observed efficiency of the disintegrating agent, and some have explored secondary attributes within the disintegrants themselves. None have succeeded, however, in advancing an explanation of disintegration that approaches a universal understanding applicable to all disintegrants. It now seems obvious that no single mechanism is applicable to all tablet disintegrants. In fact, a combination of mechanisms may be operative.

Mechanisms of Action

Water Uptake. Water uptake has been implicated as a mechanism of action for tablet disintegrants. Khan and Rhodes studied the adsorption and absorption properties of various disintegrants [64]. They concluded that the ability of particles to draw water into the porous network of a tablet (wicking) was essential for efficient disintegration. Work conducted by Mitrevej and Hollenbeck substantiates these claims [70]. A sophisticated method of determining water uptake was developed by Nogami et al. [71]. Their study further supports the theory that the rate of wetting is responsible, at least in part, for the disintegrant action.

Swelling. Perhaps the more widely accepted general mechanism of action for tablet disintegrants is swelling. Almost all disintegrants swell to some extent, and swelling has been reported quite universally in the literature [6,68,69]. Many attempts have been made to quantify swelling, but only recently have sophisticated methods been used in this area. Historically, sedimentation volumes of a slurry have been used as measures of swelling. This test is a fair appraisal of swelling capacity, but does not provide for the dynamic measurement of the swelling itself. As a result, many disintegrants studied do not show a correlation between sedimentation volumes and disintegrant efficiency. Nogami et al. developed a reliable test to measure swelling and water uptake simultaneously with the aid of two graduated columns connected by a rubber tube [71]. This apparatus was later refined by Gissinger and Stamm and by Rudnic et al. [72–74]. Figure 10 illustrates the essential features of this apparatus, which is very useful for the quantification of swelling and hydration rates for many excipients and polymers.

Both sedimentation volumes and swelling rates were evaluated by Rudnic et al., who found a poor correlation between the static test (sedimentation) and disintegrant efficiency [73,74]. In addition, they found that swelling tests, such as those reported in the foregoing, are dependent on several variables, including water transport through a gel layer and rates of hydration.

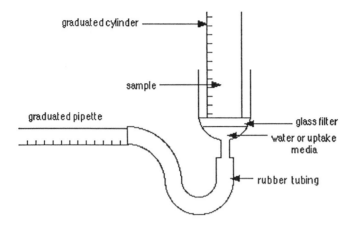

Fig. 10 Bulk swelling and water uptake apparatus.

These investigators developed methods to evaluate intrinsic swelling through the use of high-speed cinemicroscopy in conjunction with computerized image analysis. Although their method permitted more accurate and precise measure of swelling, they concluded that both bulk swelling and intrinsic swelling produced similar rank order disintegrant swelling rates.

All of the foregoing investigations have placed importance on not only the extent of swelling, but also on the rate at which swelling develops. In addition, it is important to understand that, as particles swell, there must be little or no accommodation by the tablet matrix to that swelling. If the matrix yields elasticity to the swelling, little or no force will be expended on the system, and disintegration will not take place. However, if the matrix is rigid and does not accommodate swelling, deaggregation of disintegration will occur.

Deformation. The existence of plastic deformation under the stress of tableting has been reported for many years. Evidence that disintegrant particles deformed from tablet compression was demonstrated by Hess, with the aid of scanning electron photomicrographs [61]. He found that the deformed particles returned to their normal shapes when exposed to water. Work completed by Fuhrer yielded similar results [75]. Occasionally, the swelling capacity of starch granules was improved when the granules were extensively deformed during compression. Obviously, the role of deformation and rebound under actual production conditions needs to be studied in more detail before the full effect of this phenomenon can be understood.

Particle Repulsion Theory. Another theory of tablet disintegration attempts to explain the swelling of tablets made with ''nonswellable'' starch. Ringard and Guyot-Hermann have proposed a particle repulsion theory based on the observation that particles that do not seem to swell may still disintegrate tablets [63]. In their study, they altered the dielectric constant of the disintegrating medium, in an effort to identify electric repulsive forces as the mechanism of disintegration, and concluded that water is required for tablet disintegration. These investigators espoused repulsion, secondary to wicking, as the primary mechanism of action for all tablet disintegrants.

Heat of Wetting. Matsumaru was the first to propose that the heat of wetting of disintegrant particles could be a mechanism of action [76]. He observed that starch granules exhibit slight exothermic properties when wetted and reported that this was the cause of localized stress resulting from capillary air expansion. This explanation, however, is limited to only a few types of disintegrants and cannot describe the action of most modern disintegrating agents. List and

Muazzam studied this phenomenon and also found disintegration when significant heat of wetting is generated. However, in these experiments, there is not always a corresponding decrease in disintegration time [77].

Particle Size. Physical characteristics of disintegrants, such as particle size, also have some bearing on the mechanisms of disintegration (e.g., swelling and water uptake). Several authors have attempted to relate the particle size of disintegrants to their relative efficiency. Smallenbroek et al. [78] evaluated the effect of particle size of starch grains on their ability to disintegrate tablets. They concluded that starch grains with relatively large particle size were more efficient disintegrants than the finer grades. These authors theorized that this behavior resulted from increased swelling pressure. Investigations made by Rudnic et al. [74,79] confirm these results. They also found a correlation between the rate of swelling and the amount of water uptake of sodium starch glycolate and have thus postulated that particle size plays a key role in the overall efficiency of commercial sources of this material.

Molecular Structure. In their attempts to identify the mechanism(s) of action of tablet disintegrants, researchers have recently turned their attention to the molecular structure of the disintegrants. Schwartz and Zelinskie [72] published one of the first such reports, in which they examined the two corn starch fractions amylose and amylopectin. They concluded that the linear polymer amylose was responsible for the disintegrant properties of starch, whereas the amylopectin fraction was primarily responsible for the binding properties associated with starches (Fig. 11). Because of the work by Schwartz and Zelinskie, tablet formulators can now select from a range of specialty starches that represent different ratios of amylose and amylopectin content to solve disintegrant or dissolution problems encountered in tablet formulation.

Shangraw et al. identified three major groups of compounds that have been termed superdisintegrants [6]. Many of these so-called superdisintegrants are substituted and cross-linked polymers. One group—the sodium starch glycolates—has enjoyed widespread popularity because of its exceptional ability to disintegrate tablets, as well as the relative ease with which this type of compound can be processed into tablet formulations.

Sodium starch glycolate is manufactured by cross-linking the carboxymethylating potato starch. It provides an excellent opportunity to measure the relation between molecular structure

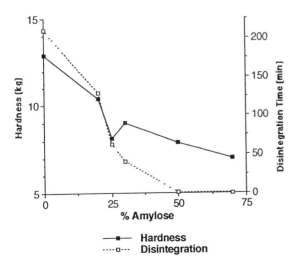

Fig. 11 Effect of amylose on the hardness and disintegration of dicalcium phosphate tablets. (From Ref. 62.)

and disintegrant efficiency, by altering the degree of substitution and the extent of cross-linking. Optimization of one sodium starch glycolate was performed with both direct compression and wet granulation methods. These studies showed that the swelling to the disintegrant particles was inversely proportional to the level of substitution [80].

Antifrictional Agents

Insoluble Lubricants. Lubricants act by interposing an intermediate layer between the tablet constituents and the die wall. The smaller the amount of stress needed to shear the material, the better its lubricant properties will be. Since they are primarily required to act at the tooling or material interface, lubricants should be incorporated in the final mixing step, after all granulation and preblending is complete. In this way, overmixing is less likely to occur.

A common mistake in the design of tablet operations is adding both the disintegrant and lubricant together in one mixing step. This causes the disintegrant to become coated with lubricant and often results in both a decrease in the disintegrant's porosity and a decrease in the efficiency of the disintegrant. Rather than add the disintegrant and lubricant simultaneously, a better approach is to add these excipients sequentially, with a disintegrant being first.

The surface area of magnesium stearate may be the most important parameter to monitor in terms of lubricant efficiency. Substantial decreases in both ejection forces and tablet hardness are noted when using brands of magnesium with larger surface areas. Lubricants with high surface areas may also be more sensitive to changes in mixing time than lubricants with low surface areas. Thus, if a particular drug or formulation is deleteriously affected by prolonged mixing of lubricants, adequate characterization for monitoring a lubricant surface area should be an integral part of product development and quality control.

Some of the more common antifrictional agents are listed in Table 10. Many of these are hydrophobic and, consequently, may affect the release of medicament. Therefore, lubricant concentration and mixing time should be kept to the absolute minimum. Lubricants may also significantly reduce the mechanical strength of the tablet (Fig. 12) [29,81]. Stearic acid and its magnesium and calcium salts are widely used, but the latter can be sufficiently alkaline to react

Table 10 Some Commonly Used Antifrictional Agents

Soluble lubricants	Insoluble lubricants
Adipic acid	Calcium, magnesium, and zinc salts of stearic acid
d,l-Leucine	Glyceryl behenate
Glyceryl triacetate	Glyceryl palmitostearate
Magnesium lauryl sulfate	Hydrogenated vegetable oils
PEG[a] 4000, 6000, and 8000	Light mineral oil
Polyoxethylene monostearates	Pafaffins
Sodium benzoate	Polytetrafluoroethylene
Sodium lauryl sulfate	Stearic acid
Sucrose monolaurate	Talc
Glidants	Waxes
Calcium silicate	Antiadherants
Fumed silicon dioxide	Most lubricants
Magnesium carbonate	Starch
Magnesium oxide	Talc
Starch	
Talc	

[a]Polyethylene glycol.

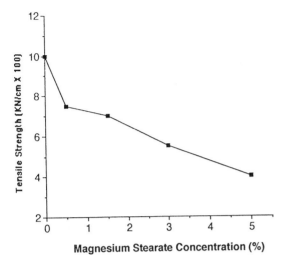

Fig. 12 Tensile strength of calcium sulfate tablets as a function of magnesium stearate concentration (solid fraction = 0.57). (From Ref. 29.)

with certain amine salts, such as aminophylline, resulting in the release of free base and discoloration of the tablet. Published formulas show levels of these lubricants between 1 and 4%, but there is evidence to show that, frequently, they can be reduced to as little as 0.25% without substantially affecting the lubrication of the system.

Liquid paraffins, particularly those of low viscosity, have been used and are said to be of value for colored tablets, and even the use of modified vegetables has been attempted. However, they appear, in general, to offer little advantage over solid lubricants, and their incorporation into the precompression mixture is more difficult, requiring solution in a volatile liquid that is then sprayed onto the unlubricated material. Because handling and Environmental Protection Agency (EPA) requirements, these materials are often rejected in the preformulation stage.

Isolated references in the literature describe the use of talc as a lubricant, but this material is better considered as a glidant [82–85]. It has several disadvantages, including its insolubility in body fluids and the abrasiveness found when using all but the fines grades of this material. Finally, it loses some of its effectiveness after compression, so that tablets containing talc cannot be readily reworked without extra quantities of it being added to the formulation.

Soluble Lubricants. Because of the association of lubricant properties with lipophilic materials (hence, poor aqueous solubility) alternative, more hydrophilic materials have been investigated. Soluble lubricants do not appear to be as efficient in lubricating tablet systems as their insoluble counterparts [82]. Increasing attention in this context is being given to a group of soluble, if less effective, lubricants that also possess surfactant qualities and are typified by the lauryl sulfates. For example, when tablets lubricated with sodium lauryl sulfate were compared with those lubricated with magnesium stearate, the tablets containing sodium lauryl sulfate exhibited a significantly higher rate of dissolution. Physical mixtures of the lubricant with stearates can lead to the best compromise in terms or lubricity, tablet strength, and disintegration. Recently, magnesium lauryl sulfate has been found to have an attractive balance of these properties; although requiring a concentration of 5% to provide the same lubricating efficiency as 2% magnesium stearate.

Some of the synthetic, soluble, waxlike polymers, typified by the high-molecular-weight polyethylene glycols (PEGs), have also been used as soluble lubricants. PEG 4000 and 6000 have been investigated, but their lubricant efficiency is less than that of magnesium stearate. In attempts to find the optimum lubricant from all standpoints, combinations of polymers, such as polyoxyethylene monostearates and polyoxethylene lauryl sulfates have undergone limited trials, with some encouraging results, but more information is required.

Glidants. Glidants are added to the formulation to improve the flow properties of the material to be fed into the die and sometimes to aid in particle rearrangement within the die during the early stages of compression [6,85]. They may act by interposing their particles between those of the other components and, thus, by virtue of their reduced adhesive tendencies, lower the overall interparticulate friction of the system. In addition, there may be adsorption of glidant into the irregularities of the other materials. It follows that, similar to lubricants, they are required at the surface of feed particles, and they should be in a fine state of division and appropriately incorporated in the mix.

Starches remain a popular glidant, in particular, those with the larger-grain sizes, such as potato starch, possibly because of their additional value as a disintegrant in the formulation. Concentrations up to 10% are common, but it should be appreciated that excess may result in exactly the opposite effect to that desired (i.e., flow properties may worsen). Talc is also widely used and has the advantage that it is superior to starches in minimizing any tendency for material to stick to the punch faces, a property sometimes classified as antiadherent. Because of its totally insoluble nature—hence, potential retardant effect of dissolution—concentration must be strictly limited and should rarely exceed 5%. In fact, the best overall compromise may be realized by using a mixture of starch and talc.

Recently, certain siliceous materials have been used successfully to induce flow. Among those quoted in the literature are pyrogenic silica, in concentrations as low as 0.25%, and hydrated sodium silioaluminate, in concentrations of about 0.75%. The former has the additional property of being able to scavenge moisture, which might otherwise contribute to restricted flow characteristics.

Antiadherents. Some material have strong adhesive properties toward the metal of the punches and dies. Although not a frictional effect, this results in material preferentially sticking to the punch faces and gives rise to tablets with rough surfaces. This effect, called picking, can also arise in formulations containing excess moisture.

Normally, the lubricants present in the tableting mass also act as antiadherents, but in the worst cases, it may be necessary to add more starch or even talc to overcome the defect. Accordingly, by judicious choice of a combination of excipients, all of these undesirable effects of the tableting process can be minimized.

Polymers for Controlled-Release Tablets

This topic is the subject of Chapter 15, but some of the materials that are used in these systems have other uses as well (see Table 5). Materials used to retard dissolution can be incorporated in the formulation on either a dry or wet basis. If they are added dry, very rarely do they improve tableting performance significantly. The most efficient use of these materials is to add a certain amount in the wet state and granulate with that system.

Table 11 lists some of the more commonly used controlled-release excipients. It can be seen from this list that the physical properties of the tablet can vary widely depending on the selection of materials to be coformulated. Great care needs to be taken in the selection of ingredients and the design of the dosage form. The design and performance of these dosage forms are the subject of Chapter 15.

Absorbents

Some tablet formulation call for the inclusion of a small amount of semisolid, or even semi-liquid, ingredient. It is highly desirable that any such component should be adsorbed onto, or

Table 11 Some Commonly Used Controlled-Release Excipients

Hydrophillic
    Acrylic acid
    Acrylic acid derivatives/esters
    Carboxymethylcellulose (CMC)
    Ethylcellulose (EC)
    Hydroxypropylcellulose (HPC)
    Hydroxypropylmethylcellulose (HPMC)
    Methylcellulose (MC)
    Polvinyl alcohol (PVA)
    Polvinylpryrrolidone (PVP)
    Polyacrylic acid (PAA)
    Polyethylene glycols (PEG)
Hydrophobic
    Carnauba wax
    Glyceryl behanate
    Glyceryl monostearate
    Hydrogenated vegetable oil
    Paraffin
    White wax
pH dependent
    Cellulose acetate phthalate (CAP)
    Hydroxypropylmethylcellulose phthalate (HPMCP)
    Polyvinyl acetate phthalate (PVAP)
    Shellac
    Zein
Surface-active
    Pluronics

---

absorbed into, one of the powders. If none of the other excipients in the formulation can act as a carrier, an absorbent may have to be included. When oily substances, such as volatile flavoring agents are involved, magnesium oxide and magnesium carbonate are suitable for this purpose. Natural earths, such as kaolin, bentonite, and Fuller's earth, have also been used and possess pronounced absorbent qualities. In general, they tend to reduce tablet hardness and may be abrasive. Therefore, fine, grit-free grade must be specified.

An absorbent may also be necessary when the formulation contains a hygroscopic ingredient, especially when absorption of moisture produces a cohesive powder that will not feed properly to the tablet press. In such instances, silicon dioxide has been of particular value.

One special problem in the tableting of volatile medicaments, such as nitroglycerin, is the loss of activity through evaporation. This effect can be reduced by fixing agents, such as Marcogol 400 or 4000 in concentrations of 85%. Alternatively, cross-linked povidone can also be used to enhance the stability of this particular drug.

Flavoring Agents

Making a formulation palatable enough to be chewed may result in enhanced availability of the drug. In addition, for patients who are unable to swallow a tablet whole (e.g., children), such a tablet may be the only reasonable alternative. Sweetening agents, such as dextrose, mannitol, saccharin, and sucrose, are widely used as flavoring agents. Perhaps the most extensively documented examples concern nitroglycerin tablets, which at one time were formulated in a chocolate base containing nonalkalized cocoa. Unfortunately, the cocoa affected the product's stability and has since been replaced by mannitol.

Flavoring agents proper are commonly volatile oils that have been dissolved in alcohol and sprayed onto the dried granules or have simply been adsorbed onto another excipient (e.g., talc). They are added immediately before compression to avoid loss through volatilization. Occasionally, they may even have some lubricating activity. If the oil normally contains terpenes, a low terpene grade is better to avoid possible deterioration in taste owing to terpene oxidation products. When an oil flavoring is prone to oxidation, it may be protected by a special type of encapsulation involving spray-drying or an aqueous emulsion containing the flavor. The emulgent used in spray-dried products may be starch or acacia gum, giving rise to the so-called dry flavors.

Pharmaceutical Colors

Colorants do not contribute to therapeutic activity, nor do they improve product bioavailability or stability. Indeed, they increase the cost and complication of the manufacturing process. Their main role is to facilitate identification and to enhance the aesthetic appearance of the product. In common with all material to be ingested by humans, solid dosage forms are severely restricted in the coloring agents that are allowable. This situation is complicated by the lack in international agreement on an approved list of colorants suitable for ingestion.

Colorants are available as either soluble dyes (i.e., giving a clear solution) or insoluble pigments that must be dispersed in the product (Table 12). There is an increasing tendency to use dyes in the form of special pigments termed "lakes." The pigment, here, is adsorbed onto some inert substrate, usually aluminum hydroxide. These pigments can be directly incorporated into tablets. It is often preferable to mix the lake with an extendor before incorporation into the tablet blend to minimize any mottling. Ordinary starch or modified starches, such as StaRx 1500, or sugars can be used for this purpose. In addition, mottling is less evident in pastel shades and colors in the center of the visible spectrum.

For tablet coating there are distinct advantages to using pigments, rather than dyes. Color development is more rapid and, hence, processing time is shorter. Since the final color is a function of the quantity of dye in the coating suspension, rather than the number of coats applied, there will be less operator influence and a better chance of achieving uniformity within

Table 12 Differences Between Lakes and Dyes

Characteristics	Lakes	Dyes
Solubility	Insoluble in most solvents	Soluble in water, propylene glycol, and glycerin
Method of coloring	By dispersion	By solution
Pure dye content	10–40%	Primary colors: 90–93%
Rate of use	0.1–0.3%	0.01–0.03%
Particle size	< 0.5 μm	12–200 mesh
Stability		
Light	Better	Good
Heat	Better	Good
Coloring strength	Not proportional to dye content	Directly proportional to pure dye content
Shades	Varies with pure dye content	Constant

Source: Warner Jenkinson Pamphlet on Lake Pigments, September 1990.

and between batches. There may also be a reduced risk of interaction between the drug and other ingredients.

All dyes, and to a smaller extent lakes, are sensitive to light to varying degrees, and their color may be affected by other ingredients in the formulation. For example, since many colorants are sodium salts of organic acids, they may react in solution with cationic drugs (e.g., antihistamines). In addition, nonionic surfactants may adversely effect color stability. This is more prevalent when using natural colorants, which also tend to have a higher degree of batch-to-batch variability. However, unlike artificial dyes, natural colors do not require the Food and Drug Administration (FDA) certification before use in drug products. Some of the more common synthetic colorants are listed in Table 13, together with their important properties.

Wetting Agents

Wetting agents have been used in tablets containing very poorly soluble drugs to enhance their rate of dissolution [86–89]. Surfactants are often chosen for this purpose, with sodium lauryl sulfate being the most common. Paradoxically, some ionic surfactants have recently been formulated with oppositely charged drugs to produce a sustained-release complex [90–92]. Thus, one might want to consider using an uncharged surfactant, such as Tween 80, that has less likelihood of interacting with charge molecules.

III. TABLET MANUFACTURE

Thus far in this chapter the emphasis has been on the materials, rather that the processes, involved in tablet manufacture, but the latter are of equal importance. The pharmaceutical industry is highly regulated and deeply concerned with Good Manufacturing Practice (GMP). In terms of equipment, this translates into preparing products in totally enclosed systems by processes that involve a minimum of handling and transfer. Irrespective of the particular production route, whether moist granulation or direct compression, the first stage is likely to involve the intimate mixing together of several powdered ingredients.

A. Powder Mixing

The successful mixing together of fine powder is acknowledged to be one of the more difficult unit operations because, unlike the situation with liquid, perfect homogeneity is practically unattainable. All that is possible is to realize a maximum degree of randomness in the arrangement of the individual components of the mix. In practice, problems also arise because of the inherent cohesiveness and resistance to movement between the individual particles. The process is further complicated, in many systems, by the presence of substantial segregative influences in the powder mix. They arise because of differences in size, shape, and density of the component particles.

It is not possible here to present a full account of the interactions between these effects; therefore, the reader is referred to standard texts dealing with this important topic [93]. However, there may be an optimum mixing time and, in such circumstances, prolonged mixing may result in an undesired product. In the special case of mixing in a lubricant, which is required at the granule surface, overmixing can be particularly detrimental.

Powder mixers vary widely in their ability to produce adequately mixed powders and the time needed to accomplish it. For intimate mixing of powders, energy must be supplied at a high enough rate to overcome the inherent resistance to differential movement between particles. Older mixers attempt to supply sufficient energy to mix the entire batch at once, but modern designs tend to be based on mechanisms for sequentially feeding a proportion of the total mix to a region of high-energy–mixing potential. Typical examples of the general design

Table 13 Some Commonly Used Pharmaceutical Colorants (Synthetic)

FD&C color	Common name	Solubility (g/100 ml at 25°C)				Stability to					
		Water	Glycerin	Propylene glycol	25% Ethanol	Light	Oxidation	pH 3	pH 5	pH 7	pH 8
Red 3[a]	Erythorosine	9.0	20.0	20.0	8.0	Poor	Fair	Insol	Insol	NNC	NNC
Red 40	Allura red AC	22.0	3.0	1.5	9.5	Good	Fair	NNC[b]	NNC	NNC	NNC
Yellow 5	Tartrazine	20.0	18.0	7.0	12.0	V. good	Fair	NNC	NNC	NNC	NNC
Yellow 6	Sunset yellow	19.0	20.0	2.2	10.0	Mod	Fair	NNC	NNC	NNC	NNC
Blue 1	Brilliant blue	20.0	20.0	20.0	20.0	Fair	Poor	S. fade[c]	V.S. fade[d]	V.S. fade	V.S. fade
Blue 2	Indigotine	1.6	1.0	0.1	1.0	V. poor	Poor	A. fade[e]	A. fade	A. fade	C. fade[f]
Green 3	Fast green	20.0	14.0	20.0	20.0	Fair	Poor	S. fade	V.S. fade	V.S. fade	S. fade

[a]Note: FD&C Red 3 lake has been delisted by FDA as of January 29, 1990.
[b]No noticeable change
[c]Slight fade
[d]Very slight fade
[e]Appreciable fade
[f]Completely fades
Source: Warner-Jenkinson Pamphlet of Certified Colors, September 1990.

of such mixers are shown diagramatically in Fig. 13. Mixing is achieved by a main impeller that feeds material to a high-speed "chopper," producing an intense shear mixing zone.

With all mixers, however, it is necessary to establish that an acceptable degree of homogeneity has been reached. Quantitative methodologies for establishing an "index of mixing" or "efficiency of mixing" are reported in the literature [94].

B. Powder Compaction

For simplicity, the physics of tablet compaction discussed here will deal with the single-punch press, in which the lower punch remains stationary. Initially, the powder is filled into the die, with the excess being swept off. When the upper punch first presses down on the powder bed, the particles rearrange themselves to achieve closer packing. As the upper punch continues to advance on the powder bed, the rearrangement becomes more difficult, and deformation of particles at points of contact beings. At first, the particle will undergo elastic deformation, which is a reversible process, but as the continual pressure is applied, the particle begins to deform irreversibly. Irreversible deformation can be due to either plastic deformation, which is a major factor attributing to the tablet's mechanical strength, or brittle fracture, which produces poor quality compacts that crumble as the tablets are ejected [25,95]. In general, as increasing pressure is applied to a compact, its porosity will be reduced.

The surface area of the individual particles themselves change during the compaction process. Initially, an increase in surface area is noted owing to fracture as compression force increases. Eventually, the surfaces are decreased by bonding and consolidation of particles at higher compression forces [25,95]. Higuchi [95] postulated that an additional increase in surface area occurs after this point, and that this effect may cause lamination of the tablet owing to extensive rebound at decompression. In other words, at the tablet punch–powder interface, there may be zones of high density during compression, but with decompression, these zones have elastic rebound and are pulled apart from the rest of the tablet that did not contain this high density.

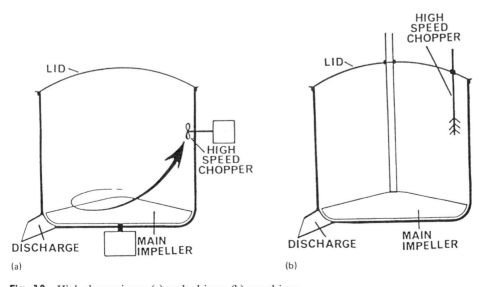

Fig. 13 High-shear mixers: (a) underdriven; (b) overdriven.

The major forces involved in the formation of a tablet compact are illustrated in Fig. 14 (note, this is a single-ended model) and are notated as follows: F_A represents the axial pressure, which is the force applied to the compact by the upper punch; F_L is the force translated to the lower punch, and F_D is the force lost to the die wall. If one remembers that, for every force, there must be an equal and opposite force, the following relationship is obvious:

$$F_A = F_L + F_D$$

F_R is the radial die wall force that develops because the powder is in a confined environment (i.e., not able to spread outward as pressure is applied down upon it because it is residing within the die). The coefficient of friction at the die wall, μ_w, is due to the shearing adhesion that occurs along the die wall as the powder is densified and compressed. The following relationship between F_D, μ_w, and F_R exists:

$$F_D = \mu_w F_R$$

The force of tablet ejection from the die F_E is a function of both μ_w and the residual die wall force, $RDWF$, that exists after decompression. As the friction decreases, one will obviously see a corresponding drop in F_E. It is important to remember here that we want F_E to be as low as possible so that minimal damage is imparted to both the tablet and the tooling.

The first applications of this technique [96] were directed to developing a more sensitive assessment of lubricant efficiency than that offered by the traditional *coefficient of lubrication*, or *R-value* (i.e., is the F_L/F_A ratio). Since then, its use has been extended to provide predictive information on formulation performance [97,98]. Lammens et al. [99] have stressed the importance of ensuring precision and accuracy in such measurements for correct interpretation of the data so produced.

Although the choice of precise tablet geometry may be more the prerogative of the marketing department than of the pharmaceutical formulation or production departments, certain general technical observations must be taken into account. Bearing in mind that most tablets are cylindrical, diameters between 3 and 12 mm are preferred, with either beveled edges or biconvex profile. Low height/diameter ratios, consistent with adequate tablet strength, are de-

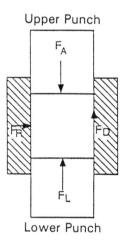

Upper Punch

Lower Punch

Fig. 14 Forces developed in the formation of a tablet compact. ▧, die wall, F_A, axial pressure applied by the upper punch; F_D, force lost to the die wall; F_R, radial die wall; F_L, force translated to the lower punch.

sirable to minimize die wall fractional effects, which can consume a significant amount of the total energy required in tableting. In addition, the internal stress differences will be minimized if a biconvex profile is selected, although the tablet's shape may affect the release of the drugs in matrix tablets [100]. The effect of punch face geometry and lubricant compression on tablet properties has been reported by Mechtersheimer and Zucker [101].

As a generalization, increasing compressional force will retard dispersion on administration; therefore, levels should be kept as low as possible, consistent with achieving acceptable mechanical properties. With some excipients, there is a critical compressional force range required to achieve minimum disintegration times. This has been demonstrated for starch-containing formulations and was thought to be linked to production of an optimum pore size distribution that allowed rapid intake of water, without providing large internal air spaces to accommodate the swelling starch grains [102]. The importance of press speed must also be taken into account, particularly if plastic deformation is thought to play a major role in tablet formation. The effect of this rate of compaction has been demonstrated quantitatively in a report by Roberts and Rowe [103].

C. Direct Compression

For obvious reasons, the possibility of compressing mixed powders into tablets without an intermediate granulating step is an attractive one. For many years, several widely used drugs, notably aspirin, have been available in forms that can be tableted without further treatment. Recently, there has been a growing impetus to develop so-called direct compression (d.c.) formulations, and the range of excipients, especially diluents, designed for this specific role has expanded dramatically.

It is possible to distinguish two types of d.c. formulations: (a) those in which a major proportion is an active ingredient, and (b) those in which the active ingredient is a minor component (i.e., < 10% of the compression weight). In the former, the inherent characteristics of the drug molecule, in particular the ability to prepare a physical form that will tablet directly, will have profound effects on the tablet's characteristics.

It may sometimes be necessary to supplement the compressional qualities of the drug, and the needs of this situation have been realized by several manufacturers of excipients. Materials that can be described as "compression aids" are now commercially available. Ideally, such adjuvants should develop mechanical strength while improving, or at least not adversely affecting, release characteristics. Among the most successful at meeting both these needs have been the microcrystalline celluloses (partially acid-hydrolyzed forms of cellulose). Several grades are available based on particle size and distribution.

Most other d.c. excipients really belong in the second category, for which the drug is present in low concentration. In such instances, the use of an inexpensive d.c. diluent is warranted. Before considering some of these, certain generalizations are worth noting, since an erroneous belief that d.c. is always a simpler formulation route seems to have arisen. For instance, many d.c. fillers, such as spray-dried lactose, should not be reworked, as this affects their compressibility. In addition, those diluents with a large particle size may give rise to mixing problems, owing to segregation, unless an optimum proportion of fine material is present. More often, we are faced with the problem of an excessively narrow particle size distribution so that flow, in general, and uniform feeding to the dies, in particular, are difficult. Sometimes batch-to-batch variations are more prevalent in soluble d.c. fillers, such as sugars. Unlike wet granulation, d.c. has little ability to mask inherent tableting deficiencies in an ingredient. In addition, there will be little possibility for prior wetting of a hydrophobic drug and subsequent dissolution enhancement, which is a proved effect of wet granulation. On the other hand, d.c. formulations are likely to be more stable, show less aging effects, and in specific cases, offer the only workable production method.

D. Wet Granulation

Although many existing products continue to be processed by a lengthy wet granulation involving blending of dry ingredients, wet massing, screening, and then tray or fluidized bed drying, there is a trend toward using machines that can carry out the entire granulation sequence in a single piece of equipment—the mixer–granulator–dryer (MGD). Additional advantages of MGD processors include reduced handling of excipients, reduced exposure of the excipients to heat, and a better opportunity to precisely control the moisture level in the granulation. Furthermore, use of such plant may reduce granulation time by factors between 5 and 10.

In one interesting study [104], granules produced by five different methods were evaluated in terms of angle of repose, hardness, density, number per unit weight, bulk–volume geometric form, and shape–volume factor. In addition, a means of comparing the efficiency of the processes in producing smooth, spherical particles was suggested. It is generally agreed that there will exist an optimum range of granule sizes for a particular formulation; therefore, certain generalizations are worthy to note here. Within limits, smaller granules will lead to higher and more uniform tablet weight, and higher tablet crushing strength, with subsequent longer disintegration time and reduced friability. The strength of granules also influences the tensile strength of the tablets prepared from them, with stronger granules generally leading to harder tablets [105].

One important finding with widespread implications arises from work by Chaudry and King [106], who demonstrated that migration of a soluble drug during moist granulation was responsible for uneven content uniformity of tablets of warfarin. A modified base (containing dibasic calcium phosphate, alginic acid, and acacia) developed from experiments to assess migration-retarding ability, enhanced tablet performance.

IV. TABLETING EQUIPMENT

A. Granulators

Mechanical aids to wet granulation have developed from the hand process of preparing a wet mass and forcing it through a screen onto trays that were placed into a convection oven where the granules were dried. As batch sizes increase, the need for bulk granulation procedures became necessary. Many of these procedures involved mixing the powdered ingredients in a special ribbon blender, which could also accomplish the wet massing process. The moistened materials were then usually granulated by forcing them through a screen, using oscillating blades of a modified comminuting mill, onto trays that were then transferred to a forced warm air oven.

More efficient high-shear–mixing machines are now commonly used to achieve both the mixing and granulating operations. Another approach to granulation is the use of fluidized bed dryers, which can now mix, granulate, and dry in the same bowl. However, the process conditions, such as drying temperature and length of the drying cycle, must be optimized, and it may be necessary to adopt a different approach to formulation than would have been used with the more traditional equipment. For example, granulating fluid will have to sprayed into the bed of mixed powder at a given temperature; its volatility and viscosity under these conditions will influence the characteristics of the final product. Additionally, similarity of the particle size distribution of the various ingredients and a minimum amount of fines are important factors to be considered. The granules produced from fluidized bed drying will tend to have lower bulk density than those from the higher-shear mixer–granulator. Sophisticated MGDs based on the high-shear principle have now been developed. The Topogranulator (Topo) and Pivotal Processor (T. K. Fielder) are typical of this new generation of single-unit, multiple function processors.

The important parameters governing the performance of fluidized beds in which all three operations are being accomplished (i.e., mixing, granulating, and drying), have been studied by Worts and Schoeffer [107]. They concluded that, in addition to the inlet air temperature, the type of binder solution and its flow rate and droplet size were critical variables that had to be controlled. These investigators also found that the residual moisture levels of granulations are a major factor contributing to the in vitro dissolution and friability of tablets made from them.

Granulation by preliminary compression, originally performed in heavy-duty tablet machines, when it was called "slugging," has been used for a small number of products for many years. This approach has been further developed in machines that are essentially roll compactors, squeezing the material to produce agglomerates. In addition, some tablet presses are equipped with precompression rollers that perform the "slugging" of the granulation before its "true" compaction (Fig. 15).

In another type of machine, the Marumeriser (Luwa), the material granulated is extruded through numerous small orifices and then rotated in a special bowl capable of spheronizing the granules. This leads to products of high-bulk density and excellent flow properties, with virtually no dustiness. The parameters controlling this process have been studied by Malinowski [108], and the effect on the tablet produced from it was evaluated.

Microwave vacuum drying of granulation has recently been developed for commercial-scale operations by companies such as T. K. Fielder, who now offer a range of such dryers. The major advantage of such systems are claimed to be fast drying at low temperatures in a dust-free environment.

B. Tablet Presses

With the exception of presses designed to produce coated or layered tablets, the development of tableting equipment has largely been one of continuing evolution. In many areas, the incentives have come from the pharmaceutical industry, rather than the tablet press manufacturers, as a result of certain trends in tableting operations. These include the desire for higher rates of production, direct compression of powders, stricter standards for cleanliness as part of an increasing awareness of GMP, and a wish to automate, or at least continuously monitor, the process. However, there is now evidence to suggest that we may be approaching inherent limits to further development of some press variables on existing lines.

At present, and in the immediate future, we may anticipate the continual vying of one manufacturer with another over relatively minor improvements, with maybe significant advances possible in the field on instrumentation, automation, and hygiene. There is the added possibility of some major design developments, but two revolutionary press designs, unveiled some time ago—the Regina (Horn) and HSR (Manesty) machines—have not been developed to the commercial stage. However, the self-cleaning press manufactured by Kikisui, has been marketed. In this machine the entire turret area can be filled with solvents and cleaned by running the press—a somewhat revolutionary approach to cleaning.

All tableting presses employ the same basic principle: they compress the granular or powdered mixture of ingredients in a die between two punches, with the die and its associated punches being called a "station of tooling." Tablet machines can be divided into two distinct categories on this basis:

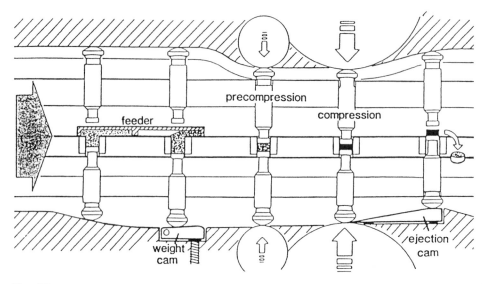

Fig. 15 Multistation press cycle.

1. Those with a single set of tooling—single-station (or single-punch) or eccentric presses
2. Those with several stations of tooling—multistation (or rotary) presses

Figures 15 and 16 provide a summary of the compression cycles for rotary and single-punch tablet presses. The formation of the tablet compact in these two types of presses differs mainly in the compaction mechanism itself, as well as the much greater speeds achieved with rotary-type presses. The single-punch basically uses a hammering-type motion (i.e., the upper punch moves down while the lower punch remains stationary), whereas rotary presses make use of an accordion-type compression (i.e., both punches move toward each other). The former find their primary use as a research and development tool, whereas the latter, having higher outputs, are used in most production operations.

Single-Station Presses

All commercial types of single-station presses have essentially the same basic operating cycle (see Fig. 16), during which filling, compression, and ejection of tablets from the die are accomplished by punch movement utilizing cam actions. Material is fed to the die from the hopper by an oscillating feed shoe; the position of the lower punch at this point determines the tablet weight. The feed shoe then moves away, and the upper punch descends into the die to compress the tablet, with the extent of this movement controlling the level of compression force. As the upper punch moves upward, the lower punch rises and in so doing ejects the tablet from the die. At this point, the feed shoe moves in and knocks the tablet out of the machine as the lower punch moves to its bottom position ready for the next press cycle.

Sizes of machines in this group vary widely from small ones capable of making tablets up to 12 mm in diameter at rates of about 80 tablets per minute (tpm) and exerting maximum forces on the order of 20 kN (Table 14), to large machines with maximum tablet diameter near 80 mm and loads up to 200 kN or more. In isolated cases, tablets can be made only on this type of machine, probably because its mode of operation gives the material a longer "dwell time" under compression. Although the table output rate form single-station presses can be

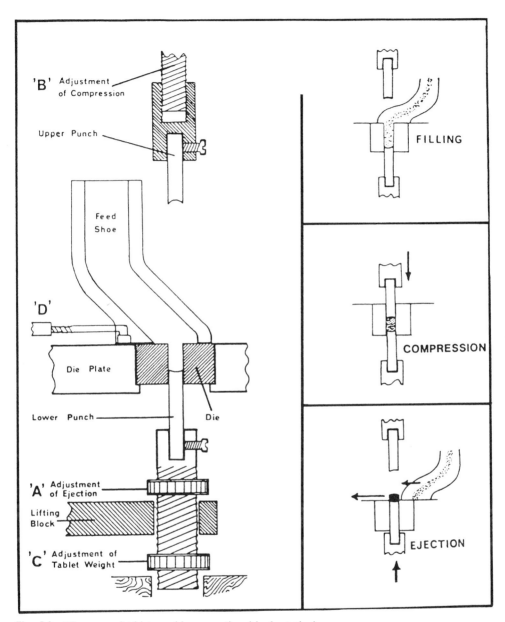

Fig. 16 Diagrams of tablet machine operation (single station).

increased by use of multitip tooling, the rotary machine remains the method of choice for production purposes.

Multistation Presses

In the multistation-type machine, the operating cycle and methods of realizing the filling, compressing, and ejection operations are different from those of single-station presses and are summarized in Table 15. More specifically, the dies and punches are mounted on a rotating turret.

Table 14 Approximate Specifications of Some Small Single-Station Laboratory Presses

| Manufacturer: | Fette | Kilian | Korsch | Manesty |
Model number:	Exacta 1	KS	EKO	F3
Maximum output (tabs/min)	75	80	100	85
Maximum tablet diameter (mm)	15	18	20	22
Maximum fill (mm)	16	16	20	17
Maximum applied force (kN)	15	25	30	40
Motor (HP)	0.75	1.1	0.5	2.0
Weight (kg)	250	300	220	476
Approximate dwell time (msec)	133	125	100	118

All operations take place simultaneously in different stations (Fig. 15). Sixteen stations were commonly used in earlier machines, with outputs between 500 and 1000 tpm and tablet diameters up to 15 mm. Presses with outputs orders of magnitude greater than this are now widely available. The dies are filled as they pass beneath a stationary feed frame, which may be fitted with paddles to aid material transfer. The die cavities are completely filled and excess ejected before compression. Compression involves the movement of both punches between compression rolls, in contrast with single-station operations, during which only the upper punch effects compression. Ejection occurs as both punches are moved away from the die on cam tracks, until the tablet is completely clear of the die, at which point it hits the edge of the feed frame and is knocked off the press. Tooling pressure may be exerted hydraulically, rather than through the use of mechanical camming actions, as with machines produced by Cortoy.

The ways in which individual manufacturers of tableting equipment have sought to achieve higher output fall into four groups:

1. Increasing the effective number of punches (i.e., multitipped types)
2. Increasing the number of stations
3. Increasing the number of points of compression
4. Increasing the rate of compression (i.e., turret speed)

Table 15 Comparative Specifications of Some High-Output Tablet Presses

| Manufacturer: | Fette | Horn | Kilian | Manesty | Stokes | Vector |
Model number:	P3000	URP/59E	RX67A	IIa	565-1	Magna
Maximum output (tabs/min)	8,250	8,850	10,000	11,100	10,000	15,000
Number of stations	55	59	67	61	65	90
Maximum fill (mm)	75	75	75	91	65	45
Maximum turret speed (rpm)	13	11	13	11	11	11
Maximum tablet diameter (mm)	18	15	16	17	17	19
Maximum fill depth (mm)	100	100	80	65	75	90
Precompression (kN)	20	Yes	Yes	Yes	Yes	Yes
Motor (HP)	7.5	7.5	7.5	7.5	7.5	7.5
Net weight (kg)	3,250	2,800	2,400	2,900	2,040	4,535
Approximate dwell time (msec)	7.3	6.8	6.0	5.4	6.0	8.0

Each of these approaches has its own particular set advantages and disadvantages. In addition, all make demands on other aspects of press design, and certain general inherent characteristics of die compaction have to be taken into account.

Generally, the high-speed machines consist of double-rotary presses in which the cycle of operation is repeated twice in one revolution of the turret carrying the tooling, although one press (Magna; Vector Corp.) has four cycles per revolution. All these machines normally have odd numbers of stations, with up to 79 in the largest presses. Double rotary presses have also been modified to produce layered tablets, whereas other machines have been adapted to produce coated tablets by a "dry" compression technique.

C. Tablet Machine Instrumentation

To produce an adequate tablet formulation, certain requirements, such as sufficient mechanical strength and desired drug release profile, must be met. At times, this may be a difficult task for the formulator to achieve, because of poor flow and compatibility characteristics of the powdered drug. This is of particular importance when one has only a small amount of active material to work with and cannot afford to make use of trial-and-error methods. The study of the physics of tablet compaction through the use of instrumented tableting machines (ITMs) enables the formulator to systematically evaluate his formula and to make any necessary changes.

ITMs provide a valuable service to all phases of tablet manufacture, from research to production and quality control [109–111]. As a research tool, ITMs allow in-depth study of the mechanism of tablet compaction by measuring the forces that develop during formation, ejection, and detachment of tablets. They can also provide clues about how materials bond, deform, and react to frictional effects. The formulator himself is able to monitor the effects of additives in the overall tableting process, as well as the effects of operation variables in the manufacture and performance of the dosage form. This markedly reduces the formulator's reliance on empiricism in formulation design. In the area of product and quality control, ITMs are able to monitor tablet weight and punch and machine wear and damage. More recently, ITMs have been used to characterize unique "typical batches" of materials so that one has a baseline for troubleshooting formulations or a basis for quality control [109–111].

Instrumented Tableting Machines in Research and Development

The tableting process involves two phenomena: (a) a reduction in the bulk volume of the tablet mass by elimination of air, which we call compression, and (b) an increase in the mechanical strength of the mass by particle–particle interactions, which is termed *consolidation*. This latter process results from utilization of the free surface energies of the particles in bond formation, referred to as "cold welding," plus intermolecular interactions by van der Waals forces, for example. The process is enhanced by generation of large areas of clean surface, which are then pressed together; such a mechanism is feasible if appreciable brittle fracture and plastic deformation can be introduced into the system. Therefore, the manner in which the various components compress will be of importance.

It is also important to appreciate that the behavior on decompression can markedly affect the characteristics of the finished tablets, because the structure must be strong enough to accommodate the recovery- and ejection-induced stresses. Indeed, tablet strength is a direct function of the number of "surviving bonds" in the finished tablet. In addition, ability to monitor ejection forces leads to valuable information on lubricant efficiency.

Analysis of Data Obtained from Instrumented Tableting Machines

Measurement of the punch and die forces plus the relative displacement of the punches can provide raw data that, when suitably processed and interpreted, facilitate the evaluation of many tabling parameters. Many of the workers first involved in instrumenting tablet presses concentrated on deriving relationships between the applied force (F_A) and the porosity (E) of the consolidating mass.

Heckel proposed that a correlation exists between yield strength and an empirically determined constant K, which is a measure of the ability of the compact to deform [28,112]. He discovered that, indeed, K is inversely proportional to yield strength. Furthermore, he derived an equation expressing the relation between the density of a compact and the compressional force applied. This relationship is based on the assumption that decreasing void space (i.e., decreasing porosity) of a compact follows a first-order rate process:

$$\frac{dD}{dP} = K(1 - D)$$

where

$$D = \text{relative density}$$
$$P = \text{pressure}$$
$$(1 - D) = \text{pore fraction}$$
$$K = \text{proportionality constant}$$

By integrating and rearranging this equation, one obtains the following linear relationship:

$$\ln\left(\frac{1}{1 - D}\right) = KP + A$$

where

$$A = \ln\left(\frac{1}{1 - D_0}\right) + B$$

and B is a measure of particle rearrangement.

It has been claimed [113] that the presentation of data in the form of Athy–Heckel plots [112,114], as illustrated in Fig. 17, facilitates assessment of the relative proportions of brittle fracture and plastic deformation present. Each set of curves within a plot represents the same formulation, with decreasing particle size fractions. If the curves remain discrete, as in Fig. 17a, one can assume that plastic deformation is the predominate mechanism, because all formulations have sufficient time to rearrange (i.e., plastic deformation is a time-dependent process). In contrast, Fig. 17b illustrates the behavior of a material that deforms primarily by brittle fracture. Here, the original particle size distribution is rapidly destroyed and the curves become superimposed. Additionally, the slope of the lines, which represents K, is approximately unity. Higher slopes (i.e., low yield pressure) are found when evaluating plastically deforming materials. Subsequent studies [115,116] have cast doubts on the universality of this claim, and conflicting data has been reported.

Applied force and displacement measurements have also been used to generate force versus displacement curves (F–D plots) [96]. Such information can be used to estimate the energy necessary to form a compact in the following manner:

$$W = F \int dD$$

where

$$W = \text{work}$$
$$F = \text{force}$$
$$D = \text{distance}$$

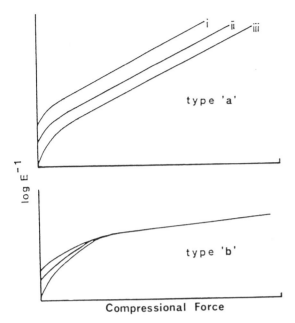

Fig. 17 Athy–Heckel plots: (a) material undergoing plastic deformation; (b) material undergoing brittle fracture.

When one plots force versus displacement, then, the area under the curve (AUC) thus represents work. In practice, the compression–decompression data take the form shown in Fig. 18. The area under the upward line represents the work done on the tableting mass during compaction, and that under the downward line arises from the fact work is done on the punch by the tablet as a result of the latter's elastic recovery on decompression.

In single-station presses, we can distinguish a further subdivision of work by considering the force transmitted to the lower punch during the compression. This will be less than that registered by the upper punch owing to frictional effects at the die wall. Three components to the total work can, therefore, be distinguished; that is, W_F, the work done in overcoming die wall friction; W_D, the work of elastic recovery; and W_N, the net work involved in forming the

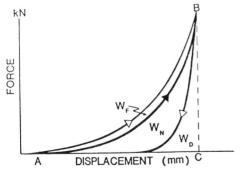

Fig. 18 A typical force–displacement curve: W_F = work done in overcoming die wall friction; W_D = work of elastic recovery; W_N = net work involved in formation of tablet compact.

tablet. Interpretation of F–D curves, on single-station presses, has proved to be particularly attractive, as demonstrated by the work of Travers and Cox [117].

Some researchers have extended the monitoring of work to the determination of the rate of doing work, or power [116] as illustrated in the following equation:

$$\frac{dW}{dt} = \text{power}$$

The rationale behind this approach is that because of the varying degrees of bond formation, different materials require a larger, or smaller, amount of energy to compress them to a given degree. In a tablet press running at a given speed, the rate of doing work, therefore, must be different and may be related to tablet strength. In addition, since plastic deformation is a time-dependent phenomenon; power, which takes time into account, may be indicative of the contribution of this mechanism.

The compressive behavior of the material is also reflected in that proportion of the axial force F_A that is transmitted radially to the die wall F_R during compression and decompression. Therefore, monitoring the ratio of these two forces during the entire machine operation can provide valuable data that we can call the *compaction profile*. As shown in Fig. 19, this normally takes the form of a hysteresis loop, the area of which is a function of the departure of the material from purely elastic behavior. Other features of the profile provide valuable guidelines to tablet strength, likely levels of lubrication required, and predominant type of deformation. The line OA is represented as a dotted line because this region is due to repacking, which can be quite variable. At point A, elastic deformation becomes dominant and continues until the yield stress, at point B, is reached. At this point, the deformation of the compact is due to plastic deformation and brittle fracture. This process continues to point C, at which time force is removed and decompression begins. From points C to D, the material is elastically recovering. If a second yield point D is reached, the material has become plastically deformed or brittley fractured. To sum up all of these processes:

Slope AB = function of elastic deformation

Slopes BC and DE = function of plastic deformation and brittle fracture

Slopes BC′ and CD = function of elastic recovery

Lines OC′ and OE = function of (residual die wall force) *RDWF*

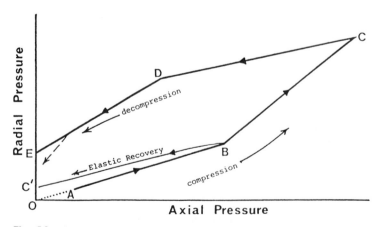

Fig. 19 Compaction profile.

One should note that BC′ represents a highly elastic material, as little plastic deformation or brittle fracture has occurred. Also, sharp differences between the slope CD and DE are indicative of weak, or failed, tablet structures. The *RDWF* estimated from these plots can provide a good indication of the ejection force. More detailed treatments of such studies are now in the open literature, to which the interested reader is referred [118–120].

Another approach to determining the contribution being made by each of the possible compression–decompression mechanisms involves monitoring the degree and rate of relaxation in tablets immediately after the point of maximum applied force has been reached. Once a powder bed exceeds a certain yield stress, it behaves as a fluid and exhibits plastic flow [121,122]. Certain investigators [122] have studied plastic flow in terms of viscous and elastic elements and have derived the following equation:

$$\ln F_t = \ln F_0 - kt$$

where

F = compressional force in viscoelastic region
t = time
k = degree of plastic flow

which can be integrated and rearranged to yield the linear equation:

$$\ln F_t = \ln F_0 - kt$$

Thus, if one plots log F versus time, one needs to calculate only the slope of the line so plotted to determine the plastic flow, k. Materials with higher values of k (i.e., more plastic flow) tend to form stronger tablets than those with low k values.

In 1972, Rees and associates [123] developed the compaction simulator which, in oversimplified terms, is a hydraulic press capable of accurately mimicking the action of any high-speed rotary press. This is an overwhelming achievement in the study of compaction physics, when one reflects on the various disadvantages associated with single-punch and rotary presses that this system can overcome. For instance, single-punch measurements, although providing a baseline for formulation development, do not accurately reflect the forces incurred at the production level. In other words, the lower punch remains stationary, rather than moving, in single-punch systems, and the dwell time, which influences the extent of plastic deformation, is significantly increased owing to the low speed of manufacture. On the other hand, rotary presses require relatively large amounts of granulation and, therefore, are inappropriate for the initial phases of formulation when only small amounts of material are available. Also, the rate of applying and removing forces varies appreciably from machine to machine, depending on the way the machine is operated, the punches used, and so on. The compaction simulator is able to compensate for all of these disadvantages, such that production level conditions can exist for small amounts of material. Additionally, the compaction simulator is capable of reproducing the variables associated with each machine and, thereby, affords the company an adequate means for transferring a formulation from one machine to another. Presently, there are only two compaction simulators in use in the United States; one at SmithKline Beecham and the other, which is used by a consortium of companies in the United States, at Rutgers College of Pharmacy. Researchers at Rutgers hope to develop a library of compression profiles for several excipients and drugs that will be made available to all those in the industry.

Several reviews of the mechanisms involved in the compaction process and interpretation of the large amount of data generated from such studies have been published [124,125]. The solution to this complex process is still incomplete and provides continuing research opportunities for the pharmaceutical scientist. The actual mechanics of instrumenting presses has also been reviewed [111], and a more recently published book [126] is devoted solely to this subject. In addition, details of individual instrumentation techniques are usually described in the papers of authors reporting data.

There has also been significant growth in the number of press manufacturers offering instrumentation packages for use with high-speed, multistation machines, to control tablet weight variation and, in some cases, divert out-of-tolerance tablets to a reject container. The more sophisticated systems result in a fully automated press operation, with essentially no operator intervention required (Fig. 20).

V. COATED TABLETS

The coating of pharmaceutical tablets may be divided conveniently into the traditional sugar, or pan-coating procedures, and contemporary techniques that include film coating and com-

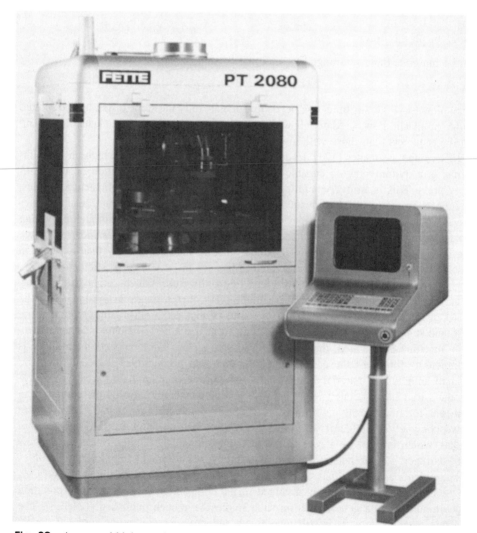

Fig. 20 Automated high-speed rotary tablet press.

pression coating. Coating methods were developed for a variety of reasons, including the need to mask an unpleasant taste or unsightly appearance of the uncoated tablet, as well as to increase patient acceptability. Protection of an ingredient from degradation effects, caused by exposure to moisture, air, and light, were further incentives. The newer techniques have extended the usefulness of coating to include the facilitation of controlled-release characteristics and the ability to coformulate inherently incompatible materials.

A. Sugar Coating

The sugar-coating process involves building up layers of coating material on the tablet cores, as they are tumbled in a revolving pan, by repetitively applying a coating solution or suspension and drying off the solvent. Traditionally, the cores were made using tooling with deep concave geometry to reduce the problems associated with producing a sufficient coat around the tablet's edge, as illustrated in Fig. 21. However, this shape may not be ideal for all products because of the inherently softer crown region exhibited in tablets manufactured from such tooling. In addition, deep concave tooling often produces tablets of poor mechanical strength [127]. Core mechanical strength, in particular friability, must be adequate enough to withstand the abrasive effects of the tumbling action while retaining the dissolution characteristics of an uncoated tablet. Large tablets, in particular, sometimes require higher compressional forces than are necessary for uncoated tablets of the same size. Care must also be exercised to minimize penetration of coating solutions into the core itself, although the coat should adhere well to the tablet surface. It will also be important to maintain a smooth, uniform surface and to provide careful control of the environment within the coating pan. Because of these requirements, the process may best be described by means of a generalized example.

In the past, the initial layers of coating (the sealing coat) were achieved by applying one or two coats of shellac. However, owing to the variability between batches of this material, PVP-stabilized types of shellac or other polymeric materials, such as cellulose acetate phthalate (CAP) and polyvinyl acetate phthalate (PVAP), are now more popular. It should be appreciated that the sealing coat must be kept to the minimum thickness consistent with providing an adequate moisture barrier—not a simple balance to achieve.

The next stage is to build up a subcoat that will provide a good bridge between the main coating and the sealed core, as well as rounding off any sharp corners. This is normally a two-

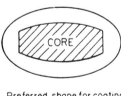

Preferred shape for coating

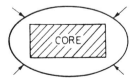

Bad choice of core geometry

Fig. 21 Tablet-coating geometry.

step procedure. The first step involves the application of a warm subcoat syrup (containing acacia or gelatin) that rapidly distributes uniformly over the tablets and, eventually, becomes partially dry and tacky. At this point, a subcoat powder (containing material such as calcium carbonate, talc, kaolin, starch, and acacia) is dusted evenly over the tablets, after which the pan is allowed to rotate until the coat is hard and dry. This subcoat cycle is usually repeated three or four times, taking care to avoid the production of rough surfaces, which would be difficult to eradicate later on, and ensuring that each coat is absolutely dry before the next is applied.

The step that follows is known as "smoothing" or "grossing." It produces the bulk of the total coating weight and involves the application of a suspension of starch, calcium carbonate, or even some of the subcoat powder, in syrup. Each application is dried and the process repeated until the desired buildup has been realized. The last few applications may be made with a syrup free from suspended powders, to produce a smooth surface. If the tablets are to be colored, colorants are normally added in these clear syrup layers. It is important that the tablet surfaces be smooth before this is attempted, otherwise uneven coloring may result. The final finishing stage is accomplished by again applying one or two layers of clear syrup, taking care not to overdry between coats and stopping while the final coat is still slightly damp. Jogging (i.e., pan stationary, apart from intermittent rotation through a small angle) is then carried out until the tablets appear dry.

The tablets are then left for several hours and are transferred to the polishing pan, which is usually of a cylindrical design with canvas side walls. The polish is a dilute wax solution (e.g., carnauba or beeswax in petroleum spirit) applied sparingly, following which the tablets are left to roll until a high luster is produced. They are then normally "racked" to allow any traces of solvent to evaporate before being sent to the inspection and packing operations.

There are as many variations in coating procedures as there are tablet coaters, and so the account given here is only a guide. Nevertheless, it illustrates the complexity and time-consuming nature of the process, and the reader will realize why efforts were made to develop alternative coatings, equipment, and methods that would permit at least some degree of automation.

B. Film Coating

Film coating has increased in popularity for various reasons. The film process is simpler and, therefore, easier to automate. It is also more rapid than sugar coating, since weight gains of only 2–6% are involved, as opposed to more than 50% with sugar coating. In addition, moisture involvement can be avoided (if absolutely necessary) through the use of nonaqueous solvents. Moreover, distinctive identification tablet markings are not obscured by film coats.

There are now many synthetic polymeric materials available to the would-be film coater, many of which meet all the requirements of a good film former. These include lack of toxicity and a suitable solubility profile for film application and, after ingestion, the ability to produce a tough, yet elastic film, even in the presence of powdered additives such as pigments. The film must be stable to heat, light, and moisture, and be free from undesirable taste or odor.

Some of the more popular materials meeting these criteria are listed in Table 16, together with some important properties. Two major groups may be distinguished: (a) those materials that are nonenteric and, for the most part, cellulose derivatives, and (b) those materials that can provide an enteric effect and are commonly esters of phthalic acid. Within both groups it is general practice to use a mixture of materials to give a film with the optimum range of properties. They may contain a plasticizer that, as the name implies, prevents the film from becoming brittle, with consequent risk of chipping [128]. Some popular choices are shown in Table 17. As they essentially function by modifying polymer-to-polymer molecular bonding,

Table 16 Some Commonly Used Film-Coating Materials

Full name	Abbreviation	Soluble in	Comments
Nonenteric			
Methylcellulose	MC	Cold water, GI fluids, organic solvents	Methocel (DOW): a useful polymer for aqueous films; low-viscosity grade best
Ethylcellulose	EC	Ethanol; other organic solvents	Ethocel (Dow): cannot be used alone as is totally insoluble in water and GI fluids; employed as a film toughener
Hydroxyethylcellulose	HEC	Water and GI fluids	Cellosize (Union Carbide): properties similar to MC, but gives clear solutions
Methylhydroxyethylcellulose	MHEC	GI fluids	Similar properties to HPMC, but less soluble in organic systems
Hydroxypropylcellulose	HPC	Cold water; GI fluids; polar organics; such as anhydrous lower alcohols	Klucel (Hercules): difficulty in handling owing to tackiness while drying
Hydroxypropylmethylcellulose	HPMC	Cold water; GI fluids; methanol/methylene chloride; alcohol/fluorohydrocarbons	Excellent film former and readily soluble throughout GIT; low-viscosity grades to be preferred, e.g., Methocel HG (Dow)
Sodium carboxymethylcellulose	Na-CMC	Water and polar organic solvents	CMC-T (Hercules): main use where presence of moisture in solvent not a problem
Povidone	PVP	Water; GI fluids; alcohol, and IPA	Plasdone (General Aniline): care needed in use owing to tackiness during drying; best used in mixtures to increase adhesion; is hygroscopic if used alone
Polyethylene	PEGs	Water; GI fluids; some organic solvents	Carbowaxes (Union Carbide): low-molecular-weight grades used mainly as film modifiers, particularly plasticizers[a]
Enteric			
Shellac		Aqueous if pH 7.0	May delay release too long; high batch-to-batch variability
Cellulose acetate	CAP	Acetone, ethyl acetate/IPA, alkalies, if pH 6.0	Dissolves in distal end of duodenum; requires presence of plasticizer, such as triacetin or castor oil; is somewhat hygroscopic (Kodak)
Polyvinyl acetate phthalate	PVAP	As above, if pH > 5.0	Dissolves along whole length of duodenum (Colorcon)
Hydroxypropylmethylcellulose phthalate	HPMCP	As above, if pH > 4.5	Dissolves in proximal end of duodenum (Shinetsu)
Polymers of methacrylic acid and its esters		Eudragit L[b] pH > 6, Eudragit S[b] pH > 7	Solubilized in alkaline media; mixtures of L and S can provide enteric coating plus sustained-release

[a]High-molecular-weight grades are less hygroscopic and give tough coating.
[b]Pharm Rohma.

Table 17   Some Commonly Used Film
Plasticizers

Phthalate esters	Propylene glycol
Citrate esters	Polyethylene glycol
Triacetin	Glycerin

the choice of plasticizer is dependent on the particular film polymer. Like so many other facets of tablet coating, this is an area for which there is no substitute for properly designed experimental trials in developing a procedure.

The nature of the solvent system may markedly influence the quality of the film [129] and, to optimize the various factors, mixed solvents are usually necessary. More specifically, the rate of evaporation and, hence, the time for the film to dry, has to be controlled within fine limits if a uniform smooth coat is to be produced. The solvent mixture must be capable of dissolving the required amount of coating material, yet give rise to a solution within a workable range of viscosity. Until relatively recently, alcohols, esters, chlorinated hydrocarbons, and ketones have been among the most frequently used types of solvents.

However, as a result of increasing regulatory pressures against undesirable solvents, there has been a pronounced trend toward aqueous film coating. Many of the same polymers can be used, but it may be necessary to employ lower-molecular-weight grades owing to their higher viscosity in aqueous systems. Alternatively, water-insoluble polymers may be dispersed as a latex (emulsion) or pseudolatex (suspension) in an aqueous medium. This approach permits a high-solids content without attendant high-viscosity problems. However, acceptable film-forming in these systems is dependent on coalescence or agglomeration. For pseudolatexes, this agglomeration requires a soft particle and, thus, a high concentration of plasticizer in the system, to ensure formation of a continuous film.

In a four-part article, Porter [130] has provided a comprehensive review of tablet-coating technology, with emphasis on contemporary practice. More specifically, a recent review [131] discusses characterization techniques for the aqueous film-coating process and provides a useful "influence matrix" between process variables and final product attributes. This matrix is summarized in Fig. 22.

Because of the need to develop a uniform color, with minimum application, the colorants used in film coating are more likely to be lakes than dyes. In lakes the colorant has been absorbed onto the surface of an insoluble substrate. This gives both opacity and brightness to the pigments, which are formulated with other materials so that they can be easily dispersed, while retaining the desired film-forming capabilities of the polymeric film former. Complete, matched-coloring systems are now available as fine powders that can be readily suspended in organic solvents or aqueous systems and, as such, provide a very convenient colorant source.

C.   Modified-Release Coatings

A coating may be applied to a tablet to modify the release pattern of the active ingredient from it. We may distinguish two general categories: enteric coating and controlled-release coating. The former are those that are insoluble in the low pH environment of the stomach, but dissolve readily on passage into the small intestine with its elevated pH. They are used to minimize irritation of the gastric mucosa by certain drugs and to protect others that are degraded by gastric juices.

The most common mode of action of enteric coating is pH-related solubility (i.e., insoluble at gastric pH, but soluble at some pH above 4.5). A list of the most widely used enteric coatings

In-Process/Final Product Characteristics / Process Variables	Picking	Peeling	Bridging	Orange Peel	Mottling	Cracking	Erosion	Overwetting	Sticking	Uneven mixing	Coating of pan	Final quality	Disintegration	Dissolution	Stability
Suspension concentration			•	•				•	•			•	•	•	•
Suspension viscosity			•	•		•		•	•			•	•	•	•
Plasticizer concentration			•			•							•	•	
Pigment homogenicity					•							•			
Film surface tension			•			•									
Spray rate	•	•		•			•	•	•			•	•	•	•
Number of guns				•				•	•		•				
Spray fan width								•	•		•				•
Atomizer air pressure				•			•	•	•		•				•
Gun distance				•				•	•						•
Spray angle				•				•	•			•	•		
Pan speed								•	•	•	•				
Baffle design								•	•	•	•				
Load size				•				•	•	•					
Tablet bed porosity				•				•	•	•					
Tablet shape	•	•		•				•	•	•	•				
Tablet hardness		•					•					•	•	•	
Tablet friability							•					•	•	•	
Inlet/outlet temperature	•	•		•				•	•			•	•	•	•
Inlet/outlet humidity	•	•		•				•	•			•	•	•	•
Air supply volume	•	•		•				•	•			•	•	•	•

Fig. 22 An influence matrix for common variables and responses in an aqueous film-coating process.

is given in Table 15. A less popular alternative has been the use of materials that are affected by the changing enzymatic activity on passage from the stomach to the small intestine. Since their performance is dependent on the digestion of the coating, the intrinsic in vivo variability of this action makes this type of enteric coating less predictable.

The second group of specialized coatings are those that produce a controlled, and usually extended, release of drug from the tablet. These are reviewed in Chapter 15.

D. Coating Equipment

Conventional coating pans are subglobular, pear-shaped or even hexagonal (Fig. 23), with a single front opening, through which materials and processing air enter and leave. Their axis is

Fig. 23 Conventional coating pan.

normally inclined at approximately 45° to the horizontal plane, and they are rotated between 25 and 40 rpm, the precise speed depending, most often, on the product involved. One modification of the normal pan has been the substitution of a cylindrical shape, rotated horizontally with regions of the walls perforated by small holes or slots. This design permits a one-way airflow through the pan, as shown in Fig. 24. This figure also illustrates the ways in which vendors have chosen to modify the basic concept. In the Accela-Cota (Thomas Engineering) and Hi-Coater (Vector Corp.) (Fig. 25), the flow of air is through the tablet bed and out through the perforated wall of the pan. In the Driacoater (Driam) the air flows from the perforated pan wall through the tablet and into the central region (i.e., countercurrent to the direction of the coating spray). The Glatt-Coater (Glatt Air Techniques) permits either co- or countercurrent airflow to suit particular products.

The traditional method of ladling the coating solutions into the rotating bed of tablets has given way to systems capable of spraying material, with or without the assistance of an air jet. More specifically, two general types are available: (a) those that rely entirely on hydraulic pressure to produce a spray when material is forced through a nozzle (airless spraying), and (b) those in which atomization of the spray is assisted by turbulent jets of air introduced into it. The latter type tends to produce a more easily controlled spray pattern and, therefore, is better for small-scale operations, although both are capable of giving the flat jet profile preferred for pan operation. Other important parameters include the distance from nozzle to the bed surface and whether continuous or intermittent spraying is used. One interesting development in this area has been the introduction of a special plough-shaped head that is immersed in the tumbling bed of tablets and through which both coating solutions and air can be delivered into the bed.

Fig. 24 Typical side-vented pan.

The film-coating process can be carried out in convential pans, although operation variables, such as speed of pan rotation, angle of pan axis, and temperature and humidity control, may be more critical. Newer pans, with one-way airflow through the tablet bed offer an even better alternative, because the pan environment can be controlled within finer limits.

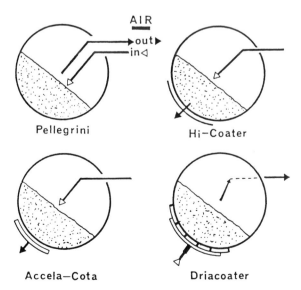

Fig. 25 Coating pan air configurations.

Coating in Fluidized Beds

Fluidized beds, in general, and the air-suspension technique patented by Wurster, in particular, now offer an attractive alternative to pan coating. The basic principle underlying their operation is to suspend the tablets in an upward moving stream of air so that they are no longer in contact with one another.

An atomizer introduces spray solution into the stream and onto the tablets, which are then carried away from the spraying region where the coating is dried by the fluidized air. In the Wurster process, the design is such that the tablets are sprayed at the bottom of the coating column and move upwards centrally, being dried as they do so. They then leave the top of the central region and return down the periphery of the column to be recycled into the coating zone, as shown in Fig. 26. The process can be controlled by careful adjustment of the air, rate of coating solution delivery, and monitoring of the exit-air temperatures.

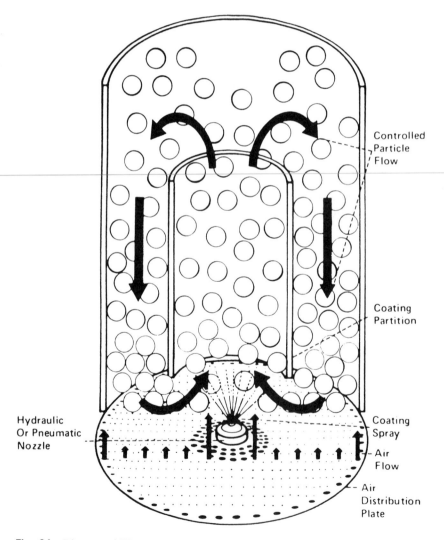

Fig. 26   Diagram of Wurster coating chamber.

Compression Coating

A method has been described [132] for compressing a coating around a tablet "core," by using specially designed presses. The process involves preliminary compression of the core formulation to give a relatively soft tablet, which is then transferred to a large die already containing some of the coating material. After centralizing the core, further coating granulation is added and the whole compressed to form the final product. From a formulation point of view, this requires a core material that develops reasonable strength at low compressional loads and a coating material in the form of fine free-flowing granules, with good binding qualities.

Perhaps the best-known commercial presses developed for this work are the Drycota (Manesty) and the Prescoter (Killian). The Drycota, illustrated in Fig. 27 consists of two rotary presses rigidly linked by a transfer station, one producing cores and then immediately coating them. The Prescoter uses cores produced on another machine and, for this reason, requires somewhat harder cores capable of withstanding the additional handling, which may result in weaker bonding between core and coating.

Incompatible drugs may be coformulated by this method, by incorporating one drug in the core and the other in the coating formulation. The possibility of having a dual-release pattern

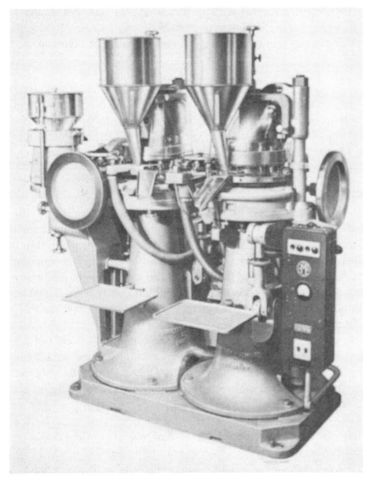

Fig. 27   Illustration of Drycota tablet press.

of the same drug (e.g., a rapidly released fraction in the coat and an extended-release component in the core) is perfectly feasible. Compression of an enteric coating around a tablet, and even a second layer of coating, have been attempted, but not widely adopted, although a machine for this purpose, the Bicota (Manesty), is available.

Layered Tablets

In a search for novelty, as much as functionality, tablets have been produced on presses capable of compressing a second (or even third) layer on top of the original material. Indeed, the standard double-rotary machines require little modification to achieve this goal. Such tableting procedures facilitate the coformulation of incompatible materials and design of complex-release patterns, as well as adding a new dimension to ease of identification. A considerable amount of expertise is needed to formulate and consistently manufacture these tablets to meet the strict regulatory requirements now demanded. Specifically designed presses to produce two- or three-layered tablets, such as the Layerpress (Thomas Engineering), are commercially available.

VI. EVALUATION OF TABLETS

Under this heading it will be convenient to divide the types of test procedures into two major categories: (a) those that are requirements in an official compendium, and (b) those that, although unofficial, are widely used in commerce. In certain circumstances, it will also be of value to consider specialized evaluative procedures that have perhaps a more academic background.

A. Official Standards

We shall largely restrict consideration to those tests that are mandatory in *The United States Pharmacopeia* (*USP*) and *The National Formulary* (*NF*), although reference to the monographs of other compendia is included where appropriate. Tests concerned with dissolution rate determinations are discussed in Chapter 20, and assay procedures are omitted as they are essentially analytical methods pertaining to a particular drug.

Uniformity of Dosage Units

The dose uniformity of tablets can be determined by two different general approaches: (a) the weight variation between a specified number of tablets, or (b) the extent of drug content uniformity. The *USP* permits the latter approach in all cases. Moreover, drug content uniformity must be measured for coated tablets as the tablet coat [which does not usually contain the active ingredient(s)] may vary significantly from tablet-to-tablet. The use of weight uniformity as a singular means of quantifying uniformity of dosage units is permitted only when the tablet is uncoated and contains 50 mg, or more, of a single active ingredient that constitutes 50%, or more, of the total tablet weight.

Most pharmacopeias include a simple weight test on a specified number of tablets that are weighed individually. The arithmetic mean weight and relative standard deviation (i.e., mean divided by standard deviation) of these tablets is then calculated. Only a specified number of test tablets may lie outside the prescribed limits. These specifications vary, depending on the type of tablet and amount of active ingredient present.

Content uniformity is a *USP* test designed to establish the homogeneity of a batch. Ten tablets are assayed individually, after which the arithmetic mean and relative standard deviation (RSD) are calculated. The *USP* criteria are met if the content uniformity lies within 85–115% of the label claim, and the RSD is not greater than 6%. Provision is included in the compendium for additional testing if one or more units fail to meet the standards.

Disintegration Testing

Determination of the time for a tablet to disintegrate when immersed in some test fluid has long been a requirement in most compendia. For many years, it was the only test available to evaluate the release of medicaments from a dosage unit. We now recognize the severe limitations of such tests in assessing this property. Hence, the introduction of dissolution rate requirements (discussed in Chapter 20).

The *USP* disintegration test is typical of most and is described in detail in a monograph of that volume. Briefly, it consists of an apparatus in which a tablet can be introduced into each of six cylindrical tubes, the lower end of which is covered by a 0.025-in.^{2} wire mesh. The tubes are then raised and lowered through a distance of 5.3–5.7 cm at a rate of 29–32 strokes per minute in a test fluid maintained at 37±2°C. Continuous agitation of the tablets is ensured by this stroking mechanism and by the presence of a specially designed plastic disk, which is free to move up and down in the tubes.

The tablets are said to have disintegrated when the particles remaining on the mesh (other than fragments of coatings) are soft and without palpable core. A maximum time for disintegration to occur is specified for each tablet, and at the end of this time, the aforementioned criteria must be met. The disintegration medium required varies, depending on the type of tablet to be tested. Apparatus meeting the official specification is available from several sources. Several modifications of the official method have been suggested in the literature, including a basket insert as an alternative to the disks [133].

The disintegration time may be markedly affected by the amount of disintegrant used, as well as the tablet process conditions. In particular, a log–linear relation between disintegration time and compressional force has been suggested by several authors [134–136]. Mufrod and Parrot have recently concluded that, although disintegration is affected by changes in compression pressure, these changes do not significantly alter the product's dissolution profile [137].

B. Unofficial Tests

Mechanical Strength

The mechanical strength of tablets is an important property of this form of drug presentation and plays a substantial role in both product development and control. It has been described by various terms, including friability [138], hardness [139], fracture resistance [140], crushing strength [141], and flexure or breaking strength [142].

Even in tablets of the simplest geometry, interpretation of this property is less straightforward than it might first appear. Some degree of anisotropy is almost certain to be present, and the ideal test conditions, employing closely defined uniform stresses, are rarely met. The mechanical strength of the tablet is primarily due to two events that occur during compression: the formation of interparticulate bonds and a reduction in porosity resulting in an increased density.

Crushing Strength. Many crushing strength testers are described in the literature [133,138–141]. Several commercial instruments are now available, and comparisons between them have been described. Among the more widely known are the Stokes (Monsanto), Strong-Cobb, Pfizer, Erweka, and Schleuniger; all hardness testers that measure the horizontal crushing strength.

In industry, mechanical strength is most often referred to as the tablet's hardness or, more precisely, its crushing strength. Brook and Marshall have described *crushing strength* as "the compressional force that, when applied diametrically to a tablet, just fractures it" [141]. In most cases, the tablet is placed on a fixed anvil, and the force is transmitted to it by a moving plunger. Many testers of this type are commercially available, including the Stokes (Monsanto),

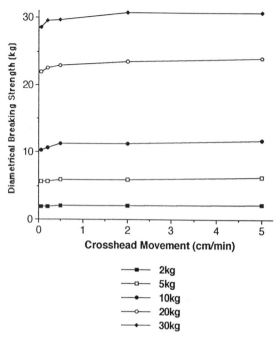

Fig. 28   Effect of loading rate on the diametrical breaking strength of tablets compressed at different levels. (From Ref. 123.)

Strong-Cobb, Pfizer, Erweka, and Schleuniger (Heberlein). Given the particular tester's design, the plunger is moved either manually or electronically. Comparisons between the different types of testers has proved that the electronic testers produce results that are much more reproducible than those obtained from the manual testers [130,141,143–145]. This is largely due to the constant rate of loading achieved with electronic testers [123] (Fig. 28).

In general, the load is applied at 90° to the longest axis (i.e., across the tablet's diameter); (Fig. 29). In such cases, the load required to break the tablet is referred to as the diametrical strength. Many testers of this type are commercially available, such that the load can also be applied across the tablet's thickness, in which case, it is referred to as flexure or breaking strength [140,142]. The tensile strength (σ) can be calculated once the load required to fracture the tablet has been determined. The precise calculation of tensile strength depends on the method used to break the tablet. When using a diametrical, or diametral, test the calculation

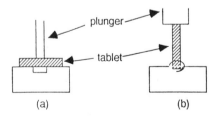

Fig. 29   Methods of evaluating tablet crushing strength: (a) bending or flexure strength; (b) diametrical compression. (From Ref. 145.)

is as follows:

$$\sigma_d = \frac{2F_d}{\pi DH}$$

where

σ_d = tensile strength
F_d = load required to fracture tablet
D = tablet diameter
H = tablet height

From a test of the tablet's flexure, tensile strength (σ_f) is calculated from the following equation:

$$\sigma_f = \frac{3F_f D'}{4DH^2}$$

where

σ_f = tensile strength
F_f = load required to fracture tablet
D = tablet diameter
D' = distance from fulcrum to fulcrum
H = tablet height

In their evaluation of flexure and diametral testing, David and Augsburger found that the tensile strength was the same, regardless of which method was used, as long as the appropriate calculation was employed [142].

Two types of inherent error may be present: (a) that associated with an incorrect zero, and (b) a scale that does not accurately indicate the actual load being applied. In some models, there is the additional problem of a variable rate of loading. Moreover, when using unpadded flat anvils, the failure may involve some compression. Thus, it is essential to realize what is being measured in a particular instrument.

The Stokes (or Monsanto) and Pfizer testers apply force through a coil spring that, after long periods of use, shows signs of fatigue and may also show some loss of load owing to frictional effects [141,145]. The rate of loading is not controlled in these types, nor in the Strong-Cobb tester, which applies force through hydraulic pressure. The scale of this machine registers air pressure, and comparison of results with other instruments, therefore, is possible only if these readings are converted to compressional load.

The Erweka and Schleuniger (or Heberlein) testers operate on a counterweight principle that eliminates fatigue, if not frictional, losses. The latter of these two devices is supplied calibrated and a "mechanical tablet" is available for periodic recalibration. More recently, testers have been introduced that measure the load being applied to the tablet by means of load cells and, therefore, facilitate direct electronic digital readout. This eliminates the two major sources of error referred to earlier and permits recording of production of hard copies of the test results.

Comparative reports in the literature confirm the necessity of employing some calibration procedure if results from testers are to be compared or reduced to actual units of force. This is particularly true when the information is being used to determine relationships between crushing strength and other tablet properties.

It can be argued that the crushing strength of a tablet is more closely related to the compressional process, and the results may not give the best indication of how the tablet will behave during handling. If this type of information is required, then the groups of instruments in the following paragraphs are more relevant.

One might anticipate that the crushing strength F of a tablet is a function of the pressure P employed during its compaction [134]. For example, the following relationship may hold true:

$$F = k \log P + k_1$$

where k and k_1 are constants.

However, deviations from this logarithmic relationship at compressional pressure values, above 150 MPa has been reported [81,135]. In addition, the crushing strength has been related to certain physical properties of the compact. For example:

$$F = kE^{-1} + k_1$$

where E, the porosity, has values between 5 and 20%. The most obvious use of crushing strength measurements has been to give indications of possible disintegration time (t_D), that is,

$$F = kt_D + k_1$$

Abrasion. Although the crushing strength of a tablet gives some indication of its mechanical robustness, it does not truly measure the ability of the tablet to withstand the handling it will encounter during processing and shipping. Test designed to assess the resistance of the surface regions to abrasion or other forms of general wear and tear may be more appropriate for this.

Many tests to assess abrasion are quoted in the literature [138,146,147]. Most measure the weight loss on subjecting tablets to a standard level of agitation for a specified time. The choice of agitation should be based on knowledge of the likely level during use or manufacture.

More specifically a certain weight of tablets W_0 is subjected to a well-defined level of agitation in a fixed-geometry, closed container for a specific time. They are then again re-weighed, W. The measure of abrasion resistance or friability B is usually expressed as a percentage loss in weight:

$$B = 100 \left[1 - \frac{W}{W_0} \right]$$

It might be advantageous to relate friability to unit time or to number of falls.

The Roche Friabilator is one of the most common methods used to test for resistance to abrasion [147]. Here, a minimum of 6 g (often 20 tablets) of dedusted and weighed tablets are placed in a 12-in.–high drum, which is then rotated for 100 revolutions. A shaped arm lifts the tablets and drops them half the height of the drum during each revolution. At the end of this operation, the tablets are removed, dedusted and reweighed. Should any tablet break up, the test is rejected. Values of B from 0.8 to 1.0% are frequently quoted as the upper level of acceptability for pharmaceutical products [147].

Indentation hardness, using modified tests based on Brinell hardness measurements, have been used by some researchers [148] to provide information on the surface hardness of tablets. In addition, these tests are capable of providing a measure of a tablet's plasticity or elasticity. For the most part, such tests have been confined to basic research applications in a few laboratories, but their value is beginning to be more widely recognized.

Porosity

The bioavailability of drugs from tablets can be markedly influenced by the rate and efficiency of the initial disintegration and dissolution process. Unfortunately, we are faced with a compromise situation: we want a strong structure before administration, but one that readily breaks down when placed in the in vivo environment. One of the major factors affecting both these properties is the structure of the tablet, in particular, its density (or porosity) and the pore structure. Study of the significance of such measurements and interpretation of the results is a relatively recent field of interest.

Determination of the porosity of a tablet presents the classic problem of defining the appropriate volume that we wish to measure. The displacement medium may be able to penetrate the most minute crevices, such as helium. Whereas other displacement media, such as mercury, are unable to enter the smallest tablet crevices and, thus, produce different porosity values. Standardization of displacement media, therefore, is necessary for comparative evaluations.

Pore Structure and Size. The relationship between applied pressure P and the diameter of the smallest circular pore penetrated d by a liquid gas is given by the equation:

$$d = \frac{4\pi\gamma \cos\beta}{P}$$

where

γ = surface tension of the liquid
β = contact angle solid and liquid

Originally, the method of porosimetry was of interest only to those involved with the high-pressure techniques associated with pore analysis. However, with the increasing availability of sophisticated porosimeters, the technique of porosimetry is being used on a frequent basis to investigate tablet structure. High-pressure mercury intrusion porosimeters are capable of assessing a wide range of pore radii. A typical example of the application of such an instrument to evaluate wet and dry techniques of precompression treatment is reported by Ganderton and Selkirk [149]. These authors found that lactose granulations resulted in a wider pore size distribution than did ungranulated lactose.

This technique has also been used in combination with nitrogen absorption to study the pore structure of some excipients, particularly microcrystalline cellulose in both the powdered and compacted state. The intraparticulate porosity of MCC is unaffected by tableting; however, the interparticular pores are gradually reduced in size [38]. Recently, this method has been used to evaluate the internal structure of tablets prepared from microcapsules [150].

Liquid Penetration. The rate at which selected liquids penetrate into tablets can be used to study their pore structure. A knowledge of the rate of liquid penetration should also provide information on the disintegration–dissolution behavior of a tablet on administration. Such investigations are capable of forming a valuable link between physicomechanical characteristics and in vivo performance.

Evaluation of Bioadhesive Tablets

With the advent of increasingly sophisticated tableted delivery systems comes the task of assessing these systems. Bioadhesive tablets, in particular, present an interesting problem to the formulator. Although such tablets are not currently marketed in the United States, they are being evaluated in many laboratories as an alternative means for providing sustained-release of drug. The sustained-release characteristics of bioadhesive tablets is afforded through their ability to adhere to the intestinal mucosa. Thus, an estimation of their adhesiveness is a key factor in their in vitro evaluation.

Ishida et al. were some of the first investigators to propose a method for investigating the adhesive properties of tablets [151]. Their method involved placing a tablet onto a membrane under constant pressure for 1 min. and then measuring the force required to remove it. Most methods published since that time involve essentially the same principle, with variations in the type of membrane used and the manner in which the adhesive force is measured [152,153]. An excellent review of these methods has been published by Duchene et al. for those interested in the precise details of such tests [154].

Jimenez-Castellanos et al. recently developed a method to measure both the adhesional and frictional forces involved in the attachment of such tablet's to mucosa. These researchers found that a good correlation existed between the maximal adhesion strength and polymer content of the tablets tested [155].

VII. RECENT NOVEL-TABLETING PROCESSES

A. Injection Molding of Tablets

An injection molding technology was developed by Keith and his associates at Zetachron. This system uses primarily polyethylene glycol as a matrix. A mixture of PEG and silicones are melted and then molded into tablets designed to physically melt at physiological temperatures. These tablets soften easily at relatively low ambient temperatures, but show remarkable performance in vivo.

B. WEB System

A system was developed at Roche Laboratories whereby a sheet (or "web") was coated with a drug–binder mixture. The solid dosage units were then punched from the web. This system was very flexible and amenable to immediate-release and sustained-release technologies. However, because of the impracticality of the system, it was scrapped in the mid-1980s and is of only historical significance.

VIII. FUTURE DEVELOPMENTS

Tablets have been a viable dosage form well before William Brockendon's patent for a tablet machine in 1843. His invention only made them easier to produce. As tablet presses and production-monitoring systems have developed, these dosage forms are the most economical of any ever developed. It will be hard to improve on their efficiency, but several attempts, such as the WEB delivery system from Roche and the injection-molding technology from Zetachron have been made recently.

Newer technologies may become available, but it is unlikely that any new tableting technology will render the old technology obsolete anytime soon. Processing technologies such as high-shear mixing and microwave drying seem to be having the most influence on processing times and efficiency. Coating technologies are becoming more controllable, but offer the most hope for efficiency improvements.

Another new development has been the application of oral absorption promoters. These materials are designed to enhance the oral bioavailability of many compounds and improve variable absorption. However, many of these compounds are hydrophobic and cause difficulty during tableting itself. The challenge for formulators is to arrive at clever solutions around the process problems, while retaining material performance.

The ultimate challenge for tablet formulators in the 21st century is to achieve a true understanding of material properties and material science. Those people that can quickly conceive a compatible, functional formulation will be irreplaceable as large companies shrink their research and development resources and the public sector demands better efficiency.

ACKNOWLEDGMENTS

The authors would like to acknowledge the leadership of the editors, as well as the assistance of Ms. Kimberly J. Ruhling and Ms. Catherine A. Grudzinski in the preparation of this manuscript.

REFERENCES

1. A. T. Florence, E. G. Salole, and J. B. Stenlake, J. Pharm Pharmacol., 26, 479 (1974).
2. P. Timmins, I. Browning, A. M. Delargy, J. W. Forrester, and H. Sen, Drug Dev. Ind. Pharm., 12, 1293 (1986).
3. N. Kaneniwa, K. Imagawa, and J. I. Ichikawa, Chem. Pharm. Bull., 36, 2531 (1988).
4. B. M. Hunter, J. Pharm. Pharmacol., 26, 58P (1974).
5. S. Vesslers, R. Boistelle, A. Delacourte, J. C. Guyot, and A. M. Guyot-Hermann, Drug Dev. Ind. Pharm., 18, 539 (1992).
6. R. F. Shangraw, J. W. Wallace, and F. M. Bowers, Pharm. Technol., 11, 136 (1987).
7. J. W. McGinity, C. T. Ku, R. Bodmeier, and M. R. Harris, Drug Dev. Ind. Pharm., 11, 891 (1985).
8. A. J. Shukla and J. C. Price, Drug Dev. Ind. Pharm., 17, 2067 (1991).
9. H. Ando, M. Ishii, M. Kayano, and H. Ozawa, Drug Dev. Ind. Pharm., 18, 453 (1992).
10. P. W. S. Heng and J. N. Staniforth, J. Pharm. Pharmacol., 40, 360 (1988).
11. J. C. Callahan, G. W. Cleary, M. Elefant, G. Kaplan, T. Tensler, and R. A. Nash, Drug Dev. Ind. Pharm., 8, 355 (1982).
12. *Handbook of Pharmaceutical Excipients.*
13. C. M. Anderson, Aust. J. Pharm., 47, S44 (1966).
14. J. W. Poole and C. K. Bahal, J. Pharm. Sci., 57, 1945 (1968).
15. O. Weis-Fogh and T. Dansk, Farm. Suppl., 11, 276 (1956).
16. H. K. Chan and E. Doelker, Drug Dev. Ind. Pharm., 11, 315 (1985).
17. S. Kopp, C. Beyer, E. Graf, F. Kubel, and E. Doelker, J. Pharm. Pharmacol., 41, 79 (1989).
18. M. Otsuka, T. Matsumoto, and N. Kaneniwa, J. Pharm. Pharmacol., 41, 667 (1989).
19. T. Matsumoto, N. Kaneniwa, S. Higuchi, and M. Otsuka, J. Pharm. Pharmacol., 43, 74 (1991).
20. C. R. Lerk and H. Vromans, Acta Pharm. Suec., 24, 60 (1987).
21. P. York, Drug Dev. Ind. Pharm., 18, 677 (1992).
22. W. I. Higuchi, P. D. Bernardo, and S. C. Mehta, J. Pharm. Sci., 56, 200 (1967).
23. E. N. Hiestand, J. Bane, and E. Strzelinski, J. Pharm. Sci., 60, 758 (1971).
24. E. N. Hiestand and C. B. Peot, J. Pharm. Sci., 63, 605 (1974).
25. E. N. Hiestand, J. E. Wells, C. B. Peot, and J. F. Ochs, J. Pharm. Sci., 66, 510 (1977).
26. E. N. Hiestand and D. Smith, Powder Technol., 38, 145 (1984).
27. E. N. Hiestand, J. Pharm. Sci., 74, 768 (1985).
28. E. N. Hiestand, Pharm. Technol., 10, 52 (1986).
29. R. O. Williams and J. W. McGinnity, Drug Dev. Ind. Pharm., 14, 1823 (1988).
30. A. J. Romero, G. Lukas, and C. T. Rhodes, Pharm. Acta Helv., 66, 34 (1991).
31. D. Q. M. Craig, C. T. Davies, J. C. Boyd, and L. B. Hakess, J. Pharm. Pharmacol., 43, 444 (1991).
32. H. G. Brittain, Drug Dev. Ind. Pharm., 15, 2083 (1989).
33. D. S. Phadke and J. L. Eichorst, Drug Dev. Ind. Pharm., 17, 901 (1991).
34. H. Bundgaard, J. Pharm. Pharmacol., 26, 535 (1974).
35. Y. K. E. Ibrahim and P. R. Olurinaola, Pharm. Acta Helv., 66, 298 (1991).
36. H. L. Avallone, Pharm. Technol., 16, 48 (1992).
37. D. C. Monkhouse and A. Maderich, Drug Dev. Ind. Pharm., 15, 2115 (1989).
38. H. Nyquist, Drug Dev. Ind. Pharm., 12, 953 (1986).
39. Z. T. Chowhan and L.-H. Chi, J. Pharm. Sci., 75, 534 (1986).
40. G. K. Bolhuis and C. F. Lerk, Pharm. Weekbl., 108, 469 (1973).
41. J. W. Wallace, J. T. Capozzi, and R. F. Shangraw, Pharm. Technol., 7, 95 (1983).
42. J. T. Carstensen and C. Ertell, Drug Dev. Ind. Pharm., 16, 1121 (1990).

43. D. H. Shah and A. S. Arambulo, Drug Dev. Ind. Pharm., 1, 495 (1974–1975).
44. K. A. Khan and C. T. Rhodes, Chemist Druggist, 159, 158 (1973).
45. N. Shah, R. Pytelewski, H. Eisen, and C. I. Jarowski, J. Pharm. Sci., 63, 339 (1974).
46. G. D. D'Alonzo and R. E. O'Connor, Drug Dev. Ind. Pharm., 16, 1931 (1990).
47. O. Shirakura, M. Yamada, M. Hashimoto, S. Ishimaru, K. Takayama, and T. Nagai, Drug Dev. Ind. Pharm., 18, 1099 (1992).
48. H. Seager, P. J. Rue, I. Burt, J. Ryder, and J. K. Warrack, Int. J. Pharm. Technol. Prod. Manuf., 2(2), 41 (1981).
49. R. M. Iyer, L. L. Augsburger, and D. M. Parikh, Drug Dev. Ind. Pharm., 19, 981 (1993).
50. Z. Chowhan, Pharm. Technol., 12, 26 (1988).
51. P. York, Int. J. Pharm., 14, 1 (1983).
52. C. F. Lerk, M. Lagas, J. P. Boelstra, and P. Broersma, J. Pharm. Sci., 66, 1480 (1977).
53. C. F. Lerk, M. Lagas, J. T. Fell, and P. Nauta. J. Pharm. Sci., 67, 935 (1978).
54. E. Shotton and G. S. Leonard, J. Pharm. Pharmacol., 24, 798 (1972).
55. M. H. Rubinstein and D. M. Bodey, J. Pharm. Pharmacol., 26, 104P (1974).
56. R. F. Shangraw, Manuf. Chem., 57, 22 (1986).
57. H. Burlinson and C. Pickering, J. Pharm. Pharmacol., 2, 630 (1950).
58. L. C. Curlin, J. Am. Pharm. Assoc. Sci. Ed., 44, 16 (1955).
59. N. R. Patel and R. E. Hopponen, J. Pharm. Sci., 55, 1065 (1966).
60. K. S. Manudhane et al., J. Pharm. Sci., 58, 616 (1969).
61. H. Hess, Pharm. Technol., 11, 54 (1987).
62. J. B. Schwartz and J. A. Zelinskie, Drug Dev. Ind. Pharm., 4, 463 (1978).
63. A. M. Guyot-Hermann and J. Ringard, Drug Dev. Ind. Pharm., 7, 155 (1981).
64. W. Lowenthal and J. H. Wood, J. Pharm. Sci., 62, 287 (1973).
65. K. A. Khan and C. T. Rhodes, J. Pharm. Sci., 64, 166 (1975).
66. K. A. Khan and C. T. Rhodes, J. Pharm. Pharmacol., 23, 261A (1971).
67. J. F. Bavitz, N. R. Bohidar, and F. A. Restaino, Drug Dev. Commun., 1, 331 (1974–1975).
68. W. Lowenthal, J. Pharm. Sci., 61, 1695 (1972).
69. J. L. Kanig and E. M. Rudnic, Pharm. Technol., 8, 50 (1984).
70. A. Mitrevej and R. G. Hollenbeck, Pharm. Technol., 6, 48 (1982).
71. H. Nogami, T. Nagai, E. Fukuoka, and T. Sonobe, Chem. Pharm. Bull., 17, 1450 (1969).
72. D. Gissinger and A. Stamm, Drug Dev. Ind. Pharm., 6, 511 (1980).
73. E. M. Rudnic, Doctoral dissertation, University of Rhode Island, 1982.
74. E. M. Rudnic, C. T. Rhodes, S. Welch, and P. Bernardo, Drug Dev. Ind. Pharm., 8, 87 (1982).
75. C. Fuhrer, Informationsdienst APV, 20, 58 (1964).
76. H. Matsumaru, Yakugaku Zashi, 79, 63 (1959).
77. P. H. List and V. A. Muazzam, Drugs Made Ger., 22, 161 (1979).
78. A. J. Smallenbroek, G. K. Bolhuis, and C. F. Lerk, Pharm. Weekbl., Sci. Ed., 3, 172 (1981).
79. E. M. Rudnic, J. M. Lausier, R. N. Chilamkurti, and C. T. Rhodes, Drug Dev. Ind. Pharm., 6, 291 (1980).
80. E. M. Rudnic, J. L. Kanig, and C. T. Rhodes, J. Pharm. Sci., 74, 647 (1985).
81. E. Shotton and C. Lewis, J. Pharm. Pharmacol., 16, 111T (1964).
82. H. C. Caldwell and W. J. Westlake, J. Pharm. Sci., 61, 984 (1972).
83. Y. Matsuda, Y. Minamida, and S.-I. Hayashi, J. Pharm. Sci., 65, 1155 (1976).
84. J. Kikuta and N. Kitamori, Drug Dev. Ind. Pharm., 11, 845 (1985).
85. S. Dawoodbhai and C. T. Rhodes, Drug Dev. Ind. Pharm., 16, 2409 (1990).
86. P. W. S. Heng, L. S. C. Wan, and T. S. H. Ang, Drug Dev. Ind. Pharm., 16, 951 (1990).
87. L. S. C. Wan and P. W. S. Heng, Pharm. Acta Helv., 62, 169 (1987).
88. L. S. C. Wan and P. W. S. Heng, Pharm. Acta Helv., 61, 157 (1986).
89. H. Schott, L. C. Kwan, and S. Feldman, J. Pharm. Sci., 71, 1038 (1982).
90. M. L. Wells and E. L. Parrott, Drug Dev. Ind. Pharm., 18, 175 (1992).
91. M. L. Wells and E. L. Parrott, Drug Dev. Ind. Pharm., 18, 265 (1992).
92. M. L. Wells and E. L. Parrott, J. Pharm. Sci., 81, 453 (1992).

93. J. T. Carstensen, *Theory of Pharmaceutical Systems*, Vol. 2, Academic Press, New York, 1973.
94. J. A. Hersey, J. Pharm. Pharmacol., 63, 1685 (1967).
95. T. Higuchi, E. Nelson, and L. W. Busse, J. Am. Pharm. Assoc. Sci. Ed., 43, 344 (1954).
96. J. Polderman and C. J. DeBlaey, Farm. Aikak., 80, 111 (1971).
97. C. J. DeBlaey and J. Polderman, Pharm. Weekbl., 105, 241 (1970).
98. C. J. DeBlaey, A. B. Weekers-Anderson, and J. Polderman, Pharm. Weekbl., 106, 893 (1971).
99. R. F. Lammens, J. Polderman, and C. J. DeBlaey, Int. J. Pharm. Technol., Prod. Manuf., 1, 26 (1979).
100. J. Cobby, M. Mayersohn, and G. C. Walker, J. Pharm. Sci., 63:725,732 (1974).
101. B. Mechtersheimer and H. Zucker, Pharm. Technol., 38 (1986).
102. T. Higuchi, L. N. Elow, and L. W. Busse, J. Am. Pharm. Assoc. Sci. Ed., 43, 688 (1954).
103. R. J. Roberts and R. C. Rowe, J. Pharm. Pharmacol., 38, 567 (1986).
104. D. E. Fonner, G. S. Banker, and J. Swarbrick, J. Pharm. Sci., 55, 181 (1966).
105. P. J. Jarosz and E. L. Parrott, J. Pharm. Sci., 72, 530 (1983).
106. I. A. Chaudry and R. E. King, J. Pharm. Sci., 61, 1121 (1972).
107. O. Worts and T. Schoefer, Arch. Pharm. Chem. Sci. Ed., 6, 1 (1978).
108. H. J. Malinowski, Ph.D. dissertation, Philadelphia College of Pharmacy, 1973.
109. J. J. Williams and D. M. Stiel, Pharm. Technol., 8, 26 (1984).
110. J. B. Schwartz, Pharm. Technol., 5, 102 (1981).
111. K. Marshall, Pharm. Technol., 7, 63 (1983).
112. R. W. Heckel, Trans. Met. Soc. AIME, 221, 671, 1001 (1961).
113. J. A. Hersey and J. E. Rees, Nature, 230, 96 (1971).
114. L. F. Athy, Bull. Am. Assoc. Petrol. Geol., 14, 1 (1930).
115. R. J. Rue and J. E. Rees, J. Pharm. Pharmacol., 30, 642 (1978).
116. M. Celik and K. Marshall, Drug Dev. Ind. Pharm., 15, 759 (1989).
117. D. N. Travers and M. Cox, Drug Dev. Ind. Pharm., 4, 157 (1978).
118. S. Leigh, J. E. Carless, and B. W. Burt, J. Pharm. Sci., 56, 888 (1967).
119. E. Shotton and B. A. Obiorah, J. Pharm. Pharmacol., 25, 37P (1973).
120. J. T. Carstensen, J.-P. Marty, F. Puisieux, and H. Fessi, J. Pharm. Sci., 70, 222 (1981).
121. E. G. Rippie and D. W. Danielson, J. Pharm. Sci., 7, 476 (1981).
122. S. T. David and L. L. Augsburger, J. Pharm. Sci., 66, 155 (1977).
123. J. E. Rees, J. A. Hersey, and E. T. Cole, J. Pharm. Pharmacol., 22, 64S (1970).
124. I. Krycer and D. G. Pope, Drug Dev. Ind. Pharm., 8, 307 (1982).
125. I. Krycer, D. G. Pope, and J. A. Hersey, Int. J. Pharm., 12, 113 (1982).
126. P. Ridgeway-Watt, *Tablet Machine Instrumentation in Pharmaceutics*, John Wiley & Sons, New York, 1988.
127. H. Seager, P. J. Rue, I. Burt, J. Ryder, J. K. Warrack, and M. Gamlen, Int. J. Pharm. Technol. Prod. Manuf., 6, 1 (1985).
128. S. M. Blaug and M. R. Gross, Drug Standards, 27, 100 (1959).
129. G. S. Banker, J. Pharm. Sci., 55, 81 (1966).
130. S. C. Porter, Drug Cosmet. Ind., 130, 46 (May); 131, 44 (June) (1981).
131. L. K. Mathur and S. J. Forbes, Pharm. Technol., 42 (1984).
132. R. C. Whitehouse, Pharm. J., 172, 85 (1954).
133. L. L. Kaplan and J. A. Kish, J. Pharm. Sci., 51, 708 (1962).
134. T. Higuchi, A. N. Rao, L. W. Busse, and J. V. Swintosky, J. Am. Pharm. Assoc. Sci. Ed., 42, 194 (1953).
135. T. Higuchi, L. N. Elowe, and L. W. Busse, J. Am. Pharm. Assoc. Sci. Ed., 43, 685 (1954).
136. K. C. Kwan, F. O. Swart, and A. M. Mattocks, J. Am. Pharm. Assoc., 46, 236 (1957).
137. Mufrod and E. L. Parrot, Drug Dev. Ind. Pharm., 16, 1081 (1990).
138. A. Nutter-Smith, Pharm. J., 163, 194 (1949).
139. A. McCallum, J. Buchter, and R. Albrecht, J. Am. Pharm. Assoc. Sci. Ed., 44, 83 (1955).
140. C. J. Endicott, W. Lowenthal, and H. M. Gross, J. Pharm. Sci., 50, 343 (1961).
141. D. B. Brook and K. Marshall, J. Pharm. Sci., 57, 481 (1968).

142. S. T. David and L. L. Augsburger, J. Pharm. Sci., 63, 933 (1974).
143. H. J. Fairchild and F. Michel, J. Pharm. Sci., 50, 966 (1961).
144. J. F. Bavitz, N. R. Bohidar, J. I. Karr, and F. A. Restaino, J. Pharm. Sci., 62, 1520 (1973).
145. F. W. Goodhardt, J. R. Draper, D. Dancz, and F. C. Ninger, J. Pharm. Sci., 62, 297 (1973).
146. H. Burlinson and C. Pickering, J. Pharm. Pharmacol., 2, 630 (1950).
147. E. G. E. Shafer, E. G. Wollish, and C. E. Engel, J. Am. Pharm. Assoc. Sci. Ed., 45, 114 (1956).
148. K. Ridgway, M. E. Aulton, and P. H. Rosser, J. Pharm. Pharmacol., 22, 70S (1970).
149. D. Ganderton and A. B. Selkirk, J. Pharm. Pharmacol., 22, 345 (1970).
150. H. Yuasa, Y. Kanaya, and K. Omata, Chem. Pharm. Bull., 38, 752 (1990).
151. M. Ishida, Y. Machida, N. Nambu, and T. Nagai, Chem. Pharm. Bull., 29, 810 (1981).
152. G. Ponchel, F. Touchard, D. Duchene, and N. A. Peppas, J. Controlled Release, 5, 129 (1987).
153. V. S. Chitnis, V. S. Malshe, and J. K. Lalla, Drug Dev. Ind. Pharm., 17, 879 (1991).
154. D. Duchene, F. Touchard, and N. A. Peppas, Drug Dev. Ind. Pharm., 14, 283 (1988).
155. M. R. Jimenez-Castellanos, H. Zia, and C. T. Rhodes, Int. J. Pharm., 89, 223 (1993).

11

Hard and Soft Shell Capsules

Larry L. Augsburger
University of Maryland School of Pharmacy, Baltimore, Maryland

I. HISTORICAL DEVELOPMENT AND ROLE AS A DOSAGE FORM

Capsules are solid dosage forms in which the drug substance is enclosed within either a hard or soft soluble shell, usually formed from gelatin. The capsule may be considered a ''container'' drug delivery system that provides a tasteless and odorless dosage form without need for a secondary coating step, as may be required for tablets. Swallowing is easy for most patients, since the shell is smooth and hydrates in the mouth, and the capsule often tends to float on swallowing in the liquid taken with it. Their availability in a wide variety of colors makes capsules aesthetically pleasing. There are numerous additional advantages to capsules as a dosage form, depending on the type of capsule employed.

Capsules may be classified as either *hard* or *soft*, depending on the nature of the shell. Soft gelatin capsules (sometimes referred to as ''softgels'') are made from a more flexible, plasticized gelatin film than hard gelatin capsules. Most capsules of either type are intended to be swallowed whole; however, some soft gelatin capsules are intended for rectal or vaginal insertion as suppositories. Most capsule products manufactured today are of the hard gelatin type. One survey [1] has estimated that the utilization of hard gelatin capsules to prepare solid dosage forms exceeds that of soft gelatin capsules by about tenfold.

The first capsule prepared from gelatin was a one-piece capsule that was patented in France by Mothes and DuBlanc in 1834 [2]. Although the shells of these early capsules were not plasticized, such capsules likely would be classified today as ''soft gelatin capsules'' on the basis of shape, contents and other features. Intended to mask the taste of certain unpleasant-tasting medication, they quickly gained popularity, primarily as a means for administering copaiba balsam, a drug popular at the time in the management of venereal disease [2]. These capsules were made one at a time by hand by dipping leather molds into a molten gelatin mixture, filled with a pipette and sealed with a drop of molten gelatin [3]. Today, soft gelatin capsules are commonly prepared from plasticized gelatin by a rotary die process in which they are formed, filled, and sealed in a single operation. With few exceptions, soft gelatin capsules are filled with solutions or suspensions of drugs in liquids that will not solubilize the gelatin shell. They are a completely sealed dosage form: the capsule cannot be opened without destroying the capsule. Because liquid contents can be metered with high-quality pumps, soft gelatin capsules are the most accurate and precise of all solid oral dosage forms. Depending on the machine tooling, a wide variety of sizes and shapes are possible. Possible shapes include spherical, oval, oblong, tube, and suppository-type; size may range from 1 to 480 minims (16.2 minims = 1 ml) [3].

Although the patent holders at first sold both filled and empty soft gelatin capsules, the sale of empty shells was discontinued after 1837 [2]. However, the demand that had been created for the empty capsules led to several attempts to overcome the patents which, in turn, resulted in the development of both the gelatin-coated pill and the hard gelatin capsule [2]. The first hard gelatin capsule was invented by J. C. Lehuby, to whom a French patent was granted in 1846 [2]. It resembled the modern hard gelatin capsule in that it consisted of two, telescoping, cap and body pieces. In Lehuby's patent, the capsule shells were made of starch or tapioca sweetened with syrup, although later additions to the patent claimed carragheen (1847) and mixtures of carragheen with gelatin (1850) [2]. The first person to describe a two-piece gelatin capsule was James Murdock, who was granted a British patent in 1848, and who is often credited as the inventor of the modern hard gelatin capsule. Since Murdock was a patent agent by profession, it has been suggested that he was actually working on behalf of Lehuby [2].

Unlike soft gelatin capsules, hard gelatin capsules are manufactured in one operation and filled in a completely separate operation. Originally, they were made by hand-dipping greased metal pinlike molds into a molten gelatin mixture, drying the resultant films, stripping them from the pins, and joining the same two pieces together [2]. Today, they are manufactured in a similar manner by means of a completely automated process. For human use, hard gelatin capsules are supplied in at least eight sizes, ranging in volumetric capacity from 0.13 to 1.37 ml. Typically, they are oblong; however, some manufacturers have made modest alterations in that shape to be distinctive.

In further contrast with soft gelatin capsules, hard gelatin capsules typically are filled with powders, granules, or pellets. Modified-release granules or pellets may be filled without crushing or compaction, thereby avoiding disruption of barrier coats or other possible adverse effects on the release mechanism. Although many manufacturers of hard capsule-filling equipment also have developed modifications to their machines that would permit the filling of liquids or semisolid matrices, there currently are few commercial examples.

Filled hard gelatin capsules are held together by interlocking bumps and grooves molded into the cap and body pieces, and the capsules are usually additionally sealed by a banding process that places a narrow strip of gelatin around the midsection of the capsule where the two pieces are joined.

Recently, hard shell capsules made from starch have become available (Capill; Capsugel, Div. Warner Lambert Co.). These consist of two, fitted cap and body pieces that are made by injection molding the glassy mass formed when starch containing 13–14% water is heated, and then dried [4]. Temperatures in the range of 140–190°C reportedly produce masses that flow satisfactorily without degradation [4]. The two parts are formed in separate molds. Unlike hard gelatin capsules that are supplied with the caps and bodies prejoined, the two parts are supplied separately. The caps and bodies do not interlock and must be sealed together at the time of filling to prevent inadvertent separation.

Additional advantages and attributes of hard and soft shell capsules are discussed in the following sections.

II. HARD GELATIN CAPSULES

A. Advantages

Hard gelatin capsules often have been assumed to have better bioavailability than tablets. Most likely, this assumption derives from the fact the gelatin shell rapidly dissolves and ruptures, which affords at least the potential for rapid release of the drug, together with the lack of utilization of a compaction process comparable with tablet compression in filling the capsules. However, capsules can be just as easily malformulated as tablets. A number of reports of bioavailability problems with capsules have been reported [5–8].

Hard shell capsules allow a degree of flexibility of formulation not obtainable with tablets: often they are easier to formulate because there is no requirement that the powders be formed into a coherent compact form that will stand up to handling. However, the problems of powder blending and homogeneity, powder fluidity, and lubrication in hard capsule filling are similar to those encountered in tablet manufacture. It is still necessary to measure out an accurate and precise volume of powder or pellets, and the ability of such dry solids to uniformly fill into a cavity (often comparable with a tablet die) is the determining factor in weight variation and, to a degree, content uniformity.

Modern filling equipment makes possible the multiple filling of diverse systems (e.g., beads or granules, tablets, powders, semisolids) in the same capsule, which offers many possibilities in dosage form design to overcome incompatibilities by separating ingredients within the same capsule, or to create modified or controlled drug delivery. Indeed, capsules are ideally suited to the dispensing of granular or bead-type modified-release products, since they may be filled without a compression process that could rupture the particles or otherwise compromise the integrity of any controlled-release coatings.

Hard gelatin capsules are uniquely suitable for blinded clinical tests and are widely used in preliminary drug studies. Bioequivalence studies of tablet formulations may be conveniently ''blinded'' by inserting the tablets into opaque capsules, often along with an inert filler powder. Even capsule products may be disguised by inserting them into larger capsules.

B. Disadvantages

From a manufacturers point of view, there perhaps is some disadvantage in the fact that the number of suppliers of shells is limited. Moreover, filling equipment is slower than tableting, although that gap has narrowed in recent years with the advent of high-speed automatic-filling machines. Generally, hard gelatin capsule products tend to be more costly to produce than tablets; however, the relative cost effectiveness of capsules and tablets must be judged on a case-by-case basis. This cost disadvantage diminishes as the cost of the active ingredient increases or when tablets must be coated [9]. Furthermore, it may be possible to avoid the cost of a granulation step by choosing encapsulation in lieu of tableting.

Highly soluble salts (e.g., iodides, bromides, or chlorides) generally should not be dispensed in hard gelatin capsules. Their rapid release may cause gastric irritation owing to the formation of a high drug concentration in localized areas. A somewhat related concern is that both hard gelatin capsules and tablets may become lodged in the esophagus, where the resulting localized high concentration of certain drugs (doxycycline, potassium chloride, indomethacin, and others) may cause damage [10]. Marvola [10] measured the force required to detach various dosage forms from an isolated pig esophagus mounted in an organ bath and found that capsules tended to adhere more strongly than tablets. However, the detachment forces were greatly reduced for both after a water rinse (to simulate drinking) or when there was a slow continuous flow of artificial saliva. In an in vivo study, Hey et al. [11] studied the esophageal transit of barium sulfate tablets and capsules radiologically in 121 healthy volunteers. The subjects' position (standing or lying down) and the volume of water taken (25 or 100 ml) during swallowing were considered. The majority (60%) of the volunteers had some difficulty in swallowing one or more of the preparations: many preparations adhered to the esophagus and began to disintegrate in the lower part of the esophagus. Delayed transit time occurred more frequently with large, round tablets than with small tablets or capsules. In contrast with tablets, patient position or the volume of water taken had less influence on the passage of capsules. Despite their findings, Hey et al. preferred not to use capsules because of their potential for esophageal adhesion. In general, it was recommended that patients should remain standing 90 sec or more after taking tablets or capsules, and that they should be swallowed with at least 100 ml water. In a study considering only the esophageal transit of barium sulfate-filled hard gelatin capsules, Channer and Virjee [12] found that 26 of 50 patients exhibited sticking; however, only 3 of these patients were aware that a capsule had lodged in their esophagus. These investigators also concluded that drugs should be taken with a drink while standing. Evans and Roberts [13] compared barium sulfate tablets and capsules and found a greater tendency for esophageal retention with tablets than with capsules. Fell [14] has pointed to the large difference in density between barium sulfate and typical pharmaceutical preparations as a complicating factor in drawing conclusions about any differences in esophageal retention between tablets and capsules.

C. The Manufacture of Hard Gelatin Capsules [15–18]

Manufacturers

Empty hard gelatin capsules are manufactured on Colton machines, which were invented about 50 years ago. It has been estimated that there are about 340 such machines worldwide [15]. There are three producers of hard shell capsules in North America (Shionogi Qualicaps, Indianapolis, IN; Capsugel Div. Warner-Lambert Co., Greenwood, SC; and Pharmaphil Corp, Windsor, Ontario).

Shell Composition

Hard gelatin shells are manufactured by a process in which stainless steel mold pins are dipped into warm gelatin solutions and the shells are formed by gelatin on the pin surfaces. Gelatin is the most important constituent of the dipping solutions, but other components may be present.

   Gelatin.   Gelatin is prepared by the hydrolysis of collagen obtained from animal connective tissue, bone, skin, and sinew. This long polypeptide chain yields, on hydrolysis, 18 amino acids, the most prevalent of which are glycine and alanine. Gelatin can vary in its chemical and physical properties, depending on the source of the collagen and the manner of extraction. There are two basic types of gelatin. Type A, which is produced by an acid hydrolysis, is manufactured mainly from pork skin. Type B gelatin, produced by alkaline hydrolysis, is manufactured mainly from animal bones. The two types can be differentiated by their isoelectric points (4.8–5.0 for type B and 7.0–9.0 for type A) and by their viscosity-building and film-forming characteristics.

   Either type of gelatin may be used, but combinations of pork skin and bone gelatin are often used to optimize shell characteristics [15,16]. Bone gelatin contributes firmness, whereas pork skin gelatin contributes plasticity and clarity.

   The physicochemical properties of gelatin of most interest to shell manufacturers are the bloom strength and viscosity. Bloom strength is an empirical gel strength measure that gives an indication of the firmness of the gel. It is measured in a Bloom Gelometer which determines the weight in grams required to depress a standard plunger a fixed distance into the surface of a 6-2/3% w/w gel under standard conditions. Those gelatins that are produced from the first extraction of the raw materials have the highest bloom strength. Bloom strengths in the range of 150–280 g are considered suitable for capsules.

The viscosity of gelatin solutions is vital to the control of the thickness of the cast film. Viscosity is measured on a standard 6-2/3% w/w solution at 60°C in a capillary pipette, and is generally in the range of 30–60 millipoise-(mP).

Colorants. Commonly, various soluble synthetic dyes ("coal tar dyes") and insoluble pigments are used. Commonly used pigments are the iron oxides.

Colorants not only play a role in identifying the product, but also may play a role in improving patient compliance. Thus, the color of a drug product may be selected in consideration of the disease state for which it is intended. For example, Buckalew and Coffield [19] found in a panel test that four colors were significantly associated with certain treatment groups (white, analgesia; lavender, hallucinogenic effects; orange or yellow, stimulants and antidepressants).

Opaquing Agents. Titanium dioxide may be included to render the shell opaque. Opaque capsules may be employed to provide protection against light or to conceal the contents.

Preservatives. When preservatives are employed, parabens are often selected.

Water. Hot, demineralized water is used in the preparation of the dipping solution. Initially, a 30–40% w/w solution of gelatin is prepared in large stainless steel tanks. Vacuum may be applied to assist in the removal of entrapped air from this viscous preparation. Portions of this stock solution are removed and mixed with any other ingredients, as required, to prepare the dipping solution. At this point, the viscosity of the dipping solution is measured and adjusted. The viscosity of this solution is critical to the control of the thickness of the capsule walls.

Shell Manufacture

The Colton machine illustrated in Fig. 1 is a fully automatic implementation of the dipping process. The steps are:

1. Dipping (Fig. 2): Pairs of stainless steel pins are dipped into the dipping solution to simultaneously form the caps and bodies. The pins are lubricated with a proprietary mold-release agent. The pins are at ambient temperature (about 22°C); whereas the dipping solution is maintained at a temperature of about 50°C in a heated, jacketed dipping pan. The length of time to cast the film has been reported to be about 12 sec, with larger capsules requiring longer dipping times [16].
2. Rotation: After dipping, the pins withdrawn from the dipping solution, and as they are done so, they are elevated and rotated 2-1/2 times until they are facing upward. This rotation helps to distribute the gelatin over the pins uniformly and to avoid the formation of a bead at the capsule ends. After rotation, the film is set by a blast of cool air.
3. Drying: The racks of gelatin-coated pins then pass into a series of four drying ovens. Drying is done mainly by dehumidification by passing large volumes of dry air over the pins. A temperature elevation of only a few degrees is permissible to prevent film melting. Drying must not be too rapid to prevent "case hardening." Overdrying must be avoided, as this could cause films to split on the pins from shrinkage or at least make them too brittle for the later trimming operation. Underdrying will leave the films too pliable or sticky for subsequent operations.
4. Stripping: A series of bronze jaws (softer than stainless steel) strip the cap and body portions of the capsules from the pins.

Fig. 1 View of a hard gelatin capsule manufacturing machine. (Courtesy of Elanco Qualicaps, formerly a Division of Eli Lilly and Co., Indianapolis, IN)

5. Trimming (Fig. 3): The stripped cap and body portions are delivered to collets in which they are firmly held. As the collets rotate, knives are brought against the shells to trim them to the required length.
6. Joining (Fig. 4): The cap and body portions are aligned concentrically in channels and the two portions are slowly pushed together.

The entire cycle takes about 45 min; however, about two-thirds of this time is required for the drying step alone.

Sorting

The moisture content of the capsules as they are ejected from the machine will be in the range of 15–18% w/w. Additional adjustment of moisture content toward the final desired specification will occur during the sorting step. During sorting, the capsules passing on a lighted moving conveyor are examined visually by inspectors. Any defective capsules spotted are thus manually removed. Defects are generally classified according to their nature and potential to cause problems in use. The most serious of these are those that could cause stoppage of a filling machine, such as imperfect cuts, dented capsules, or those with holes. Other defects may cause problems on use, such as capsules with splits, long bodies, or grease inside. Many less important, cosmetic faults that detract only from appearance also may occur (small bubbles, specks in the film, marks on the cut edge, and such).

Fig. 2 Dipping of pins in the manufacture of hard gelatin capsules. (Courtesy of Elanco Qualicaps, formerly a Division of Eli Lilly and Co., Indianapolis, IN)

Printing

In general, capsules are printed before filling. Empty capsules can be handled faster than filled capsules and, should there be any loss or damage to the capsules during printing, no active ingredients would be involved [15]. Generally, printing is done on offset rotary presses having throughput capabilities as high as three-quarter million capsules per hour [15]. Available equipment can print either axially along the length of capsules or radially around their circumference.

Sizes and Shapes

For human use, empty gelatin capsules are manufactured in eight sizes, ranging from 000 (the largest) to 5 (the smallest). The volumes and approximate capacities for the traditional eight sizes are listed in Table 1.

The largest size normally acceptable to patients is a No. 0. Size 0 hard gelatin capsules having an elongated body (size 0E) also are available that provide greater fill capacity without an increase in diameter. Three larger sizes are available for veterinary use: 10, 11, and 12, having capacities of about 30, 15, and 7.5 g, respectively. Although the standard shape of capsules is the traditional, symmetrical bullet shape, some manufacturers have employed distinctive proprietary shapes. Lilly's Pulvule is designed with a characteristic body section that tapers to a bluntly pointed end. Smith Kline Beacham's Spansule capsules exhibited a characteristic taper at both the cap and body ends.

Fig. 3 Trimming the newly cast and dried shells to proper length. (Courtesy of Elanco Qualicaps, formerly a Division of Eli Lilly and Co., Indianapolis, IN)

Sealing and Self-Locking Closures

Positive closures help prevent the inadvertent separation of filled capsules during shipping and handling. Such safeguards have become particularly important with the advent of high-speed filling and packaging equipment. This problem is particularly acute in the filling of noncompacted, bead or granular formulations.

Hard gelatin capsules are made self-locking by forming indentations or grooves on the inside of the cap and body portions. When fully engaged, a positive interlock is created between the cap and body portions. Indentations formed farther down on the cap provide a prelock that keeps the empty capsules joined during shipping and handling, yet allows ready separation for filling. Examples of self-locking capsules include Posilok (Shionogi Qualicaps), Loxit (Pharmaphil), and Coni-Snap (Capsugel, Div. Warner-Lambert Co.). The rim of the body portion of Coni-Snap capsules is tapered to help guide the cap onto the body. In high-speed automatic capsule-filling machines, this feature can reduce or eliminate snagging or splitting of capsules. The Coni-Snap principle and prelock feature are illustrated in Fig. 5.

The Coni-Snap Supro capsule (Fig. 6) is similar to the Coni-Snap relative to locking mechanism and tapered body edge, but differs in that it is short and squat, and the cap overlaps the body to a greater degree [21].

Hard gelatin capsules may be made hermetically sealed by *banding* (i.e., layering down a film of gelatin, often distinctively colored, around the seam of the cap and body). Parke Davis' Kapseal is a typical example. In one newer process, Quali-Seal (Shionogi Qualicaps), two thin

Fig. 4 Joining caps and bodies. (Courtesy of Elanco Qualicaps, formerly a Division of Eli Lilly and Co., Indianapolis, IN)

layers are applied, one on top of the other. Banding currently is the single most commonly used sealing technique. Banded capsules can provide an effective barrier to atmospheric oxygen [22].

Spot welding was once commonly used to lock the cap and body sections of bead-filled capsules together. In the thermal method, two hot metal jaws are brought into contact with the area where the cap overlaps the filled body [23].

Table 1 Capsule Capacities

Size	Volume	Fill weight (g) at 0.8 g/cm^3 powder density
000	1.37	1.096
00	0.95	0.760
0	0.68	0.544
1	0.50	0.400
2	0.37	0.296
3	0.30	0.240
4	0.21	0.168
5	0.13	0.104

Source: Ref. 20.

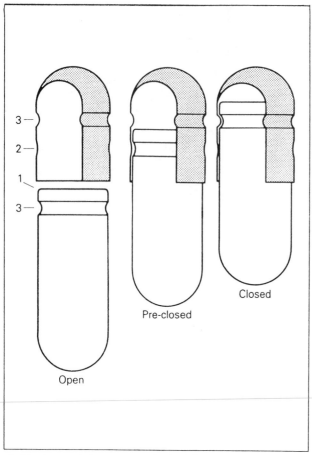

1 The tapered rim prevents faulty joins
2 These indentations prevent the pre-closed capsule
   from opening too early
3 These groovs lock the two halves together after filling
   (SNAP-FIT ® principle)

Fig. 5 Coni-Snap mechanically locking capsule showing prelock feature. (Courtesy of Capsugel, a Division of Warner-Lambert Co., Greenwood, SC)

Capsules are sealed and somewhat reshaped in the Etaseal process (Capsule Technology International, LTD., Canada). This thermal welding process forms an indented ring around the waist of the capsule where the cap overlaps the body.

Capsugel had proposed a low-temperature thermal method of hermetically sealing hard gelatin capsules [23]. The process involved immersion of the capsules for a fraction of a second in an hydroalcoholic solvent, followed by rapid removal. Excess solvent is then drained off, leaving traces in the overlapping area of the cap and body (held by capillary forces). Finally, the capsules are dried with warm air. A more recent adaptation of this approach involves the spraying of a mist of the hydroalcoholic solution onto the inner cap surface immediately before closure in filling machines. Such a process is used by Capsugel to seal Capill starch capsules together.

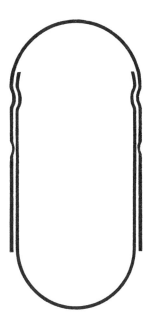

Fig. 6 Coni-Snap Supro. (Courtesy of Capsugel, a Division of Warner-Lambert Co., Greenwood, SC)

In the wake of several incidents of tampering with over-the-counter (OTC) capsules, sometimes with fatal consequences, much thought was given on how to make capsules safer [23]. Attention was focused on sealing techniques as possible means of enhancing the safety of capsules by making them *tamper-evident* (i.e., so that they could not be tampered with without destroying the capsule or at least causing obvious disfigurement).

Storage, Packaging, and Stability Considerations

Finished capsules normally contain an equilibrium moisture content of 13–16%. This moisture is critical to the physical properties of the shells, since at lower moisture contents (< 12%), shells become too brittle; at higher moisture contents (> 18%) they become too soft [24,25]. It is best to avoid extremes of temperature and to maintain a relative humidity of 40–60% when handling and storing capsules.

The bulk of the moisture in capsule shells is physically bound, and it can readily transfer between the shell and its contents, depending on their relative hygroscopicity [26,27]. The removal of moisture from the shell could be sufficient to cause splitting or cracking, as has been reported for the deliquescent material, potassium acetate [28]. Sodium cromoglycate has been reported to act as a ''sink'' for moisture, in that moisture was continuously removed from hard gelatin shells, especially at higher temperatures [29]. Conditions that favor the transfer of moisture to powder contents may lead to caking and retarded disintegration or other stability problems. It may be useful to preequilibrate the shell and its contents to the same relative humidity within the acceptable range [30,31].

One issue that is receiving current attention is the loss of water solubility of shells, apparently as a result of sufficient exposure to high humidity and temperature or to exposure to trace aldehydes [32]. Such capsules develop a ''skin'' or pellicle during dissolution testing, exhibit retarded dissolution, and may fail to meet the *U.S. Pharmacopea* (*USP*) drug dissolution specifications. This insolubilization of gelatin capsules has been attributed to ''gelatin cross-linking.'' In one example, photoinstability compounded by humidity has been suggested as the explanation for the retarded dissolution of model compounds from hard gelatin capsules containing certified dyes, particularly when FD & C Red No. 3 was incorporated in both the cap and the shell [33,34]. The problem also has been attributed to the presence of trace aldehydes in excipients [35], as well as to the liberation of furfural from the rayon stuffing in bottles [32]. These results point to the need for appropriate storage conditions and moisture-tight packaging, as well as to the need to exclude aldehydes. The issue is not new, nor is it a capsule issue per se; rather, it is a gelatin issue. The loss of water solubility on exposure of gelatin to elevated temperature and humidity was reported in 1968 to be "particularly disadvantageous in the case of gelatin desserts" [36]. The phenomenon also has been reported to occur with gelatin-coated acetaminophen tablets [37]. The inclusion of gastric enzymes in dissolution media tends to negate these effects [34,37]; thus, the phenomenon may have little physiological significance [38,39]. Should the latter prove to be the case, it would seem appropriate to modify official capsule dissolution tests to provide for the use of enzymes if dissolution fails using the simpler nonenzymatic media.

D. The Filling of Hard Gelatin Capsules

The several types of filling machines in use in the pharmaceutical industry have in common the following operations:

1. Rectification: The empty capsules are oriented so that all point the same direction (i.e., body-end downward). In general, the capsules pass one-at-a-time through a channel just wide enough to provide a frictional grip at the cap end. A specially designed blade pushes against the capsule and causes it to rotate about its cap end as a fulcrum. After two pushes (one horizontally and one vertically downward), the capsules will always be aligned body-end downward, regardless of which end entered the channel first.
2. Separation of caps from bodies: This process also depends on the difference in diameters between cap and body portions. Here, the rectified capsules are delivered body-end first into the upper portion of split bushings or split filling rings. A vacuum applied from below pulls the bodies down into the lower portion of the split bushing. The diameter of the caps is too large to allow them to follow the bodies into the lower bushing portion. The split bushings are then separated to expose the bodies for filling.
3. Dosing of fill material: Various methods are employed, as described later.
4. Replacement of caps and ejection of filled capsules: The cap and body bushing portions are rejoined. Pins are used to push the filled bodies up into the caps for closure, and to push the closed capsules out of the bushings. Compressed air also may be used to eject the capsules.

These machines may be either semiautomatic or fully automatic. Semiautomatic machines such as the Capsugel Type 8 machines require an operator to be in attendance at all times. Depending on the skill of the operator, the formulation, and the size capsule being filled, these machines are capable of filling as many as 120,000–160,000 capsules in an 8-hr shift. This output contrasts sharply with the output of fully automatic machines, some models of which are rated to fill that many capsules in 1 hr. Some representative automatic capsule-filling machines are listed in Table 2. These machines may be classified as either intermittent- or continuous-motion machines. Intermittent machines exhibit an interrupted-filling sequence, as indexing turntables must stop at various stations to execute the basic operations described earlier. Continuous-motion machines execute these functions in a continuous cycle. The elimination of the need to decelerate and accelerate from one station to the next makes greater machine speeds possible with continuous-motion machines [40]. Although capsule-filling ma-

Table 2 Representative Automatic Capsule-Filling Machines

Make/model	Dosing principle	Motion[a]	Rated capacity[b] (capsules/hr)
Hofliger-Karg[c]	Tamping/dosing disk		
GKF 400		I	24,000
GKF 1200		I	48,000
GKF 3000		I	180,000
Zanasi[d]	Tamping/dosator		
Zanasi 6		I	6,000
Zanasi 40		I	40,000
Matic 120		C	120,000
MG2[e]	Tamping/dosator		
Futura		C	36,000
G38/N		C	60,000
G100		C	100,000
Dott. Bonapace & Co.[f]	Tamping/dosator		
RC530		I	30,000
Harro Hofliger[g]	Tamping/dosing		
KFM III	disk	I	18,000

[a]I, intermittant; C, continuous.
[b]Based on manufacturer's, distributor's literature; some capacities are listed as approximate.
[c]Robert Bosch Corp., Packaging Machinery Div., 121 Corporate Blvd., South Plainfield, NJ 07080.
[d]IMA North America, Inc., 107 Ardmore St., Fairfield, CT 06430.
[e]MG America, Inc., 31 Kulick Rd., Fairfield, NJ 07004.
[f]Sitco, 7E Easy St., Bound Brook, NJ 08805.
[g]M. O. Industries, 53 South Jefferson Rd., Whippany, NJ 07981.

chines may vary widely in their engineering design, the main difference among them, from a formulation point of view, is the means by which the formulation is dosed into the capsules.

Powder Filling

Capsule-filling equipment has been the subject of several reviews [18,40–44], and for powder filling, four main dosing methods may be identified:

Auger Fill Principle. At one time, nearly all capsules were filled by means of semiautomatic equipment, wherein the powder is driven into the capsule bodies by means of a rotating auger. This type of filling machine is exemplified by the Capsugel Type 8 machines. A Type 8 machine is illustrated in Fig. 7. The empty capsule bodies are held in a filling ring that rotates on a turntable under the powder hopper. The fill of the capsules is primarily volumetric. Because the auger mounted in the hopper rotates at a constant rate, the delivery of powder to the capsules tends to be at a constant rate. Consequently, the major control over fill weight is the rate of rotation of the filling ring under the hopper. Faster rates produce lighter fill weights, because bodies have a shorter dwell-time under the hopper. Ito et al. [45] compared an experimental flat-blade auger with an original screw auger and found that the screw auger provided greater fill weight (30–60% more for a test lactose formulation) and smaller coefficients of weight variation (up to 50% smaller at the two fastest ring speeds). The formulation requirements of this type of machine have been the subject of only a limited number of reports. In general, the flow properties of the powder blend should be adequate to assure a uniform flow

(a)

(b)

(c)

(d)

Fig. 7 Type 8 semiautomatic capsule-filling machine. (a) "Sandwich" of cap and body rings positioned under rectifier to receive empty capsules. Vacuum is pulled from beneath the rings to separate caps from bodies. (b) Body ring is positioned under foot of powder hopper for filling. (c) After filling the bodies, the cap and body rings are rejoined and positioned in front of pegs. A stop plate is swung down in back of rings to prevent capsule expulsion as the pneumatically driven pegs push the bodies to engage the caps. (d) The plate is swung aside and the pegs are used to eject the closed capsules.

rate from the hopper. Glidants may be helpful. Ito et al. [45] studied the glidant effect of a colloidal silica using a Capsugel Type 8 filling machine. They found that there was an optimum concentration for minimum weight variation (approximately 0.5% for lactose capsules; approximately 1% for corn starch capsules). With the Elanco Type 8 machine, Reier et al. [46] reported that the presence of 3% talc reduces weight variation, compared with 0% talc in a multivariate study involving several fillers. These investigators analyzed their data by multiple stepwise regression analysis and concluded that the mean fill weight was dependent on machine speed, capsule size, and on the formulation-specific volume, in that order. Weight variation was found to be a function of machine speed, specific volume, flowability, and the presence of glidant, but independent of capsule size.

Lubricants, such as magnesium stearate and stearic acid, are also required. These facilitate the passage of the filling ring under the foot of the powder hopper and help prevent the adherence of certain materials to the auger.

Vibratory Fill Principle. The Osaka machines (Fig. 8) employ a vibratory feed mechanism [47,48]. In this machine, the capsule body passes under a feed frame that holds the powder in

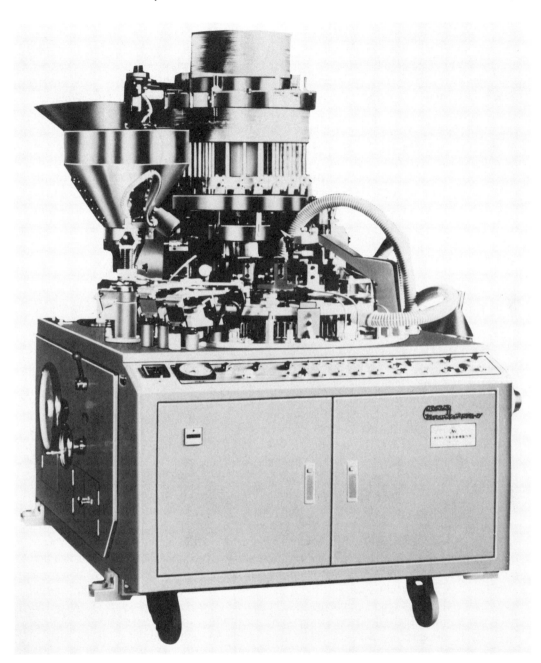

Fig. 8 Osaka model R-180 automatic capsule-filling machine. (Courtesy of Sharpley-Stokes Division, Pennwalt Corp., Warminster, PA)

the filling section. In the powder, a perforated resin plate is positioned that is connected to a vibrator. The powder bed tends to be fluidized by the vibration of the plate, and this assists the powder to flow into the bodies through holes in the resin plate [48]. The fill weight is controlled by the vibrators and by setting the position of the body under the feed frame. Much like the fill mechanism of a tablet press, there is overfill and then adjustment with scrape-off of the excess material as the capsule bodies pass under the feed frame. The capsule bodies are supported on pins in holes bored through a disk plate. While they pass under the feed area, the pins may be set to drop the bodies to below the level of the disk, thereby causing "overfill." However, before their passage is completed under the feed frame, the capsules are eventually pushed up so their upper edges become level with the surface of the disk plate. When this occurs, the excess powder is forced out and eventually scraped off by the trailing edge of the feed frame. This process affords some light compression of the powder against the resin plates and offers the opportunity to modify the fill weight. Weight variation has been related to the formulation flow properties. Kurihara and Ichikawa [47] reported that the fill weight variation with Model OCF-120 was more closely related to the minimum orifice diameter than to the angle of repose. Apparently, the minimum orifice diameter is a better analogy of the flowing of powder into capsule bodies than the static angle of repose. No studies of the formulation requirements for this machine have been reported; however, typical stearate lubricants may be indicated to prevent the binding of push rods and guides.

Piston-Tamp Principle. Piston-tamp machines are fully automatic fillers in which pistons tamp the individual doses of powders into plugs (sometimes referred to as "slugs"), which often resemble soft tablets in consistency, and eject the plugs into the empty capsule bodies. There are two types of piston-tamp fillers: dosator machines and dosing-disk machines.

1. DOSING-DISK MACHINES.   This type of machine is exemplified by the Hofliger-Karg GKF models and the Harro-Hofliger KFM models (Fig. 9). The dosing-disk–filling principle has been described [49,50] and is illustrated in Fig. 10. The dosing-disk, which forms the base of the dosing or filling chamber, has a number of holes bored through it. A solid brass "stop" plate slides along the bottom of the dosing disk to close off these holes, thereby forming openings similar to the die cavities of a tablet press. The powder is maintained at a relatively constant level over the dosing disk. Five sets of pistons (Hofliger-Karg machines) compress the powder into the cavities to form plugs. The cavities are indexed under each of the five sets of pistons so that each plug is compressed five times per cycle. After the five tamps, any excess powder is scraped off as the dosing disk indexes to position the plugs over empty capsule bodies where they are ejected by transfer pistons. The dose is controlled by the thickness of the dosing disk (i.e., cavity depth), the powder depth, and the tamping pressure. The flow of powder from the hopper to the disk is auger-assisted. A capacitance probe senses the powder level and activates an auger feed if the level falls to below the preset level. The powder is distributed over the dosing disk by the centrifugal action of the indexing rotation of the disk. Baffles are provided to help maintain a uniform powder level. However, while working with a Hofliger-Karg model 330, Shah et al. [50] noted that a uniform powder bed height was not maintained at the first tamping station because of its nearness to the scrape-off device. Kurihara and Ichikawa [47] reported that variation in fill weight was closely related to the angle of repose of the formulation; however, a minimum point appeared in the plots of the angle of repose versus coefficient of variation of filling weight. Apparently, at higher angles of repose, the powders did not have sufficient mobility to distribute well under the acceleration of the intermittent-indexing motion. At lower angles of repose, the powder was apparently too fluid to maintain a uniform bed. However, the investigators did not appear to make use of powder compression through tamping, and this complicates the interpretation of their results.

These machines generally require that formulations be adequately lubricated for efficient plug ejection, to prevent filming on pistons, and to reduce friction between any sliding components that may come into contact with powder. A degree of compactibility is important, as coherent plugs appear to be desirable for clean, efficient transfer at ejection. However, there may be less dependence on formulation cohesiveness than exists for dosator machines [42].

Fig. 9 Hofliger Karg model GKF 1500 automatic capsule-filling machine. (Courtesy of Robert Bosch Corp., Packaging Machinery Division, South Plainfield, NJ)

The Harro-Hofliger machine is similar to Hofliger-Karg machines, except that it employs only three tamping stations. However, at each station, the powder in the dosing cavities is tamped twice before rotating a quarter turn to the next station. One other difference is that the powder in the filling chamber is constantly agitated to help in the maintenance of a uniform powder bed depth.

2. DOSATOR MACHINES. The dosator machines are exemplified by the Zanasi, MG2, Dott. Bonapace, and Macofar machines. Examples of Zanasi and MG2 machines are pictured in Figs. 11 and 12. Figure 13 illustrates the basic dosator principle. The dosator principle has been previously described [51,52]. The dosator consists of a cylindrical dosing tube fitted with a moveable piston. The end of the tube is open, and the position of the piston is preset to a

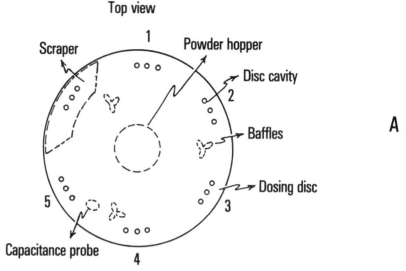

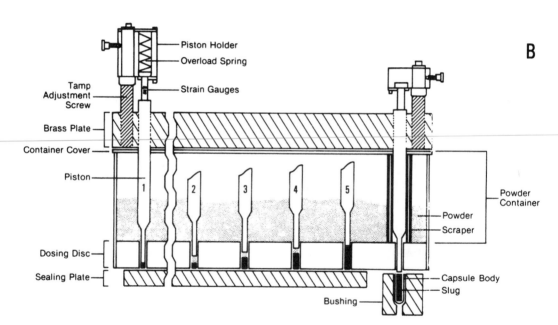

Fig. 10 Illustration of the dosing–disk-filling principle: (A) view looking down on the dosing disk; (B) side view (projected) showing progressive plug formation. Note the placement of strain gauges on the piston to measure tamping and plug ejection forces (see text). (From Ref. 37.)

particular height to define a volume (again, comparable with a tablet press die cavity) that would contain the desired dose of powder. In operation, the dosator is plunged down into a powder bed maintained at a constant preset level by agitators and scrapers. The powder bed height is generally greater than the piston height. Powder enters the open end and is slightly compressed against the piston (termed ''precompression'' [51]). The piston then gives a tamping blow, thus forming the powder into a plug. The dosator, bearing the plug, is withdrawn from the powder hopper and is moved over to the empty capsule body where the piston is

Fig. 11 Zanasi Matic 90 automatic capsule-filling machine. (Courtesy of IMA North America, Inc., Fairfield, CT)

Fig. 12 MG2 Futura automatic capsule-filling machine. (Courtesy of MG America, Inc., Fairfield, NJ)

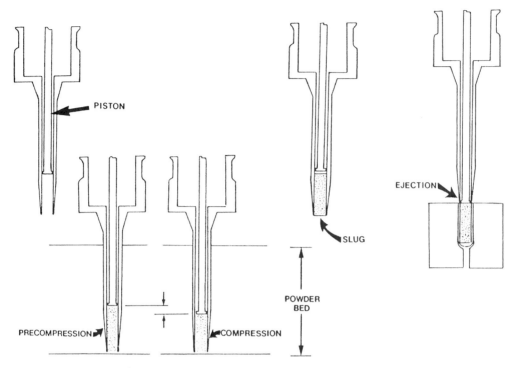

Fig. 13 Dosator-filling principle. (From Ref. 48.)

pushed downward to eject the plug. In certain machines, such as the Macofar machines, the body bushing is rotated into position under the dosator to receive the ejected plug [53]. The primary control over fill weight (for a given set of tooling) is the initial piston height in the dosing tube. A secondary control of weight is the height of the powder bed into which the dosator dips.

In one of the earliest reports evaluating the Zanasi machine, Stoyle [52] suggests that formulations should have the following characteristics for successful filling:

1. *Fluidity* is important for powder feed from the reservoir to the dipping bed and also to permit efficient closing in of the hole left by the dosator.
2. A degree of *compactibility* is important to prevent loss of material from the end of the plug during transport to the capsule shell.
3. *Lubricity* is needed to permit easy and efficient ejection of the plug.
4. It was suggested that formulations have a *moderate bulk density*. Low-bulk density materials or those that contain entrapped air will not consolidate well, and capping similar to what occurs in tableting may result.

More recent studies have further examined the relation between formulation *flow properties and weight variation* on Zanasi machines. For example, Irwin et al. [54] compared the weight variation of capsules filled on a Zanasi LZ-64 machine from formulations composed of different diluents and lubricants. The formulations had different flow properties, as judged in a recording flow meter. Generally, the better the rate of flow, the more uniform was the capsule fill weight. Chowhan and Chow [55] compared the powder consolidation ratio with the coefficient of variation (relative standard deviation) of capsule weight and found a linear relationship for a test formulation containing 5 or 15% drug, 10% starch, 0.5% magnesium stearate, and lactose q.s. The capsules were filled on a Zanasi machine. Powder flow characteristics were inferred from the volume reduction (consolidation) that occurs when a series of loads are applied to the surface of the loosely packed powder bed in cylindrical containers. The powder consolidation ratio was the intercept of the plot of

$$\log \frac{V_0 - V}{V} \text{ vs. } \frac{P}{P_0}$$

where V_0 = initial powder volume, V = powder volume at a given surface pressure, P = surface pressure and P_0 = 1 kg/cm^2. Further work to assess the usefulness and limitations of this approach appears warranted.

The effect of *machine variables on fill weight* and its uniformity were evaluated by Miyake et al. [56] using a Zanasi Z-25. In general, they found that the filling mechanism was a compaction process. The following relationship was found to apply:

$$r = -a(i) \log Pr + b(i)$$

where r = density ratio, $a(i)$ and $b(i)$ are constants, and Pr = compression ratio = $(H - L)/L$, where H = powder bed height and L = piston height (within the dosator).

The quantitative retention of powder within the dosator during transfer from the powder bed to the capsule shell is essential to a successful filling operation. Jolliffe et al. [57,58] have studied this theoretically by the application of hopper design theory. The retention of powder during transfer requires the formation of a stable powder arch at the dosator outlet, and this depends on the angle of wall friction. Generally, theory predicts that cohesive materials will be retained with minimal compressive stress on rough dosator walls and that smoother walls provide the best conditions for retaining more freely flowing powders.

Nonpowder Filling

Modern automatic capsule filling machines offer enormous flexibility in terms of what can be filled into hard shell capsules. In addition to powder dosing, filling devices also are available that can feed beads or pellets, microtablets, tablets, and liquid or pasty materials into capsules. Often, these can be installed at different filling stations of the same machine such that capsules may be dosed from several different filling devices as they pass by each station before closure and ejection. Such arrangements could, for example, permit the dosing of several different tablets, different batches of beads (perhaps immediate-release and modified-release beads) or combinations of tablets, powder plugs, and beads into the same capsule.

Beads, pellets, and such may be poured directly into the capsule body by gravity-feed devices that rely on the free-flowing nature of such materials. In this approach, capsules are filled to their volumetric capacity, and partial fills for multiple dosing are not possible. Modern automatic-filling machines circumvent this issue by employing various indirect-filling methods (i.e., the required quantity of beads, granules, and such, are first fed to a separate, volumetric metering chamber, and then the measured volume of material is transferred to the capsule body. The metering chamber is usually filled by gravity (e.g., Hofliger-Karg); however, in certain machines (e.g., Zanasi), the chamber is a modified dosator that draws and holds the beads into its open end by means of vacuum. In general, the dose is determined by the size of the metering chamber. If blends are being dispensed, the uniformity of the dose dispensed depends on the size and shape of the granules or pellets, since differences in these properties may cause pellet segregation. The development of electrostatic charges on beads or pellets also may cause separation of individual beads as well as problems in flowing and transferring from chambers.

Typically, tablets are fed to the bodies through a tube and are simply released in the required number as the body passes beneath. Pumpable, liquid fills are dosed by conventional liquid-dispensing devices.

E. Instrumentation of Capsule-Filling Machines and Their Role in Formulation Development

A major development in pharmaceutical technology has been the application of instrumentation techniques to tablet presses. The ability to monitor the forces that develop during the compaction, ejection, and detachment of tablets has brought about new insights into the physics of compaction, facilitated formulation development, and provided a means for the in-process control of tablet weight in manufacturing [59,60]. Usually, automatic capsule filling is carried out on dosator or dosing-disk machines, which resemble tableting in that there are compression and ejection events. Given this similarity to tableting and the benefits that have accrued from instrumented tablet machines, it was only logical that similar instrumentation techniques be applied to these capsule-filling machines. Although both types of machines have been instrumented, most reports have been concerned with dosator machines [61].

Cole and May [62,63] were the first investigators to report the instrumentation of an automatic capsule filling machine. They bonded strain gauges to the piston of a Zanasi LZ-64 dosator. Because of dosator rotation, this machine required modification by installation of a planetary gear system to prevent the continuous twisting of the output cable during operation. Their work demonstrated for the first time that compression and ejection forces could be recorded during plug formation. They reported (a) an initial compaction force as the plug was being formed by the dosator dipping into the powder bed, (b) a partial retention of this force during passage to the ejection station, and (c) an ejection force as the plug was pushed out of the dosator.

Small and Augsburger [51] also reported on the instrumentation of the same model Zanasi with strain gauges. Twisting of the output cable was avoided by connecting it to a low-noise mercury contact swivel mounted over the capsule hopper. This was a simpler arrangement than that employed by Cole and May [62] in that it permitted electrical contact to be maintained during experimental runs without the need for a planetary gear system nor any other machine modification. Figures 14 and 15 illustrate their instrumented piston and the mounting of the mercury swivel.

In contrast with Cole and May [62], Small and Augsburger reported a two-stage plug formation trace: (a) a precompression force which occurs when the dosator dips into the powder bed and (b) compression of the powder by the tamping of the piston at the bottom of dosator travel in the powder bed. Apparently, the earlier workers did not make use of the piston compression feature of the Zanasi filling principle. Like Cole and May, Small and Augsburger also reported retention and ejection forces. The retention force, which apparently is a result of elastic recovery of the plug against the piston, was observed by Small and Augsburger only when running unlubricated materials under certain conditions. This phenomenon was not observed in any lubricated runs, apparently because the lubricant permits the plug to more readily slip to relieve any residual pressure [51]. It is interesting that both teams of investigators reported instances of drag on the piston as it returns to the original position after ejection. This was manifested by a negative force (i.e., a trace below the baseline) during retraction of the piston. Sample traces appear in Fig. 16.

Small and Augsburger [64] later reported a detailed study of the formulation lubrication requirements of the Zanasi LZ-64. Three fillers were studied (microcrystalline cellulose, pregelatinized starch, and anhydrous lactose). Powder bed height, piston height, compression force,

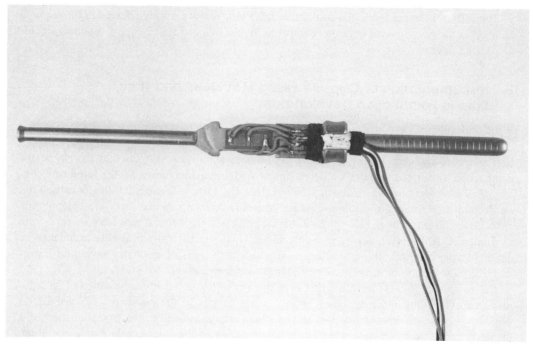

Fig. 14 Strain gauges bonded to Zanasi piston. (From Ref. 38.)

Fig. 15 Instrumented Zanasi LZ-64 showing mercury swivel for signal removal. (A) Dosator containing strain-gauged piston; (B) mercury swivel. (From Ref. 38.)

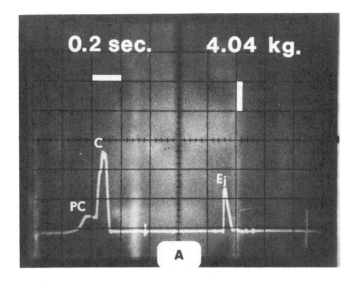

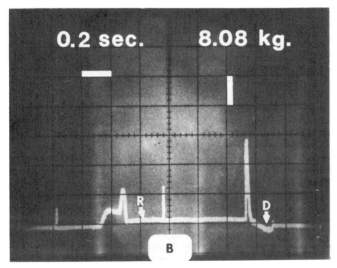

Fig. 16 Typical force–time trace from an instrumented Zanasi LZ-64 automatic capsule-filling machine. PC, precompression resulting from dipping of dosator into the powder bed; C, compression resulting from actual piston tamping; R, retention force; Ej, ejection; D, drag force developing during retraction of piston. (From Ref. 38.)

and lubricant type and concentration were varied to determine their effects on ejection force. In general, anhydrous lactose exhibited higher lubrication requirements than either pregelatinized starch or microcrystalline cellulose. Comparing several concentrations of magnesium stearate, minimum ejection forces were recorded at 1% with anhydrous lactose, at 0.5% with microcrystalline cellulose, and at 0.1% with pregelatinized starch. Magnesium lauryl sulfate compared favorably with magnesium stearate in the starch filler, but was not as efficient as magnesium stearate for the other two fillers. Also, the magnitude of the ejection force was affected by machine-operating variables. After precompression, ejection force increased with the compression force. However, at a given compression force, ejection force also increased with an increase in either the piston height or the powder bed height. Figure 17 is typical. These results suggest the possibility of manipulating machine-operating variables to reduce formulation lubricant requirements.

Mehta and Augsburger [65] later reported the mounting of a linear variable displacement transducer on the previously instrumented Zanasi LZ-64 machine [51] to allow the measurement of piston displacement during compression and ejection. The work of ejection, calculated from force–displacement profiles was different for several formulations having comparable peak ejection forces [66].

Following the approach of Small and Augsburger [51], Greenberg [67] used strain gauges to instrument a larger Zanasi machine (model AZ-60). This intermittent-motion machine employs three groups of eight dosators. Two instrumented pistons were installed in two dosators in one group. The system was unique in that a high quality ten-pole slip ring was used to avoid twisting of the cables. More recently, Botzolakis [68] described the successful replacement of the previously reported mercury swivel with a ten-pole gold-contact slip ring assembly.

Piezoelectric transducers have also been used to instrument automatic capsule-filling machines. Mony et al. [69] instrumented a Zanasi RV/59 by fitting a piezoelectric load cell to the upper end of a piston. This system can register a force only when the upper end of the piston is in actual contact with the compression or ejection knobs. Although this instrumentation provides a measure of overall compression and ejection forces, it does not permit the detection of precompression, retention, or piston retraction drag forces. Moreover, this instrumentation adds the force required to compress the piston retraction spring to any forces measured. No attempt to correct their data for this variable was reported. Rowley et al. [70] reported the mounting of a small piezoelectric load cell to the ejection knob of a Zanasi LZ-64 machine to monitor ejection force. This approach suffers from the same disadvantages as that of Mony et al. [69]. However, these latter investigators did report subtracting out the force required to compress the dosator spring from their measurements. This correction was obtained by making a "blank" run with an empty dosator.

Jolliffe et al. [72] reported the instrumentation of an MG2 capsule-filling simulator. The simulator employs the filling turret of a model G-36 machine and a drive mechanism that allows the normal up and down motion of the dosators, but without the usual turret rotation. One dosator was employed, which was instrumented by bonding strain gauges to the piston. Additionally, displacement transducers were fitted to permit registration of piston movement relative to the dosator and dosator movement relative to the turret. With this device, Jolliffe and Newton [72] studied the effects of changes in the compression ratio on fill weight variation and the resultant compression and ejection stresses for four different-sized fractions of lactose. In general, the ranges of compression ratio over which uniform weights could be obtained with minimum tamping pressure was far greater for finely divided, cohesive powders than for the coarser, freely flowing sized fractions. Fine, cohesive powders have greater void volumes and, therefore, are capable of greater volume reduction than free-flowing powders.

Shah et al. [50] reported the instrumentation of an Hofliger-Karg model 330 filling machine. Two pistons were instrumented using strain gauges to enable simultaneous monitoring of either two of the tamping stations or one tamping station and ejection (see Figs. 10 and 18). This preliminary study revealed the complexity of the interaction of the various tamping stations on the final fill weight. By using additional instrumented pistons and microprocessor-controlled data acquisition techniques, Shah et al. [73] later evaluated seven compaction parameters and concluded that, aside from station 1, all tamping stations and all piston positions within a station contribute equally to plug formation. The nearness of station 1 to the scrape-off bar results in nonuniform powder bed height and a high degree of compression force variability. Model calculations suggesting that fill weight could be achieved with only three tamps were supported by experiments in which fill weight was determined as a function of tamping force and the number of tamps for typical lubricated fillers. The effect of tamping force and multiple tamping on drug dissolution also was investigated using this equipment [74]. Cropp et al. [75] later installed displacement transducers on the machine previously instrumented by Shah et al. [73] to further study the multiple-tamping effect and to assess the role of overload spring tension on fill weights obtained.

Instrumentation has also been developed to measure the mechanical strength of plugs. Greenberg [67] was the first investigator to report the measurement of plug "hardness." A pneumatically driven piston, moving at a controlled rate, was brought against the plug held in a narrow channel. A ring indicator registered the highest force developed as the plug fails. Hardness values were generally under 0.1 N. Later, Mehta and Augsburger [76] measured the maximum bending resistance of plugs in a three-point flexure test. Here the plug was supported at each end. A blunt-edged blade, mounted on the moving head of a bench type tensile strength tester, was lowered at a slow, controlled rate against the unsupported midpoint of the plug.

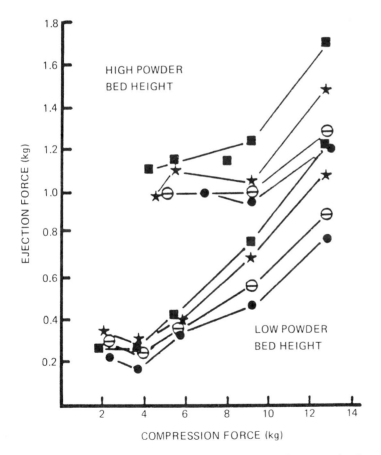

Fig. 17 Effect of powder bed height, piston height, and compression force on plug ejection force in an instrumented Zanasi LZ-64 automatic capsule filling machine (pregelatinized starch lubricated with 0.005% magnesium stearate). Note that the first point of each curve is precompression. Piston height (mm): ■, 15; ★, 14; ⊖, 13; ●, 12. Powder bed height (mm); heavy line, 30; light line, 50. (From Ref. 51.)

Fig. 18 Strain-gauged pistons mounted in a Hofliger Karg model 330 automatic capsule filling machine. (From Ref. 37.)

The maximum force required for plug failure was measured by a push–pull transducer. Values of up to 1 N were reported.

F. General Considerations in the Design of Hard Gelatin Capsule Powder Formulations and Choice of Excipients

Powder formulations for encapsulation should be developed in consideration of the particular filling principle involved. The requirements on the formulation imposed by the filling process, such as lubricity, compactibility, and fluidity are not only essential to a successful filling operation, but also may be expected to influence drug release from the capsules. Indeed, the various filling principles themselves may be expected to influence drug release. This seems particularly evident for those machines that form compressed plugs.

When immersed in a dissolution fluid at 37°C, hard gelatin capsules can be seen to rupture first at the shoulders of the cap and body where the gelatin shell is thinnest [77,78]. As the dissolution fluid penetrates the capsule contents, the powder mass begins to disintegrate and deaggregate from the ends to expose drug particles for dissolution. The efficiency by which the drug will be released will depend on the wettability of the powder mass, how rapidly the dissolution fluid penetrates the powder, the rate of disintegration and deaggregation of the contents, and the nature of the primary drug particles. These processes, in turn, can be significantly affected by the design of the formulation and the mode of filling. Such factors as the amount and choice of fillers and lubricants, the inclusion of disintegrants or surfactants, and the degree of plug compaction can have a profound effect on drug release.

Active Ingredient

The amount and type of active ingredient influences capsule size and the nature and amount of excipients to be used in the formulation. Although there are a growing number of exceptions, drugs having doses less than 10 mg are seldom formulated into capsules. These can usually be as easily formulated into tablets that are more economical. Thus, the active ingredient often tends to make up a high percentage of the contents of a capsule; much more so than is usual for tablets.

The dissolution of the drug in gastrointestinal fluids must occur before absorption can occur. Drugs having high water solubility generally exhibit few formulation problems. For drugs of low water solubility, the absorption rate may be governed by the dissolution rate. In such circumstances, if dissolution occurs too slowly, absorption efficiency may suffer. Drug stability in gastrointestinal fluids is another concern for slowly dissolving drugs, which can affect their bioavailability. Drugs of low water solubility are usually micronized to increase the dissolution rate. Particle size reduction increases the surface area per unit weight of the drug, thereby increasing the surface area available from which dissolution can occur. For instance, Fincher et al. [79] studied the different particle size fractions of sulfathiazole administered in capsules to dogs, and found that the smallest particle size gave the highest blood level. Also, Bastami and Groves [80] reported that reducing the particle size of sodium phenytoin improved the dissolution rate from capsules containing 100 mg of the drug and 150 mg lactose. There are, however, practical limitations to this approach. Micronized particles with high surface/mass ratios may tend to aggregate, owing to surface cohesive interaction, thereby reducing the surface area effectively available for dissolution. Newton and Rowley [81] found that at equivalent bed porosities, larger particle size fractions of a poorly soluble drug, ethinamate, gave better dissolution from capsules of the pure drug than from those of smaller particle sizes. They attributed this result to the smaller particle size fractions having reduced effective surface area for dissolution, owing to aggregation. The compaction of fine particles into capsules also reduces the bed permeability and generally retards dissolution [81].

From a manufacturing point of view, a compromise may have to be struck between small particle size and good flow properties. Small particles, in general, are more poorly flowing than larger particles. Surface cohesive and frictional interactions, which oppose flow properties, are more important in smaller particle size powders because of their larger specific surface areas. One possible way to both reduce the effects of aggregation of fine particles and enhance flow properties is granulation. When micronized ethinamate was granulated in a simple moist process with isopropanol, bed permeability and drug dissolution from capsules were greatly enhanced compared with the micronized powder [81].

Fillers

Fillers (diluents) are often needed to increase the bulk of the formulation. The most common capsule diluents are starch, lactose, and dicalcium phosphate. Modifications of these materials for direct-compression tableting, such as pregelatinized starch (Starch 1500, Colorcon Inc., West Point, PA) or spray-processed lactose (Fast-Flo, lactose, Foremost, Div. Wisconsin Dairies, Baraboo, WI) or unmilled dicalcium phosphate dihydrate (Ditab, Rhone-Poulenc Basic Chemicals Co., Shelton, CT; Emcompress, Mendell, A Penwest Co., Patterson, NY) can also be used. These substances improve flow and compactibility while maintaining the basic properties of the original materials.

Formulations intended to be run on dosator machines may sometimes benefit from the greater compactibility of microcrystalline cellulose (Avicel, FMC Corp., Food and Pharmaceutical Products Div., Philadelphia, PA), particularly when drug dosage is large. In these machines it is essential to prevent powder loss from the end of the cylinder during the transfer from the powder bed to ejection into the capsule body. The failure to have a cohesive plug may also cause a "blow off" of powder as the plug is ejected into the shell. As previously pointed out, a degree of compactibility is also important in formulations for dosing-disk machines.

From a drug dissolution point of view, formulators may need to consider the solubility of both the filler and the drug. For instance, Newton et al. [82] demonstrated that the dissolution of poorly soluble ethinamate from capsules improved greatly when the concentration of lactose in the formulation was increased to 50%. However, with the soluble drug, chloramphenicol, Withey and Mainville [83] found that the inclusion of 80% lactose in the formulation severely retarded drug dissolution from capsules; there was little or no effect on dissolution when up to 50% lactose was included. It was suggested that dissolution of the lactose occurs more rapidly and that chloramphenicol dissolution is retarded because of the high concentration of lactose already in solution. The effect of the filler on bioavailability was graphically illustrated [84] when Australian physicians noted an increase in the number of patients exhibiting phenytoin toxicity while using a particular sodium phenytoin capsule product. This occurrence coincided with the manufacturer changing the filler from calcium sulfate to lactose and was the result of increased bioavailability when lactose was the filler. Here, the effect may not be solely due to the greater solubility of lactose. Bastami and Groves [80] reported that the in vitro dissolution of phenytoin may not be complete in the presence of calcium sulfate and suggested the formation of an insoluble calcium salt of the drug. It has also been reported that lactose, at a concentration of 50%, enhanced the dissolution of phenobarbital from capsules, but had no effect on the dissolution of the water-soluble sodium salt [85].

Corn starch at 50% slowed the dissolution of sodium phenobarbital and improved the dissolution of the free acid; however, the effect in either case was dependent on the moisture content of the starch [85]. The t_{50} (time required for 50% drug dissolution) for 50:50 phenobarbital/corn starch capsules decreased from 28 min to 9 min as the starch moisture content increased from 1.2% w/w to 9.5% w/w. This compares with a t_{50} of 25 min for the drug alone. Drug dissolution also improved in the 50:50 sodium phenobarbital/corn starch mixtures with increased moisture; however, even at the highest moisture content (13.5%) t_{50} was still about double that of the drug alone (4.9 vs. 2.5 min).

Glidants

Glidants are used to improve the fluidity of powders. They are fine particles that appear to coat the particles of the bulk powder and enhance fluidity by one or more of several possible mechanisms [86,87]: (a) reducing roughness by filling surface irregularities, (b) reducing attractive forces by physically separating the host particles, (c) modifying electrostatic charges, (d) acting as moisture scavengers, and (e) serving as ball bearings between host particles. Usually, there is an optimum concentration for flow, generally less than 1% and typically 0.25–0.50%. The optimum concentration may be related to the concentration just needed to coat the host particles [88]. Exceeding this concentration usually will result in either no further improvement in flow and, even, a worsening of flow. Glidants include the colloidal silicas, corn starch, talc, and magnesium stearate.

Lubricants

Capsule formulations usually require lubricants just as do tablet formulations. Lubricants ease the ejection of plugs, reduce filming on pistons and adhesion of powder to metal surfaces, and reduce friction between sliding surfaces in contact with powder. The same lubricants are used in both tablet and capsule formulations: magnesium stearate and stearic acid are typical.

Increasing the concentration of hydrophobic lubricants, such as magnesium stearate, is generally understood to retard drug release by making formulations more hydrophobic [82,89–91]. However, exceptions to that rule are possible. Stewart et al. [92] reported that the effect of magnesium stearate concentration on the dissolution of a model low-dose drug, riboflavin, from capsules was dependent in some manner on the type of filler. Soluble fillers exhibited the anticipated prolonged times with increasing lubricant levels. However, the trends with insoluble fillers were less predictable. In some cases, insoluble fillers were only slightly affected by the concentration of magnesium stearate. For others, such as microcrystalline cellulose, there appeared to be an ideal concentration of lubricant at which the dissolution rate was maximized. These capsules were filled using an isolated Zanasi dosator fitted to a moveable crosshead.

In a follow-up of this work, Mehta and Augsburger [76] suggested that the mechanical strength of plugs produced in a dosator may be reduced by the amount of lubricant used and that this could have a beneficial effect on drug dissolution. As previously described, plug hardness was assessed by measuring their breaking load in a three-point flexure test. With use of hydrochlorothiazide as the tracer drug for dissolution, these investigators compared the time for 60% of the drug content to dissolve (T_{60}) and plug hardness for two fillers lubricated with 0.05–0.75% magnesium stearate and compressed at the same 22-kg compression force. With microcrystalline cellulose, T_{60} decreased from 55 to 12 min. Paralleling this was a dramatic decrease in plug hardness from 84 to about 2.0 g. With lactose, T_{60} increased with the lubricant level from 12 to 18 min, whereas plug hardness decreased slightly, although not significantly ($p = 0.05$), from 18 to 13 g. For the microcrystalline cellulose, it was suggested that the increase in hydrophobicity owing to increased lubricant concentration initially was more than offset by reduced plug hardness, which probably enhances moisture penetration and promotes deaggregation in the dissolution medium.

This dual effect of magnesium stearate has also been noted in a study of the dissolution of rifampin from hard gelatin capsules [93].

Disintegrants

Although tablet disintegrants are being used in some capsule formulations, until recently, the role they play in capsules has been a relatively unexplored area. The few studies that had been reported produced only mixed results and usually involved hand-filled capsules [90,94,95, for example]. Capsules filled by methods that afford little compression of contents (e.g., auger method) are much looser than tablets, and there is little structure for disintegrants to swell against to effect disintegration. However, the advent in recent years of filling machines that actually compress capsule contents, together with the development of newer disintegrants that have superior swelling or moisture-absorbing properties, appear to warrant serious consideration of disintegrants in modern capsule formulations. These newer disintegrants, which have been called ''super disintegrants'' [96,97], include croscarmellose sodium, type A (AcDiSol, FMC Corp., Food and Pharmaceutical Products, Philadelphia, PA), sodium starch glycolate (Primojel, Generichem Corp., Little Falls, NJ; Explotab, Mendell, A Penwest Co., Patterson, NY), and crospovidone (Polyplasdone XL, ISP Corp., Wayne, NJ). For instance, Botzolakis et al. [98] compared various levels of these newer disintegrants against 10% starch and 0% disintegrant as controls in dicalcium phosphate-based capsules filled on an instrumented Zanasi LZ-64 at a uniform compression force. In most cases, the dissolution rate of hydrochlorothiazide was dramatically enhanced (Fig. 19). Disintegrant efficiency was concentration-dependent. Although the typical use levels of these disintegrants in tablets is 2–4%, the most effective disintegrants required 4–6% for fast dissolution. The importance of drug solubility or magnesium stearate level was also clear. When magnesium stearate was reduced (from 1 to 0.5%)

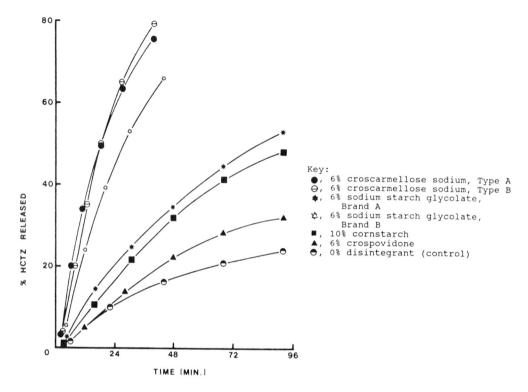

Fig. 19 Effect of disintegrants on hydrochlorothiazide dissolution from hard gelatin capsules (filler, dicalcium phosphate; lubricant, 1% magnesium stearate). (From Ref. 82.)

or when a more soluble drug (acetaminophen) was substituted for hydrochlorothiazide, less croscarmellose was needed to exert a similar effect on dissolution (Figs. 20 and 21) [98].

In a later study [99], the effect of disintegrants on hydrochlorothiazide dissolution from both soluble (anhydrous lactose) and "insoluble" (dicalcium phosphate) fillers was compared for different lubricant levels and tamping forces (instrumented Zanasi LZ-64 machine). Statistical analysis of this multivariable study revealed that all main factors and their interactions were significant. However, by averaging the results for each factor over all conditions, the relative magnitude of each main factor could be assessed, as in Figs. 22 and 23. Although the disintegrants were effective in promoting drug dissolution from both fillers, the effect was much less dramatic with lactose. This finding is not surprising, since the lactose-based capsule without disintegrant is already a fast-releasing formulation. Soluble fillers tend to dissolve, rather than to disintegrate. A beneficial effect of increasing the tamping force also was much more evident with the dicalcium phosphate-based capsules. As compression force increases, plug porosity may decrease, possibly making more effective the swelling action of disintegrants. Again, the retardant effect on dissolution of the hydrophobic lubricant is evident; however, it is apparent that the soluble lactose-based formulation is much less profoundly affected.

Surfactants

Surfactants may be included in capsule formulations to increase the wetting of the powder mass and enhance drug dissolution. The "waterproofing" effect of hydrophobic lubricants may be offset by the use of surfactants. Numerous studies have reported the beneficial effects of surfactants on disintegration and deaggregation or drug dissolution [80,82,89,95]. Botzolakis

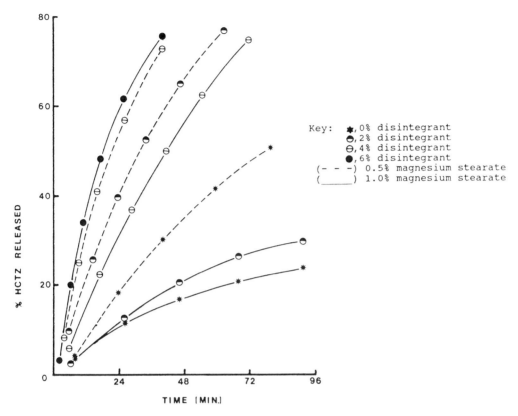

Fig. 20 Effect of lubricant level on hydrochlorothiazide dissolution from hard gelatin capsules (filler, dicalcium phosphate; disintegrant, croscarmellose sodium, type A). (From Ref. 82.)

[68] demonstrated enhanced liquid uptake into capsule plugs owing to surfactants. The most common surfactants employed in capsule formulations are sodium lauryl sulfate and sodium docusate. Levels of 0.1–0.5% are usually sufficient to overcome wetting problems. Figure 24 illustrates the dramatic effect surfactants can have on dissolution.

Hydrophilization

Another approach to improving the wettability of poorly soluble drugs is to treat the drug with a solution of a hydrophilic polymer. Lerk et al. [100] reported that both wettability of the powder and the rate of dissolution of hexobarbital from hard gelatin capsules could be greatly enhanced if the drug was treated with methylcellulose or hydroxyethylcellulose. In this process, called *hydrophilization*, a solution of the hydrophilic polymer was spread onto the drug in a high-shear mixer and the resultant mixture dried and screened. No benefit accrued when the drug and polymer were merely dry blended. No other excipients were included, and the capsules were loosely packed by hand. Lerk et al. [101] later treated phenytoin by hydrophilization with methylcellulose and compared the pure and treated drug compressed into plugs at 120 N to simulate a tamping machine. The plugs were manually filled into hard gelatin capsules. The treated phenytoin dissolved and absorbed (in humans) considerably faster than the untreated drug. However, no lubricants or fillers, were included in these capsules, as may be required for actual filling. The beneficial effect of hydrophilization on disintegration of benylate from hard gelatin capsules was demonstrated in vivo in humans by external scintigraphy [102].

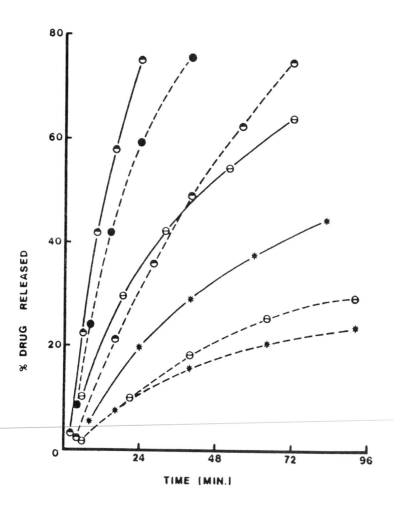

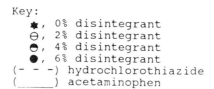

Fig. 21   Effect of croscarmellose sodium (type A) on drug dissolution from hard gelatin capsules (filler, dicalcium phosphate; lubricant, 1% magnesium stearate). (From Ref. 82.)

III.   SOFT GELATIN CAPSULES [3,103–108]

A.   Advantages

Several advantages of soft gelatin capsules derive from the fact that the encapsulation process requires that the drug be a liquid or at least dissolved, solubilized, or suspended in a liquid vehicle. Since the liquid fill is metered into individual capsules by a positive-displacement pump, a much higher degree of reproducibility is achieved than is possible with powder or

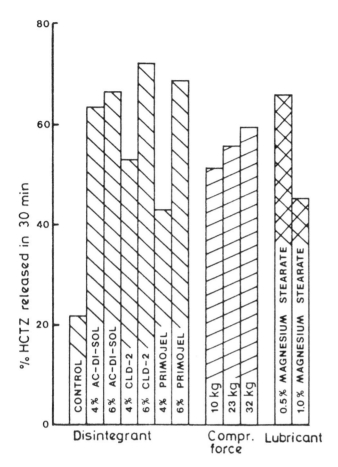

Fig. 22 Averaged effect of disintegrant, lubricant, and compression force on hydrochlorothiazide dissolution from dicalcium phosphate-based capsules. (From. Ref. 83.)

granule feed in the manufacture of tablets and hard gelatin capsule products. Moreover, a higher degree of homogeneity is possible in liquid systems than can be achieved in powder blends. A content uniformity of ± 3% has been reported [3] for soft gelatin capsules manufactured in a rotary die process.

Another advantage that derives from the liquid nature of the fill is rapid release of the contents with potential enhanced bioavailability. The proper choice of vehicle may promote rapid dispersion of capsule contents and drug dissolution. Hom and Miskal [107,109] compared the in vitro dissolution rates of 20 drugs from soft gelatin capsules and tablets. The drugs were either dissolved in polyethylene glycol 400 or suspended in polyols or nonionic surfactants. In all cases, more rapid dissolution occurred from the capsules. Several in vivo studies have demonstrated beneficial effects on the bioavailability of drugs administered in soft gelatin capsules [105,108].

For example, single-dose studies in humans comparing the sedative temazepam as a powder-filled hard gelatin capsule and as a polyethylene glycol solution in soft gelatin capsules revealed higher and earlier peak plasma levels, although there was no significant difference in their total availabilities [110]. In another example, digoxin dissolved in a vehicle consisting of polyethylene glycol 400, ethanol, propylene glycol, and water, and filled in soft gelatin capsules,

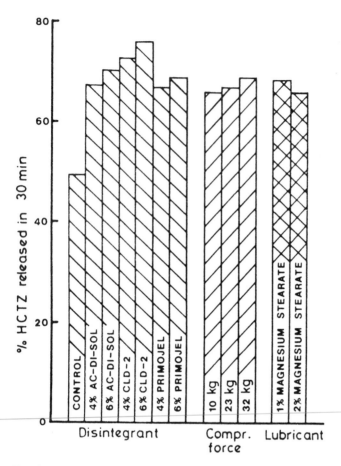

Fig. 23 Averaged effect of disintegrant, lubricant, and compression force on hydrochlorothiazide dissolution from anhydrous lactose-based capsules. (From Ref. 83.)

produced higher mean plasma levels in humans during the first 7 hr after administration than either an aqueous solution or commercial tablets [111]. In addition, the areas under the 14-hr plasma concentration curves were also greater for the soft gelatin capsule, compared with the solution or tablets.

Soft gelatin capsules are hermetically sealed as a natural consequence of the manufacturing process. Thus, this dosage form is uniquely suited for liquids and volatile drugs. Many drugs subject to atmospheric oxidation may also be formulated satisfactorily in this dosage form. Hom et al. [112] have shown that the soft gelatin shell can be an effective barrier to oxygen.

Soft gelatin capsules are available in a wide variety of sizes and shapes. Specialty packages in tube form (ophthalmics, ointments) or bead form (various cosmetics) are possible [3].

B. Disadvantages

One disadvantage of soft gelatin capsules is that such products must be contracted out to a limited number of firms having the necessary filling equipment and expertise. Materials must be shipped to the soft gelatin capsule facility and products must be shipped back to the phar-

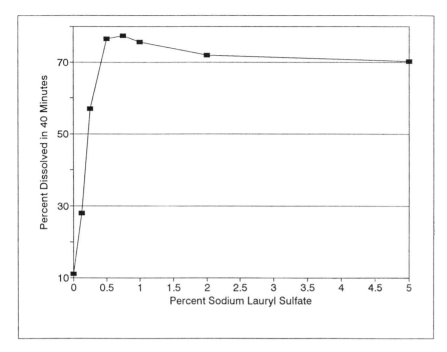

Fig. 24 The effect of percentage of sodium lauryl sulfate in the formulation on the dissolution of hydrochlorothiazide from capsules containing dicalcium phosphate as filler, 0.75% magnesium stearate, and filled on a dosator machine with 300-N–tamping force (*USP* method 2, 900 ml 0.1N HCl, 50 rpm). [Redrawn from Ref. 68.]

maceutical manufacturer for final packaging and distribution. Additional quality control measures may be required.

Soft gelatin capsules are not an inexpensive dosage form, particularly when compared with direct compression tablets [3].

There is a more intimate contact between the shell and its liquid contents than exists with dry-filled hard gelatin capsules, which increases the possibility of interactions. For instance, chloral hydrate formulated with an oily vehicle exerts a proteolytic effect on the gelatin shell; however, the effect is greatly reduced when the oily vehicle is replaced with polyethylene glycol [3].

Drugs can migrate from an oily vehicle into the shell, and this has been related to their water solubility and partition coefficient between water and the nonpolar solvent [113]. Studying 4-hydroxybenzoic acid as an encapsulated solution, Armstrong et al. [113] found that most transfer to the shell occurred during drying subsequent to manufacture; however, after completion of drying, only a combined 89% of the original amount of solute could be found in the shell and contents. It was thus considered that some of the solute may migrate to the outer shell surface where it can be lost by erosion or washing [113].

The possible migration of a drug into the shell must be considered in the packaging of topical products in soft gelatin tubelike capsules, as this could affect drug concentration in the ointment, as applied [3]. For other products, such as oral capsules or suppository capsules, both the shell and the contents must be considered in judging drug content when migration occurs. It is interesting that drug in the shell may provide for an initial dissolution of drug before shell rupture. When shell migration was negligible, there was a definite lag time in drug dissolution from encapsulated oily solution, indicating that rupture of the shell had to occur before solute release [113]. The larger the proportion of drug in the shell, the more overall drug release was dependent on the dissolution of the shell.

C. Composition of the Shell [3]

Similar to hard gelatin shells, the basic component of soft gelatin shells is gelatin; however, the shell has been plasticized by the addition of glycerin, sorbitol, or propylene glycol. Other components may include dyes, opacifiers, preservatives, and flavors. The ratio of dry plasticizer to dry gelatin determines the "hardness" of the shell and can vary from 0.3–1.0 for a very hard shell to 1.0–1.8 for a very soft shell. Up to 5% sugar may be included to give a "chewable" quality to the shell. The basic gelatin formulation from which the plasticized films are cast most usually consists of 1 part gelatin, 1 part water, and 0.4–0.6 part plasticizer. The residual shell moisture content of finished capsules will be in the range of 6–10%.

One physical parameter of finished capsules measured in stability studies is "softness." Whereas this parameter traditionally has been measured subjectively, Vemuri [114] has reported the objective measurement of this property by compression of individual capsules between the platens of a physical-testing machine.

D. Formulation of Soft Gelatin Capsules

The formulation for soft gelatin capsules [3,103] involves liquid, rather than powder technology. Materials are generally formulated to produce the smallest possible capsule consistent with maximum stability, therapeutic effectiveness, and manufacture efficiency [3].

Soft gelatin capsules contain a single liquid, a combination of miscible liquids, a solution of a drug in a liquid, or a suspension of a drug in a liquid. The liquids are limited to those that do not have an adverse effect on the gelatin walls. The pH of the liquid can be between 2.5 and 7.5. Liquids with more acid pHs would tend to cause leakage by hydrolysis of the gelatin. Both liquids with pHs higher than 7.5 and aldehydes decrease shell solubility by tanning the gelatin. Emulsions cannot be filled because inevitably water will be released that will affect the shell. Bauer and Dortunc [115] have proposed nonaqueous emulsions for both soft and hard gelatin capsules. In general, the emulsions were composed of a hydrophilic liquid, such as polyethylene glycol 400, and a triglyceride oil as the lipophilic liquid. Other liquids that cannot be encapsulated include water (greater than 5% of contents) and low molecular weight alcohols, such as ethyl alcohol. The types of vehicles used in soft gelatin capsules fall into two main groups.

1. Water-immiscible, volatile, or more likely nonvolatile liquids, such as vegetable oils, aromatic and aliphatic hydrocarbons (mineral oil), medium-chain triglycerides, and acetylated glycerides.
2. Water-miscible, nonvolatile liquids, such as low molecular weight polyethylene glycol (PEG-400 and 600) have come into use more recently because of their ability to mix with water readily and accelerate dissolution of dissolved or suspended drugs.

All liquids used for filling must flow by gravity at a temperature of 35°C or less. The sealing temperature of gelatin films is 37°–40°C. The limiting particle size for suspensions is that which can be handled by the pump. It is common practice to micronize (colloid mill) all materials during the preparation of the suspension. Typical suspending agents for oily bases and concentration of base are beeswax (5%), paraffin wax (5%), and animal stearates (1–6%). Aluminum monostearate and ethylcellulose are used with volatile organic liquids such as butylchloride and tetrachloroethylene. Suspending agents for nonoily bases include PEG 4000 and 6000 (1–5%), solid nonionics (10%), or solid glycol esters (10%). Little work has been published on the physical properties of materials for filling into soft gelatin capsules.

E. Manufacture of Soft Gelatin Capsules [3,104]

Plate Process

The oldest commercial process, this semiautomatic, batch process has been supplanted by more modern, continuous processes. Equipment for the plate process is no longer available. In general, the process involved (a) placing the upper half of a plasticized gelatin sheet over a die plate containing numerous die pockets, (b) application of vacuum to draw the sheet into the die pockets, (c) filling the pockets with liquor or paste, (d) folding the lower half of the gelatin sheet back over the filled pockets, and (e) inserting the "sandwich" under a die press where the capsules are formed and cut out.

Rotary Die Process

The first continuous process is the rotary die process that was invented in 1933 by R. P. Scherer. Aside from its being a continuous process, the rotary die process reduced manufacturing losses to a negligible level and content variation to ± 1–3% range, both major problems with earlier processes. In this process, the die cavities are machined into the outer surfaces of two rollers (i.e., die rolls). The die pockets on the left-hand roller form the left side of the capsule; the die pockets on the right-hand roller form the right side of the die capsule. The die pockets on the two rollers match as the rollers rotate. Two plasticized gelatin ribbons (prepared in the machine) are continuously and simultaneously fed with the liquid or paste fill between the rollers of the rotary die mechanism. The forceful injection of the feed material between the two ribbons causes the gelatin to swell into the left- and right-hand die pockets as they converge. As the die rolls rotate, the convergence of the matching die pockets seals and cuts out the filled capsules. The process is illustrated in Fig. 25 and an actual machine is pictured in Fig. 26.

Accogel Process

A continuous process for the manufacture of soft gelatin capsules filled with powders or granules was developed by Lederle Laboratories in 1949. In general, this is another rotary process involving (a) a measuring roll, (b) a die roll, and (c) a sealing roll. The measuring roll rotates directly over the die roll, and the pockets in the two rolls are aligned with each other. The powder or granular fill material is held in the pockets of the measuring roll under vacuum. A plasticized sheet is drawn into the die pockets of the die roll under vacuum. As the measuring roll and die rolls rotate, the measured doses are transferred to the gelatin-linked pockets of the die roll. The continued rotation of the filled die converges with the rotating sealing roll where a second gelatin sheet is applied to form the other half of the capsule. Pressure developed between the die roll and sealing roll seals and cuts out the capsules.

Bubble Method

The Globex Mark II Capsulator (Kinematics and Controls Corp., Deer Park, NY) produces truly seamless, one-piece soft gelatin capsules by a "bubble method." A concentric tube dispenser simultaneously discharges the molten gelatin from the outer annulus and the liquid content from the inner tube. By means of a pulsating pump mechanism, the liquids are discharged from the concentric tube orifice into a chilled-oil column as droplets that consist of a liquid medicament core within a molten gelatin envelop. The droplets assume a spherical shape under surface tension forces and the gelatin congeals on cooling. The finished capsules then must be degreased and dried.

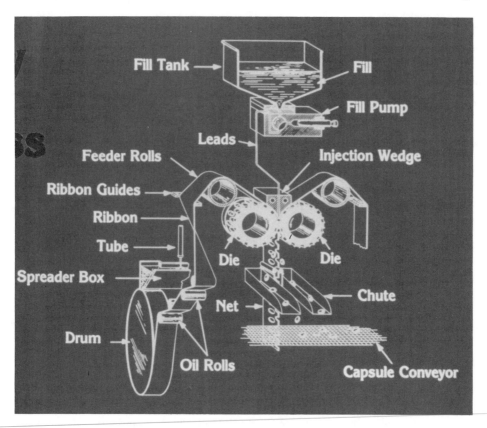

Fig. 25 Illustration of the rotary die process. (Courtesy of R. P. Scherer North America, Clearwater, FL)

IV. SOFT/LIQUID-FILLED HARD GELATIN CAPSULES

Perhaps the most important reason soft gelatin capsules became the standard for liquid-filled capsules was the inability to prevent leakage from hard gelatin capsules. The advent of such sealing techniques as banding and of self-locking hard gelatin capsules, together with the development of high–resting-state viscosity fills, has now made liquid/semisolid-filled hard gelatin capsules a feasible dosage form [118]. Currently, the only commercial example in this country is Vancocin HCl (Eli Lilly and Co., Indianapolis, IN).

Such a dosage form essentially offers the advantages of soft gelatin capsules. In addition, since most hard shell-filling machine manufacturers have developed models capable of filling liquids and semisolids, this technology can be brought in-house, thus avoiding the necessity of having to contract the work outside to a specialty house [118]. As with soft gelatin capsules, any materials filled into hard capsules must not dissolve, alter, or otherwise adversely affect the integrity of the shell. Generally, the fill material must be pumpable.

Three formulation strategies based on having a high resting-state viscosity after filling have been described [116,118]:

1. Thixotropic formulations: Such systems exhibit shear thinning when agitated and thus are pumpable. Yet, when agitation stops, the system rapidly establishes a gel structure, thereby avoiding leakage.

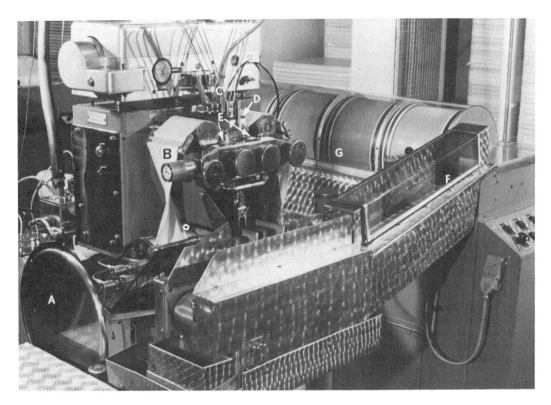

Fig. 26 Rotary die machine. (A) Ribbon casting drum; (B) ribbon; (C) filling leads; (D) injection mechanism; (E) rotary die; (F) capsule wash; (G) infrared dryer. (Courtesy of R. P. Scherer North America, Clearwater, FL)

2. Thermal-setting formulations: For these, excipients are used that are liquid at filling temperatures, but that gel or solidify in the capsule to prevent leakage.
3. Mixed thermal–Thixotropic systems: Improved resistance to leakage may be realized for low- or moderate-melting–point systems. Above its melting point, the liquid can still be immobile because of thixotropy.

A high resting-state viscosity is not a formulation prerequisite if capsules are banded and the contents have a viscosity of about 100 cP to prevent leakage from the mechanically interlocked capsules before banding.

Materials that may be used as carriers for the drug cover a wide range of chemical classes and melting points. These include vegetable oils, hydrogenated vegetable oils, various fats such as carnauba wax and cocoa butter, and polyethylene glycols (molecular weights 200–20,000). For thixotropic systems, the liquid excipient is often thickened with colloidal silicas. In one example [118], the liquid-active clofibrate was formulated as a thermal-setting system by adding 30% (based on total weight) polyethylene glycol 20,000. In another example [118], vitamin E was filled as a thixotropic system by adding approximately 6% each (based on total weight) beeswax and fumed silicon dioxide. Powdered drugs may be dissolved or suspended in thixotropic or thermal-setting systems.

In general, the more lipophilic the contents, the slower the release rate. Thus, by selecting excipients with varying hydrophilic–lipophilic balance (HLB), varying release rates may be achieved [118].

REFERENCES

1. R. Delaney, Surveying consumer preferences, Pharm. Executive, 2(3), 34 (1982).
2. B. E. Jones and T. D. Turner, A century of commercial hard gelatin capsules, Pharm. J., 213, 614 (1974).
3. J. P. Stanley, Capsules II. Soft gelatin capsules, in *The Theory and Practice of Industrial Pharmacy*, 2nd ed. (L. Lachman, H. A. Leiberman, and J. L. Kanig, eds.), Lea & Febiger, Philadelphia, 1976, pp. 404–420.
4. L. Eith, R. F. T. Stepto, F. Wittwer, and I. Tomka, Injection-moulded drug-delivery system, Manuf. Chem., 58, 21 (1987).
5. L. M. Mortada, F. A. Ismail, and S. A. Khalil, Correlation of urinary execretion with in vitro dissolution using four dissolution methods for ampicillin capsules, Drug Dev. Ind. Pharm. 11, 101 (1985).
6. K. Arnold, N. Gerber, and G. Levy, Absorption and dissolution studies on sodium diphenylhydantoin capsules, Can. J. Pharm. Sci., 5(4), 89 (1970).
7. A. J. Aguiar, L. M. Wheeler, S. Fusari, and J. E. Zelmar, Evaluation of physical and pharmaceutical factors involved in drug release and availability from chloramphenicol capsules, J. Pharm. Sci., 57, 1844 (1968).
8. A. J. Glazco, A. W. Kinkel, W. C. Alegnani, and E. L. Holmes, An evaluation of the absorption characteristics of different chloramphenicol preparations in normal human subjects, Clin. Pharmacol. Ther., 9, 472 (1968).
9. A study of physician attitudes toward capsules and other pharmaceutical product forms, Elanco Products Co., Div. Eli Lilly & Co., Indianapolis, IN, 1971, E1-0004.
10. M. Marvola, Adherence of drug products to the oesophagus, Pharm. Int., 3, 294 (1982).
11. H. Hey, F. Jorgensen, H. Hasselbach, and T. Wamberg, Oesophageal transit of six commonly used tablets and capsules, Br. Med. J., 285, 1717 (1982).
12. K. S. Channer and J. Virjee, Effect of posture and drink volume on the swallowing of capsules, Br. Med. J., 285, 1702 (1982).
13. T. Evans and G. M. Roberts, Where do all the tablets go? Lancet, 2, 1237 (1976).
14. J. T. Fell, Esophageal transit of tablets and capsules, Am. J. Hosp. Pharm., 40, 946 (1983).
15. G. W. Martyn, Jr., Production history of hard gelatin capsules from molding through filling, paper presented at Symposium on Modern Capsule Manufacturing, Society of Manufacturing Engineers, Philadelphia, PA, January 31, 1978.
16. L. C. Lappas, The manufacture of hard gelatin capsules, paper presented to Research and Development Section, The American Drug Manufacturers Association, Atlantic City, NJ, October 8, 1954.
17. G. W. Martyn, Jr., The people computer interface in a capsule molding operation, Drug Dev. Commun., 1, 39 (1974–75).
18. B. E. Jones, Hard gelatin capsules and the pharmaceutical formulator, Pharm. Technol., 9(9), 106 (1985).
19. L. W. Buckalew and K. E. Coffield, An investigation of drug expectancy as a function of capsule color and size and preparation form, J. Clin. Psychopharmacol., 2, 245 (1982).
20. General specifications for capsugel hard gelatin capsules, Bulletin BAS-114a-E, USA, Capsugel, Div. Warner-Lambert Co., Greenwood, SC, 1985.
21. "Coni-Snap—the hard gelatin capsule with the advantages that matter, Bulletin BAS-112-E, USA, Capsugel, Div. Warner-Lambert Co., Greenwood, SC, 1982.
22. R. Shah and L. L. Augsburger, Oxygen permeation in banded and non-banded hard gelatin capsules, Pharm. Res., 6, S-55 (1989).
23. F. Wittwer, New developments in hermetic sealing of hard gelatin capsules, Pharm. Manuf., 2(6), 24 (1985).
24. D. Scott, R. Shah, and L. L. Augsburger, A comparative evaluation of the mechanical strength of sealed and unsealed hard gelatin capsules, Int. J. Pharm., 84, 49 (1992).
25. K. S. Murthy and I. Ghebre-Sellassie, Current perspectives on the dissolution stability of solid oral dosage forms, J. Pharm. Sci., 82, 113 (1993).

26. K. Ito, S.-I. Kaga, and Y. Takeya, Studies on hard gelatin capsules I. Water vapor transfer between capsules and powders, Chem. Pharm. Bull., 17, 1134 (1969).
27. W. A. Strickland, Jr. and M. Moss, Water vapor sorption and diffusion through hard gelatin capsules, J. Pharm. Sci., 51, 1002 (1962).
28. Incompatibilities in prescriptions IV. The use of inert powders in capsules to prevent liquefaction due to deliquescence, J. Am. Pharm. Assoc. (Sci. Ed.), 29, 136 (1940).
29. J. H. Bell, N. A. Stevenson, and J. E. Taylor, A moisture transfer effect in hard gelatin capsules of sodium cromoglycate, J. Pharm. Pharmacol., 25(Suppl.), 96P (1973).
30. M. J. Kontny and C. A. Mulski, Gelatin capsule brittleness as a function of relative humidity at room temperature, Int. J. Pharm., 54, 79 (1989).
31. G. Zographi, G. P. Grandolfi, M. J. Kontny, and D. W. Mendenhall, Prediction of moisture transfer in mixtures of solids: Transfer via the vapor phase, Int. J. Pharm., 42, 77 (1988).
32. J. R. Schwier, G. G. Cooke, K. J. Hartauer, and L. Yu, Rayon: A source of furfural—a reactive aldehyde capable of insolubilizing gelatin capsules, Pharm. Technol., 17(5), 78 (1993).
33. K. S. Murthy, N. A. Enders, and M. B. Fawzi, Dissolution stability of hard-shell products, Part I: The effect of exaggerated storage conditions, Pharm. Technol., 13(3), 72 (1989).
34. K. S. Murthy, R. G. Reisch, Jr., and M. B. Fawzi, Dissolution stability of hard-shell capsule products, Part II: The effect of dissolution test conditions on in vitro drug release, Pharm. Technol., 13(6), 53 (1989).
35. H. Mohamad, R. Renoux, S. Aiache, and J.-M. Aiache, Etude de la stabilite biopharmaceutique des medicaments application a des gelules de chlorohydrate de tetracycline I. Etude in vitro, STP Pharma, 2, 531 (1986).
36. E. M. Marks, D. Tourtellotte, and A. Andus, The phenomenon of gelatin insolubility, Food Technol., 22, 1433 (1968).
37. T. Dahl, I. L. T. Sue, and A. Yum, The effect of pancreatin on the dissolution performance of gelatin-coated tablets exposed to high-humidity conditions, Pharm. Res., 8, 412 (1991).
38. H. Mohamad, J.-M. Aiche, R. Renoux, P. Mougin, and J.-P. Kantelip, Etude de la stabilite biopharmaceutique des medicaments application a des gelules de chlorohydrate de tetracycline IV. Etude complimentaire in vivo, STP Pharma, 3, 407 (1986).
39. M. Dey, R. Enever, M. Krami, D. G. Prue, D. Smith, and R. Weierstall, The dissolution and bioavailability of etodolac from capsules exposed to conditions of high relative humidity and temperatures, Pharm. Res., 10, 1295 (1993).
40. G. Cole, Capsule filling, Chem. Eng. (Lond.), 382, 473 (1982).
41. L. L. Augsburger, Powdered dosage forms, in Sprowl's American Pharmacy, 7th Ed. (L. W. Dittert, ed.), J. B. Lippincott, Philadelphia, 1974, pp. 301–343.
42. H. Clement and H. G. Marquart, The mechanical processing of hard gelatin capsules, News Sheet 3/70, Capsugel, A. G., CH-4000, Basel, Switzerland.
43. K. Ridgway and J. A. B. Callow, Capsule-filling machinery, Pharm. J., 212, 281 (1973).
44. V. Hostetler and J. Q. Bellard, Capsules I. Hard capsules, in The Theory and Practice of Industrial Pharmacy, 2nd Ed. (L. Lachman, H. A. Lieberman, and J. L. Kanig, eds.), Lea & Febiger, Philadelphia, 1976, pp. 389–404.
45. K. Ito, S.-I. Kaga, and Y. Takeya, Studies on hard gelatin capsules II. The capsule filling of powders and effects of glidant by ring filling machine-method, Chem. Pharm. Bull., 17, 1138 (1969).
46. G. Reier, R. Cohn, S. Rock, and F. Wagenblast, Evaluation of factors affecting the encapsulation of powders in hard gelatin capsules I. Semi-automatic capsule machines, J. Pharm. Sci., 57, 660 (1968).
47. K. Kurihara and I. Ichikawa, Effect of powder flowability on capsule filling weight variation, Chem. Pharm. Bull., 26, 1250 (1978).
48. Osaka R-180 Brochure, Osaka Automatic Machine Mfg. Co., Osaka 591, Japan (Sharpley-Stokes, Div. Pennwalt Corp., Warminster, PA 18944, distributors).
49. GKF, filling and sealing machine for hard gelatin capsules, Hofliger–Karg, Brochure HK/GKF/4/82-2E, Robert Bosch Corp., Packaging Machinery Div., South Plainfield, NJ.
50. K. B. Shah, L. L. Augsburger, L. E. Small, and G. P. Polli, Instrumentation of a dosing disc automatic capsule filling machine, Pharm. Technol., 7(4), 42 (1983).

51. L. E. Small and L. L. Augsburger, Instrumentation of an automatic capsule filling machine, J. Pharm. Sci., 66, 504 (1977).

52. L. E. Stoyle, Jr., Evaluation of the Zanasi automatic capsule machine, paper presented to the Industrial Pharmacy Section, A.Ph.A. 113th Annual Meeting, Dallas, TX, April 1966.

53. Macofar Mod. MT 13-1 and 13-2 Brochure, Macofar s.a.s., Bologna, Italy.

54. G. M. Irwin, G. J. Dodson, and L. J. Ravin, Encapsulation of clomacran phosphate I. Effect of flowability of powder blends, lot-to-lot variability, and concentration of active ingredient on weight variation of capsules filled on an automatic capsule filling machine, J. Pharm. Sci., 59, 547 (1970).

55. Z. T. Chowhan and Y. P. Chow, Powder flow studies I. Powder consolidation ratio and its relationship to capsule-filling weight variation, Int. J. Pharm., 4, 317 (1980).

56. Y. Miyake, A. Shimoda, T. Jasu, M. Furukawa, K. Nesuji, and K. Hoshi, Packing properties of pharmaceutical powders into hard gelatin capsules, Yakuzaigaku, 34, 32 (1974).

57. I. G. Jolliffe and J. M. Newton, Powder retention within a capsule dosator nozzle, J. Pharm. Pharmacol., 30(Suppl.), 41P (1978).

58. I. G. Jolliffe, J. M. Newton, and J. K. Walters, A theoretical approach to optimizing capsule filling by a dosator nozzle, J. Pharm. Pharmacol., 31(Suppl.), 70P (1979).

59. J. B. Schwartz, The instrumented tablet press: Uses in research and production, Pharm. Technol., 5(9), 102 (1981).

60. K. Marshall, Instrumentation of tablet and capsule filling machines, Pharm. Technol., 7(3), 68 (1983).

61. L. L. Augsburger, Instrumented capsule filling machines: Development and application, Pharm. Technol., 6(9), 111 (1982).

62. G. C. Cole and G. May, Instrumentation of a hard shell encapsulation machine, J. Pharm. Pharmacol., 24(Suppl.), 122P (1972).

63. G. C. Cole and G. May, The instrumentation of a Zanasi LZ/64 capsule filling machine, J. Pharm. Pharmacol., 27, 353 (1975).

64. L. E. Small and L. L. Augsburger, Aspects of the lubrication requirements for an automatic capsule-filling machine, Drug Dev. Ind. Pharm., 4, 345 (1978).

65. A. M. Mehta and L. L. Augsburger, Simultaneous measurement of force and displacement in an automatic capsule filling machine, Int. J. Pharm., 4, 347 (1980).

66. A. M. Mehta and L. L. Augsburger, Quantitative evaluation of force development curves in an automatic capsule filling machine, paper presented to the IPT Section, A.Ph.A. Academy of Pharmaceutical Sciences, 128th Annual A.Ph.A. Meeting, St. Louis, MO, March–April 1981.

67. R. Greenberg, Effects of AZ-60 filling machine dosator settings upon slug hardness and dissolution of capsules, Proc. 88th National Meeting, Am. Inst. Chem. Engrs., Session 11, Philadelphia, PA, June 8–12, 1980 (Fiche 29).

68. J. E. Botzolakis, Studies on the mechanism of disintegrant action in encapsulated dosage forms, Ph.D. Thesis, University of Maryland (1985).

69. C. Mony, C. Sambeat, and G. Cousin, Interet des measures de pression dans la formulation et le remplissage des gelules, Newsheet 1977, Capsugel A. G., Basel Switzerland.

70. D. J. Rowley, R. Hendry, M. D. Ward, and P. Timmins, The instrumentation of an automatic capsule filling machine for formulation design studies, paper presented at 3rd International Conference on Powder Technology, Paris, 1983.

71. I. G. Jolliffe, J. M. Newton, and D. Cooper, The design and use of an instrumented mG2 capsule filling machine simulator, J. Pharm. Pharmacol., 34, 230 (1982).

72. I. G. Jolliffe and J. M. Newton, An investigation of the relationship between particle size and compression during capsule filling with an instrumented mG2 simulator, J. Pharm. Pharmacol., 34, 415 (1982).

73. K. B. Shah, L. L. Augsburger, and K. Marshall, An investigation of some factors influencing plug formation and fill weight in a dosing disk-type automatic capsule-filling machine, J. Pharm. Sci., 75, 291 (1986).

74. K. B. Shah, L. L. Augsburger, and K. Marshall, Multiple tamping effects on drug dissolution from capsules filled on a dosing-disk type automatic capsule filling machine, J. Pharm. Sci., 76, 639 (1987).

75. J. W. Cropp, L. L. Augsburger, and K. Marshall, Simultaneous monitoring of tamping force and piston displacement (F–D) on an Hofliger-Karg capsule filling machine, Int. J. Pharm., 71, 127 (1991).

76. A. M. Mehta and L. L. Augsburger, A preliminary study of the effect of slug hardness on drug dissolution from hard gelatin capsules filled on an automatic capsule filling machine, Int. J. Pharm., 7, 327 (1981).

77. A. Ludwig and M. Van Ooteghem, Disintegration of hard gelatin capsules, Part 2: Disintegration mechanism of hard gelatin capsules investigated with a stereoscopic microscope, Pharm. Ind., 42, 405 (1980).

78. A. Ludwig and M. Van Ooteghem, Disintegration of hard gelatin capsules, Part 5: The influence of the composition of the test solution on disintegration of hard gelatin capsules, Pharm. Ind., 43, 188 (1981).

79. J. H. Fincher, J. G. Adams, and M. H. Beal, Effect of particle size on gastrointestinal absorption of sulfisoxazole in dogs, J. Pharm. Sci., 54, 704 (1963).

80. S. M. Bastami and M. J. Groves, Some factors influencing the in vitro release of phenytoin from formulations, Int. J. Pharm., 1, 151 (1978).

81. J. M. Newton and G. Rowley, On the release of drug from hard gelatin capsules, J. Pharm. Pharmacol., 22, 1635 (1970).

82. J. M. Newton, G. Rowley, and J. F. V. Tornblom, The effect of additives on the release of drug from hard gelatin capsules, J. Pharm. Pharmacol., 23, 452 (1971).

83. R. J. Whithey and C. A. Mainville, A critical analysis of a capsule dissolution test, J. Pharm. Sci., 58, 1120 (1969).

84. Diphenylhydantoin, Clin. Alert, 299, Dec. 8 (1970).

85. P. York, Studies of the effect of powder moisture content on drug release from hard gelatin capsules, Drug Dev. Ind. Pharm., 6, 605 (1980).

86. L. L. Augsburger and R. F. Shangraw, Effect of glidants in tableting, J. Pharm. Sci., 55, 418 (1966).

87. P. York, Application of powder failure testing equipment in assessing effect of glidants on flowability of cohesive pharmaceutical powders, J. Pharm. Sci., 64, 1216 (1975).

88. H. M. Sadek, J. L. Olsen, H. L. Smith, and S. Onay, A systematic approach to glidant selection, Pharm. Technol., 6(2), 43 (1982).

89. J. M. Newton, G. Rowley, and J. F. V. Tornblom, Further studies on the effect of additives on the release of drug from hard gelatin capsules, J. Pharm. Pharmacol., 23, 156S (1971).

90. J. C. Samyn and W. Y. Jung, In vitro dissolution from several experimental capsules, J. Pharm. Sci., 59, 169 (1970).

91. K. S. Murthy and J. C. Samyn, Effect of shear mixing on in vitro drug release of capsule formulations containing lubricants, J. Pharm. Sci., 66, 1215 (1977).

92. A. G. Stewart, D. J. W. Grant, and J. M. Newton, The release of a model low-dose drug (riboflavine) from hard gelatin capsule formulations, J. Pharm. Pharmacol., 31, 1 (1979).

93. H. Nakagwu, Effects of particle size of rifampicin and addition of magnesium stearate in release of rifampicin from hard gelatin capsules, Yakugaku Zasshi, 100, 1111 (1980).

94. P. T. Shah and W. E. Moore, Dissolution behavior of commercial tablets extemporaneously converted to capsules, J. Pharm. Sci., 59, 1034 (1970).

95. F. W. Goodhart, R. H. McCoy, and F. C. Ninger, New in vitro disintegration and dissolution test method for tablets and capsules, J. Pharm. Sci., 62, 304 (1973).

96. R. F. Shangraw, A. Mitrevej, and M. Shah, A new era of tablet disintegrants, Pharm. Technol., 4(10), 49 (1980).

97. R. F. Shangraw, J. W. Wallace, and F. M. Bowers, Morphology and functionality of tablet excipients for direct compression: Part II, Pharm. Technol., 5(10), 44 (1981).

98. J. E. Botzolakis, L. E. Small, and L. L. Augsburger, Effect of disintegrants on drug dissolution from capsules filled on a dosator-type automatic capsule-filling machine, Int. J. Pharm., 12, 341 (1982).

99. J. E. Botzolakis and L. L. Augsburger, The role of disintegrants in hard-gelatin capsules, J. Pharm. Pharmacol., 37, 77 (1984).

100. C. F. Lerk, M. Lagas, J. T. Fell, and P. Nauta, Effect of hydrophilization of hydrophobic drugs on release rate from capsules, J. Pharm. Sci., 67, 935 (1978).

101. C. F. Lerk, M. Lagas, L. Lie-A-Huen, P. Broersma, and K. Zuurman, in vitro and in vivo availability of hydrophilized phenytoin from capsules, J. Pharm. Sci., 68, 634 (1979).
102. M. Lagas, H. J. C. deWit, M. G. Woldring, D. A. Piers, and C. F. Lerk, Technetium labelled disintegration of capsules in the human stomach, Pharm. Acta Helv., 55, 114 (1980).
103. W. R. Ebert, Soft elastic gelatin capsules: A unique dosage form, Pharm. Technol., 1(10), 44 (1977).
104. G. Muller, Methods and machines for making gelatin capsules, Manuf. Chem., 32, 63 (1961).
105. I. R. Berry, Improving bioavailability with soft gelatin capsules, Drug Cosmet. Ind., 133(3), 32 (1983).
106. I. R. Berry, One-piece, soft gelatin capsules for pharmaceutical products, Pharm. Eng., 5(5), 15 (1985).
107. F. S. Hom and J. J. Miskel, Enhanced drug dissolution rates for a series of drugs as a function of dosage form design, Lex Sci., 8(1), 18 (1971).
108. H. Seager, Soft gelatin capsules: A solution to many tableting problems, Pharm. Technol., 9(9), 84 (1985).
109. F. S. Hom and J. J. Miskel, Oral dosage form design and its influence on dissolution rates for a series of drugs, J. Pharm. Sci., 59, 827 (1970).
110. L. J. Fuccella, G. Bolcioni, V. Tamassia, L. Ferario, and G. Tognoni, human pharmacokinetics and bioavailability of temazepam administered in soft gelatin capsules, Eur. J. Clin. Pharmacol., 12, 383 (1977).
111. B. F. Johnson, C. Bye, G. Jones, and G. A. Sabey, A completely absorbed oral preparation of digoxin, Clin. Pharmacol. Ther., 19, 746 (1976).
112. F. S. Hom, S. A. Veresh, and W. R. Ebert, Soft gelatin capsules II: Oxygen permeability study of capsule shells, J. Pharm. Sci., 64, 851 (1975).
113. N. A. Armstrong, K. C. James, and W. K. L. Pugh, Drug migration into soft gelatin capsule shells and its effect on the in-vitro availability, J. Pharm. Pharmacol., 36, 361 (1984).
114. S. Vemuri, Measurement of soft elastic gelatin capsule firmness with a universal testing machine, Drug Dev. Ind. Pharm., 10, 409 (1984).
115. K. H. Bauer and B. Dortunc, Non-aqueous emulsions as vehicles for capsule filling, Drug Dev. Ind. Pharm., 10, 699 (1984).
116. D. Francois and B. E. Jones, Making the hard capsule with the soft center, Manuf. Chem. Aerosol News, 50(3), 37 (1979).
117. S. E. Walker, J. A. Ganley, K. Bedford, and T. Eaves, The filling of molten and thixotropic formulations into hard gelatin capsules, J. Pharm. Pharmacol., 32, 389 (1980).
118. The hard capsule with the soft center, Elanco Products Co., Div. Eli Lilly and Co., Indianapolis, IN.

GENERAL REFERENCES

Augsburger L. L., Powdered dosage forms, in Sprowl's American Pharmacy, 7th Ed. (L. W. Dittert, ed.), J. B. Lippincott, Philadelphia, 1974, pp. 301–343.
Hostetler, V. and J. Q. Bellard, Capsules I. Hard capsules, in The Theory and Practice of Industrial Pharmacy, 2nd Ed. (L. Lachman, H. A. Lieberman, and J. L. Kanig, eds.), Lea & Febiger, Philadelphia, 1976, pp. 389–404.
King, R. E. and J. B. Schwartz, Oral solid dosage forms, in Remington's Pharmaceutical Sciences, 17th Ed. (A. R. Gennaro, ed. and chairman), Mack Publishing, Easton, PA, 1985, pp. 1603–1632.
Marshall, K., Solid oral dosage forms, in Modern Pharmaceutics (G. S. Banker and C. T. Rhodes, eds.), Marcel Dekker, New York, 1979, pp. 359–427.
Stanley, J. P., Capsules II. Soft gelatin capsules, in Theory and Practice of Industrial Pharmacy, 2nd Ed. (H. A. Lieberman and J. L. Kanig, eds.), Lea & Febiger, Philadelphia, 1976, pp. 404–420.

12

Parenteral Products

James C. Boylan
Abbott Laboratories, Abbott Park, Illinois

Alan L. Fites
Fites Consulting, Greenwood, Indiana

Steven L. Nail
Purdue University, West Lafayette, Indiana

I. INTRODUCTION

The first official injection (morphine) appeared in the *British Pharmacopoeia* (*BP*) of 1867. It was not until 1898 when cocaine was added to the *BP* that sterilization was attempted. In this country, the first official injections may be found in the *National Formulary* (*NF*), published in 1926. Monographs were included for seven sterile glass-sealed ampoules. The *United States Pharmacopeia* (*USP*) published in the same year contained a chapter on sterilization, but no monographs for ampoules. The current *USP* contains monographs for over 400 injectable products [1].

Parenteral administration of drugs by intravenous (IV), intramuscular (IM), or subcutaneous (SC) routes is now an established and essential part of medical practice. Advantages for parenterally administered drugs include the following: rapid onset; predictable effect; predictable and nearly complete bioavailability; and avoidance of the gastrointestinal tract (GIT), and hence, the problems of variable absorption, drug inactivation, and GI distress. In addition, the parenteral route provides reliable drug administration in very ill or comatose patients.

The pharmaceutical industry directs considerable effort toward maximizing the usefulness and reliability of oral dosage forms in an effort to minimize the need for parenteral administration. Factors that contribute to this include certain disadvantages of the parenteral route, including the frequent pain and discomfort of injections, with all the psychological fears associated with "the needle," plus the realization that an incorrect drug or dose is often harder or impossible to counteract when it has been given parenterally (particularly intravenously), rather than orally.

In recent years, parenteral dosage forms, especially IV forms, have gained immensely in use. The reasons for this growth are many and varied, but they can be summed up as (a) new and better parenteral administration techniques; (b) new forms of nutritional therapy, such as intravenous lipids, amino acids, and trace metals; (c) the need for simultaneous administration of multiple drugs in hospitalized patients receiving IV therapy, (d) the extension of parenteral therapy into the home; and (e) an increasing number of drugs that can be administered only by a parenteral route.

Many important drugs are available only as parenteral dosage forms. Notable among these are biotechnology drugs; insulin; several cephalosporin antibiotic products; and drugs such as heparin, protamine, and glucagon. In addition, other drugs, such as lidocaine hydrochloride and many anticancer products, are used principally as parenterals. The reasons that certain drugs are administered largely or exclusively by the parenteral route are very inefficient or unreliable absorption from the GIT, destruction or inactivation in the GIT, extensive mucosal or first-pass metabolism following oral administration, or clinical need in particular medical situations for rapid, assured high blood and tissue levels.

Along with this astounding growth in the use of parenteral medications, the hospital pharmacist has become a very knowledgeable, key individual in most hospitals, having responsibility for hospital-wide IV admixture programs, parenteral unit-dose packaging, and often central surgical supply. By choice, by expertise, and by responsibility, the pharmacist has accumulated the greatest fund of information about parenteral drugs—not only their clinical use, but also their stability, incompatibilities, methods of handling and admixture, and proper packaging. More and more, nurses and physicians are looking to the pharmacist for guidance on parenteral products.

To support the institutional pharmacist in preparing IV admixtures (which typically involves adding one or more drugs to large-volume parenteral fluids), equipment manufacturers have designed laminar flow units, electromechanical compounding units, transfer devices, and filters specifically adaptable to a variety of hospital programs.

The nurse and physician have certainly not been forgotten either. A wide spectrum of IV and IM administration devices and aids have been made available in recent years for bedside use. Many innovative practitioners have made suggestions to industry that have resulted in product or technique improvements, particularly in IV therapy. The use of parenteral products is growing at a very significant rate in nonhospital settings, such as outpatient surgical centers and homes. The reduction in costs associated with outpatient and home care therapy, coupled with advances in drugs, dosage forms, and delivery systems, has caused a major change in the administration of parenteral products [2].

II. ROUTES OF PARENTERAL ADMINISTRATION

The routes of parenteral administration of drugs are (a) subcutaneous, (b) intramuscular, and (c) intravenous; other more specialized routes are (d) intrathecal, (e) intracisternal, (f) intra-arterial, (g) intraspinal, (h) intraepidural, and (i) intradermal. The intradermal route is not typically used to achieve systemic drug effects. The similarities and differences of the routes or their definitions are highlighted in Table 1. The major routes will be discussed separately.

A. The Subcutaneous Route

Lying immediately under the skin is a layer of fat, the superficial fascia (see Fig. 1 in Chapter 8), that lends itself to safe administration of a great variety of drugs, including vaccines, insulin, scopolamine, and epinephrine. Subcutaneous (SC; also SQ or sub-Q) injections are usually administered in volumes up to 2 ml using a ½- to 1-in. 22-gauge (or smaller) needle. Care must be taken to ensure that the needle is not in a vein. This is done by lightly pulling back on the syringe plunger (aspiration) before making the injection. If the needle is inadvertently located in a vein, blood will appear in the syringe and the injection should not be made. The injection site may be massaged after injection to facilitate drug absorption. Drugs given by this

Table 1 Various Parenteral Routes of Drug Administration

Routes	Usual volume (ml)	Needle commonly used	Formulation constraints	Types of medication administered
Primary parenteral routes				
Small-volume parenterals				
Subcutaneous	2	⅝ in., 23 gauge	Need not be isotonic	Insulin, vaccines
Intramuscular	2	1½ in., 22 gauge	Can be solutions, emulsions, oils, or suspensions, isotonic preferably	Nearly all drug classes
Intravenous	50	Veinpuncture 1½ in., 22 gauge	Solutions and some emulsions	Nearly all drug classes
Large-volume parenterals	100 and larger (infusion unit)	Venoclysis 1½ in., 19 gauge	Solutions and some emulsions	Nearly all drug classes (see precautionary notes in text)
Other parenteral routes				
Intra-arterial: directly into an artery (immediate action sought in peripheral area)	2–20	20–22 gauge	Solutions and some emulsions	Radiopaque media, antineoplastics, antibiotics
Intrathecal (intraspinal; into spinal canal)	1–4	24–28 gauge	Must be isotonic	Local anesthetics, analgesics; neurolytic agents
Intraepidural (into epidural space near spinal column)	6–30	5 in., 16–18 gauge	Must be isotonic	Local anesthetics, narcotics, α_2-agonists, steroids
Intracisternal: directly into caudal region of the brain between the cerebellum and the medulla oblongata			Must be isotonic	
Intra-articular: directly into a joint, usually for a local effect there, as for steroid anti-inflammatory action in arthritis	2–20	1.5–2 in., 18–22 gauge	Must be isotonic	Morphine, local anesthetics, steroids, NSAIDs, antibiotics
Intracardial: directly into the heart when life is threatened (epinephrine stimulation in severe heart attack)	0.2–1	5 in., 22 gauge		Cardiotonic drugs, calcium
Intrapleural: directly into the pleural cavity or a lung (also used for fluid withdrawal)	2–30	2–5 in., 16–22 gauge		Local anesthetics, narcotics, chemotherapeutic agents
Diagnostic testing				
Intradermal	10	⅝ in., 26 gauge	Should be isotonic	Diagnostic agents

route will have a slower onset of action than by the IM or IV routes, and total absorption may also be less.

Sometimes dextrose or electrolyte solutions are given subcutaneously in amounts from 250 to 1000 ml. This technique, called hypodermoclysis, is used when veins are unavailable or difficult to use for further medication. Irritation of the tissue is a danger with this technique. Administration of the enzyme hyaluronidase can help by increasing absorption and decreasing tissue distention. Irritating drugs and vasoconstrictors can lead to abscesses, necrosis, or inflammation when given subcutaneously. Body sites suitable for SC administration include most portions of the arms and legs plus the abdomen. When daily or frequent administration is required, the injection site can and should be continuously changed or rotated, especially by diabetic patients self-administering insulin.

B. The Intramuscular Route

The IM route of administration is second only to the IV route in rapidity of onset of systemic action. Injections are made into the striated muscle fibers that lie beneath the subcutaneous layer. The principal sites of injection are the gluteal (buttocks), deltoid (upper arm), and vastus lateralis (lateral thigh) muscles. The usual volumes injected range from 1.0 to 3.0 ml, with volumes up to 10.0 ml sometimes being given (in divided doses) in the gluteal or thigh areas (see Table 1). Again, it is important to aspirate before injecting to ensure that the drug will not be administered intravenously. Needles used in administering IM injections range from 1 to $1\frac{1}{2}$ in. and 19 to 22 gauge, the most common being $1\frac{1}{2}$ in. and 22 gauge.

The major clinical problem arising from IM injections is muscle or neural damage, the injury normally resulting from faulty technique, rather than the medication.

Most injectable products can be given intramuscularly. As a result, there are numerous dosage forms available for this route of administration: solutions, oil-in-water (o/w) or water-in-oil (w/o) emulsions, suspensions (aqueous or oily base), colloidal suspensions, and reconstitutable powders. Those product forms in which the drug is not fully dissolved generally result in slower, more gradual drug absorption, a slower onset of action, and sometimes longer-lasting drug effects.

Intramuscularly administered products typically form a ''depot'' in the muscle mass from which the drug is slowly absorbed. The peak drug concentration is usually seen within 1–2 hr. Factors affecting the drug-release rate from an IM depot include the compactness of the depot (the less compact and more diffuse, the faster the release), the rheology of the product (affects compactness), concentration and particle size of drug in the vehicle, nature of the solvent or vehicle, volume of the injection, tonicity of the product, and physical form of the product.

C. The Intravenous Route

Intravenous medication is injected directly into a vein either to obtain an extremely rapid and predictable response or to avoid irritation of other tissues. This route of administration also provides maximum availability and assurance in delivering the drug to the site of action. However, a major danger of this route of administration is that the rapidity of absorption makes antidoting very difficult, if not impossible, in most instances. Care must also be used to avoid too rapid a drug administration by the IV route because irritation or an excessive drug concentration at the target organ (drug shock) can occur. The duration of drug activity is dependent on the initial dose and the distribution, metabolism, and excretion properties (pharmacokinetics) of the drug. For most drugs, the biological half-life is independent of the initial dose, because the elimination process is first-order. Thus, an intravenous drug with a short half-life would not provide a sustained blood level. The usual method of administration for drugs with short half-lives is to use continuous IV drip. Intravenous injections (vein puncture) normally range from 1 to 100 ml and are given with either a 20- or 22-gauge $1\frac{1}{2}$-in. needle, with an injection rate of 1 ml/10 sec for volumes up to 5 ml and 1 ml/20 sec for volumes over 5 ml. Only drugs in aqueous or hydroalcoholic solutions are to be given by the IV route.

Large proximal veins, such as those located inside the forearm, are most commonly used for IV administration. Because of the rapid dilution in the circulating blood and the general insensitivity of the venous wall to pain, the IV route may be used to administer drugs that would be too irritating or caustic to give by other routes (e.g., nitrogen mustards), provided that proper dosing procedures are employed. The risk of thrombosis is increased when extremity sites such as the wrist or ankle are used for injection sites, or when potentially irritating IV products are used, with the risk further increasing in patients with impaired circulation.

The IV infusion of large volumes of fluids (100–1000 ml) has become increasingly popular (Figs. 1 and 2). This technique, called *venoclysis*, utilizes products known as large-volume parenterals (LVPs). It is used to supply electrolytes and nutrients, to restore blood volume, to prevent tissue dehydration, and to dilute toxic materials already present in body fluids. Various parenteral drug solutions may often be conveniently added to the LVP products as they are being administered (Figs. 3 and 4), or before administration, to provide continuous and prolonged drug therapy. Such drug additions to LVP has become very common in hospitals. Combining parenteral dosage forms for administration as a unit product is known as *IV admixtures*. Pharmacists practicing such IV additive product preparation must be very knowledgeable to avoid physical and chemical incompatibilities in the modified LVP, creation of any therapeutic incompatibilities with other drugs being given parenterally or by any other route, or loss of sterility or addition of extraneous matter.

Commonly administered large-volume parenterals include such products as sodium chloride injection [USP] (0.9% saline), which replenish fluids and electrolytes, and 5% dextrose injection [USP], which provides fluid plus nutrition (calories) or various combinations of dextrose and saline. In addition, numerous other nutrient and ionic solutions are available for clinical use, the most popular of which are solutions of essential amino acids or lipid emulsions. These solutions are modified to be hypertonic, isotonic, or hypotonic to aid in maintaining both fluid, nutritional, and electrolyte balance in a particular patient according to need. Indwelling needles or catheters are required in LVP administration. Care must be taken to avoid local or systemic infections or thrombophlebitis owing to faulty injection or administration technique.

D. Other Parenteral Routes

Other more specialized parenteral routes are listed and described briefly in Table 1. The intra-arterial route involves injecting a drug directly into an artery. This technique is not simple and may require a surgical procedure to reach the artery. It is important that the artery not be missed, since serious nerve damage can occur to the nerves lying close to arteries. Doses given by this route should be minimal and given gradually, since, once injected, the drug effect cannot be neutralized. As shown in Table 1, the intra-arterial route is used to administer radiopaque contrast media for viewing an organ, such as the heart or kidney, or to perfuse an antineoplastic agent at the highest possible concentration to the target organ.

The intrathecal route is employed to administer a drug directly into the cerebrospinal fluid at any level of the cerebrospinal axis. This route is used when it is not possible to achieve sufficiently high plasma levels to accomplish adequate diffusion and penetration into the cerebrospinal fluid. This is not the same route used to achieve spinal anesthesia, for which the drug is injected within the dural membrane surrounding the spinal cord, or in extradural or

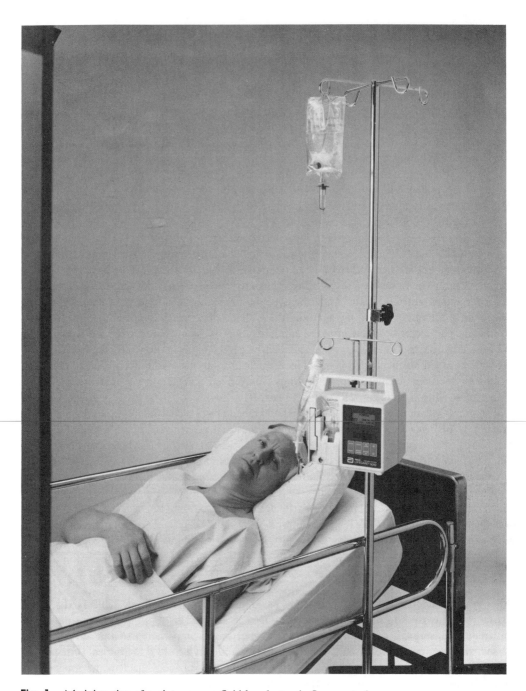

Fig. 1 Administration of an intravenous fluid by electronic flow control.

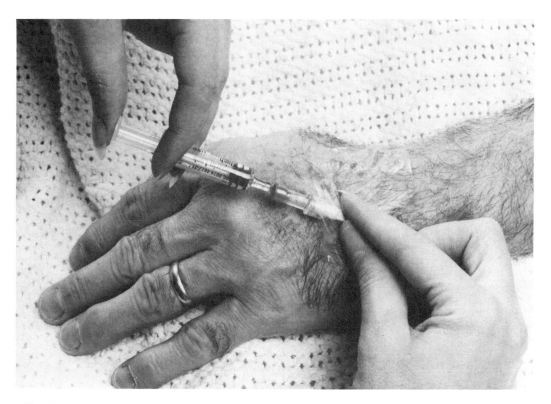

Fig. 2 Direct intravenous administration using gum rubber injection site.

epidural anesthesia (caudal or sacral anesthesia), for which the drug is deposited outside the dural membrane and within the body spinal caudal canals. Parenteral products administered by the intrathecal, intraspinal, and intracisternal routes must be especially carefully formulated, with ingredients of the highest purity because of the sensitivity of nerve tissue.

Intradermal (ID) administration involves injection into the skin layer (see Fig. 3 in Chapter 8). Examples of drugs administered by this route are allergy test materials. Since intradermal drugs are normally given for diagnostic purposes, it is important that the product per se be nonirritating. Volumes are normally given at 0.05 ml/dose and the solutions are isotonic. Intradermal medication is usually administered with a $1/2$- or $5/8$-in., 25- or 26-gauge needle, inserted at an angle nearly parallel to the skin surface. Absorption is slow and limited from this site, since the blood vessels are extremely small, although the area is highly vascular. The site should not be massaged after the injection of allergy test materials. Skin testing includes not only allergens, such as pollens or dust, but also microorganisms, as in the tuberculin or histoplasmin skin tests.

III. SPECIALIZED LARGE-VOLUME PARENTERAL AND STERILE SOLUTIONS

Large-volume parenterals designed to provide fluid (water), calories (glucose solutions), electrolytes (saline solutions), or combinations of these materials have been described. Several other specialized LVP and sterile solutions are also used in medicine and will be described

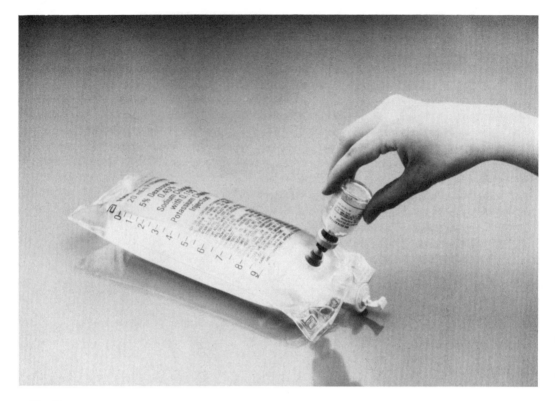

Fig. 3 Addition of intravenous medication directly to primary intravenous solution container.

here, even though two product classes (peritoneal dialysis and irrigating solutions) are not parenteral products.

A. Hyperalimentation Solutions

Parenteral hyperalimentation involves administration of large amounts of nutrients (e.g., carbohydrates, amino acids, and vitamins) to maintain a patient who is unable to take food orally, for several weeks, at caloric intake levels of 4000 cal/day or more. Earlier methods of parenteral alimentation, which involved IV administration, were not typically able to maintain patients without a weight loss and gradual deterioration in physical condition. Parenteral hyperalimentation involves continuous administration of the nutrient solution into the superior vena cava by an indwelling catheter. Available hyperalimentation solutions vary in various amino acids, vitamins, minerals, and electrolytes. The method permits administration of lifesaving or life-sustaining nutrients to comatose patients or to patients undergoing treatment for esophageal obstruction, GI diseases (including cancer), ulcerative colitis, and other disease states.

B. Peritoneal Dialysis Solutions

The sterile peritoneal dialysis solutions are infused continuously into the abdominal cavity, bathing the peritoneum (the semipermeable membrane covering the viscera of the abdominal cavity), and are then continuously withdrawn. The purpose of peritoneal dialysis is to remove toxic substances from the body, or to aid and accelerate the excretion function normal to the kidneys. The process is employed to counteract some forms of drug or chemical toxicity as

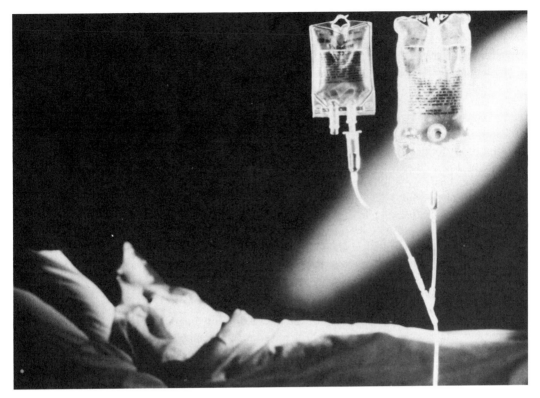

Fig. 4 ''Piggybacking'' of a small-volume intravenous fluid into the primary large-volume intravenous solution.

well as to treat acute renal insufficiency. Peritoneal dialysis solutions contain glucose and have an ionic content similar to normal extracellular fluid. Toxins or metabolites diffuse into the circulating dialysis fluid through the peritoneum and are removed. At the same time, excess fluid is removed from the patient if the glucose content renders the dialysis solution hyperosmotic. An antibiotic is often added to these solutions as a prophylactic measure.

C. Irrigating Solutions

Irrigating solutions are intended to irrigate, flush, and aid in cleansing body cavities and wounds. Although certain IV solutions, such as normal saline, may be used as irrigating solutions, solutions designed as irrigating solutions should not be used parenterally. Since irrigating solutions used in treatment of serious wounds infuse into the bloodstream to some degree, they must be sterile, pyrogen-free, and made and handled with the same care as parenteral solutions.

IV. PHYSICOCHEMICAL FACTORS AND COMPONENTS

Physicochemical properties of the active drug and the components used in a parenteral dosage form can significantly affect the availability of the drug substance. Other factors that influence drug availability are physiological (biological conditions and disease state of the patient), the route of administration, and the type of dosage form. Intramuscular and subcutaneous routes of parenteral administration require drug absorption before blood or cerebrospinal fluid levels can be achieved. The rate at which the drug is absorbed has a significant influence on the concentration of the drug in the blood. With an IM suspension, drug dissolution is usually the rate-limiting step in the absorption of the drug at the injection site [3]. The absorption of the drug following IM administration is greatly influenced by the physicochemical properties of the drug.

Components that are incorporated into parenteral dosage forms may have very rigid specifications and standards. Because of these requirements, extensive analytical and toxicological testing is performed to ensure that a chemical is acceptable. Thorough toxicity testing is required of a new drug or other component not previously approved for parenteral use, and the accumulated data are evaluated. Testing may be done on the individual components as well as the final dosage forms. Given such explicit qualities as purity, safety, and lack of (or minimum) pharmacological effect required of parenteral additives, the formulator is restricted to a very few materials as excipients, preservatives, suspending agents, and surfactants. Because of the very extensive pharmacological and toxicological data required to obtain approval for any new additive, most formulators continue to depend on materials of known acceptability.

A. The Active Drug

A thorough evaluation of properties of the active drug or drugs is essential in developing a stable and safe parenteral dosage form. The physical and chemical factors that may significantly affect the development of a parenteral dosage form are discussed in Chapter 7 and by Motola and Agharkar [4]. Important properties include solubility and rate of solution. Factors that influence solubility include particle size; salt, ester, or other chemical form; solution pH; polymorphism; purity; and hydrate formulation.

Crystal Characteristics

Control of the crystallization process to obtain a consistent and uniform crystal form, habit, density, and size distribution is particularly critical for drug substances to be utilized in suspensions or powders. For example, when the crystallization of sterile ceftazidime pentahydrate was modified to significantly increase the density to reduce the volume of the fill dose, the rate of dissolution increased significantly. Many dry solid parenteral products, such as the cephalosporins, are prepared by sterile crystallization techniques.

To obtain a uniform product from lot to lot, strict adherence to the procedures developed for a particular crystallization must be followed, including control of pH, rates of addition, solvent concentrations and purity, temperature, and mixing rates. Each crystallization procedure has to be designed to ensure sterility and minimize particulate contamination. Changes, such as using absolute ethyl alcohol instead of 95% ethanol during the washing procedure, can destroy the crystalline structure if the material being crystallized is a hydrate structure.

Drugs that associate with water to produce crystalline forms are called hydrates. Water content of the hydrate forms of sodium cefazolin as a function of relative humidity is seen in Fig. 5. As shown in Fig. 5, the sesquihydrate is the most stable structure when exposed to extreme humidity conditions [5]. This figure also reveals the importance of choosing the proper combination of hydrate and humidity conditions when designing a manufacturing process or facility.

Chemical Modifications

Improvement of the properties of a drug may be achieved by the chemical modification of the parent drug. The preparation of an ester, salt, or employed other modification of the parent structure may be employed with parenteral drugs to increase stability, alter drug solubility,

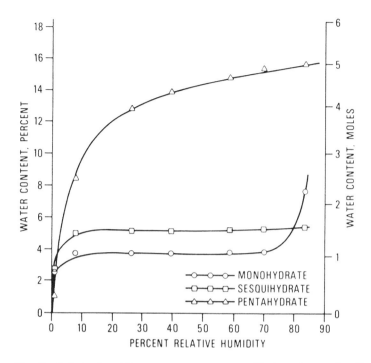

Fig. 5 Relative humidity versus water content of hydrate forms of sodium cefazolin. (From Ref. 5.)

enhance depot action, ease formulation difficulties, and possibly, decrease pain on injection. The molecularly modified drug that converts back to the active parent structure is defined as a *prodrug*. This conversion usually occurs within the body system or, for some drugs that are formulated as dry powders, occurs on reconstitution. The preparation of prodrugs is becoming a common practice with many types of drugs. Examples of antibiotic prodrugs include benzathine penicillin, procaine penicillin, metronidazole phosphate, and chloramphenicol sodium succinate.

The preparation of salts of organic compounds is one of the most important tools available to the formulator. Compounds for both IM and IV solutions require high solubility so that the drug may be incorporated into small volumes for IM administration and also be acceptable for IV use. Sodium and potassium salts of weak acids and hydrochloride and sulfate salts of weak bases are widely used in parenterals requiring highly soluble compounds, based on their overall safety and history of clinical acceptance.

If a drug's solubility is to be reduced to enhance stability or to prepare a suspension, the formulator may prepare water-insoluble salts. A classic example is procaine penicillin G, the decreased solubility (7 mg/ml) of which, when compared with the very soluble penicillin G potassium, is utilized to prepare stable parenteral suspensions. Another alternative to preparing an insoluble drug is to use the parent acidic or basic drug and to buffer the pH of the suspension in the range of minimum solubility.

Polymorphism

The literature lists numerous examples of polymorphism; that is, the existence of several crystal forms of a given chemical that exhibit different physical properties [6]. The conversion of one polymorph to another may cause a significant change in the physical properties of the drug.

Studies of polymorphs in recent years have pointed out the effects of polymorphism on solubility and, more specifically, on dissolution rates. The aspect of polymorphism that is of concern to the parenteral formulator is basically one of product stability [7]. Substances that form polymorphs must be evaluated so that the form used is stable in that particular solvent system. Physical stresses that occur during suspension manufacture may also give rise to changes in crystal form [8].

pH and pK_a

Profiles of pH versus solubility and pH versus stability are needed for solution and suspension formulations to help assure physical and chemical stability as well as to maximize or minimize solubility. This information is also valuable for predicting the compatability of drugs with various infusion fluids.

In summary, the physical and chemical data that should be obtained on the drug substance include the following:

Molecular structure and weight
Melting point
Thermal profile
Particle size and shape
Hygroscopicity potential
Ionization constant
Light stability
Optical activity
pH solubility profile
pH stability profile
Polymorphism potential
Solvate formation

B. Added Substances in Parenteral Formulations

To provide efficacious, safe, and elegant parenteral dosage forms, added substances must frequently be incorporated into the formula to maintain pharmaceutical stability, control product attributes, ensure sterility, or aid in parenteral administration. These substances include antioxidants, antimicrobial agents, buffers, bulking materials, chelating agents, inert gases, solubilizing agents, protectants, and substances for adjusting toxicity. In parenteral product development work, any additive to a formulation must be justified by a clear purpose and function. In addition, every attempt should be made to choose added substances that are accepted by regulatory agencies throughout the world, since most pharmaceutical development is international in scope.

Some of the added substances most commonly used are listed in Table 2. Pharmacists involved in IV additive programs must be aware of the types of additives that may be present in the products being combined.

Antioxidants

Salts of sulfur dioxide, including bisulfite, metasulfite, and sulfite, are the most common antioxidants used in aqueous parenterals. These antioxidants maintain product stability by being preferentially oxidized and gradually consumed over the shelf life of the product. Irrespective of which salt is added to the solution, the antioxidant moiety depends on the final concentration of the thio compound and the final pH of the formulation [9]. While undergoing oxidation

Table 2 Classes and Examples of Parenteral Additives

Additive class	Examples of parenteral additives	Usual concentration (%)
Antimicrobial	Benzalkonium chloride	0.01
	Benzyl alcohol	1–2
	Chlorobutanol	0.25–0.5
	Metacresol	0.1–0.3
	Butyl *p*-hydroxybenzoate	0.015
	Methyl *p*-hydroxybenzoate	0.1–0.2
	Propyl *p*-hydroxybenzoate	0.2
	Phenol	0.25–0.5
	Thimerosal	0.01
Antioxidants	Ascorbic acid	0.01–0.05
	Cysteine	0.1–0.5
	Monothioglycerol	0.1–1.0
	Sodium bisulfite	0.1–1.0
	Sodium metabisulfite	0.1–1.0
	Tocopherols	0.05–0.5
Buffers	Acetates	1–2
	Citrates	1–5
	Phosphates	0.8–2.0
Bulking agents	Lactose	1–8
	Mannitol	1–10
	Sorbitol	1–10
	Glycine	1–2
Chelating agents	Salts of ethylenediaminetetraacetic acid (EDTA)	0.01–0.05
Protectants	Sucrose	2–5
	Lactose	2–5
	Maltose	2–5
	Human serum albumin	0.5–2
Solubilizing agents	Ethyl alcohol	1–50
	Glycerin	1–50
	Polyethylene glycol	1–50
	Propylene glycol	1–50
	Lecithin	0.5–2.0
Surfactants	Polyoxyethylene	0.1–0.5
	Sorbitan monooleate	0.05–0.25
Tonicity-adjusting agents	Dextrose	4–5
	Sodium chloride	0.5–0.9

reactions, the sulfites may be converted to sulfate and other species. Sulfites can also react with certain drug molecules (e.g., epinephrine).

Sulfite levels are determined by the reactivity of the drug, the type of container (glass seal versus rubber stopper), single- or multiple-dose use, container headspace, use of inert gas purge, and the expiration dating period to be employed. Upper limits for sulfite levels are specified in most pharmacopeias; for example, the *USP* allows 3.2 mg of sodium bisulfite per millimeter of solution, whereas the French pharmacopeia (*Pharmacopée Française*) allows only 1.6

mg/ml. Allowances on upper limits are made for concentrated drugs that are diluted extensively before use. An oxygen-sensitive product to be used in France might require a smaller container (less headspace) or a glass seal ampoule to maintain product stability because of the reduced bisulfite level.

Sulfites have been reported to precipitate an allergic reaction in some asthmatics. If possible, alternative antioxidants should be considered or the product should be manufactured and packaged in a manner such as to eliminate or minimize the concentration of bisulfite required. Deoxygenation of the makeup water, maintaining the solution under a nitrogen atmosphere throughout the manufacturing process, and purging the filled vials with an inert gas could significantly reduce the amount of antioxidant required.

Antimicrobial Agents

A suitable preservative system is required in all multiple-dose parenteral products to inhibit the growth of microorganisms accidentally introduced during withdrawal of individual doses. Preservatives may be added to single-dose parenteral products that are not terminally sterilized as a sterility assurance measure; that is, to prevent growth of any microorganisms that could be introduced if there were any inadvertent breach of asepsis during filling operations. However, the inclusion of a preservative in single-dose parenteral products must be weighed against the need to develop formulations that are acceptable to regulatory bodies worldwide. Inclusion of a preservative can be a difficult challenge here, given the wide range of viewpoints concerning which preservatives are acceptable and when it is appropriate to include them in a formulation. Partly because of this, there is a trend in parenteral product development to eliminate preservatives wherever it is practical to do so. This may require added measures in manufacturing to improve sterility assurance—such as using barrier technology to provide positive separation of personnel from product during aseptic filling and transfer steps.

The formulation scientist must be aware of interactions between preservatives and other components of a formulation that could compromise the efficacy of the preservative. For example, proteins can bind thimerosal, reducing preservative efficacy. Partitioning of preservative into a micellar phase or an oil phase (in an emulsion) can also reduce the effective concentration of preservative available for bactericidal or bacteriostatic action. Preservative efficacy testing should be done on the proposed formulation to assure an effective preservative concentration.

Several investigators have published research on incompatibilities of preservatives with rubber closures and other packaging components, particularly polymeric materials [10]. Again, challenging the product with selected microorganisms to measure bacteriostatic or bactericidal activity is necessary, including evaluation of efficacy as a function of time throughout the anticipated shelf life of the product.

More subtle effects of preservatives on injectable formulations are possible. Formulation of insulin is an illustrative case study. Insulin is usually formulated as a multiple-dose vial, since individual dosage varies among patients. Preservation of zinc insulin with phenol causes physical instability of the suspension, whereas methylparaben does not. However, the presence of phenol is required for obtaining protamine insulin crystals [8].

Buffers

Many drugs require a certain pH range to maintain product stability. As discussed previously, drug solubility may also be strongly dependent on the pH of the solution. An important aid to the formulator is the information contained in a graph of the solubility profile of the drug as a function of pH (Fig. 6). The product can then be buffered to approach maximum or minimum solubility, whichever is desired.

Parenteral products should be formulated to possess sufficient buffer capacity to maintain proper product pH. Factors that influence pH include product degradation, container and stopper

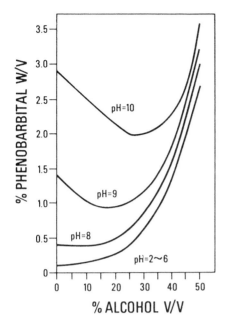

Fig. 6 Interdependence of pH and alcohol concentration on the solubility of phenobarbital. (From Ref. 11.)

effects, diffusion of gases through the closure, and the effect of gases in the product or in the headspace. However, the buffer capacity of a formulation must be readily overcome by the biological fluids; thus, the concentration and ratios of buffer ingredients must be carefully selected.

Buffer systems for parenterals consist of either a weak base and the salt of a weak base or a weak acid and the salt of a weak acid. Buffer systems commonly used for injectable products are acetates, citrates, and phosphates (see Table 2). Amino acids are receiving increased use as buffers, especially for polypeptide injectables.

Chelating Agents

Chelating agents are added to complex and, thereby, inactivate metals such as copper, iron, and zinc that generally catalyze oxidative degradation of drug molecules. Sources of metal contamination include raw material impurities; solvents, such as water, rubber stoppers and containers; and equipment employed in the manufacturing process [12]. The most widely used chelating agents are ethylenediaminetetraacetic acid (edetic acid; EDTA) derivatives and salts. Japan does not allow the use of these particular chelating agents in any parenteral products. Citric and tartaric acids are also employed as chelating agents.

Inert Gases

Another means of enhancing the product integrity of oxygen-sensitive medicaments is by displacing the air in the solution with nitrogen or argon. This technique may be made more effective by first purging with nitrogen or boiling the water to reduce dissolved oxygen. The container is also purged with nitrogen or argon before filling and may also be topped off with the gas before sealing.

Glass-seal ampoules provide the most impervious barrier for gas transmission. A butyl rubber stock is used with rubber-stoppered products that are sensitive to oxygen because it provides better resistance to gas permeation than other rubber stocks.

Solubilizing Agents and Surfactants

Drug solubility can be increased by the use of solubilizing agents, such as those listed in Table 2, and by nonaqueous solvents or mixed solvent systems, to be discussed shortly. When using solubilizing agents, the formulator must consider their effect on the safety and stability of the drug.

A surfactant is a surface-active agent that is used to disperse a water-insoluble drug as a colloidal dispersion. Surfactants are used for wetting and to prevent crystal growth in a suspension. Surfactants are used quite extensively in parenteral suspensions for wetting powders and to provide acceptable syringability. They are also used in emulsions and for solubilizing steroids and fat-soluble vitamins.

Tonicity Adjustment Agents

It is important that injectable solutions that are to be given intravenously are isotonic, or nearly so. Because of osmotic pressure changes and the resultant exchange of ionic species across red blood cell membranes, nonisotonic solutions, particularly if given in quantities larger than 100 ml, can cause hemolysis or crenation of red blood cells (owing to hypotonic or hypertonic solutions, respectively). Dextrose and sodium chloride or potassium chloride are commonly used to achieve isotonicity in a parenteral formula.

Protectants

A protectant is a substance that is added to a formulation to protect against loss of activity caused by some stress that is introduced by the manufacturing process or to prevent loss of active ingredients by adsorption to process equipment or to primary packaging materials. Protectants are used primarily in protein and liposomal formulations. For example, cryoprotectants and lyoprotectants are used to inhibit loss of integrity of the active substance resulting from freezing and drying, respectively. Compounds that provide cryoprotection are not necessarily the same as those that provide lyoprotection. For example, polyethylene glycol protects lactate dehydrogenase and phosphofructokinase from damage by freezing, but does not protect either protein from damage by freeze-drying. Compounds such as glucose and trehalose are effective lyoprotectants for both proteins [13]. Effective cryo- and lyoprotectants must be determined on a case-by-case basis, but sugars and polyhydroxy compounds are usually the best candidate compounds. These same types of compounds also tend to markedly improve the stability of proteins against inactivation by thermal denaturation.

Another type of protectant is used to prevent loss of active substance—again, usually a protein and usually present at a very low concentration—by adsorption to materials or equipment in the manufacturing process or to components of the primary package. In manufacturing, particular attention should be given to adsorption of the active entity to filters (especially nylon) and to silicone tubing used for transfer operations. For packaging materials, rubber closures and other polymeric materials should be examined carefully for adsorptive potential. The same consideration applies to infusion equipment, particularly considering that most materials in modern IV infusion therapy are polymeric.

Human serum albumin (HSA) is commonly used as a protectant against adsorptive loss. HSA is present at higher concentration than the active substance, and is preferentially adsorbed, coating the surface of interest and preventing adsorption of the drug. For example, insulin is subject to adsorptive loss to hydrophobic materials. Addition of 0.1–1.0% HSA has been reported to prevent adsorptive loss [8].

C. Vehicles

Aqueous Vehicles

"Water for injection" (WFI) is the most widely used solvent for parenteral preparations. The requirements for WFI are generally the same throughout the world. Companies involved in international markets must be assured that their products comply with the applicable standards. The most common means of obtaining WFI is by the distillation of deionized water. Water for injection must be prepared and stored in a manner to ensure purity and freedom from pyrogens.

Microorganisms, dissolved organic and inorganic substances, and foreign particles are the most common contaminants found in water. New purification methods and systems are continually being investigated to improve the quality of water for parenteral use. Inorganic compounds are commonly removed by distillation, reverse osmosis, deionization, or a combination of these processes. Membrane and depth filters are used to remove particulate contaminants, and charcoal beds may be used to remove organic materials. Filtration, chilling or heating, or recirculation of water are used to reduce microbial growth and to prevent pyrogen formation that will occur in a static deionization system. To inhibit microbial growth, WFI must be stored at either 5°C or 60–90°C if it is to be held for over 24 hr.

The USP also lists sterile water for injection and bacteriostatic water for injection, which unlike WFI, must be sterile. Higher levels of solids are allowed in these vehicles because of the possible leaching of glass constituents into the product during high-temperature sterilization and subsequent storage. Bacteriostatic water for injection must not be placed in containers larger than 30 ml. This is to prevent the administration of large quantities of bacteriostatic agents (such as phenol) that could become toxic if large volumes of solution were administered. Other aqueous vehicles that may be used in place of sterile water for injection or bacteriostatic water for injection for reconstitution or administering drugs include 5% dextrose, 0.9% sodium chloride, and a variety of other electrolyte and nutrient solutions, as noted earlier.

Nonaqueous and Mixed Vehicles

A nonaqueous solvent or a mixed aqueous–nonaqueous solvent system may be necessary to stabilize drugs, such as the barbiturates, that are readily hydrolyzed by water, or to improve solubility (e.g., digotoxin). Nonaqueous solvents must be carefully screened and tested to ensure that they exhibit no pharmacological action, are nontoxic and nonirritating, and are compatible and stable with all ingredients of a formulation.

A major class of nonaqueous solvents is the fixed oils. The USP [1] recognizes the use of fixed oils as parenteral vehicles and lists their requirements. The most commonly used oils are corn oil, cottonseed oil, peanut oil, and sesame oil. Because fixed oils can be quite irritating when injected and may cause sensitivity reactions in some patients, the oil used in the product must be stated on the label.

Sesame oil is the preferred oil for most of the official injections in oil. This is because it is the most stable (except in light) and, thus, will usually meet the official requirements. Fixed oils must never be administered intravenously and are, in fact, restricted to IM use.

The USP usually does not specify an oil, but states that a suitable vegetable oil can be used. The main use of such oils is with the steroids, with which they yield products that produce a sustained-release effect. Sesame oil has also been used to obtain slow release of fluphenazine esters given intramuscularly [4]. Excessive unsaturation of an oil can produce tissue irritation. The use of injections in oil has diminished somewhat in preference to aqueous suspensions, which generally have less irritating and sensitizing properties. Benzyl benzoate may be used to enhance steroid solubility in oils if desired.

Water-miscible solvents are widely used in parenterals to enhance drug solubility and to serve as stabilizers. The more common solvents include glycerin, ethyl alcohol, propylene glycol, and polyethylene glycol 300. A common example of an injectable product formulated with nonaqueous solvents is IV Valium, which contains 40% propylene glycol and 10% ethanol. Mixed-solvent systems do not exhibit many of the disadvantages observed with the fixed oils, but may also be irritating or increase toxicity, especially when present in large amounts or in high concentrations. A solution containing a high percentage of ethanol will produce pain on injection.

The formulator should be aware of the potential of nonaqueous solvents in preparing a solubilized or stable product that may not have been otherwise possible. The reader is directed to comprehensive reviews of nonaqueous solvents for additional information [15,16].

V. DOSAGE FORMS

A. Solutions

The most common of all injectable products are solutions. Solutions of drugs suitable for parenteral administration are referred to as *injections*. Although usually aqueous, they may be mixtures of water with glycols, alcohol, or other nonaqueous solvents. Many injectable solutions are manufactured by dissolving the drug and a preservative, adjusting the pH, sterile filtering the resultant solution through a 0.22-μm–membrane filter and, when possible, autoclaving the final product. Most solutions have a viscosity and surface tension very similar to water, although streptomycin sulfate injection and ascorbic acid injection, for example, are quite viscous.

Sterile filtration, with subsequent aseptic filling, is common because of the heat sensitivity of most drugs. Those drug solutions that can withstand heat should be terminally autoclave-sterilized after filling, since this better assures product and package sterility.

Large-volume parenterals (LVPs) and small-volume parenterals (SVPs) containing no antimicrobial agent should be terminally sterilized. It is standard practice to include an antimicrobial agent in SVPs that cannot be terminally sterilized or are intended for multiple-dose use. The general exceptions are products that pass the *USP* Antimicrobial Preservative Effectiveness Test [1] because of the preservative effect of the active ingredient, vehicle, pH, or a combination of these. For example, some barbiturate products have a pH of 9–10 and a vehicle that includes glycol and alcohol.

Injections and infusion fluids must be manufactured in a manner that will minimize or eliminate haze and color. Parenteral solutions are generally filtered through 0.22-μm–membrane filters to achieve sterility and remove particulate matter. Prefiltration through a coarser filter is often necessary to maintain adequate flow rates, or to prevent clogging of the filters during large-scale manufacturing. A talc or carbon filtration aid (or other filter aids) may also be necessary. If talc is used, it should be pretreated with a dilute acid solution to remove surface alkali and metals.

The formulator must be aware of the potential for binding when filtering protein solutions. Because of the cost–availability of most protein materials, a membrane should be used that minimizes protein adsorption to the membrane surface. Typical filter media that minimize this binding include hydrophilic polyvinylidene difluoride and hydroxyl-modified hydrophilic polyamide membranes [17]. Filter suppliers will evaluate the compatibility of the drug product with their membrane media and also validate the selected membrane.

The total fluid volume that must be filled into a unit parenteral container is typically greater than the volume that would contain the exact labeled dose. The fill volume is dependent on the viscosity of the solution and the retention of the solution by the container and stopper. The *USP* [1] provides a procedure for calculating the fill dose that is necessary to ensure the delivery of the stated dose. It also provides a table of excess volumes that are usually sufficient to permit withdrawal and administration of the labeled volume.

B. Suspensions

One of the most difficult parenteral dosage forms to formulate is a suspension. It requires a delicate balance of variables to formulate a product that is easy to fill, ships without caking or settling, and ejects through an 18- to 21-gauge needle through its shelf life. To achieve these properties it is necessary to select and carefully maintain particle size distribution, zeta potential, and rheological parameters, as well as the manufacturing steps that control wettability and surface tension. The requirements for, limitations in, and differences between the design of injectable suspensions and other suspensions have been previously summarized [18,19].

A formula for an injectable suspension might consist of the active ingredient suspended in an aqueous vehicle containing an antimicrobial preservative, a surfactant for wetting, a dispersing or suspending agent, and perhaps a buffer or salt.

Two basic methods are used to prepare parenteral suspensions: (a) sterile vehicle and powder are combined aseptically, or (b) sterile solutions are combined and the crystals formed in situ. Examples of these procedures may be illustrated using sterile penicillin G procaine suspension (*USP*) and sterile testosterone suspension (*USP*).

In the first example, procaine penicillin, an aqueous vehicle containing the soluble components (such as lecithin, sodium citrate, povidone, and polyoxyethylene sorbitan monooleate) is filtered through a 0.22-µm–membrane filter, heat sterilized, and transferred into a presterilized mixing–filling tank. The sterile antibiotic powder, which has previously been produced by freeze-drying, sterile crystallization, or spray-drying, is gradually added to the sterile solution aseptically while mixing. After all tests have been completed on the bulk material, it is aseptically filled.

An example of the second method of parenteral suspension preparation is testosterone suspension. Here, the vehicle is prepared and sterile-filtered. The testosterone is dissolved separately in acetone and sterile-filtered. The testosterone–acetone solution is aseptically added to the sterile vehicle, causing the testosterone to crystallize. The resulting suspension is then diluted with sterile vehicle, mixed, the crystals allowed to settle, and the supernatant solution siphoned off. This procedure is repeated several times until all the acetone has been removed. The suspension is then brought to volume and filled in the normal manner.

The critical nature of the flow properties of parenteral suspensions becomes apparent when one remembers that these products are frequently administered through 1¼-in. or longer needles having internal diameters in the range of only 300–600 µm. In addition, microscopic examination shows a very rough interior needle surface, further hindering flow. The flow properties of parenteral suspensions are usually characterized on the basis of syringeability or injectability. The term *syringeability* refers to the handling characteristics of a suspension while drawing it into and manipulating it in a syringe. Syringeability includes characteristics such as ease of withdrawal from the container into the syringe, clogging and foaming tendencies, and accuracy of dose measurement. The term *injectability* refers to the properties of the suspension during injection; it includes such factors as pressure or force required for injection, evenness of flow, aspiration qualities, and freedom from clogging. The syringeability and injectability characteristics of a suspension are closely related to viscosity and to particle characteristics.

C.   Emulsions

An emulsion is a heterogeneous dispersion of one immiscible liquid in another. This inherently unstable system is made possible through the use of an emulsifying agent, which prevents coalescence of the dispersed droplets. Parenteral emulsions are rare because it is necessary (and difficult) to achieve stable droplets of less than 1 μm to prevent emboli in the blood vessels, and it is not usually necessary to achieve an emulsion for drug administration.

Parenteral emulsions have been used for several purposes, including (a) water-in-oil emulsions of allergenic extracts (given subcutaneously), (b) oil-in-water sustained-release depot preparations (given intramuscularly), and (c) oil-in-water nutrient emulsions (given intravenously). Formulation options are severely restricted through a very limited selection of stabilizers and emulsifiers, primarily owing to the dual constraints of autoclave sterilization and parenteral injection. Additionally, unwanted physiological effects (e.g., pyrogenic reaction and hemolysis) have further limited the use of intravenous emulsions.

An increasingly popular class of intravenous emulsions is fat emulsions. These preparations have been available in Europe for over 20 years and in the United States since 1975. Fat is transported in the bloodstream as small droplets called chylomicra. Chylomicra are 0.5- to 1.0-μm spheres consisting of a central core of triglycerides and an outer layer of phospholipids. Intravenous fat emulsions usually contain 10% oil, although they may range up to 20% (Table 3). These emulsions yield triglycerides that provide essential fatty acids and calories during total parenteral nutrition of patients who are unable to absorb nutrients through the gastrointestinal tract. The products commercially available in the United States range from 0.1 to 0.5 μm and have a pH of 5.5–8 (blood plasma has a pH of 7.4). Glycerol and glucose are added to make the product isotonic.

D.   Dry Powders

Many drugs are too unstable—either physically or chemically—in an aqueous medium to allow formulation as a solution, suspension, or emulsion. Instead, the drug is formulated as a dry powder that is reconstituted by addition of water before administration. The reconstituted product is usually an aqueous solution; however, occasionally it may be an aqueous suspension (for example, ampicillin trihydrate and spectinomycin hydrochloride are sterile powders that are reconstituted to form a sterile suspension).

Dry powders for reconstitution as an injectable product may be produced by several methods—filling the product into vials as a liquid and freeze-drying, aseptic crystallization followed by powder filling, and spray-drying followed by powder filling. A brief discussion of each follows.

Freeze-Drying

The most common form of sterile powder is a *freeze-dried* or *lyophilized*, powder. The advantages of freeze-drying are that (a) water can be removed at low temperature, avoiding damage to heat-sensitive materials; (b) if freeze-drying is done properly, the dried product has a high specific surface area, which facilitates rapid, complete rehydration (or "reconstitution") of the solid; and (c) from an operations point of view, freeze-dried dosage forms allow drug to be filled into vials as a solution. This makes control of the quantity filled into each vial more precise than filling drug into vials as a powder. In addition, since drug is filled as a solution, there is minimal concern with airborne particulate matter and potential cross-contamination as is the problem with powder filling.

Despite the advantages of freeze-drying, there are some limitations that must be kept in mind.

Table 3 Intravenous Fat Emulsions

Component	Intralipid[a]		Liposyn II[b]		Infonutrol[c]	Lipofundin[d]	Lipihysan[e]	
(g/100 ml)	10%	20%	10%	20%	15%	10%	10%	15%
Soybean oil	10	20	5	10				
Safflower oil			5	10				
Cottonseed oil					15	10	10	15
Egg phospholipids	1.2	1.2	1.2	1.2				
Soybean phospholipids					1.2	1.2		
Soybean lecithin							1.5	2
Glycerol	2.25	2.25	2.5	2.5				
Glucose					4			
Sorbitol						5	5	5
Pluronic F-68					0.3			
DL-α-Tocopherol							0.05	0.05
Water for injections q.s. ad	100 ml	100 ml	100 ml	100 ml	100 ml	100 ml	100 ml	

[a]Kabi-Vitrum A. G., Stockhold, Sweden.
[b]Abbott Laboratories, North Chicago, IL.
[c]Astra-Hewlett, Södertäye, Sweden
[d]Braun, Melsunger, West Germany
[e]Egic, L'Equilibre Biologique S. A., Loiret, France

1. Some proteins are damaged by freezing, freeze-drying, or both. Although the damage can often be minimized by using protective agents in the formulation, the problem is still substantial.
2. Often the stability of a drug in the solid state depends on its physical state (i.e., crystalline or amorphous [20]). If freeze-drying produces an amorphous solid, and the amorphous form is not stable, then freeze-drying will not provide an acceptable product.
3. Freeze-drying is a relatively expensive drying operation. Although this is not an issue for many high-cost drug products, it may become an issue for more cost-sensitive pharmaceutical products.

In freeze-drying, a solution is filled into vials, a special slotted stopper is partially inserted into the neck of the vial (Fig. 7), and trays of filled vials are transferred to the freeze-dryer (Fig. 8). The solution is frozen by circulation of a fluid, such as silicone oil, at a temperature in the range of −35° to about −45°C through internal channels in the shelf assembly. When the product has solidified sufficiently, the pressure in the freeze-dry chamber is reduced to a pressure less that the vapor pressure of ice at the temperature of the product, and heat is applied to the product by increasing the temperature of the circulating fluid. Under these conditions, water is removed by *sublimation* of ice, or a phase change from the solid state directly to the vapor state without the appearance of an intermediate liquid phase. The phase diagram in Fig. 9 illustrates the difference between freeze-drying and conventional drying methods, during which drying takes place by a phase change from the liquid state to the vapor state. Freeze-drying takes place below the triple point of water, at which solid, liquid, and vapor all coexist in equilibrium. As freeze-drying proceeds, a receding boundary can be observed in the vial as the thickness of the frozen layer decreases. This phase is called *primary drying*, during which ice is removed by direct sublimation through open channels created by prior sublimation of ice. After primary drying, additional drying is necessary to remove any water that did not

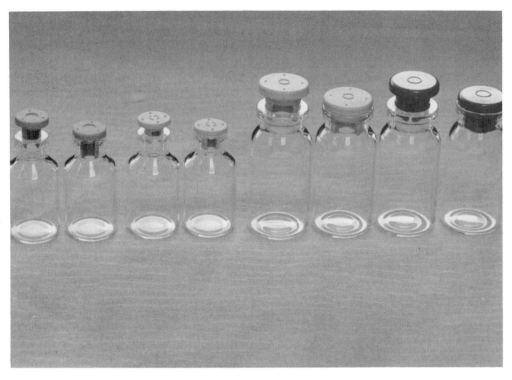

Fig. 7 Vials typically used for lyophilization showing special slotted stopper.

freeze during the freezing process, but instead remained associated with the solute. This is called *secondary drying* and consists of water removal by diffusion and desorption of water from the partially dried solid phase. The phases of a typical freeze-dry cycle—freezing, primary drying, and secondary drying—are illustrated by means of a plot of shelf temperature, chamber pressure, and product temperature in Fig. 10.

The most important objective in developing a freeze-dried product is to assure that critical quality attributes are met initially and throughout the shelf life of the product. Examples of critical quality attributes are recovery of original chemical or biological activity after reconstitution, rapid and complete dissolution, appropriate residual moisture level, and acceptable cake appearance. In addition, process conditions should be chosen to maximize process efficiency; that is, those conditions that minimize drying time without adversely affecting product quality. The driving force for sublimation is the vapor pressure of ice, and the vapor pressure of ice is highly temperature-dependent as shown below:

Temperature (°C)	Vapor pressure (mm Hg)
−40	0.096
−30	0.286
−20	0.776
−10	1.950
0	4.579

Therefore, freeze-drying should be carried out at the highest allowable product temperature that maintains the appropriate attributes of a freeze-dried product. This temperature depends

Fig. 8 Filled vials being transferred to freeze dryer.

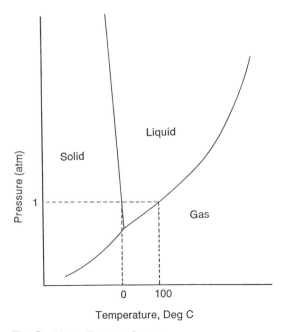

Fig. 9 Phase diagram of water.

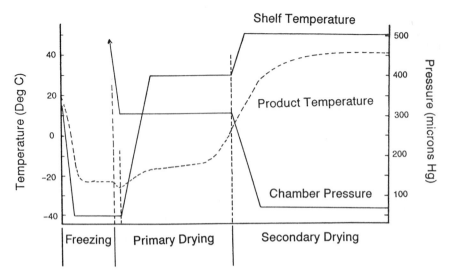

Fig. 10 Process variables during a representative freeze-dry cycle.

on the nature of the formulation. Process development and validation requires characterizing the physical state of the solute, or solutes, that result from the freezing process and identifying a maximum allowable product temperature for the primary drying process [21,22].

The term *eutectic temperature* is often misused in reference to freeze-drying. A eutectic phase—an intimate mixture of ice and crystals of solute that melts as if it were a single, pure compound—is present only if the solute crystallizes when the solution is frozen. Eutectic melting can often be detected by a thermal analysis technique, such as differential scanning calorimetry DSC [23,24]. An example of a eutectic system is neutral glycine in water. The presence of a eutectic phase is indicated by a melting endotherm in the DSC thermogram of the solution (Fig. 11) in addition to the melting endotherm for ice. In this example, the theoretical maximum allowable product temperature during primary drying is the eutectic melting temperature at $-3.5°C$. In practice, the product temperature should be maintained a few degrees below this temperature to assure that melting does not occur during the process. Examples of some other common solutes that form eutectics, along with the eutectic temperature, are shown below (24):

Solute	Eutectic temperature (°C)
Calcium chloride	-51.0
Citric acid	-12.2
Mannitol	-1.0
Potassium chloride	-10.7
Sodium carbonate	-18.0
Sodium chloride	-21.5
Sodium phosphate, dibasic	-0.5

However, many solutes do not crystallize during the freezing process, but instead, remain amorphous. Examples include sugars, such as sucrose, lactose, maltose, and many polymers. In this case, no eutectic phase is formed. Instead, the freeze concentrate becomes more concentrated and more viscous as the temperature is lowered and ice crystals grow. This process

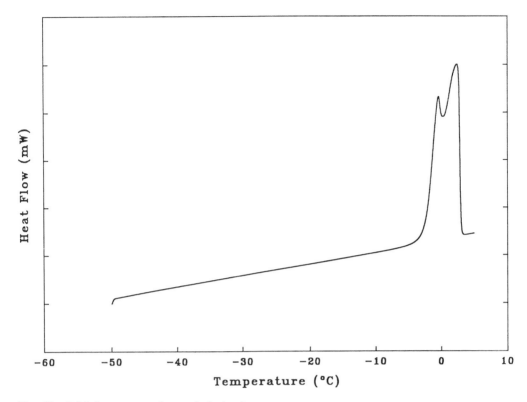

Fig. 11 DSC thermogram of neutral glycine in water.

continues until a temperature is reached at which the viscosity of the freeze concentrate increases dramatically with only a small change in temperature, and ice crystal growth ceases on a practical time scale. This temperature is a *glass transition temperature*, and is an important characteristic of amorphous systems. Below the glass transition temperature, the freeze concentrate exists as a rigid glass. Above the glass transition temperature, the freeze concentrate behaves as a viscous liquid. The significance of the glass transition temperature of the freeze concentrate (commonly referred to as Tg') is that it is closely related to the *collapse temperature* in freeze-drying. If drying is carried out above the collapse temperature, the freeze concentrate will flow and lose the microstructure established by freezing, once the supporting structure of ice crystals is removed. Collapse can be observed in a variety of forms, from a slight shrinkage of the dried cake (where the cake has pulled away from the wall of the vial) to total loss of any cake structure.

The glass transition of solutes that remain amorphous during and after the freezing process can often be seen in the DSC thermogram as a shift in the baseline toward higher heat capacity. This is illustrated in the DSC thermogram of sucrose solution in Fig. 12, in which the glass transition is observed at $-34°C$. Glass transition (Tg') values of some other solutes common to freeze drying are shown below (24):

Solute	Glass transition (Tg')
Dextran	-9
Gelatin	-10

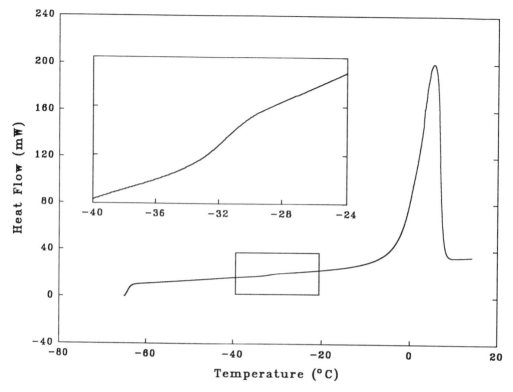

Fig. 12 DSC thermogram of sucrose solution.

Glucose	−43
Lactose	−32
Maltose	−32
Polyvinylpyrrolidone	−24
Sorbitol	−48

Some solutes may first form a metastable amorphous phase initially on freezing and then crystallize when the material is heated. This is the basis for *thermal treatment*, or *annealling*, in the freeze-dry process, during which the product is frozen to perhaps −40°C, the product is then heated to some temperature below the melting point of ice, held for a few hours, then cooled before starting the drying process. Some cephalosporins are known to crystallize by thermal treatment [20], and mannitol is a common excipient which frequently forms a metastable amorphous phase that crystallizes with subsequent heating [26].

In general, crystallization of the solute is desirable in terms of freeze-drying properties, as well as quality attributes of the final product, for several reasons. First, when the solute crystallizes, nearly all of the water is present as ice in the frozen matrix, and it is removed by direct sublimation during primary drying. Therefore, there is little water to be removed by secondary drying. This improves process efficiency, since water removed during secondary drying must be removed primarily by the process of diffusion, rather than by bulk flow. Second, eutectic temperatures are usually higher than collapse temperatures, which allows higher product temperatures and more efficient drying. Eutectic temperatures of most organic compounds are in the range of −1° to about −12°C, whereas collapse temperatures commonly are −30°C or lower. Third, the chemical and physical stability of a compound in crystalline form is generally better than that of the same compound in an amorphous form [20,27]. This can be a critical aspect of determining the feasibility of a freeze-dried dosage form.

An understanding of the effect of formulation on freeze-drying behavior is important to the pharmaceutical scientist involved in the development of freeze-dried products. Mixtures of components should be expected to behave differently from single-component systems. For example, a compound that crystallizes readily from aqueous solution when it is the only solute present may not crystallize at all when other solutes are present. For solutes that remain amorphous on freezing, the glass transition temperature is affected by the presence of other solutes. Subtle variations in the composition of the formulation, such as changes in ionic strength or pH, may have a significant effect on the physical chemistry of the freezing and freeze-drying processes.

Many drugs are present in a dose too small to form a well-defined freeze-dried cake, and must be formulated with a *bulking agent*, the purpose of which is to provide a dried matrix in which the active ingredient is dispersed. Common bulking agents are mannitol, lactose, glycine, and mixtures of these compounds. Buffers are commonly used, such as sodium or potassium phosphate, acetate, citrate, or tris-hydroxymethylaminomethane (THAM). Formulations of proteins, liposomes, or cells generally require the presence of a *protectant*, or a substance that protects the active compound from damage by freezing, by drying, or both. Disaccharides, such as sucrose, lactose, and maltose, are, in general, the most effective protectants [28]. Trehalose, a disaccharide of glucose, is also a well-known protectant, but is not currently used in freeze-dried products for use in humans.

In addition to the effects of formulation factors on freeze-drying behavior, it is important for the pharmaceutical scientist to understand basic principles of heat and mass transfer in freeze-drying [29,30]. Because of the high heat input required for sublimation—670 calories/g—transfer of heat from the heated shelf to the sublimation front is often the rate-limiting step in the coupled heat and mass transfer process. There are three basic mechanisms for heat transfer: conduction, convection, and radiation. *Conduction* is the transfer of heat by molecular motion between differential volume elements of a material. *Convection* is the transfer of heat by bulk flow of a fluid, either from density differences (natural convection) or because an external force is applied (forced convection). Because of the relatively low pressures used in freeze-drying, convection is probably not a large contributing factor in heat transfer. Heat transfer by *thermal radiation* arises when a substance, because of thermal excitation, emits radiation in an amount determined by its absolute temperature. Of these mechanisms, heat transfer by conduction is the most important. Heat transfer by conduction takes place through a series of resistances—the bottom of the vial, the frozen layer of product, the metal tray (if used), and through the vapor phase caused by lack of good thermal contact between the vial and the shelf. The thermal conductivity of the vapor phase at the pressures used in freeze-drying is dependent on pressure in the chamber. Therefore, to maintain consistent drying conditions from batch to batch, it is as important to control the chamber pressure as it is to control shelf temperature [31]. In addition, changes in the geometry of the system that will affect heat transfer will also affect process consistency. Examples include changing from molded to tubing vials, changing the depth of fill in the vials, and changing from trays with metal bottoms to those without bottoms. Thermal radiation is a small, but significant, contributor to the total quantity of heat transferred to the product. This can be a significant issue in scale-up of cycles from pilot dryers to production-scale equipment.

Mass transfer in freeze-drying refers to the transfer of water vapor from the sublimation front through open channels in the partially dried layer, created by prior sublimation of ice, through the headspace of the vial, past the lyostopper, through the chamber, to the condenser.

The reader is referred to basic studies of mass transfer in freeze-drying by Pikal and co-workers for in-depth treatment of the theoretical and practical aspects of mass transfer [30,32]. Briefly, the rate-limiting step in mass transfer is transfer of water vapor through the partially dried matrix of solids. Resistance of the dried layer increases in a more or less continuous fashion as the depth of the dried layer increases, and the resistance also increases with the concentration of solids in the dried layer. Other factors can also affect the resistance of the dried layer, such as the method of freezing; faster freezing tends to create a higher resistance in the dried layer.

Mass transfer of the ''unfrozen'' water through a glassy phase during secondary drying occurs slower than bulk flow of water vapor by direct sublimation, since no open channels are present in the glassy phase. The high resistance of the solid material to mass transfer is why secondary drying can be the most time-consuming phase of the freeze-dry cycle for amorphous solutes containing a large percentage of unfrozen water. According to studies reported by Pikal, shelf temperature is the most critical process variable, affecting the rate of secondary drying and final moisture level [32]. Chamber pressure had no measurable influence on secondary drying kinetics.

The quantity of residual water is frequently a critical product characteristic relative to chemical and physical stability of freeze-dried products, particularly amorphous solids. Water acts as a plasticizer of the solid material, lowering the glass transition temperature. A low glass transition temperature, relative to the storage temperature, can result in physical instability, such as cake shrinkage or collapse, or accelerated rates of chemical reactions leading to instability. Often, a small change in moisture content can result in a large change in the glass transition temperature; therefore, careful consideration of appropriate limits on residual moisture is often an important part of the product development process.

Aseptic Crystallization and Dry Powder Filling

Aseptic crystallization is primarily used for manufacture of sterile aqueous suspensions. However, if the physical form of the drug is critical to quality of the final product, better control over physical form can be attained by aseptic crystallization because a large variety of organic solvents can be used to control the crystallization process. In aseptic crystallization, the drug is dissolved in a suitable solvent and sterile filtered through an appropriate membrane filter. A second solvent—a sterilely filtered nonsolvent for the drug—is then added at a controlled rate, causing crystallization and precipitation of the drug. The crystals are collected on a funnel, washed if necessary, and dried by vacuum drying. After drying, it may be necessary to mill or blend the drug crystals. The powder is then transferred to dry–powder-filling equipment and filled into vials. Although simple in principle, there are obvious drawbacks to this approach. Batch-to-batch variability in crystal habit and crystal size and the resulting variability in physical properties can be troublesome for consistent product quality. Maintenance of asepsis between sterile crystallization and filling of the powder is a challenge during material handling, and will usually result in decreased sterility assurance. Also, since the drug is filled into vials as a powder, maintenance of fill weight uniformity is generally more troublesome than when filling with a liquid.

Spray-Drying

A solution of drug is sterile filtered and metered into the drying chamber, where it passes through an atomizer that creates an aerosol of small droplets of liquid (Fig. 13). The aerosol comes into contact with a stream of hot sterile gas—usually air. The solvent evaporates quickly, allowing drug to be collected as a powder in the form of uniform hollow spheres. The powder is then filled into vials using conventional powder-filling equipment. Spray-drying may be more economical than freeze-drying, but challenges in the use of this technique include sterile fil-

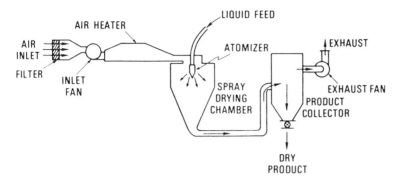

Fig. 13 Schematic drawing of spray-dryer.

tration of very large volumes of air, constructing and maintaining a spray dryer that can be readily sterilized, aseptic transfer of powder from the spray dryer to the powder-filling line, and precise control of the drying conditions to prevent overheating of the product while providing adequate drying. Probably because of these limitations, the technique is not widely used.

E. Protein Formulations

The first biotechnology-derived therapeutic agent to be approved by FDA in the United States was human insulin (Humulin, Eli Lilly) in 1982. The number of such products has grown steadily since then, and now accounts for roughly 3 billion dollars in sales in the United States. This growth is expected to continue during the next decade. Because biotechnology-derived pharmaceuticals are generally proteins and glycoproteins, they require special consideration in formulation and processing.

Special problems with formulation and processing of proteins arise from the hierarchy of structural organization of proteins. In addition to primary structure (the amino acid sequence), proteins have secondary structure (interactions between peptide bonds, resulting in helical or sheetlike structures), tertiary structure (folding of chain segments into a precise three-dimensional arrangement), and in some, quaternary structure (association of individual protein subunits into a multimeric structure). Disruption of this higher-order structure can lead to loss of the biologically active, or native, conformation which, in turn, causes physical instability and may accelerate reactions that are characteristic of chemical instability of proteins.

Loss of the native conformation of a protein generally exposes hydrophobic amino acid residues that are normally buried on the inside of the self-associated structure and are shielded from the aqueous environment. This leads to association between the exposed hydrophobic residues of neighboring proteins (aggregation), or between these exposed residues and hydrophobic surfaces that the protein may encounter either in the manufacturing process or in the primary package.

Processing variables, usually are not critical for traditional low molecular weight drugs, may be critical for protein formulations. For example, vigorous agitation of a protein solution can cause foaming, generating a large air–water interface that is an excellent site for denaturation, aggregation, and perhaps precipitation of protein. Loss of protein by adsorption to surfaces, such as tubing and filters used in manufacturing, can result in subpotent product. Other potentially critical factors in maintenance of the native structure during processing include temperature, pH, the presence of organic solvents, and ionic strength of the formulation.

Disruption of the native structure of a protein can also contribute to chemical instability by accelerating the rates of a variety of degradation routes, including deamidation, hydrolysis, oxidation, disulfide exchange, β-elimination, and racemization.

Formulation strategies for stabilization of proteins commonly include additives such as other proteins (e.g., serum albumin), amino acids, and surfactants to minimize adsorption to surfaces (see Sec. IV.B). Modification of protein structure to enhance stability by genetic engineering may also be feasible, as well as chemical modification such as formation of a conjugate with polyethylene glycol.

Most proteins are not sufficiently stable in aqueous solution to allow formulation as a sterile solution. Instead, the protein is freeze-dried and reconstituted before use. Development of a freeze-dried protein formulation often requires special attention to the details of the freezing process (potential pH shifts and ionic strength increase with freezing) as well as to potential loss of activity with drying. Formulation additives, such as sugars and polyhydroxy compounds, are often useful as cryoprotectants and lyoprotectants (see also Sec. IV.B). Residual moisture may also be critical to the stability of the dried preparation [33].

F. Novel Formulations

A summary of sustained- and controlled-release parenteral dosage forms is included in Chapter 16. This subject is also covered extensively by Chien [34].

Concepts in drug delivery that have received increasing attention include drug carrier systems, implants, intravenous infusers, and implantable infusion pumps. Carrier systems include microspheres, liposomes, monoclonal antibodies, and emulsions. Drugs are incorporated into these systems to increase the duration of drug action and to provide selective delivery of the drug to a specific target site or organ. Implants are used for the same reason. Unwanted side effects and adverse reactions are usually reduced because of selective delivery, which also results in a lower concentration of drug required to achieve the desired therapeutic effect. Infusion pumps provide a delivery system with uniform, continuous flow. A specific dose of a drug, such as insulin, may be administered to a patient on a continual or intermittent basis.

VI. PACKAGING

Container components for parenteral products must be considered an integral part of the product because they can dramatically affect product stability, potency, toxicity, and safety. Parenteral dosage forms, especially solutions, usually require more detailed evaluation of packaging components for product compatability and stability than do other pharmaceutical dosage forms. Common container components in direct contact with the product include various types of glass, rubber, plastic, and stainless steel (needles)—all of which may react with the drug. Maintenance of microbiological purity and product stability, adaptability to production operations and inspections, resistance to breakage and leakage, and convenience of clinical use are factors that must be evaluated when selecting the container.

Parenteral packaging includes ampoules (glass-seal), rubber-stoppered vials and bottles, plastic bags and bottles, glass and plastic syringes, and syringe–vial combinations. Glass containers have traditionally achieved widespread acceptability for parenteral products because of their relative inertness. In recent years, hospital preference for unit-dose and clinical convenience has resulted in an increase in products packaged in disposable syringes and the development of polyvinyl chloride, polyester, and polyolefin plastic containers for IV fluids. Package systems, such as the dual chamber plastic container and Add-Vantage, have been developed for

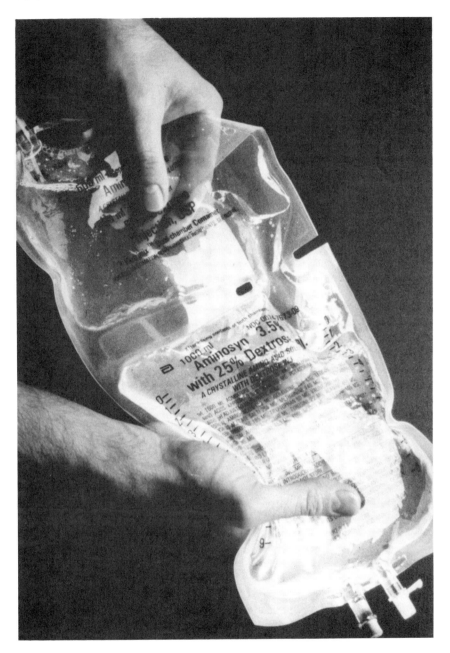

Fig. 14 Dual-chamber Nutri-Mix container for amino acid and dextrose solutions.

combining unstable mixtures of drugs and solutions (Figs. 14 and 15). Several antibiotics that are unstable in solution are now available as a frozen product in a plastic container. These systems are designed for convenience and cost efficiency as well as minimizing the potential of contamination when preparing the admixture. Parenteral packaging materials are discussed in Chapter 18.

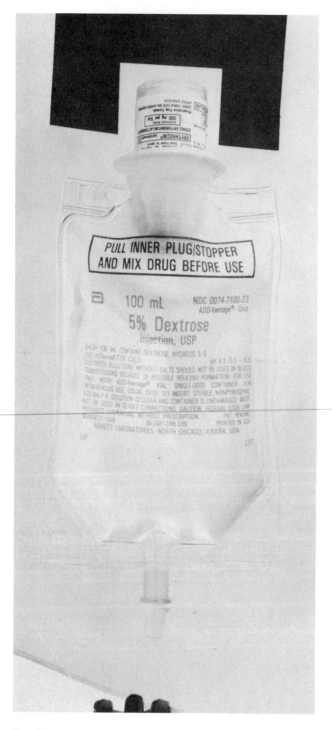

Fig. 15 Add-Vantage drug-diluent admixture system.

VII. STABILITY

A formal stability program is needed to assure that all critical attributes of any drug product are maintained throughout the shelf life of the product. A validated stability-indicating assay is essential to measure chemical or biological activity, and acceptance criteria should be established before initiating stability studies. Particular attention should be given to developing a detailed protocol for a stability study before preparing stability samples, including assays to be performed, storage conditions, and sampling intervals.

In general, expiration dating is based on the estimated time required for the active compound to reach 90–95% of labeled potency at the specified storage temperature. However, other considerations may limit the shelf life of the product. For example, the shelf life of products containing a preservative may be determined by adsorption of preservative to a rubber closure or another elastomeric component of the container–closure system. The drug substance itself may be subject to physical instability such as adsorption. The stability program should include placing enough units at the specified storage conditions to allow inspection of a statistically valid number of units to verify acceptable appearance of the product, such as the development of haze or discreet particulate matter in solution products, as well as to check for discoloration or any other physical attribute that would result in unacceptable pharmaceutical elegance. Formulation pH is often a critical attribute that must be monitored during a stability study, since pH may be affected both by chemical reactions in solution or by interactions between the formulation and the container–closure system.

Sterile powders may require special attention to identify which tests are required to assure adequate physical and chemical stability. The stability of many dried products is often sensitive to small differences in the amount of residual water present, requiring monitoring of residual moisture by Karl Fisher titration or loss on drying. This is particularly important for protein formulations. Special efforts may be needed to minimize the residual moisture in rubber stoppers, since water vapor can transfer from the closure to the powder during prolonged storage. Reconstitution time—the amount of time required after addition of diluent until all solids are dissolved—should be measured routinely. For freeze-dried products, cake shrinkage with time is not uncommon. This may be accompanied by discoloration, increased reconstitution time, or crystallization of one or more component of the formulation. The physical state of the drug—crystalline or amorphous—has an important influence on stability, particularly for cephalosporins. Periodic examination of stability samples by x-ray diffraction may be valuable to identify changes in physical state of either drug or excipients that could influence critical quality attributes. For some solid dosage forms subject to oxidative degradation, it is critical to exclude oxygen from the vial headspace. The headspace of selected vials should be analyzed periodically for oxygen. Many freeze-dried powders are stoppered under vacuum or an inert gas. Testing selected vials during the stability study for presence of vacuum in the headspace of the vial is a useful method of verifying container–closure integrity.

Sterile suspensions can be challenging for physical stability, and this should be reflected in the stability protocol. Examples of physical stability issues for suspensions include (a) caking, which causes poor resuspendability; (b) changes in the particle size distribution, particularly growth of large crystals of drug, which can cause poor syringeability; and (c) polymorphic transformations, which can result in changes in dissolution characteristics and, therefore, the bioavailability of the drug.

For parenteral emulsions, the formulation scientist must be particularly aware of changes in particle size distribution of the oil phase. Droplet coalescence results in increased droplet size. As a general rule, average droplet size should be less than 1 μm. Droplet sizes more than about 6 μm can cause adverse effects.

VIII. STERILIZATION METHODS

Five sterilization processes are described in the *USP* [1]: steam, dry-heat, filtration, gas, and ionizing radiation. All are commonly used for parenteral products, except gas and ionizing radiation, which are widely used for devices and surgical materials. To assist in the selection of the sterilization method, certain basic information and data must be gathered. This includes determining (a) the nature and amount of product bioburden, and (b) whether the product and container–closure system will have a predominately moist or dry environment during sterilization. Both of these factors are of critical importance in determining the conditions (time and temperature) of any sterilization method chosen.

The natural bioburden in a well-maintained pharmaceutical parenteral manufacturing plant is quite low, often to the point that it is difficult to isolate and propagate plant bioburden for sterilization studies. Nevertheless, it is still important to characterize the microbiological bioburden in the process and then monitor it at regular intervals.

For sterilization purposes, microorganisms can be categorized into three general categories: (a) easy to kill with either dry or moist heat; (b) susceptible to moist heat, but resistant to dry heat (e.g., *Bacillus subtilis*); or (c) resistant to moist heat but susceptible to dry heat (e.g., *Clostridium sporogenes*). Organisms such as *B. Subtilis* and *C. sporogenes* are often used as biological indicators because they are spore formers of known heat resistance. When used in a known concentration, they will be killed at a reproducible rate. In this manner, when a product has a low bioburden, biological indicator organism(s) can be used at a concentration of 1×10^3 in kill studies to simulate 10^6 kills of natural (environmental) bioburden. Processing and design of container–closure systems for individual products must be reviewed carefully to ascertain whether moist or dry conditions predominate, particularly in difficult-to-reach inner portions of closures. A good review of the use of biological indicators in validating parenteral container–closure systems may be found in Ref. 35.

The *USP* also recommends the use of biological indicators, whenever possible, to monitor all sterilization methods except sterile filtration. Biological indicators are generally of two types. If a product to be sterilized is a liquid, microorganisms are added directly to carefully identified representative samples of the product. When this is not practical, as with solids or equipment to be sterilized, the culture is added to strips of filter paper. The organism chosen varies with the method of sterilization.

Sterilization tests are performed to verify that an adequate sterilization process has been carried out. Validation of the sterilization cycle also gives assurance of process. Sterility is not assured simply because a product passes the *USP* sterility test. As outlined in the *USP*, the sterility test is described in considerable detail, including procedures for sampling, general conditions of the test, and specific procedures for testing solids and liquids. In addition, guidelines for the design of an aseptic work environment are outlined in some detail. Sample limitations, plus the impossibility of cultivating and testing all viable microorganisms that may be present, affect the reliability of sterility tests. Because of these problems, it is necessary to monitor and test sterilization equipment continuously. Reference 35 provides a good review of validation of sterile products.

A. Sterilization by Steam

When drug solutions and containers can withstand autoclaving conditions, this method is preferred to other sterilization methods because moist heat sterilizes quickly and inexpensively. However, judgment must be exercised and experiments run to ensure that the solution and container are permeable to steam. Oils and tightly closed containers, for example, are not normally sterilizable by steam.

Autoclave steam sterilization is a well-established and widely used procedure. Normally, steam enters through the top of the chamber (Fig. 16). Being lighter than air, it remains at the top of the chamber, but steadily and continuously drives the air out of the chamber through the bottom vent throughout the sterilization cycle. The velocity of steam entering the autoclave, the efficiency of water separation from incoming steam, the size of the drain, and the amount of vacuum applied, all are examples of factors that must be controlled to obtain efficient and reproducible steam sterilization in an autoclave. A thorough description of steam sterilization variables and theory can be found in Ref. 36.

With the widespread use of flexible packaging for LVP products, the use of steam sterilization has increased. Compared with the traditional LVP glass bottles closed with rubber stoppers, flexible LVP plastic containers (polyvinyl chloride, polyester, or polyolefin) offer autoclaving advantages. Specifically, (a) a larger surface area is available for heating per unit volume of liquid; (b) if held in a ''flattened'' position during sterilization, the heat penetration depth required is reduced, resulting in a more uniform thermal mapping of the contents; and (c) shorter heat-up and cool-down periods are required. The net effect is to allow a much shorter sterilization cycle for LVP products packaged in flexible containers, thus exposing the product to less heat, less potential for degradation, and reduced manufacturing costs.

B. Sterilization by Dry Heat

Dry heat is widely used to sterilize glassware and equipment parts in manufacturing areas for parenteral products. It has good penetration power and is not as corrosive as steam. However, heat-up time is slow necessitating long sterilization periods at high temperatures. It is important to allow sufficient circulation around the materials to be sterilized. Metal cans are often used to contain the parts or containers that are to be sterilized.

The two principal methods of dry-heat sterilization are infrared and convection hot air. Infrared rays will sterilize only surfaces. Sterilization of interior portions must rely on conduction. Convection hot-air sterilizers are normally heated electrically and are of two types: gravity or mechanical. In gravity convection units, a fan is used to promote uniformity of heat distribution throughout the chamber.

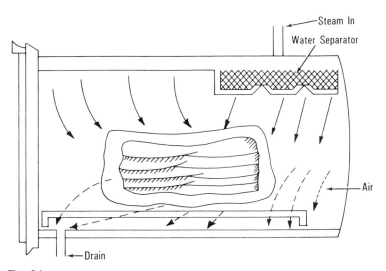

Fig. 16 Gravity displacement steam sterilizer.

Dry-heat processes kill microorganisms primarily through oxidation. The amount of moisture available to assist sterilization in dry-heat units varies considerably at different locations within the chamber and at different time intervals within the cycle. Also, the amount of heat available, its diffusion, and the environment at the spore–air interface all influence the microorganism kill rate. Consequently, cycles tend to be longer and hotter than would be expected from calculations, to ensure that varying conditions do not invalidate a run. In general, convection dry-heat sterilization cycles are run above 160°C [37].

C. Sterilization by Ethylene Oxide

Ethylene oxide (ETO), a colorless gas, is widely used as a sterilant in hospitals and industry for items that cannot be sterilized by steam. It is often diluted with carbon dioxide, or sometimes fluorocarbons, to overcome its flammable and explosive nature. The mechanism by which ETO kills microorganisms is by alkylation of various reactive groups in the spore or vegetative cell. One of the more resistant organisms to ETO is *B. subtilis* var. *niger* (*globigii*). It is the *USP* biological indicator for monitoring the effectiveness of ETO sterilization cycles. Several factors are important in determining whether ETO is effective as a sterilizing gas: gas concentration, temperature, humidity, spore water content, and substrate for the microorganisms. Ethylene oxide should be present at a concentration of about 500 ml/liter for maximum effectiveness. Once gas concentration is not a limiting factor, the inactivation rate of spores by ETO doubles for each 10°C rise in temperature. Relative humidity plays an important role, in that the sensitivity of spores to ETO largely depends on the water content of the spore.

A "typical" ETO sterilization cycle is shown in Fig. 17. As discussed at the beginning of this section, it is important to determine and monitor the bioburden level of the product entering the sterilizer. Also, the load configuration in the sterilizer is important in achieving uniform and reliable sterilization. Unfortunately, commercially available biological indicators used in ETO sterilization are often unreliable. Hopefully, progress will be made in this field in the years ahead.

Unlike other methods, it is necessary to posttreat the product, either through vacuum purging or by allowing the product to remain at ambient conditions for a time, to allow removal of residual ETO and ethylene chlorhydrin/ethylene glycol by-products before use by the consumer.

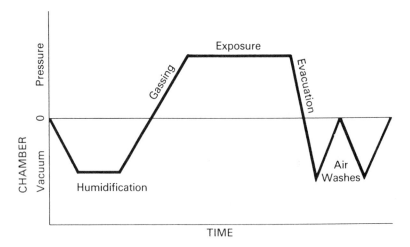

Fig. 17 Ethylene oxide sterilization cycle.

In addition, in 1984 OSHA lowered the maximum permissible operator 8-hr exposure level from 50 ppm in air to 1 ppm (on a time-weighted average) [38].

D. Sterilization by Filtration

It has been only in the past 20 years that filters have become sufficiently reliable to use them on a wide scale to sterilize injectable solutions. Even now, it is prudent to use filtration to sterilize only those products that can not be terminally sterilized.

Filters are of two basic types, depth and membrane. Depth filters rely on a combination of tortuous pathway and adsorption to retain particles or microorganisms. They are made from materials such as diatomaceous earth, inorganic fibers, natural fibers, and porcelain. They carry a nominal rating; that is, a particle size above which a certain percentage of particles is retained. The major advantage of depth filters is their ability to retain large quantities of particles, including many below the nominal rating of the particular filter. Disadvantages of depth filters include grow-through and reproduction of microorganisms, tendency of the filter components to slough during line surges, and retention of some liquid in the filter. Membrane filters rely on sieving and, to a lesser degree, absorption to prevent particles from passing. Although all pores in a membrane filter are not of the same size, nevertheless, the filter can retain all particles larger than the stated size.

Similar to depth filters, membrane filters are made from a variety of materials, although filters made from cellulose ester derivatives are by far the most common. The advantages of membrane filters include no retention of product, no media migration, and efficiency independent of flow-rate pressure differential. The major disadvantages are low capacity before clogging and the need to prewash the filters to remove surfactants. Given the advantages and disadvantages of each type of filter, when large quantities of liquids are to be sterile filtered, such as in industrial applications, it is very common to use a relatively coarse depth filter (1–5 μm) to remove the great majority of particles and, subsequently, use a membrane filter to remove the remaining particles and microorganisms down to a predetermined size (0.22 μm).

Filter cartridges are used for filtering large volumes of solution or more viscous products because of the large surface area that is available through the pleated design. Hydrophobic filters are also available for sterile filtering of gases and solvents [39].

IX. CLINICAL CONSIDERATIONS IN PARENTERAL PRODUCT DESIGN

The formulator must take into consideration all the factors that will improve the clinical acceptability and safety of a product. To minimize tissue damage and irritation and to reduce hemolysis of blood cells and prevent electrolyte imbalance, isotonic solutions should be formulated, if possible. This is not always feasible as a result of the high concentrations of drugs used and the low volumes required for some injections; the wide variety of dose regimens and methods of administration; or because of product stability considerations. Historically, there has been concern over the osmolarity or tonicity of IV fluids. There has also been interest in the osmolarity of other parenteral dosage forms. As mentioned previously, sodium or potassium chloride and dextrose are commonly added to adjust hypotonic solutions.

The effect of isotonicity on reducing pain on injection is somewhat vague, although it may at least reduce tissue irritation. Pain on injection may occur during and immediately following the injection, or it may be a delayed or prolonged type of pain that becomes more severe after subsequent injections. The actual cause of the pain is often unknown and will vary significantly among patients and according to the product. In some cases, pain may be reduced by minor formulation changes, such as adjusting tonicity and pH, or by adding on anesthetic agent, such as benzyl alcohol or lidocaine hydrochloride. In other cases, pain is more inherent to the drug, and pain reduction is more difficult or impossible to resolve. Pain, soreness, and tissue inflammation are often encountered in parenteral suspensions, especially those containing a high amount of solids.

Thrombophlebitis, which is an inflammation of the venous walls, may occur during IV administration and may be related to the drugs being infused, the administration techniques, the duration of the infusion, and the tonicity, and possibly, the pH of the infusion fluid [40]. It has been difficult to define the relative importance of each because of the interplay of all these variables. Some drugs do cause a more significant amount of phlebitis than others. Perhaps the most important factor is administration technique. Proper cleaning of the injection site, the type of cannula employed, strict adherence to aseptic techniques when one is preparing and administering the fluid, and the duration of the infusion are factors that influence the degree of phlebitis that may occur.

The formulator should be aware of the types of clinical use of a drug when designing the dosage forms. Specific examples are pediatric dosage forms and unit dosage forms—including disposable syringes and special packages for hospital, office, or home administration. Hospital packages can take several forms, depending, for example, on whether the package is to be unit-dose, reconstituted by a nurse, bulk dispensed in the pharmacy, or administered as a secondary ''piggyback'' IV container.

Drugs that affect tissue properties, particularly blood flow at the absorption site, may be used to control the rate of absorption. Reduced drug absorption may be achieved physiologically with an IM preparation by incorporating epinephrine, which causes a local constriction of blood vessels at the site of injection. Increased muscular activity may enhance drug absorption because of increased drug flow.

When preparing preparations for IV and IM use, the formulator must be aware of the effect of added substances when unusually large doses of the drug are administered. Although the USP limits the use of some added substances (Table 4), these types of problems cannot always be anticipated. The USP urges special care when administering over 5 ml [1]. When effects do become apparent, the formulator should consider additional dosage sizes or formulation changes. Sometimes during the life of a drug product, new uses and larger doses make the original formula unsatisfactory. When this happens, a new dosage form should be designed and the appropriate cautionary statements placed on the respective labels. The ''precautions,

Table 4 Maximum Amounts of Added Substances Permitted in USP Injectable Products

Substance	Maximum (%)
Mercury compounds	0.01
Cationic surfactants	0.01
Chlorobutanol	0.5
Cresol	0.5
Phenol	0.5
Sulfur dioxide	0.2
or sodium bisulfite equivalent	0.2
or sodium sulfite equivalent	0.2

problems, hazards, and complications associated with parenteral drug administration'' are discussed extensively by Duma and Akers [41].

All parenteral dosage forms must be sterile, pyrogen-free and essentially free from particles that can be detected by visual inspection. Sources of particulate matter include the drug substance, other components or the vehicle; the manufacturing process, which includes the personnel, environment, equipment; and the packaging components. The *USP* provides methods and standards for a microscopic and an electronic liquid-borne particle counting system for large-volume and small-volume parenterals.

The preparation of a new drug substance or dosage form for evaluation in clinical trials must meet the same regulatory requirements and controls as a marketed product. The cGMP requirements for clinical trial products are outlined by the FDA and are discussed in Chapter 21.

The formulator of a new product must consider the manufacturing process to be used for full-scale production of the product. Many new product failures or deficiencies occur because of the inability to resolve or foresee production-related problems, rather than poor product development per se. Therefore, the scientist who has formulated the product must be involved in the development of its manufacturing process and testing. For example, at scale-up, it was found that there was a loss of preservative in the portion of solution/suspension that was in prolonged contact with the transfer tubing during ''down'' time. This was not observed on a small-scale process with less down time and minimum tubing exposure.

The use of clinical trial manufacture as trial production runs provides valuable information and experience for evaluating the scale-up of the formulation and the process. If validated bulk-drug substance is available during the latter stages of the clinical studies (Phase III), this provides an ideal time to validate the manufacturing process. By using this approach, stability data are included in the validation package, thereby reducing the total amount of stability studies. Also, the validated lots can be used in clinical studies. This also permits a thorough evaluation of the process and specifications before submission of the NDA or the registration document.

A. Toxicity Studies

In toxicity studies, acute toxicity tests are usually carried out in the rat, mouse, cat, and dog. Subacute toxicity studies for IM products are performed by giving SC injections to rats and IM injections to dogs. In IV studies the rat tail vein or a front leg is used. Deliberate overdosing usually ''washes out'' metabolism differences between species. In dogs it is common to give an IV dose five times that intended for humans. In rats this is increased to ten times.

Irritation studies are done in rabbits. Each rabbit serves as its own control. The concentration selected for the irritation studies is that intended for humans.

B. Clinical Evaluation

Clinical evaluation of the dosage form is the most expensive and critical phase of product development. All that has been done before this point has been done in an effort to ensure a safe and reliable product for the clinician.

A drug company normally assigns one of its staff physicians as ''monitoring'' physician for the clinical trial (CT) program. The monitoring physician has the key role in the conduct of the CT program (Fig. 18). He or she coordinates the establishment of clinical protocols, the awarding of grants, the gathering and ''in-house'' evaluation of clinical data, and preparation of the FDA submission.

The monitoring physician must first establish what the clinical protocol is going to be. With injectable products, this involves both a clinical pharmacology safety test and a dose-ranging

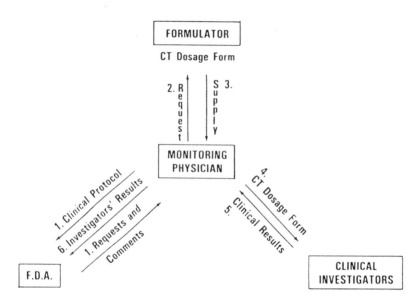

Fig. 18   Clinical trial scheme.

study in humans. These studies are normally a single IM injection in several patients. Depending on the drug, clinical studies may proceed eventually to controlled double-blind studies. Care must be exercised to involve a sufficient number of patients to make the studies statistically meaningful. If it is intended that treatment of several clinical conditions is to be claimed for the product, each must be evaluated separated. Throughout the course of the studies, there is a continuing dialogue between the FDA and the monitoring physician.

When the clinical program has been approved by a peer review committee and filed with the FDA, the monitoring physician requests a sufficient amount of material from the formulator to initiate the clinical program. Before the dosage form is released to the custody of the monitoring physician, the new drug substance and the formulated product must be thoroughly evaluated to ensure proper potency, purity, and safety. Stability studies must also be initiated so that, if the product becomes subpotent or physically unstable during the course of the clinical trial, it can be recalled before any harm can result to the patients in the study.

After release of CT material, the monitoring physician supplies it to the clinical investigators with whom the clinical program has previously been discussed. As the clinical investigators use the product, they begin sending in reports to the monitoring physician, who evaluates them and sends them to the FDA. Although the concept shown in Fig. 18 is oversimplified, it does convey the principal framework under which the clinical trials are conducted.

X.   QUALITY ASSURANCE

The terms *quality assurance* and *quality control* are sometimes used interchangeably, but there is an important difference in meaning. Quality control generally refers to testing of raw materials, packaging components, and final product for conformance to established requirements. Quality assurance is a term that includes quality control, but has broader meaning to also include written operating procedures, personnel training, record keeping, and facility design and monitoring. The philosophy of a quality assurance program is to build quality into the product, rather than to rely only on final product testing to cull out defective product.

Although principles of quality control and quality assurance are important to all pharmaceutical dosage forms, they are especially critical when considering the unique attributes of parenteral dosage forms–sterility, absence of pyrogens, and freedom from extraneous particulate matter. Quality control is generally divided into three areas: (a) raw materials, (b) in-process controls, and (c) product specifications. However, numerous attributes for a product have to be considered throughout all phases of development, evaluation, production, and storage to guarantee good quality.

The factors necessary to achieve quality in a product during the developmental stage have been discussed. The formulator of a new product must consider the manufacturing process to be used for full-scale production of the product. Many new product failures or deficiencies occur because of inability to resolve or foresee production-related problems, rather than to poor product development per se. Therefore, the scientist involved in the development of a product must be involved in development of its manufacturing process and testing. Standards must be carefully established for all raw materials and packaging components used in the product so that the quality of the product will be maintained. Trial production runs should be performed on a new product for stability testing and process evaluation.

A. Regulatory and Compendial Requirements

The manufacture and sale of parenteral products is regulated by federal and state laws, as well as by the *USP*. Federal drug regulations are discussed in detail in Chapter 21. The *USP* provides specifications, test procedures, standards, and so on. for parenteral products and their packaging components. In addition to individual monographs, the *USP* limits the use of certain additives (see Table 4), limits the size of multiple-dose containers to 30 ml, and requires a suitable preservative to be added to containers intended for multiple use.

The current Good Manufacturing Practice (cGMP) regulations are guidelines that the FDA requires a pharmaceutical manufacturer to meet. Compliance with the cGMPs is a prerequisite for the approval of NDAs, INDs, and antibiotic forms. General areas in which GMP guidelines must be established and adhered to include:

Organization and personnel
Buildings and facilities
Control of drug components, packaging, and materials
Production and process control
Equipment
Packaging and labeling control
Holding and distribution
Laboratory control
Records and reports

Parenteral formulation and the preparation of parenterals for clinical trial use obviously require adherence to cGMPs. A development group that generates CT materials should have guidelines and written procedures covering such areas as equipment (validation, maintenance, and cleaning), environment (monitoring and cross-contamination), instruments (maintenance and monitoring), housekeeping, documentation, training, and material handling and storage. Sterilization methods, aseptic processing, and filling techniques and methods, all must be validated to assure product sterility and quality. *Validation* is the process of proving that a process or equipment does what it is supposed to do within the established limits. All individuals performing an aseptic process must periodically pass a test to verify their aseptic technique.

B. Monitoring Programs

Process Facilities

Continual evaluation of manufacturing processes are necessary to maintain "good manufacturing practices." Facilities, buildings, and equipment used in the production of parenteral products must be specially designed for this purpose. Factors to be considered when designing a new plant include environmental conditions, work flow, equipment, choice of materials, personnel, organization, process, documentation, production hygiene, and process controls [42,43]. Thorough planning and engineering of a parenteral facility will not only help maintain the quality of the manufactured products, but will simplify cleanup and maintenance requirements. Contamination of a product is minimized by maintaining a clean facility.

Production Areas

Production areas can be separated into seven general classes: cleanup area, preparation area for packaging materials, preparation area for drug products, sterilization facilities, aseptic filling and processing areas, sorting and product holding areas, and a labeling section.

The exact identity of all packaging components, the bulk and filled product, labels, and so on, must be carefully maintained. The production ticket must be written so that it is easily understood and followed by the appropriate production personnel. All procedures should be clearly outlined and limits established for all operators, (e.g., "Heat water to 35–45°C" or "Autoclave sterilize for 15–20 min at 121–124°C).

All production processes, such as ampoule washing and sterilization, solution filtration, equipment setup and operation, sorting, and freeze-drier cleaning and operation, should be covered in detail in a procedure manual to ensure that all operations are understood as well as carried out properly and uniformly. Cleaning, sterilization, sterile filtration, filling, and aseptic processing operations must be validated.

Personnel

People are the principal source of contamination in clean room operations. All personnel involved throughout the development and production of a parenteral product must be aware of those factors that influence the overall quality of a product as well as those factors on which they directly impinge. It is of particular importance that production personnel be properly trained so that human error is minimized. They should be made aware of the use of the products with which they are involved and the importance of following all procedures, especially proper aseptic techniques. Procedures must be set up to check and verify that the product is being manufactured as intended. After manufacture of a batch, production tickets must be carefully checked, sterilization charts examined, and labels verified for correctness and count.

Environmental Monitoring

Control of environmental factors is important to product quality. Air quality and air movement, care and maintenance of the area, and personnel movement and attire are of particular importance.

The air quality in preparation and aseptic areas can be one of the greatest sources of product contamination. However, this problem can be minimized by use of the effective equipment currently available to provide clean air essentially free from microorganisms and dirt particles. Depth-type filters, electrostatic filters, and dehumidification systems are used to remove the major portion of the airborne contaminants. Air for aseptic areas is then passed through high-efficiency particulate air (HEPA) filters, which remove 99.97% of all particles 0.3 μm or larger. To prevent outside air from entering aseptic areas, a positive pressure is maintained relative to corridors.

A laminar flow enclosure provides a means for environmental control of a confined area for aseptic use. Laminar flow units utilize HEPA filters, with the uniform movement of air along parallel lines. The air movement may be in a horizontal or vertical direction and may involve a confined area, such as a workbench, or an entire room. Laminar flow modules are suspended above filling lines, vial- and stopper-washing equipment, and other processes to provide an aseptic and particulate-free environment.

Regardless of the methods used to obtain a clean air environment, unless the parenteral operator is made completely aware of the limits of laminar flow, uses careful, planned movements, and is wearing proper clothing, he or she can be a source of product contamination. Operator movement within aseptic rooms should be minimized. The rooms must be disinfected regularly and thoroughly before setting up for aseptic operation.

Commonly used environmental monitoring techniques include the following:

Passive Air Sampling. Petri dishes containing microbiological growth media are placed in aseptic areas for specified lengths of time, the ''settling plates'' are then incubated and colonies are counted and identified. This is a qualitative test, since there is no way of knowing the volume of air represented by a given number of colonies.

Active Air Sampling. Active air sampling provides quantitative data because air at a known flow rate is impacted on a strip of nutrient media, followed by incubation of the nutrient strips and enumeration of colonies. Common active air sampling instruments include the slit-to-agar impact sampler and the centrifugal (Reuter) sampler.

Air Classification Measurement. Electronic airborne particle monitoring instruments count and size particulate matter in the sampled air with no consideration of whether the particles are viable or nonviable. Air classification is defined as the number of particles per cubic foot of air that are larger than 0.5 μm in diameter. Climet and HIAC-Royco are common instruments for airborne particulate monitoring).

Surface Monitoring. Contact (or Rodac) plates of growth media are applied to surfaces such as bench tops, walls, and personnel, then incubated. Colony-forming units (CFUs) are counted and identified.

Differential Pressure Measurement. Differential manometers are instruments that measure the difference in pressure between two adjacent rooms. Cleaner environments must have a higher pressure than adjacent less clean environments to prevent flow of relatively dirty air into the cleaner environment. This differential pressure must be monitored and controlled.

C. Product Testing and Evaluation

Quality control testing and evaluation is involved primarily with incoming raw materials, the manufacturing process, and the final product. Testing of incoming raw materials includes routine testing on all drugs, chemicals, and packaging materials.

Process controls include daily testing of water for injection (USP), conformation of fill doses and yields, checking and approving intermediate production tickets, and checking label identity and count. Finished product control includes all the tests necessary to ensure the potency, purity, and identity of the product. Parenteral products require additional tests, which include those for sterility, pyrogens, clarity, and particulate analysis; and for glass-sealed ampoules, leaker testing.

Sterility Testing

The purpose of a sterility test is to determine the probable sterility of a specific batch. The *USP* lists the procedural details for sterility testing and the sample sizes required [1]. The *USP* official tests are the direct (or culture tube inoculation) method and the membrane filtration method.

The interpretation of sterility results is divided into two stages by the *USP* relative to the type of sterility failure if one occurs. If sterility failure of the test samples occurred because of improper aseptic technique or as a fault of the test itself, stage 1 may be repeated with the same sample size. Sample size is doubled in a stage 2 testing, which is performed if microbial growth is observed in stage 1 and there is no reason that the test was invalid. The only absolute method to guarantee the sterility of a batch would be to test every vial or ampoule.

There is a probability of non-product-related contamination in the order of 10^{-3} when performing the sterility test because of the aseptic manipulations necessary to carry out the procedure. This level (10^{-3}) is comparable with the overall efficiency of an aseptic operation. Confidence in the sterility test is dependent on the fact that the batch has been subjected to a sterilization procedure of proved effectiveness. Records of all sterility tests must be maintained, as well as temperature recordings and records from autoclaves, ovens, or other equipment used during the manufacturing process. All sterilizing equipment must be validated to ensure that the proper temperatures are obtained for the necessary time period. These validations are obtained by the use of thermocouples, chemical and biological indicators, sealed ampoules containing culture medium with a suspension of heat-resistant spores, and detailed sterility testing.

Pyrogen Testing

Pyrogenic substances are primarily lipid polysaccharide products of the metabolism of microorganisms; they may be soluble, insoluble, or colloidal. Pyrogens produced by gram-negative bacilli are generally the most potent. Minute amounts of pyrogens produce a wide variety of reactions in both animals and humans, including fever, leukopenia, and alterations in blood coagulation. Large doses can induce shock and eventually death.

Pyrogens readily contaminate biological materials because of their ability to withstand autoclaving as well as to pass through many filters. Several techniques are used to remove them from injectable products. The ideal situation is one in which there are no pyrogens present in the starting materials. This is achieved by strict control of the cleanliness of equipment and containers, distillation of water, and limited processing times. In general, pyrogens may be destroyed by being subjected to prolonged heating. Other pyrogen-removal techniques, which are generally less effective or applicable, include filtration, absorption or adsorption, chemical (oxidation), aging, or a combination of these.

One pyrogen test is a qualitative biological test based on the fever response of rabbits. If a pyrogenic substance is injected into the vein of a rabbit, a temperature elevation will occur within 3 hr. Many irritative medical agents will also cause a fever.

A preferred method for the detection of pyrogens is the limulus amebocyte lysate (LAL) test. A test sample is incubated with amebocyte lysate from the blood of the horseshoe crab, *Limulus polyphemus*. A pyrogenic substance will cause a gel to form. This is a result of the clottable protein from the amebocyte cells reacting with the endotoxins. This test is more sensitive, more rapid, and easier to perform than the rabbit test.

Leaker Testing and Sealing Verification

Ampoules that have been sealed by fusion must be tested to ensure that a hermetic seal was obtained. The leaker test is performed by immersing the ampoules in a dye solution, such as 1% methylene blue, and applying at least 25 in. (ca. 64 cm) of vacuum for a minimum of 15 min. The vacuum on the tank is then released as rapidly as possible to put maximum stress on weak seals. Next, the ampoules are washed. Defective ampoules will contain blue solution.

Another means of testing for leakers is a high-frequency spark test system developed by the Nikka Densok Company of Japan, which detects the presence of pinholes in ampoules.

Some advantages of this system include higher inspection accuracy, higher processing speed, and eliminating the possibility of product contamination [44].

Bottles and vials are not subjected to such a vacuum test because of the flexibility of the rubber closure. However, bottles that are sealed under vacuum may be tested for vacuum by striking the base of the bottle sharply with the heel of the hand to produce a "water hammer" sound. Another test is the spark test, in which a probe is applied outside the bottle. When it reaches the air space of the bottle, a spark discharge occurs if the headspace is evacuated.

The microbiological integrity of various packages, such as vials and stoppers, disposable syringes, and plastic containers, should be determined. A microbiological challenge test is performed by filling the package with a sterile medium and then exposing the sealed container to one of the following tests that is appropriate for the package system: (a) static-aerosol challenge, (b) static-immersion challenge, (c) static-ambient challenge, or (d) dynamic-immersion challenge. The static-immersion challenge test is used commonly with new package combinations. The sealed containers are periodically challenged by immersion into a suspension of challenge organisms. Storing the containers at 5° or 40–50°C, or both, before immersion provides additional stress.

Clarity Testing and Particulate Analysis

Clarity is defined as the state or quality of being clear or transparent to the eye. Clarity is a relative term subject to the visual acuity, training, and experience of the sorter. Clarity specifications are not given in the *USP*, other than to state that all injections be subjected to visual inspection.

Particulate matter is defined in the *USP* as extraneous, mobile, undissolved substances, other than gas bubbles, unintentionally present in parenteral solutions. Test methods and limits for particulates are stated in the *USP* for large-volume injections and small-volume injections.

The development of sorting standards is the responsibility of the manufacturer. Parenteral solutions are sorted for foreign particles, such as glass, fibers, precipitate, and floaters. The sorter also checks for any container deficiency and improper dose volume when feasible. All products containing clear solutions should be inspected against a black and sometimes a white background using a special light source. Although manual visual inspection is the most common means of inspection, electronic particle detection equipment and computer-controlled electro-optic systems are replacing manual inspection and use a light source or camera, or both, positioned behind, above, or below the units being inspected.

Instruments that measure scattered light, such as the Photo-Nephelometer (Coleman Instruments, Oak Brook, IL) are used to evaluate and set clarity standards for parenteral preparations. It is not possible to establish an overall standard value for all products (e.g., 30 nephelos) because the value itself is relative and influenced by many factors, including concentration, aging, stopper extracts, and the solubility characteristics of the raw materials. Nephelometer readings are insensitive to contamination by large (visible) particulates.

The significance of particulate contamination in all parenteral preparations and devices has received much attention. Although it has not been established that particles can cause toxic effects, the pharmaceutical industry, the medical profession, hospital pharmacists, and the FDA, all realize the importance of reducing particulate levels in all parenteral products and devices.

Sources of particulate matter include the raw materials, processing and filling equipment, the container, and environmental contamination. Several methods have been developed for identifying the source of particulates in a product so that they may be eliminated or reduced. The most effective method is that of collecting the particulates on a membrane filter and identifying and counting them microscopically. However, this method is time-consuming and not adaptable to in-line inspection. Several video image projection methods for in-line detection of particles have been developed that provide potential for mechanizing inspection. Electronic particulate counters have been applied to parenterals because of the rapidity at which they do particulate analysis. Their main disadvantages are the lack of differentiation of various types of particulates including liquids such as silicones, and the fact that particle size is measured differently from microscopic analysis. The *USP* tests for particulate matter in injections utilizes both the microscopic and light obscuration methods [1].

Labeling

The package and, in particular, the labeling for parenteral dosage forms are integral and critical parts of the product. The labeling must be legible and clearly identify the drug, its concentration, handling or storage conditions, and any special precautions. The dose or concentration must be prominently displayed when other concentrations of the same drug are marketed. Proper labeling is difficult with the space limitation dictated by small containers used for many parenteral products. Smaller containers have become increasingly popular because of the unit-dose concept.

REFERENCES

1. *The United States Pharmacopeia*, 23rd Ed., U. S. Pharmacopeial Convention, Rockville, MD, 1995.
2. S. Turco, *Sterile Dosage Forms*, 4th ed., Lea & Febiger, Philadelphia, 1994.
3. S. Feldman, Bull. Parenter. Drug. Assoc., 28, 53 (1974).
4. S. Motola and S. N. Agharkar, *Preformulation Research of Parenteral Medications*, 2nd Ed., *Pharmaceutical Dosage Forms: Parenteral Medications*, Vol. 1 (K. E. Avis, H. A. Lieberman, and L. Lachman, eds.), Marcel Dekker, New York, 1992, pp. 115–172.
5. G. Engel and R. Pfeiffer, unpublished data, Eli Lilly & Co., Indianapolis, Ind.
6. J. K. Haleblian, J. Pharm. Sci., 64, 1269 (1975).
7. M. J. Pikal, A. L. Lukes, J. E. Lang, and K. J. Gaines, Pharm. Sci., 67, 767, (1978).
8. J. Brage, *Galenics of Insulin*, Springer-Verlag, New York, 1982, p. 41.
9. L. C. Schroeter, J. Pharm. Sci., 50, 891 (1961).
10. L. Lachman, P. B. Sheth, and T. Urbanyi, J. Pharm. Sci., 53, 311 (1964).
11. K. S. Lin, J. Anschel, and C. J. Swartz, Bull. Parenter. Drug Assoc., 25, 40 (1971).
12. S. Motola and C. Clawans, Bull. Parenter. Drug Assoc., 26, 163 (1972).
13. J. F. Carpenter, S. J. Prestrelski, and T. Arakawa, Arch. Biochem. Biophys., 303, 456–464, (1993).
14. J. Freyfuss, J. M. Shaw, and J. J. Ross, Jr., J. Pharm. Sci., 65, 1310 (1976).
15. A. J. Spiegel and M. M. Noseworthy, J. Pharm. Sci., 52, 917, (1963).
16. S. L. Hem, D. R. Bright, G. S. Banker, and J. P. Pogue, Drug. Dev. Commun., 1, 471 (1974–75).
17. M. J. Akers, A. L. Fites, R. L. Robinson, J. Parenter. Sci. Technol., 41, 88 (1987).
18. J. C. Boylan, Bull. Parenter. Drug Assoc., 19, 98 (1965).
19. J. C. Boylan and R. L. Robison, J. Pharm. Sci., 57, 1796 (1968).
20. M. J. Pikal, A. L. Lukes, J. E. Lang, and K. Gaines, J. Pharm. Sci., 67, 767–773, (1978).
21. A. P. MacKenzie, Dev. Biol. Stand. 36, 51–67 (1977).
22. F. Franks, Cryo Lett. 11, 93–110 (1990).
23. S. L. Nail and L. A. Gatlin, in *Pharmaceutical Dosage Forms: Parenteral Medications*, 2nd Ed., Vol. 2 (K. Avis, H. Lieberman, and L. Lachman, eds.), Marcel Dekker, New York, 1993.
24. L. M. Her and S. L. Nail, Pharm. Res., 11, 54–59, (1994).
25. L. A. Gatlin and P. P. DeLuca, J. Parenter. Drug Assoc., 34, 398–408 (1980).
26. N. A. Williams, Y. Lee, G. P. Polli, and T. A. Jennings, J. Parenter. Sci. Technol., 40, 135–41 (1986).
27. T. Oguchi, Chem. Pharm. Bull., 37, 3088–91 (1989).
28. J. Crowe, L. Crowe, and J. Carpenter, Cryobiology, 27, 219–231 (1990).
29. M. J. Pikal, S. Shah, D. Senior, and J. E. Lang, J. Pharm. Sci., 72, 635–650 (1983).
30. M. J. Pikal, M. L. Roy, and S. Shah, J. Pharm. Sci., 73, 1224–1237 (1984).

31. S. L. Nail, J. Parenter. Drug Assoc., 34, 358–368 (1980).

32. M. J. Pikal, S. Shah, M. L. Roy, and R. Putman, Int. J. Pharm., 60, 203–217 (1990).

33. T. J. Ahern and M. J. Manning, *Stability of Protein Pharmaceuticals*, Part A, *Chemical and Physical Pathways of Protein Degradation*, Plenum Press, New York, 1992.

34. Y. W. Chien, *Novel Drug Delivery Systems: Fundamentals, Developmental Concepts, Biomedical Assessments*, Marcel Dekker, New York, 1982, pp. 219–292.

35. M. J. Akers and N. R. Anderson, *Pharmaceutical Process Validation* (B. T. Loftus and R. A. Nash, eds.), Marcel Dekker, New York, 1984, pp. 29–97.

36. E. J. Leuthner, *Autoclaves and Autoclaving, Encyclopedia of Pharmaceutical Technology*, Vol. 1 (J. Swarbrick and J. Boylan, eds.), Marcel Dekker, New York, 1988, pp. 393–414.

37. F. M. Groves and M. J. Groves, *Dry Heat Sterilization and Depyrogenation, Encyclopedia of Pharmaceutical Technology*, Vol. 4 (J. Swarbrick and J. Boylan, eds.), Marcel Dekker, New York, 1991, pp. 447–484.

38. R. R. Reich and D. J. Burgess, *Ethylene Oxide Sterilization, Encyclopedia of Pharmaceutical Technology*, Vol. 5 (J. Swarbrick, and J. Boylan, eds.), Marcel Dekker, New York, 1992, pp. 315–336.

39. T. H. Meltzer, *Filters and Filtration, Encyclopedia of Pharmaceutical Technology*, Vol. 6 (J. Swarbrick and J. Boylan, eds.), Marcel Dekker, New York, 1992, pp. 51–91.

40. S. J. Turco, Bull. Parenter. Drug Assoc., 28, 197 (1974).

41. R. J. Duma and M. J. Akers, *Pharmaceutical Dosage Forms: Parenteral Medications*, Vol. I (K. E. Avis, L. Lachman, and H. A. Lieberman, eds.), Marcel Dekker, New York, 1984, p. 35.

42. H. E. Hempel, Bull Parenter. Drug Assoc., 30, 88 (1976).

43. R. A. Blackmer, J. Parenter. Sci. Technol., 38, 183 (1984).

44. M. J. Akers, *Parenteral Quality Control: Sterility, Pyrogen, Particulate, and Package Integrity Testing*, Vol. 1 (J. R. Robinson, ed.), Marcel Dekker, New York, 1985, pp. 207–209.

13

Design and Evaluation of Ophthalmic Pharmaceutical Products

Gerald Hecht, Robert E. Roehrs, John C. Lang, Denise P. Rodeheaver, and Masood A. Chowhan
Alcon Laboratories, Inc., Fort Worth, Texas

I. INTRODUCTION

Any modern text on drug product design and evaluation must place into perspective the unique nature of the ophthalmic dosage form in general. More specifically, it must consider that bodily organ which, probably better than any other, serves as a model structure for the evaluation of drug activity—the eye. In no other organ can a practitioner, without surgical or mechanical intervention, so well observe the activity of the drug being administered. With such modern instrumentation as the biomicroscope (Fig. 1), the specular microscope (Fig. 2) and the various intraocular pressure-measuring devices, the ophthalmologist can readily view most of the ocular structures from the cornea to the retina and, in so doing, detect sign of ocular or systemic disease long before sight-threatening or certain general health-threatening disease states become intractable. With this specialized type of instrumentation, the practitioner can view the activity of the drug product on the entire eye or, for those products administered to the internal structure of the eye, the activity or effect of the drug product on a cell or, a group of cells.

Although *official* ophthalmic solutions, suspensions and ointments [the *United States Pharmacopeia* (USPXXII) and the *National Formulary* (NF XVII) list 134 in all] can be compounded extemporaneously by the community pharmacist with the required expertise, equipment, and care, the time and expense in doing so usually encourage the pharmacist to seek commercially compounded and packaged products for his or her prescription needs. This is so because the *USP* requires that "Ophthalmic solutions are sterile solutions, essentially free from foreign particles, suitably compounded and packaged for installation into the eye," or "Ophthalmic suspensions are sterile liquid preparations containing solid particles dispersed in a liquid vehicle intended for application to the eye" [1]. Ophthalmic suspensions are required to be made with the insoluble drug in a micronized form to prevent irritation or scratching of the cornea. Ophthalmic ointments are (to be) manufactured from presterilized ingredients under rigidly aseptic conditions and must meet the requirements of the *USP* under Sterility Tests ⟨71⟩. If the specific ingredients used in the formulation do not lend themselves to routine

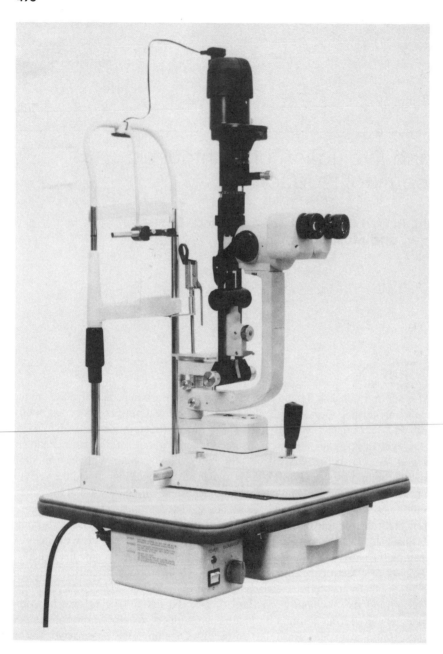

Fig. 1 Topcon biomicroscope.

sterilization techniques, ingredients that meet the sterility requirements described under Sterility Tests ⟨71⟩, along with aseptic manufacture, may be employed. Ophthalmic ointments must contain a suitable substance or mixture of substances to prevent growth of, or to destroy, microorganisms accidentally introduced when the container is opened during use, unless otherwise directed in the individual monograph, or unless the formula itself is bacteriostatic. The finished ointment must be free from large particles and must meet the requirements for *Leakage*

Outflow Reservoir

Thermistor Gauge

Pressure Transducer

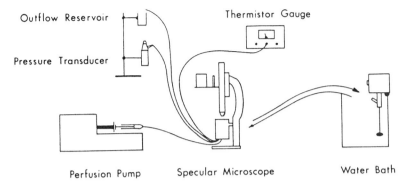

Perfusion Pump   Specular Microscope   Water Bath

Fig. 2   Specular microscope setup for in vitro evaluation of effect of drugs on ocular tissue.

and for *Metal Particles* under Ophthalmic Ointments ⟨771⟩. The *USP* goes on further to specify requirements of the immediate container for ointments and ointment bases. The *USP* also states that "Preparation of an ophthalmic solution and suspension requires careful consideration of such factors as the inherent toxicity of the drug itself, isotonicity values, the need for buffering agents, the need for a preservative (and, if needed, its selection), sterilization and proper packaging."

Behind the relatively straightforward compositional nature of ophthalmic solutions, suspensions, and ointments, however, lie many of the same physicochemical parameters that affect drug stability, safety, and efficacy, as they do for most other drug products. Additionally, specialized dosage forms, such as parenteral-type ophthalmic solutions for intraocular, subtenons, and retrobulbar use; suspension of insoluble substances, such as dexamethasone or fluorometholone; and solids for reconstitution, such as acetylcholine chloride and epinephrine bitartrate for ophthalmic solution, all present the drug product designer with composition and manufacturing procedural challenges second to none in the development of pharmaceuticals.

Like most other products in the medical armamentarium, ophthalmic products are currently undergoing a process termed *optimization*. New modes of delivering a drug to the eye are being actively explored, ranging from a solid, hydrophobic or hydrophilic device that is inserted into the ophthalmic cul-de-sac, to conventionally applied dosage forms that, owing to their formulation characteristics, markedly increase the drug residence time in the fornix of the eye, thereby providing drug for absorption for prolonged periods and reducing the frequency that a given drug product must be administered.

Inasmuch as products for the diagnosis and treatment of ocular disease cover the spectrum of practically all dosage forms and, thus, require the same pharmaceutical sciences for their development, in this chapter we discuss the entire scope of considerations involved in the development of ophthalmic products, ranging from regulatory and compendial requirements, through physicochemical, safety, and efficacy considerations, to a discussion of each type of dosage form currently used by the medical practitioner.

The final consideration, but by no means a minor one, is the design of contact lens care products for the correction of visual acuity. These products are currently regulated by the Food and Drug Administration (FDA) and the official compendia. They include polymethyl methacrylate lenses as well as the currently approved hydrophilic and oxygen-permeable contact lenses and their associated lens care products that are now classified as devices by the FDA, requiring licensure before sale.

II. HISTORICAL BACKGROUND

"If a physician performed a major operation on a seignior [a nobleman] with a bronze lancet and has saved the seigniors's life, or he opened the eye socket of seignior with a bronze lancet and has saved the seignior's eye, he shall receive ten shekels of silver." But if the physician in so doing "has caused the seignior's death, or has destroyed the seignior's eye, they shall cut off his hand." The foregoing excerpts are from two of 282 laws of King Hammurabi's Code, engraved about 100 B.C., in a block of polished black igneous stone about 2.7 m high, now permanently preserved at the Louvre [2].

Mention is made of the Code of Hammurabi only to place in human history that period when reference to eye medicines or poultices was beginning to appear. The Sumerians, in southern Mesopotamia, are considered to be the first to record their history, beginning about 3100 B.C. The Egyptians used copper compounds, such as malachite and chrysocalla, as green eye makeup with, no doubt, some beneficial effect against infection, owing to the antibacterial properties of copper [3].

The standard wound salve of the Smith Papyrus (approximately 1700 B.C.)—grease, honey, and lint—probably served as one of the earliest ointments or ointment bases for the treatment of eye disease or wounds. The Greeks expanded on this basic salve to arrive at a typical enaimon (enheme), a drug for fresh wounds, which might have contained copper, lead, or alum, in addition of myrrh and frankincense [4]. The use of the aromatic substance myrrh in the form of sticks, blocks, or probes has been documented and attributed to the Romans and Greeks. Such sticks were called *collyria* (sing.: *collyrium*) and were dissolved in water, milk, or egg white for use as eyedrops. The Latin word *collyrium* is a derivative of the Greek word, *kollyrien* (in turn derived from *kollyra*, a roll of coarse bread), meaning a glutinous paste made from wheat and water that was rolled into thin cones, rods, or blocks. Often the physician's name was inscribed on these bodies [5]. Pliney the Elder [ca. A.D. 23–79] advocated the use of egg whites to "cool" inflamed eyes, and lycium, one of the most popular of the plant extracts of India, was recommended for use especially for "eye troubles" [6].

After having placed the origin of at least two dosage forms (solution and ointment) for treating disorders or wounds of the eye between approximately the first and second millennium B.C., we can readily reflect on the progress that the designers of dosage forms for eye products have made down through the ages—until relatively recently, little or none. Over the past two decades, however, we have begun to see new concepts emerging, some receiving the enthusiastic support of the ophthalmologist and optometrist, whereas others, not as fortunate, have been relegated to status of little-used novelties.

III. ANATOMY OF THE EYE AND ADNEXA

No attempt will be made in this chapter to present an in-depth discussion of the anatomy of the eye and adnexa, as this subject has been adequately covered by previous authors in the pharmaceutical literature [7,8] and in recent texts on ocular anatomy. For the purpose of this discussion, however, an anatomical cross section of the human eye is presented (Fig. 3) to identify specific tissues, their function, and their involvement in selected ophthalmic disease states, as well as to locate specific sites of drug administration and action. In this discussion, consideration will primarily be given to drugs applied topically; that is, onto the cornea or conjunctiva or into the palpebral fornices. Additionally, drugs are administered by parenteral-type dosage forms subconjunctivally, into the anterior and posterior chambers, the vitreous chamber, and Tenon's capsule, or by retrobulbar injection. Because some of the dosage forms described may be considered as adjunctive to ophthalmic surgical procedures, those procedures

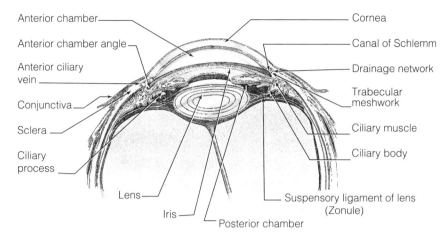

Anterior chamber

Anterior chamber angle

Anterior ciliary vein

Conjunctiva

Sclera

Ciliary process

Lens

Iris

Posterior chamber

Cornea

Canal of Schlemm

Drainage network

Trabecular meshwork

Ciliary muscle

Ciliary body

Suspensory ligament of lens (Zonule)

Fig. 3 Anatomical cross section of the human eye. (Courtesy of Alcon Laboratories, Inc.)

and the concomitant use of the drug are described in Sec. VIII.D. For orientation, readers are encouraged to familiarize themselves with the following anatomical structures of the eye (Table 1), some of which are shown in Fig. 3.

IV. PHARMACOLOGY AND THERAPEUTICS OF OPHTHALMIC MEDICATION

It is not the purpose of a text on the design and evaluation of drugs to present an in-depth review of the pharmacology of ophthalmic drugs. For this purpose the reader is referred to one of the highly regarded treatments of this subject [9–11]. However, since this topic is not commonly covered in pharmacy school curricula, a brief treatment is presented here. For the most part, drugs used in the eye fall into one of several categories, including miotics, mydriatics

Table 1 Anatomical Structures of the Eye

Conjunctiva	Iris
Inferior conjunctival sac	Uvea
Superior conjunctival sac	Posterior chamber
	Zonuies of Zinn
Cornea	Lens
Epithelium	Vitreous humor
Bowman's membrane	
Stroma (substantia propria)	Tenon's capsule
Descemet's membrane	Retina
Endothelium	Ciliary body (zone)
	Meibomian glands
Anterior chamber	
Angle of anterior chamber	Posterior chamber
Schlemm's canal	Vitreous chamber
Spaces of fontana	
Retina	

(with or without cycloplegic activity), cycloplegics, anti-inflammatories, anti-infectives (including antibiotics, antivirals, and antibacterials), antiglaucoma drugs, surgical adjuncts, diagnostics, and a category of drugs for miscellaneous uses. The intended ophthalmic use will dictate the general drug category and dosage form chosen. However, the extent of activity, required onset and duration times, and various other disease state–patient considerations will more precisely indicate what drug or combination of drugs are to be used, including their dosage form and route of administration. For example, the practitioner will, with knowledge of certain contraindications, use mydriatic drugs specifically for their pupillary and accommodative effects, both in the process of refraction and in the management of iridocyclitis, iritis, accommodative exotropia, and so on. Atropine, homatropine, scopolamine, tropicamide, and cyclopentolate are examples of parasympathomimetic drugs possessing mydriatic and cycloplegic activity, whereas phenylephrine and epinephrine are examples of sympathomimetic drugs possessing only mydriatic activity.

Drugs that may be chosen for use in the management of glaucoma may be topically applied miotics, such as pilocarpine hydrochloride or nitrate, carbachol, echothiophate iodide, or demecarium bromide; new drugs (e.g., β-adrenergic blocking agents, epinephrine prodrugs) devoid of pupillary effect, such as betaxolol hydrochloride, timolol maleate, bunolol hydrochloride, or dipivefrin hydrochloride; or, they may be orally administered drugs to present an osmotic effect that will lower intraocular pressure, such as 50% glycerin or 45% isosorbide. Other drugs administered orally to lower intraocular pressure are the carbonic anhydrase inhibitors acetazolamide and methazolamide. Furthermore, the miotic drugs may be chosen to reverse the effect of mydriatics after refraction or during surgical procedures such as cataract removal. There is now available an antimydriatic drug devoid of pupillary activity, dapiprazole hydrochloride, which is gaining importance in the reversal of the effect of mydriatics.

Depending on the location of ocular inflammation, a specific corticosteroid in a specific dosage form may be chosen. For instance, a corticosteroid of high potency, such as prednisolone acetate, fluorometholone, or dexamethasone, may be chosen for deep-seated inflammation of the uveal tract. Further treatment of such inflammation may take the form of subtenon injections or oral (systemic) administration of selected corticosteroids, depending on the indication and the dosage forms available. For inflammation of a more superficial nature the lower strengths of prednisolone acetate or the lower-potency corticosteroids, such as hydrocortisone or medrysone, will usually be chosen.

Drugs used for the treatment of ocular infection will generally be chosen based on the presumptive diagnosis of the causative agent by the ophthalmologist. Laboratory confirmation by microbial culture and identification is routinely conducted concurrently with the initiation of therapy. This is generally necessary because of the severity and sight-threatening nature of some types of infections. For example, if a patient has a foreign body lodged in the cornea originating from a potentially contaminated environment, the physician may choose to begin treatment of the eye, after foreign-body removal, with a single or combination antibiotic, such as gentamicin, tobramycin, chloramphenicol, and a neomycin–polymyxin combination. This is considered appropriate, since an infection with *Pseudomonas aeruginosa* can destroy a cornea in 24–48 hr, generally the time it takes to identify an infectious agent. Less fulminating, but no less dangerous, are infections caused by various staphylococcal and streptococcal organisms. For superficial bacterial infections of the conjunctiva and eyelids, sulfonamides, such as sodium sulfacetamide, are usually prescribed, as are yellow mercuric oxide and mild silver protein.

For fungal and viral infections, there are very few agents that the ophthalmologist can prescribe. For instance, idoxuridine, a selective metabolic inhibitor, has been shown to be useful against herpes simplex virus infection of the cornea. For the trachoma virus and the viruses

TRIC (the single largest cause of blindness worldwide)],
strated satisfactory activity, and the secondary bacterial
aged by conventional antibiotics, such as tetracycline,
he trachoma virus itself seems to be somewhat suscep-
to 6 weeks of treatment three times per day are required

atment of fungal keratitis. The antifungal antibiotic drugs
ective to varying degrees in superficial fungal infection,
facetamide [13,14]. For both of these drugs iontophoresis
iduces enhanced activities.
re primarily irrigating solutions, solutions of proteolytic
ataract removal, intraocular lens placement, vitrectomy,
egrity. These drugs are considered true parenteral dosage
which are discussed in greater detail elsewhere in this

fluorescein, are administered topically or intravenously to
is as corneal abrasions or ulceration and various retinop-
ost widely used diagnostic agent in the practice of oph-
gal has also been used topically, although to a far lesser
h is available as well-preserved alkaline solutions in con-
[15,16], as fluorescein-impregnated absorbent sterile paper
inally sterilized intravenous injections in concentrations

iesthetics are routinely used by the eye care specialist in
i and for various relatively simple surgical procedures. The
ne, in concentrations ranging from 1 to 4% [19]. More
such as tetracaine hydrochloride and proparacaine hydro-
rugs of choice in these procedures.
complex nature, lidocaine hydrochloride and similar local
iave been used [20]. The foregoing overview has presented
rug use. One additional class of drugs that merits brief
I for the treatment of various dry eye syndromes. The most
sicca, involves diminished secretion of a mucuslike agent
:osaminoglycans) that serves to coat the corneal epithelium
rmly attracts water molecules, resulting in even hydration
secretion of this substance causes dry spots to develop on
hydration, which can lead to ulceration, scarring, corneal
pharmaceutical products are available (Hypotears, Tears
ic high molecular weight polymers that serve to temporarily
nting the aforementioned dehydration and affording the dry
previously unavailable [22,23].

IDERATIONS

manufactured under conditions validated to render it sterile
testing is conducted on each lot of ophthalmic product by
the *USP* and validated in each manufacturer's laboratory.

Sterile preparations in special containers for individual use on a single patient must be made available. This availability is especially critical for every hospital, office, or other installations where accidentally or surgically traumatized eyes are treated.

The *USP* recognizes six methods of achieving a sterile product: (a) steam sterilization, (b) dry-heat sterilization, (c) gas sterilization, (d) sterilization by ionizing radiation, (e) sterilization by filtration, and (f) aseptic processing [25]. For ophthalmic products packaged in plastic containers, the practice in the manufacture of modern pharmaceuticals for the eye, a combination of two or more of these six methods is routinely used. For example, for a sterile ophthalmic suspension, bottles, dropper-tips, and caps may be sterilized by ethylene oxide or gamma radiation; the suspended solid may be sterilized by dry heat, gamma radiation, or ethylene oxide; and the aqueous portion of the composition may be sterilized by filtration.

The compounding is completed under aseptic conditions, and the product is filled into the previously sterilized containers, again under aseptic conditions. One can see by the complexity of these types of manufacturing procedures, that much care and attention to detail must be maintained by the manufacturer. This sterile manufacturing procedure must then be validated to prove that no more than 3 containers in a lot of 3000 containers (0.1%) are nonsterile. Ultimately, it is the manufacturer's responsibility to ensure the safety and efficacy of the manufacturing process and the absence of any adverse effect on the product, such as the possible formation of substances toxic to the eye, an everpresent possibility with gas sterilization or when using ionizing radiation. For ophthalmic products sterilized by terminal sterilization (e.g. steam under pressure), the sterilization cycle must be validated to ensure sterility at a probability of 10^6 or greater.

Currently, the *British Pharmacopoeia* suggests five methods of sterilization: (a) sterilization by autoclaving, (b) dry-heat sterilization, usually to $> 60°C$, (c) sterilization by filtration, (d) ionizing radiation (electron accelerator or gamma radiation), and (e) ethylene oxide. During the manufacture of an ophthalmic product, sterility may be checked while the finished product is in its bulk form before filling. It is then also tested on a random sampling basis in the finished package. Suggested guidelines for the number of samples are dependent on whether or not sterilization has taken place in the sealed final container. Class A products are those sterilized in bulk form and filled aseptically into sterile final containers without further sterilization. Class B products are those sterilized in sealed final containers. Class B is further subdivided according to method of sterilization: type 1 comprises those products sterilized by steam under pressure; type 2 comprises those products sterilized by any other means. Class A products require a minimum random sample number of no fewer than 30 items from each filling operation. Class B products require varying sample sizes, depending on whether the sterilization occurs in a chamber or by a continuous process. This generally ranges from 5 to 30 units per lot, depending on conditions of sterilization.

B. Ocular Toxicity and Irritation

Assessment of ocular irritation potential of ophthalmic solutions represents an extremely important step in the development of both over-the-counter (OTC) and prescription pharmaceuticals. Excellent reviews have appeared in recent years [26–28,64] on the evaluation procedures used. Significant advances have been made, resulting in greater reliability, reproducibility, and predictability. The historical evaluation of these procedures can be traced through the literature [29–38], as can an understanding of the mechanisms of ocular response to irritants, based on examination of the conjunctiva [39–42], the cornea [30,43–45] or the iris [30,46,87].

Albino rabbits are currently used to test the ocular toxicity and irritation of ophthalmic formulations. Several articles relate to the use of rabbits as predictors for human responses.

The rabbit has obvious advantages associated with its use. It is readily available, docile, easily handled, relatively inexpensive, easy to maintain, has a large eye, both the corneal surface and the bulbar conjunctival areas are large and easily observed, and the iris is unpigmented, allowing ready observation of the iridal vessels [27]. The primary differences between rabbits and humans in ophthalmic studies relate to decreased tearing in rabbits, decreased blinking rate, loosely attached eyelids, presence of a nictating membrane [41,47–49], differences in the structure of Bowman's membrane, and a slower reepithelialization of the rabbit cornea [30]. The primate has also gained in popularity as an ocular model for the evaluation of drugs and chemicals [47,48].

Various governmental agencies have published guidelines for eye irritancy studies [49,50]. These guidelines are directed toward ophthalmic formulations, chemicals, cosmetics, extractables from ophthalmic containers, and other materials that may intentionally or accidentally contact the eye during use. It is the manufacturer's responsibility to determine those specific studies appropriate to test the safety of the ophthalmic formulation. The *USP* presents guidelines for a 72-hr ocular irritation test in rabbits using saline and cottonseed oil extracts of plastic containers used for packaging ophthalmic products. Containers are cleaned and sterilized as in the final packaged product and extracted following submersion in saline and cottonseed oil. Topical ocular instillation of the extracts and blanks in rabbits is completed and ocular changes examined. If ocular changes between extracts and blanks are similar, the plastic part passes and is rated as satisfactory.

As a part of the Federal Hazardous Substances Act (FHSA), a modified Draize test was adopted [51–53] as the official method for eye irritancy evaluation [54]. Although a proposed method change to include ocular irritation tests as part of the FHSA methods was made in 1972 [55], it was determined that the proposed method performed no better [56] than those already used.

It has been stated that the best way to determine the degree of irritation or differences between test materials may not be the FHSA or Draize methods, as these are pass–fail procedures [56]. Possibly, a better judgment for irritancy is based on degree, frequency, and duration of ocular changes. These changes are graded by examination, and a provision allows slit-lamp (biomicroscope) examination or fluorescein staining of the cornea, or both [27,52,53].

Current guidelines for toxicity evaluation of ophthalmic formulations involve both single and multiple applications. These guidelines [27] are not for contact lens solutions. The multiple applications extend over a 21-day period and involve both irritation and systemic toxicological studies.

As mentioned previously (and discussed in detail in Sec. IX), contact lens products have specific guidelines for testing accessory solutions used with contact lens materials (other than polymethyl methacrylate). These solutions are viewed as new drug devices and require testing with the contact lenses with which they are to be used. These testing guidelines are in addition to the guidelines necessary for ophthalmic solutions. The tests include a 21-day ocular study in rabbits, employing various types (group I, II, III, or IV) of the lenses as they are to be used clinically, including solutions that may be used with the lens as well as other tests [27].

During the application of the various guidelines for both ophthalmic and contact lens products, ocular examination and biomicroscopic examination on rabbit eyes are completed with objective reproducible grading [27,77] for conjunctival congestion, conjunctival swelling, conjunctival discharge, aqueous (humor) flare, iris involvement, corneal cloudiness, severity, area of corneal opacity or cloudiness, pannus, and intensity of fluorescein staining.

In addition to in vivo testing of ophthalmic preparations, primarily in rabbit eyes and, secondarily, in primate eye, numerous in vitro methods have developed over the past few years as alternatives to in vivo ocular testing [57–66]. In vivo methods that incorporate new technology and reduced numbers of animals have also been developed. Particular attention has recently been given to evaluation of preservative effect on corneal penetration [67,68], cytotoxicity [69–73], and effects on wound healing [74–76]. Several manufacturers are currently using in vitro toxicity tests in the development of ophthalmic solutions.

C. Preservation and Preservatives

In 1953 the FDA required that all ophthalmic solutions be manufactured sterile [78]. Preservatives are included as a major component of multiple-dose eye solutions for the primary purpose of maintaining sterility of the product after opening and during use unless prepared sterile in a unit-dose package. The use of the popular plastic eyedrop container has reduced, but not completely eliminated, the chances of inadvertent contamination. There can be a ''suckback'' of an unreleased drop when pressure on the bottle is released. If the tip is allowed to touch a nonsterile surface, contamination could be introduced. Therefore, it is important that the pharmacist instruct the patient on the proper use of an ophthalmic-dispensing container to minimize the hazards of contamination. The contamination hazard is magnified in the busy clinical practice of the eye care professional where numerous diagnostic solutions of cycloplegics, mydriatics, and dyes are used in many patients from the same container. The cross-contamination hazard can be eliminated by the use of packages containing small volumes and designed for single application only (i.e., unit-dose). However, these single-use packages still contain (as a large-scale manufacturing necessity) an amount in excess of the several drops (0.05–0.20 ml) to be used. Unfortunately, there is the tendency to use the entire contents and, thereby, reintroduce the contamination hazards and defeat the purpose of this special packaging. The *USP* outlines a test procedure for antimicrobial effectiveness and how to interpret the results. This test is not a mandatory requirement of the *USP* or the FDA [79], but it is used by the manufacturers to guide them in developing adequately preserved products. This testing of formulas is carried out as a part of the formulation development sequence. Cultures of *Candida albicans, Apergillus niger, Escherichia coli, Pseudomonas aeruginosa,* and *Staphylococcus aureus* are used: a standardized inoculum with organism counts of 100,000–1 million/ml for each microorganism is prepared and tested against the preserved formula. The inoculated tubes or containers are incubated at 20° or 25°C for 28 days, with examination at days 7, 14, 21, and 28. The preservative is effective in the product if (a) the concentrations of viable bacteria are reduced to no more than 0.1% of the initial concentrations by day 14, (b) the concentrations of viable yeasts and molds remain at or below the initial concentrations during the first 14 days, and (c) the concentration of each test microorganism remains at or below these designated levels during the remainder of the 28-day test period. Importantly, most ophthalmic product manufacturers use this as a minimum standard and attempt to formulate their products with a safety margin for preservation.

Considerable emphasis in the ophthalmic literature is placed on the effectiveness of preservatives against *Pseudomonas* spp. because of the reports of the loss of eyes through corneal ulcerations from eye solutions contaminated with *P. aeruginosa*. This organism is not the most prevalent cause of bacterial eye infections, even though it is a common inhabitant of human skin, but it is the most opportunistic and virulent. *Staphylococcus aureus* is responsible for most bacterial infections of the eye. The eye seems to be remarkably resistant to infection when the corneal epithelium is intact. When there is a corneal epithelial abrasion, organisms can enter freely and *P. aeruginosa* can grow readily in the cornea and rapidly produce an ulceration and loss of vision. This microorganism has been found as a contaminant in a number of studies on sterility of ophthalmic solutions, particularly in sodium fluorescein solutions used to detect corneal epithelial damage. The chances for serious infections and cross-contamination are greatly enhanced by multiple use of this dye solution—a danger that has led to the practice by the ophthalmologist of using sterile disposable applicator strips of fluorescein.

An additional test procedure employed by one manufacturer is the preservative evaluation "cidal" test. A formulation is tested against 5–14 microorganisms, including gram-negative and gram-positive bacteria, fungi, and yeasts in a standardized inoculum. Cidal times (no growth) are measured for each organism within 24, 48, and 72 hr of contact.

One area of ophthalmic products for which stricter microbiological guidelines have recently been imposed is in the area of flexible hydrophilic contact lens accessory products. Specific guidelines have been devised by the FDA for this area of ophthalmic products and differ primarily in the necessary kill rate, depending on whether or not the solution is to be used as a cleaning or rinsing product or as a flexible hydrophilic lens disinfectant. These microbiological guidelines have evolved and changed to the current position during the last decade. They are discussed in Sec. IX of this chapter.

The need for and testing of preservatives in ophthalmic products was described previously. There are a group of eyedrops in which the use of preservatives is prohibited. These products are those that are used at the time of eye surgery. In this type of usage, there is the risk of the solution contacting the internal eye tissues causing toxicity. If sufficient concentration of the preservative is in contact with the cornea endothelium for a sufficient time period; the cells can become damaged to the point of causing clouding of the cornea and possible loss of vision. These products should be packaged in sterile, unit-of-use containers without preservatives.

Choice of Preservative

Although this section is directed to ophthalmic products, it is largely applicable to parenteral and even nonsterile products (solutions, emulsions, and suspensions). The choice of preservative is limited to only a few chemicals that have been found, over the years, to be safe and effective for this purpose. These are benzalkonium chloride, Polyquad, thimerosal, methyl- and propylparaben, phenylethanol, chlorhexidine, and polyaminopropyl biguanide. The chelating agent, disodium edetate (EDTA), is sometimes used to increase the activity against certain *Pseudomonas* strains, particularly with benzalkonium chloride. Chlorhexidine as the hydrochloride, acetate, or gluconate salts is widely used in the United Kingdom and Australia, but was not used in the United States until 1976, and then only in a soft contact lens disinfection solution. This limited choice of preservative agents is further narrowed when the requirements of chemical and physical stability and compatibility are considered for a particular formulation and package, or a contact lens material. Many times it is necessary to design the formula to fit the requirements of the preservative system chosen.

Several guidelines are available in the literature for the pharmacist who must extemporaneously prepare an ophthalmic solution. The *USP* contains a section on ophthalmic solutions, as do several standard compoundial textbooks. Since the pharmacist does not have the facilities to test the product, he or she should dispense only small quantities, with an expiration date of no more than 30 days. Refrigeration of the product should also be required as a precautionary measure. To reduce the potentially largest source of microbial contamination, only sterile purified water should be used in compounding ophthalmic solutions. Sterile water for injection, USP, from unopened IV bottles or vials is the highest-quality water available to the pharmacist. Prepackaged sterile water with bacteriostatic agents should *not* be used.

Benzalkonium Chloride. The most widely used preservative is benzalkonium chloride, used in combination with disodium edetate. The official benzalkonium chloride is a quarternary ammonium compound defined in *USP* XXII as an alkylbenzyldimethylammonium chloride mixture with alkyl chains ranging from C_8 to C_{16}. This compound's popularity is because, despite its compatibility limitations, it has generally been shown to be the most effective and rapid-acting preservative and has an excellent chemical stability profile. It is stable over a wide pH range and does not degrade, even under excessively hot storage conditions. It has pronounced surface-active properties, and its activity can be reduced by adsorption. It is cationic, and this, unfortunately, leads to a number of incompatibilities with large negatively charged molecules, through production of a salt of lower solubility and possibly precipitation. For example, it cannot be used with nitrates, salicylate, anionic soaps, and large anionic drugs, such as sodium sulfacetamide and sodium fluorescein. It is usually advisable to design the formula to avoid these incompatible anions, rather than to substitute a less effective preservative where this is feasible. There are a number of lists of incompatibilities of benzalkonium chloride in the literature, which may be helpful, but they should not be relied on entirely. Compatibility is determined by the total environment in which the drug molecules exists (i.e., the total product formula). The pharmaceutical manufacturer can sometimes design around what appears to be an incompatibility, whereas the extemporaneous compounder may not have this option, or, more importantly, the ability to test the final product for its stability, safety, and efficacy.

The usual concentration of benzalkonium chloride used in eyedrops is 0.01%, with a range of 0.004–0.02% [80]. The uptake of benzalkonium chloride itself into ocular tissues is limited [82]; however, lower concentrations of benzalkonium chloride have enhanced corneal penetration of compounds [67,81,83].

Richards [84], Mullen et al. [85], and the American College of Toxicology [86] have summarized the literature of benzalkonium chloride. The conclusion drawn was that benzalkonium chloride, up to 0.02%, has been well substantiated as being suitable for use in topical ophthalmic solutions when the conditions of its use are properly controlled. McDonald [88] studied various concentrations in rabbit eyes with several dosing regimens, the most severe being a 1-day acute regimen in which a 0.05-ml dose was instilled in the cul-de-sac at 20 min intervals for 6 consecutive hours. Ocular changes were graded by macroscopic and biomicroscopic (slit-lamp) examination. A dose–response pattern for conjunctival congestion, swelling, discharge, and iritis was noted. He concluded that up to 0.02% benzalkonium chloride is a permissible level in ophthalmic solutions.

Numerous articles have appeared in the literature describing antibacterial activity studies of benzalkonium chloride alone and in comparison with other preservatives. Many of the articles give conflicting results, which is not surprising, considering the many different test methods, formulas, and criteria used to arrive at these diverse conclusions. Adequate data are available to permit us to conclude that the manufacturer can rely only on the test results for each particular product, using the *USP* (or similar) validated test to decide on which preservative system to use to achieve a satisfactory composition.

Some strains of *P. aeruginosa* are resistant to benzalkonium chloride and, in fact, can be grown in concentrated solution of this agent. This has caused great concern because of the virulent nature of this organism in ocular infections, as discussed previously. Thus, it was an important finding in 1958 that the acquired resistance could be eliminated by the presence of ethylenediaminetetraacetic acid (edetic acid; EDTA) in the formulation. This action of EDTA has been correlated with its ability to chelate divalent cations. The use of disodium EDTA, where it is compatible, is recommended in concentrations up to 0.1%.

Another quaternary ammonium germicide, benzethonium chloride, has been used in several opthalmic solutions. It has the advantage of not being a chemical mixture; however, it does not possess the bactericidal effectiveness of benzalkonium chloride and is subject to the same incompatibility limitations.

Organic Mercurials. When benzalkonium chloride cannot be used in a particular formulation, as with pilocarpine nitrate, eserine salicylate, or fluorescein sodium (because of potential anion–cation association), one of three organic mercurials, phenylmercuric nitrate, phenylmercuric acetate, and thimerosal had, until recent years, been used. Because of environmental concerns, however, the use of organic mercurials has fallen into disfavor, and its use must be rigorously defended based on the premise that no suitable alternate exists.

In those situations for which the use of an organic mercurial is the only avenue available, the usual concentration range for the phenylmercuric compounds is 0.002–0.004% and for thimerosal it is 0.02–0.1%. Although they can be used effectively in some products, the mercurials are relatively weak and slow in their antimicrobial activity. The organic mercurials are generally restricted to use in neutral to alkaline solutions; however, they have been used successfully in slightly acid formulations. The phenyl mercuric ion can react with halide ions to form salts of lower solubility and a reduction in their effectiveness. Thimerosal has a greater solubility and is relatively more stable than the phenylmercuric compounds and has not been shown to deposit in the lens of the eye. The latter phenomenon has been observed with the phenylmercury compounds.

These organic mercurials have been brought under the umbrella of the mercury toxicity alarm that occurred in the early 1970s, although they have not been implicated directly. The FDA has issued a regulation restricting the mercurial preservative content of cosmetics applied around the eye. Several countries have banned them entirely.

Ocular sensitization to thimerosal has been well documented over the years [89–95]. Although thimerosal had, at one time, been referred to as the preservative of choice for soft contact lens care products [96–98], its use has largely been supplanted by the polymeric quaternary ammonium preservatives such as Polyquat.

Since the organic mercurials offer an alternative to the quaternary ammonium preservatives, and since adequate preservation of ophthalmic solutions is essential, the benefit-to-risk must be considered before a ban is imposed on their use.

Chlorobutanol. This aromatic alcohol has been an effective preservative and is used in several ophthalmic products. Over the years, it has proved to be a relatively safe preservative for ophthalmic products [102]. In one test system, this preservative was considered to be the only safe preservative for ophthalmic products using cytolytic activity as the criteron [60] and has produced minimal effects in other tests [73,99]. In addition to its relatively slower activity, it has a number of formulation and packaging limitations. It possesses adequate stability when stored at room temperature in an acidic solution, usually about pH 5 or below. If autoclaved for 20–30 min at a pH of 5, it will decompose about 30%. The hydrolytic decomposition of chlorobutanol produces HCl, resulting in a decreasing pH as a function of time. As a result, the hydrolysis rate also decreases. Use of chlorobutanol usually dictates the use of glass packaging, since the volatile compound readily permeates popular polyolefin plastic ophthalmic containers. Chlorobutanol is generally used at a concentration of 0.5%. Its maximum water solubility is only about 0.7% at room temperature, and it is slow to dissolve. Heat can be used to increase its dissolution rate, but will also cause some decomposition and loss from sublimation. Concentrations as low as 0.125% have shown antimicrobial activity under the proper conditions.

Methyl- and Propylparaben. These esters of p-hydroxybenzoic acid have been used primarily to prevent mold growth but in higher concentrations they do possess some weak antibacterial activity. Their effective use is limited by their low aqueous solubility and by reports of their causing stinging and burning sensations in the eye. They bind to a number of nonionic surfactants and polymers, thereby reducing their bioactivity. They are used in combination, with the methyl ester at 0.03–0.1% and the propyl ester at 0.01–0.02%.

Phenylethyl Alcohol. This substituted alcohol has been used at 0.5% concentration, but in addition to its weak activity, it has several limitations. It is volatile and will lose activity by permeation through a plastic package. It has limited water solubility, can be "salted out" of solution, and can produce burning and stinging sensations in the eye. It has been recommended primarily for use in combination preservative systems.

Polyquad. This preservative is relatively new to ophthalmic preparations and is a quaternary ammonium germicide. Its advantage over other quaternary ammoniums seems to be in its inability to penetrate ocular tissues, especially the cornea. At clinically effective preservative levels, Polyquad is approximately ten times less toxic than benzalkonium chloride [61,100]. Various in vitro tests and a chronic in vivo evaluation substantiate the safety of this compound [101]. This preservative has been extremely useful for soft contact lens solutions because of its inability to be adsorbed onto or absorbed into the lens, and its practically nonexistent sensitization potential.

Chlorhexidine. Chlorhexidine, a bisbiguanide, has been demonstrated to be somewhat less toxic than benzalkonium chloride and thimerosal at clinically relevant concentrations [61,63,69,103,104]. This work was confirmed in a series of in vitro and in vivo experiments [100,105–107].

Polyaminopropyl Biguanide. This preservative is also relatively new to ophthalmic formulations, and has been used primarily in contact lens solutions. At the concentrations used in these solutions, polyaminopropyl biguanide has a low toxicity potential [108,109].

VI. ABSORPTION OF DRUGS IN THE EYE

From the perspective of drug design, several requirements need be satisfied by topical ophthalmic therapeutic agents. The drug must be (a) both biochemically and pharmacologically potent, (b) nontoxic to both ocular and systemic tissues, (c) sufficiently stable that neither significant loss in potency from diminished availability nor little increase in toxicity from byproducts of degradation arises, (d) targetable either to tissues and location of primary disease-state etiology or to sites responsible for symptomatic response, and (e) sufficiently compatible with the dosage form, and with the tissues exposed to it, to achieve an effective pharmacokinetic tissue profile.

Often the demand for such a complement of properties requires a hierarchical strategy in which only the broadest possible limits are satisfied for those fourth and fifth design requirements. For example, topical administration assists in limiting toxicity while improving targeting and pharmacokinetic response. The requirements for effective absorption of such topical ophthalmic medications correspond with the physical, chemical, and transport characteristics of the drug primarily designed to satisfy the more stringent biological and pharmacological criteria. Simple guidelines can be appreciated readily by examining the factors affecting absorption of an antiglaucoma agent administered in the conventional manner as drops into the cul-de-sac [110].

The first factor affecting drug availability is loss of drug from the palpebral fissure. This takes place by spillage of drug from the eye and its removal by the nasolacrimal drainage. The normal volume of tears in the human eye is estimated to be approximately 7 μl, and if blinking occurs, the human eye can accommodate a volume of up to 30 μl, without spillage from the palpebral fissure. With an estimated drop volume of 50 μl, 70% of the administered volume of 2 drops can be seen to be expelled from the eye by overflow. If blinking occurs, the residual volume of 10 μl indicates that 90% of the administered volume of 2 drops will be expelled within the first several minutes [111,112].

Drainage from an administered drop is a second factor. Drainage of the drop through the nasolacrimal system into the gastrointestinal tract begins immediately on instillation. This takes place when either reflex tearing or the dosage form causes the volume of fluid in the palpebral tissue to exceed the normal lacrimal volume of $7-10$ µl. Reference to Fig. 4 indicates the pathway for this drainage. The excess fluid volume enters the superior and inferior lacrimal puncta, moves down the canalicula into the lacrimal sac, and continues into the gastrointestinal tract. It is due to this mechanism that significant systemic effects for certain potent ophthalmic medications have been reported [113–115]. This also is the mechanism by which a patient may occasionally sense a bitter or salty taste, typical of therapeutic ammonium salts, following the use of eye drops. The influence of drop size on bioavailability has been investigated thoroughly for conventional formulations and is significant [116,117]. Even for nonconventional viscoelastic formulations, drop volume can be expected to influence efficacy and needs to be optimized [118] The clinical significance of drainage is so well recognized that manual nasolacrimal occlusion has been recommended as a means of improving the therapeutic index of antiglaucoma medications [119].

A third mechanism competing for drug absorption into the eye is superficial absorption of drug into the palpebral and bulbar conjunctiva, with generally concomitant rapid removal from ocular tissues by the peripheral blood flow. For example, the extensive vascularity of the uvea underlies the bulbar conjunctiva, a mucous membrane, and the sclera, the white part of the eye and a tough covering, to which it is attached anteriorly [120]. Binding of drug to either external sites (e.g., by mucins) or internal tissues (e.g., sclera) can be detrimental to efficcy.

In competition with the three foregoing forms of therapeutically *in*effective drug removal from the palpebral fissure is the transcorneal absorption of drug, often the route most effective in bringing drug to the anterior portion of the eye. Although transport of hydrophilic and macromolecular drugs has been reported to occur by limbal or scleral routes, often this is at rates significantly reduced from those expected for transcorneal transport of conventional, mod-

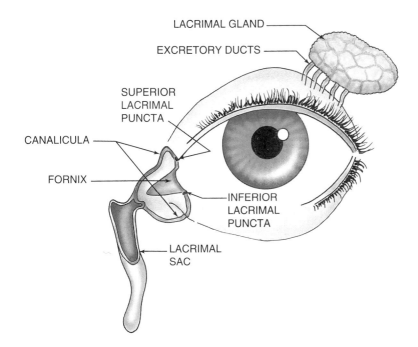

Fig. 4 Anatomical view of the lids and lacrimal system. (Courtesy of Alcon Laboratories, Inc.)

estly lipophilic agents of low molecular weight [121–123]. Even here, transmembrane transport is a significant requirement for availability.

Loss of drug from a precornea volume has been investigated both in vivo and in vitro. These studies relate to both design of dosage forms as well as investigations of transport, bioavailability and pharmacokinetics. Simultaneous release profiles of drugs and adjuvant from in vitro cells in volumes characteristic for the eye can be correlated simply with exposure for transmembrane transport [124]. An example of release profiles and the influence of dosage form is shown in Fig. 5. Simple hydrodynamic analysis of the in vitro mechanism idicates that the elution concentration, in the absence of absorption, is a linear kinetic process, with a release profile that scales as the ratio of the tear production to tear volume, V/V_R, specifically:

$$N_T(t) = N_1\left(1 - \exp\left(\frac{-\dot{V}_t}{V_R}\right)\right) \tag{1}$$

where

$\quad V_R$ = volume of the reservoir
$\quad V$ = flow rate through reservoir (alternatively, Q_T)
$\quad N_1$ = amount of drug in reservoir at time zero
$N_T(t)$ = time dependent total amount eluted from volume

and where the complementary amount, the amount of drug in the reservoir, is defined as

$$N_c(t) = N_1 - N_T(t) = \text{amount contained in reservoir} \tag{2}$$

characteristic of the stirred-tank chemical reactor models [125–127]. Combining these containment profiles, $N_c(t)$, with diffusional transmembrane transport, yields expected tissue profiles the rates of which are dictated by both containment profile and tissue affinities, and the magnitude of which is often dominated by transmembrane flux. A pictorial representation of the processes is shown in Fig. 6. Pharmacokinetic modeling with this scheme has been successful in fitting the aqueous humor levels of pilocarpine following topical administration (Fig. 7) [116]. Although this type of data fitting has been quite successful, there has not been a

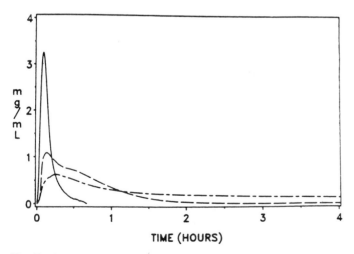

Fig. 5 Comparison of time–release profiles from three different preparations of betaxolol: (——) drug solution representing marketed product; (---) suspension formulation; (— · —) gel formulation.

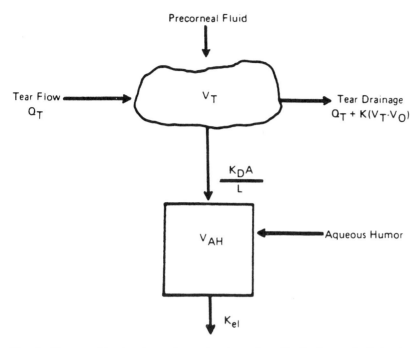

Fig. 6 Pharmacokinetic scheme for ocular absorption, distribution, and elimination.

sufficient number of systematic studies to determine the role of all the molecular and physiological properties influencing various pharmacokinetic parameters. The pharmacokinetic consequences of these competing transport processes has been reviewed recently [123]. Analysis of such data using Green's function solutions, for responses to unit impulse, which can be integrated as a means of generating responses to more complicated dosage regimens, appears to offer promise [129,130].

Ex vivo studies of transcorneal transport can be used to establish the characteristics of passive diffusional motion, the conventional means by which drugs reach internal ocular tissues. Although such analysis neglects the complications of tear flow, tear drainage, nonproductive membrane absorption, elimination from the aqueous humor, and so forth, the corneal transport results are crucial and can be grafted into the primarily hydrodynamic effects subsequently.

Modifications [131] of the classical ex vivo experiments of transport across excised, but metabolizing, rabbit corneas [132–135] have provided useful information both about targeting of similar molecules from the same pharmacological class [136], as well as confirmation of the balance of different anatomical pathways for accession [137]. Systemic interpretation of such results has followed from analysis of the anatomical structure of the barrier membrane and correlation of transmembrane transport with physicochemical properties of the therapeutic agents.

The cornea is a transparent, avascular tissue and, with the adherent precorneal tear film, is the first refracting surface operant in the process of sight. It is composed of three general layers: a multilayered, lipid-rich epithelium, a well-hydrated and lipid-poor stroma, and a lipid-rich endothelium of one-cell layer thickness. Differential studies of the relative lipid densities for these three corneal layers have shown that the densities of lipid in epithelium and endothelium are approximately 40 times as large as that in the stroma [138], although more recent

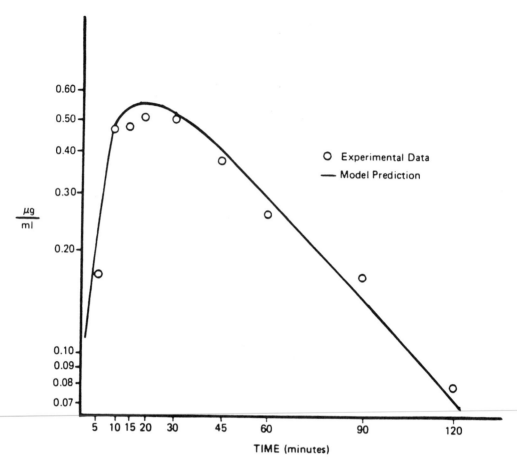

Fig. 7 Comparison of predicted and experimental aqueous humor concentrations following topical administration of pilocarpine.

studies suggest the disparity may be less [139,140]. This can be a primary physiological factor influencing drug penetration through the cornea and into the aqueous humor. For a topically administered drug to traverse an intact cornea and to appear in the aqueous humor, it must possess dual or differential solubility.

One of the key parameters for correlating molecular structure and chemical properties with bioavailability has been transcorneal flux, or alternatively, the corneal permeability coefficient. The epithelium is modeled as a lipid barrier (possibly, with a limited number of aqueous "pores" that, for this physical model, serve as the equivalent of the extracellular space in a more physiological description) and the stroma as an aqueous barrier (Figure 8). The endothelium is very thin and porous compared with the epithelium [133] and can be ignored in the analysis, although mathematically it can be included as part of the lipid barrier. Diffusion through bilayer membranes of various structures has been modeled for some time [141] and adapted to ophthalmic applications more recently [142,143]. For a series of molecules of similar size, it was shown that the permeability increases with octanol/water distribution (or partition) coefficient until a plateau is reached. Modeling of this type of data has led to the earlier statement that drugs need to be both oil- and water-soluble. If pores are not included in the

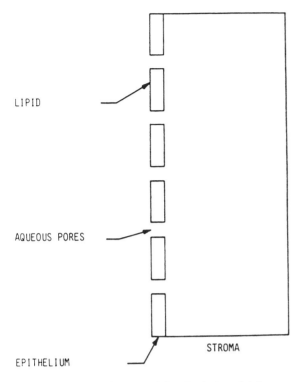

Fig. 8 Schematic diagram of the physical model for transcorneal permeation; features are not to scale.

analysis, the steady-state corneal flux, J_s, can be written as:

$$J_s = \frac{PC_w}{\dfrac{Pl_s}{D_s} + \dfrac{l_e}{D_e}} \tag{3}$$

where

C_w = concentration of drug in donor phase
l_s, l_e = stromal and epithelial thickness, respectively
D_s, D_e = corresponding diffusion coefficients
P = oil-to-water distribution coefficient

The permeability coefficient K_{per} is just the flux divided by C_w. It is apparent that the permeability coefficient is linear with P for small distribution coefficients and constant for large P. Thus, for small P the epithelium is the barrier, and for large P the stroma is the barrier. A fit for steroid permeability is shown in Fig. 9, where the regression analysis gave $D_e = 1.4 \times 10^{-9}$ cm²/sec and $D_s = 2.0 \times 10^{-6}$ cm²/sec for $l_e = 4 \times 10^{-3}$ cm and $l_s = 3.6 \times 10^{-2}$ cm [144]. These values for the diffusion coefficients are reasonable compared with those of aqueous gels and lipid membranes.

A simple estimate of the diffusion coefficients can be approximated from examining the effects of molecular size on transport through a continuum for which there is an energy cost of displacing solvent. Since the molecular weight dependence of the diffusion coefficients for polymers obeys a power law equation [145], a similar form was chosen for the corneal barriers.

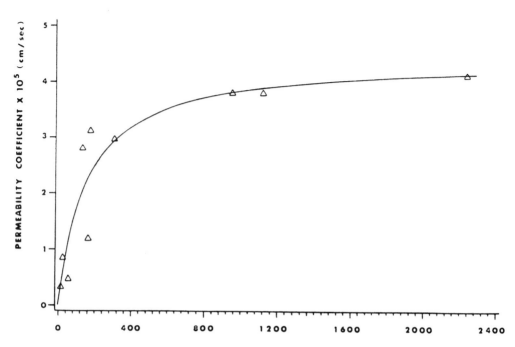

Fig. 9 Predicted versus experimental values for corneal steroid permeability as a function of partition coefficient.

That is, the molecular weight (*M*) dependence of the diffusion coefficients was written as:

$$D_e = D_e^{(0)}M^\alpha$$
$$D_s = D_s^{(0)}M^\gamma \tag{4}$$

Using regression analysis on a data set of about 50 different molecules it was found that $\alpha = -4.4$, $\gamma = -0.5$, $D_e^{(0)} = 12$ cm^2/sec, and $D_s^{(0)} = 2.5 \times 10^{-5}$ cm^2/sec [136]. A graphic representation of the effect of relative molecular mass (*M*$_r$) and distribution coefficient on corneal permeability is shown in Fig. 10. One observes a rapid reduction in permeability coefficient with decreasing *P* and increasing *M*$_r$. The addition of pores to the model, a mathematical construct, is necessary to account for permeability of polar molecules, such as mannitol and cromolyn. These would also be required for correlating effects of compounds, such as benzalkonium chloride, which may compromise the epithelial barrier by increasing the volume of the extracellular space.

Another perspective provided by this model is the effect of three physicochemical parameters; solubility, distribution coefficient, and molecular mass on transcorneal flux. All of these properties can be influenced by molecular design. These influences are illustrated in Fig. 10 in which the logarithm of the flux is plotted as a function of solubility and distribution coefficient for two different *M*$_r$. Several features of the model are depicted and these qualitative, or semiquantitative, aspects presumably encompass the principles of corneal permeation.

Inferred from this model is the relative independence of the effects of solubility and partitioning. For each property there is a characteristic threshold above which the log of the flux increases more slowly than below it; and the value of the threshold for one variable is not very dependent on the value of the other variable. This tabletop perspective has led to the name *mesa model*. The relative independence signifies that neither property can totally com-

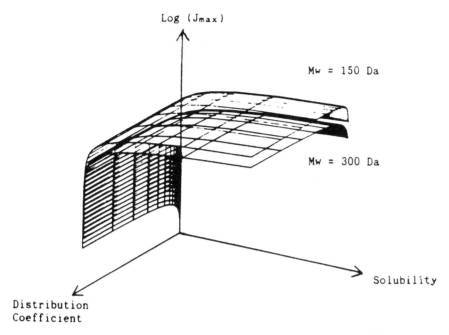

Fig. 10 This figure illustrates the "mesa" response resulting for the diffusion model. Two plateau functions corresponding to different M_r are shown.

pensate for a deficiency in the other. This is not to say that these properties are independent of one another in a chemical sense, quite the contrary. However, in the hypothetical sense that if one property were varied independently of the other, then the consequences on flux are relatively independent. Clearly molecular mass dependence for relatively low molecular mass agents can be significant.

The rough brush stroke agreement between model and experiment is illustrated by the results shown in Fig. 11, for which the correspondence of theoretical with experimental permeability coefficients for the compounds listed in Table 2, β-adrenergic blockers studied by Lee [146,147] and Schoenwald [135] are potted. The calculated values utilized the physical model with pores [144]. Characteristic of correlations of this type is the slope's value, less than 1. The origin of these small values of calculated permeability coefficients is unknown. A reasonable conjecture, however, is that the estimated diffusion coefficients (i.e., the laws presented in Eq. 4) on which the permeability is based, are not quite correct for the drugs in different ocular environments. The predictability of the model is useful both for providing approximate values and for distinguishing departure from simple diffusional transport. Also apparent on comparison of the last two figures (Figs. 10 and 11) is the significance of solubility, since it is the value of C_w that controls flux by orders of magnitude.

As more is learned about accession of drugs into the eye, it is becoming more obvious that passive diffusion through the cornea is not the only pathway likely to be exploited for future delivery of drugs. A drug is known either to bind to, or to be taken up by and accumulated in, epithelial cells. Interesting work is emerging in the areas of facilitated transport in which enhancers are used to diminish diffusional barriers temporarily [148,149], and areas of active transport in which drug conjugates can be employed for transport of larger molecules [131].

In conclusion, it should be apparent from this cursory summary that, although the major features influencing drug absorption are well known, implementation of a delivery strategy is

CALCULATED vs MEASURED PERMEABILITY COEFFICIENTS

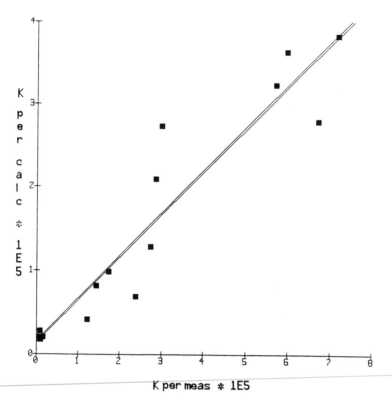

Fig. 11 Plot of data in Table 2, the theoretically computed permeability coefficient versus that measured. The larger influence on flux is often the range of solubility, which can increase the range in flux by several orders of magnitude.

Table 2 Permeability of β-Adrenergic Blockers

Compound	Chemical formula	Molecular weight	Distribution coefficient	K_{per}^{calc} *1E5 cm/sec	K_{per}^{meas} *1E5 cm/sec
Acebutolol	$C_{18}H_{28}N_2O_4$	336.43	1.58	0.28	0.085
Alprenolol	$C_{15}H_{23}NO_2$	249.34	8.92	2.09	2.86
Atenolol	$C_{14}H_{22}N_2O_3$	266.34	0.03	0.20	0.068
Betaxolol	$C_{18}H_{29}NO_3$	307.44	60.3	2.72	3.02
Bevantolol	$C_{20}H_{27}NO_4$	345.44	154.8	2.78	6.76
Bufuralol	$C_{16}H_{23}NO_2$	261.36	204.2	3.82	7.24
Labetolol	$C_{19}H_{24}N_2O_3$	328.41	7.77	0.82	1.43
Levobunolol	$C_{17}H_{25}NO_3$	291.39	5.25	0.99	1.74
Metoprolol	$C_{15}H_{25}NO_3$	267.38	1.91	0.69	2.40
Nadolol	$C_{17}H_{27}NO_4$	309.42	0.15	0.17	0.10
Oxprenolol	$C_{15}H_{23}NO_3$	265.34	4.90	1.28	2.75
Penbutolol	$C_{18}H_{29}NO_2$	291.44	338.84	3.62	6.03
Propranolol	$C_{16}H_{21}NO_2$	259.34	41.69	3.22	5.75
Sotalol	$C_{12}H_{20}N_2O_3S$	272.36	0.06	0.20	0.16
Timolol	$C_{13}H_{24}N_4O_3S$	316.42	2.19	0.42	1.23

highly specific for any compound, and many variables need to be adjusted for their significant influence on absorption, and more importantly on bioavailability. In addition, from the suggestion of the role of ocular hydrodynamics, delivery strategies also will include considerations of dosage form and its effect on local, systemic, and target pharmacokinetics.

VII. MANUFACTURING CONSIDERATIONS

Because the official compendia require all topically administered ophthalmic medication to be sterile, the manufacturer of such medication must consider all the current concepts in the manufacture of sterile pharmaceuticals in designing a manufacturing procedure for sterile *ophthalmic* pharmaceutical products.

It is quite rare that the composition or the packaging of the ophthalmic pharmaceutical will lend itself to terminal sterilization, the simplest form of manufacture of sterile products. Only a few of those drugs formulated in simple aqueous vehicles are stable to normal autoclaving temperatures and times (121°C for 20–30 min). Such drug products must be packaged in glass or other heat–deformation-resistant packaging and thus can be sterilized in this manner.

Most ophthalmic products, however, do not fall into the foregoing category. In general, the active principle is not particularly stable to heat, either physically or chemically. Moreover, to impart viscosity, aqueous products are generally formulated with the inclusion of high molecular weight polymers, which may, similarly, be adversely affected by heat. Finally, the convenience of plastic dispensing bottles has led to the use of modern polyolefins that resist heat deformation and, with a proper sterilization cycle, are now being used for some ophthalmic pharmaceutical products.

Because of these product sensitivities, most ophthalmic pharmaceutical products are aseptically manufactured and filled into previously sterilized containers in aseptic environments using aseptic filling-and-capping techniques. This is the case for ophthalmic solutions, suspensions, and ointments, and rather specialized technology is involved in their manufacture.

In general, however, the manufacture of sterile ophthalmic pharmaceutical products requires special attention to environment, manufacturing techniques, raw materials (including packaging components), and equipment. These will be dealt with individually herein for those products that cannot be terminally sterilized. For those that can, comments in Secs. VII.A (environment), VII.C (raw materials, including packaging components), and VII.D (equipment) apply as well.

A. Environment

Aside from drug safety, stability, and efficacy, the major design criteria of an ophthalmic pharmaceutical product are the additional safety criteria of sterility, preservative efficacy, and freedom from extraneous foreign particulate matter. Current U.S. standards for Good Manufacturing Practices (GMP) provide for the use of specially designed, environmentally controlled areas for the manufacture of sterile large- and small-volume injections for terminal sterilization. These environmentally controlled areas must meet the requirements of class 100,000 space in all areas where open containers and closures are *not* exposed, or where product filling-and-capping operations are *not* taking place. The latter areas must meet the requirements of class 100 space [150]. As defined in Federal Standard 209, class 100,000 and 100 space contain not more than 100,000 or 100 particles, respectively, per cubic foot of air of a diameter of 0.5 μm or larger. Often these design criteria are coupled with laminar airflow concepts [151,152]. This specification deals with total particle counts and does not differentiate between viable and nonviable particles. Federal Standard 209 was promulgated as a ''hardware'' or mechanical specification for the aerospace industry and has found applications in the pharmaceutical industry as a tool for the design of aseptic and particle-free environments. Class 100,000 conditions can be achieved in the conventionally designed cleanroom where proper filtration of air supply and adequate turnover rates are provided. Class 100 conditions over open containers can be achieved with properly sized HEPA (high-efficiency particulate air) filtered laminar airflow sources. Depending on the product need and funds available, some aseptic pharmaceutical manufacturing environments have been designed totally to class 100 laminar-flow specifications, although during actual product manufacture the generation of particulate matter by equipment, product, and (most importantly), people, may cause these environments to demonstrate particulate matter levels two or more orders of magnitude greater than design. It is for this reason that specialists in the design of pharmaceutical manufacturing and hospital operating room environments are beginning to view these environments not from the standpoint of total particles per cubic foot of space alone, but from the standpoint of the ratio of viable to nonviable particles [153].

Such environmental concepts as mass air transfer may lead to meaningful specifications for the space in which a nonterminally sterilized product can be manufactured with a high level of confidence [154].

When dealing with the environment in which a sterile product is manufactured, the materials used for construction of the facility, as well as personnel attire, training, conduct in the space; the entrance and egress of personnel, equipment, and packaging, and the product, all bear heavily on the assurance of product sterility and minimization of extraneous particulate matter.

Walls, ceilings, and floors should be constructed of materials that are hard, nonchipping or nonflaking, smooth, and unaffected by surface-cleaning agents and disinfectants. All lights and windows should be flush-mounted in walls and ceilings for ease of cleaning and disinfection. Ultraviolet lamps may be provided in recessed, flush-mounted fixtures to maintain surface disinfection; however, their use may be difficult to validate, and may lead cleanroom personnel into a false sense of security toward cleanroom techniques. Separate entrances for personnel and equipment should be provided through specially designed air locks that are maintained at a negative pressure relative to the aseptic manufacturing area and at a positive pressure relative to nonenvironmentally controlled area. Equipment should be designed for simplicity of operation and should be constructed for ease of disassembly, cleaning, and sterilization or sanitization.

The importance of personnel training and behavior cannot be overemphasized in the maintenance of an acceptable environment for the manufacture of sterile ophthalmic products or sterile pharmaceutical agents in general. Personnel must be trained in the proper mode of gowning with sterile, nonshedding garments, and also in the proper techniques and conduct for aseptic manufacturing. The Parenteral Drug Association can be contacted at their offices in Bethesda, Maryland for a listing of training films on this subject. For the maximum in personnel comfort and to minimize sloughing of epidermal cells and hair, a cool working environment should be maintained, with relative humidities controlled to between 40 and 60%.

B. Manufacturing Techniques

In general, aqueous ophthalmic solutions are manufactured by methods that call for the dissolution of the active ingredient and all or a portion of the excipients into all or a portion of the water, and the sterilization of this solution by heat or by sterilizing filtration through sterile depth or membrane filter media into a sterile receptacle. If incomplete at this point, this sterile solution is then mixed with the additional required sterile components, such as previously sterilized solutions of viscosity-imparting agents, preservatives, and so on, and the batch is brought to final volume with additional sterile water.

Aqueous suspensions are handled in much the same manner, except that before bringing the batch to final volume with additional sterile water, the solid that is to be suspended is previously rendered sterile by heat, by exposure to ethylene oxide, or ionizing radiation (gamma or electrons), or by dissolution in an appropriate solvent, sterile filtration, and aseptic crystallization. The sterile solid is then added to the batch, either directly or by first dispersing the solid in a small portion of the batch. After adequate dispersion, the batch is brought to final volume with sterile water. Because the eye is sensitive to particles not much larger than 25 μm in diameter, proper raw material specifications for particle size of the solid to be dispersed must be established and verified on each lot of raw material and final product.

When an ophthalmic ointment is manufactured, all raw material components must be rendered sterile before compounding, unless the ointment contains an aqueous fraction that can be sterilized by heat, filtration, or ionizing radiation. The ointment base is sterilized by heat, and appropriately filtered while molten to remove extraneous foreign particulate matter. It is then placed into a sterile steam jacketed kettle to maintain the ointment in a molten state under aseptic conditions, and the previously sterilized active ingredients(s) and excipients are added aseptically. While still molten, the entire ointment may be passed through a previously sterilized colloid mill for adequate dispersion of the insoluble components.

After the product is compounded in an aseptic manner, it is filled into previously sterilized containers. Commonly employed methods of sterilization of packaging components include exposure to heat, ethylene oxide gas, and ^{60}Co (gamma) irradiation. When a product is to be used in conjunction with ophthalmic surgical procedures and must enter the aseptic operating area, the exterior of the primary container must be rendered sterile by the manufacturer and maintained sterile with appropriate packaging. This may be accomplished by aseptic packaging or by exposure of the completely packaged product to ethylene oxide gas, ionizing radiation, or heat.

C. Raw Materials

All raw materials used in the compounding of ophthalmic pharmaceutical products must be of the highest quality available. Complete raw material specifications for each component must be established and verified for each lot purchased. When raw materials are rendered sterile before compounding, the reactivity of the raw material with the sterilizing medium must be completely evaluated, and the sterilization must be validated to demonstrate its capability of sterilizing raw materials contaminated with large numbers (10^5–10^7) of microorganisms that have been demonstrated to be most resistant to the mode of sterilization appropriate for that raw material. As mentioned previously, for raw material components that will enter the eye as a suspension in an appropriate vehicle, particle size must be carefully controlled both before use in the product and as a finished product specification.

As for most sterile (and nonsterile) aqueous pharmaceuticals, the largest portion of the composition is water. At present, *USP* XXII allows the use of "purified water" as a pharmaceutical aid for all official aqueous products, with the exception of preparations intended for parenteral administration [155]. For preparations intended for parenteral administration, *USP* XXII requires the use of water for injection (WFI), sterile water for injection, or bacteriostatic water for injection as a pharmaceutical aid. Because some pharmaceutical manufacturers produce a line of parenteral ophthalmic drugs and devices (large-volume and small-volume irrigating and "tissue-sparing" solutions) as well as topical ophthalmic drugs, the provision of WFI manufacturing capability is being designed into new and existing facilities to meet this requirement. Some manufacturers have made the decision to compound *all* ophthalmic drugs from WFI, thus employing the highest grade of this raw material economically available to the pharmaceutical industry. In doing so, systems must be designed to meet all the requirements for WFI currently listed in the *USP* [156] and the guidelines listed for such systems by the FDA in their Good Manufacturing Practices guidelines for large- and small-volume parenterals [157]. Briefly, these proposals call for the generation of water by distillation or by reverse osmosis and its storage and circulation at relatively high temperatures of up to 80°C (or, alternatively, its disposal every 24 hr), in all stainless steel equipment of the highest-attainable, corrosion-resistant quality.

D. Equipment

The design of equipment for use in controlled environment areas follows similar principles, whether for general injectable manufacturing or for the manufacture of sterile ophthalmic pharmaceuticals. All tanks, valves, pumps, and piping must be of the best available grade of corrosion-resistant stainless steel. In general, stainless steel type 304 or 316 is preferable. All product-contact surfaces should be finished either mechanically or by electropolishing to provide a surface as free as possible from scratches or defects that could serve as a nidus for the commencement of corrosion [158]. Care should be taken in the design of such equipment to provide adequate means of cleaning and sanitization. For equipment that will reside in aseptic-filling areas, such as filling-and-capping machines, care should be taken in their design to yield equipment as free as possible from particle-generating mechanisms. Wherever possible, belt- or chain-drive concepts should be avoided in favor of sealed gear or hydraulic mechanisms. Additionally, equipment bulk should be held to an absolute minimum directly over open containers during filling-and-capping operations to minimize introduction of equipment-generated particulate matter and to minimize creation of air turbulence, particularly when laminar flow is used to control the immediate environment around the filling–capping operation.

In the design of equipment for the manufacture of sterile ophthalmic (and nonophthalmic) pharmaceuticals, manufacturers and equipment suppliers are turning to the relatively advanced technology in use in the dairy and aerospace industries, where such concepts as CIP (clean-in-place), COP (clean-out-of-place), automatic heliarc welding, and electropolishing have been in use for several years. As a guide here, the reader is referred to the so-called 3A Standards of the dairy industry issued by the U. S. Public Health Service [159].

VIII. CLASSES OF OPHTHALMIC PRODUCTS

A. Topical Eyedrops

Administration and Dosage

Although many methods of instilling drugs to the eye have been experimented with, the use of eyedrops remains the major method of administration for the topical ocular route. The usual method of self-administration is to place the eyedrop from a dropper or dropper bottle into the lower cul-de-sac by pulling down the eyelid, tilting the head backward, and looking at the ceiling after the tip is pointed close to the sac, and applying a slight pressure to the rubber bulb or plastic bottle to allow a single drop to form and fall into the eye. Most people become quite adept at this method with some practice and may develop their own modifications. However, elderly, arthritic, low-vision, and glaucoma patients often have difficulty in self-administration and may require another person to instill the drops.

The pharmacist should instruct patients to keep in mind the following considerations in administering drops to help improve the accuracy and consistency of dosage and to prevent contamination potential: be sure that the hands are clean; do not touch the dropper tip to the eye, surrounding tissue, or any surface; prevent squeezing or fluttering of lids, which causes blinking; place the drop in the conjunctival sac, not on the globe; close the lids for several moments after installation. The administration of eyedrops to young children can be a difficult task. A way to simplify the task involves the parent sitting on the floor or a flat surface and placing the child's head firmly between the parents thighs and crossing legs over the child's lower trunk and legs. The parent's hands are then free to lower the eyelid and administer the drops.

Eyedrops are one of the few dosage forms that are not administered by exact volume or weight dosage, yet this seemingly imprecise method of dosing is quite well established and accepted by ophthalmologists. The volume of a drop is dependent on the physicochemical properties of the formulation, particularly surface tension; the design and geometry of the dispensing orifice; and the angle at which the dispenser is held in relation to the receiving surface. The manufacturer of ophthalmic products controls the tolerances necessary for the dosage form and dispensing container to provide a uniform drop size. How precise does the actual dose have to be? As noted earlier, the normal tear volume is about 7 μl, and with blinking about 10 μl can be retained in the eye. There are approximately 1.2 μl of tears produced per minute, for about a 16% volume replacement per minute. Commercial ophthalmic droppers deliver drops from about 30 to 50 μl/drop. Therefore, the volumes delivered normally are more than threefold in excess of that which the eye can hold, and the fluid that does remain in the eye is continuously being removed until the normal tear volume is attained. It can be seen, then, that the use of more than 1 drop/dose must take into account the fluid volume and dynamics of the lacrimal system of the eye. If the effect of multiple drops is desired, they should be administered 1 drop at a time with a 3- to 5-min interval in between. Some doctors may prescribe more than 1 drop/dose to ensure that the patients retains at least 1 drop in the eye.

Dosage Forms

Solutions.   The two major physical forms of eyedrops are aqueous solutions and suspensions. Nearly all the major ophthalmic therapeutic agents are water-soluble or can be formulated as water-soluble salts. A homogeneous solution offers the potential of greater assurance of uniformity of dosage and bioavailability and simplifies large-scale manufacture. The selection of the appropriate salt form depends on its solubility; the therapeutic concentrations required; the ocular toxicity; the effect of pH, tonicity, and buffer capacity; its compatibility with the total formulation; and the intensity of any possible stinging or burning sensations produced (i.e., discomfort reactions). The most common salt forms used are the hydrochloride, sulfate, nitrate, and phosphate. Salicylate, hydrobromide, and bitartrate salts are also used. For drugs that are acidic, such as the sulfonamides, sodium and diethanolamine salts are used. The effect that choice of salt form can have on resulting product properties is exemplified by the epinephrine solutions available, as shown in Table 3. The bitartrate form is a 1:1 salt, and the free carboxyl group acts as a strong buffer resisting neutralization by the tears, causing considerable stinging. The borate form results in a solution with lower buffer capacity, a more nearly physiological pH, and better patient tolerance; however, it is less stable than the other

Table 3   Effect of Salt Form on Product Properties

Salt form	Discomfort reaction	pH range	Buffer capacity
Epinephrine hydrochloride	Mild to moderate stinging	2.5–4.5	Medium
Epinephrine bitartrate	Moderate to severe stinging	3–4	High
Epinephrine borate	Only occasional mild stinging	5.5–7.5	Low

two salts. The hydrochloride salt combines better stability than the borate with acceptable patient tolerance.

Suspensions.   If the drug is not sufficiently soluble, it can be formulated as a suspension. A suspension may also be desired to improve stability, bioavailability, or efficacy. The major topical ophthalmic suspensions are the steroid anti-inflammatory agents; prednisolone acetate, dexamethasone, fluorometholone, and medrysone. Water-soluble salts of prednisolone phosphate and dexamethasone phosphate are available; however, they have a lower steroid potency and are poorly absorbed.

An ophthalmic suspension should use the drug in a microfine form, usually 95% or more of the particles have a diameter of 10 μm or less. This is necessary so that the particles do not cause irritation of the sensitive ocular tissues and to help ensure that a uniform dosage is delivered to the eye. Since a suspension is made up of solid particles, it is at least theoretically possible that they may provide a short-lived reservoir in the cul-de-sac for slightly prolonged activity. However, it appears that this is not so, since the drug particles are extremely small, and with the rapid tear volume turnover rate, they are washed out of the eye relatively quickly.

Powders for Reconstitution.   Several ophthalic drugs are prepared as sterile powders for reconstitution by the pharmacist before dispensing to the patient. These include α-chymotrypsin and echothiophate iodide. The sterile powder is usually manufactured by lyophilization and is packaged separately from the diluent, and a sterile dropper assembly is provided. In powder form these drugs have a much longer shelf life than that of their solution forms. The pharmacist should use only the diluent provided with the products, since it has been developed to maintain the optimum potency and preservation of the reconstituted solution.

Inactive Ingredients in Topical Drops

The therapeutically inactive ingredients in ophthalmic solution and suspension dosage forms are necessary to perform one or more of the following functions: adjunct tonicity, buffer and adjust pH, stabilize the active ingredients against decomposition, increase solubility, impart viscosity, and act as the solvent. The use of unnecessary ingredients is to be avoided, and the use of ingredients solely to impart a color, odor, or flavor is prohibited.

The choice of the particular inactive ingredient and concentration is based not only on physical and chemical compatibility, but also on biocompatibility with the sensitive and delicate ocular tissues. Because of the latter requirement, the use of inactive ingredients is greatly restricted in ophthalmic dosage forms.

Tonicity and Tonicity-Adjusting Agents.   In the past a great deal of emphasis was placed on teaching the pharmacist to correctly adjust an ophthalmic solution to be isotonic (i.e., exert an osmotic pressure equal to that of tear fluids, generally agreed to be equal to 0.9% NaCl). In compounding an eye solution, it is more important to consider the sterility, stability, and preservative aspects, and not jeopardize these aspects to obtain a precisely isotonic solution. A range of 0.5–2.0% NaCl equivalency does not cause a marked pain response, and a range of about 0.7–1.5% should be acceptable to most persons. Manufacturers are in a much better position to make a precise adjustment, and thus their products will be close to isotonic, since they are in a competitive situation and are interested in a high percentage of patient acceptance for their products. In certain instances, the therapeutic concentration of the drug will necessitate using what might otherwise be considered an unacceptable tonicity. This is the case for sodium sulfacetamide, for which the isotonic concentration is about 3.5%, but the drug is used in 10–30% concentrations. Fortunately, the eye seems to tolerate hypertonic solutions better than hypotonic ones. Various textbooks deal with the subject of precise tonicity calculations and determination. Several articles [e.g. Ref. 160] have recommended practical methods of obtaining an acceptable tonicity in extemporaneous compounding. Tonicity-adjusting ingredients usually used include NaCl, KCl, buffer salts, dextrose, glycerin, and propylene glycol.

pH Adjustment and Buffers. The pH and buffering of an ophthalmic solution is probably of equal importance to proper preservation, since the stability of most commonly used ophthalic drugs is largely controlled by the pH of their environment. Manufacturers place particular emphasis on this aspect, since economics indicate that they produce products with long shelf lives that will retain their labeled potency and product characteristics under the many and varied storage conditions outside the makers' control. The pharmacist and wholesaler must become familiar with the labeled storage directions for each product and see that they are properly stored. Particular attention should be paid to those products requiring refrigeration. The stability of nearly all products can be enhanced by refrigeration, except a few in which a decrease in solubility and precipitation might occur. Freezing of ophthalmic products, particularly suspensions, should be avoided. A freeze–thaw cycle can induce particle growth or crystallization of a suspension and increase the chances of causing ocular irritation and loss of dosage uniformity. Glass-packaged liquid products may break owing to the volume expansion of the solution when it freezes. It is especially important that the pharmacist fully advise the patient on proper storage and use of ophthalmic products to ensure their integrity and their safe and efficacious use.

In addition to stability effects, pH adjustment can influence the comfort, safety, and activity of the product. Comfort can be described as the subjective response of the patient after installation of the product in the cul-de-sac (i.e., whether it causes a pain response such as stinging or burning). Eye irritation is normally accompanied by an increase in tear fluid secretion (a defense mechanism) to aid in the restoration of normal physiological conditions. Accordingly in addition to the discomfort encountered, products that produce irritation will tend to be flushed from the eye and, hence, a more rapid loss of medication may occur, with a probable reduction in the therapeutic response.

Ideally, every product would be buffered to a pH of 7.4, which is considered the normal physiological pH of tear fluid. The argument for this concept is that the product would be comfortable and possibly have optimum therapeutic activity. Various experiments, primarily in rabbits, have shown an enhanced effect when the pH was increased, owing to the solution containing a higher concentration of the nonionized lipid-soluble drug base, which is the species that can more rapidly penetrate the corneal epithelial barrier. This would not be true if the drug were an acidic moiety. The tears have some buffer capacity of their own, and it is believed that they can neutralize the pH of an instilled solution if the quantity of solution is not excessive and if the solution does not have a strong resistance to neutralization. Pilocarpine activity is apparently the same whether applied from vehicles with nearly physiological pH values or from more acidic vehicles, provided the latter are not strongly buffered [161]. A pH difference of 6.6 versus 4.2 produced a statistically insignificant difference in pilocarpine miosis [162]. The pH values of ophthalmic solutions are adjusted to a range at which an acceptable shelf life stability of at least 2 years can be achieved, and if necessary, they are buffered to remain within this range. If buffers are required, their capacity is controlled to be as low as possible, thus enabling the tears to bring the pH of the eye back to the physiological range. Since the buffer capacity is determined by buffer concentration, the effect of buffers on tonicity must also be taken into account and is another reason that ophthalmic products are usually only lightly buffered.

The pH value is not the sole contributing factor to discomfort of some ophthalmic solutions. It is possible to have a product with a low pH and little buffer capacity that is more comfortable than a similar product with a higher pH and a strong buffer capacity. Epinephrine hydrochloride and dipivefrin hydrochloride solutions, used for treatment of glaucoma, have a pH of about 3, yet they have acceptable comfort such that they can be used daily for many years. The same pH solution of epinephrine bitartrate has an intrinsically higher buffer capacity and will produce much more discomfort.

Stabilizers. Stabilizers are ingredients added to a formula to decrease the decomposition of the active ingredients. Antioxidants are the principal stabilizers added to some ophthalmic solutions, primarily those containing epinephrine and other oxidizable drugs. Sodium bisulfite or metabisulfite are used in concentration up to 0.3% in epinephrine hydrochloride and bitartrate solutions. Epinephrine borate solutions have a pH in the range 5.5–7.5 and offer a more difficult challenge to formulators who seek to prevent oxidation. Several patented antioxidant systems have been developed specifically for this compound. These consist of ascorbic acid and acetylcysteine, and sodium bisulfite and 8-hydroxyquinoline. Isoascorbic acid is also an effective antioxidant for this drug. Sodium thiosulfate is used with sodium sulfacetamide solutions.

Surfactants. The use of surfactants is greatly restricted in formulating ophthalmic solutions. The order of surfactant toxicity is anionic > cationic >> nonionic. Several nonionic surfactants are used in relatively low concentrations to aid in dispensing steroids in suspensions and to achieve or to improve solution clarity. Those principally used are polysorbate 20 and 80, tyloxapol, and polyoxyl 40 stearate. The lowest concentration possible is used to perform the desired function. Their effect on preservative efficacy and their possible binding by macromolecules must be taken into account, as well as their effect on ocular irritation. The use of surfactants as cosolvents for an ophthalmic solution of chloramphenicol has been described [163]. This composition includes polyoxyl 40 stearate and polyethylene glycol to solubilize 0.5% chloramphenicol. These surfactants–cosolvents provide a clear aqueous solution of chloramphenicol and a stabilization of the antibiotic in aqueous solution.

Viscosity-Imparting Agents. Polyvinyl alcohol, methylcellulose, hydroxypropyl methylcellulose, hydroxyethylcellulose, and one of the several high molecular weight cross-linked polymers of acrylic acid, known as Carbomers [162], are commonly used to increase the viscosity of ophthalmic solutions and suspensions. Although they reduce surface tension significantly, their primary benefit is to increase the ocular contact time, thereby decreasing the drainage rate and increasing drug bioavailability. A secondary benefit of the polymer solutions is a lubricating effect that is largely subjective, but noticeable to many patients. One disadvantage to the use of the polymers is their tendency to dry to a film on the eyelids and eyelashes; however, this can be easily removed by wiping with a damp tissue.

Numerous studies have shown that increasing the viscosity of ophthalmic products increases contact time and pharmacological effect, but there is a plateau reached after which further increases in viscosity produce only slight or no increases in effect. Blaugh and Canada [163] using methylcellulose solutions found increased contact time in rabbits up to 25 cP (centipoise) and a leveling off at 55 cP. Lynn and Jones [164] studied the rate of lacrimal excretion in humans using a dye solution in methylcellulose concentration from 0.25 to 2.5%, corresponding to viscosities of 6–30,000 cP, the latter being a thick gel. The results were as shown in Table 4.

Chrai and Robinson [165] conducted studies in rabbits and found that, over a range of 1.0–12.5 cP viscosity, there is a threefold decrease in the drainage rate constant and a further threefold decrease over the viscosity range of 12.5–100 cP. This decrease in drainage rate increased the concentration of drug in the precorneal tear film at zero time and subsequent time periods, which resulted in a higher aqueous humor drug concentration. The magnitude of aqueous humor drug concentration increase was smaller than the increase in viscosity, about 1.7 times, for the range 1.0–12.5 cP, and only a further 1.2-fold increase at 100 cP.

Since direct determination of ophthalmic bioavailability in humans is not possible without endangering the eye, investigators have used fluorescein to study factors affecting bioavail-

Table 4 Effect of Viscosity on Product Contact Time

Methylcellulose concentration (%)	Time to dye appearance through nasolacrymal duct (sec)
0.0	60
0.25	90
0.50	140
1.00	210
2.50	255

ability in the eye, because its penetration can be quantitated in humans through the use of a slit-lamp fluorophotometer. Adler [166], using this technology, found only small increases in dye penetration over a wide range of viscosities. The use of fluorescein data to extrapolate vehicle effects to ophthalmic drugs in general would be questionable owing to the large differences in chemical structure, properties, and permeability existing between fluorescein and most ophthalmic drugs.

The major commercial viscous vehicles are hydroxypropyl methylcellulose (Isopto) and polyvinyl alcohol (Liquifilm). Isopto products most often use 0.5% of the cellulosic and range from 10 to 30 cP in viscosity. Liquifilm products have viscosities of about 4–6 cP and use 1.4% polymer.

Vehicles. Ophthalmic drops are, with few exceptions, aqueous fluids using purified water *USP* as the solvent. Water for injection is not required as it is in parenterals. Purified water meeting *USP* standards may be obtained by distillation, deionization, or reverse osmosis.

Oils have been used as vehicles for several topical eyedrop products that are extremely sensitive to moisture. Tetracycline HCl is an antibiotic that is stable for only a few days in aqueous solution. It is supplied as a 1% sterile suspension with Plastibase 50W and light liquid petrolatum. White petrolatum and its combination with liquid petrolatum to obtain a proper consistency is routinely used as the vehicle for ophthalmic ointments.

When oils are used as vehicles in ophthalmic fluids they must be of the highest purity. Vegetable oils such as olive oil, castor oil, and sesame oil have been used for extemporaneous compounding. These oils are subject to rancidity and, therefore, must be used carefully. Some commercial oils, such as peanut oil, contain stabilizers that could be irritating. The purest grade of oil, such as that used for parenteral products would be advisable for ophthalmics.

Packaging

Eyedrops have been packaged almost entirely in plastic dropper bottles since the introduction of the Drop-Tainer plastic dispenser in the 1950s. A few products still remain in glass dropper bottles because of special stability considerations. The main advantage of the Drop-Tainer and similarly designed plastic dropper bottles are convenience of use by the patient, decreased contamination potential, lower weight, and lower cost. The plastic bottle has the dispensing tip as an integral part of the package. The patient simply removes the cap and turns the bottle upside down and squeezes gently to form a single drop that falls into the eye. The dispensing tip can be designed to deliver only 1 drop or a stream of fluid for irrigation, depending on the pressure applied. When used properly, the solution remaining in the bottle is only minimally exposed to airborne contaminants during administration; thus, it will maintain very low to nonexistent microbial content as compared with the old-style glass bottle with separate dropper assembly.

The plastic bottle and dispensing tip is made of low-density polyethylene (LDPE) resin, which provides the necessary flexibility and inertness. Because these components are in contact with the product during its self life, they must be carefully chosen and tested for their suitability for ophthalmic use. In addition to stability studies on the product in the container over a range of normal and accelerated temperatures, the plastic resins must pass the *USP* biological and chemical tests for suitability. The LDPE resins used are compatible with a very wide range of drugs and formulation components. Their one disadvantage is their sorption and permeability characteristics. Volatile ingredients; such as the preservatives chlorobutanol and phenylethyl alcohol, can migrate into the plastic and eventually permeate through the walls of the container. The sorption and permeation can be detected by stability studies if it is significant. If the permeating component is a preservative, a repeat test of the preservative effectiveness with time will determine if the loss is significant. If necessary, a safe and reasonable excess of the permeable component may be added to balance the loss over the shelf life. Another means of overcoming permeation effects is to employ a secondary package, such as a peel-apart blister or pouch composed of nonpermeable materials (e.g., aluminum foil or vinyl). The plastic dropper bottles are also permeable to water, and weight loss by water vapor transmission has a decreasing significance as the size of the bottle increases. The consequences of water vapor transmission must be taken into consideration when assessing the stability of a product, and appropriate corrections for loss of water on the analysis of components must be applied.

The LDPE resins are translucent, and if the drug is light-sensitive, additional package protection may be required. This can be achieved by using a resin containing an opacifying agent such as titanium dioxide, by placing an opaque sleeve over the exterior of the container, or by placing the bottle in a cardboard carton. Extremely light-sensitive drugs, such as epinephrine and proparacaine, may require a combination of these protective measures. Colorants, other than titanium dioxide, are rarely used in plastic ophthalmic containers; however, the use of colorants is common for the cap for a very important purpose. Red is used to denote a mydriatic drug, such as atropine, and green a miotic drug, like pilocarpine. This is an aid to the physician and the dispensing pharmacist to prevent potentially serious mistakes. The pharmacist should dispense the ophthalmic product only in the original unopened container. A tamper-evidence feature, such as a cellulose or metal band around the cap and bottle neck is provided by the manufacturer.

The LDPE resin used for the bottle and the dispensing tip cannot be autoclaved, and they are sterilized either by ^{60}Co gamma irradiation or ethylene oxide. The cap is designed such that when it is screwed tightly onto the bottle, it mates with the dispensing tip and forms a seal. The cap is usually made of a harder resin than the bottle, such as polystyrene or polypropylene, and is also sterilized by gamma radiation or ethylene oxide gas exposure.

A special plastic ophthalmic package has been introduced that uses a special grade of polypropylene that is resistant to deformation at autoclave temperatures. With this specialized packaging, the bottle can be filled, the dispensing tip and cap applied, and the entire product sterilized by steam under pressure at 121°C.

The glass dropper bottle is still used for products that are extremely sensitive to oxygen or contain permeable components that are not sufficiently stable in plastic. Powders for reconstitution also use glass containers, owing to their heat-transfer characteristics that are necessary during the freeze-drying process. The glass used should be USP type I for maximum compatibility with the sterilization process and the product. The glass container is made sterile by dry-heat or steam autoclave sterilization. Amber glass is used for light-resistance and is superior to green glass. A sterile dropper assembly is usually supplied separately. It is usually gas-sterilized in a blister composed of vinyl and Tyvek, a fused, porous polypropylene material. The dropper assembly is made of a glass or LDPE plastic pipette and a rubber dropper bulb. The manufacturer carefully tests the appropriate plastic and rubber materials suitable for use with the product; therefore, they should be dispensed with the product. The pharmacist should aseptically place the dropper assembly in the product before dispensing it and instruct the patient on precautions to be used to prevent contamination.

B. Semisolid Dosage Forms: Ophthalmic Ointments and Gels

Formulation

The principle semisolid dosage form used in ophthalmology is an anhydrous ointment with a petrolatum base. The ointment vehicle is usually a mixture of mineral oil and white petrolatum. The mineral oil is added to reduce the melting point and modify the consistency. The principal advantages of the petrolatum-based ointments are their blandness and their anhydrous and inert nature, which make them suitable vehicles for moisture-sensitive drugs. Ophthalmic ointments containing antibiotics are used quite frequently following operative procedures, and their safety is supported by the experience of a noted eye surgeon [167] who, in over 20,000 postsurgical patients, saw no side effects secondary to the ointment use. No impediment to epithelial or stromal wound healing was exhibited by currently used ophthalmic ointments tested by Hanna et al. [168]. The same investigators have reported that, even if these ointments were entrapped in the anterior chamber and did not exceed 5% of the volume, little or no reaction was caused [169]. Granulomatous reactions requiring surgical excision have been reported secondary to therapeutic injection of ointment into the lacrimal sac [170].

The chief disadvantages of the use of ophthalmic ointments are their greasy nature and the blurring of vision they produce. They are most often used as adjunctive nighttime therapy, with eyedrops administered during the day. The nighttime use obviates the difficulties produced by blurring of vision and is stated to prolong ocular retention when compared with drops. Ointments are used almost exclusively as vehicles for antibiotics, sulfonamides, antifungals, and anti-inflammatories. The petrolatum vehicle is also used as an ocular lubricant following surgery or to treat various dry eye syndromes. Anesthesiologists may prescribe the ointment vehicle for the nonophthalmic surgical patients to prevent severe and painful dry eye conditions that could develop during prolonged surgeries.

The anhydrous petrolatum base may be made more miscible with water through the use of an anhydrous liquid lanolin derivative. Drugs can be incorporated into such a base in aqueous solution if desired. Polyoxyl 40 stearate and polyethylene glycol 300 are used in an anti-infective ointment to solublize the active principle in the base so that the ointment can be sterilized by aseptic filtration. The cosmetic-type bases, such as the oil-in-water (o/w) emulsion bases popular in dermatology, should not be used in the eye, nor should liquid emulsions, owing to the ocular irritation produced by the soaps and surfactants used to form the emulsion.

In an attempt to formulate an anhydrous, but water-soluble, semisolid base for potential ophthalmic use, five bases were studied and reported on [171]. The nonaqueous portion of the base was either glycerin or polyethylene glycols in high concentrations. The matrix used to form the phases included silica, Gantez AN-139, and Carbopol 940. Eye irritation results were not reported, but the authors have studied representative bases from the research report and found them to be quite irritating in rabbit eyes. The irritation is believed to be primarily due to the high concentration of the polyols used as vehicles.

An aqueous semisolid gel base has been developed that provides significantly longer residence time in the cul-de-sac and increases drug bioavailability and, thereby, may prolong the therapeutic level in the eye. The gel contains a high molecular weight, cross-linked, polymer to provide the high viscosity and optimum rheological properties for prolonged ocular retention. Only a relatively low concentration of polymer is required, so that the gel base is more than 95% water.

Schoenwald et al. [172] have demonstrated the unique ocular retention properties of this polymeric gel base in rabbits, in which the miotic effect of pilocarpine was significantly prolonged. The use of other polymers, such as cellulosic gums, polyvinyl alcohol, and polyacrylamides at comparable apparent viscosities, did not provide a significant prolonged effect. The prolonged effect of pilocarpine has also been demonstrated in human clinical trials, in which a single application of 4% pilocarpine HCl-containing carbomer gel at bedtime, provided a 24-hr duration of reduced intraocular pressure (IOP), compared with the usually required q.i.d. dosing for pilocarpine solution [173]. As a result, some glaucoma patients can now use pilocarpine in this aqueous gel base (Pilopine HS Gel), dosing only once a day at bedtime to control their IOP without the significant vision disturbance experienced during the day for the use of conventional pilocarpine eyedrops. The gel is applied in a small strip in the lower conjunctival sac from an ophthalmic ointment tube.

The carbomer polymeric gel base itself has been used successfully to treat moderate to severe cases of dry eye (keratoconjunctivitis sicca) [174]. The dry eye syndrome is usually characterized by a deficiency of tear production and, therefore, requires frequent instillation of aqueous artificial tear eyedrops to keep the corneal epithelium moist. The gel base applied in a small amount provides a prolonged lubrication to the external ocular tissues, and some patients have reduced the frequency of dosing to control their symptoms to three times a day or fewer.

Sterility and Preservation

Since October 1973, FDA regulations require that all U. S. ophthalmic ointments be sterile. This legal requirement was a result of several surveys on microbial contamination of ophthalmic ointments and followed reports in Sweden and the United Kingdom of severe eye infections resulting from use of nonsterile ointments. In its survey published in 1973, the FDA found that of 82 batches of ophthalmic ointments tested from 27 manufacturers, 16 batches were contaminated, including 8 antibiotic-containing ointments. The contamination levels were low and were principally molds and yeasts [175]. The time lag in imposition of a legal requirement for sterility of ointments compared with solutions and suspensions was due to the absence of a reliable sterility test for the petrolatum-based ointments until isopropyl myristate was employed to dissolve these ointments and allow improved recovery of viable microorganisms by membrane filtration. Manufacturers found that, in fact, many of their ointments were sterile, but revised their manufacturing procedures to increase the assurance of sterility.

A suitable substance or mixture of substances to prevent the growth of microorganisms must be added to ophthalmic ointments that are packaged in multiple-use containers, regardless of the method of sterilization employed, unless otherwise directed in the individual monograph, or unless the formula itself is bacteriostatic [USP XXII, p. 1594]. Schwartz [176] has commented that a sterile ointment cannot become excessively contaminated by ordinary use because of its consistency and the fact that in a nonaqueous medium microorganisms merely survive, but do not multiply. An official test for effectiveness of a preservative in a nonaqueous medium has not been devised, thus making rational selection by the formulator an even more difficult task. A survey of U. S. ophthalmic ointments finds that chlorobutanol and methyl- and propylparaben are the most often used when an ointment contains a preservative, and they are used in the same concentration as in aqueous systems. The USP preservative test, if used, must be evaluated with a full understanding of its shortcomings for nonaqueous media [163].

Packaging

Ophthalmic ointments are packaged in small collapsible tin tubes, usually holding 3.5 g. The pure tin tube is compatible with a wide range of drugs in petrolatum-based ointments. Aluminum tubes have been considered and may eventually be used because of their lower cost and as an alternative should the supply of tin become a problem. Until internal coating technology for these tubes advances, the aluminum tube will be a secondary packaging choice. Plastic tubes made from flexible LDPE resins have also been considered as an alternative material, but do not collapse and tend to suck-back the ointment. Plastic tubes recently introduced as containers for toothpaste have been investigated and may offer the best alternative to tin. These tubes are laminates of plastic and various materials, such as paper, foil, and so on. A tube can be designed by selection of the laminate materials and their arrangement and thickness to provide the necessary compatibility, stability, and barrier properties. The various types of metal tubes are sealed using an adhesive coating covering only the inner edges of the bottom of the open tube to form the crimp, which does not contact the product. Laminated tubes are usually heat-sealed. The crimp usually contains the lot code and expiration date. Filled tubes may be tested for leakers by storing them in a horizontal position in an oven at 60°C for at least 8 hr. No leakage should be evidenced except for a minute quantity that could only come from within the crimp of the tube or the end of the cap. The screw cap is made of polyethylene or polypropylene. Polypropylene must be used for autoclave sterilization, but either material may be used when the tubes are gas sterilized. A tamper-evident feature is required for sterile ophthalmic ointments, and may be accomplished by sealing the tube or the carton holding the tube such that the contents cannot be used without providing visible evidence of destruction of the seal. The Teledyne Wirz tube used by most manufacturers has a flange on the cap that is visible only after the tube has been opened the first time.

The tube can be a source of metal particles and must be cleaned carefully before sterilization. The *USP* contains a test procedure and limits the level of metal particles in ophthalmic ointments. The total number of metal particles detected under 30 times magnification that are 50 μm or larger in any dimension is counted. The requirements are met if the total number of such particles counted in ten tubes is not more than 50, and if not more than one tube is found to contain more than 8 such particles.

C. Solid Dosage Forms: Ocular Inserts

In earlier times, it has been reported that lamellae or disks of glycerinated gelatin were used to supply drugs to the eye by insertion beneath the eyelid. The aqueous tear fluids dissolved the lamella and released the drug for absorption. The medical literature also describes a sterile paper strip impregnated with drug for insertion in the eye. These appear to have been the first attempts at designing a sustained-release dosage form.

Nonerodible Ocular Inserts

In 1975, the first controlled-release *topical* dosage form was marketed in the United States by the Alza Corporation. Zaffaroni [177] describes the Alza therapeutic system as a drug-containing device or dosage form that administers a drug or drugs at programmed rates, at a specific body site, for a prescribed time period to provide continuous control of drug therapy and to maintain this control over extended periods. Therapeutic systems for uterine delivery of progesterone, transdermal delivery of scopolamine, and oral delivery of systemic drugs have also been developed.

The Ocusert Pilo-20 and Pilo-40 system is an elliptical membrane that is soft and flexible and designed to be placed in the inferior cul-de-sac between the sclera and the eyelid and to release pilocarpine continuously at a steady rate for 7 days. The design of the dosage form is described by Alza in terms of an open-looped therapeutic system, having three major components: (a) the drug, (b) a drug delivery module, and (c) a platform. In the Ocusert Pilo-20 and Pilo-40 systems, the drug delivery module consists of (a) a drug reservoir, pilocarpine (free base), and a carrier material, alginic acid; (b) a rate controller, ethylene vinyl acetate (EVA) copolymer membrane; (c) an energy source, the concentration of pilocarpine in the reservoir; and (d) a delivery portal, the copolymer membrane. The platform component for the pilocarpine Ocusert consists of the EVA copolymer membranes, which serve as the housing, and an annular ring of the membrane impregnated with titanium dioxide that forms a white border for visibility. The laminate structure of the Ocusert is seen in Fig. 12. The free-base form of pilocarpine is used, since it exhibits both hydrophilic and lipophilic characteristics. Use of the extremely water-soluble salts of pilocarpine would have necessitated the use of a hydrophilic membrane which, if it osmotically imbibed an excessive amount of water, would cause a significant decline in the release rate with time. Use of the free base allowed a choice of more hydrophobic membranes that are relatively impermeable to water; accordingly the release rate is independent of the environment in which it is placed. EVA, the hydrophobic copolymer chosen, was found to be very compatible with the sensitive ocular tissues [178], an important feature.

The pilocarpine Ocusert is seen by Alza to offer a number of theoretical advantages over drop therapy for the glaucoma patient. The Ocusert exposes a patient to only one-fourth to one-eighth the amount of pilocarpine, compared with drop therapy. This could lead to reduced local side effects and toxicity. It provides continuous around-the-clock control of IOP, whereas drops used four times a day can permit periods where the IOP might rise. Additionally, the Ocusert provides for more patient convenience and improved compliance, as the dose needs to be administered only once per week. However, the clinical experience seems to indicate that the Ocusert has a compliance problem of its own (i.e., retention in the eye for the full 7 days). The patient must check periodically to see that the unit is still in place, particularly in the morning on arising. Replacement of a contaminated unit with a fresh one can increase the price differential of the already expensive Ocusert therapy compared with inexpensive drop or once-a-day gel therapy.

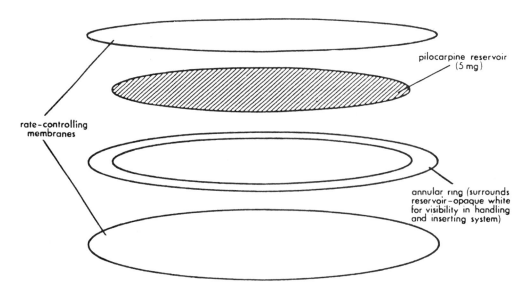

pilocarpine reservoir (5 mg)

rate-controlling membranes

annular ring (surrounds reservoir–opaque white for visibility in handling and inserting system)

Pilo-20 Ocusert®:13.4 mm x 5.7 mm x 0.3 mm, weight = 19 mg

Fig. 12 Elements and dimensions of pilocarpine Ocusert system.

Erodible Ocular Inserts

Since polymers have been added to solutions to increase their conjunctival retention time, it is not surprising that similar solutions have been dried to form films of the polymer drug system. These films have been inserted in the cul-de-sac and have been reported to increase retention time, increase the penetration of the drug, and prolong its effect. Considerable experimentation in humans has been reported in the former USSR with soluble ophthalmic film inserts made from polymers of polyacrylamide, ethylacrylate, and vinylpyrollidone, containing pilocarpine, antibiotics, and other ophthalmic drugs [179]. A solid disk of pilocarpine alginate has provided a greater degree of miosis and prolonged activity compared with aqueous solutions of the hydrochloride and alginate salts [180]. These solid inserts absorb the aqueous tear fluid and gradually erode or disintegrate. The drug is slowly leached from the matrix, but a constant zero-order delivery, the kinetics that best prolong delivery to the eye, is not achieved. They quickly lose their solid integrity and are squeezed out of the eye with eye movement and blinking. If they are made more hydrophobic, they become irritating when placed in the eye. These erodible inserts possess an advantage over the nonerodible inserts described previously in that they do not have to be removed at the end of their useful dosing cycle. Several patents have been issued to the Alza Corporation for bioerodible ocular inserts that do possess this advantage and are designed to release the drug by controlled erosion of the polymer matrix, rather than by leaching from a hydrophilic matrix [181,182]. These offer the possibility for zero-order delivery, and since the release mechanism is not diffusion, they may be applicable to high molecular weight drugs that are not highly water-soluble. A polypeptide matrix with a release rate of 10 μg/hr of steroid acetate was as efficacious as steroid acetate eyedrops in treating a severe conjunctivitis in rabbits [183].

Lacrisert is a sterile ophthalmic insert that is used in the treatment of moderate to severe dry eye syndrome and is usually used in patients when an adequate trial of artificial tear solutions has not provided symptomatic relief. The insert is composed of 5 mg of hydroxyproplycellulose in a rod-shaped form of about 1.27 mm diameter by about 3.5 mm long. No preservative is used, since it is essentially anhydrous. The cellulose rod is placed in the lower conjunctival sac and first imbibes water from the tears and forms a gellike mass after several hours, which gradually erodes as the polymer dissolves. This action thickens the tear film and provides increased lubrication, which can provide symptomatic relief for dry eye states. It is usually used once or twice daily.

D. Intraocular Dosage Forms

A special class of ophthalmic drugs and irrigating solutions requiring the application of parenteral dosage form technology in their design and manufacture comprises those that are introduced into the interior structure of the eye during surgery. These drugs have been developed to meet the needs of the new and rapidly advancing techniques in use today by the ophthalmologist–surgeon.

In this discussion, the reader is referred to the anatomical representation presented in Fig. 3. Particular attention to the anterior chamber, the iris, the lens, the suspensory ligaments (zonules) of the lens, the cornea, and the angle of the anterior chamber will help pinpoint the site of product administration and the purpose it plays in the particular procedure in which it is used.

To illustrate the varied use of intraocular dosage forms by the ophthalmologist–surgeon, it will be helpful to review, in general terms, the procedures that are involved in a typical intraocular surgical procedure such as the removal of cataracts. The cataract is an opacity of the crystalline lens of the eye that interferes with its refractive role and gradually obliterates sight in the affected eye. Surgery for the removal of the cataract, or the opacified lens, has been practiced for many years, and has restored the sight of millions of people who otherwise would be totally or partially sightless owing to these lens opacities.

Presurgical therapy to maintain pupil dilation and enhance the surgeon's ability to remove the cataract and insert a plastic intraocular lens replacement (IOL) is widely practiced. During surgery, prostaglandins are released as a result of the tissue trauma, and one of their actions is to produce miosis as well as inflammatory effects. The nonsteroidal anti-inflammatory drugs (NSAID) such as suprofen and indomethacin are potent inhibitors of prostaglandin biosynthesis. Clinical trials have demonstrated that topical ophthalmic use of these drugs just before surgery, as an adjunct to the normal mydriatic agents, can prolong the desired degree of pupil dilation. The NSAIDs have also been used after cataract surgery to prevent or reduce a serious inflammatory complication of ocular surgery that can reduce visual function and acuity.

After the surgical patient is prepared and anesthetized, the surgeon–ophthalmologist makes a small incision in the corneal limbus. This usually results in the immediate loss of the aqueous humor from the anterior chamber. The limbus is the junction of the cornea and the sclera. Sutures are loosely affixed and the anterior chamber is now accessible to the surgeon. To expose the opacified lens that lies behind the iris, 1 or 2 drops of a mydriatic agent will have been applied before surgery to dilate the iris. Such drugs as 10% phenylephrine hydrochloride or 1% atropine sulfate may be used for this purpose. To gain further access to the cataractous lens, the surgeon may perform a procedure called an iridectomy. In this procedure, a portion of the iris is removed, thereby providing a pathway for the insertion of an intraocular lens.

The missing iris segment can also be seen in Fig. 13. When the procedure calls for the complete removal of the opacified lens, the lens, which is held firmly in place by the zonulas of Zinn, the fiberlike strands that connect to lens to the ciliary body, a solution of the proteolytic enzyme chymotrypsin is introduced into the posterior chamber. After a few minutes, the proteolytic enzyme will soften and weaken the zonules. The lens is then gripped by forceps or adhered to a cryosurgical instrument and gently rocked to free it from the zonulas. The lens

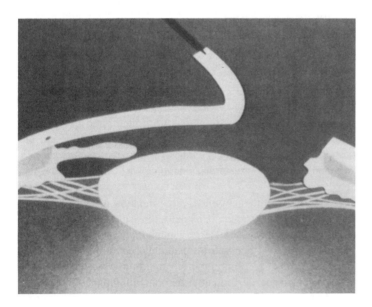

Fig. 13 Preparation of eye for removal of opacified lens—corneal flap with sutures and iridectomy shown. (Courtesy of Alcon Laboratories, Inc.)

is then extracted through the dilated or surgically enlarged iris as depicted in Fig. 14. In the present state-of-the-art, newer procedures and surgical devices are available to avoid iridectomy, disruption of the zonulas, and leave the lens capsule essentially intact. The lens capsule is the relatively tough sac that holds the gellike substance of the lens. More recently, laser treatment before surgery readily allows removal of approximately one-half of the anterior capsule. The opacified, gellike cortex can then be macerated, irrigated, and aspirated out of the lens capsule, thus making way for the implantation of an intraocular lens into this same, now empty, lens capsule.

Throughout this entire procedure, the surgical field is irrigated periodically by specially formulated irrigation solutions to keep the delicate tissues of the anterior segments of the eye from desiccation and to clear the surgical field of blood and debris. Most ophthalmic surgeons also use a viscous solution of sodium hyaluronate alone or in combination with chondroitin sulfate as an aid to help maintain anterior chamber depth and visibility and to minimize interaction between tissues or surgical trauma to them during the procedure. It is also used to provide a hydrophilic coating for the plastic intraocular lens before and during insertion. These polymers are naturally occurring high molecular weight linear polysaccharides. They form viscoelastic solutions in water, which give them unique properties for use in ocular surgery. For this use, they must be specially purified to remove pyrogenic, inflammatory, and antigenic components. The molecular weight fraction must be selected and carefully controlled to provide the desired degree of viscoelasticity, without excessive viscosity, which can detract from their handling characteristics and could result in prolonged retention in the anterior chamber, where the polymers can block aqueous outflow and cause a dangerous increase in intraocular pressure. These solutions must be refrigerated until just before use.

The plastic IOL that the surgeon implants as an artificial lens is made primarily of the same polymethyl methacrylate (PMMA) as that used for hard contact lenses. However, other poly-

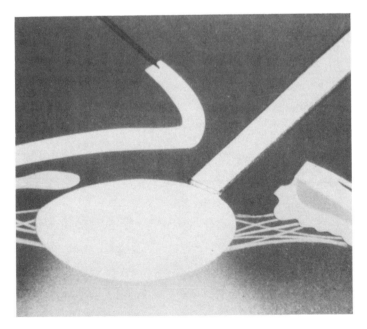

Fig. 14 Delivery of the cataract opacified crystalline lens by cryosurgical technique. (Courtesy of Alcon Laboratories, Inc.)

meric materials such as silicone and the hydroxyethyl methacrylic acid (HEMA) polymer used for soft contact lenses are also being investigated. Having completed the extractive phase of this surgical procedure, and placed an intraocular lens, if this is indicated, the surgeon will now prepare the eye for closure. The first step in this process is the constriction of the iris by the administration of a miotic drug, such as carbachol or acetylcholine chloride, directly into the anterior chamber. This is done to protect the eye against prolapse of the face of the vitreous humor into the anterior chamber and to pull the iris out of the angle of the anterior chamber during suturing of the cornea. Several sutures are then applied to the small limbal incision and, before securing the incision, the surgeon commonly re-forms the anterior chamber with a small quantity of the specially formulated irrigating solution previously referred to. Figure 15 shows a completed cataract removal. Postoperatively, the surgeon may administer a drop or two or a small amount of an ophthalmic solution or ophthalmic ointment of an antibiotic such as tobramycin, ciprofloxacin, or a neomycin–polymyxin combination and then the eye is bandaged.

The foregoing description briefly illustrates a procedure used for the removal of cataracts, which may require up to 2 hr for completion. During this procedure, the use of at least four different intraocular pharmaceutical products was cited, all having the same requirements of extremely high quality, equaling those applicable for parenteral products in general. The newer surgical procedures of irrigation and aspiration may require the use of volumes of the specially formulated irrigating solutions of up to 1 liter [184].

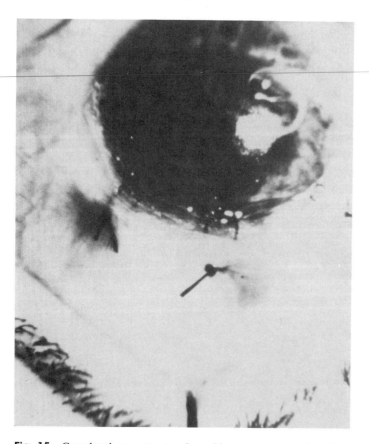

Fig. 15 Completed cataract extraction with cornea sutured closed.

For those products that are to be injected intravenously and are to be used for diagnostic ophthalmology, formulation considerations would be practically identical with those for any IV small-volume injectable fluid. Subconjunctival and subtenons therapy, which are occasionally used for the treatment of disorders of the posterior segments of the eye would also fall into this category [185,186]. Additional discussion, however, is in order for solutions instilled intraocularly.

Inasmuch as systemic toxicity is seldom encountered in the use of intraocularly administered drug products, the prime concern in the design of such products is that of tissue compatibility. This is particularly important when one recognizes that a product instilled into the anterior chamber may remain in contact with the anterior segment tissue, including the very delicate corneal endothelium for up to 2 hr [187]. For such drug products, it is mandatory to design specific ex vivo testing protocols that amount to continuous infusion of the specific product composition, both freshly made and aged, into the anterior chamber of excised rabbit eyes for prolonged periods. Judgments for product–tissue compatibility can then be made by observing the corneal endothelium by specular microscopy [188] and histopathology.

In vitro these materials can also be evaluated against specific cell lines in tissue culture. Here, again, corneal endothelial tissue is of prime interest. As tissue culture technology progresses, cell lines for the other tissues in the anterior segment of the eye will be established and will become useful in tissue compatibility testing as well. In considering the design of a drug for intraocular use, simplicity of composition is the key. In general, the fewer the ingredients, the less likelihood of tissue incompatibility. This is not to say that a simple solution of drug in water is optimal. Indeed, a simple isotonic solution of sodium chloride is toxic to human corneal epithelial, endothelial, iris, and conjunctival cells, whereas a solution properly balanced with various organic and inorganic ions and nutrients is nontoxic to these cells in vitro and in vivo [188–191]. Figures 16–18 serve to illustrate the foregoing. In these electron photomicrographs of human corneal endothelium, the effect of solution composition on tissue integrity is illustrated. Figure 16 illustrates human corneal endothelial tissue after corneal perfusion of 3 hr with lactated Ringer's solution, while Fig. 17 illustrates the same tissue perfused for 3 hr with Ringer's solution containing glutathione, adenosine, and bicarbonate. In the former, cell darkening and swelling are in evidence, whereas in the latter, normal cell confluence is retained. Figure 18 shows the same tissue after 3 hr perfusion with a solution devoid of ingredients essential for normal cell confluence. The discontinuity of cell structure is quite evident.

If this type of information is used as a criterion for designing an intraocular pharmaceutical product, the formulator can begin to focus on the development of suitable irrigation–aspiration solution products. For example, the preservative agents commonly used in topical ophthalmic preparations are not compatible with the tissues of the anterior segments of the eye and in several cell lines in tissue culture [192]. The *USP* recognizes this problem and specifically warns against their use in intraocular solutions [193,194].

The question of drug stabilizers and solubilizers and the question of toxicity of these products are closely related. Drug stabilizers, such as antioxidants and chelating agents, must be used with care and in absolutely minimal quantities, when necessary. Occasionally, it may seem desirable to solubilize an otherwise sparingly soluble ingredient. Whereas this may be a practical consideration in some injectables, only aqueous solutions should be employed intraocularly. Furthermore, only fairly low concentrations of typical cosolvents such as glycerin and propylene glycol can be employed because of their osmotic effect on the surrounding tissues. If a hyperosmotic solution were to be instilled into the anterior chamber, some transient desiccation of the tissues may occur. A solution that is hypotonic, however, may cause edema of these and could lead to corneal clouding. There appears to be little or no experience with these or other common cosolvents in products of this type.

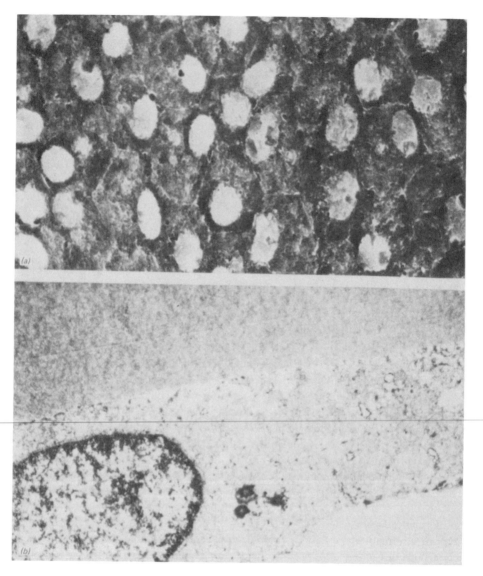

Fig. 16 Human corneal endothelium 3-hr perfusion with lactated Ringer's solution: (a) scanning electron microscopy (2100×); (b) transmission electron microscopy (9100×). (Courtesy of H. Edelhauser.)

Another formulation variable to consider is that of pH and buffer capacity. Inasmuch as the anterior chamber fluid (aqueous humor) contains essentially the same buffering system as the blood, products of a pH somewhat dissimilar to the physiological range of 7.0–7.4 are converted to this range by the buffering capacity of the aqueous humor, if it is in sufficient volume relative to the solution introduced. Often, however, aqueous humor is lost in the procedure being used; consequently, drug products should be formulated as close as possible to this physiological range. If at all possible, the use of buffering agents should be avoided.

To those engaged in the design, development, and manufacture of solutions for intraocular use, particularly for instillation into the anterior chamber, the question of particulate matter is of great importance. Although the total effect of particulate inclusion in the anterior chamber

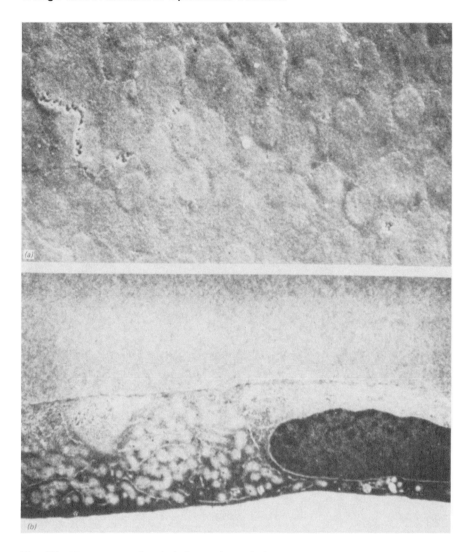

Fig. 17 Human corneal endothelium 3-hr perfusion with glutathione bicarbonated Ringer's solution: (a) scanning electron microscope (1950×); (b) transmission electron microscopy (8450×). (Courtesy of H. Edelhauser.)

is not completely known, some possible results of this inclusion have been postulated [194]. Certain amounts of iritis and uveitis might be expected, as well as the production of granulomas similar to the type reported for pulmonary tissue as a result of particulates in large-volume parenterals. At least as important, however, is the possibility that particulate matter can block the canals of Schlemm, which provide the outflow mechanism for the aqueous humor. If this should occur to any great extent, the continued undiminished production of aqueous humor could lead to a rapid increase in intraocular pressure and the onset of an acute attack of glaucoma.

No small factor in the utility and success of pharmaceutical products designed for instillation into the anterior chamber are those considerations having to do with packaging. Standard injectable packaging, while constituting the majority of packaging systems for drug products

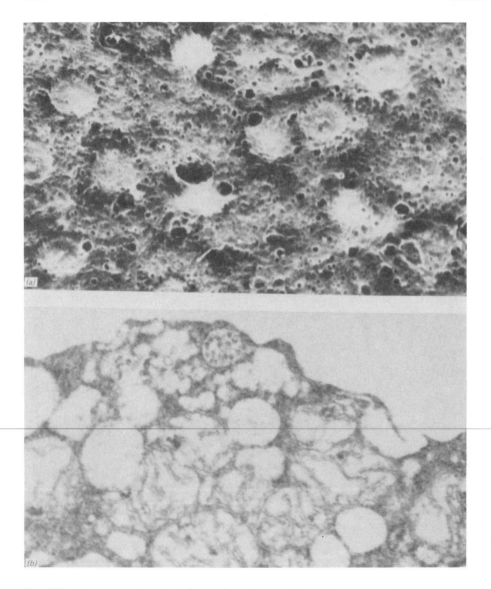

Fig. 18 Human corneal endothelium 3-hr perfusion with solution devoid of essential nutrients: (a) scanning electron microscope (2100×); (b) transmission microscopy (9100×). (Courtesy of H. Edelhauser.)

of this type, is beginning to yield to more specialized packaging designed to circumvent some of the recognized difficulties inherent in the use of ampoules and single- or multiple-dose vials. Since particulate contamination is of serious concern, the glass fragments produced in glass ampoule fracture and the elastomer particles generated during stopper penetration should be kept in mind. Very specialized stopper design, cleaning procedures, and lubrication should be considered when this type of packaging is used.

In addition to the foregoing, the question of packaging sterility must be posed. Since these products are generally used in conjunction with surgical procedures, the need for sterility of the exterior of the primary package exists. Figure 19 illustrates products for intraocular use

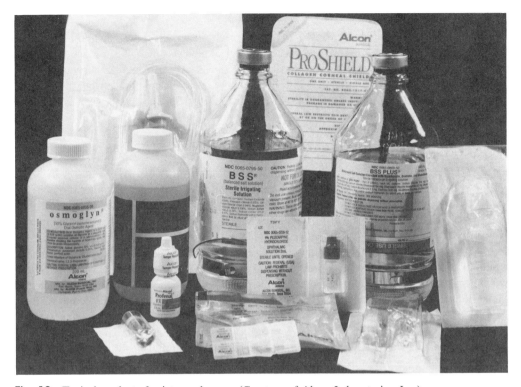

Fig. 19 Typical products for intraocular use. (Courtesy of Alcon Laboratories, Inc.)

that encompass the aforementioned design considerations and, in addition, owing to their specialized packaging, offer the surgeon a large latitude for control of flow as well as the convenience of a package adaptable to the proper types of irrigating cannulae. Since these drugs are administered through an incision, it is obvious that needle penetration through the tissue need not take place. Indeed, the sharp points of standard hypodermic needles are contraindicated in that they could conceivably damage some of the sensitive tissues in the anterior segment of the eye.

IX. CONTACT LENS CARE PRODUCTS

Contact lenses are optical devices that are either fabricated from preformed polymers or polymerized during lens manufacture. The main purpose of contact lenses is to correct defective vision. For this application, they are called *cosmetic* lenses. Contact lenses used medically for the treatment of certain corneal diseases are called bandage lenses.

A. Evolution of Contact Lenses

In 1508, Leonardo da Vinci conceived the concept of the contact lens. It was not until 1887 that scleral contact lenses were fabricated by Dr. A. E. Fick, a physician in Zurich; F. A. Mueller, a maker of prosthetic eyes in Germany; and Dr. E. Kalt, a physician in France. Muller, Obrig, and Gyorry fabricated contact lenses made from polymethyl methacrylate (PMMA) in the late 1930s. K. Tuohy filed the patent for contact lens design in 1948 which were made of PMMA material [195]. Although they were safe and effective, these lenses were uniformly uncomfortable, thus suppressing their potential growth for contact lens wear. Lenses made from polyhydroxyethyl methacrylate (HEMA), the so-called soft lenses or hydrophilic lenses were introduced in 1970. Since then, significant technological advances have been made in the lens materials, lens fabrication, and lens designs [196]. Consequently, a phenomenal growth in lens wearers necessitated the need and development of lens care products. Today, rigid lenses made from materials polymerized with PMMA and in combination with various siloxanes and fluorocarbons are available to meet the broadest needs of lens wearers.

B. Composition of Contact Lenses

Contact lenses are broadly classified as PMMA, rigid gas-permeable (RGP), and soft hydrogel (HEMA) lenses. Dyes may be added during polymerization or after fabrication to improve lens handling or to change the color of the lens wearer's eyes. Lenses made from numerous polymers are available today [197]. In soft hydrogel lenses, HEMA is a commonly used monomer. However, to avoid infringement of existing patents, many comonomers, as for example methyl methacrylic acid or a blend of comonomers, are used. Comonomers produce changes in the water content or ionic nature of lenses that is significantly different from HEMA lenses. For example, addition of acrylic acid in HEMA increases the water content and ionic nature of lenses. Some lenses are made from n-vinylpyrrolidone and have high water contents. Such lenses have pore sizes that are much larger than low water content lenses. Cross-linkers, such as ethyleneglycol dimethacrylate, and initiators as, for example, benzyl peroxide, in appropriate amounts are added for polymerization and to achieve desirable physical and chemical properties. Table 5 gives a list of monomers, comonomers and cross-linkers along with their effects on polymer properties. In 1985, FDA published a classification for soft hydrophilic lenses based on their water content and ionic nature. Grouping for soft lenses and their generic names are listed in Table 6. Adequate levels of oxygen are necessary to maintain normal corneal metabolism [198]. Lenses that are poorly designed and worn overnight deprive the cornea of oxygen, causing edema [199]. Contact lenses made from PMMA materials are virtually impermeable to gases [200]. The PMMA lenses are also inflexible, causing discomfort in a large percentage of individuals while the lens is worn. During the 1980s, lenses that were somewhat flexible and permeable to oxygen were introduced. These lenses were made from either cellulose acetate butyrate (CAB), or silicone elastomer. Although comfortable and flexible, such lenses accumulated lipids, were nonwettable, and adhered to the cornea. Several reports detailing difficulty in removing CAB and silicone lenses appeared in the published literature [201]. Lenses made from fluorocarbons, and in various combinations of fluorocarbon, silicone, methyl methacrylate, and acrylic acid are currently available. Desired properties of these lenses include flexibility, wettability, and gas transmissibility. Grouping for rigid gas-permeable lenses was published by the FDA in 1989. Table 7 provides the generic name and oxygen permeability of rigid gas permeable (RGP) lens materials [202].

C. Complications of Contact Lens Wear and the Need for Care Products

Lens design, user compliance with manufacturers' instructions, hygiene, environmental conditions, poor fit, lens materials, and tear chemistry are the major causes of lens wear complications. Complications owing to lens design, compliance, hygiene, environmental conditions, and poor fit are beyond the scope of this chapter and are not critical to an understanding of the concepts required for the development of care products. However, knowledge of tear chemistry is important in understanding the complex chemical processes between tear components and contact lenses. The tear film can be broadly divided into three distinct layers: lipids,

Table 5 Commonly Used Monomers, Comonomers, and Cross-linkers in Contact Lens Polymers

Name	Abbreviation	Lens properties
Acrylic acid	AA	Flexibility Hydrophilicity pH sensitivity; acidic Reactivity; ionically interacts with positively charged tear components Wettability
Butyl methacrylate	BMA	Softness Flexibility Hydrophobicity; attracts lipids Wettability Gas transmisssibility
Cellulose acetate butyrate	CAB	Clarity Wettability Gas transmissibility Physical stability
Dimethyl siloxane	DMS	Hydrophobicity Wettability Gas transmissibility Physical stability
Diphenyl siloxane	DPS	Hydrophobicity Wettability Gas transmissibility Physical stability
Ethonyethyl methacrylate	EOEMA	Flexibility Softness Hydrophobicity Wettability Gas transmissibility
Ethylene glycol dimethacrylate	EGDME	Hydrophilicity Wettability
Glyceryl methacrylate	GMA	Wettability Gas transmissibility Hydrophilicity Machinability
Hydroxyethyl methacrylate	HEMA	Flexibility Wettability Gas transmissibility Softness Machinability
Methacrylic acid	MA	Hardness Hydrophilicity
Methyl methacrylate	MMA	Hardness Machinability Wettability Gas transmissibility Hydrophobicity
Methylphenyl siloxane	MPS	Hydrophobicity Gas transmissibility
Methyl vinyl siloxane	MVS	Hydrophobicity Gas transmissibility
N-Vinyl pyrrolidone	NVP	Hydrophilicity Wettability Machinability Color, clarity
Siloxanyl methacrylate	SMA	Hardness Wettability Gas transmissibility

Table 6 FDA Grouping for Soft Hydrophilic Lenses and Generic Names

Group 1 Low water (< 50% H_2O) nonionic polymers	Group 2 High water (> 50% H_2O) nonionic polymers	Group 3 Low water (< 50% H_2O) ionic polymers	Group 4 High water (> 50% H_2O) ionic polymers
Tefilcon (38%)	Lidofilcon A (70%) Lidofilcon B (79%)	Etafilcon (43%)	Bufilcon A (55%)
Tetrafilcon A (43%)	Surfilcon (74%)	Bufilcon A (45%)	Perfilcon (71%)
Grofilcon (39%)	Vilifilcon A (55%)	Deltafilcon A (43%)	Etafilcon A (58%)
Dimefilcon A (38%)	Scafilcon A (71%)	Dronifilcon A (47%)	Ocufilcon C (55%)
Hefilcon A (43%)	Xylofilcon A (67%)	Phenifilcon A (38%)	Phenfilcon A (55%)
		Ocufilcon (44%)	Tetrafilcon B (58%)
Hefilcon B (43%)		Mafilcon (33%)	Methafilcon (55%)
Phenifilcon A (30%)			Vifilcon A (55%)
Isofilcon (36%)			
Polymacon (38%)			
Mafilcon (33%)			

aqueous, and mucin [203]. Each layer of the tear film performs a specific function. The mucin layer spreads and coats the hydrophobic corneal cells and extends into the aqueous layer. The aqueous layer contains 98% water and 2% solids. Solids in this layer are predominantly the electrolytes (Na^+, K^+, C^{2+}, Mg^{2+}, Cl^-, and HCO_3^-), nonelectrolytes (urea and glucose), and proteins. Major proteins in the tear film are presented in Table 8.

The lipid layer, which consists of cholesterol esters, phospholipids, and triglycerides, prevents and regulates aqueous evaporation from the tear film.

Table 7 FDA Grouping of Hydrophobic Hard and Rigid Gas-Permeable Lenses

Lens materials	Generic name	D_k
Cellulose acetate butyrate	Cabufocon A	> 150
	Powfocon A	> 150
	Powfocon B	> 150
t-Butylstyrene	Aufocon A	> 150
Silicone	Elastofilcon A	> 150
	Dimofocon	> 150
	Silafilcon A	> 150
t-Butylstyrene–silicone acrylate	Pentasilcon P	120
Fluoroacrylate	Fluorofocon A	100
Fluoro silicone acrylate	Itafluorofocon A	60
	Porflufocon A	30–92
Silicone acrylate	Pasifocon A	14
	Pasifocon B	16
	Pasifocon C	45
	Itafocon A	14
	Itafocon B	26
	Nefocon A	20
	Telefocon A	15–45
	Amefocon A	40

Table 8 Major Proteins of the Tear Film

Name	Total protein (%)	Function
Lysozyme	30–40	Antimicrobial, collagenase regulator
Lactoferin	2–3	Bacteriostatic, anti-inflammatory
Albumin	30–40	Anti-inflammatory
Immunoglobins	0.1	Immunological, anti-inflammatory

Components of the tear attaches to contact lenses by electrostatic and van der Waals forces, and build up to form deposits. Deposits on the surface and in the lens matrix may result in reduced visual acuity, irritation, and in some instances, serious ocular complications. The composition of deposits vary because of the complexity of an individual's ocular physiology–pathology. Lysozyme is a major component of soft lens deposits, especially found on high-water–content ionic lenses [204]. Calcium [205] and lipds [206] are infrequent components of deposits, occurring as inorganic salts, organic salts, or as an element of mixed deposits, or as a combination thereof [207,208].

Lenses are exposed to a broad spectrum of microbes during normal wear and handling and become contaminated relatively quickly. Failure to effectively remove microorganisms from lenses can cause ocular infections. Ocular infections, particularly those caused by pathogenic microbes, such as *P. aeruginosa*, can lead to loss of the infected eye if left untreated.

D. Types of Lens Care Products

Contact lens care products can be divided into three categories: cleaners, disinfectants, and lubricants. Improperly cleaned lenses can cause discomfort, irritation, decrease in visual acuity, and giant papillary conjunctivitis (GPC). This latter condition often requires discontinuation of lens wear, at least until the symptoms clear. Deposits can also accumulate preservatives from lens care products and produce toxicity, and can act as a matrix for microorganism attachment to the lens [209]. Thus, cleaning is one of the most important steps in successful lens wear. It helps in removal of surface debris, tear components, and contaminating microorganisms, resulting in safety and efficacy of lens-wearing experience [210].

Daily cleaners and weekly cleaners are employed to clean deposits that accumulate on lenses during normal wear. A list of cleaning agents commonly used in daily cleaners is provided in Table 9. Single cleaning agents or combinations of cleaning agents may be used in a cleaner. Surfactant(s), solvent(s), and complexing agent(s) chosen for cleaner formulations must be capable of solubilizing lens deposits and must have low irritation potential. They must be easily rinsed, leaving very low or nondetectable residue levels on the lens. Many problems that contact lens wearers experience with their lenses are the results of incomplete deposit(s) removal [211]. Nonionic and amphoteric surfactants are commonly used in daily cleaner products. Because of their toxicity to the cornea and binding to the lenses, anionic and cationic surfactants are avoided. Solvents capable of solubilizing lens deposits without altering the lenses polymer properties should be carefully selected. Complexing agents, such as citrates, are included in daily cleaner formulations [212]. They counter the binding of positively charged proteins to the lenses and render the proteins more soluble in the media by ion-pair or salt formation.

Table 9 Cleaning Agents Commonly Used in Daily Cleaners

Class	Trade name	Chemical name
Abrasive particles	Nylon 11	11-Aminoundecanoic acid
	Silica	Silicon dioxide
Complexing agents	Citric acid	2-Hydroxy-1,2,3-propane tri-carboxylic acid
Solvents	Isopropyl alcohol	2-Propanol
	Propylene glycol	1,2-Propanediol
	Hexalmethyene glycol	1,6-Hexanediol
Surfactants (nonionic)	Tween 21	Polysorbate 21
	Tween 80	Polysorbate 80
	Tyloxapol	4-(1,3,3-Tetramethylbutyl)-phenol polymer with formaldehyde and oxirane
	Pluronic	Poloxamer
	Tetronic	Poloxamine
Surfactants (ionic)	Miracare	Cocoamphocarboxy-glycinate

Mechanical force is a key aspect in the cleaning process. For daily cleaning, mechanical force is generally provided through the rubbing action of the fingers over the lens during the actual cleaning process. Cleaning lenses by rubbing typically removes 1.7 ± 0.5 logs of microorganisms, rinsing the lens removes 1.9 ± 0.5 logs of microorganism, and cleaning and rinsing the lens removes 3.7 ± 0.5 logs of microorganisms of a typical 10^6 colony-forming units (CFU)/ml challenge [213]. Abrasive particles are included in products to enhance the mechanical force applied to the lens during the cleaning process [214]. The abrasive properties are evaluated by testing the hardness of the included abrasive particles. Particles that have Rockwell hardness lower than the hardness of the lens polymers are typically used. If the hardness of abrasive particles is higher than the hardness of the lens polymer, it is possible that the lens would be damaged. Some contact lenses are reported to require special treatment. Abrasive particles may alter surface treatment effects even when their hardness is lower than that of the lens polymer.

Enzymatic cleaners contain enzymes derived from animals, plants, or microorganisms. Plant- and microorganism-derived enzymes cause sensitization problems in many lens wearers [215]. A list of commonly used enzymes is provided in Table 10. All of these enzymes are effective in removing deposits from the contact lens surface [216]. They are biochemical catalysts that are specific for catalyzing certain chemical reactions. Those that aid in removing contact lens debris are protease (protein-specific enzyme), lipase (lipid-specific enzyme), and amylase (poly-

Table 10 Enzymes Commonly Used in Weekly Cleaners

Name	Origin	Active against	Active at pH
Pancreatin	Animal (porcine)		
Proteases		Proteins	7.0
Lipase		Lipids	8.0
Amylase		Carbohydrates	6.7–7.2
Papain	Plant (papaya)	Proteins	5.0
Subtilisin A	Microorganisms	Proteins	8–10
Subtilisin B	Microorganisms	Proteins	8–10

saccharide-specific enzyme). Such enzymes act by attacking substrate molecules, such as protein, lipid, and mucin, on the lens and catalyzing their breakdown into smaller molecular units. This process yields fragments that are readily removed by mechanical force and rinsing.

In the past, only tablet dosage forms of enzymatic cleaners were available. They required soaking lenses in solutions prepared from a tablet for a period of 15 min to more than 2 hr before disinfecting the lenses. Although this process provided sufficient time for cleaning, it was a cumbersome process and required multiple steps. Complicated or cumbersome processes inevitably lead to poor user compliance. The newer products are either in a tablet or a solution product form. Simultaneous cleaning and disinfection can be achieved, which reduces care time and the need for multiple steps [217].

Contact lenses are contaminated with microorganisms during lens handling and lens wear. They must be disinfected to prevent ocular infections, especially from pathogenic microorganisms. The two disinfection methods used are thermal and chemical. In thermal disinfection systems, lenses are placed in a preserved or unpreserved solution in a lens case. The lens case is then heated in a device to kill the microorganisms. The current FDA requirement for thermal disinfection requires heating at a minimum of 80°C for 10 min. The unpreserved salines are either packaged in a unit-dose or an aerosol container, and they do have some antimicrobial activity [218]. Preservatives must be used in salines packaged in nonaerosol multidose containers. The type and names of preservatives and antimicrobial disinfectants commonly used in lens care products are listed in Table 11. Thimerosal and sorbic acid are commonly used preservatives in these products; however, concerns over the sensitization potential and discoloration of lenses has led to the introduction of new and safer molecules like Polyquad (a polymerically bound quaternary ammonium compound) and Dymed. Specifically, Polyquad is resistant to absorption into the lenses; thus, it may not diffuse out of the lens into the eye, leading to corneal toxicity, an inherent problem associated with nonpolymerically bound quaternary ammonium compounds. The FDA and the *USP* have specific preservative effectiveness standards that these products must meet. The FDA standards detailing the method were pub-

Table 11 Antimicrobial Agents Commonly Used in Lens Care Products

Class	Generic name	Molecular weight	Used in lens type	
			Soft	RGP, PMMA
Acids	Benzoic acid	122	No	Yes
	Boric acid	62	Yes	Yes
	Sorbic acid	112	Yes	Yes
Alcohols	Benzyl alcohol	108	No	Yes
	Phenyl ethyl alcohol	122	No	Yes
Biguanides	Chlorhexidine	505	Yes	Yes
	Polyaminopropyl biguanide	~1200	Yes	Yes
Mercurial	Thimerosal	404	Yes	Yes
	Phenylmercuric nitrate	634	Yes	Yes
Oxidizing	Hydrogen peroxide	34	Yes	No
	Sodium dichloroisocyanurate	220	Yes	No
	Halazone	270	Yes	No
Quaternary	Tris(2-hydroxyethyl) tallow ammonium chloride	≈424	Yes	No
	Benzalkonium chloride	≈363	No	Yes
	Benzethonium chloride	448	No	Yes
	Polyquaternium-1	≈6000	Yes	Yes

lished in 1985 [219]. Oxidizing agents and nonoxidizing chemical disinfectants that are non-toxic at product concentrations are used to chemically disinfect lenses. Mostly, hydrogen peroxide is used as an oxidizing agent [220]. It is used in concentrations of 0.6–3.0%. Peroxides are very toxic to the cornea of the eye. After the disinfection cycle, and before placing the lens in the eye, hydrogen peroxide must be completely neutralized by reducing agents, catalase or transition metals, such as platinum.

An ideal chemical-disinfecting agent would have the following properties: (a) it should be nonirritating nonsensitizing, and nontoxic in tests for cytotoxicity; (b) it should have an adequate antimicrobial spectrum and be able to kill ocular pathogens during a short lens-soaking period; (c) it should not bind to the lens surface; and (d) it should be compatible with the lens

Table 12 Types of Tests and Requirements
Proposed by the FDA for Product Development

I. Chemistry/manufacturing
    A. Solution/container descriptions
    B. Solution stability testing
    C. Lens group selection for solution testing
II. Toxicology
    A. Solution testing
        1. Acute oral toxicity assessment
        2. Acute systemic toxicity assessment
        3. Acute ocular irritation and cytotoxicity assessment
        4. Sensitization/allergic response assessment
            a. Preservative uptake and release test
            b. Guinea pig maximization testing
    B. Container/accessory testing
        1. In vitro testing
        2. Systemic toxicity testing
        3. Primary ocular irritation testing
III. Microbiology
    A. Sterilization of the solution by the manufacturer
        1. Validation of the sterilization cycle
        2. *USP* sterility tests
        3. *USP* type perservative effectiveness test
        4. *USP* microbial limits test
    B. Shelf life-testing requirements
        1. Shelf life sterility
        2. Shelf life preservative effectiveness
        3. Extension of shelf life protocol
    C. Disinfection of the lens
        1. Chemical disinfection systems
            a. Contribution of elements test
            b. D-value determinations
            c. Multi-item microbial challenge test
        2. Thermal (heat) disinfection system
            Same as for chemical disinfection system
IV. Clinical
    A. Patient characteristics
    B. Number of eyes duration and number of investigators
    C. Initial patient visit parameters

---

and not cause lens discoloration or alter the tint of colored contact lenses. Polyquad and Dymed have most of these characteristics. They have been recently introduced in the marketplace and are performing to expectations.

Contact lens wearers may experience increasing awareness of their lenses during the day owing to ocular dryness [221]. With some lens materials, this increase in awareness may arise from a decrease in the wettability of the lens surface. Dehydration of the lens or accumulation of debris on the lens surface can cause similar symptoms [222]. The lens wearer may achieve relief from these symptoms with periodic administration of lubrication of rewetting drops. These solutions contain polymers or surfactants that enhance the wettability of the surface, facilitate the spreading of tears, and improve the stability of the tear film. They may also provide cushioning and lubrication actions, thereby reducing the frictional force between the eyelids and the lens. Some products are specifically designed to rehydrate the lens. These products are unpreserved and packaged in a unit-dose. However, a preservative is required for a multidose product.

E. Summary

Generally, contact lens products are sterile solutions or suspensions. Formulators for these products must have the training for the technologies practiced in the development of sterile pharmaceutical products, such as injectable and large-volume intravenous fluids. The products must be effective and compatible with a wide range of lens materials. Components of the formulations should not accumulate in the lens or change the lens properties. They must be adequately preserved and be well tolerated by the sensitive ocular tissues. The products should also be simple to use to assure good compliance on the part of lens wearers. Additionally, they should be developed following the guidelines enumerated in Table 12.

ACKNOWLEDGMENT

The authors thank Ms. Terry Praznik for her assistance in the library sciences that has led to the extensive bibliography on which this chapter is based.

REFERENCES

1. *The United States Pharmacopeia* (USP), U.S. Pharmacopeial Convention, Rockville, MD, 1990, pp. 1692,1693.
2. G. Majno, *The Healing Hand—Man and Wound in the Ancient World*, Harvard University Press, Cambridge, MA, 1975, pp. 43–45.
3. G. Majno, *The Healing Hand—Man and Wound in the Ancient World*, Harvard University Press, Cambridge, MA, 1975, pp. 112–114.
4. G. Majno, *The Healing Hand—Man and Wound in the Ancient World*, Harvard University Press, Cambridge, MA, 1975, p. 154.
5. G. Majno, *The Healing Hand—Man and Wound in the Ancient World*, Harvard University Press, Cambridge, MA, 1975, pp. 216,359.
6. G. Majno, *The Healing Hand—Man and Wound in the Ancient World*, Harvard University Press, Cambridge, MA, 1975, pp. 348,377.
7. D. L. Deardorf, *Remington's Pharmaceutical Sciences*, 14th Ed., Mack Publishing, Easton, PA, 1970, pp. 1545–1548.
8. S. Riegelman and D. L. Sorby, *Dispensing of Medication*, 7th Ed., Mack Publishing, Easton, PA, 1971, pp. 880–884.
9. W. H. Havener, *Ocular Pharmacology*, 2nd Ed., C. V. Mosby, St. Louis, MO, 1970.
10. B. Smith, *Handbook of Ocular Pharmacology*, Publication Sciences Group, Action, MA, 1974.

11. P. Ellis and D. L. Smith, *Ocular Therapeutics and Pharmacology*, 3rd Ed., C. V. Mosby, St. Louis, MO, 1969.
12. J. T. Grayston, S. P. Wang, R. L. Woolridge, and P. B. Johnson, JAMA, 172, 602 (1962).
13. J. L. Byers, M. G. Holland, and J. H. Allen, Am. J. Ophthalmol., 49, 267 (1960).
14. W. D. Gingrich, JAMA, 179, 602 (1962).
15. S. Mishima and D. M. Maurice, Invest. Ophthalmol., 1, 794 (1962).
16. D. M. Maurice, Invest. Ophthalmol., 6, 464 (1967).
17. S. J. Kimura, Am. J. Ophthalmol., 34, 446 (1951).
18. A. Wessing, *Fluorescein Angiography of the Retina* (transl. by G. K. von Noorden), C. V. Mosby, St. Louis, MO, 1969.
19. K. Koller, Arch. Ophthalmol., 13, 404 (1884).
20. Council on Drugs, New drugs and developments in therapeutics, JAMA, 183, 178 (1963).
21. M. A. Lemp, C. H. Dohlman, and F. J. Holly, Ann. Ophthalmol., 2, 258 (1970).
22. M. A. Lemp and E. S. Szymanski, Arch. Ophthalmol., 93, 134 (1975).
23. M. A. Lemp, Scientific Exhibit, American Academy of Ophthalmology and Otolaryngology, Dallas, Sept. 1975.
24. *USP XXII*, U. S. Pharmacopeal Convention, Rockville, MD, 1990, ⟨71⟩ p. 1483.
25. *USP XXII*, U.S. Pharmacopeal Convention, Rockville, MD, 1990, ⟨1211⟩ p. 1706.
26. F. N. Marzulli and M. E. Simon, Am J. Optom., 48, 61 (1971).
27. R. B. Hackett and T. O. McDonald, Eye Irritation, in *Dermatotoxicology*, 4th Ed. (F. N. Marzulli and H. L. Maibach, eds.), Hemisphere Publishing, Washington, DC, 1991, pp. 749–815.
28. P. K. Basu, J. Toxicol. Cutan. Ocul. Toxicol., 2, 205–227 (1984).
29. J. S. Friedenwald, W. F. Hughes, Jr., and H. Hermann, Arch. Ophthalmol., 31, 379 (1944).
30. C. Carpenter and H. Symth, Am J. Ophthalmol., 29, 1363 (1946).
31. L. W. Hazelton, Proc. Sci. Sect. Toilet Goods Assoc., 17, 490 (1973).
32. L. M. Carter, G. Duncan, and G. K. Rennie, Exp. Eye Res., 17, 5 (1952).
33. J. H. Kay and J. C. Calandra, J. Soc. Cosmet. Chem., 13, 281 (1962).
34. K. L. Russell and S. G. Hoch, Proc. Sci. Sect. Toilet Goods Assoc., 37, 27 (1962).
35. I. Gaunt and K. H. Harper, J. Soc. Cosmet. Chem., 15, 290 (1964).
36. S. P. Battista and E. S. McSweeney, J. Soc. Cosmet. Chem, 16, 119 (1965).
37. J. H. Becklet, Am. Perfum. Cosmet., 80, 51 (1965).
38. C. T. Bonfield and R. A. Scala, Proc. Sci. Sect. Toilet Goods Assoc., 43, 34 (1965).
39. E. V. Buehler and E. A. Newmann, Toxicol. Appl. Pharmacol., 6, 701 (1964).
40. C. H. Dohlman, Invest. Ophthalmol., 10, 376 (1971).
41. M. J. Hogan and L. E. Zimmerman, in *Ophthalmic Pathology: An Atlas and Textbook*, 2nd Ed., W. B. Saunders, Philadelphia, 1962.
42. R. R. Phister, Invest. Ophthalmol, 12, 654 (1973).
43. D. M. Maurice, *The Eye*, Vol. 1 (H. Davson, ed.), Academic Press, New York, 1969, pp. 489–600.
44. J. H. Prince, C. D. Diesen, I. Eglitis, and G. L. Ruskell, *Anatomy and Histology of the Eye and Orbit in Domestic Animals*, Charles C. Thomas, Springfield, IL, 1960.
45. H. Davson, in *The Eye*, Vol. 1 (H. Davson, ed.), Academic Press, New York, 1969, pp. 217–218.
46. B. S. Fine and M. Yanoff, *Ocular Histology: A Text and Atlas*, Harper & Row, New York, 1972.
47. W. R. Green, J. B. Sullivan, R. M. Hehir, L. G. Scharpf, and A. W. Dickinson, *A Systemic Comparison of Chemically Induced Eye Injury in the Albino Rabbit and Rhesus Monkey*, Soap and Detergent Association, New York, 1978, pp. 405–415.
48. E. V. Buehler and E. A. Newmann, A comparison of eye irritation in monkeys and rabbits, Toxicol. Appl. Pharmacol., 6, 701–710 (1964).
49. Committee for the Revision of NAS Publication 1138, National Research Council, *Principles and Procedures for Evaluating the Toxicity of Household Substances*, National Academy of Sciences, Washington, DC, 1977, pp. 41–56.
50. Interagency Regulatory Liaison Group, Testing Standards and Guidelines Work Group, *Recommended Guidelines for Acute Eye Irritation Testing*, Jan. 1981.
51. J. H. Draize, Food Drug Cosmet. Law J., 10, 722 (1955).

52. J. H. Draize and E. A. Kelley, Proc. Sci. Sect. Toilet Goods Assoc., 17, 1 (1959).
53. J. H. Draize, J. Pharmacol. Exp. Ther., 82, 377 (1944).
54. Food Drug Cosmet. Law Rep., 233, 8311; 440, 8313; 476, 8310.
55. Federal Register, 37, 8534 (1972).
56. W. R. Markland, *Norda Briefs*, 470 (1975).
57. C. Shopsis, E. Borenfreund, J. Walberg, and D. M. Stark, *Alternative Methods in Toxicology*, Vol. 2 (A.M. Goldberg, ed.) Mary Ann Liebert, New York, 1984, pp. 103–114.
58. E. Borenfreund and O. Borrero, Cell Biol. Toxicol., 1, 55–65 (1984).
59. C. Shopsis and S. Sathe, Toxicology, 29, 195–206 (1984).
60. R. Neville, P. Dennis, D. Sens, and R. Crouch, Curr. Eye Res., 5, 367 (1986).
61. M. E. Stern, H. F. Edelhauser, and J. W. Hiddemen, Methods of Evaluation of Corneal Epithelial and Endothelial Toxicity of Soft Contact Lens Preservatives. Presented at Contact Lens International Congress, Las Vegas, Nevada, March, 1985.
62. H. F. Edelhauser, M. E. Antione, H. J. Pederson, J. W. Hiddemen, and R. G. Harris, J. Toxicol. Cutan. Ocul. Toxicol., 2(1), 7 (1983).
63. S. J. Krebs, M. E. Stern, J. W. Hiddemen, and H. F. Edelhauser, CLAO J., 10(1), 35 (1984).
64. H. E. Seifried, J. Toxicol. Cutan. Ocul. Toxicol., 5, 89–114 (1986).
65. D. M. Stark, C. Shopsis, E. Borenfreund, and J. Walberg, *Alternative Methods of Toxicology*, Vol. 1, *Product Safety Evaluation*, Mary Ann Liebert, New York, 1983, pp. 127–204.
66. J. M. Frazier, *Dermatotoxicology*, 4th Ed. (F. N. Marzulli and H. L. Maibach, eds.), Hemisphere Publishing, Washington, DC, pp. 817 (1991).
67. N. L. Burstein, Invest. Ophthalmol. Vis. Sci., 25, 1453–1457 (1984).
68. D. Maurice and T. Singh, A permeability test for acute corneal toxicity, Toxicol. Lett., 31, 125–130 (1986).
69. N. L. Burstein, Invest. Ophthalmol. Vis. Sci., 7, 308–313 (1980).
70. R. R. Pfister and N. Burstein, Invest. Ophthalmol. Vis. Sci., 15, 246–249 (1976).
71 A. M. Tonjum, Acta Ophthalmol., 53, 358–366 (1975).
72. H. Ichijima, W. M. Petroll, J. V. Jester, and H. D. Cavanagh, Cornea, 11, 221–225 (1992).
73. P. S. Imperia, H. M. Lazarus, R. E. Botti, Jr., and J. H. Lass, J. Toxicol. Cutan. Ocul. Toxicol., 5, 309–317 (1986).
74. H. B. Collins and B. E. Grabsch, Am. J. Optom. Physiol. Opt., 59, 215–222 (1982).
75. J. Ubels, J. Toxicol Cutan. Ocul. Toxicol., 1, 133–145 (1982).
76. B. J. Tripathi and R. C. Tripathi, Lens Eye Toxicol. Res., 6, 395–403 (1987).
77. H. A. Baldwin, T. O. McDonald, and C. H. Beasley, J. Soc. Cosmet. Chem., 25, 181 (1973).
78. Federal Register, 18, 351 (1953).
79. C. W. Bruch, Drug Cosmet. Ind., 118(6), 51 (1976).
80. K. Green and J. M. Chapman, J. Toxicol. Cutan. Ocul. Toxicol., 5, 133–142 (1986).
81. C. Thode and H. Kilp, Fortschr. Ophthalmol., 79, 125–127 (1982).
82. K. Green, J. Chapman, L. Cheeks, and R. Clayton, Conc. Toxicol., 4, 126–132 (1987).
83. A. R. Gassett, Y. Ishii, H. E. Kaufman, and T. Miller, Am. J. Ophthalmol., 78, 98–105 (1975).
84. R. M. E. Richards, Aust, J. Pharm. Sci., 55, S86, S96 (1967).
85. W. Mullen, W. Shepherd, and J. Labovitz, Surv. Ophthalmol., 17, 469 (1973).
86. J. Am. Coll. Toxicol., 8, 589–625 (1989).
87. W. H. Havener, *Ocular Pharmacology*, C. V. Mosby, St. Louis, MO, 1966.
88. T. O. McDonald, Technical Report, Alcon Laboratories Inc., August 1975.
89. P. S. Binder, D. Rasmussen, and M. Gordon, Arch. Ophthalmol., 99, 87 (1981).
90. A. Tosti and G. Tosti, Contact Dermatitis, 18, 268–273 (1988).
91. E. Shaw, Contact Intraocul. Lens Med. J., 6, 273 (1980).
92. J. Molinari, R. Nash, and D. Badham, Int. Contact Lens Clin., 9, 323 (1982).
93. E. L. Gaul, J. Invest. Dermatol., 31, 91 (1958).
94. F. A. Ellis and H. M. Robinson, Arch., Fermatol. Syphilol., 46, 425 (1941).
95. R. E. Reisman, J. Allergy, 43, 245 (1969).
96. D. MacKeen, Contact Lens J., 7, 14 (1978).

97. R. C. Meyer and L. B. Cohn, J. Pharm. Sci., 67, 1636 (1978).
98. W. R. Baily, Contact Lens Soc. Am. J., 6, 33 (1972).
99. E. M. Salonen, A. Vaheri, T. Tervo, and R. Beuerman, J. Toxicol. Cutan. Ocul. Toxicol., 10, 157–166 (1991).
100. B. J. Tripathi, R. C. Tripathi, and P. K. Susmitha. Lens Eye Toxicol. Res., 9, 361–375 (1992).
101. D. P. Rodeheaver, R. B. Hackett, and J. W. Hiddemen, *Safety of the Ophthalmic Preservative, Polyquad*, in press, 1994.
102. W. M. Grant, *Toxicology of the Eye*, Charles C. Thomas, Springfield, IL, 1974, p. 264.
103. N. L. Burstein, Invest. Ophthalmol. Vis. Sci., 19, 308 (1980).
104. A. M. Tonjum, Acta Ophthalmol., 53, 358 (1975).
105. J. A. Dormans and J. J. Van Logten, Toxicol. Appl. Pharmacol., 62, 251 (1982).
106. A. R. Gassett and Y. Ishii, Can. J. Ophthalmol., 10, 98 (1975).
107. K. Green, V. Livingston, K. Bowman, and D. S. Hull, Arch. Ophthalmol., 19, 1273, (1980).
108. C. G. Begley, P. J. Waggoner, G. S. Hafner, T. Tokarski, R. E. Meetz, and W. H. Wheeler, Opt. Vis. Sci., 68, 189–197 (1991).
109. K. Green, R. E. Johnson, J. M. Chapman, E. Nelson, and L. Cheeks, Lens Eye Toxicol. Res., 6, 37–41 (1989).
110. J. D. Mullins and G. Hecht, Ophthalmic preparations, in *Remington's Pharmaceutical Sciences*, Vol. 18 (A. R. Genaro, Ed.), Mack Publishing, Easton, PA, 1990, pp. 1581–1595.
111. R. A. Moses, *Adler's Physiology of the Eye*, 5th Ed., C. V. Mosby, St. Louis, MO, 1970, p. 49.
112. S. S. Chrai, M. C. Makoid, S. P. Eriksen, and J. R. Robinson, J. Pharm Sci., 63, 333 (1974).
113. D. I. Weiss and R. D. Schaffer, Arch. Ophthalmol., 68, 727 (1962).
114. F. T. Fraunfelder and S. M. Meyer, *Drug-Induced Ocular Side Effects and Drug Interactions*, Lea & Febiger, Philadelphia, (1989), pp. 442–487.
115. B. C. P. Polak, Drugs used in ocular treatment, in *Meyler's Side Effects of Drugs* (M. N. G. Dukes, Ed.), Elsevier, New York, 1988, pp. 988–998.
116. T. F. Patton, in *Ophthalmic Drug Delivery Systems* (J. R. Robinson, Ed.), Academy of Pharmaceutical Sciences, American Pharmaceutical Association, Washington, DC, 1980, pp. 23–54.
117. J. C. Keister, E. R. Cooper, P. J. Missel, J. C. Lang, and D. F. Hager, J. Pharm. Sci., 80, 50 (1991).
118. A. Rozier, C. Mazuel, J. Grove, and B. Plazonnet, Int. J. Pharm., 57, 163 (1989).
119. T. J. Zimmerman, M. Sharir, G. F. Nardin, and M. Fuqua, Am. J. Ophthalmol., 114, 1–13 (1992).
120. A. Durward, The skin and the sensory organs, in *Cunningham's Testbook of Anatomy* (G. J. Romanes, Ed.), Oxford University Press, London, 1964, p. 796.
121. O. A. Candia, Invest. Ophthalmol. Vis. Sci. 33, 2575 (1992).
122. A. J. Huang, S. C. Tseng, and K. R. Kenyon, Invest. Ophthalmol. Vis. Sci. 30, 684 (1989).
123. N. Narawane, Oxidative and hormonal control of horseradish peroxidase transcytosis across the pigmented rabbit conjunctiva, Ph. D. Thesis, University of Southern California (1993).
124. L. E. Stevens, P. J. Missel, and J. C. Lang, Anal. Chem. 64, 715 (1992).
125. R. H. Perry and C. H. Chilton, *Chemical Engineer's Handbook*, 5th Ed., McGraw-Hill, New York, 1973, Sec. 4.
126. J. M. Smith, *Chemical Engineering Kinetics*, 3rd Ed., McGraw-Hill, New York, 1981, Chap. 3.
127. C. G. Hill, *An Introduction to Chemical Engineering Kinetics and Reactor Design*, John Wiley & Sons, New York, 1977.
128. A. K. Mitra, *Ophthalmic Drug Delivery*, Marcel Dekker, New York, 1993.
129. P. Veng-Pedersen and W. R. Gillespie, J. Pharm. Sci., 77, 39 (1988).
130. W. R. Gillespie, P. Veng-Pedersen, E. J. Antal, and J. P. Phillips, J. Pharm. Sci., 77, 48 (1988).
131. E. Hayakawa, D.-S. Chien, K. Inagaki, A. Yamamoto, W. Wang, and V. H. L. Lee, Pharm. Res., 9, 769 (1992).
132. H. F. Edelhauser, J. R. Hoffert, and P. O. Fromm, Invest. Ophth., 4, 290 (1965).
133. D. M. Maurice and S. Mishima, *Ocular Pharmacokinetics* (M. L. Sears, Ed.), Springer-Verlag, Berlin, 1984, pp. 19–116.
134. R. D. Schoenwald and R. L. Ward, J. Pharm. Sci., 67, 786 (1978).
135. R. D. Schoenwald and H. S. Huang, J. Pharm. Sci., 72, 1266 (1983).

136. R. D. Schoenwald, Clin. Pharmacokinet, 18, 255 (1990).

137. D.-S. Chien, J. J. Homsy, C. Gluchowski, and D. D.-S. Tang-liu, Curr. Eye Res., 9 1051 (1990).

138. D. M. Maurice and M. V. Riley, The cornea, in *Biochemistry of the Eye* (C. N. Graymore, Ed.), Academic Press, New York, 1970, Chap. 1.

139. E. R. Berman, *Biochemistry of the Eye*, Plenum Press, New York, 1991.

140. H. E. P. Bazan, private communication.

141. G. L. Flynn, S. H. Yalkowsky, and T. J. Roseman, J. Pharm. Sci, 63, 479 (1974).

142. E. R. Cooper and G. Kasting, J. Controlled Release, 6, 23 (1987).

143. G. Hecht, R. E. Roehrs, E. R. Cooper, J. W. Hiddemen, F. F. Van Duzee, in *Modern Pharmaceutics*, 2nd Ed. (G. S. Banker and C. T. Rhodes, Eds.), Marcel Dekker, New York, 1990, Chap. 14.

144. E. R. Cooper, *Optimization of Transport and Biological Response with Epithelial Barriers in Biological and Synthetic Membranes.* Alan R. Liss, New York, 1989, pp. 249–260.

145. R. W. Baker and H. K. Lonsdale, *Controlled Release: Mechanisms and Rates* (A. C. Tanquary and R. E. Lacy, Eds.), Plenum Press, New York, 1974, pp. 15–71.

146. W. Wang, H. Sasaki, D.-S. Chien, and V. H. Lee, Curr. Eye Res. 10, 57 (1991).

147. P. Ashton, S. K. Podder, and V H. L. Lee, Pharm Res., 8, 1166 (1991).

148. J. Liaw and J. R. Robinson, Ocular penetration enhancers, in *Ocular Drug Delivery Systems* (A. K. Mitra Ed.), Marcel Dekker, 1993, pp. 369–381.

149. Y. Rojanaskul, J. Liaw, and J. R. Robinson, Int. J. Pharm., 66, 133 (1990).

150. Clean Room and Work Station Requirements, Controlled Environment, Sec. 1-5 Federal Standard 209, Office of Technical Services, U. S. Department of Commerce, Washington, DC, Dec. 16, 1963.

151. P. R. Austin and S. W. Timmerman, *Design and Operation of Clean Rooms*, Business News Publishers, Detroit, MI, 1965.

152. P. R. Austin, *Clean Rooms of the World.* Ann Arbor Science Publishers, Ann Arbor, MI, 1967.

153. K. R. Goodard, Air filtration of microbial particles, Publication 953, U. S. Public Health Service, Washington, DC, 1967.

154. K. R. Goddard, Bull. Parenter, Drug Assoc., 23, 699 (1969).

155. *USP*, U. S. Pharmacopeial Convention, Rockville, MD, 1990, p. 1456.

156. *USP*, U. S. Pharmacopeial Convention, Rockville, MD, 1990, p. 1457.

157. Federal Register, 41, 106, June 1, 1976.

158. T. L. Grimes, D. E. Fonner, J. C. Griffin, and L. R. Rathburn, Bull. Parenter. Drug Assoc., 29, 64 (1975).

159. E-3A Accepted Practices for Permanently Installed Sanitary Product Pipeline and Cleaning Systems, Serial E-60500, U. S. Public Health Service, Washington, DC.

160. D. E. Cadwallader, Am. J. Hosp. Pharm., 24, 33 (1967).

161. F. G. Kronfeld and J. E. McDonald, J. Am. Pharm. Assoc. (Sci. Ed.), 42, 333 (1951).

162. S. Riegelman and D. G. Vaughn, J. Am. Pharm. Assoc. (Pract. Pharm. Ed.), 19, 474 (1958).

163. S. M. Blaugh and A. T. Canada, Am. J. Hosp. Pharm., 22, 662 (1965).

164. M. L. Linn and L. T. Jones, Am J. Ophthalmol., 65, 76 (1968).

165. S S. Chrai and J. R. Robinson, J. Pharm. Sci., 63, 1218 (1974).

166. C. A. Adler, D. M. Maurice, and M. E. Patterson, Exp. Eye Res., 11, 34 (1971).

167. R. Castroviejo, Arch. Ophthalmol., 74, 143 (1965).

168 C. Hanna, F. T. Fraunfelder, M. Cable, and R. E. Hardberger, Am. J. Ophthalmol., 76, 193 (1973).

169. F. T. Fraunfelder, C. Hanna, M. Cable, and R. E. Hardberger, Am J. Ophthalmol., 76, 475 (1973).

170. R. Mouly, Ann. Chir. Plast., 17, 61 (1972).

171. D. W. Newton, C. H. Becker, and G. Torosian, J. Pharm. Sci., 62, 1538 (1973).

172. R. L. Schoenwald, et al., J. Pharm. Sci., 67, 1280 (1978).

173. W. F. March, et al., Arch. Ophthalmol., 100, 1270 (1982).

174. H. M. Liebowitz, et al., Ophthalmology, 91, 1199 (1984).

175. F. W. Bowman, E. W. Knoll, M. White, and P. Mislivic, J. Pharm. Sci., 61, 532 (1972).

176. T. W. Schwartz, Am. Perfum. Cosmet., 86, 39 (1971).

177. A. Zaffaroni, *Proc. 31st International Congress on Pharmaceutical Science*, Washington, DC, 1971.

178. J. W. Shell and R. W. Baker, Ann. Ophthalmol., 7, 1037 (1975).

179. S. Lerman and B. Reininger, Can. J. Ophthalmol., 6, 14 (1971).
180. Y. F. Maichuk, Invest. Ophthalmol., 14, 87 (1975).
181. S. P. Loucas and H. M. Haddad, J. Pharm. Sci., 61, 985 (1972).
182. J. Hiller and R. W. Baker (To Alza Corporation.), U. S. Patent 3,811,444, 1974.
183. A. Michaels (To Alza Corporation), U. S. Patent 3,867,519, 1975.
184. N. Keller, A. M. Longwell, and S. A. Biros, Arch. Ophthalmol., 94, 644 (1976).
185. H. F. Edelhauser, Arch. Ophthalmol., 93, 649 (1975).
186. W. H. Havener, *Ocular Pharmacology*, 2nd Ed., C. V. Mosby, St. Louis, MO, 1970, p. 27.
187. J. W. Shell and R. W. Baker, Ann. Ophthalmol., 7, 1637 (1975).
188. W. M. Grant, *Toxicology of the Eye*, Charles C. Thomas, Springfield, IL, 1974, p. 259.
189. B. E. McCarey, H. F. Edelhauser, and D. L. Van Horn, Invest. Ophthalmol., 12, 410 (1973).
190. D. L. Merrill, T. C. Fleming, and L. J. Girard, Am. J. Ophthalmol., 49, 895 (1960).
191. L. J. Girard, *Proceedings International Congress on Ophthalmology*, Brussels, Sept. 1958.
192. H. F. Edelhauser, D. L. Van Horn, R. W. Scholtz, and R. A. Hyndiuk, Am J. Ophthalmol., 81, 473 (1976).
193. S. E. Herrell and D. Heilman, Am. J. Med. Sci., 206, 221 (1943).
194. *USP*, U. S. Pharmacopeial Convention, Rockville, MD, 1990, p. 1693.
195. N. S. Jaffee, Bull. Parenter. Drug Assoc., 24, 218 (1970).
196. N. J. Baily, Contact Lens Spectrum, 2(7), 6–31 (1987).
197. K. J. Randeri, R. P. Quintana, and M. A. Chowhan, Lens care products in *Encyclopedia of Pharmaceutical Technology*, Vol. 8 (James Swarbrick and James C. Boylan, eds.), Marcel Dekker, New York, 1993, pp. 361–402.
198. P. R. Kastl., M. J.. Refojo, and O. H. Dabezies, Jr., Review of polymerization for the contact lens fitter, in *Contact Lenses: The CLAO Guide to Basic Science and Clinical Practice*, 2nd Ed., Vol. 1 (O. H. Dabizies, Jr., ed.), Little, Brown & Co., Boston, 1989, pp. 6.21–6.24.
199. K. A. Polse and R. B. Mandell, Arch Ophthalmol., 84, 505–508 (1970).
200. P. S. Binder, Ophthalmology, 86, 1093 (1978).
201. I. Fatt and R. M. Hill, Am. J. Optom. Arch. Am. Acad. Optom., 47, 50 (1970).
202. I. Fatt, Contacto, 23(1), 6 (1979).
203. Food and Drug Administration, *Guidance Document for Class III Contact Lenses*, United States Food and Drug Administration, Silver Springs, MD, 1989.
204. N. J. Van Haeringen, Clinical biochemistry of tear, Surv. Ophthalmol., 26(2), 84–96 (1981).
205. E. J. Castillo, J. L. Koenintg, and J. M. Anderson, Biomaterials, 7, 89–96 (1986).
206. M. Ruben, Br. J. Ophthalmol., 59, 141 (1975).
207. D. E. Hart, Int. Contact Lens Clin. 11, 358–360 (1984).
208. C. G. Begley and P. J. Waggoner, J. Am. Optom. Assoc., 62, 208–214 (1991).
209. R. C. Tripathi, B. J. Tripathi, and C. B. Millard, CLAO J., 14, 23–32 (1988).
210. M. J. Miller, L. A., Wilson, and D. G. Ahrean, J. Clin. Microbiol., 16, 513–517 (1988).
211. M. Chowhan, T. Bilbault, R. P. Quintana, and R. A. Rosenthal, Contactologia, 15, 190–195 (1993).
213. R. Jacob, Int. Contact Lens Clin., 15, 317–325 (1988).
214. D. Holsky, J. Am. Optom. Assoc., 55, 205–211 (1993).
215. M. M. Hom and M. Pickford, Int. Eyecare, 2, 325–326 (1986).
216. R. L. Davis, Int. Contact Lens Clin., 10, 277–284 (1983).
217. C. G. Begley, S. Paraguia, and C. Sporm, An analysis of contact lens enzyme cleaner, J. Am. Optom. Asssoc., 61, 190–193 (1990).
218. N. Tarrantino, R. C. Courtney, L. A. Lesswell, D. Keno, and I. Frank, Int. Contact Lens Clin., 15, 25–32 (1988).
219. R. D. Houlsby, M. Ghajar, and G. Chavez, J. Am. Optom. Assoc., 59, 184–188 (1988).
220. C. B. Anger, K. Ambrus, J. Stocker, S. Kapadia, and L. Thomas, Spectrum, 9, 46–51 (1990).
221. Food and Drug Administration, *Draft Testing Guidelines for Class III Soft (Hydrophilic) Contact Lens Solutions*, U. S. Food and Drug Administration, Silver Springs, MD, 1985.
222. N. A. Brennan and N. Efron, Optom. Vis. Sci., 66, 834–838 (1989).
223. N. Efron, T. R. Goldwig, and N. A. Brennan, CLAO J., 17, 114–119 (1991).

14

Pharmaceutical Aerosols

John J. Sciarra
Sciarra Laboratories, Inc., Hicksville, New York

I. INTRODUCTION

A. History

The first modern-day pressurized aerosol dosage form was developed in the early 1950s and was introduced as Medihaler Epi by Riker Laboratories (now 3M Pharmaceuticals). These products were formulated as a solution and as a suspension. In solution form, the products consisted of epinephrine (as the hydrochloride) dissolved in hydroalcohol and a mixture of chlorofluorocarbon (CFCs) propellants, consisting of dichlorodifluoromethane (Propellant 12) and dichlorotetrafluoroethane (Propellant 114). A suspension system was formulated with epinephrine bitartrate, suspended with the aid of a dispersing agent, such as sorbitan trioleate, in a mixture of dichlorofluoromethane, dichlorotetrafluoroethane, and trichloromonofluoromethane (Propellant 12/114/11). These products were packaged in small plastic-coated glass bottles and stainless steel containers (30 ml or less), respectively, and fitted with a metered-dose valve that dispensed about 50 μl of total product. The dose of released drug, solvent, or dispersing agent, and propellant would then be inhaled through a specially designed mouthpiece (oral adapter) by the patient.

From these beginnings, one has seen the expansion of this concept to include other drugs, such as isoproterenol hydrochloride or sulfate, albuterol, metaproterenol sulfate, isoetharine mesylate or hydrochloride, ipratropium bromide, terbutaline sulfate, pirbuterol acetate, salmeterol xinafoate and bitolterol mesylate. Steroids, including flunisolide, triamcinolone acetonide, dexamethasone sodium phosphate, and beclomethasone dipropionate have also been delivered by means of a pressurized aerosol system. Cromolyn sodium and nedocromil sodium have also been used. These products have collectively been referred to as "metered-dose inhalers" (MDIs) and are intended for oral inhalation whereby the particles of drug can be deposited in the pulmonary airways.

Pressurized aerosol systems have also been developed for absorption through the mucous membranes of the oral cavity (sublingual or buccal). One such product includes nitroglycerin as the active ingredient and Propellants 12/114 and is intended for administration onto or under the tongue. Additional aerosol products have been developed for administration intranasally and have been especially useful in the treatment of allergic rhinitis. These drugs include dexamethasone sodium phosphate, beclomethasone dipropionate, budesonide, and triamcinolone acetonide, and all have been formulated as a suspension system.

The aerosol dosage form has also been used for topical pharmaceutical applications. Before 1980, these aerosols utilized a CFC, hydrocarbon, or compressed gas as the propellant, and over the years have been used for a variety of different applications, resulting in an increased demand for these products. Ranging from topical anti-infective sprays, dermatological sprays or foams, to local anesthetics, and rectal and vaginal foams, the topical aerosol dosage form has taken its rightful place alongside ointments, creams, lotions, solutions, and the rest of the dosage forms. These aerosol products have been formulated and dispensed in a variety of ways, including wet sprays, dry sprays, foams, powders, creams, and ointments.

According to the Clean Air Act of 1990 as well as regulations of the Food and Drug Administration (FDA) and Environmental Protection Agency (EPA), the use of CFCs in nonessential products (except for certain exempted products, which include metered-dose inhalers and vaginal foams) has been prohibited. All topical pharmaceuticals have been reformulated using alternative acceptable propellants. This will be discussed in greater detail in a later section of this chapter.

B. Definitions

Aerosols are those products that depend on the power of a compressed or liquefied gas to expel the contents from the container. Terms such as "pressure pak" and "pressurized package" are also used to describe this type of product. The product can be dispensed as a fine or wet spray, a foam, or a semisolid stream. Other systems classify aerosols according to the particle size of the emitted spray. Space sprays generally produce particles no larger than 50 μm, whereas those larger than this are classified as wet or coarse sprays.

All of the existing aerosol products can be described adequately by the foregoing definition. Pharmaceutical aerosols are included in the general definition; however, since the principles and considerations involved in the formulation of aerosol products used for a therapeutic response are quite different from other aerosols, it is advantageous from a discussion as well as a formulation viewpoint to distinguish between these types of aerosol products.

Pharmaceutical aerosols may be defined as aerosol products containing therapeutically active ingredients dissolved, suspended, or emulsified in a propellant or a mixture of solvent and propellant, and intended for topical administration; for administration into one of the body cavities, such as the ear, rectum, and vagina; or intended for administration orally or nasally as fine solid particles or liquid mists through the pulmonary airways, nasal passages, or oral cavity (buccal or sublingual).

C. Advantages of the Aerosol Dosage Form

General

1. The product is more convenient to use, since it is in a compact unit and can be applied or administered easily and quickly. No other ancillary equipment is needed, such as applicators, brush, atomizer, or nebulizer.
2. There is no danger of contamination of the product from outside foreign matter. Moisture, bacteria, and so on, generally cannot enter the container as long as adequate pressure is maintained within the container. The aerosol dosage form is essentially

"tamper-proof," a most important consideration as a result of the problems associated with the deliberate tampering of capsules and other products.

3. Through use of metered valves, doses of drugs can be delivered accurately, effectively, and efficiently.

Metered Dose Aerosols (Oral, Inhalation, Nasal)

1. Drugs generally administered parenterally may be given by inhalation or intranasally. These drugs include vaccines, antiviral compounds, hormones, such as insulin, and other medicinal agents. Not only is this method of administration more acceptable to the patient, but the need for elaborate manufacturing precautions to maintain sterility are no longer required.
2. There generally is a rapid onset of action, avoidance of degradation of the drug in the gastrointestinal tract, and circumvention of the first-pass effect.
3. Generally, a lower dosage of drug can be used that will minimize adverse and side effects.
4. Through use of metered valves, the dose can be tailored to suit individual needs and is desirable for use with drugs that are prescribed on a "use when necessary" basis.
5. Inhalation, nasal, or sublingual route of administration can be used as an alternative route when the therapeutic agent may chemically or physically interact with other medicinals needed concurrently.
6. The aerosol dosage form is considered a feasible alternative when the drug entity exhibits erratic pharmacokinetic behavior with oral or parenteral administration.

Topical Aerosols (Dermal, Vaginal, and Rectal)

1. Sprays and foams applied topically to the skin reduce or eliminate the irritation brought about by mechanical application of medication to abraded areas of the skin.
2. The aerosol dosage form is generally considered to be more efficient, as there is no waste or messiness such as accompanies the use of cotton swabs, applicators, and so on.
3. For rectal and vaginal aerosols, the product can be delivered as an expanding foam to ensure direct contact of the drug with the rectal or vaginal mucosa.
4. Through use of metered-dose valves, an accurately measured amount of product can be dispensed to the vagina or rectum.
5. Disposal applicators can be used to ensure cleanliness and lack of contamination of the unused medication.

Although almost any drug can be formulated for topical administration, those drugs for use orally or nasally must possess the following characteristics: (a) be nonirritating to the respiratory airways or the nasal mucosa; (b) be reasonably soluble in respiratory and nasal fluids (drugs having poor solubility characteristics are more likely to be irritating); (c) be therapeutically effective at a relatively low dose; (d) should exhibit passive transport (absorption) through respiratory membranes (carrier and active transport are possible, but this area is not yet clearly defined); (e) have minimal local or topical nasal mucosa activity (unless specifically used for this purpose); and (f) be stable and compatible in intranasal vehicles and have a pH between 5.5 and 7.5.

II. OPERATION OF THE AEROSOL PACKAGE

The mode of operation of an aerosol package is dependent on the type of propellant used. Aerosols are different from other products in that they are packaged under pressure; that is, there is energy in the container to push the contents out. This can be accomplished in many different ways. If all that is desired is to push the material out of the container, energy can be supplied in the form of a liquefied or compressed gas. If the product characteristics are to change on dispensing, additional energy in the form of a liquefied gas or mechanical breakup systems are required.

All aerosol products consist of a volatile and nonvolatile portion. The volatile portion consists of pure propellant or a mixture of pure propellant and volatile solvents, such as ethanol. The nonvolatile portion is generally the active ingredient, solvents, and dispersing agents.

An aerosol system consists of a product concentrate, propellant, container, and suitable valve. The concentrate can be of the solution, dispersion, emulsion, or semisolid type, while the propellant can be either a liquefied or a compressed gas. Depending on the nature of the propellant and product concentrate, as well as the combination of these two components with the appropriate valve and actuator, the product can be dispensed as a spray, foam, or semisolid.

There are various ways in which the aerosol components can be brought together to formulate a finished pharmaceutical aerosol product. The product concentrate, which may be a solid or a liquid, can be dissolved or mixed with the propellant so that a true solution is formed. A cosolvent may be added if the propellant and concentrate are immiscible, although this is not always possible. The concentrate can also contain insoluble solids that can be dispersed throughout the propellant. It may also be possible to disperse the propellant throughout the powder. If the propellant and concentrate phases are immiscible, it may also be possible to bring them together by means of an emulsifying agent or surfactant, resulting in the formulation of an emulsion. This emulsion can be dispensed as a foam or as a spray, depending on the nature of each phase and the design of the valve and actuator. Additional systems that serve simply to push the contents out of the container or else to separate the propellant from the concentrate have also been used for pharmaceutical aerosols.

A. Liquefied Gas Systems

Liquefied gases are commonly used to supply the energy required for proper functioning of the aerosol container. Liquefied gases, as the name indicates, are materials that at room temperature and atmospheric pressure exist in the gaseous or vapor state and are capable of being liquefied at relatively low pressures or temperatures. Although most gases can be liquefied, only those that can be liquefied at a relatively low pressure or at a temperature close to room temperature are useful as aerosol propellants. Chlorofluorocarbon (CFCs) and hydrocarbons liquefy at relatively low pressures (15–80 psig) and are present in the liquid state. They function as aerosol propellants in that, because of their relatively low vapor pressure, they can be maintained in the liquefied state in a fairly inexpensive, thin-walled metal container that is capable of withstanding the pressures normally used for aerosol products.

Depending on whether the product concentrate is dissolved, suspended, or emulsified with the propellant, the final product will be discharged as a spray or foam. The same material, formulated differently, can be discharged in one case as a spray and in another as a stream or foam. The final form is dependent on the product concentrate and the valve. Both CFCs and hydrocarbons function in a similar manner, but there is a difference in the properties of the two groups. Such characteristics as density, solubility, pressure, heat exchange, and flammability may limit the use of some of them. As a result of the ban on the use of CFCs, other liquefied gases are available for use and will be discussed in a later portion of this chapter.

B. Compressed Gas Systems

An aerosol utilizing a compressed gas as the propellant operates essentially as a pressure package in that the pressure of the gas forces the product from the container in its original state. Under limited conditions, a spray or foam-type product may be achieved. A compressed gas propellant is essentially a gas that has been placed in a container under pressure, for example, nitrogen at 90 psig. Because of the difference between atmospheric pressure and 90 psig, as soon as the valve is opened the gas forces the contents out the valve. The gas can be nitrogen, nitrous oxide, or carbon dioxide. Only the product is discharged; the gas remains behind, occupying the head space. The gas will then expand because of the increase in volume. As it expands, the pressure will drop from the original level of 90 psig to 80, 70, or 60 psig. The drop in pressure is related to the amount of material discharged and may be calculated from the gas laws. Since there is no propellant in the liquid phase (unless some solubility is present), the pressure decreases as the contents are used.

C. Barrier Packs

Although barrier packs have been available in the past, their use has increased since they can effectively use some of the environmentally acceptable propellants. Sprays, foams, and semisolids can be employed in this system. Since the system is based on the separation of propellant and product, the propellant serves only to push the contents out of the container. These systems consist essentially of an inner bag (barrier) placed within a regular container (generally metal). The product is sealed within the inner bag. The propellant is added to the outer can and occupies the space between the inner bag and outer container. This propellant provides the necessary pressure to expel the product in the desired form.

III. AEROSOL PACKAGING COMPONENTS

A. Propellants

The propellant supplies the push needed to expel the contents from the container in the desired form. A *propellant* has been defined as a "liquefied gas with a vapor pressure greater than atmospheric pressure at a temperature of 40°C (105°F)." The propellant is said to be the "heart" of the aerosol and serves to supply the power or push to expel the product and to deliver it in the proper form. The propellant, together with the valve, must deliver the product to the site of action in a form in which it can be used. The propellant actually performs the work normally performed by the person using a nonaerosol product. There are several chemical compounds used as aerosol propellants. Of greatest interest for oral, inhalation, and nasal aerosols is the group of halogenated hydrocarbons derived from methane and ethane, such as dichlorodifluoromethane (Propellant 12), trichloromonofluoromethane (Propellant 11), and dichlorotetrafluoroethane (Propellant 114). These compounds are relatively nontoxic, inert, and are nonflammable. They are nonpolar and are miscible with most nonpolar solvents. For topical aerosols, the hydrocarbons—butane, isobutane, and pentane—are most useful. This has come about because of the ban placed on the use of CFCs, which are believed to destroy the "ozone" in the atmosphere, resulting in an increase in the incidence of skin cancer, which is largely caused by the increased exposure of the body to the ultraviolet rays of the sun.

The rules and regulations promulgated by the Environmental Protection Agency, Consumer Product Safety Commission, Food and Drug Administration and recently, the Clean Air Act of 1990, relative to the use of fluorocarbons as propellants has resulted in the need to reformulate all topical aerosols, with only a few exceptions. Only aerosols for inhalation, oral and nasal use, and contraceptive foams have been exempted. All aerosol products manufactured or packaged on or after December 15, 1978, must comply with the regulations. To develop products that would comply with these regulations, a newer technology was required that reexamined propellants, valves, and containers, as well as other means of dispensing products as a spray. Hydrocarbons have now replaced the CFCs, except for those used orally, nasally, or in certain topical pharmaceutical aerosols.

Chlorofluorocarbons

Chlorofluorocarbons are the propellant of choice for oral, inhalation, and nasal aerosols. Table 1 indicates selected properties of those fluorocarbons that are used in the formulation of pharmaceutical aerosols. The most important and commonly used propellants for oral, inhalation, and nasal aerosols are of the fluorinated, chlorinated hydrocarbon type and include trichloromonofluoromethane (P-11), dichlorodifluoromethane (P-12), and dichlorotetrafluoroethane (P-114). These three propellants represent the bulk of propellants used in MDI and nasal aerosol formulations. Because of their relatively low-order toxicity, nonflammability, and nonreactive properties, they have been readily accepted for use as propellants.

Nomenclature. A numbering system to identify specific compounds has been developed so that the chemical structure of the compound can be determined from the number. The system consists of three digits: when only two are present, the first digit from the right is the number of fluorine (F) atoms; the second digit from the right is one *more* than the number of hydrogen (H) atoms; while the third digit from the right is one *less* than the number of carbon (C) atoms present. The number of chlorine (Cl) atoms in the compound is obtained by subtracting the sum of the fluorine and hydrogen atoms from the total number of atoms required to saturate the compound (e.g., four for the methane series and six for the ethane series).

Several examples follow:

Propellant 11 consists of one atom of fluorine, no hydrogen, and is a one-carbon chain; therefore, it must contain three atoms of chlorine, and its formula is $CFCl_3$.

Propellant 114 comprises

Two carbon atoms
Four flourine atoms
No hydrogen atoms
Two chlorine atoms

The formula would be $CClF_2CClF_2$.

When isomers are possible, the most symmetrical compound is given the designated number, and all other isomers are assigned a letter, such as a, b, c, . . . , in descending order of symmetry.

Propellant 134a comprises

Two carbon atoms
Four fluorine atoms
Two hydrogen atoms
Two chlorine atoms

The formula would be CF_3CH_2F and *not* $CF_2CH_2F_2$ (symmetrical and known as Propellant 134).

Physical Properties. The CFCs exist as liquids at relatively low temperature or at high pressure and are known as liquefied gases. Their boiling point ranges from about 24°C for Propellant 11 to about −30°C for Propellant 12. Related to the boiling point is the vapor pressure of the propellant. *Vapor pressure* is defined as the pressure that exists when there is an equilibrium between the molecules that exist in the vapor and liquid states. Vapor pressure is dependent on temperature and is independent of quantity. That is, the vapor pressure of a pure material is the same for 1 g or 1 ton of the compound. The vapor pressure ranges from about 13.4 psia for Propellant 11 to about 85 psia for Propellant 12. Vapor pressures between these values may be obtained by blending Propellant 11 with Propellant 12 and Propellant 12

Table 1 Properties of Chlorofluorocarbon Propellants[a]

Propellant	Chemical name	Boiling point (°F)	Vapor pressure at 70°F (psig)	Liquid density at 70°F (g/ml)	Water solubility at 70°F (% w/w)
11	Trichloromonofluoromethane	74.8	−1.3	1.485	0.009
12	Dichlorodifluoromethane	−21.6	70.3	1.325	0.008
114	Dichlorotetrafluoroethane	38.8	12.9	1.468	0.007

[a] Available as Dymel from E.I. du Pont de Nemours and Company, Inc., and as Genetron from Allied Signal Corporation.

with Propellant 114, and Propellant 12, 114, and 11. The vapor pressure of a mixture of propellants can be determined using Raoult's law, which states that the vapor pressure of a system consisting of two or more components is equal to the sum of the mole fractions of each component multiplied by the vapor pressure of the pure compound. That is,

$$p_a = \frac{n_a}{n_a + n_b} P_a^0$$

$$p_b = \frac{n_b}{n_a + n_b} P_b^0$$

where p_a and p_b are the partial pressures of components a and b, respectfully, n_a and n_b are the mole fraction of components a and b, and P_a^0 and P_b^0 are the vapor pressure of the pure compound. The total vapor pressure of the system is then obtained by adding p_a and p_b.

Another important property of the propellants is their density. They are heavier than water and have a density of approximately 1.2–1.3 g/ml. Density is important from a formulation viewpoint when one considers the relationship between the weight and volume of the propellant.

Chemical Properties. Chemically, the propellants are inert, but they may be subject to hydrolysis (Propellant 11). In this case, Propellant 11 in the presence of water will form hydrochloric acid, which will increase the danger of corrosion to the container, and for topical products such dermatological foams may be irritating to the skin. When water is present in the formulation, Propellant 12 or a mixture of Propellants 12 and 114 is used. It should be emphasized that these propellants cannot be used for topical aerosols unless the product is essential, cannot be made without CFCs, and is specifically exempted for use with the product.

Hydrocarbons

Nomenclature. The hydrocarbons are designated by their vapor pressure, as illustrated in Table 2. Blends of hydrocarbons will produce different vapor pressures, depending upon the vapor pressure of each component. For example, a commonly used blend for aerosol foams is known as A-46. A typical blend consists of propane (19.7%), isobutane (77.3%), and n-butane (3.0%).

Physical and Chemical Properties. Hydrocarbons are used in topical pharmaceutical aerosols. They are preferred for use as a propellant because of their environmental acceptance and their lower cost. However, they are flammable and explosive. Propane, butane, and isobutane are generally used for this purpose. Isobutane is used alone or in combination with propane. These are virtually nontoxic and nonreactive. Since they do not contain any halogens, hydrolysis does not take place, making these propellants useful for water-based aerosols.

Table 2 Physiochemical Properties of Hydrocarbons

Property	Propane	Isobutane	Butane	Propane/isobutane
Designation	A-108	A-31	A-17	A-46
Vapor pressure at 70°F (psig)	108	31	17	46
Boiling point (°F)	−46	9	28	
Density (g/ml)	0.509	0.564	0.585	0.556
Flash point (°F)	−156	−120	−100	

Table 2 illustrates some of the physicochemical properties of these hydrocarbons. They can be blended with one another and with other liquefied gases to obtain the desired vapor pressure, density, and decreased flammability. They can be blended with Propellant 22 to produce a nonflammable product or one with less flammability than the hydrocarbon propellant alone. Propellants 142b and 152a can also be used to reduce the flammability of the overall propellant blend and of the product. By using the appropriate aerosol valve with a vapor tap, the flame extension of the product can be reduced substantially and, in many cases, it may be possible to classify the product as ''nonflammable.'' This can be noted from Table 3. Figure 1 illustrates the vapor pressure of the hydrocarbon propellants.

Hydrochlorofluorocarbons and Hydrofluorocarbons

Nomenclature. Propellants in this category are designated as indicated under CFCs. They differ from CFCs in that they may not contain chlorine and have one or more hydrogen atoms. These compounds break down in the atmosphere at a faster rate than the CFCs, resulting in a lower ozone-destroying effect. Although they currently are environmentally acceptable, some of them are scheduled for phase-out sometime during the second decade of the 21st century.

Physical and Chemical Properties. Table 4 indicates some of the essential properties that make these compounds useful as alternative propellants for topical pharmaceuticals; Fig. 2 illustrates their vapor pressures. Dimethyl ether has also been suggested for use with topical aerosols. Although it is a good solvent, its use is limited because of its flammability. It also shows a greater solubility in water than the other propellants, as noted in Table 4.

Although Propellants 142b and 152a have been available for many years, they have not been used to any great extent, even though they do possess desirable properties. They are now being used for topical aerosols. The fact that they have greater miscibility with water than the other propellants and possess greater solvent power for many drugs makes them likely candidates for pharmaceutical aerosols. Although they are slightly more flammable than the other propellants (except Propellant 22), the extent of increased flammability is not considered to be a disadvantage.

Table 3 Flammability of Propellants

Propellant[a]	Status
Propellant 22	Nonflammable
Dimethyl ether and blends with 22, 142b, 152a, and hydrocarbons	Flammable
Propellant 142b and blends with 22, 152a, and hydrocarbons	Flammable
Propellant 152a, and blends with 22, 142b, and hydrocarbons	Flammable
Hydrocarbons and blends with 22, 142a, and 152b	Flammable
Propellant 11	Nonflammable
Propellant 12	Nonflammable
Propellant 114	Nonflammable

[a]Propellants 142b and 152a are sometimes referred to as 142 and 152 without the ''letter'' designation.

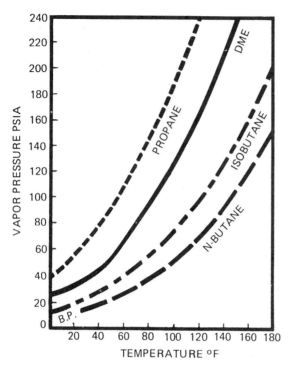

Fig. 1   Effect of temperature on vapor pressure of hydrocarbon propellants.

Compressed Gases

Compressed gases are used as aerosol propellants, since they are inexpensive, nontoxic, and nonreactive. Nitrogen is used, primarily with barrier packs, as the propellant that pushes the product out of the container. It has also been used to produce sprays. Further uses will be indicated under the formulation section of this chapter.

IV.   VALVES AND ACTUATORS

The aerosol valve must be capable of delivering the product in the desired form. An aerosol valve must be multifunctional, and in addition to being able to be easily turned on and off it must change the product into the proper form. For example, a liquid preparation desired for inhalation therapy must be delivered as a fine spray of given particle size, whereas an emulsion developed for topical application must be dispensed as a foam. As such, aerosol valves are available for dispensing aerosol products as sprays, foams, and semisolid streams. Of importance to pharmaceutical aerosols is control of dosage. Metered valves have been developed that are capable of dispensing given quantities of medication.

A.   Spray Valves

Spray valves have been used for a large number of aerosol pharmaceutical products. Depending on the formulation and the design of the valve and actuator, the particle size of the emitted spray can be varied. Fine-to-coarse wet sprays can be obtained, including foams. A valve consists of several subcomponent parts and a series of small orifices. The aerosol solution will

Table 4 Properties of Hydrofluorocarbons, Hydrochlorofluorocarbons, and Dimethyl Ether Propellants

Propellant	Chemical name	Boiling point (°F)	Vapor pressure at 70°F (psig)	Liquid density at 70°F (g/ml)	Water solubility at 70°F (% w/w)
134a	Trifluoromonofluoroethane[a]	−15.0	71.1	1.21	
227	Heptafluoropropane[c]	1.4	43.0[b]	1.41	
	Dimethyl ether[a]	−13.0	63.0	0.66	34.0
142b	Monochlorodifluoroethane[a]	14.4	29.1	1.119	0.054
152a	Difluoroethane[a]	−12.5	61.7	0.911	0.17
22	Monochlorodifluoromethane[a]	−41.4	121.4	1.209	0.11

[a] Available as Freon or Dymel from E.I. du Pont de Nemours and Company, Inc., and as Genetron from Allied Signal Corporation.
[b] At 68°F.
[c] Available from Great Lakes Chemical Corporation and Hoechst-Celanese Corporation.

pass through these orifices, which open into chambers allowing for expansion of the product and subsequent dispersion into the proper particle size.

With the introduction of the ''vapor tap'' valve a greater degree of freedom was possible in formulating aerosols. This valve is basically a standard valve, except that a small hole has been placed into the valve housing. This allows the escape of vaporized propellant along with the dispensed product and produces a spray with a greater degree of dispersion. This is used primarily with aqueous and hydroalcoholic products, which contain large amounts of water and less than 50% propellant. In addition, the vapor tap allows use of the product in either the upright or upside-down position, since the product can now be dispensed through this ''vapor tap'' orifice as well as through the dip tube.

B. Metered Valves

Metered valves fitted with a 20-mm ferrule are used with aluminum and glass containers for all oral, inhalation, and nasal aerosols. The metered valve should deliver a measured amount of product accurately, and the amount should be reproducible not only for each dose delivered from the same package, but from package to package. Two basic types of metered valves are available, one for inverted use and the other for upright use. Generally, valves for upright use contain a thin capillary dip tube, whereas those for inverted use do not contain a dip tube. Figures 3 and 4 illustrate both types of valves and are typical of those commercially available.

An integral part of these valves is the metering chamber, which is directly responsible for the delivery of the desired amount of therapeutic agent. The size of the metering can be varied

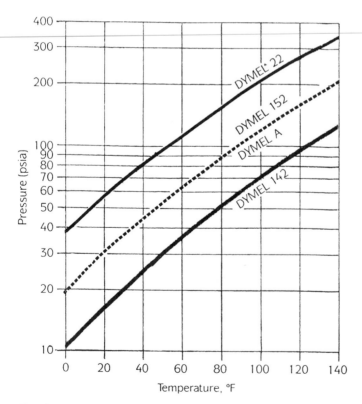

Fig. 2 Vapor pressure of Dymel propellants.

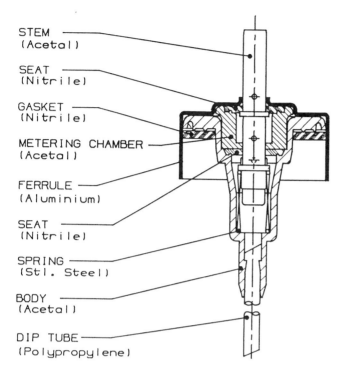

STEM
(Acetal)

SEAT
(Nitrile)

GASKET
(Nitrile)

METERING CHAMBER
(Acetal)

FERRULE
(Aluminium)

SEAT
(Nitrile)

SPRING
(Stl. Steel)

BODY
(Acetal)

DIP TUBE
(Polypropylene)

Fig. 3 Metering valve—upright. (Courtesy of Bespak Inc.)

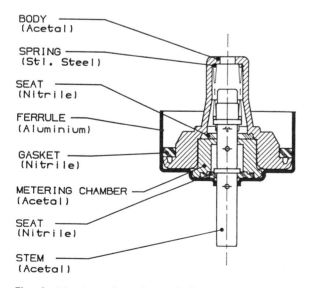

BODY
(Acetal)

SPRING
(Stl. Steel)

SEAT
(Nitrile)

FERRULE
(Aluminium)

GASKET
(Nitrile)

METERING CHAMBER
(Acetal)

SEAT
(Nitrile)

STEM
(Acetal)

Fig. 4 Metering valve—inverted. (Courtesy of Bespak Inc.)

so that from about 25 to 150 μl of product can be delivered per actuation. Most of the products now commercially available use dosages in the range of 50–75 μl. The chamber is sealed by the metering gasket and the stem gasket. In the actuated position, the stem gasket will allow the contents of the metering chamber to be dispensed, while the metering gasket will seal off any additional product from entering the chamber. In this manner, the chamber is always filled and ready to deliver the desired amount of therapeutic agent.

These valves should retain their prime over fairly long periods. However, it is possible for the material in the chamber to return slowly to the main body of the product in the event the container is stored upright (for those used in the inverted position). The degree to which this can occur varies with the construction of the valve and the length of time between actuations of the valves.

To overcome "loss of prime," some valves have been fitted with a drain tank, as seen in Fig. 5. A recently available valve allows retention of the drug in the metering chamber for substantially longer periods than the conventional capillary action/surface tension retention. This is due to the incorporation of a siphon feature seen in Fig. 6.

Both types of valves are currently used in the commercially available oral inhalation aerosols. During the developmental stage, the compatibility of the valves should be determined with the exact formulation to be used so that the accuracy of the metered dose developed relative to doses delivered from the same container of product and from different containers can be determined. There should be no interaction between the various valve subcomponents and the formulation. If distortion or elongation of some of the plastic subcomponents occurs, this may result in leakage, inaccurate dosage, or decomposition of the active ingredients.

There have also been instances whereby the therapeutic agent was adsorbed onto the various plastic components, resulting in a low dosage of therapeutic agent. Therefore, one must determine not only the total weight of the product dispensed per dose, but the actual amount of active ingredient in each dose. Some test procedures use the results obtained by taking ten doses of material and determining the amount of active ingredient present, whereas others determine the amount present in one dose. When possible and when the analytical procedure is capable of detecting these fairly small amounts of active ingredient present per dose, the

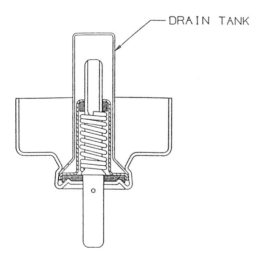

Fig. 5   Cutaway diagram of a metered-dose aerosol valve (a) with and (b) without drain tank shown with stem down, or firing orientation. (Courtesy of Riker Laboratories, Inc./3M, Minneapolis, MN.)

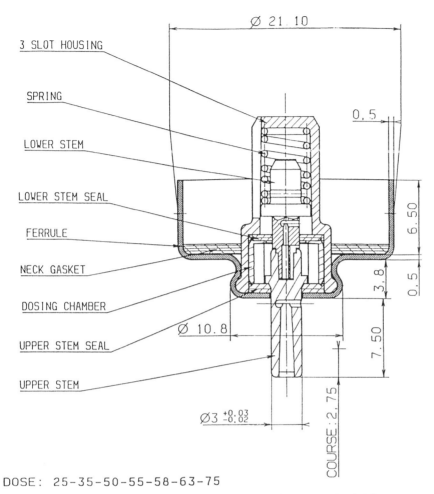

3 SLOT HOUSING

SPRING

LOWER STEM

LOWER STEM SEAL

FERRULE

NECK GASKET

DOSING CHAMBER

UPPER STEM SEAL

UPPER STEM

Ø 21.10

0.5

6.50

3.8

0.5

7.50

2.75 COURSE:

Ø 10.8

Ø3 $^{+0.03}_{-0.02}$

DOSE: 25-35-50-55-58-63-75

Fig. 6 Metering valve with increased retention of dose. (Courtesy of Valois.)

latter procedure should be used. The average of ten doses may fail to reveal problems of variations in each of the doses dispensed. It is possible to have good reproducibility on the basis of ten doses and not on a unit-dose basis. These valves have been used on most of the aerosol inhalation products both in the United States and other countries of the world.

V. AEROSOL CONTAINERS

Various containers have been used for pharmaceutical aerosols. Because of aesthetics and excellent compatibility with drugs, glass, stainless steel, and aluminum containers have found widespread use in the pharmaceutical industry.

A. Plastic-Coated Glass Bottles

Plastic-coated glass bottles ranging in size from 15 to 30 ml have been used primarily with solution aerosols, although there is no technical or scientific rationale for this, other than that one can note the amount of material left in the container by holding the bottle in the path of a strong light. Glass bottles are not recommended for suspension aerosols owing to the visibility of the suspended particles, which may present an aesthetic problem. Glass, being inert, has always been preferred for use with all types of pharmaceuticals, although, with the advent and introduction of many newer materials of construction, it has been replaced in certain instances with these materials.

The plastic coating on the glass containers serves to protect from flying glass in the event of glass shattering when the container is accidentally dropped. Additionally, the plastic coating around the neck of the container serves to absorb some of the shock from the crimping operation and decreases the danger of breaking around the neck area.

All commercially available bottles have a 20-mm neck finish and adapt easily to all the metered aerosol valves presently available. In addition, the plastic coating also serves as an ultraviolet light absorber so that the contents are protected from the deleterious effects of light. These plastic coatings are available in a clear finish or in various colors. The advantage of plastic-coated glass lies in its excellent compatibility with pharmaceuticals and its ability to permit one to view the level of contents remaining in the container.

B. Aluminum Containers

Aluminum is used as the material of construction for most other oral aerosols. This material is extremely light weight and is also essentially inert, although aluminum will react with certain solvents and chemicals. Even though aluminum can be used without an internal organic coating for certain aerosol formulations (especially those that contain only active ingredient and propellant), many containers are available that have an internal coating made from an epon or epoxy-type resin. The coating formulations are generally confidential, but many of the container manufacturers have on file with the FDA a Drug Master File (DMF) which contains all pertinent information about the exact formulation, safety evaluations, inertness, and so on, of the coating material. The aluminum container may also be anodized to form a stable coating of aluminum oxide.

These aluminum containers are also made with a 20-mm opening to receive the generally available metered valves. However, for various reasons, a variety of openings ranging from 15 to 20 mm are also available for special and customized applications. The most generally used opening, however, remains the 20 mm. Aluminum containers are manufactured from a ''slug'' of aluminum and are seamless; therefore, there is virtually no danger of leakage occurring.

C. Tin-Plate Containers

The three-piece tinplated steel container finds use in topical pharmaceutical aerosols. To decrease the compatibility problems, an internal organic coating has been used. This coating generally consists of a film made of vinyl or epoxy resins. These materials are resistant to many of the pharmaceutical ingredients and can be successfully formulated with topical pharmaceuticals.

D. Barrier Packs

Although the greater number of aerosol pharmaceutical products in use today employ the typical aerosol systems, other systems that separate product from propellant are also available. These systems include barrier packs and are of several types.

The piston system, consisting of a ''free'' piston fitted into a two-piece aluminum container, can be used for viscous and semisolid products. It can be used to package pharmaceutical ointments and creams. The top of the piston (generally polyethylene) is contoured to fit the top of the valve so that when the piston is pushed to the top, the product will be expelled and, since the piston fits snugly against the top of the valve, all of the material would be dispensed. In use, the product is filled through the 1-in. opening of the metal container and occupies the space above the piston. The valve is then sealed into place. After the gas is added through a small hole in the bottom of the can, the opening is sealed with a rubber plug and pressurized with nitrogen to about 90 psig, a hydrofluorocarbon, or with about 15 g of isobutane or pentane/isobutane.

Another type of container consists of an accordion-pleated, collapsible plastic bag fitted into a standard three-piece, tin-plated can. The product is placed within the bag, and the propellant is added through the bottom of the container, which is fitted with a one-way valve. Since the product is placed in a plastic bag, there is no contact between the product and the container walls. The compatibility of the product with the plastic bag must be considered. Several different materials are available and have been used successfully. This system has been used for postfoaming gels and is useful for viscous pharmaceuticals. About 15 g of hydrocarbon propellant is used in this system.

Other barrier-type systems include various laminated bags into which the product is placed. Some types consist of a latex membrane that expands as the product is filled into the bag. When the valve is opened, the latex collapses and forces the product out of the bag. Depending on the type of valve, the product is dispensed as a spray, foam, or semisolid.

VI. FORMULATION OF PHARMACEUTICAL AEROSOLS

A. Solution Aerosols

Solution aerosols consist of therapeutically active ingredients in pure propellant, or a mixture of propellant and solvents. The solvent is used to dissolve the active ingredients or to retard the evaporation of the propellant. Solution aerosols are relatively easy to formulate, provided the ingredients are soluble in the propellant or propellant–solvent system. However, the propellants are nonpolar and, in most cases, are poor solvents for some of the commonly used medicinal agents. Through use of a solvent that is miscible with the propellant, one can achieve varying degrees of solubility. Unfortunately, one is limited to the number of solvents that can be used for this purpose, owing to toxicity considerations. Ethyl alcohol has found greatest use for this purpose, although solvents such as polyethylene glycols, dipropylene glycol, ethyl acetate, hexylene glycol, acetone, glycol ethers, and so on, can be used. However, consideration must be given to their toxicity, especially for aerosols used orally and nasally.

Solutions have been used to formulate foot preparations, local anesthetics, spray-on-protective films, anti-inflammatory preparations, and aerosols for oral and nasal applications. They are generally formulated to consist of from 50 to 90% propellant and from 10 to 50% of active ingredients and cosolvent for topical aerosol and up to 99.5% propellant for oral and nasal aerosols. The greater the amount of propellant present, the greater will be the degree of dispersion and the finer the spray. As the concentration of propellant decreases, the wetness of the spray will increase. Generally, inhalation aerosols contain higher proportions of propellant compared with topical sprays. This is due to the higher degree of atomization required by inhalation aerosols and the relatively low concentration of active ingredient.

When the valve of a solution aerosol is depressed, a mixture of active ingredients, solvents, and propellants are emitted into the atmosphere. As the liquid propellant hits the warm surrounding air, it tends to vaporize and, in so doing, breaks up the active ingredients and solvents into fine particles. Depending on the size of the particles, they remain suspended in air for relatively long periods. The particles of the spray can vary from as small as 5–10 μm or less for inhalation aerosols to as much as 50–100 μm for topicals.

Table 5 Prototype Formulation for Oral,
Inhalation, and Nasal Aerosol Solutions

Active ingredient	Dissolved in system
Solvent	Ethyl alcohol
	Glycol
	Water
	Solublizer (surfactant)
	Liposome
Antioxidant	Ascorbic acid
Flavor	Aromatic oils
Propellant(s)[a]	12/11
	12/114
	12
	12/114/11

[a]Other combinations of these propellants can also be used
to obtain desired solubility.

Characteristic of the liquefied gas aerosols is that the pressure within the container remains constant through the life of the product. As the product is utilized, the vapor in the headspace expands, and there is a temporary drop in pressure. However, some of the propellant in the liquid state will vaporize and restore the original pressure. This is quite different from compressed gas aerosols, for which the pressure will decrease during use of the product.

A prototype formulation for pressurized aerosol solutions for oral or nasal use is illustrated in Table 5 while Table 6 illustrates a prototype formulation for a topical aerosol solution.

In developing solution-type pharmaceutical aerosols, one must consider the following: (a) effect of solvent–propellant blends on the solubility and stability of the active ingredient, (b) particle size and the surface tension of the droplet, (c) irritating properties of the various additives when used both orally and topically (antioxidants, preservatives, or other), (d) esoph-

Table 6 Prototype Formulation for Topical Aerosol
Solutions

Active ingredient	Dissolved in system
Solvents	Ethyl and isopropyl alcohol
	Glycols
	Isopropyl esters
	Surfactants
Antioxidants	Ascorbic acid
Preservative	Methyl- and propylparaben
Propellant(s)[a]	Isobutane
	Propane/butane
	Propane/isobutane
	Propellant 22
	Propellant 152a/142b
	Propellant 22/142b
	Dimethyl ether

[a]Other combinations of these propellants can also be used to obtain
desired solubility and flammability characteristics.

ageal irritability, and (e) for oral and nasal aerosols, the solubilizing agents must be readily metabolized and nonocclusive or retentive by the pulmonary mucosa.

B. Suspension System

For those substances that are insoluble in the propellant or mixture of propellant and solvent, or in those cases for which a cosolvent is not desirable, the therapeutically active ingredient can be suspended in the propellant vehicle. When the valve is depressed, the suspension is emitted followed by rapid vaporization of the propellant, leaving behind the finally dispersed active ingredients. This system has been used successfully to dispense antiasthmatic aerosols, steroids, antibiotics, and similar ingredients. However, the formulation of this type of aerosol is not without difficulty. Problems arise in caking, agglomeration, particle size growth, and clogging of the valve. Valves with a vapor tap are now available that can be used to dispense powders. Some of the more important factors that must be considered in formulating this type of system include the following:

Moisture Content of Ingredients

Moisture content is a most important aspect for dispersion aerosols. The moisture content of the entire preparation must be kept below 200–300 ppm. All active and inactive ingredients, solvents, and propellants must be essentially anhydrous or made anhydrous by drying before use. The propellants can be dried by passing them through desiccants, and the other ingredients can be dried by the usual methods.

Particle Size

The initial particle size of the insoluble ingredients should be in the micrometer range, generally from 1 to 5 μm and not more than 10 μm, depending on the amount of powder to be dispensed. The use of a Jet Pulverizer or ball mill can be used to reduce the particle size. When considering solids for topical aerosols, all powders must pass through a No. 325 sieve (40–50 μm).

Solubility of Active Ingredients

That derivative of the active ingredient that has minimum solubility in propellant and solvents should be selected. It is the slight solubility of the active ingredients in the propellants and solvents that results in particle size growth. However, one must also consider solubility from the therapeutic viewpoint in that the drug must also have sufficient solubility in body fluids. For example, epinephrine bitartrate is used as a suspension in an aerosol for inhalation. It has minimum solubility in the propellant system, as compared with the hydrochloride and sufficient solubility in fluids surrounding the lung areas to be therapeutically effective. It should be soluble in respiratory secretions at a reasonable rate, or it may act as an irritant. Polymorphs of the drug that have high thermodynamic activity should be avoided.

Surfactant or Dispersing Agents

Final consideration should be given to the use of a surfactant or dispersing agent. For topical aerosols, isopropyl myristate (which is not a surfactant) has been used primarily for its lubricating properties. Mineral oil has also been used in a similar manner. Surfactants that have been suggested, especially for oral, inhalation, and nasal use, include several of the polysorbate and sorbitan esters, especially sorbitan trioleate, and others, including lecithin derivatives, oleyl alcohol, and ethanol. The surfactant should be nontoxic, biodegradable, and minimally irritating to the respiratory airways, the oropharynx mucosa, and the nasal mucosal lining, in the case of intranasal preparations.

Table 7 Prototype Formulation for Aerosol Suspensions
for Inhalation Use

Active ingredient(s)	Micronized
Dispersing agent or surfactant	Sorbitan trioleate, lecithin, and lecithin derivatives
	Oleyl alcohol
	Ethyl alcohol
Propellant(s)	12/11
	12/114
	12/114/11
	12

Table 7 illustrates a prototype formulation for a dispersion aerosol for use by inhalation, Table 8 includes a typical topical pharmaceutical aerosol suspension, and Table 9 shows a starting formulation for a nasal aerosol.

C. Emulsion Aerosols

Emulsion aerosols consist of active ingredients, aqueous or a nonaqueous vehicle, surfactant, and propellant. Depending on the choice of ingredients, the product can be emitted as a stable or quick-breaking foam. Approximately 7–10% of propellant is used in conjunction with 90–93% of emulsion concentrate. The propellant is generally considered to be part of the immiscible phase and, as such, can be in the internal or external phase. When the propellant is included in the internal phase, a typical foam is emitted. When the propellant is in the external phase, the product is dispensed as a spray. If the product is dispensed as a foam, varying consistency of the foam may be obtained without any material becoming airborne from the escape of fine particulate matter. This will decrease the danger of inhaling particles intended for topical administration. The pressure will be approximately 40 psig, depending on the propellant used. Depending on the formulation, either a stable foam, such as would be expected in shave cream, or a quick-breaking foam will occur. A quick-breaking foam will be dispensed as a foam, but will collapse in a relatively short time. This type is not desirable in a shave cream, but may be advantageous in preparations where, for example, one may wish to have the material dispensed in a foam so that it can be applied to a limited area of the skin and collapse. One chief advantage of a foam system over a spray system is that the area with which

Table 8 Prototype Formulation for Topical Aerosol
Suspensions

Active ingredient(s)	Pass through a 325-mesh screen
Dispersing agents	Isopropyl myristate
	Mineral oil
	Sorbitan esters
	Polysorbates
	Glycol ethers and derivatives
Propellant(s)	12/11; 12/114 (only if exempted)
	Hydrocarbons
	142b, 152a, 22
	Dimethyl ether

Table 9 Prototype Formulation for Nasal Aerosol
Suspensions or Solutions

Active ingredients	Solubilized or suspended
Antioxidants	Ascorbic acid
	Bisulfites
Preservatives	Benzalkonium chloride
Adjuvants	Phosphate buffer (pH 6.5)
	Sodium chloride (tonicity)
	Water
	Emulsifying agent or surfactant
Propellant(s)	12 ⎫
	12/11 ⎬ (nonaqueous only)
	12/114 ⎭
	Nitrogen

the product can come into contact is controlled and can be "pinpointed." The most stable foams [whether they be water–oil (w/o), or oil–water (o/w), in terms of physical appearance] are achieved when the surfactants employed exhibit only minimum solubility in both the formulation's aqueous and organic phases. It should be noted, however, that, although the surfactant must be poorly soluble in both phases, it should be easily wetted by each phase to permit its uniform dispersion.

It has been demonstrated that surfactants possessing these qualities on product actuation will concentrate at the interface between the propellant and the aqueous phase, forming a thin film commonly referred to as the "lamellae." It is the specific composition of this lamellae that dictates the structural strength and general characteristics of the foam. The composition of the lamallae may consist of simply a single surfactant, a blend of surfactants or, in some more complex situations, a mixture of surfactants combined with additional agents present in the foam formulation that are molecularly associated with, or adsorbed to, the surfactant(s). The latter dealing with particulate adsorption is commonly referred to as particulate foam stabilization. As a general rule, thick and tightly layered lamellae generally produce very structured foams having high yield values (capable of supporting their own weight). Foam concentrates (emulsions) generally consist of a fatty acid saponified with triethanolamine and diluted with water. Other ingredients, including perfumes, are then incorporated to impart such properties as softening, wetting, cooling, and smoothness. The active ingredients are incorporated into this base. The formulation is then pressurized with propellant, forming the internal phase of the emulsion. After actuation, the liquefied propellant, which is partially emulsified in the concentrate, vaporizes and forms a relatively stiff foam matrix.

Of the three types of surfactants, the anionic have been most widely used in foam formulations. In recent years, however, the use of nonionic surface-active agents has grown very rapidly in the pharmaceutical industry. The nonionics are seen by pharmaceutical manufacturers as inherently less likely (because they are not charged) to present compatibility problems. Popular nonionic surfactants presently used in aerosol products include the polyoxyethylene (POE) fatty esters, polyoxyethylene sorbitan esters, alkyl phenoxy ethanols, fatty acid esters, and alkanolamides. A prototype formulation of an emulsion product is illustrated in Table 10.

It is also possible to formulate edible foams to be used to dispense cough remedies, calcium supplements, antacids, vitamins, and many other similar products. These systems should be readily acceptable to children and to the geriatric population, who may have difficulty in swallowing other types of dosage forms. These generally consist of the active ingredients

Table 10 Prototype Formulation for Topical Aerosol Emulsions (Foam)

Active ingredient	Solubilized in fatty acid, vegetable oil, glycol
Emulsifying agents	Fatty acid soaps (triethanolamine stearate)
	Polyoxyethylene sorbitan esters
	Emulsifiable waxes
	Surfactants
Other modifiers	Emollients
	Lubricants
	Preservatives
	Perfumes, etc.
Propellant(s)	12/114 (only if exempted)
	Hydrocarbons
	22/152a
	22/142b
	152a/142b
	Dimethyl ether

dissolved or suspended in a vegetable oil and emulsified with a food-grade emulsifier, such as glyceryl monostearate. Through use of nitrogen, nitrous oxide, carbon dioxide, or a hydrocarbon as the propellant, a relatively stable foam may be obtained. Additional studies are needed to develop a suitable metering valve or device to control the dosage accurately. Recently, two such valves became available from Lablabo (Valois of America, Inc., Greenwich, Conn.) and Bespak, Inc. (Cary, North Carolina).

D. Semisolid Aerosols

The semisolid aerosol preparations are formulated in the usual manner and depend on a compressed gas, such as nitrogen, to push the contents from the package. Viscosity plays an important role in dispensing this type of product. It can be used with ointmentlike products. Foam dispensing is also possible through use of a soluble compressed gas (nitrous oxide or carbon dioxide, or a mixture of both) together with an emulsion concentrate. A mixture of the two gases is desirable because of their collective solubility characteristics and physical and chemical properties. In addition, carbon dioxide tends to have an acid reaction, which may or may not present compatibility problems when this gas is used alone. Nitrogen is also used as the propellant for metered-dose liquids dispensed with a metered valve. These valves are capable of delivering from 0.25 to about 5 ml of solution each time the valve is depressed.

VII. ORAL, INHALATION, AND NASAL AEROSOLS

These aerosols are similar in many respects and can best be discussed as a group. Oral aerosols are generally used sublingually (buccal) and at the present time are used to dispense nitroglycerin. Inhalation aerosols are also referred to as ''oral aerosols,'' but they are intended to be inhaled through the respiratory system for either local pulmonary activity or by absorption and systemic activity. Intranasal aerosols are administered into the nasal passages and can act either locally or systemically.

A. Oral Aerosols (Buccal)

Oral aerosols are formulated as a solution of the active ingredient with, or without, cosolvents in a mixture of CFC propellants. Generally, Propellants 12/11 or 12/114, or mixtures of all three propellants, are used for this type of aerosol. These aerosols in formulation are similar to mouth fresheners, which have been available for many years. The product is emitted through a metered-dose valve and sprayed directly under the tongue, where it is quickly absorbed. Nitroglycerin is available as this type of aerosol. Nitrolingual Spray contains 0.4 mg of nitroglycerin per metered-dose. There are 200 metered-doses packaged in an aluminum container fitted with a metered valve. These preparations are dispensed as a fairly wet spray to produce relatively large particles.

B. Inhalation Aerosols

The largest group of oral aerosol products, the inhalation aerosols, are formulated either as a solution or a suspension-type aerosol. Aluminum containers ranging in size from 15 to 30 ml have been used together with a metered valve, delivering about 35–75 μl/dose. These valves are available up to a dose of 150 μl/dose. Plastic-coated glass has also been used. The commonly used propellants include Propellants 12/11, 12/114, and 12/114/11.

The ultimate deposition of materials actuated through the valve into the respiratory system is of utmost importance, since this will determine the therapeutic activity of the product. Accurate assessment of drug deposition profiles in terms of both the quantity of drug reaching the respiratory airways and its depth of penetration are critical parameters in evaluating the biopharmaceutics (the delivery of drug to site of absorption) of inhalation aerosol products. The major factors influencing the ultimate deposition of inhalation aerosols include the product's formulation, the design of components (specifically the valve and oral adaptor), the administrative skills and techniques of the product user, and the anatomical and physiological status of the respiratory system. The interdependence of one of these factors on the other cannot be overemphasized. Included among formulation factors are the physicochemical characteristics of the active ingredients, the formulation's particle size and shape, the type and concentration of surface-active agent used, and to some extent, the vapor pressure of propellants. In terms of physiochemical properties, the lipoidal solubility and pulmonary absorption rates of the active ingredient are of utmost importance. Another physicochemical factor governing the biopharmaceutics of a drug is its dissolution characteristics in pulmonary fluids. Drugs having a rapid dissolution rate in pulmonary fluids predictably produce much more intense and rapid onset of action, having a shorter duration than that of their corresponding, less soluble, derivatives. Therapeutic agents that exhibit a very poor solubility in pulmonary fluids are to be avoided, since they are likely to serve as irritants and precipitate bronchial spasms.

Particle size and shape also play an important role in the drug's deposition pattern in the respiratory passageways. It is imperative to keep approximately 90% or better of the particles in inhalation products between 0.5 and 10 μm to maximize their delivery and deposition in respiratory fluids. Most workers agree that particles in the size range of 3–6 μm are most useful. Particles of this size have been demonstrated to deposit in the lung by gravitational sedimentation, inertial impaction, or by diffusion into terminal alveoli by brownian motion.

The selection of the appropriate surface-active agent (required in most pressurized inhalation suspension aerosols) is another important consideration, since the surfactant will influence droplet evaporation, particle size, and overall hydrophobicity of the particles reaching the respiratory passageways and pulmonary fluids.

Component designs, specifically the valve and oral adaptor also alter the particle size and the penetration and deposition of drugs into the lungs.

One component that has undergone enormous modification in the last few years to improve drug delivery is the oral adaptor. Most adaptors were short and rather simplistic to minimize possible holdup of material in it. The holdup in the short-stem adaptors averages anywhere from 5 to 20%. Recently, however, numerous customized adaptors, having specific designs and dimensions, have entered the marketplace. Many refer to these newer, rather large, adaptors as tube spacers.

Interest in the larger adaptors can be attributed to any one or more of the following reasons. The large actuator designs permit a complete evaporation of propellant, reducing initial droplet velocity and particle size. This reduction of particle size should improve depth of drug penetration into the lungs, while a lower initial velocity is expected to decrease product impaction to the back of the esophagus (whiplash effect).

The MDI has been modified to improve the drug delivery of aerosolized particles into the nasal passageways and respiratory airways. Some of these modifications have included the introduction of tube spacers, breath-activated actuators, and portable plastic reservoirs with inhalation aerosols. During the late 1970s and early 1980s, there were a number of in vivo and in vitro studies evaluating the differences between conventional adapters and the expanded-chamber adapters referred to as "spacers or tube spacers," shown in Fig. 7. Currently, many conventional short-term MDIs deliver at best only about 10–15% of the dose actuated into the respiratory airways. The balance of the dose is either lost to the inner surface of the adaptor (approximately 10%) or is deposited by inertial impaction in the oropharynx area (80%). The latter leads to swallowing and possible systemic absorption of the therapeutic agent(s). To reduce this fraction lost to the oropharynx and swallowed, several tube spacers of various geometric shapes and dimensions were considered, since they should, at least in theory, minimize some of the effects produced through inertial compactions, which contribute significantly to this problem.

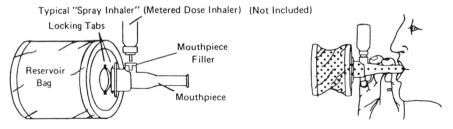

Fig. 7   InspirEase unit. (Courtesy of Key Pharmaceuticals, Inc., a Division of Schering Corporation.)

One such delivery system now available consists of a 700-ml collapsible plastic bag into which the aerosol can be injected, as shown in Fig. 7. The system has been marketed as InspirEase by Key Pharmaceuticals (Schering Corp.). InspirEase is designed to allow patients more time to breathe in the medication after actuating than did the conventional MDI and eliminated some of the loss of medication associated with too rapid propulsion or inhalation of aerosolized drugs. The InspirEase device consists of a collapsible reservoir bag and a special mouthpiece fitted with a reed placed inside the mouthpiece that produces a warning sound when patients are inhaling too quickly. Manufacturers of InspirEase claim that, in addition to eliminating the coordinating problems associated with the use of conventional MDIs, the device also encourages proper breathing patterns during administration. Other tube spacers are available from Forest Pharmaceuticals, Inc. (Aerochamber) and from Schering Corporation (Inhal-Aid). Figure 8 illustrates some of the more conventional oral adaptors.

An MDI fitted with a breath-activated oral adaptor has recently been made available as Maxair (3M Pharmaceuticals) which contains pirbuterol acetate. It is said that this device increased the efficiency of the MDI from 50 to 91%.

C. Nasal Aerosols

The use of nasal aerosols is a most promising area of development, and several aerosols are now available for nasal administration. Three such products that have gained widespread acceptance within the medical profession were those using the steroid intranasal anti-inflammatory agents beclomethasone dipropionate, (marketed as Vancenase by Schering Corporation and Beconase by Glaxo Inc.), triamcinolone acetonide (marketed as Nasacort by Rhone-Poulenc Rorer, Inc.), dexamethasone sodium phosphate (marketed as Decadron Phosphate Turbinaire by Merck) and budesonide (marketed as Rhinocort by Astra, Inc.). These are potent anti-inflammatory steroids indicated for the relief of the symptoms of perennial rhinitis. Delivery

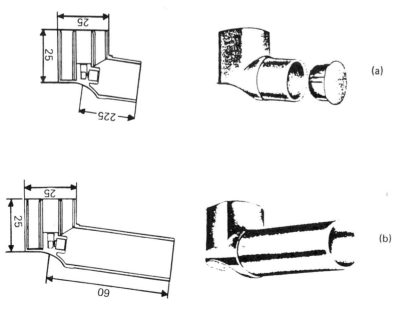

Fig. 8 Oral adaptors for inhalation aerosols: (a) short version; (b) long version. (Courtesy of Valois of America, Inc., Greenwich, CT.)

of steroids, such as beclomethasone dipropionate, triamcinolone acetonide, and dexamethasone sodium phosphate directly to the nasal mucosa by this aerosol system provides continuous relief of nasal congestion, sneezing, running nose, and itching associated with hay fever and nasal allergies, without the problems associated with the use of antihistamines and decongestant, such as drowsiness, rebound congestion, and cardiovascular stimulation.

Insulin is currently under study, along with other agents, for administration as a nasal aerosol. Other proteins and peptides are also under study using the nasal route of administration.

VIII. MANUFACTURERING AND PACKAGING THE AEROSOL DOSAGE FORM

Two basic methods are available for the manufacture and packaging of aerosols. Both a cold- and a pressure-filling process are available. Metered-dose inhalers and nasal aerosols generally can be filled by either process. Topical aerosols are filled by the pressure process, although, for the most part, they can be packaged by either method, except for foams and products containing water, which will freeze or solidify at the low temperature required for cold fill.

A. Cold-Fill Process

As the name indicates, this process is dependent on lowering the temperature of the concentrate (either a solution or suspension) to temperatures below room temperature (generally $-30°$ to $-60°C$) so that the propellant remains liquefied. A cold solution or dispersion of the concentrate is then added to the chilled container. The propellant is added, a valve is crimped in place, followed by passing the container into a water bath (approximately 55°C). This test checks to ensure there is no leakage or distortion of the container (leakage in the case of plastic-coated glass bottles).

This process was originally the process-of-choice for all aerosols containing a metered-dose valve, as equipment was not available at the time to fill propellant through the valve. With the introduction of pressure-fill equipment, it is possible to package MDIs using the pressure process.

B. Pressure-Fill Process

The pressure-fill process is carried out at essentially room temperature. The concentrate (solution or suspension) may be chilled slightly (to approximately 15–20°C) to reduce vaporization of any volatile solvent or Propellant 11. This is added to an open container, a valve is crimped in place, and the propellant is added, under pressure, through the valve. The filled container is then passed through the water bath.

C. Single-Stage–Filling System

Either the cold-fill method or the pressure-fill method can be used for this system. In the cold fill the concentrate is made in an open tank and the propellants are kept liquid by lowering the temperature to about −30 to −60°C. The propellants are added in the liquid state to the concentrate in the tank. The tank will be fitted with a cover to minimize loss of propellant; however, at this low temperature, any loss by vaporization is negligible. The total product is added to an open container and the valve is crimped in place.

The same process can be carried out at room temperature, except the product must be made in a pressure tank capable of withstanding internal pressures of at least 125–150 psi. By having the pressure tank fitted with a cooling jacket, the internal temperature can be lowered so that the product can be made at a lower pressure.

Pressure filling is preferred over the cold-fill process, and it is possible to package aerosols fitted with at metered-dose valve. This process is fast and can fill about 160 cans a minute using rotary packaging equipment having multiple-filling heads. Less propellant escapes into the atmosphere, and pressure filling is useful for those drugs that are moisture-sensitive, such as albuterol.

IX. TESTING OF AEROSOLS

Metered-dose inhalers are probably one of the most tested pharmaceutical product. Approximately 15–17 different tests are performed on the product during the research and development stage. These include delivery of drug through the oral adaptor, delivery of total product through the valve, amount of drug retained in the oral adaptor, particle size distribution, number of doses per container, leakage rate, pressure, total can contents, microbial limits, degradation products, or others.

Additionally, there are some specific tests indicated in the *USP/NF* monograph. Some of these tests, such as particle size, are carried out with a sample taken from the beginning, middle, and end of the can to ensure uniform distribution of the active ingredient.

X. FUTURE DEVELOPMENTS

During the past few years, there has been a resurgence of interest in therapeutic aerosols for oral, inhalation, nasal, and topical use (including rectal and vaginal). This increasing interest within the pharmaceutical industry, can probably be best attributed to any one or more of the following:

A rethinking and reevaluation of the lung and related respiratory passageways as more than just a "target organ" for pulmonary and respiratory therapy

A need for alternative routes of administration for those drugs that are poorly or erratically absorbed orally or parenterally

A need for greater understanding and more finite assessment of the biopharmaceutics and pharmacokinetics of drugs aerosolized into respiratory airways

Technological advances in the hardware (e.g., valves, actuators, and tube spacers) of metered-dose inhalers (MDIs)

Increased recognition of the need to improve administrative techniques and, more importantly, patient compliance (breath-activated oral adaptors)

Increased emphasis within the pharmaceutical industry on new and novel delivery systems (e.g., transdermal delivery, intranasal delivery, osmotic pumps)

Several of the metered-dose aerosols' (albuterol, metaproterenol sulfate, ipatropium bromide) patent protection has expired, making these products available for generic development.

Basic technoogy in aserosol formulation is being restudied relative to the development rationale for inhalation aerosol drug dosage forms

Given these observations and the activity in this field, it is reasonable to predict that a large number of these products will become available to the patient before the start of the next decade, and certainly before the end of the twentieth century.

BIBLIOGRAPHY

Hickey, A. J., *Pharmaceutical Inhalation Aerosol Technology*, Marcel Dekker, New York, 1992.

Hollenbeck, R. G., and T. H. Wiser, Inhalation drug-delivery systems, in *Pharmaceutics and Pharmacy Practice* (G. S. Banker and R. K. Chambers, eds.), J. B. Lippincott, Philadelphia, 1982, pp. 353–395.

Johnsen, M. A., *The Aerosol Handbook*, 2nd Ed., Wayne Dorland Company, Mendham, NJ, 1982.

Nasr, M. M., Single-puff particle-size analysis of albuterol metered-dose inhalers (MDIs) by high pressure chromatography with electrochemical detection (HPLC-EC), Pharm. Res., 10, 1381–1384 (1993).

Sanders, P. A., *Handbook of Aerosol Technology*, 2nd Ed., Robert E. Krieger Publishing Company, Malabar, FL, 1987.

Sciarra, J. J., Aerosols, in *Remington's Pharmaceutical Sciences*, 18th Ed. (A. R. Gennaro, ed. and chairman), Mack, Easton, PA, (in press).

Sciarra, J. J., and A. J. Cutie, Pharmaceutical aerosols, in *The Theory and Practice of Industrial Pharmacy*, 3rd Ed. (L. Lachman, H. A. Lieberman, and J. L. Kanig, eds.), Lea & Febiger, Philadelphia, 1986, pp. 589–618.

Sciarra, J. J., and L. Stoller, *The Science and Technology of Aerosol Packaging*, John Wiley & Sons, New York, 1974.

15

Sustained- and Controlled-Release Drug Delivery Systems

Gwen M. Jantzen and Joseph R. Robinson
School of Pharmacy, University of Wisconsin, Madison, Wisconsin

I. INTRODUCTION

Over the past 30 years, as the expense and complications involved in marketing new drug entities have increased, with concomitant recognition of the therapeutic advantages of controlled drug delivery, greater attention has been focused on development of sustained- or controlled-release drug delivery systems. There are several reasons for the attractiveness of these dosage forms. It is generally recognized that for many disease states, a substantial number of therapeutically effective compounds already exist. The effectiveness of these drugs, however, is often limited by side effects or the necessity to administer the compound in a clinical setting. The goal in designing sustained- or controlled-delivery systems is to reduce the frequency of dosing or to increase effectiveness of the drug by localization at the site of action, reducing the dose required, or providing uniform drug delivery.

If one were to imagine the ideal drug delivery system, two prerequisites would be required. First, it would be a single dose for the duration of treatment, whether it be for days or weeks, as with infection, or for the lifetime of the patient, as in hypertension or diabetes. Second, it should deliver the active entity directly to the site of action, thereby minimizing or eliminating side effects. This may necessitate delivery to specific receptors, or to localization to cells or to specific areas of the body.

It is obvious that this imaginary delivery system will have changing requirements for different disease states and different drugs. Thus, we wish to deliver the therapeutic agent to a specific site, for a specific time. In other words, the objective is to achieve both spatial and temporal placement of drug. Currently, it is possible to only partially achieve both of these goals, with most drug delivery systems.

In this chapter, we present the theory involved in developing sustained- and controlled-release delivery systems and applications of these systems as therapeutic devices. Although suspensions, emulsions, and compressed tablets may demonstrate sustaining effects within the body, compared with solution forms of the drug, they are not considered to be sustaining and are not discussed in this chapter. These systems classically release drug for a relatively short period, and their release rates are strongly influenced by environmental conditions.

II. TERMINOLOGY

In the past, many of the terms used to refer to therapeutic systems of controlled and sustained release have been used in an inconsistent and confusing manner. Although descriptive terms such as "timed release" and "prolonged release" gives excellent manufacturer identification, they can be confusing to health care practitioners. For purposes of this chapter, sustained release and controlled release will represent separate delivery processes. *Sustained release* constitutes any dosage form that provides medication over an extended time. *Controlled release*, however, denotes that the system is able to provide some actual therapeutic control, whether this be of a temporal nature, spatial nature, or both. In other words, the system attempts to control drug concentrations in the target issue. This correctly suggests that there are sustained-release systems that cannot be considered controlled-delivery systems.

In general, the goal of a sustained-release dosage form is to maintain therapeutic blood or tissue levels of the drug for an extended period. This is usually accomplished by attempting to obtain *zero-order* release from the dosage form. Zero-order release constitutes drug release from the dosage form that is independent of the amount of drug in the delivery system (i.e., a constant release rate). Sustained-release systems generally do not attain this type of release and usually try to mimic zero-order release by providing drug in a slow first-order fashion (i.e., concentration-dependent). Systems that are designated as prolonged release can also be considered as attempts at achieving sustained-release delivery. Repeat-action tablets are an alternative method of sustained release in which multiple doses of a drug are contained within a dosage form, and each dose is released at a periodic interval. Delayed-release systems, in contrast, may not be sustaining, since often the function of these dosage forms is to maintain the drug within the dosage form for some time before release. Commonly, the release rate of drug is not altered and does not result in sustained delivery once drug release has begun. Enteric-coated tablets are an example of this type of dosage form.

Controlled-release, although resulting in a zero-order delivery system, may also incorporate methods to promote localization of the drug at an active site. In some cases, a controlled-release system will not be sustaining, but will be concerned strictly with localization of the drug. *Site-specific* systems and *targeted-delivery* systems are the descriptive terms used to denote this type of delivery control.

The ideal of providing an exact amount of drug at the site of action for a precise time period is usually approximated by most systems. This approximation is achieved by creating a constant concentration in the body or an organ over an extended time; in other words, the amount of drug entering the system is equivalent to the amount removed from the system. All forms of metabolism and excretion are included in the removal process: urinary excretion, enterohepatic recycling, sweat, fecal, and so on. Since, for most drugs, these elimination processes are first-order, it can be said that at a certain blood level, the drug will have a specific rate of elimination. The idea is to delivery drug at this exact rate for an extended period. This is represented mathematically as

$$\text{Rate in} = \text{rate out} = k_{\text{elim}} \times C_d \times V_d$$

where C_d is the desired drug level, V_d is the volume of distribution, and k_{elim} the rate constant for drug elimination from the body. Often such exacting delivery rates prove to be difficult to achieve by administration routes other than intravenous infusion. Noninvasive routes (e.g., oral) are obviously preferred.

Figure 1 shows comparative blood level profiles obtained from administration of conventional, controlled-, and sustained-release dosage forms. The conventional tablet or capsule provides only a single and transient burst of drug. A pharmacological effect is seen as long as the amount of drug is within the therapeutic range. Problems occur when the peak concentration is above or below this range, especially for drugs with narrow therapeutic windows. The slow first-order release obtained by a sustained-release preparation is generally achieved by slowing the release of drug from a dosage form. In some cases, this is accomplished by a continuous release process; however, systems that release small bursts of drug over a prolonged period can mimic the continuous-release system.

III. ORAL SYSTEMS

Historically, the oral route of administration has been used the most for both conventional and novel drug delivery systems. There are many obvious reasons for this, not the least of which would include acceptance by the patient and ease of administration. The types of sustained- and controlled-release systems employed for oral administration include virtually every currently known theoretical mechanism for such application. This is because there is more flexibility in dosage design, since constraints, such as sterility and potential damage at the site of administration, are minimized. Because of this, it is convenient to discuss the different types of dosage forms by using those developed for oral administration as initial examples.

With most orally administered drugs, targeting is not a primary concern, and it is usually intended for drugs to permeate to the general circulation and perfuse to other body tissues (the obvious exception being medications intended for local gastrointestinal tissue treatment). For this reason, most systems employed are of the sustained-release variety. It is assumed that increasing concentration at the absorption site will increase the rate of absorption and, therefore, increase circulating blood levels which, in turn, promotes greater concentrations of drug at the site of action. If toxicity is not an issue, therapeutic levels can thus be extended. In essence, drug delivery by these systems usually depends on release from some type of dosage form, permeation through the biological milieu, and absorption through an epithelial membrane to

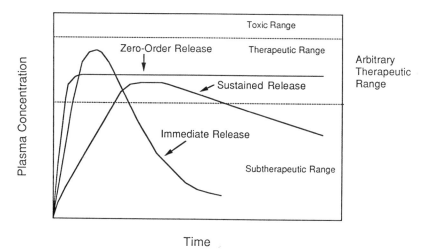

Fig. 1 Drug level versus time profile showing differences between zero-order controlled release, slow first-order sustained release, and release from a conventional tablet or capsule.

the blood. There are a variety of both physicochemical and biological factors that come into play in the design of such systems.

A. Biological Factors Influencing Oral Sustained-Release Dosage Form Design

Biological Half-Life

The usual goal of an oral sustained-release product is to maintain therapeutic blood levels over an extended period. To this, drug must enter the circulation at approximately the same rate at which it is eliminated. The elimination rate is quantitatively described by the half-life ($t_{1/2}$). Each drug has its own characteristic elimination rate, which is the sum of all elimination processes, including metabolism, urinary excretion, and all other processes that permanently remove drug from the bloodstream.

Therapeutic compounds with short half-lives are excellent candidates for sustained-release preparations, since this can reduce dosing frequency. However, this is limited, in that drugs with very short half-lives may require excessively large amounts of drug in each dosage unit to maintain sustained effects, forcing the dosage form itself to become limitingly large. In general, drugs with half-lives shorter than 2 hr, such as furosemide or levodopa [1], are poor candidates for sustained-release preparations. Compounds with long half-lives, more than 8 hr, are also generally not used in sustaining forms, since their effect is already sustained. Digoxin, warfarin, and phenytoin are some examples [1]. Furthermore, the transit time of most dosage forms in the gastrointestinal (GI) tract (i.e., mouth to ileocecal junction) is 8–12 hr, making it difficult to increase the absorptive phase of administration beyond this time frame. Occasionally, absorption from the colon may allow continued drug delivery for up to 24 hr.

Absorption

The characteristics of absorption of a drug can greatly affect its suitability as a sustained-release product. Since the purpose of forming a sustained-release product is to place control on the delivery system, it is necessary that the rate of release is much slower than the rate of absorption. If we assume that the transit time of most drugs and devices in the absorptive areas of the GI tract is about 8–12 hr, the maximum half-life for absorption should be approximately 3–4 hr; otherwise, the device will pass out of the potential absorptive regions before drug release is complete. This corresponds to a minimum apparent absorption rate constant of 0.17–0.23 hr^{-1} to give 80–95% over this time period [3]. The absorption rate constant is an apparent rate constant, and should, in actuality, be the release rate constant of the drug from the dosage form. Compounds that demonstrate true lower absorption rate constants will probably be poor candidates for sustaining systems.

The foregoing calculations assume that absorption for the therapeutic agent occurs at a relatively uniform rate over the entire length of the small intestine. For many compounds, this is not true. If a drug is absorbed by active transport, or transport is limited to a specific region of the intestine, sustained-release preparations may be disadvantageous to absorption. Absorption of ferrous sulfate, for example, is maximal in the upper jejunum and duodenum, and sustained-release mechanisms that do not release drug before passing out of this region are not beneficial [5].

One method to provide sustaining mechanisms of delivery for compounds such as these has been to try to maintain them within the stomach. This allows slow release of the drug, which then travels to the absorptive site. These methods have been developed as a consequence of the observation that coadministration of food results in a sustaining effect [6]. Although administration of food can create highly variable effects, there have been methods devised to circumvent this problem. One such attempt is to formulate low-density pellets, capsules [7], or tablets [8]. These float on top of the gastric juice, delaying their transfer out of the stomach [9]. The increase in gastric retention results in higher blood levels for *p*-aminobenzoic acid, a drug with a limited GI absorption range [10], however, drugs that have widespread absorption in the intestinal system would likely not benefit from an increase in emptying time [11].

Another approach is that of bioadhesive materials. The principle is to administer a device with adhesive polymers having an affinity for the gastric surface, most probably the mucin coat [12]. Bioadhesives have demonstrated utility in the mouth, eye, and vagina, with a number of commercially available products. To date, use of bioadhesives in oral drug delivery is a theoretical possibility, but no promising leads have been published.

An alternative to GI retention for drugs with poor absorption characteristics, is to use chemical penetration enhancers. Membrane modification through chemical enhancers has been very well demonstrated for a variety of tissues in the body, including the gastrointestinal tract. Concern about this approach is the potential toxicity that may arise when protective membranes are altered. Although there are numerous safety studies for oral products containing surfactants, which are known penetration enhancers, there has not been a definitive safety study in humans using an agent that is specifically present in the formulations as a penetration enhancer.

Metabolism

Drugs that are significantly metabolized before absorption, either in the lumen or tissue of the intestine, can show decreased bioavailability from slower-releasing dosage forms. Most intestinal wall enzyme systems are saturable. As the drug is released at a slower rate to these regions, less total drug is presented to the enzymatic process during a specific period, allowing more complete conversion of the drug to its metabolite. For example, aloprenolol was more extensively metabolized in the intestinal wall when given as a sustained-release preparation [13]. High concentrations of dopa-decarboxylase in the intestinal wall will result in a similar effect for levodopa [14]. If levodopa is formulated in a dosage form with a drug compound that can inhibit the dopa-decarboxylase enzyme, the amount of levodopa available for absorption increases and can sustain its therapeutic effects. Formulation of these enzymatically susceptible compounds as prodrugs is another viable solution.

B. Physicochemical Factors Influencing Oral Sustained-Release Dosage Form Design

Dose Size

For orally administered systems, there is an upper limit to the bulk size of the dose to be administered. In general, a single dose of 0.5–1.0 g is considered maximal for a conventional dosage form [15]. This also holds for sustained-release dosage forms. Those compounds that require a large dosing size can sometimes be given in multiple amounts or formulated into liquid systems. Another consideration is the margin of safety involved in administration of large amounts of a drug with a narrow therapeutic range.

Ionization, pK$_a$, and Aqueous Solubility

Most drugs are weak acids or bases. Since the unchanged form of a drug preferentially permeates across lipid membranes, it is important to note the relationship between the pK$_a$ of the compound and the absorptive environment. It would seem, untuitively, that presenting the drug in an uncharged form is advantageous for drug permeation. Unfortunately, the situation is made more complex by the fact that the drug's aqueous solubility will generally be decreased by conversion to an uncharged form. Delivery systems that are dependent on diffusion or dissolution will likewise be dependent on the solubility of drug in the aqueous media. Considering that these dosage forms must function in an environment of changing pH, the stomach being acidic and the small intestine more neutral, the effect of pH on the release processes must be defined. For many compounds, the site of maximum absorption will also be the area in which the drug is the least soluble. As an example, consider a drug for which the highest solubility is in the stomach and is uncharged in the intestine. For conventional dosage forms, the drug can generally fully dissolve in the stomach and then be absorbed in the alkaline pH of the intestine. For dissolution- or diffusion-sustaining forms, much of the drug will arrive in the small intestine in solid form, meaning that the solubility of the drug may change several orders of magnitude during its release.

Compounds with very low solubility (less than 0.01 mg/ml) are inherently sustained, since their release over the time course of a dosage form in the GI tract will be limited by dissolution of the drug. Examples of drugs that are limited in absorption by their dissolution rate are digoxin [16], griseofulvin [17], and salicylamide [18]. The lower limit for the solubility of a drug to be formulated in a sustained-release system has been reported to be 0.1 mg/ml [19], so it is obvious that the solubility of the compound will limit the choice of mechanism to be employed in a sustained delivery system. Diffusional systems will be poor choices for slightly soluble drugs, since the driving force for diffusion, which is the drug's concentration in solution, will be low.

Partition Coefficient

When a drug is administered to the GI tract it must cross a variety of biological membranes to produce a therapeutic effect in another area of the body. It is common to consider that these membranes are lipidic; therefore, the partition coefficient of oil-soluble drugs becomes important in determining the effectiveness of membrane barrier penetration. *Partition coefficient* is generally defined as the ratio of the fraction of drug in an oil phase to that of an adjacent aqueous phase. Accordingly, compounds with a relatively high partition coefficient are predominantly lipid-soluble and, consequently, have very low aqueous solubility. Furthermore, these compounds can usually persist in the body for long periods, because they can localize in the lipid membranes of cells. Phenothiazines are representative of this type of compound [20]. Compounds with very low partition coefficients will have difficulty penetrating membranes, resulting in poor bioavailability. Furthermore, partitioning effects apply equally to diffusion through polymer membranes. The choice of diffusion-limiting membranes must largely depend on the partitioning characteristics of the drug.

Stability

Orally administered drugs can be subject to both acid–base hydrolysis and enzymatic degradation. Degradation will proceed at a reduced rate for drugs in the solid state; therefore, this is the preferred composition of delivery for problem cases. For drugs that are unstable in the stomach, systems that prolong delivery over the entire course of transit in the GI tract are beneficial; likewise, for systems that delay release until the dosage form reaches the small intestine. Compounds that are unstable in the small intestine may demonstrate decreased bioavailability when administered from a sustaining dosage form. This is because more drug is delivered in the small intestine and, hence, is subject to degradation. Propantheline [21] and probanthine [22] are representative examples of such drugs.

C. Oral Sustained- and Controlled-Release Products

Because of their relative ease of production and cost, compared with other methods of sustained or controlled delivery, dissolution and diffusion-controlled systems have classically been of primary importance in oral delivery of medication. Dissolution systems have been some of the oldest and most successful oral systems in early attempts to market sustaining products.

D. Dissolution-Controlled Systems

It seems inherently obvious that a drug with a slow dissolution rate will demonstrate sustaining properties, since the release of drug will be limited by the rate of dissolution. This being true, sustained-release preparations of drugs could be made by decreasing their rate of dissolution. The approaches to achieve this include preparing appropriate salts or derivatives, coating the drug with a slowly dissolving material, or incorporating it into a tablet with a slowly dissolving carrier. Representative products using dissolution-controlled systems are listed in Tables 1 and 2.

Dissolution-controlled systems can be made to be sustaining in several different ways. By alternating layers of drug with rate-controlling coats, as shown in Fig. 2, a pulsed delivery can be achieved. If the outer layer is a quickly releasing bolus of drug, initial levels of drug in the body can be quickly established with pulsed intervals following. Although this is not a true controlled-release system, the biological effects can be similar. An alternative method is to administer the drug as a group of beads that have coatings of different thicknesses. This is also shown in Fig. 3. Since the beads have different coating thicknesses, their release will occur in a progressive manner. Those with the thinnest layers will provide the initial dose. The maintenance of drug levels at later times will be achieved from those with thicker coatings. This is the principle of the Spansule capsule marketed by SmithKline Beecham.

This dissolution process can be considered to be diffusion-layer controlled. This is best explained by considering the rate of diffusion from the solid surface to the bulk solution through an unstirred liquid film as the rate-determining step. This dissolution process at steady state is described by the Noyes–Whitney equation:

$$\frac{dC}{dt} = k_D A(C_s - C) = \frac{D}{h} A(C_s - C) \tag{1}$$

where

dC/dt = dissolution rate
k_D = dissolution rate constant
D = diffusion coefficient
C_s = saturation solubility of the solid
C = concentration of solute in the bulk solution

It can be seen that the dissolution rate constant k_D is equivalent to the diffusion coefficient divided by the thickness of the diffusion layer (D/h).

Equation (1) predicts that the rate of release can be constant only if the following parameters are constant: (a) surface area, (b) diffusion coefficient, (c) diffusion layer thickness, and (d) concentration difference. These parameters, however, are not easily maintained constant, especially surface area. For spherical particles, the change in surface area can be related to the weight of the particle; that is, under the assumption of sink conditions, Eq. (1) can be rewritten as the cube-root dissolution equation:

$$W_0^{1/3} - W^{1/3} = k_D t \tag{2}$$

where k_D is the cube-root dissolution rate constant, W_0 and W are the initial weight and the weight of the amount remaining at time t, respectively.

Table 1 Encapsulated Dissolution Products

Product	Active ingredient(s)	Manufacturer
Ornade Spansules	Phenylpropanolamine hydrochloride, chlorpheniramine maleate	Smith Kline Beecham
Thorazine Spansules	Chlorpromazine hydrochloride	Smith Kline Beecham
Contac 12-Hour capsules	Phenylpropanolamine hydrochloride, chlorpheniramine maleate, atropine sulfate, scopolamine hydrobromide, hyoscyamine sulfate	Smith Kline Consumer Products
Artane Sequels	Trihexyphenidyl hydrochloride	Lederle
Diamox Sequels	Acetazolamide	Lederle
Nicobid Temples	Nicotinic acid	Rorer
Pentritol Temples	Pentaerythritol tetranitrate	Rorer
Chlor-Trimeton Repetabs	Chlorpheniramine maleate	Schering
Demazin Repetabs	Chlorpheniramine maleate, phenylephrine hydrochloride	Schering
Polaramine Repetabs	Dexchlorpheniramine maleate	Schering

E. Diffusional Systems

Diffusion systems are characterized by the release rate of a drug being dependent on its diffusion through an inert membrane barrier. Usually, this barrier is an insoluble polymer. In general, two types or subclasses of diffusional systems are recognized: reservoir devices and matrix devices. These will be considered separately.

Reservoir Devices

Reservoir devices, as the name implies, are characterized by a core of drug, the reservoir, surrounded by a polymeric membrane. The nature of the membrane determines the rate of

Table 2 Matrix Dissolution Products

Product (tablets)	Active ingredient(s)	Manufacturer
Dimetane Extentabs	Brompheniramine maleate	Robins
Dimetapp Extentabs	Brompheniramine maleate, phenylephrine hydrochloride, phenylpropanolamine hydrochloride	Robins
Donnatal Extentabs	Phenobarbital, hyoscyamine sulfate, atropine sulfate, scopolamine hydrobromide	Robins
Quinidex Extentabs	Quinidine sulfate	Robins
Mestinon Timespans	Pyridostigmine bromide	ICN
Tenuate Dospan	Diethylpropion hydrochloride	Merrel
Disophrol Chronotabs	Dexbrompheniramine maleate, pseudoepherine sulfate	Schering

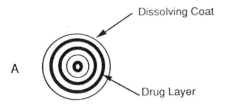

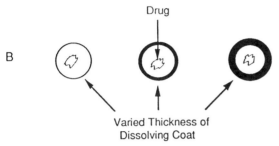

Fig. 2 Two types of dissolution-controlled, pulsed delivery systems: (A) single bead-type device with alternating drug and rate-controlling layers; (B) beads containing drug with differing thickness of dissolving coats.

release of drug from the system. A schematic description of this process is given in Fig. 4, and characteristics of the system are listed in Table 3.

The process of diffusion is generally described by a series of equations that were first detailed by Fick [23]. The first of these states that the amount of drug passing across a unit area is proportional to the concentration difference across that plane. The equation is given as

$$J = -D \frac{dC}{dX} \tag{3}$$

where the flux J, given in units of amount/area−time, D is the diffusion coefficient of the drug in the membrane in units of area/time. This is a reflection of the drug molecule's ability to diffuse through the solvent and is dependent on such factors as molecular size and charge.

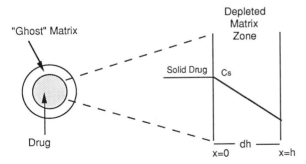

Fig. 3 Schematic representation of a matrix release system. C_s is the saturation concentration of drug controlling the concentration gradient over the distance h, of the remaining ghost matrix.

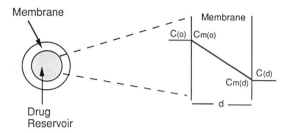

Membrane

Drug
Reservoir

Fig. 4 Schematic representation of a reservoir diffusional device. $C_{m(0)}$ and $C_{m(d)}$ represent concentrations of drug at inside surfaces of the membrane and $C_{(0)}$ and $C_{(d)}$ represent concentrations in the adjacent regions. (From Ref. 29.)

This coefficient may be dependent on concentration [24]; hence, its designation as a coefficient and not a constant, although for the purpose of designing a pharmaceutical system it is usually considered a constant [25]. dC/dX represents the rate of change in concentration C relative to a distance X in the membrane.

It is useful to make the assumption that a drug on either side of the membrane is in equilibrium with its respective membrane surface. There is, then, an equilibrium between the membrane surfaces and their bathing solutions as shown in Fig. 4. This being so, the concentration just inside the membrane surface can be related to the concentration in the adjacent region by the following expressions:

$$K = \frac{C_{m(0)}}{C_{(d)}} \quad \text{at } x = 0 \tag{4}$$

$$K = \frac{C_{m(d)}}{C_{(d)}} \quad \text{at } x = d \tag{5}$$

where K is the partition coefficient. This coefficient denotes the ratio of drug concentration in the membrane to that in the bathing medium at equilibrium. In general, a hydrophilic molecule will partition favorably to the medium, whereas a hydrophobic compound will preferentially partition to the polymer. C_m is the concentration of drug on the inside surface of the membrane, $C_{m(d)}$ the concentration on the outside surface, and d the thickness of the diffusion layer, the diffusional path length.

Assuming that D and K are constant, Eq. (3) can be integrated and simplified to give

$$J = \frac{DK\Delta C}{d} \tag{6}$$

Table 3 Characteristics of Reservoir Diffusional Systems

Description	Drug core surrounded by polymer membrane that controls release rate
Advantages	Zero-order delivery is possible
	Release rate variable with polymer type
Disadvantages	System must be physically removed from implant sites
	Difficult to deliver high-molecular-weight compounds
	Generally increased cost per dosage unit
	Potential toxicity if system fails

Where ΔC is the concentration difference across the membrane. The other variables are as defined previously. Drug release will vary, depending on the geometry of the system. The simplest system to consider is that of a slab, where drug release is from only one surface, as shown in Fig. 5. In this case, Eq. (6) can be written as

$$\frac{dM_t}{dt} = \frac{ADK\Delta C}{d} \tag{7}$$

where M_t is the mass of drug released after time t, dM_t/dt the steady-state release rate at time t, and A the surface area of the device. Equations of a similar form can be written for other geometries, such as spheres or cylinders [26].

Since the left side of Eq. (7) represents the release rate of the system, a true controlled-release system with a zero-order release rate can be possibly only if all of the variables on the right side of Eq. (7) remain constant. A constant effective area of diffusion, diffusional path length, concentration difference, and diffusion coefficient are required to obtain a release rate that is constant. These systems often fail to deliver at a constant rate, since it is especially difficult to maintain all these parameters constant. The use of a solid drug core reservoir results in a constant effective concentration, that of the solubility of the drug. Often, however, the polymer may be affected by the bathing medium. Swelling or contraction of the polymer membrane causes a change in the diffusional path length of the diffusion coefficient of the drug through the barrier. For example, if the polymer swells, the diffusion path length will increase. The ability of the drug to diffuse through the membrane, however, will increase. This is because the diffusion coefficient of the drug in the bathing medium, which has perfused the polymer during swelling, will be greater than in the unswelled polymer.

Although the partition coefficient is expected to remain constant, its magnitude is important. Since this coefficient represents the concentration of drug in the membrane relative to that in the core, an excessively high partition coefficient will allow quick depletion of the core and an ineffective delivery system. For effective diffusional systems, the partition coefficient should be less than unity. If the value of this coefficient is greater than 1, the surrounding polymer does not represent a barrier, and drug release becomes first-order.

Although diffusional systems can provide constant release at steady state, they will demonstrate initial release rates, which may be faster or slower. This depends on the device [27]. For reservoir devices, a system that is used relatively soon after construction will demonstrate a large time in release, since it will take time for the drug to diffuse from the reservoir to the membrane surface. On the other hand, systems that are stored will demonstrate a burst effect, since, on standing, the membrane becomes saturated with available drug. The magnitude of these effects is dependent on the diffusing distance (i.e., the membrane thickness). Figure 6 gives examples of this phenomenon. This plot shows the approach to steady-state release for

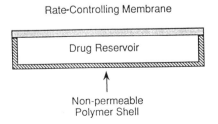

Fig. 5 Diagrammatic representation of the slab configuration of a reservoir diffusional system.

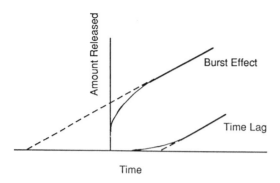

Fig. 6 Plot showing the approach to steady state for a reservoir device that has been stored for an extended period (the burst effect curve) and for a device that has been freshly made (the lag time curve). (From Ref. 29.)

a typical reservoir device that has been stored (burst effect), and for a device that has been freshly made (time lag).

Reservoir diffusional systems have several advantages over conventional dosage forms. They can offer zero-order release of drug, the kinetics of which can be controlled by changing the characteristics of the polymer to meet the particular drug and therapy conditions. The inherent disadvantages are that, unless the polymer used is soluble, the system must somehow be removed from the body after the drug has been released. This is an important dosage form consideration with implantable systems. A silicone elastomer reservoir has been used to orally deliver iodine, through the water supply, to large populations suffering from deficiency [28]. For a system such as this, the nonerodible device poses no significant problem; however, the appearance of the drug-depleted matrix in the stool can often alarm a naive patient.

Another important point to consider is that, in general, the amount of drug contained in the reservoir is far greater than the usual dose needed, since the dosage form is designed to sustain delivery over many dosing intervals. Any error in production or any accidental damage to the dosage form that would directly expose the reservoir core could expose the patient to a potentially toxic dose of drug. This becomes important when designing these dosage forms for drugs with narrow therapeutic ranges or high toxicity. Table 4 gives a representative listing of available products employing reservoir diffusion systems.

Matrix Devices

A matrix device, as the name implies, consists of drug dispersed homogeneously throughout a polymer matrix as represented in Fig. 7. In the model, drug in the outside layer exposed to

Table 4 Reservoir Diffusional Products

Product	Active ingredient(s)	Manufacturer
Duotrate	Pentaerythritol tetranitrate	Jones
Nico-400	Nicotinic acid	Jones
Nitro-Bid	Nitroglycerin	Marion
Cerespan	Papaverine hydrochloride	Rhône-Poulenc Rorer
Nitrospan	Nitroglycerin	Rorer
Measurin	Acetylsalicylic acid	Sterling Winthrop

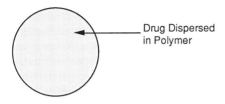

Time = 0

Time = t

Fig. 7   Matrix diffusional system before drug release (time = 0) and after partial drug release (time = *t*).

the bathing solution is dissolved first and then diffuses out of the matrix. This process continues with the interface between the bathing solution and the solid drug moving toward the interior. Obviously, for this system to be diffusion-controlled, the rate of dissolution of drug particles within the matrix must be much faster that the diffusion rate of dissolved drug leaving the matrix. Derivation of the mathematical model to describe this system involves the following assumptions [29,30]: (a) a pseudo-steady state is maintained during drug release, (b) the diameter of the drug particles is less than the average distance of drug diffusion through the matrix, (c) the bathing solution provides sink conditions at all times, (d) the diffusion coefficient of drug in the matrix remains constant (i.e., no change occurs in the characteristics of the polymer matrix).

The next equations, which describe the rate of release of drugs dispersed in an inert matrix system, have been derived by Higuchi [29]. The following equation can be written based on Fig. 3:

$$\frac{dM}{dh} = C_0 \, dh - \frac{C_s}{2} \tag{8}$$

where

dM = change in the amount of drug released per unit area
dh = change in the thickness of the zone of matrix that has been depleted of drug
C_0 = total amount of drug in a unit volume of the matrix
C_s = saturated concentration of the drug within the matrix.

From diffusion theory,

$$dM = \frac{D_m C_s}{h} \, dt \tag{9}$$

where D_m is the diffusion coefficient in the matrix. Equating Eqs. (8) and (9), integrating, and solving for h gives

$$M = [C_s D_m (2C_0 - C_s)t]^{1/2} \qquad (10)$$

When the amount of drug is in excess of the saturation concentration, that is, $C_0 \gg C_s$

$$M = (2C_s D_m C_0 t)^{1/2} \qquad (11)$$

which indicates that the amount of drug released is a function of the square root of time. In a similar manner, the drug release from a porous or granular matrix can be described by

$$M = \left[D_s C_a \frac{p}{T} (2C_0 - pC_a)t \right]^{1/2} \qquad (12)$$

where

 p = porosity of the matrix
 T = tortuosity
 C_a = solubility of the drug in the release medium
 D_s = diffusion coefficient in the release medium

This system is slightly different from the previous matrix system in that the drug is able to pass out of the matrix through fluid-filled channels and does not pass through the polymer directly.

For purposes of data treatment, Eq. (11) or (12) can be reduced to

$$M = kt^{1/2} \qquad (13)$$

where k is a constant, so that a plot of amount of drug released versus the square root of time will be linear, if the release of drug from the matrix is diffusion-controlled. If this is the case, then, by the Higuchi model, one may control the release of drug from a homogeneous matrix system by varying the following parameters [31–35]: (a) initial concentration of drug in the matrix, (b) porosity, (c) tortuosity, (d) polymer system forming the matrix, and (e) solubility of the drug.

Matrix systems offer several advantages. They are, in general, easy to make and can be made to release high-molecular-weight compounds. Since the drug is dispersed in the matrix system, accidental leakage of the total drug component is less likely to occur, although, occasionally, cracking of the matrix material can cause unwanted release. The primary disadvantages of this system are that the remaining matrix "ghost" must be removed after the drug has been released. Also, the release rates generated are not zero-order, since the rate varies with the square root of time. A substantial sustained effect, however, can be produced through the use of very slow release rates, which in many applications are indistinguishable from zero-order. The characteristics of the system are summarized in Table 5, and a representative listing of available products is given in Table 6.

Table 5 Characteristics of Matrix Diffusion Systems

Description	Homogeneous dispersion of solid drug in a polymer mix
Advantages	Easier to produce than reservoir devices
	Can deliver high-molecular-weight compounds
Disadvantages	Cannot obtain zero-order release
	Removal of remaining matrix is necessary for implanted systems

Table 6 Matrix Diffusional Products

Product (tablets)	Active ingredient(s)	Manufacturer
Desoxyn-Gradumet	Methamphetamine hydrochloride	Abbott
Fero-Gradumet	Ferrous sulfate	Abbott
Tral Filmtab	Hexocyclium methylsulfate	Abbott
PBZ-SR	Tripelennamine	Geigy
Procan SR	Procainamide hydrochloride	Parke-Davis
Choledyl SA	Oxtriphylline	Parke-Davis

F. Bioerodible and Combination Diffusion and Dissolution Systems

Strictly speaking, therapeutic systems will never be dependent on dissolution only or diffusion only. However, in the foregoing systems, the predominant mechanism allows easy mathematical description. In practice, the dominant mechanism for release will overshadow other processes enough to allow classification as either dissolution rate-limited or diffusion-controlled. Bioerodible devices, however, constitute a group of systems for which mathematical descriptions of release characteristics can be quite complex. Characteristics of this type of system are listed in Table 7. A typical system is shown in Fig. 8. The mechanism of release from simple erodible slabs, cylinders, and spheres has been described [36]. A simple expression describing release from all three of these erodible devices is

$$\frac{M_t}{M} = 1 - \left(1 - \frac{k_0 t}{C_0 a}\right)^n \tag{14}$$

where $n = 3$ for a sphere, $n = 2$ for a cylinder, and $n = 1$ for a slab. The radius of a sphere, or cylinder, or the half-height of a slab is represented by a. M_t is the mass of a drug release at time t and M is the mass released at infinite time. As a further complication, these systems can combine diffusion and dissolution of both the matrix material and the drug. Drug not only can diffuse out of the dosage form, as with some previously described matrix systems, but the matrix itself undergoes a dissolution process. The complexity of the system arises from the fact that, as the polymer dissolves, the diffusional path length for the drug may change. This usually results in a moving-boundary diffusion system. Zero-order release can occur only if surface erosion occurs and surface area does not change with time. The inherent advantage of such a system is that the bioerodible property of the matrix does not result in a ghost matrix. The disadvantages of these matrix systems are that release kinetics are often hard to control, since many factors affecting both the drug and the polymer must be considered.

Table 7 Characteristics of Bioerodible Matrix Systems

Description	A homogeneous dispersion of drug in an erodible matrix
Advantages	All the advantages of matrix dissolution system
	Removal from implant sites is not necessary
Disadvantages	Difficult to control kinetics owing to multiple processes of release
	Potential toxicity of degraded polymer must be considered

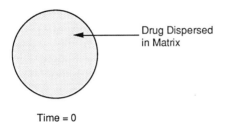

Time = 0

Time = t

Fig. 8 Representation of a bioerodible matrix system. Drug is dispersed in the matrix before release at time = 0. At time = t, partial release by drug diffusion or matrix erosion has occurred.

Another method for the preparation of bioerodible systems is to attach the drug directly to the polymer by a chemical bond [37]. Generally, the drug is released from the polymer by hydrolysis or enzymatic reaction. This makes control of the rate of release somewhat easier. Another advantage of the system is the ability to achieve very high drug loading, since the amount of drug placed in the system is limited only by the available sites on the carrier.

A third type, which in this case utilizes a combination of diffusion and dissolution, is that of a swelling-controlled matrix [38]. Here the drug is dissolved in the polymer, but instead of an insoluble or eroding polymer, as in previous systems, swelling of the polymer occurs. This allows entrance of water, which causes dissolution of the drug and diffusion out of the swollen matrix. In these systems the release rate is highly dependent on the polymer-swelling rate, drug solubility, and the amount of soluble fraction in the matrix [39]. This system usually minimizes burst effects, since polymer swelling must occur before drug release.

G.   Osmotically Controlled Systems

In these systems, osmotic pressure provides the driving force to generate controlled release of drug. Consider a semipermeable membrane that is permeable to water, but not to drug. A tablet containing a core of drug surrounded by such a membrane is shown in Fig. 9. When this device is exposed to water or any body fluid, water will flow into the tablet owing to the osmotic pressure difference. The rate of flow, dV/dt, of water into the device can be represented as

$$\frac{dV}{dt} = \frac{Ak}{h(\Delta\Pi - \Delta P)} \tag{15}$$

where

   k = membrane permeability
   A = area of the membrane

h = membrane thickness
$\Delta\Pi$ = osmotic pressure difference
ΔP = hydrostatic pressure difference

These systems generally appear in two different forms, as depicted in Fig. 9. The first contains the drug as a solid core together with electrolyte, which is dissolved by the incoming water. The electrolyte provides the high osmotic pressure difference. The second system contains the drug in solution in an impermeable membrane within the device. The electrolyte surrounds the bag. Both systems have single or multiple holes bored through the membrane to allow drug release. In the first example, high osmotic pressure can be relieved only by pumping solution, containing drug, out of the hole. Similarly, in the second example, the high osmotic pressure causes compression of the inner membrane, and drug is pumped out through the hole.

In the system with the bag, or if the hole is large enough in either system, the hydrostatic difference becomes negligible, and Eq. (15) becomes

$$\frac{dV}{dt} = \frac{Ak}{h(\Delta\Pi)} \tag{16}$$

indicating that the flow rate of water into the tablet is governed by permeability, area, and thickness of the membrane. The rate of drug leaving the orifice, dM/dt, is equivalent to the

Type A

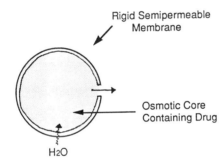

Type B

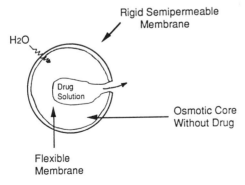

Fig. 9 Diagrammatic representation of two types of osmotically controlled systems. Type A contains an osmotic core with drug. Type B contains the drug solution in a flexible bag, with the osmotic core surrounding.

flow rate of incoming water multiplied by the solution concentration of drug, C_s, within the device:

$$\frac{dM}{dt} = \frac{dV}{dt} C_s \tag{17}$$

Osmotic systems have application in pharmacological studies, implantation therapies, and oral drug delivery.

In systems with solid drug dispersed with electrolyte, the size or number of bored hole(s) are the rate-limiting factors for release of drug. Quality control of the manufacture of these systems must be exceptional, since any variation in boring of the hole, accomplished with a laser drill, can have a substantial effect on release characteristics. Most of the orally administered osmotic systems are of this variety. A variation on this theme is an osmotic system of similar design without a hole. The building osmotic pressure causes the tablet to burst, causing all the drug to be rapidly released [40]. This design is useful for drugs that are difficult to formulate in tablet or capsule form.

These osmotic systems are advantageous in that they can deliver large volumes, and some are refillable. Most important, the release of drug is in theory independent of the drug's properties [41,42]. This allows one dosage form design to be used for almost any drug. Disadvantages are that the systems are relatively expensive and, for certain applications, require implantation. For drugs that are unstable in solution, these systems may be inappropriate because the drug remains in solution form for extended periods before release. System characteristics are summarized in Table 8.

H. Ion-Exchange Systems

Ion-exchange systems generally use resins composed of water-insoluble cross-linked polymers. These polymers contain salt-forming functional groups in repeating positions on the polymer chain. The drug is bound to the resin and released by exchanging with appropriately charged ions in contact with the ion-exchange groups.

$$\text{Resin}^+\text{-drug}^- + X^- \rightarrow \text{resin}^+\text{-}X^- + \text{drug}^-$$

conversely,

$$\text{Resin}^-\text{-drug}^+ + Y^+ \rightarrow \text{resin}^-\text{-}Y^+ + \text{drug}^+$$

where X^- and Y^+ are ions in the GI tract. The free drug then diffuses out of the resin. The drug–resin complex is prepared either by repeated exposure of the resin to the drug in a chromatography column, or by prolonged contact in solution.

Table 8 Characteristics of Osmotically Controlled Devices

Description	Drug surrounded by semipermeable membrane and release governed by osmotic pressure
Advantages	Zero-order release is obtainable
	Reformulation is not required for different drugs
	Release of drug independent of the environment of the system
Disadvantages	Systems can be much more expensive than conventional counterparts
	Quality control is more extensive than most conventional tablets

The rate of drug diffusing out of the resin is controlled by the area of diffusion, diffusional path length, and rigidity of the resin, which is a function of the amount of cross-linking agent used to prepare the resin.

This system is advantageous for drugs that are highly susceptible to degradation by enzymatic processes, since it offers a protective mechanism by temporarily altering the substrate. This approach to sustained release, however, has the limitation that the release rate is proportional to the concentration of the ions present in the area of administration. Although the ionic concentration of the GI tract remains rather constant with limits [15], the release rate of drug can be affected by variability in diet, water intake, and individual intestinal content. A representative listing of ion-exchange products is given in Table 9.

An improvement in this system is to coat the ion-exchange resin with a hydrophobic rate-limiting polymer, such as ethylcellulose or waxes [43]. These systems rely on the polymer coat to govern the rate of drug availability.

IV. TARGETED DELIVERY SYSTEMS

Targeted systems represent the next level in "state-of-the-art" controlled drug delivery systems. These systems address the problem of spatial placement of therapeutic compounds. Since the site of drug action is the target of these systems, oral administration is generally not used as a method of delivery.

A. Liposomes

Liposomes have been, and continue to be, of considerable interest in drug delivery systems. A schematic diagram of their production is shown in Fig. 10. Liposomes are normally composed of phospholipids that spontaneously form multilamellar, concentric, bilayer vesicles, with layers of aqueous media separating the lipid layers. These systems, commonly referred to as multilamellar vesicles (MLVs), have diameters in the range of 1–5 μm. Sonication of MLVs results in the production of small unilamellar vesicles (SUVs), with diameters in the range 0.02–0.08 μm. These vesicles are a single, lipid outer layer, with an aqueous inner core. Large unilamellar vesicles (LUVs) can also be made by evaporation under reduced pressure, resulting in liposomes with a diameter of 0.1–1 μm. Further extrusion of LUVs through a membrane filter will also result in SUVs.

To use liposomes as delivery systems, drug is added during the formation process. Hydrophilic compounds usually reside in the aqueous portion of the vesicle, whereas hydrophobic species tend to remain in the lipid proteins. The physical characteristics and stability of liposomal preparations depend on pH, ionic strength, the presence of divalent cations, and the nature of the phospholipids and additives used [44–46].

In general, these vesicle systems demonstrate low permeability to ionic and polar substances, but this varies greatly with liposome composition. Those made with positively charged phos-

Table 9 Ion-Exchange Products

Product	Active ingredient(s)	Manufacturer
Biphetamine capsules	Amphetamine, dextroamphetamine	Fisons
Tussionex suspension	Hydrocodone, chlorpheniramine	Fisons
Ionamin capsules	Phenteramine	Pennwalt
Delsym solution	Dextromethorphan hydrobromide	McNeil

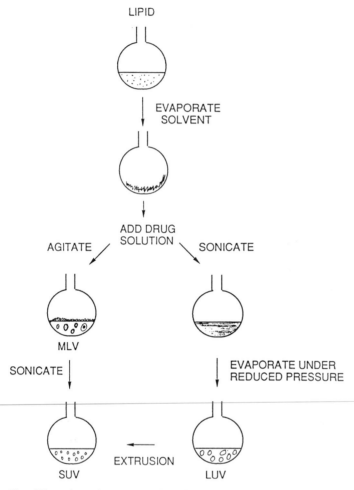

Fig. 10 Schematic representation of a procedure for the production of liposomes.

pholipids are impermeable to cations, whereas negatively charged liposomes are permeable to cations, and both types are readily permeated by anions [47]. The degree of saturation or the length of the phospholipid fatty acid chain will also greatly affect the solute permeability of the liposomes [48]. An increase in temperature can also alter permeability [49], by causing the lipids to undergo a phase transition to a less-ordered, more fluid configuration. Again, the transition is characteristic for differing types of lipids. This has been employed in a unique-targeting approach, by creating an environment of local hyperthermia, the liposomes are encouraged to release their encapsulated cargo in that specified area, for example, a capillary bed [50,51].

Some proteins, such as those found in serum, are able to deform, penetrate the bilayer, or remove lipid components, resulting in changes in liposome permeability [52]. Many additives, such as cholesterol, are able to inhibit this effect, stabilizing the membrane structure of the vesicle and limiting cargo leakage [53]. This is achieved by allowing closer lipid packing [54]. The fact that impurities, such as cholesterol or free fatty acids [55], can dramatically change the permeability and surface charge of liposomes points to the necessity for strict controls on the quality and purity of lipids used in liposomal preparation.

Liposomes that remain impermeable to their contents cannot release these compounds without interaction with cells. This cellular interaction occurs by three different mechanism (Fig. 11) [56]. Of these, fusion and adsorption usually involve drug leakage, whereas effective drug delivery results from endocytosis.

1. *Fusion of the liposome with the cell membrane.* For this, the lipid portion of the vesicle becomes part of the cell wall.
2. *Adsorption to the cell wall.* For this, transfer of liposome content must be by diffusion through the lipids of the liposome and the cell membrane.
3. *Endocytosis of the vesicle by the cell.* The entire liposomal contents are made available to the cell.

The advantageous effects of liposomal carrier systems include protection of compounds from metabolism or degradation, as well as enhanced cellular uptake. Liposome-mediated delivery of cytotoxic drugs to cells in culture has resulted in improved potency [57,58]. Prolonged release of encapsulated cargo has also been demonstrated [59,60]. More recently, liposomes with extended circulation half-lives and dose-independent pharmacokinetics (Stealth liposomes) [61], have shown promise in delivery of drugs that are normally very rapidly degraded.

Liposomes, however, also have inherent disadvantages in the areas of stability and uniformity of production. Once a system has demonstrated merit for treatment of a particular disease state, the following must be determined before a formulation is acceptable for marketing and human use: (a) lipid purity and stability; (b) drug stability and leakage from the vesicles; (c) lipid–drug cargo interaction; and (d) control of vesicle size and drug-loading efficiency for large-batch production.

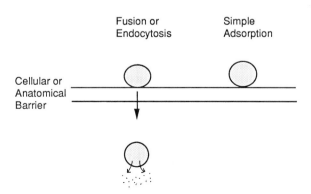

Fig. 11 Schematic representation of liposome interactions at a membrane surface.

B. Prodrugs

A *prodrug* is a compound resulting from chemical modification of a biologically active compound that will liberate the active form in vivo by enzymatic or hydrolytic cleavage. The primary purpose in forming a prodrug is to modify the physicochemical properties of the drug, usually to alter the membrane permeability of the parent compound. This change in physicochemical properties of the drug influences the ultimate localization of the drug. There are various reasons for formulating a prodrug system. If the parent compound is insoluble, this can be modified [62]. If it is easily degraded, modification can protect the parent compound from enzymatic or hydrolytic attack. Modifications can also reduce side effects, such as GI irritation [63]. Several drugs are now marketed in the form of a prodrug; for example, sulindac, a nonsteroidal anti-inflammatory agent, and numerous angiotension-converting enzyme (ACE) inhibitors. The necessary conversion of prodrug to parent can occur by a variety of reactions, the most common being hydrolytic cleavage [64]. The prodrug ester forms of a hydroxyl or carboxyl group of the parent compound can be readily cleaved by blood esterase. Other activation processes may include biochemical reduction or oxidation. However the conversion occurs, to achieve sustained drug action, the rate of conversion from prodrug to active compound should not be too high [65]. Site-specific, controlled delivery is achieved by the antiviral prodrug acyclovir, being converted to active form by a virus-specific enzyme [66]. Sustained release of steroid prodrugs, especially progestagens and progestagen–estrogen combinations, have seen a substantial amount of clinical experience, both as a means of birth control and as symptomatic menopausal treatment [67].

The concept of the double prodrug (proprodrugs), may allow more controlled delivery of various prodrug compounds [68]. For example, if a prodrug that shows site-specific activation, but has poor transport properties or stability problems, it could be converted to a proprodrug that transported better or is more stable (Fig. 12). Prodrug systems have been taken even further by including as prodrugs, polymer prodrugs, in which a drug is covalently linked to a polymer backbone. This type of system could encompass a staggering number of possibilities. Encouraging results have been shown with mitomycin [69,70], for example.

The most serious disadvantage to the prodrug approach to controlled–sustained delivery is that extensive development must be undertaken to find the correct chemical modification for a specific drug. Additionally, once a prodrug is formed, it is a new drug entity and, therefore, requires extensive and costly studies to determine safety and efficacy.

C. Nanoparticles

Nanoparticles are solid colloidal particles ranging in size from 10 to 1000 nm. They can be used as drug carriers, with the drug encapsulated, dissolved, adsorbed, or covalently attached

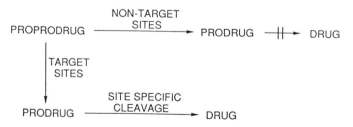

Fig. 12 Illustration of prodrug and proprodrug concept.

[71,72]. The small size of the nanoparticles permits administration by intravenous injection, and also permits their passage through capillaries that remove larger particles. They are usually taken up by the liver, spleen, and lungs [73,74].

Preparation of nanoparticles can be by a variety of different ways. The most important and frequently used is emulsion polymerization; others include interfacial polymerization, solvent evaporation, and desolvation of natural proteins. The materials used to prepare nanoparticles are also numerous, but most commonly they are polymers such as polyalkylcyanoacrylate, polymethylmethacrylate, polybutylcyanoacrylate, or from albumin or gelatin. Distribution patterns of the particles in the body can vary depending on their size, composition, and surface charge [75–77]. In particular, nanoparticles of polycyanoacrylate have been found to accumulate in certain tumors [78,79].

There are several possible ways that the drug cargo can be incorporated into nanoparticles. They may be bound by polymerization of the nanoparticles in the presence of drug solution, or by absorption of the drug onto prepolymerized nanoparticles. The drug will be dispersed in the particle's polymer matrix [80], or adsorbed to the surface, depending on its affinity to the polymer. Drugs used for nanoparticle delivery have been, for the most part, cytotoxic agents; dactinomycin (actinomycin D) [81], 5-fluorouracil [82,83], doxorubicin [84,85], and methotrexate [86]. But also for delivery of bioactive peptides and proteins, for example, growth hormone-releasing factor [87,88].

Nanoparticles show great promise as devices for the controlled release of drugs, provided that the choice of material for nanoparticle formation is made with the appropriated considerations of the drug cargo, administration route, and the desired site of action.

D. Resealed Erythrocytes

When red blood cells are placed in hypotonic media, they swell, which causes rupturing of the membrane and formation of pores. These pores allow free exchange of intra- and extracellular components. Readjustment of the solution tonicity to isotonic allows resealing of the membrane. This technique usually allows encapsulation of up to 25% of the drug or enzyme in solution [89]. In addition to this method, called the preswell dilution technique, there are other ways to form drug-loaded erythrocytes. In the dialysis technique, the red blood cells are placed in dialysis tubes that are immersed in a hypotonic medium. This results in retention of cytoplasmic components when the cells are resealed. Another method involves subjecting the cells to an intense electric field, causing pores to form, which again, can be resealed after drug uptake.

The potential advantages of loaded red blood cells as delivery systems are [90]:

1. They are biodegradable and nonimmunogenic.
2. They can be modified to change their resident circulation time, depending on their surface (cells with little membrane damage can circulate for prolonged periods).
3. Entrapped drug is shielded from immunological detection and external enzymatic degradation.
4. The system is relatively independent of the physicochemical properties of the drug (i.e., it does not require chemical modifications).

In general, normally aging erythrocytes and slightly damaged cells are sequestered in the spleen, whereas those heavily damaged or modified are removed from circulation by the liver [91]. This, along with a short storage life of about 2 weeks [92], constitutes the major drawbacks of using resealed erythrocytes as drug carriers.

E. Antibody-Targeted Systems

An alternative drug delivery system makes use of macromolecular attachment for delivery using immunoglobulins as the macromolecule. The obvious advantage of this system is that it can be targeted to the site of the antibody specificity. Although this usually does not provide much of a sustaining mechanism, the problem of spatial placement is addressed. The advantage in this is that far less drug is used, and side effects can be reduced substantially.

Drugs are linked, covalently or noncovalently, to the antibody [93], or placed in vesicles such as liposomes or microspheres, and the antibody used to target the liposome [94,95] (Fig. 13). Covalent attachment is generally not very efficient and also diminishes the antigen-binding capacity [96,97]. There are only a few functional groups available per antibody that can be used for chemical coupling without affecting the antibody's binding activity. If conjugation is done through an intermediate carrier molecule, one can increase the drug/antibody ratio [98,99]. Such intermediates have included dextran or poly-L-glutamic acid [100–102].

There are many drugs that have been conjugated to antibodies or their fragments, a few are daunomycin [103], cyclosporine [104], platinum [105], chlorambucil [106], and vindesine [107]. When choosing a drug for this type of delivery, one must consider many things [108], such as whether the drug is active extra- or intracellularly, if it must be cleaved from the antibody to be active, and the strength and method of coupling.

Immunoliposomes—liposomes loaded with drug cargo that have been surface-conjugated to antibodies, or antibody fragments—have also been investigated by a number of researchers. Linkage of antibody to a liposome can be covalent or noncovalent. Spacers are used for covalent binding, or the antibody is modified by attaching an "anchor" group [109] for noncovalent coupling. The anchor group, which is hydrophobic, inserts into the bilayer of the liposome, "anchoring" the antibody to the vesicle. Numerous antibody–liposome combinations have been looked at, delivering both drugs [110–112], and genetic material [113–115].

The obvious advantage to antibody-targeted systems is that through the use of monoclonal antibodies, which recognize only the tumor antigen, side effects of cytotoxic chemicals on the rest of the body could be greatly reduced. These systems represent a novel and currently high-interest research area of drug delivery. Their potential value in the delivery of compounds to directed targets has generated considerable interest.

V. DENTAL SYSTEMS

Controlled and sustained drug delivery has recently begun to make an impression in the area of treatment of dental diseases. Many researchers have demonstrated that controlled delivery of antimicrobial agents, such as chlorhexidine [118–120], ofloxacin [121–123], and metroni-

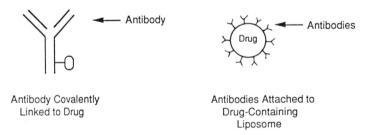

Antibody Covalently
Linked to Drug

Antibodies Attached to
Drug-Containing
Liposome

Fig. 13 Diagrammatic representation of two types of antibody-targeted systems. Drug is either covalently linked directly to the antibody, or is contained in liposomes that are targeted by attached antibodies.

dazole [124] can effectively treat and prevent periodontitis. The incidence of dental caries and formation of plaque can also be reduced by controlled delivery of fluoride [125,126]. Delivery systems used are film-forming solutions [119,120], polymeric inserts [122], implants, and patches. Since dental disease are usually chronic, sustained release of therapeutic agents in the oral cavity would obviously be desirable.

VI. OCULAR SYSTEMS

The eye is unique in its therapeutic challenges. An efficient mechanism, that of tears and tear drainage, which quickly eliminates drug solution, makes topical delivery to the eye somewhat different from most other areas of the body [127]. Usually less than 10% of a topically applied dose is absorbed into the eye, leaving the rest of the dose to potentially absorb into the bloodstream [128], resulting in unwanted side effects. The goal of most controlled delivery systems is to maintain the drug in the precorneal area and allow its diffusion across the cornea. Suspensions and ointments, although able to provide some sustaining effect, do not offer the amount of control desired [129,130]. Polymeric matrices can often significantly reduce drainage [131], but other newer methods of controlled drug delivery can also be used.

The application of ocular therapy generally includes glaucoma, artificial tears, and anticancer drugs for intraocular malignancies. The sustained release of artificial tears has been achieved by a hydroxypropylcellulose polymer insert [132]. However, the best-known application of diffusional therapy in the eye, Ocusert-Pilo, the device shown in Fig. 14, is a relatively simple structure with two rate-controlling membranes surrounding the drug reservoir containing pilocarpine. Thus, a thin, flexible lamellar ellipse is created and serves as a model reservoir device. The unit is placed in the eye and resides in the lower cul-de-sac, just below the cornea. Since the device itself remains in the eye, the drug is released into the tear film.

The advantages of such a device are that it can control intraocular pressure for up to a week [133]. Control is achieved with less drug and fewer side effects, since the release of drug is zero-order. The system is more convenient, since application is weekly, as opposed to the four times a day dosing for pilocarpine solutions. This greatly improves patient compliance and assures round-the-clock medication, which is of great importance for glaucoma treatment. The main disadvantage of the system is that it is often difficult to retain in the eye, and can be of some discomfort.

Another method of delivery of drug to the anterior segment of the eye, which has proved successful, is that of prodrug administration [134]. Since the corneal surface presents an effective lipoidal barrier, especially to hydrophilic compounds, it seems reasonable that a prodrug that is more lipophilic than the parent drug will be more successful in penetrating this barrier.

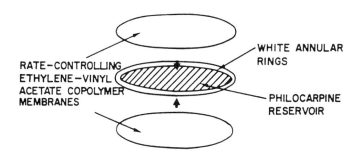

Fig. 14 Schematic diagram of the Ocusert intraocular device for release of pilocarpine.

One drug that has been formulated in this manner is dipivalyl epinephrine (Dividephrine), a dipivalyl ester of epinephrine. Epinephrine itself is poorly absorbed owing to its polar characteristics and is highly metabolized. The prodrug form is approximately ten times as effective at crossing the cornea and produces substantially higher aqueous humor levels [134,135]. For another prodrug, phenylephrine pivalate, there is some possibility that the prodrug itself is therapeutically active [136,137]. Many other drugs have been derivatized for prodrug ocular delivery: timolol [138,139], nadolol [140], pilocarpine [141,142], prostaglandin $F_{2\alpha}$ [143,144], terbutaline [145], acyclovir [146], vidarabine [147], and idoxuridine [148,149].

New sustained-release technologies are gaining in ocular delivery, as in other routes. Liposomes as drug carriers have achieved enhanced ocular delivery of certain drugs [150]; antibiotics [151–153], and peptides [154]. Biodegradable matrix drug delivery to the anterior segment has also been studied [155,156]. Prolonged delivery of pilocarpine can be achieved with a polymeric dispersion [157]. Implantation of polymers containing endotoxin for neovascularization [158], gancyclovir [159], 5-flurouracil [160], and injections of doxorubicin (Adriamycin) [161] have also resulted in sustained delivery. However, topical ocular delivery is preferred considerably over implants and injections.

VII. TRANSDERMAL SYSTEMS

The transdermal route of drug administration offers several advantages over other methods of delivery. For some cases, oral delivery may be contraindicated, or the drug may be poorly absorbed. This would also include situations for which the drug undergoes a substantial first-pass effect [162], and systemic therapy is desired.

The skin, although presenting a barrier to most drug absorption, provides a very large surface area for diffusion. Below the barrier of the stratum corneum is an extensive network of capillaries. Since the venous return from these capillary beds does not flow directly to the liver, compounds are not exposed to these enzymes during absorption [162]. A most notable example of such a drug is nitroglycerin, which as been administered both sublingually and transdermally to avoid first-pass metabolism. Other drugs that have seen success in controlled transdermal delivery are testosterone [163], fentanyl [164,165], bupranolol [166], and clonidine.

Transdermal controlled-release systems can be used to deliver drugs with short biological half-lives and can maintain plasma levels of very potent drugs within a narrow therapeutic range for prolonged periods. Should problems occur with the system, or a change in the status of the patient require modification of therapy, the system is readily accessible and easily removed.

One of the primary disadvantages to this method of delivery is that drugs requiring high blood levels to achieve an effect are difficult to load into a transdermal system owing to the large amount of material required. These systems would naturally be contraindicated if the drug or vehicle caused irritation to the skin. Also, various factors affecting the skin, such as age, physical condition, and device location, can change the reliability of the system's ability to deliver medication in a controlled manner. In other words, both the drug and the nature of the skin can affect the system design.

Current controlled transdermal-release systems can be classified into four types, as follows, with a representative product and manufacturer.

1. Membrane permeation-controlled system in which the drug permeation is controlled by a polymeric membrane.
    Transderm-Scop (scopolamine; Ciba-Geigy)

2. Adhesive dispersion-type system is similar to the foregoing but lacks the polymer membrane, instead the drug is dispersed into an adhesive polymer.
    Deponit (nitroglycerin; Wyeth)
3. Matrix diffusion-controlled system in which the drug is homogeneously dispersed in a hydrophilic polymer, diffusion from the matrix controls release rate.
    Nitrodur (nitroglycerin; Key)
4. Microreservoir dissolution-controlled system in which microscopic spheres of drug reservoir are dispersed in a polymer matrix.
    Nitrodisc (nitroglycerin; Searle)

Most marketed systems are of the polymeric membrane-controlled type, representative of these is Transderm-Scop. This product, shown in Fig. 15, is designed to deliver scopolamine over a period of days, without the side effects commonly encountered when the drug is administered orally [167]. The system consists of a reservoir containing the drug dispersed in a separate phase within a highly permeable matrix. This is laminated between the rate-controlling microporous membrane and an external backing that is impermeable to drug and moisture. The pores of the rate-controlling membrane are filled with a fluid that is highly permeable to scopolamine. This allows delivery of the drug to be controlled by diffusion through the device and skin. Control is achieved because, at equilibrium, the membrane is rate-limiting for drug permeation. To initiate an immediate effect, a priming dose is contained in a gel on the membrane side of the device.

Another drug that is popular for controlled transdermal release is nitroglycerin. Conventionally, this drug is administered sublingually, although the duration of action by this route is quite short. This is acceptable for acute anginal attacks, but not for prophylactic treatment. Oral administration has the disadvantage that large fractions of the dose are lost to first-pass metabolism in the liver. Topical ointments have long been used for prophylactic treatment of angina, but their duration is only 4–8 hr and, in addition, are not aesthetically acceptable. The transdermal nitroglycerin devices employ a variety of systems to provide 24-hr delivery.

VIII. VAGINAL AND UTERINE SYSTEMS

Sustained- and controlled-release devices for drug delivery in the vaginal and uterine areas are most often for the delivery of contraceptive steroid hormones. The advantages in administration by this route—prolonged release, minimal systemic side effects, and an increase in bioavailability—allow for less total drug than with an oral dose. First-pass metabolism that inactivates many steroid hormones can be avoided [168,169].

One such application is the medicated vaginal ring [170]. Therapeutic levels of medroxyprogesterone have been achieved at a total dose that was one-sixth the required oral dose [171]. Ring delivery devices have several problems that have limited their use; vaginal wall erosion and ring expulsion, to name a few. Microcapsules have also recently been useful for vaginal and cervical delivery [172]. Local progesterone release from this dosage form can alter cervical mucus to interfere with sperm migration [173]. Other steroids have also attained sustained delivery by an intracervical system [174]. The sustained release of progesterone from various polymers given vaginally have also been found useful in cervical ripening and induction of labor [175–177].

A more common contraceptive device is the intrauterine device (IUD). The first intrauterine devices used were of the undedicated type. These have received increased attention since the use of polyethylene plastics and silicone rubbers [178–180]. These materials had the ability to resume their shape following distortion. Because they are unmedicated, these IUDs cannot

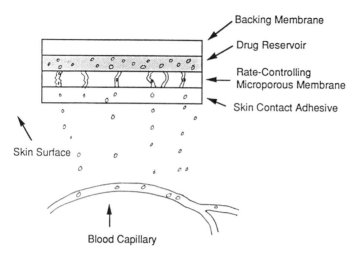

Backing Membrane

Drug Reservoir

Rate-Controlling
Microporous Membrane

Skin Contact Adhesive

Skin Surface

Blood Capillary

Fig. 15   Schematic diagram of a transdermal device for the delivery of scopolamine.

be classified as sustained-release products. It is believed their mechanism of action is due to local endometrial responses, both cellular and cytosecretory [181]. Initial investigations of these devices led to the conclusions that the larger the device, the more effective it was in preventing pregnancy. Large devices, however, increased the possibility of uterine cramps, bleeding, and expulsion of the device.

Efforts to improve IUDs have led to the use of medicated devices. Two types of agents are generally used, contraceptive metals and steroid hormones. The metal device is exemplified by the CU-7, a polypropylene plastic device in the shape of the number 7. Copper is released by a combination of ionization and chelation from a copper wire wrapped around the vertical limb. This system is effective for up to 40 months.

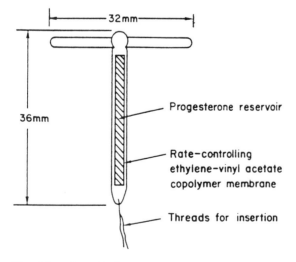

32mm

36mm

Progesterone reservoir

Rate-controlling
ethylene-vinyl acetate
copolymer membrane

Threads for insertion

Fig. 16   Schematic diagram of the Progestasert intrauterine device for the release of progesterone.

The hormone-releasing devices have a closer resemblance to standard methods of sustained release because they involve the release of a steroid compound by diffusion [182,183]. The Progestasert, a reservoir system, is diagrammed in Fig. 16. Progesterone, the active ingredient, is dispersed in the inner reservoir, surrounded by an ethylene/vinyl acetate copolymer membrane. The release of progesterone from this system is maintained almost constant for 1 year. The effects of release are local, with none of the systematic side effects observed with orally administered contraceptives [184–191].

IX. INJECTIONS AND IMPLANTS

One of the most obvious ways to provide sustained-release medication is to place the drug in a delivery system and inject or implant the system into the body tissue. The concept of these delivery methods is not new, but the technology applied is contemporary. Administration of these systems often requires surgical implantation or specialized injection devices. The fact that these systems are in constant contact with exposed tissue components places certain requirements on the systems and their polymer composition.

In general, the materials used must be biocompatible; that is the polymers themselves must not cause irritation at the implantation site, or promote infection or sterile abscess. The most common polymers used are hydrogels, silicones, and biodegradable materials [192]. Hydrogels have the advantageous property of being able to retain large amounts of water within their structure without dissolving [193]. This high aqueous content makes them very compatible with living tissues, but disadvantageously, allows low-molecular-weight substances to diffuse out quickly. Cross-linking agents can be used to reduce this diffusional loss and to provide structural rigidity, but this can increase the frictional irritation of the hydrogel with its surrounding tissue.

Subcutaneous implantation is currently one of the most utilized routes to investigate the potential of sustained-delivery systems. This is because favorable absorption sites are available, and removal of the device can be accomplished at any time. Surgery is often required, and in itself can be considered a disadvantage, as well as the fact that once implanted, the delivery rate of the drug is usually fixed until the device is removed. The development of implants has a long history, starting initially with investigations on implanted silicone devices. The most notable new implantable product is Norplant, a contraceptive device releasing levonorgestrel for up to 5 years [194]. This product is implanted subdermally and requires only a local anesthetic. A variety of other drugs have also been used, including thyroid hormones, steroids, cardiovascular agents [195–198], insulin [199], and nerve growth factor [200].

Sustained-release injections, subcutaneous and intramuscular, have been investigated in a variety of different formulations [201,202]. Injections of degradable microspheres have efficiently prolonged delivery of numerous drugs [203–206], even antigenic substances and vaccines to produce immunity [207,208].

Some implantation devices have extended well beyond the classic diffusional systems and have included not only bioerodible devices, but implantable therapeutic systems that can be activated. There are devices activated by change in osmotic pressure to deliver insulin [209], morphine release trigger by vapor pressure [210], and pellets activated by magnetism to release their encapsulated drug load [211]. Such external control of an embedded device would eliminate many of the disadvantages of most implanted delivery systems.

X. CONCLUSIONS

The space limitations of a written work such as this do not permit a complete discourse on all of the sustained and controlled mechanisms available for possible drug delivery. Such a chapter would be quite expansive, so that instead, an attempt has been made to cover as much as possible the major and currently marketed types. Current research in this area involves many more novel systems, many of which have strong therapeutic potential. The future of this area is limited only by the imagination of those who chose to become involved in this field.

REFERENCES

1. V. H. Lee and J. R. Robinson, in *Sustained and Controlled Release Drug Delivery Systems* (J. R. Robinson, ed.), Marcel Dekker, New York, 1978, pp. 71–121.
2. M. E. Jacobson, Postgrad. Med., 49, 181, (1971).
3. B. Beerman, K. Helstrom, and A. Rosen, Clin. Pharmacol. Ther., 13, 212 (1972).
4. A. B. Morrison, C. B. Perusse, and J. A. Campbell, N. Engl. J. Med., 263, 115 (1960).
5. E. J. Middleton, E. Nagy, and A. B. Morrison, N. Engl. J. Med., 274, 136 (1966).
6. P. G. Welling and R. H. Barbhaiya, J. Pharm. Sci., 71, 32 (1982).
7. B. C. Thanoo, M. C. Sunny, and A. Jayakrishnan, J. Pharm. Pharmacol., 45, 21 (1993).
8. W. L. Xu, X. D. Tu, and Z. D. Lu, Acta Pharm. Sin., 26, 541 (1991).
9. S. Watanabe, M. Kayano, Y. Ishino, and K. Miyao, U.S. Patent 3,976,764, 1976.
10. S. Watanabe, M. Ichikawa, and Y. Miyake, J. Pharm. Sci., 80, 1062 (1991).
11. S. Watanabe, M. Ichikawa, T. Kato, M. Kawahara, and M. Kayano, J. Pharm. Sci., 80, 1153 (1991).
12. S. S. Leung and J. R. Robinson, J. Controlled Release, 5, 223 (1988).
13. R. Johansson, C. G. Regardh, and J. Sjogren, Acta Pharm. Suec., 8, 59 (1971).
14. A. C. Woods, G. A. Glaubiger, and T. N. Chase, Lancet, 1, 1391 (1973).
15. S. Eriksen, in *The Theory and Practice of Industrial Pharmacy* (L. Lachman, H. A. Lieberman, and J. L. Kanig, eds.), Lea & Febiger, Philadelphia, 1970, p. 408.
16. V. Manninen, K. Ojala, and P. Reisell, Lancet, 2, 922 (1972).
17. J. G. Wagner, P. G. Welling, K. O. Lee, and J. E. Walker, J. Pharm. Sci., 69, 666 (1971).
18. T. R. Bates, D. A. Lambert, and W. H. Jones, J. Pharm. Sci., 58, 1488 (1969).
19. J. H. Fincher, J. Pharm. Sci., 57, 1825 (1968).
20. N. P. Salzman and B. B. Brodie, J. Pharmacol. Exp. Ther., 118, 46 (1956).
21. B. Beerman, K. Helstrom, and A. Rosen, Clin. Pharmacol. Ther., 13, 212 (1972).
22. W. H. Bachrach, Am. J. Dig. Dis., 3, 743 (1958).
23. A. Fick, Poggendorffs Ann., 94, 59 (1885).
24. R. M. Barrier, Discuss. Faraday Soc., 21, 138 (1956).
25. G. L. Flynn, S. H. Yalkowsky, and T. J. Roseman, J. Pharm. Sci., 63, 479 (1974).
26. J. Crank, *The Mathematics of Diffusion*, Oxford University Press, Oxford, 1956.
27. R. W. Baker and H. K. Lonsdale, in *Controlled Release of Biologically Active Agents* (A. C. Tanquary and R. E. Lacey, eds.), Plenum Press, New York, 1974, p. 15.
28. A. Fisch, E. Pichard, T. Prazuck, R. Sebbag, G. Torres, G. Gernez, and M. Gentilini, Am. J. Public Health, 83, 540 (1993).
29. T. Higuchi, J. Pharm. Sci., 50, 874 (1961).
30. G. L. Flynn, S. H. Yalkowsky, and T. J. Roseman, J. Pharm. Sci., 63, 479 (1974).
31. S. J. Desai, P. Singh, A. P. Simonelli, and W. I. Higuchi, J. Pharm. Sci., 55, 1224 (1966).
32. S. J. Desai, A. P. Simonelli, and W. I. Higuchi, J. Pharm. Sci., 54, 1459 (1965).
33. S. J. Desai, P. Singh, A. P. Simonelli, and W. I. Higuchi, J. Pharm. Sci., 55, 1230 (1966).
34. S. J. Desai, P. Singh, A. P. Simonelli, and W. I. Higuchi, J. Pharm. Sci., 55, 1235 (1966).
35. H. Lapidus and N. G. Lordi, J. Pharm. Sci., 55, 840 (1966).
36. H. B. Hopfenberg, in *Controlled Release Polymeric Formulations* (D. R. Paul and F. W. Harris, eds.), American Chemical Society, Washington, DC, 1976, p. 26.
37. E. Goldberg, In *Polymeric Delivery Systems, Midland Macromolecular Symposium* (R. J. Kostelnek, ed.), Gordon and Breach, New York, 1978, p. 227.
38. H. B. Hopfenberg and K. C. Hsu, Polym. Eng. Sci., 18, 1186 (1978).

39. H. Nakagami and M. Nada, Drug Design Discovery, 8, 103 (1991).

40. R. W. Baker, U.S. Patent 3,952,741, 1976.

41. W. Bayne, V. Place, F. Theeuwes, J. D. Rogers, R. B. Lee, R. O. Davies, and K. C. Kwan, J. Clin. Pharmacol. Ther., 32, 270 (1982).

42. R. Theeuwes, J. Pharm. Sci., 64, 1987 (1975).

43. S. Motycka and J. G. Naira, J. Pharm. Sci., 67, 500 (1978).

44. G. Gregoriadis, in *Drug Carriers in Biology and Medicine* (G. Gregoriadis, ed.), Academic Press, London, 1979.

45. H. K. Kimelberg and G. G. Meyhem, CRC Crit. Rev. Toxicol., p. 25 (Dec. 1978).

46. G. Gregoriadis, C. Kirby, P. Large, A. Meehan, and J. Senior, in *Targeting of Drugs* (G. Gregoriadis, J. Senior, and A. Trouet, eds.), Plenum Press, New York, 1982, p. 155.

47. D. Chapman in *Liposome Technology*, Vol. 1 (G. Gregoriadis, ed.), CRC Press, Boca Raton, FL, 1984.

48. J. de Gien, J. G. Mandersloot, and L. L. M. van Deenen, Biochim. Biophys. Acta, 150, 166 (1968).

49. B. D. Ladbrooke and D. Chapman, Chem. Phys. Lipid, 3, 304 (1969).

50. R. L. Magin and J. N. Weinstein, in *Liposome Technology*, Vol. 3, (G. Gregoriadis, ed.), CRC Press, Boca Raton, FL, 1984.

51. J. N. Weinstein, R. L. Magin, R. L. Cysyk, and D. S. Zaharko, Cancer Res., 40, 1388 (1980).

52. J. D. Morrissett, R. L. Jackson, and A. M. Gotto, Biochim. Biophys. Acta, 472, 93 (1977).

53. F. Szoka and D. Papahadjopoulos, Annu. Rev. Biophys. Bioeng., 9, 467 (1980).

54. R. A. Damel, S. C. Kinsky, C. B. Kinsky, and L. L. M. Van Deenen, Biochim. Biophys. Acta, 150, 655 (1968).

55. H. L. Kantor and J. H. Prestegard, Biochemistry, 14, 1790 (1975).

56. R. E. Pagano and J. N. Weinstein, Annu. Rev. Biophys. Bioeng., 7, 435 (1978).

57. T. D. Heath, N. G. Lopez, J. R. Piper, J. A. Montgomery, W. H. Stern, and D. Papahadjopoulos, Biochim. Biophys. Acta, 862, 72 (1986).

58. T. D. Heath and C. S. Brown, J. Liposome Res., 1, 303, (1989).

59. J. Vaage and E. Mayhew, Int. J. Cancer, 47, 582 (1991).

60. H. A. Titulaer, W. M. Eling, D. J. Crommelin, P. A. Peeters, and J. Zuiderma, J. Pharm. Pharmacol., 42, 529 (1990).

61. T. M. Allen, T. Mehra, C. Hansen, and Y. C. Chin, Cancer Res. 52, 2431 (1992).

62. G. L. Amidon, G. D. Leesman, and R. L. Elliot, J. Pharm. Sci., 69, 1363 (1980).

63. G. W. Carter, P. R. Young, L. R. Swett, and G. Y. Paris, Agents Actions, 10, 240 (1980).

64. J. Bungaard, in *Bioreversible Carriers in Drug Design, Theory and Application* (E. B. Roche, ed.), Pergammon Press, New York, 1987.

65. H. Bungaard, in *Design of Prodrugs* (H. Bungaard, ed.), Elsevier, Amsterdam, 1985.

66. G. B. Elion, J. A. Fyfe, L. Beauchamp, P. A. Furman, P. De Miranda, and H. J. Schaeffer, Proc. Natl. Acad. Sci. USA, 74, 5716 (1977).

67. A. A. Sinkula, in *Design of Prodrugs* (H. Bungaard, ed.), Elsevier, Amsterdam, 1985.

68. A. Bundgaard, Adv. Drug Deliv. Rev., 3, 39 (1989).

69. T. Kojima, M. Hashida, S. Muranishi, and H. Sezaki, J. Pharm. and Pharmacol., 32, 30 (1980).

70. A. Kato, Y. Takakura, M. Hashida, T. Kimura, and H. Sezaki, Chem. Pharm. Bull. 30, 2951 (1982).

71. J. Kreuter and W. Liehl, J. Pharm. Sci., 70, 367 (1981).

72. J. Kreuter, Pharm. Acta Helv., 58, 196 (1983).

73. R. C. Oppenhem, in *Drug Delivery Systems* (R. L. Juliano, ed.), Oxford University Press, New York, 1982, p. 182.

74. J. Kreuter, Pharm. Acta Helv., 58, 217 (1983).

75. L. Illum and S. S. Davis, FEBS Lett., 167, 79 (1984).

76. D. Leu, B. Manthey, J. Kreuter, P. Speiser, and P. P. DeLuca, J. Pharm. Sci., 73, 1433 (1984).

77. S. D. Troster, U. Mueller, and J. Kreuter, Int. J. Pharm., 61, 85 (1991).

78. L. Grislain, P. Couvreur, V. Lenaerts, M. Roland, D. Deprez-Decampeneere, and P. Speiser, Int. J. Pharm., 15, 335 (1983).

79. E. M. Gipps, R. Arshady, J. Kreuter, P. Groscurth, and P. P. Speiser, J. Pharm. Sci., 75, 256 (1986).

80. T. Harmia, P. Speiser, and J. Kreuter, J. Microencapsul., 3, 3 (1986).
81. F. Brasseur, P. Couvreur, B. Kante, L. Deckers-Passau, M. Roland, C. Deckers, and P. Speiser, Eur. J. Cancer, 16, 1441 (1980).
82. J. Kreuter and H. R. Hartmann, Oncology, 40, 363 (1983).
83. K. Sugibayashi, M. Akimoto, Y. Morimoto, T. Nadai, and Y. Kato, J. Pharmicobiodyn., 2, 350 (1979).
84. P. Couvreur, B. Kante, L. Grislain, M. Roland, and P. Speiser, J. Pharm. Sci., 71, 790 (1982).
85. Y. Morimoto, K. Sugibayashi, and Y. Kato, Chem. Pharm. Bull., 29, 1433 (1981).
86. J. Kreuter, in *Drug Targeting* (P. Buri and A. Gumma, eds.), Elsevier, Amsterdam, 1985.
87. J. L. Grangier, M. Puygrenier, J. V. Hautier, and P. Couvreur, J. Controlled Deliv., 15, 3 (1991).
88. J. C. Gautier, J. L. Grangier, A. Barbier, P. Dupont, D. Dussossoy, G. Pastor, and P. Couvreur, J. Controlled Release, 20, 67 (1992).
89. E. Pitt, C. M. Johnson, D. A. Lewix, D. A. Jenner, and R. E. Offord, Biochem. Pharmacol., 32, 3359 (1983).
90. G. Ihler, in *Drug Carriers in Biology and Medicine* (G. Gregoriadis, ed.), Academic Press, London, 1979, p. 287.
91. R. A. Cooper, in *Hematology*, 2nd Ed. (W. J. Williams, E. Beutler, A. J. Erslev, and R. W. Rundles, eds.), McGraw-Hill, New York, 1977, p. 216.
92. D. A. Lewis and H. O. Alpar, Int. J. Pharm., 22, 137 (1984).
93. C. A. Scheinberg and M. Strand, Cancer Res., 42, 44 (1982).
94. G. Gregoriadis, Drugs, 24, 261 (1982).
95. L. D. Lesserman, J. Barbet, F. Kourilsky, and J. N. Weinstein, Nature, 288, 602 (1980).
96. P. N. Kulkarni, P. H. Blair, and T. Ghose, Fed. Proc., 40, 642 (1982).
97. G. F. Rowland, R. G. Simmons, J. R. F. Corvalan, R. W. Baldwin, J. P. Browns, M. J. Embelton, C. H. J. Ford, K. E. Hillstorm, I. Hellstrom, J. T. Kemshead, C. E. Newmans, and X. S. Woodhouse, in *Proteins of the Biological Fluids*, Proceedings Colloquim 30 (H. Peters, ed.), Pergammon Press, New York, 1983.
98. J. M. Whiteley, Ann. N.Y. Acad. Sci., 79, 621 (1982).
99. M. Muirhead, P. J. Martin, B. Torok-Storb, J. W. Uhr, and S. Vitelia, Blood, 62, 327 (1983).
100. G. F. Rowland, in *Targeted Drugs* (E. P. Goldberg, ed.), Wiley-Interscience, New York, 1983.
101. M. B. Primm, J. A. Jones, M. R. Price, J. G. Middle, M. J. Embleton, and R. W. Baldwin, Cancer Immunol. Immunother., 12, 125 (1982).
102. Z. Brich, S. Ravel, T. Kissel, J. Fritsch, and A. Schoffmann, J. Controlled Release, 19, 245 (1992).
103. I. Tsukada, W. K. Bishop, N. Hibe, H. Hirai, E. Hurwitz, and M. Sela, Proc. Natl. Acad. Sci. USA, 79, 621 (1982).
104. B. Rihoua, A. Jegorov, J. Strohalm, W. Matha, P. Rossmann, L. Fornusek, and K. Ulbrich, J. Controlled Release, 19, 25 (1992).
105. F. Hurwitz, R. Kashi, and M. Wilcheck, JNCI, 69, 47 (1982).
106. L. G. Bernier, M. Page, R. C. Gaudreault, and L. P. Joly, Br. J. Cancer, 49, 245 (1984).
107. M. J. Embleton, G. F. Rowland, R. S. Simmonds, E. Jacobs, C. H. Marsden, and R. W. Baldwin, Br. J. Cancer, 47, 43 (1983).
108. J. B. Cannon and H. W. Hui, in *Targeted Therapeutic Systems*, (P. Tyle and B. P. Ram, eds.), Marcel Dekker, NY, 1990.
109. S. Wright and L. Huang, Adv. Drug Deliv. Rev., 3, 343 (1989).
110. M. Udayachander, A. Meenakshi, R. Muthiah, and M. Sivanandham, Int. J. Radiol. Oncol. Biol. Phys., 13, 1713 (1987).
111. A. K. Agrawal, A. Singhal, and C. M. Gupta, Biochim. Biophys. Res. Commun., 148, 357 (1987).
112. T. Tadakuma, in *Medical Applications of Liposomes* (K. Yagi, ed.), Japan Scientific Society Press, Tokyo, 1986.
113. C. Y. Wang and L. Huang, Proc. Natl. Acad. Sci. USA, 84, 7851 (1987).
114. P. Machy and L. D. Lesernman, Biochim. Biophys. Acta, 730, 313 (1983).
115. K. K. Matthay, T. D. Heath, and D. Papahadjopoulos, Cancer Res., 44, 1850 (1984).

116. D. Papahadjopoulous, T. Heath, F. Martin, F. Fraley, and R. Straubinger, in *Targeting of Drugs* (G. Gregoriadis, J. Senior, and A. Trouet, eds.), Plenum Press, New York, 1982.
117. W. Magee, H. Croneberger, and D. E. Thor, Cancer Res., 38, 1173 (1978).
118. F. Cervone, L. Tronstad, and B. Hammond, Endodont. Dent. Traumatol., 6, 33 (1990).
119. D. Steinberg, N. Friedman, A. Soskolne, and M. N. Sela, J. Periodontol., 61, 393 (1990).
120. A. Kozlorsky, A. Sintor, Y. Zubery, and H. Tal, J. Dent. Res., 71, 1577 (1992).
121. K. Higashi, K. Morisaki, S. Hayashi, M. Kitamura, N. Fujimoto, S. Kimura, S. Ebisu, and H. Okada, J. Perdont. Res., 25, 1 (1990).
122. H. Yamagami, A. Takomori, T. Sakamotok, and H. Okada, J. Periodontol., 63, 2 (1992).
123. S. Kimura, H. Toda, Y. Shimabukuro, M. Kitamura, N. Fujimoto, Y. Miki, and H. Okada, J. Periodont. Res., 26, 33 (1991).
124. J. P. Fiorellini and D. W. Paguette, Curr. Opin. Dent., 2, 63 (1992).
125. K. S. Aithal, A. R. Aroor, Y. B. Pathak, and J. Uchil, Indian J. Dent. Res., 2, 174 (1990).
126. S. Tamburic, G. Vuleta, M. Gajic, R. Stevanovic, Slomatoloski Glasnik Srbije, 37, 307 (1990).
127. M. C. Makoid and J. R. Robinson, J. Pharm. Sci., 68, 435 (1978).
128. H. Benson, Arch. Ophthalmol., 91, 313 (1974).
129. J. W. Seig and J. R. Robinson, J. Pharm. Sci., 68, 724 (1979).
130. R. D. Schoenwald and P. Stewart, J. Pharm. Sci., 69, 391 (1980).
131. J. W. Shell, Surv. Ophthalmol., 26, 207 (1982).
132. D. W. Lamberts, D. P. Langston, and W. Chu, Ophthalmol., 85, 794 (1978).
133. J. W. Shell and R. W. Baker, Ann. Ophthalmol., 6, 1037 (1974).
134. J. A. Anderson, W. L. Davis, and C. Wei, Invest. Ophthalmol. Visual Sci., 19, 817 (1980).
135. A. I. Mandell, F. Stentz, and A. E. Kitabchi, Ophthalmol., 85, 268 (1978).
136. J. S. Mindel, S. T. Shaikewite, and S. M. Podos, Arch. Ophthalmol., 98, 2220 (1980).
137. D. S. Chien and R. D. Schoenwald, Pharm. Res., 7, 476 (1990).
138. D. S. Chien, H. Bundgaard, A. Buur, and V. H. L. Lee, J. Ocular Pharmacol., 4, 137 (1988).
139. H. Bundgaard, A. Buur, S. C. Chang, and V. H. L. Lee, Int. J. Pharm., 33, 15 (1986).
140. E. Duzman, C. C. Chen, J. Anderson, M. Blumenthal, and L. Twizer, Arch. Ophthalmol., 100, 1916 (1982).
141. H. Bundgaard, E. Falch, C. Larsen, and T. J. Mikkelson, J. Pharm. Sci., 75, 775 (1986).
142. G. L. Mosher, H. Bundgaard, E. Falch, C. Larsen, and T. J. Mikkelson, Int. J. Pharm., 39, 113 (1986).
143. L. Z. Bito and R. A. Baroody, Exp. Eye Res., 44, 217 (1984).
144. O. Camber, P. Edman, and L. I. Olsson, Int. J. Pharm., 37, 27 (1987).
145. T. L. Phipps, D. E. Potter, and J. M. Rowland, J. Ocular Pharmacol., 2, 109 (1986).
146. P. C. Maudgal, K. D. Clercq, J. Descamps, and L. Missotten, Arch. Ophthalmol., 102, 140 (1984).
147. D. Pavan-Langston, R. D. North, P. A. Geary, and A. Kinkel, Arch. Ophthalmol., 94, 1585 (1976).
148. M. M. Narurkar and A. K. Mitra, Pharm. Res., 6, 887 (1989).
149. M. M. Narurkar and A. K. Mitra, Pharm. Res., 5, 734 (1988).
150. V. H. L. Lee, P. T. Ureaa, R. E. Smith, and D. J. Schanzlin, Surv. Ophthalmol., 29, 335 (1985).
151. M. W. Fountain, A. S. Janoff, M. J. Ostro, M. C. Popescu, and A. L. Weiner, 10th Int. Symp. Contr. Rel. Bioactive Mat., 1983.
152. K. Singh and M. Mezei, Int. J. Pharm., 19, 263 (1984).
153. A. G. Palestine, R. B. Nussenblatt, M. V. W. Bergamini, I. E. Bolcsak, and W. T. Robinson, Invest. Ophthalmol. Vis. Sci., 27 (3 suppl.), 112 (1986).
154. M. Barza, J. Baum, and F. Szoka, Invest. Ophthalmol. Vis. Sci., 25, 486 (1984).
155. G. M. Grass, J. C. Cob, and M. C. Makoid, J. Pharm. Sci., 75, 618 (1984).
156. J. W. Shell, Sur. Ophthalmol. 29, 117 (1984).
157. S. P. Vyas, S. Ramchandraiah, C. P. Jain, S. K. Jain, J. Microencapsul., 9, 347 (1992).
158. W. W. Li, G. Grayson, J. Folkman, and P. A. D'Amore, Invest. Ophthalmol. Vis. Sci., 32, 2906 (1991).
159. R. Anand, S. D. Nightingale, R. H. Fish, T. J. Smith, and P. Ashtor, Arch. Ophthalmol., 111, 223 (1993).

160. D. L. Blondford, T. J. Smith, J. D. Brown, P. A. Pearson, and P. Ashtor, Invest. Ophthalmol. Vis. Sci., 32, 2906 (1991).

161. J. Kimura, Y. Ogura, T. Moritera, Y. Honda, R. Wada, and S. H. Hyon, Invest. Ophthalmol. Vis. Sci., 33, 3436 (1992).

162. S. K. Chandrasekaran, W. Bayne, and J. E. Shaw, J. Pharm. Sci., 67, 1370 (1978).

163. M. Bals-Pratsch, Y. D. Yoon, U. A. Knuth, and E. Nieschlag, Lancet, 2, 943 (1986).

164. D. J. R. Duthie, D. J. Rowbotham, R. Wyld, P. D. Henderson, and W. S. Nimmo, Br. J. Anaesth., 60, 614 (1988).

165. F. O. Holley and C. van Steennis, Br. J. Anaesth., 60, 608 (1988).

166. A. Wellstein, H. Kuppers, H. F. Pitscher, and D. Palm, Eur. J. Clin. Pharmacol., 31, 419 (1986).

167. S. K. Chandrasekaran, H. Benson, and J. Urquhart, in *Sustained and Controlled Release Delivery Systems* (J. R. Robinson, ed.), Marcel Dekker, New York, 1978, p. 578.

168. D. P. Benziger and J. Edelson, Drug Metab. Rev., 14, 137 (1983).

169. D. J. Back, A. Breckenridge, M. E. Orme, and M. A. Shaw, Lancet, 1(8369), 171 (1984).

170. E. Diczfalusy and B. M. Landgren, in *Long-Acting Contraceptive Delivery Systems* (G. I. Zatuchni, A. J. Sobero, J. J. Speidel, and J. J. Sciarra, eds.), Harper and Row, New York, 1984.

171. Y. W. Chien, in *Drug Delivery Systems* (R. L. Juliano, ed.), Oxford University Press, New York, 1982, p. 42.

172. G. A. Digenis, M. Jay, R. M. Beihn, T. R. Rice, and L. R. Beck, in *Long-Acting Contraceptive Delivery Systems* (G. I. Zatuchni, A. J. Sobero, J. J. Speidell, and J. J. Sciarra, eds.), Harper and Row, New York, 1984.

173. N. S. Mason, D. V. S. Gupta, D. W. Liller, R. S. Youngquist, and R. S. Sparks, in *Biomedical Applications of Microencapsulation* (R. Lim, ed.), CRC Press, Boca Raton, 1984.

174. P. Lahteenmaki, H. Kurunmaki, T. Luukkainer, P. O. A. Lahteenmaki, K. Ratsula, and J. Toivonen, in *Biomedical Applications of Microencapsulation* (R. Lim, ed.), CRC Press, Boca Raton, 1984.

175. T. A. Johnson, I. A. Green, R. W. Killy, and A. A. Calder, Brit. J. Ob. Gyn., 99, 877 (1992).

176. F. R. Witter, L. E. Rocco, and T. R. Johnson, Am. J. Obstet. Gynecol., 99, 877 (1992).

177. A. V. Taylor, J. Boland, and I. Z. MacKenzie, Prostaglandins, 40, 89 (1990).

178. W. Oppinheimer, Am. J. Obstet. Gynecol., 78, 446 (1959).

179. A. Ishihama, Yokahama Med. Bull., 10, 89 (1959).

180. A. Ishihama, T. Kagabu, T. Iamai, and M. Shimal, Acta Cytol., 14, 35 (1970).

181. M. Rowland, *Response to Contraception*, W. B. Saunders, Philadelphia, 1973, p. 111.

182. *Progestasert Product Monograph*, Alza Corporation, Palo Alto, CA.

183. V. A. Place and B. B. Pharriss, J. Reprod. Med., 13, 66 (1974).

184. R. Aznar-Ramos, B. B. Pharriss, and J. Martinez-Mamautou, Fertil. Steril., 25, 308 (1974).

185. B. D. Kulkarni, T. D. Avila, B. B. Pharriss, and A. Scommegna, Contraception, 8, 299 (1973).

186. U. Leone, Int. J. Fertil., 19, 17 (1974).

187. A. Scommegna, Obstet. Gynecol., 43, 769 (1974).

188. A. Zaffaroni, Acta Endocrinol., 75 (Suppl. 185), 423 (1974).

189. V. A. Place and B. B. Pharriss, J. Reprod. Med., 13, 66 (1974).

190. A. Rosado, J. J. Hicks, R. Aznar, and E. Mercado, Contraception, 9, 39 (1974).

191. B. Seohadri, Y. Gibor, and A. Scommegna, Am. J. Obstet. Gynecol., 109, 536 (1971).

192. R. Langer, Chem. Eng. Commun., 6, 1 (1980).

193. T. Tanaka, Sci. Am., 224, (1) 124, 138 (1981).

194. C. M. Klaisle and S. Wysocke, Clin. Issues Perinat. Wom. Health, 3, 267 (1992).

195. H. M. Creque, R. L. Langer, and T. Folkham, Diabetes, 29, 39 (1980).

196. Y. W. Chien and E. P. K. Lau, J. Pharm. Sci., 65, 488 (1976).

197. A. S. Lifichez, Fertil. Steril., 21, 426 (1970).

198. A. Cuadros, A. Brinson, and K. Sundaram, Contraception, 2, 29 (1970).

199. P. Y. Wans, Biomaterials, 12, 57 (1991).

200. E. M. Powell, M. R. Sobarzo, and W. M. Saltzman, Brain Res., 515, 309 (1990).

201. J. Skarda, J. Slaba, P. Krejci, and I. Mikular, Physiol. Res., 41, 151 (1992).

202. F. M. Kahan and J. D. Rogers, Chemotherapy, 2, 21 (1991).

203. S. Li, M. Lepage, Y. Merand, A. Belanger, F. Labrie, Breast Cancer Res. Treat., 24, 127 (1993).

204. T. Hashimoto, T. Wada, N. Fukuda, and A. Nagoaka, J. Pharm. Pharmacol., 45, 94 (1993).

205. R. A. Burns, K. Vitale, and L. M. Sanders, J. Microencapsul., 7, 397 (1990).

206. N. S. Jones, M. G. Glenn, L. A. Orloff, and M. R. Mayberg, Arch. Otolaryngol., 116, 779 (1990).

207. D. T. O'Hagan, D. Ragman, J. P. McGee, H. Jeffrey, M. C. Davies, P. Williams, S. S. Davis, and S. J. Challacombe, Immunology, 73, 239 (1991).

208. D. T. O'Hagan, H. Jeffrey, M. J. Roberts, J. P. McGee, and S. S. Davis, Vaccine, 9, 768 (1991).

209. Y. W. Chein, in Novel Drug Delivery Systems (T. J. Roseman and S. Z. Mansdorf, eds.), Marcel Dekker, New York, 1982.

210. Implantable pump for morphine, Am. Pharm., NS24, 9, 20 (1984).

211. D. S. T. Hsieh and R. Langer, in Controlled Release Delivery Systems (T. J. Roseman and S. Z. Mansdorf, eds.), Marcel Dekker, New York, 1983.

16

Target-Oriented Drug Delivery Systems

Vijay Kumar and Gilbert S. Banker
University of Iowa, Iowa City, Iowa

I. INTRODUCTION

Although the idea of drug targeting to a specific site in the body was first introduced almost a century ago by Ehrlich [1], it is only in recent years that the field has emerged as an important area of research. This long silence in the field can be attributed to the inadequate understanding of various diseases; the lack of a detailed description, at the cellular–molecular level, of how drugs are processed; and the difficulties in identifying and producing carrier molecules specific to the targeted organs, cells, or tissues. The recent advent of recombinant DNA technology and progress in biochemical pharmacology and molecular biology have provided not only a clearer elucidation of pathogenesis of many diseases and identification of various types of surface cell receptors, but also enabled the production of several new classes of highly potent protein and peptide drugs (e.g., homo- and heterologous peptidergic mediators and sequence-specific oligonucleotides) [2]. For these new drugs, and for some conventional drugs (e.g., antineoplastic agents) that have narrow therapeutic windows and require localization to a particular site in the body, it is essential that they be delivered to their target sites intact, in adequate concentrations, and in an efficient, safe, convenient, and cost-effective manner. Most drug therapies currently available provide little, if any, target specificity. The selective delivery of drugs to their pharmacological receptors should not only increase the therapeutic effectiveness, but should limit side effects.

In this chapter, an overview of various target-specific drug delivery systems, including discussion of biological events and processes that influence drug targeting, is presented. To avoid an unwieldy bibliography, only selected book references, review articles, and papers are cited.

II. RATIONALE FOR TARGETED DRUG DELIVERY

Most drugs, after administration in a conventional immediate- or controlled-release dosage form, freely traverse throughout the body, typically leading to uptake by cells, tissues, or organs other than where their pharmacological receptors are located. Figure 1 illustrates the distribution, metabolism, and elimination of drugs that may occur by natural pathways. This lack of target specificity, illustrated in Fig. 1, for the most part, can be attributed to the formidable barriers that the body presents to a drug. For example, a drug taken orally (most drugs are administered by this route, if possible) must withstand large fluctuations in pH as it travels along the gastrointestinal (GI) tract, as well as resist the onslaught of the enzymes that digest our food, and the metabolism by microflora that live there. To be systemically active, the drug must then be absorbed from the GI tract into the blood, before it passes its region of absorption in the tract. Once in the blood, it needs to survive inactivation by metabolism and extraction (first-pass effects). To produce its therapeutic effect(s), the drug must then selectively access and interact with its pharmacological receptor(s). The concentration of drug at the active site must be adequate. Administration of drugs by parenteral routes avoids the GI-associated problems, but deactivation and metabolism of the drug and dose-related toxicity are frequently observed. Furthermore, it can not be assured that, of all the paths the drug may take following administration, one of them will lead the drug to its desired destination in adequate concentrations. There are many diseases that are poorly accessible to drugs. These include rheumatoid arthritis, diseases of the central nervous system, some cancers, and intractable bacterial, fungal, and parasitic infections. The treatment of these diseases often require high doses and frequent administration of drugs, which can lead to toxic manifestations, inappropriate pharmacodisposition, untoward metabolism, and other deleterious effects.

Numerous other reasons enumerating why it is preferable to direct drugs to their sites of action are listed in Table 1 [3]. Thus, a target-oriented drug delivery system must supply drug selectively to its site(s) of action(s) in a manner that provides maximum therapeutic activity (through controlled and predetermined drug-release kinetics), prevents degradation or inactivation during transit to the target sites, and protects the body from adverse reactions because of inappropriate disposition. For drugs that have a low therapeutic index (ratio of toxic dose to therapeutic dose), targeted drug delivery must provide an effective treatment at a relatively

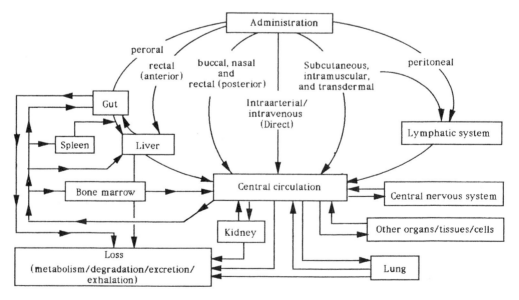

Fig. 1 A schematic representation of drug disposition in the body.

Table 1 Reasons for Site-Specific Delivery of Drugs

Pharmaceutical
  Drug instability as delivered from
    conventional formulation
  Solubility
Biopharmaceutical
  Low absorption
  High membrane binding
  Biological instability
Pharmacokinetic and pharmacodynamic
  Short half-life
  Large volume of distribution
  Low specificity
Clinical
  Low therapeutic index
  Anatomical or cellular barriers
Commercial
  Drug presentation

Source: Ref. 3.

low drug concentration. Other requirements for target-oriented drug delivery include that (a) the delivery system should be biochemically inert (nontoxic), nonimmunogenic, and physically and chemically stable in vivo and in vitro; (b) the carrier must be biodegradable, or readily eliminated without any problem; and (c) the preparation of the delivery system must be reproducible, cost-effective, and reasonably simple.

III. BIOLOGICAL PROCESSES AND EVENTS INVOLVED IN DRUG TARGETING

Drug targeting has been classified into three types: (a) First-order targeting—this describes delivery to a discrete organ or tissue; (b) second-order targeting—this represents targeting to a specific cell type(s) within a tissue or organ (e.g., tumor cells versus normal cells and hepatocytic cells versus Kupffer cells); and (c) third-order targeting—this implies delivery to a specific intracellular compartment in the target cells (e.g., lysosomes) [4]. Basically, there are three approaches for drug targeting. The first approach involves the use of biologically active agents that are both potent and selective to a particular site in the body (magic bullet approach of Ehrlich). The second approach involves the preparation of pharmacologically inert forms of active drugs that when they reach the active sites become activated by a chemical or enzymatic reaction (prodrug approach). The third approach utilizes a biologically inert macromolecular carrier system that directs a drug to a specific site in the body where it is accumulated and effects its response (magic gun or missile approach). Regardless of the approach, the therapeutic efficacy of targeted drug delivery systems depends on the timely availability of the drug in active form at the target site(s) and its intrinsic pharmacological activity. The intrinsic pharmacokinetic properties of the free drug should be the same, irrespective of whether or not it is introduced into the body attached to a carrier. Figure 2 shows a schematic representation of possible anatomical and physiological pathways that a drug may follow to reach its target site(s) [5]. As shown in this figure, a drug can selectively access to, and interact with, its

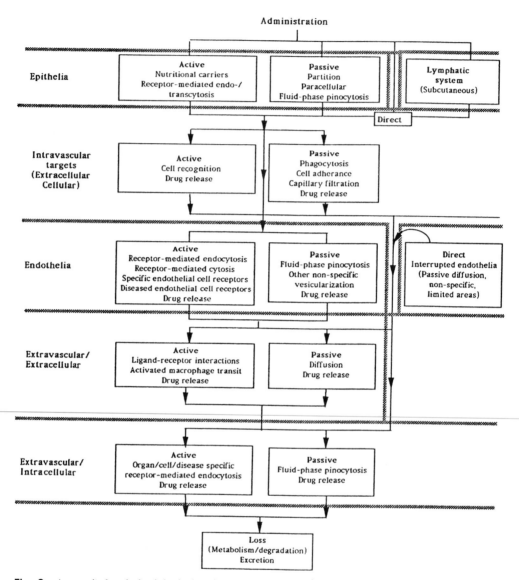

Fig. 2 Anatomical and physiological pathways for site-specific delivery. (From Ref. 5.)

pharmacological receptors, either passively or by active processes. Passive processes rely on the normal distribution pattern of a drug–drug-carrier system, whereas the active routes use cell receptor-recognizing ligand(s) or antibodies (''homing'' or ''vector'' devices) to access specific cells, tissues, or organs in the body. Various biological processes and events that govern drug targeting are discussed in the following sections.

A. Cellular Uptake and Processing

Following administration, a drug frequently passes through various cells, membranes, and organs to reach its target site(s). Various passive and active processes or mechanisms by which the drug can achieve this are shown in Fig. 2 [5]. These pathways offer opportunities for cell selection and access by targeted drug delivery.

Low-molecular-weight drugs can enter into, or pass through, various cells by simple diffusion processes. Targeted drug delivery systems often comprise macromolecular assemblies, and are unable to enter into cells by such simple processes. Instead, they are captured by a process called endocytosis. *Endocytosis* is defined as a phenomenon that involves internalization of the plasma membrane, with concomitant engulfment of the extracellular material (particulate or fluid). This process can be constitutive or nonconstitutive. Other methods of gaining access to cells include passive diffusion, membrane fusion, and binding to either specific or nonspecific regions of the cell.

Endocytosis is divided into two types: phagocytosis and pinocytosis (Fig. 3). The former refers to the capture of particulate matter, whereas the latter represents engulfment of fluids. Phagocytosis is carried out by specialized cells of the mononuclear phagocyte system (MPS), called phagocytes. It is mediated by the absorption of specific blood components [e.g., immunoglobulin (Ig) G, complement C3b, and fibronectin], called opsonins, and relevant receptors located on macrophages. The extent to which a drug is opsonized, and by what plasma protein, depends on the size and surface characteristics of the particles. This, in turn, determines the engulfment mechanism. For example, red blood cells treated with glutaraldehyde are opsonized by IgG and rapidly phagocytosed by the Fc receptor. In contrast, cells treated with *n*-ethylmaleimide are opsonized by C3b factor and are engulfed with a minimal membrane–receptor contact. Changes in the glycoprotein levels of patients may lead to variations in the opsonization of administered particles and, consequently, their ultimate distribution in the body [6]. Particles with higher hydrophilic surface characteristics tend to undergo opsonization to a lesser extent and, as a result, exhibit decreased phagocytic uptake [7–10]. This has direct

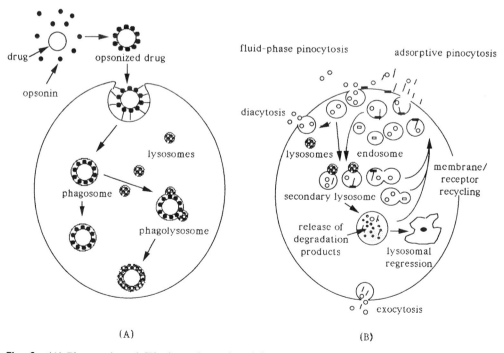

Fig. 3 (A) Phagocytic and (B) pinocytic uptake of drugs.

implications in targeting microparticulate drugs to cells other than those of the reticuloendo-thelial system (RES), because the longer the drug stays in the central circulation, the greater the chances of uptake by other cells. Nonspecific phagocytic uptake of particles, triggered by particle size and hydrophilic coatings [11], and mediated by membrane components [12], has also been reported.

Following ingestion, the phagocytic vacuole (or phagosome) fuses with one or more lyso-somes to form phagolysosomes (or secondary lysosomes); (see Fig. 3A). It is here that the digestion of particles by lysosomal acid hydrolases (e.g., proteinases, glycosidases, nucleases, phospholipases, phosphatases, and sulfatases) occurs, making the drug available to exert its therapeutic effect. The internal pH of lysosomes is between 4.5 and 5.5.

Compared with phagocytosis, pinocytosis appears to be a universal phenomenon in all cells, including phagocytes. Unlike phagocytosis, which is mediated by the serum opsonin, pinocy-tosis does not require any external stimulus. Pinocytosis is divided into two types: fluid-phase pinocytosis and adsorptive pinocytosis (see Fig. 3B). Fluid-phase pinocytosis is a nonspecific, continuous process, and it is believed to be useful as a general process for transporting mac-romolecular constructs through epithelia, some endothelia, and into various blood cells. Ad-sorptive pinocytosis, in contrast, refers to internalization of macromolecules that bind to the cell surface membrane. If the macromolecule adheres to a general cell surface site, then uptake is referred to as simply nonspecific pinocytosis. However, if it binds to a specific cell receptor site, then the process is called receptor-mediated pinocytosis. Before membrane internalization, the pinocytic substrate often patches into domains or areas of the membrane called *coated pits*. Coated pits have a cytoplasmic coat consisting of clathrin and other proteins. Once internalized, the pinocytic vesicles can interact among themselves, or with vesicles of other intracellular origins, such as endosomes and lysosomes. Endosomes are rich in pinocytic receptors and contain an active ATP-powered proton pump (not identical with that present in lysosomes) that maintains the internal pH between 5.0 and 5.5. The mild internal acidic pH condition induces dissociation of the receptor–drug carrier complex, freeing the receptor for recycling. Endo-somes also serve as a sorting station to route internalized substrates to their appropriate intra-cellular locations. Internalized substrates that remain intact in the endosome are usually then transferred to the lysosome (transcellular transfer) where their digestion by acid hydrolases continues. In some cells (e.g., endothelial cells), the endosomes, instead of transferring their contents to the lysosome, release them outside the cell. This process is termed diacytosis or retroendocytosis, and can achieve a vectorial translocation of substances through an otherwise impenetrable barrier of cells [13]. In cells, such as secretory polymeric IgG in the neonatal gut, polymeric IgA in hepatocyte, and low-density lipoprotein (LDL) in endothelia, the sec-ondary lysosome transports its content to the other side of the membrane by a process called transcytosis. The secondary lysosome can also regress to form residual bodies that continue to retain nondegraded macromolecules.

Nonspecific pinocytic uptake appears to be dependent on the size (molecular weight and configuration), charge, and hydrophobicity of the pinocytic substrates. Polycation macromol-ecules have increased pinocytic uptake in rat yolk sacs and rat peritoneal macrophages cultured in vitro, compared with the neutral and anionic macromolecules [14–16]. The rate of pinocytic uptake in different cells also increases with an increase in the size and hydrophobicity of the substrates. The molecular size of the pinocytic substrate is detrimental to the movement of macromolecules from one compartment to another.

The receptor-mediated form of endocytic uptake has been identified for a wide variety of physiological ligands, such as metabolites, hormones, immunoglobulins, and pathogens (e.g., virus and bacterial and plant toxins). Several endosomotropic receptors identified in cells are listed in Table 2.

Table 2 Distributions of Some Endosomotropic Receptors (Various Species)

Cell	Receptor for
Hepatocytes	Galactose, low density lipoprotein, polymeric IgA
Macrophages	Galactose (particles), mannose–fucose, acetylated LDL, alpha$_2$-macroglobulin–protease complex (AMPC)
Leukocytes	Chemotactic peptide, complement C3b, IgA
Basophils, mast cells	IgE
Cardiac, lung, diaphragm endothelia	Albumin
Fibroblasts	Transferrin, epidermal growth factor, LDL, mannose-6-phosphate, transcobalamine II, AMPC, mannose
Mammary acinar	Growth factor
Enterocytes	Maternal IgG, dimeric IgA, transcobalamine–B$_{12}$/intrinsic factor
Blood–brain endothelia	Transferrin, insulin

Source: Ref. 5.

Compared with phagocytosis, fluid-phase pinocytic capture of molecules is relatively slower, being directly proportional to the concentration of macromolecules in the extracellular fluid. It is also dependent on the size (molecular weight) of macromolecules; lower-molecular-weight fractions are captured faster than the higher-molecular-weight fractions. The magnitude of the rate of capture by adsorptive pinocytosis is higher than fluid-phase pinocytosis and relates to the nature of substrate–membrane interactions.

B. Transport Across the Epithelial Barrier

The oral, buccal, nasal, vaginal, and rectal cavities, all are internally lined with one or more layers of epithelial cells. Depending on the position and function in the body, epithelial cells can be of varied forms, ranging from simple columnar, to cuboidal, and to squamous types. Irrespective of their morphological differences, these cells are extremely cohesive. The lateral membrane of these cells exhibits several specialized features that form intercellular junctions (tight junction, zonula adherens, and gap junction), that serve not only as sites for adhesion, but also as seals to prevent flow of materials through the intercellular spaces (paracellular pathway), and to provide a mechanism for intercellular communication. The strong intercellular cohesion is partly due to the binding action of the glycoproteins, which are an integral part of the plasma membrane, and of a small amount of the intercellular proteoglycan. Calcium ions also play a role in maintaining this cohesion. Below the epithelial cells is a layer of connective tissue called lamina propria, which is bound to epithelium by the basal lamina. The latter also connects epithelium to other neighboring structures. The luminal side of the epithelium is covered with a more or less coherent, sticky layer of mucus. This is the layer that first interacts with foreign materials (e.g., food, drugs, bacterial organisms, and chemicals). It contains large glycoproteins (mucins), water electrolytes, slough epithelial cells, enzymes, bacteria and bacterial products, and various other materials, depending on the source and location of the mucus. Mucin, which is synthesized by goblet cells, or by special exocrine cells, acini, constitutes about 5% of the total weight of mucus. The structure of mucin consists of a polypeptide backbone with oligosaccharide side chains. Each oligosaccharide chain contains eight to ten monosaccharide residues of molecular weight of 320–4500, and has sialic acid or L-fucose as the terminal group. The oligosaccharide side chains are covalently linked to hydroxyamino acids, serine, and threonine residues along the polypeptide backbone.

The absorption of low-molecular-weight drugs from oral, buccal, nasal, vaginal, and rectal cavities is well established. Various transport processes used frequently by drugs to cross the epithelial barrier lining these cavities include passive diffusion, carrier-mediated transfer systems, and selective and nonselective endocytosis. Additionally, polar materials also can diffuse through the tight junctions of epithelial cells (the paracellular route). However, there is now evidence to suggest that macromolecules (particulate and soluble), including peptides and proteins, can also reach the systemic circulation, albeit in small amounts, following administration through these routes. This may have far-reaching consequences in certain therapies, such as immune reactions and hormone replacement treatment. Both passive and active transport pathways are energy-dependent processes, and they may occur simultaneously. Passive transport is usually higher in damaged mucosa, whereas active transport depends on the structural integrity of epithelial cells.

Harris [17] reported that nasal administration of biopharmaceuticals (polypeptides) results in bioavailabilities of the order 1–20% of administered dose, depending on the molecular weight and physiochemical properties. It is widely accepted that (macro)molecules with an M_r of less than 10 kDa in size can be absorbed from the nasal epithelium into the systemic circulation in sufficient amounts without the need for added materials, except for bioadhesives [18]. Larger molecules, such as proteins [e.g., interferon, granulocyte colony-stimulating factor (G-CSF), human growth hormone], however, require both a penetration enhancer (e.g., bile salts and surfactants) and bioadhesives. Since all the dose passes through one tissue, these flux enhancers may cause deleterious effects to the nasal mucosa and muciliary function. Thus, caution must be exercised in using them. Recently, cyclodextrin [19] and phospholipids [20] have been reported to significantly increase the absorption of macromolecules, without causing any damage to the nasal mucosal membrane. The phospholipid approach is particularly attractive, in that they are biocompatible and bioresorbable, and thus, pose no threat of toxicity.

The transport of macromolecules across intestinal epithelium may occur by cellular vesicular processes by either fluid-phase pinocytosis or specialized (receptor-mediated) endocytic processes [21]. Matsuno et al. [22] reported that spheres of 20-nm diameter, when given orally to the suckling mice, pass through the epithelial layer, and become localized in the omentum, the Kupffer cells of the lumen, the mesenteric lymph nodes, and even the thymic cortex. Recent studies with polyalkylcyanoacrylate nanocapsules of a size smaller than 300 nm suggest that particles can also pass intact through the intestinal barrier by the paracellular route [23]. The M cells found in the Peyer's patch have also been suggested to transport particles that exist within the epithelium membrane. These are specialized absorptive cells, and are known to absorb and transport indigenous bacteria (i.e., *Vibrio cholerea*); macromolecules, such as ferritin and horseradish peroxidase; viruses; and carbon particles, from the lumen of the intestine to submucosal lymphoid tissue [21,24]. Further transport of absorbed materials to the systemic circulation through lymph fluid and by lymphocytes has been suggested as possible. An increase in the lymph flow or a decrease in the blood supply could make lymphatic uptake of particles important [24]. Since Peyer's patches are more prevalent and larger in young individuals, and drastically decrease with increasing age, the transport by this route is of significance only in younger individuals [25].

Various factors influencing the absorption of drugs, including peptides, from the GI tract have recently been reviewed [26,27]. Table 3 lists some of the parameters discussed in these reviews, including important physiological and biochemical variables affecting drug absorption, such as pH, enzymes, surface area, segment length, microflora, and transit time. A variety of

Table 3 Factors Influencing the Absorption of Drugs

	Average length diameter (cm)	Average surface area (m^2)	Average pH (range)	Enzymes and others	Mean transit times	Microflora per gram content
Mouth cavity	15–20/10	0.07	6.4 (5.8–7.1)	Ptyalin, maltose, mucin	At will	
Esophagus	25/2.5	0.02	5.6		9–15 sec	
Stomach	20/15	0.11	1.5 (1.0–3.5)	Pepsin, lipase, rennin, HCl	0.5–4.5 hr	
Duodenum	25/5	0.09	6.9 (6.5–7.6)	Bile, trypsin, chymotrypsin, amylase, maltase, lipase, nuclease, peptidases		<10^3
Jejunum	300/5	60	6.9 (6.3–7.3)	Erepsin, amylase, maltase, lactase, sucrase, peptidases	1–4 hr	
Ileum	300/5	60	7.6 (6.9–7.9)	Lipase, nuclease, enterokinase, nucleotidase, peptidases		10^5–10^7
Cecum	10–30/7	0.05	7.7 (7.5–8.0)		4–16 hr	
Colon	150/5	0.25	7.95 (7.9–8.0)			10^{10}–10^{13}
Rectum	15–19/ 2.5	0.015	7.7 (7.5–8.0)		2–8 hr	

Source: Ref. 27.

penetration enhancers have been used to improve intestinal absorption of peptides and other macromolecular drugs. These include chelators (e.g., ethylenediaminetetraacetic acid, citric acid, salicylates, N-acetyl derivatives of collagen, and enamines); natural, semisynthetic, and synthetic surfactants (e.g., bile salts, derivatives of fusidic acid, sodium lauryl sulfate, polyoxyethylene-9-laurylether, and polyoxyethylene-20-cetylether); fatty acids and their derivatives (e.g., sodium caprate, sodium laurate, oleic acid, monoolein, and acrylcarntines); and a variety of mixed micelle solutions [28,29]. The different regions of the GI tract show different sensitivity to the penetration enhancers. The following order of sensitivity is suggested: rectum > colon > small intestine > stomach. Present evidence, however, suggests that oral administration of peptides and proteins results in less than 1% bioavailability [30].

There is very little evidence to suggest that soluble or particulate macromolecules can be transported across the buccal mucosa [31]. More work is needed to determine whether this route could be of any benefit in drug targeting.

The absorption of drugs from rectal [32] and vaginal [33] cavities has been studied in detail recently. Muranishi et al. [34] have shown that a significant increase in the absorption and lymphatic uptake of soluble and colloidal macromolecules can be achieved by pretreating the rectal mucosal membrane with lipid–nonionic surfactant mixed micelles. They found no evidence of serious damage of the mucosal membrane. Davis [30] suggested that vaginal cavity could be of benefit for certain pharmaceuticals, such as calcitonin, used for the treatment of postmenopausal osteoporosis.

C. Extravasation

Many diseases are known that result from the dysfunction of cells located outside the cardiovascular system. Thus, for a drug to exert its therapeutic effects, it must egress from the central circulation and interact with its extravascular–extracellular or extravascular–intracellular target(s). This process of transvascular exchange is called *extravasation*, and it is governed by the permeability of blood capillary walls. The main biological features that control permeability of capillaries include the structure of the capillary wall, under normal and (patho)physiological conditions, and the rate of blood and lymph supply. Physicochemical factors that are of profound importance in extravasation of compounds are molecular size, shape, charge, and hydrophilic–lipophilic balance (HLB) characteristics.

The structure of the blood capillary wall is complex and varies in different organs and tissues. It consists of a single layer of endothelial cells joined together by intercellular junctions. Each endothelial cell, on an average, is 20–40 μm long, 10–15 μm wide, and 0.1–0.5 μm thick, and contains 10,000–15,000 of uniform, spherical vesicles called plasmalemmal vesicles. These vesicles range in size between 60 and 80 nm in diameter. About 70% of these vesicles open on the luminal side of the endothelial surface, and the remaining open within the cytoplasm. Plasmalemmal vesicles are believed to be involved in the pinocytic transport of substances across the endothelium. The transition time of pinocytic vesicles across the cell is about 1 sec. Fusion of plasmalemmal vesicles leads to the formation of transendothelial channels. The endothelial cells are covered, on the luminal side, with a thick (10- to 20-nm) layer of a glycosaminoglycan coating. This layer continues into the plasmalemmal vesicles and into transendothelial channels, and is believed to be involved in cell adhesion, the stabilization of receptors, cellular protection, and in the regulation of extravasation. It also provides many microdomains of differing charge or charge density on the endothelial cell surface. On the external side, the endothelium is supported by a 5- to 8-nm–thick membrane called the basal lamina. Below the basal lamina a layer of connective tissues is present, called adventitia. The connective tissues surround the basal lamina as well as blending externally with the surrounding fibroaerolar tissues.

Depending on the morphology and continuity of the endothelial layer and the basement membrane, blood capillaries are divided into three types: continuous, fenestrated, and sinusoidal. The distribution of these capillaries in the body and their characteristics are presented in Table 4, and a schematic representation of the differences in their structures is shown in Fig. 4 [35]. Continuous capillaries are common and widely distributed in the body. They exhibit tight interendothelial junctions and an uninterrupted basement membrane. Fenestrated capillaries show interendothelium gaps of 20–80 nm at irregular intervals. These gaps have a thin membrane, which is believed to be derived from the basal membrane. Sinusoidal capillaries show interendothelial gaps of up to 150 nm. Depending on the tissue or organ, the basal membrane in sinusoidal capillaries is either absent (e.g., in liver) or present as a discontinuous membrane (e.g., in spleen and bone marrow). Sinusoidal capillaries are also wider in diameter, have an irregular lumen, and their wall is very thin. Furthermore, they have hardly any connecting tissues between the endothelial cells and the cells in which they are located. This area is occupied by a variety of cells, including highly active phagocytic cells.

There are also numerous important variations in the microvasculature bed (i.e., arterioles, capillaries, and venules) that affect permeability. For example, venular portions of the capillaries have thin endothelial cells (170 nm), with frequent interendothelial discontinuities. About

Table 4 Distribution and Characteristics of Endothelium in Various Tissues

Tissue	Characteristics
Continuous endothelium	
Connective tissue, muscle (skeletal and smooth), heart, pancreas, brain, lung, gonads, mesentery	Tight junctions (up to 2 nm) with continuous basement membrane; extravasation mainly by vesicular trafficking
Discontinuous endothelium	
Fenestrated	
Kidney glomeruli, GI tract mucosa, exocrine and endocrine glands, certain tumors, pertibular capillaries, choroid plexus, pancreas, intestinal wall	Interruptions (20–80 nm) between cell junctions; membrane thickness 4–6 nm; basement membrane continuous
Sinusoidal	
Liver, spleen, red bone marrow, suprarenal and parathyroid glands, certain tumors, carotid and coccygeal bodies	Junctions up to 150 nm, basement membrane absent in liver and discontinuous in spleen and bone marrow

30% of venular junctions are believed to have gaps of about 6 nm. Arterioles, in contrast, have endothelial cells that are linked by the tight junctions and communicating junctions, whereas the capillary endothelium contains only occluding junctions. Communicating gaps are small and rare in muscular venules and are absent in capillaries and pericytic venules. Endothelial cells in capillaries have more vesicles than those in arterioles ($1000/\mu m^3$ versus $190/\mu m^3$). The

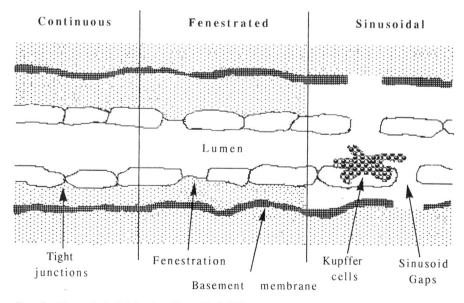

Fig. 4 The endothelial barrier. (From Ref. 35.)

intercellular sealing is strong in arterioles, well developed in capillaries, and particularly loose in venules. Furthermore, capillaries and postcapillaries venules have more transendothelial channels.

The transport of macromolecules across endothelium has been recently reviewed [36, 37]. Macromolecules can traverse the normal endothelium by passive processes, such as nonspecific fluid-phase transcapillary pinocytosis and passage through interendothelial junctions, gaps, or fenestrae, or by receptor-mediated transport systems. Passive extravasation is affected by the regional differences in the capillaries structure, the disease state of the tissue or organ, the number and size of the microvascular surface area, and physicochemical characteristics of the macromolecules. In general, the transfer of macromolecules across endothelium decreases progressively with an increase in the molecular size. It is widely recognized that low-molecular-mass solutes and a large number of macromolecules, up to 30 nm in diameter, can cross the endothelium under certain normal and pathophysiological conditions [24]. For proteins, the threshold restricting free passage through the glomerular endothelium is at a molecular mass of 60–70 kDa. Molecules with a molecular mass larger than 70 kDa are predominantly retained in the blood until they are degraded and excreted. Certain hydrophilic polymers, such as polyvinylpyrrolidone (PVP), dextran, polyethylene glycol (PEG), N-(2-hydroxypropyl)methacrylamide (HPMA), exhibit much greater hydrodynamic radii, compared with proteins of the same molecular weight, and consequently, the threshold molecular weight restricting glomerular filtration is lower than for proteins (25,000 for PVP; 50,000 for dextran; and 45,000 for HPMA) [38].

Because of the presence of anionic sites on the endothelium and on the glycocalyx layer, anionic macromolecules show a significantly slower rate of extravasation compared with neutral and cationic macromolecules. Kern and Swanson [39] found a threefold increase in the permeability of the pulmonary vascular system to cationic albumin, compared with native albumin of the same molecular weight and hydrodynamic radius.

Regional differences in the capillary structure and the number and size of the microvascular surface area determine the flux of macromolecules in the interstitium [24]. For example, organs such as the lung, with very large surface areas, will have a proportionately large total permeability and, consequently, a high extravasation. Renal endothelia has a thick basement membrane and contains anionic groups and heparin sulfate proteoglycan in the basement membrane. Thus, extravasation through this membrane will largely depend on the molecular charge, shape, size, and lipophilic–hydrophilic balance characteristics of the macromolecules. Intestinal endothelium, although fenestrated, is highly restrictive to passage of macromolecules. The absolue rate of extravasation varies considerably from one region to another within the alimentary canal. There is a large change in permeation of solute macromolecules with a radius of less than 6 nm, and no decrease in the permeation occurs for molecules of radius between 6 and 13.5 nm [40]. The lung endothelium, which is nonfenestrated and has vesicles of 50–100 nm, is more selective to the passage of macromolecules; the lymph/plasma ratio was decreased from a value of 0.7 to 0.25 when the molecular radius of the macromolecule increased from 3.7 to 11.0 nm [41]. Skeletal muscle, adipose tissue, liver, and myocardial endothelia, all show extravasation as a function of macromolecular size. The endothelium of the brain is the tightest of all endothelia in the body. It is formed by the continuous, nonfenestrated endothelial cells, and shows virtually no pinocytic activity. There are, however, certain regions of the brain (e.g., choroid plexus) that have fenestrated endothelium. Macromolecules, such as horseradish peroxidase, reach the cerebrospinal fluid by this route. Also, certain pathophysiological conditions, such as osmotic shocks, thermal injury, arterial hypertension, air or fat embolism, hypovolemia, and traumatic injury, causes transcapillary leakage and onset of pinocytic activity. This may have some implications in extravasation of macromolecules across the blood–brain barrier.

The changes in the permeability of capillaries as a result of inflammation is believed to be due to the effect of histamine, bradykinin, and a variety of other mediators [36]. The latter act directly on the capillary venule and endothelial vessel wall, effecting a rapid interaction between venular endothelial cells and circulating neutrophils. Damaged capillaries, in general, usually show increased openings (ranging in size between 80 and 140 nm) in the endothelium and, hence, increased transport activity. Macromolecules of up to 3000 kDa are capable of extravasation from blood vessels within experimental solid tumor, whereas molecules between 70–150 kDa extravasate mainly from the vascular plexus induced around solid tumors. It has been recently suggested that inflamed tissues also show changes in the glycocalyx layer, which cause increased vesicular trafficking and, consequently, increased extravasation of blood-borne materials. The metabolic changes, which are mediated through a reduced oxygen concentration, an increased carbon dioxide concentration, and a local increase in pH owing to accumulation of various metabolites, also affect extravasation.

Soluble macromolecules permeate the endothelial barrier more readily than particulate macromolecules. The rate of movement of fluid across the endothelium appears to be directly related to the difference between the hydrostatic and osmotic forces.

Receptor-mediated transport systems include both the fluid-phase and the constitutive and nonconstitutive endocytosis or transcytosis. It appears that particles smaller than 40 nm in diameter are able to enter these pathways. Ghitescu et al. [42], using 5-nm gold–albumin particles, showed that particles are first absorbed onto the specific binding sites of the endothelia examined (i.e., in lung, heart, and diaphragm), then transported in transcytotic vesicles across the endothelium by receptor-mediated transcytosis and, to a lesser extent, by fluid-phase processes. Low-density lipoproteins pass through sinusoids, enter the space of Disse, and then are processed into the liver hepatocytes, after interaction with the apolipoprotein ligands on the surface of the hepatocytes. Studies indicate that particles can be directed to other cells in the liver by altering the surface with ligands specific for the plasma membrane of those cells. Table 2 lists various receptors and the cells that have them.

D. Lymphatic Uptake

Following extravasation, the drug molecules can either reabsorb into the bloodstream directly by the enlarged postcapillary interendothelial cell pores found in most tissues [43], or enter into the lymphatic system and then return with the lymph (a constituent of the interstitial fluid) to the blood circulation [41] (Fig. 5). Also, drugs administered by subcutaneous, intramuscular, transdermal, and peritoneal routes can reach the systemic circulation by the lymphatic system (see Fig. 1). A schematic representation of the integration of lymph and blood circulation is shown in Fig. 6 [44]. As shown in Fig. 6, the lymphatic system originates in tissues as a network of fine capillaries. These capillaries coalesce regionally to form large vessels (referred to as afferent vessels), which extend centrally to one or more lymph nodes. The (efferent) ducts from the centrally located lymph nodes unite and form the major lymph trunks (e.g., intestinal, cervical, and thoracic ducts), which finally coalesce with the venous supply at the root of the neck.

Similar to blood capillaries, the lymphatic capillaries consist of a single layer of endothelial cells joined together by intercellular junctions. The diameter of small pores is 12 nm, whereas large pores range between 50 and 70 nm. The rate of formation of lymph depends on the hydrostatic pressure of blood and the permeability of the capillary wall. As blood enters the arterial end of the capillary, the hydrostatic pressure increases and, consequently, extravasation of water, electrolytes, and other blood-borne substances (e.g., proteins), occurs. By the time blood reaches the venular end of the capillary, the hydrostatic pressure drops, and some water and other low-molecular mass (less than 10,000 Da) substances are reabsorbed. However, there

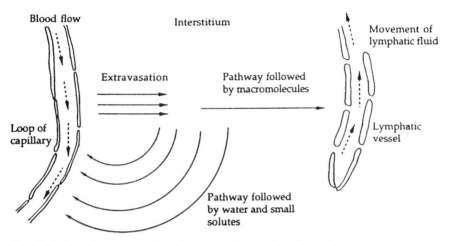

Fig. 5 Schematic representation of extravasation and lymphatic drainage. (From Ref. 38.)

is a net excess of extravasation over reabsorption, which results in accumulation of excess lymph in the tissues. This accumulation of excess fluid causes an increase in the interstitial pressure, which forces the lymph to enter the lymphatic system. The larger lymphatic vessels contain bicuspid valves, which prevent the retrograde flow of the lymph, while a coat of circular smooth muscle propels the lymph to flow centrally at a rate proportion to its rate of formation [45]. Following absorption in the peripheral capillary bed, the lymph is transported (by large

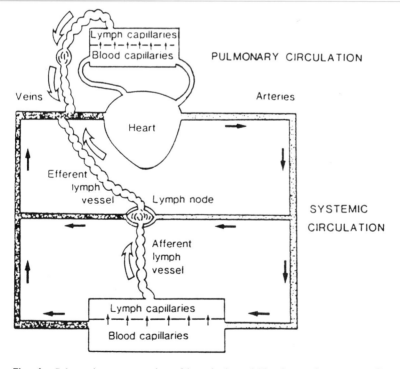

Fig. 6 Schematic representation of lymphatic and blood vascular systems. (From Ref. 44.)

lymph capillaries) to the regional lymph node where lymphocytes are added. The lymph is then taken to the next node up the chain and, finally, into the great vein.

Factors known to influence the clearance of drugs from interstitial sites, following extravasation or parenteral interstitial or transepithelial administration, include size and surface characteristics of particles, formulation medium, the composition and pH of the interstitial fluid, and disease within the interstitium. Studies indicate that soluble macromolecules smaller than 30 nm can enter the lymphatic system, whereas particulate materials larger than 50 nm are retained in the interstitial sites and serve as a sustained-release depot. The use of lipids or an oil in the formulation and the presence of negative surface charge all appear to facilitate the absorption of particles into the lymphatic system.

Solid tumors, in general, lack lymphatic drainage; therefore, macromolecular drugs that enter tumor interstitium by extravasation remain there. This mechanism is commonly referred to in terms of a tumor ''enhanced permeability and retention'' or ''EPR'' effect. The trapped drug in the tumor interstitium may then be released, either intra- or extracellularly, by tumor-associated proteolytic enzymes. The released drug is then able to penetrate readily through cell membranes and reach its intracellular targets. Possible exploitation of this phenomenon in selective tumor therapy has been discussed in detail by Seymour [38]. The direct delivery of drugs into lymphatics has also been proposed as a potential approach to kill malignant lymphoid cells located in lymph nodes.

IV. PHARMACOKINETIC AND PHARMACODYNAMIC CONSIDERATIONS

The human body can be considered to be made up of a series of anatomically discrete compartments connected to each other through blood circulation and by physiological and biochemical links. When a drug is administered, it is readily distributed to various compartments by blood. The relative amounts of drug available at the target (response compartment) and nontarget (toxicity compartment) sites determine the therapeutic effect and toxicities relative to that effect. In conventional therapy, the natural distribution characteristics of the drug determine the ratio of therapeutic response to the toxic effects.

Targeted drug delivery systems are designed to maximize therapeutic response by delivering drug selectively to its pharmacological site(s). There are several factors that determine the availability of drug at the target site [46,47]. These include the rates of (a) input of targeted drug into the body plasma, (b) distribution of targeted drug to the active site, (c) release of active drug from the targeted drug at the site of action, (d) removal (elimination) of targeted drug and free drug from the target site, (e) diffusion or transport of targeted drug and free drug from the active site to nontarget sites, and (f) blood and lymph flow to and from the target site. A three-compartment pharmacokinetic model used by Boddy et al. [46] to describe these processes is schematically shown in Fig. 7. The release of free drug from the targeted drug delivery system may occur either passively or by an active mechanism mediated by an internal or external stimulus (e.g., pH, temperature, and enzymes). Thus, the rate of release of drug varies, depending on the mechanism involved. In enzyme-mediated reactions, the rate of release of free drug depends on the activity and concentration of the enzyme involved, whereas in reactions that are selective, but nonspecific to the local chemical characteristics of the active site, the rate of release of free drug depends on the concentration of the targeted drug available at that site.

Thus, if the rate of distribution of targeted drug delivery to active site(s) is slow, or if the rate of elimination of targeted drug from the active site is faster than the rate of delivery to

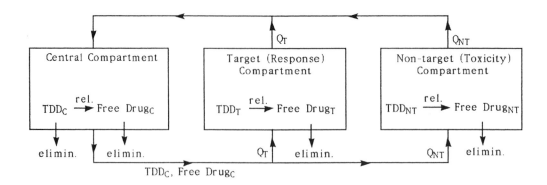

(TDD = Targeted Drug Delivery)

Fig. 7 A three-compartment pharmacokinetic model describing targeted drug delivery. (From Ref. 46.)

this site, then a sufficient amount of drug may not be available at the target site to produce the desired pharmacological effect [46]. The rate of distribution of targeted drug delivery to the target site depends on the rate of blood flow, whereas the removal of targeted delivery and free drug from the active site depends on the permeability of the endothelial barrier, on the rates of flow of blood and lymph to and from the target site, and on the rate of release of free drug. The upper limit of distribution of targeted drug or free drug to a specific tissue or organ in the body is provided by the product of blood flow and the concentration of the species in the blood. The inability of macromolecules or charged species to cross membrane barriers can limit their access to, and removal from, the target sites. The binding of the targeted drug delivery system or drug within the target site reduces the concentration available for removal. For drugs that are active only in the free form, the binding may also reduce the effective amount required to produce the desired therapeutic effect. Levy [48], on the basis of the assumption that the targeted drug delivery product will be transported from the target site to the rest of the body (which acts at least initially as an infinite sink), by diffusion, convection, or transport processes, concluded that (a) drug elimination from the target site will frequently be much more rapid than drug elimination from the body as a whole; (b) the duration of action of a targeted bolus dose will often be much shorter than the duration of action of a conventionally administered bolus dose and, consequently, the rate of drug administration to maintain a constant pharmacologic effect will need to be much higher for targeted than for conventionally administered drug; and (c) changes in the biotransformation and excretion kinetics or of other processes (e.g., the liver perfusion rate) that determine the systemic clearance of a drug by the body will have no effect on the kinetics of elimination of targeted drug from the site of action.

According to Levy [48], if there is a large difference between the rates of drug elimination from the active site and from the body, if the ratio of the effective dose at the active site and in body plasma is small, and if elimination at the target site does not represent biotransformation, then the amount of drug in the body will gradually accumulate if the targeted drug delivery system is continuously administered. As a result the pharmacological effect will gradually increase. However, this will also lead to the loss of drug-targeting selectivity because the amount of drug in the body plasma will continue to rise. Thus, to maintain the selectivity of drug targeting, the delivery system must be designed to require a very low continuous input relative to the rate of elimination of drug from the body [48].

In conventional delivery, the pharmacological response to a drug is assumed to be linearly related to the drug concentration in the plasma. This relationship between concentration and effect is much more complex in targeted drug delivery. It can vary in different organs or tissues, depending on access, retention (maintenance of adequate levels of targeted delivery and free drug at the active site), and timing of release of drug within that site.

The various approaches that have been used to quantitate targeted drug delivery systems have been recently reviewed by Gupta and Hung [49]. These authors suggested that the overall drug-targeting efficiency (T_e^*), which represents selectivity of a delivery system for the target tissue (T) compared with n nontarget (NT) tissues, can be reliably calculated according to the following expression:

$$\%T_e^* = \frac{(AQU_0^\infty)_T \times 100}{\sum_{i=1}^n (AQU_0^\infty)_{NT}}$$

Where (AUQ_0^∞) is the area under the amount of drug (Q) in a tissue versus time curve. Q can be obtained, at any time t, by the relationship $Q = CV$ (or W), where C is the concentration of drug at time t and V and W are the volume and weight, respectively, of that tissue.

V. TARGETED DRUG DELIVERY SYSTEMS

As noted in Sec. III, three strategies have been used to achieve drug targeting. These include use of site-specific, pharmacologically active molecules (magic bullet approach); preparation of pharmacologically inert agents that are activated only at the active site (prodrugs); and use of biologically inert carrier systems that selectively direct drugs to a specific site in the body (magic gun/missile, or drug-carrier approach). In this section, prodrugs and drug-carrier delivery systems are discussed in detail.

A. Prodrugs

A *prodrug* is pharmacologically inert form of an active drug that must undergo transformation to the parent compound in vivo by either a chemical or an enzymatic reaction to exert its therapeutic effects. The theory and practice of prodrugs have been reviewed by Notari [50]. Stella and Himmelstein [51,52] have critically reviewed the use of prodrugs in site-specific delivery. For a prodrug to be useful in site-specific delivery, it must exhibit adequate access to its pharmacological receptor(s). Also, the enzyme or chemical agent responsible for reactivating the prodrug should show major activity only at the target site. Furthermore, the enzyme should be in adequate supply to produce the required level of drug to manifest its pharmacological effect. Finally, the active drug produced in situ must be retained at the target site and not leak out into the systemic circulation, which could lead to adverse effects. Thus, prodrugs are designed to alter the absorption, distribution, and metabolism of the parent compound and, thereby, to increase its beneficial effects and decrease its toxicity. Prodrugs are also used to overcome formulation problems and to avoid an unpleasant taste or odor of the parent compounds.

Table 5 lists some commonly used types of prodrugs and their methods of regeneration [53]. Many of these prodrugs are simple esters, and can be reactivated in vivo by an esterase enzyme. Prodrugs containing an amide bond can be regenerated by peptidases, but their use in vivo has had varying degree of success. The chemically reconvertible prodrugs frequently lack selectivity of activation at the target sites and, thus, offer little opportunity for drug targeting [a detailed discussion of prodrugs listed in Table 5 can be found in Refs. 54 and 55].

Table 5 Prodrug Modifications and Method of Regeneration

Drug	Prodrug	Regeneration method
R-OH (alcohols and phenols)	Alkyl esters and half esters	Enzyme
	Phosphate and sulfate esters	Enzyme
	Sulfoacetyl, dialkyl aminoacyl	Enzyme
	Acyloxyalkyl ethers and thioethers	Enzyme
	Carbamates	Enzyme
R-COOH	Alkyl and glyceryl esters	Enzyme
	Acyloxyalkyl and lactonyl esters	Enzyme
	Alkoxycarbonyloxyalkyl esters	Enzyme
	(2-Oxo-1,3-dioxolenyl)alkyl esters	Enzyme
	Amides and amino acid derivatives	Enzyme
RNH_2, R_2NH, and R_3N	Enamines, Schiff bases, Mannich bases and oxalzolidines	Chemical
	Amides and peptides	Enzyme
	Hydroxymethyl derivatives	Chemical
	Hydroxymethyl esters	Enzyme
	Soft quaternary ammonium slats	Enzyme
	Carbamates	Enzyme
R-CHO and >C=O	Enol esters	Enzyme
	Thiazolidines and oxazolidines	Chemical
R-C(O)-NH$_2$ and imides	Hydroxymethyl derivatives	Chemical
	Hydroxymethyl esters such as acetate and phosphates	Enzyme
	Mannich bases	Chemical

Source: Ref. 53.

There are numerous reports of prodrugs in the literature that show improved drug effects. Prodrugs that have shown some measures of success for site-specific delivery include L-3,4-dihydroxyphenylalanine (L-dopa) to the brain [56], dipivaloyl derivative of epinephrine to the eye [57], γ-glutamyl-L-dopa to the kidney [58], β-D-glucoside dexamethasone and prednisolone derivatives to the colon [59], thiamine-tetrahydrofuryldisulfide [60] to the red blood cells, and various amino acid derivatives of antitumor agents such as daunorubicin [61,62], acivicin [63], doxorubicin [63], and phenylenediamine [63] to tumor cells.

The selective delivery of drugs to the brain has been, and continues to be, one of the greatest challenges. Only highly lipid-soluble drugs can cross the blood–brain barrier. Prodrugs with high lipid solubility can be used, but they may show increased partitioning to other tissues and, thereby, cause adverse reactions. For example, L-dopa, the precursor of dopamine, when administered orally, readily partitions throughout the body, including the brain. Its conversion to dopamine in the corpus striatum produces the therapeutic response, whereas its conversion in the peripheral tissues results in many undesirable side effects. Although many of these side effects can be overcome by additional administration of an inhibitor of aromatic amino acid decarboxylase, such as carbidopa (this does not penetrate into the brain and thereby permits the conversion of L-dopa to the dopamine in the brain, but prevents its transformation in the peripheral tissues), the direct delivery of dopamine to the brain constitutes an attractive alternative. One approach that can be used is the prodrug carrier system developed by Bodor and Simpkins (Fig. 8) [64]. This approach is based on the observation that certain dihydropyridines

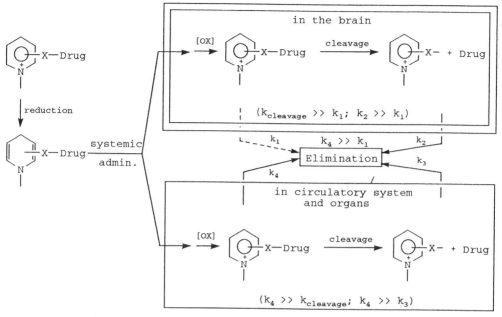

Fig. 8 Schematic representation of dihydropyridine–pyridinium redox delivery system. (From Ref. 66.)

fairly readily enter the brain, where they are oxidized to the corresponding quaternary salts. The latter, owing to difficulty in crossing the blood–brain barrier, remain in the brain. The formation of quaternary salts in the peripheral tissues, on the contrary, rapidly accelerates their removal by renal or biliary mechanisms. This results in a significant buildup of the quaternary salt concentration in the brain, and a significant reduction in the systemic toxicity. Chemical or enzymatic hydrolysis of the quaternary salt (in the brain) then slowly releases the drug in the cerebrospinal fluid, allowing the therapeutic concentration to be maintained over some period. Examples of drugs that have been investigated using this approach include pralidoxime iodide (2-pyridine aldoxime methyl iodide) [65], phenylethylamine [66], dopamine [67], 3'-azido-2',3'-dideoxyuridine (Azddu) [68], and 3'-azido-3'-deoxythymidine (AZT; zidovudine) [68–70], and sex hormones [71].

B. Drug–Carrier Delivery Systems

Drug–carrier delivery systems employ biologically inert macromolecules to direct a drug to its target site in the body. These are divided into two types: particulate and soluble macromolecular. Depending on the carrier system, the drug can be either molecularly entrapped within the carrier matrix or covalently linked to the carrier molecules. The major advantage of drug-carrier delivery systems is that the distribution of drugs in the body depends on the physicochemical properties of the carriers not those of drugs. This implies that targeting can be manipulated by choosing an appropriate carrier, or by alterations in the physicochemical properties of the carrier. There are, however, several other factors that must be considered in the pharmaceutical development and clinical use of both soluble macromolecular and particulate biotechnical and synthetic site-specific systems. These are listed in Table 6.

Targeting with drug–carrier systems can be divided into three types: passive, active, and physical [38,72,73]. Passive targeting relies on the natural distribution pattern of the drug–

Table 6 Considerations in the Pharmaceutical Development and Clinical Use of Both Soluble Macromolecular and Particulate Biotechnical and Site-Specific Drug Delivery Systems

Specification/activity	Specification/activity

I. Pharmaceutical development
  A. Production
    1. Purity
    2. Evaluation of novel production safety hazard (e.g., sparkling with particulate)
  B. Characterization
    1. Identity
    2. Conformation
    3. Size
    4. Size distribution
    5. Charge, aggregation
    6. Density
    7. Surface configuration, homogeneity of attachment moieties:
      Polymers
      Ligands
      Spacers
  C. In vitro functionality
    1. Drug loading efficiency
    2. Drug release
    3. Retention of recognition characteristics
  D. Stability
    1. Characteristics of the breakdown products
      In storage formulation
      In biological fluids
    2. Parameters to be assessed on storage:
      Chemical stability
      Character
      In vitro functionality
      Sterility and functionality
      Colloidal character (e.g., aggregation, size, charge)
      Surface properties (including conformation and epitopic character)
II. Safety pharmacology (non human)
  A. General considerations
    1. General safety
    2. Sterility and pyrogenicity
      Major organ function tests
      Acute and subacute toxicity studies
  B. Potential novel toxicities
    1. MPS uptake
    2. Uptake in specialized immune
    3. Depression/exhaustion of MPS
      Bone marrow
      Bacterial and viral infections
      Immunological depression
      Hemorrhagic and endotoxin shock
      Altered drug response
	4. Low level activation of MPS
	      Interleukin-1
	      Amyloidosis
	      Hyperplastic liver foci
	      Altered stem cell kinetics
	      Altered drug metabolism
	      Altered response to drugs
	  C. Biotechnics: For biotechnics (and specifically monoclonal antibodies), factors affecting safety include
	    1. Hybridoma background:
	      Murine–murine
	      Human–human
	      Interspecies (chimerics)
	    2. Contaminants
	      General safety
	      Pyrogens
	      Sterile
	      Free of hazardous viruses
	      Free of detectable DNA
	    3. Interaction with the host
	      Immunogenicity
	      Cross-reactivity
	      Hypersensitivity to foreign epitopes
	      Anti-idiotypic response to normal cells
	      Immune complex disease
	      Potential MPS toxicity
	  D. Specific specificity
	    1. Issues include altered drug disposition/ metabolism, and the need for a tier assessment of safety to include knowledge on
	      Pathology of lymphoid tissues
	      Antibody and cell-mediated immunity
	      Host cell resistance
	      Phagocytic cell function
	      Immune/immunotoxicity reaction
	      Testing
	      Antigen specificity
	      Complement binding
	III. Metabolism: Issues here include species-specific metabolism (related to novel pattern of drug release at receptor sites), and possible use of novel paracrine- and endocrine-like peptidergic mediators. These could manifest themselves in novel:
	    Dose response
	    Absorption sites/rates

Table 6 Continued

Specification/activity	Specification/activity
Bioavailabilities (at receptor)	New routes of administration:
Organ, tissue, cell disposition	transmucosal; specific regional uptake
Disease-dependent release	in gastrointestinal tract
Excretion routes/rates	New pattern of drug release: bolus/first-
IV. Efficacy: Major considerations in the clinical	order/pulsatile; feedback control;
development of a site-specific system,	disease-related release of drug
relating to effect, utility, and efficacy could	(Analytical techniques will need to encompass the
include:	identification of very low levels of site-specific
Novel pharmacokinetic and disposition	systems and their degradation and metabolic
Novel modalities of cell/tissue/receptor	products.)
exposure	V. Extended nonclinical development: These
Utilization of novel cellular transport	parallel activities will include
processes	Reproductive toxicology
Species-specific drugs and delivery	Chronic toxicology
modalities	Selection of market formulation
Novel drug interactions	Definition of marketed specifications
Novel drug metabolism	Development of implementation of
Local versus general distribution	market-related scale-up
Biphasic drug action	Confirmation of specification following
Chronopharmacology	scale-up

Source: Ref. 24.

drug–carrier system. For example, particles of 5 μm or smaller are readily removed from the blood by macrophages of the RES when administered systemically. This natural defense mechanism of the RES thus provides an opportunity to target drug, encapsulated in or conjugated to an appropriate carrier system, to macrophages. Mechanical filtration of large carriers by capillary blockage can also be exploited to target drugs to the lungs by the venous supply and to other organs through the appropriate arterial supply. By controlling the rate of drug release, one can achieve the desired therapeutic action in the targeted organ. Passive targeting also includes delivery of drug–carrier systems directly to a discrete compartment in the body (e.g., different regions of the GI tract, eye, nose, knee joints, lungs, vagina, rectum, respiratory tract, or other). This offers the opportunity for the treatment of diseases that require a persistent and sustained presentation of drug at that site.

Active targeting employs a deliberately modified drug–drug–carrier molecule capable of recognizing and interacting with a specific cell, tissue, or organ in the body. Modifications of the carrier system may include a change in the molecular size, alteration of the surface properties, incorporation of antigen-specific antibodies, or attachment of cell receptor-specific ligands [73].

Physical targeting refers to delivery systems that release a drug only when exposed to a specific microenvironment, such as a change in pH or temperature, or the use of an external magnetic field [73].

A detailed discussion of particulate and soluble macromolecular delivery systems is presented in the following sections.

Particulate Drug Delivery Systems

The concept of using particles to deliver drugs to selected sites in the body originated from their use as radiodiagnostic agents in medicine in the investigation of the RES (liver, spleen, bone marrow, and lymph nodes), gastrointestinal examination, and so on. Particles ranging in sizes from 20 to up to 300 μm have been proposed for drug targeting. Because of the small size of the particles, particulate drug delivery systems can be introduced directly into the central circulation by intra-articular or intravenous injection, or delivered to a given body compartment, for example, by injection into a joint or by administration by an aerosol to the lungs and nose. Subcutaneous and intraperitoneal administration routes have also been used to deliver drugs to the lymphatic system and regional lymph nodes. Some suggested uses of particulate drug delivery systems in drug targeting are presented in Table 7 [24].

Particulate drug delivery systems can be monolithic (i.e., containing an intimate mixture of drug and the core material), capsular (in which the drug is surrounded by the carrier material), or emulsion (in which the drug is dispersed in a suspension of the carrier material) types. The biofate (passive targeting) of particulate drug delivery systems depends on the size and shape,

Table 7 Some Uses for Particulate Drug Delivery Systems

Target site/purpose (Particle size)	Disease/therapy
Direct administration to discrete compartments (0.05–100 μm)	
Eye	Infection
Lung	Allergy
Joints	Arthritis
Gastrointestinal tract	Crohn's disease, immunization
Intralesional	Tumor
Bladder	Infection
Cerebral ventricles	Infection
Interstitial administration (0.005–100 μm)	
Subcutaneous	Lymph node targeting (e.g., some cancers)
Intramuscular	Depot for anesthetics, proteins
Intravascular targets	
Diseased macrophages (0.1–1.0 μm)	Parasitic, fungal, viral, enzyme storage disease; autoimmune diseases; gene therapy
Other blood cells (0.1–1.0 μm)	Cancerous; platelets; gene therapy (bone marrow erythroblasts); immune cells (vaccination/adjuvant); antivirals
Circulating depot (0.1–1.0 μm)	Anti-infectives; antileukemics; antithrombotics; antivirals; release of polypeptides and protein drugs
Capillary filtration (> 1.0 μm)	Cancer; emphysema; thrombi; drug acting on local endothelia
Extravascular targets	
Macrophages activation (0.1–1.0 μm)	Abnormal cells (e.g., cancerous and virally infected)
Discontinuous endothelia (< 0.15 μm)	
Basement membrane	Spleen
Parenchymal cells	Liver
Diseased endothelia (< 0.5 ? μm)	Rheumatoid arthritis, malignant hypertension, myocardial infarct, transluminal angioplasty
Ex vivo (> 0.5–50 μm)	
Cells	Cell targeting (e.g., for gene therapy)

Source: Ref. 24.

charge, and surface hydrophobicity of the particles. A relation between particle size and bio-logical targeting, after intravascular injection, is schematically depicted in Fig. 9 [73]. After intravenous administration, particles larger than 7 μm are normally mechanically filtered by the smallest capillaries of the lungs (particles of 15 μm have been homogeneously distributed throughout the lung, whereas particles of 137 μm exhibited a more peripheral distribution [74]), and particles smaller than 7 μm in diameter (between 2 and 7 μm) may pass the smallest lung capillary beds and be entrapped in the capillary network of the liver and spleen. Larger particles can also be injected intra-arterially. Here, particles will be retained in the first capillary bed encountered (first-order targeting). For examples, administration into the mesenteric, portal, or renal artery leads to complete entrapment of particles in gut, liver, or kidneys, respectively. For organs that bear solid tumors, this approach may lead to localization in the tumor cells. Particles in a size between 0.05 and 2 μm are rapidly cleared from the bloodstream by mac-rophages (primarily by the Kupffer cells of the liver) of the RES after intravenous, intra-arterial, or intraperitoneal administration. Extraction of particles by macrophages of the RES can be 90% or greater, with a half-life of less than 1 min. This natural targeting to the liver offers opportunities for the treatment of tropical diseases (leishmaniasis) and fungal infections (can-didiasis). Because of the dominant role of Kupffer cells, other cells of the RES play a small role in removing particles from the blood. Since the fenestrae of the liver endothelium have a diameter of 0.1 μm, particles smaller than 0.1 μm can pass through the sieve plates of the sinusoid and become localized in the spleen and bone marrow.

Negatively charged particles are more rapidly cleared from the blood than are neutral and positive ones [75]. The clearance rate of particles by the reticuloendothelial system is inversely related to the load of the particles; that is, the rate of clearance of a larger dose of micropar-ticulate is slower than the smaller dose [76].

The targeting of drugs to sites in the body other than RES (e.g., parenchymal cells or tumor cells of the liver or monocytes in the blood) has been extensively studied. In vitro studies show that this can be achieved by linking particles to monoclonal antibodies [77–80] or to cell-specific ligands (e.g., desialyated fetuin [81], glycoproteins [82], native immunoglobins [83], or heat-aggregated immunoglobins [84], or by alterations of the particles' surface characteristics (e.g., by using bioadhesives [85] or nonionic surfactants [7]) so that they are not recognized by the RES as being foreign (active targeting) [73]. The changes in the surface properties prevent particles from adhering to the macrophages and, consequently, their endocytosis. For

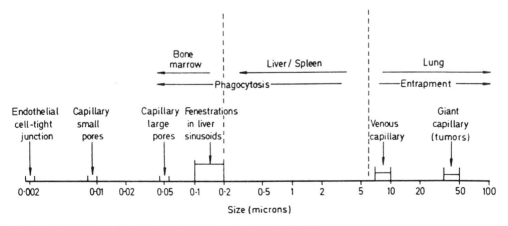

Fig. 9 The vascular bed and passive targeting. (From Ref. 73.)

example, coating particles with high-molecular-weight copolymers, consisting of a long hydrophobic chain (e.g., poloxmers), to anchor the polymers on the surface of the particles, and two or more hydrophilic chains that prevent endocytosis by steric stabilization, has significantly reduced uptake by the RES and, as a result, the particles are distributed to other parts of the body where they do not normally localize (Table 8) [7].

Other approaches used to avoid RES uptake of particles include incorporation of ferromagnetic materials, such as carboxyl-iron powder and Fe_3O_4 [86]; formulation of particles in oils [87]; and the use of an appropriate carrier (e.g., phospholipids) capable of degrading and releasing the drug into the surroundings with slight changes in the temperature [88] or in pH [89] conditions. The approach involving magnetic materials has been successfully used in humans in the therapy of carcinoma of the prostate and bladder [90]. It allows maneuvering of the delivery system with external magnetic fields. The particle-in-oil approach has been used by Sezaki et al. [87] in the treatment of VXZ carcinoma in rabbits and cystic hygroma in pediatric patients. They reported a significant increase in the survival rate of the rabbits and the management of 22 cases of pediatric cystic hygroma when treated with bleomycin–gelatin microsphere–oil emulsion. Cystic hygroma are benign tumors and are difficult to remove completely by surgery. Free bleomycin had no effect on these tumors.

The various particulate drug carrier systems that have been investigated can be grouped into the following classes:

Microparticles and Nanoparticles. Colloidal particles ranging in size between 10 and 1000 nm (1 μm) are known as nanoparticles, whereas particles larger than 1 μm, but small enough not to sediment when suspended in water (but large enough to scatter incoming light), are called microparticles. A partial list of various natural and synthetic materials used in the preparation of microparticles and nanoparticles is presented in Table 9 [3]. Also included in the list are the sizes of particles investigated, the names of the active agents entrapped or proposed for entrapment, and the intended or suggested use in drug targeting.

The most commonly practiced methods to prepare microparticles and nanoparticles are emulsion, micelle, and interfacial polymerization, and coacervation. The emulsion polymerization method involves heating a mixture of monomer and active agent(s) in an aqueous or nonaqueous phase that contains an initiator, a surfactant (employed usually in excess of its critical micelle concentration), and a stabilizer. Vigorous agitation is employed during the emulsion formation to produce particles smaller than 100 μm, usually below 1 μm. The smaller particle size assures good tissue tolerance, uptake, and transfer, and causes no foreign body reaction. Examples of carrier systems prepared by the emulsion polymerization approach include poly(methyl methacrylate) nanoparticles, which exhibit excellent adjuvant properties for vaccines [91], and polyalkylcyanoacrylate nanoparticles [92], which are biodegradable. The main advantage of emulsion polymerization is that higher-molecular-weight polymers are usu-

Table 8 Uptake of Polystyrene Microsphere (60 nm in Diameter) in the Organ of the Rat Following IV Injection

System	Liver	Spleen	Blood[a]	Femurs
Uncoated control	47.4[b] + 2.6 (7)	1.05 + 0.65 (4)	3.7 + 0.28 (7)	0.059 + 0.002 (3)
Coated with poloxamer 338	3.5 + 1.1 (2)	0.39 + 0.015 (2)	39.2 + 5.4 (2)	0.142 + 0.037 (2)
Coated/uncoated ratio	0.073	0.36	10.6	2.4

[a]Blood volume taken as 6.5 ml/100 g body weight.
[b]Percentage uptake at 1 hr.
Source: Ref. 7.

Table 9 Particulate Drug Delivery Systems

Matrix material	Diameter (μm)	Intended or suggested use	Actual or suggested active molecules
Chylomicron	0.1–0.5		Factor VIII
Low-density lipoprotein	0.017–0.025	Delivery to neoplastic cells	Methotrexate, anticancer agents
High-density lipoprotein	0.007–0.012	Delivery to adrenals and ovaries after intravenous injection	Gram-negative lipopolysaccharide
Polyalkylcyanoacrylate	0.2	Lysosomtropic after intravenous injection	Antimitotics (e.g., daunorubicin, actinomycin-D, doxorubicin)
	0.213	Blood glucose regulators after intra-articular injection	Insulin, triamcinolone diacetate
Ferromagnetic			
Poly(isobutylcyanoacrylate)	0.22	Biodegradable TDS with extracorporeal guidance	[³H]Dactinomycin
Poly(methylmethacrylate)	0.1–1.0	Vaccine therapy	Vaccines
Polyamide	60–120	Oral for saccharose tolerance	L-Invertase
Polyacrylamide	0.3, 18, 36	Intraperitonial, intravenous for acute leukemia	L-Asparaginase
	0.25–3	Intramuscular/subcutaneous for reduction in enzyme antibody effects	L-Asparaginase
	0.7	Lysosomotropic after intravenous injection to treat enzyme deficiencies (e.g., adult Gaucher's disease)	Enzymes
Polyacryldextran	25–75	Biodegradable TDS for protein delivery	Proteins (e.g., L-asparaginase)
DL-polylactic acid	125	Subcutaneous delivery of local anesthetics	Dibucaine; tetracaine
	38–297	Subdermal antimalarial implant	Quinazoline analogs
Sulfonic ion-exchange resin		Oral, anthelmintic	Levamisole
Carnauba	30–800	Chemoembolism TDS for intra-arterial delivery of cytostatics	5-Fluorouracil; CCNU; methotrexate
Ethylcellulose	225 (mean)	Arterial chemoembolism for delivery of cytostatics to kidney and liver	Mitomycin
Ferromagnetic ethylcellulose	307	Extracorporeal guidance to tumors	Mitomycin
Modified cellulose	40–160	Nondegradable (model) parenteral TDS for delivery to lungs	Methotrexate
Gelatin	0.301	Intra-articular injection	Triamcinolone diacetate
	1.6, 1.9	Intralymphatic delivery (lymphotropic), as microspheres-in-oil emulsion for delivery of cytostatics	5-Fluorouracil
	0.28	Delivery to liver and spleen after intravenous injection	Bleomycin; water-soluble drugs

Table 9 Continued

Matrix material	Diameter (μm)	Intended or suggested use	Actual or suggested active molecules
Gelatin core with dextran conjugate of drug	15	Lung, after intravenous injection	Mitomycin–dextran conjugate
Dextran cross-linked functionalized by carboxymethylation	10–30	Intratumor direct delivery	Doxorubicin; mitomycin
Self-forming microspheres polymercaptol	0.8	Oral and hemoperfusion for treatment of heavy metal poisioning	Mercaptol
Insulin	0.2	Oral and intramuscular delivery for treatment of diabetes	Insulin
Hemoglobin	5–60	Oxygen transport function	Hemoglobin
Polystyrene	3–25	Percorneal retention studies	
Agarose		Injection into tumor tissue	Mitomycin
Starch (Spherex)	40	Intra-arterial administration of cytostatics	5-Fluorouracil, hepatic BCNU, renal antinomycin-D
Ferromagnetic starch	1–50	Extracorporeal maneuvering to tumor tissue after IV injection	Ethanolamine and albumin as model compounds

Albumin	0.1–1	IV immunosuppressives delivery: infestation of the RES (e.g., histoplasmosis, typhoid)	[^{14}C]Mercaptopurine-8-hydrate
	0.169	Intra-articular injection	Triamcinolone diacetate
	0.66	Intravenous delivery to liver (RES)	[^{3}H]5-Fluorouracil
	10–30	Intrarenal delivery to tumors	Doxorubicin, bleomycine, 5-fluorouracil, methotrexate
	10	Intravenous delivery to lungs of antiallergic compounds	Sodium cromoglicate (Cromolyn sodium)
	7, 15	Treatment of emphysema (IV)	Leukocyte elastase inhibitor
	10–200	Intramuscular or subcutaneous depot	Norgestrone; progesterone
	All sizes	Variously: intra-arterial, intravenous	Antiasthmatics; analgesics, bronchodialators; narcotics; mucolytics; antibacterials; antituberculars; hypoglycemics; steroids; antitumor agents; amino acids
	10–40	Supplementation of drug therapy using internal radiation by intra-arterial delivery of radiolabeled microspheres	Yttrium 90
Ferromagnetic albumin	1–2	Delivery to tumors by extracorporeal guidance	Doxorubicin
	1–2; 2–4; 3–7	Intravenous delivery to lung; renal artery delivery	Doxorubicin
	1	Probe for neurological function	Mylein basic protein
	7	Immunoglobulin incorporation using staphylococcal protein A	

Source: Ref. 2.

ally formed at a faster rate and a lower temperature. A major disadvantage, however, is that product cannot be readily freed from the residual monomers. Micellar polymerization differs from emulsion polymerization only in that the monomers and active agent(s) are contained within the micelle formed by a suitable concentration of a surfactant before the polymerization is commenced. This allows very little, if any, increase in particle size as polymerization proceeds.

In interfacial polymerization, monomers react at the interface of two immiscible liquid phases to produce a film that encapsulates the dispersed phase. The process involves an initial emulsification step in which an aqueous phase, containing a reactive monomer and a core material, is dispersed in a nonaqueous continuous phase. This is then followed by the addition of a second monomer to the continuous phase. Monomers in the two phases then diffuse and polymerize at the interface to form a thin film. The degree of polymerization depends on the concentration of monomers, the temperature of the system, and the composition of the liquid phases.

The coacervation approach uses heating or chemical denaturation and desolvation of natural proteins or carbohydrates. As high as 85% of water-soluble drugs can be entrapped within a protein matrix by freeze-drying the emulsion prepared in this manner. For water-insoluble drugs, a microsuspension–emulsion procedure has been suggested as a method of choice to achieve high drug payloads.

Most products investigated to date are designed as sterile, freeze-dried, free-flowing powders, usually containing 0.1% w/w of a nonionic surfactant to assist redispersion in saline. These can be administered either systemically or by an intramuscular route. As is obvious from Table 7, the major use of microspheres and nanospheres, including magnetic microspheres, has been in tumor therapy. Microsphere-in-oil emulsions have been used as lysosomotropic systems. After subcutaneous administration, the oily product is readily taken up by the lymphatic system, not by the cardiovascular [87]. Recently, biodegradable polymeric nanospheres, with potential applications in medical imaging, gene therapy, and drug targeting to specific cells or tissues, have been developed [93]. These nanospheres have a polymer core in which a drug is dispersed. The core is, in turn, covalently linked to a polyethylene glycol coating that prevents rapid clearance of the particles from the body. The drug releases by diffusion through the coating or as the nanospheres break down. The nanospheres can carry high doses of drug or agent (up to 45% of particle weight), with entrapment efficiency of over 95%. Antibodies or other protein ligands can also be attached to nanospheres. The nanospheres can be freeze-dried and reconstituted.

Various factors that influence the release of drugs from particulate carriers are listed in Table 10. Drugs can be released by diffusion or by surface erosion, disintegration, hydration, or breakdown (by a chemical or an enzymatic reaction) of the particles. The release of drugs from microspheres follows a biphasic pattern; that is, an initial fast release followed by a slower first-order release (Fig. 10) [94]. The higher the solubility of a drug in water, the greater will be the release rate from the microspheres. Yapel [95] reported that, for epinephrine, the release from (albumin) microspheres becomes monophasic when the drug load exceeds 30%, by weight. The release (of drug) is also dependent on the degree of cross-linking (or heat denaturation) of the polymer and the size of the particles. At least over some initial range, the higher the cross-linking, the greater the water uptake characteristics of the polymer and, consequently, the slower the release rates. This suggests that the release of a drug from albumin microspheres with extended cross-linking is primarily due to the hydration of the polymer, rather than degradation. Figure 10 also shows the effect of microsphere size on the release rate. It is obvious that the release rate decreases as the size of microspheres decreases, suggesting that the release of cromolyn sodium (sodium cromoglicate) from albumin microspheres

Table 10   Factors Affecting the Release of Drugs from Particulate Carriers

Drug
      Position in the particle
      Molecular weight
      Physicochemical properties
      Concentration
      Drug–carrier interaction
      Diffusion; desorption from surface; ion-exchange
Particles
      Type and amount of matric material
      Size and density of the particle
      Capsular or monolithic
      Extent and nature of any cross-linking; denaturation of polymerization
      Presence of adjuvants
      Surface erosion; particle diffusion and leaching
      Total disintegration of particles
Environment
      Hydrogen ion concentration
      Polarity
      Ionic strength
      Presence of enzymes
      Temperature
      Microwave
      Magnetism
      Light

Source: Ref. 3.

is a diffusion-controlled process. Sezaki et al. [87] reported that incorporation of a gelatin-mitomycin drug conjugate, instead of free mitomycin into gelatin microspheres, leads to a monophasic drug release, similar to the rate of hydrolytic cleavage of the conjugate linkage. In magnetic microspheres, the release rate of drug apparently depends on the strength of the magnetic field, the frequency of oscillation [96], the shape of the embedded magnet [97], and the composition of the polymer [98–100].

Liposomes.   Liposomes are versatile, efficient, and probably the most extensively studied class of carrier systems. They have been used experimentally in all areas of medicine. A comprehensive review of preparation, analysis, drug entrapment, and interactions with the biological milieu, including drug targeting, can be found in a compendium entitled, *Liposome Technology*, edited by Gregoriadis [101,102]. There are also several books [103–106], book chapters [107–111], and review articles [112–119] that cover various aspects of liposome technology. A general discussion of liposomes as a dispersed system, including preparation, characterization and uses in pharmacy, is presented in Chapter 9 of this book.

The important attributes of liposomes as a drug carrier are (a) they are biologically inert and completely biodegradable; (b) they pose no concerns of toxicity, antigenicity, or pyrogenicity because phospholipids are natural components of all cell membranes; (c) they can be prepared in various sizes, compositions, surface charges, and so forth, depending on the requirements of a given application; (d) they can be used to entrap or encapsulate a wide variety of hydrophilic and lipophilic drugs, including enzymes, hormones, vitamins, antibiotics, and cytostatic agents; (e) drugs entrapped in liposomes are physically separated from the environment and, thus, are less susceptible to degradation or deactivation by the action of external

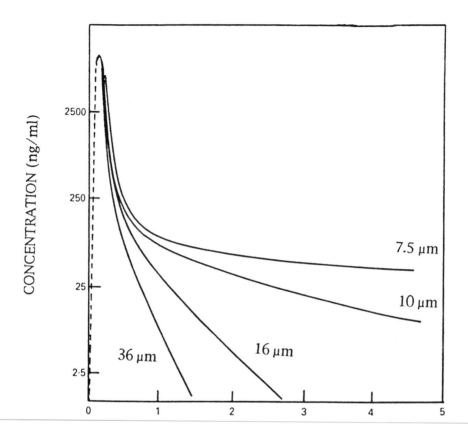

TIME (HOURS)

Fig. 10 Release of cromolyn sodium (sodium cromoglicate) from human serum albumin microspheres prepared using a water–oil emulsion technique with 5% glutaraldehyde as cross-linking agent. Dissolution medium: pH 7 phosphate buffer. (From Ref. 94.)

media (e.g., enzymes or inhibitors); and (f) liposome-entrapped drugs offer new possibilities for drug targeting, since entrapped drugs follow the fate of liposomes and are released only at the site of liposome destruction.

The drug-loading capacities of liposomes depend on the properties of the drug, the phospholipids, and other additives. Typically, hydrophobic drugs are solubilized with lipid(s) in an organic solvent that is then dried and subsequently hydrated to yield liposomal drug complexes. The loading of hydrophilic drugs is limited by their aqueous solubility and on the entrapment efficiency of the formulation method. By submitting the liposome–drug solution to several freeze-thaw cycles, the entrapment efficiency of water-soluble drugs can be increased from 1–5% to 50–80%. However, this is possible only at high lipid concentrations [120]. The loading of cationic (or anionic) drugs can be significantly improved by using liposomes containing negatively (or positively) charged lipids. Gabizon [121] reported that the entrapment efficiency can be increased from 10% for neutral liposomes to 60% for charged liposomes. Another approach to increase the entrapment efficiency is the use of transmembrane pH gradients. For example, liposomes with low internal pH will have a higher loading of cationic drugs, whereas the high intraliposomal pH will enable increased entrapment of negatively charged (anionic drugs) and of amphiphilic drugs [122]. Recently, a new dehydration–rehydration approach has been developed that allows entrapment of drug in the range of 50–85%. This method is easy to scale up, and the liposome vesicles containing drug, including protein, can be freeze-dried and reconstituted with saline solution without affecting the entrapped drug [113]. Proteins, sugar residues, and antibodies can also be incorporated into liposomes. The various approaches used include physical adsorption, incorporation during liposomal preparation, and covalent binding (direct or through a spacer) to the active drug or an inert additive (e.g., polymer) incorporated into the liposomal membrane [119,123,124].

After intravenous administration, liposomes are rapidly removed from the blood, primarily by cells of the reticuloendothelial system, and foremost by the liver (Kupffer cells). The half-lives of liposomes in the blood may range from a few minutes to many hours, depending on the nature and compositions of the lipids, surface properties, and size of the liposomal vesicles. In general, smaller unilamellar vesicles (SUVs) show much longer half-lives in the blood than multilamellar vesicles (MLVs) and large unilamellar vesicles (LUVs). Negatively charged liposomes are cleared more rapidly from the circulation than the corresponding neutral or positively charged liposomes. Also, the uptake by the spleen is greater for negatively charged liposomes than for positive or neutral liposomes. The SUVs can penetrate 0.1 μm fenestrations located in the endothelium of discontinuous (sinusoidal) capillaries lining the liver, spleen, and bone marrow [125], and reach the underlying parenchymal cells. The endothelium of the hepatic sinusoid contains openings larger than 0.1 μm in diameter, and this may allow penetration of MLVs and LUVs. An increase in the liposome dose causes a relative decrease in liver uptake and, consequently, an increase in blood levels and, to some extent, in spleen and bone marrow uptake [125]. Prolongation of the blood clearance times of the liposomes by blocking the RES uptake may increase the likelihood of liposomes to interact with vascular endothelial cells and circulating blood cells.

Irrespective of the size, liposomes, when injected intraperitoneally, partially accumulate in the liver and spleen. It has been suggested that the transport of liposomes from the peritoneal cavity to the systemic circulation, and eventually to tissues, occurs by lymphatic pathways. Local injection of larger liposomes leads to quantitative accumulation at the site of injection. The slow disintegration of the carrier then releases the drug, which diffuses into the blood circulation. Smaller liposomes, on the contrary, enter the lymph nodes and blood circulation and, eventually, accumulate in the liver and spleen.

Since the RES is the natural target, liposomes have been extensively investigated as carriers for the treatment of liver and RES organ diseases (passive targeting). Belchetz et al. [126] reported that liposome-entrapped glucocerebroside, when administered intravenously in patients suffering with Gaucher's disease, reduced the liver size significantly. The effect is attributed to the penetration ability of the liposomal drug into the cells. The native enzyme gave no effects because of its inability to penetrate the cells. A similar finding was reported in patients, suffering with glycogenosis type II disease, following administration of amyloglycosidase entrapped in liposomes [127]. Encapsulation of antimonial drugs within liposomes has increased the efficacy by 800- to 1000-fold, compared with the free drug, against experimental visceral and cutaneous leishmaniasis in rats [128–130]. Bakker-Woudenberg et al. [131] found about a 90-fold increase in the therapeutic efficacy following administration of liposome-encapsulated ampicillin, compared with free ampicillin, against *Listeria monocytogenes* infection.

Liposomes have also been used for delivering immunomodulating agents to macrophages. Macrophages are immunologically competent, extravascular cells that contribute to the host defense mechanisms. Activated macrophages are capable of selectively killing tumor cells, thereby leaving normal cells unharmed. Fidler et al. [132] have shown that lymphokines (mac-

rophage-activating factor, interferon), muramyl dipeptide, and a lipophilic derivative of muramyl tripeptide, encapsulated within liposomes are highly effective in activating antitumor
functions in rodent and human macrophages in vitro and in the mouse and the rat in vivo.
Dose–response measurements show that liposome-encapsulated preparations of these agents
induce maximum levels of macrophage activation at a significantly lower dose than needed for
equivalent activation by the nonencapsulated preparation [133–135]. Roerdnik et al. [108]
reported a 50–60% increase in the tumoricidal activity of muramyl dipeptide when encapsulated within liposomes, compared with free drug, against B-16 melanoma cells in vitro. The
free drug gave a maximum activity of 30% cytotoxicity versus a 250- to 1000-fold increase
in potentiation of muramyl-induced cytotoxicity as a result of encapsulation within liposomes.
Fidler et al. [136,137] and Sone et al. [138] have found that encapsulation of more than one
agent within the same liposome produces synergistic activation of macrophages in vitro and in
vivo. The activation of macrophages, in general, requires phagocytosis of the liposome, followed
by a lag period of 4–8 h before tumoricidal activity is expressed [134]. No participation of
macrophages surface receptors is required. This suggests that tumoricidal activity of macrophages
results from the interaction of immunomodulating agents with intracellular targets [139].

Liposomes also serve as a carrier for a variety of antineoplastic drugs. Mayhew and Rustum
[140] demonstrated that liposomes containing doxorubicin (Adriamycin) are 100 times more
effective, compared with free drug, against the liver metastasis of the M5076 tumor. Liposomal
encapsulation of amphotericin B, a potent, but extremely toxic, antifungal drug, also resulted
in much reduced toxicity, while it maintained potency [141]. Rosenberg et al. [142] and Burkhanov et al. [143] have reported that liposomes prepared using autologous phospholipids obtained from tumor cells are taken up by the tumor cells two to six times better than the control
egg lecithin liposomes.

Liposomes containing specific-targeting molecules, such as tumor-specific antibodies or cell
receptor-specific ligands (e.g., glycolipids, lipoproteins, and aminosugars), have been prepared
to provide liposomes with increased direct transport properties [119]. These cell-specific–
targeting molecules can be either adsorbed on or covalently attached, directly or by a spacer,
to the outer surface of the liposomal mambrane. The user of spacers enables binding of considerable quantities of targeting molecules, without affecting its specific-binding properties or
the integrity of the liposomes.

Temperature- and pH-sensitive liposomes have been investigated for targeting drugs to primary tumors and metastases or sites of infection and inflammation [72]. The basis for the
temperature-sensitive drug delivery is that at elevated temperatures, above the gel to liquid-
crystalline phase transition temperature (Tc), the permeability of liposomes markedly increases,
causing the release of the entrapped drug. The release rate depends on the temperature and the
action of the serum components, principally the lipoproteins. Weinstein et al. [144] investigated
the effect of heating on incorporation of [^{3}H]methotrexate, administered in the free form and
encapsulated in 7:3 (w/w) dipalmitoyl and distearoyl phosphatidylcholine liposomal vesicles,
in L1210 tumors implanted in the hind feet of mice. They found about a 14% increase in
[^{3}H]methotrexate incorporation from the liposomal form, compared with the free drug, after
heating. This approach has been extended to a bladder transitional cell carcinoma, implanted
in the hind legs of C^{3}H/Bi mice [145], and for delivery of cisplatin (cis-diamminedichloropla-
tinum) selectively to tumors [146].

The pH-sensitive liposomes consist of mixtures of several saturated egg phosphatidylcho-
lines and several N-acylamino acids. The release of drug is suggested to be a function of
acid–base equilibrium effected by the interaction between ionizable amino acids and N-acylamino acid headgroups of the liposomes. There appears to be a close relation between Tc and
pH effect [72].

Liposomes also offer potential for use as carriers to transfer genetic materials to cells. Nicolau et al. [147] reported that a recombinant plasmid containing the rat preproinsulin I gene, encapsulated in large liposomes, when injected intravenously resulted in the transient expression of this gene in the liver and spleen of the recipient animals. A significant fraction of the expressed hormone was in physiologically active form. Recently, liposomes have also been used to block the initial binding of human immunodeficiency virus (HIV) to host cells [113]. This binding takes place between a glycoprotein (gp 120) on the virus coat and the CD4 receptor on the surface of T-helper lymphocytes and other cells. Antiviral drugs, such as zalcitabine (2′,3′-dideoxycytidine)-5′-triphosphate [148] and zidovudine (AZT) [149], have also been incorporated into liposomes and studied for their antiviral activities.

Niosomes. Niosomes are globular submicroscopic vesicles composed of nonionic surfactants. They can be formed by techniques analogous to those used to prepare liposomes [150]. To predict whether the surfactant being used will produce micelles or bilayer niosome vesicles, an arbitrary critical packing parameter (CPP), $v/a \cdot l$, (where v and l are specific volume and length of the hydrophilic portion of the surfactant, and a is the area of the hydrophobic segment of the surfactant), can be used [151]. A CPP value of 0.5 or less is considered to favor the formation of micelles, whereas a value between 0.5 and 1.0 indicates the formation of vesicles. The various types of nonionic surfactants used to prepare niosomes include polyglycerol alkylethers [152], glucosyl dialkylethers [153], crown ethers [154], and polyoxyethylene alkylethers [154]. Similar to liposomes, niosome vesicles can be unilamellar, oligolamellar, or multilamellar. A variety of lipid additives, such as cholesterol, can be incorporated in the niosome bilayer. Incorporation of cholesterol in the niosome bilayer enhances the stability of niosomes against destabilizing effects of plasma and serum proteins, and decreases the permeability of the vesicle to entrapped solute [155]. Niosomes are osmotically active, and require no special conditions for handling and storage.

Niosomes have been investigated as a drug carrier in experimental cancer chemotherapy and in murine visceral leishmaniasis [155]. The structures of nonionic surfactants (polyglycerol-based) employed to prepare niosomes used in these studies are shown in Fig. 11. Surfactants I and II contain an ether linkage, whereas surfactant III has an ester linkage. The latter can be degraded in vivo by the enzymes called esterases. When compared with free drug, niosomal

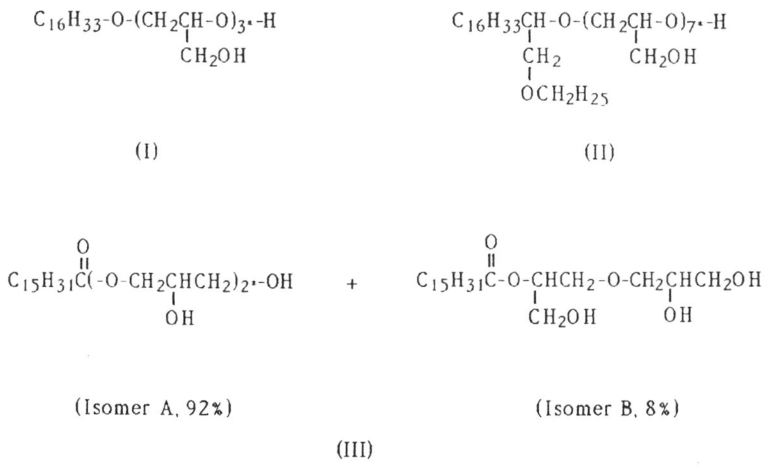

$$C_{16}H_{33}\text{-O-}(CH_2CH\text{-O})_{3*}\text{-H}$$
$$\underset{|}{\phantom{C_{16}H_{33}\text{-O-}(CH_2}}CH_2OH$$

(I)

$$C_{16}H_{33}CH\text{-O-}(CH_2CH\text{-O})_{7*}\text{-H}$$
$$\underset{|}{CH_2}\qquad \underset{|}{CH_2OH}$$
$$OCH_2H_{25}$$

(II)

$$\overset{O}{\overset{\|}{C_{15}H_{31}C}}(\text{-O-}CH_2CHCH_2)_{2*}\text{-OH}$$
$$\underset{|}{OH}$$

(Isomer A, 92%)

$+$

$$\overset{O}{\overset{\|}{C_{15}H_{31}C}}\text{-O-}CHCH_2\text{-O-}CH_2CHCH_2OH$$
$$\underset{|}{CH_2OH}\qquad \underset{|}{OH}$$

(Isomer B, 8%)

(III)

Fig. 11 Structures of three polyglycerol-based nonionic surfactants used in the preparation of niosomes (* represents an average number value of glycerol units).

forms of methotrexate, prepared using nonionic surfactant I, cholesterol, and dicetylphosphate, after intravenous administration by the tail vein in mice, exhibited prolonged lifetimes in the plasma and produced increased methotrexate levels in the liver and the brain. In addition, encapsulation within niosomes caused a reduction in the metabolism and urinary and fecal excretion of methotrexate [156,157]. Polysorbate 80, a nonionic surfactant that does not form niosomes, when coadministered with free methotrexate, provided reduced efficacy, compared with methotrexate encapsulated in niosomes [157]. This suggests that it is essential for surfactants to have a vesicular structure to effect enhanced targeting of drugs.

The delivery of doxorubicin to S180 sarcoma (tumor) in mice using niosomes as a carrier has been studied by Rogerson [158]. Much higher tumor drug levels were reported with niosomes prepared using nonionic surfactant I and 50% cholesterol than with free drug or drug encapsulated in cholesterol-free niosomes. The initial serum drug concentrations were higher following administration of free drug, but between 2 and 6 h after administration, the concentrations dropped and were lower than those observed using niosomal drugs, suggesting a rapid metabolism or distribution of free drug from the vascular system.

Niosomes, (prepared using surfactant I and surfactant I, II, or III and 30% cholesterol), containing stibogluconate have been as effective as the corresponding liposomal drugs in the visceral leishmaniasis model. Free drug showed reduced efficacy [159].

Lipoproteins. A lipoprotein is an endogenous macromolecule consisting of an inner apolar core of cholesteryl esters and triglycerides surrounded by a monolayer of phospholipid embedded with cholesterol and apoproteins. The functions of lipoproteins are to transport lipids and to mediate lipid metabolism. There are four main types of lipoproteins (classified based on their floatation rates in salt solutions): chylomicrons, very low-density lipoprotein (VLDL), low-density lipoprotein (LDL), and high-density lipoprotein (HDL). These differ in size, molecular weight, and density, and have different lipid, protein, and apoprotein compositions (Table 11). The apoproteins are important determinants in the metabolism of lipoproteins—they serve as ligands for lipoprotein receptors and as mediators in lipoproteins interconversion by enzymes.

Table 11 Physicochemical Properties and Composition of Human Lipoproteins

Lipoproteins	Chylomicrons	VLDL	LDL	HDL
Density (g/ml)	< 0.950	< 1.006	1.019–1.063	1.063–1.210
Size (nm)	80–1000	30–90	20–25	8–12
Molecular weight; M_r	10^9	10×10^6	2.3×10^6	300,000
Composition (%)				
Phospholipids	3–6	15–20	18–24	26–32
Cholesterol	1–3	4–8	6–8	3–5
Cholesterol esters	2–4	16–22	45–50	15–20
Triglycerides	80–95	45–65	4–8	2–7
Protein	1–2	6–10	18–22	45–55
Apoproteins[a]				
Major	A-I, A-IV, B-48, C-I, C-II, C-III	B-100, C-I, C-II, C-III, E	B-100	A-I, A-II, E
Minor	A-II, E	A-I, A-II, Ed	C-I, C-II, C-III	C-I, C-II, C-III, D, E

[a]Designations A, B, and C represent heterogeneous groups of apoproteins, each containing more than one polypeptide.
Source: Refs. 160, 231.

A schematic representation of the metabolism of lipoproteins is shown in Fig. 12 [160]. Chylomicrons are synthesized and secreted by the small intestine. They are hydrolyzed in blood by the enzyme lipoprotein lipase (LPL; found on the endothelial surfaces of the blood capillaries) to produce chylomicron remnants, which are then removed from the circulation by specific remnant receptors located on parenchymal liver cells. The VLDLs are secreted by the liver. Following their secretion in blood, VLDLs undergo metabolism in a way analogous to chylomicron. The resulting VLDL remnants are either removed by the hepatic remnant receptors or further metabolized to LDL, the major cholesterol-carrying lipoprotein in humans. The LDLs are removed from the circulation mainly by the liver and, to some extent, by peripheral cells. The HDL apoproteins are synthesized and secreted by the liver and the intestine. These apoproteins combine with the phospholipids, cholesterol, and protein components, produced as side products during in vivo conversion of chylomicrons and VLDL to the corresponding remnants by the lipoprotein lipase enzyme, and form the nascent HDL intermediate. The later is then converted to HDL by the enzyme lecithin cholesterylacyltransferase (LCAT). Depending on the types of apoproteins present, HDL can be removed from the circulation by LDL-specific receptors, remnant receptors, or by specific high-affinity HDL-binding sites present on adipocytes, adrenocortical cells, and various liver cells. The function of HDL is to transport cholesterol from peripheral tissues to the liver. It also serves as the site of plasma cholesterol esterification [161].

Lipoproteins have been suggested as potential drug carriers because (a) they are natural macromolecules and thus pose no threats of any anti-immunogenic response; (b) unlike other particulate carriers, lipoproteins are not rapidly cleared from the circulation by the reticuloendothelial system; (c) the cellular uptake of lipoproteins is by high-affinity receptors; (d) the inner core of a lipoprotein, which comprises triglycerides and cholesterol, provides an ideal

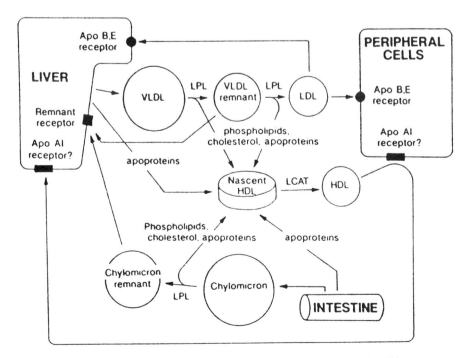

Fig. 12 Schematic representation of lipoprotein metabolism. (From Ref. 160.)

domain for transporting highly lipophilic drugs, whereas amphiphilic drugs can be incorporated in the outer phospholipid coat of the core; (e) drugs incorporated in the core are protected from the environment during transport, and the environment is protected from the drug; and (f) drugs located in the core do not affect the specificity of the ligand(s) present on the surface of the particle for binding to various cells.

Several methods are known to entrap or incorporate drugs into lipoproteins. The three most commonly practiced procedures include [160]: (a) direct addition of an aqueous solution of a drug to the lipoprotein; (b) transfer of a drug from a solid surface (e.g., the wall of a glass tube, glass beads, or small siliceous earth crystals) to the lipoprotein; and (c) delipidation of lipoprotein with sodium desoxycholate or an organic solvent, followed by reconstitution with drug–phospholipid microemulsion or drug alone.

The use of LDL and other lipoproteins in drug-targeting has been recently reviewed [160, 162]. Damle et al. [163] have shown that radiopharmaceuticals, such as iopanoic acid, a cholecystographic agent, could be incorporated in chylomicrons remnants by esterification with cholesterol, and used for liver imaging. About 87% of the chylomicron remnant-loaded iopanoic acid accumulated in the liver within 0.5 h after administration, compared with 31% accumulated using a saline solution containing the same amount of the drug. The LDLs have been used as a carrier to selectively deliver chemotherapeutic agents to neoplastic cells. The rationale is that tumor cells, compared with normal cells, express higher amounts of LDL receptors and, thus, can be selectively targeted with LDL. Thus, Samadi-Baboli et al. [164] have shown that LDL loaded with 9-methoxyellipticin incorporated in an emulsion containing dimyristoylphosphatidylcholine and cholesteryl oleate, exhibited much higher activity than free drug against L1210 and P388 murine leukemia cells in vitro. The eradication of the L1210 cells by the drug–LDL complex occurred exclusively by an LDL receptor mechanism. The LDL–drug complex showed higher cytotoxicity against cells preincubated with lipoprotein-deficient serum than those incubated in fetal serum, confirming that higher LDL expression on the cells leads to higher uptake of LDL. A recent study [165] indicated that acrylophenone antineoplastic molecules when incorporated within LDL can be delivered selectively to cancer cells without being entrapped into other blood proteins and cleared by the reticuloendothelial cells.

Kempen et al. [166] synthesized a water-soluble cholesteryl-containing trigalactoside, (I, Tris-Gal-Chol), which when incorporated in lipoproteins,

(I), Structure of Tris-Gal-Chol

allows the utilization of active receptors for galactose-terminated macromolecules as a trigger for the uptake of lipoproteins. The effect of increasing concentrations of Tris-Gal-Chol on the removal of LDL and HDL from serum and their quantitative recovery in the liver is shown in Fig. 13. These data show that lipoproteins containing Tris-Gal-Chol can be used as a liver-specific drug carrier system.

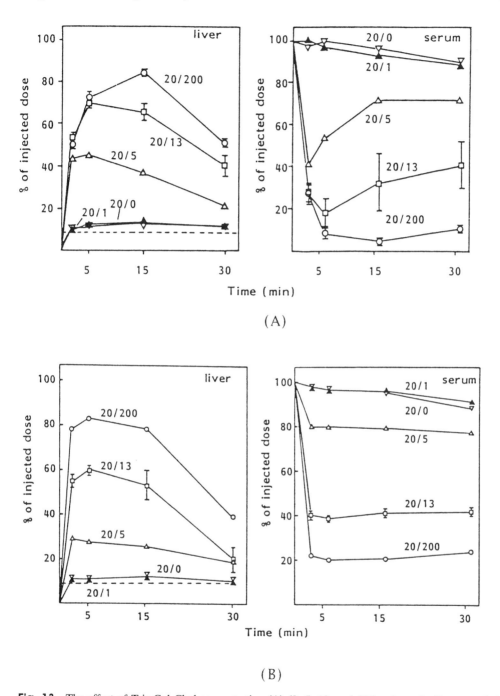

Fig. 13 The effect of Tris-Gal-Chol concentration (A) (0, 5, 13, and 200 μg) on the liver association and serum decay of ^{125}I-LDL and (B) ^{125}I-HDL (20 μg). The dotted line represents the maximal contribution of the serum value to the liver uptake. (From Ref. 160.)

The LDL and HDL have also been chemically modified to provide new recognition markers so that they can be selectively targeted to various types of cells in the liver [160,162]. Lactosylated LDL and HDL, which contain D-galactose residues as a ligand, can be prepared by incubating the corresponding protein with lactose (D-galactosyl-D-glucose) and sodium cyanoborohydride. Incubation of LDL with acetic anhydride produces the acetylated LDL. When injected intravenously in rats, both lactosylated LDL and HDL and acetylated LDL are rapidly cleared from the circulation by the liver (Table 12). Lactosylated LDL is specifically taken up by the Kupffer cells, whereas lactosylated HDL is mainly cleared by the parenchymal cells. Acetylated LDL shows higher accumulations in endothelial and parenchymal cells than in Kupffer cells. Thus, lactosylated HDL can be used to deliver antiviral drugs to parenchymal cells, whereas lactosylated LDL may serve as a carrier for immunomodulators, antivirals, and antiparasitic drugs to Kupffer and parenchymal cells. Acetylated LDL, on the other hand, is suitable for targeting anti-infective drugs to parenchymal and endothelial liver cells. Both Kupffer and endothelial cells have been recently implicated in HIV infections [167].

Activated Carbon (Charcoal).   Activated carbon is commonly used as an adsorbent. It has a microporous structure, and possesses a large surface area for adsorption. Drugs or chemicals adsorbed on the activated carbon particles exist in dynamic equilibrium with nonadsorbed drugs. The aqueous suspension forms of activated carbon, available commercially under the trade names Actidose with Sorbitol and Actidose-Aqua (208 mg/ml), have been accepted for oral use in humans to remove toxic substances.

Activated carbon, when injected into tissues, is taken into the lymphatic capillaries and becomes localized in the regional nodes. When administered into a cancerous pleural or abdominal cavity, activated carbon was adsorbed onto cancer and serosal surfaces. Takahashi [168] reported that mitomycin binds reversibly to activated carbon with desorption rates of 90 ± 4.8% and 107.2 ± 7.3% in saline and Ringer's solutions, respectively. Activated carbon adsorbs 20–500 times more mitomycin than that considered to be effective for cancer cells. The acute toxicity, evaluated in vivo in non–tumor-bearing and tumor-bearing humans and animals, showed an increase in the median lethal dose (LD_{50}) value with an increase in the amount of activated carbon. When administered peritoneally in non–tumor-bearing rats, activated carbon–mitomycin particles (equivalent to 5.0 mg of mitomycin) deposited in the omentum and peritoneum cavities and adsorbed into the lymphatic system. In tumor-bearing rats, particles were observed in lymph nodes of the omentum, paar aorta, perirenal, and thoracic angle 10 min after administration. There were no particles in the lymph nodes displaced by tumor tissues at a terminal stage, suggesting that the administered activated charcoal–mitomycin particles are delivered to distant lymph nodes with metastasis through the lymph vessels. A graphic presentation of the mitomycin concentration of ascites in Donryu male rats bearing ascites hepatoma AH130, following intraperitoneal administration of mitomycin-activated char-

Table 12 Distribution of Acetylated LDL, Lactosylated LDL, and Lactosylated HDL Over Liver Cell Types

	Percentage of total liver uptake ($n = 3$)		
	Acetylated LDL	Lactosylated LDL	Lactosylated HDL
Parenchymal cells	38.8 ± 5.8	31.8 ± 4.9	98.1 ± 0.6
Kupffer cells	7.4 ± 1.7	57.1 ± 1.9	1.0 ± 0.5
Endothelial cells	53.8 ± 5.7	11.1 ± 3.2	0.9 ± 0.2

Source: Ref. 160.

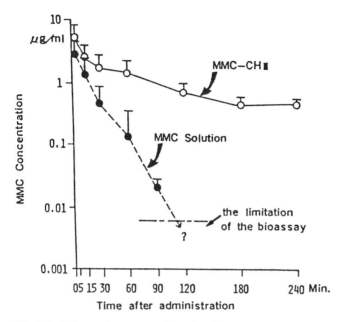

Fig. 14 Mitomycin concentration of ascites in rats bearing AH-130 after intraperitoneal administration (dose: 2 mg/kg in saline solution; open circles, mitomycin adsorbed on activated carbon; and solid circles, free mitomycin; $n = 5$). (From Ref. 168.)

coal and free mitomycin in saline solutions, is shown in Fig. 14. Table 13 lists the concentration of mitomycin in tumors and body organs. In rabbits, mitomycin was delivered only to lymphatic system [168].

Studies in dogs indicated that mitomycin-activated carbon, when injected into the canine gastric wall, is taken up mostly by the lymphatic system and transported rapidly to the regional lymph nodes, with retained activity of the drug [168]. A similar observation was made by Ito et al. [169] following administration of activated charcoal–mitomycin suspension into the sub-

Table 13 Mitomycin Concentration (mg/g) in Tumors and Organs After Intraperitoneal Administration of Mitomycin-Activated Carbon and Mitomycin in Saline Solutions (dose: 2 mg/kg).

	Mitomycin solution					Mitomycin-activated carbon				
Time after injection (min)	5	30	60	120	240	5	30	60	120	240
Tumor of greater omentum						0.95	0.55	0.09	0.13	0.013
Tumor of mesentrium						0.06	0.35	0.014	0.082	0.037
Tumor of epidermis		a				0.20	0.61	0.08	0.036	0.046
Spleen						0.01	0.029	0.10	0.14	0.082
Retroperitoneum						0.15	0.89	0.67	0.066	0.096
Lung	0.09	a								
Heart		a								

[a]Below the limitation of the bioassay (< 0.005 mg/g).
Source: Ref. 168.

serosal space. In both studies, compared with a mitomycin solution, a significant inhibition of lymph node metastasis was noted with mitomycin-activated charcoal. In humans, both early stomach cancers and advanced cancers of Borrmann I type either decreased by more than 50% in size or completely disappeared following the local injection of mitomycin–activated charcoal. No significant effect was noted in advanced cancers with Borrmann II, III, and IV types. High drug activity was demonstrated in regional lymph nodes for a prolonged period. The drug in the peripheral blood was scarce. Patients tolerated about a five times larger dose of the anticancer agent when presented in the adsorbed (on activated charcoal) form than in free form [168].

It appears that activated carbon might be a potential carrier for lymphatic delivery, or to peritoneal or pleural cavities, the most common sites in cancer metastasis. Minimal side effects are expected, since constant low concentrations of drug are maintained in the general circulation.

Cellular Carriers. Erythrocytes, leukocytes, platelets, islets, hepatocytes, and fibroblasts, all have been suggested as potential carriers for drugs and biological substances. They can be used to provide slow-release of entrapped drugs in the circulatory system, to deliver drugs to a specific site in the body, as cellular transplants to provide missing enzymes and hormones (in enzyme–hormone replacement therapy), or as endogenous cells to synthesize and secret molecules that affect the metabolism and function of other cells. Because these carriers are actual cells, they produce little or no antigenic response, and when old or damaged, they, like normal cells, are removed from the circulation by macrophages. Another important feature of these carriers is that, once loaded with drug, they can be stored at 4°C for several hours to several days, depending on the storage medium and the entrapment method used.

Since erythrocytes, platelets, and leukocytes have received the greatest attention, the discussion that follows will be limited to these carriers. Fibroblasts [170] and hepatocytes [171] have been specifically used as viable sources to deliver missing enzymes in the management of enzyme deficiency diseases, whereas islets are useful as a cellular transplant to produce insulin [172,173].

ERYTHROCYTES. Erythrocytes are biconcave disk-shaped (with pits or depressions in the center on both sides) blood cells, the primary function of which is to transport hemoglobin, the oxygen-carrying protein. The biconcave shape of the erythrocyte provides a large surface/volume ratio and, thereby, facilitates exchange of oxygen. The average diameter of erythrocytes is 7.5 μm, and the thickness at the rim is 2.6 μm and in the center about 0.8 μm. The normal concentration of erythrocytes in blood is approximately 3.9–5.5 million/μl in women and 4.1–6 million/μl in men. The total life span of erythrocytes in blood is 120 days.

The erythrocyte cell is surrounded by a membrane called plasmalemma, which consists of equal weights of lipid (major components: phospholipids, cholesterol, and glycolipids) and protein (glycophorin, ankyrin, and protein 4.1) components. The lipid portion of the membrane exists as a bilayer. The preponderant phospholipid constituents of the lipid bilayer membrane include phosphatidylcholine, phosphatidylethanolamine, phosphatidylserine, phosphatidylinositol, and sphingomylein. The polar headgroups of the outer lipid layer face out to the extracellular fluid, whereas those of the inner lipid layer face inward to the cells. The hydrocarbon portions of the lipid lie adjacent to each other in the middle. Glycolipids are present only as a minor component. About half of the protein present in the plasmalemma span the lipid bilayer. Erythrocytes show a net negative surface charge (owing to the presence of carboxylic groups of sialic acids), which prevents erythrocytes from agglutinating in the presence of IgG. In addition to lipids and proteins, some membranes contain carbohydrates, which are responsible for some of their surface antigenic properties.

Erythrocytes have been suggested as potential carriers for a number of biologically active substances, including drugs, nucleic acids, and enzymes [174,175]. They can be used as storage depots for sustained-drug release or potentially be modified to permit targeting to specific cell types in the blood (e.g., direct targeting to cells in leukemia [176]). Although constrained to move within blood vessels, erythrocytes can exit from blood vessels in tissues [177], potentiating their use as carriers in treating inflammations. Examples of drugs used for entrapment within erythrocytes and their suggested action against disease or their site-specificity in the body is presented in Table 14. A number of enzymes such as β-glucuronidase, β-fructofuranosidase, β-galactosidase, glutaminase, urease, hexosaminidase B, uricase, neuraminidase, thymidine kinase, and hypoxanthine-guanine phosphoribosyltransferase have also been entrapped within erythrocytes, and used for treating enzyme deficiency [175]. Erythrocytes are removed from the circulation by the RES, especially by cells located in the spleen and the liver.

Several methods have been used to incorporate exogenous substances into erythrocytes [175]. These include hypotonic hemolysis, dielectric breakdown, endocytosis, and entrapment

Table 14 Examples of Erythrocyte-Targeted Drug Delivery Systems

Agents	Entrapment method[*]	Suggested action against target/disease
As circulating carriers		
L-Asparaginase	a	Leukemia
Indolyl-3-alkane-α-hydroxylase	a	Sarcomas/carcinomas
Arginase	a, b	Hyperargininemia
δ-Aminolevulinate dehydratase	a	Porphyrias
Factor IX hemophilia	b	Hemophilia B
Desferroxamine	a	Excess body iron storage
Ara-C and phosphorylated Ara-C	a, b	Histiocytic medullary reticulosis/leukemia
Primaguine phosphate	c	Casual prophylaxis and radical cure of malaria
Insulin	b	Diabetes mallitus
Dideoxycytidine-5′-phosphate		AIDS
As targeted drug carriers		
Carboplatin		Liver
Rubomycin		Tumor P388
Daunomycin	a, d	Leukemia
Actinomycin D–DNA	a	Tumor
Doxorubicin		Liver tumor
Adriamycin	b	Liver and lung tumors
Methotrexate	a, b, e	Liver tumor
Daunorubicin	f	Leukemia
Bleomycin	a	RES
Gentamicin	b	RES
Antibodies	b	Diphtheria toxin; T cells
Diflubenzuron	b	Blood-sucking flies
Soybean trypsin inhib.	b	Blood-sucking flies
Imidocarb	b	Babesiasis
β-Glucoceribrosidase	a, b	Gaucher's disease
Meglumine antimoniate	g	Leishmaniasis
Pentamidine		Leishmaniasis
Heparin	b	Thromboembolism

[*]a, dilution; b, dialysis; c, endocytosis; d, amphotericin B; e, dielectric breakdown; f, coupling reaction; g, preswell.

without hemolysis, using amphotericin B. Of these, hypotonic hemolysis is most commonly used. It involves placing the cell in a hypotonic solution containing drug or chemical substance to be incorporated. As a result the cell swells. When the internal (osmotic) pressure exceeds a critical value, the cell ruptures and releases its content in the external medium. Substances present in the external medium enter into the cell at this time and become entrapped after the erythrocyte membrane is resealed. Resealing of erythrocyte membrane can be achieved by incubating the entrapped cells at 37°C under isotonic conditions. This is the fastest and simplest method and works efficiently for encapsulation of low-molecular weight (<130,000) substances. The major disadvantages of the method, however, are (a) it requires a relatively large amount of the starting material (owing to the large extracellular volume compared with the small intercellular volume); (b) a substantial percentage of the cellular content of erythrocyte enzymes and hemoglobin is lost during lysis; (c) a small change in ionic strength of the external medium may cause a significant alteration in the structure of the cell membrane (owing to the loss of membrane polypeptides) and, consequently, a decrease in the life span of the erythrocytes.

Several modifications in the foregoing method (hypotonic hemolysis) have been made to circumvent many of the disadvantages just noted. For example, the loss of enzymes and hemoglobin during lysis can be reduced by using hypotonic solution and resealing hemolysates prepared by the lysis and dialysis of other erythrocytes, or by preswelling the erythrocytes in 0.6% NaCl solution before subjecting them to hypotonic hemolysis. The morphological and structural integrity of the resealed erythrocytes can be preserved by hemolyzing and resealing erythrocyte cells in a dialysis tube. The dialysis approach allows a slow and gradual decrease in the ionic strengths and, consequently, preserves the elasticity of erythrocyte membrane. Other advantages of the dialysis method are (a) entrapped materials do not leak out to an appreciable extent; (b) the use of a high hematocrit during dialysis allows more efficient encapsulation; and (c) erythrocytes with appreciably higher drug percentages can be prepared. Several factors can influence the optimal encapsulation of the drug and the integrity of the erythrocytes and need to be properly optimized, including the composition of the buffer solution, centrifugation speed, osmolarity of the hypotonic buffers, the hematocrit, the temperature during hemolysis and resealing, the time of resealing, and the nature of the lysis procedure.

The dielectric breakdown method involves application of an electric pulse to a suspension of erythrocytes in isotonic or slightly hypotonic solution. Hemolysis is typically achieved at 0–4°C for $\frac{1}{2}$–1 h. It occurs as a result of dielectric breakdown of the cell membrane which, in turn, causes a reversible change in the permeability of the membrane. An increase in the temperature to 37°C intitiates the resealing process, and the original membrane resistance and impermeability of the erythrocytes are restored within minutes at this temperature. A major advantage of this method is that the size of the pore during lysis and, hence, the permeability of the membrane, can be controlled by appropriate manipulations of the electric pulse intensity, pulse duration, or the ionic strength of the pulsation medium. The electric breakdown technique has also been used to prepare "magnetic" erythrocytes [178].

The loading of drugs in erythrocytes by endocytosis typically involves incubating the intact or resealed erythrocytes with the substance to be entrapped for 30 min at 37°C in the presence of varying amounts of membrane-active agents, such as primaquine or chlorpromazine, in a buffer solution. Drugs can also be loaded in erythrocytes without lysing them first [179]. This can be achieved by first incubating the erythrocytes with amphotericin B in an isotonic medium for 30 min at 37°C and then with the material to be encapsulated for another 30 min at 37°C. This method causes no observable structural changes in the erythrocyte cell membrane. Other methods that have been used include chemical-induced isotonic lysis of cells [180] and the use of anesthetic agents, such as halothane [181]. The latter method avoids the use of both isotonic and hypotonic solutions.

PLATELETS. Platelets are nonnucleated discoid or elliptical cells that originate from the fragmentation of giant polyploid megakaryocytes located in the bone marrow. The average diameter of the platelet is 1.5 μm. Each platelet is surrounded by a trilaminar membrane, and its cytoplasm contains a dense body (delta granule), a surface-connected canalicular system, microchondrion, alpha granules, a lysosome, peroxisomes, glycogen, and a dense tubular system. The normal platelet count ranges from 1.5 to 4.0 × 10^{10}/dl of blood. Once they enter the blood, platelets have a total life span of about 10 days. Although the primary role of platelets is in controlling hemorrhage, they are also involved in immune reactions, inflammation, maintenance of vascular wall integrity, and in the evolution of vascular diseases (e.g., vasculitis and atherosclerosis) [182,183].

Platelets have been used as a carrier for several biological substances and drugs useful in the management of various hematological diseases [182]. Platelets can accumulate drugs by selective active transport. Certain drugs, such as angiotensin, hydrocortisone, imipramines, vinca alkaloids (vinblastine and vincristine), and many other drugs, are known to bind platelets. The first clinical trial of platelet-loaded vinca alkaloids in the treatment of idiopathic thrombocytopenic purpura, an autoimmune disorder characterized by decreased platelets counts, was reported by Ahn et al. [184]. The platelets were loaded with vinblastine and vincristine, separately, by incubation at 37°C in the dark for an hour. Following removal of the unbound drug, the platelets were resuspended in donor's plasma and then infused to patients over a period of 30 min. They found that patients treated with platelet-loaded alkaloids, compared with free drugs, required fewer treatments to provide a long-lasting remission from the disorder, without any form of maintenance therapy. Despite its success, the technique has several limitations, such as tedious preparation and high cost for their production and, thus, is restricted to those patients refractory of readily available therapies. The use of vinca-loaded platelets in the management of autoimmune hemolytic anemia [185], an autoimmune disease of red cells in which they become sensitized with autoantibodies of IgG, and are cleared by macrophages, and of various malignant disorders of the mononuclear phagocyte system (e.g., malignant histiocytosis, Rosai Dorfman syndrome, hairy cell leukemia, monocytic leukemia, familial erythrophagocytic lymphohistiocytosis, and other platelet phagocytosing tumors) [186–188], has also been reported.

LEUKOCYTES. Leukocytes are white blood cells involved in the cellular and humoral defense of the organism against foreign material. They are grouped into two classes: polymorphonuclear leukocytes, which comprise neutrophils, eosinophils, and basophils, and mononuclear leukocytes that include lymphocytes and monocytes. Of these, neutrophils and lymphocytes have been suggested as potential cellular carriers.

Neutrophils constitute about 60–80% of the total blood leukocyte level. They are spherical, with a diameter ranging between 12 and 15 μm. The total life span is about 6–7 h. They are known to carry a wide range of digestive enzymes and carrier proteins. Unlike erythrocytes, which are constrained to move within blood vessels, neutrophils can leave the capillaries and accumulate in large number at localized areas of disease. This property and their ready availability in large numbers (the average production per day in adults is 1 × 10^{11}) and in highly pure form, make them very attractive as a natural drug carrier. Segal and co-workers [189] reported that neutrophils containing ^{111}In-oxinate complex as a radiolabel marker, when administered intravenously in patients with abscesses and inflammation, preferentially accumulated at the diseased sites. Gainey and McDougall [190] successfully used this approach for the detection of acute inflammation in children and adolescents. Neutrophils could also be used as carriers for drugs that are effective in the treatment of pyrogenic infections, including diseases such as ulcerative colitis, acute arthritis, and other infections.

Sioux and Teissie [191] loaded propidium iodide in 70% leukocytes in whole blood using the dielectric breakdown method. The entrapped drug showed a half-life of longer than 4 h at 4 and 37°C. When compared with the nonpulsed cells, leukocytes loaded with the drug showed ten times more accumulation in the inflammation area than in control areas.

Lymphocytes are of two sizes: smaller lymphocytes, with a diameter of 6–8 μm; and larger lymphocytes, with a diameter up to 18 μm. They are found not only in blood, but in lymph and in every tissue of the body. Larger lymphocytes are believed to be cells that differentiate into T and B lymphocytes when activated by specific antigens. The life span of lymphocytes may vary from a few days to many years. Lymphocytes have been suggested as potential carriers for transporting macromolecules, particularly DNA, to other cells. Low-molecular-weight exogenous substances can be introduced into lymphocytes by electrical breakdown methods [192,193].

Soluble Macromolecular Drug Delivery Systems

Soluble macromolecules of both natural and synthetic origin have been used as drug carriers. When compared with the particulate carriers, soluble macromolecules (a) encounter fewer barriers to their movement around the body and can enter into many organs by transport across capillary endothelium or (in the liver) by passage through the fenestration connecting the sinusoidal lumen to the space of Disse; (b) penetrate the cells by pinocytosis, which is a phenomenon universal to all cells and that, unlike phagocytosis, does not require an external stimulus; and (c) can be found in the blood many hours after their introduction (particulate carriers, in contrast, are rapidly cleared from the blood by the RES). The fate of soluble macromolecules in animals and humans, with special reference to the transfer of polymers from one body compartment to another, has been reviewed by Drobnik and Rypacek [43].

Macromolecular drugs, in general, can be divided into four types: (a) polymeric drugs—these represent macromolecules that themselves display pharmacological activity, and polymers that contain therapeutically active groups as an integral part of the main chain; (b) macromolecule-drug analogs—these are derivatives of drugs that require no separation from the macromolecule to fulfill their therapeutic actions; (c) macromolecular prodrugs (or macromolecule–drug conjugates)—these represent drugs that must be detached from the macromolecule at the target site(s) to exert their therapeutic effects; and (d) noncovalently linked macromolecule–drug complexes. In this section, only macromolecule–drug conjugates are discussed.

A general configuration of an ideal macromolecular–drug conjugate is shown in Fig. 15 [194]. The drug can be attached to the macromolecule (or polymer chain) either directly or by a spacer, and may present as pendant groups or as a terminal (not shown in Fig. 15) group. The macromolecule–drug conjugate may also contain a homing or vector device (e.g., antibodies or receptor-specific ligands) to achieve selective access to, and interaction with, the target cell, and a moiety for controlling physical and chemical properties of the conjugate. The use of a spacer arm to attach a drug to a macromolecule enhances both configuration and drug–receptor interaction efficiencies.

Depending on the chemical functional groups present on the drug and on the macromolecule, a variety of methods can be used to prepare covalently linked macromolecule–drug conjugates. The most commonly used coupling reactions are shown in Fig. 16. Proteinaceous compounds can be conjugated using free thiol groups of cysteine, whereas the coupling between carboxylic-containing macromolecules and amine drugs, or vice versa, is achieved by the carbodiimide method. Polysaccharides can be attached to amine drugs by an dialdehyde intermediate, prepared using periodates. The covalently linked antibody–toxin conjugates are typically prepared by using the N-succinimidyl-3-(2-pyridyldithio)propionate (SDPD) reagent. It is important that the coupling reaction being used does not adversely affect the therapeutic activity of the drug.

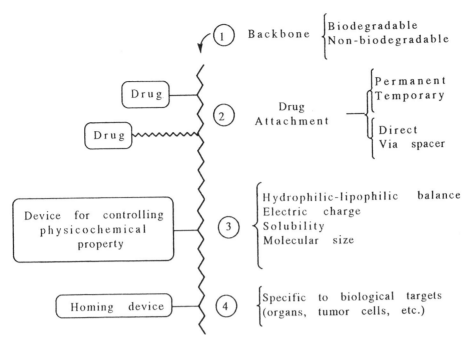

Fig. 15 A schematic representation of an ideal macromolecule–drug conjugate. (From Ref. 194.)

The fate of macromolecular drug conjugates in vivo depends on their distribution and elimination properties. In general, the plasma half-life of a macromolecule–drug conjugate depends on its molecular weight, ionic nature, configuration, and interaction tendencies in the physiological milieu. Although RES is the natural target, other organs in the body can be targeted as well. Because many tumors possess vasculature that is hyperpermeable to macromolecules, soluble macromolecules also serve as potential drug carriers in the treatment of cancers. Present evidence suggests that the greatest levels of tumor accumulation are achieved using macromolecules that carry a negative charge. Several excellent reviews that describe the distribution of soluble macromolecules in biological systems have been published [24,36,38,43,195].

The choice of a macromolecular carrier depends on the intended clinical objectives and the nature of therapeutic agents being used. In general, the properties of an ideal soluble carrier system include the following [194,196]: (a) The carrier and its degradation products must be biodegradable (or at least, should not show accumulation in the body), be nontoxic, and non-antigenic, and not alter the antigenicity of the therapeutic agents being transported. (b) The carrier must have an adequate drug-loading capacity; that is, the carrier must have functional groups for chemical fixation of the therapeutic agent. (c) The carrier must remain soluble in water when loaded with drug. (d) The molecular mass of the carrier should be large enough to permit glomerular filtration, but small enough to reach all cell types. (e) The carrier–drug conjugate must retain the specificity of the original carrier, and must maintain the original activity of the therapeutic agent until it reaches the targeted site(s). (f) The carrier–drug conjugate must be stable in body fluids, but should slowly degrade in extracellular compartments or in the lysosomes. (g) For lysosomotropic drug delivery, the macromolecule should not interfere with the pinosome formation at the cell surface, and subsequent intracellular fusion events. Furthermore, the macromolecule–drug linkage must be sensitive to acid hydrolysis or degradation by specific lysosomal enzymes.

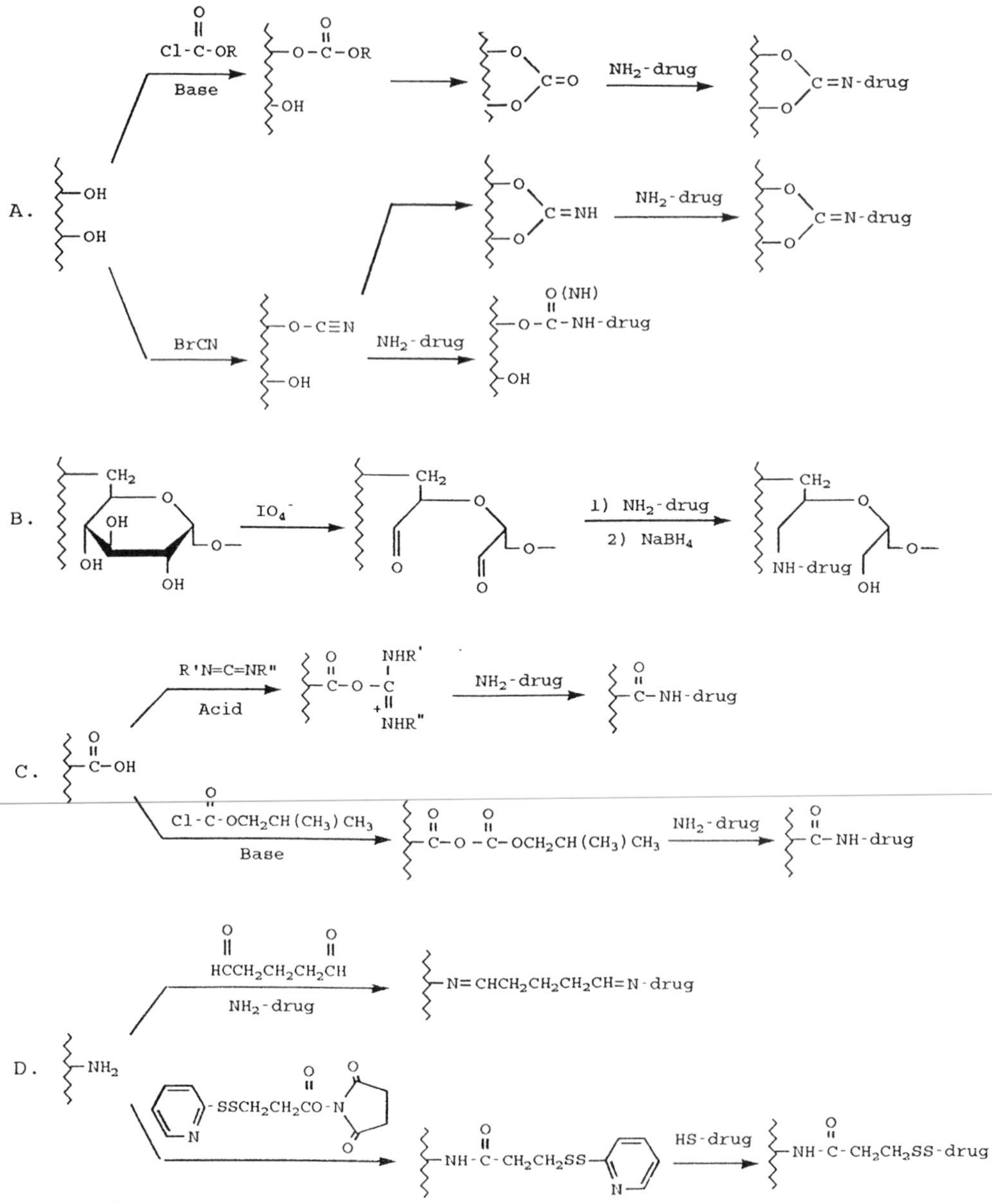

Fig. 16 Commonly used coupling methods for the preparation of macromolecule–drug conjugates. Functional groups of the drug and macromolecule are interchangeable with each other.

The various natural and synthetic soluble macromolecules investigated are listed in Table 15. Natural polymers are monodisperse (i.e., same chain lengths), have rigid structures, and are generally more biodegradable. The rigid structures may facilitate interactions between the determinant groups on the natural polymer and the binding region of immunoglobulin. Synthetic polymers, by contrast, are less immunogenic and can be tailor-made to predetermined specifications (i.e., molecular size, charge, hydrophobicity, and their capacity for drug-loading

Table 15 Suggested Soluble Macromolecular Drug-Delivery Systems and Their Uses

Carrier	Targets/diseases/therapy
Proteins	
Antibodies; antibody fragments (e.g., collagen-specific drug–toxin conjugates)	Injured sites of blood vessels walls; tumor cells
Albumin drug–conjugates; glycoproteins	Hepatocyte-specific agents (infections disease especially viruses)
Lipoproteins	Liver; cancers of ovaries and gonads
Lectins	General carrier/recognition ligands
Hormones (toxin–drug–hormone conjugates)	Tumors
Dextrans (e.g., enzyme–drug–dextran conjugates)	Tumors
Deoxyribonucleic acid (lysosomotropic carrier)	Cancer cells
Synthetic polymers	
Poly-L-lysine and polyglutamic acid	Carrier for targeting to cancer
Poly-L-aspartic acid	Hydrolyzable targeting carrier for cancer
Polypeptide–mustard conjugates	Lung targeting; tumor targeting
HPMA	Lysosomotropic carrier for cytotoxics
Pyran copolymers	Lysosomotropic carrier for cytotoxics

Source: Ref. 24.

can be optimized). Synthetic polymers are also easier and cheaper to produce in large quantity and high purity, and the chemistry to load drugs is much less laborious. In addition, they are more robust and, thus, are more stable during manipulation and storage. A detailed discussion of some natural and synthetic soluble carrier systems follows.

Antibodies. Antibodies are circulating plasma proteins of the globulin group. They are produced by plasma cells that arise as a result of differentiation and proliferation of B lymphocytes. Antibodies interact specifically with antigenic determinants (molecular domains of the antigens) that elicit their formation. In humans, there are five main types of antibodies, commonly designated as IgG, IgA, IgM, IgE, and IgD. Of these, IgG is the most abundant. It constitutes 75% of the serum immunoglobulins and is the only immunoglobulin that crosses the placental barrier and is incorporated in the circulatory systems of the fetus, thereby protecting newborn from infection. The basic structure of the immunoglobulin molecule consists of two identical heavy (long) chains, each with a molecular weight of 50,000, and two light (short) chains with a molecular weight of 23,000 (Fig. 17). These chains are held together by noncovalent forces as well as by disulfide linkage. Each chain also contains interchain disulfide linkages. In addition, each chain contains a region of constant amino acid sequence and a region of variable amino acid sequence. Antibodies belonging to the same class share the constant region in their heavy chains and may have kappa- or lambda-type light chains. The ends of the constant region (carboxyl ends) of the heavy chains form the Fc region, which is responsible for binding to Fc receptors present on many cells. The specificity of a particular antibody is determined by amino acid sequences of the variable region that are similar in the light and heavy chains. The ends of the variable region, referred to as amino (NH_2) terminals, serve as the binding sites for antigens.

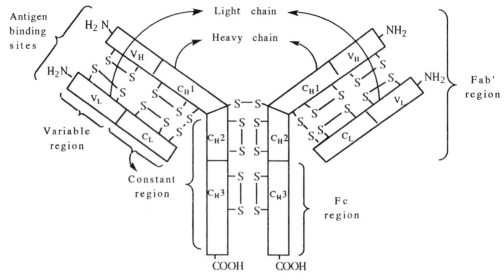

Fig. 17 A schematic representation of structure of an antibody.

The use of antibodies for active targeting of drugs to specific cell types in vivo has long been recognized [197–199]. They have been extensively explored as carriers in cancer diagnosis and in targeting drugs, toxins, and other therapeutic agents to tumor cells. Monoclonal antibodies of defined class and antigen specificity can be obtained in a highly purified form, and in virtually unlimited amounts, by immunizing mice with human tumor cells, followed by hybridizing their spleen cells with myeloma cells and, subsequently, screening the hybridoma cells for the formation of antibodies that bind only to immunizing tumor cells. The clinical usefulness of a given antibody in drug-targeting, however, depends on its degree of cross-reaction with normal tissues and its affinity to target antigens. The presence of free antigens in the general circulation is also an important determinant in drug-targeting, because an antibody specific to these antigens would combine with it prematurely, thereby lowering site-specificity. To improve specificity of action and to aid penetration into the target site (e.g., tumor cells), antibodies have been broken enzymatically to Fab′ fragments (see Fig. 17) and used in drug-targeting.

The various cytotoxic drugs that have been investigated using antibodies as a carrier include alkylating agents (e.g., chlorambucil [200], trenimon [201], and *p*-phenylenediamine mustard [202]); antimetabolites (e.g., methotrexate [203,204]); antitumor drugs (e.g., daunomycin [205, 206]); and toxic proteins (e.g., A-chains of diphtheria toxin and ricin). Depending on the nature of the drug, the antibody–drug conjugates can be prepared by a carbodiimide [206], mixed anhydride, periodate, or glutaraldehyde [207]-mediated coupling reaction (see Fig. 16). Diphtheria toxin has been linked to antibodies using either a homobifunctional (e.g., glutaraldehyde) or heterobifunctional (e.g., toluene diisocyanate) agent, or through disulfide groups by using the *N*-succinimidyl-3-(2-pyridyldithio)propionate (SPDP) reagent. Antibodies, in general, are fairly robust molecules and are little affected by chemical manipulations during conjugate synthesis. However, a method that works well with one monoclonal may not necessarily produce the same (intended) conjugated product with another monoclonal antibody. Rowland et al. [208] found that four monoclonals, recognizing different human tumor-associated antigens, when conjugated using an active azide derivative of a vinca alkaloid, produced vindsine–

monoclonal antibody conjugates with a drug/antibody conjugation ratio varying from 3:1 to 10:1, depending on the antibody used. The percentage activity of the vindsine–monoclonal conjugate coupled at 6:1 was 98%, whereas the conjugate with a drug/antibody ratio of 4:1 showed only 2% activity.

Samokhin et al. [209] demonstrated that biotinated red blood cells complexed with biotin–antihuman collagen antibodies (by an avidin cross-linking molecule) bind efficiently and specifically to collagen-coated surfaces, and thus can be used to deliver drugs to injured sites of blood vessel walls. Biotinylated antibodies specific to tumor-associated ganglio-*N*-triosylceramide (GalNAcβ1→ 4GalNAcβ1→ 4Glcβ1→ ceramide) have been prepared and found very effective in killing tumor cells that express ganglio-*N*-triosylceramide [210].

Antibodies have also been adsorbed or covalently linked to a variety of other carrier systems, such as liposomes [119], erythrocytes [211,212], and microcapsules and nanospheres, to selectively target them to a specific site in the body.

Deoxyribonucleic Acid. Deoxyribonucleic acid (DNA) has been suggested as a potential carrier for drugs that show strong affinity for DNA and that can be released from DNA by a simple equilibrium process, or after digestion of DNA by extra- or intra-cellular enzymes. It has been extensively investigated in chemotherapy of a number of malignant disorders, because (a) DNA complexed with drugs acts as a lysosomotropic drug; (b) DNA itself is a potent inducer of pinocytosis and is easily degraded by lysosomal hydrolases; and (c) compared with normal cells, tumor cells exhibit higher endocytic activity. To avoid clearance by the RES, DNA–drug complexes are usually administered intraperitoneally [213].

The DNA forms stable complexes with doxorubicin (Adriamycin; ADR) and daunorubicin (DNR). Doxorubicin and DNR, although structurally similar, show distinctly different properties: ADR is more toxic and active than DNR in the treatment of various human solid tumors; the apparent binding affinity of ADR to DNA is about 1.8 times higher than that of DNR to DNA. Trouet et al. [214] found ADR–DNA complex to be more active than ADR, DNR, or DNR–DNA in subcutaneously inoculated leukemic mice, whereas DNR–DNA complex showed the highest activity against the intravenously inoculated leukemic cells. Similar results have been reported in other tumor models. A recent pharmacokinetics study in patients revealed an enhanced uptake into leukemic cells and a reduction in distribution volume, cardiotoxicity, and rates of plasma clearance following administration of DNA conjugates than with the free drug [215].

DNA has also been tested as a carrier for ethidium bromide, a drug used in the treatment of protozoal diseases. When compared with free drug, the DNA-bound drug showed decreased toxicity and higher therapeutic efficacy in mice infected with *Trypanosoma cruzi* [216].

A major advantage of DNA as a carrier is that no chemical synthesis or manipulations are needed to obtain DNA–drug complex. The efficacy of DNA–drug complex depends on its stability in the blood stream, endocytic behavior of normal and tumor cells, presence of extracellular deoxyribonuclease activity in the tumor tissues, and capillary barriers that separate normal and tumor cells from the blood stream.

Low-Molecular-Weight Proteins. Low-molecular-weight proteins, such as lysozyme, have been recently suggested as a potential carrier for targeting drugs to the kidney [217]. The rationale is that these endogenous proteins, when administered intravenously, rapidly and extensively accumulate in the proximal tubule cells of the kidney [218]. Following glomerular filtration, these proteins are endocytosed into these cells and are taken to lysosomes where their hydrolysis into amino acids occurs by the lysosomal enzymes, causing the release of free drug. The released drug can act either locally or be transferred into the tubular lumen of the kidney.

Glycoproteins. Glycoproteins are endogenous substances consisting of a polypeptide backbone, with oligosaccharides as side chains. Each oligosaccharide chain contains monosaccharide residues and has sialic acid as the terminal group. The exact sequence of carbohydrate groups varies in different glycoproteins. The sialic acid groups from the terminal carbohydrate residues can be removed by treatment with the enzyme neuraminidase. When the sialic acid groups are removed, the resulting asialoglycoproteins are readily recognized and cleared by certain cells of the liver, depending on the sugar residues exposed. Hepatocytes recognize and internalize mainly galactose- and *N*-acetylgalactosamine-terminated glycoproteins, whereas Kupffer cells and endothelial cells recognize fucose-, mannose-, or *N*-acetylglucosamine-terminated glycoproteins. Kupffer cells also endocytose particulate material to which galactose groups are attached. Thus, asialoglycoproteins covalently linked to drug, protein, or diagnostic agent, can be used as hepatocyte-specific carriers [219].

Recently, a variety of glycoconjugates (also known as neoglycoproteins), such as mannosylated albumin, lactosylated albumin, galactosylated poly(L-lysine) and Ficoll (a polycarbohydrate), and galactosylated poly(hydroxymethylacrylamide) [219–221] have been prepared. These, like glycoproteins, are readily recognized by various liver cell types. A partial list of drug–substrate used to target various liver cell types with glycoproteins and glycoconjugates as carriers is presented in Table 16. The rationale for using asialoglycoproteins as a carrier for antiviral drugs is that the sugar residues (e.g., mannose, fucose, and galactose) on the carrier are capable of recognizing lectin receptors on the T lymphocytes and, thus, enhance targeting of antiviral drugs.

Albumin. Albumin is available in highly pure and uniform form, and exhibits low toxicity and good biological stability. It has been used as a carrier for methotrexate and a variety of antiviral drugs [amantadine, floxuridine (5-fluorodeoxyuridine), and cytarabine (cytosine arabinoside)] to treat macrophage tumors and infections caused by DNA viruses growing in macrophages. Heavily modified albumins are known to be readily endocytosed by cells of the macrophage system. After penetration into the cell, albumin is digested by lysosomal enzymes, causing the release of free drug. The albumin (bovine and murine)–methotrexate conjugates exhibit higher and more prolonged serum levels and decreased excretion rates than the free drug. When administered intraperitoneally into L1210 tumor-bearing mice, the methotrexate–albumin conjugate produced increased localization of methotrexate in the ascitic fluid and elevated intracellular methotrexate levels [222]. It was also more effective, compared with free drug, in reducing the number of metastases in female (C57BL/6 × DBA/2)F mice bearing subcutaneously implanted Lewis lung carcinoma [223]. Whiteley et al. [224] examined the mechanism of the therapeutic activity of the conjugate using the methotrexate–transport-resistant strain of Reuber hepatoma H35 cells. They found no suppression of the tumor growth, suggesting that the observed increased therapeutic activity of the conjugate occurs either through the transport of free drug following its extracellular release from the albumin, or the conjugate invokes a specific transport system to penetrate the cell.

The conjugates of antiviral drugs have been investigated to treat *ectromelia*, a virus known to infect Kupffer cells of the liver [225]. Albumins containing a variety of carbohydrate residues (e.g., lactosyl, galactosyl, and mannosyl) as terminal groups have also been chemically prepared and used as carriers to selectively deliver antitumor and antiviral drugs to the liver cells [219–221].

Hormones. Hormones, such as human placental lactogen, human chorionic gonadotropin, epidermal growth factor, and melanotropin, all have been suggested as carriers to deliver toxins (e.g., ricin and diphtheria toxin) and antineoplastic agents (e.g., daunomycin) to cancer cells. The rationale is that cancer cells frequently possess receptors for hormones; thus, drugs co-

Table 16 Targeting of Drugs to the Liver with Glycoproteins and Glycoconjugates

Drug/substrate	Glycoprotein carrier–glycoconjugates	Species
Antiviral agents		
Ara–Amp, acyclovir	Lactosaminated HSA	Woodchuck [266], Rat [267]
Ara–Amp	Galactosylated poly-L-lysine	Mouse [268]
PMEA	Mannosylated poly-L-lysine	Mouse [269]
Ara C	Galactose–mannose carboxymethyldextran	Mouse
N-Acetylmuramyl peptide	Mannosylated BSA	Mouse [270]
Antiparasitic agents		
Alloprinol riboside	Mannosylated poly-L-lysine	Mouse [271]
Antineoplastic agents		
Mitomycin	Asialofetuin	Rat [272]
Agents affecting lipid metabolism		
LDL	Lactosaminated antiapolipoprotein B antibodies	Rat [273]
HDL	tris-Galactosyl cholesterol	Rat [273]
Antitoxicants/cytoprotective agents		
Uridine monophosphate	Polylysine asialofetuin	Rat in vivo [274]
Naproxen	Galactosyl–mannosyl HSA	Rat [275]
Superoxide dismutase	Mannosylated form	Rat [267]

Source: Courtesy of Professor D. K. F. Meijer, University Center of Pharmacy, University of Groningen, Groningen, The Netherlands. Other examples of drug targeting with glycoproteins and neoglycoproteins can be found in Ref. 219.

valently linked to hormones can be selectively targeted to these cells. Other reasons include (a) hormones are easy to obtain in a chemically pure form; (b) they are not bound by Fc receptors on macrophages; and (c) they cause no allergic reactions since hormones from different species are structurally similar [226].

Dextran and Other Polysaccharides. Dextrans (Fig. 18) are colloidal, hydrophilic, and water-soluble substances produced from sugar by microorganisms of the *Lactobacillaceae* family or by cell-free system containing dextran sucrase. Pharmaceutical and commercial dextrans of different molecular weights are produced by controlled hydrolysis and repeated fractionation of native dextran. Dextrans are inert in biological fluids, and do not affect cell viability. Colloidal dextrans of 40,000, 70,000, and 110,000 molecular weight can be used in the prevention of thromboembolic diseases in high-risk surgical patients [227]. The immunogenicity of dextran increases with increasing molecular weight and branching. Dextrans of molecular weight equal to or smaller than 45,000 are completely excreted in the urine within 48 h following intravenous administration, whereas those with a molecular weight higher than 45,000 remain in the blood for a long period. However, they are eventually cleared by the RES cells. Dextrans are slowly hydrolyzed *in vivo* into soluble sugars, (the major products being isomaltose and isomaltotriose), by dextranase, an enzyme found in the intestinal mucosa and RES cells [228]. Despite their removal, administration of large doses of the dextran can cause a storage problem.

Dextrans have been used as carriers for a variety of drugs and enzymes, including antimicrobial agents (e.g., ampicillin, kanamycin, and tetracycline), cytostatic agents [e.g., mitomycin, methotrexate, bleomycin, daunorubicin, daunomycin, and cytanabine (cytosine-1-β-D-arabinoside)], cardiac drugs (e.g., alprenolol and procainamide), peptides and proteins (e.g., asparaginase, carboxypeptidase, α- and β-amylase, and insulin), and enzyme inhibitors (e.g., pancreatic trypsin inhibitor) [229]. Drugs or enzymes can be coupled to dextrans either directly, or through a spacer (e.g., γ-aminobutyric acid, ε-aminocaproic acid, and ω-caprylic acid), by activated intermediates produced from reactions with alkali metal periodates, cyanogen bromide, carbodiimide, or chloroformate (see Fig. 16), depending on the nature of the drug and

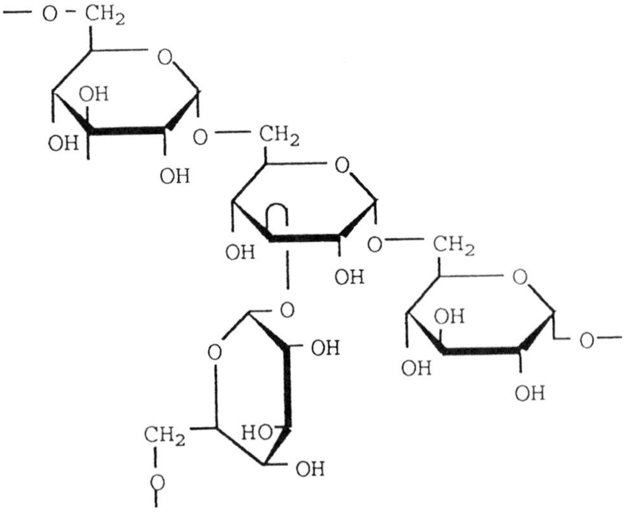

Fig. 18 Structure of dextran.

carrier molecules. The direct esterification reaction can be used to link organic acids to dextrans. Drugs coupled to dextran by hydrolyzable covalent linkages are more efficacious than free drugs or drug–dextran conjugates containing nonhydrolyzable covalent linkages. Studies also revealed that direct substitution of drugs to antibody (4–6 mol of a drug per mole of antibody) causes a significant loss of the antibody activity, whereas antibody–dextran conjugates of drugs, with and without a spacer, show increased stability, longer duration of activity, reduced toxicity, and enhanced cell selection activity (drug-targeting). The antitumor activity of dextran–mitomycin conjugates against B16 melanoma has been shown to be molecular weight dependent [230]. Modifications of dextran–drug conjugates by an appropriate choice of linkage has permitted the synthesis of a variety of cationic and anionic forms. In vivo studies in mice bearing subcutaneous sarcoma 180 showed that cationic forms clear much more quickly from the bloodstream than the anionic forms, accumulating primarily in liver and spleen, and little uptake into tumor tissues. Anionic conjugates with molecular mass ranging between 10,000 and 50,000 kDa, on the contrary, show greater tumor accumulation (following intravenous administration) and, thereby, are more effective in the treatment of a subcutaneous tumor. This probably occurs as a result of prolonged residence time of anionic conjugates in the bloodstream.

Other polysaccharides that have been suggested as potential carriers in drug targeting are chondroitin sulfate, heparin sulfate, dermatin sulfate, hyaluronic acid, and keratin sulfate. All are highly water-soluble, linear polyanions composed of 1:1 mixture of uronic acid and hexosamine (Fig. 19). These are commonly called glycosaminoglycans (GAGs), and are generally biocompatible, nonimmunogenic, and biodegradable. They contain carboxylic and sulfate groups, in addition to primary and secondary hydroxyl groups, and thus provide additional sites for drug attachment. Since polyanions interact with cell membranes, it is proposed that the use of GAGs may facilitate transport to drugs into the interior of the cells [231]. *O*-Palmitoylamylpectin, another polysaccharide, has been used to coat liposomes to alter their distribution in vivo.

Synthetic Polymers. Synthetic polymers are versatile, and offer promise for both targeting and extracellular–intracellular drug delivery. Of the many soluble synthetic polymers known, poly(amino acids) [poly(L-lysine), poly(L-aspartic acid), and poly(glutamic acid)], poly(hydroxypropylmethacrylamide) copolymers (polyHPMA), and maleic anhydride copolymers have been investigated extensively, particularly in the treatment of cancers. A brief discussion of these materials is presented.

POLY(AMINO ACIDS). Both anionic [e.g., poly(L-aspartic acid) and poly(glutamic acid)] and cationic [e.g., poly(L-lysine)] poly(amino acids) have been suggested as potential drug carriers. Poly(L-lysine) is a homopolymer consisting of repeating units of L-lysine. It exhibits some affinity for cancer cells, and possesses antimicrobial and antiviral properties. It also shows some activity against murine tumors. The covalently linked methotrexate conjugate of poly(L-lysine), prepared by the carbodiimide method, penetrated the Chinese hamster ovary cells faster and was more effective, than free drug [232]. Poly(D-lysine)–methotrexate conjugate, in contrast, had no effect, because it is resistant to degradation by intracellular enzymes. In vitro growth inhibition studies revealed that the poly(L-lysine)–methotrexate conjugate was more effective against five cell lines of human solid tumors than five cell lines of lymphocytes [233].

Poly(amino acids) can be covalently linked to daunorubicin by a nucleophilic substitution reaction of the 14-bromo derivative of the drug [234]. This method avoids alteration or modification of the amino sugar moiety of the drug. When compared with free drug, the poly(L-aspartic acid)–daunorubicin conjugate prepared by this method was less toxic to HeLa cells in vitro, but more effective against P388 leukemia, Gross leukemia, and MS-2 sarcoma in in

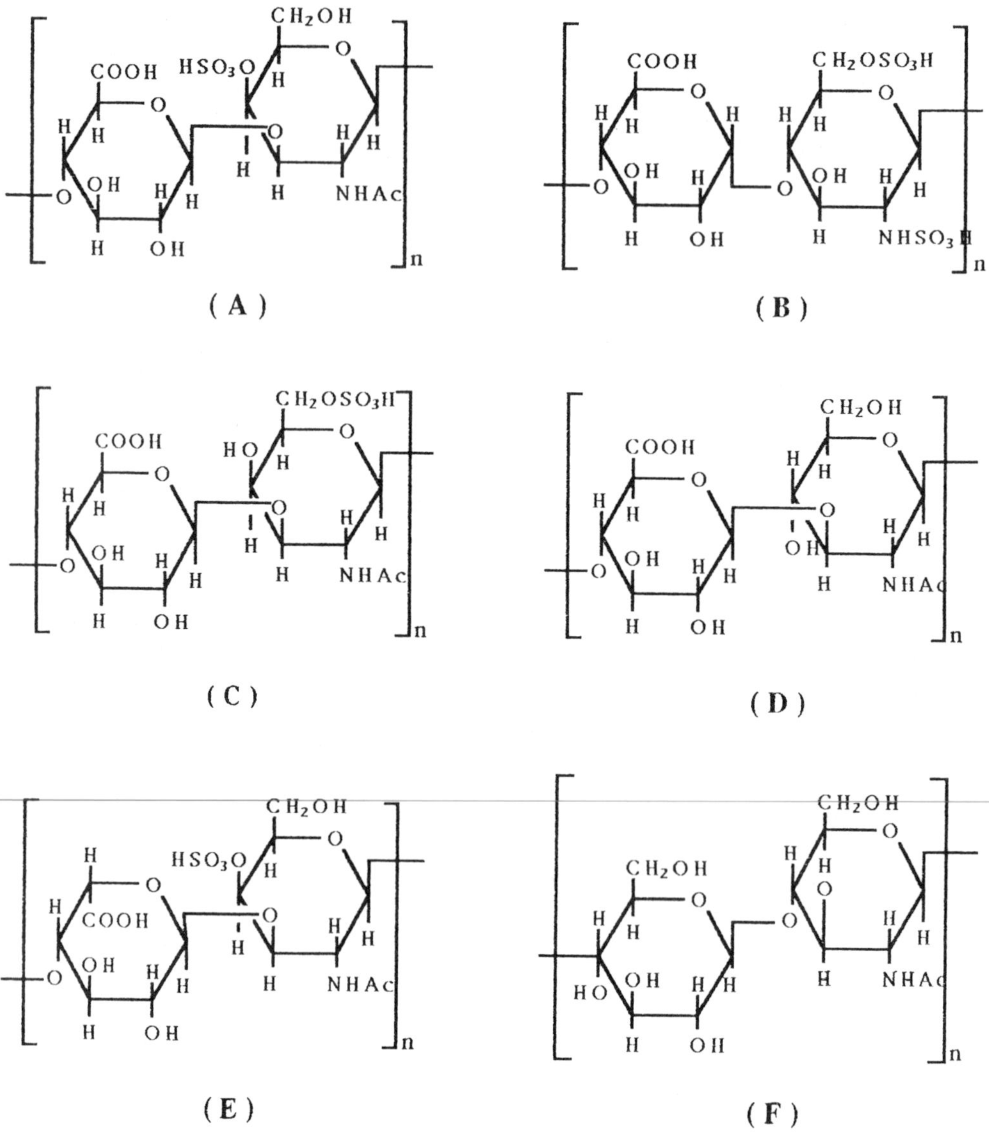

Fig. 19 Structures of glycosaminoglycans: (A) chondroitin-4-sulfate; (B) heparin sulfate; (C) dermatin sulfate; (D) hyaluronic acid; (E) chondroitin-6-sulfate; (F) keratin sulfate.

vivo. The corresponding conjugate of poly(L-lysine) showed markedly reduced activity overall. This was attributed to the more stable amide linkage in poly(L-lysine) conjugate compared with that in the poly(L-aspartic acid) conjugate [234].

Hurwitz et al. [235] used a hydrazide derivative of poly(glutamic acid) as a carrier for daunamycin. This was less toxic than free drug against mouse lymphoma in vitro, but was as effective, or more effective, against the same lymphoma in vivo.

Poly(L-lysine) has also been suggested as a carrier for pepstatin, a specific inhibitor of the lysosomal proteinase cathepsin D, responsible for causing muscle-wasting diseases, such as muscular dystrophy [236].

POLY(HYDROXYPROPYLMETHACRYLAMIDE) COPOLYMER. Poly(hydroxypropylmethacrylamide) copolymer (polyHPMA; (Fig. 20A) is a nonbiodegradable, nonimmunogenic, biocompatible polymer. It has been extensively investigated as a plasma expander and as a carrier for a variety of anthracycline antibiotic–antitumor agents (e.g., doxorubicin and daunorubicin), alkylating agents (e.g., sarcolysin and melphalan), chlorin e^6 (phytochlorin) and mesalamine (5-amino-salicylic acid) [237]. It is cleared from the circulation, depending on the molecular weight. The molecular weight threshold permitting the renal excretion was 45,000. In general, the lower the molecular weight, the faster the clearance rate. Larger polymers are more susceptible to capture by the RES cells, and have little chance to pass the basement membrane and other natural barriers.

PolyHPMA has been linked to drugs by oligopeptidyl linkages (e.g., Gly-Phe-Leu-Gly) [195]. The latter degrades under the influence of lysosomal enzymes, causing the release of free drug. The digestibility of oligopeptides by the lysosomal enzymes depends on the length and detailed structure of the oligopeptide sequence and follows the order: tetrapeptide > tri-peptide > dipeptide. It is important that the oligopeptide linkage used to attach drug to the polymer must not be susceptible to degradation during transit in the bloodstream. The Gly-Phe-Leu-Gly tetrapeptide sequence fulfills this and the intralysosomal digestibility criteria.

Targeting to a specific cell has been accomplished by attaching appropriate cell receptor-specific ligands to polyHPMA. Thus, polyHPMA derivatized with glycylglycylgalactosamine (Gly-Gly-GalNH$_2$), when injected into rat bloodstream, is very efficiently removed by the liver parenchmal cells and taken into the lysosomes. Only 5–10% hydroxypropyl residues need to be substituted to achieve maximal targeting [196]. Studies have shown that the presence of low levels of side chains decrease the pinocytic uptake which, in turn, causes targeting residues to achieve maximum targeting. More complex-targeting systems, such as the use of melanocyte-stimulating hormone [238] and antibodies [80], have also been investigated.

To obviate the accumulation of polyHPMA in the body (polyHPMA copolymers are not biodegradable), low-molecular-weight fractions (small enough to pass through the glomerular

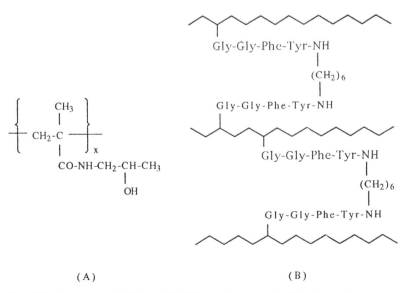

(A)   (B)

Fig. 20 Structures of (A) polyHMPA copolymer and (B) schematic representation of cross-linked polyHMPA copolymer.

filtrate) of polyHPMA have been cross-linked by diamine-containing oligopeptidyl sequences and used as drug carriers (see Fig. 20B) [195]. Similar to non–cross-linked polyHPMA, these are also cleared from the circulation in a molecular weight-dependent fashion. In vivo studies in rats revealed the appearance of lower-molecular-weight polymer chains in the urine 8–24 h after ther administration (intravenously).

Recently, polyHPMA copolymers containing galactosamin–chlorin e_6 (phytochlorin) and anti-Thy 1.2 antibody–chlorin e_6 have been developed as targetable polymeric photoactivable drugs [239]. When compared with polyHPMA–chlorin-e_6, polyHPMA–galactosamin-chlorin e_6 conjugate was more active against human hepatoma cell line (PLC/PRF/5; Alexander cells) in vitro. The polyHPMA–anti-Thy 1.2 antibody–chlorin e_6 conjugate was prepared in two ways—one was linked by N^ϵ-amino groups of lysine residues and the other by oxidized carbohydrate. Both, showed higher activity toward mouse splenocytes than the nontargeted conjugate. The polyHPMA–anti-Thy 1.2 antibody–chlorin e_6 conjugate linked by oxidized carbohydrate was the most active in its photodynamic effect on the viability of splenocytes and the suppression of the primary antibody response of mouse splenocytes toward sheep red blood cells in vitro.

MALEIC ANHYDRIDE COPOLYMERS.   Examples of this class of drug carriers include styrene–maleic anhydride (SMA) and divinyl ether–maleic anhydride (DIVEMA) copolymers (Fig. 21). Both SMA and DIVEMA have been suggested as potential carriers for neocarzinostatin (NCS), a naturally occurring antitumor antibiotic, and other antitumor agents. NCS is more potent than many conventional antitumor drugs (e.g., mitomycin and doxorubicin). It is composed of a protein (molecular weight 12,000) and a low molecular weight cytotoxic chromophore. Following release from the protein, the chromophore enters neighboring cells and interacts with the DNA and, eventually, kills the cells. Despite its higher antitumor activity, NCS can not be used clinically because (a) it is nonspecific in its action, and (b) it is rapidly cleared ($t_{1/2}$ = 1.8 min) by urinary excretion following intravenous administration.

Maeda et al. [240] synthesized styrene–maleic anhydride conjugates of NCS by reacting the two amino groups (one belonging to Ala-1 and the other to Lys-20 residues) on NCS, with an anhydride group of SMA at a pH of 8.5 for 10–12 hr. Compared with free NCS, the SMA–NCS conjugate was retained longer ($t_{1/2}$ = 18 min in mice) in the blood circulation and, consequently, promoted accumulation of NCS in peripheral tumor tissues (by up to eightfold in the solid sarcoma 180; Fig. 22). The SMA–NCS conjugate was also effective against lym-

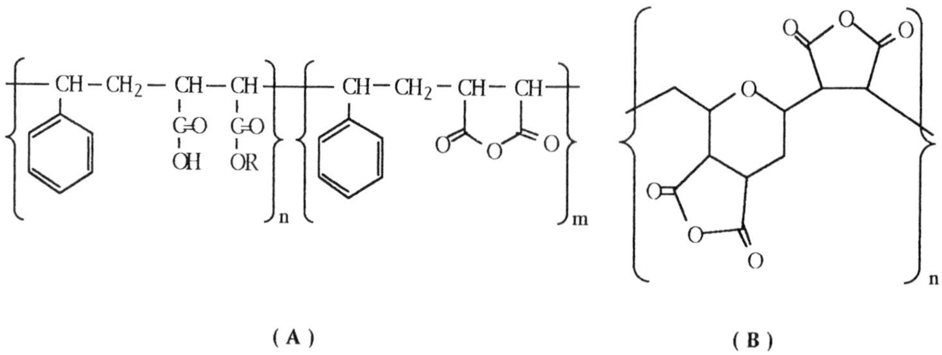

(A)   (B)

(n, m ≥ 1; R = alkyl group)

Fig. 21   Structures of (A) SMA–NCS and (B) DIVEMA.

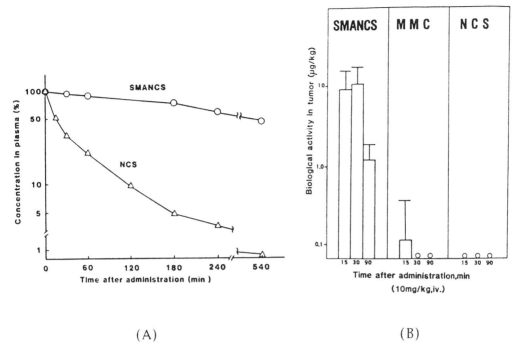

Fig. 22 (A) Plasma concentration of SMA-NCS and NCS in human after an intravenous bolus injection. (B) Intratumor concentration of SMA-NCS, NCS, and mitomycin (MMC). SMA–NCS exhibits a much higher and more prolonged tumor concentration than MMC and NCS. All drugs were given as an intravenous bolus at 10 mg/kg to rabbits bearing VX-2 tumor in the liver (assayed by antibacterial activity). (From Ref. 244.)

phatic metastases [240]. Various vasoactive agents can be used to modulate the rates of extravasation and, consequently, penetration of SMA–NCS in tumor tissue [241]. A multiinstitutional Phase II clinical trial of SMA–NCS, formulated in an oily lipid contrast medium, Lipiodol (an ethyl ester of iodinated poppyseed oil; 37%, w/w in iodine content; used clinically as a lymphographic agent), was completed in Japan in 1990. The tumor/blood ratio was greater than 2500 following intra-arterial administration of this formulation. The results of the first pilot study (total number of patients 44) revealed detectable tumor shrinkage in 95% of the cases [242]. Of 21 patients, 9 showed a 40–99% reduction in tumor mass within 1–5 months, and extensive necrosis was seen in biopsy specimens. Mean survival rate was greater than 18 months for the treated patients, compared with 3.7 months for the controls. In a parallel study, 24 patients, with tumors other than hepatoma, were used, and SMA–NCS was administered by various arterial routes [243]. The results showed a regression of the tumors in 6 (of 9) patients with metastatic liver cancer, 4 (of 4) with adenocarcinoma of the lung, and 1 (of 3) patient with unresectable gallbladder tumors.

SMA–NCS can also be used to activate macrophages, natural cells, and T cells, and to induce interferon gamma production [244].

The DIVEMA–NCS conjugate, on the contrary, exhibited cytotoxic activity (on a molar basis) in vitro against eight cell lines and bone marrow cells similar to that observed with free NCS [245]. In vivo toxicity data indicated about a 1.7-fold higher LD_{50} value for the conjugate than for NCS. Studies also revealed a lower distribution of the conjugate than that for NCS in bone marrow and spleen cells and, to some extent, in other organs. The biological activity of DIVEMA–NCS in plasma was about 2.2 times higher than that of NCS. Because of reduced acute toxicity, DIVEMA–NCS showed ten times higher antitumor activity than NCS (after 5 min of injection) at a high dose. At lower doses, there was no difference in the antitumor activities of DIVEMA–NCS and free NCS. The relatively low antitumor activity of DIVEMA–NCS, compared with SMA–NCS, has been attributed to its rapid clearance from the circulation [245].

The DIVEMA copolymer has also been covalently linked to methotrexate [246]. This polymer spontaneously released methotrexate from the polymer backbone by hydrolysis. The DIVEMA–methotrexate conjugate (molecular weight 22,000) was more effective against L1210 leukemia and Lewis lung carcinoma at optimal equivalent doses than either DIVEMA or NCS alone or the combination of DIVEMA and NCS. As such, DIVEMA also shows various biological properties. It is a well-known interferon inducer and shows activity against several solid tumors and virus. It also possesses antibacterial, antifungal, anticoagulant, and anti-inflammatory properties, and is capable of stimulating macrophages activation. Biological studies revealed that the toxicity of DIVEMA increases with an increase in the molecular weight. Low-molecular weight pyrans stimulate phagocytosis, whereas high-molecular weight copolymers decrease the rate of uptake. The inhibition of polymer metabolism also increases with an increase in the molecular weight [247].

VI. TARGETING IN THE GASTROINTESTINAL TRACT

The alimentary canal has been, and continues to be, the preferred route for drug administration for systemic drug action. Thus, there is a growing interest in developing oral dosage forms that can be targeted in the gastrointestinal (GI) tract to either (a) exert a therapeutic effect at a specific site in the GI tract [see Ref. 26 for diseases of the GI tract], or (b) allow systemic absorption of drugs or prodrugs utilizing a specific region of the alimentary canal, without being affected by the GI fluid, pH fluctuations, enzymes, and microflora, while traveling along the GI tract. The various anatomical, physiological, physicochemical, and biochemical features of various regions of the alimentary canal that influence drug targeting, and possible formulation approaches that can be used, have recently been reviewed by Ritschel [27].

Depending on the site in the GI tract where drug release is sought, a variety of approaches can be used. Bioadhesive polymers are used to prepare adhesive tablets and films for use in the buccal cavity and other regions of the alimentary canal [248]. These polymers adhere to biological tissue for an extended period, thereby providing increased local therapeutic effect, or prolonged maintenance of therapeutic amounts of drug in the blood. Examples of bioadhesive polymers commonly used include hydroxypropylcellulose, polyacrylic acid (Carbopol), and sodium carboxymethylcellulose. Enteric polymers (e.g., cellulose acetate phthalate, hydroxypropylcellulose acetate phthalate, polyvinyl acetate phthalate, methacrylate–methacrylic acid copolymers, styrol maleic acid copolymers, and others), which remain insoluble in the stomach, but dissolve at higher pH of the intestine, are used to deliver drugs to the small intestine. Enteric coating also prevents drugs from degradation by gastric fluid and enzymes, and protects the gastric mucosa from irritating properties of the drugs. Recently, Klokkers-Bethke and Fisher [249] developed a multiple-unit drug delivery system to target the lower parts of the intestine; one formulation releasing the drug in the lower part of the small bowel and the second one in the colon. The delivery to the stomach can be achieved using hydrodynamically balanced (floating) dosage forms. Because of lower bulk density, such dosage forms stay buoyant on the stomach, thereby resisting gastric emptying. Drug delivery systems that contain inflatable chambers (these become gas-filled at body temperature), or solids (e.g., carbonates and bicarbonates) that form gas when in contact with gastric fluid, also stay buoyant, and are used to target drugs in the stomach. The successful delivery of drugs to the colon, either for local treatment, such as ulcerative colitis and irritable bowel syndrome, or for the systemic absorption of drugs that are not well absorbed from the other regions of the GI tract, can be achieved by coating drugs with azopolymers [250]. The latter forms an impermeable film that is resistant to proteolytic digestion in the stomach and small intestine. In the colon, the azopolymer reduces into the corresponding amines by the indigenous microflora; consequently, the film breaks, causing the release of free drug. Targeting in the colon may be of significance to polypeptide delivery, because there are no digestive enzymes in the colon, and the duration of residence is longer in the colon than in other regions of the GI tract (see Table 3). Targeting in the colon is also feasible by time–pulsatile systems. Rectal delivery, depending on the position of the dosage form in the rectum, can be used to target either the systemic circulation or the liver. For systemic targeting, the dosage form should be placed and located directly behind the internal rectal sphincter, whereas targeting to the liver requires the dosage form to be placed in the ampulla recti (about 12–15 cm up the rectum).

Targeting to esophageal mucosa and other regions of the GI tract can also be achieved using bioadhesive magnetic granules [251]. The various parameters that are detrimental in targeting to a specific site using bioadhesive magnetic granules include the composition of the formulation, the amount of the magnetic material in the granules, and the magnitude of the magnetic field. This approach can be used for local chemotherapy of esophageal cancer and for other diseases in the alimentary canal [251].

Although most drugs are absorbed from the intestine by the blood capillary network in the villi, they can be taken up by the lymphatic system (an integral and necessary part of the vascular system, the function of which is to collect extra tissue fluid and returned it to the vascular compartment), particularly by M cells that reside in the Payer's patch regions of the intestine. The Payer's patches have also been implicated in the regulation of secretory immune response. Wachsmann et al. [252] reported that an antigenic material encapsulated within a liposome, when administered perorally, is taken up by these cells and exhibited better saliva and serum IgA (primary and secondary) immune response than from a simple solution of antigen (Fig. 23). In an attempt to demonstrate the use of microparticles as potential oral immunological adjuvants, O'Hagan et al. entrapped ovalbumin, a model, but poor, immunogen, into polyacrylamide (2.55 μm diameter) [253], poly(butyl-2-cyanoacrylate) 3 μm and 100 nm) [254], and poly(D, L-lactide–co-glycolide) [255–257] particles, and found significantly elevated secretory IgA and systemic IgG antibody responses in rats, compared with soluble antigen, following oral administration. Although additional work is needed, these studies show that microparticles can be used as potential oral adjuvants for the induction of long-term immune responses.

VII. MECHANICAL PUMPS

The mechanical pump approach employs miniature mechanical devices, such as implantable and portable infusion pumps and percutaneous infusion catheters, to deliver drugs into appropriate blood vessels or to a discrete site in the body. When compared with the conventional drug therapy, these devices offer several advantages: (a) the rate of the drug infusion can be better controlled; (b) relatively large volumes of (relatively) dilute drugs can be administered; (c) the drug dose can be readily changed, stopped, or alternated with other drugs, or a placebo, when required; and (d) the drug can be directed into a vascular site or body cavity using the drug delivery cannula (e.g., hepatic arterial chemotherapy, intrathecal morphine infusion for

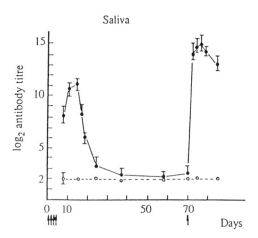

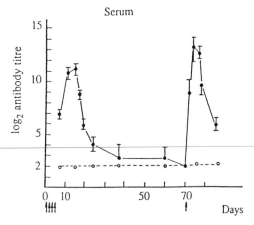

Fig. 23 Saliva and serum IgA (primary and secondary) response following orally administered soluble antigen *Streptococcus mutans* cell wall extract (open circles, soluble antigen; solid circles, liposome-encapsulated material) (phosphatidylcholine, phosphatidic acid, cholesterol). (From Ref. 252).

pain control, intraventricular and intra-articular treatment of central nervous system tumors, intravenous infusions of heparin in thrombotic disorders, and intravenous infusions of insulin in type II diabetes). Several excellent review articles describing design and applications of infusion and implantable pumps in drug therapy, particularly insulin therapy, have been published [258–264]. A diagram of the INFUSAID Model 400 Implantable Pump produced by Infusaid Inc (Norwood, MA) is shown in Fig. 24. Drugs such as floxuridine (5-fluorodeoxyuridine) [265] and zidovudine (3′-azido-2′,3′-dideoxythymidine; AZT) [265] can also be delivered using an implantable pump.

VIII. SUMMARY

The field of targeted drug delivery has grown rapidly in the last two decades. Several delivery systems based on passive-, active-, and physical-targeting strategies have been explored. The two approaches that have received the most attention include prodrugs and polymer–macro-

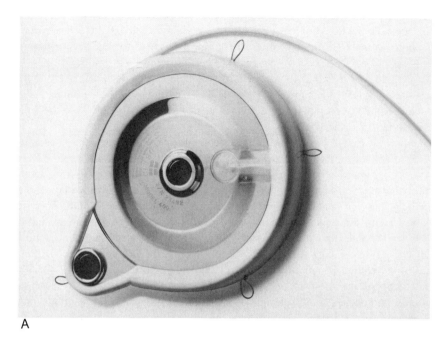

A

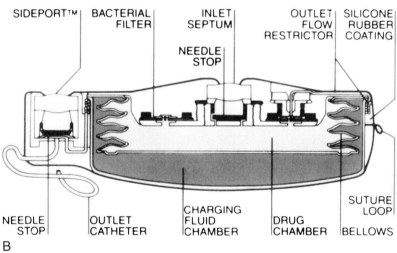

B

SIDEPORT™ | BACTERIAL FILTER | INLET SEPTUM | OUTLET FLOW RESTRICTOR | SILICONE RUBBER COATING

NEEDLE STOP

NEEDLE STOP | OUTLET CATHETER | CHARGING FLUID CHAMBER | DRUG CHAMBER | SUTURE LOOP | BELLOWS

Fig. 24 (A) A top view and (B) a cross-sectional illustration of the INFUSAID Model 400 Implantable pump. (Courtesy of Infusaid Inc., Norwood, MA.)

molecular-carried drug delivery systems. Prodrugs are pharmacologically inert forms of the active drug that must be converted to their active form (i.e., parent drug) by either a chemical or an enzymatic reaction at the site of action. As our understanding of the active sites becomes clearer, this approach should lead to the production of drugs that have a targeting moiety built into the structure. Currently, several examples exist that show promise for site-specific drug delivery.

Polymer–macromolecular-carried drug delivery systems are of two types: particulate and soluble macromolecular. Particulate drug delivery systems, owing to their rapid clearance from the central circulation by the RES system, offer the greatest promise for use in combating diseases (e.g., tumors) of the RES system. Several of these systems (e.g., glucocerebroside glucosidase-encapsulated liposomes for the treatment of type I Gaucher's disease) are currently undergoing clinical trials. Various strategies to avoid uptake of particles by the RES have been developed. This may provide opportunities to deliver drugs to cellular targets within the vasculature and to sites other than the RES.

Soluble macromolecular (natural and synthetic) systems are frequently used as lysosomotropic agents. Because of their ability to extravasate, they have been extensively explored for treating cancer and other remotely located diseases. The recent advent of the hybridoma technology and the progress made in identifying target-specific antibodies and ligands that enable ready target selectivity, have provided additional impetus to design and develop site-specific delivery systems. Several of these systems have been proved very effective in animals, and it remains to be seen how these results will be translated in clinical trials (many thorough clinical trials of several of these systems, e.g., SAMCNS in Japan and polyHMPA copolymer–anthracycline conjugates in the United Kingdom, have been completed or are in progress).

As our understanding of the drug action and pathogenesis of various diseases becomes clearer, more rational approaches to the design of therapeutic systems with functions that selectively target the disease, or deliver the drug to its intended site of action, with no or with reduced side effects, will emerge. The advent of the control of gene expression has already provided several new classes of biopharmaceuticals, including peptidergic mediators and sequence-specific oligonucleotides, and it is important that they be delivered to their sites of action exclusively.

Also discussed in this chapter is a brief account of targeting in the GI tract and localized chemotherapy using mechanical pumps.

REFERENCES

1. P. Ehrlich, John Wiley & Sons, London, 1906, Chap. 36.
2. E. Tomlinson, J. Pharm. Pharmacol., 44 (Suppl.1), 147 (1992).
3. E. Tomlinson, Int. J. Pharm. Technol. Prod. Manuf., 4, 49 (1983).
4. K. J. Widder, A. E. Senyei, and D. F. Ranney, Adv. Pharmacol. Chemother., 16, 213 (1979).
5. E. Tomlinson, (Patho)physiology and the temporal and spatial aspects of drug delivery, in *Site-Specific Drug Delivery. Cell Biology, Medical, and Pharmaceutical Aspects* (E. Tomlinson and S. S. Davis, eds.), John Wiley & Sons, Chichester, 1986, p. 1.
6. C. J. Vanoss, C. F. Gillman, P. M. Bronson, and J. R. Border, Immunol. Commun., 3, 329 (1974).
7. S. S. Davis, S. J. Douglas, L. Illum, P. D. E. Jones, E. Mak, and R. H. Muller, Targeting of colloidal carriers and the role of surface properties, in *Targeting of Drugs with Synthetic Systems* (G. Gregoriadis, J. Senior, and G. Poste, eds.), Plenum Press, New York, 1986, p. 123.
8. L. Illum, S. S. Davis, and P. D. E. Jones, Polym. Preprints, 27, 25 (1986).
9. L. Illum and S. S. Davis, Life Sci., 40, 1553 (1987).
10. K. Saito, J. Ando, M. Yoshida, M. Haga, and Y. Kato, Chem. Pharm. Bull., 36, 4187 (1988).
11. S. M. Moghimi, C. J. Porter, I. S. Muir, L. Illum, and S. S. Davis, Biochem. Biophys. Res. Commun., 177, 861 (1991).
12. M. J. Hsu and R. L. Juliano, Biochim. Biophys. Acta, 720, 411 (1982).
13. C. R. Hopkins, Site-specific drug delivery—cellular opportunities and challenges, in *Site-specific Drug Delivery. Cell Biology, Medical and Pharmaceutical Aspects* (E. Tomlinson and S. S. Davis, eds.), John Wiley & Sons, Chichester, 1986, p. 27.
14. R. Duncan, H. C. Cable, F. Rypacek, J. Drobnik, and J. B. Lloyd, Biochem. Soc. Trans., 12, 1064 (1984).

15. R. Duncan, H. C. Cable, P. Rejmanove, and J. Kopecek, Biochim. Biophys. Acta, 799, 1 (1984).
16. T. Kooistra, A. Duursma, J. M. W. Bouma, and M. Gruber, Biochim. Biophys. Acta, 631, 439 (1980).
17. A. S. Harris, Biopharmaceutical aspects of the intranasal administration of peptides, in *Delivery Systems for Peptides* (S. S. Davis, L. Illum, and E. Tomlinson, eds.), Plenum Press, New York, 1986, p. 191.
18. L. Illum, N. F. Farraj, H. Critichley , and S. S. Davis, Int. J. Pharm., 46, 261 (1988).
19. W. A. J. J. Hermans, M. J. M. Deurloo, S. G. Romeyn, J. C. Verhoef, and F. W. H. M. Merkus, Pharm. Res., 7, 500 (1990).
20. A. N. Fisher, N. F. Farraj, D. T. O'Hagan, I. Jabbal-Gill, B. R. Johansen, S. S. Davis, and L. Illum, Int. J. Pharm., 74, 147 (1991).
21. A. Pusztai, Adv. Drug Deliv. Rev., 3, 215 (1989).
22. K. Matsuno, T. Schaffner, H. A. Gerbel, C. Ruchti, M. W. Hess, and H. Cottier, J. Reticuloendothel. Soc., 33, 263 (1983).
23. C. Damge, M. Aprahamian, G. Balboin, A. Hoetzel, V. Andrieu, and J. P. Devissaguet, Int. J. Pharm., 36, 121 (1987).
24. E. Tomlinson, Adv. Drug Deliv. Rev., 1, 87 (1987).
25. W. A. Ritschel, Methods Find. Exp. Clin. Pharmacol., 13, 205 (1991).
26. J. B. Dressman, P. Bas, W. A. Ritschel, D. R. Friend, A. Rubinstein, and E. Zhiv, J. Pharm. Sci., 82, 857 (1993).
27. W. A. Ritschel, Method. Find. Exp. Clin. Pharmacol., 13, 313 (1991).
28. V. H. L. Lee and A. Yamamoto, Adv. Drug Deliv. Rev., 4, 171 (1990).
29. S. Muranishi, Crit. Rev. Ther. Drug Carrier Syst., 7, 1 (1990).
30. S. S. Davis, J. Pharm. Pharmacol., 44 (Suppl. 1), 186 (1992).
31. H. P. Merkle, R. Anders, J. Sandow, and W. Schurr, Drug delivery of peptides: The buccal route, in *Delivery Systems for Peptide Drugs* (S. S. Davis, L. Illum, and E. Tomlinson, eds.), Plenum Press, New York, 1986, p. 159.
32. W. A. Ritschel, G. B. Ritschel, B. E. C. Ritschel, and P. W. Lucker, Methods Find. Exp. Clin. Pharmacol., 10, 645 (1988).
33. J. L. Richardson, P. S. Minhans, N. W. Thomas, and L. Illum, Int. J. Pharm., 16, 29 (1989).
34. S. S. Muranishi, K. Takada, H. Yoshikawa, and M. Murakami, Enhanced absorption and lymphatic transport of macromolecules via the rectal route, in *Delivery Systems for Peptide Drugs* (S. S. Davis, L. Illum, and E. Tomlinson, eds.), Plenum Press, New York, 1986, p. 177.
35. S. S. Davis and L. Illum, Colloidal delivery systems—opportunity and challenges, in *Site-Specific Drug Delivery. Cell Biology, Medical and Pharmaceutical Aspects* (E. Tomlinson and S. S. Davis, eds.), John Wiley & Sons, Chichester, 1986, p. 93.
36. W. J. Joyner and D. F. Kern, Adv. Drug Deliv. Rev., 4, 319 (1990).
37. K. Patrak and P. Goddard, Adv. Drug Deliv. Rev., 3, 191 (1989).
38. L. W. Seymour, Crit. Rev. Ther. Drug Carrier Syst., 19, 135 (1992).
39. D. F. Kern and J. A. Swanson, FASEB J., 3, A1390 (1989).
40. G. J. Arturson and K. Granath, Clin. Chim. Acta, 37, 309 (1972).
41. A. E. Taylor and D. N. Granger, Exchange of macromolecules across the microcirculation, in *Handbook of Physiology*, 6, *Microcirculation*, Part 1 (E. M. Renkin and C. C. Michel, eds.), Waverly Press, Baltimore, 1984, p. 467.
42. L. Ghitescu, A. Fixman, M. Simionescu, and N. Simionescu, J. Cell. Biol., 102, 1304 (1986).
43. J. Drobnik and F. Rypacek, Adv. Polym. Sci., 57, 2 (1984).
44. J. M. Yoffey and F. C. Courtice, *Lymphatics, Lymph and the Lymphomyeloid Complex*, Academic Press, London, 1970.
45. J. G. Hall, The lymphatic system in drug targeting: An overview, in *Targeting of Drugs. Anatomical and Physiological Considerations* (G. Gregoriadis and G. Poste, eds.), Plenum Press, New York, 1985, P. 15.
46. A. Boddy and L. Aarons, Adv. Drug Deliv. Rev., 3, 155 (1989).
47. A. Boddy, L. Aarons, and K. Petrak, Pharm. Res., 6, 367 (1989).

48. G. Levy, Pharm. Res., 4, 3 (1987).
49. P. K. Gupta and C. T. Hung, Int. J. Pharm., 56, 217 (1989).
50. R. E. Notari, Prodrugs kinetics: Theory and practice, in *Optimization of Drug Delivery* (H. Bundgaard, A. B. Hansen, and H. Koford, eds.), Munksgaard, Copenhagan, 1982, p. 117.
51. V. J. Stella and K. J. Himmelstein, J. Med. Chem., 23, 1275 (1980).
52. V. J. Stella and K. J. Himmelstein, Critique of prodrugs and site-specific drug delivery, in *Optimization of Drug Delivery, Alfred Benzon Symposium 17* (H. Bundsgaard, A. B. Hansen, and H. Kofod, eds.), Munksgaard, Copenhagen, 1982, p. 134.
53. C. R. Gardner and J. Alexander, Prodrug approaches to drug targeting: past accomplishments and future potential, in *Drug Targeting* (P. Buri and A. Gumma, eds.), Elsevier Science Publishers, Amsterdam, 1985, p. 145.
54. V. J. Stella, Prodrugs: An overview and definition, in *Prodrugs in Novel Drug Delivery Systems* (T. Higuchi and V. Stella, eds.), American Chemical Society, Washington, DC, 1975, p. 1.
55. E. B. Roche, ed., *Design of Biopharmaceutical Properties Through Prodrugs and Analogs*, American Pharmaceutical Association, Washington, DC, 1977.
56. G. C. Cotzias, M. H. VanWoert, and L. M. Schiffer, N. Engl. J. Med., 276, 374 (1967).
57. A. Hussain and J. E. Truelove, J. Pharm. Sci., 65, 1510 (1976).
58. S. Wilk, H. Mizoguchi, and M. Orlowski, J. Pharmacol. Exp. Ther., 206, 227 (1978).
59. D. R. Friend and G. W. Chang, J. Med. Chem., 28, 51 (1985).
60. V. J. Stella, T. J. Mikkelson, and J. D. Pipkin, Prodrugs: The control of drug delivery via bioreversible chemical modification, in *Drug Delivery Systems: Characteristics and Biomedical Applications* (R. L. Juliano, ed.), Oxford University Press, New York, 1980, p. 112.
61. R. Baurain, M. Masquelier, D. D.-D. Campeneere, and A. Trouet, J. Med. Chem., 23, 1171 (1980).
62. M. Masquelier, R. Baurain, and A. Trouet, J. Med. Chem., 23, 1166 (1980).
63. P. K. Chakravarty, P. L. Carl, M. J. Weber, and J. A. Katzenellenbogen, J. Med. Chem., 26, 638 (1983).
64. N. Bodor and J. W. Simpkins, Science, 221, 65 (1983).
65. E. Shek, T. Higuchi, and N. Bodor, J. Med. Chem., 19, 113 (1976).
66. N. Bodor, Methods Enzymol., 112, 381 (1985).
67. N. Bodor and H. H. Farag, J. Med. Chem., 26, 528 (1983).
68. C. K. Chu, V. S. Bhadti, K. J. Doshi, J. T. Etse, J. M. Gallo, F. D. Boudinot, and R. F. Schinazi, J. Med. Chem., 33, 2188 (1990).
69. E. Palomino, Drugs Future, 15, 361 (1990).
70. M. E. Brewster, W. Anderson, and N. Bodor, J. Pharm. Sci., 80, 843 (1991).
71. N. Bodor and H. H. Farag, J. Pharm. Sci., 73, 385 (1984).
72. M. B. Yatvin, T. C. Cree, and I. M. Tegmo-Larsson, Theoretical and practical considerations in preparing liposomes for the purpose of releasing drug in response to changes in temperature and pH, in *Liposome Technology* (G. Gregoriaids, ed.), CRC Press, Boca Raton, FL, 1984, p. 157.
73. L. Illum and S. S. Davis, Passive and active targeting using colloidal carrier systems, in *Drug Targeting* (P. Buri and A. Gumma, eds.), Elsevier Science Publishers, Amsterdam, 1985, p. 65.
74. C. Chiles, L. W. Hedlund, R. J. Kubek, C. Harris, D. C. Sullivan, J. C. Tsai, and C. E. Putman, Invest. Radiol, 21, 618 (1986).
75. C. K. Kim, M. K. Lee, J. H. Han, and B. J. Lee, Int. J. Pharm., 108, 21 (1994).
76. B. M. Altura and T. M. Saba, *Pathophysiology of the Reticuloendothelial system*, Raven Press, New York, 1981.
77. I. Ahmad and T. M. Allen, Cancer Res., 52, 4817 (1992).
78. E. Hurwitz, A. Adler, D. Shouval, J. R. Takahashi, J. R. Wands, and M. Sela, Cancer Immunol. Immunother., 35, 86 (1992).
79. K. Affleck and M. J. Embleton, Br. J. Cancer, 65, 838 (1992).
80. L. W. Seymour, A. Al-Shamkhani, P. A. Flanagan, V. Subr, K. Ulbrich, J. Cassidy, and R. Duncan, Select. Cancer Ther., 7, 59 (1991).
81. G. Gregoriadis and E. D. Neerunjun, Biochem. Biophys. Res. Commun., 65, 537 (1975).
82. R. I. Juliano and D. Stamp, Nature, 261, 235 (1976).

83. G. Gregoriadis, N. Meehan, and M. M. Mah, Biochem. J., 200, 203 (1981).
84. G. Weissman, D. Bloomgarden, R. Kaplan, C. Cohen, S. Hoffstein, T. Collins, A. Gotlieb, and D. Nagle, Proc. Natl. Acad. Sci. USA, 72, 88 (1975).
85. M. R. Kaplan, E. Calef, T. Bercovivi, and C. Gitler, Biochim. Biophys. Acta, 728, 112 (1983).
86. K. J. Widder and A. E. Senyei, Magnetic microsphere: A vehicle for selective targeting of organs, in *Methods of Drug Delivery* (G. M. Ihler, ed.), Pergamon Press, New York, 1986, p. 39.
87. H. Sezaki, M. Hashida, and S. Muranishi, Gelatin microspheres as carriers for antineoplastic agents, in *Optimization of Drug Delivery* (H. Bundgaard, A. B. Hansen, and H. Kofod, eds.), Munksgaard, Copenhagen, 1982, p. 316.
88. M. B. Yatvin, J. N. Weinstein, W. H. Dennis, and R. Blumenthal, Science, 202, 1290 (1978).
89. M. B. Yatvin, W. Kreutz, B. A. Horowitz, and M. Shinitzky, Science, 210, 1253 (1980).
90. T. Kato, Encapsulated drugs in targeted cancer therapy, in *Controlled Drug Delivery* (S. D. Bruck, ed.), CRC Press, Boca Raton, FL, 1983, p. 189.
91. J. Kreuter and E. Liehl, J. Pharm. Sci., 70, 367 (1981).
92. P. Couvreur, Crit. Rev. Ther. Drug Carrier Syst., 5, 1 (1988).
93. R. Langer, Science, 263, 1600 (1994).
94. E. Tomlinson and J. J. Burger, Monolithic albumin particles as drug carriers, in *Polymers in Controlled Drug Delivery* (L. Illum and S. S. Davis, eds.), Wright, Bristol, 1987, p. 25.
95. A. F. Yapel, U.S. Patent 4,147,747, 1979.
96. E. Edelman, J. Brown, J. Taylor, and R. Langer, J. Biomed. Mater. Res., 21, 339 (1987).
97. M. McCarthy, D. Soong, and E. Edelman, J. Controlled Release, 1, 143 (1984).
98. H. Hsu and R. Langer, J. Biomed. Mater. Res., 19, 445 (1985).
99. J. Kotsr, K. E. Noecker, and R. Langer, J. Biomed. Mater. Res., 19, 935 (1985).
100. P. K. Gupta and C. K. Hung, Int. J. Pharm., 59, 57 (1990).
101. G. Gregoriadis, ed., *Liposome Technology* Vol. 1–3, CRC Press, Boca Raton, FL, 1984.
102. G. Gregoriadis, ed., *Liposome Technology*, 2nd Ed., Vol. 1–3, CRC Press, Boca Raton, FL, 1993.
103. P. R. C. New, ed., *Liposome–A Practical Approach*, IRL Press, Oxford, 1990.
104. K. Yagi, ed., *Medical Applications of Liposomes*, Japan Scientific Press and Karger, Tokyo and Basel, 1986.
105. K. H. Schmidt, ed., *Liposomes as Drug Carriers*, Georg Thieme Verlag, Stuttgart, 1986.
106. G. Gregoriadis and A. C. Allison, eds., *Liposome in Biological Systems*, John Wiley & Sons, Chichester, 1980.
107. K. Maruyama, A. Mori, S. J. Kennel, M. V. B. Waalkes, G. L. Scherphof, and L. Huang, ACS Symp. Ser., 469, 275 (1990)
108. F. H. Roerdnik, T. Daemen, I. A. J. M. Bakker-Woundenberg, G. Storm, D. J. A. Crommelin, and G. L. Scherphof, Therapeutic utility of liposomes, in *Drug Delivery Systems. Fundamental and Techniques* (P. Johnson and J. G. L. Jones, eds.), Ellis Horwood, Chichester, 1987, p. 67.
109. F. Roerdink, J. Regts, T. Daemen, I. Bakker-Woundenberg, and G. Scherphof, Liposomes as drug carriers to liver macrophages, in *Targeting of Drugs with Synthetic Systems* (G. Gregoriadis, J. Senior, and G. Poste, eds.), Plenum Press, New York, 1985, p. 193.
110. G. Gregoriadis, J. Senior, B. Wolff, and C. Kirby, Fate of liposomes in vivo: Control leading to targeting, in *Receptor-Mediated Targeting of Drugs* (G. Gregoriadis, G. Poste, J. Senior, and A. Trouet, eds.), Plenum Press, New York, 1984, p. 243.
111. G. Lopez-Berestein, R. L. Juliano, K. Mehta, R. Mehta, T. McQueen, and R. L. Hopfer, Liposomes in antimicrobial therapy, in *Targeting of Drugs with Synthetic Systems* (NATO ASI- Series. Series A, Life Sciences; V. 113) (G. Gregoriadis, J. Senior, and G. Poste, eds.), Plenum Press, New York, 1985, p. 193.
112. S. M. Sugarman and R. Peres-Solar, Crit. Rev. Oncol. Hematol., 12, 231 (1992).
113. O. Alpar, Pharm. J., 246, 172 (1991).
114. G. Gregoriadis and A. T. Florence, Cancer Cells, 3, 144 (1991).
115. G. Gregoriadis, J. Antimicrob. Chemother., 28, 39 (1991).
116. J. Liliemark, Eur. J. Surg. Suppl., 561, 49 (1991).
117. F. Sozaka, Jr., Biotechnol. Appl. Biochem., 12, 496 (1990).

118. G. Lopez-Berstein, Antimicrob. Agents Chemother., 31, 675 (1987).
119. V. P. Torchilin, Crit. Rev. Ther. Drug Carrier Syst., 2, 65 (1985).
120. M. J. Hope, M. B. Bally, G. Webb, et al., Biochim. Biophys. Acta, 812, 210 (1985).
121. A. Gabizon, A. Dagan, D. Goren, Y. Barenholz, and Z. Funks, Cancer Res., 42, 4734 (1982).
122. L. D. Mayer, M. B. Bally, M. J. Hope, and P. R. Cullis, Biochim. Biophys. Acta, 816, 294 (1985).
123. V. Weissig, J. Lasch, and G. Gregoriadis, Pharmazie, 46, 56 (1991).
124. J. Senior and G. Gregoriadis, Biochim. Biophys. Acta, 1002, 58 (1989).
125. G. Poste, R. Kirsh, and T. Koestler, eds., *Liposome Technology. Targeted Drug Delivery and Biological Interactions*, Vol. 3, CRC Press, Boca Raton, FL, 1984.
126. P. E. Belchetz, I. P. Braidman, J. C. W. Crawly, and G. Gregoriadis, Lancet, 2, 116 (1977).
127. D. A. Tyrrel, B. E. Ryman, B. R. Keeton, and V. Dubowitz, Br. Med. J., 2, 88 (1976).
128. R. R. C. New, M. L. Chance, S. C. Thomas, and W. Peters, Nature, 272, 55 (1978).
129. C. R. Alving, E. A. Steck, J. W. L. Chapman, V. B. Waits, L. D. Hendricks, J. G. M. Swartz, and W. L. Hanson, Proc. Natl. Acad. Sci. USA, 75, 2959 (1978).
130. C. D. V. Black, G. J. Watson, and R. J. Ward, Trans. R. Soc. Trop. Med. Hyg., 71, 550 (1977).
131. I. A. J. M. Bakker-Woudenberg, A. F. Lokerse, F. H. Roerdnik, D. Regts, and M. F. Michel, J. Infect. Dis., 151, 917 (1985).
132. I. J. Fidler, The generation of tumoricidal activity in macrophages for the treatment of established mestastases, in *Cancer Invasion and Metastasis: Biologic and Therapeutic Aspects* (G. L. Nicolson and L. Milas, eds.), Raven Press, New York, 1984, p. 421.
133. S. Sone, S. Matsura, M. Ogawara, and E. Tsubura, J. Immunol., 132, 2105 (1984).
134. I. J. Fidler, S. Sone, W. E. Fogler, D. Smith, D. G. Graun, L. Tarcsay, R. J. Gisler, and A. J. Schroit, J. Biol. Response Mod., 1, 43 (1982).
135. I. J. Fidler, A. Raz, W. E. Fogler, L. C. Hoyler, and G. Poste, Cancer Res., 41, 495 (1981).
136. L. Saiki, S. Sone, W. E. Fogler, E. S. Kleinerman, G. Lopez-Berestein, and I. J. Fidler, Cancer Res., 45, 6188 (1985).
137. L. Saiki and I. J. Fidler, J. Immunol., 135, 684 (1985).
138. S. Sone, P. Tandon, T. Utsugi, and M. Ogawara, Int. J. Cancer, 38, 495 (1986).
139. I. J. Fidler and A. J. Schroit, J. Immunol., 133, 515 (1984).
140. E. Mayhew and Y. Rustum, Biol. Cell, 47, 81 (1983).
141. R. R. C. New, M. L. Chance, and S. Heath, Antimicrob. Agents Chemother., 8, 371 (1981).
142. O. A. Rosenberg, V. Y. Berkreneva, L. V. Loshakova, S. P. Rezvaya, E. F. Davidenkova, and K. P. Hanson, Vopr. Onkol., 29, 56 (1983).
143. S. A. Burkhanov, V. A. Kosykh, V. S. Repin, T. S. Saatov, and V. P. Torchilin, Int. J. Pharm., 46, 31 (1988).
144. J. N. Weinstein, R. L. Magin, R. L. Cysyk, and D. S. Zaharko, Cancer Res., 40, 1388 (1980).
145. J. R. Tacker and R. U. Anderson, J. Urol., 127, 1211 (1982).
146. M. B. Yatvin, H. Muhensipen, W. Porschen, J. N. Weinstein, and L. F. Feinendegen, Cancer Res., 41, 1602 (1981).
147. C. Nicolau, A. L. Pape, P. Soriano, F. Fargette, and M. F. Juhel, Proc. Natl. Acad. Sci. USA, 80, 1068 (1983).
148. J. Szebeni, S. M. Wahl, G. V. Betageri, L. M. Wahl, S. Gartner, M. Popovic, R. J. Parker, C. D. V. Black, and J. N. Weinstein, AIDS Res. Hum. Retroviruses, 6, 791 (1990).
149. N. C. Philips, E. Skamene, and C. Tsoukas, J. AIDS, 4, 959 (1991).
150. A. J. Baillie, A. T. Florence, L. H. Muirhead, and A. Rogerson, J. Pharm. Pharmacol., 37, 863 (1985).
151. J. N. Israelachvili, S. Marcelja, and R. G. Horn, Q. Rev. Biophys. 13, 121 (1980).
152. C. A. Hunter, T. F. Dolan, G. H. Coombs, and A. J. Baillie, J. Pharm. Pharmacol., 40, 161 (1988).
153. H. Kiwada, H. Nimura, Y. Fujisaki, S. Yamada, and K. Kato, Chem. Pharm. Bull., 33, 753 (1985).
154. L. E. Echegoyen, J. C. Hernandez, A. E. Kaifer, G. W. Gokel, and L. Echegoyen, J. Chem. Soc. Chem. Commun., 863 (1988).
155. A. J. Baillie, Niosomes: A putative drug carrier system, in *Targeting of Drugs. Anatomical and Physiological Considerations* (G. Gregoriadis and G. Poste, eds.), NATO Series. Series A. Life Science, Vol. 155, Plenum Press, New York, 1988, p. 143.

156. M. N. Azmin, A. T. Florence, R. M. Handjani-Vila, J. F. B. Stuart, G. Vanlerberghe, and J. S. Whittaker, J. Pharm. Pharmacol., 3, 237 (1985).

157. M. N. Azmin, A. T. Florence, R. M. Handjani-Vila, J. F. B. Stuart, G. Vanlerberghe, and J. S. Whittaker, J. Microencapsul., 3, 95 (1986).

158. A Rogerson, Ph.D. thesis, University of Strathclyde, Glasgow, 1986; A. Rogerson, J. Cummings, and A. T. Florence, J. Microencapsul., 4, 321 (1987); A Rogerson, J. Cummings, N. Willmott, and A. T. Florence, J. Pharm. Pharmacol., 40, 337 (1988); D. J. Kerr, A. Rogerson, G. J. Morrison, A. T. Florence, S. B. Kaye, Br. Jr. Cancer, 58, 432 (1988).

159. A. J. Baillie, G. H. Coombs, T. F. Dolan, and J. Laurie, J. Pharm. Pharmacol., 38, 502 (1986).

160. M. K. Bijsterbosch and T. J. C. V. Berkel, Adv. Drug Deliv. Rev., 5, 231 (1990).

161. S. Eisenberg, J. Lipid Res., 25, 1017 (1984).

162. P. C. D. Smidt and T. J. C. V. Berkel, Crit. Rev. Ther. Drug Carrier Syst., 7, 99 (1990).

163. N. S. Damle, R. H. Seevers, S. W. Schwendner, and R. E. Counsell, J. Pharm. Sci., 72, 898 (1983).

164. M. Samadi-Baboli, G. Favre, E. Blancy, and G. Soula, Eur. J. Cancer Clin. Oncol., 25, 233 (1989).

165. S. Lestavel-Delattre, F. Martin-Nizard, V. Clavey, et al., Cancer Res., 52, 3629 (1992).

166. H. J. M. Kempen, C. Hoes, J. H. V. Boom, H. H. Spanjer, J. D. Lange, A. Langendoen, and T. J. C. V. Berkel, J. Med. Chem., 27, 1306 (1984).

167. J. Y. Scoazec and G. Feldman, Hepatology, 10, 627 (1989).

168. T. Takahashi, Crit. Rev. Ther. Drug Carrier Syst., 2, 245 (1985).

169. K. Ito, K. Kiriyama, T. Watanabe, M. Yamauchi, S. Akiyama, K. Kondou, and H. Takagi, ASAIO Trans., 36, M199 (1990).

170. M. F. Dean, H. Muir, P. F. Benson, L. R. Button, J. R. Batcheolor, and M. Bewica, Nature, 257, 609 (1975).

171. A. J. Matas, D. E. R. Sutherland, M. W. Steffes, R. L. Simmons, and J. S. Najarian, Science, 192, 892 (1976).

172. R. Baum, Chem. Eng. News, 72, 4 (1994).

173. R. Younoszai, R. L. Sorenson, and A. W. Lindal, Diabites, 19, (Suppl.), 406 (1971).

174. U. Spraundel, Res. Exp. Med., 190, 267 (1990).

175. G. M. Ihler and H. C. Tsang, Crit. Rev. Ther. Drug Carrier Syst., 1, 155 (1984).

176. U. Benatti, E. Zocchi, M. Tonetti, et al., Pharmacol. Res. 21, (Suppl. 2), 27 (1989).

177. H. J. Leu, A. Wenner, and M. A. Spycher, VASA, 10, 17 (1981).

178. U. Zimmermann, Dtsch. Apoth. Ztg., 122, 1170 (1982).

179. T. Kitao and K. Hattori, Cancer Res., 40, 1351 (1980).

180. M. M. Billah, J. B. Finean, R. Coleman, and R. H. Mitchell, Biochim. Biophys. Acta, 433, 54 (1976).

181. P. S. Lin, D. F. H. Wallach, R. B. Mikkelsen, and R. Schmidt-Ulrich, Biochim, Biophys. Acta, 401, 73 (1975).

182. L. Gordon and A. J. Milner, Blood platelets as multifunctional cells, in *Platelets in Biology and Pathology* (J. L. Gordon, ed.), Elsevier/North Holland Biomedical Press, Amsterdam, 1976, p. 3.

183. J. M. Weiss, N. Engl. J. Med., 293, 531 (1975).

184. Y. S. Ahn, J. J. Byrnes, and W. J. Harrington, N. Engl. J. Med., 298, 1101 (1978).

185. Y. S. Ahn, W. J. Harrington, J. J. Byrnes, L. Pall, and J. McCranie, JAMA, 249, 2189 (1983).

186. Y. S. Ahn, W. J. Harrington, J. J. Byrnes, A. S. Collin, M. L. Cayer, J. McCranie, and L. M. Pall, Blood, 58, 134 (1981).

187. N. S. Penney, Y. S. Ahn, and E. C. McKinney, Cancer, 49, 1944 (1982).

188. S. Y. Woo, R. S. Klappenbach, G. M. McCullars, D. M. Kerwin, G. Rowden, and L. F. Sinks, Cancer, 46, 2566 (1986).

189. A. W. Segal, M. L. Thakur, R. N. Arnot, and J. P. Lavender, Lancet, 2, 1056 (1976).

190. M. A. Gainey and I. R. McDougall, Clin. Nucl. Med., 9, 71 (1984).

191. S. Sixou and J. Teissie, Biochem. Biophys. Res. Commun., 186, 860 (1992).

192. G. M. Ihler, R. H. Glew, and F. W. Schnure, Proc. Natl. Acad. Sci. USA, 70, 2663 (1973).

193. U. Zimmermann, J. Vienken, and G. Pilwat, Bioelectrochem. Bioeng., 7, 332 (1980).

194. H. Sezaki and M. Hashida, Crit. Rev. Ther. Drug Carrier Syst., 1, 1 (1984).

195. J. Kopecek and R. Duncan, Poly(*N*-(2-hydroxypropyl)methacrylamide) macromolecules as drug carrier systems, in *Polymers in Controlled Drug Delivery* (L. Illum and S. S. Davis, eds.), Wright, Bristol, 1987, p. 152.

196. J. B. Lloyd, Soluble polymers as targetable drug carriers, in *Drug Delivery Systems. Fundamental and Techniques* (P. Johnson and J. G. Llyod, eds.), Ellis Horwood and VCH Verlagsgesellschaft, Chichester and Weinheim, 1987, p. 95.

197. D. C. Blakey, Acta Oncol., 31, 91 (1992).

198. K. E. Hellstrom, I. Hellstrom, and G. E. Goodman, Antibodies for drug delivery, in *Controlled Drug Delivery: Fundamental and Applications* (J. R. Robinson and V. H. Lee, eds.), Marcel Dekker, New York, 1987, p. 623.

199. K. Sikora, Monoclonal antibodies and drug targeting in cancer, in *Targeting of Drugs. Anatomical and Physiological Considerations* (G. Gregoriadis and G. Poste, eds.), Plenum Press, New York, 1987, p. 69.

200. A. Tai, A. H. Blair, and T. Ghose, Eur. J. Cancer, 15, 1357 (1979).

201. T. Ghose, J. Tai, A. Guclu, R. R. Ramam, and A. H. Blair, Cancer Immunol. Immunother., 13, 185 (1982).

202. G. F. Rowland, G. J. O'Neil, and D. A. L. Davies, Chemotherapy, 8, 11 (1977).

203. M. V. Pimm, R. A. Robins, M. J. Embleton, E. Jacobs, A. J. Markham, A. Charleston, and R. W. Baldwin, Br. J. Cancer, 61, 508 (1990).

204. M. K. Ghosh, D. O. Kildsig, and A. K. Mitra, Drug Des. Deliv., 4, 13 (1989).

205. L. Diang, J. Samuel, G. D. MacLean, A. A. Noujaim, E. Diener, and B. M. Longenecker, Cancer Immunol. Immunother., 32, 105 (1990).

206. F. Hudecz, H. Ross, M. R. Price, and R. W. Baldwin, Bioconjug. Chem., 1, 197 (1990).

207. M. Page, D. Thibeault, C. Noel, and L. Dumas, Anticancer Res., 10, 353 (1990).

208. G. F. Rowland, R. G. Simmonds, J. R. F. Corvalan, et al., Protides Biol. Fluids, 30, 375 (1983).

209. G. P. Samokhin, M. D. Smirnov, V. R. Muzykantove, S. P. Domogatsky, and V. N. Smirnov, FEBS Lett., 154, 259 (1983).

210. D. L. Urdal and S. Hakomori, J. Biol. Chem., 255, 10509 (1980).

211. H. G. Eichler, S. Gasic, K. Bauer, A. Korn, and S. Bacher, Clin. Pharmacol. Ther., 40, 300 (1986).

212. M. A. Glukhova, S. P. Domogatsky, A. E. Kabakov, V. R. Muzykantov, O. I. Ornatsky, D. V. Sakharov, M. G. Frid, and V. N. Smirnov, FEBS Lett., 198, 155 (1986).

213. A. Trouet, D. D. Campeneere, R. Burain, M. Huybrechts, and A. Zeneberch, Desoxyribonucleic acid as carrier of antitumor drugs, in *Drug Carriers in Biology and Medicine* (G. Gregoriadis, ed.), Academic Press, London, 1979, p. 87.

214. A. Trouet, Carriers for bioactive materials, in *Polymeric Delivery Systems* (R. J. Kostelnik, ed.), Gordon & Breach, New York, 1978, p. 157.

215. C. Paul, J. Lliemark, U. Tidefelt, G. Gahrton, and C. Peterson, Ther. Drug Monitor., 11, 140 (1989).

216. A Trouet, J. M. Jadin, and F. V. Hoof, Lysosomotropic chemotherapy in protozoal diseases, in *Biochemistry of Parasites and Host–Parasites Relationships* (H. van den Bossche, ed.), North-Holland, Amsterdam, 1976, p. 519.

217. E. J. Franssen, R. G. van Amsterdam, J. Visser, F. Moolenaar, D. D. Zeeuw, and D. K. F. Meijer, Pharm. Res., 8, 1223 (1991).

218. T. Maack, V. Johnson, S. T. Kau, J. Figueiredo, and D. Siguelm, Kidney Int., 16, 251 (1979).

219. D. K. F. Meijer and P. V. D. Sluijs, Pharm. Res., 6, 105 (1989).

220. G. Molema, R. W. Jansen, R. Pauwels, E. Clerco, and D. K. F. Meijer, Biochem. Pharmacol., 40, 2603 (1990).

221. A. C. Roche, P. Midoux, V. Pimpaneau, E. Negre, R. Mayer, and M. Monsigny, Res. Virol., 141, 243 (1990).

222. B. C. F. Chu and J. M. Whiteley, Mol. Pharmacol., 13, 80 (1977).

223. B. C. F. Chu and J. M. Whiteley, JNCI, 62, (1979).

224. J. M. Whiteley, Z. Nimec, and J. Galivan, Mol. Pharmacol., 19, 505 (1981).

225. L. Fiume, C. Busi, and A. Mattioli, FEBS Lett., 153, 6 (1983).

226. J. M. Varga and N. Asato, Hormones as drug carriers, in *Targeted Drugs* (E. P. Goldberg, ed.), John Wiley & Sons, New York, 1983, p. 73.

227. A. D. Ross and D. M. Angaran, Drug Intell. Clin. Pharm., 18, 202 (1984).

228. L. Moleteni, Dextran as drug carriers, in *Drug Carriers in Biology and Medicine* (G. Gregoriadis, ed.), Academic Press, New York, 1979, p. 25.

229. E. Schacht, Polysaccharide macromolecules as drug carriers, in *Polymers in Controlled Drug Delivery* (L. Illum and S. S. Davis, eds.), Wright, Bristol, 1987, p. 131.

230. S. Matsumoto, Y. Arase, Y. Takakura, M. Hashida, and H. Sezaki, Chem. Pharm. Bull., 33, 2941 (1985).

231. D. R. Friend and S. Pangburn, Med. Res. Rev., 7, 53 (1987).

232. W. C. Shen and H. J. P. Ryser, Mol. Pharmacol., 16, 614 (1979).

233. B. C. F. Chu and S. B. Howell, Biochem. Pharmacol., 30, 2545 (1981).

234. F. Zunino, G. Savi, F. Giuliani, R. Gambetta, R. Supinio, S. Tinelli, and G. Pezzoni, Eur. J. Cancer Clin. Oncol., 20, 421 (1984).

235. E. Hurwitz, M. Wilchek, and J. Pitha, J. Appl. Biochem., 2, 25 (1980).

236. P. Campbell, G. Glover, and J. M. Gunn, Arch. Biochem. Biophys., 203, 676 (1980).

237. J. Koecek, J. Controlled Release, 11, 279 (1990).

238. L. W. Seymour, K. O'Hare, R. Duncan, J. Strohalm, and K. Ulbrich, Br. J. Cancer, 63, 882 (1991).

239. J. Kopecek, B. Rihova, and N. L. Krinick, J. Controlled Release, 16, 137 (1991).

240. H. Maeda and Y. Matsumura, New tactics and basic mechanisms of targeting chemotherapy in solid tumors, in *Cancer Chemotherapy: Challenge for the Future* (K. Kimura, ed.), Excerpta Medica, Tokyo, 1989, p. 42.

241. Y. Matsumra, K. K., T. Yamamoto, and H. Maeda, Jpn. J. Cancer Res., 47, 852 (1988).

242. T. Konno and H. Maeda, Targeting chemotherapy of hepatocellular carcinoma: Arterial administration of SMANCS/Lipiodol, in *Neoplasm of the Liver* (K. Okada and K. G. Ishak, eds.), Springer-Verlag, New York, 1987, p. 343.

243. T. Konno, Targeting anticancer chemotherapy for primary and secondary liver cancer using arterially administered oily anticancer agents, in *Cancer Chemotherapy: Challenges for the Future* (K. Kimura, ed.), Excerpta Medica, Tokyo, 1987, p. 287.

244. H. Maeda, Adv. Drug Deliv. Rev., 6, 181 (1991).

245. H. Yamamoto, T. Miki, T. Oda, T. Hirano, Y. Sera, M. Akagi, and H. Maeda, Eur. J. Cancer, 26, 253 (1990).

246. M. Przybylski, E. Fell, H. Ringsdorf, and D. S. Zaharko, Makromol. Chem., 179, 1719 (1978).

247. D. S. Breslow, E. I. Edwards, and N. R. Newburg, Nature 246, 160 (1973).

248. V. Lenaerts and R. Gurny, eds., *Bioadhesive Drug Delivery Systems*, CRC Press, Boca Raton, FL, 1990.

249. K. Klokkers-Bethke and W. Fischer, J. Controlled Release, 15, 105 (1991).

250. M. Saffran, G. S. Kumar, C. Savariar, J. C. Burnham, F. Williams, and D. Neckers, Science, 233, 1081 (1986).

251. R. Ito, Y. Machida, T. Sannan, and T. Nagai, Int. J. Pharm., 61, 109 (1990).

252. D. Wachsmann, J. P. Klein, M. Scholler, and R. M. Frank, Immunology, 54, 189 (1985).

253. D. T. O'Hagan, K. Palin, S. S. Davis, P. Arthursson, and I. Sjoholm, Vaccine, 7, 421 (1989).

254. D. T. O'Hagan, K. J. Palin, and S. S. Davis, Vaccine, 7, 213 (1989).

255. D. T. O'Hagan, D. Rahman, J. P. McGee, H. Jeffery, M. C. Davies, P. Williams, S. S. Davis, and S. J. Challacombe, Immunology, 73, 239 (1989).

256. S. J. Challacombe, D. Rahman, H. Jeffery, S. S. Davis, and D. T. O'Hagan, Immunology, 76, 164 (1992).

257. D. T. O'Hagen, D. Rahman, H. Jeffery, S. Sharif, and S. J. Challacombe, Int. J. Pharm., 108, 133 (1994).

258. B. D. Wigness, F. D. Dorman, T. D. Rhode, and H. Buchwald, ASAIO J., 38, M454 (1992).

259. H. Buchwald and T. D. Rhode, ASAIO J., 38, 772 (1992).

260. J. L. Salem, P. Micossi, F. L. Dumm, and D. M. Nathan, Diabetes Care, 15, 877 (1992).

261. J. L. Salem, Horm. Metab. Res. 24 (Suppl.), 144 (1990).

262. J. L. Salem and M. A. Charles, Diabetes Care, 13, 955 (1990).
263. P. J. Blackshear and T. D. Rhode, Horiz. Biochem. Biophys., 9, 293 (1989).
264. P. J. Blackshear, Implantable pumps for insulin delivery: Current clinical status, in *Drug Delivery Systems, Fundamentals and Techniques* (P. Johnson and J. G. Lloyd-Jones, eds.), Ellis Horwood, Chichester, 1987, p. 139.
265. J. M. Gallo, Y. Sanzgiri, E. W. Howerth, T. S. Winco, J. Wilson, J. Johnston, R. Tackett, and S. C. Budsberg, J. Pharm. Sci., 81, 11 (1992).
266. A. Ponzetto, L. Fiume, B. Forzani, et al., Hepatology, 14, 16 (1991).
267. R. W. Jensen, J. K. Kruijt, T. J. V. Berkel, and D. K. Meijer, Hepatology, 18, 146 (1993).
268. L. Fiume, G. D. Stefano, C. Busi, and A. Mattioli, Biochem. Pharmacol., 47, 643 (1994).
269. P. Midoux, E. Negre, A. C. Roche, et al., Biochem. Biophys. Res. Commun., 167, 1044 (1990).
270. S. Kuchler, M. N. Graff, S. Gobaille, G. Vincendon, A. C. Roche, J. P. Delaunoy, M. Monsigny, and J. P. Zanetta, Neurochem. Int., 24, 43 (1994).
271. E. Negre, M. L. Chance, S. Y. Hanboula, M. Monsigny, A. C. Roche, R. M. Mayer, and M. Hommel, Antimicrob. Agents Chemother., 36, 2228 (1992).
272. Y. Kaneo, T. Tanaka, and S. Iguchi, Chem. Pharm. Bull., 39, 999 (1991).
273. T. J. V. Berkel, J. K. Kruijt, P. C. D. Smidt, and M. K. Bijsterbosch, Targeted Diagn. Ther., 5, 225 (1991).
274. V. Keegan-Rogers and G. Y. Wu, Cancer Chemother. Pharmacol., 26, 93 (1990).
275. E. J. Franssen, R. W. Jansen, M. Vaalburg, and D. K. Meijer, Biochem. Pharmacol., 45, 1215 (1993).

17

Packaging of Pharmaceutical Dosage Forms

Donald C. Liebe
Schering-Plough Research Institute, Kenilworth, New Jersey

I. INTRODUCTION

A. Scope of This Chapter

Not many years ago the appropriate packaging of pharmaceuticals involved nothing more than carrying out stability studies in the desired package. The package, usually made of glass, was viewed as relatively inert; hence, it received little attention. This situation has changed radically with the worldwide emergence of drug regulations, the increasing sophistication and variety of dosage forms, and the development of new, synthetic packaging materials. Today the pharmacist must be keenly aware of a wide range of packaging issues that relate directly to the stability and acceptability of the dosage form. For example, to optimize shelf life, the industrial pharmacist must understand the interrelationships of container material properties, while the retail pharmacist must not compromise that shelf life through inadvertent underpackaging.

Traditionally, a review of pharmaceutical packaging focuses on materials science, acquainting the reader with the various types of materials used to package drug products, discussing relevant properties, and perhaps, offering some cautions to observe. This review is a departure from that tradition. Emphasis will be placed on acquainting the reader, assumed to be an aspiring or a practicing pharmacist or a pharmaceutical scientist, with information that should be part of his or her working knowledge. A packaging engineer entering the pharmaceutical industry will also find these materials useful. More esoteric or detailed reviews on the specific types of package materials can be found elsewhere in the specialized technical literature. It is my desire to focus the reader's attention on the essential questions related to the pharmaceutical package and the direction to take to gain answers to those questions.

The emphasis in this chapter will be on the appropriate packaging of major dosage forms, which includes the emergence in recent years of many new and novel forms. But these have not altered the fundamental approach to identifying the most suitable primary package materials for a particular product. Therefore, in this chapter an expanded discussion will be given to solid oral, liquid parenteral, and aerosol dosage forms because they serve as prototypes in reviewing the logic, approach, and potential pitfalls in most package evaluations. It is then possible to apply the same approach to any dosage form.

The last three decades have witnessed a sustained growth in awareness of pharmaceutical packaging. The attention has been given in many levels of detail, from research articles, to book chapters, to monographs and textbooks, to dedicated conferences, to annual conferences [1,2]. At least three factors greatly stimulated this interest, including the desire to improve the effectiveness of the dosage through increasingly sophisticated packaging forms, the increasing costs of package materials (to the point that the package may cost more then the contents), and a greater focus on packaging in the drug regulations. The regulations call attention to the importance of the package as related to product function and comment on the role of packaging in achieving a clean, safe, and efficacious product. Therefore, it is instructive to understand the key elements of the regulations as they pertain to the package system.

B. Package Definition and Function

In this chapter attention will be centered on the primary package system; that is, those package components and subcomponents that actually come into contact with the product, or those that may have a direct effect on the product shelf life. The other, outer packaging forms are referred to as the secondary and tertiary packaging and include such items as cartons, corrugated shippers, and pallets. Those types of components are covered in standard texts on packaging [3] and, with the possible exception of labeling, do not require any special consideration when applied to pharmaceuticals.

Historically, pharmaceutical packaging meant the effective containment of pharmaceutical dosage forms such that, at any time point before the expiration date of the drug product, a safe and efficacious dosage form was available. Ideally, no interaction occurred between product and package, a situation that was confirmed by the formulator through stability studies. However, a virtual revolution has occurred within pharmaceutical technology because of the intricacy and complexity both in dosage forms and in packaging forms. Although the traditional situation, as just sketched, still occurs, the package and dosage have commonly become so intertwined that the drug product must be defined in terms of both packaging and formulation. More specifically, examples such as prefilled syringes, transdermal patches, metered-dose inhalers, and nasal sprays, all contain formulations the quantity and successful delivery of which are strongly dependent on the proper function of the package system.

Therefore, a contemporary definition of *pharmaceutical packaging is that combination of components necessary to contain, preserve, protect, and deliver a safe, efficacious drug product.*

This definition is restricted, however, to the *primary* package components, that is, the components and subcomponents that come into direct contact with the product during storage and delivery. Functionally, the packaging of pharmaceuticals involves *containment* as well as *protection* from potentially damaging environmental factors such as moisture, oxygen, temperature, and light. The packaging system must also safely *transport* the product through distribution channels to the pharmacist or physician. Ultimately, in many, if not most situations, the packaging system may *deliver* the drug product to the end user, the patient. Here, the package is an integral part of the dosage because the package controls or affects the quantity of drug delivered: that is, the dosage. Therefore, today the pharmacist must be critically aware of a wide variety of packaging issues that relate directly to the stability and acceptability of the dosage form. During development, this array of relationships of container material and properties must be understood and managed to optimize shelf life, whereas at the other end of the development chain, the retail pharmacist must not compromise that shelf life through improper or underpackaging.

C. Regulatory Requirements in Pharmaceutical Packaging

Although there have been many technological advances, perhaps the most dramatic changes in the pharmaceutical industry in recent decades are due to the effects of regulations. Although emphasis in this chapter is on regulations enacted in the United States by the Food and Drug Administration (FDA), most other countries with major markets around the world apply similar regulations, especially as related to packaging [4]. Because of growing governmental regulation, an understanding of regulations on a global basis is necessary to properly set the scope for a packaging system. Furthermore, because there is a ''harmonization'' that is occurring among regulating bodies, properly designed developmental studies carried out for submission to one government agency should be acceptable for submission to that of another nation.

On the other hand, and in contrast with the simplifying effect of harmonization, additional requirements of the United States regulatory body (FDA) as perpetrated through published guidelines, the *United States Pharmacopeia (USP)*, and policy statements given in lectures and seminars have significantly increased the complexity and depth of a development plan.

Paragraph 211.94 of the Current Good Manufacturing Practice (CGMP) regulations [5], entitled ''Drug Product Containers and Closures,'' provides the most relevant statements to packaging (Table 1). These regulations, to a great extent, form the charter of the packaging function in many modern pharmaceutical organizations. From a historical perspective, however, they brought good science to management's attention, and they forced the development project team to consider the relation of the package to the product in very concrete terms. For example, paragraph (a) specifically deals with interactions that might occur between the dosage form and the container or closure. Testing must be carried out that demonstrates that the primary package components neither add to nor take from the drug. Chemical inertness must be demonstrated. Beyond the appropriateness of the package by the foregoing criteria, paragraph (b) requires that adequate protection be assured from environmental factors in the distribution system and in subsequent storage. Paragraphs (c) and (d) provide for adequate quality control procedures for both incoming materials and in-process handling. The most detailed testing for

Table 1 Paragraph 211.94 of the Current Good Manufacturing Practice Regulations for Finished Pharmaceuticals

---

Drug Product Containers and Closures

    (a) Drug product containers and closures shall not be reactive, additive, or absorptive so as to alter the safety, identity, quality, or purity of the drug beyond the official or established requirements.

    (b) Container closure systems shall provide adequate protection against foreseeable external factors in storage and use that can cause deterioration or contamination of the drug product.

    (c) Drug product containers and closures shall be clean and, where indicated by the nature of the drug, sterilized and processed to remove pyrogenic properties to assure that they are suitable for their intended use.

    (d) Standards or specifications, methods of testing, and, where indicated, methods of cleaning, sterilizing, and processing to remove pyrogenic properties shall be written and followed for drug product containers and closures.

Source: Ref. 5.

in-process handling applies to parenteral drug product packaging and is covered in paragraph (c). The receipt and preparation of package materials must meet specifications designed to assure the required quality level.

There are other sections of the CGMPs that apply to packaging. For example, Subpart E, "Control of Components and Drug Product Containers and Closures," deals with receipt, storage, testing, approval, usage, retesting, and rejection of containers and closures. Subpart G, "Packaging and Labeling Control," covers other packaging requirements, including those on tamper-evident and child-resistant packaging. Clearly, the CGMPs have much to say to the packaging scientist and to the industrial or retail pharmacist concerning the package.

The other regulatory document that has a great effect on the work of the pharmaceutical packaging scientist is the *United States Pharmacopeia/National Formulary* (*USP/NF*) [6]. Compendia from other nations, such as Great Britain, Japan, and Europe, must also be considered for a worldwide product development and launch. Because the *USP/NF* is specifically mentioned in the Federal Food, Drug, and Cosmetic Act, especially as related to drugs recognized in the *USP/NF*, the many procedures and tests presented there have the legal force of law. For example, for solid dosage forms, the sections on container tests are quite explicit. Similarly, for parenteral products, evaluation of elastomeric closures by the *USP/NF* procedures is required. As specific dosage form packaging is discussed in this chapter, the relevant portions of the *USP/NF* will be reviewed.

In summary, because of the emergence of strong regulatory concerns, each component must have adequate prior testing to assure the appropriateness of the chosen packaging system. Subsequently, shelf tests (to assure the expiration dating) and specifications (to assure continuing quality of the components) are required for the package system in which the product will be marketed. The tests and procedures outlined in *USP/NF* serve as the starting point for the development of a satisfactory package. Adequate documentation should be maintained over the course of the developmental process so that the package choice can be defended. The information gathered over the course of the package development must be effectively summarized, such that it is available for regulatory review. To aid the packaging scientist in the documentation-gathering process, the FDA has developed guidelines for submitting support documentation for the packaging of human drugs and biologics [7].

D. Nature of Package Evaluations

When developing a new product, each dosage form requires a package evaluation study to obtain a desirable expiration date. The traditional method of simply carrying out stability studies on all possible permutations and combinations of package and product variables is too costly in materials, resources, and, most importantly, time. It is essential to reach the most cost-effective packaging as quickly as possible so that applicable stability studies can begin at the earliest stage of a project.

A package evaluation is performed to investigate the physicochemical interactions that might occur between the product and the package. The ideal package would be completely inert relative to the product and would provide maximum shelf life as well. In the real world, however, interactions often occur. Therefore, the evaluation is designed to identify, characterize, and monitor these interactions to achieve a safe, unadulterated, stable, and efficacious product. The ultimate goal of the evaluation is to eliminate or control any interactions that are discovered so that they are rendered innocuous. Whereas most studies are concerned with interactions between the primary package components and the product, in cases where the secondary package influences product stability, it must also be included in the evaluation. Such a situation might be a light-sensitive product that relies on the carton to give light protection and achieve a desirable shelf life.

Protocols for characterizing various packaging forms are covered in the *USP/NF*, which can serve as a starting point, if not the entire methodology, to adequately qualify a package. For many packaging systems the *USP/NF* also establishes standards for the results of the evaluation. These protocols are grouped into two categories: characterization of the package material or container and evaluation of the package and product as a system. Included in the former are physicochemical and occasionally biological procedures to evaluate glass and plastic bottles, metal closures, elastomeric closures, flexible and blister materials, syringe components, and aerosol packaging. In the second category of guidelines, the procedures take the form of accelerated, short-term tests and long-term stability studies (i.e., characterization of the effect of the package component on the product). Because official stability studies usually require validated analytical support, they are a very costly way to reach a packaging decision. Details of the accelerated test methods are discussed in the respective sections of this chapter. Nevertheless, the *USP/NF* is an important reference for designing most accelerated methods.

The end result of the package evaluation should be a well-written technical or development report summarizing the approach taken, the results obtained, and the conclusion. In addition, copies of final specifications and quality control tests for incoming components should be incorporated in the regulatory submission to support the package system chosen.

E.   Package Specifications and Quality Assurance

An important result of a package investigation is the specification for each of the packaging components selected. The more effective the package evaluation, the more effective each specification will be in assuring the quality of the product. This is true for a simple product stability assessment of a solid dosage form, or for a more complex system, such as a particle-free, small-volume parenteral product in a glass vial with an elastomeric closure [8]. The final specification has at least two purposes: to be used as the guide to test the quality of an incoming lot of materials, and to serve as the tool for purchasing the component. The specification should include detailed material descriptions and dimensions, which serve as the target results of prescribed quality test methods. In addition, the sample plan necessary to obtain reliable characteristics of the lot of materials can also be covered in the specification. Thus, the detailed specification is a tool for effective evaluation, inspection, and release on incoming package goods. Many purchasing agents share such a specification with the vendor to avoid any confusion about expectation of component quality on arrival at the drug manufacturing plant. The specification can also serve as a purchasing document to clearly communicate to suppliers the level of quality required, as well as the liability should that quality not be met.

An alternative approach is for the purchasing and quality assurance functions to use the package evaluation test results and final specification as parts of a larger, foundational document from which an abstracted version is used for their specific requirements. Then, not only will the most appropriate tests have been elucidated, thereby saving much time and energy by eliminating useless tests, but an effective sampling plan will also have emerged. This last point cannot be overemphasized. It is impossible to estimate the millions of dollars being wasted in many quality control departments (for both the component manufacturer and the user), who allow their staff to carry out useless measurements that will essentially never identify a noncompliant lot. On the other hand, a thorough review of the well-executed developmental protocol to evaluate the best package material will often uncover a single, most important test for characterization of the true quality of the lot.

When the packaging is also a delivery system, such as a spray pump or a metered-dose inhaler, a package performance specification should also be developed. This specification should incorporate in-process and final product tests to demonstrate proper container function. Then, a decision must be made about whether these tests should be carried out during development only or after manufacturing commences. In the quality and regulatory climate of the last decade, this decision is frequently made to continue with the testing on the manufactured goods.

II. PACKAGING OF SOLID ORAL DOSAGE FORMS

A. Scope

The purpose of this section is to review the key factors involved in the selection of the package components for solid dosage forms, which includes tablets, capsules (both hard and soft gelatin types), cachets, powders, and granules. At first the packaging of such solid dosage forms may appear trivial and uninteresting, but the emergence of newer types of coatings, such as those developed for controlled-release or enteric delivery, or the development of complex solid matrices to achieve a variety of desired behaviors, including sustained-release or controlled dissolution rates, makes the determination of appropriate package components a more challenging task.

When carrying out the necessary evaluations for the selection of both multi- and unit-dose container systems, five key *technical factors* must be considered. These include product compatibility, satisfactory protection from environmental conditions, package and product integrity through the distribution system, ease of accessibility of product to the user, and resistance to children and tamperers. Beyond these factors there are other *user-dependent* and *business factors* related to design and cost of materials, respectively. For example, packaging forms available for solids, in addition to bottles, include blister and strip packaging, which are used principally for unit-dose packaging, and composite and metal cans, which are used for powders, granules, tablets, and capsules. Optimization of these technical, design, and economic factors is a major challenge, because they often conflict with one another.

Historically, the developer of a solid dosage form carried out stability tests in glass bottles that had metal screw-cap closures. If, through preformulation studies or other experience, moisture or light sensitivity was identified, a desiccant was added and the bottle used was amber glass. In fact, an amber bottle was probably used initially as a safety factor to avoid redoing the study.

Recent developments in formulation technology and shelf life simulation potentially permit quicker, clearer answers to the questions on how much package protection is required to achieve a desired shelf life. Today, with the development of modern plastics and other synthetic materials, a plastic bottle is much more likely to be chosen for a multidose container, because of its lighter weight, resistance to breakage, and lower unit cost. Table 2 summarizes some of the key economic factors to consider when choosing a plastic or glass container. Even though plastic does have economic advantages, commonly used plastics such as high- or low-density polyethylene (HDPE or LDPE), polypropylene (PP), and polyethylene terephthalate (PET) do not possess the level of barrier provided by glass (Table 3). Clearly, it is essential to understand the nature and behavior of the dosage form and the packaging form. Then, the package decision will be reliable and cost-effective.

Beyond multidose packaging in glass and plastic bottles, there are various other packaging forms available for solids. These include blister and strip packaging, used principally for unit-dose packaging, and composite and metal cans for powders, granules, tablets, and capsules. In

Table 2 Summary of Economic Factors in Comparing Glass with Plastic

	Glass		Plastic	
Container material and type	Type I	Type II	HDPE	Oriented PET
Mold cost (4-cavity tool)	$15,000–$25,000	$15,000–$25,000	$65,000–$70,000	$95,000–110,000
Bottle costs[a] ($ each)				
30-ml Boston round	0.191	0.131	0.110	0.116
60-ml Boston round	0.356	0.282	0.152	0.287
120-ml Boston round	0.513	0.310	0.161	0.294
Container weight (g)				
30 ml	54	54	6.0	7.0
60 ml	85	85	11.1	12.5
120 ml	116	116	16.0	18.0
Density (g/ml)	2.2–2.6	2.2–2.6	0.95	1.29–1.40
Chance of breakage	High	High	Low	Low

[a]Bottle costs are for orders of 250,000 units.
Source: Personal communication with plastic and glass bottle suppliers.

the remaining portions of this section these major packaging forms are discussed in greater detail, and the features needed to make the proper package selection are considered.

B. Packaging, Stability, and Shelf Life

Before discussing these features, it is appropriate to first assess the criterion by which adequate protection is judged, namely expiry. Expiry is the period of time over which the specific dosage form will adequately retain its potency, purity, bioavailability, and relevant physical properties under nominal storage conditions. Frequently referred to as shelf life, expiry varies with the specific product and often equates to the time for a product to degrade from 100% to 90% of its labeled potency. A broader definition is the time to reach 90% potency from a larger initial potency. Thus, if acceptable, shelf life can be doubled by preparing tablets with an initial

Table 3 Barrier Properties of Typical Plastic Bottle Materials

Property	Permeability constant $[(cm^3 \, mm)/(sec \, cm^2 \, cm \, Hg)]$			
Permeating gas molecule	H_2O	O_2	CO_2	N_2
Temperature (°C)	25	30	30	30
Type of plastic				
Low-density polyethylene	800	55	352	19
High-density polyethylene	130	10.6	35	2.7
Polypropylene	680	23	92	4
Polyvinyl chloride (unplasticized)	1,560	1.2	10	0.4
Polyvinylidene dichloride (Saran)	14	0.053	0.29	0.0094
Polystyrene	12,000	11	88	2.9
Polyester terephthalate	1,300	0.22	1.53	0.05

Source: Ref. 9 (p. 283, 317).

potency of 110% of the specified dosage. The range of acceptable potency in the foregoing argument is often more restricted with improved analytical methodology and a deeper understanding of the degradation profile. For example, an acceptable potency range might be 100 ± 5% or even tighter.

When maximizing shelf life through such formulation approaches, the formulator wants assurance that the packaging will not shorten the shelf life of the product. Thus a careful analysis of the relation between package properties and shelf life is important. Prediction of shelf life of a packaged product is a topic of pharmaceutical packaging that has been actively and innovatively pursued for at least two reasons. First, the time-honored approach to determining shelf life employs accelerated stability studies at elevated temperatures, along with parallel confirmatory real-time studies. Such an approach is costly in materials, possesses higher risk because of inadequate background information, and perhaps more important, is time-consuming. Second, it is generally felt that drug products are frequently "overpackaged," that is, the package offers protection of the product significantly beyond that required for the desired expiration date. This is a natural outcome when inadequate background information exists and an attempt to minimize the downside risk is made. This process will again increase the total cost of the product.

The objective of package stability studies, then, is to match package performance with needed protection, using cost-effective package materials, and to do so in the shortest time possible. The various shelf life prediction models that have been recently developed are designed to address these concerns. To develop a reasonably accurate shelf life model, it is necessary to have some knowledge of the product, the package, and the environment. Synthesizing information from these three sources and obtaining shelf life predictions began primarily in the field of food science. An excellent introduction and overview to this subject can be found in the monograph of Cairns et al. [10]. More recently, others have amplified [11–13] and applied [14–21] these earlier theories to pharmaceuticals to obtain very realistic estimates of shelf life.

A major driving force for attempting to predict shelf life is the potential inapplicability of accelerated testing for a number of drug products. The rules and assumptions for using the Arrhenius equation on the temperature dependence of a degradation reaction often break down for a solid dosage form (see Chapter 7). Consider the situation of a coated tablet in a plastic bottle, in which the various polymeric materials in both product and package might each pass through temperature-dependent transitions, (e.g., glass transitions or thermal melts). Such thermal effects cause changes in the actual physical state of the polymers and, as a result, seriously impair the reliability and predictability of the accelerated test method. These kinds of transitions render it improper to use the Arrhenius equation for rate constant versus temperature data because the mechanism of degradation under the accelerated conditions is different from that under ambient storage conditions. In the extreme, a similar argument can be made for even the pure solid phase of drug, when, as has often been discovered [22], the pure drug alone may exist in any of several temperature-dependent morphological forms. Then, not only does the crystalline structural form change, but also, conceivably, the rate and mechanism of degradation.

Conversely, an alternative shelf life prediction approach [23,24] characterized the properties of the components of the system under ambient conditions. By measuring the time rate of change of these properties with sufficient precision, a mathematical model was applied, and a shelf life was predicted. In the traditional stability study, the rate of degradation of the active ingredient is measured by a stability-indicating method, which usually involves the measurement of a small change in a relatively large quantity. For example, when following the degradation of 100 mg of active drug in a tablet, a 1% degradation represents a 1.0 mg loss in the active agent. The loss will correspond to a 1% reduction in the measured property being used to quantitate the active drug. Furthermore, variability of content uniformity, the imprecision and inaccuracies of the assay method, and other sources of error may require that a number of tablets be assayed to identify whether or not a 1% decrease in potency is statistically significant. In contrast, Gilbert and colleagues [23–25] showed that by following the appearance of a degradation product, sensitivity is vastly improved. In the foregoing example, the appearance of the degradation product corresponding to the 1% loss of active drug represents a 100% increase (from 0 to 1 mg) in the parameter quantitating a degradation product. This increase in signal/noise ratio may allow the investigator to eliminate working under accelerated conditions and, instead, determine degradation rates in real-time under normal conditions.

Workers at Michigan State University [18–21] took a different approach in shelf life prediction and employed the food-packaging concepts [11,12] to characterize the behavior of drug products at constant temperature as a function of equilibrium moisture content (EMC). Results were obtained based on each of three fundamental considerations that relate to the stability of the drug product; that is, the product, the package, and the environment. The EMC of the product was determined as a function of relative humidity (Fig. 1). Then a physical or chemical property related to dosage form stability was measured as a function of relative humidity or moisture content (Fig. 2). Given these data, a critical moisture content was defined as that beyond which the drug product no longer met specification and, therefore; had reached expiration. Separately, a third set of results, the package moisture transmission rate, was characterized under known environmental conditions (e.g., 25°C and 50%RH). From these results a mathematical model was applied to predict a shelf life under given environmental conditions. Figures 3a and b illustrate the synthesis of these three results into a mathematical prediction. Fulcoly [19] and Wang [20] carried this method to a high level of sophistication, and the predictions appeared excellent. Furthermore, both Wang and a Japanese group [14–17] extended certain elements of this approach by allowing for variability in environmental conditions to simulate more closely the actual conditions that the package may encounter during its lifetime.

One caution when applying this "food" model to predict shelf life is the underlying assumption that once a product reaches the critical moisture content it no longer meets the specification. The shelf life calculation is then reduced to time-dependent factors that are determined by the package and the environment. (See Gyeszly [26] for a thorough discussion of these assumptions.) Once the key package parameters (permeability constant, wall thickness), the environmental conditions (relative humidity, temperature), and the critical moisture content of the product are known, the model produces a prediction of shelf life. But to apply this model properly, it must be established that stability is, in fact, dependent on moisture content.

A second limitation of the model is that it is "static"; that is, it is based on a relation between the aforementioned parameters concerning package, environment, and product. Once a parameter that characterizes the dosage form, such as hardness, moisture content, dissolution rate, or chemical assay, is known or measured, it is usually taken to be unchanging or static and, accordingly, descriptive of the state of the dosage form. In fact, such properties characterize the drug product at a point in time, but they do not necessarily predict the behavior in the future. The most complete model must correlate package performance to the degradation rate of the drug product or the rate of change of other critical dosage form properties that relate to stability. This would then lead to a dynamic model wherein the kinetics of degradation, in terms of the reaction rate constants and package permeation rates, are correlated.

This analysis does not render the existing models unusable, because there are situations during which the active drug is extremely inert, but the physicomechanical properties of the dosage form change with, and are dependent on, moisture content. Under such circumstances this approach appears to work well to predict shelf life.

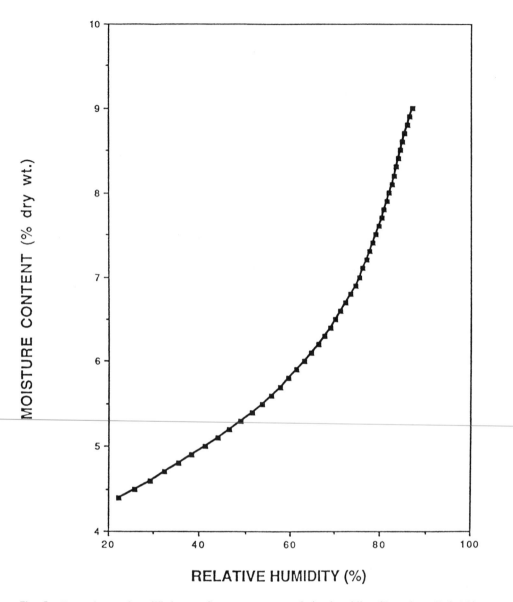

Fig. 1 Dependence of equilibrium moisture content on relative humidity. (Data from Ref. 14.)

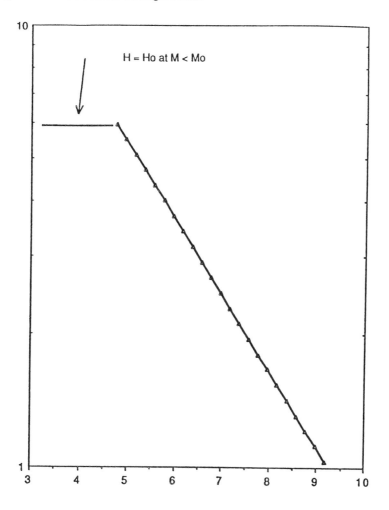

Fig. 2 Example of the dependence of a drug property (hardness, assay, etc.) on moisture content. As shown in Fig. 1, moisture content depends on relative humidity. Critical moisture content (CMC) is the moisture content at which the hardness (or other property) is outside the specification range. (Data from Ref. 14.)

C. Containers and Closures

The regulatory requirements, along with the guidelines of the *USP/NF* [6] give direction to the pharmacist concerning repackaging of solid dosage forms in multidose as well as unit-dose containers. A pharmacist should repackage a product in a container equivalent to that employed by the manufacturer of the original product. Containers should be clean and provide adequate protection to assure the identity, strength, quality, and purity of the drug product for its shelf life. The drug manufacturers are required to carry out tests that demonstrate that the specified packaging meets these standards.

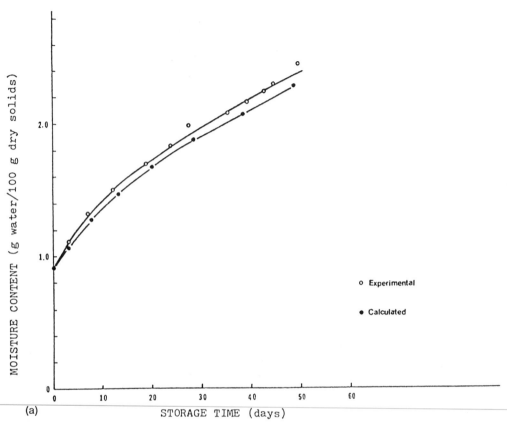

Fig. 3 (a) Prediction vs. observation of moisture content as a function of storage time at 22°C, 73% RH for a multivitamin tablet in a PVC blister.

For solid dosage forms, the packaging usually consists of a glass or plastic bottle with a closure that might range from a plastic-lined metal screw cap to an unlined plastic cap. Color of the container will depend on the specific properties of the drug. A fibrous ''coil'' is then usually added to prevent damage during shipment. The major reason for selecting a glass bottle is that it provides the closest approach to an absolute barrier in bottle form. The only potential source of gas transmission is the joint between the closure and the bottle neck. From a materials perspective, glass and bottle technologies are well known and adequately covered in the *USP/NF*, in packaging texts [3], and in technical encyclopedias [21].

Key properties of glass bottles include type of glass, shape, total volume (also known as overflow capacity), neck finish, color, and light transmission. The glass type commonly used is NP, a soda-lime glass for nonparenteral products (i.e., for products intended for oral or topical use). The most common color, when used, is amber. The *USP/NF* provides a test method to characterize the degree of light transmission for type NP containers. Broken pieces of the container with nominal wall thickness are placed in a spectrophotometer, and the light transmission is measured. The light transmission must not exceed 10% at any wavelength in the range 290–450 nm. The remaining properties are part of industry standards and should be included in the specification, along with a drawing, which is usually provided by the supplier.

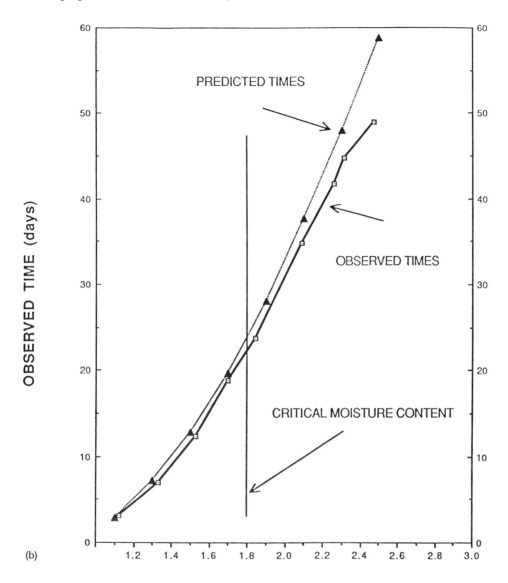

Fig. 3 Continued (b) Predicted and observed time to reach equilibrium moisture level under storage conditions of 22°C, 73% RH for a multivitamin tablet in a PVC blister. CMC was 1.8 g water per 100 g of dry tablet weight. Shelf life is estimated as 22 days in this PVC blister. Any parameter that affects the MVTR will alter the shelf life. (From Ref. 20.)

Specification of plastic bottles is more complex because the normally accepted requirement in a submission is to have stability data in the exact container-closure system. Prudence dictates that multiple bottle sizes and alternative resin sources be qualified. Fortunately, the *USP/NF* provides guidelines and test protocols for determining equivalent resins. For solid dosage pharmaceuticals the most common resin is high-density polyethylene (HDPE), although the methodology and principles involved in the testing protocols apply equally well to other common molding resins, such as low-density polyethylene (LDPE), polypropylene (PP), polyvinyl chloride (PVC), or polyethylene terephthalate (PET).

A problem that often unexpectedly arises is the discontinuance of a resin qualified for use in a package component. Because of resource constraints and pressure to develop a product in the shortest possible time, a bottle and closure may be single-sourced for both resin vendor and molder. Nevertheless, resin manufacturers do make changes for several reasons. Real examples include a change in method of catalysis, discontinuance of a process because of its environmental impact, or a natural disaster such as fire, explosion, or flooding. The simple fact is that resin suppliers cannot be relied on to provide a specified resin forever. Thus, the astute industrial pharmacist will have a solid dosage product qualified in a resin type (e.g., HDPE) supplied by a number of alternative manufacturers. An example of the need for multiple sourcing was vividly demonstrated to the author when, in April 1981, a major petroleum corporation announced that it would no longer supply a particular HDPE resin after June 1982. This announcement became more critical when that company unexpectedly announced that it would discontinue its resin business completely and close its plant by the end of January 1982. This corporation was the primary source of HDPE resin supplied to many molders of pharmaceutical bottles. To be single-sourced at that time and unable to qualify alternative resins could conceivably have shut down products packaged in that resin. To avoid this pitfall, an effective strategy is to develop the equivalent of a drug master file on plastic bottles and their resins. The file would contain a database demonstrating equivalence of several resins. The benefit of such a strategy is that it allows maximum flexibility to both formulator and purchasing agent. The file should also contain letters from the resin manufacturers and the bottle fabricators stating that all ingredients used, including additives and processing agents, conform to the requirements in the applicable sections of the Code of Federal Regulations, Title 21.

Although there may be less risk involved for a plastic bottle in combination with a solid dosage product than other dosage forms, such as liquid parenterals, it is still necessary to demonstrate equivalence between packaging materials [4]. Plastic containers with the same polymer type, while generally appearing identical, may have slightly different compositions. For HDPE, molecular composition, shape, and size may vary depending on the polymerization process itself. Branching and cross-linking will be quite different if the polymerization takes place from monomers of ethene, butene, or hexene. The resulting structures offer many property variations [28]. Numerous additives may also be incorporated into the resin to improve selected properties in processing or end use. Compounds that may be added to a resin include antioxidants, antistatic agents (i.e., antistats), colorants, mold lubricants, and more. Polyethylene is stabilized against oxidation that might occur during manufacturing and processing and, thus, prevent color change and stress cracking. Certain phenolic compounds are common antioxidants for HDPE since they meet requirements of low volatility, high molecular weight, and broad FDA approval [29]. Hydrocarbon polymers are usually nonconductive and easily acquire, but do not readily lose electrostatic charge. This charge may build up during molding or on the filling line and, since the charge is not easily dissipated, antistats are added [30]. These antistats serve to make the surface of the plastic more conductive and, hence, prevent charge buildup. The nonionic antistats such as glycerin monoesters of fatty acids and ethoxylated tertiary amines are common for polyolefins, especially since they are well accepted for food packaging. The most common colorant used in pharmaceutical grade HDPE is titanium dioxide (TiO_2). The TiO_2 renders the resin opaque white and has been reported to be chemically inert [31]. Other compounds that may be added to plastic containers include impact modifiers, lubricants, plasticizers, and stabilizers. With this many possible added ingredients, there is potential for variation between resins and resin suppliers, and the necessity for establishing equivalence is understood.

The protocol and standards for the tests to demonstrate resin equivalence are found in the current *USP/NF* [6] and include three categories of tests: chemical, spectral, and moisture barrier. The chemical tests are based on an analysis of extractions of fresh plastic bottle pieces in water, ethanol, and hexane. These tests are designed to given a quantitative assessment of the extractable substances present in a particular resin. The extracts, obtained by incubation of a specified surface area of plastic, in the appropriate solvent, at a given temperature, are then tested for nonvolatile residue, which gives the weight of substances extracted into each of the foregoing test solvents. When one considers that HDPE is a long-chain hydrocarbon, and the nature of the additives, it would be expected that more compounds would be extracted into a nonpolar solvent. This is reflected in the *USP/NF* tolerances of 100, 75, and 15 mg, respectively, for hexane, ethanol, and water. The water extract is also tested for residue on ignition, heavy metals, and buffering capacity. Leachable heavy metals could be a source of concern, since metals might react with a drug product or be toxic. This is more serious for a liquid product, but a high level of heavy metals in the extract may indicate an improper catalytic stage during polyethylene production. The buffering capacity serves to check for extraction of organic acids and bases.

The spectral analyses include multiple internal reflectance infrared spectroscopy (MIR-IR), light transmission in the ultraviolet and visible regions, and thermal analysis by differential scanning calorimetry (DSC), differential thermal analysis (DTA), or thermogravimetric analysis (TGA). The *USP/NF* is able to supply reference standards for both HDPE and LDPE, and the corresponding spectra of these standards then serve as a comparison with those of the test samples. The *USP/NF* specifies that conformance is required within certain stated tolerances. Specifically, the MIR-IR spectra of the test sample must give major absorption bands only at the same wavelengths as the reference standards when the spectra are similarly determined over the range $3500-600$ cm^{-1}. The thermograms must give endotherms and exotherms that match the standard within 6°C for HDPE and 8°C for LDPE. Light transmission standards for plastics, which are the same as those for glass, must be met only when light protection is required for the product.

The final set of standards is based on the water vapor permeation rate. Bottles that contain either a desiccant or a nonabsorbent material, such as glass beads, are sealed and weighed. The weight change is recorded as samples are stored for 2 weeks in a chamber of known environmental conditions: 20 ± 2°C and $75 \pm 3\%$ RH. Although this method is highly useful to compare resins, there are two aspects of the experimental design that should be reviewed by the *USP/NF*. First, the number of test samples used (those with desiccant fill) is *ten*, whereas the number of reference samples used (those with glass beads) is *two*. Since the final result involves subtraction of the average value of the weight change obtained for the reference samples from the weight change of each individual test sample, it can be seen that there is statistically unequal confidence between the average weight change for test and reference samples. Thus, the chance for error in the final result is increased. The final result would have greater statistical validity if the distribution of samples were, for example, *six* test and *six* reference. The second concern with the method is that the time interval for gathering of data distorts the results, such that the true permeation rate might not be measured. Because permeation initially involves dissolving into and diffusing across the plastic of the container wall followed by steady-state permeation, there is a lag period during which very little or no weight change occurs. This phenomenon is well known (see Chapter 2, Ref. 10), with the result that the *USP/NF* value is biased on the low side (Fig. 4). On the other hand, the *USP/NF* method is superior to most automated permeation measurements for two reasons: variability in molding (and, hence, variability in container walls) and the time interval for measurement. In the former case the resulting bottle variability leads to a relatively large variation in individual bottle permeation rate. Thus, a large number of test samples are appropriate. In the latter case a measurement of only a few hours to 1 day is far too brief to account for lag time. A protocol that requires incubation for 4 weeks, with measurement every week, and calculation of the permeation rate from the linear portion of the curve, would yield more reliable results.

In addition to the *USP/NF* criteria, choice of bottle size, expressed as overflow volume, is one more parameter that must be specified for the package system. An optimum bottle fill includes some headspace (i.e., the volume above the fill level). The exact headspace volume varies somewhat, depending on bottle design and the overflow capacity of the bottle. The volume of fill should be considered the untapped bulk volume of the tablets or capsules. For large-volume bottles (> 1000 ml) the headspace should be in the range of 5–10%. For a wide-mouthed bottle, the larger figure will be appropriate, whereas the reverse is true for a narrow-mouthed bottle. Because neck finish does not decrease as rapidly as volume, a headspace of 20–25% can be expected for smaller bottles (~100 ml). Table 4 summarizes typical headspace targets.

For solid dosage forms the remaining headspace volume is then filled with a pharmaceutical coil. The coil can be omitted only when the tablets or capsules are undamaged during shipping. The coil material can be chosen from cotton, rayon, and polyester. Although cotton was historically the material of choice, it is not used as much today. Rayon is now the most commonly used material because of the consistency and high purity achievable. However, both cotton and rayon can cause problems because both are known to retain moisture. Rayon, for example, can contain up to 13% moisture when exposed to 21°C, 65% RH conditions [32]. I have measured moisture levels as high as 16% at 40°C, 80% RH. Not only is the moisture level an issue, but the rate of moisture regain of dried rayon is also quite high. If rayon is heated in a warm oven to a condition of near dryness, it will regain the moisture within a matter of a few hours unless

Table 4   Fill Target and Headspace for Bottles[a]

Shape[b]	Nominal fill (ml)	Overflow capacity (ml)	Headspace volume (ml)	Headspace (% overflow)
Wide-mouth square	22	26	20	24.5
	45	44	34	22.7
	60	68	54	21.2
	120	124	99	20.2
	290	294	255	13.3
Blake	180	198	161	18.7
	500	563	507	9.9
	750	834	753	9.7
	1050	1128	1059	6.1

[a]Fill target is set at the bottle height where the curve of the shoulder breaks from the vertical line of the bottle sidewall.
[b]Bottles were stock containers from a single molder for each shape.
Source: Ref. 1.

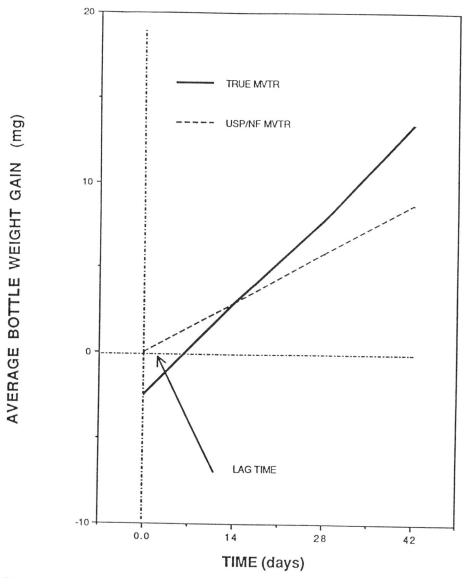

Fig. 4 Comparison of measured MVTR as compared with *USP/NF* method. The *USP/NF* method does not take into account the lag time and, therefore, underestimates MVTR. (From Ref. 1.)

hermetically packaged. Therefore packaging a rayon coil with a moisture-sensitive product in a summer, tropical environment such as Puerto Rico should be avoided, unless the atmospheric moisture level of the packaging line and the rayon are tightly controlled. Otherwise, the container will experience a "greenhouse" effect, wherein the atmosphere in the bottle becomes highly concentrated in water vapor released from the coil. Although it is the most expensive of the coils, for moisture-sensitive products, polyester is the material of choice. It has negligble moisture content and shows essentially no capacity to hold moisture.

Another component that may be added to a bottled product is a desiccant. If a rayon or cotton coil must be used for a moisture-sensitive product, the use of a desiccant counteracts the added moisture brought to the package by the coil. However, even though the desiccant may be thought of as creating a dry atmosphere in the closed bottle, that addition must be done cautiously. Once the desiccant is enclosed in a bottle and the ambient moisture absorbed, the desiccant will pull additional moisture into the container, especially for plastic bottles. Also, the physicomechanical properties of the tablet may deteriorate significantly if the atmosphere overdries the tablet. This could then cause the tablet to become excessively friable (causing easy breakage) or hard (causing poor disintegration or slower dissolution). When desiccants are used, the acceptable compounds (from a regulatory perspective) are calcium chloride and silica gel. The choice is usually made based on moisture uptake capacity, the degree of dryness required, and cost.

The final packaging component to consider for multidose solid dosage forms is the closure. The attention to closures, because of child resistance (CR) and tamper evidence (TE), has brought new life to what was a relatively stable industry. Tamper evidence should not be confused with tamper "resistance," which may be impossible to achieve in the face of malicious intent. In fact Iwaskiewicz and others at Michigan State University [33–35] showed that truly tamperproof packaging does not exist. Virtually all packaging forms can be tampered with and not detected if sufficient intent and knowledge is present. Thus, the general approach focuses on TE packaging, which helps the consumer recognize packaging that has been previously opened.

Although the closure may be a small part of the package system, it is by no means the least important. Several features of the closure must be considered to maintain the degree of security needed to achieve the desired shelf life. The principal components of the closure include (beginning from the outside and moving to the product, Fig. 5) the cap material, the liner, and the inner seal. Typically, the closure is a threaded screw-on type made of metal, such as tinplate, or of plastic, such as polypropylene (PP), HDPE, or LDPE. More important than the closure material is the liner. A typical liner is made to aid in the formation of a hermetic seal between the cap and the bottle. The liner is usually a multilayer structure consisting of a pulp or paperboard backing, a foil layer, and an inner layer of an acceptable plastic such as LDPE or PVC or even PVDC (polyvinylidene dichloride; i.e., Saran) for a high moisture barrier. The liner may be either glued to the cap or held in place by a friction fit. When the cap is fabricated it may also include an inner seal, which is added for a more positive seal surface and to provide TE. For glass bottles the inner seal is transferred to the bottle by means of an adhesive that is coated on the bottle neck or present on the inner seal as a pressure-sensitive adhesive. The same materials and methods apply to plastic bottles. For these, however, the inner seal can actually be heat-sealed to the mouth of the plastic bottle. If the inner seal structure contains aluminum foil, an extremely good hermetic seal as well as a reasonably effective TE feature (required for most over-the-counter drug products) can be produced. Of special interest is the requirement that the material on the inner surface, which may contact the product, consists of a plastic and additives that meet regulatory standards. The backing material must also comply with pertinent regulations, even though it does not come into direct contact with the product

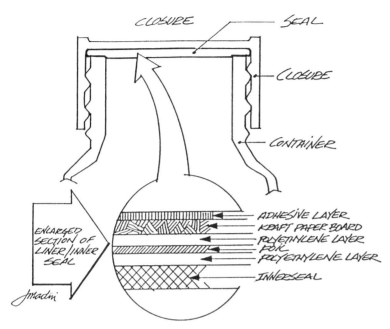

Fig. 5 Typical closure structure consisting of closure (plastic or metal), liner (usually a laminated paper structure, such as paper/foil/polyethylene) and an innerseal, which is transferred to the bottle mouth by adhesive or heat seal.

[36]. An alternative to this multicomponent closure is the linerless closure. Typically made of the same plastics mentioned earlier (PP, HDPE, LDPE), these caps are well suited when the highest level of hermetic seal is not required. Under such circumstances the plastic cap may not even be a screw-on, but rather, a lug, or press-on-type.

D. Unit-Dose Packaging

Unit-dose packaging has become very popular because of the convenience to the pharmacist, the physician, and the patient. Unit-dose packaging is also commonly used in hospital settings. The convenience in these applications has been catalyzed by the requirements of the more sophisticated dosage forms developed in recent years, such as effervescent or other very moisture-sensitive products. In addition, with many sustained-release solid dosage forms, the dosage regimen is greatly simplified. Whereas many tablets were previously taken at least four times daily, now a single tablet or capsule, taken every 12 or 24 hr, is prescribed. Therefore, the prescription may consist of far fewer dosages, in which case dispensing by unit-dose packaging is ideal.

Unit-dose packaging of solid dosage forms offers the potential for better patient compliance to the dosage regimen, a major problem for physician, pharmacist, and patient. An excellent example for which compliance is important and has been made simpler through special packaging is the oral contraceptive. Another application has come through hospital pharmacies, in which drug products are repackaged in blister or strip packages. This process has become so popular that it receives special mention in the *USP/NF* [6]. According to the *USP/NF*, any repackaging of official dosage forms into single-unit or unit-dose packages for dispensing pursuant to prescription must comply with guidelines and regulations for labeling and storage

Table 5 *USP/NF* Classification of Permeability of Unit-Dose Packages

Category	Results of *USP/NF* moisture permeation test for unit-dose packaging by method I (individual blisters)
Class A	Not more than 1 of 10 containers exceeds 0.5 mg per day in moisture permeation and none exceeds 1 mg per day.
Class B	Not more than 1 of 10 containers exceeds 5.0 mg per day in moisture permeation and none exceeds 10 mg per day.
Class C	Not more than 1 of 10 containers exceeds 20 mg per day in moisture permeation and none exceeds 40 mg per day.
Class D	Containers meet none of the requirements above
	Results of *USP/NF* moisture permeation test for unit-dose packaging by method II (blister packs)
Class A	No pack tested exceeds 0.5 mg per day in average blister moisture permeation rate.
Class B	No pack tested exceeds 5.0 mg per day in average blister moisture permeation rate.
Class C	No pack tested exceeds 20.0 mg per day in average blister moisture permeation rate.
Class D	The packs tested meet none of the above average blister moisture permeation rate requirements

Source: Ref. 6, p. 1789.

(i.e., resistance to temperature and humidity conditions). For example, the repackager cannot extend the expiration date and must ascertain the suitability of the chosen packaging material by carrying out a water-vapor permeation test similar to that of plastic bottles. It is curious that the flaw in methodology discussed for plastic bottles is not present in this protocol. In this case ten samples and ten controls are used. The results of the moisture uptake test allow classification of the package into one of four categories, as presented in Table 5.

The convenience of unit-dose packaging has been made all the more possible with major developments in packaging machinery. Several equipment makers have developed very sophisticated means of bringing the drug to the packaging line, feeding the drug into the package, detecting the presence of the drug in the package, and then sealing, labeling, and cartoning the package—all automatically, at a rapid rate, and with total quality control. At the other end of the cost scale these same equipment makers have developed inexpensive, small-scale machines that permit easy repackaging in hospital pharmacies.

There are two major classes of unit-dose packaging that are commonly used today: blister packaging and strip packaging. Blister packaging involves the formation of a plastic see-through bubble by a thermoforming process. The tablet is placed in this blister and a lidding material is heat sealed over the blister (Fig. 6). The lidding stock contains any label printing for the dosage. Typically, the lidding material is a thin foil coated on the inner side with a heat-sealable material. Thus, the tablet or blister is obtained simply by pushing it through the lid material.

Unit-dose blister packaging is considered one of the more reliable forms of TE packaging. Not only is tampering easier to detect, but it is also more difficult to tamper with a large number of dosages. Therefore, there is a disincentive to tamper with blister packaging. In addition, if the lidding material is appropriately designed, for example with a layer of mylar

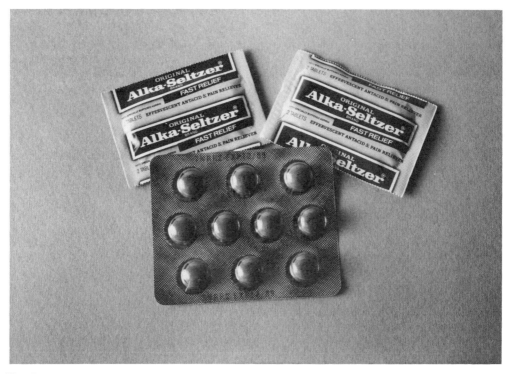

Fig. 6 Examples of unit-dose packaging: blister packaging and foil-strip packaging.

over the foil, the opening of the blister is not done by the normal procedure just discussed. Instead, the plastic layer portion of the lidding material must first be torn or peeled away to allow the dosage to be pushed through the foil. Thus the relatively rare situation is obtained of high-quality child resistance and tamper evidence in the same package.

There are several alternative materials from which to choose to form blisters. Chemical and material properties and methods of production are thoroughly covered in an excellent recent review [37]. Table 6 summarizes some of the key barrier properties of the more common materials. Factors to consider when choosing a material are barrier properties to moisture,

Table 6 Moisture Barrier Properties of Blister Films and the Effect of Blister Formation

Film structure	WVTR (mg/m^2 day^{-1}, at 40°C, 80% RH)		Percentage increase from blister formation
	Sheet	Blister	
Polyvinyl chloride (PVC)	3500	6500	86
PVC/polyvinylidene dichloride	500	1000	100
PVC/Aclar	200	250	25

Source: Ref. 1.

oxygen, and light, as well as cost. For example, a lamination of PVC and Aclar provides almost a tenfold improvement in moisture barrier protection at a 40% price increase. The effect of processing on barrier properties must also be considered. Table 6 shows that, whereas a blister formed of PVC was significantly more permeable than the material in sheet form, a PVC–Aclar blister was relatively unchanged. Blister materials can be fabricated with amber dyes to create a light-absorbent material for protection from light. Such light inhibition should be quantitated to assure adequate protection. The *USP/NF* standards for light protection by bottles are a useful guideline.

The other common form of unit-dose packaging is strip packaging, which involved two films or laminates being joined, usually by a heat seal, around the tablet or capsule. This is actually the same process as pouch packaging, but on a smaller scale. The two layers can range from an inexpensive paper/polyethylene structure, which offers very little protection, except possibly for dust, to a foil structure, which offers perhaps the ultimate in dosage protection: a nearly hermetic seal and complete light protection. If this latter packaging operation is carried out under nitrogen or carbon dioxide, even the gas space around the drug product can be controlled. A recent review has thoroughly covered both materials and processes used for this form of packaging [37].

III. PACKAGING OF PARENTERAL AND OPHTHALMIC DOSAGE FORMS

A. Scope

The packaging of parenteral and ophthalmic dosage forms presents a major scientific challenge to the industrial pharmacist or packaging scientist, because the package, the product, and the package–product interactions must all be characterized in greater detail than for solid or non-sterile liquid dosage forms. The reason for this is apparent: the consequences of incorrect judgment are more likely to have a negative, even life-threatening impact on the health of the patient. Because it is possible to bypass most, if not all, of the body's defense mechanisms with an injectable dosage form, one must have correspondingly greater assurance that extraneous chemicals are not added and that active chemical entities are not diminished or degraded in the formulated drug product as a result of the chosen packaging. In this section the major factors involved in package material selection are reviewed. Emphasis is placed on testing that reveals package–product interactions and on minimizing any unfavorable interactions that might occur. The variety of package materials and packaging forms used for parenterals has greatly expanded in recent years through the application of plastics, prefilled syringes, new types of elastomers, and other packaging innovations and conveniences. Thus, the variety and type of potential interactions have been greatly multiplied, requiring more sophisticated and sensitive analytical methods to assure that the package materials are as inert and as protective as possible.

Even though an evaluation typically beings with characterization of each primary package component by the *USP/NF* protocol—that is, a determination of the effect of water and other solvents on the package component as measured by physicochemical and biological indicators—the effect of the package component on product stability must also be determined. Other tests, related to the specific type of drug product and its known environmental sensitivities (light, air, or other) should also be carried out. For each major type of parenteral product, the investigator should consider not only the potential for these package–product interactions, but also the effect that the drug product manufacturing process will have on the materials and on the interactions. For example, if the product is given terminal sterilization, the effect of the method of sterilization must be evaluated. Finally, specific regulatory requirements, including adequate stability studies and extractables studies, should be covered with appropriate test methods.

Each of the aforementioned concerns leads to protocols that should be evaluated in detail for the specific type or form of the product. These forms include solids, such as freeze-dried and powder-filled, and liquids, such as solutions and emulsions in the form of small- and large-volume parenterals and ophthalmics. Because many of the issues that arise are similar, regardless of whether the product is solid, liquid, or emulsion, we will focus on the most common parenteral packaging, the sterile liquid dosage form. Furthermore, although reference is made to parenteral products, it should be understood that, where appropriate, ophthalmic products are included in the discussion.

A valuable first resource with which to plan the packaging evaluation is, as with solid dosage forms, the *USP/NF* [6]. Detailed protocols are given there that can be applied to both the drug product and its packaging components. Several other references are available to guide the packaging scientist to effective methods for carrying out evaluations for parenteral products, including reviews [38–41] and bulletins from vendors [42] and professional societies, especially the Parenteral Drug Association [43–45].

B. Regulatory Requirements

From a regulatory perspective, the focus of the package qualification program must be on demonstration that the proper tests were carried out and that satisfactory results were obtained. All pertinent documentation should be included in a master file or development report that accompanies the regulatory submission. The content of the documentation must confirm that the chosen sterile product packaging form satisfactorily meets the stated objectives and standards. This means that results from package evaluation studies on the specified components are presented and include results from the basic *USP/NF* test protocols and from any tests specifically designed for the particular product–package combination. An FDA representative once stated that the *USP/NF* ''container/closure test results on plastic containers and rubber or plastic closures are acceptable'' [46]. However, since that time the testing protocols have become more rigorous. For example, it is now common to carry out studies that assay for extractables in the product formulation. This topic is of such concern to the agency and to the pharmaceutical industry that an entire conference was devoted to identification and analysis of extractables derived from packaging materials and components [47]. From an understanding of the chemistry of the type of packaging material involved (elastomer, plastic, glass), analytical tests for the presence of specific contaminants, such as polynuclear aromatics (PNAs), nitrosamines, and mercaptobenzothiazole (MBT), are often carried out [47]. Letters referencing vendor drug master files (DMF) should also be present in the summary documentation for each primary package component and material. There must be assurance not only that the final package interacts minimally with the product, but that all ingredients used in fabricating the component are acceptable. This is particularly true for elastomeric components, for which the degree of polymerization and cross-linking are known to vary considerably.

Stability data should be present or referenced that demonstrate the long-term safety and adequacy of the package components. This portion of the file is most important because slow permeation and leaching phenomena can cause the chemical composition of the product to change substantially over a period of 6–12 months. There are additives from closures and metal ions from glass that can have a dramatic effect on some drug products. It has been suggested, and is common practice, that all containers of liquids, including small- and large-volume parenterals, should be stored inverted and upright to assess the effect of long-term contact with the closure [46].

Since the container integrity often involves different materials at the closure container interface, both the short- and long-term adequacy of the seal must be demonstrated. Results proving satisfactory seal integrity should be gathered at reasonable intervals throughout the stability studies and included in the file. The principles involved in stability of a parenteral product are somewhat different from those for nonsterile oral products. Whereas in the latter case a continual degradation of drug product to a specified lower limit is permissible (as long as the degradation products are not toxic), for parenteral products, absolute sterility must be present even at expiry. A 95% sterile product is not an acceptable concept. Therefore, the test protocols and presentation of results in the master file must be correspondingly rigorous in demonstrating supportive long-term container integrity.

C. Containers

The containers used for parenteral products encompass a variety of materials (Figs. 7a and b). Quite commonly, and indeed more so than with oral solid dosage forms, parenteral and ophthalmic dosage forms use glass containers. On the other hand, plastic containers have gained wider acceptance. For parenteral products, especially liquids, the interaction potential between product and container is significant because the product is in continuous, intimate physical contact with the container. Also to be evaluated are the enhancing potential and destructive effects of the component cleaning and sterilization processes.

Glass

Even though it is one of the oldest container materials known, glass is still the most common packaging material for liquid parenteral products because of its inherent stability and impermeability. Glass-packaging components are made by one of two processes: they are either blown in molds, or formed from glass tubing. Containers made by these two methods of manufacture are readily distinguishable (see Fig. 7b): the blown container has a seam line running from the top finish to the bottom of the sidewall and possibly an embossed marking on the bottom indicating the mold number and the glass molder. The tubing container, on the other hand, has smooth, seamless sidewalls and no bottom markings. Tubing glass can be used for such package forms as ampoules, prefilled syringes, and vials. Blown glass is used mainly for vials and bottles.

Even though glass is considered to be a very inert and stable packaging form, there are many examples of the leaching and corrosion of glass surfaces in contact with water or other aqueous solutions [27,48]. In addition, the mechanical handling of the glassware, at both the glass manufacturer and the bottler, can damage the surface externally and internally. As a result, glass containers manufactured for food and drug purposes frequently receive some type of surface treatment. These treatments are carried out by the glass manufacturer in the vicinity of the annealing oven, also known as the lehr. At the front end of the lehr, the glass temperature is considerably higher than the back end; hence, treatments applied before the annealing lehr are called "hot-end" treatments, whereas those applied after the lehr are called "cold-end" treatments. The treatments achieve any of three purposes: increased durability of the glass during handling, increased resistance to chemical corrosion after filling with product, and increased lubricity of the glass on the production lines. Furthermore, some of the treatments are temporary and removable before product filling.

During one type of hot-end process, the hot glass passes through a vapor hood only seconds after the component is formed. Inside the hood a vaporized metal compound, usually tin or titanium chloride, is blown on the outside of the containers to create a thin film of metal oxide

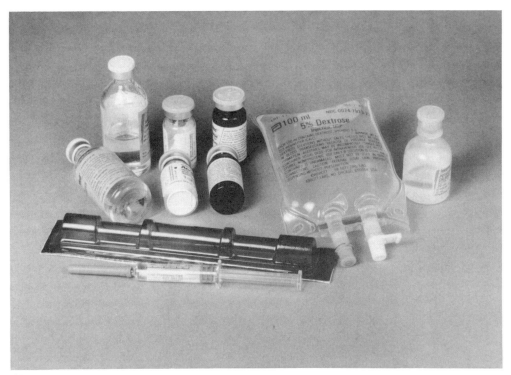

(a)

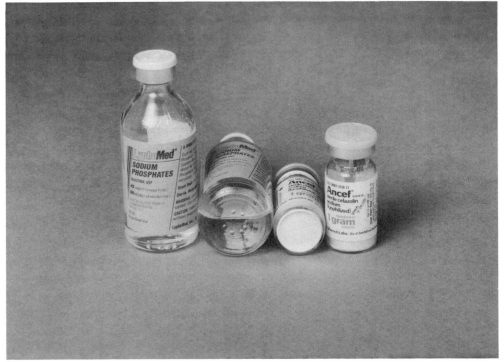

(b)

Fig. 7 (a) Examples of parenteral packaging. Materials shown include plastic, rubber, and glass. Amber blister is designed to give adequate light protection for the prefilled syringe. (b) Closer view of glass packaging, illustrating the two types of processes, blown glass and tubing vials. The tubing vial has no visible markings on the bottom, whereas the molded vial has a parting line and other mold markings.

bonded to the glass surface. This hot-end treatment then serves as a base coat or primer for the cold-end treatment applied at the other end of the annealing lehr. There are four lubricants commonly used for the cold-end treatment: a stearate spray, an oleic acid vapor, a polyethylene spray, or a Surlyn (a brand of ionomer) spray or roller coat. Vendor literature [49] provides advice on the best choice for a particular application. For example, as shown in Table 7, the polyethylene coating provides poor label adhesion, but excellent abrasion resistance, whereas the Surlyn coating gives good label adhesion, but only fair abrasion protection. This pair of treatments, at the hot and cold ends, respectively, appear to be synergistic by providing the retention of bottle strength far greater than when used alone.

Another hot-end treatment, applied internally to enhance the chemical resistance of the inner surface of the glass, is available. Here a small quantity of a dilute solution (e.g., ammonium sulfate) is sprayed into the container as it enters the annealing lehr. The temperature, in the range 1000–1200°F (550–650°C) induces the following possible chemical reactions [50]:

$$(NH_4)_2SO_4 \rightarrow (NH_4)HSO_4 + NH_3$$

$$Na_2O + (NH_4)HSO_4 \rightarrow Na_2SO_4 + NH_3 + H_2O$$

The result of these reactions is the conversion of alkali oxides (e.g., sodium) in the surface layer of the glass to the sulfate salt (sodium sulfate), which later appears as a grayish bloom on the inner surface of the ware. Since sodium sulfate is highly soluble in water, the salt is flushed out in the normal wash cycle as the glass is being prepared for filling. An alternative internal treatment to increase chemical resistance is the fluoride treatment. Again the strategy is to add a compound that forms soluble compounds with undesirable, leachable metals in the glass, such as calcium, barium, manganese, and sodium, among others. These fluoride salts are easily water-washable when preparing the glass for filling.

The glass companies have provided extensive data to customers and to the FDA to demonstrate that these treatments are acceptable for packaging food and drugs. Nevertheless, the packaging scientist must be aware of the existence and application of these treatments, since it is quite possible that the surface will have been treated and not mentioned on the vendor specification. Although safety is not an issue as a result of the treatment and the chemicals used to treat the surface, there may be consequences either on product stability or in the formation of particulates. Therefore, specifications should be explicit on what treatment, if any, is acceptable and present.

The glass usable for pharmaceutical (including parenteral) packaging has been classified by the *USP/NF* into types based on the capability of resisting hydrolytic attack (Table 8). However,

Table 7 Evaluation and Ratings of Cold End Exterior Glass Treatments

Description	Results			
Type of coating	Stearate	Oleic acid	Polyethylene	Surlyn
Method of application	Spray	Vapor	Spray	Spray/roll on
Lubricity				
Dry	Good	Good	Excellent	Fair
Wet	Poor	Good	Excellent	Fair
Label adhesion				
Dry	Poor	Good	Poor	Good
Rinsed	Excellent	Good	Poor	Good
Abrasion protection	Fair	Good	Excellent	Fair

Source: Ref. 49.

Table 8 *USP/NF* Glass Classifications

Type of glass	General description	Type of test[a]	Overflow capacity	Volume of acid (ml)[b]
I	Highly resistant borosilicate glass	Powdered glass	All sizes	1.0
II	Treated soda–lime glass	Water attack	100 ml or less	0.7
			Over 100 ml	0.2
III	Soda–lime glass	Powdered glass	All sizes	8.5
NP	General-purpose soda–lime glass	Powdered glass	All sizes	15.0

[a]Detailed procedure is given in Ref. 6.
[b]Volume of 0.020 N sulfuric acid to titrate to a methyl red endpoint.
Source: Ref. 6.

careful examination of the published literature reveals that there are many more "types" of glass than listed in the *USP/NF* [27,50]. Table 9 lists the chemical composition of several common type I and type III glasses. Type II glass, which has the same composition as type III glass, receives an internal surface treatment at the glass fabrication plant that renders the glass more inert than type III. As can be seen in Table 9, there is a significant variation in composition, even within a given type. This is in part related to the requirements of the two manufacturing processes mentioned earlier. The chemical composition is designed to facilitate that process. Glass manufacturers then designate the glass with a code that reveals both composition and processing method, which clearly reveals more information about the glass than the *USP/NF* nomenclature.

Because of the chemical differences in glass within a given type, it is important to take these compositional factors into account when choosing and testing glass containers. It may even be possible that particle-free product can develop insoluble precipitates with time. For example, if a product contains anions such as phosphate or sulfate, these anions can combine with certain divalent trace metals leached from the glass after a time to form insoluble particulates (e.g., calcium phosphate). A good method to use to assess such long-term effects is to force the reaction by multiple autoclavings. Each autoclaving is followed by a particulate inspection by a trained expert. Work done under the author's direction clearly demonstrated that the time to observe particulates could be accelerated by autoclaving and that the rate of appearance of particulates varied significantly with the glass composition.

Another potentially serious problem of glass containers is the ability to absorb active drug on the surface and, thereby, reduce potency. A well-known example of this phenomenon is insulin [51]. Any peptide or nonpolar product should be considered as a potential adsorbent. This problem might be solved by a postmanufacture treatment with special silicates that render the glass surface nonwettable and reduce adsorption.

A property of glass that has been considered useful is its processibility; that is, its ability to withstand cleaning, filling, sterilizing, and other processing and handling methods. Again caution is advised as studies [52] have shown that glass may not be as "clean" as originally thought. The mechanical shocks and other "stresses" alluded to earlier have led to the observation of glass flaking [53]. This is a potentially serious problem if glass particles are generated that are small enough to pass through a needle, into the syringe, and then be injected.

In summary, even the relatively inert glass container cannot simply be taken off the shelf to package a parenteral product. Today, high standards (especially for parenteral products), advanced tests methods, and well-developed protocols are available to discriminate among glass types and find the glass best suited for a particular application. We have seen that performance,

Table 9 Chemical Composition of Glasses Used for Pharmaceuticals

Vendor glass type	R-6	CA-2	800	Amber 900	KG-33	KG-35	N-51A	Amber RN-3	Amber 203	NSV400	NS-51	NS-33	NSV	Amber NSV500
USP/NF glass type	III	III	III	I	I	I	I	I	I	I	I	I	I	I
Process T = tubing M = molding	T	M	M	M	T	T	T	M	T	M	T	T	T	M
Oxide														
SiO_2	68	73	72	73	80	69	71	67	69	70	73	81	73	66
B_2O_3	2	0	0.5	0.5	13	13	11	9	10	10	10	13	10	9
Al_2O_3	3	3	2	2	3	6	7	6	6	6	6	2	6	7
Fe_2O_3	0	0	0	0.2	0	0	0	1	1	0	0	0	0	1
ZnO	0	0	0	0	0	0	0	0	0	0.5	0	0	0	0.5
TiO_2	0	0	0	0	0	0	0	0	3	0	0	0	0	0
MnO	0	0	0	0	0	0	0	6	0	0	0	0	0	6
BaO	2	0	0	0	0	2	2	1	2	2	2	0	2	1
CaO	5	10	8	9	<0.1	1	1	2	1	1	1	0	0.5	0.5
MgO	4	0	3	1	0	0	0	0	0	0.5	0	0	0	0
Na_2O	15	13	14	14	4	8	6	7	6	9	6	4	7	8
K_2O	1	1	0	0	<0.1	1	2	1	2	1	1	0	1	1

Source: Ref. 45.

both physical and chemical properties, can be optimized with a thorough understanding of the glass formulation and its method of fabrication and processing, including surface treatments. With these variables in mind, one can select and develop the most appropriate glass container, whether it be a vial, ampoule, bottle, or prefilled syringe.

Plastic

The application of plastic packaging to parenterals and ophthalmics has continued to broaden. Modified resin and new resin development, improved methods of molding and assembly of components, and added dimensional precision have opened the door for many new and novel packaging forms to be developed. Included for both large- and small-volume parenterals are plastic bags, plastic vials, form–fill–seal containers, as well as other forms of plastic components, such as plastic syringes, tubing for infusions, and filters.

The two most pressing problems that plastic packaging must overcome include full characterization of the effects of direct product–package contact and the ability to carry out effective product inspection. Other additional considerations for component qualification include material purity, degree of clarity, material stability, stress cracking, effect of light and high temperature, and potential for interactions, permeability, and leaching of extractables. Thus, the emergence of plastic packaging means that many of these technical problems have been solved along the way.

The strategy for selecting plastic materials, similar to that for glass should begin with protocols based on *USP/NF* chapters ⟨661⟩ Containers, ⟨87⟩ Biological Reactivity Tests, in vitro, and ⟨88⟩ Biological Reactivity Tests, in vivo [6]. If the results of these tests meet the *USP/NF* standards, general suitability and safety can be claimed. However, beyond these results, testing should be completed to evaluate for (a) product–package compatibility (preferably with an accelerated protocol), (b) extractables into product, (c) functionality testing (if appropriate), and (d) stability testing. In qualifying plastic components, one must be particularly aware of the potential for drug–plastic interactions [54]. Leaching of extractables should be approached with highly sensitive methodology because of the potential existence of small quantities of unpolymerized monomers or other additives. Monomeric components, such as vinyl chloride, acrylonitrile, styrene, and benzene, that have known human toxicity and that may be present even after polymerization is complete, must be eliminated from the converted plastic to assure adequate safety.

A given plastic material may arise as a parenteral packaging candidate on the basis that the plastic is used for food packaging. This implies the existence of a solid body of safety and physicochemical data supporting the use as proposed. However, risk is added that injectables and ophthalmics may bypass the normal body detoxification system, as compared with food ingestion. Thus, whereas the same plastic may be suitable for food packaging, greater caution must be taken for this kind of drug packaging; therefore, adequate data must be generated. An additional factor to be considered for plastic packaging of parenterals and ophthalmics is the ability to effectively inspect filled containers and demonstrate acceptably low levels of particulate formation [55]. Despite these challenges, plastics continue to grow as a form of parenteral packaging.

D. Closures

With the exception of ampoules, all the glass and many of the plastic parenteral product containers that have been discussed require a "closure." That closure is usually made of a rubber elastomeric material. These same elastomers are also used with prefilled syringes for plungers, needle shields, and barrel tips. It has already been shown that a significant amount of evaluative testing must occur for the other packaging forms of parenteral products, but in general, these are usually less substantial when compared with the level of rigor necessary for approval of rubber components. The reason for such detailed testing is the complexity in composition and manufacture of elastomeric components.

The purpose of this section is to present the basics of an elastomer formulation and to describe a responsible testing program to assure appropriate component selection for the product–package system. A rubber stopper will be used as an example of a generalized elastomeric component as the manufacturing process is reviewed. This will aid in comprehending the potential problems and, therefore, the reasons for thorough test protocols. There are hundreds of references and several excellent reviews that go into greater depth in discussing the nature of stopper formulations and their evaluation as used in the pharmaceutical industry, and which can provide further insight and detail in assembling an appropriate development plan [1,56].

The process whereby the rubber compound cross-links or cures is known as vulcanization. This process is responsible for the change in properties from individual ingredients to a blended, cured elastomer. Each ingredient is present for some specific function, either in the vulcanization process or to achieve a certain final property of the elastomer. Typically, a stopper formulation consists of a recipe of some 6–12 ingredients, some of which may be toxic in the free state. The stoppers are made by a batch process wherein ingredients are added in a prescribed amount and sequence and blended. Examples of the types of chemicals used in rubber formulations are shown in Table 10. Not every type of chemical is found in all formulations, and the amounts of the ingredients vary widely, but some typical formulations are summarized in Table 11.

An important point relative to elastomeric packaging components is that no single rubber formulation will fulfill the requirements of all possible applications. Therefore, most often, the formulation is designed for a specific functional purpose [58]. Because of this uniqueness and specificity, stopper manufacturers do not release details of their formula to customers, which accentuates the need to do adequate testing of an elastomer for a given product both before approval, and after approval on an ongoing basis.

Typically, rubber stoppers are manufactured by first milling the elastomer, which softens because of the friction of milling. The other ingredients are blended together by a kneading process, with the vulcanizer and accelerator compounds added last. The blend is then extruded and cut into pellets that are vulcanized under elevated temperature and pressure in a mold of the desired shape. This curing process causes the formation of three-dimensional cross-linked networks of the rubber polymeric chains. Furthermore, the original tackiness of the rubber is lost as it changes to a relatively inert matrix that should be resistant to attack by solvent, light, heat, and oxidation.

Key control parameters in this process include cure time and amounts of ingredients. Unfortunately, not all desired final properties of the stopper reach their optimum at the same time, or with the same ratio of compounds. Ingredients are not simply added in prescribed stoichiometric amounts and allowed to react until completion, as one would, for example, in a classic organic chemistry reaction. Therefore, some choice must be made, based on the end use of the stopper, as to which of these changing properties to optimize when the vulcanization reaction is halted. An additional complication is that, unlike glass, which remains extremely stable once processed, a rubber formulation will continue to ''ripen'' (i.e., undergo further chemical reaction, albeit slowly) long after the formal curing process has ended.

The thorough qualification of a rubber stopper, therefore, must take into account the following factors: basic properties of the cured elastomer (physical, mechanical, chemical, and biological); compatability with the drug product; effect of drug processing (washing, lubrication, sterilization); ability to effectively contain the product (seal integrity, stability); subsequent

Table 10 Main Ingredients of a Stopper Formulation

Functional ingredient	Chemical type or example
Elastomer	Polyisoprene
	Ethylene propylene/dicyclopentadiene copolymer
	Styrene/butadiene copolymer (SBR)
	Polybutadiene
Vulcanizing material	
Vulcanizing agent	Hexamethylenediamine carbamate
	Sulfur
Accelerator	Copper dimethyldithiocarbamate
	2-Mercaptobenzothiazole
	Zinc 2-mercaptobenzothiazole
	Ziram (zinc dimethyldithiocarbamate)
Retarder	Cyanoguanidine
	Phthalic anhydride
	Salicylic acid
Activator	Diethylamine
	Fatty acids
	Glycerin
	Glyceryl monostearate
	Magnesium stearate
	Stannous chloride
	Triethanolamine
	Zinc oxide
	Zinc stearate
Antioxidant	Butylated hydroxytoluene (BHT)
	2,2'-Methylenebis(4-methyl-6-t-butylphenol)
Plasticizer	Butyl stearate
	Castor oil
	Dibutyl or dioctyl phthalate
	Epoxidized linseed oil
	Lanolin
	Mineral wax
	Petrolatum
	Polypropylene glycol
	Terpene resins
Filler	Aluminum or magnesium silicate (clay)
	Barium sulfate
	Calcium carbonate
	Calcium silicate (talc)
	Carbon black
	Silica
	Titanium dioxide
	Zinc oxide or sulfide
Colorant	Chromium oxide
	Iron oxide
	Titanium dioxide
	Zinc chromate
Lubricant	Paraffinic wax
	Polyethylene
	Silicone oil
	Sodium stearate
Emulsifier	Fatty acid salts
	Sodium lauryl sulfate

Source: Ref. 43.

Table 11 Examples of Rubber Component Formulations

Elastomer type	Ingredient list	Purpose	Relative amount (arbitrary units)
Natural rubber	Natural rubber	Elastomer	100.0
	Zinc oxide	Activator	5.0
	Stearic acid	Activator	2.0
	N-t-Butyl-2-benzothiazole sulfenamide	Accelerator	0.7
	HAF black	Reinforcing agent	35.0
	Sulfur	Curing agent	2.25
Ethylene propylene diene monomer (EPDM)	EPDM	Elastomer	100.0
	Zinc oxide	Activator	5.0
	FEF black	Reinforcing/pigment	70.0
	SRF black	Reinforcing/pigment	35.0
	Naphthenic oil	Extending oil	100.0
	Paraffin wax	Dispersant mold release	1.0
	Stearic acid	Activator	1.0
	Mercaptobenzothiazole (MBT)	Accelerator	0.8
	Tetramethyl thiuram disulfide (TMTDS)	Curing agent	0.8
	Thiazole/carbamate	Accelerator	0.8
	Sulfur	Curing agent	0.8
Butyl rubber	Butyl rubber	Elastomer	100.0
	GPF black	Reinforcing agent	70.0
	Parafinnic process oil	Extender	25.0
	Zinc oxide	Activator	5.0
	Sulfur	Curing agent	2.0
	TMTDS	Accelerator	1.0
	MBT	Accelerator	0.5

Source: Ref. 57.

handling of the package (needle penetration force, coring, resealability); and lot-to-lot varia-bility. There have been some excellent reviews [1,38–44,56,58,59] to aid the investigator in optimizing component selection, specification development, and incoming quality assurance.

As for many of the components discussed earlier the *USP/NF* offers a good starting point for test protocols. Both biological and physicochemical test procedures are specified in the compendium. The former include acute systemic toxicity and intracutaneous reactivity tests on saline solution, cottonseed oil, and drug product solution extracts of the stopper, at a volume of 1 ml of extractant per 1.25 cm^2 of stopper surface area. Other biological tests useful in qualifying an elastomer formulation include pyrogen and hemolysis tests. The physicochemical tests are done on extracts in purified water, drug product vehicle (where applicable), and iso-propyl alcohol, at a surface area/volume ratio of 0.5 (cm^2/ml). The tests run on the extracts include quantification of turbidity, reducing agents, heavy metals, pH change, and total extract-ables. I have found that increasing this ratio fourfold, to 2.0 cm^2/ml, provides better discrim-ination when a battery of stoppers are being initially screened for optimum compatability with a particular drug product formulation. Once an optimum elastomer has been chosen, the full *USP/NF* protocol should be repeated on several different lots of stopper samples to assure the consistency of the stoppers and of the results. Other physicochemical tests that are highly recommended [43] to characterize the extracts include the spectrophotometry protocol outlined in Fig. 8. Such a thorough investigation of all possible extractables gives good assurance that subsequent problems will be avoided when production begins.

IV.   PACKAGING OF SEMISOLIDS AND TOPICALS

A.   Scope

Broadly defined, the packaging of semisolids consists of the packaging of oral liquids (solu-tions, syrups, emulsions, and suspensions) and products used for topical applications, such as ointments, gels, emulsions, suspensions, creams, and suppositories. Because of the emphasis on topical drug delivery, a brief review of transdermal device packaging is also part of this section. Most often the semisolids used for topical purposes are packaged in collapsible tubes, jars, glass and plastic bottles, with optional dispensing closures, which may include flip-top caps, dropper assemblies, medium-viscosity pumps (as used for products such as hand and body lotions), and high-viscosity pumps (as used for tooth gels). Pouches of assorted materials and shapes are commonly used for single-dose and sample dispensing.

Each packaging form must be carefully evaluated in light of the dosage form to give guid-ance for the direction of the package qualification. For example, as summarized in the previous section, packaging of sterile liquids for injection requires a very thorough test protocol. The packaging of most semisolids involves test protocols that are closer to those discussed for solids. The main technical issues involve adequate demonstration that interaction between con-tact surface and product are not significant and that product containment and protection is adequate. The challenge to the package developer for these types of dosage forms is the need to devise tests that will be effective in discriminating among material choices on an accelerated basis so that real-time stability studies can be initiated at the earliest point in the project. When properly done, such protocols will yield reliable results that shorten developmental time, reduce project costs, and minimize material costs.

B.   Liquids and Semisolids

Bottles and Jars

Packaging of semisolids and oral liquids in bottles involves test protocols similar to those for solids. The standards for selecting liquid packaging are also similar to those for solid oral

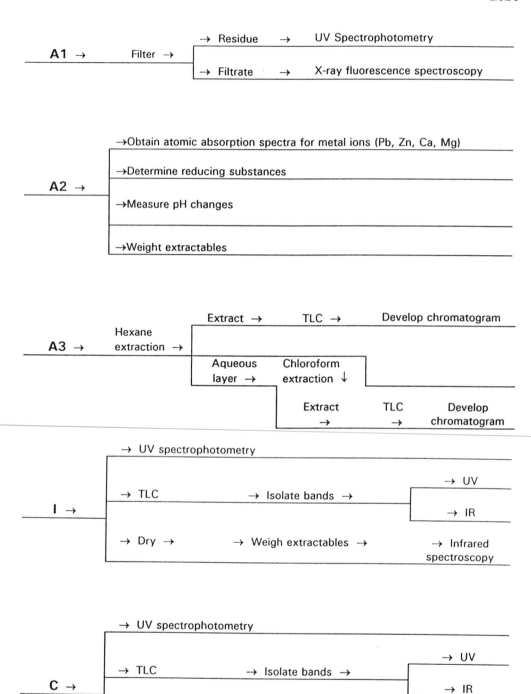

Fig. 8 Analytical scheme for characterizing rubber component extracts. A1, A2, and A3 are extracts prepared by 2-hr autoclaving of stoppers in water. I is an isopropanol extract obtained by 30-min refluxing of stoppers in isopropranol. C is a chloroform extract prepared by 24-hr Soxhlet extraction of diced stoppers. (From Ref. 43.)

dosage forms. In fact, the recent *USP/NF* XXIII contains sections dedicated to dispensing liquid (oral) dosage forms [6; Chap. ⟨661⟩, p. 1784]. That is, use type NP glass bottles or plastic bottles of resin that restricts permeation to an acceptable rate. The *USP/NF* [6; Chap. ⟨661⟩, p. 1781] specifically mentions that "Containers of Type NP glass are intended for packaging nonparenteral articles, i.e., those intended for oral or topical use." One caution to applying this standard is that the glass chosen should be compatible with the formulation on a long-term basis. For example, as mentioned earlier in the discussion of parenteral packaging, a phosphate-buffered solution packaged in a glass that has a high level of leachable divalent metals can form insoluble phosphates. Although this reaction may not produce a health-related contaminant, as it would for parenterals, the appearance of a precipitate might create questions concerning the purity of the dosage form. When plastic bottles are used, the *USP/NF* standards for moisture should be applied. Here, however, weight loss to a dry environment should be characterized instead of weight gain in a moisture cabinet. The *USP/NF* extraction tests should also be performed on any plastic material considered for these dosage forms, and if possible, an additional extraction profile into the product formulation or product placebo (drug vehicle without the active) should be carried out. As with all other plastic components, it is important to qualify alternative resins and vendors to allow maximum flexibility in resin and supplier sources.

Closures and Pumps

One must also assess the degree of interaction, if any, with the closure and, if present, the liner. A property that should be obtained is the "leakage" rate; that is, the rate of moisture loss (obtained as weight loss) of the bottle, when the closure has been applied to the container at the same torque as that obtained on the production line. Leakage reflects moisture loss through the closure–container juncture, rather than through the bottle material itself, which is permeation. The control sample for this test is an identically prepared container, sealed (e.g., with a foil laminate; see *USP/NF* [6; Chap. ⟨661⟩, p. 1786 and Chap. ⟨671⟩, p. 1788]), such that the only moisture loss possible is through the walls. Other considerations for evaluation of product compatibility with the closure include stress cracking for plastic closures and corrosion for metal closures. When a liner is used, food-grade materials compatible with the drug product must be used.

Generally, the flip-open dispensing closure is a linerless design of either polypropylene or high-density polyethylene. These dispensing closures should be tested and selected by the same protocols as those just discussed.

Another form of dispensing closure for liquids or semisolids is the pump. Three classes of pumps are most prominent [1]. First are spray pumps that produce a spray or mist from liquid products. Common products include throat and nasal sprays. These will be discussed in more detail in the following section on aerosol and spray packaging. The second type of pump is the dispenser pump for discharging viscous liquids and creams. Hand and body lotions are one type of product that use these closures. Both spray and dispenser pumps consist of several subcomponents that come into contact with the product. Most prominent are the metal ball and spring, the rubber gaskets and seals, and the different plastics for the other parts. Clearly, a thorough compatibility study should be carried out, involving individual subcomponents as well as assembled pumps, before final approval of the pump system is given. The third type of pump is the somewhat newer high-viscosity dispenser used initially for dentifrice compounds and more recently for ointments and skin creams. These units operate like a displacement pump wherein the user forces out product while depressing a lever, and the resulting vacuum in the container pulls a piston up the product-containing tube body. The high-viscosity dispensers are usually entirely plastic in product contact areas, but one must be aware that the product is exposed to some degree to the environment once the package is opened, whereas in the fore-going dispenser pumps, only the product left in the actuator button is exposed. At this end of the packaging spectrum, the OTC marketplace is prominent and other convenient and inter-esting packaging forms dedicated to consumer convenience have been developed. A variety of such systems came to the dentrifice marketplace in the last decade as the perception arose of user discontent with the collapsible tube. The penalty for such a change was cost. More re-cently, however, examination of the toothpaste shelves demonstrates a return to the functional, and less costly, collapsible tube.

Another form of dispensing closure is the dropper assembly, consisting of a dropper bulb, a pipette, and the closure itself. Once again these subcomponents might be made of, and usually comprise, different materials. For example, the bulb may be one of a variety of elastomeric rubbers, the pipette may be glass or polyolefin, and the cap might be PP or HDPE. If the finished drug product is manufactured with the dropper indwelling, appropriate weight loss and extraction studies should be carefully carried out.

Still another form of dispensing closure system is the dropper tip. The materials involved may differ from the bottle and appropriate compatibility studies should be carried out.

Unit-Dose and Pouch Packaging

A frequently used application of pouches for packaging of semisolids is to create unit-dose or sample packages. Pouches of assorted materials and shapes are also often used for single-dose dispensing at pharmacies. Owing to structural and material similarities, the same types of chemical interactions may take place in pouches as in tubes, especially laminated tubes, for which the chemical nature of the product will usually dictate which material is best used.

For unit-dose packaging of liquids and semisolids the same guidelines and requirements should be followed as listed under solids [6; see Chap. ⟨661⟩]. This is true whether the unit-dose packaging takes place when the product is manufactured or when it is repackaged by the local pharmacist, such as in a larger hospital pharmacy. A typical protocol might include weight loss, burst force, product analysis, and analysis of the package material in contact with the product. The package analysis should be done both instrumentally (e.g., spectrophotometrically or colorimetrically) and visually. When possible, a reasonable elevation of temperature, for example to within the range 40–70°C, may be appropriate to accelerate any deteriorative reactions. Weight loss and burst force studies monitor the resistance of the heat seal to product attack with time. The product and package analyses will detect migration into the product or into the plastic of the package, respectively. Because of the potential for metal interaction and the necessity for a heat seal, with pouches an organic coating over the metal is present. If there is moisture or any volatile chemical in the product, the investigator must carefully observe the adhesion of the organic lining to the metal substrate. In time, smaller molecules diffuse across the lining, and if corrosion or liner attack occurs, loss of adhesion to the metal would result.

Frequently, the experience of the package supplier can yield material candidates with a very good probability of success, provided that sufficient detail concerning the formulation is pro-vided. Ideally, stability tests can then be initiated on samples prepared at the earliest stage of the project, preferably on production equipment or pilot plant equipment that closely duplicates the production equipment.

Suppositories

Suppositories are made by a process quite similar to form–fill–seal unit-dose packaging. Typ-ically, a foil or other rigid thermoforming plastic is formed into longitudinal halves of a tor-pedolike shape that are sealed together, except for a fill opening at one end. This form becomes the mold into which melted product is added. The open end is then sealed and sent through a cooling tunnel to solidify the product. The major packaging concerns are product compatibility with the heat-sealed layers of the forming materials. A good starting point is to use food-grade foil or thermoforming plastics that have known resistance to the type of chemicals in the suppository formulation. Accelerating conditions, especially elevated temperature, are difficult for this dosage form because the product will melt to the liquid state. If no interaction is seen after incubation at elevated temperatures, however, one can assume that satisfactory compatibility will be found in room-temperature stability studies. Confirming room-temperature studies should be carried out concurrently.

Collapsible Tubes

Four major classes of collapsible tubes are used in packaging semisolids. These include unlined metal (aluminum or tin), lined metal (same metal with an epoxy or other organic coating), laminated tubes, and plastic tubes. Choice of the most appropriate tube is usually based on chemical compatibility issues. If compatibility is not an issue, then a cost-driven decision can be made.

For collapsible tubes, both chemical and mechanical integrity must be assured. Thus, a typical protocol should include weight loss, burst force, product analysis, and analysis of the package material in contact with the product. The package analysis should be done both instrumentally (e.g., spectrophotometrically or colorimetrically) and visually. When possible, an elevated temperature, for example, in the range 40–70°C, may be appropriate to accelerate any deteriorative interactions. Weight loss and burst force studies monitor the resistance of the crimp or heat seal to product attack. The product and package analyses will detect migration into the product or into the plastic of the package, respectively. Because of the potential for metal interaction with tubes, an organic coating or plastic laminate is usually present. As with pouches, the investigator must carefully observe the adhesion of the organic coating to the metal substrate. In time the smaller molecules diffuse across the lining, and if corrosion or liner attack occurs, the liner loses adhesion to the metal. This shows the importance of characterizing package integrity as a function of time and temperature.

As with pouch material, for tubes the experience of the package vendor can yield material candidates with a high probability of success, provided there is sufficient understanding of the formulation. Ideally, stability samples should be prepared at the earliest stage of the project on production equipment or pilot plant equipment that duplicates the production equipment.

C. Transdermal Devices

The past decade has witnessed a great deal of excitement and activity in the drug industry concerning transdermal drug delivery systems. The dosage form itself has been subject of numerous textbooks and reviews [60], but the package is so intimately connected with the product that the package materials can be thought of as part of the product (Fig. 9).

The introduction to the market place of transdermally delivered nitroglycerin by CIBA-Geigy, Key Pharmaceuticals, and G. D. Searle greatly expanded, and perhaps even changed, the usefulness and applicability of that very old drug. From a packaging view the accomplishment was significant because of the difficulty in containing the active principle, nitroglycerin. Given its high vapor pressure and inherent instability, nitroglycerin had been sold largely as tablets in glass containers. The problem of finding appropriate containers and closures was exacerbated by the discovery that nitroglycerin is highly soluble in such polymers as polyvinyl chloride and low-density polyethylene. In a study of package material interactions with nitroglycerin, many components, including cotton stuffing and vinyl cap liners, had an adverse effect on nitroglycerin stability [61]. A strong dependence of drug stability on both temperature and type of packaging material was also demonstrated [62]. The authors made the interesting observation that every condition that limited the free gas exchange between the drug product

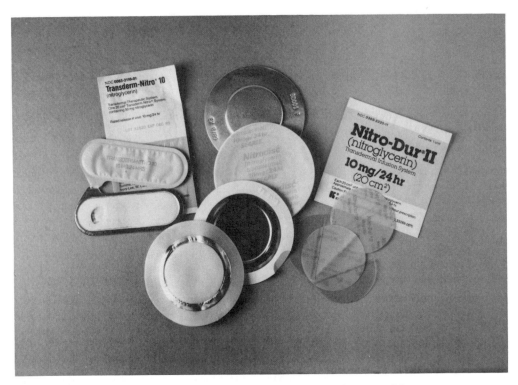

Fig. 9   Nitroglycerin transdermal products, illustrating numerous unique material usages.

and the environment had a favorable effect on stability and, therefore, recommended the development of suitable heat-resistant packaging forms to minimize gas exchange, as well as absorption into plastic, rubber, or other materials. Subsequently, unit-dose packaging of nitroglycerin tablets was characterized and a large variation in affinity of the drug for the packaging material was observed [63]. When properly unit-dose packaged, tablets maintained acceptable potency for 2 years at room temperature. This discovery of proper containment was essential to development of the nitroglycerin transdermal patch [64]. Transdermal patches typically include a backing to prevent drug release or skin transpiration when applied to the body, an adhesive applied to a suitable substrate to hold the patch adequately to the body, a release liner to cover the adhesive until the dosage is used, and possibly a rate-controlling release membrane between the reservoir and the adhesive. Joint effort of nonpharmaceutical disciplines, such as physical scientists, plastic film converters, materials experts, and process engineers, in addition to the industrial pharmacist, were all required to bring transdermal delivery into reality from an interesting laboratory curiosity.

V. PACKAGING OF AEROSOLS

A. Scope

Aerosol technology, in the broadest sense, involves the generation of aerosolized particles, and thus includes aerosol and spray technology as well as aspects of powder science. Furthermore, the packaging of aerosol products represents a supreme example of the merging of both dosage form and packaging form. While focusing on the product formulation, by dosage form we mean the delivery of droplets or particles from a container system. By packaging form, we mean the container system (i.e., the hardware), designed and engineered to produce those droplets or particles in the desired size range and, most often, in the specified amount. Given this general definition, the aerosol should not be limited simply to the traditional pressurized packaging. With the advent of modern pump systems, spray pumps that generate aerosolized particles are also to be considered in this dosage–packaging form. In addition, recent efforts to bypass the use of propellants have led to the introduction of inhalers that deliver a respirable, dry powdered form of drug. This, too, is a delivery of aerosolized particles. Thus, *aerosol* packaging is taken to include pressurized containers with continuous or metered-dose valves, spray pumps, and dry powder inhalers. In each of these examples, the packaging form is expected to deliver a predetermined amount of drug within a specified particle size range.

The aerosol is perhaps the most complex and demanding packaging form used for pharmaceuticals. It is the ultimate example in which drug performance is directly linked to package performance. Consider a metered-dose inhaler, where product dosage and package performance are completely interdependent. To achieve consistent product delivery, the formulation must be homogeneous as the product is dispensed over the entire usage of an individual canister of product. In addition, to provide the equivalent of content uniformity, the delivery volume must be consistent. That is, drug concentration in each burst must be constant as the patient uses up the product. Since most inhalation products are suspensions of finely ground, insoluble drug particles in a suspending delivery medium, one can understand the significant demands placed on both formulation and package.

When the package is also a drug delivery system, all facets of a package evaluation must be carefully considered for suitability. Both compatibility and component function are of more significance than most, if not all, other packaging systems. Well-designed compatibility studies must be carried out to test for all potential interactions of the valve, pump, or device subcomponents with the product formulation. Numerous materials, including metals, plastics, and elastomers may be involved. For example, steel, aluminum, acetal, nylon, polypropylene, polyethylene, nitrile, and butyl rubber, all may be used to produce components and subcomponents in one packaging system, and all may come into direct contact with product. Needless to say, the potential for interaction is high and, therefore, must be minimized with astute selection of materials and product ingredients.

The most recent issue of the *USP/NF* [19] contains a new, completely rewritten chapter on aerosol containers [see Chapter ⟨601⟩]. The chapter gives protocols, test procedures and guidelines on delivery rate, leak testing, pressure testing, metered dosing, and propellants. In fact, this latest revision ties together all forms of inhalation products into the one aerosol chapter.

Although there have been a number of general texts on the aerosol and aerosol technology [64,65], the aerosol as a package and delivery system for pharmaceuticals is covered in depth in this volume (see Chapter 16) and in several other recent reviews [66–68]. Therefore, this overview will focus on pointing the reader to key issues and appropriate literature references.

B. Pressurized Containers: Metered-Dose Inhalers

Until recently, pressurized aerosol packaging as a drug delivery system was rather limited in its application, possibly because industrial experience and expertise gave more attention to household, personal care, and other consumer products [65]. Nevertheless, popularity of metered-dose inhalers has grown rapidly in recent years as an application of pharmaceutical aerosols because of major advantages that are available that cannot be found in other forms of packaging. For example, the product cannot be tampered or contaminated, the atmosphere in the container can be kept oxygen-free, sterility can be maintained, and delivery can be controlled. With the development of compartmented aerosol containers [64,65], such as the Sepro of U.S. Can Company, which consists of a special plastic bag to hold the product inside the can without contacting the propellant, and the piston can, which consists of a plastic barrier piston that is forced up the walls of the can by the expanding propellant below the piston (much as a plunger in a syringe), propellant is kept segregated from the formulation, allowing the possibility of more convenient and less expensive delivery than the mechanical pump systems.

The metered-dose inhaler (MDI) represents the largest volume of pressurized packaging of pharmaceuticals [66]. Used primarily for delivery of drug product to the lungs, the MDI consists of three packaging components, an aluminum canister, a metering valve, and an actuator, and sometimes a fourth (i.e., a spacer), which is thought to facilitate improved delivery of the respirable fraction of the drug product. Each of these components must be developed with appropriate engineering inputs. For example, the canister must be able to withstand certain pressures without deforming. The valve must deliver a constant-volume dosage over the course of container emptying, and the actuator must provide proper breakup to achieve the desired particle size. All of these issues are dependent on the nature of the formulation, and appropriate material choices are involved in the selection of compatible subcomponents.

Details of the theory and operation of MDIs and other pressurized aerosol packaging of pharmaceuticals are covered in several standard reference books [64,65] and reviews [66,67] and in Chapter 16 of this text, where a discussion of propellant principles is also covered. The only additional observation is to note the influence of the Montreal Protocol as drug companies seek to replace chlorofluorocarbon (CFC) with the new hydrofluoroalkane (HFA) propellants, which do not deplete the upper ozone layer. Several existing products will be relaunched in the next few years with these new propellants.

The other new aspect of MDI technology to be aware of is the application of spacers to improve the delivery of the respiratory fraction to the lungs. The spacer serves to reduce the effective velocity of emerging particles from the actuator exit orifice. As this is done, the user is thought to be able to more easily inhale the medicament, as there is less chance for the dosage to move down the throat. More data will undoubtedly be forthcoming to demonstrate or disprove this hypothesis.

C. Nonpressurized Containers: Spray Pumps

Pumps can be thought of as a different form of closure for liquid or semisolid packaging. The container portion of the package system should be evaluated by the methods mentioned under oral liquids or solids. The component of most concern is the pump itself, which, for pharmaceuticals, is usually a metering device for the drug formulation. Of the three classes of pumps mentioned earlier, the most commonly used for pharmaceuticals, especially within the context of aerosol and spray technology, is the spray pump used for intranasal drug delivery. As in the metered-dose inhaler, the packaging delivery system is as critical in drug function as the active drug because the amount and presentation of product is controlled by the spray pump mechanism. With many newer drug products, especially those of peptide origin, nasal delivery has become a method of drug delivery of great interest. The dosage form offers rapid absorption, quick onset of activity as well as sustained delivery, avoidance of the first-pass effect, ease of administration, and usefulness for prolonged medication [68].

As with MDIs, spray pumps bring a high level of engineering to a packaging component. Several subcomponents made from a variety of materials, ranging from steel springs, to elastomeric seals, to assorted thermoplastic resin parts, all combine to produce a consistent dosage. There is an aspect to spray pumps that is even more complicated and difficult to execute than for aerosols; namely, the effect of the user and other design elements on the final output. Specifically, a key aspect to producing a spray of desired droplet size is the design of the swirl chamber of the nozzle (or actuator) [70]. The specific design has a dramatic effect on both average droplet size and the droplet size distribution. In addition, the pressure differential between the compression chamber in the pump and the atmosphere directly affects the breakup of the product liquid into droplets. Although efforts are made to engineer out this dependence on user application force, there is little doubt that variability is created as a result of the hand and finger pressure applied by the user. Therefore traits of the user can have as much effect on the dosage as any variability in the pump subcomponents and its construction. In spite of this challenge, the nasal spray offers exciting opportunities for new ways to deliver drug products that might otherwise be too inconvenient (e.g., as a replacement for injections) or technically impossible (e.g., as in poor oral uptake).

D. Nonpressurized Containers: Dry Power Inhalers

Without question one of the most actively pursued new packaging forms is the dry powder inhaler (DPI). Motivated by the Montreal Protocol and the desire to move away from propellants, drug companies are pursuing the DPI because it offers an alternative for rapid and convenient delivery to the lungs. As with the MDI and the spray pump, the DPI is a merger of dosage form and delivery device. In this case the dosage form is a dry powder of pure drug or of pure drug with excipient prepared to the proper size range, otherwise known as the respirable fraction. Several excellent reviews of the commercial devices and the developing products have recently appeared [68,71]. Shepherd [71] documents this rapid expansion through the number of related patents that have been issued each year. Before 1980 a total of only 3 relevant patents on multidose DPIs were issued, whereas in 1991 alone 26 patents were issued. He also points out that, although many other forms of devices have been patented over the last 20 years, not many have been effectively commercialized. This is a strong indication of the difficulty that has arisen in developing a device with the consistency and reliability of the MDI. This is partly due to the large number of subcomponents that are assembled in the final device. Some units consist of over 20 separate subcomponents that must be consistently and correctly manufactured and assembled. Nevertheless, strong demands are being placed on the DPI to meet the standards of current inhalation products. In fact the DPI is included in a proposed revision [72] of *USP/NF* Chapter ⟨601⟩, which would be retitled "Aerosols, Metered-Dose Inhalers, and Dry Powder Inhalers." Under the proposed standards, DPIs would be required to deliver drug product with the same tolerances as MDIs. In spite of, or perhaps because of, these challenges, numerous drug and device companies continue to patent and to develop unabated.

VI. CONCLUDING REMARKS

From this review on the packaging of pharmaceutical dosage forms, clearly the investigator who is responsible for determining the packaging must have a broad background in a variety of disciplines. A solid foundation in the areas of chemistry, regulatory affairs, pharmaceutical dosage forms, materials science, and package engineering is important in making the most astute packaging choices. The synthesis of these disciplines allows for a number of possible ways to economize the process of drug development. Cost savings may include both direct and indirect costs, through, for example, improved accuracy of the data, shorter project timing, reduced project costs (through fewer stability studies), and reduced packaging material costs.

The successful pharmaceutical development packaging scientist requires expertise in both the packaging science and applicable government regulations to assemble a successful submission. With the ever-increasing sophistication and technical depth needed for the new packaging and dosage forms discussed in this review, expertise has necessarily become increasingly focused in the industry. That focus is seen in the experience with greater frequency of stand-alone package development groups.

The horizon reveals that the future will be even more demanding on the pharmaceutical packaging scientist. With the advent of other new dosage forms, new materials, and new drug classes (such as those produced through bioengineering), the package may be the single, most important factor in delivering the drug product and, ultimately, in bringing the product to the marketplace. The pharmaceutical house that can effectively stabilize and deliver drugs in new and novel ways, such as has been done through transdermal delivery or dry-powder inhalers, will not only maintain, but will create, a larger presence in the marketplace.

The future is bright for packaging, in that ever-greater stress will be placed on what has been coined the "delivery system"; that is, the mechanical components that contain and convey the product to the patient in the most convenient and effective manner. In the future this concept of delivery system and all of its nuances will discriminate between those products that will succeed and those that will fail.

This strategy applies not only to new and novel drugs, but to the increasing number of products that either go off-patent or are changed from ethical to over-the-counter drugs. In a marketplace where efficacy has already been demonstrated and is not the discriminating issue among similar products, a delivery system that brings *convenience* at a *lower cost* to the users (which include physician, pharmacist, and patient) will be a major factor for success.

REFERENCES

1. D. C. Liebe, Packaging of pharmaceutical dosage forms, in *Modern Pharmaceutics*, 2nd Ed. (G. S. Banker and C. T. Rhodes, eds.), Marcel Dekker, New York, 1990, pp. 695–740.
2. D. C. Liebe, Pharmaceutical packaging, in *Encyclopedia of Pharmaceutical Technology*, Vol. 12 (J. Swarbrick and J. C. Boylan, eds.), Marcel Dekker, New York, 1995, pp. 1–28.
3. J. F. Hanlon, *Handbook of Package Engineering*, 2nd Ed., McGraw-Hill, New York, 1984.
4. J. Cooper, *Plastic Containers for Pharmaceuticals—Testing and Control*, WHO Offset Publ. 4, Geneva, 1974.
5. *Code of Federal Regulations*, Title 21, *Food and Drugs*, Part 211, Current good manufacturing practice for finished pharmaceuticals (Apr. 1, 1992), p. 88.
6. *The United States Pharmacopeia/National Formulary*, 23rd Ed., U.S. Pharmacopeia Convention, Rockville, MD, 1994.
7. U.S. Department of Health and Human Services, Public Health Services, Food and Drug Administration, *Guidelines for Submitting Documentation for Packaging Human Drugs and Biologics*, Feb. 1987.
8. E. A. Leonard, *Packaging: Specifications, Purchasing, and Quality Control*, 3rd Ed., Marcel Dekker, New York, 1987.
9. R. J. Ashley, Permeability and plastics packaging, in *Polymer Permeability* (J. Comyn, ed.), Elsevier, Amsterdam, 1985, pp. 269–308.
10. J. A. Cairns, C. R. Oswin, and F. A. Paine, *Packaging for Climatic Protection*, Butterworth, London, 1974.
11. P. Labuza, S. Mizrahi, and M. Karel, Mathematical models for optimization of flexible film packaging of foods for storage, Trans. ASAE, 15, 150 (1972).
12. I. Sagvy and M. Karel, Modeling of quality deterioration during food processing and storage, Food Technol., 13, 715 (1980).

13. N. A. Peppas and G. S. Sekhon, Mathematical analysis of transport properties of polymer films for food packaging. IV. Prediction of shelf-life of food packages using Halsey sorption isotherms, SPE Tech. Pap., 26, 681 (1980).

14. K. Nakabayashi, T. Shimamoto, and H. Mima, Stability of packaged solid dosage forms. I. Shelf-life prediction for package tablets liable to moisture damage, Chem. Pharm. Bull., 28, 1090 (1980).

15. K. Nakabayashi, T. Shimamoto, and H. Mima, Stability of packaged solid dosage forms. II. Shelf-life prediction for packaged sugar-coated tablets liable to moisture and heat damage, Chem. Pharm. Bull., 28, 1099 (1980).

16. K. Nakabayashi, T. Tsuchida, and H. Mima, Stability of packaged solid dosage forms. IV. Shelf-life prediction for packaged aspirin aluminum tablets under the influence of moisture and heat, Chem. Pharm. Bull., 29, 2027 (1981).

17. K. Nakabayashi, T. Shimamoto, H. Mima, and J. Okada, Stability of packaged solid dosage forms. V. Prediction of the effect of aging on the disintegration of packaged tablets influenced by moisture and heat, Chem. Pharm. Bull., 29, 2051, (1981).

18. M. D. Kentala, H. E. Lockhart, J. R. Giacin, and R. Adams, Computer-aided simulation of quality degradation of oral solid drugs following repackaging, Pharm. Technol., 6, 46 (1982).

19. J. S. Fulcoly, Predicting the moisture uptake of tablets packaged in semipermeable blister packages and stored under static conditions of temperature and humidity, M.S. thesis, School of Packaging, Michigan State University, 1984.

20. M. J. Wang, Prediction of moisture content of a packaged moisture sensitive pharmaceutical product stored under fluctuating temperature and humidity environments, M.S. thesis, School of Packaging, Michigan State University, 1985.

21. J. R. Giacin, C. L. Pires, and H. E. Lockhart, Predicting packaged product shelf life: experimental and mathematical models, Pharm. Technol., 15(9), 98 (1991).

22. J. Haleblain and W. McCrone, Pharmaceutical applications of polymorphism, J. Pharm. Sci., 58, 911 (1969).

23. S. G. Gilbert, Prediction of product–package shelf life, Arden House Conference on Pharmaceutical Packaging, Feb. 1982.

24. I. Jagnandan, The use of inverse gas chromatography in accelerated stability testing for products affected by temperature and humidity, Ph.D. thesis, Rutgers University, 1980.

25. S. G. Gilbert, J. Miltz, and J. R. Giacin, Transport considerations of potential migrants from food packaging materials. J. Food Process. Preserv., 4, 27 (1980).

26. S. Gyeszly, Shelf-life simulation, Package Eng., 25, 70 (1980).

27. R. P. Abendroth, Glass as a packaging material for pharmaceuticals, in Encyclopedia of Pharmaceutical Technology, Vol. 7 (J. Swarbrick and J. C. Boylan, eds.), Marcel Dekker, New York, 1993, pp. 79–99.

28. J. B. Riley, High and low density polyethylene, Mod. Plast. Encycl., 58, 58 (1981).

29. J. D. Capolupo and T. M. Chucta, Antioxidants, Mod. Plast. Encycl., 63, 112 (1986).

30. J. D. van Drumpt, Antistatic agents, Mod. Plast. Encycl., 63, 117 (1986).

31. R. C. Schiek, Colorants, Mod. Plast. Encycl., 63, 118 (1986).

32. R. L. Mitchell and G. C. David, Rayon. In Kirk-Othmer Encyclopedia of Chemical Technology, 2nd Ed., Vol. 17, John Wiley & Sons, New York, 1968, pp. 168–209.

33. R. A. Iwaskiewicz, Prospects for Tamper Evidency in Packaging, M.S. thesis, School of Packaging, Michigan State University, 1991.

34. H. E. Lockhart, Tamper-evident packaging—is it really? in Food Product–Package Compatibility (J. I. Gray, B. R. Harte, and J. Miltz, eds.), Technomic, Lancaster, PA, 1986, p. 270.

35. J. Sneden, Testing of tamper-resistant packaging, M.S. thesis, Michigan State University, 1983.

36. J. J. Maloney and R. Ullrich, Closures, liners and seals, Packag. Encycl., 32, 113 (1987).

37. S. W. Shalaby and B. L. Williams, Films and sheets for packaging, in Encyclopedia of Pharmaceutical Technology, Vol. 6 (J. Swarbrick and J. C. Boylan, eds.), Marcel Dekker, New York, 1993, pp. 29–49.

38. A. I. Kay, The selection and evaluation of package components for parenteral products, Pharm. Technol., 6, 54 (1982).

39. S. J. Borchert, G. A. Kelley, and E. A. Hardwidge, A program for identification testing of package materials, Pharm. Technol., 7, 72 (1983).

40. S. Turco and M. Akers, Parenterals: Large volume, in *Encyclopedia of Pharmaceutical Technology*, Vol. 11 (J. Swarbrick and J. C. Boylan, eds.), Marcel Dekker, New York, 1995.

41. S. Turco and M. Akers, Parenterals: small volume, in *Enclopedia of Pharmaceutical Technology*, Vol. 11 (J. Swarbrick and J. C. Boylan, eds.), Marcel Dekker, New York, 1995.

42. R. C. Hughes and G. H. Hopkins, Pharmaceutical applications for West Company elastomeric closures, Tech. Rep. 11, The West Company, Phoenixville, PA, July 1, 1965.

43. Extractables from elastomeric closures: Analytical procedures for functional group characterization/identification, Tech. Methods Bull. 1, Parenteral Drug Association, Philadelphia, 1980.

44. Elastomeric closures: evaluation of significant performance and identity characteristics, Tech. Methods Bull. 2, Parenteral Drug Association, Philadelphia, 1982.

45. Glass containers for small volume parenteral products: Factors for selection and test methods for identification, Tech. Methods Bull. 3, Parenteral Drug Association, Philadelphia, 1982.

46. R. M. Patel, Stability and the FDA guidelines, address presented at the 13th Annual Industrial Pharmacy Management Conference, Madison, WI, Oct. 13, 1980.

47. Proceedings of the AAPS Workshop on Pharmaceutical Packaging: Issues and Challenges for the 90's, April 22–23, 1993, Arlington, VA.

48. W. J. Passl and E. Renshaw, Chemical resistance of glass bottles for intravenous infusions, Pharm. Ind., 44, 955 (1982).

49. *Lubricative Bottle Coatings*, technical bulletin of Wheaton Glass Company, Millville, NJ, 1981.

50. S. V. Sanga, Review of glass types available for packaging parenteral solutions, J. Parenter. Drug Assoc., 33, 61 (1979).

51. F. P. Miltrano and D. W. Newton, Factors affecting insulin adherence to type I glass bottles, Am. J. Hosp. Pharm., 39, 1491 (1982).

52. J. Anschel, Processing of glass containers for parenteral products, J. Test. Eval., 5, 58 (1977).

53. R. V. Kasubick, E. P. Mariani, and E. C. Shinal, Investigations on glass delamination with parenteral solutions, *Pharm. Tech. Conference '82 Proceedings*, 1982, p. 577.

54. J. Autian, Plastics and medication, in *Dispensing of Medication*, 7th Ed. (E. W. Martin, ed.), Mack Publishing, Easton, PA, 1971, p. 652.

55. J. A. Uotila and N. T. Santasalo, New concepts in the manufacturing and sterilization of LVP's in plastic bottles, J. Pharm. Sci. Technol., 35, 170 (1981).

56. K. E. Avis and E. J. Smith, Elastomeric parenteral closures, in *Encyclopedia of Pharmaceutical Technology*, Vol. 5 (J. Swarbrick and J. C. Boylan, eds.), Marcel Dekker, New York, 1993, pp. 73–88.

57. J. J. Farley, Rubber: Physical, chemical, and functional aspects, Arden House Conference on Pharmaceutical Packaging, Feb. 1982.

58. G. H. Hopkins, Elastomeric closures for pharmaceutical packaging. J. Pharm. Sci., 54, 138 (1965).

59. G. H. Hopkins, Prescreening closure formulations for specific applications, Bull. Parenter. Drug Assoc., 29, 278 (1975).

60. Y. W. Chien, *Transdermal Controlled Systemic Medications*, Marcel Dekker, New York, 1987.

61. S. A. Fusari, Nitroglycerin sublingual tablets. I. Stability of conventional tablets, J. Pharm. Sci., 62, 122 (1973).

62. K. Thomas and R. Gröning, Influence of packaging and dosage form on the stability of solid nitroglycerin preparation, Pharm. Ind., 36, 876 (1974).

63. M. J. Pikal, D. A. Bibler, and B. Rutherford, Polymer sorption of nitroglycerin and stability of molded nitroglycerin tablets in unit-dose packaging, J. Pharm. Sci., 66, 1293 (1977).

64. P. A. Sanders, *Handbook of Aerosol Technology*, 2nd Ed., Van Nostrand, Reinhold, New York, 1979.

65. M. A. Johnsen, *The Aerosol Handbook*, 2nd Ed., Wayne Dorland, Mendham, NJ, 1982.

66. J. J. Sciarra and A. J. Cutie, Pharmaceutical aerosols, in *Modern Pharmaceutics*, 2nd Ed. (G. S. Banker and C. T. Rhodes, eds.), Marcel Dekker, New York, 1990, pp. 605–634.

67. G. W. Hallworth, Metered-dose inhalers: Pressurized systems, in *Encyclopedia of Pharmaceutical Technology*, Vol. 9 (J. Swarbrick and J. C. Boylan, eds.), Marcel Dekker, New York, 1993, pp. 299–329.

68. P. Atkins, Metered-dose inhalers: Nonpressurized systems, in *Encyclopedia of Pharmaceutical Technology*, Vol. 9 (J. Swarbrick and J. C. Boylan, eds.), Marcel Dekker, New York, 1993, pp. 287–298.

69. K. S. E. Su, Intranasal drug delivery, in *Encyclopedia of Pharmaceutical Technology*, Vol. 7 (J. Swarbrick and J. C. Boylan, eds.), Marcel Dekker, New York, 1993, pp. 175–201.

70. A. H. Lefebvre, *Atomization and Sprays*, Hemisphere Publishing, New York, 1989.

71. M. T. Shepherd, Dry powder delivery devices for asthma, in *Packag. Technol. Sci.*, 7, 215–227 (1994).

72. Recommendations of the USP Advisory Panel on Aerosols on the USP General Chapters on Aerosols ⟨601⟩ and Uniformity of Dosage Units ⟨905⟩, Pharm. Forum, 20, 7477–7505 (1994).

Optimization Techniques in Pharmaceutical Formulation and Processing

Joseph B. Schwartz
Philadelphia College of Pharmacy and Science, Philadelphia, Pennsylvania

Robert E. O'Connor*
R. W. Johnson Pharmaceutical Research Institute, Raritan, New Jersey

I. INTRODUCTION

A significant portion of this book is devoted to the concepts involved in formulating drug products in their various forms. Physical, chemical, and biological properties all must be given due consideration in the selection of components and processing steps for that dosage form. The final product must be one that meets not only the requirements placed on it from a bioavailability standpoint, but also the practical mass production criteria of process and product reproducibility. In the current regulatory climate, formulation and process justification is a requirement for preapproval inspections for all new drug applications. In fact, a development report for both formulation and process are reviewed during these inspections. It is in the best interest of the pharmaceutical scientist to understand the theoretical formulation and target processing parameters, as well as the ranges for each excipient and processing parameter. Optimization techniques provide both a depth of understanding and an ability to explore and defend ranges for formulation and processing factors. With a rational approach to the selection of the several excipients and manufacturing steps for a given product, one qualitatively selects a formulation. It is at this point that optimization can become a useful tool to quantitate a formulation that has been qualitatively determined. Optimization is not a screening technique.

The word *optimize* is defined as follows: to make as perfect, effective, or functional as possible [1]. The last phrase, "as possible," leads one immediately into the area of decision-making, since one might ask: (a) perfect by whose definition; (b) for what characteristics; and (c) under what conditions? The term *optimization* is often used in pharmacy relative to formulation and to processing; and one will even find it in the literature referring to *any* study of the formula. In developmental projects, one generally experiments by a series of logical

---

Current affiliation: Whitehall-Robins Healthcare, Hammonton, New Jersey

steps, carefully controlling the variables and changing one at a time, until a satisfactory system is produced. If the experimenter had sufficient help or sufficient time, he or she would eventually perfect the formulation, but under the circumstances the "best" one is often simply the last one prepared. It is satisfactory, but how close is it to the optimum, and how does the experimenter know?

No matter how rationally designed, the trial-and-error method can be improved on. It is the purpose of this chapter to discuss the general principles behind the techniques of optimization and, then, to review the specific techniques that have been applied to pharmaceutical systems.

II. OPTIMIZATION PARAMETERS

A. Problem Types

There are two general types of optimization problems: the constrained and the unconstrained. Constraints are those restrictions placed on the system by physical limitations or perhaps by simple practicality (e.g., economic considerations). In unconstrained optimization problems there are no restrictions. For a given pharmaceutical system one might say: make the hardest tablet possible. The constrained problem, on the other hand, would be stated: make the hardest tablet possible, but it must disintegrate in less than 15 min.

Within the realm of physical reality, and most important in pharmaceutical systems, the unconstrained optimization problem is almost nonexistent. There are always restrictions that the formulator wishes to place or must place on a system, and in pharmaceuticals, many of these restrictions are competing. For example, it is unreasonable to assume, as just described, that the hardest tablet possible would also have the lowest compression and ejection forces and the fastest disintegration time and dissolution profile. It is sometimes necessary to trade off properties; that is, to sacrifice one characteristic for another. Thus, the primary objective may not be to optimize absolutely (i.e., a maxima or minima), but to realize an overall preselected or desired result for each characteristic or parameter. Drug products are often developed by an effective compromise between competing characteristics to achieve the best formulation and process within a given set of restrictions.

An additional complication in pharmacy is that formulations are not usually simple systems. They often contain many ingredients and variables which may interact with one another to produce unexpected, if not unexplainable, results.

B. Variables

The development of a pharmaceutical formulation and the associated process usually involves several variables. Mathematically, they can be divided into two groups. The independent variables are the formulation and process variables directly under the control of the formulator. These might include the level of a given ingredient or the mixing time for a given process step. The dependent variables are the responses or the characteristics of the in-progress material or the resulting drug delivery system. These are a direct result of any change in the formulation or process.

The more variables one has in a given system, the more complicated becomes the job of optimization. But regardless of the number of variables, there will be a relationship between a given response and the independent variables. Once this relationship is known for a given response, it defines a response surface, such as that represented in Fig. 1. It is this surface that must be evaluated to find the values of the independent variables, X_1 and X_2, which give the most desirable level of the response, Y. Any number of independent variables can be consid-

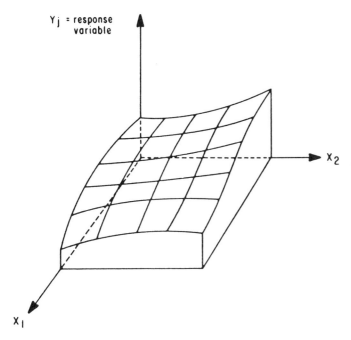

Fig. 1 Response surface representing the relationship between the independent variables, X_1 and X_2, and the dependent variable Y.

ered; representing more than two becomes graphically impossible, but mathematically only more complicated.

III. CLASSIC OPTIMIZATION

Classic optimization techniques result from application of calculus to the basic problem of finding the maximum or minimum of a function. The techniques themselves have limited application, but might be useful for problems that are not too complex and do not involve more than a few variables. The concept, however, is important.

The curve in Fig. 2 might represent the relation between a response, Y, and a single independent variable X, in a hypothetical system, and since we can see the whole curve, we can pick out the highest point or lowest, the maximum or minimum. Use of calculus, however, makes the task of plotting the data or equation unnecessary. If the relationship, that is, the equation for Y as a function of X is available [Eq. (1)],

$$Y = f(X) \tag{1}$$

we can take the first derivative, set it equal to zero, and solve for X to obtain the maximum or minimum. For many functions of X, there will be more than one solution when the first derivative is set equal to zero. The various solutions may all be maxima or minima, or a mixture of both.

There are also techniques to determine whether we are dealing with a maximum or a minimum, that is, by use of the second derivative. And there are techniques to determine whether we simply have *a* maximum (one of several local peaks) or *the* maximum. Such approaches are covered in elementary calculus texts and are well presented relative to optimization in a review by Cooper and Steinberg [2].

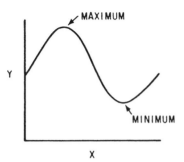

Fig. 2 Graphic location of optimum (maximum or minimum).

When the relationship for the response Y is given as a function of two independent variables, X_1 and X_2,

$$Y = f(X_1, X_2) \tag{2}$$

the problem is slightly more involved. Graphically, there are contour plots (Fig. 3) on which the axes represent the two independent variables, X_1 and X_2, and the contours (analogous to elevations, as on a contour map) represent a specific level of Y. Again, we can select an optimum graphically. Mathematically, appropriate manipulations with partial derivatives of the function can locate the necessary pair of X values for the optimum.

The situation with multiple variables (any more than two) becomes graphically impossible. It is still possible by mathematics, but very involved, making use of partial derivatives, matrices, determinants, and so on. The reader is referred to optimization texts for further details. Because of the complications involved and because the classic calculus methods apply basically to unconstrained problems, more practical methods are generally used.

IV. STATISTICAL DESIGN

The techniques most widely used for optimization may be divided into two general categories: one, in which experimentation continues as the optimization study proceeds; and second, in which the experimentation is completed before the optimization takes place. The first type is

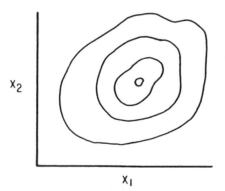

Fig. 3 Contour plot. Contours represent values of the dependent variable Y.

represented by evolutionary operations (EVOP) and the simplex methods, and the second by the more classic mathematical and search methods. (Each of these is discussed in Sec. V.)

For the techniques of the second type, it is necessary that the relation between any dependent variable and the one or more independent variables be known. To obtain the necessary relationships, there are two possible approaches: the theoretical and the empirical.

If the formulator knows a priori the theoretical equation for the formulation properties of interest, no experimentation is necessary. However, much of the work in pharmaceutics has been in the pursuit of such relationships and, to our knowledge, most have not been determined. Therefore, it remains the task of the formulator to generate the relationships between the variables for the particular formulation and process.

In a text on experimental design, Davis states [3]:

> Theoretically, the behavior of chemical reactions, or for that matter the behavior of any system, is governed by ascertainable laws, and it should be possible to determine optimum conditions by applying such laws. In practice, however, the underlying mechanisms of the system are frequently so complicated that an empirical approach is necessary.

To apply the empirical or experimental approach for a system with a single independent variable, the formulator experiments at several levels, measures the property of interest, and obtains a relationship; usually by simple regression analysis, or by the least squares method. In general, however, there is more than one important variable, so the experimenter must enter into the realm of "statistical design of experiments and multiple linear regression analysis." Statistical design and multiple linear regression analysis are separate and rather large fields and, again, the reader is referred to appropriate texts [3–5,40]. The concept of interest to the pharmacist planning to utilize optimization techniques is that there are methods available for selecting one's experimental points so that (a) the entire area of interest is covered or considered, and (b) analysis of the results will allow separation of variables (i.e., statistical analysis can be performed, which allows the experimenter to know which variable caused a specific result).

One of the most widely used experimental plans is that of the factorial design, or some variation of it (two of the techniques in the following section utilize it). By multiple regression techniques, the relationships between variables, then, are generated from experimental data, and the resulting equations are the basis of the optimization. These equations define the response surface for the system under investigation.

V. APPLIED OPTIMIZATION METHODS

There are many methods that can be, and have been, used for optimization, classic and otherwise. These techniques are well documented in the literature of several fields. Deming and King [6] presented a general flowchart (Fig. 4) that can be used to describe general optimization techniques. The effect on a real system of changing some input (some factor or variable) is observed directly at the output (one measures some property), and that set of real data is used to develop mathematical models. The responses from the predictive models are then used for optimization. The first two methods discussed here, however, omit the mathematical-modeling step; optimization is based on output from the real system.

A. Evolutionary Operations

One of the most widely used methods of experimental optimization in fields other than pharmaceutical technology is the *evolutionary operation*, or EVOP. This technique is especially well suited to a production situation. The basic philosophy is that the production procedure

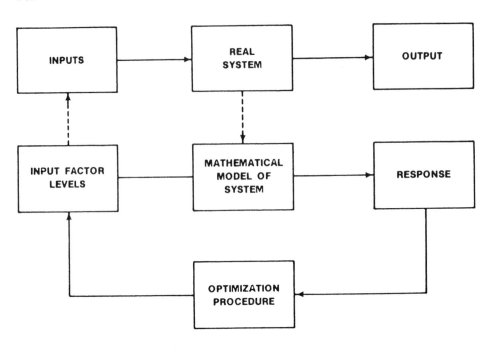

Fig. 4 Flowchart for optimization.

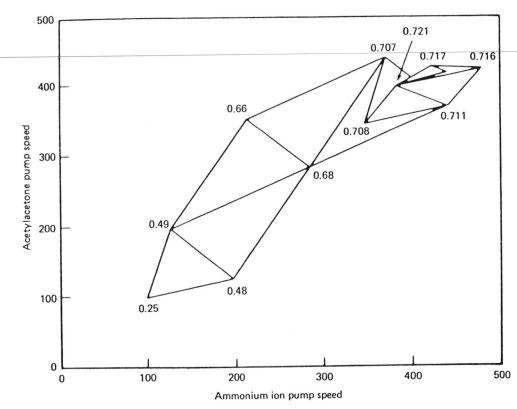

Fig. 5 The simplex approach to optimization. Response is spectrophotometric reading at a given wavelength. (From Ref. 6.)

(formulation and process) is allowed to evolve to the optimum by careful planning and constant repetition. The process is run in a way such that it both produces a product that meets all specifications and (at the same time) generates information on product improvement.

By this method the experimenter makes a very small change in the formulation or process, but makes it so many times (i.e., repeats the experiment so many times) that he or she can determine statistically whether the product has improved. If it has, the experimenter makes another change in the same direction, many times, and notes the results. This continues until further changes do not improve the product or perhaps become detrimental. The experimenter then has the optimum—the peak.

In an industrial process, this large number of experiments is usually not a problem, since the process will be run over and over again. The application of this technique to tablets has been advocated by Rubinstein [7]. It has also been applied to an inspection system for parenterals [8].

In most pharmaceutical situations, however, there is often insufficient latitude in the formula or process to allow the necessary experimentation. The pharmaceutical industry is subject to regulatory constraints that make EVOP impossible to employ in validated production processes and, therefore, impractical and expensive to use. Moreover, EVOP is not a substitute for good laboratory-scale investigation, and because of the necessarily small changes utilized, is not particularly suitable to the laboratory. In pharmaceutical development, more efficient methods are desired.

B. The Simplex Method

The simplex approach to the optimum is also an experimental method and has been applied more widely to pharmaceutical systems. Originally proposed by Spendley et al. [9], the technique has even wider appeal in areas other than formulation and processing. A particularly good example to illustrate the principle is the application to the development of an analytical method (a continuous flow analyzer) by Deming and King [6].

A *simplex* is a geometric figure that has one more point than the number of factors. So, for two factors or independent variables, the simplex is represented by a triangle. Once the shape of a simplex has been determined, the method can employ a simplex of fixed size or of variable sizes that are determined by comparing the magnitudes of the responses after each successive calculation. Figure 5 represents the set of simplex movements to the optimum conditions using a variable size technique.

The two independent variables (the axes) show the pump speeds for the two reagents required in the analysis reaction. The initial simplex is represented by the lowest triangle; the vertices represent the spectrophotometric response. The strategy is to move toward a better response by moving away from the worst response. Since the worst response is 0.25, conditions are selected at the vortex, 0.68 and, indeed, improvement is obtained. One can follow the experimental path to the optimum, 0.721.

For pharmaceutical formulations, the simplex method was used by Shek et al. [10] to search for an optimum capsule formula. This report also describes the necessary techniques of reflection, expansion, and contraction for the appropriate geometric figures. The same laboratories applied this method to study a solubility problem involving butoconazole nitrate in a multicomponent system [11].

Bindschaedler and Gurny [12] published an adaptation of the simplex technique to a TI-59 calculator and applied it successfully to a direct compression tablet of acetaminophen (paracetamol). Janeczek [13] applied the approach to a liquid system (a pharmaceutical solution) and was able to optimize physical stability. In a later article, again related to analytical techniques, Deming points out that when complete knowledge of the response is not initially available, the simplex method is probably the most appropriate type [14]. Although not presented here, there are sets of rules for the selection of the sequential vertices in the procedure, and the reader planning to carry out this type of procedure should consult appropriate references.

C. The Lagrangian Method

This optimization method, which represents the mathematical techniques, is an extension of the classic method, and was the first, to our knowledge, to be applied to a pharmaceutical formulation and processing problem. Fonner et al. [15] chose to apply this method to a tablet formulation and to consider two independent variables. The active ingredient, phenylpropanolamine HCl, was kept at a constant level, and the levels of disintegrant (corn starch) and lubricant (stearic acid) were selected as the independent variables, X_1 and X_2. The dependent variables include tablet hardness, friability, volume, in vitro release rate, and urinary excretion rate in human subjects.

This technique requires that the experimentation be completed before optimization so that mathematical models can be generated. The experimental design here was a full 3^2 factorial and, as shown in Table 1, nine formulations were prepared. Polynomial models relating the response variables to the independent variables were generated by a backward stepwise regression analysis program. The analyses were performed on a polynomial of the form

$$y = B_0 + B_1X_1 + B_2X_2 + B_3X_1^2 + B_4X_2^2 + B_5X_1X_2 + B_6X_1X_2^2 + B_7X_1^2X_2 + B_8X_1^2X_2^2 \quad (3)$$

and the terms were retained or eliminated according to standard stepwise regression techniques. In Eq. (3), y represents any given response and B_i represents the regression coefficient for the various terms containing levels of the independent variables. One equation is generated for each response or dependent variable.

A graphic technique may be obtained from the polynomial equations, as represented in Fig. 6. Figure 6a shows the contours for tablet hardness as the levels of the independent variables are changed. Figure 6b shows similar contours for the dissolution response, $t_{50\%}$. If the requirements on the final tablet are that hardness be 8–10 kg and $t_{50\%}$ be 20–33 min, the

Table 1 Tablet Formulations

Formulation No.	Ingredient per tablet (mg)			
	Phenylpropanolamine HCl	Dicalcium phosphate · 2H$_2$O	Starch	Stearic acid
1	50	326	4 (1%)	20 (5%)
2	50	246	84 (21%)	20
3	50	166	164 (41%)	20
4	50	246	4	100 (25%)
5	50	166	84	100
6	50	86	164	100
7	50	166	4	180 (45%)
8	50	86	84	180
9	50	6	164	180

Source: Ref. 15.

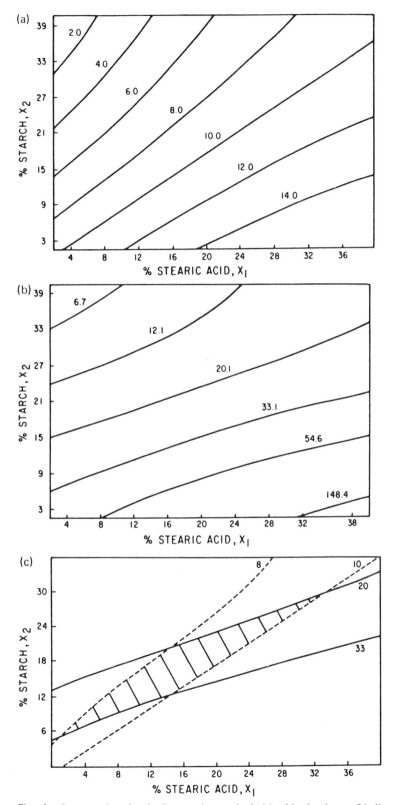

Fig. 6 Contour plots for the Lagrangian method: (a) tablet hardness; (b) dissolution ($t_{50\%}$); (c) feasible solution space indicated by crosshatched area. (From Ref. 15.)

feasible solution space is indicated in Fig. 6c. This has been obtained by superimposing Fig. 6a and b, and several different combinations of X_1 and X_2 will suffice.

Slightly different constraints are used to illustrate the mathematical technique. In this example, the constrained optimization problem is to locate levels of stearic acid (X_1) and starch (X_2) that minimize the time of in vitro release (y_2) such that the average tablet volume (y_4) did not exceed 9.422 cm^2 and the average friability (y_3) did not exceed 2.72%.

To apply the Lagrangian method, this problem must be expressed mathematically as follows: Minimize

$$y_2 = F_2(X_1, X_2) \tag{4}$$

such that

$$y_3 = f_3(X_1, X_2) \leq 2.72 \tag{5}$$

$$y_4 = F_4(X_1, X_2) \leq 0.422 \tag{6}$$

and

$$5 \leq X_1 \leq 45 \tag{7}$$

$$1 \leq X_2 \leq 41 \tag{8}$$

Equations (7) and (8) serve to keep the solution within the experimental range.

The foregoing inequality constraints must be converted to equality constraints before the operation begins, and this is done by introducing a slack variable q, for each. The several equations are then combined into a Lagrange function F, and this necessitates the introduction of a Lagrange multiplier, λ, for each constraint.

Then, following the appropriate steps (i.e., partial differentiation of the Lagrange function) and solving the resulting set of six simultaneous equations, values are obtained for the appropriate levels of X_1 and X_2 to yield an optimum in vitro time of 17.9 min ($t_{50\%}$). The solution to a constrained optimization program may depend heavily on the constraints applied to the secondary objectives.

A technique called *sensitivity analysis* can provide information so that the formulator can further trade off one property for another. For sensitivity analysis the formulator solves the constrained optimization problem for systematic changes in the secondary objectives. For example, the foregoing problem restricted tablet friability, y_3, to a maximum of 2.72%. Figure 7

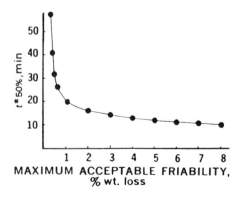

Fig. 7 Optimum in vitro $t_{50\%}$ release rate as a function of restrictions on tablet friability. (From Ref. 15.)

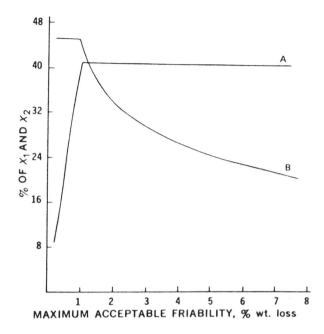

Fig. 8 Optimizing values of stearic acid and starch as a function of restrictions on tablet friability. A, percent starch; B, percent stearic acid. (From Ref. 15.)

illustrates the in vitro release profile as this constraint is tightened or relaxed and demonstrates that substantial improvement in the $t_{50\%}$ can be obtained up to about 1–2%. Subsequently, the plots of the independent variables, X_1 and X_2, can be obtained as shown in Fig. 8. Thus the formulator is provided with the solution (the formulation) as he changes the friability restriction.

The several steps in the Lagrangian method can be summarized as follows.

1. Determine objective function.
2. Determine constraints.
3. Change inequality constraints to equality constraints.
4. Form the Lagrange function, F:
   a. One Lagrange multiplier λ for each constraint
   b. One slack variable q for each inequality constraint
5. Partially differentiate the Lagrange function for each variable and set derivatives equal to zero.
6. Solve the set of simultaneous equations.
7. Substitute the resulting values into the objective functions.

Although many steps in the procedure may be carried out by computer, the application requires significant mathematical input from the person involved.

Buck et al. [16] have expanded on the previous work and have proposed that the statistical design technique can be incorporated into an overall management philosophy for proposed product design. The authors discuss four phases in this philosophy, which are defined as (a) a preliminary planning phase, (b) an experimental phase, (c) an analytical phase, and (d) a verification phase. They include case studies of a tablet design and a suspension design to

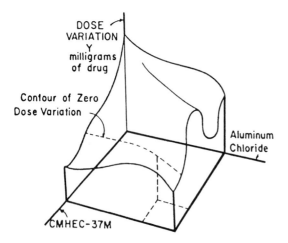

Fig. 9 Response surface concept and results of the second case study. (From Ref. 16.)

illustrate the efficient and effective procedures that might be applied. Representation of such analysis and the available solution space is shown for the suspension in Figs. 9 and 10.

D. Search Methods

In contrast with the mathematical optimization methods, search methods do not require continuity or differentiability of the function—only that it be computable. In these methods the response surfaces, as defined by the appropriate equations, are searched by various methods to find the combination of independent variables yielding the optimum.

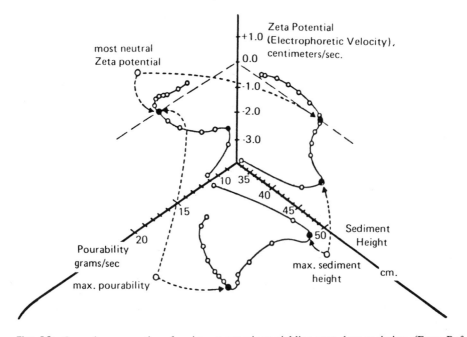

Fig. 10 Secondary properties of various suspensions yielding zero dose variation. (From Ref. 16.)

Table 2 Formulation Variables
(Independent)

X_1	Diluent ratio
X_2	Compressional force
X_3	Disintegrant level
X_4	Binder level
X_5	Lubricant level

Although the Lagrangian method was able to handle several responses or dependent variables, it was generally limited to two independent variables. A search method of optimization was also applied to a pharmaceutical system and was reported by Schwartz et al. [17]. It takes five independent variables into account and is computer-assisted. It was proposed that the procedure described could be set up such that persons unfamiliar with the mathematics of optimization and with no previous computer experience could carry out an optimization study.

The system selected here was also a tablet formulation. The five independent variables or formulation factors selected for this study are shown in Table 2. The dependent variables are listed in Table 3. Since each dependent variable is considered separately, any number could have been included.

The experimental design used was a modified factorial and is shown in Table 4. The fact that there are five independent variables dictates that a total of 27 experiments or formulations be prepared. This design is known as a five-factor, orthogonal, central, composite, second-order design [3]. The first 16 formulations represent a half-factorial design for five factors at two levels, resulting in $\frac{1}{2} \times 2^5 = 16$ trials. The two levels are represented by $+1$ and -1, analogous to the high and low values in any two-level factorial design. For the remaining trials, three additional levels were selected: zero represents a base level midway between the aforementioned levels, and the levels noted as 1.547 represent extreme (or axial) values.

The translation of the statistical design into physical units is shown in Table 5. Again the formulations were prepared and the responses measured. The data were subjected to statistical analysis, followed by multiple regression analysis. This is an important step. One is not looking for the best of the 27 formulations, but the "global best." The type of predictor equation used with this type of design is a second-order polynomial of the following form:

$$Y = a_0 + a_1X_1 + \cdots + a_5X_5 + a_{11}X_1^2 + \cdots + a_{55}X_5^2 + a_{12}X_1X_2 + a_{13}X_1X_3 + \cdots \qquad (9)$$
$$+ a_{45}X_4X_5$$

Table 3 Response Variables
(Dependent)

Y_1	Disintegration time
Y_2	Hardness
Y_3	Dissolution
Y_4	Friability
Y_5	Weight uniformity
Y_6	Thickness
Y_7	Porosity
Y_8	Mean pore diameter

Table 4 Experimental Design

Trial	X_1	X_2	X_3	X_4	X_5
	— Factor Level in Experimental Units				
1	−1	−1	−1	−1	1
2	1	−1	−1	−1	−1
3	−1	1	−1	−1	−1
4	1	1	−1	−1	1
5	−1	−1	1	−1	−1
6	1	−1	1	−1	1
7	−1	1	1	−1	1
8	1	1	1	−1	−1
9	−1	−1	−1	1	−1
10	1	−1	−1	1	1
11	−1	1	−1	1	1
12	1	1	−1	1	−1
13	−1	−1	1	1	1
14	1	−1	1	1	−1
15	−1	1	1	1	−1
16	1	1	1	1	1
17	−1.547	0	0	0	0
18	1.547	0	0	0	0
19	0	−1.547	0	0	0
20	0	1.547	0	0	0
21	0	0	−1.547	0	0
22	0	0	1.547	0	0
23	0	0	0	−1.547	0
24	0	0	0	1.547	0
25	0	0	0	0	−1.547
26	0	0	0	0	1.547
27	0	0	0	0	0

Source: Adapted from Ref. 17.

Table 5 Experimental Conditions

Factor	−1.547 eu	−1 eu	Base 0	+1 eu	+1.547 eu
X_1 = Calcium phosphate/ lactose ratio (1 eu = 10 mg)	24.5/55.5	30/50	40/40	50/30	55.5/24.5
X_2 = Compression pressure (1 eu = 0.5 ton)	0.25	0.5	1	1.5	1.75
X_3 = Corn starch disintegrant (1 eu = 1 mg)	2.5	3	4	5	5.5
X_4 = Granulating gelatin (1 eu = 0.5 mg)	0.2	0.5	1	1.5	1.8
X_5 = Magnesium stearate (1 eu = 0.5 mg)	0.2	0.5	1	1.5	1.8

Source: Ref. 17.

where Y is the level of a given response, a_{ij} the regression coefficients for second-order polynomial, and X_i the level of the independent variable. The full equation has 21 terms, and one such equation is generated for each response variable. The usefulness of the equation is evaluated by the R^2 value, or the index of determination, which is an indication of the fit. In most cases the fit was satisfactory, and the equations were used. One possible disadvantage of the procedure as it is set up is that not all pharmaceutical responses will fit a second-order regression model. In fact, further analysis was attempted, and the results indicated that one of the responses was adequately described by a modified third-order model (interaction terms were eliminated.) However, a significant advantage of the digital system utilized is that it can be modified to accept other mathematical models—another order polynomial, any other empirical relationship, or a mathematical model based on first principles.

For the optimization itself, two major steps were used: the feasibility search and the grid search. The feasibility program is used to locate a set of response constraints that are just at the limit of possibility. One selects the several values for the responses of interest (i.e., the responses one wishes to constrain) and a search of the response surface is made to determine whether a solution is feasible. For example, the constraints in Table 6 were fed into the computer and were relaxed one at a time until a solution was found. The first feasible solution was found at disintegration time = 5 min, hardness = 10 kg, and dissolution = 100% at 50 min. This program is designed so that it stops after the first possibility; it is not a full search. The formulation obtained may be one of many possibilities satisfying the constraints.

The next step, the grid search, is essentially a brute-force method in which the experimental range is divided into a grid of specific size and methodically searched. The method is called an exhaustive grid search. From an input of the desired criteria, the program prints out all points (formulations) that satisfy the constraints.

The purpose of the preliminary step of the feasibility program is simply to limit the number of solutions in the grid search. In addition to providing a printout of each formulation, the grid search program also gives the corresponding values for the responses. At this point, the experimenter can trade off one response for another, and the fewer possibilities there are, the easier the job. Thus, the best or most acceptable formulation is selected from the grid search printout to complete the optimization.

The two steps just discussed require that one or more responses be constrained, and a question may arise of which ones to select. The formulator may have certain basic constraints,

Table 6 Specifications for Feasibility Search

Variable	Constraint	Experimental range[a]
Disintegration time (min)	1 (1)[b]	1.33–30.87
	3 (2)	
	5 (3)	
Hardness (kg)	12 (1)[b]	3.82–11.60
	10 (2)	
	8 (3)	
Dissolution (% at 50 min)	100 (1)[b]	13.30–89.10
	90 (2)	
	80 (3)	

[a]It is possible to request values for a response that are more desirable than any data obtained in the set of 27 experiments.
[b](1) = first choice.

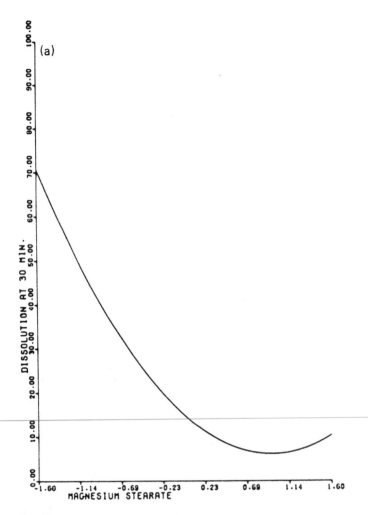

Fig. 11   Computer-generated plots for a single variable.

such as a minimum hardness value, but it is, nevertheless, important to know which property or properties can be used to distinguish between the available choices. Generally, this is done by an educated guess, based on experience with the system and with pharmaceutical systems in general.

However, there is a mathematical method of selecting those variables that best distinguish between formulations—those variables that change most drastically from one formulation to another and that should be the criteria on which one selects constraints. A multivariate statistical technique called *principal component analysis* (PCA) can effectively be used to answer these questions. PCA utilizes a variance–covariance matrix for the responses involved to determine their interrelationships. It has been applied successfully to this same tablet system by Bohidar et al. [18].

In addition to the programs to select the optimum discussed previously, graphic approaches are also available and graphic output is provided by a plotter from computer tapes. The output includes plots of a given response as a function of a single variable (Fig. 11) or as a function of all five variables (Fig. 12). The abscissa for both types is produced in experimental units,

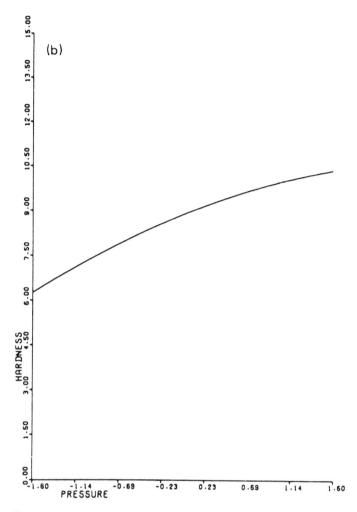

Fig. 11 Continued.

rather than physical units, so that it extends from −1.547 to +1.547 (see Table 5). Use of the experimental units allows the superpositioning of the single plots (see Fig. 11) to obtain the composite plots (see Fig. 12).

An infinite number of these plots is possible, since for each curve represented, four of the five variables must remain constant at some level. This is analogous to a partial derivative situation, and the slope of any one graph does indeed represent a partial derivative of the response for one of the independent variables. It will change, depending on the level of the other four variables.

Contour plots (Fig. 13) are also generated in the same manner. The specific response is noted on the graph and, again, the three fixed variables must be held at some desired level. For the contour plots shown, both axes are in experimental units (eu). This technique is automated so that a formulator with no previous computer experience and no familiarity with the mathematics of optimization can follow the steps necessary to complete such a study. Those steps may be summarized as follows.

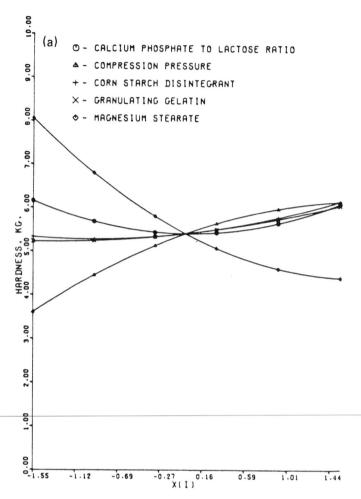

Fig. 12 Computer-generated composite plots.

1. Select a system.
2. Select variables:
   a. Independent
   b. Dependent
3. Perform experiments and test product.
4. Submit data for statistical and regression analysis.
5. Set specifications for feasibility program.
6. Select constraints for grid search.
7. Evaluate grid search printout.
8. Request and evaluate.
   a. "Partial derivative" plots, single or composite
   b. Contour plots

The last step, which concerns the graphic techniques, may be requested at any time after the regression analysis has been performed and will probably be appropriate at several different stages of a project.

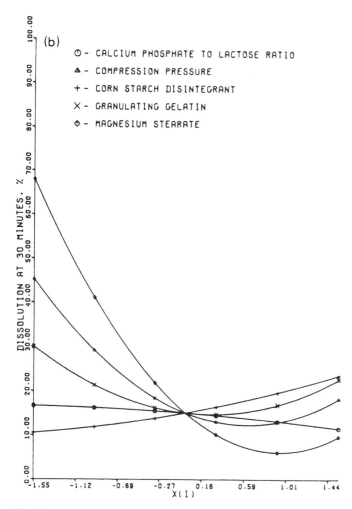

(b)

⊙ – CALCIUM PHOSPHATE TO LACTOSE RATIO
△ – COMPRESSION PRESSURE
+ – CORN STARCH DISINTEGRANT
× – GRANULATING GELATIN
◇ – MAGNESIUM STEARATE

Fig. 12 Continued.

The key to successful application of the experimental optimization techniques is based on adequate experimental design. A system based on this experimental design (see Table 4), but utilizing a special analog computer for analysis, was presented by Claxton [19] as the Firestone Computer/Optimizer.

This approach demonstrates that use of only a part of this procedure will represent a step forward over the trial-and-error method of formula and process modification. It is not always necessary to carry these studies to completion. For example, once the designed experimentation has been completed, one might be able to accomplish the task simply by analyzing the graphs; therefore, further mathematical treatment or search programs will not be necessary. Some of the examples in the following section illustrate this fact.

E. Canonical Analysis

Canonical analysis, or canonical reduction, is a technique used to reduce a second-order regression equation, such as Eq. (9), to an equation consisting of a constant and squared terms,

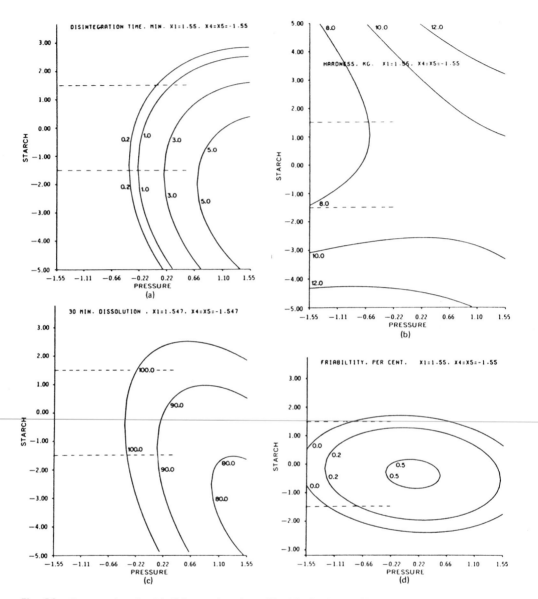

Fig. 13 Contour plots for (a) disintegration time; (b) tablet hardness; (c) dissolution response (%); (d) tablet friability as a function of disintegrant level and compressional force. Dashed lines on ordinate denote limits of experimental range (−1.547 to +1.547 eu; see the text).

as follows:

$$Y = Y_0 + \lambda_1 W_1^2 + \lambda_2 W_2^2 + \lambda_3 W_3^2 + \cdots \tag{10}$$

The technique allows immediate interpretation of the regression equation by including the linear and interaction (cross-product) terms in the constant term (Y_0 or stationary point), thus simplifying the subsequent evaluation of the canonical form of the regression equation. The first report of canonical analysis in the statistical literature was by Box and Wilson [37] for determining optimal conditions in chemical reactions. Canonical analysis, or canonical reduction,

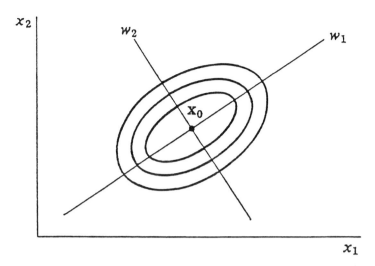

Fig. 14 Two-dimensional representation of the rigid rotation and translation involved in canonical analysis.

was described as an efficient method to explore an empirical response surface to suggest areas for further experimentation. In canonical analysis or canonical reduction, second-order regression equations are reduced to a simpler form by a rigid rotation and translation of the response surface axes in multidimensional space, as shown in Fig. 14 for a two-dimension system. This mathematical technique, which makes use of eigenvalues and eigenvectors, is based in matrix algebra and is described in textbooks on response surface methodology [38,39].

A reported application of canonical analysis involved a novel combination of the canonical form of the regression equation with a computer-aided grid search technique to optimize controlled drug release from a pellet system prepared by extrusion and spheronization [28,29]. Formulation factors were used as independent variables and in vitro dissolution was the main response, or dependent variable. Both a minimum and a maximum drug release rate was predicted and verified by preparation and testing of the predicted formulations. Excellent agreement between the predicted values and the actual values was evident for the four-component pellet system in this study.

VI. OTHER APPLICATIONS

In the last few years, optimization techniques have become more widely used in the pharmaceutical industry. Some of these have appeared in the literature, but a far greater number remain as "in-house" information, using the same techniques indicated in this chapter, but with modifications and computer programs specific to the particular company. An excellent review of the application of optimization techniques in the pharmaceutical sciences was published in 1981 [20]. This covers not only formulation and processing, but also analysis, clinical chemistry, and medicinal chemistry.

Designed experimentation, involving mostly some type or modification of factorial design, has been used to study many different types of formulation problems. These include a pharmaceutical suspension [21], a controlled-release tablet formulation [22], and a tablet-coating operation [23]. In the latter case, Dincer and Ozdurmus studied an enteric film coating and utilized the steepest descent graphic method to select the optimum.

Adaptation of the modified factorial techniques to desk-top computers has also been accomplished [24,25]. Down et al. [25] have presented this concept and have applied the programs to a tablet problem. The statistics involved are presented in some detail. A similar design was also used to study a high-performance liquid chromatography (HPLC) analysis [26]. In an unusual application, optimization techniques were even used to study the formulation of a culture medium in the field of virology [27].

Other applications of the previously described optimization techniques are beginning to appear regularly in the pharmaceutical literature. A recent literature search in *Chemical Abstracts* on process optimization in pharmaceuticals yielded 17 articles in the 1990–1993 timeframe. An additional 18 articles were found between 1985 and 1990 for the same narrow subject. This simple literature search indicates a resurgence in the use of optimization techniques in the pharmaceutical industry. In addition, these same techniques have been applied not only to the physical properties of a tablet formulation, but also to the biological properties and the in vivo performance of the product [30,31]. In addition to the usual tablet properties, the authors studied the pharmacokinetic parameters (a) time of the peak plasma concentration, (b) lag time, (c) absorption rate constant, and (d) elimination rate constant. The graphs in Fig. 15 show that for the drug hydrochlorothiazide, the time of the plasma peak and the absorption rate constant could, indeed, be controlled by the formulation and processing variables involved.

VII. COMPUTERS AND SYSTEMS

It is obvious that the use of computers will facilitate the data analysis steps in the procedures discussed and will be needed for any mathematical analysis or search methods. In fact, a textbook has appeared describing the practical application of computer-aided optimization and provides direction for the implementation of these techniques to formulation [41]. Most of the examples presented have made use of computers in some way, and a few were completely performed by computer.

Several companies have adapted these experimental analysis techniques to computer software, but have kept the programs in-house. Representatives of a few have, in fact, presented data at various conferences [24,32–34]. However, there are several commercially available programs that may be bought or licensed and several courses in experimentation address this subject.

The interested reader might be alerted to courses offered by the American Chemical Society, Dupont Corporation, or the Foremost Corporation. Some of these programs offer the use of statistical or response surface software. Specific computer packages are also available through Statistical Analysis Systems (SAS), IBM, and RS/Discover (RS1), which are designed for mainframe computers. The number of software packages available for standard desk-top office computers is large and is expected to increase. The following software packages, ECHIP, XStat, and Design Expert, are commonly used in the pharmaceutical industry, but these titles do not provide a complete list of available programs.

VIII. CONCLUDING REMARKS

As the list of applications illustrates, the techniques of optimization are not limited to tablets or even to solids. Any dosage form and any process should be amenable to this type of experimentation and analysis. From the most simple formulation to the most complicated one, there are ingredient levels and processing steps that can be varied, and any information on the result of such variation should be useful to the formulator.

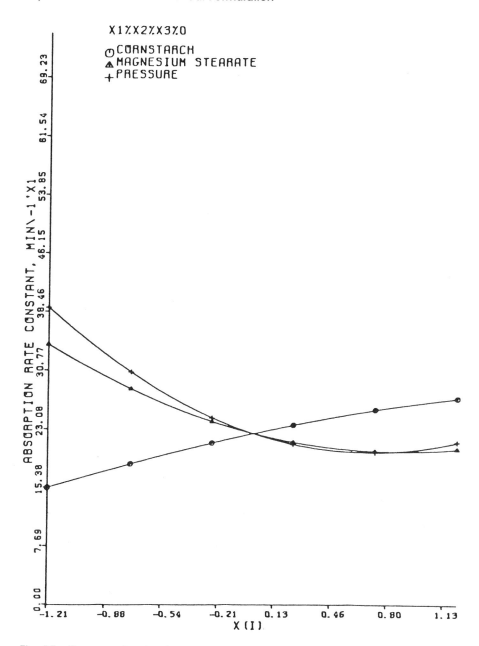

Fig. 15 Computer plots for absorption rate constant and time of plasma peak. (From Ref. 31.)

Properly designed experimentation and subsequent analysis can lead not only to the optimum or most desirable product and process, but, if carried far enough, can shed light on the mechanism by which the independent variables affect the product properties. There are appropriate statistical techniques involving the use of selective regression analysis by which such analyses can be carried out [35].

By appropriate analysis and generation of model (regression) equations (which are continuous), the formulator is able to select not the best of the formulations experimentally prepared,

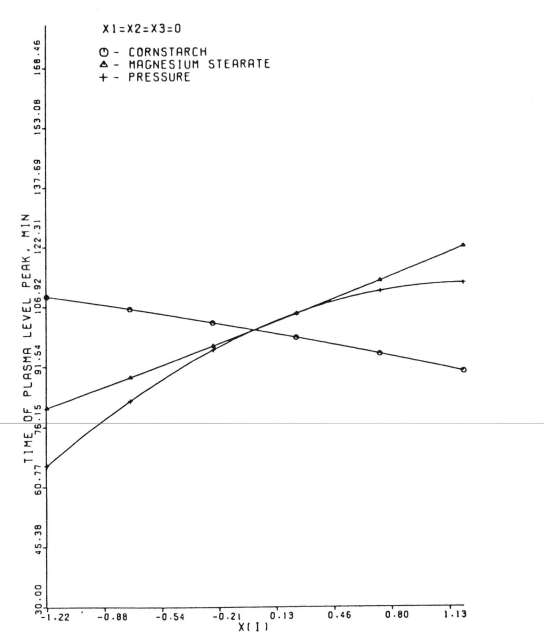

Fig. 15 Continued.

but the best within experimental range; the optimum may be a combination of ingredients that the formulator has never prepared (and might never think to prepare).

These techniques of optimization can be useful, even if selecting the optimum is not the primary objective. The formulator may have no intention of drastically changing a given formulation. Many times a very small change in processing or ingredient level can dramatically improve a particular property. The use of such information in "troubleshooting" situations has been demonstrated [36].

The independent variables have been, or should have been, selected by the formulator, and there is no substitute for experience. Experience with the system or with pharmaceutical systems in general can guide the formulator to select those variables most likely to have an effect on those levels which are most practical. The results of an optimization study, especially the graphic output, can give direction for product improvement—no matter why the improvement is necessary or desirable.

Once experimental data have been collected and relationships generated by regression analysis (or even derived from first principles), the formulator has many options available for subsequent analysis. These need not be restricted to mathematical techniques or to elaborate computerized systems.

A side benefit of this designed type of experimentation is its potential usefulness in product and process validation. The subject of validation is of great interest to those in the operations area, but if approached rationally, validation must begin in the product development phase. The designs usually selected lend themselves to the concept of processing limits and "challenge." The resulting data can be applied to scale-up, can aid in the transfer of information to the operations area, and should be the basis of the protocol design for validation.

The emphasis, once again, is that appropriate statistical design is an important consideration. For a formulator planning such a study, it should be noted that the independent variables can be anything that he or she can quantitate and control; and the dependent variables can be anything that he or she can quantitate. From the data resulting from the required number of experiments, one is able to generate a mathematical model to which the appropriate optimization technique is applied (e.g., graphic, mathematical, or the search method).

The final conclusion is the ultimate benefit: The more the formulator knows about a system, and the better that he or she can define it, the more closely it can be controlled.

REFERENCES

1. *Webster's New Collegiate Dictionary*, G. & C. Merriam, Springfield, MA, 1974.
2. L. Cooper and D. Steinberg, *Introduction to Methods of Optimization*, W. B. Saunders, Philadelphia, 1970.
3. O. L. Davis (ed.), *The Design and Analysis of Industrial Experiments*, Macmillan (Hafner Press), New York, 1967. [2nd Ed., Longman, Harlow, Essex, England, 1978.]
4. W. G. Cochran and G. M. Cox, *Experimental Designs*, John Wiley & Sons, New York, 1957, Chap. 8A.
5. G. E. P. Box, W. G. Hunter, and J. S. Hunter, *Statistics for Experimenters: An Introduction of Design, Data Analysis and Model Building*, John Wiley & Sons, New York, 1978.
6. S. N. Deming and P. G. King, Res./Dev., p. 22, May (1974).
7. M. H. Rubinstein, Manuf. Chem. Aerosol News, p. 30, Aug. (1974).
8. T. N. DiGaetano, Bull. Parenter. Drug Assoc., 29, 183 (1975).
9. W. Spendley, G. R. Hext, and F. R. Himsworth, Technometrics, 4, 441 (1962).
10. E. Shek, M. Ghani, and R. Jones, J. Pharm. Sci., 69, 1135 (1980).
11. S. T. Anik and L. Sukumar, J. Pharm. Sci., 70, 897 (1981).
12. C. Bindschaedler and R. Gurny, Pharm. Acta Helv., 57(9), 251 (1982).
13. D. Janeczek, M. S. thesis, Philadelphia College of Pharmacy and Science (1979).
14. S. L. Morgan and S. N. Deming, Anal. Chem., 46, 1170 (1974).
15. D. E. Fonner, Jr., J. R. Buck, and G. S. Banker, J. Pharm. Sci., 59, 1587 (1970).
16. J. R. Buck, G. E. Peck, and G. S. Banker, Drug Dev. Commun., 1, 89 (1974–75).
17. J. B. Schwartz, J. R. Flamholz, and R. H. Press, J. Pharm. Sci., 62, 1165 (1973).
18. N. R. Bohidar, F. A. Restaino, and J. B. Schwartz, J. Pharm. Sci., 64, 966 (1975).

19. W. E. Claxton, paper presented to the Industrial Pharmaceutical Technology Section, American Pharmaceutical Association, Academy of Pharmaceutical Sciences, San Francisco meeting, Mar. 1971.
20. D. A. Doornbos, Pharm. Weekbl. Sci. Ed., 3, 33 (1981).
21. J. L. O'Neill and R. E. Dempski, through J. Soc. Cosmet. Chem., 32, 287 (Sept./Oct. 1981).
22. M. Harris, J. B. Schwartz, and J. McGinity, Drug Dev. Ind. Pharm., (in press).
23. S. Dincer and S. Ozdurmus, J. Pharm. Sci., 66, 1070 (1977).
24. D. R. Savello, K. J. Koenig, K. R. Nelson, J. B. Schwartz, and C. G. Thiel, paper presented to the American Pharmaceutical Association, Academy of Pharmaceutical Sciences, Atlanta meeting, Nov. 1975.
25. G. R. B. Down, R. A. Miller, S. K. Chopra, and J. F. Millar, Drug Dev. Ind. Pharm., 6, 311 (1980).
26. M. L. Cotton and G. R. B. Down, J. Chromatogr., 259, 17 (1983).
27. E. M. Scattergood, J. B. Schwartz, M. O. Villarejos, W. J. McAleer, and M. R. Hilleman, Drug Dev. Ind. Pharm., 9, 745 (1983).
28. R. E. O'Connor, N. R. Bohidar, and J. B. Schwartz, Optimization by Canonical Analysis Applied to Controlled Release Pellets, Abstracts of the 1st National AAPA Meeting, Washington, D.C., Nov. 2–6, 1986.
29. R. E. O'Connor, The Drug Release Mechanism and Optimization of a Microcrystalline Cellulose Pellet System, Ph.D. dissertation, Philadelphia College of Pharmacy & Science, 1987.
30. A. N. Karabelas, R. W. Mendes, and J. B. Schwartz, paper presented to the Industrial Pharmaceutical Technology Section, American Pharmaceutical Assocation, Academy of Pharmaceutical Sciences, San Antonio meeting, Nov. 1980.
31. J. B. Schwartz, J. Soc. Cosmet. Chem., 32, 287, Sept/Oct. (1981).
32. A. R. Lewis and R. Poska, paper presented to the Midwest Regional IPT meeting, American Pharmaceutical Association, Academy of Pharmaceutical Sciences, May 1981.
33. D. R. Savello, Workshop presentation, American Pharmaceutical Association, Academy of Pharmaceutical Sciences, Las Vegas meeting, April 1982.
34. J. E. Brown, paper presented at RxPo Conference, New York, June 1984.
35. N. R. Bohidar, F. A. Restaino, and J. B. Schwartz, Drug Dev. Ind. Pharm., 5, 1975 (1979).
36. J. B. Schwartz, J. R. Flamholz, and R. H. Press, J. Pharm. Sci., 62, 1518 (1973).
37. G. E. P. Box and K. B. Wilson, J. R. Statist. Soc. Ser. B, (Methodological), 13, 1 (1954).
38. G. E. P. Box and N. R. Draper, *Empirical Model Building and Response Surfaces*, John Wiley & Sons, New York, 1987.
39. R. H. Myers, *Response Surface Methodology*, Allyn & Bacon, Boston, 1971.
40. D. G. Kleinbaum, L. L. Kupper, and K. E. Muller, *Applied Regression Analysis and Other Multivariate Methods*, PWS-Kent, Boston, 1988.
41. A. H. Bohl (ed.), *Computer-Aided Formulation*, VCH, New York, 1990.

19

Food and Drug Laws that Affect Drug Product Design, Manufacture, and Distribution

Garnet E. Peck
Purdue University, West Lafayette, Indiana

I. IMPACT OF THE FOOD AND DRUG LAWS

A. Historical Perspective: Effect of the 1906 Act on Drug Product Distribution

The establishment of a set of laws to control the purity of the food and drink offered to the public can be traced back in history to 1202, when King John of England issued the first English food law, which included the prohibition of the adulteration of bread with such ingredients as ground peas or beans [1]. The earliest history of food and drug laws in the United States occurred as follows:

1784   Massachusetts enacted the first general food law in the United States.
1848   The Import Drugs Act—first federal statute to ensure the quality of drugs—was passed when quinine used by American troops in Mexico to treat malaria was found to be adulterated.
1850   California passed a pure food and drink law 1 year after the gold rush.
1879–1905   During these 25 years, more than 100 food and drug bills were introduced into Congress.
1879   Chief Chemist Peter Collier, Division of Chemistry, U.S. Department of Agriculture, began investigating food and drug adulteration. The following year he recommended enactment of a national food and drug law.

Harvey Wiley, who became the chief chemist of the U.S. Bureau of Chemistry in 1883, is generally considered to be the father of the original U.S. pure food and drug laws. His life's work was concerned with food and drug adulteration. He not only championed the first Federal Food, Drug, and Cosmetic Act, but can also be considered the federal government's first strong consumer advocate [2]. In the eighteenth and nineteenth centuries in the United States, drugs of very dubious quality, as well as many quack or patent remedies that ranged from valueless, to harmful, to addicting, were distributed and sold totally without control.

In 1902 Congress passed the Biologics Control Act, which licensed and regulated the interstate sale of serums and vaccines used to prevent and treat diseases in humans. The effect of this would seem obvious because, after much discussion before the Committee on Interstate and Foreign Commerce, the final result was passage of the Federal Food, Drug and Cosmetic Act of 1906 [2]. This bill was signed into law by President Theodore Roosevelt on June 30, 1906 [3]. The Meat Inspection Act was also passed that same day. This was the result of shocking disclosures of unsanitary conditions of meat-packing plants. It is obvious that numerous problems did exist in the food and drug supply of the United States before the twentieth century. They were not simply centered around dangerous preservatives and dyes in, and adulteration of, foods and alcoholic beverages, as well as totally unfounded claims and worthless to dangerous patent medicines, but included problems in the total distribution and control of the safety, purity, and quality of all foods and drugs supplied to the public.

The next important drug-related law was the Caustic Poison Act of 1927, which required warning labels and antidotes on ten dangerous or corrosive substances to be sold for household use [1]. As we reflect on the history of the food and drug laws, we find that changes came about because of accidents or improper use of substances in drug products or active agents being used without adequate testing. Selected and probably rare episodes of negligence on the part of a few pharmaceutical manufacturers led to new drug laws or amended laws. Both the 1938 law and the 1962 amendments, were triggered in this manner. In 1937, a new wonder drug, sulfanilamide, was formulated into an elixir that contained ethylene glycol, rather than propylene glycol, with the result that at least 107 people were killed [4]. The resulting public outcry led, at least in part, to the major revision of the Federal Food, Drug, and Cosmetic Act the following year. This 1938 act contained the following new provisions [5]:

Extended coverage to cosmetics and devices
Required predistribution clearance for safety of new drugs
Eliminated Sherley Amendment requirement to prove intent to defraud in drug-misbranding cases
Provided for tolerances for unavoidable poisonous substances
Authorized standards of identity, quality, and fill of container for foods
Authorized factory inspections
Added the remedy of court injunction to previous remedies of seizure and persecution

This revision concerned itself not only with new drug substances, but extended to cosmetics and devices.

Trade associations and the compendia also played a role in drug product control and quality in the nineteenth and twentieth centuries. The *United States Pharmacopeia* (*USP*) and *National Formulary* (*NF*) established standards of potency and purity for the most commonly used drugs and drug products. The first *USP* was issued in 1833 and the first *NF* was published in 1887 [6]. In the early 1900s two trade associations existed that were to benefit the industry and attempt to bring them together on common causes. These two trade associations were the American Drug Manufacturers Association and the American Pharmaceutical Manufacturers Association. In 1924 a committee was formed between these two organizations, which was known as the Contact Committee. This committee's purpose was to examine the need of the pharmaceutical industry for standards, tolerances, and test methodology for tablet and parenteral products [7]. This Contact Committee later formed the quality control section of the Pharmaceutical Manufacturers Association, which was the union of both original associations concerned with pharmaceutical manufacturing. This represented the industry's own attempts to better control the manufacturer of drug products by establishing better standards and test methods. Standards were issued in a loose-leaf publication entitled *Pharmaceutical Standards Including Tolerances and Methods of Analyses*, first issued in 1924 and revised over the following 2 years to be more closely linked to the official compendia, the *USP* and the *NF*. The Contact Committee served as the sounding board for drug product quality for a number of years.

In 1940, the Food and Drug Administration (FDA) was transferred from the U.S. Department of Agriculture to the Federal Security Agency, after which the First Commissioner of Food and Drugs was named [1]. In 1945 the Food, Drug, and Cosmetic Act was amended to require the certification of the safety and efficacy of penicillin. The act has since been amended many times over the years to include all antibiotics and antibiotic products. The next major revision came in 1951 in the form of the Durham–Humphrey Amendment, which required the labeling of prescription items: "Legend drugs may not be dispensed without a prescription from a physician, dentist, or other designated practitioner." In 1955, it was recommended that the FDA expand its facilities to improve its educational and informational programs. Citizen input provided some of the impetus to have this done. In 1960 the color additive amendments were enacted, which allowed FDA to establish by regulations the conditions of safe use of color additives in foods, drugs, and cosmetics, and to require manufacturers to perform the necessary studies to establish safety. As happened previously following a serious drug toxicity episode, new legislation passed in 1962, known as the Kefauver–Harris Drug Amendments, required a much greater degree of safety assurance and strengthened the new drug clearance procedures. This legislation was a result of the use of thalidomide in Western Europe, Canada, and to the limited extent the United States, on a clinical trial. This particular drug substance produced malformed babies when taken by pregnant mothers. However, it should be noted that the Pharmaceutical Manufacturers Association was also very active during this time in trying to improve the law. For example, in May 1961, they issued a document entitled *Principles of Control of Quality in the Drug Industry*. This combined the thoughts of many of the manufacturers and their concern over improving drug product quality [7]. From this point on, drugs and drug products for distribution in the United States were required to show clear evidence of safety, effectiveness, and high quality, and to be manufactured by prescribed procedures, as regulated by law, but also as intended by the major drug manufacturers.

B. Functions and Organization of the Federal Food and Drug Administration

The FDA was created to administer the Federal Food, Drug, and Cosmetic Acts and their various amendments; currently, this agency functions within the U.S. Department of Health and Human Services. Figure 1 outlines the major divisions of the FDA. As shown in the figure, the Commissioner of the FDA has reporting to him or her a number of associate commissioners who are responsible for various centers which, in turn, have responsibility for the various types of products that currently come under FDA regulations. The FDA also has several offices that are concerned with various functions of the agency. The Office of Compliance, for example, is concerned with inspection of manufacturing facilities and assurance that they are adhering to that section of the regulations entitled Current Good Manufacturing Practices (GMPs). These manufacturing practices are summarized later in this chapter. Figure 2 gives a detailed outline of the Center of Drugs and Biologics, which is the FDA division having the greatest relevance to most pharmacists in their professional practice. This center is divided into six major parts, plus an office for planning and evaluation. Many of these component parts are self-explanatory. The Center of Drug Evaluation and Research in the FDA division responsible for the approval of all new human prescription drugs marketed in the United States; the reevaluation of older

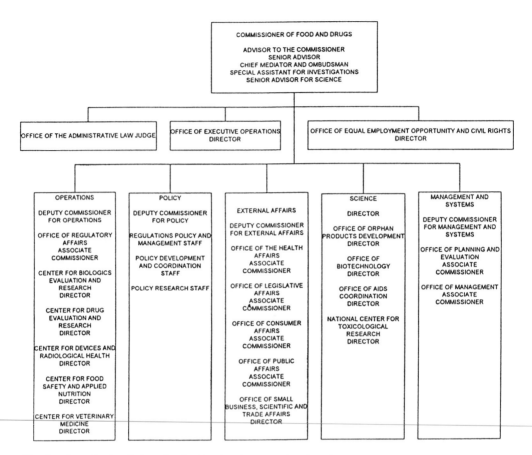

Fig. 1 Organizational chart for the Food and Drug Administration.

drugs; evaluation, approval, and control of over-the-counter (OTC) drugs (i.e., drugs sold without prescription); and basically is concerned with the regulation of the manufacture, distribution, and sale of all drugs involved in interstate commerce. The Center for Medical Devices and Radiological Health (see Fig. 1) has become more prominent in the regulatory area. As new regulations are promulgated to govern medical devices and diagnostic aids, this center will have an increasing impact on pharmacy practice, especially in the hospital field.

Figure 3 describes the functional units of the Center of Veterinary Medicine. Many pharmacy practitioners are also involved in the drugs controlled by this unit of the FDA and may have occasion to communicate with it.

II. LAWS GOVERNING EVALUATION OF NEW DRUG PRODUCTS

A. Claimed Investigational Exemption for a New Drug

When the Federal Food, Drug, and Cosmetic Act of 1938 was passed, a new era of drug product development began. It was the beginning of a requirement for the preclearance of a drug product before its marketing. The act required the assurance of safety and stated minimum requirements for manufacturing and quality control. It required only 60 days for review by the

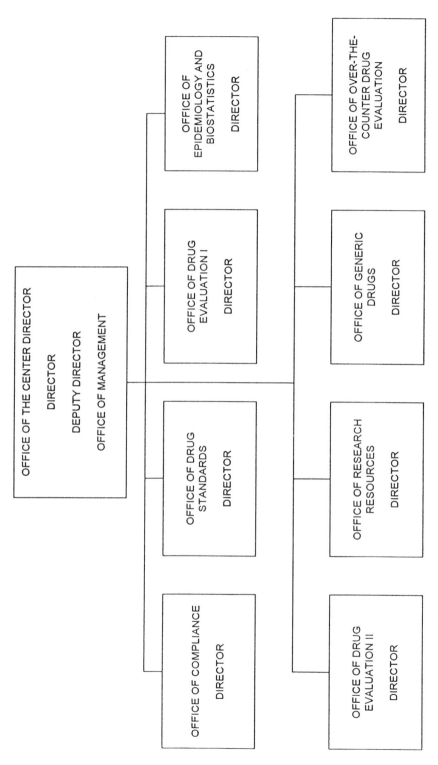

Fig. 2 Organizational chart for the Center of Drug Evaluation and Research, FDA.

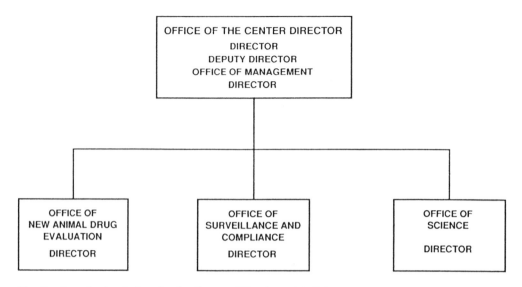

Fig. 3 Organizational chart for the Center of Veterinary Medicine, FDA.

FDA before the distribution of any new drug product [8]. The 1962 Kefauver–Harris Amendments to the act required greatly increased information concerning the safety, effectiveness, manufacture, and controls of drug products [1]. For the first time it was also necessary for a drug firm or other organization interested in a new drug substance to inform the FDA of the desire for the testing of the drug in humans. The document to initiate a human study involving a new drug (which includes not only new drug substances, but also new dosage forms for existing drugs) is known officially as a Notice of Claimed Investigational Exemption for a New Drug (Form FD 1571; also often called the IND). These regulations appear in Section 505 of the act [9]. A "new drug" is one not presently recognized by experts in the field of clinical pharmacology to be safe and effective based on currently available clinical evidence. All the definitions for a new drug appear in the following regulations for a new drug entity [10]:

1. *New-drug substance* means any substance that when used in the manufacture, processing, or packing of a drug causes that drug to be a new drug, but does not include intermediates used in the synthesis of such substances.
2. The newness of a drug may arise by reason (among other reasons) of:

   a. The newness for drug use of any substance which composes such drug, in whole or in part, whether it be an active substance or a menstruum, excipient, carrier, coating, or other component.

   b. The newness for a drug use of a combination of two or more substances, none of which is a new drug.

   c. The newness for drug use of the proportion of a substance in a combination, even though such combination containing such substance in other proportion is not a new drug.

   d. The newness of use of such drug in diagnosing, curing, mitigating, treating, or prevention a disease, or to affect a structure or function of the body, even though such drug is not a new drug when used in another disease or to affect another structure or function of the body.

e. The newness of a dosage, or method of duration of administration or application, or other condition of use prescribed, recommended, or suggested in the labeling of such drug, even though such drug when used in other dosage, or other method or duration of administration or application, or different condition, is not a new drug.

To determine whether the filing of an IND (and human testing) is reasonable for a new drug substance, it is first necessary to conduct preclincal studies, which include:

1. Acute toxicity (to determine the safety of the drug in animals)
2. Preformulation studies covering the physical and chemical characterization of the new drug substance
3. Pharmacological screening and evaluation (to verify that a new drug substance has some potential for therapeutic activity)

Once such studies have been completed successfully, work may proceed to process the IND. Pharmacists working in major hospitals may become involved in the distribution, controlled use, or keeping the records of drugs or drug products being tested under an IND. The necessary information needed to prepare the IND is outlined in Table 1. The IND is to contain information on appropriate prior animal studies for safety evaluation, any available clinical data, adequate drug identification and manufacturing instructions, and a detailed outline of the proposed clinical study, routes of administration, approximate number of patients to be used, and an estimate of the length of treatment and an environmental impact statement may also be needed.

The clinical evaluation of a new drug is divided into three phases. The first phase of human testing considers what chemical actions a drug has, how it is absorbed by the body, and what is a safe dosage range [10]. Normally, fewer than ten patients are used as subjects. This phase is to describe the human pharmacology of the drug and the preferred route of administration. Phase II studies involve the dosing of a limited number of patients for treatment or prevention of the disease of interest. This step evaluates the effectiveness of the drug. The amount of animal toxicity testing will increase if the drug appears to be a favorable candidate for extended study. Phase III studies are the next step. The drug is assessed for its safety, effectiveness, and

Table 1 Components of a Notice of Claimed Investigational Exemption for a New Drug (IND)

1. Best available descriptive name of the drug product and how it is to be administered.
2. Complete list of components of the drug product
3. Quantitative composition of the drug product
4. Description of the source and preparation of any new drug substance used as a component
5. Description of the methods, facilities, and controls used to prepare and distribute the drug product
6. Information available from preclinical investigations and any clinical studies that may have been conducted
7. Copies of any labels or labeling to be used in the study
8. Scientific training and experience to qualify the investigators
9. Complete information of each investigator, including training and experience
10. Outline of any phase or phases of the planned clinical investigation
11. Assurance that the Food and Drug Administration will be notified if the studies are discontinued
12. Assurance that investigators will be notified if a new drug application is approved or the studies discontinued
13. Explanation of why a drug product must be sold during a study
14. Agreement not to start a study for 30 days before a notice of receipt of the IND by the Food and Drug Administration

most desirable dosage for the disease to be treated and the results are verified in a large number of patients.

The sponsor is required to obtain statements from the clinical investigators describing his or her qualifications to take part in the proposed studies. Reference to the *review committee* in this document is to the institutional human review committee of the hospital or other facility where the study will be undertaken. Note that both documents require that the investigator certify that all subjects (patients), whether used as controls or test subjects, be notified that they are receiving drugs being used for investigational purposes, and that their consents are obtained. After the IND has been filed with the FDA and the date of receipt noted, a 30-day waiting period is needed before the initiation of the studies by the sponsor. Following this period, and if no objections have been raised by the FDA, the drug may be shipped to the clinical investigators, as long as the label bears the statement "Caution: New Drug–Limited by Federal (or United States) Law to Investigational Use." The sponsor must maintain records of distribution of the new drug for 2 years after approval of the drug for use; or if not approved, the records must be retained for 2 years after the last shipment and delivery to investigators. An additional important point of the 1962 amendment was the requirement that the FDA be informed if and when a study was stopped and for what reasons. Before this time, no record was made of any clinical study until the time came to apply for a new drug release.

B. New Drug Application

Once a new drug substance, or anything that is by definition a new drug, has been through Phase III (i.e., the clinical trial) and the appropriate data accumulated, a document is prepared entitled *New Drug Application* (*NDA Form 356H*). The data must normally be obtained under a specific IND that has permitted the gathering of the following information: acute and chronic toxicity, extended clinical pharmacology, full-scale clinical evaluation, complete product design, complete package design, and complete labeling requirements.

Because of today's myriad regulations, the NDA submission has become a compilation of information that could be compared in size with any one of the well-known encyclopedias. Before 1962, only a few volumes were necessary for an NDA submission [11]. Today it is not uncommon for this justification to consist of as many as 200 volumes of information (see Table 2). It is important for the pharmacy and medical practitioner to have some sensitivity to the many requirements and high cost placed on manufacturers who market new drugs (especially new drug substances) today.

Some of the points of the NDA that are quite significant will not be reviewed. Concerning the statement on labeling, it is necessary for manufacturers to abstract from their clinical summaries that which is both recommendef for the description of the use of the drugs and that which is necessary for the warning or cautions needed for the administration of the new products. The labeling, which includes the package insert, must assist all professionals in proper administration and use of new drugs. The labeling is also approved by the FDA to ensure that only those prescribed uses verified in the clinical studies are recommended [12]. The professional prescribing the drug may not legally use the substance for any other use. The only use for which the drug can be administered is that described in the package insert or other accompanying literature. Other points that are of great significance to the manufacturer and, in turn, to the physician and pharmacist are the methods used, the facilities and controls used, and the overall processing and packaging of any new drug substance. The NDA requires full descriptions of all of these to ensure safe and effective marketed drug substances. These descriptions are so detailed in the original NDA that even drying temperatures for tablets prepared by wet granulation must be closely adhered to in all subsequent manufactured lots. The NDA will

Table 2 Components of a New Drug Application (Drugs for Human Use)

1. Table of contents
2. Summary
3. Evaluation of safety and effectiveness
4. Copies of the label and all other labeling to be used for the drug
5. A statement as to whether the drug is (or is not) limited in its labeling and by this application to use under the professional supervision of a practitioner licensed by law to administer it
6. A full list of the articles used as components of the drug
7. A full statement of the composition of the drug
8. A full description of the methods used in, and the facilities and controls used for, the manufacture, processing, and packing of the drug
9. Samples of the drug and articles used as components
10. Full reports of preclinical investigation that have been made to show whether or not the drug is safe for use and effective use
11. List of investigators
12. Full reports of clinical investigations that have been made to show whether or not the drug is safe for use and effective in use
13. If this is a supplemental application, full information on each proposed change concerning any statement made in the approved application

contain descriptions of the exact equipment and processes, and the specifications of all drugs and excipients that are used for manufacture of the product. At a later date, if these are modified, the manufacturer may be required to submit a supplement to the NDA, which verifies that no quality features of the product (such as bioavailability) have been altered.

Many products have specific packaging requirements owing to inherent problems in their physical, chemical, and possibly biological stability. For this reason detailed information must be given for the packaging of any new drug covered by an NDA. The pharmacist or physician should be aware of the extensive research that goes into package selection to satisfy the FDA. Any repackaging of a drug product should, as a rule, attempt to place the product in a package that is at least as protective as the one from which it was removed.

Another important section of the NDA is a full report of all preclinical investigations of the new drug substance or new drug. These investigations ensure safety and efficacy of the new drug. No animal study, large or small, should be unreported. There have been instances in the past in which some animal data were not reported and problems resulted later when the drug was administered to humans. Ideally, the more animal data from chronic studies that are gathered, the better perception one will have for the drug's safety in humans. This information can be as important as the clinical work in patients to demonstrate the efficacy of the drug.

Of ultimate importance are the full reports of the clinical studies in humans and their results. These data will be treated statistically for their validity. The number of studies for a specific compound or combination of compounds will vary with the type of drug being tested, as well as the number of tests needed to appraise relative or absolute safety and to clearly demonstrate efficacy. The basic requirement is the proof of safety and efficacy of the product being submitted under the NDA system. A drug that does not contribute to therapy, as, for example, a new antihistamine that does not demonstrate greater safety or efficacy, or both, compared with drugs already on the market, will have a difficult or impossible time achieving approval.

After the new drug application has been compiled, it is forwarded to the FDA, where it receives a specific identification number and is submitted to review. However, in recent years, extensions of this time limit have been the rule, rather than the exception for the FDA to complete its review. These extensions are generally for a specific number of days or months. Once reviewed by the FDA, a response is returned to the sponsor that generally contains numerous questions concerning the application. The sponsor will then attempt to satisfy the questions raised and resubmit the application for review by the FDA. This process may be repeated for from a few months to several years, until the agency is convinced that the drug is, in fact, safe and effective. Once so proved, permission is granted to the sponsor to manufacture and distribute the product to the public. Currently, the period from the time of synthesis of a compound, or its isolation, to its release for marketing is generally some 10 years or longer.

C. Drug Efficacy Study

In 1966, the Commissioner of the FDA approached the Division of Medical Sciences of the National Academy of Sciences–National Research Council (NAS–NRC) for assistance in reviewing drugs that had been marketed under NDAs approved during the period 1938–1962 [13]. There were approximately 7000 drug formulations in this category, and of these about 4000, which contained 300 distinctly different medicinal agents, were still being marketed. Since these drugs had been approved under the old law, which did not include the new requirement of the 1962 amendments that the drug clearly demonstrate effectiveness, James Goddard, FDA Commissioner, sought the help of the NAS–NRC establish review committees to undertake this formidable review task. To proceed in this undertaking, a Policy Advisory Committee was first formed whose members needed to become acquainted with major medical, legal, and industrial problems in the field of drugs. Early in the development of the study, it was recognized that a system of evaluating panels would have to be established. Twenty-seven panels were originally formed. Later, additional panels were added as the need arose. Some of the panels formed concerned themselves with drugs used in allergy, anesthesiology, antineoplastic and cardiovascular therapy, dermatology, and other specific fields of medicine or drug therapy. Guidelines were established by the Policy Advisory Committee for the review activities of the panels, and appropriate individuals representing various scientific and professional disciplines were appointed to the panels. By October 1966, the review program was formerly organized and in operation. The type of products reviewed included single-drug entities, as well as products containing two or more active ingredients. The extensive duplication of products was most evident in the antibiotic, antihistimine, and antihypertensive areas.

One of the major problems in initiating the study was to attempt to establish a categorical rating of effectiveness. The FDA developed four effectiveness categories: (a) effective, (b) probably effective, (c) possibly effective, and (d) ineffective. In a little more than 2 years, a large amount of data was collected on more than 80% of the drugs currently on the market. The results of the NAS–NRS study and the effectiveness categorization of various products were periodically published by the FDA in the *Federal Register* (the official publication of the federal government for all notices, proposed rules, and regulations). In many cases, time was given for response by the companies affected by the results of this study. Some of the results were (a) the immediate removal of products currently marketed; (b) the recognition that certain clinical studies were required to verify or establish effectiveness under the new law and according to past product claims; and (c) the recognition that directions or labeling of certain products were poorly organized, repetitive, outdated, evasive, and oriented to the promotion of products.

As one broad positive impact of this study, it was recognized that package inserts needed to be brought up to modern standards of accurate and objective drug information. Several products were voluntarily removed from the market place by manufacturers who were unable or unwilling to establish the necessary proofs of efficacy. The number of recalls occurring during this evaluation should not be thought of as a poor scientific approach, but rather, an updating of knowledge. The industry had not fallen down as far as its drug product quality efforts were concerned, but these recalls were based strictly on the latest in drug rationale and utilization. Some of the types of products that ceased to exist following this review were antibiotics containing troches and lozenges for the treatment of minor sore throat; antibiotic, antihistamine, and analgetic/antipyretic combinations for treatment of colds and influenza; and many other drug combination products. This review was undoubtedly beneficial to the public, in that more stringent principles of rational therapeutics were applied to the products meeting the standards of the review.

D. Abbreviated New Drug Application

A detailed description of the concept of an Abbreviated New Drug Application (ANDA) was published in the *Federal Register* of June 20, 1975 [14]. The ANDA represents a form of new drug application in which certain information is not needed because previously acquired data has been filed with the FDA. In the ANDA procedure, the FDA's intent is to minimize duplication of effort in preparing applications for drugs about which some of the needed information is already available, while assuring that the new product will be equivalent to established marketed products. Normally, the information allowed to be omitted relates to preclinical and clinical studies pertaining to the safety and effectiveness of the active ingredient(s) that have been on the market for many years. Since many of the items are established generic drugs, it is unnecessary to reverify their efficacy. It may simply be necessary to have assurance that the manufacturing procedures, specifications, and labeling are adequate. In other cases, it will be necessary to demonstrate the bioequivalence of the new product relative to currently available standard products, or to the original NDA-approved product.

Unlike the NDA—which requires submission of well-controlled clinical studies to demonstrate effectiveness, data to show safety, and detailed description of the manufacturing and packaging of a drug as well as stability data—the ANDA requires only the following: a description of the components and composition of the dosage form to be marketed; brief statements that identify the place where the drug is to be manufactured; the name of the supplier of the active ingredients; assurance that the drug will comply with appropriate specifications; an outline of the methods and facilities used in the manufacture and packing; certification that the drug will be manufactured in compliance with current GMP, as defined by regulation; labeling; and—when so specified—data adequate to assure the drug's biological availability [15].

In selecting drugs suitable for the ANDA, one considers the characteristics of the drug, method of use, method of manufacture and packaging, stability, extent of use, safety and potency, and past history. Difficulties or known problems in any of these areas usually prevent the drug from being included in the abbreviated new-drug procedure category. For example, the limited information included in an abbreviated application would not be sufficient to permit the FDA to conclude that a high-potency drug that requires extremely careful handling during manufacture is suitable for marketing. The same is true of a drug for which the container plays a critical part in its administration, for instance, a metered aerosol. The purpose of the ANDA procedure is thus to eliminate unnecessary and costly animal and human experimentation, to assist the manufacturers in attempting to market duplication drug products, and to make all drug substances not covered by patents readily available to the consumer in a competitive market.

In 1984, an amendment to the Food, Drug and Cosmetic Act was passed and is entitled Drug Price Competition and Patent Term Restoration Act of 1984 [19]. This revised the ANDA procedure and permitted the approval of post-1982 drugs once the patent of the innovator had expired. This has also become known as the Waxman–Hatch Amendment to the act. The result of this amendment has allowed many generic products to be placed in the market, with a minimum amount of clinical effort, but with the establishment of the physical and chemical equivalency of the marketed materials [20]. These are generic equivalent materials of previously patented drug products. The amount of bioequivalency study necessary will vary with the particular drug product. This act was also tied into the patent restoration act, which allows an extended time to be given to innovator firms for new drug substances. Although this amendment allows numerous individuals to get into generic manufacturing of drugs, the act is specific about the requirements, and only those who can document the ANDA adequately will be permitted to distribute the drugs in the marketplace. This is also an important application of the CGMPs and their importance to drug products and drug product distribution.

III. LAWS GOVERNING PREPARATION AND DISTRIBUTION OF EXISTING PRODUCTS

A. Current Good Manufacturing Practices

On June 20, 1963, the FDA published new regulations governing ''current good manufacturing practices'' in the ''manufacturing, processing, packaging, and holding of finished pharmaceuticals'' [16]. These regulations represented an official step in the interpretation of GMP and quality control functions of a pharmaceutical operation. They also recognized the potential for innovation and made allowances for such developments as automatic, mechanical, or electronic equipment to be used in the production of drugs, provided adequate controls are available to verify the production operation.

The section on current GMP of the Code of Federal Regulations (revised as of April 1, 1985) is divided into two parts [17]. The first is Part 210: *Current Good Manufacturing Practices in Manufacturing Processing, Packaging, or Holding of Drugs: General*. This section defines or refers to other sections of the law for definitions. It includes definitions of what a ''component'' is, what a ''batch'' is, what the term ''lot'' means, what the use of lot number or control number is, what an ''active ingredient'' means as to its application in a dosage form for either humans or animals, what the term ''inactive ingredient'' means, what the term ''materials approval unit'' means, and what the term ''strength'' means. It is very important that in this last item some basic concepts of concentration of the drug substance have been clearly defined. The examples used include weight/weight; weight/volume, or unit-dose/volume basis, or it is accepted that the therapeutic activity of the drug substance is indicated based on laboratory and controlled clinical data. The second section, Part 211, is entitled *Current Good Manufacturing Practices for Finished Pharmaceuticals*. This section is subdivided as follows [18]:

Subpart A: General Provisions
Subpart B: Organization and Personnel
Subpart C: Buildings and Facilities
Subpart D: Equipment
Subpart E: Control of Components and Drug Product Containers and Closures
Subpart F: Production and Process Controls
Subpart G: Packaging and Labeling Control
Subpart H: Holding and Distribution

Subpart I: Laboratory Controls
Subpart J: Records and Reports
Subpart K: Returned and Salvaged Drug Products

To help understand the CGMPs, the following general comments will be made concerning several of the subparts of the regulations. It is hoped that by giving these explanations, the reader will be able to better understand where the regulations fit into the control of the manufacture of those products that are being dispensed and distributed to the consumer. For more details of these regulations, the reader is to refer to Section 21, Food and Drug Code of Federal Regulations, Part 211. The descriptions of the sections of importance are as follows.

Section 211.1: Scope of the Regulations

For all pharmaceutical products, it must first be assured that the drugs will be prepared under conditions that will assure them to be safe and to have the identity and strength as stated by the manufacturer. The regulations require a complete description of all that is involved with the manufacture of the dosage form.

Section 211.25: Personnel

In this section, the personnel responsible for directing the manufacture and control of the drugs and drug products are described. They must be of a sufficient number and have an educational background, which may be of a formal nature or of a training nature, or a combination of the two, so that they can make appropriate decisions on the safety and identity of the products being produced by the manufacturer or agency. It is imperative that they understand all manufacturing and control procedures to be performed that will ensure the preparation of quality products, and that they clearly understand their respective functions.

It is further stated that any individuals who have an apparent illness or open lesions that could in any way affect the safety and quality of drugs shall be excluded from direct contact with the pharmaceutical products being prepared. It may be necessary to assign these individuals temporarily to other duties until they are in satisfactory health. It is imperative that the employees be instructed to report any of the foregoing conditions to their supervisors. The supervisory personnel have additional responsibility to ensure that this requirement is met.

Section 211.42 to .58: Buildings

It is evident that buildings used for the manufacture, processing, packaging, labeling, or holding of drug products should be clean and kept in an orderly manner. The law, however, is much more specific, and states that space provisions must be made and gives certain general instructions on space utilization, as for the following:

1. The appropriate place of equipment and materials to minimize the risk of mix-ups between different drugs and all that is required to prepare a product (should also provide for the minimizing of cross-contamination)
2. The receipt, storage, and holding of all substances for a pharmaceutical product before release for use
3. The holding and storage of rejected materials
4. The storage of all items needed in pharmaceutical production
5. The obvious space for manufacturing and processing operations
6. The packaging and labeling operations to be done in close proximity to other packaging and labeling operations
7. The storage of finished products to be very well provided for
8. The control and production laboratory operations to be considered as important as any other space commitment

The law also specifies the type of lighting, ventilation, and screening required to control the environment of the facility, together with other steps to assure that certain areas shall be protected from microbial contamination or dust, or be maintained at special humidity and temperature conditions. These requirements are in order to:

1. Minimize contamination of products. This contamination may come from other products being made either in the same building or nearby.
2. Endeavor to minimize microbial contamination from one area to another, regardless of the dosage form being prepared. We now see greater emphasis of this on nonsterile products as well as products that are intended to be sterile.
3. Provide storage conditions that will maintain raw materials and finished products in a stable form.

Other general provisions are as follows:

1. Locker facilities must be provided for the employees such that they can prepare themselves adequately for the duties to be performed in their specific work areas.
2. Suitable water must be available, continuously under pressure from a system free of defects, so that any water used has not been contaminated in any way. There shall also be a good system of drains within the entire operation.
3. When needed, suitable housing shall be provided for the care of any laboratory animals needed.
4. There must be a safe and sanitary procedure for the disposal of sewage, trash, and other materials that could cause contamination of the building or the immediate surroundings. It should be noted that in 1967 the Pharmaceutical Manufacturers Association established sound sanitary rules and regulations for all the member firms.

Section 211.63 to .72: Equipment

"Equipment used for the manufacture, processing, packaging, labeling, holding, testing, or control of drugs shall be maintained in a clean and orderly manner and shall be of suitable desing, size, construction, and location to facilitate cleaning, maintenance, and operation for its intended purpose." The equipment should:

1. Be so constructed that all surfaces that come into contact with a drug product shall not be reactive, additive, or absorptive
2. Be so constructed that any substances required for operation of the equipment, such as lubricants or coolants, do not contact drug products
3. Be so constructed and installed to facilitate adjustment, disassembly cleaning, and maintenance to assure the reliability of control procedures, uniformity of production, and exclusion from the drugs of contaminants from previous and current operations
4. Be of suitable type, size, and accuracy for any testing, measuring, mixing, weighing, or other processing or storage operations

Section 211.100 to .115: Production and Process Control

"Production and control procedures shall include all reasonable precautions, indicating the following, to assure that the drugs produced have the safety, identity, strength, quality, and purity they purport to possess."

1. Each significant step in the process, such as the selection, weighing, and measuring of components, the addition of ingredients during the process, weighing and measuring during various stages of the processing, and the determination of the finished yield shall be performed by a competent and responsible person and checked by a second

competent and responsible person. The written record of the significant steps in the process shall be identified by the person performing these tests and by the person charged with checking these steps.

2. It is necessary that all containers, production lines, and equipment be properly identified to ensure batch integrity and the stage or location during processing.

3. To minimize and essentially eliminate contamination, and to eliminate mix-ups, equipment, utensils, and containers are to be thoroughly cleaned and adequately stored. Reference to previous use must also be removed from these items.

4. If products are to be of a sterile nature or their requirements are such that these should be free of objectionable microorganisms, precautions must be taken to aid in this final goal.

5. All precautions that might contribute to cross-contamination of drugs during manufacture or storage must be taken.

6. In-process controls, such as the checking of weights and disintegration times of tablets, satisfactory mixing, appropriate suspension preparation, or clarity of solutions must be conducted to ensure appropriate product content uniformity and performance.

7. A sufficient number of samples of various dosage forms should be tested to see if they meet product specifications.

8. This section requires the review and approval of all production and control records before the release or distribution of any batch. If any unexplainable discrepancy cannot be resolved, a thorough investigation is required, terminating in a written report of the investigation.

9. This section concerns itself with returned goods. It should be obvious to all handling drug products that it is absolutely necessary to be able to identify and evaluate products that may be either returned to stock or reprocessed. If neither of these is capable of being done, the material should be destroyed. There is no room for uncertainty.

10. This section concerns itself with asbestos-containing or other fiber-releasing filters. Current technology, as well as current feelings on these items, accept the fact that such release of fibers should not happen, especially for parenteral injections for humans; hence, the need for different types of filter media. However, it would be well to concern ourselves with any filter media that release fibers, and a recommendation is that "a fiber is defined as any particle with length at least three times greater than its width." As new filter media become available, we anticipate that this difficulty will be eliminated.

This section specifically concerns itself with filters that must be used that do release fibers. It is noted that an additional filter with a maximum pore size of either 0.2 or 0.45 μm must also be used to finish the filtration. This is an absolute requirement with asbestos filters used because of the total concept of safety and effectiveness of the drug. This section simply set the time limit for instituting good filtration procedures for parenteral products.

Section 221.80 to .94: Control of Components and Drug Product Containers and Closures

The section on components deals with all materials that may be used in the manufacture, processing, and packaging of drug products, plus materials used for maintenance of the building and equipment. They must be stored and handled in a "safe, sanitary, and orderly manner." These precautions are needed to prevent mix-ups and cross-contamination of drugs and drug products. All items should be held until they have been sampled and tested according to the company's specifications and not released until the tests have been completed. In this section it is required that

1. A check should be made of the container of a component to see whether it has been damaged or contaminated. This is done by visual inspection.
2. Samples should be taken that are representative of each lot. Samples of components that may be susceptible to insect infestation or similar contamination should be carefully examined.
3. Further sampling should be carried out to ensure potency, lack of microbiological contamination, identity, and the like. It should be noted that because of the above, approved components must therefore be stored to prevent contamination, and that the oldest stock should be used first, and that rejected components should be so identified for appropriate disposal.
4. Records should be maintained to ensure the identity, the supplier, the lot number, the date of receipt, the decision whether to accept or reject, and an inventory scheme for each component.

A specific statement then is given that requires that a reserve sample be retained for all required tests and that it should be available for at least 2 years after distribution of the last drug lot that utilized the components or 1 year after a stated expiration date of the finished product.

Section 211.122 to .137: Product Containers and Their Components

This section requires written procedures for identifying and storing of all items involved with packaging. It also describes how labels are to be prepared and stored. Strict regulation is enforced concerning labeling and how this material is issued within a pharmaceutical operation or a repackaging operation. This should be noted, since there are times when a large pharmacy may be involved with relabeling. Reference should be made to these regulations for guidance in the design of an operation such as repackaging. This section also discusses the expiration dates required for drug products so that they meet the standards of identity, strength, quality, and purity at the time of use.

Section 211.160 to .176: Laboratory Controls

The backbone of the laboratory control system rests on the philosophy used to establish sound and appropriate specifications, standards, and test procedures to ensure drug product quality. Some of the components of this section of the law are as follows:

1. The laboratory should maintain the master records for all lots of incoming raw material as well as finished product. Samples should be retained by the laboratory for immediate identification or future reference, which constitutes a reserve sampling system.
2. The master records should also indicate the sampling and test procedures involved with in-process control and sampling procedures for finished drug products.
3. Provisions should be made for checking the identity and strength of drug products for all active ingredients and, where applicable, sterility, pyrogenicity, and minimal contamination of ophthalmic ointments by "foreign particles and harsh or abrasive substances." It is also important that for sustained-release products a satisfactory laboratory method be available to ensure that release specifications are met.
4. This section also concerns itself with the ability to review laboratory test procedures and laboratory instruments used. It is stressed that reserve samples of marketed products must be retained in their original containers for at least 2 years after the final distribution of a lot or at least 1 year after the drug's expiration date. In this same light, laboratory data for each batch or lot of drug should be retained for these time periods.
5. This section requires that animals used for quality control purposes must be maintained in the proper manner and records kept for their identity and for their use.

6. An important element was added to this section several years ago which concerned the contamination of products by penicillin, especially when a firm manufacturers both penicillin-containing products and non−penicillin-containing products. It is established in the regulation that there may not be more than 0.05 unit of penicillin G per single dose of any drug when it is to be injected, or no more than 0.5 unit of penicillin G for drugs taken orally. This has led many companies to establish separate facilities for the preparation of penicillin-containing products.

Because of the success of the CGMP laws, a new line of thought was developed that involved the subject of good laboratory practices. On December 22, 1978, the regulations entitled *Non-clinical Laboratory Studies* were issued in the *Federal Register* [21]. These regulations, which became known as Good Laboratory Practice, are involved in the evaluation of nonclinical laboratory studies.

The purpose of the regulations was to ensure proper operation of laboratories that generated data to support either INDs or NDAs. In particular, they addressed animal studies, including animal care and animal accountability. They also involved the equipment used to do the various procedures for analysis and the maintenance and calibration of the equipment. In general, these regulations now cover the operation of the laboratories and how the data are collected and stored. This ensures that good procedures are followed during various studies.

B. Over-the-Counter Human Drugs

In the Code of Federal Regulations, Title 21, Part 330 outlines the general information for the regulations covering over-the-counter (OTC) human drug products that are generally recognized as safe and effective and not misbranded. Any product that fails to conform to the conditions outlined for the products would be in regulatory violation. In this section, monographs have been prepared for various classes of drugs and the conditions that should be met for their distribution. It is noted that the products must be manufactured in compliance with CGMPs.

This section and those that immediately follow outline what has been approved for use and the necessary warnings for these over-the-counter drugs. The categories that have been designated in the OTC area are antacids, laxatives, antidiarrhea products, emetics, antiemetics, antiperspirants, sunburn prevention and treatment products, vitamin−mineral products, antimicrobial products, dandruff products, oral hygiene products, hemorrhoidal products, hematinics, bronchodilator and antiasthmatic products, analgesics, sedatives and sleep aids, stimulants, antitussives, allergy treatment products, cold remedies, antirheumatic products, ophthalmic products, contraceptive products, miscellaneous dermatological products, dentifrices and dental products, and miscellaneous. The last category covers all OTC products that do not fall into any specific therapeutic category.

Materials that are under the OTC classification must be reviewed by the appropriate FDA division to evaluate elements concerning labeling, quantities of active ingredient, animal safety data, human safety data, and other elements that would support the marketing of the appropriate products.

Following this section of the regulations, other parts are given for specific categories, such as Part 331—*Antacid Products of Over-the-Counter Human Use*. In the monograph, such items are addressed as active ingredients, tests and procedures, and labeling. Very specific information is given on how materials are categorized and what tests are used for their evaluation. A manufacturer of these particular types of products would have to comply with all points as stated in the regulations. This is to ensure that appropriate products are in the marketplace and that they have been satisfactorily labeled. Depending on the monograph, appropriate warning labels are necessary to ensure safety for the consuming public. This might include, "avoid alcoholic beverages while taking this product," "do not take this product if you are taking sedatives of tranquilizers without first consulting your doctor and so forth," that would appear in the particular monograph. Other appropriate precautions would be noted and would be required on the labeling. It is very important when working with the patient that, if it is necessary to draw attention to caution labels, it should be done at the time of presenting production information to the consumer.

C. The FDA Recall System

One of the important functions of the FDA is responsibility to the American consumer to ensure that defective or hazardous products will be immediately removed from distribution and the marketplace. The principal legal procedure that is followed is to remove from commerce any product that is deemed unsafe through seizure. When the recall system was first initiated, it did not have a well-defined mechanism for either seizure or voluntary removal of products found defective from the marketplace. This lack of ability to recall products was evident in the late 1930s when the famous "elixir of sulfanilamide," which killed more than 100 people (see Sec. I) had to be found and returned to the manufacturer [6]. This situation initiated considerations of a better recall system. Also, there was no clear legal statement on a recall system; it was simply an accepted concept in the food and drug industry. The recall procedure increased in usage during the 1960s, and in 1967 the FDA started to publish a weekly recall list. In 1971 the FDA refined the procedure and stated that two types of recalls should be defined [22]:

1. Products that were an immediate threat to public safety
2. Products that may have or could be of potential hazard

On further investigation by an internal FDA task force, the following recall definitions were established in 1973, which seemed to better define the types of product defects that currently are being found. The recalls are now divided into the following three classes.

Class I recalls are those that have been judged to present a serious threat to the health of the consumer. Examples of this type of recall include brewers' yeast tablets (product contaminated with Salmonella), defective pacemakers (electronically unsafe), and numerous cases of mislabeling of potential drug substances, such as the belladona alkaloids [20].

Class II recalls are those in which the use of or exposure to a product found in violation of the law may cause a temporary health problem that is reversible, or in which the situation would not cause serious adverse health consequences. Examples of this type of recall would include uncertainty of the sterility of an injectable product, Salmonella contamination of various types of oral dosage forms, inadequate directions, for use, and improper buffering of solution for injection [20].

Class III recalls are those that involve violations of the law, but for which a health hazard is remote. Examples in this case include insect parts or droppings in flour, the marketing of a product without an approved NDA, unacceptable disintegration or dissolution of tablets, or swollen cans of certain food products [23].

To obtain weekly information on drug recalls, the practitioner should follow various professional association journals or newsletters for this information [i.e., Ref. 23]. It should be noted that it is possible to receive information on both drug recalls and new drug approvals by following the USP DI monthly supplements [24]. This is an excellent reference also for drug information for health professionals and patients. It is important to be aware of drug recalls as well as to know the new drug products entering the marketplace, or what new generic companies are preparing various prescription items.

Every practicing pharmacist has a responsibility to follow the recall lists to ensure that all listed products are immediately removed from the distribution system. This should be a concern of all practitioners and to be taken on as a direct obligation within his or her professional capabilities.

D. Tamper-Resistant Packaging

On November 5, 1982, regulations were issued in the *Federal Register* concerning the use of tamper-resistant packaging for OTC drugs, which included certain human drugs, cosmetic products, contact lens solutions, and tablets. This was due to the unfortunate poisoning that took place in Chicago beginning on September 30, 1982, and that was followed by a number of cases in the ensuing weeks [26].

A *tamper-resistant package* is defined as "one having an indicator or barrier to entry which, if breached or missing, can reasonably be expected to provide visible evidence to consumers that tampering has occurred." Because of the tragedy of 1982, this evidence on over-the-counter products was very important to tablets, capsules, certain liquids, and other substances that would be detrimental to the public if tampered with. Unfortunately, it did not stop, and during 1985 further tamper cases became well known within this country. Further efforts were made to improve packaging to help prevent this unfortunate happening.

Packages considered to be tamper-resistant are defined as follows: film grappers; blister or strip packs; bubble packs; shrink seals and bands; foil, paper, or plastic pouches; bottle seals; tape seals; breakable caps; sealed tube; sealed carton; and aerosol containers. These and other designs are currently being used in an attempt to produce tamper-resistant packaging for the consumer. Products must contain a label indicating to the consumer the existence of a specific tamper-resistant mechanism. Tamper resistance for OTC products is an aim of the industry to protect the public. It must be understood that tamper-proofing is impossible. Only through proper education of the consumer can we ensure adequate package protection. The pharmacist in contact with the public has an opportunity to educate them further in proper use of OTC products.

IV. FURTHER CONSIDERATIONS

This chapter has attempted to summarize current federal regulations that affect the manufacture and distribution of pharmaceutical products. There are other U.S. agencies involved with the manufacture of pharmaceutical products, and these include the Occupational Safety and Health Administration (OSHA) and the Environmental Protection Agency (EPA). For the clinical practitioner, there are those laws involving the cost of drug distribution and selection, which include Maximum Allowable Costs (MAC) and other federally sponsored funding programs. Thus, we see that both the pharmaceutical manufacturer and the clinical pharmacist must be continuously observant of changes in federal and state regulations.

REFERENCES

1. *Milestones in U.S. Food and Drug Law History*, Publ. (FDA) 73-1018, U.S. Dept. of Health, Education, and Welfare, Food and Drug Administration, Washington, DC, 1972.
2. H. W. Wiley and Harvey W. Wiley, *The History of a Crime Against the Food Law*, Washington, DC 1929, p. 1.

3. H. W. Wiley and Harvey W. Wiley, *The History of a Crime Against the Food Law*, Washington, DC, 1929, p. 56.

4. H. A. Toulmin, *A Treatise on the Law of Foods, Drugs, and Cosmetics*, 2nd Ed., W. H. Anderson, 1963, p. 13.

5. H. A. Toulmin, *A Treatise on the Law of Foods, Drugs, and Cosmetics*, 2nd Ed., W. H. Anderson, 1963, p. 19.

6. A. Hecht, FDA Consumer, p. 25 (Oct. 1977).

7. *Fifty Years of Quality Assurance in the Pharmaceutical Industry*, Pharmaceutical Manufacturers Association, Washington, DC, 1974.

8. C. Kumkumian, Manufacturing and controls: Guidelines for IND's and NDA's, paper presented at the 15th Annual International Industrial Pharmacy Conference, University of Texas, Austin, TX, 1976.

9. Federal Food, Drug, and Cosmetic Act as Amended, Aug. 1991.

10. *Code of Federal Regulations*, Title 21, *Food and Drugs*, Part 310.3, Apr. 1, 1992.

11. M. Pernarowski and M. Darrach (eds.), *The Development and Control of New Drug Products*, Evergreen Press, Vancouver, BC, Canada, 1972, pp. 118–120.

12. *Code of Federal Regulations*, Title 21, *Food and Drugs*, Part 312, pp. 62–97, Apr. 1, 1992.

13. *Drug Efficacy Study: A Report to the Commissioner of Food and Drugs*, National Academy of Sciences, National Research Council, Washington, DC, 1969.

14. Fed. Regist., 40, 26142 (1975).

15. L. Geismer, FDA Pap., Dec. 1970 to Jan. 1971 (1971).

16. Kefauver–Harris Amendments, *Public Law (PL) 87-781*, 87th Congress, Oct. 10, 1962.

17. *Code of Federal Regulations*, Title 21, *Food and Drugs*, Part 210, Apr. 1, 1992.

18. *Code of Federal Regulations*, Title 21, *Food and Drugs*, Part 211, Apr. 1, 1992.

19. Drug price competition and patent term restoration act of 1984, *Public 98-417*, 98 Stat. 1585-1605, Sept. 24, 1984.

20. W. M. Troetel, How new drugs win FDA approval, U.S. Pharm. (Nov. 1986).

21. Fed. Regist., 43, 59985 (1978).

22. Statement of FDA recall policy, by A. M. Schmidt (Commissioner, Food and Drug Administration), Sept. 21, 1973.

23. Pharm. Wk., American Pharmaceutical Association, Washington, D.C.

24. *USP DI and Supplements*, United States Pharmacopeial Convention, 1993.

25. Fed. Regist., 47, 50441 (1982).

20

European Aspects of the Regulation of Drug Products

Brian R. Matthews
Alcon Laboratories, Croydon, England

I. INTRODUCTION

A. Europe: Why Is It Important?

It costs an enormous amount of time and money to develop drug products. So if the return on that investment may be optimized, it is essential that a new product is sold in as many markets as possible as quickly as possible. The world's pharmaceuticals market covers the globe, but it is a fact of economic life that the majority of sales come from three regions: The United States and Canada, Japan, and Europe.

An illustration of the importance of these three regions is given in Figs. 1 and 2, which show the turnover of the whole of the chemical industries and production figures for the pharmaceuticals industries, respectively, between 1980 and 1990 for the United States (US), Japan, and the European Community (EC). The significance of the increasing size of the EC's pharmaceutical industry is obvious from Fig. 2. The levels of imports and exports of pharmaceuticals to and from the EC are shown in Fig. 3, which clearly shows that there is a significant positive balance of trade in this sector as far as the EC is concerned.

The size of the pharmaceuticals markets in the three regions is shown in Fig. 4 for the period 1984–1989. The EC has shown a steady growth during this period: the US market rather more stops and starts (and the impact of the change of administration in the United States in 1992 has still to be shown). Figure 5 indicates that between 1984 and 1989 the US share of the world pharmaceuticals market had fallen from 42 to 37%, whereas that of Japan had increased from 23 to 26%. The EC share had increased from 35 to 37% in that period. Some recent data on the annual amount spent per person on pharmaceuticals in certain countries is shown in Fig. 6: Japan is the leader here at £198, with the U.S. at £110, and European countries showing a wide variety (from £64, in the United Kingdom [UK] to £112, in France).

773

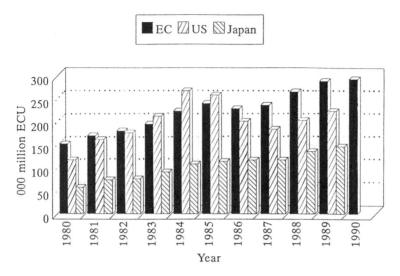

Fig. 1   Chemical industries' turnover. (From CEFC: cited in *Panorama of EC Industries*, pp. 8–13.)

So what is the message from all these figures? The message is that the EC is one of the major pharmaceuticals markets in the world, and its requirements are important when it comes to developing new pharmaceutical products if these are to gain approval and maximize the return on the investment in them.

B.   European Cooperation

The EC consists of an alliance of 15 countries: Austria, Belgium, Denmark, Finland, France, the unified Germany, Greece, Ireland, Italy, Luxembourg, Netherlands, Spain, Sweden, Portu-

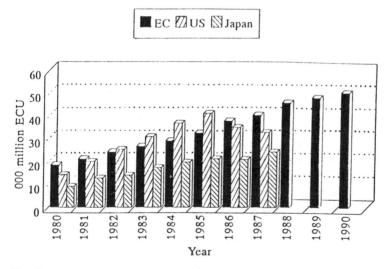

Fig. 2   Pharmaceutical industries' production. (From EFPIA/Eurostat: cited in *Panorama of EC Industries*, pp. 8–55.)

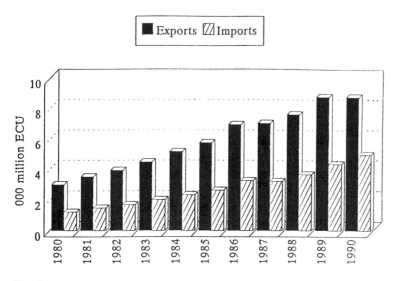

Fig. 3 Pharmaceuticals—EC trade. (From EFPIA: cited in *Panorama of EC Industries*, pp. 8–55.)

gal, and the United Kingdom. Table 1 gives some basic information on these countries. The total population of the EC (then consisting of 12 countries–Austria, Finland, and Sweden joining in 1995) was about 341 million in 1988. This compares with 246.3 million for the United States and 122.6 million for Japan. The total industrial output in the same year was 1758.7 billion dollars for the EC, 1250 billion dollars for the United States and 1155 billion dollars for Japan. In addition to the links between the EC member states, there is an economic relationship between the EC and the European Free Trade Association (EFTA) member states (Iceland, Liechtenstein, Norway, and Switzerland; Table 2).

The alliance between the EC and EFTA has resulted in a joint population of 373 million and the total industrial output was 2013 billion dollars at 1988 levels. This alliance was for-

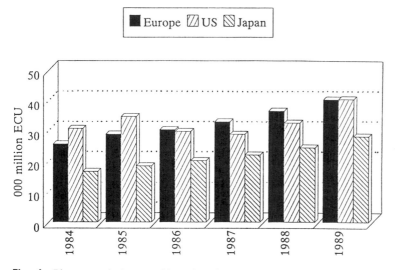

Fig. 4 Pharmaceuticals—world market. (From *Panorama of EC Industries*, pp. 5–58.)

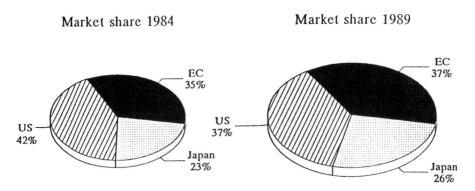

Fig. 5   World market share in pharmaceuticals. (From *Panorama of EC Industries*, pp. 5–58.)

malized as the European Economic Area, or EEA, before the full membership of the EC was expanded in 1995. The importance of the EEA was recognized in the EC Committee of the American Chamber of Commerce in Belgium's publication *EC Information handbook 1992* as the creation of the largest economic common market in the world; as an important trial run for those countries that want to join the EC; and reinforcing links with Eastern Europe necessary to stimulate the economic and political reform taking place.

European countries have, in the course of the last few centuries, not shown an overwhelming propensity to get on well together. After World War II serious attempts were made to increase the amount of cooperation between countries.

In 1948, the Organization for European Economic Cooperation (OEEC) was formed—largely to administer and implement Marshall Plan aid. This also led to the liberalization of trade between the concerned member states with the establishment of the European Payments Union in 1950. The OEEC developed in the Organization for Economic Cooperation and Development (OECD) in 1960. The OECD now has 24 members. Belgium, Netherlands, and Luxembourg established the BENELUX Customs Union in 1948.

The Council of Europe was formed in 1949 and now consists of 26 members and has developed about 140 conventions in the areas of cultural and social affairs, health, environmental protection, safeguarding natural resources, and so forth. Among its achievements have been the establishment of the first parliamentary assembly at European level, the adoption of

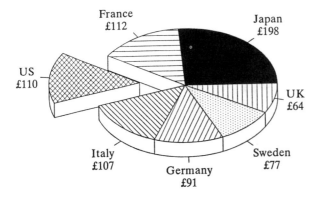

Fig. 6   Pharmaceuticals: spending per capita. (From *Script*, June 4, 1993.)

Table 1 Information on the Member States of the European Community

Country	Population (millions) (1994)	GDP ($ bn) (1994)	GDP with purchasing power parity (1994)	Industrial output ($ bn) (1988)	Weighted votes
Austria	7.9	175	79.9	50.63	4
Belgium	10	210	79.1	47	5
Denmark	5.1	134	80.8	31	3
Finland	5.06	116	72.9	35.5	3
France	57.3	1279	83.3	305	10
Germany	80.5	1846	89.3	545	10
Greece	10.4	75	34.6	15.9	5
Ireland	3.5	43	51.6	12.0	3
Italy	57.8	1187	77.0	286	10
Luxembourg	0.39	13.7	(79.0)[a]	2.19	2
Netherlands	15.1	312	76	76.48	5
Portugal	9.8	73	42.7	16.6	5
Spain	39.1	548	57.3	127	8
Sweden	8.7	233	79.0	75.17	4
United Kingdom	57.7	1025	73.8	295	10
(United States)	(246.33)	(4881)	(100)	(1250)	
(Japan)	(122.61)	(2859)	(71.5)	(1155)	

[a]Latest available.

Sources: The Economist Pocket World in Figures (1991); *The Economist Pocket Europe* (1994); *The Economist Book of Vital World Statistics* (1990).

the European Convention for the Protection of Human Rights and Fundamental Freedoms, and the Establishment of the European Court and Commission for Human Rights. In the area of health care one of the notable developments is the convention on the elaboration of an *European Pharmacopoeia.*

The Treaty of Paris was signed by France, Italy, the Federal Republic of Germany, Belgium, Luxembourg, and Netherlands on 18 April 1951 and established the European Coal and Steel Community (ECSC). This came into effect in May 1953. This established many of the principles later adopted or modified in the formation of the European Economic Community (EEC) and the European Atomic Energy Community (Euratom).

Table 2 Information on the Member States of the European Free Trade Community

Country	Population (millions) (1988)	GDP ($ bn) (1988)	GDP with purchasing power parity	Industrial output ($ bn) (1988)
Iceland (IS)	0.26	n/a	(79.0)	n/a
Liechtenstein	0.028	n/a	n/a	n/a
Norway (NO)	4.3	110	77.6	29
Switzerland (CH)	6.7	249	98.4	63

Sources: The Economist Pocket World in Figures (1991); *The Economist Pocket Europe* (1994); *The Economist Book of Vital World Statistics* (1990).

The EEC was formed as a result of the signing of the Treaty of Rome by France, Italy, the Federal Republic of Germany, Belgium, Luxembourg, and Netherlands on 25 March 1957. Its purpose was to establish a common market with free movement between the member states of persons, labor, services, goods, and capital. Other purposes of the Community include harmonization of laws to ensure that there are no competitive distortions; harmonization of indirect taxation; and the development of common policies. There is also a possibility of protection against nonmembers by means of a common external tariff (customs barrier).

The EEC came into being on 1 January 1958, and the customs union was effected on 1 July 1968. A unified system of taxation, value-added tax, was introduced in 1967, and this was used in part to finance the Community. The Community was enlarged in 1973 by the United Kingdom, Denmark, and Ireland joining; again in 1981 when Greece joined; and again in 1986 when Spain and Portugal joined. Austria, Finland, and Sweden joined in 1995. There are pending applications to join at the time of writing from Turkey (submitted in 1987), Cyprus, and Malta (1990), and a special expression of interest in joining at a later date by Hungary. Special agreements have been signed with Hungary, Poland, the Czech Republic, Slovakia, Bulgaria, and Rumania.

Direct elections to the European Parliament was introduced in 1979. The European Monetary System was introduced in 1979. The Single European Act was adopted in February 1986, with the expressed aim of establishing a single European market by the beginning of 1993. This was achieved in part by the introduction of a new system for implementing EC policies and objectives that incorporated a system for adopting necessary legislation concerned with the single market using weighted majority voting (the number of votes available to each member state being related to its population), rather than requiring unanimity in votes up to the end of 1994. Until the enlargement of the Community in 1995 for a particular measure to be adopted it needed to have 54 of the 76 available votes cast in its favor. Special provisions have been introduced following the enlargement of the European Union in 1995. The distribution of votes is shown in Table 1.

The ECSC, the EEC, and Euratom have, in effect, been merged into the European Community (EC) since the mid-1960s. The EC has common policies in several areas, including commercial, competition, agriculture, transport, fisheries, regional development, energy, research and technological development, and social matters. The ''European Union'' was introduced following the ratification of the Maastricht Treaty.

One attempt at European union in the area of defense, the European Defence Community (consisting of France, Italy, the Federal Republic of Germany, Belgium, Luxembourg, and Netherlands) failed. However the Western European Union (WEU) was established in 1954 by Belgium, the Federal Republic of Germany, France, Italy, Luxembourg, Netherlands, Portugal, Spain, and the United Kingdom. This is the only purely European Defence Organization. The North Atlantic Treaty Organization (NATO) has 16 member states including the United States and Canada.

This short account of postwar European cooperation has introduced a number of asssociations of countries, and for the sake of clarity Table 3 indicates which countries belong to which of the groups.

Relative to the system of controls applied to pharmaceuticals in Europe, two of the foregoing are of significance: the Council of Europe for the *European Pharmacopoeia* and the EC for the system of controls applied under the terms of a series of pharmaceuticals Directives.

C. The Development of Legal Controls in the European Community

The EC has various institutions that are involved in the legislative process. The key ones are as follows.

Table 3 Membership of Various International Organizations

Country	OECD	EC	EFTA	NATO	WEU	Council of Europe	EEA
Australia	✓						
Austria	✓	✓				✓	✓
Belgium	✓	✓		✓	✓	✓	✓
Canada	✓			✓			
Cyprus						✓	
Czech Republic/Slovakia						✓	
Denmark	✓	✓		✓		✓	✓
Finland	✓	✓				✓	✓
France	✓	✓		✓	✓	✓	✓
Germany	✓	✓		✓	✓	✓	✓
Greece	✓	✓		✓		✓	✓
Hungary						✓	
Iceland	✓		✓	✓		✓	✓
Ireland	✓	✓				✓	✓
Italy	✓	✓		✓	✓	✓	✓
Japan	✓						
Liechtenstein			✓				✓
Luxembourg	✓	✓		✓	✓	✓	✓
Malta						✓	
Netherlands	✓	✓		✓	✓	✓	✓
New Zealand	✓						
Norway	✓		✓	✓		✓	✓
Poland						✓	
Portugal	✓	✓		✓		✓	✓
San Marino						✓	
Spain	✓	✓		✓		✓	✓
Sweden	✓	✓				✓	✓
Switzerland	✓		✓			✓	✓
Turkey	✓			✓		✓	✓
United Kingdom	✓	✓		✓	✓	✓	✓
United States	✓						

Key: OECD, Organization for Economic Cooperation and Development; EC, European Communities; EFTA, European Free Trade Association; NATO, North Atlantic Treaty Organization; WEU, Western European Union; EEA, European Economic Area.

1. The Commission of the European Communities, consisting of 20 members nominated by the member states for a renewable period of 4 years (but who take an oath to represent EC, rather than national interests) and its 23 Directorates-General (a civil service that has the responsibility of proposing measures to implement Community policies and, then, has the responsibility for monitoring that those measures are brought into the laws of the member states and that the principles involved are followed);

2. The Councils of Ministers (political representatives of each of the governments of the member states; the membership changes according to the items under discussion. The Trade

Ministers are responsible for matters relating to pharmaceuticals and medical devices, since these are items under the single-market legislation. The Councils of Ministers can consider proposals on legislation submitted to them by the Commission (but cannot make proposals of their own) and take into account the comments of the European Parliament and the Economic and Social Committee, among others, when adopting the measures using either a unanimous approval procedure or weighted majority voting (see Table 1), depending on the article of the Treaty of Rome under which the proposals were put forward. The more routine level of contact between the Commission and the member states is at the level of a Committee of Permanent Representative ("ambassadors") (COREPER) and of working parties of experts (often civil servants) from the member states.

3. The European Parliament, a directly-elected body consisting of 567 members in December 1994 (i.e., before the inclusion of Austria, Finland, and Sweden: the total when these countries' Members of Parliament [MEPs] are elected will be 624), which is consulted at least once and sometimes twice or more on legislation proposals by the Councils of Ministers, depending on whether the legislation was referred to them under the consultation procedure, the cooperation procedure, or the codecision procedure (i.e., depending on the article of the Treaty of Rome under which the proposal has been put forward). The groupings within the Parliament are politically organized, rather than by country. Parliamentary work is organized around a series of more than 19 committees and subcommittees.

4. The Economic and Social Committee, consisting of 222 members from January 1995, nominated by the governments of the member states and representing employers, workers, and miscellaneous activities, such as agriculture, transport, trade, small- and medium-sized businesses, the liberal professions, and consumers. Work is undertaken in a series of nine sections. This committee has an advisory role and is consulted on all proposals for Community legislation;

5. The European Court of Justice, which consists of 13 judges and 6 advocates-general organized into a series of chambers, to whom matters relating to the application and interpretation of community law can be referred and whose decision is final. The judges and the advocates-general are nominated by the member states and are appointed for renewable six-year terms of office. There is also a Court of First Instance.

The numbers of members of some of these bodies elected or nominated by the member states are indicated in Table 4.

The legislative process involves the Commission identifying a measure that is necessary, in its opinion, to help satisfy the requirements of a Community policy. It can seek advice of outside bodies—including European-level trade associations (known as "economic interest groups") and consumer interests. Having taken advice, the Commission services (i.e., the civil servants) will draft a proposed measure—which may be a draft Regulation (which once adopted is directly binding on all those addressed without the need to transpose the measure into national legislation) or a draft Directive (which once adopted will be binding as to effect on all those addressed, but that the member states, for example, can enact into their national laws as they consider appropriate). The draft Directive may be one that needs to be adopted by a Council of Ministers of one that the Commission is empowered under specific legislation to adopt after consultation. Nonbinding measures available to the Commission include Recommendations and Opinions. Decisions are binding; Opinions are not.

After a proposed measure is drafted, it is published in one of the series of the *Official Journal of the European Communities*—the C-series is used for proposals and the L-series for final legislation. This opens the opportunity for general discussion and input on the proposals. The appropriate *Council of Ministers* will consider the document and refer it to the *Economic*

Table 4   Numbers of Nominees or Elected Persons From the Member States at Certain European Community Institutions

Country	Commission of the European Communities	European Parliament	Economic and Social Committee
Austria	1	(20)[a]	12
Belgium	1	25	12
Denmark	1	16	9
Germany	2	99	24
Greece	1	25	12
Spain	2	64	21
Finland	1	(16)[a]	9
France	2	87	24
Ireland	1	15	9
Italy	2	87	24
Luxembourg	1	6	6
Netherlands	1	31	12
Sweden	1	(21)[a]	12
Portugal	1	25	12
United Kingdom	2	87	24

[a]The elections for these members of the European Parliament had not been held at the time of writing.
Source: Certain information taken from *Vacher's European Companion and Consultants' Register.*

and Social Committee and to the *European Parliament* for its only reading or for its first reading (depending on which article of the Treaty of Rome it is proposed under). Proposals concerned with the completion of the single market were formerly referred to the Parliament twice under the cooperation procedure, but are now referred under the codecision procedure (with additional consultation stages and the possibility of arbitration should the Council of Ministers and the Parliament disagree).

The comments of these two bodies are considered by the Commission and an expert working party of representatives of the member states. Consultations with the appropriate Community level economic interest groups and other interested parties continue, and national parliaments are consulted.

Revisions to the original propoosals may be accepted (or not) by the Commission or the Council, and a revised document submitted to the Council of Ministers once it has reached a nearly finalized version and has been agreed by COREPER. Once a common position has been reached, the document may be referred back to the European Parliament for a second reading (and so on), under the terms of the cooperation or codecision procedures.

Once the Council of Ministers has adopted the final document, taking into account (but not necessarily accepting) any alterations or amendments proposed by the European Parliament, the proposal becomes law. The text has to be translated into each of the Community languages (of which there are eleven) and published in the *Official Journal*.

The measure will indicate to whom it is addressed. The date of implementation (and of any transitional arrangements) will be indicated in the text published in the *Official Journal*. In the case of a Regulation the implementation date is 20 days after publication if there is no other date in the published text.

II. THE PHARMACEUTICAL LAWS OF THE EUROPEAN COMMUNITY

A. Introduction

The introductory section of this chapter was intended to give background information on the economic importance of the EC, and a general overview of the legislative procedures by which EC-level laws are adopted. In this section the framework of European law as it related to pharmaceuticals will be discussed.

It was recognized at an early date that the market in pharmaceuticals represented one for which there were nonfiscal barriers to trade between the member states, and it was decided to enact Community legislation to try to achieve a single market in the pharmaceuticals sector. The first (enabling) legislation relating to marketing authorizations for pharmaceuticals for human use was Directive 65/65/EEC, which was adopted by the Council on 26 January 1965 and published in the *Official Journal* on 9 February 1965. Another key piece of legislation relating to manufacturing authorizations was Directive 75/319/EEC, adopted by the Council on 20 May 1975 and published in the *Official Journal* on 9 June 1975. The effects of these and the later Directives, Recommendations, guidelines and the *Notice to Applicants* will be discussed later in more detail. A list of the Directives current at January 1995 and other legislation affecting human pharmaceuticals is given in Table 5.

In addition to the harmonization of controls on human pharmaceuticals, the EC has also harmonized the system of controls applying to veterinary pharmaceuticals. The enabling Directive for veterinary products is Council Directive 81/851/EEC, adopted on 28 September 1981 and published in the *Official Journal* on 6 November 1981. Council Directive 81/852/EEC (also adopted and published on the same dates) together with Directive 81/851/EEC also control maximum residue limits of veterinary medicinal products in foodstuffs of animal origin. A list of the Directives current in January 1995 and other legislation applying to veterinary products is given in Table 6.

B. The Systems for Control of Pharmaceuticals in the European Community

National Systems

National systems of control for pharmaceuticals exist in all of the member states of the EC. To market a product, a separate product license/marketing authorization is required and must be applied for locally or through centralized procedures. National language requirements may be applied and national fees are payable.

Because the systems of control developed separately, there is a considerable variation in their design in the different countries. Some authorities use in-house assessors (reviewers) for the examination of all of the data submitted in the applications for a marketing authorization. Others use external assessors for some of the work. Others again contract out all of the work. Appeal mechanisms available to applicants who are refused a marketing authorization in the terms for which one was applied differ between different countries. Information on these aspects of product registration cannot be dealt with in detail here because of space limitations, and readers are referred to *Pharmaceutical Product Licensing Requirements for Europe*, which has a detailed chapter on the United Kingdom system and information in a second chapter on the systems in the other member states.

It is possible to apply to each individual country for a marketing authorization, and this will be the subject of normal national processing times. The times taken to grant authorizations will vary between member states, and this may affect the strategy chosen to gain EC-wide

Table 5 The Legal Basis for Controls on Pharmaceuticals for Human Use in the European Community

Directive number	Subject	Date adopted	Date published
65/65/EEC	Medicinal products	26 Jan. 1965	9 Feb. 1965
2309/93[a]	EMEA	22 July 1993	24 Aug. 1994
75/319/EEC	Medicinal products	20 May 1975	9 June 1975
75/318/EEC	Standards, protocols	20 May 1975	9 June 1975
91/507/EEC[b]	Standards, protocols	19 July 1991	26 Sept. 1991
83/570/EEC	Medicinal products	26 Oct. 1983	28 Nov. 1983
87/19/EEC	Standards, protocols	22 Dec. 1986	17 Jan. 1987
87/21/EEC	Medicinal products	22 Dec. 1986	17 Jan. 1987
93/39/EEC	Medicinal products	14 June 1993	24 Aug. 1993
87/22/EEC	High-tech/biotech	22 Dec. 1986	17 Jan. 1987
93/41/EEC	Repeal of 87/22/EEC	14 June 1993	24 Aug. 1993
89/341/EEC	Medicinal products	3 May 1989	25 May 1989
89/342/EEC	Immunologicals	3 May 1989	25 May 1989
89/343/EEC	Radiopharmaceuticals	3 May 1989	25 May 1989
89/381/EEC	Human blood products	14 June 1989	28 June 1989
92/25/EEC	Wholesaling	31 Mar. 1992	30 Apr. 1992
92/26/EEC	Supply classification	31 Mar. 1992	30 Apr. 1992
92/27/EEC	Labeling, leaflets	31 Mar. 1992	30 Apr. 1992
92/28/EEC	Advertizing	31 Mar. 1992	30 Apr. 1992
92/73/EEC	Homoeopathics		
91/356/EEC	GMP	13 June 1991	17 July 1991
89/105/EEC	Pricing/reimbursement	21 Dec. 1988	11 Feb. 1989
(Unnumbered)[d]	Pricing/reimbursement		4 Dec. 1986
78/25/EEC	Colors	12 Dec. 1977	14 Jan. 1978
90/219/EEC	Genetically modified organisms containment	23 Apr. 1990	8 May 1990
91/448/EEC[c]	Classifications GMOs	29 July 1991	28 Aug. 1991
90/220/EEC	Deliberate release GMOs	23 Apr. 1990	8 May 1990
90/679/EEC	Biological agents exposure	26 Nov. 1990	31 Dec. 1990
87/18/EEC	GLP	18 Dec. 1986	17 Jan. 1987
88/320/EEC	Inspection of GLP	9 June 1988	11 June 1988
90/18/EEC[b]	Inspection of GLP	18 Dec. 1989	13 Jan. 1990
86/609/EEC	Protection of animals	24 Nov. 1986	18 Dec. 1986
85/374/EEC	Defective product liability	25 July 1985	7 Aug. 1985
83/189/EEC	Technical standards	28 Mar. 1983	26 Apr. 1983
88/182/EEC	Technical standards	22 Mar. 1988	26 Mar. 1988
75/320/EEC[e]	Pharmaceutical Committee	20 May 1975	9 June 1975
(Unnumbered)[d]	Parallel imports		6 May 1982

Council Directives, except:
[a]Council Regulation
[b]Commission Directive
[c]Commission Decision
[d]Commission Communication
[e]Council Decision

Source: *Rules Governing Medicinal Products in the European Community*, Vol. I: *The Rules Governing Medicinal Products for Human Use in the European Community.*

Table 6 The Legal Basis for the Control of Pharmaceuticals for Veterinary Use in the European Community

Directive number	Subject	Date adopted	Date approved
81/851/EEC	Veterinary medicines	28 Sept. 1981	6 Nov. 1981
81/852/EEC	Standards, protocols	28 Sept. 1981	6 Nov. 1981
87/20/EEC	Standards, protocols	22 Dec. 1986	17 Jan. 1987
90/676/EEC	Veterinary medicines	13 Dec. 1990	31 Dec. 1990
90/677/EEC	Veterinary immunologicals	13 Dec. 1990	31 Dec. 1990
87/22/EEC	High-tech/biotech	22 Dec. 1986	17 Jan. 1987
78/25/EEC	Colors	12 Dec. 1977	14 Jan. 1978
87/18/EEC	GLP	18 Dec. 1986	17 Jan. 1987
88/320/EEC	Inspection of GLP	9 June 1988	11 June 1988
88/609/EEC	Protection of animals	24 Nov. 1988	18 Dec. 1988
83/189/EEC	Technical standards	28 Mar. 1983	26 Apr. 1983
88/182/EEC	Technical standards	22 Mar. 1988	26 Mar. 1988
Regulation 2377/90	Maximum residue limits	26 June 1990	

Source: The Rules Governing Medicinal Products in The European Community: Vol. V: *Veterinary Medicinal Products*; and Vol. VI: *Establishment by the European Community of Maximum Residue Limits (MRLs) for Residues of Veterinary Medicinal Products in Foodstuffs of Animal Origin.*
Additional information is included in the Table.

approvals. Up-to-date average-processing times should be checked (using information published in *Scrip*, for example) when an application is under consideration or preparation. The German and the Spanish authorities have a reputation for slow processing of applications, for example.

The granting of a marketing authorization does not always result in immediate access to the market. In some countries, there are separate negotiations for terms for inclusion in national social security–health service-prescribing lists or reimbursement terms. These arrangements can cause a considerable delay in marketing an otherwise approved product. However, the Commission of the European Communities failed in recent attempts to influence these practices—which for many could be seen as a barrier to trade. However, the Commission has encouraged price competition by the use of generic products and parallel imports.

There are wide differences in the organizational arrangements of the individual ministries or agencies responsible for the control of pharmaceutical products in the different countries. Again, space does not allow for expansion of these aspects in this chapter, but readers are referred to the *Regulatory Affairs Journal*, which regularly includes organization charts for the different institutions. This journal is also a useful source for information on necessary procedures to be followed for when undertaking clinical trials in the different countries—this is one area for which there has, as yet, been no attempt at harmonization of the national requirements, although the Commission has indicated that legislative proposals will be forthcoming in this area shortly.

Supranational Procedures

It would appear, then, that there is little harmonization of the control systems for national applications. 'Is there any harmonization at all?' you may well be asking. The answer is 'yes.' There has been harmonization of the data requirements and the application format, and two supranational procedures have been introduced to permit the application for a marketing authorization in several member states at once. These procedures follow:

1.   The *multistate procedure*, introduced under Directive 83/570/EEC, in which an authorization first has to be granted in one of the member states, using its national procedures, and this is then used as the basis for applications to two or more additional member states. When the predecessor of the multistate procedure was first introduced under Directive 75/319/EEC it required the first authorization to form the basis for applications to five or more other member states.

2.   The *concertation procedure*, which was compulsory for some products of biotechnology and which was optional for other biotechnology products and for certain defined classes of high-technology or otherwise innovative products, was introduced under Directive 87/22/EEC. It was possible for applications thought to be of interest at a Community level to be referred to the procedure by a member state. Under the terms of Directive 87/21/EEC some protection was provided for innovative products against second applicants for a period of 10 years after the grant of the marketing authorization (for concertation procedure applications) or 6 to 10 years (depending on the member state) for other innovative applications.

The aim of the multistate procedure was to allow applicants to gain access to new markets based on mutually recognized prior approvals. Applications made under this procedure had a defined timetable for the stages of the assessment (and subsequent provision of additional information or appeals). The company could select who acted as rapporteur for the procedure by the choice of the first country to which the application was submitted or by choosing from among those countries where an authorization was held.

Provision was made within the procedure for the reference to an arbitration mechanism of cases when there was disagreement among the different authorities—the consideration of the application by the Committee for Proprietary Medicinal Products (CPMP). Any deficiencies identified in the application had to be transmitted to the rapporteur (and thence to the applicant) within the agreed period of the assessment, and this could be very useful in minimizing the amount of resource that needed to be applied to the same application at different times.

In practice almost all of the 400 or so applications submitted through the procedure were referred to the CPMP. Since the outcome of the reference to the CPMP was a nonbinding Opinion, there was no certainty that an Opinion in favor of an application would result in approvals in all of the targeted member states. In such circumstances, however, the applicant company had the right of access to any available national appeal procedures.

The potential advantages of using the multistate procedure included

1.   In theory at least, some degree of mutual recognition of authorizations granted in other member states
2.   A defined timetable for the processing of the application that could have been shorter than that applied to national applications—although processing time should not necessarily be equated with time to the market, since reimbursement and pricing hurdles also had to be overcome in some countries
3.   A reduction in the amount of contact needed with individual national authorities (although local contacts could be useful in easing the procedure at some points)
4.   The availability of all reasoned objections within a defined time scale (although some member states were often late in providing their comments)
5.   The accelerated review of applications in countries outside the EC where the Product Evaluation Report scheme could be brought into play.

Potential disadvantages of using the system have been identified:

1.   The potential for member states who are not party to the process for a specific application becoming involved at the CPMP

2. The reduction in the amount of informal dialogue possible with the national authority staff, especially where additional questions were raised after the submission of the company's response to the consolidated list of objections
3. The variation in the depth and detail of the assessments between the member states
4. The need to have available sufficient resource to deal with the consolidated list of questions within the available time
5. The lack of a binding Decision from the CPMP.

The concertation procedure involved the application being assessed by a rapporteur member state (chosen by the applicant), which then submitted an assessment report to the other concerned member states. The assessment report and a consolidated list of queries from the member states was considered by the CPMP, with the possibility of an appeal to the CPMP or its working parties when additional information or other input from the applicant was required, and an Opinion being delivered. This Opinion was sent to the member states, but was nonbinding. Again, any decision by a member state to decline to grant a marketing authorization was subject to any available national appeal mechanisms.

Many of the foregoing comments relating to the multistate application route also relate to the concertation application route.

In both of these procedures the rapporteur had a variety of important functions, including

1. Advising the potential applicant on the suitability of a particular product for the procedure
2. Seeking the advice of the CPMP on the admissibility of a particular product into the high-technology concertation procedure
3. Initiating the preparation of an assessment report once an application had been accepted into the procedure
4. Discussing the reasoned objections with the applicant and helping the company to decide whether a written representation or a hearing would be the best way of dealing with them
5. Advising the applicant of any outstanding points one week before a hearing before the CPMP, and confirming these on the day before the hearing
6. Following up any problems encountered at national level after the delivery of the CPMP Opinion.

The nonobligatory nature of the CPMP Opinion meant that national appeal procedures sometimes needed to be followed after the end of the CPMP processes. It has been reported in the literature that member states could reintroduce points at this stage that had been dropped from consideration during the CPMP procedures. Completely new points have also been introduced immediately before the CPMP has considered an appeal.

More information on the multistate and concertation procedures is included in *Pharmaceutical Product Licensing Requirements for Europe*. The first issue of the *Drug Information Journal* for 1993 included a series of 16 articles on experience of the industry and the regulatory authorities with the multistate and concertation procedures.

The Future (or New) System

The Commission of the European Communities proposed the introduction of a new system of controls for the EC: The *Future System*, now more correctly termed the *New System*. This represents an evolution from the preexisting system of national applications, multistate applications, and concertation applications.

It is intended that *national applications* will continue after the operational date of the new scheme—1 January 1995, at least for the mutual recognition procedure—but ultimately, this is intended to be available for products that will be marketed in only a single member state. As soon as an application for the same product is made in a second (or subsequent) member state, the mutual recognition procedure may be initiated (although for an interim period the use of parallel national applications is not excluded).

Reference into the *mutual recognition procedure* will occur whether the application is made to only one additional country or to several member states, if the second or subsequent member states are unable to agree with the assessment report and terms of the originating member state's approval. The mutual recognition procedure, therefore, is a development of the multistate procedure. It allows for the reassessment of the original application by a rapporteur (possibly with the assistance of a corapporteur) if the concerned member states are unable to reach agreement among themselves. The application will then be referred to the CPMP, as established under the new procedure, under the auspices of a European Medicines Evaluation Agency (EMEA), which was also set up under the terms of the new system.

Having completed the appeal procedure—which will be before the CPMP—an Opinion will be arrived at. The significant difference with the New System is that there will be a mechanism by which the Opinion can be converted into a binding Commission Decision. Space does not permit the proposed mechanism to be discussed in detail in this chapter, however.

In addition there is a *centralized procedure*, which is a development of the concertation procedure. This is applicable to many biotechnology and high technology products (including all new drugs)—either compulsorily or at the choice of the applicant—and includes the registration fee being paid to the EMEA. The assessment of these applications will be undertaken by a team assembled by a member of the CPMP appointed as the rapporteur (and a corapporteur). Applicants will have an opportunity to give an oral explanation concerning any problem areas. The outcome of the first round of consideration will be a CPMP Opinion, which will then enter the procedure to convert the Opinion into a Commission Decision. A single successful application under the centralized procedure will result in an EC-wide product authorization.

Under the New System, the final choice of rapporteur rests with the CPMP—particularly in the centralized procedure and in the appeal stages of the mutual recognition procedure—but account will be taken of any preferences expressed by the applicant. It should be borne in mind that the decision will be binding on all involved parties. This could mean that existing authorizations in the EC will need to be modified.

Data Requirements for Marketing Authorization Approval

The basic data requirements are laid out in the Annex to Directive 91/507/EEC for human products and in the Annex to Directive 81/852/EEC for veterinary products. The former are discussed more fully later in this chapter.

Application Format

The application format is described in detail in the *Notice to Applicants*, the full title of which is *Notice to Applicants for Marketing Authorizations for Medicinal Products for Human Use in the Member States of the European Community*, published as Volume II of the *Rules Governing Medicinal Products in the European Community* (for human products); and *Notice to Applicants for Marketing Authorizations for Veterinary Medicinal Products in the Member States of the European Community*, published as part of Volume V of the *Rules Governing Medicinal Products . . . Veterinary Medicinal Products*.

For the sake of brevity the remainder of this chapter will concentrate on human medicinal products. The following sections will be concerned with the format of an application and with a general outline of the data requirements for an EC application.

C. Basis for Approval of Marketing Authorizations for Pharmaceuticals for Human Use in the European Community

No medicinal product may be placed on the EC market unless it has an authorization issued by the relevant competent authority in the member state concerned or an authorization issued by the Commission following a centralized procedure. There is a requirement in the pharmaceutical Directives that an application be processed by the competent authority within 120 days of the receipt of a valid application, subject to the clock being stopped if additional information is requested of the applicant. The grounds for assessment are limited to the quality, safety, and efficacy of the product concerned, although from 1995 an environmental risk assessment was also required in certain applications.

Grounds for refusal, revocation, or suspension of marketing authorizations are limited to the following:

1. Failure to provide the necessary particulars, documents and Expert Reports
2. Products intended for use as contraceptives where the laws of the member state concerned prohibit such use
3. If the product is harmful in normal conditions of use
4. If the therapeutic efficacy is lacking, or if it is insufficiently substantiated by the applicant
5. If the product is not of the declared qualitative or quantitative composition
6. If the labeling and package leaflet do not comply with the requirements of the relevant Directives
7. If the controls on the finished product or ingredients and the controls carried out during manufacture have not been carried out or other obligations arising from the grant of the manufacturing authorization have not been fulfilled.

Any decision to refuse, suspend, or revoke an authorization has to be notified to the applicant or authorization holder with the reasons for the decision. Any available appeal rights have to be advised at the same time. Decisions to revoke an authorization have to be published in an official journal in the country concerned.

Once granted, an authorization is valid for 5 years, and applications for renewals must be submitted 3 months (or more, depending on the member state concerned) before the expiration of the original authorization. The grant of an authorization has to be the subject of a suitable notice published in an official journal in the member state concerned.

The granting of a marketing authorization does not affect the civil and criminal liability of the manufacturer or the person placing the product on the EC market.

When applications for authorizations or for renewals are made the pharmaceutical directives require that due account be taken of technical and scientific progress and that any necessary changes be effected to the control methods.

D. The Format of an Application for a Marketing Authorization in the European Community

An application for a marketing authorization to any member state of the EC has to be in the format prescribed in the *Notice to Applicants*. A number of other countries will accept the same format, including Australia, South Africa, the EFTA/EEA member states, and Canada.

The documents and particulars required in an application are defined in Directive 65/65/EEC. Some aspects are also detailed in guidelines and other documents. In completing the application form and writing the Summary of Product Characteristics (SPC) due account should be taken of guidance note III/3593/91-EN *Guidance on Standard Terms for Marketing Authorization Applications*, which lists permitted terms for routes of administration, dosage forms, and containers.

The application for a marketing authorization is required to be accompanied by the following:

1. *An application form* (the content of which is described in EC guideline III/3038/91-EN *Note for Guidance: EC Application Format*, approved by the CPMP in March 1992), which identifies the nature of the application, the name of the product, the ingredients (qualitatively and quantitatively; active ingredients and excipients), the pharmacotherapeutic classification, the pharmaceutical form and strength, the route of administration, the container and closure details, the supply classification proposed, the applicant's details (and other relevant persons' details including the manufacturer—and the name of the responsible qualified person—and the storage–distribution sites, and any contract companies involved), and details of any marketing authorizations (granted or pending or rejected–withdrawn or suspended–revoked). In the case of multistate and concertation applications the rapporteur country has to be identified and the concerned member states named. This administrative data forms Part I A of the submission.

2. A brief description of the *method of manufacture.* This forms part of Part II of the dossier.

3. *Therapeutic indications, contraindications and side effects.* This information is included in the SPC.

4. *Posology*: This also forms part of the SPC.

5. *Pharmaceutical form; route and method of administration*: This is included in the SPC.

6. Expected *shelf life*: This forms part of the Summary of Product Characteristics and also Part II of the dossier.

7. Description of the *control methods* used by the manufacturer, including qualitative and quantitative analyses, sterility tests, and so forth. This forms part of Part II of the dossier.

8. *Results of physicochemical, biological, and microbiological tests; pharmacological and toxicological tests; and clinical tests* unless the two later categories can be omitted. The grounds for omission of those sections and the contents of all sections will be discussed later. These data form Parts II, III, and IV of the dossier.

9. A *summary of product characteristics* (SPC) is also required. The sequence given for the SPC in Directive 65/65/EEC has been amended in guidance note III/9163/90-EN *Note for Guidance: Summary of Product Characteristics* (adopted by the CPMP in October 1991). The recommended sequence is now:
   a. Name of the medicinal product
   b. Qualitative and quantitative composition
   c. Pharmaceutical form
   d. Clinical particulars: therapeutic indications; posology and method of administration; contraindications; special warnings and special precautions for use; interaction with other medicaments and other forms of interaction; pregnancy and lactation; effects on ability to drive and operate machines; undesirable effects; overdose
   e. Pharmacological effects: pharmacodynamics; pharmacokinetics; preclinical safety data

   f. Pharmaceutical particulars: List of excipients; incompatibilities; shelf life; spe-
       cial precautions for storage; nature and contents of container; instructions for
       use and handling; name/style/permanent address or registered place of business
       of the holder of the marketing authorization.

   The SPC forms Part I B of the application dosssier. It is an important document
and needs to be drafted carefully. Not only will it be used by the assessor as a key to
the basis for the review of the application, but the final version of the document after
the grant of the marketing authorization will control the permitted claims that may be
made for the product in advertising and promotion, and the information made available
to prescribers and suppliers of medicinal products has to be closely based on it. A
copy of the original approved SPC has to accompany an application for a supplemen-
tary patient certificate—which can extend the patient life of a pharmaceutical in the
EC.

10. *Expert reports* prepared by appropriately qualified and experienced persons and
     including:
     a. A product profile
     b. *Expert Reports* consisting of summary tabulations/formats, written summaries,
         and critical commentaries for each of the pharmaceutical, the preclinical, and
         the clinical sections. These documents form Part I C of the application.
11. The *applicant's signature* and the *date of the application* must be included. The ap-
     plication is invalid unless it bears a signature.

E. Data Requirements for an European Community Marketing Authorization Application

Part II, III, and IV of the application contain the technical and scientific and clinical data to
support the application. This is required to address the following topics:

Part II

   II A: Composition of the product (formula, container, clinical trial formulae, development
       pharmaceutics)
   II B: Method of preparation (manufacturing formula, manufacturing method, validation of
       the manufacturing method)
   II C: Control of the starting materials (specification and routine tests; data on the active
       ingredients, other ingredients, and packaging components)
   II D: Control tests on intermediate products
   II E: Control tests on the finished product
   II F: Stability tests
   II Q: Bioavailability or bioequivalence (with relevant references to part IV) and any other
       information

Part III

   III A: Single dose toxicity
   III B: Repeated dose toxicity (subacute, chronic)
   III C: Reproduction studies (fertility/general reproductive performance; embryotoxicity; peri-
       and postnatal toxicity)
   III D: Mutagenic potential (in vitro, in vivo)
   III E: Oncogenic/carcinogenic potential

III F: Pharmacodynamics (effects relating to the proposed indications; general pharmacodynamics; drug interactions)

III G: Pharmacokinetics (single-dose, repeated-dose; distribution in normal and pregnant
   animals; biotransformation)

III H: Local tolerance/toxicity

III Q: Other information

From January 1995 it is also necessary to include an environmental risk assessment as part
of Part III for products containing new drugs and certain other materials.

Part IV

IV A: Clinical pharmacology (pharmacodynamics, pharmacokinetics)

IV B: Clinical experience (clinical trials; postmarketing studies; published and unpublished
   studies)

IV Q: Other information.

Information on the data requirements is included in the Annex to Directive 91/507/EEC,
but this is further expanded on in a series of Guidance Notes issued at various times by the
CPMP. A list of the guidelines available in 1994 is included in Table 7.

The fundamental requirement is that the applicant should submit with the application sufficient information to establish the quality, safety, and efficacy of the medicinal product, taking
into account its intended uses. The basis of the assessment by the regulatory agency will be
the Directive requirements and any applicable guidelines current at the time of the application,
and the claims included by the applicant in the SPC.

Part II: Chemical, Pharmaceutical and Biological Testing

Particular emphasis is placed in the need for adequate *validation data for analytical methods*
at the start of the section on Part II data requirements in the Annex to Directive 91/507/EEC.
There is a similar emphasis in the section on preclinical testing and clinical studies, especially
on the analytical methods used to support safety data and pharmacokinetics. All test procedures
should be acceptable by the standards current at the time of the application. It is necessary to
describe them in sufficient detail for the national agency to carry them out for verification or
other purposes. In particular, detailed descriptions should be provided of any noncommercial
equipment. Some, but not all, registration authorities in the EC will undertake tests on samples
of the product before or after the product approval.

Part II A. Part II A of the dossier should include information on the *composition of the
product*: its active ingredient(s), the excipients, and those constituents (such as capsule shells)
that are intended to be ingested when the dosage form (= drug product) is taken. *Clinical trial
formulae* should be stated where these differed from the formulation to be marketed.

The titles given to ingredients should be the name at the head of a *European Pharmacopoeia*
monograph or the name at the head of a monograph in the pharmacopoeia of a member state
of the EC (or an approved synonym of one of these), or the International Nonproprietary Name
(INN). These may be accompanied by other nonproprietary names. Failing this the exact scientific name may be used. For coloring matter the E-number assigned should be indicated.

The *quantitative content of the dosage form* should be stated in an appropriate way.

The development pharmaceutics section is included in Part II A, also. This should include
an explanation of the choice of the composition and container. Information should be included
on the intended function of the excipients and any excess of an ingredient (overage) explained
and justified. Particular attention should be paid to a justification for the use of any novel or
unusual excipients. Compatibility of (active or inactive) ingredients may need to be discussed.

Table 7 Guidelines Adopted by the Committee for Proprietary Medicinal Products

Subject	Where published[a]
Development pharmaceutics and process validation	Vol. III
Chemistry of active ingredients	Vol. III
Stability tests	Vol. III
Quality of herbal remedies	Vol. III
Analytical validation	Add 1990
European DMF procedure	Add 1990
Radiopharmaceuticals	Add 1992
Radiopharmaceuticals based on monoclonal antibodies	Add 1992
Amendments to the Notice to Applicants for Radiopharmaceutical applications	
Adaptation of the pharmaceutical Expert Report to Radiopharmaceuticals	
Requirements relating to active ingredients	Add 1992
Use of ionizing radiation in the manufacture of medicinal products	Add 1992
Specifications and control tests on the finished product	Add 1992
CPMP list of allowed terms for the pharmaceutical dosage form, route of administration, container, closure, and administration device	
Investigation of chiral active ingredients	
Stability testing of new drug substances and products (ICH)	
Limitation to the use of ethylene oxide in the manufacture of medicinal products	
Manufacture of products derived from human blood or human plasma	
Manufacture of investigational medicinal products	
Quality of prolonged release oral solid dosage forms	
Use of the European DMF procedure. A practical guide to implementation by the chemical industry (active ingredient manufacturers) and the pharmaceutical industry (marketing authorization applicant)	
Replacement of CFCs in metered-dose inhalation products	
Definition of a new active substance	
Plastic primary packaging materials	
Excipients in the dossier for application for marketing authorization of a medicinal product	
Validation of analytical procedures (ICH)	
Biotechnology headings for the Notice to Applicants Part II	
Production/quality control: derived by recombinant DNA technology	Vol. III
Production/quality control: murine monoclonal antibodies	Vol. III
Preclinical biological safety testing: biotechnology products	Vol. III
Production/quality control: cytokines derived by biotechnology	Add 1990
Production/quality control: human monoclonal antibodies	Add 1990
Validation of virus removal and inactivation procedures	Add 1992
Harmonization: influenza vaccines	May 1992
Medicinal products derived from human blood and plasma	May 1992
Minimizing risk of transmission of agents causing spongiform encephalopathies by medicinal products	May 1992
Validation of virus removal/inactivation procedures: choice of viruses; and priority setting (two documents)	
Allergen products	May 1992
Control authority batch release of vaccines and blood products (several products)	
Plasma pool testing	
Single-dose toxicity	Vol. III
Repeated-dose toxicity	Vol. III
Reproduction studies	Vol. III
Testing of medicinal products for their mutagenic potential	Vol. III
Carcinogenic potential	Vol. III
Pharmacokinetics and metabolic studies in the safety evaluation of new drugs in animals	Vol. III
Nonclinical local tolerance testing of medicinal products	May 1992
Guideline to detection of toxicity to reproduction for medicinal products (ICH)	
Pharmacokinetics: guidance for repeated dose tissue distribution studies (ICH)	

Table 7 Continued

Subject	Where published[a]
Toxicokinetics: the assessment of systemic exposure in toxicity studies (ICH)	
Carcinogenicity: guidance for dose selection for carcinogenicity studies (ICH)	
Recommended basis for the conduct of clinical trials	Vol. III
Clinical investigation: children	Vol. III
Clinical investigation: the elderly	Vol. III
Pharmacokinetics studies in humans	Vol. III
Clinical testing: drugs for long-term use	Vol. III
Fixed-combination products	Vol. III
Good Clinical Practice	Add 1990
Clinical testing: prolonged-action forms (extended-action forms)	Add 1990
Investigation: bioavailability and bioequivalence	May 1992
Good Clinical Practices: guideline for essential documents (ICH)	
Studies in support of special populations: geriatrics (ICH)	
The extent of population exposure to assess clinical safety for drugs intended for long-term treatment of non–life-threatening conditions (ICH)	
Clinical safety data management: definitions and standards for expedited reporting	
Clinical investigation; oral contraceptives	Vol. III
Clinical investigation: drugs for chronic peripheral vascular diseases	Vol. III
Nonsteroidal anti-inflammatory compounds: chronic disorders	Vol. III
Antiepileptic/anticonvulsant drugs	Vol. III
Antianginal drugs	Vol. III
Medicinal products used in the treatment of cardiac failure	Vol. III
Antiarrhythmic medicinal products	Vol. III
Antidepressant medicinal products	Vol. III
Evaluation of anticancer medicinal products in humans	Add 1990
Medicinal products for the treatment of epileptic disorders	Add 1990
Clinical investigation of hypnotic medicinal products	Add 1992
Guideline for the SPC of β-adrenergic blocking agents	
Guideline for the SPC of benzodiazepines	
Guideline for the SPC of angiotensoin-converting enzyme inhibitors	
User information on oral contraceptives	Vol. III
Data sheets: antimicrobial drugs	Vol. III
(Revised): antibacterial medicinal products	(Add 1990)
OECD Principles of Good Laboratory Practice	Vol. III

[a]Key:

Vol. III = *The Rules Governing Medicinal Products in the European Community* Volume III: *Guidelines on the Quality, Safety and Efficacy of Medicinal Products for Human Use;*

Add 1990 = Volume III *Addendum July 1990;*

Add 1992 = Volume III *Addendum No. 2*, published May 1992.

Items without a publication reference had not been officially published by the Commission at the time of writing, but were available through trade associations, the MCA EuroDirect service, and others.

In addition, the following guidelines were incorporated in Volume II A of the *Rules Governing Medicinal Products*, the *Notice to Applicants* issued by the Commission of the European Communities in December 1992 (in edited form):

Procedure for CPMP Guidelines

Abridged applications

EC application format

Summary of Product Characteristics

Dossier check-in procedure

Operational procedures for the rapporteur and the concerned member states in the Multistate Procedure

Operational procedures for the rapporteur and the concerned member states in the Concertation Procedure.

Other guidelines include

Variations to a marketing authorization

Guideline on the Assessment Report (and explanatory memorandum on Expert Reports in the application dossier)

Any factors likely to influence the bioavailability or the acceptability for use of the finished product should be addressed. Information should also be included to show that the formulation has been optimized.

Part II B. The *method of manufacture* and the *manufacturing formula* of the finished product should be described in Part II B. The various stages of the manufacturing process should be described in sufficient detail to allow an assessment to be made of any potential adverse changes induced in the ingredients. If continuous production processes are used, there should be information on the measures taken to ensure homogeneity of the resulting product. In-process control points should be indicated, and the nature of the tests and limits applied should be stated.

Adequate *validation data* should be submitted for the manufacturing process, particularly when nonstandard or critical processes are used. In particular, attention should be paid to sterilization procedures or aseptic techniques employed. Environment- and equipment-related issues may need to be addressed, depending on the nature of the processes applied.

Part II C. Part II C is concerned with the *controls applied to the starting materials.* These include active ingredients, excipients, and container components.

The amount of information that is required to support the use of a particular ingredient in an application depends on the availability of an acceptable and relevant pharmacopeial monograph. In this context, priority is given to monographs in the *European Pharmacopoeia*, which are expected to be applied if relevant. Failing this priority will be given to the pharmacopeia of a member state of the EC. Other pharmacopeial specifications may be accepted in the absence of these. For a pharmacopeial monograph to be acceptable, it must be shown that the route of synthesis used for the ingredient does not leave impurities not mentioned in the monograph, and that the monograph is otherwise adequate to control the quality of the ingredient. When it is necessary to use a foreign pharmacopeia, the monograph (and any other relevant information) should be translated before submission. The routine tests applied to the material should be stated, and any nonroutine tests and the frequency with which they are applied should also be given.

If there is no pharmacopeial monograph for an ingredient, it is necessary to provide information on the following points:

1. The name of the substance.
2. A definition comparable with the statements included in the *European Pharmacopoeia.*
3. Methods capable of identifying the substance.
4. Purity tests in relation to known impurities (individually named and total) and predictable, but unnamed impurities (individually and in total), especialy those that may have a harmful effect, that may affect stability, or that may distort analytical results. The application should also include a detailed discussion of potential and actual impurities found.
5. Any special storage requirements and retest intervals.
6. A description of the manufacture of the material, with a structural diagram indicating the stereochemistry (if any) of the material. Annotated flow charts of the synthetic process are particularly useful. Factors such as scale and yield, and details such as the reactants, reagents, solvents, catalysts, reaction conditions, isolation of intermediates and the finished substance, and the final purification procedure should be described.
7. Developmental chemistry data should be provided to demonstrate the chemical structure of the compound.
8. Physicochemical characteristics of the compound should be demonstrated.
9. Analytical development should be described.

10. A specification for the ingredient should be provided, fully characterizing the physical characteristics, identity, purity, and potency of the material (with stated methods and limits).

Batch analyses should be provided for several (ideally five) recent batches of the material prepared by the proposed synthetic route and at the intended scale of manufacture. Data should also be provided on any reference materials.

Coloring matter should meet the requirements of Directive 78/25/EEC. Permitted coloring agents are: curcumin, lactoflavin/riboflavin, tartrazine, quinoline yellow, orange yellow S/sunset yellow FCF, cochineal carminic acid, azorubine carmoisine, amaranth, cochineal red A/ponceau 4R, erythrosine, patent blue V, indigotin/indigo carmine, copper complexes of chlorophylls and chlorophyllins, acid brilliant green/lissamine green, caramel, brilliant black BN/black PN, charcoal, listed carotenoids, listed xanthophylls, beetroot red/betanin, anthocyanins, calcium carbonate, titanium dioxide, iron oxides and hydroxides, aluminium, silver, and gold.

It is possible to submit a Drug Master File (DMF) under the European Drug Master File Procedure. This does not reduce the data requirements, but allows active ingredient manufacturers to submit some information to the regulatory authorities on a confidential basis if necessary. However, the DMF itself is not approved—it is the product containing that ingredient that will be the subject of approval. European DMFs are different from FDA DMFs.

Additional information may be required on any materials of human, animal, or vegetable origin.

Part II D.   Part II D of the application concerns any tests applied at *intermediate stages of the manufacture* of the finished product.

Part II E.   Part II E includes details of the *control tests applied to the finished product.* Due account should be taken of the requirements of the general monographs of the *European Pharmacopoeia* applicable to the dosage form concerned.

The tests that might be expected in a finished product specification include the following:

1. Identity tests for the active ingredient and any essential excipients (e.g., antioxidants and antimicrobial preservatives): An identity test is also required for any colors present in the finished product.
2. Assays for active ingredients and essential excipients: Upper and lower limits would normally be expected. At the time of release the assay limit for an active ingredient should normally be the nominal content (allowance being made for any justifiable overages) $\pm$ 5%, unless wider limits can be adequately justified.
3. Controls on impurities and degradation products.
4. Sterility tests, pyrogenicity/endotoxins tests, other biological safety tests, as appropriate.
5. Dissolution tests for solid dosage forms taken orally, unless their absence can be justified.

Part II F.   In Part II F of the application the *stability studies* on the active ingredient and the dosage form should be described, with numerical results for all tests included in the drug substance specification and the finished product specification, respectively, and any additional tests that may be indicative of the overall stabiity of the material. The provisions of the ICH guideline on stability testing requirements should be taken into account.

Bioavailability or bioequivalence data should be discussed in Part II Q.

Part III: Pharmacological and Toxicological Tests

The tests described in this section of the application should have been carried out under Good Laboratory Practice conditions. The studies should be designed to demonstrate the potential toxicity of the ingredient and any potentially dangerous or undesirable erffects that may occur during use in humans. The pharmacological properties in relation to the proposed use in humans should also be reported. Mathematical and statistical procedures should be applied to the results as appropriate.

Systemic absorption of products intended for topical application should be investigated. Provided that negligible systemic absorption occurs, then certain tests (e.g., repeated dose systemic toxicity, fetal toxicity, and reproductive toxicity) may be omitted.

Part III A.   Part III A includes reports on the toxicity investigations. *Single-dose toxicity studies* should be carried out in both sexes of two or more mammalian species of known strain, unless the use of a single strain or sex can be justified. Two or more routes of administration should be used, one of which should be that intended for use in humans. The formulation used for administration should be as bland as possible.

Systemic exposure to the test substance should be ensured. Signs of toxicity (including local toxicity) should be recorded for at least 7 days and usually for 14 days, provided that the animals are not exposed to unnecessary suffering. The full spectrum of toxicity signs should be present and the mode of death determined. Information on the dose–effect relationship should be gathered, but high-precision quantitative evaluation of this or of lethal dose are not required. All animals should be autopsied, including those dying during the observation period.

When a combination of active substances is used checks should be made to see whether enhanced or novel toxic effects are produced by the combination compared with the separate ingredients. The impurities in the test substance should be of the same pattern as in the material to be given to humans, or additional data may be required.

Part III B.   Subchronic and chronic *repeated-dose toxicity studies* are required and are included in Part III B. The former should extend over 2–4 weeks or up to 3 months; the latter will last for a period determined by the intended duration of clinical use in humans, but normally for 3–6 months for products intended for use in humans in repeated doses over longer than 30 days. Two mammalian species should be used, one of which should be a nonrodent. Both sexes should be included. The species selected should have metabolic and pharmacokinetic characteristics as close as possible to human. The pharmacodynamic effects and target organs should be as close as possible to human.

The route(s) of administration should consider the intended therapeutic use, taking into account the possibility of systemic absorption. The dosage and frequency of administration should be such that target organs and harmful secondary effects can be established. It is usually necessary to undertake dosing on 7 days/week throughout the study.

The evaluation of toxic effects should be based on observations on behavior, growth, hematology, and biochemistry tests, and autopsy and histology reports.

Part III C.   Part III C includes *reproductive function*, which should be investigated if any test suggests harmful effects on progeny or impairment of male or female reproductive function. The omission of the tests should be justified (e.g. because the product will be used only in men or in women who are not of childbearing potential).

Any effect on mating behavior, or any effect leading to fetal loss, abnormality or damage to offspring in later life, should be revealed by the studies. Embryotoxicity studies should be undertaken in two species, one other than a rodent. For fertility and perinatal studies at least one species (preferably one having a drug metabolism similar to human) should be used. One of the species should have been used in the long-term toxicity studies. Adequate numbers of animals may be as follows: for embryotoxicity studies, 20 pregnant female rodents and 12 pregnant female nonrodents; for fertility studies, 24 male and 24 female rodents; and for perinatal studies, 12 pregnant females of whichever species used.

Three dose levels are normally expected, the highest producing signs of maternal toxicity. Dosing should continue throughout the period of embryogenesis in embryotoxicity studies. For fertility studies, male and female animals should have been dosed for a sufficient time before the proposed mating to demonstrate any effects on gametogenesis. Perinatal study dosing should extend throughout the period of gestation, from the end of organogenesis to parturition, and extend to lactation and weaning. The level of exposure of the fetus to the drug should be determined as far as technically possible.

Part III D.   In vivo and in vitro *mutagenic potential studies* should be reported in Part III D for all new active substances. The tests selected should be capable of detecting the main classes of genetic damage: gene or point mutation, chromosomal mutation, and, if possible, genomic mutation. At least one in vivo test should be included.

A system that is based on the use of four categories of validated test is recommended, although alternative strategies may be used if justified: tests for gene mutations in bacteria; tests for chromosomal aberrations in mammalian cells in vitro; gene mutation tests carried out in eukaryotes; and in vivo tests for genetic damage. There is no test in this panel that detects genomic mutation, because at the time of writing the guideline on mutagenicity tests there were no sufficiently validated procedures.

The studies are intended to demonstrate whether the compound under test is mutagenic; a separate question is whether it represents a genetic hazard to humans. The significance of positive and negative test results should be assessed, not according to their number, but according to their nature and significance. The overall risk/benefit assessment of a mutagen should take into account the results of the mutagenicity tests; the pharmacokinetics, metabolism, and toxicity profile of the compound; together with the intended patient population and extent of use for the product.

Part III E.   Carcinogenic–oncogenic potential studies are required when the substance has a close chemical analogy to a known carcinogen or cocarcinogen; when suspicious changes have occurred during long-term toxicity tests; when suspicious results have been obtained in mutagenicity studies or short-term carcinogenicity studies; or when the maternal is to be administered to humans regularly over a prolonged period. These data should be included in Part III E. Due account should be taken of the intended use of the medicinal product: for example, if it is to be used only in patients with a limited life expectancy (shorter than would be expected to reveal any carcinogenic potential in humans) it might not be necessary to undertake the tests.

Two species should normally be used for carcinogenicity tests. When possible, the metabolic handling of the drug in the animals should be similar to that in humans. Species or strains with high incidence of spontaneous tumor formation should be avoided. Positive controls are not normally required, but the spontaneous formation of tumors in the strains used should be recorded.

The clinical route of administration should be used in the studies and evidence provided for absorption of the test substance. Dosing should be on a daily basis. Three dose levels should be used: the highest one should demonstrate signs of minimal toxic effect (e.g., 10% weight loss, growth failure, or other) and target organ toxicity. Studies should last for 24 months in rats and 18 months in mice. In rats, mice, and hamsters, there should be two groups of 50 animals of each sex in the control treated with the vehicle, and 50 animals per sex per treatment group. A uniform diet should be used.

A full autopsy should be undertaken on all animals dying during the course of the study or killed because of their poor condition. All surviving animals should be sacrificed at the end of the study, and a full autopsy should be conducted on each animal. Hematological and biochemical investigations may also be useful in the interpretation of any lesions found.

The results should be reported for each group and each sex separately. The data should be analyzed with consideration of the incidence of tumor-bearing animals, the total incidence of tumors, the incidence of tumors affecting a specific tissue, the incidence of malignant tumors, and the latent period to tumor appearance. The effects seen in the three treatment groups and the control group(s) should be compared.

Part III F. Pharmacodynamics studies should investigate the therapeutic basis for the use of the drug (with results expressed quantitatively using time-effect curves, dose-response curves, and such). Comparative data with substances of known activity should be used as far as possible. In addition, the general pharmacological properties of the test material should also be described. Any modifications of the effects of the drug on repeated administration should be indicated. Drug interaction studies should also be reported.

For combination products, data will be required on the combination and possibly on the individual components (e.g., when a novel ingredient is present). If the combination is based on a pharmacological theory, then data should be provided to demonstrate the interactions that might give rise to therapeutic benefit. If the combination is based on therapeutic experiments, the investigations should include an attempt to demonstrate the expected effects in animals, with the importance of collateral effects being examined.

Part III G. Part III G of the dossier includes the *pharmacokinetic data*, which should cover absorption, distribution, metabolism, and the excretion. Studies on the pharmacokinetics after a single dose and after repeated doses; the distribution of the drug in normal and pregnant animals; and the effect on biotransformation are required for all pharmacologically active ingredients. Biotransformation data are necessary for all products from which the information helps to establish the human dose. They are also required for chemotherapeutic substances, such as antibiotics, and for substances that depend on nonpharmacodynamic effects. If the toxicity tests or therapeutic experiments justify it, pharmacokinetic studies may be omitted for new combinations of known substances.

Part III H. This includes *local toxicity (tolerance) studies*: These should distinguish between the mechanical effects of administering the substance and physicochemical effects on the one hand, and toxicological and pharmacological effects on the other. The studies should investigate tolerance for both active ingredient and excipients.

Part III Q. This section contains any additional information.

Part IV: Clinical Documentation

Before clinical trials are initiated there should be adequate knowledge relating to the animal pharmacology and toxicology of the compound under investigation. Relevant information should be included in an Investigator's Brochure. All clinical studies should be undertaken under Good Clinical Practice. Due account should be taken of the current revision of the Declaration of Helsinki, with freely given consent from all trial subjects.

The opinion of an appropriate ethics committee (similar to an institutional review board, but with some differences depending to some extent on the country in which the trial is to be undertaken) is required before the initiation of trials in patients and possibly volunteers, depending on the member state concerned. The organization, conduct, data collection, documentation, and verification procedures for the trial should be established before it commences.

Appropriate arrangements need to be made for archival storage of the information gathered in the trial: patient identification codes should be retained by the investigator for at least 15 years after the trial had been completed or discontinued; patient files and other source data should be retained for the maximum period of time permitted by the institution involved; and the owner of the data should retain the trial protocol, standard operating procedures, written options on protocol and procedures, the investigator's brochure, case report forms for each subject, the final report, and audit certificates (if available).

For each clinical trial, the data should be presented in sufficient detail to permit objective judgments to be made of the protocol, statistical design, and methods used. The conditions under which the trial was performed and managed should be stated and available audit certificates provided, with a list of the investigators and where the study was undertaken, as well as details of the investigational product and the case report forms from the study subjects. The final report should be supplied and should be signed by the investigator(s).

Part of the information may be omitted by prior arrangement with the national authority(ies) concerned with the application. The omission of any data should be explained. However, any such omitted information should be available for submission to the authorities at short notice if requested.

For each trial the summary of clinical observations should include:

1. The number and sex of the patients treated
2. The selection and distribution of the investigational and the control groups
3. The numbers of subjects withdrawn from the study, and the reasons for their withdrawal
4. The nature of the control groups (no treatment, placebo, comparator medicinal product, comparator treatment)
5. The frequency of observed side effects
6. Details of at-risk patients (e.g., the elderly, menstruating or pregnant women, children, those with pathological or physiological conditions requiring special consideration)
7. Evaluation criteria for efficacy and the results when assessed by those criteria
8. Statistical evaluation and the variable factors involved.

The investigator should express an opinion on the safety of the product under normal conditions of use. Compatibility, efficacy, and any other useful information relating to indications and contraindications, dosage, average duration of treatment, any special precautions to be taken during treatment, clinical symptoms of overdosage, and so on, should be included. Observations should be incuded on any signs of habituation, any interactions with concomitantly administered medicinal products, criteria for excluding patients from trials, and any deaths occurring during the trial or follow-up period. Any modification of the pharmacological effects of the product under investigation after repeated administration should be reported.

Due account should be taken of the general guidelines on clinical data requirements, and those guidelines relating to relevant clinical areas.

Part IV A

1. CLINICAL PHARMACOLOGY.   The following should be addressed in Part IV A:

Pharmacodynamic action relating to efficacy and any other pharmacodynamic action
Dose–response relationship
Time course of effect
Justification for the dosage
Justification for the conditions of administration
If possible, the mode of action.

2. PHARMACOKINETICS IN HUMANS.   The adsorption (rate and extent), distribution, metabolism, and excretion should be reported, together with any clinically significant aspects (e.g., any patient groups that might be assumed to be at risk on the basis of the kinetic data). Results should be reported for any healthy volunteer studies, patients, and at-risk groups. Differences between the results in humans and in experimental animals should be highlighted.

3.   INTERACTIONS.   Joint administration tests should be reported with the test product and other medicinal products with which it is likely to be used concomitantly. The clinical signif-icance of interactions with other medicinal products, alcohol, caffeine, tobacco, or nicotine should be described and discussed. Suitable statements should be included in the draft SPC.

4.   BIOAVAILABILITY/BIOEQUIVALENCE.   Bioavailability should be assessed for products with a narrow therapeutic index, or when previous tests have indicated that there are anomalies in the pharmacokinetic properties, such as variable absorption. Bioequivalence studies may be needed to support an abridged application made under Directive 65/65/EEC Article 4(8)(a).

Part IV B: Clinical Efficacy and Safety.   As far as possible clinical studies should be of randomized, controlled design. The inclusion of a large number of patients is not an alternative to an adequately designed and conducted study. Blinding should be used to avoid bias. Un-substantiated statements of clinical efficacy or safety are not admissible.

The statistical methods to be used in the analyses should be described in the protocol, including the level of significance to be used. Numbers of patients and the reasons for their inclusion in the trial should be stated, and the power of the study should be estimated. The method used to avoid bias (e.g., randomization) should be stated. Data are generally more valuable if they are derived from several independent competent investigators.

All adverse events (including abnormal laboratory values) should be individually reported and discussed. The overall experience of adverse events and their nature, seriousness, and frequency should be discussed. Relative safety should be considered, taking into account the disease being treated, available alternative therapies, characteristics of patient subgroups, and the preclinical toxicology and pharmacology data. Recommendations intended to reduce the incidence of adverse events should be included.

The Expert Report should include a discussion of the pertinence of the various clinical trials to the evaluation of the clinical efficacy and safety of the product.

Postmarketing Experience

When the product has been authorized in other countries, information should be included on the adverse event profile obtained. The experience with other products containing the same ingredient(s) should also be reported.

Submission of Applications in Special Circumstances

It is possible to submit an application that does not include comprehensive data on quality, safety, and efficacy in certain circumstances. Such instances may relate to conditions that are encountered so rarely that the applicant cannot be expected to provide comprehensive data; or if in the present state of scientific knowledge such data cannot be provided; or if it would be unethical to collect such information. Marketing authorizations granted in such cases will have certain conditions (e.g., the completion of a stated program of additional studies within an agreed timescale, the results from which will be used as the basis for a reassessment of the clinical safety and efficacy of the product; the supply of the product only on prescription (possibly with additional restrictions on the administration of the product; e.g., under strict medical supervision, in a hospital); and inclusion in the package leaflet of information for the practitioner indicating which aspects of the available data are incomplete.

Justification for the Omission of Pharmacotoxicology Data and Clinical Data

It was indicated earlier that an application usually has to include supporting pharmaceutical, pharmacotoxicological, and clinical data. In addition to the special provisions indicated in the foregoing paragraph, it is also possible to justify the submission of applications without phar-macotoxicological or clinical data in another circumstance: the abridged or the ''hybrid'' application. The applicant needs to justify the omission of those data in the relevant Expert Reports.

Admissible grounds are as follows:

1. That the product is ''essentially similar'' to a product authorized in the member state concerned and that the holder of the original authorization has agreed to allow the second applicant access to the data on which it was approved
2. By means of detailed references to the published literature showing that the constituents have a well-established medicinal use with recognized efficacy and an acceptable level of safety
3. That the new product is ''essentially similar'' to a product that has been approved in the EC for not less than 6 years or 10 years, depending on the member state concerned (and 10 years for products covered by the Annex to Directive 87/22/ECC, i.e., biotechnology and high-technology products) and that is marketed in the member states to which the abridged or hybrid application is submitted.

In all these cases, the intended therapeutic use should be the same as that of the original product, and the dose and route of administration must not have changed (but see later). In this context *essentially similar* means:

1. That the product has the same qualitative and quantitative composition relative to the active ingredient
2. That the pharmaceutical form is the same (although the concept of essential similarity is also extended to different oral dosage forms intended for immediate release)
3. That, when appropriate, bioavailability studies have been undertaken.

The impurity and related substances profile and the decomposition products arising during storage should be clearly indicated to allow appropriate assessment of the efficacy and safety of the new product. Bioequivalence should be established and the need for bioavailability or bioequivalence studies, or both, addressed in the Expert Reports.

It is also possible to submit a hybrid application (i.e., one that incorporates some preclinical and/or clinical data as well as the necessary quality data) if the two products are not strictly essentially similar. This might apply

1. When a different salt or ester of the active ingredient is to be used (when evidence will be required to demonstrate that there has been no change to the pharmacokinetics or pharmacodynamics or toxicity that could affect the safety or efficacy profile)
2. When a new therapeutic use is claimed (requiring clinical safety and efficacy data and, in some cases, preclinical data too)
3. When a new oral dosage form for immediate release is proposed (when bioequivalence data will be needed)
4. When a different route of administration is intended or a modified release product is proposed for the same route of administration (requiring clinical safety and efficacy data as well as pharmacokinetics data, local toxicity data, or other preclinical data may also be required)
5. When a different dosing schedule is proposed; for example, a reduction in the number of units or doses (requiring bioavailability data); or a change in the frequency, amount, dose, or daily dose (requiring clinical safety and efficacy data and pharmacokinetics data)
6. When a different strength of the product is proposed for use by the same route, using the same dosage form and the same dose schedule (requiring bioequivalence data)

7. When the product is suprabioavailable compared with the original (which may need only bioavailability studies, or may need clinical safety and efficacy data, depending on the circumstances)
8. When the new product is locally acting and the equivalence is difficult to establish other than by using a pharmacodynamic or therapeutic endpoint (when it may be possible to justify the omission of preclinical and clinical data in the Expert Reports).

F. Controls on Advertising of Medicinal Products

Directive 92/28/EEC introduced harmonized requirements on advertising of medicinal products throughout the EC. Advertising in this context includes any activity intended to promote the prescription, supply, sale, or consumption of medicinal products, and in particular advertising to the general public or to persons qualified to prescribe or supply the products; visits by representatives to those qualified to prescribe or supply medicines, and the supply of samples is also controlled. Sponsorship of promotional meetings and scientific congresses are also controlled, and the supply of other than insubstantial gifts connected with the practice of pharmacy or medicine is prohibited.

No unauthorized medicinal product may be advertised. All advertising materials are required to be in line with the SPC.

Advertising to the general public is not allowed for the following products:

1. Those that are available only on prescription
2. Those that contain psychotropic or narcotic substances by the United Nations definitions
3. Those that require the intervention of a medical practitioner for diagnosis or monitoring of treatment
4. Those for the treatment of therapeutic indications, such as tuberculosis, sexually transmitted diseases, other serious infectious diseases, cancer and tumoral diseases, chronic insomnia, diabetes, or other metabolic diseases
5. Those for which the cost may be reimbursed (when required by the member state).

An exemption is made in connection with approved vaccination campaigns.

The direct supply of medicinal products to the public for promotional purposes is not permitted. Any advertisement to the public is required to clearly set out that the message is an advertisement and that the product concerned is a medicine; and should include the name of the product, the name of the active ingredient (if only one is present), information necessary for the correct use of the product, and an invitation to read the instructions on the package leaflet or the outer packaging as appropriate. When intended solely as a reminder, member states may authorize advertisements to include only the name of the product.

The following may not appear in an advertisement aimed at the general public:

1. Any impression that a medical consultation or surgical operation is unnecessary or offer of a diagnosis or suggestion of treatment by mail
2. Any suggestion that the effects of taking the medicine are guaranteed, unaccompanied by side effects, or are better than or equivalent to another treatment or medicinal product
3. Suggestions that the patient's health can be enhanced by taking the medicine, or that their health can be affected by not taking the medicine (other than in the case of approved vaccination campaigns)
4. Exclusive targeting at children
5. References to recommendations by scientists, health professionals, or celebrities that may encourage consumption of the medicines

6. Suggestions that the safety or efficacy of the product is due to its being "natural"
7. Any information that may lead to an erroneous diagnosis by the inclusion of a case history
8. Claims of recovery in improper, misleading, or alarming terms
9. Pictorial representations of changes in the human body caused by disease or injury or by the action of a medicinal product on the body or its parts
10. Suggestion that the product is a foodstuff, cosmetic, or other consumer article
11. Any mention that the product has a marketing authorization.

Advertising to *health professionals* is required to be in line with the SPC, accurate, up to date, verifiable, sufficiently complete to allow the readers to form their own opinion on the therapeutic value of the product; and to include the supply classification for the product and, if required by the member state, the selling or indicative price and conditions of reimbursement by social security bodies. Provision is made for the permitted issue of a reminder including only the name of the product.

Medical sales representatives are required to have a sufficient knowledge of the products being promoted and are to give each person visited (or have available for distribution) the SPC for each product promoted and details of the price or reimbursement conditions. The representatives are also required to report to the company's scientific services any information on adverse events reported to them by the persons they visit.

Free samples of the products are to be supplied only on an exceptional basis and only to persons qualified to prescribe them and on their written request. A limited number of packs of a given product may be provided to each prescriber in any one year, and each has to be the same as the smallest marketing pack (identified as a free sample), accompanied by a copy of the SPC. Samples of psychotropic or narcotic products are not allowed. Different member states have applied different limits on the number of samples of other products that can be supplied to a particular person in a given year.

Provisions are included in the directive for monitoring of advertisements:

G. Labeling of Medicinal Products and Requirements for Leaflets

Directive 92/27/EEC introduced harmonized requirements for labeling and patient leaflets for medicinal products.

Labeling

The information is required to be easily legible, comprehensible, and indelible. National language requirements apply. The labeling and leaflets may include more than one language provided that the information in each is identical. Pictograms or symbols may be used.

The information required includes the following:

The name of the product
The name of the active ingredient (where only one)
The pharmaceutical form and strength (if more than one is available)
Qualitative and quantitative statements concerning the active ingredient content per dose
The pharmaceutical form and the contents by number, weight, or volume
A list of excipients included in a guideline (not finalized at the time of writing), except for ophthalmic and parenteral products for which all excipients have to be listed
The method or route of administration
If the product is for self-medication, instructions for use
A warning that the product must be kept out of the reach of children

Any special warnings or any special storage precautions

The batch reference number, and the expiry date in plain language

Any special precautions for the disposal of unused product or waste materials derived from it

The marketing authorization number

The name and address of the person responsible for placing the product on the market and of the manufacturer.

For blister packs, the immediate packaging needs to contain only the name of the product, the name of the marketing authorization holder, the batch expiry date, and the batch number, provided that the full information appears on the outer packaging.

For small immediate containers on which the full information cannot be included, the label may contain only the following information: the name/strength/route of administration, method of administration, expiry date, batch number, and contents by weight or volume.

Member states may require that the following information is included in the labeling:

The price and reimbursement status of the product

The legal supply classification

Some form of identification and authentication.

Guidelines are under consideration for special warnings for certain categories of product, information related to self-medication, legibility, methods of identification and confirmation of authenticity, and the list of excipients, which must appear on the label (and the way in which these are to be identified). These had not been finalized at the time of writing.

Leaflets

A user package leaflet is required for all products except those for which all the required information appears on the packaging.

The contents of the package leaflet must be in accordance with the SPC. It is required to be legible, to be in the national language(s) of the member states in which the product is marketed, and to be in a specified sequence. No promotional information is to be included. Symbols and pictograms may be used.

The sequence of the information is as follows:

1. For *identification of the product*: name of the product, name of the active ingredient if only one, pharmaceutical form and strength; a full statement of active ingredients (qualitative and quantitative) and excipients (qualitative); the pharmaceutical form and the contents by number, weight, or volume; the pharmacotherapeutic group or activity in terms understandable by the patients; name and address of the marketing authorization holder and of the manufacturer

2. *Therapeutic indications* unless the competent authorities decide that this information may be omitted if its dissemination may have serious disadvantages for the patient

3. *Information necessary before taking the product*: contraindications, precautions for use, forms of interaction with other medicinal products or alcohol, tobacco, foodstuffs, and so on, which may affect the action of the product; special warnings (e.g., special categories of patient: children, pregnant or breastfeeding women, the elderly, persons with specific pathological conditions, effects on driving or operating machinery, excipients)

4. *Instructions for proper use:* dosage; method and route of administration; frequency of administration; duration of treatment if this should be limited; action to be taken in the case of overdose; what should be done if one (or more) doses is missed; any indications for risks associated with withdrawal of the product

5. *Undesirable effects*: the effects, the action to be taken, and an express invitation to report to a doctor or pharmacist any undesirable effect not included in the list in the leaflet
6. A reference to the labeled *expiry date*, with a warning not to use the product after this date, with special storage precautions if necessary.

H.   The Role of Experts and Expert Reports

The requirement for Expert Reports to form part of the dossier was introduced in Directive 75/319/EEC. Three sections to these reports are required: on the chemical, pharmaceutical and biological documentation; on the pharmacotoxicological documentation; and on the clinical documentation. Each report should contain a product profile; a tabulated or graphic (and *accurate*) summary of the data, with adequate cross-referencing to the main part of the submission in which the original data may be found (compulsory); a written summary of the data, and a critical assessment. Each part of the report should be signed by a suitably qualified and experienced person. The nationality of the Expert is not relevant: their background and experience is.

The pharmaceutical Expert Report should contain a critical assessment of not more than 10 pages in addition to the tabulated summaries. The pharmacotoxicological Expert Report should include a critical summary of not more than 25 pages, in addition to the tabulations and the written summary. The medical Expert Report should include a critical summary of not more than 25 pages, in addition to the tabulations and written summaries. Each Expert Report should be prefaced by a product profile (1–2 pages), including brief extracts from the SPC as well as the following key points: the type of application; chemical and pharmacokinetic properties of the drug; indications; and a list of postmarketing surveillance undertaken and any authorizations already granted.

Failure to provide the Expert Report is grounds for refusal of a marketing authorization.

The provision of a well-written Expert Report is helpful to the applicant since it will often be used as the basis for the assessment report that has to be prepared if the product is referred through the CPMP procedures, or in connection with the Product Evaluation Report scheme.

I.   Manufacturing Authorizations

Manufacturers (including assemblers) in the EC are required to hold an authorization under the terms of Directive 75/319/EEC. To be acceptable for import into the EC, the member states must be satisfied that products manufactured in third countries have been made under conditions at least equivalent to those required inside the EC.

To obtain a manufacturing authorization, the applicant must submit details of the products and pharmaceutical forms and the place where they are to be manufactured, imported, or controlled; and indicate that suitable and sufficient premises, technical equipment, and control facilities are available for the manufacture, control, and storage of medicinal products. In addition, the applicant must have available the services of at least one Qualified Person. Applications for manufacturing authorizations are required to be processed within 90 days of the receipt of a valid application. The time limits are suspended if additional information is requested.

Manufacturing authorizations may be refused, suspended, or revoked, if it is found that the necessary conditions for the grant of an authorization are not met. This will occur if the information supplied with the application is found to be no longer accurate, if the manufacturer is found not to be meeting obligations imposed as part of the authorization, or if import procedures are not applied. Any decision to refuse, suspend, or revoke an authorization must be accompanied by detailed reasons for the decision and a notification of the remedies available under the national laws of the member state concerned, if any (and their time scale.)

Products imported from third countries (i.e., countries outside the EC) are required to be tested at import into the EC (but are required to be tested only once) unless exemptions from some or all of the requirements have been negotiated. The special relaxations in these cases are now negotiated by the Commission of the European Commuities on behalf of the EC.

Manufacturers' facilities (and also the premises of laboratories carrying out necessary checks on the part of the authorization holder) are subject to repeated inspections. This includes manufacturers outside the EC, although special arrangements (using mutual recognition of local inspections) are likely to apply for EFTA member states under the EEA.

After each inspection, the competent authority is required to make a report on whether or not the manufacturing facility complies with the EC principles and guidelines on Good Manufacturing Practices (GMP). A copy of the report is made available to the manufacturing authorization holder.

On reasoned request, member states are required to provide copies of inspection reports to the competent authorities of other member states. There are provisions for dealing with disputes between the member states on issues arising from manufacturing authorizations.

Manufacturing authorization holders are required to

1. Have the services of appropriate staff for both manufacture and control purposes
2. Dispose of medicinal products only as permitted by the member state
3. Give prior notice to the relevant competent authority of any changes that will affect the manufacturing authorization and to inform them immediately if the qualified person is replaced unexpectedly
4. Allow access to the premises at any time to agents of the competent authority
5. Enable the qualified person to undertake the duties of the post
6. Comply with the requirements of the Guide to Good Manufacturing Practice for medicinal products.

The *Qualified Person* is responsible for confirming that products manufactured within the EC have been manufactured and checked in accordance with the requirements of the marketing authorization and, for products coming from outside the EC, that each production batch has undergone the necessary import testing or is accompanied by a control report signed by a Qualified Person (unless special derogations apply).

Minimum requirements are defined for the Qualified Person (Directive 75/319/EEC, Article 23). These requirements include

1. Possession of a diploma, certificate, or other evidence of formal qualifications awarded after completion of a university course of study, or the equivalent extending over not less than 4 years of theoretical and practical study in pharmacy, medicine, veterinary medicine, chemistry, pharmaceutical chemistry, or biology; or a minimum of $3\frac{1}{2}$ years of such study if the course is followed by a period of at least 6 months in a pharmacy open to the public as part of a 12-month period of practical training, corroborated by an examination taken at university level.
2. The course should have included theoretical and practical study of the following topics: applied physics; general and inorganic chemistry; organic chemistry; analytical chemistry; pharmaceutical chemistry (including analysis of medicinal products); general and applied medicinal biochemistry; physiology; microbiology; pharmacology; pharmaceutical technology; toxicology; and medical aspects of pharmacognosy.

3. Practical experience of at least 2 years duration in one or more authorized manufacturing facilities with involvement in analysis of medicinal products or ingredients. (This may be reduced in certain circumstances.)

J. Principles and Guidelines of Good Pharmaceutical Manufacturing Practice

Directive 91/356/EEC applies the principles of Good Manufacturing Practice (GMP) to pharmaceuticals for human use. It includes definitions for terms such as *pharmaceutical quality assurance* (the sum total of organized arrangements made with the object of ensuring the medicinal products are of the quality required for their intented use), and *Good Manufacturing Practice* (the part of quality assurance that ensures that the products are consistently produced and controlled to the quality standards appropriate to their intended use).

The Directive cross-refers to the *Rules Governing Medicinal Products in the European Community*, Volume IV, *Good Manufacturing Practice for Medicinal Products*. This, in effect, makes compliance with the guidelines on GMP an obligation under the Directive. The provisions of the GMP guide are to be taken into account when an inspection is undertaken for or by a member state of the EC. Included in the guide are the following sections: quality management; personnel; premises and equipment; documentation; production; quality control; contract manufacture and analysis; complaints and product recall; and self-inspection.

There are also a series of annexes that cover special aspects of pharmaceuticals manufacture: sterile medicinal products; biological medicinal products for human use; veterinary medicinal products, other than immunologicals; medicinal gases; herbal medicinal products; sampling of starting and packaging materials; manufacture of liquids, creams, and ointments; manufacture of pressurized metered-dose aerosol preparations for inhalation; computerized systems; and the use of ionizing radiation in the manufacture of medicinal products. (Note that there is also a CPMP guidelines on the use of ionizing radiation in the manufacture of medicinal products, included in the *Addendum 1992* to *Volume III* of the *Rules Governing Medicinal Products in the European Community*.)

A recently adopted additional annex is that on manufacture of investigational medicinal products, available as document 111/3004/91-EN Final. Since there is no mandatory Community-wide system of controls on clinical trials, this guideline cannot be considered to be mandatory at the time of writing.

Detailed discussion of these various provisions is precluded by space limitations.

BIBLIOGRAPHY

CPMP Guidelines adopted but not published at the time of writing:
    Guidance on Standard Terms for Marketing Authorization Applications
    Note for Guidance: EC Application Format
    Note for Guidance: Summary of Product Characteristics
(Available through European pharmaceutical trade associations, the Pharmaceutical Manufacturers Association, or the Japanese Pharmaceutical Manufacturers Association).
Drug Information Journal, 27(1), pp 1–107, (1993).
The Economist Book of Vital World Statistics, The Economist Books, Hutchinson: London, 1990, ISBN 0 09 174652 3.
The Economist Pocket Europe, Hamish Hamilton in association with The Economist, London, 1994, ISBN 0 241 00280 X.
The Economist Pocket World in Figures, The Economist Books, London, 1991, ISBN 0 099 174913 1.

EC Information Handbook 1992, EC Committee of the American Chamber of Commerce in Belgium, 1992.

European Pharmacopoeia, 2nd ed., Maisonneuve, Ste Ruffine, France.

Official Journal of the European Communities (published in two main series—C and L—daily by the Office for Official Publications of the European Communities, Luxembourg; available on subscription, or by purchase of single issues, through agencies for the Commission of the European Communities).

Panorama of EC Industries 1991–1992, Commission of the European Communities, Office for Official Publications of the European Communities, Luxembourg, 1991, ISBN 92 826 3103 6.

Pharmaceutical Product Licensing Requirements for Europe, A. C. Cartwright and B. R. Matthews, eds., Ellis Horwood, Chichester, 1991, ISBN 0 13 662883 4.

The Regulatory Affairs Journal (monthly journal, with quarterly sister publication *The Regulatory Affairs Journal (Devices)*, published by The Regulatory Affairs Journal, Ltd., Bagshot, UK).

Rules Governing Medicinal Products in the European Community

    Volume I: *The Rules Governing Medicinal Products for Human Use in the European Community*, September 1991 revision, ISBN 92 826 3166 4

    Volume II: *Notice to Applicants for Marketing Authorizations for Medicinal Products for Human Use in the Member States of the European Community*, 1989, ISBN 92 825 9503 X.

    Volume III: *Guidelines on the Quality, Safety and Efficacy of Medicinal Products for Human Use*, 1989, ISBN 92 825 9619 2.

    Volume III: *Addendum, July 1990*, ISBN 92 826 0421 7.

    Volume III: *Addendum No. 2*, 1992, ISBN 92 826 4550 9.

    Volume IV: *Good Manufacturing Practice for Medicinal Products*, 1992, ISBN 92 826 3180 X.

    Volume V: *Veterinary Medicinal Products*, 1989, ISBN 92 825 9643 5. (Note that this volume includes the *Notice to Applicants for Marketing Authorizations for Veterinary Medicinal Products in the Member States of the European Community*.)

    Volume VI: *Establishment by the European Community of Maximum Residue Limits (MRLS) for Residues of Veterinary Medicinal Products in Foodstuffs of Animal Origin*, 1991 ISBN 92 826 3173 7.

    Office for Official Publications of the European Communities, Luxembourg.

Scrip (newsletter published twice weekly by PJB Publications, UK).

Vachers' European Companion and Consultants' Register (E. Gunn ed.), Vachers Publications: Berkhamsted, November 1994 (A quarterly listing of diplomatic, political, and commercial information).

21

Pediatric and Geriatric Aspects of Pharmaceutics

Michele Danish
St. Joseph Hospital, Providence, Rhode Island

Mary Kathryn Kottke
AutoImmune, Inc., Lexington, Massachusetts

I. INTRODUCTION

A. Classification of Age Groups

Definition of Pediatric Age Groups

The rapid physical maturation that occurs between birth and adulthood is well known. Logically, it would be anticipated that these changes would result in altered responses to xenobiotics. Yet only 25% of drugs marketed in the United States have been evaluated as safe and effective in children. Most drug doses are prescribed for children based on a fractionation of the adult dose, or by weight.

In its continuing effort to improve the safety and efficacy of drugs in the pediatric population, the U.S. Food and Drug Administration (FDA) has defined five subgroups of this population by age. Each subgroup is not homogeneous, but does contain similar characteristics that are considered milestones in growth and development. The FDA classifications are listed in Table 1.

Age classifications do not provide an all-inclusive method for establishing pediatric doses. Given current knowledge, the most accurate pediatric doses are determined by both weight and age. Dosing based on surface area has no practical advantages for pediatric patients [1].

The classifications listed in Table 1 will be used throughout this chapter. Caution must be exercised when referring to observations in the neonatal period, however, since premature or very low birth weight neonates often have drug responses and disposition significantly different from those of full-term neonates. These differences will be clearly identified in the text.

Definitions of Elderly

Old age has been defined as the "advanced years of life when strength and vigor decline" [2]. Although this definition appears to be quite ambiguous, one realizes that it is necessarily so because the aging process itself is subject to a large amount of interindividual variation

Table 1 FDA Classification by Age of the Pediatric
Population

Age group	Description
Intrauterine	Conception to birth
Neonate	Birth to 1 mo
Infant	1 mo to 2 yr
Child	2 yr to onset of puberty
Adolescent	Onset of puberty to adult

[3,4]. This means that when comparing studies of a "young adult" population with studies of an "elderly" population, one will find the variance to be much greater in the elderly group. For instance, many readers probably know of a 65-year-old who "doesn't look a day past 50," whereas others may have made the acquaintance of a 50-year-old who is exceedingly frail.

Chronological Age. Because of this wide variation within the elderly population, it is difficult to devise a "catch-all" age one must attain to be considered elderly. Within the government, there also appears to be a problem of consistently defining this age group. Table 2, which lists a variety of federally funded programs and their corresponding age criteria, illustrates this point quite well. Table 3 lists what seems to be one of the better classification systems that has been devised [5]. As well as being the system currently being used by the U.S. Bureau of the Census, it is often implemented in studies that specifically deal with elderly populations [6,7].

Biological and Functional Age. To compensate for this wide variation noted among older populations, researchers have attempted to assess age in terms of biological or functional age. This basically involves combining a variety of factors together, such as physiological, psychological, and intellectual parameters, to develop a *functional age* that would serve as the older counterpart of *developmental age* that is used when assessing neonatal development [3,4,8]. The derivation of biological and functional age is quite complex and, although many interesting approaches to this problem have been studied, no one derivation has been universally accepted [3,8].

II. THE PEDIATRIC POPULATION

A. The Effects of Maturational Changes on Drug Disposition

Traditionally, drug dosing in infants and children has been based on age or weight ratio reduction of the adult dose. Data compiled in the past 20 years on maturational changes in organ function have allowed us to gain a much greater appreciation of the differences in drug disposition in pediatrics when compared with adults. Numerous review articles have addressed

Table 2 Eligibility, by Age for Federally Funded Programs

Age criteria	Program
60	Older Americans Act Title VII
62	Housing and urban development
65	Medicare Program Title XVII
70	Mandatory retirement age

Table 3 Commonly Utilized Age Classification
System

Age	Category
65–74	Young–old
75–84	Middle–old
85 +	Old–old

Source: Ref. 5.

in detail the current state of the art concerning clinical pharmacokinetics in neonates, infants, and children [9–14]. In this chapter, we will highlight the changes that are pertinent in the development of pediatric dosage forms.

Absorption

Age affects the capacity of all the physiological functions of the gastrointestinal tract. Gastric acid output follows a biphasic pattern in neonates, with the lowest gastric acid output observed between 10 and 30 days. Gastric acid output approaches adult values by 3 months of age. Acid output, on a kilogram basis, is similar to adult levels by 24 months of age [15]. Enteral feedings influence gastric acid secretion [16]. This suggests that hospitalized neonates receiving only parenteral nutrition will be relatively achlorhydric. This hypothesis is supported by reports of increased absorption of acid-labile drugs in the newborn period [10].

Gastric-emptying time and intestinal transit time are erratic in neonates. Gastric-emptying time is influenced by gestational and postnatal age and the type and frequency of feeding [17]. Although there is some controversy over the postnatal age at which adult patterns of gastric emptying are attained, it is generally accepted that adult values are reached by 6–8 months of age. There have been reports of erratic absorption of sustained-release products in children up to 6 years of age [18]. The possible role of gastric emptying on this clinical finding is unknown.

The first-pass effect has not been evaluated in infants and children. The maturational rate of metabolic pathways would be directly related to the oral bioavailability of a drug subject to first-pass effect. Drugs that undergo glucuronidation during enterohepatic recirculation may have altered systemic availability in children up to approximately 3 years of age because of delayed maturation of conjugation.

Pancreatic enzyme activity is very low in premature neonates. Lipase activity increases 20-fold in the first 9 months of life. Since concentration of bile salts is also low, it is anticipated that lipid-soluble drugs would be poorly absorbed in early infancy.

Colonization and metabolic activity of gastrointestinal bacterial flora do not approach adult values until 2–4 years of age [19]. This has resulted in increased bioavailability of digoxin in infants and young children.

Drugs absorbed by active transport mechanisms appear to have a delayed rate, but not extent of absorption, in the neonatal period [20].

Clinical studies and case reports suggest highly variable absorption patterns of drugs in neonates and infants [21,22]. There have been insufficient studies comparing children and adults to determine if there is an adult pattern of absorption in these age groups. There has been a case report in which carbamazepine, a drug with known solubility problems, was mal-absorbed in a child [23]. A rapid gastrointestinal transit time may have contributed to the incomplete absorption of the tablets. Selection of a more readily bioavailability dosage form, such as chewable tablets and liquids should be promoted for pediatric patients.

Distribution

Significant changes occur in the percentage of total body water, albumin, alpha-acid glycoprotein concentration, and fat composition from the neonatal period through adulthood [24]. The most rapid changes occur during the first month of life [25,26]. The healthy older infant and child have been the subjects of very few drug distribution studies. The influence of body composition on drug distribution and drug response needs further exploration.

Metabolism

Drug metabolic activity cannot be reliably predicted from gestational or postnatal age. Perinatal exposure to inducing agents, nutritional status, and hormonal changes, all play a role in the determination of metabolic activity. An added consideration is that those neonates who require medications are often subject to other medical and surgical interventions that may influence drug disposition by the liver [27].

In general, the phase I reactions, such as oxidation and n-demethylation are delayed in the neonate, but are fully operational at or above adult levels by 4–6 months of age in the full-term neonate [28–30]. Conjugation pathways, such as glucuronidation, do not approach adult values until 3 or 4 years of age. Sulfation activity does appear to reach adult levels in early infancy. For drugs that are subject to metabolism by both pathways, such as acetaminophen, the efficient activity of the sulfation pathway allows infants and children to compensate for low glucuronidation ability [31]. The elimination of other compounds in which sulfation (e.g., chloramphenicol) is not an alternative pathway are subject to prolonged elimination half-lives [32].

Infants and children older than 1 year of age are usually considered to be very efficient metabolizers of drugs and may actually require larger doses than those predicted by weight adjustment of adult doses, or shorter dosing intervals [33]. On the basis of metabolic activity, sustained-release formulations would appear to be ideal for children 1–10 years old, if bioavailability issues prove not to be problematic. The ability to clear drugs in critically ill children may be severely compromised; therefore, dosing in this subgroup of patients will require very careful titration [34].

Metabolic activity declines with the onset of adolescence. After puberty, adolescents metabolize drugs at a rate similar to adults [35].

Renal Excretion

The renal excretion of drugs depends on glomerular filtration, tubular secretion, and tubular absorption. A twofold increase in glomerular filtration occurs in the first 14 days of life [36]. The glomerular filtration rate continues to increase rapidly in the neonatal period and reaches a rate of about 86 ml/min per 1.73 m^2 by 3 months of age. Children 3–13 years of age have an average clearance of 134 ml/min per 1.73 m^2 [37]. Tubular secretion approaches adult values between 2 and 6 months [11]. There is more variability observed in maturation of tubular reabsorption capacity. This is likely linked to fluctuations in urinary pH in the neonatal period [38].

Summary

It is evident from the foregoing discussion that the greatest effects of maturation on drug disposition are observed in the first 6 months of life. However, individual variation in maturation in the first 3 years necessitates individual monitoring in both the ill neonate and the young child. Pediatric-dosing formulations that readily provide for small increments or decreases in single doses would greatly facilitate dosing in this age group.

B. Pharmacodynamic Differences Observed in Pediatric Patients

It has been generally assumed that therapeutic serum drug concentration ranges based on data obtained in adults were applicable to children. However, in many instances, when drug response is studied in children, differences in drug distribution and metabolism and in receptor sensitivity has rendered this assumption invalid [39].

Receptor Sensitivity and Response

Investigators have shown that children have lower brain/serum phenobarbital ratios than adults and that the ratio increases with gestational age [40,41]. This data clearly supports clinical findings that infants and children require higher serum concentrations of phenobarbital to maintain seizure control. Conversely, a lower therapeutic range for children has been identified for phenytoin, cyclosporine, and digoxin [42].

Age does not significantly affect plasma concentrations or disposition of ibuprofen; however, investigators have found that the onset of antipyresis and maximum antipyretic effect is greater in children younger than 1 year old than in children older than 6 years [43]. The authors hypothesized that this accelerated response was due to the greater relative body surface area of the young child.

Adverse Drug Reactions

Adverse effects to drugs differ in both type and incidence in the pediatric population. Because of immature metabolic pathways, infants and children may have metabolic patterns different from those of adults. This at least partially explains why neonates require lower theophylline serum concentrations for the treatment of neonatal apnea and why the incidence of hepatotoxicity following acetaminophen overdose is much lower in young children than adults [44,45]. For other drugs, such as chloramphenicol and codeine, the immaturity of metabolic pathways leads to the accumulation of excessive serum concentrations and toxicity [46,47].

Differences in receptor sensitivity have been offered to explain the spectrum of unexpected drug responses observed in children. Neonates and young children are at increased risk to experience paradoxical central nervous system (CNS) stimulation following antihistamine administration. Symptoms observed in pediatric cases of acute overdose include hallucinations, excitation, and seizures. A physiological explanation for this reaction has not been identified. Antihistamines should not be included in over-the-counter (OTC) cough and cold products recommended for infants and young children.

Resistance to drug toxic effects has also been observed in children. The incidence of aminoglycoside toxicity has been reported to be much lower in infants and children than in adults [39,48]. Diminished tissue sensitivity has been suggested as an explanation.

The high reported incidence of gastrointestinal abnormalities in children receiving nonsteroidal agents was also unexpected [49]. It was anticipated that they would be more resistant to gastrointestinal injury than adults.

The incidence of reported adverse drug reaction (ADR) hospital admissions in pediatrics ranges from 1.8 to 3.2%. This is lower than the incidence observed in the general population and may be the result of differences in organ function, or of a reduced level of drug exposure in children [50]. In a survey of outpatients, the incidence of adverse events in children was similar to adults [51]. A significant number of the adverse event reports were related to excessive responses to OTC drug products; this implies that inappropriate dosing may contribute to the high frequency of adverse effects. More precise dosing instructions on the labeling of OTC products intended for use in the pediatric population may alleviate this problem.

C. Drug Delivery System and Compliance Issues

Inactive Ingredients

Over 1300 excipients have been approved by the FDA for use as "inactive ingredients" in drug products. The FDA Division of Drug Information Resources compiles a list of all inactive ingredients in approved prescription drug products in the *Inactive Ingredient Guide*. The FDA requires the listing of excipients in pharmaceuticals other than oral. The labeling of inactive ingredients in oral drug products is voluntary. The reported incidence of adverse drug reactions to excipients is much lower than the incidence reported with active drug. This may be due to several factors, including the generally safe nature of the excipients, the low concentration found in single doses, or to the lack of identification of an excipient as a causative agent.

Selecting excipients for incorporation into a drug product for human consumption is complex. In selecting excipients for drug products intended for use in the pediatric population, additional cautions must be taken. Several subgroups of the pediatric population who are likely to be hospitalized and receive both oral and parenteral drug products have been identified as being particularly susceptible to excipient reactions. Serious events have been reported in low birth weight neonates and infants, asthmatics, and diabetics. Reactions have ranged from dermatitis to seizures and death [52,53]. Table 4 lists the excipients that have been reported to cause adverse events in the pediatric population.

Many of these reactions are related to the quantity of excipient found in a dosage form. Benzyl alcohol, propylene glycol, lactose, and polysorbates, all are associated with dose-related toxic reactions [52–54]. Large-volume parenterals containing 1.5% benzyl alcohol as a preservative have caused metabolic acidosis, cardiovascular collapse, and death in low birth weight premature neonates and infants. The cumulative dose of benzyl alcohol ranged from 99 to 234 mg/kg per day in these patients [55,56]. Dose-related adverse effects to excipients are of particular concern in the preterm, low birth weight infant because of the known immaturity of hepatic and renal function in this population (see Sec. II. A).

Table 4 Excipients Reported to Cause Adverse Reactions in Children

Name	Route of administration	Reaction	Ref.
Azo dyes (tartrazine)	Oral	Urticaria, bronchoconstriction, angioedema	100 101
Benzalkonium Cl	Oral, inhalation, topical	Bronchoconstriction	102 103
Benzyl alcohol	IV	Metabolic acidosis, neurotoxicity	55 56
Lactose	Oral	Lactose intolerance, prolonged diarrhea	104 105
Parabens	Oral, IV	Hypersensitivity	106
Polysorbate 80 and 20/tocopherol	IV	Liver and kidney failure in low birth weight infants	107
Propyl gallate	Oral	Methemoglobinemia	108
Propylene glycol	Oral, IV, topical	Seizures, hyperosmolality, contact dermatitis	57 109
Thiomersal	Topical, otic, IM	Hypersensitivity, mercury toxicity	110 111 112

Dose-related reversible CNS effects have also been reported in children receiving long-term therapy in which propylene glycol was a cosolvent [57].

Dyes. Dose does not appear to be a factor in patient reaction to dyes. The mandatory labeling of the azo dye tartrazine (FD&C yellow No. 5) in OTC and prescription medications [58] has focused the attention of pharmaceutical manufacturers and the consumer on the potential danger of dyes in susceptible individuals.

Hypersensitivity reactions have been reported with several azo dyes, particularly FD&C No. 5 and No. 6. Tartrazine-induced bronchoconstriction is commonly considered a cross-reaction to aspirin in sensitive asthmatics, although urticaria has been reported in other patient populations [59]. In a double-blind challenge involving aspirin-sensitive asthmatics, hypersensitivity to dyes was only 2% [60]; however, numerous case reports involving azo dyes suggest caution when using a drug containing an azo dye in asthmatics [61]. Non-azo dyes are considered to be weak sensitizers.

The association of dye content with hyperactivity in children remains controversial and unproved.

Sweeteners. Sweeteners are commonly included in pediatric formulations to increase palatability.

Sucrose is still the most popular sweetener. Chewable tablets may contain 20–60% sucrose, and liquid preparations may contain up to 85% sucrose. In a recent survey of sweetener content of 107 pediatric antibiotic liquid preparations, only four were sucrose-free [62].

The sucrose content of oral liquids may cause significant problems when these products are prescribed for long term therapy (asthma, seizure control, recurrent infections). Oral liquid preparations can represent a substantial carbohydrate load to children with labile diabetes, particularly if a child is ingesting more than one liquid medication with a high-sugar content.

A wider problem exists with the possible role of liquid medications in dental caries formation [63]. The extent of acid production in the oral cavity is closely related to caries formation. In a study of liquid medication, investigators have observed that medications with sucrose concentrations higher than 15% were able to significantly lower pH; there was an inverse relation between sucrose content and a decrease in oral cavity pH [64]. In a comparison of sorbitol and sucrose-sweetened liquid iron preparations, only sucrose-containing products produced a significant decrease in oral cavity pH [65].

Viscous formulations with a high-sucrose content are especially prone to contribute to caries formation because of their prolonged contact time in the oral cavity.

Sorbitol is a polyhydric alcohol, with a high caloric content (4 calories/g). In a survey of 129 oral liquid dosage forms stocked at a large university teaching hospital, 42% contained sorbitol [66]. The sorbitol concentration in the identified products varied from 3.5 to 72% w/v (0.175–3.6 g/ml).

Sorbitol is an appropriate substitute for children whose sucrose intake must be restricted. Although it is prepared by hydrogenation of glucose or corn syrup, sorbitol does not require insulin for metabolism [67]. It is considered to have very low cariogenic potential because it is not fermented by salivary bacteria. Sorbitol is a hyperosmotic laxative, and diarrhea has been reported in children receiving only 9 g/day [68]. In patients with sorbitol intolerance, abdominal cramping and diarrhea may occur with even lower daily ingestion. When the active ingredient produces similar adverse GI effects, it is difficult to identify the causative agent. Acetaminophen elixir, theophylline elixir, valproic acid syrup, cimetidine and hydralazine solutions contain sorbitol; all have been reported to cause osmotic diarrhea in children.

Special consideration for sorbitol content is necessary for diabetic patients. Diabetics tend to have altered GI motility and ingest sorbitol from many food sources in addition to medications.

Both solid and liquid dosage forms may contain saccharin. Saccharin is a nonnutritive sweetening agent, 300 times as sweet as sucrose. In a survey of sweetener content of pediatric medications, seven out of nine chewable tablets contained saccharin (0.45–8.0 mg/tablet) and sucrose, or mannitol. Seventy-four of the 150 liquid preparations investigated contained saccharin (1.25–33 mg/5 ml) [62].

It is recommended that daily saccharin intake be maintained below 1 g because of a risk of bladder cancer. A lifetime daily diet containing 5–7.5% saccharin has induced bladder tumors in rats [69]. However, it is probable that saccharin is only a very weak carcinogen in humans. The amount contained in pharmaceutical preparations is well below the recommended maximum human daily intake.

Because of the high incidence of lactose intolerance in the general population, lactose is not recommended as a sweetener for pediatric preparations [70].

Vehicle Selection

Ethanol has long been employed as a solvent in pharmaceuticals. Since it also acts as a preservative and flavoring agent, it is second only to water for use in liquid preparations. Investigators have also suggested that it may enhance the oral absorption of the active ingredient [71].

Hepatic metabolism of ethanol involves a nonlinear saturable pathway. Young children have only a limited ability to metabolize and, thereby, detoxify ethanol. Ethanol intoxication has been recorded in children with blood levels as low as 25 mg/dl. Alcohol has a volume of distribution of approximately 0.65 L/kg. Ingestion of 20 ml of a 10% alcohol solution will produce a blood level of 25 mg/dl in a 30 lb (13.6 kg) child. In 1984, the American Academy of Pediatrics (AAP) Committee on Drugs recommended that pharmaceutical formulations intended for use in children should not produce ethanol blood levels > 25 mg/dl after a single dose.

In general, manufacturers have voluntarily complied with the recommendations of the AAP and have reformulated many of their pediatric preparations. In 1992, the Nonprescription Drug Manufacturers Association established voluntary limits for alcohol content of OTC products [72].

The voluntary limits are

1. A maximum of 10% alcohol in products for adults and teens, 12 years old and over.
2. A maximum of 5% alcohol in products intended for children 6–12 years of age.
3. Less than 0.5% alcohol content for products intended for children younger than 6 years of age.

Compliance with these guidelines is requested by 1994. Currently, alcohol content of some OTC products is higher than 25% [73].

Extemporaneous production of pediatric dosage forms is commonly undertaken in hospitals throughout the United States. Without the sophisticated formulation capabilities of pharmaceutical manufacturers, alcohol-based vehicles have been recommended for extemporaneous preparation of liquid dosage forms [74]. There is a critical need to conduct research studies that will assist the pharmacist in replacing current formulations with stable, alcohol-free preparations [75].

Administration Considerations in Dosage Form Development

Oral Administration.   Oral administration is the preferred route of administration. There is a general consensus among pediatricians and parents that children younger than 5 years of age have great difficulty with, or are unable to swallow, a solid oral dosage form. Manufacturers, therefore, have developed liquid formulations for many of the commonly used pediatric products. The liquid dosage form, however, is not free of problems. Liquid products are often unstable and have short expiration dates; accurate measurement and administration of the prescribed dose is also a problem, especially in infants.

Chewable tablets and sprinkle capsule formulations have been very well received by both patients and their parents for use in children with full dentition (older than 3 years) [76–78]. This is potentially a very fruitful area for future research and development.

Rectal Administration. The administration of drugs by a solid rectal dosage form (i.e., suppositories) results in a wide variability in the rate and extent of absorption in children [79]. This fact, coupled with the inflexibility of a fixed dose, makes this a route that should not be promoted for pediatric patients. There has been at least one death involving a 7-month-old infant, that can be directly attributed to the use of a solid rectal dosage form of a therapeutic dose of morphine [80].

Rectal administration of a drug is considered unacceptable by adolescents and various ethnic groups.

Transdermal Administration. The development of the stratum corneum is complete at birth and is considered to have permeability similar to that of adults, except in preterm infants [81]. Preterm neonates and infants have an underdeveloped epidermal barrier and are subject to excessive absorption of potentially toxic ingredients from topically applied products.

No transdermal products have been marketed for use in the pediatric population. The development of transdermal products in pediatric doses could be very beneficial for children who are unable to tolerate oral medications.

Parenteral Administration. Absorption of medication following an intramuscular injection is often erratic in neonates owing to their small muscle mass and an inadequate perfusion of the intramuscular site [82]. In a study of infants and children 28 days to 6 years of age, the intramuscular administration of chloramphenicol succinate produced serum levels that were not significantly different from intravenous administration [83]. However, the bioavailability of most drugs administered intramuscularly has not been evaluated in the pediatric population. In addition to bioavailability issues, there are other concerns specific to pediatrics with the intramuscular administration of drugs. The volume of solution injected is directly related to the degree of pain and discomfort associated with an IM injection. Manufacturer's recommendations for reconstitution of IM products often result in a final volume that is excessive for a single injection site in a child's smaller muscle mass, thereby requiring multiple injections and a significant degree of discomfort for the patient. If a smaller volume is used for reconstitution, the problems of drug solubility and high osmotic load at the site of injection must be addressed [84]. The inclusion of a local anesthetic, such as lidocaine, as part of the reconstituted product is often necessary [84,85].

In a report from the Boston Collaborative Drug Surveillance Program, pediatric nurses have reported a much higher frequency of complications from IM injections than that observed in the adult population. Twenty-three percent of pediatric nurses surveyed had observed complications (local pain, abscess, hematoma) versus a rate of 0.4% reported in adult patients [86]. Serious complications, such as paralysis from infiltration of the sciatic nerve, quadriceps myofibrosis, and accidental intra-arterial injection, are usually the result of the difficulty in placement of an intramuscular injection in children.

The major problem with the intravenous route in children is dosing errors. Because of the unavailability of stock solutions prepared for pediatric doses, errors in dilution of an adult stock solution have resulted in 10- to 20-fold errors in administered doses [87,88]. A secondary problem is the maintenance of patent intravenous lines in infants and nonsedated children.

Pulmonary Drug Delivery. Endotracheal drug delivery is a very effective method of administering emergency medications (i.e., epinephrine, atropine, lidocaine, naloxone) to children when an intravenous line is not available. To optimize drug delivery to the distal portions of the airway, the drug must be administered rapidly, using an adequate volume of diluent: 5–10 ml in young children; 10–20 ml for adolescents [89].

Pressurized inhalation products have also been very successfully employed in the pediatric population to provide a drug directly to the desired site of action, the lung. These products are designed to deliver a unit dose at high velocity, with small particle size, the ideal conditions for drug delivery to distal airways [90].

The use of the aerosol route for delivery of antibiotics for pulmonary infections remains controversial. Most pediatric studies have been conducted in children with cystic fibrosis. In these patients, distribution of the antibiotic to the desired tissue site is impeded because of the viscosity of the sputum in patients with acute exacerbations of their pulmonary infections [91,92]. Long-term studies suggest there may be some preventive benefits of aerosolized antibiotics in children with cystic fibrosis who have *Pseudomonas aeruginosa* colonizing their lungs, but are not acutely ill [93,94].

The potential for development of bacterial resistance from airborne antibiotic particles preclude the use of aerosolized antibiotics, particularly the aminoglycosides, in institutional settings.

Systemic treatment through the respiratory tract needs further study to determine its usefulness.

Compliance Issues: Taste Preference and Palatability

Written and verbal education of parents, midtherapy reminders, and special packaging, all have been employed to improve compliance to prescribed dosing regimens for pediatric patients [95]. Two factors make taste preference and palatability critical considerations in pediatric compliance.

The dosage forms most commonly employed for pediatric formulations are liquids and chewable tablets. A perceived unpleasant taste is much more evident with these dosage forms than when a drug is administered as a solid oral dosage form, such as a tablet or capsule. Second, it is widely believed that children younger than the age of 6 years have more acute taste perception than older children and adults. Taste buds and olfactory receptors are fully developed in early infancy. Loss of taste perception accompanies the aging process.

Smell, taste, texture, and aftertaste, therefore, are important factors in the development of pediatric dosage forms. In a study of six brands of OTC chewable vitamins, flavor type and intensity, soft texture, and short aftertaste were critical factors in product preference. The flavor and texture attributes of the best-selling product were significantly different from the other brands [96].

There are at least 26 different flavorings used in pediatric antimicrobial preparations [70]. Cherry is the most commonly used flavoring, although a blind taste comparison found that other flavorings, such as orange, strawberry, and bubble gum, are well accepted in pediatric antimicrobial suspensions [97].

In many circumstances, it may be difficult to mask the unpleasant taste of the active ingredient. Regardless of flavoring used, parents consistently report that children prefer cephalosporin products to penicillin suspensions [97].

D. Future Directions

Effective and safe drug therapy for newborns, infants, and children depends on knowledge of pediatric pharmacokinetics and pharmacodynamics and knowledge of the drug formulation and delivery issues that are specific to this population.

The FDA has recently issued guidelines recommending that pediatric safety and efficacy studies be completed before marketing of a new drug that would be widely used in children [98]. In a 1990 survey conducted by the Pharmaceutical Manufacturer's Association, 114 drugs and vaccines were being tested by 56 companies for pediatric use [99].

There has been tremendous progress in identification of pharmacokinetic and pharmaco- dynamic parameters in chronically ill infants and children. Pharmacokinetic and pharmacody- namic issues in the acutely ill child have been sporadically addressed and deserve further evaluation.

The critical void in pediatric drug therapy is now in effective drug delivery systems. There have been some inroads made in the manufacturing of pediatric dosing systems, particularly with OTC cough, cold, and analgesic products. There needs to be a redirection of the focus in drug formulation toward pediatric dosage forms with known stability and bioavailability that can be easily and accurately administered to infants and children.

III. THE ELDERLY POPULATION

A. Diminution of Physical Function and Its Effects on Drug Disposition

Within the medical community, it has been acknowledged that elderly patients often respond to drug therapy differently from their younger counterparts. Aside from alteration of various pharmacokinetic and pharmacodynamic processes, elderly patients tend to suffer from a number of chronic conditions and, thus, have more complex dosage regimens. Additionally, there are a variety of physical limitations prevalent among the elderly that may hinder their ability to self-administer medication.

Most of those involved in health care administration agree that elderly patients are the primary consumers of drug products. The actual extent to which this occurs is shown quite clearly in Fig. 1. This figure gives an analysis of the "drug mentions" in the United States by age group (the term *drug mention* referes to those medications that have been "prescribed, recommended or given in any medical setting by a private physician") [113]. As is shown in this figure, those older than the age of 65, henceforth "the elderly," account for more than 28% of the drug mentions in the United States. Thus, although the elderly constitute only 12.5% of our population, they are the biggest consumers of drug-related products [113,114].

The effects of aging on drug disposition is one aspect of drug-taking behavior among the elderly that has been researched by individuals within both the medical and pharmaceutical fields. Before discussing the actual changes that occur with aging, two points must be stressed. First, because of wide variation among older individuals, it is very difficult to quantify the extent of changes that occur within this population. Second, most of these changes are related to the fact that, with increasing age, there is an overall decrease in the capacity of homeostatic mechanisms to respond to physiological changes.

Pharmacokinetics

During the past decade, numerous articles reviewing the effects of aging on pharmacokinetic processes (i.e., absorption, distribution, metabolism, and elimination) have been published [115–124]. An outline of the observations made in these reports is supplied in Table 5. The absorption process is the only process that will be covered in depth in this chapter, as this is the process that can most easily be manipulated through formulation techniques.

First of all, there is a decrease in gastric secretion that causes the elevated pH that has been noted in elderly patients [116–127]. This condition is commonly referred to as hypochlorhydria or, in severe cases, achlorhydria and may be the result of atrophic gastritis [123,126–128]. It

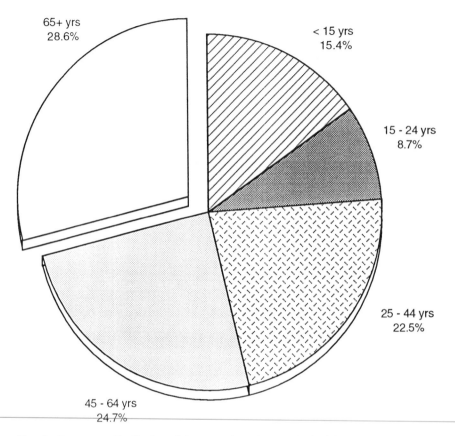

Fig. 1 Percentage distribution of drug mentions, by age: United States, 1985.

may have a substantial effect on a formulation if an enteric-coated product or a weakly acidic or weakly basic drug is being considered. In the former case, the increased pH may cause the contents of the formulation to be prematurely released in the stomach, rather than in the small intestine and may lead to excessive gastrointestinal (GI) irritation. The elevated pH that exists within the stomach may also result in incomplete absorption of weakly acidic compounds from the stomach and decreased rate of absorption of poorly soluble weak bases. The reduced gastric blood flow that has been noted in elderly patients may hinder the rate of absorption [116,117,123,126,127]. In most instances, this decrease in the rate of absorption does not necessarily cause a decrease in the extent of absorption. In fact, only those compounds that are actively absorbed (e.g., riboflavin) have a decreased rate of absorption [126,128]. Drugs that are degraded in the acidic environment of the stomach (e.g., penicillin), however, may actually have an increased extent of absorption in elderly patients because less acid is available to degrade the drug before absorption.

There appears to be an ongoing dispute over whether or not gastric-emptying rate (GER) and GI motility are affected by the aging process [116,117,124,126,129–134]. Most studies tend to suggest that there is, indeed, a decrease in GER as the body ages. As GER is the primary physiological determinant of the rate of absorption of solid oral dosage forms, one can see that a decrease in GER may result in a subsequent decreased rate of absorption, particularly when coupled with the compromised blood flow to the GI track also noted in

Table 5 Changes in Pharmacokinetic Processes That Are Observed With Aging

Process	Changes	Effects
Absorption	↓ Intestinal blood flow ↑ Gastric pH ↓ Active absorption ↓ GER?	↓ Rate of absorption
Distribution	↓ Cardiac output ↑ Fat/lean body mass ↓ Serum albumin conc.	↓ V_d water soluble drugs ↑ V_d lipid soluble drugs ↑ V_d protein bound drugs
Metabolism	↓ Hepatic blood flow ↓ Liver size ↓ Phase I metabolism ↑ Incidence liver dysfunction	↑ $t_{1/2}$ hepatically extracted drugs
Elimination	↓ Renal blood flow ↓ GFR ↓ ARTS ↓ No. functioning nephrons	↑ $t_{1/2}$ renally excreted drugs

Abbreviations: GER, gastric-emptying rate; V_d, volume of distribution; $t_{1/2}$, half-life; GFR, glomerular filtration rate; ARTS, active renal tubular secretion.

elderly patients. Additionally, unpredictable GER has a significant influence on extended-release formulations, as it becomes difficult to predict whether or not acceptable blood levels will be obtained [123]. To circumvent the possible problems that may arise from decrease in GER, a liquid or readily disintegrating formulation may be used.

Pharmacodynamics

Although there are several reviews assessing the changes in pharmacodynamics that are prevalent among the elderly population, this area has not been as widely studied as those changes occurring in pharmacokinetic processes [115–117,119]. In Table 6 some of the major changes that have been evaluated in elderly patients are listed.

The decrease in the ability of the aging body to respond to baroreflexive stimuli can result in very serious consequences for elderly patients [115–117]. Because of this decrease in sensitivity and the decreased cardiac output witnessed in elderly patients, they are predisposed to the effects of orthostatic hypotension that can occur when one is taking antihypertensive medication (e.g., prazosin). Indeed, the fact that elderly persons are prone to accidental falls may be due to this change in sensitivity [115–117].

Table 6 Pharmacodynamic Changes Observed with Aging

Decrease baroreflex sensitivity
Decrease β_1-receptor response
Decrease α_2-receptor response
Increase sensitivity to barbiturates
Decrease glucose tolerance

Decreases in β_1-receptor response were investigated when it was found that elderly patients taking β-adrenergic blockers (e.g., propranolol) were experiencing the ADRs associated with these medications, but they were not obtaining the proper therapeutic response (i.e., decrease in heart rate) [115–117]. Whether the exact mechanism for this decreased response is due to a decrease in affinity or a decrease in the number of receptors has yet to be conclusively determined [115].

The changes in α_2-receptor response observed in some elderly patients have not yet been found to have any clinical significance. Theoretically, this decrease should result in an increase in the amount of norepinephrine being released from nerve terminals, but this has not yet been demonstrated [115,116].

The incidence of diabetes and decreased glucose tolerance among the elderly is well documented [116,135–137]. Because of this, formulators should make every attempt to avoid using any sugar-containing excipients in their production processes.

Absorption Within the Oral Cavity

When dealing with oral dosage forms, it is important to study the various changes occurring within the oral cavity, particularly if a buccal or sublingual formulation is being considered. Table 7 lists the changes within the oral cavity that have thus far been elucidated [124,127,138–144]. It is very important to note that there is a decrease in the capillary blood supply to the oral mucosa. This may make it difficult to predict accurately the absorption rates that wil occur when using sublingual and buccal formulations in the elderly age group. Additional changes occurring in and about the oral cavity will be discussed at length in another section of this review.

Percutaneous Absorption

With the increasing acceptance of transdermal formulations by the pharmaceutical industry and the trend toward an aging population that is occurring in our nation, it is vital that the effects of aging on percutaneous absorption be evaluated. Certainly, elderly patients are the primary users of such drug delivery systems (e.g., Transderm Nitro; Nitro Dur II), so the need for assessment of percutaneous absorption in the elderly should be emphasized. In light of this, it is surprising to find that there have been relatively few studies published that specifically address percutaneous absorption in the elderly [145,146]. Table 8 provides an outline of changes in characteristics of the skin that occur with aging [145–148].

Researchers assessing the various factors surrounding percutaneous absorption (e.g., permeation and clearance) have theorized that, although there is an increase in the rate of per-

Table 7 Changes in and About the Oral Cavity Observed with Aging

Mucosa	Drier
	Increase susceptibility to injury
	Decrease capillary blood supply
Muscle	Decrease bulk and tone
	Decrease masticatory efficiency
Salivary Glands	Decrease resting secretory rate
	Increase viscosity of saliva
	Decrease enzyme activity of saliva
Miscellaneous	Decrease number of taste buds
	Increase dysfunction and cancer

Table 8 Changes in Skin Characteristics Observed with Aging

Dry skin
Loss of elasticity
Impaired wound healing
Deletion and derangement of small blood vessels
Increase permeation to water and some chemicals
Decrease clearance into blood stream
Decrease absorption?

meation through aging skin, substances that permeate through the skin have a slower rate of removal into the general circulation and, thus, distribution may be incomplete [145,146]. Unfortunately, there appear to be few published reports addressing this phenomenon. Studies that do specifically evaluate percutaneous absorption in the elderly have used only one compound, testosterone, in their analyses [145,146]. Therefore, before formulating drugs for transdermal delivery in the elderly, changes in percutaneous absorption that occur on aging should be assessed further.

Physical Limitations

Table 9 lists the top ten chronic conditions prevailing in the elderly population [136]. Many of these conditions severely limit the range of activities that one can perform (Fig. 2) [135,136]. Indeed, researchers have studied, in depth, the extent to which age limits one's activities of daily living (ADL) [114,135,137,149,150]. Moreover, some of these conditions, such as arthritis and impaired vision, impinge on the patient's ability to accurately self-administer medication.

Dexterity. Dexterity may be impaired in the elderly for a variety of reasons, such as the following: (a) over 45% of the elderly suffer from some form of arthritis [136]; (b) many elderly experience tremors associated with parkinsonism or other neurological disorders; and (c) frailty and weakness are prevalent in many elderly patients. In fact, the NIA has been conducting a comprehensive study to assess all of the characteristics common among the elderly. This study reveals that at least 13% of the elderly have some difficulty in handling small objects (e.g., tablets) [149]. Another government study reports that more than 4% of the elderly

Table 9 Top Ten Chronic Conditions in the United States, 1986–1988

Rank	All persons	65–74	75+
1	Chronic sinusitis	Arthritis	Arthritis
2	Orthopedic impairment	Hypertension	Hypertension
3	Arthritis	Heart disease	Hearing impairment
4	Hypertension	Hearing impairment	Heart disease
5	Allergies	Orthopedic impairment	Cataracts
6	Hearing impairment	Chronic sinusitis	Orthopedic impairment
7	Chronic bronchitis	Diabetes	Chronic sinusitis
8	Hemorrhoids	Cataracts	Visual impairment
9	Asthma	Tinnitus	Hardening of arteries
10	Visual impairment	Hemorrhoids	Constipation

Source: Ref. 136.

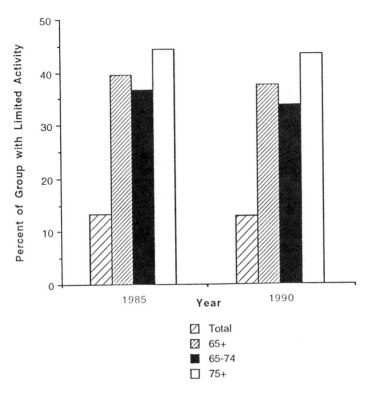

Fig. 2 Limitation of activity caused by chronic conditions: United States, 1985 and 1990.

experience difficulty preparing their own meals and, therefore, are likely to encounter problems when self-administering medication [137].

Vision. Many people experience visual decline as they age (Fig. 3) [135]. Impaired vision may also hinder one's ability to self-administer medication. Listed in Table 10 are some of the effects that are associated with impaired vision in the elderly. Some of the processes of self-administration on which impaired vision may impinge are as follows: (a) the ability to accurately measure liquids; (b) the ability to correctly read instructions; and (c) the ability to differentiate between various types of medications (both the labeling of these drugs and their physical characteristics) [137,151,152,153].

Swallowing and Chewing. In addition to those changes occurring in the oral cavity (see Table 6), there are other factors that may inhibit an elderly patient's ability to both swallow and chew. For instance, xerostomia, or dry mouth, is a condition that is prevalent among older people. Xerostomia may be caused by any one of the following conditions: (a) elderly patients often do not consume adequate amounts of liquid and are thus dehydrated; (b) many elderly patients "mouth breathe" because of asthma or other respiratory diseases; and (c) elderly patients often take medications having anticholinergic side effects (e.g., antidepressants and neuroleptics) [142,143,154–156]. Patients experiencing xerostomia often have difficulty swallowing tablets or capsules because they tend to adhere to the esophageal mucosa when it is dry [124,127,157–161]. In addition, esophageal lesions are common among the elderly and may affect a patients ability to swallow owing to inhibition of peristalsis by the weakened esophageal musculature [127,129]. The ability of elderly patients to chew is also compromised [127,138,139,144], perhaps as a result of the decreased bulk and tone of the oral musculature

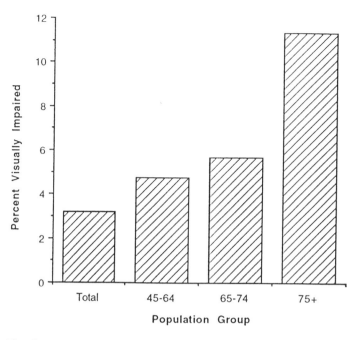

Fig. 3 Incidence of visual impairment: United States, 1991.

as one ages [144]. Additionally, it has been estimated that 50% of all elderly persons in the United States are fully edentulous (i.e., toothless) [127,139]. The absence of teeth not only hinders one's ability to chew, but also changes the bacterial flora within the oral cavity from predominantly anaerobic to aerobic [127].

B. Dosing Considerations Determined by Alterations in Physiology with Aging

The pharmacokinetics of each compound should be determined when one is deciding which drug candidates to use in designing formulations for the elderly (see Table 3). For instance, some medications have an increased half-life within the bodies of older adults, either because these drugs undergo extensive hepatic metabolism (e.g., diazepam, verapamil, and pentazocine), or because they are excreted primarily by the kidneys (e.g., lithium, aminoglycosides, and digoxin) [114–120]. In addition, drugs that are highly protein-bound (e.g., warfarin) may be the cause of serious adverse reactions among elderly patients because of the decreased concentration of serum albumin in these patients and the subsequent rise in circulating "free"

Table 10 Some Visual Declines Observed with Aging

General acuity
Peripheral vision
Ability to see in low-light levels
Ability to see highly reflective surfaces in bright light
Ability to discriminate between colors
Ability to adapt to darkness

drug [114–120]. So, if the pharmacokinetic behavior of the drug is known to change in elderly patients, it may be wise to avoid such drugs or to adjust the dosage accordingly. A guide has been recently published that lists a number of suggestions for dosing regimens in the geriatric patient [162].

C. Drug Delivery Systems and Compliance Issues

The changes experienced in aging may affect a patient's ability to use some of the drug delivery systems existing today. It should be kept in mind, however, that within the context of this review, a drug delivery system is not merely a novel dosage form. It is the dosage form with its container, labeling, *and* any other information supplied with the medication to the user.

Dosage Form Preferences

Solid oral dosage forms, particularly tablets, are the preferred type of formulation in the United States. Not only are these products widely accepted by consumers, but they are also relatively cheaper to develop and manufacture than oral liquids or suspensions, parenterals, or suppositories. Figure 4 shows, quite clearly, that even the elderly primarily make use of solid oral dosage forms [162].

Oral Dosage Forms. 1. CHEWABLE TABLETS.   It has already been noted that most elderly patients experience a decrease in their ability to chew efficiently [125,137,138,143]. Therefore, by virtue of their design, chewable tablets are not often recommended for use by elderly patients (particularly those who are edentulous) [153–155,163,164]. Most chewable formulations also rely on an adequate amount of chewing action to obtain full release of their ingredients (e.g., chewing promotes the foaming action provided by some chewable antacid products). So, aside from being difficult for the elderly patient to use, full benefit of a chewable

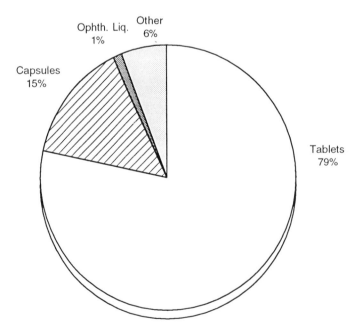

Fig. 4 Dosage forms commonly used by elderly patients.

dosage form may not be achieved by these patients. Additionally, the use of chewable tablets by denture-wearers may cause local irritation in the oral cavity [155].

2. SUBLINGUAL AND BUCCAL TABLETS.   Although sublingual and buccal tablets (e.g., nitroglycerin or isosorbide) are used by many elderly patients, few, if any, researchers have determined the effects that aging may have on the bioavailability of these dosage forms [136–142]. Patient acceptance of these types of formulations must also be considered. Elderly patients who suffer from dry mouth conditions may find sublingual or buccal tablets irritation and may refuse to use such medications. Another problem may involve demented or contrary patients who may feel foreign objects inside their mouths and pull the tablets out before the active ingredients have been released from the formulation.

3. CAPSULES.   A number of scientists have investigated the effect of formulation on esophageal transit and have found that capsules tend to adhere to the esophageal mucosa more often than any other type of oral dosage form [157–160]. Moreover, owing to the conditions of xerostomia and hindered swallowing that are prevalent among the elderly, mucosal adherence of capsules in these patients may be more pronounced [127,125,141,153–155]. In light of these observations, the use of capsules by older patients may not be advisable. This is an extremely important consideration if the drug to be delivered is one that is known to cause esophageal ulceration (e.g., tetracycline or aspirin) [159].

4. LIQUIDS AND SUSPENSIONS.   Most liquid formulations are not packaged in unit-dosage form. Therefore, before administration, the proper amount of medication to be taken for each dose must be measured. This additional requirement may compound any difficulties a patient may have in following a prescribed schedule. Patients suffering from visual impairment, arthritis, or tremors associated with neurologic disorders are particularly likely to become frustrated with this type of formulation. Visual impairments make it difficult, if not impossible, for many elderly patients to measure the prescribed amounts of medication accurately. Impaired dexterity, owing to tremors or arthritis, may have effects on a patient's ability to hold both a spoon and a bottle at the same time while pouring out the desired amount of liquid.

Additional difficulties are encountered by elderly patients if a medication is in the form of a suspension. Problems my occur because a patient cannot see, or disregards, the words "Shake Well" on the label or is not able to exert the amount of agitation necessary to provide a uniform suspension. Certainly, unevenly distributed amounts of active ingredients throughout a suspension may result in serious consequences for a patient, either in terms of under- or overdosing.

Transdermal Delivery Systems.   Although transdermal drug delivery may offer a means of increasing compliance among elderly patients, it has yet to be determined whether or not changes exist between the bioavailability of such products when administered to elderly patients versus young adults. Indeed, there may be a decrease in the absorption of compounds that are transdermally delivered to elderly patients [143–146]. Still, some may prefer the transdermal design and may opt to perform all the preliminary investigations necessary to quantify percutaneous bioavailability in older patients. If this route is chosen, one should keep in mind that transdermal products formulated around these changes may be applicable to only the older patients for whom the product have been designed. Such products may have differing release characteristics in the rest of the population. Therefore, it is possible that these products, which will probably require lengthy preliminary studies, will be able to be used effectively by only a limited portion of the population.

Parenteral Dosage Forms and Invasive Devices.   Parenteral and invasive devices provide the distinct advantage of the delivery of medication directly into the bloodstream or at the site of action. Additionally, these methods result in certain patient compliance because, in most cases, an individual other than the patient is responsible for the administration of medication by these means. Unfortunately, this attribute is counteracted by numerous problems that are illustrated in Table 11.

In this table, it can be recognized that there are problems inherent in these types of formulations from both the patient's and the manufacturer's point of view. This is why most pharmaceutical companies make attempts to avoid these types of dosage forms, if at all possible. However, with the advent of biotechnological products, which often do not lend themselves to "conventional" dosage formulations, parenteral and invasive measures may be the only answer.

Alternative Delivery Systems

Although elderly patients seem to experience difficulties with those drug delivery systems listed in the foregoing, other systems are currently available that may be more suited to the needs of these patients. Although most of these systems are not specifically designed for the geriatric community, they may offer aid to this group of patients. This may be accomplished through the use of dosage formulations or packaging designs that are easier to handle or by supplying patients with devices or information that will enable them to better follow their prescribed dosing regimen.

Compliance Aids. Many pharmacists and physicians have recognized that most elderly patients have complicated dosing schedules and need some sort of reminder that will help them keep track of their prescribed regimens [115,154,155,160,164–173]. This can be achieved through several methods. First, various types of packages, such as Dosett Trays, Calendar-Paks, and Patient Med Paks, can be prepared by pharmacists [167,169,174–176]. These packages are usually devised so that all of the medications that have been prescribed to be taken at a specified time are packed together. For instance, all the medications that are to be taken before breakfast are placed in the same container, and all those to be taken 1 h after breakfast are placed in another container. Labeling for each container should specify the time at which its contents are to be taken, as well as a list of each individual medication it holds. In an effort to promote this type of packaging, the *United States Pharmacopeia* (*USP*) has set guidelines for pharmacists to follow when preparing Patient Med Paks [176]. In this manner, the *USP* has provided the pharmacist with a means to help out those patients with complex-dosing schedules and still comply with official-labeling criteria (i.e., it is not standard practice to supply more than one drug in a single container). Another way in which pharmacists and physicians can help patients regulate their dosing schedules is by supplying them with drug-reminder cards [115,167,171,172]. The concept for the cards is essentially the same as that for

Table 11   Possible Problems Associated with Parenteral Delivery

---

Cost to patient and manufacturer
Patient discomfort
Risk of infection
Administration by trained personnel required
Limitations on particle size
Sterilization necessary
Prone to chemical, mechanical, and microbial instability
Complex manufacturing processes
Cumbersome and fragile packaging

---

the labels on the packaging just described (i.e., time of day and all medications to be taken at that time). Various modifications of this design can be made, such as including the physical characteristics of each medication or providing stickers to be placed on the card and on its corresponding container (i.e., one sticker is placed on the prescription bottle and the other sticker is placed next to the medication's name on the reminder card) [171,172].

Unfortunately, both the drug reminder cards and packaging methods described here must be made for each patient on an individual basis, and some people are not willing to take the extra time needed to prepare these systems.

Oral Dosage Forms. The advantage of oral dosage formulations have already been discussed. It appears that this route of delivery is preferred by physician, patients, and manufacturers alike. The relatively low cost of oral dosage forms makes them a particularly attractive means of drug delivery for those patients, such as the elderly, who may be economically depressed [177]. These dosage forms are also comparatively easier to formulate, package, and ship than other types of dosage forms [179]. Moreover, changes in pharmacokinetic parameters among the elderly have been assessed only in those formulations that are oral or parenteral in nature [115–125]. Therefore, it is appropriate to focus on oral dosage formulations for drug delivery in the elderly.

1. GRANULES. Granules are one type of oral formulation of which use among the elderly is warranted [163,164]. This type of dosage form not only circumvents the difficulty in swallowing that may be encountered by older patients, but it also provides the patient with a certain amount of rehydration. As the elderly are often dehydrated, this is a feature that should not be overlooked. More importantly, medications that have been dispersed in a liquid are not likely to be affected by changes in gastric-emptying rate that may occur in older patients.

Problems may still arise because granules may be supplied in either unit-dose packages or in bulk containers. If unit-dose packages are used, patients with impaired manual dexterity may have difficulty opening the packets. With bulk containers, most of the handling problems associated with administering liquid formulations, that have been discussed, can occur. However, bulk containers do offer the advantage of dosage flexibility that cannot be realized with other solid dosage formulations.

2. COATED TABLETS. Investigators studying the effects of dosage formulation on esophageal transit have concluded that coated tablets are less likely to adhere to the esophageal mucosa than other solid dosage forms (e.g., uncoated tablets or capsules) [155,157–160]. In addition, this effect may be complemented by the use of oval-shaped tablets, or what are commonly referred to as "caplets" [159–161]. This type of tablet offers advantages over uncoated tablets and capsules, especially in those patients who have difficulty swallowing. Still, it is imperative that physicians and pharmacists instruct patients to take their medications with a full glass of water because esophageal adherence is still possible in those patients who are dehydrated.

3. EFFERVESCENT TABLETS. Effervescent tablets are another means of supplying medications to the elderly. This type of formulation provides the patient with an easy-to-swallow product that is aesthetically pleasing (i.e., forms a clear solution, rather than a cloudy suspension). However, pharmaceutical chemists are well aware of the problems that exist when preparing effervescent formulations. These problems may be partly solved by certain advances in phamaceutical technology that allow direct compression of all excipients necessary to form such tablets [180–182]. Although this makes the manufacture of such products easier, stability problems still exist because these formulations must be adequately protected against moisture. As with granules, the type of packaging required for effervescent tablets can be a problem for

those patients with impaired manual dexterity. Moreover, and perhaps more importantly, the high sodium content necessary to manufacture effervescent products may have serious implications when used by patients with hypertension or congestive heart failure.

4. DISPERSIBLE OR SOLUBLE TABLETS.   The trend toward formulation of dispersible tablets is evident in Europe [183]. For example, it is understood that all tablets marketed in the Netherlands must have the ability to form an adequate dispersion when placed in water. In England, some nonsteroidal anti-inflammatory drugs (NSAIDs), which are extensively used by arthritic patients, are not marketed as dispersible tablets.

A challenge faced by formulators designing dispersible tablets is the ability to develop a formulation that rapidly disintegrates and is able to withstand shipping processes. In addition, this type of tablet should form a uniform and somewhat stable suspension when dispersed in water. An interesting answer to this challenge is the design of a "porous tablet" [184], in which a volatilizable solid (e.g., urethane on ammonium bicarbonate) is added to a standard, directly compressible formulation. After the tablets have been compressed, the volatilizable solid is sublimed off by a freeze-drying or heating process. Water easily penetrates through the pores and promotes rapid disintegration of the tablets produced in this manner. Thus, these tablets are able both to maintain their mechanical strength and to provide rapid disintegration or dispersion of product. As with granules and effervescent tablets, dispersible tablets offer the patient a dosage form that is both potable and easy to swallow.

5. TILTABS.   Tiltab tablets represent one of the few dosage formulations that has been developed expressly to meet the needs of patients with impaired dexterity [185]. Marketed by Smith, Kline & French Laboratories, Ltd. in several European countries, the novelty of the tiltab design is its irregular shape that prevents it from lying flat. Apparently, tablets manufactured in this fashion are easier to handle by those with impaired dexterity. Moreover, these tablets are readily identifiable by patients so that differentiation from other medication is facilitated. Other innovations like this are needed for drug delivery systems with the particular needs of the geriatric patient in mind.

6. CONCENTRATED ORAL SOLUTIONS.   Presentation of a drug may be made in the form of a concentrated solution that allows the entire dose to be held within a volume of less than 5 ml (e.g., Intensol Concentrated Oral Solutions, Roxanne). This opens up another means of providing medication to the aged, infants, or any other patients experiencing difficulties swallowing. Such preparations can be mixed with food or drink. Taste and poor solubility are problems that may set limits on the number of successful formulations that can be prepared in this way. Also, small errors in the measurement of such preparations represent large errors in dosing.

Taste Preferences in Oral Dosage Forms

Changes in elderly patients' abilities to taste various substances do not necessarily affect the ease or difficulty of administration of medications, but these changes do have an effect on the patients' acceptance of a product. For instance, although it may be easier for patients to swallow liquid medications, they may find the taste or smell of the product so objectionable that they will refuse to take any medication prepared in this manner. Indeed, even some solid dosage forms carry with them objectionable tastes or odors that result in limitation of patient acceptance. Although there have been few studies assessing elderly patients' taste preferences, these reports have indicated that differences in taste preference and perception do exist between elderly and young adult populations [127,134,138,186,187]. It has been determined that, although the number of taste buds declines with age, thresholds for certain tastes are affected, whereas others are not. Unfortunately, reports of taste threshold changes among the elderly are contradictory, and it is difficult to ascertain what changes really do occur [127,134,138,186,187]. Current reports claim that these changes in taste thresholds are not due to the aging process, per se, but to medications that the patient may be taking [138,187]. For example, it appears that an increased concentration of sour compounds must exist for it to be detected by patients taking medications [187].

Package and Label Design

One of the most important aspects of drug delivery design for the elderly is the presentation of the package and its label. If the patient is unable to open a package or cannot read a label properly, even the best dosage formulation design will be unsuccessful [176]. For prescription medications the package design is difficult to control, because the container supplied by the manufacturer is not necessarily the container in which the medicine will be dispensed by the pharmacist. But, pharmacist selection of special packaging is a prospect open to most drugs whenever elderly–friendly packaging is in hand. The OTC products, in partial contrast, provide a manufacturer with the opportunity to make substantial changes in the design of packages and their labels. When developing a design, it is important to always keep in mind that impaired dexterity and visual decline are prevalent among the elderly. Listed in Table 12 are some suggestions that may be useful when designing a product's label [152,153,165,166,188,189].

In terms of the package itself, it is difficult to devise a package that is both childproof and tamperproof and still able to be opened easily by someone with impaired dexterity. It has been suggested that packaging a medication in unit-dose Calendar-Paks may increase patient compliance [8]. The problem is that other studies have shown that most elderly patients encounter difficulties when attempting to open this type of packaging (blister packaging) [167,175]. Additionally, this type of packaging (i.e., C-Paks and the like) for OTC products may be unacceptable to the Federal Trade Commission and FDA, as it promotes the daily use of the product. So, it is apparent that different types of packaging are needed, depending on whether a drug is for OTC or prescription-only use.

IV. PHARMACEUTICAL FACTORS IN ORAL DOSAGE FORMULATIONS FOR SELECT POPULATIONS

Before the start of any formal laboratory work, the characteristics of the drug and excipients to be used must be considered. When performing this evaluation, it is necessary to keep in mind all of the pharmacokinetic and physical changes experienced by the elderly and pediatric population.

Table 12 Suggestions for Labels Designed To Be Used by Elderly Patients

---

Avoid pastels
Use matte surfaces to minimize glare
Light colors on dark background are more visible than dark
  colors on light background
Use distinct spacing between letters
Increase height and thickness of letters
Use additional labels that explain purpose of medication[a]

---

[a]This type of label is required on drugs dispensed in Denmark.

A. Particle Size

The relevance of particle size on solid dosage formulations has been determined by a number of individuals [154,190–196]. As is illustrated in Table 13, particle size affects the solid dosage formulation in numerous ways. For instance, particle size can be a profound effect on the dissolution of a formulation within the GI tract. This most notably characterized by the Noyes–Whitney equation in which

$$\frac{dM}{dt} = \frac{DS}{h(C_s - C)}$$

where

M = mass of drug dissolved
t = time
D = diffusion coefficient of drug
S = effective surface area of drug particles
h = stationary layer thickness
C_s = concentration of solution at saturation
C = concentration of solute at time t

As decreases in particle size produce increases in surface area S, it becomes evident that particle size reduction provides an increase in the rate of dissolution (dM/dt). When one remembers that elderly patients are likely to have decreased GER, one can see that increased dissolution rates are particularly desirable in these patients because once the formulation is in solution, GER is no longer able to limit significantly the rate of absorption [123,131,164].

Particle size impinges on aspirin-induced GI blood loss [195,196]; gastrointestinal microbleeding decreases as the particle size of the formulation declines [195,196]. In theory, this occurs because the duration in which the aspirin particles are in contact with the GI mucosa is shortened as a result of the decreased particle size of the formulation and, subsequently, the increased rate of dissolution [190,195,196]. Clearly then, given the effects on dissolution and GI erosion, formulations with smaller particle size are desired. However, if the particle size is too small, problems may arise in the handling of the powder during manufacture, and its flow properties may be impeded, leading to poor dosage uniformity [195,196].

B. pK$_a$ and Stability

As the acid content within the GI tract is known to change as the body ages, it is also important to evaluate the dissociation constant(s) (K_a) of the drug(s) that are to be used. From the Henderson–Hasselbach equation of weak acids where

$$pH = pK_a + \log\left(\frac{[salt]}{[acid]}\right)$$

Table 13 Processes in Solid Dosage Formulation
That Are Affected by Particle Size

Disintegration
Dissolution
Flowability
Compressibility
(Gastrointestinal bleeding)

it is apparent that changes in pH cause the proportions of drug that are ionized (salt) and nonionized (acid) to change. Since most compounds dissolve more readily when they are in the ionized form, it can be theorized that conversion of drugs with poor solubility to their salt form facilitates their rate of dissolution. Indeed, Hoener and Benet have modified the Noyes–Whitney equation so that it includes the effects of pH and pK_a on the rate of dissolution [197]. For weak bases, this equation is defined as

$$\frac{dM}{dt} = \left(\frac{DS}{h}\right) \left[C_s\left(1 + \frac{[H^+]}{K_a}\right) - C_g \right]$$

where

C_s = concentration of solution at saturation
C_g = concentration of drug in the GI tract
$[H^+]$ = H^+ concentration in the GI tract
K_a = dissociation constant of drug

This further serves to illustrate the fact that, as pH increases, poorly soluble, weak bases are more likely to exist in the nonionized form and, thus, have a decreased rate of dissolution. Since gastric pH is elevated in most elderly patients, products containing weak bases that are meant to be used in this population, should be formulated with these changes in mind.

Although not exclusively relevant to the design of drug delivery systems for the elderly or pediatric population, the stability of drug(s) and excipient(s) used within a formulation may have a dramatic effect on the final product. If there are instabilities inherent in the materials used in the formulations, these problems, most often, are translated to the final product. Instability may be physical, chemical, or microbial. When one is dealing with solid dosage forms, physical and chemical instabilities often occur. Instability may be due to decomposition of an active ingredient through hydrolysis or oxidation, or it may be a result of incompatibilities that exist between a drug and the excipients being used. The use of materials that are polymorphic (i.e., have more than one type of molecular orientation in the solid state) may also be the cause of instability within a formulation (e.g., dissolution and absorption rates differ among the various polymorphs of a material). Additionally, if drug–excipient interactions are not carefully evaluated, tablet hardening and an accompanying decreased rate of dissolution may result as the tablets age (e.g., dibasic calcium phosphate and ascorbic acid).

C. Disintegration

To provide the rapid rate of disintegration that is required to achieve efficient dispersion from tablets, the addition of a disintegrating agent is necessary. Such agents enable a formulator to produce a tablet that will quickly disintegrate when placed in a liquid. In addition, the use of suspending agents or surfactants may be desired so that more stable suspensions can be formed. In choosing these agents, it should be remembered that most elderly patients are unable to handle large loads of sodium. Therefore, compounds such as sodium starch glycolate may be inappropriate for use by these patients.

D. Compressibility and Flow

To produce tablets that are uniform in weight and content and exhibit a certain degree of mechanical strength, one needs a mixture of powders with (a) good flow properties; (b) a minimum tendency for segregation, and (c) the ability to be compressed. To achieve these ends, granulation with other excipients is often necessary.

Segregation can be minimized by ensuring that all particles within a mixture have approximately the same size and density [191–194]. The flow of particles, however, is a bit more complex. On the other hand, mixtures of small granule size (800 μm) have the propensity to produce tablets with minimal dosage variation. But, if the particle size becomes too small (<14 μm), flow through an orifice (e.g., a tablet press hopper) becomes impaired because the cohesive forces between the particles are of the same magnitude as the gravitational forces being imparted on the powder bed. The restricted flow may be improved by the addition of glidants, such as magnesium stearate or talc, but most of these agents are hydrophobic and may impair the tablet's ability to disperse. Superfine, high molecular weight polyethylene glycols (PEGs), which are water-soluble, have been proposed for use as lubricant–glidant in tablet formulation [180]. The PEGs may then provide the formulator with an agent that is both lubricating and hydrophilic and, as such, may be a viable choice for use in the formulation of dispersible tablets.

Another factor that impinges on tablet formation is the ability, or inability, of powders to be compressed. The ability of materials to be compressed may be due to the following: (a) compression force, (b) particle size, and (c) deformation processes. Obviously, by increasing the force of compression, one can, in theory, increase the mechanical strength of the compact. There is some evidence suggesting that the strength of the tablet may be increased by using granulations of smaller particle size [191–194]. Here again, this increased strength must be balanced with the ability of the powder bed to flow evenly through a hopper. Also, for suitable compacts to be formed, a material must exhibit a certain amount of plastic (i.e., permanent) deformation [198]. Powders that undergo more elastic than plastic deformation will lose their structure after ejection from the tablet die. Therefore, if one is using a material that does not undergo a significant amount of plastic deformation, additional processing steps are necessary so that a stable structure can be produced.

It becomes evident that a variety of factors need to be considered when designing tablet formulations. In addition, certain compromises between these parameters must be made, because it is nearly impossible to meet all specifications with the same process variables.

E. Salt and Sucrose Content

The increased incidence of glucose intolerance, congestive heart failure, and hypertension among elderly patients make them particularly sensitive to levels of sucrose and sodium. Those involved in the manufacture of antacids are well apprised of this as they are required to carry a precautionary statement on the level of those products that contain more than 5 mEq of sodium per dose [155,156,164].

Additionally, when reviewing the various liquid preparations that are available in the OTC market today, one can see that many of these products are now sugar- or sodium-free or both. Indeed, decreased sodium and sugar levels are beneficial to the entire population. This would seem to suggest that every effort should be made to keep sodium and sucrose contents to an absolute minimum in those products that are intended for use in older patients.

REFERENCES

1. G. Maxwell, Pediatric drug dosing, Drugs, 37,113–115 (1989).
2. D. B. Guralnik (ed.), *Webster's New World Dictionary of the American Language*, Simon & Schuster, New York, 1979.
3. R. C. Adelman, Definition of biological aging. In *Second Conference on the Epidemiology of Aging* (S. G. Haynes and M. Feinleib, eds.), U.S. Government Printing Office, Washington, DC, 1980.

4. P. T. Costa and R. R. McCrae, Functional age: A conceptual and empirical critique, in *Second Conference on the Epidemiology of Aging* (S. G. Haynes and M. Feinleib, eds.), U.S. Government Printing Office, Washington, DC, 23–32 (1980).

5. G. L. Maddox, ed., *The Encyclopedia of Aging*, Springer Publishing, New York, 1987.

6. R. N. Butler, Current definitions of aging, in *Second Conference on the Epidemiology of Aging* (S. G. Haynes and M. Feinleib, eds.), U.S. Government Printing Office, Washington, DC, 7–13 (1980).

7. R. Temple, The clinical investigation of drugs for use by the elderly, J. Geriatr. Drug Ther., 2, 33–44 (1988).

8. E. D. Sumner, General considerations, in *Handbook of Geriatric Drug Therapy for Health Care Professionals*, Lea & Febiger, Philadelphia, 1–9 (1983).

9. S. Yaffe and M. Danish, Problems of drug administration in the pediatric patient, Drug Metab. Rev. 8, 303–318 (1978).

10. E. Assael, Pharmacokinetics and drug distribution during postnatal development, Pharmacol. Ther., 18, 159–197 (1982).

11. C. Stewart and E. Hampton, Effect of maturation on drug disposition in pediatric patients, Clin. Pharm., 6, 548–64 (1987).

12. M. Reed and J. Besunder, Developmental pharmacology: Ontogenic basis of drug disposition, Pediatr. Clin. North Am., 36, 1053–1074 (1989).

13. G. Kearns and M. Reed, Clinical pharmacokinetics in infants and children, Clin. Pharmacokinet., 17(Suppl. 1), 29–67 (1989).

14. M. Danish and S. Rosenbaum, Dosing considerations for non-prescription drugs in infants and children, Clin. Res. Regul. Affairs, 9, 89–102 (1992).

15. J. Deren, Development of structure and function of the fetal and newborn stomach, Am. J. Clin. Nutr., 24, 144–159 (1971).

16. P. Hyman, E. Feldman, and M. Ament, Effect of external feeding on the maintenance of gastric acid secretory function, Gastroenterology, 84, 341–345 (1983).

17. M. Gupta and Y. Brans, Gastric retention in neonates, Pediatrics, 62, 26–29 (1978).

18. L. Hendeles, R. Iafrate, and M. Weinberger, A clinical and pharmacokinetic basis for the selection and use of slow release theophylline products, Clin. Pharmacokinet., 9, 95–135 (1984).

19. L. Linday, J. Dobkin, and T. Wang, Digoxin inactivation by the gut flora in infancy and childhood, Pediatrics, 79, 544–548 (1987).

20. W. Jusko, G. Levy, and S. Yaffe, Effect of age on intestinal absorption of riboflavin in humans, J. Pharm. Sci., 59, 487–490 (1970).

21. G. Heimann, Enteral absorption and bioavailability in children related to age, Eur. J. Clin. Pharmacol., 18, 43–50 (1980).

22. L. Pedersen-Bjergaard and K. Petersen, Oral absorption of pivampicillin and ampicillin in young children, Clin. Pharmacokinet., 2, 451–456 (1977).

23. J. Gilman, M. Duchown, T. Resnick, and E. Hershorin, Carbamazepine malabsorption: A case report, Pediatrics, 82, 518–519 (1988).

24. B. Friss-Hansen, Body water compartments in children during growth and related changes in body composition, Pediatrics, 28, 169–181 (1961).

25. R. Heimler, B. Doumas, B. Jendrzeczal, P. Nemeth, R. Hoffman, and L. Nelin, Relationship between nutrition, weight change and fluid compartments in preterm infants during the first week of life, J. Pediatr., 122, 110–114 (1993).

26. L. Notarianni, Plasma protein binding of drugs in pregnancy and in neonates, Clin. Pharmacokinet., 18, 20–36 (1990).

27. I. Gauntlett, D. Fisher, R. Hertzka, E. Kuhls, M. Spellman, and C. Rudolph, Pharmacokinetics of fentanyl in neonatal humans and lambs: Effects of age, Anesthesiology, 69, 683–687 (1988).

28. J. Aranda, S. MacLeod, K. Renton, and N. Eade, Hepatic microsomal drug oxidation and electron transport in newborn infants, J. Pediatr., 85, 534–542 (1974).

29. J. Rosen, M. Danish, M. Ragni, C. Lopez Saccar, S. Yaffe, and H. Lecks, Theophylline pharmacokinetics in the young infant, Pediatrics, 64, 248–251 (1979).

30. O. Carrier, G. Pons, E. Rey, M. Richard, C. Moran, J. Badoual, and G. Olive, Maturation of caffeine metabolic pathways in infancy, Clin. Pharmacol. Ther., 44, 145–151 (1988).

31. R. Miller, R. Roberts, and L. Fischer, Acetaminophen elimination kinetics in neonates, children, and adults, Clin. Pharmacol. Ther., 19, 284–294 (1976).

32. J. Glazer, M. Danish, S. Plotkin, and S. Yaffe, Disposition of chloramphenicol in low birth weight infant, Pediatrics, 66, 573–578 (1980).

33. J. Thompson, D. Bloedow, and F. Leffert, Pharmacokinetics of intravenous chlorpheniramine in children, J. Pharm. Sci., 70, 1284–1286 (1981).

34. D. Fisher, P. Schwartz, and A. Davis, Pharmacokinetics of exogenous epinephrine in critically ill children, Crit. Care Med., 21, 111–117 (1993).

35. I. Matsuda, A. Higashi, and N. Inotsume, Physiologic and metabolic aspects of anticonvulsants, Pediatr. Clin. North Am., 36, 1099–1111 (1989).

36. J. Guignard, A. Torrado, O. Da Cunha, and E. Gautier, Glomerular filtration rate in the first three weeks of life, J. Pediatr., 87, 268–272 (1975).

37. G. Schwartz, L. Feld, and D. Langford, A simple estimate of glomerular filtration rate in full term infants during the first year of life, J. Pediatr., 104, 849–854 (1984).

38. R. McCance, Renal function in early life, Physiol. Rev., 28, 331–348 (1948).

39. M. Nahata, Progress in pediatric drug therapy, Drug Intell. Clin. Pharm., 20, 388–390 (1986).

40. M. Painter, C. Pippenger, C. Wasterlein, M. Barmada, and W. Pitlick, Phenobarbital and phenytoin in neonatal seizures: Metabolism and tissue distribution, Neurology, 31, 1107–1112 (1981).

41. S. Onishi, O. Yoshiki, Y. Nishimura, S. Itoh, and K. Itobe, Distribution of phenobarbital in serum, brain and other organs from pediatric patients, Dev. Pharmacol. Ther., 7, 153–159 (1984).

42. C. Hayes, V. Butler, and W. Gersony, Serum digoxin studies in infants and children, Pediatrics, 52, 561–568 (1973).

43. R. Kauffman and M. Nelson, Effect of age on ibuprofen pharmacokinetics and antipyretic response, J. Pediatr., 121, 969–973 (1992).

44. M. Boutroy, P. Vert, R. Royer, P. Monin, and M. Royer-Morrot, Caffeine, a metabolite of theophylline during treatment of apnea in the premature infant, J. Pediatr., 94, 996–998 (1979).

45. Scientists explore effects of drugs on children, adults. World Pharm. Stand. Rev., pp. 8–9 (1990).

46. R. Steele and G. Kearns, Antimicrobial therapy for pediatric patients, Pediatr. Clin. North Am., 36, 1321–1349 (1989).

47. S. Segal and Committee on Drugs, Use of codeine and dextromethorphan containing cough syrups in pediatrics, Pediatrics, 62, 118–121 (1978).

48. T. Finitzo-Hieber, G. McCracken, R. Roeser, D. Allen, D. Chrane, and J. Morrow, Ototoxicity in neonates treated with gentamicin and kanamycin: Results of a four year controlled follow-up study, Pediatrics, 63, 445–450 (1979).

49. A. Muhlberg, C. Linz, E. Bern, L. Tucker, M. Verhave, and R. Grand, Identification of nonsteroidal antiinflammatory drug induced gastrointestinal injury in children with juvenile rheumatoid arthritis, J. Pediatr., 122, 647–649 (1993).

50. T. Einarson, Drug related hospital admissions, Ann. Pharmacother., 27, 832–840 (1993).

51. C. Woods, M. Rylance, R. Cullen, and G. Rylance, Adverse reactions to drugs in children, Br. Med. J., 294, 869–870 (1987).

52. L. Golightly, S. Smolinske, M. Bennet, E. Sutherland, and B. Rumack, Pharmaceutical excipients (Part I), Med. Toxicol., 3, 128–165 (1988).

53. L. Golightly, S. Smolinske, M. Bennet, E. Sutherland, and B. Rumack, Pharmaceutical excipients (Part II), Med. Toxicol., 3, 209–240 (1988).

54. A. Pruitt, Committee on Drugs, ''Interactive'' ingredients in pharmaceutical products, Pediatrics, 76, 635–643 (1985).

55. P. Menon, B. Thach, C. Smith, M. Landt, J. Robert, R. Hillman, and L. Hillman, Benzyl alcohol toxicity in a neonatal intensive care unit, Am. J. Perinatol., 1, 288–292 (1984).

56. G. Little and A. Pruitt, Benzyl alcohol: Toxic agent in neonatal units, Pediatrics, 72, 356–358 (1983).

57. K. Arulanantham and M. Genel, Central nervous system toxicity associated with ingestion of propylene glycol, J. Pediatr. 93, 515–516 (1978).

58. Yellow No. 5 (tartrazine) drug labeling, FDA Drug Bull., 9, 18 (1979).

59. R. Pohl, R. Balon, R. Berchow, and V. Yeragani, Allergy to tartrazine in antidepressants, Am. J. Psychiatry, 144, 237–238 (1987).

60. R. Weber, M. Hoffman, D. Raine, and H. Nelson, Incidence of bronchoconstriction due to aspirin, azo dyes, non-azo dyes and preservatives in a population of perennial asthmatics, J. Allergy Clin. Immunol., 64, 32–37 (1979).

61. R. Buswell and M. Lefkowitz, Oral bronchodilators containing tartrazine, JAMA, 235, 1111 (1976).

62. E. Hill, C. Flaitz, and G. Frost, Sweetener content of common pediatric oral liquid medications, Am. J. Hosp. Pract., 45, 135–142 (1988).

63. L. Shaw and H. Glenwright, The role of medications in dental caries formation: Need for sugar-free medication for children, Pediatrician, 16, 153–155 (1989).

64. R. Feigal, M. Jensen, and C. Mensing, Dental caries potential of liquid medications, Pediatrics, 68, 416–419 (1981).

65. P. Lokken, J. Birkeland, and E. Sannes, pH changes in dental plaque caused by sweetened iron containing medicine, Scand. J. Dent. Res. 83, 279–283 (1975).

66. D. Lutomski, M. Gora, S. Wright, and J. Martin, Sorbitol content of selected oral liquids, Ann. Pharmacother., 27, 269–274 (1993).

67. W. Dills, Sugar alcohols as bulk sweeteners, Annu. Rev. Nutri., 9, 161–186 (1989).

68. M. Veerman, Excipients in valproic acid syrup may cause diarrhea: A case report, DICP Ann. Pharmacother., 24, 832–833 (1990).

69. D. Arnold, C. Moodie, and H. Grice, Long term toxicity of *ortho*-toluene sulfonamide and sodium saccharin in the rat, Toxicol. Appl. Pharmacol., 52, 113–152 (1980).

70. A. Kumar, M. Weatherly, and D. Beaman, Sweeteners, flavorings and dyes in antibiotic preparations, Pediatrics, 87, 352–360 (1991).

71. R. Koysooko and G. Levy, Effect of ethanol on intestinal absorption of theophylline, J. Pharm. Sci., 63, 829–834 (1974).

72. NDMA statement on alcohol content of OTC drugs, Nonprescrip. Drug Manuf. Assoc. Newslett., (Dec. 18, 1992).

73. E. Feldman ed., *Handbook of Nonprescription Drugs*, 9th Ed., APHA Washington, DC, 1990.

74. P. Rappaport, Extemporaneous dosage preparations for pediatrics, U.S. Pharm., 9, H-1–H-12 (1984).

75. M. Nahata, Personal communication, (1993).

76. C. Cornaggia, S. Gianetti, D. Battino, T. Granato, A. Romeo, F. Viani, and G. Limido, Comparative pharmacokinetic study of chewable and conventional carbamazepine in children, Epilepsia, 34, 158–160 (1993).

77. J. Cloyd, R. Kriel, C. Jones-Saete, B. Ong, J. Jancik, and R. Remmel, Comparison of sprinkle versus syrup formulations of valproate for bioavailability, tolerance and preference, J. Pediatr., 120, 634–638 (1992).

78. C. Starr, New delivery system is ideal for children, the elderly, Drug Top., p. 28 (Jan. 22, 1990).

79. M. Nowak, B. Brundhofer, and M. Gibaldi, Rectal absorption from aspirin suppositories in children and adults, Pediatrics, 54, 23–26 (1974).

80. G. Gourlay and R. Boas, Fatal outcome with use of rectal morphine for postoperative pain control in an infant, Br. Med. J., 304, 766–767 (1992).

81. R. Ghadially and N. Shear, Topical therapy and percutaneous absorption, in *Pediatric Pharmacology*, 2nd Ed. (S. Yaffe and J. Aranda, eds.), W. B. Saunders, Philadelphia, 1992.

82. J. Paisley, A. Smith, and D. Smith, Gentamicin in newborn infants, Am. J. Dis. Child., 126, 473–477 (1973).

83. F. Shann, M. Linnemann, A. Mackenzie, J. Barker, M. Gratten, and N. Crinis, Absorption of chloramphenicol sodium succinate after intramusclar administration in children, N. Engl. J. Med., 313, 410–414 (1985).

84. J. Bradley, L. Compogiannis, W. Murray, M. Acosta, and G. Tsu, Pharmacokinetics and safety of intramuscular injection of concentrated ceftriaxone in children, Clin. Phar., 11, 961–964 (1992).

85. D. Doyle, Personal communication, 1992.

86. A. McIvor, M. Paluzzi, and M. Meguid, Intramuscular injection abscess—past lessons relearned, N. Engl. J. Med., 324, 1897–1898 (1991).

87. R. Hard, Pharmacists work on pediatric dosage problems. Hospitals, 66 (Oct. 20, 1992).

88. G. Koren, Z. Baryilay, and Greenwald, Tenfold errors in drug administration for children, Pediatrics, 77, 848–849 (1986).

89. C. Johnston, Endotracheal drug delivery, Pediatr. Emerg. Care, 8, 94–97 (1992).

90. A. Hickey, Factors influencing aerosol desposition in inertial impactors and their effect on particle size characterization, Pharm. Tech., pp. 118–120 (Sept. 1990).

91. B. Sagger and D. Lawson, Some observations on the penetration of antibiotics through mucus in vitro, J. Clin. Pathol., 19, 313–317 (1966).

92. B. Saggers and D. Lawson, In vivo penetration of antibiotics into sputum in cystic fibrosis, Arch. Dis. Child., 43, 404–409 (1968).

93. M. Gibaldi, Understanding and treating some genetic diseases, Ann. Pharmacother., 26, 1589–1594 (1990).

94. L. Jew and L. Hart, Inhaled aminoglycosides in cystic fibrosis, DICP Ann. Pharmacother., 24, 711–712 (1990).

95. J. Finney, P. Friman, M. Rapoff, and E. Chrisopherson, Improving compliance with antibiotic regimens for otitis media, Am. J. Dis. Child., 139, 89–95 (1985).

96. N. Mantick and C. Jantz, Children's OTC pharmaceuticals: Sensory directed flavor formulation, Profile Attribute Analysis, Arthur D. Little (1991).

97. M. Ruff, D. Schotik, J. Bass, and J. Vincent, Antimicrobial drug suspensions: A blind comparison of taste of fourteen common pediatric drugs, Pediatr. Infect. Dis. J., 10, 30–33 (1991).

98. Institute of Medicine, Report of a Workshop—Drug Development and the Pediatric Population, National Academy Press, Washington, DC, 1991.

99. G. Mossinghoff, 114 pediatric drugs, vaccines in testing, new medicines, Pharm. Manuf. Assoc., Fall (1990).

100. D. Hariparsad, N. Wilson, and C. Dixon, Oral tartrazine challenge in childhood asthma: Effect on bronchial reactivity, Clin. Allergy, 14, 81–85 (1984).

101. F. Chafee and G. Settipane, Asthma caused by FD & C approved dyes, J. Allergy, 40, 65–72 (1967).

102. R. Clarke, Exacerbation of asthma after nebulized beclomethasone diproprionate, Lancet, 2, 574–575 (1986).

103. C. Beasley, P. Rafferty, and S. Holgate, Benzalkonium chloride and bronchoconstriction, Lancet, 2, 1227 (1986).

104. D. Paige, E. Leonardo, and J. Nakasima, Response of lactose intolerant children to different lactose levels, Am. J. Clin. Nutr., 25, 467–469 (1972).

105. J. Lieb and D. Kazienko, Lactose filler as a cause of "drug-induced" diarrhea, N. Engl. J. Med., 299, 314 (1978).

106. Y. Kaminer, A. Apter, S. Tyano, E. Livni, and H. Wysenbeek, Delayed hypersensitivity reaction to orally administered methylparaben, Clin. Pharm., 1, 469–470 (1982).

107. W. Balistreri, M. Farrell, and K. Bove, Lessons from E-Ferol tragedy, Pediatrics, 78, 503–506 (1986).

108. M. Nitzan, B. Volovitz, and E. Topper, Infantile methemoglobinemia caused by food additives, Clin. Toxicol., 15, 273–280 (1979).

109. G. Angelini and C. Meneghini, Contact allergy from propylene glycol, Contact Dermatol., 7, 197–198 (1981).

110. H. Moller, Merthiolate allergy: A nationwide iatrogenic sensitization, Acta Dermatol. Venereol, 57, 509–517 (1977).

111. J. Royhans, P. Walson, G. Wood, and W. MacDonald, Mercury toxicity following merthiolate ear irrigations, J. Pediatr., 104, 311–313 (1984).

112. M. Haeney, G. Carter, W. Yeoman, and R. Thompson, Long term parenteral exposure to mercury in patients with hypogammaglobulinaemia, Br. Med. J., 2, 12–14 (1979).

113. H. Koch and D. A. Knapp, Highlights of drug utilization in office practice: National Ambulatory Medical Care Survey, 1985, Adv. Data, 134 (1987).

114. National Center for Health Statistics, *Health, United States, 1991*, Public Health Service, Hyattsville, MD (1992).

115. J. Roberts and N. Turner, Pharmacodynamic basis for altered drug action in the elderly, Clin. Geriatr. Medic., 4, 127–149 (1988).

116. R. E. Vestal, Drug use in the elderly: A review of problems and special considerations, Drugs, 16, 358–382 (1978).

117. M. L. Rocci, P. H. Blases, and W. B. Abrams, Geriatric clinical pharmacology, Cardiol. Clin., 4, 213–225 (1986).

118. J. A. Cromarty, Medicines for the elderly, Pharm. J., 235, 511–514 (1985).

119. F. Pucino, C. L. Beck, R. L. Seifert, G. L. Strommen, P. A. Sheldon, and I. L. Silbergleit, *Pharmacogeriatr*. Pharmacother. 5, 314–326 (1985).

120. J. Crooks, K. O'Malley, and H. Stevenson, Pharmacokinetics in the elderly, Clin. Pharmacokinet., 1, 280–296 (1976).

121. W. F. Kean, W. W. Buchanan, Pharmacokinetics of NSAID with special reference to the elderly, Singapore Med. J., 28, 383–389 (1987).

122. S. D. Black, M. J. Denham, R. M. Acheson, V. W. M. Drury, J. G. Evans, A. N. Exton-Smith, C. F. George, M. Hamilton, D. A. Heath, H. M. Hodkinson, T. Rawlins, T. E. D. Arie, I. G. J. R. Evans, M. Rowland, J. P. Kerr, E. S. Snell, B. Wade, D. G. Williams, G. M. G. Tibbs, and H. Irons, Medications for the elderly, J. R. Coll. Physicians Lond., 18, 7–17 (1984).

123. M. Mayersohn, Drug disposition in the elderly, in *Pharmacy Practice for the Geriatric Patient* (F. B. Penta, et al., eds.), American Association of Colleges of Pharmacy, North Carolina, 1986, pp. 9.5–9.11.

124. A. M. M. Shepherd, Physiological changes with aging—relevance to drug study design, in *Drug Studies in the Elderly: Methodological Concerns* (N. R. Cutler and P. K. Narang, eds.), Plenum Publishing, New York, 1986, pp. 50–54.

125. C. M. Castleden, C. N. Volans, and K. Raymond, The effect of ageing on drug absorption from the gut, Age and Ageing, 6, 138–143 (1977).

126. P. P. Gerbino and C. J. Wordell, Gastrointestinal disorders, in *Pharmacy Practice for the Geriatric Patient* (F. B. Penta, et al., eds.), American Association of Colleges of Pharmacy, North Carolina, 1986, pp. 20.1–20.24.

127. T. W. Sheely, The gastrointestinal system and the elderly, in *Contemporary Geriatric Medicine*, Vol. 2 (S. R. Gambert, ed.), Plenum Publishing, New York, 1986.

128. M. C. Geokas and B. J. Haverback. The aging gastrointestinal tract, Am. J Surg. 117, 881–891, (1969).

129. P. R. Holt, Gastrointestinal drugs in the elderly, Am. J. Gastroenterol., 81, 403–411 (1986).

130. S. Anuras and V. Loeing-Baucke, Gastrointestinal motility in the elderly, J. Am. Geriatr. Soc., 32, 386–390 (1984).

131. J. G. Moore, C. Tweedy, P. E. Christian, and F. L. Datz, Effect of age on gastric emptying of liquid-solid meals in man, Dig. Dis. Sci. 28, 340–344 (1983).

132. M. A. Evans, E. J. Triggs, M. Cheung, G. A. Broe, and H. Creasey, Gastric emptying rate in the elderly: Implications for drug therapy, J. Am. Geriatr. Soc., 24, 201–205 (1981).

133. M. A. Evans, G. A. Broe, E. J. Triggs, M. Cheung, H. Creasy, and P. D. Paull, Gastric emptying rate and the systemic availability of levodopa in the elderly parkinsonian patient, Neurology, 31, 1288–1294 (1981).

134. P. Mojaverian and P. H. Vlasses, Effects of gender, posture, and age on gastric residence time of an indigestible solid: Pharmaceutical considerations, Pharm. Res., 5, 639–644 (1988).

135. P. F. Adams and V. Benson, Current estimates from the National Health Interview Survey, 1991, National Center for Health Statistics, Vital Health Stat., 10(184) (1993).

136. J. G. Collins, Prevalence of selected chronic conditions, United States, 1986–88, National Center for Health Statistics, Vital Health Stat., 10(182) (1993).

137. J. Cornoni-Huntley, D. B. Brock, A. M. Ostfeld, J. O. Taylor, and R. B. Wallace (eds.), *Established Populations for Epidemiologic Studies for the Elderly*, U.S. Government Printing Office, Washington, DC, 1986.

138. D. B. Ferguson, An overview of physiological changes in the aging mouth, Front. Oral Physiol., 6, 1–6, (1987).

139. J. C. Ofstehage and K. Magilvy, Oral health and aging, Geriatr. Nurs., 7, 238–241 (1986).

140. S. Kamen and L. B. Kamen, Aging and oral function, in *Contemporary Geriatric Medicine*, Vol. 2. (S. R. Gambert, ed.), Plenum Publishing, New York, 1986.

141. G. M. Ritchie, Mouth and dentition, in *Practical Geriatric Medicine* (A. N. Exton-Smith and M. E. Weksler, eds.), Churchill Livingstone, New York, 1985.

142. H. Heeneman and D. H. Brown, Senescent changes in and about the oral cavity and pharynx, J. Otolaryngol., 15, 214–216 (1986).

143. H. Ben-Aryeh, D. Miron, I. Berdicevsky, R. Szargel, and D. Gutman, Xerostomia in the elderly: Prevalence, diagnosis, complications and treatment, Gerodontology, 4, 77–82 (1985).

144. B. J. Baum and L. Bodner, Aging and oral motor function: Evidence for altered performance among older persons, J. Dent. Res., 62, 2–6 (1983).

145. E. Christophers and A. M. Kligman, Percutaneous adsorption in aged skin, in *Advances in Biology of the Skin*, Vol. 6 (W. Montagna, ed.), Permagon Press, New York, 1964, pp. 163–175.

146. C. J. Behl, N. H. Bellantone, and G. L. Flynn, Influence of age on percutaneous absorption of drug substance, in *Transdermal Delivery of Drugs*, Vol. 2 (A. F. Kyodonieus and B. Berner, eds.), CRC Press, Boca Raton, FL, 1988, pp. 109–132.

147. A. M. Kligman, Perspectives and problems in cutaneous gerontology, J. Invest. Dermatol., 73, 39–46 (1979).

148. C. H. Daly and G. F. Odland, Age-related changes in the mechanical properties of human skin, J. Invest. Dermatol., 73, 83–87 (1979).

149. S. Katz and T. D. Downs, Progress in the development of an index of ADL, Gerontologist, 10, 20–30 (1970).

150. L. G. Branch and S. Katz, A prospective study of functional status among community elderly, Am. J. Public Health, 74, 266–268 (1984).

151. C. C. Maloney, Identifying and treating the client with sensory loss, Phys. Occup. Ther. Geriatr., 5, 31–41 (1987).

152. W. Kosnik, L. Winsolow, D. Kline, K. Rasinski, and R. Sekuler, Visual changes in daily life throughout adulthood, J. Gerontol., 43, P63–P70 (1988).

153. J. Cerella, Age-related decline in extrafoveal letter perception, J. Gerontol., 40, 727–736 (1985).

154. P. P. Lamy, Over-the-counter medication: The drug interactions we overlook, J. Am. Geriatr. Soc. 30(Suppl), S69–S75 (1982).

155. P. P. Lamy, Appropriate and inappropriate drug use, in *Pharmacy Practice for the Geriatric Patient* (F. B. Penta, et al., eds.), American Association of Colleges of Pharmacy, North Carolina, 1985, pp. 13.1–13.27.

156. C. C. Fuselier, General principles of drug prescribing, in *Pharmacy Practice for the Geriatric Patient* (F. B. Penta, et al., eds.), American Association of Colleges of Pharmacy, North Carolina, 1985, pp. 8.1–8.28.

157. M. K. Kottke, G. Stetsko, S. R. Rosenbaum, and C. T. Rhodes, Problems encountered by the elderly in the use of conventional dosage forms, J. Geriatr. Drug Ther., 5, 77–92 (1990).

158. M. Marvola, Adherence of drug products to the oesophagus, Int. J. Pharm., 3, 294–296 (1984).

159. K. S. Channer and J. P. Virjee, The effect of formulation on oesophageal transit, J. Pharm. Pharmacol., 37, 126–129 (1985).

160. M. Marvola and M. Rajaniemi, Effect of dosage form and formulation factors on adherence of drugs to the esophagus, J. Pharm. Sci., 72, 1034–1036 (1983).

161. H. Hey and J. Jorgensen, Oesophageal transit of six commonly used tablets and capsules, Br. Med. J., 235, 1717–1719 (1982).

162. T. Semla, J. Breizer, and M. Higbee, *Geriatric Dosage Handbook*, American Pharmaceutical Association (1993).

163. R. B. Wallace, Drug utilization in the rural elderly: Perspectives from a population study, in *Geriatric Drug Use—Clinical and Social Perspectives* (S. R. Moore and R. W. Teal, eds.), Permagon Press, New York, 1985, pp. 78–85.

164. R. G. Hollenbeck and P. P. Lamy, Dosage form considerations in clinical trials involving elderly patients, in *Drug Studies in the Elderly: Methodological Concerns* (N. R. Cutler and P. K. Narang, eds.), Plenum Publishing, New York, 1986, pp. 335–353.

165. J. E. Finchman, Over-the-counter drug use and misuse by the ambulatory elderly: A review of the literature, J. Geriatr. Drug Ther. 1, 3–21 (1986).

166. J. Williamson, R. G. Smith, and L. E. Burley, Drugs and safer prescribing, in *Primary Care of the Elderly: A Practical Approach*, IOP Publishing, London, 1987.

167. W. Simonson, Compliance to drug therapy, in *Medications and the Elderly: A Guide to Promoting Proper Use*, Aspen Publishers, Baltimore, 1984, pp. 70–79.

168. P. P. Lamy, The future is not what it used to be, M. Pharm., 63, 10–14 (1987).

169. J. L. Richardson, Perspectives on compliance with drug regimens among the elderly, J. Compliance Health Care, 1, 33–45 (1986).

170. B. Wade and A. Bowling, Appropriate use of drugs by elderly people, J. Adv. Nurs., 11, 47–55 (1986).

171. E. D. Sumner, Compliance with drug therapy, in *Handbook of Geriatric Drug Therapy for Health Care Professionals*, Lea & Febiger, Philadelphia, 1983, pp. 43–52.

172. R. B. Hallworth and L. A. Goldberg, Geriatric patients' understanding of labelling of medicines, Br. J. Pharm. Pract., 6, 6–14 (1984).

173. R. B. Hallworth and L. A. Goldberg, Geriatric patients' understanding of labelling of medicines, Part 2, Br. J. Pharm. Pract., 6, 42–48 (1984).

174. B. S. M. Wong and D. C. Norman, Evaluation of a novel medication aid, the Calendar Blister-Pak, and its effect on drug compliance in a geriatric outpatient clinic, J. Am. Geriatr. Soc., 35, 21–26 (1987).

175. S. Keram and M. E. Williams, Quantifying the case of difficulties older persons experience opening medical containers, J. Am. Geriatr. Soc., 36, 198–201 (1988).

176. J. R. Davidson, Presentation and packaging of drugs for the elderly, J. Hosp. Pharm., 31, 180–184 (1973).

177. United States Pharmacopeial Convention, *Fourth Supplement of the United States Pharmacopeia—National Formulary*, United States Pharmacopeial Convention, Rockville, MD, 1988, pp. 2249–2250.

178. U.S. Bureau of the Census, Demographic and socioeconomic aspects of aging in the United States, *Current Population Reports, Series P-23, No. 138*, U.S. Government Printing Office, Washington, DC, 1984.

179. G. S. Banker and N. R. Anderson, (1985).

180. J. Tsumara, Process for the preparation of water-soluble tablets, U.S. Patent 3,692,896, 1972.

181. G. Crivellaro, and F. Oldani, Soluble tablets, U.S. Patent 3,819,824, 1974.

182. L. G. Daunora, Water soluble tablet, U.S. Patent 4,347,235, 1982.

183. T. Martin, Tablet dispersion as alternative to mixtures, N.Z. Pharm. 7(Jul), 34–35 (1987).

184. H. Heinemann and W. Rothe, Preparation of porous tablets, U.S. Patent 3,885,026, 1975.

185. G. D. Tovey, The development of the Tiltab Tablets, Pharm. J., 239, 363–364 (1987).

186. C. Murphy, Aging and chemosensory perception, Front. Oral Physiol., 6, 135–150 (1987).

187. M. E. Spitzer, Taste acuity in institutionalized and non-institutionalized elderly men, J. Gerontol., 43, P71–P74 (1988).

188. B. A. Cooper, A model for implementing color contrast in the environment of the elderly, Am. J. Occup. Ther., 39, 253–258 (1985).

189. G. Zuccollo and H. Liddell, The elderly and the medication label: Doing it better, Age Ageing, 14, 371–376 (1985).

190. R. J. Michocki, What to tell patients about over-the-counter drugs, Geriatrics, 37, 113–124 (1982).

191. P. Timmins, I. Browning, A. M. Delargy, J. W. Forrester, and H. Sen, Effect of active raw material variability on tablet production, Drug Dev. Ind. Pharm., 12, 1293–1307 (1986).

192. N. Kaneniwa, K. Imagaw, and J.-I. Ichikawa, The effect of particle size on the compaction properties and compaction mechanism of sulfadimethoxine and sulfaphenazole, Chem. Pharm. Bull., 36, 2531–2537 (1988).

193. B. M. Hunter, The effect of the specific surface area of primidone on its tableting properties, J. Pharm. Pharmacol., 26, 58P (1974).

194. S. Vesslers, R. Boistelle, A. Delacourte, J. C. Guyot, and A. M. Guyot-Hermann, Influence of structure and size of crystalline aggregates on their compression ability, Drug Dev. Ind. Pharm., 18, 539–560 (1992).

195. J. R. Leonards and G. Levy, Biopharmaceutical aspects of aspirin-induced blood loss in man, J. Pharm. Sci., 58, 1277–1279 (1969).

196. G. M. Phillips, B. T. Palermo, Physical form as a determinant of effect of buffered acetylsalicylate formulations on GI microbleeding, J. Pharm. Sci., 66, 124–126 (1977).

197. B. Hoener and L. Z. Benet, Factors influencing drug absorption and drug availability, in *Modern Pharmaceutics* (G. S. Banker and C. T. Rhodes, eds.), Marcel Dekker, New York, 1979, pp. 143–182.

198. R. W. Heckel, An analysis of powder compaction phenomena, Trans. Metallurg. Soc. AIME, 221, 1001–1008 (1961).

22

Biotechnology-Based Pharmaceuticals

S. Kathy Edmond Rouan
SmithKline Beecham, King of Prussia, Pennsylvania

I. INTRODUCTION

During the last two decades, two principal biotechnologies have had a significant influence on drug discovery and development in the pharmaceutical industry. Recombinant DNA and hybridoma techniques have provided means of producing sufficient quantities of purified macromolecules suitable for clinical evaluation. Table 1 lists approved therapeutic agents derived from these biotechnologies. In addition to approved biopharmaceuticals, there are currently over 140 such products in various stages of clinical development within the United States alone. Numerous recent articles further record and reflect the influence of biotechnology-based medicines on modern pharmaceuticals [1–5]. This chapter will discuss the basic concepts behind these relatively novel technologies, the significant tasks and challenges encountered in their product development to provide commercial medicines, and some of the future prospects for biotechnology-based pharmaceuticals.

II. BACKGROUND

The use of therapeutic proteins to replace or supplement endogenous protein molecules has been a long-established treatment for diseases such as diabetes, growth hormone deficiency, and hemophilia. However, treatment was often limited by immunological responses to heterologous protein molecules, contamination of proteins derived from complex natural sources, and the difficulty and expense of obtaining useful quantities of materials of human and animal origin. The first successful treatment of diabetic patients with animal-derived insulin was described more than 70 years ago, by Banting and Best [6]. Insulin was classically derived from bovine or porcine sources, so immunological responses to the heterologous protein were common. Before the availability of recombinantly derived human growth hormone (hGH), approximately 50 cadaver pituitary glands were required to treat a single growth hormone-deficient child for just 1 year [7]. However, one of the reasons for the withdrawal of pituitary-derived

Table 1 Biotechnology-Based Pharmaceuticals

Trade name	Product type	Manufacturer	Abbreviated indication	Approval date (U.S.)
Actimmune	Interferon gamma-1b	Genentech	Chronic granulomatous disease	Dec. 1990
Activase	Alteplase (t-PA)	Genentech	Acute myocardial infarction	Nov. 1987
			Acute pulmonary embolism	June 1990
Alferon N	Interferon alfa-n3	Interferon Sciences	Genital warts	Oct. 1989
Engerix-B	Hepatitis B vaccine	SmithKline Beecham	Hepatitis B	Sept. 1989
Epogen	Epoetin alfa	Amgen	Anemia of chronic renal failure	June 1989
PROCRIT	Epoetin alfa	Ortho Biotech	Anemia of chronic renal failure	Dec. 1990
Humatrope	Somatropin for injection	Eli Lilly	hGH deficiency in children	Mar. 1987
Humulin	Human insulin	Eli Lilly	Diabetes	Oct. 1982
Intron A	Interferon alfa-2b	Schering-Plough	Hairy cell leukemia	June 1986
			Genital warts	June 1988
			AIDS-related Kaposi's sarcoma	Nov. 1988
			Non-A, non-B hepatitis	Feb. 1991
			Chronic hepatitis B	July 1992
Leukine	Sargramostim (GM-CSF)	Immunex	Autologous bone marrow transplant	Mar. 1991
			Delayed/failed bone marrow graft	Dec. 1991
Prokine	Sargramostim (GM-CSF)	Hoechst-Roussel	Autologous bone marrow transplant	Mar. 1991
Neupogen	Filgrastim (rG-CSF)	Amgen	Chemotherapy-induced neutrotropenia	Feb. 1991
Orthoclone OKT3	Muromonab-CD3	Ortho Biotech	Acute kidney transplant rejection	June 1986
Protropin	Somatrem for injection	Genentech	hGH deficiency in children	Oct. 1985
Recombivax HB	Hepatitis B vaccine, MSD	Merck	Hepatitis B prevention	July 1986
Roferon-A	Interferon alfa-2a	Hoffmann-La Roche	Hairy cell leukemia	June 1986
			AIDS-related Kaposi's sarcoma	Nov. 1988
Proleukin	Interleukin-2	Chiron	Metastatic renal cell carcinoma	May 1992
Recombinate	Antihemophilic factor	Baxter/Genetics Inst.	Factor VIII deficiency	Dec. 1992

Source: Refs. 2, 157, 158.

hGH from the market was its implication in deaths from Creutzfeldt–Jakob disease, a slow virus that may have contaminated some preparations of the material [8]. The tragic consequences of the treatment of hemophiliacs with clotting factors derived from acquired immunodeficiency syndrome- (AIDS) or hepatitis-infected blood stand as further harsh testimony to the risks involved in isolating therapeutic proteins from their natural source. Significantly, therefore, recombinant DNA and hybridoma techniques are capable of providing molecules of a well-defined chemical composition and producing them in cell culture media that can be carefully controlled. In short, the critical advance these techniques offer is that supply and purity no longer impede the development and clinical usefulness of therapeutic protein molecules.

Recombinant DNA technology, often used synonymously with genetic engineering, involves the isolation of cellular DNA fragments that code for proteins of therapeutic interest [see Refs. 9 and 10 for reviews]. The DNA fragments are inserted into cellular hosts that, by normal replication, make multiple copies of the original sequence (Fig. 1). This amplification of the original sequence enables the production of useful quantities of protein in the cell culture medium. By using established biochemical purification techniques, the protein may be isolated in a highly purified form.

The development of antibody-producing hybridomas was first described in the Nobel Prize-winning work of Kohler and Milstein [11]. Hybridoma technology involves the fusion of a cell that produces an antibody of therapeutic interest with a continually dividing cell line, to produce a hybrid cell that has inherited specific antibody production from one parent and the immortality of the other parent [see Ref. 12 for a review]. Importantly, the specificity of the antibody may be predetermined by obtaining the lymphocyte (antibody-producing cell) from an animal immunized with an antigen of interest. Classically, the source of the parental cell lines is murine. The clinical administration of the resultant murine antibody elicits a human immune response with the production of human antimouse antibody (HAMA) and a short serum half-life of the administered antibody. To circumvent this limitation, the original techniques of Kohler and Milstein [11] have been substantially adapted, and genetic-engineering tools are now used to produce monoclonal antibodies with greater sequence homology than human antibodies. The most immunogenic portion of the antibody is the species-conserved constant region. Therefore, several laboratories have used recombinant DNA technology to construct chimeric rodent–human monoclonal antibodies by attaching human constant regions to the rodent variable regions [13]. A more sophisticated approach to obtain ''humanized'' antibodies involves grafting the rodent hypervariable complementarity-determining regions (CDRs) onto human variable framework regions [14–16]. More recently, ''primatized'' antibodies have been developed that combine cynomolgus monkey variable region genes with human constant region fragments [17]. The newest phase of monoclonal antibody technology avoids the need for cell fusion partners altogether and will likely provide the ultimate goal: true human antibody homologues [18,19].

Clearly, the tools are at hand to provide a steady stream of novel macromolecular pharmacological agents. The challenge for the pharmaceutical industry lies in the development of the purified macromolecule into a stable, pharmaceutically elegant, medicinal product. The products of biotechnology are typically proteins, herein distinguished arbitrarily from peptides as molecules having in excess of 30 amino acid residues. The established use of insulin and other macromolecules has heralded some of the difficulties in formulating therapeutic proteins. Despite this experience, the emergence of the large range of products of biotechnology has lead to intense investigation and expansion in the fields of protein chemistry, analytical evaluation of macromolecules, and the formulation strategies to support the biotechnology pipeline. Unlike the more traditional, low molecular weight drugs, protein therapeutics pose unique

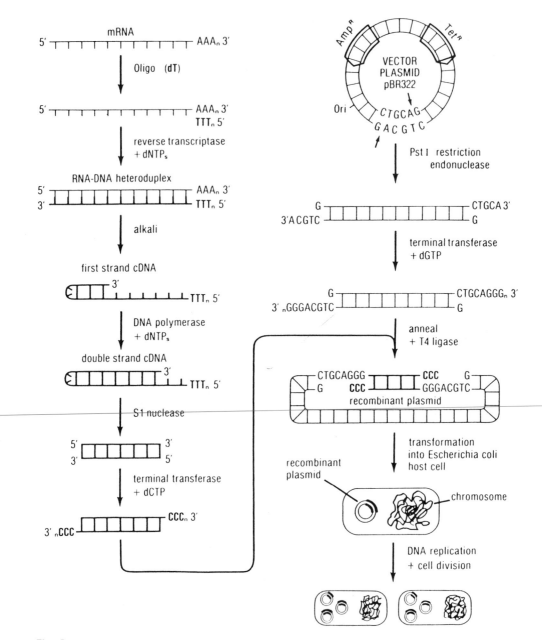

Fig. 1 Schematic outline of procedures employed in the synthesis of a cDNA gene copy from a poly-adenylated mRNA template, insertion of the cDNA into a bacterial plasmid vector by a homopolymer tailing strategy, and cloning of the recombinant plasmid in an *Escherichia coli* host.

problems in their pharmaceutical development owing to the need to stablilize multiple orders of chemical structure. Therefore, successful development relies on a knowledge of the processes capable of destabilizing proteins and, subsequently, the application of pharmaceutically acceptable means of preventing or retarding these processes.

III. PROTEIN STRUCTURE

The essential distinction between the approaches used to formulate and evaluate proteins, compared with conventional low molecular weight drugs lies in the need to maintain several levels of protein structure and the unique chemical and physical properties that these higher-order structures convey. Proteins are condensation polymers of amino acids, joined by peptide bonds. The levels of protein architecture are typically described in terms of the four orders of structure [20,21] depicted in Fig. 2. The primary structure refers to the sequence of amino acids and the location of any disulfide bonds. Secondary structure is derived from the steric relations of amino acid residues that are close to one another. The α-helix and β-pleated sheet are examples of periodic secondary structure. Tertiary structure refers to the overall three-dimensional architecture of the polypeptide chain. Proteins under physiological conditions assume their distinctive tertiary structure (native conformation) of minimum free energy, which is a prerequisite for their biological function [22]. Proteins that contain more than one polypeptide chain display an additional level of structural organization; namely, the quaternary structure, which refers to the way in which the chains are packed together. In addition to the structural arrangement of

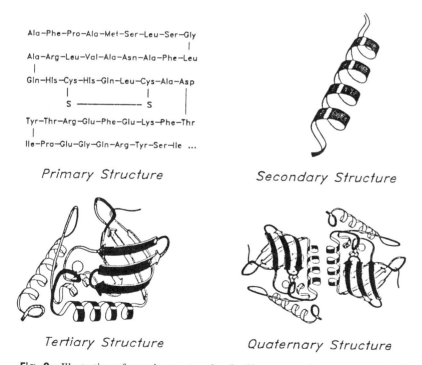

Fig. 2   Illustration of protein structure levels. Shown are primary structure (amino acid sequence), secondary structure (local order of protein chain, α-helix shown as an example), tertiary structure (assembly of secondary structure elements), and quaternary structure (relationship of different protein chain in multisubunit protein). (From Ref. 69.)

the polypeptide chains, another structural feature of many recombinant proteins is the attachment of oligosaccharide groups by means of glycosidic linkages. The glycosylation pattern of many proteins is important for them to exert their biological effect, or it may influence the biodistribution of a protein after administration [23–25].

The native, biologically active form of a protein molecule is held together by a delicate balance of noncovalent forces: hydrophobic, ionic, van der Waals interactions, and hydrogen bonds. In addition, disulfide linkages are covalent bonds that form between sulfur-containing amino acid residues and, thus, may contribute substantially to maintaining conformation in proteins that contain two or more cysteine residues. It has been confirmed by X-ray structure analysis that most water-soluble proteins may be grossly described as a hydrophobic core of nonpolar amino acid groups, surrounded by a hydrophilic shell of polar-solvated amino acids [22]. With exposure to certain denaturants or adverse environmental conditions, the noncovalent forces are weakened, and then broken apart, leading to the unfolding and consequent inactivation of the protein. Typically, the native structure exhibits only marginal stability that is easily upset by even subtle environmental changes in pressure, temperature, pH, ionic strength, or a combination thereof. The free energy of denaturation of globular proteins rarely exceeds 15 kcal mol^{-1} [22]. The complete or partial unfolding of the protein is usually fully reversible after removal from the antagonistic agent. However, this reversible unfolding event is the precursor to irreversible covalent and noncovalent reactions that lead to irreversible protein denaturation. The term *denaturation* is applied to both reversible and irreversible disruption of the native, biologically active conformation. Denaturation involves changes in noncovalent interactions, such as hydrogen bonding, hydrophobic interactions, and electrostatic forces. Although noncovalent processes can be involved in both irreversible and reversible denaturation, chemical reactions involving covalent bond breakage or rearrangement are irreversible inactivation events.

IV. MECHANISMS AND CAUSES OF PROTEIN DESTABILIZATION

The complex hierarchy of native protein structure may be disrupted by multiple possible destabilizing mechanisms. As has been described in the foregoing, these processes may disrupt noncovalent forces of interaction, or may involve covalent bond breakage or formation. A summary of the processes involved in the irreversible inactivation of proteins is illustrated in Fig. 3 and described briefly in the following section. Detailed discussions of mechanisms of protein destabilization processes are provided in several review articles [see Refs. 26–30].

A. Covalent Protein Destabilization

As with conventional smaller drug molecules, the chemical reactions involved in protein destabilization may be classified as those involving hydrolysis, oxidation, and racemization. However, within these categories of chemical reaction, specific reaction mechanisms are characteristic of polypeptide and protein molecules. Disulfide bond cleavage and exchange are further reaction mechanisms that specifically affect proteins. A striking feature of protein destabilization is that several different reaction mechanisms may proceed simultaneously. This was demonstrated by Ahern and Klibanov [31] in an elegant series of experiments that showed that the mechanism of inactivation of hen egg-white lysozyme at 100°C was highly dependent on the solution pH. At pH 4 and 6, inactivation was largely due to deamidation of asparagine residues. At pH 8, however, inactivation was associated with the combined contributions of noncovalent conformational processes, destruction of disulfide bonds, and deamidation.

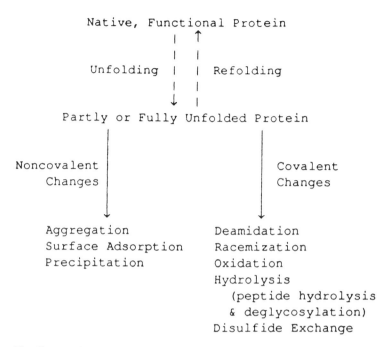

Fig. 3 Mechanisms involved in irreversible degradation of protein pharmaceuticals.

Because of the multiple degradation pathways that may take place at elevated temperature, protein stability monitoring data may not conform to the Arrhenius relationship, and the maximum temperature selected for accelerated stability studies must be carefully selected. Gu et al. [32] described the different mechanisms of inactivation of interleukin-1β (IL-1β) in solution above and below 39°C. In this example, the multiple mechanisms precluded the prediction of formulation shelf-life from accelerated temperature data. In contrast, by working at 40°C and lower, Pearlman and Nguyen [33] were able to successfully extrapolate data from stability studies of tissue plasminogen activator down to 5°C.

Chemical degradative processes in proteins and the relative incidence of each of the possible mechanisms depend on the nature of the protein, temperature, pH, ionic strength, oxygen tension, and the solutes present. Clearly, hydrogen ion concentration of the protein environment strongly influences hydrolytic reactions. However, pH also plays a significant role in driving other chemical processes that may exhibit a distinctive pH, as well as a temperature-dependent, rate-of-reaction profile. Proteins are generally most stable and least soluble at their isoelectric point (pI). At the pH corresponding to the protein's pI, the net charge of all the ionizable species is zero, with diminished potential for charge repulsion and protein unfolding.

Hydrolytic Reactions

The primary hydrolytic reactions in protein degradation are peptide bond hydrolysis and deamidation. Loss of oligosaccharide moieties through hydrolysis of glycosidic bonds may also influence protein activity [23–25].

Deamidation is the hydrolysis of the side-chain amide on glutamine and asparagine. Asparagine residues are more susceptible to deamidation than glutamine residues. Deadmidation introduces a new negative charge that may influence tertiary structure and stability. By using model peptides, Geiger and Clarke [34] demonstrated that deamidation under physiological conditions proceeds essentially completely through an imide intermediate. They further concluded that deamidation can be accompanied by simultaneous isomerization of the peptide linkage and racemization at the α-carbon, and that these mechanisms are linked by a common succinimide intermediate. The rate of deamidation could be significantly affected by the nature of the amino acid residue adjacent to the asparagine. The most labile asparagine residues in smaller peptides are those followed by a glycine residue. Asparagine–serine sequences are the next most labile sites of deamidation. However, in globular proteins, the location of the susceptible residue in the folded conformation may be more important in controlling the deamidation rate [35]. Certain proteins therefore, will not undergo deamidation unless they have been denatured. Tyler-Cross and Schirch [36] provide a recent and detailed examination of the effects of amino acid sequence, buffer, and ionic strength on the rate and mechanism of deamidation of asparagine-residue small peptides.

Peptide bond hydrolysis occurs readily under strongly acidic conditions, or by a combination of milder pH and elevated temperature. Although complete acid hydrolysis of protein into its amino acids is obtained under extreme conditions (6 M HCl, 24 hr, 110°C), shorter exposures, under less acidic conditions, show preferred peptide hydrolysis at aspartic acid residues. Aspartyl–prolyl linkages are especially vulnerable [28,37].

Oxidation

Oxidation of amino acids, especially those with aromatic side chains, as well as methionine, cysteine, and cystine residues, can occur with a variety of oxidants. Molecular oxygen, hydrogen peroxide, and oxygen radicals, all are known oxidants of proteins [26]. Oxidation of methiamine residues to their corresponding sulfoxides is associated with loss of biological activity for many peptide hormones and proteins [28,38]. The thiol group of cysteine can be oxidized in steps successively to a sulfenic acid (RSOH), a disulfide (RSSH), a sulfinic acid (RSO2H), and finally a sulfonic acid (RSO3H) depending on reaction conditions. The factors that influence the rate of reaction include the temperature, pH, buffer medium used, the type of catalyst, and the oxygen tension. Oxidation of thiols occurs readily at basic pH in the presence of transition metal ions such as Cu^{2+}. As with deamidation reactions, the steric arrangement of the protein may influence the oxidative reaction and its sequelae. Where oxidized thiol groups are exposed on the protein surface, intermolecular disulfide bonds may subsequently form, leading to protein aggregation (Fig. 4).

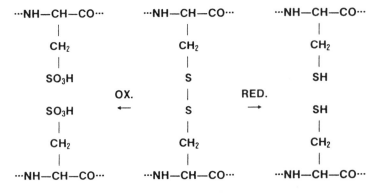

Fig. 4 Oxidation and reduction of disulfide bridge. (From Ref. 66.)

Racemization

Racemization of the native L-amino acid in peptides and proteins to the D-enantiomer generally results from base-catalyzed removal of the α-proton to produce a negatively charged planar carbanion. Return of the proton to the carbanion intermediate, through reaction with water, produces an enantiomeric mixture.

Rates of racemization depend on the particular amino acid and are influenced by temperature, pH, ionic strength, and metal ion chelation [39]. Aspartic acid and serine residues are the most prone to racemization. An electron-withdrawing group in the side chain of the amino acid, as in serine, for example, stabilizes the carbanion intermediate which, in turn, accelerates the rate of racemization. Geiger and Clarke [34] have provided data that suggest that intermediate succinimide formation plays a major role in racemization at aspartyl and asparaginyl residues. The latter results from stabilization of the carbanion through charge resonance involving the nitrogen atom and α- and β-carbonyl groups of the imide intermediate. Racemization of amino acids in proteins can generate nonmetabolizable forms of amino acids (D-enantiomers) or create peptide bonds inaccessible to proteolytic enzymes [28].

Disulfide Bond Exchange

Disulfide bonds provide covalent structural stabilization in proteins, so cleavage and subsequent rearrangement of these bonds can alter the tertiary structure, thereby affecting biological activity. Disulfide exchange is catalyzed by thiols, which can arise by initial hydrolytic cleavage of disulfides, or by β-elimination in neutral or alkaline media (Fig. 5) [28,40,41]. Zale and Klibanov [40] showed that, at pH 6 and 8, thermal inactivation of ribonuclease at 90°C is caused primarily by disulfide exchange. This process was inhibited by thiol scavengers, such as N-ethylmaleimide, p-(chloromercuri)benzoate, and copper ion, and accelerated in the presence of thiols, such as with the addition of cysteine. Volkin and Klibanov [42] demonstrated β-elimination of disulfide, with the generation of free thiols, occurring at 100°C in more than a dozen proteins (Fig. 6). The rates were generally accelerated under alkaline conditions. The widespread formation of covalent insulin dimers (CIDs) in insulin formulations as a result of disulfide interchange has also been reported [43].

B. Noncovalent Protein Destabilization

Three major categories of irreversible protein inactivation occur as a result of perturbation of the noncovalent forces that maintain the three-dimensional native state of proteins. Aggregation,

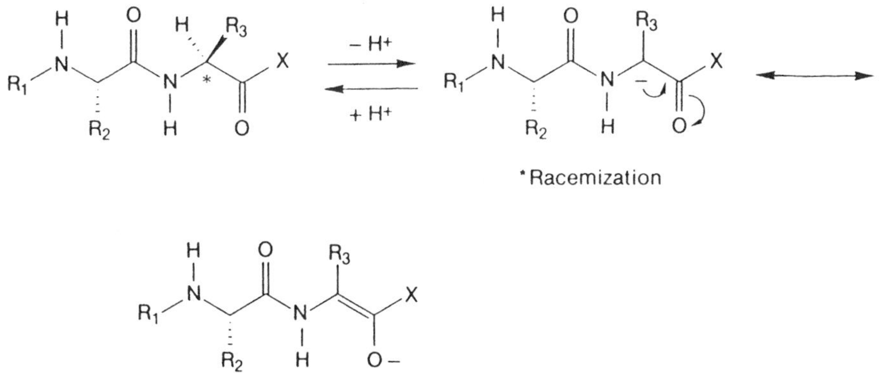

Fig. 5 Racemization of peptides by enolization.

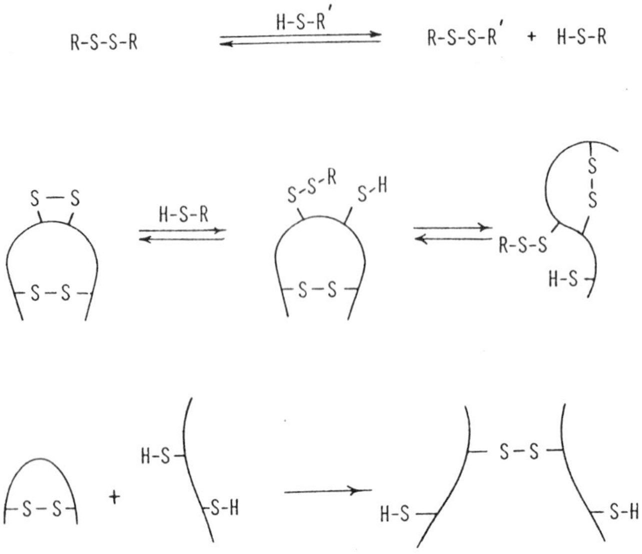

Fig. 6 Cystine disulfide exchange. (From Ref. 46.)

macroscopic precipitation, and surface adsorption are characteristic and, often, very troubling phenomena encountered during the formulation development of proteins and polypeptides.

Noncovalent electrostatic forces, hydrogen bonds, hydrophobic interactions and protein hydration may be altered as a result of thermal or pH effects. The irreversible inactivation of proteins proceeds following initial reversible unfolding of the native state. With an increase in temperature, a protein molecule in solution will undergo a transition from the native to unfolded state at the melting temperature (T_m). The T_m is defined as the temperature at which 50% of the molecules are unfolded. The pH of the solution influences the net charge on the protein, depending on its pI. Therefore, solution pH may affect electrostatic interactions, also referred to as salt bridges, which form between amino acids with ionic side groups. At extremes of pH, the net charge of the protein increases with greater charge repulsion, leading to protein unfolding.

Protein conformation is also markedly affected by the concentration and type of ionic species present. In solution, individual salt effects can be either stabilizing or denaturing [26,27]. These effects correspond to the Hofmeister lyotropic series:

$$SO_4^- > HPO_4^2 > OAc^- > F^- > \text{citrate} > Cl^-$$
$$> NO_3^- > I^- > CNS^-, ClO_4^- \ (CH_3)_4N^+ > NH_4^+ > K^+, Na^+ > Mg^{2+} > Ca^{2+} > Ba^{2+}$$

Anions and cations to the left of the series are the most stabilizing and reduce the solubility of hydrophobic groups (''salting-out'') on the protein molecule by increasing the ionic strength of the solution. Anions and cations to the right of the series are destabilizing and are known to denature proteins. These ions bind extensively to charged groups of proteins, causing an increase in observed solubility or salting-in.

Mechanical forces, such as shearing, shaking, and pressure, may also denature proteins. Shaking proteins may lead to inactivation owing to an increase in the area of the gas–liquid interface [26]. At the interface, the protein unfolds and maximizes exposure of hydrophobic residues to the air. Surface denaturation may also occur at the protein–container interface and has been observed following adsorption of proteins to filter materials [44]. The effects of shear on proteins have been reviewed [45]. High pressure also induces structural changes in proteins [46,47].

Aggregation and Precipitation

Aggregation of proteins is a microscopic process of protein molecule association. The aggregates may be dimers or larger oligomers that remain in solution, yet may affect the observed biological activity. *Precipitation* refers to the formation of visible proteinaceous particles, which may reduce potency in addition to altering the appearance of a formulation and its performance in various infusion devices.

Irreversible aggregation may follow unfolding, as a result of incorrect refolding of the protein. Protein unfolding exposes the interior hydrophobic region to the solvent, usually water. The structuring of hydrophilic solvent molecules around the hydrophobic protein regions is thermodynamically unfavorable. This drives intermolecular reactions between exposed hydrophobic regions, leading to aggregation. Although a two-state equilibrium has been proposed as a general rule in protein unfolding, several examples of intermediate conformational states have been described [48,49], and these may play a significant role in the formation of protein aggregates and subsequent precipitation. The conformational intermediates have considerable secondary structure, but lack tertiary structural interactions. Aggregation results from association of exposed hydrophobic regions on the monomeric intermediates. Protein concentration may influence the rate of formation of the associated intermediate species [48].

Mulkerrin and Wetzel [50] have investigated the effect of pH on the irreversible thermal denaturation of human interferon gamma by aggregation. This study demonstrates the dependence of aggregation on thermal unfolding and on pH. The aggregation rate versus pH profile resembles a titration curve, with the half maximal rate at pH 5.7. The effect of pH was related to the protonation state of histidine residues in the protein molecule. Deprotonation of histidine may reduce the solubility of the thermally unfolded state, rendering it more susceptible to aggregation.

The shaking of protein solutions may lead to aggregation and precipitation as a result of several mechanisms, such as air oxidation, denaturation at the interface, adsorption to the vessel, or mechanical stress. These possibilities were systematically examined for solutions of human fibroblast interferon [51]. In this example, mechanical stress was identified as the causative factor in the inactivation. The proposed mechanism of inactivation by mechanical stress was through orientation of the asymmetric protein in the shear field, which promotes association of aligned molecules. After association, disulfide bonds may form between reactive thiol groups in the unfolded molecules [51,52].

Insulin aggregation and precipitation has been an impediment to the development of implantable devices for insulin delivery and has been noted by several investigators working with conventional insulin infusion devices [53–56]. The potential causes of the observed aggregation and precipitation are thermal effects, mechanical stress, the nature of the materials in contact with the insulin solution, formulation factors, and the purity of the insulin preparation.

Surface Adsorption

Adsorption of proteins and peptides to the surfaces of the immediate container, filter, or materials of the infusion system can be substantial when the initial concentration of the protein in solution is low and the relative loss of drug to adsorption is consequently high. Andrade and Hlady [57] provide a comprehensive review of the principles of protein adsorption at the solid–liquid interface. As with protein aggregation, surface adsorption results from hydrophobic and electrostatic interaction and, therefore, will depend on the conformational state of the protein, the pH, and ionic strength of the solution, as well as the nature of the exposed surface. The interaction between a protein and a surface increases with increasing hydrophobicity of the surface, and increases with increasing hydrophobicity of the protein. The tendency for the protein to undergo conformational change after surface adsorption is a time-dependent, protein- and interface-specific process.

As membrane filtration is the only currently acceptable method of sterilizing protein phar- maceuticals, the adsorption and inactivation of proteins on membranes is of particular concern during formulation development. Pitt [58] examined nonspecific protein binding of polymeric microporous membranes typically used in sterilization by membrane filtration. Nitrocellulose and nylon membranes had extremely high protein adsorption, followed by polysulfone, cellu- lose diacetate, and hydrophilic polyvinylidene fluoride membranes. In a subsequent study by Truskey et al. [44], protein conformational changes after filtration were observed by CD spec- troscopy, particularly with nylon and polysulfone membrane filters. The conformational changes were related to the tendency of the membrane to adsorb the protein, although the precise mechanism was unclear. The effect of pressure and shear were not considered to be destabilizing factors, as the pressure levels and shear stresses under the filtration conditions used were lower than those typically associated with protein denaturation [45,46].

V. METHODS USED TO EVALUATE PROTEIN PHARMACEUTICALS

As is evident from the discussion of protein destabilization, protein formulation development must be supported by multiple analytical approaches to capture a complete profile of the chemi- cal and physical stability of prototype formulations. In addition to stability-indicating meth- odologies, analytical tools are available that assist in predicting the optimal excipients for stabilizing particular proteins in solution or in the solid state. Methods used to assess physical and chemical protein formulation attributes are described briefly in the following sections. A combination of these methods is typically employed during formulation development and sta- bility evaluation of protein pharmaceuticals.

A. Liquid Chromatography

As with the smaller drug molecules, chromatography is a powerful tool used to assess the purity and degradation profile of proteins. The most common chromatographic methods used in protein analyses are reverse-phase–high-pressure liquid chromatography (RP–HPLC), ion- exchange chromatography (IEC), and size-exclusion chromatography (SEC).

The RP–HPLC technique relies on a relatively nonpolar stationary phase in conjunction with an aqueous-based, polar mobile phase. For large protein molecues, C-4 or C-8 alkanes are bonded to silica-based or polymeric supports as the stationary phase. Stationary phases should possess wide pore sizes (diameter about 300 Å and greater) to permit the adequate diffusion of proteins with a relative molecular mass (M_r) of 50,000 Da and greater into the matrix [59]. Mobile phases typically consist of aqueous acetonitrile gradients, often with an ion-pairing agent, such as trifluoroacetic acid, constant at 0.1%. Phosphate or Tris buffers may be used to adjust the pH for optimal separation.

Use of RP-HPLC can denature some proteins owing the organic solvents, hydrophobic interactions, and low pHs typically employed for separation [27,59]. Nevertheless, the tech- nique has proved particularly useful for several proteins. It has been used extensively to char- acterize insulin formulations [60,61]. Kroeff and Chance [61] describe the use of RP-HPLC as a definitive identity test for insulin, as a means of monitoring insulin-related substances, and as a method for evaluating stability samples. A correlation was established between the assay value obtained by RP-HPLC and that obtained using the conventional potency assay. In the latter assay, the hypoglycemic activity of the insulin was determined in rabbits relative to a standard preparation. In fact, an official compendial RP-HPLC method is now described in the *United States Pharmacopeia* (*USP*) as an acceptable alternative to the rabbit bioassay for certain insulin preparations. Geigert et al. [62] used RP-HPLC as part of the stability evaluation of a recombinant human interferon beta formulation. The RP-HPLC methods used to characterize malaria protein antigens [63], interleukin-2, muteins [64], and human growth hormone have been described [65].

Ion-exchange chromatography is based on the selective retention of proteins relative to electrostatic interaction with charged groups, reacting with either cationic or anionic species in the protein molecule, covalently bonded to solid supports of porous material, such as silica gel. Elution is achieved with a gradient mobile phase of increasing ionic strength of the counterion. Chromatofocusing is an extension of ion-exchange chromatography that separates proteins by elution in a pH gradient on the basis of differences in the pI, rather than an ionic strength gradient. An example of the usefulness of the latter technique is found in the separation of recombinant IL-1β from an *N*-terminally methionylated form [66].

Size-exclusion chromatography, sometimes referred to as gel permeation chromatography, separates proteins on the basis of the molecular size. This technique, therefore, can yield information on the levels of aggregation and fragmentation in a protein formulation. Watson and Kenney [67] describe the use of high-performance size-exclusion chromatography to examine the aggregation of interferon gamma, and interleukin-2 after storage at elevated temperature, after mechanical agitation, and following rapid freeze-thaw.

B. Optical Spectroscopy

Quantitation of the manner in which proteins absorb, emit, and scatter light provides valuable information about the amount of protein present in the sample, the protein's conformation, and the tendency of the protein to aggregate. Optical spectroscopic techniques used to evaluate protein pharmaceuticals include ultraviolet (UV) and visible absorption spectroscopy, optical rotatory dispersion (ORD) and circular dichroism (CD), fluorescence, and infrared (IR) and Raman spectroscopy. Detailed theoretical discussions of these techniques are provided in the review by Cantor and Timasheff [68]. Havel [69] provides an overview of the application of these techniques to the investigation of protein structure.

Ultraviolet absorbance is routinely used to determine the molar concentration of proteins in solution by applying the Beer–Lambert law. Absorption of proteins in the 230- to 300-nm range is determined by the aromatic side chains of tyrosine (λ_{max} = 274 nm), trytophan (λ_{max} = 280 nm), and phenylalanine (λ_{max} = 257 nm). The molar absorbtivity of phenylalanine is, however, an order of magnitude less than tryptophan and tyrosine so it does not contribute significantly to the observed spectrum. As the difference in the absorption spectra of native and unfolded protein molecules is generally small, difference spectra can be generated by difference spectroscopy and, thereby, provide a convenient means of monitoring conformational changes in a protein [59,70]. When protein aggregates are present in solution, a sloping baseline in the 310- to 400-nm range results from light scattering. This may bias the absorption measured at 280 nm; consequently, the latter figure should be corrected to account for any light scattering [70].

Light scattering may be useful to measure the number of protein aggregates in a formulation. The intensity of the scattered light can be expressed as a function of the number of centers of scattering per unit volume [71]. Brems [72] measured turbidity at 450 nm as part of an evaluation of the solubility of different folding conformers of bovine growth hormone. Mulkerrin and Wetzel [50] also used light-scattering measurements, at a nonabsorbing wavelength, to study the effects of pH on the thermal denaturation of alpha interferons. One of the limitations is that the turbidity is only linearly related to the extent of aggregation when the particle size is small compared with the wavelength of light [71,73]. When the formulation contains large macroscopic precipitate, a visual ranking system, or the use of light microscopy, may be more valuable in assessing the physical stability of the formulation [53,56,73].

Of the visible spectroscopic techniques, CD spectroscopy has seen the most rapid and dramatic growth. The far-UV circular dichroism spectrum of a protein is a direct reflection of its secondary structure [74–76]. An asymmetrical molecule, such as a protein macromolecule, exhibits circular dichroism because it absorbs circularly polarized light of one rotation differently from circularly polarized light of the other rotation. Therefore, the technique is useful in determining changes in secondary structure as a function of stability, thermal treatment, or freeze-thaw. Brems et al. [72] used far-UV CD to study the helical structure of isolated fragments of bovine growth hormone. The amount of helix was found to be dependent on pH and peptide concentration. Johnson [75] and Manning [77] provide excellent reviews of the technique, with further examples of its application.

C. Electrophoresis

Electrophoretic techniques are based on the differences in the electrically induced migration of a protein in a sievelike gel, depending on the molecule's size or net charge. The most widely used electrophoretic techniques in protein analysis are reducing and nonreducing sodium dodecyl sulfate–polyacrylamide gel electrophoresis (SDS–PAGE) [78–80] and isoelectric focusing (IEF) [81,82]. More recently, capillary zone electrophoresis has generated considerable interest as a complementary technique in protein formulation analysis [83–85].

The SDS–PAGE technique involves denaturation of the sample with the SDS anionic detergent, followed by electrophoresis through the polyacrylamide support. Differential migration through the gel, followed by detection using either Coomassie or silver staining of the protein, reveals the presence of protein molecules of different molecular size, owing to their aggregation or fragmentation. A comparison of reduced versus nonreduced samples can be used to identify covalently cross-linked aggregation. Isoelectric focusing uses a pH gradient to separate molecules, based on their isoelectric points. As the isoelectric point is dependent on the charged functional groups in the amino acid sequence, the migration will be influenced by such processes as deamidation, or the degree of glycosylation, of the molecule. In capillary electrophoresis (CE) the electric field drives separation of charged molecules across a small-bore fused-silica capillary. The advantages of CE that are currently being explored are its high resolution, detection sensitivity, relatively short analytic time, and ease of automation [85].

D. Immunoassays and Biological Activity Assays

The ultimate test of the stability of the native protein conformation is maintenance of biological activity. Therefore, biological activity assays are included in the stability monitoring of protein formulations, and regulatory guidelines comment on the need for an assessment of biological potency for products of recombinant DNA and hybridoma technology. To substitute a physicochemical method for the biological potency assay, a correlation must exist between the two tests. As described earlier, such correlation has been established for certain insulin preparations [61,86]. For many other pharmaceutical proteins, a test of biological potency is included in the product specifications and stability testing. The design of functional assessment assays is highly protein-specific, and their adequacy in assessing the medically relevant in vivo activity is often very difficult to determine. Chayen [87] provides a useful discussion of the distinction drawn between analytical assays and functional assays. Functional assays to determine potency may be animal model assays (e.g., hypoglycemic response in rabbits with insulin) or cytochemical assays (e.g., the cytopathological effect assay for interferon and the clot lysis assay used to assess the potency of tissue plasminogen activator).

If the required in vivo response is binding of the protein to a target receptor or antigen, as in monoclonal antibodies used for diagnostic imaging, immunoassays provide a direct and relevant measure of the intended function. In other instances, specific protein-binding is the primary step that triggers the series of events culminating in the observed biological response. Immunoassays provide convenient in vitro tests to confirm maintenance of the specific protein-binding site through its interaction with a specific antibody or antigen; therefore, they are frequently used to monitor protein conformational stability. However, their primary limitation needs to be considered in that they may not provide a measure of true biological activity because antigenically similar proteins may be detected that do not have the same in vivo functional activity. The specificity of the antigen–antibody interaction is, therefore, a significant factor in the interpretation of immunoassay data. In the quality control of protein formulations, immunoassays may also be used to quantify proteins compared with a known amount of standards, or to detect specific protein contaminants, such as host cell proteins. Detection of specific protein-binding to the target is typically achieved through the use of enzyme- or radioisotope-labeled antibodies. Several examples and detailed reviews of these techniques are available for the interested reader [88–93].

E. Other Techniques

In addition to the foregoing methods, a host of other techniques may be valuable in supporting excipient selection and evaluating destabilizing mechanisms. In particular, the traditional role of differential scanning calorimetry (DSC) in the study of thermal unfolding of proteins [94] is being expanded and is now widely used to evaluate the relative ability of excipients to stabilize proteins to thermal denaturation [95–98]. Nuclear magnetic resonance spectroscopy in solution can also yield information about three-dimensional protein structure and, thereby, the nature of structural changes induced by environmental conditions [99]. Levine et al. [100] describe the use of surface tension measurements in formulation design. Surface tension measurements of monoclonal antibody solutions were predictive of susceptibility to surface denaturation, leading to aggregation and precipitation. Protein solutions exhibiting lower surface tension were more susceptible to protein denaturation. Furthermore, those surface-active excipients with the greatest surface activity provided the best stabilization from shaking-induced aggregation and precipitation. Carbohydrate analyses are reviewed by Parekh and Patel [23] and may provide information on the relevance of deglycosylation in protein destabilization. Enzymic digests of denatured proteins, may also be prepared, which are then analyzed by chromatographic separation and identification of fragments, to elucidate the mechanism of destabilization. Pearlman and Nguyen [101] describe the use of tryptic digestion of human growth hormone dimer, followed by HPLC and mass spectrometry to reveal the formation of an oxidized methionine when the formulation is exposed to intense fluorescent light.

VI. FORMULATION APPROACHES TO PROTEIN STABILIZATION

Once satisfied that suitable methods are in place to characterize covalent and noncovalent changes in the protein, and with a fundamental understanding of potential pathways of inactivation, one can critically evaluate the effect of the formulation environment on protein stabilization. A variety of approaches exist for stabilizing proteins (e.g., chemical modification, immobilization, and site-directed mutagenesis [22,102,103]); however, for the purposes of developing a stable parenteral pharmaceutical, the principal formulation strategy is to stabilize the protein using clinically acceptable additives or through the use of suitable pharmaceutical-processing technologies. From an ease of use and processing perspective, a solution formulation is preferred and is typically the primary goal in developing a parenteral formulation. However, this approach may be limited by inadequate stability of a protein in solution under refrigerated or ambient storage conditions. Therefore, methods of dehydrating proteins, such as by lyophilization (also referred to as freeze-drying), to retard the degradative reactions that may occur in solution are frequently evaluated. Table 2 lists examples of formulations for currently approved biotechnology-based pharmaceuticals. The principles used in developing these types of products are described in the following.

A. Protein Stabilization in Solution Using Additives

As for other pharmaceuticals, the protein formulation program may begin with an assessment of the effect of pH and ionic strength on the stability and solubility of the protein. The potential influence of pH on the degradation process has been discussed in Section IV. Buffer selection for the formulation is guided by the pH range of interest and the suitability of the buffer for use in a medicinal product (GRAS status). The effect of different salts and their concentration on protein stability has also been reviewed in Section IV. In addition to affecting protein stability, salt concentration may have a profound effect on protein solubility and aggregation. Schein [104] provides a useful overview of the means of stabilizing proteins against aggregation and of methods to determine, predict, and increase solubility. Pearlman and Nguyen [33] report that, for human growth hormone, a sodium phosphate buffer concentration of about 5 mM produced less aggregation of the protein, as measured by light scattering, compared with solutions of 2.5, 10, and 20 mM. These authors further highlight the formulation development of tissue plasminogen activator. For this agent, the solubility of the protein at the optimally stable pH was insufficient for the therapeutic application. A positively charged amino acid, arginine, was included in the formulation to increase the protein's solubility at the desired pH range. Although ionic surfactants are often associated with denaturation of proteins [26], the nonionic surfactant polysorbate 80 has been included in several marketed formulations and serves to inhibit protein aggregation. The mechanism may be the greater tendency of the surfactant molecules to align themselves at the liquid–air interface, so excluding the protein from the interface and inhibiting surface denaturation. Other recent strategies used to manipulate protein solubility include the use of cyclodextrins. Simpkins [105] reports the use of 2-hydroxypropyl-β-cyclodextrin to solubilize ovine growth hormone and to prevent shaking-induced precipitation.

As a native, properly folded protein aggregates less than an unfolded, denatured one, solution additives that are known to stabilize the native proteins in solution may inhibit aggregation and enhance solubility. A diverse range of chemical additives are known to stabilize proteins in solution. These include salts, polyols, amino acids, and various polymers. Timasheff and his colleagues have provided an extensive examination of the effects of solvent additives on protein stability [106]. The unifying mechanism for protein stabilization by these cosolvents is related to their preferential exclusion from the protein surface. With the cosolvent preferentially excluded, the protein surface is preferentially hydrated, and key structural elements remain in their native conformation.

The mechanism of cosolvent exclusion may be divided into two classes. In the first class, interactions are determined by the properties of the solvent and, in the second class, interactions depend on the chemical nature of the protein surface. In the first class of cosolvents, steric exclusion of the larger cosolvent molecule, compared with the smaller water molecule, can account for the preferential hydration, as occurs with polyethylene glycol. Also in the first class of cosolvents, additives such as sugars, amino acids, and many salts, may act by increasing the surface tension of water, which leads to preferential hydration at the protein–solvent interface. For the second class of cosolvents, solvophobicity and repulsion from charges on the protein surface may account for the additives preferential exclusion. Here, the contact with the water–cosolvent mixture is thermodynamically less favorable than the protein–water contact, forcing the cosolvent away from the surface into the bulk solvent. Glycerol and other polyols, such as sorbitol and mannitol, belong to this category of stabilizers.

Despite these mechanisms of exclusion, other forces may serve to attract a particular cosolvent to the surface of the molecule. The net effect of the additive as a stabilizing or destabilizing agent, therefore, depends on the balance between the exclusion from the protein surface (stabilizing) and the propensity of the additive to bind to the protein (destabilizing) by electrostatic or hydrophobic interactions. Electrostatic interactions that promote binding of an additive may be highly influenced by the concentration of the additive, as for many salts. Table 3 [106] summarizes these interactions for various additives. For additives in class I, preferential exclusion predominates. For class II additives, the stabilization may depend on protein charge and concentration and charge. Class III cosolvents predominantly bind to the unfolded denatured form of the protein through interaction with hydrophobic groups. As many of the additives described in the work of Timasheff [106] are also clinically safe for inclusion in a parenteral formulation, these concepts provide a rationale basis for excipient selection during formulation development.

Globular proteins are known to act as polymeric stabilizers of protein structure in solution. Wang and Hanson [29] review the mechanisms of protein stabilization by serum albumin, and it has been included in marketed protein pharmaceuticals, as shown in Table 3. The ability of albumin to adsorb at surfaces, and so inhibit adsorption of a low concentration of the therapeutic protein, provides one rationale for the inclusion of albumin in the formulation. However, there are a number of compelling reasons not to include human serum albumin in a formulation [101,107]. The addition of a protein excipient decreases the specific activity and may confound analytical methods aimed at evaluating the therapeutic protein. Human serum albumin is a product derived from human blood and so carries with it the potential to introduce protein impurities or pathogens. Its inclusion in a formulation, therefore, counteracts one of the primary advantages of the biotechnology-based pharmaceuticals: purity.

B. Protein Stabilization in the Dried Solid State

In situations for which the strategies described in the foregoing fail to produce solution formulations that are chemically and physically stable for 1 year or more, retarding degradation by using low–temperature-drying processes is often required. Lyophilization and spray-drying are two such processes, with a long history of use in the pharmaceutical and food industries for stabilization of otherwise easily degraded substances. Both technologies may be used to dehydrate heat-sensitive molecules and, thereby, inhibit the degradative reactions that may be observed when proteins are formulated in solution. Lyophilization is currently the more common technique; however, there is increasing interest in the use of spray-drying, owing to the spherical physical nature of the spray-dried powder and its potential usefulness in protein drug delivery. A comprehensive discussion of the lyophilization process is provided in Chapter 13 of this volume, and so only a brief account of considerations relevant to protein lyophilization is given in the following. Freeze-drying and formulation of biological products was also the

Table 2 Composition of FDA-Approved Biotechnology-Based Pharmaceuticals

Product	Dosage form	Storage	Composition
Actimmune (interferon gamma-1b)	Solution for IV or SC injection	2–8°C Stable up to 12 hr at room temperature. Do not freeze or shake.	Each 0.5 ml; 100 µg interferon gamma-1b, 20 mg mannitol, 0.36 mg sodium succinate, 0.5 mg polysorbate 20 in WFI
Activase (tissue plasminogen activator)	Lyophilized product with WFI as diluent	Stable up to 8 hr at room temperature after reconstitution. Light sensitive	20 mg vial: 0.7-g L-Arg, 0.2-g H$_3$PO$_4$, <1.6 mg polysorbate 80 50 mg vial: 1.7-g L-Arg, 0.5-g H$_3$PO$_4$, <4 mg polysorbate 80
Alferon N (interferon alfa-n3)	Solution for intralesional injection	2–8°C Do not freeze or shake.	Each ml contains: 5 million IU interferon alfa-n3 8.0 mg NaCl, 1.74 mg Na$_2$HPO$_4$, 0.20 KH$_2$PO$_4$, 0.20 mg KCl 3.3 mg phenol as preservative, 1 mg human albumin as stabilizer
Engerix B (hepatitis B vaccine)	Suspension for IM injection in single-dose vials	2–8°C Do not freeze.	Each ml: 20 µg hepatitis B surface antigen adsorbed on 0.5 mg Al as Al(OH)$_3$, 1:20,000 thimerosal (preservative), 9 mg NaCl, 0.98 mg disodium potassium dihydrate, 0.71 mg dihydrogen phosphate dihydrate
Epogen (erythropoietin)	Solution for IV or SC injection	2–8°C	Each vial: 2,000–10,000 IU protein, 2.5 mg HSA, 5.8 mg sodium citrate, 5.8 mg NaCl, 0.06 mg citric acid in WFI
PROCRIT (erythropoietin)	Solution for IV or SC injection in single-use vials	2–8°C Do not freeze or shake.	Each vial (1 ml): 2,000–10,000 units epoetin alfa in pH 6.9 ± 0.3 isotonic citrate buffer, 2.5 mg HSA, 5.8 mg sodium citrate, 5.8 mg NaCl, 0.06 mg citric acid in WFI
Humatrope (human growth hormone)	Lyophilized product with diluent (0.3% m-cresol and 1.7% glycerin)	Stable for 14 days at 2–8°C after reconstitution	Each vial: 5 mg protein with 1.13 mg Na$_2$HPO$_4$, 25 mg glycine, 25 mg mannitol
Humulin (insulin)	Solution or suspension	Stable within expiration date, preferably under refrigeration.	Each vial—with or without Na$_2$HPO$_4$ as the buffer

Product	Dosage form	Storage/stability	Composition
Intron A (interferon alfa-2b)	Lyophilized product with bacteriostatic WFI as diluent	Stable for 1 mo at 2–8°C after reconstitution	Each vial: 5 mg protein with 9 mg Na_2HPO_4, 2.25 mg NaH_2PO_4 (pH 7), 43 mg NaCl, 1.0 mg Tween 80
Leukine (sargramostim GM-CSF)	Lyophilized powder for infusion following reconstitution with 1 mg WFI, USP	2–8°C. Do not freeze or shake. Use within 6 hr of reconstitution/dilution for IV infusion (as is unpreserved).	Each vial: 250 or 500 μg sargramostim, 40 mg mannitol, 10 mg sucrose, 1.2 mg tromethamine, pH 7.4 ± 0.3
Prokine (sargramostim)	Lyophilized powder for infusion following reconstitution with 1 mg WFI, USP	2–8°C. Do not freeze or shake. Use within 6 hr of reconstitution/dilution for IV infusion (as is unpreserved).	Each vial: 500 μg sargramostim, 40 mg mannitol, 10 mg sucrose, 1.2 mg tromethamine, pH 7.4 ± 0.3
Neupogen (filgrastim)	Solution for IV or SC injection in single-use vial	2–8°C. Do not freeze or shake. Do not leave at room temperature for longer than 6 hr	Each ml: 300 μg filgrastim, 0.59 mg acetate, 50.0 mg mannitol, 0.004% Tween 80, 0.035 mg sodium, WFI qs 1 ml, pH 4.0
Orthoclone OKT3 (murine Monoab-CD3)	Solution	2–8°C	Each 5 ml: 0.015–0.24 mg protein, 20 mg glycine, 2.3 mg Na_2HPO_4, 0.55 mg NaH_2PO_4, 1.0 mg HSA
Protropin (human growth hormone)	Lyophilized product with bacteriostatic WFI as diluent	Stable for 14 days at 2–8°C after reconstitution	5-mg vial: 1.6 mg Na_2HPO_4, 0.1 mg NaH_2PO_4 (pH 7), 40 mg mannitol; 10-mg vial: 3.2 mg Na_2HPO_4, 0.2 mg NaH_2PO_4 (pH 7), 80 mg mannitol
Recombivax HB (hepatitis B vaccine)	Suspension	2–8°C	Each ml: 10 μg hepatitis B surface antigen adsorbed onto 0.5 mg Al provided as $Al(OH)_3$
Roferon (interferon alfa-2a)	Solution (3, 6, 36 million IU) or lyophile (18 million IU) per vial	Stable for 1 month at 2–8°C after reconstitution	Each ml contains: 3, 6, 36 million IU protein with: 9 mg NaCl, 5 mg HSA and 3 mg phenol

WFI, water for injection; HSA, human serum albumin.
Source: Ref. 159.

Table 3 Classification of Cosolvent Interactions with Proteins

Cosolvent	Mechanism of exclusion	Mechanism of binding	Expected activity
Class I			
Sugars (sucrose, glucose, mannose)	Surface tension increase	Inert	Good stabilizers of globular proteins and assembled organelles
Some amino acids (glycine, alanine, glutamic, and aspartic acids)	Surface tension increase	Weak binding	Stabilizers of globular proteins
Salting-out salts (Na_2SO_4, NaCl, $MgSO_4$)	Surface tension increase	Weak binding	Good stabilizers of globular proteins and precipitants of native and denatured proteins
Glycerol, polyols (sorbitol, mannitol)	Solvophobicity	Affinity for polar regions	Stabilizers of globular proteins and assembled organelles, decreasing for proteins of high polarity
Class II			
Weakly interacting salts ($MgCl_2$, NaCl, Gn_2SO_4)	Surface tension increase	To charged groups or peptide bonds	Stabilization dependent on protein charge and salt concentrations
Arginine-HCl, lysine-HCl	Surface tension increase	To peptide bonds and negative charges	Stabilization dependent on protein charge and amino acid concentration
Valine (possibly other nonpolar amino acids)	Surface tension increase	To hydrophobic regions	Weak stabilization
Class III			
PEG	Steric exclusion	To hydrophobic regions	Good precipitants; stabilizers of native structure at low temperature, of unfolded structure at high temperature; stabilizers and solubilizers of hydrophobic patches or domains in proteins
MPD	Repulsion from charges	To hydrophobic regions	

Source: Ref. 106.

topic of a recent symposium organized by the Center for Biologics Evaluation and Research of the U. S. Food and Drug Administration. The published proceedings of this symposium provide a wealth of information on this subject, as applied to biotechnology-based pharmaceuticals [108].

Lyophilization and Protein Formulation Development

Lyophilized protein formulation development is characterized by two primary concerns: One focus is excipient selection to optimize long-term stability of the protein in the dried state. Simultaneously, the effect of the excipient on the processing parameters of the lyophilization cycle, such as maximum and minimum drying temperatures and cycle length, and on the appearance of the lyophilized powder cake need to be considered.

Relative to additive selection for protein stabilization, it must be recognized that, although lyophilization is intended to stabilize the protein in a dried state, the freezing and drying processes to which a protein is subjected during lyophilization may actually be a source of inactivation. The phenomenon of freeze denaturation of proteins has been widely described in the literature [109–113]. The denaturing effect of freezing is related to the gradual concentration of solutes surrounding the protein molecule as water is removed by the phase change from a liquid to solid ice crystals. The high concentration of salts, potential shifts in pH because of crystallization of buffer components, or of temperature sensitivity of the buffer pK_a, and the limiting solubility of the protein itself as it concentrates, can lead to protein inactivation by one or several of the mechanisms described earlier. In selecting a buffer for a lyophilized formulation, the effect of temperature on the pH and the buffer solubility needs to be considered. Tris buffer, for example, exhibits a 0.028 unit shift in pK_a per degree centigrade [114]. The greater solubility of potassium compared with sodium phosphate buffer salts may also lead to greater protection by the former buffer from the denaturing effects of pH changes on freezing [27]. The recovery of protein activity after freezing can be influenced by the addition of certain protective additives, including sugars, polyols, amino acids, and certain salts [112,115,116]. The nature of these cryoprotective compounds and their mechanism of action have been examined by Carpenter and colleagues [116,117]. The types of compounds that provide cryoprotection are essentially those that have been shown by Timasheff [106] to be preferentially excluded from the protein–solute interface in aqueous solution [117,118]. By analogy, therefore, it is proposed that cryoprotective agents act by preferential exclusion from the protein surface, making it thermodynamically unfavorable for the protein to unfold.

In contrast with the wide range of compounds that stabilize proteins in solution and on freezing, Carpenter et al. [119–121] have shown that only certain carbohydrates (e.g., disaccharides) can preserve phosphofructokinase activity during either freeze-drying or air-drying. This suggests a fundamental difference in the stabilizing mechanism between the dried state and the solution or frozen state [122]. The mechanism suggested for stabilization of dried proteins is the binding of the additive to the dried protein to act as a ''water-substitute'' after removal of the hydration shell. As water-substitutes, sugars, such as trehalose and lactose, can serve to partially satisfy hydrogen-bonding requirements of polar groups in dried proteins [117]. In support of this theory, Fourier transform IR spectroscopy was used to demonstrate that hydrogen bonding occurs between proteins and stabilizing carbohydrates and that solute binding is required for preservation of labile proteins during drying [121].

Often the unit-dose amount of protein is low; thus, each vial may contain very low amounts of total solid. After the lyophilization process and the removal of water, the vial may appear to contain very little product and no lyophilized plug will form. Therefore, bulking agents are used to enhance the appearance of the lyophilized product. Mannitol is a widely used bulking agent that produces an elegant lyophilized cake. It has the further advantages that it may also exert a cryoprotective effect, and its concentration in a formulation can be adjusted to achieve an isoosmotic solution on reconstitution. Other polyhydric alcohols, such as sorbitol, and sugars, such as sucrose, dextrose, and dextrans, are also used as bulking agents.

The thermal behavior of the formulation may be affected by the solution components which, in turn, may influence the processing parameters used during lyophilization. Accordingly, the effect of additives on the thermal behavior must be considered to avoid the need for excessively long process-drying times, or freezing temperatures that cannot be achieved using typical production-scale lyophilizers. The thermal properties of importance for lyophilization cycle development include the freezing temperature of the formulation, the temperature at which the frozen formulation will melt or collapse, and the temperature above which the product will rapidly degrade. Differential-scanning calorimetry is a particularly useful technique for characterization and optimization of the freeze-drying processes and formulations. The application of this technique in lyophilization cycle and formulation development has been described by several authors [96,123–127].

The moisture level and physical character of the lyophilized cake (amorphous or crystalline) compatible with long-term stability also need to be considered during the development of the lyophilization cycle. The formation of a crystalline, rather than an amorphous, cake is affected by temperature cycling during lyophilization and the nature of the excipients in the formulation [125,128,129]. The final moisture content of an amorphous, lyophilized cake is generally higher owing to reduced transport of water vapor from the cake during drying. Increasing moisture reduces the physical stability of the lyophilized cake and may lead to collapse of the cake with storage [96,127]. In addition to affecting the physical stability of the lyophilized product, the moisture level may affect the physicochemical stability of the protein itself [130]. The transfer of water vapor from the stopper to the product and its adverse effect on product stability has recently been described [131,132].

Spray-Drying of Protein Pharmaceuticals

Spray-drying provides an alternative to freeze-drying proteins as a process capable of drying thermally labile materials. It has been used extensively in processing foods. The application of spray-drying in the pharmaceutical industry was the topic of a recent review article [133]. In the spray-drying process, a liquid-feed stream is first atomized for maximal air spray contact. The particles are then dried in the air stream in seconds, owing to the high surface area contact with the drying gas.

The major advantage of spray-drying, compared with lyophilization, is the particle size and shape of the final dried powder. Spray-drying can produce spherical particles that have good flow properties, and the process can be adapted to produce particles of a range of sizes, dependent on the application. These process features may be particularly useful in the development of novel protein formulations for controlled-release or for noninvasive routes of drug delivery. A recent European patent application [134] describes the use of spray-drying and spray-cooling to produce sustained-release microspheres of biologically active proteins for parenteral administration. Bovine growth hormone was prepared in the size range suitable for incorporation into fat and wax microspheres by spray-drying. Sustained-release microspheres of bovine growth hormone prepared in this manner were used to increase weight gain and milk production in dairy cattle. The spray-drying process also has clear potential as a means of providing protein particles for inhalation. Vidgren et al. have published a series of papers comparing spray-dried and mechanically micronized cromolyn sodium (sodium cromoglycate) [135–137]. These studies demonstrate the use of spray-drying to generate particles of an appropriate size, which were shown with an in vitro model, to be suitable for delivery to the therapeutically important alveolar region of the lung.

As described earlier, disaccharide molecules protect proteins during drying [120]; hence, they may be effective stabilizers during the spray-drying process. Labrude et al. [138] found that sucrose had a protective effect on both the spray-drying and lyophilization of oxyhemoglobin. Spray-drying is often limited by the high product losses experienced when operating on a small scale and the difficulties in achieving low-moisture levels compatible with long-term physical and chemical stability. However, as reported recently by Broadhead et al., careful control of the process variables and the selection of suitable stabilizing additives can lead to optimized yields of fully active protein particles of 3–5 μm and with moisture levels of 5–6% [139]. Clearly, with the heightened interest and requirement for novel protein formulations and alternative drug delivery systems, spray-drying and other similar low-temperature–processing technologies are likely to receive greater attention.

VII. REGULATORY ASPECTS OF BIOTECHNOLOGY-BASED PHARMACEUTICALS

Although the regulatory climate is evolving in concert with the developments in biotechnology, at the time of this writing several general guidelines can be given about regulatory considerations for biotechnology medicines. Many, although not all, of these products are classified as "biologics." As such, these products require licensing under the Public Health Service Act (Sec. 351) and must comply with the regulations set forth in the Code of Federal Regulations, Title 21, Parts 600–680. Regulatory control and review of biotechnology-derived biologics, except for those considered to be medical devices, is administered by the Center for Biologics Evaluation and Research (CBER) of the U.S. Food and Drug Administration (FDA). Regulatory product approval for biologics is based on the submission of a product license application (PLA) to CBER, in contrast with a New Drug Application (NDA) submitted for products to be reviewed by the Center for Drugs Evaluation and Research (CDER).

Biologics licenses issued by CBER include approval of a specific series of production steps and in-process control tests, as well as end-product specifications that must be met on a lot-by-lot basis. Each lot of a licensed biologic is approved for distribution when it has been determined that the lot meets the specific control requirements set forth in the product license. Therefore, samples of each production lot, as well as documentation, are submitted to the FDA to show that all the applicable tests have been performed and the results of these tests. The FDA will review these data and may select certain tests that are repeated at the FDA laboratories: they often repeat the test for potency. The preferred tests involve comparison of the product with a potency standard. The standard may be obtained from the manufacturer or prepared by the FDA. The reference standards for biological products are currently held by CBER and are generally supplied only to the manufacturer of the licensed product.

Two forms of license are required for the manufacture of biologics. First, an establishment license is required. An establishment license is issued only after inspection of the establishment by an inspector from the FDA has determined that the establishment complies with applicable standards prescribed in the regulations in Subchapter F, Title 21, CFR 600.10–600.15. A product license is required for every biologic manufactured at the licensed facility.

To provide guidance to manufacturers of biologics, the FDA has developed a series of *Points to Consider* . . . documents that are designed to convey the current consensus of CBER relating to biologic product development and testing. These documents are not guidelines that carry the force of law, but provide details of criteria that CBER expects producers of biologics to consider in their product development and license applications. To respond to developments in this rapidly changing field, the FDA expects these documents to remain as dated ''drafts'' that will be updated as needed using input from industry, academia, and other regulatory agencies both from within the United States and abroad. The relevant *Points to Consider Documents* . . . are available from CBER (Contact the Congressional, Consumer and International Affairs Staff, HFB-142 Suite 109, Metro Park North III, 5600 Fishers Lane, Rockville, MD, 20857, USA.) and include: *Points to Consider in the Characterization of Cell Lines Used to Produce Biologicals* (1992), *Points to Consider in the Production and Testing of New Drugs and Biologicals Produced by Recombinant DNA Technology* (1985), and *Points to Consider in the Manufacture and Testing of Monoclonal Antibodies for Human Use* (1992) [140].

As stated earlier, not all of the products of biotechnology are regulated by CBER. A recent publication, which considers the development of public standards for licensed biologics by the *United States Pharmacopeia*, also provides a useful outline of the types of products that will be assigned to either CBER or CDER [140].

VIII. DEVELOPMENTS IN PROTEIN DRUG DELIVERY

As pharmaceutical scientists gain experience and tackle the primary challenges of developing stable parenteral formulations of proteins, the horizons continue to expand and novel delivery systems and alternative routes of administration are being sought. The interest in protein drug delivery is reflected by the wealth of literature that covers this topic [141–152]. The interested reader is referred to these articles for more extensive information. Some of the more promising recent developments are summarized in the following.

Clinical experience with therapeutic proteins has highlighted the need for extending the plasma half-life of many high-clearance proteins. In other cases, a pulsatile, non–zero-order delivery system may be necessary [153]. Therefore, approaches to alter the pharmacokinetic profile of proteins, have been explored. Plasma half-life extension may be obtained through chemically modifying the molecule to inhibit its pharmacological clearance, or by controlling the rate at which the protein is delivered to the bloodstream. Of the chemical-modifying technologies explored to date, polyethylene glycol (PEG) modification appears to be the most promising [148]. The PEG modification increases the plasma half-life of protein molecules through various mechanisms. The PEG-conjugated proteins may be too large for glomerular filtration, or may sterically hinder the protein's interaction with cellular receptors required for metabolism and excretion. This approach has been adopted for adenosine deaminase (ADA): PEG–ADA, for use in ADA-deficient severe combined immunodeficiency syndrome, is the first such product to gain FDA approval.

To achieve controlled protein delivery, the use of biodegradable microspheres is being actively pursued [149–151,154,155]. In particular, biodegradable lactic–glycolic acid copolymer-based microspheres have proved useful for the controlled delivery of several polypeptides and proteins. The first FDA-approved system for controlled release of a peptide was an injectable poly(lactide–coglycolide) microsphere formulation of leuprolide acetate. This formulation provides controlled release of the peptide over 30 days for the treatment of prostate cancer. Although promising, many problems remain, and the more general use of formulations of this type for other biopharmaceuticals is often limited by the instability of the molecule within the acid environment created in vivo as the polmer degrades [144].

The delivery of protein drugs by a noninvasive route is certainly a prominent goal within the pharmaceutical industry. Such an achievement would greatly increase patient compliance and expand the usefulness and market for these agents. However, despite the intense desire, the challenges are formidable and, perhaps, reflected by the fact that, although insulin has been used clinically for more than 70 years, it is still given exclusively by daily injections. Possible noninvasive routes for delivery of proteins include nasal, buccal, rectal, vaginal, transdermal, ocular, oral, and pulmonary. For each route of delivery, there are two potential bariers to absorption: permeability and enzymatic barriers. All of the potential routes have received some attention; however, the pulmonary and nasal routes appear to hold the greatest promise. The oral route would be by far the most popular route, yet, despite extensive investigation, current strategies to prevent degradation and poor absorption in the gastrointestinal tract have proved of limited value for macromolecular proteins [141,152].

The nasal route possesses higher permeability and presents less of an enzymatic barrier than does the oral route. The nasal route has been successful for a number of polypeptide drugs. Nasal formulations for luteinizing human-releasing hormone (LH-RH) analogues—desmopressin, oxytocin, and calcitonin—have reached the marketplace. Notably, however, this route of delivery has not been as successful for larger proteins with M_r greater than 10 kDa and may be associated with local irritation and toxicity with long-term administration [144,156]. Illum and Davis [142] describe recent approaches used to enhance the nasal delivery of insulin through the use of absorption enhancers or bioadhesive microspheres.

The pulmonary route of protein drug delivery has recently received increased attention. Three therapeutic peptides (leuprolide: nine amino acids; insulin: 51 amino acids; and growth hormone: 192 amino acids) were reported to be absorbed in biologically active form from the lungs, with bioavailabilities of 10–25% [143]. These values exceed those reported for nasal delivery of insulin and growth hormone, in the absence of permeation enhancers. Current challenges in pulmonary protein delivery include assessment of the safety of long-term administration, the molecular size limitation of pulmonary absorption and strategies for enhancing permeation, and formulation approaches capable of delivering suitable doses of stable proteins to the vast absorptive surface presented by the lung.

ACKNOWLEDGMENT

I gratefully acknowledge the invaluable assistance of Ann C. Dailey in the preparation of this chapter.

REFERENCES

1. J. Sterling, The next decade of biotechnology—where are we going? J. Parenter. Sci. Technol. 44: 63–66 (1990).
2. Pharmaceutical Manufacturer's Association, Products in the pipeline, Biotechnology, 9, 947–949 (1991).
3. W. Sadee, Protein drugs: A revolution in therapy? Pharm. Res., 3, 3–6 (1986).
4. W. J. Black, Drug products of recombinant DNA technology, Am. J. Hosp. Pharm. 46, 1834–1844 (1989).
5. L. P. Gage, Biopharmaceuticals: Drugs of the future, Am. J. Pharm. Educ., 50, 368–370 (1986).
6. F. G. Banting and C. H. Best, Pancreatic extracts, J. Lab. Clin. Med., 7, 464–472 (1922).
7. J. D. Baxter, Recombinant DNA and medical progress, Hosp. Pract., 15, 57–67 (1980).
8. P. Brown, C. C. Gajdusek, C. J. Gibbs, and D. N. Asher, Potential epidemic of Creutzfeldt-Jakob disease from human growth hormone therapy, N. Engl. J. Med., 313, 728 (1985).
9. J. D. Watson, J. Tooze, and D. T. Kurtz, *Recombinant DNA: A Short Course*, W. H. Freeman and Co., New York (1983).
10. R. L. Rodriguez and R. C. Tait, *Recombinant DNA Techniques: An Introduction*, Addison-Wesley Publishing, Reading, MA, 1983.
11. G. Kohler and C. Milstein, Continuous cultures of fused cells secreting antibody of predefined specificity, Nature, 256, 495–497 (1975).
12. J. W. Goding, *Monoclonal Antibodies: Principal and Practice*, Academic Press, Orlando, FL, 1983.

13. S. L. Morrison and V. T. Oi, Genetically engineered antibody molecules, Adv. Immunol., 44 65–92 (1989).

14. P. T. Jones, P. H. Dear, J. Foote, M. S. Neuberger, and G. Winter, Replacing the complementarity determining regions in a human antibody with those from a mouse, Nature, 321, 1534–1536 (1986).

15. C. Queen, W. P. Schneider, H. E. Selick, P. W. Payne, N. F. Landolphi, J. F. Duncan, N. M. Avdalovic, M. Levitt, R. P. Junghans, and T. A. Waldman, A humanized antibody that binds to the interleukin 2 receptor, Proc. Natl. Acad. Sci. USA, 86, 10029–10033 (1989).

16. L. Reichmann, L. M. Clarke, H. Waldman, and G. Winter, Reshaping human antibodies for therapy, Nature, 332, 323–327 (1988).

17. R. Newman, J. Alberts, D. Anderson, K. Carner, C. Heard, F. Norton, R. Raab, M. Reff, S. Shuey, and N. Hanna, Primatization of recombinant antibodies for immunotherapy of human diseases: A macaque/human chimeric antibody against human CD4, Biotechnology, 10, 1455–1460 (1992).

18. R. E. Hawkins, B. L. Meirion, and S. J. Russell, Adapting antibodies for clinical use, Br. Med. J., 305, 1348–1352 (1992).

19. R. A. Lerner, A. S. Kang, J. D. Bain, D. R. Burton, and C. F. Barbas, Antibodies without immunization, Science, 258, 1313–1314 (1992).

20. L. Stryer, *Biochemistry*, W. H. Freeman, San Francisco, 1981.

21. F. Franks, *Characterization of Proteins*, Humana Press, Clifton, NJ, 1988.

22. R. D. Schmid, Stabilized soluble enzymes, Adv. Biochem. Eng., 12, 42–118 (1979).

23. R. B. Parekh and T. P. Patel, Comparing the glycosylation patterns of recombinant glycoproteins, *TIBTech*, 10 (Aug.), 276–280 (1992).

24. K. Kaushansky, Structure–function relationships of the hematopoietic growth factors, Proteins Struct. Funct. Genet. 12, 1–9 (1992).

25. R. J. Leatherbarrow, T. W. Rademacher, R. A. Dwek, J. M. Woof, A. Clark, R. Burton, N. Richardson, and A. Feinstein, Effector functions of a monoclonal aglycosylated mouse IgG2a: Binding and activation of complement component C1 and interaction with human monocyte Fc receptor, Mol. Immunol., 22, 407–415 (1985).

26. D. B. Volkin and A. M. Klibanov, Minimizing protein inactivation, in *Protein Function: A Practical Approach* (T. E. Creighton, ed.), IRL Press, Oxford, 1989, pp. 1–24.

27. T. Chen, Formulation concerns of protein drugs, Drug Dev. Ind. Pharm., 18, 1311–1354 (1992).

28. M. C. Manning, K. Patel, and R. T. Borchardt, Stability of protein pharmaceuticals, Pharm. Res., 6, 903–913 (1989).

29. Y. J. Wang and M. A. Hanson, Parenteral formulations of proteins and peptides: Stability and stabilizers, J. Parenter. Sci. Technol., 42 (Suppl. 2), S2–S26 (1988).

30. T. J. Ahern and M. C. Manning (eds.), *Stability of Protein Pharmaceuticals Part A: Chemical and Physical Pathways of Protein Stabilization*, Plenum Press, New York, 1992.

31. T. J. Ahern and A. M. Klibanov, The mechanism of irreversible enzyme inactivation at 100°C, Science, 288, 1280–1284 (1985).

32. L. C. Gu, E. A. Erdos, H. Chiang, T. Calderwood, K. Tsai, G. C. Visor, J. Duffy, W. C. Hsu, and L. C. Foster, Stability of interleukin 1β (IL-1β) in aqueous solution: Analytical methods, kinetics, products and solution formulation implications, Pharm. Res. 8, 485–490 (1992).

33. R. Perlman and T. H. Nguyen, Formulation strategies for recombinant proteins: Human growth hormone and tissue plasminogen activator, in *Therapeutic Peptides and Proteins: Formulation, Delivery and Targeting* (D. Marshak and D. Liu, eds.), Cold Spring Harbor Laboratory, Cold Spring Harbor, NY, 1989, pp. 23–31.

34. T. Geiger and S. Clarke, Deamidation, isomerization and racemization at asparaginyl and aspartyl residues in peptides, J. Biol. Chem., 262, 785–794 (1987).

35. S. J. Wearne and T. E. Creighton, Effect of protein conformation on rate of deamidation: Ribonuclease A, Proteins Struct. Funct. Genet., 5, 8–12 (1989).

36. R. Tyler-Cross and V. Schirch, Effects of amino acid sequence, buffers and ionic strength on the rate and mechanism of deamidation of asparagine residues in small peptides, J. Biol. Chem., 266, 22549–22556 (1991).

37. A. S. Inglis, Cleavage at aspartic acid, Methods Enzymol., 91, 324–332 (1983).

38. M. J. Pikal, K. M. Dellerman, M. L. Roy, and R. M. Riggin, The effects of formulation variables on the stability of freeze-dried human growth hormone, Pharm. Res., 8, 427–436 (1991).

39. J. L. Bada, In vivo racemization in mammalian proteins, Methods Enzymol., 106, 98–115 (1984).

40. S. E. Zale and A. M. Klibanov, Why does ribonuclease irreversibly inactivate at high temperatures? Biochemistry, 25, 5432–5444 (1986).

41. T. E. Creighton, Disulphide bonds and protein stability, *Bioessays*, 8, 57–63 (1988).

42. D. B. Volkin and A. M. Klibanov, Thermal destruction processes in proteins involving cystine residues, J. Biol. Chem., 262, 2945–2950 (1987).

43. J. Brange, S. Havelund, and P. Hougaard, Chemical stability of insulin 2: Formation of higher molecular weight transformation products during storage of pharmaceutical preparations, Pharm. Res., 9, 727–734 (1992).

44. G. A. Truskey, R. Gabler, A. Dileo, and T. Manter, The effect of membrane filtration upon protein conformation, J. Parenter. Sci. Technol., 41, 180–193 (1987).

45. S. E. Charm and B. L. Wong, Shear effects on enzymes, Enzyme Microbiol. Technol., 3, 111 (1981).

46. K. Heremans, High pressure effects on proteins and other biomolecules, Annu. Rev. Biophys. Bioeng., 11:1 (1982).

47. R. Buchet, D. Carrier, P. T. T. Wond, I. Jona, and A. Martonosi, Pressure effects on sarcoplasmic reticulum: A Fourier transform infrared spectroscopy study, Biochim. Biophys. Acta, 1023, 107 (1990).

48. D. N. Brems, Solubility of different folding conformers of bovine growth hormone, Biochemistry, 27, 44541–44546 (1988).

49. E. Zerovnik, R. Jerala, L. Kroon-Zitko, R. H. Pain, and V. Turk., Intermediates in denaturation of a small globular protein, recombinant human Stefin B, J. Biol. Chem., 13, 9041–9046 (1992).

50. M. G. Mulkerrin and R. Wetzel, pH dependence of the reversible and irreversible thermal denaturation of gamma interferons, Biochemistry, 28, 6556–6561 (1989).

51. T. Cartwright, O. Senussi, and M. D. Grady, The mechanism of the inactivation of human fibroblast interferon by mechanical stress, J. Gen. Virol., 36, 317–321 (1977).

52. J. J. Sedmark and S. E. Grossberg, Stabilization of interferons, Tex. Rep. Biol. Med., 35, 198–204 (1977).

53. R. Quinn and J. D. Andrade, Minimizing the aggregation of neutral insulin solutions, J. Pharm. Sci., 72, 1472–1473 (1983).

54. J. Brange, S. Havelund, P. E. Hansen, L. Langkjaer, E. Sorensen, and P. Hildebrandt, Formulation of physically stable neutral insulin solutions for continuous infusion by delivery systems, in *Hormone Drugs*, U.S. Pharmacopeial Convention, Rockville, MD, 1982, pp. 96–105.

55. J. R. Brennan, S. P. Gebhart, and W. G. Blackard, Pump-induced insulin aggregation: A problem with the biostator, Diabetes, 34, 353–359 (1985).

56. W. D. Lougheed, A. M. Albisser, H. M. Martindale, J. C. Chow, and J. R. Clement, Physical stability of insulin formulations, Diabetes, 32, 424–432 (1983).

57. J. D. Andrade and V. Hlady, Protein adsorption and materials biocompatibility: A tutorial review and suggested hypotheses, Adv. Polymer Sci., 79, 1–63 (1986).

58. A. M. Pitt, The nonspecific protein binding of polymeric microporous membranes, J. Parenter. Sci. Technol., 41, 110–113 (1987).

59. R. Pearlman and T. H. Nguyen, Analysis of protein drugs, in *Peptide and Protein Drug Delivery* (Vincent H. Lee, ed.), Marcel Dekker, New York, 1990, pp. 247–301.

60. B. S. Welinder and F. H. Andresen, Characterization of insulin and insulin-like drugs by high performance liquid chromotography, in *Hormone Drugs*, U.S. Pharmacopeial Convention, Rockville, MD, 1982, pp. 163–177.

61. E. P. Kroeff and R. E. Chance, Application of high performance liquid chromatography for the analysis of insulins, in *Hormone Drugs*, U.S. Pharmacopeial Convention, Rockville, MD, 1982, pp. 148–162.

62. J. Geigert, B. M. Panschar, S. Fong, H. N. Huston, D. E. Wong, D. Y. Wong, C. Tafaro, and M. Pemberton, The long-term stability of recombinant (Serine-17) human interferon-β, J. Interferon Res., 8, 539–547 (1988).

63. K. Benedek, B. Hughes, M. B. Seaman, and J. K. Swadesh, Reversed-phase liquid chromatography and sodium dodecyl sulfate polyacrylamide gel electrophoresis characteristics of recombinant derived malaria antigen, J. Chromatogr., 444, 191–202 (1988).

64. M. Kunitani, P. Hirtzer, D. Johnson, R. Halenbeck, A. Boosman, and K. Koths, Reversed-phase chromotography of interleukin-2 muteins, J. Chromotogr., 359, 391–402 (1986).

65. R. M. Riggin, G. K. Dorulla, and D. J. Miner, A reversed-phase high performance liquid chromatographic method for characterization of biosynthetic human growth hormone, Anal. Biochem., 167, 199–209 (1987).

66. P. T. Wingfield and P. Graber, Chromatofocusing of N-terminally processed forms of proteins— isolation and characterization of two forms of interleukin-1β and of bovine growth hormone, J. Chromotogr. 387, 291–300 (1987).

67. E. Watson, and W. C. Kenney, High performance size exclusion chromotography of recombinant derived proteins and aggregated species, J. Chromotogr., 436, 289–298 (1988).

68. C. R. Cantor and S. N. Timasheff, Optical spectroscopy of proteins, in The Proteins, Vol. 5, 3rd Ed. (H. Neurath, ed.), Academic Press, New York, 1982, pp. 145–306.

69. H. A. Havel, R. S. Chao, R. J. Haskell, and T. J. Thamann, Investigation of protein structure with optical spectroscopy: Bovine growth hormone, Anal. Chem., 61, 642–650 (1989).

70. F. X. Schmid, Spectral methods of characterizing protein conformation and conformational changes, in Protein Structure—A Practical Approach (T. E. Creighton, ed.), IRL Press, Oxford, 1989, pp. 251–285.

71. M. Bier, Light scattering measurements, Methods Enzymol., 4, 147–166 (1957).

72. D. N. Brems, S. M. Plaisted, E. W. Kauffman, M. Lund, and S. R. Lehrman, Helical formation in isolated fragments of bovine growth hormone, Biochemistry, 26, 7774–7778 (1987).

73. H. Sonntag and K. Strenge, The surface coagulation of proteins during shaking, J. Colloid Interface Sci., 32, 162–165 (1970).

74. J. P. Hennessey and W. C. Johnson, Information content in the circular dichroism of proteins, Biochemistry, 20, 1085–1094 (1981).

75. W. C. Johnson, Circular dichroism and its empirical application to biopolymers, Methods Biochem. Anal., 31, 61–163 (1985).

76. W. C. Johnson, Secondary structure of proteins through circular dichroism spectroscopy, Annu. Rev. Biophys. Chem., 17, 145–166 (1988).

77. M. C. Manning, Underlying assumptions in the estimation of the secondary structure content in protein by circular dichroism spectroscopy—a critical review, J. Pharm. Biomed. Anal., 7, 1103–1119 (1989).

78. Y. P. See and G. Jackowski, Estimating molecular weights of polypeptides by SDS gel electrophoresis, in Protein Structure—A Practical Approach (T. Creighton, ed.), IRL Press, Oxford, 1989, pp. 1–21.

79. D. P. Goldenberg, Analysis of protein conformation by gel electrophoresis, in Protein Structure— A Practical Approach (T. Creighton, ed.), IRL Press, Oxford, 1989, pp. 225–250.

80. U. K. Laemmli, Cleavage of structural proteins during the assembly of the head of bacteriophage T4, Nature, 227, 680–685 (1970).

81. P. G. Righetti, Isoelectric focusing of proteins in conventional and immobilized pH gradients, in Protein Structure—A Practical Approach (T. Creighton, ed.), IRL Press, Oxford, 1989, pp. 23–63.

82. D. R. Hoffman, Studies of the structure and synthesis of immunoglobulins by isoelectric focusing, in Biological and Biomedical Application of Isoelectric Focusing (N. Catsinpoolas and J. Drysdale, eds.), Plenum Press, New York, 1977, pp. 121–153.

83. N. A. Guzman, H. Ali, J. Moschera, K. Iqbal, and A. W. Malick, Assessment of capillary electrophoresis in pharmaceutical applications: Analysis and quantification of a recombinant cytokine in an injectable dosage form, J. Chromatogr., 559, 307–315 (1991).

84. W. G. Kuhr and C. A. Monnig, Capillary electrophoresis, Anal. Chem., 64, R389–R407 (1992).

85. J. P. Landers, R. P. Oda, T. C. Spelsberg, J. A. Nolan, and K. J. Ulfelder, Capillary electrophoresis: A powerful microanalytical technique for biologically active molecules, Biotechniques, 14, 98–111 (1993).

86. M. Pingel, A. Volund, E. Sorensen, and A. R. Sorensen, Assessment of insulin potency by biological and chemical methods, in *Hormone Drugs*, U.S. Pharmacopeial Convention, Rockville, MD, 1982, pp. 200–207.

87. J. Chayen, Cytochemical bioassay and its potential place in compendial definitions: A method that offers sensitivity as well as specificity, in *Hormone Drugs*, U.S. Pharmacopeial Convention, Rockville, MD, 1982, pp. 48–58.

88. T. Porstmann and S. T. Kiessig, Enzyme immunoassay techniques: An overview, J. Immunol. Methods, 150, 5–21 (1992).

89. H. van Vunakis and J. J. Langone, eds., Methods Enzymol., 70 (1980).

90. S. Petska, B. Kelder, and S. J. Tarnowski, Procedures for measurement of interferon dimers and higher oligomers by radioimmunoassay, Methods Enzymol., 119, 588–593 (1986).

91. J. M. Teal and D. C. Benjamin, Antibody as an immunological probe for studying refolding of bovine serum albumin, J. Biol. Chem., 4603–4608 (1976).

92. L. Sjodin, K. Holmberg, I. Stadenberg, and E. Viitanen, Quantitation of insulin by radioreceptorassay, in *Hormone Drugs*, U.S. Pharmacopeial Convention, Rockville, MD., 1982, pp. 192–200.

93. B. Friguet, L. Djavadi-Ohaniance, and M. E. Goldberg, Immunochemical analysis of protein conformation, in *Protein Structure—A Practical Approach* (T. Creighton, ed.), IRL Press, Oxford, 1989, pp. 287–310.

94. J. M. Sturtevant, Biochemical applications of differential scanning calorimetry, Annu. Rev. Phys. Chem., 38, 463–488 (1987).

95. D. Dollimore, Thermal analysis, Anal. Chem., 64, R147–R153 (1992).

96. M. P. W. M. te Booy, R. A. Ruiter, and A. L. J. de Meere, Evaluation of the physical stability of freeze dried sucrose containing formulations by differential scanning calorimetry, Pharm. Res., 9, 109–114 (1992).

97. L. R. Maneri, A. R. Farid, P. J. Smialkowski, M. B. Seaman, J. Baldoni, and T. D. Sokoloski, Preformulation of proteins using high sensitivity differential scanning calorimetry (DSC), Pharm. Res., 8, S-48 (1991).

98. R. Williams, R. Northey, and J. Schrier, Evaluation of RHM–CSF liquid stability by DSC, SDS–PAGE and HPLC analysis, Pharm. Res., 8, S48 (1991).

99. K. Wuthrich, Six years of protein structure determination by NMR spectroscopy: What have we learned? in *Protein Conformation* (Ciba Foundation Symposium 161), John Wiley & Sons, Chichester, 1991, pp. 136–149.

100. H. L. Levine, T. C. Ransohoff, R. T. Kawahata, and W. C. McGregor, The use of surface tension measurements in the design of antibody-based product formulations, J. Parenter. Sci. Technol., 45, 160–165 (1991).

101. R. Pearlman and T. Nguyen, Pharmaceutics of protein drugs, J. Pharm. Pharmacol., 44 (Suppl. 1), 178–185 (1992).

102. V. Mozhaev, I. V. Berezin, and K. Martinek, Structure stability relationship in proteins: Fundamental tasks and strategies for the development of stabilized enzyme catalysts for biotechnology, CRC Crit. Rev. Biochem., 23, 235–281 (1988).

103. C. O'Fagain, and R. O'Kennedy, Functionally stabilized proteins—a review, Biotechnol. Adv., 9, 351–409 (1991).

104. C. H. Schein, Solubility as a function of protein structure and solvent components, Biotechnology, 8, 308–317 (1990).

105. J. W. Simpkins, Solubilization of ovine growth hormone with 2-hydroxypropyl-β-cyclodextrin, J. Parenter. Sci. Technol., 45, 266–269 (1991).

106. S. N. Timasheff, Stabilization of protein structure by solvent additives, in *Stability of Protein Pharmaceuticals*, Part B: *In Vivo Pathways of Degradation and Strategies for Protein Stabilization* (T. J. Ahern and M. C. Manning, eds.), Plenum Press, New York, 1992, pp. 265–285.

107. M. A. Hanson and S. K. E. Rouan, Introduction to formulation of protein pharmaceuticals, in *Stability of Protein Pharmaceuticals*, Part B: *In Vivo Pathways of Degradation and Strategies for Protein Stabilization* (T. J. Ahern and M. C. Manning, eds.), Plenum Press, 1992, New York, pp. 209–233.

108. J. C. May and F. Brown (eds.), *Developments in Biological Standardization: Biological Product Freeze-Drying and Formulation*, Vol. 74, S. Karger, Basel, 1991.
109. O. P. Chilson, L. A. Costello, and N. O. Kaplan, Effects of freezing on enzymes, Fed. Proc. 24 (Suppl. 15), S55–S65 (1965).
110. F. Franks, R. H. M. Hatley, and H. L. Friedman, The thermodynamics of protein stability: Cold destabilization as a general phenomenon, Biophys. Chem., 31, 307–315 (1988).
111. P. E. Bock and C. Frieden, Another look at the cold lability of enzymes, Trends Biochem. Sci. 3, 100–103 (1978).
112. T. Tamiya, N. Okahashi, R. Sakuma, T. Aoyama, T. Akahane, and J. J. Matsumoto, Freeze denaturation of enzymes and its prevention with additives, Cryobiology, 22, 446–456 (1985).
113. P. L. Privalov, Cold denaturation of proteins, CRC Crit. Rev. Biochem. Mol. Biol., 25, 281–305 (1990).
114. J. S. Blanchard, Buffers for enzymes, Methods Enzymol., 104, 404–414 (1984).
115. G. F. Doebbler, Cryoprotective compounds—review and discussion of structure and function, Cryobiology, 3, 2–11 (1966).
116. J. F. Carpenter and J. H. Crowe, The mechanism of cryoprotection of proteins by solutes, Cryobiology, 25, 244–255 (1988).
117. J. F. Carpenter, T. Arakawa, and J. H. Crowe, Interactions of stabilizing additives with proteins during freeze-thawing and freeze-drying, Dev. Biol. Stand., 74, 225–239 (1991).
118. T. Arakawa, Y. Kita, and J. F. Carpenter, Protein–solvent interactions in pharmaceutical formulations, Pharm. Res., 8, 285–291 (1991).
119. J. F. Carpenter, L. M. Crowe, and J. H. Crowe, Stabilization of phosphofructokinase with sugars during freeze-drying: Characterization of enhanced protection in the presence of divalent cations, Biochim. Biophys. Acta, 923, 109–115 (1987).
120. J. F. Carpenter, B. Martin, L. M. Crowe, and J. H. Crowe, Stabilization of phosphofructokinase during air-drying with sugars and sugar/transition metal mixtures, Cryobiology, 24, 455–464 (1987).
121. J. F. Carpenter and J. H. Crowe, An infrared spectroscopic study of the interactions of carbohydrates with dried proteins, Biochemistry, 28, 3916–3922 (1989).
122. J. H. Crowe, J. F. Carpenter, L. M. Crowe, and T. J. Anchordoguy, Are freezing and dehydration similar stress vectors? A comparison of modes of interaction of stabilizing solutes with biomolecules, Cryobiology, 27, 219–231 (1990).
123. B. S. Chang and C. S. Randall, Use of subambient thermal analysis to optimise protein lyophilization, Cryobiology, 29, 632–656 (1992).
124. R. H. M. Hatley, The effective use of differential scanning calorimetry in the optimisation of freeze-drying processes and formulations, Dev. Biol. Stand., 74, 105–122 (1991).
125. L. Gatlin and P. DeLuca, A study of the phase transitions in frozen antibiotic solutions by differential scanning calorimetry, J. Parenter. Drug Asso., 34, 398–408 (1980).
126. R. H. M. Hatley and F. Franks, Applications of DSC in the development of improved freeze-drying processes for labile biologicals, J. Thermal Anal., 37, 1905–1914 (1991).
127. Y. Roos and M. Karel, Differential scanning calorimetry study of phase transitions affecting the quality of dehydrated materials, Biotechnol. Prog., 6, 159–163 (1990).
128. D. R. Macfarlane, Devitrification in glass-forming aqueous solutions, Cryobiology, 23, 230–244 (1986).
129. D. J. Korey and J. B. Schwartz, Effects of excipients on the crystallization of pharmaceutical compounds during lyophilization, J. Parenter. Sci. Technol., 43, 80–83 (1989).
130. M. J. Hageman, The role of moisture in protein stability, Drug Dev. Ind. Pharm., 14, 2047–1070 (1988).
131. M. J. Pikal and S. Shah, Moisture transfer from stopper to product and resulting stability implications, Dev. Biol. Stand., 74, 165–179 (1991).
132. J. P. Earle, P. S. Bennet, K. A. Larson, and R. Shaw, The effects of stopper drying on moisture levels of *Haemophilus influenzae* conjugate vaccine, Dev. Biol. Stand., 74, 203–210 (1991).
133. J. Broadhead, S. K. Edmond Rouan, and C. T. Rhodes, The spray drying of pharmaceuticals, Drug Dev. Ind. Pharm., 18, 1169–1206 (1992).

134. W. Steber, R. Fishbein, and S. M. Cady, Compositions for parenteral administration and their use, European Patent Application A1 0257 368, 1988.
135. M. T. Vidgren, P. A. Vidgren, and T. P. Paronen, Comparison of physical and inhalation properties of spray-dried and mechanically micronized disodium cromoglycate, Int. J. Pharm., 35, 139–144 (1987).
136. M. Vidgren, P. Vidgren, J. Uotila, and P. Paronen, In vitro inhalation of disodium cromoglycate powders using two dosage forms, Acta Pharm. Fenn., 97, 187–15 (1988).
137. P. Vidgren, M. Vidgren, and P. Paronen, Physical stability and inhalation behaviour of mechanically micronized and spray dried disodium cromoglycate in different humidities, Acta Pharm. Fenn., 98, 71–78 (1989).
138. P. Labrude, M. Rasolomana, C. Vigneron, C./ Thirion, and B. Chaillot, Protective effect of sucrose on spray drying of oxyhemoglobin, J. Pharm. Sci., 78, 223–229 (1989).
139. J. Broadhead, S. K. E. Rouan, I. Hau, K. Marshall, and C. T. Rhodes, The effect of process variables on the properties of a spray dried enzyme, Pharm. Res., 9, S-78 (1992).
140. U.S.P. Rationale for the Development of Public Standards for Biological Products Licensed by the Food and Drug Administration's Center for Biologics, Evaluation and Research (FDA-CBER), Pharmacopeial Forum, 18, 3754–3761 (1992).
141. P. L. Smith, D. A. Wall, C. Gochoco, G. Wilson, Routes of delivery: Case studies—(5) Oral absorption of peptides and proteins, Adv. Drug Deliv. Rev., 8, 253–290 (1992).
142. L. Illum and S. S. Davis, Intranasal insulin, Clin. Pharmacokinet., 23, 30–41 (1992).
143. J. S. Patton and R. M. Platz, Routes of drug delivery: Case studies (2) pulmonary delivery of peptides and proteins for systemic action, Adv. Drug Deliv. Rev., 8, 179–196 (1992).
144. S. S. Davis, Delivery systems for biopharmaceuticals, J. Pharm. Pharmacol., 44 (Suppl. 1), 186–190 (1992).
145. X. H. Zhou and A. Li Wan Po, Peptide and protein drugs: II Non-parenteral routes of delivery, Int. J. Pharm., 75, 117–130 (1991).
146. L. L. Wearley, Recent progress in protein and peptide delivery by noninvasive routes, Crit. Rev. Ther. Drug Carrier Syst., 8, 331–394 (1991).
147. V. H. Lee, ed., *Peptide and Protein Drug Delivery: Advances in Parenteral Sciences/4*, Marcel Dekker, New York, 1990.
148. C. Delgado, G. E. Francis, and D. Fisher, The uses and properties of PEG-linked proteins, Crit. Rev. Ther. Drug Carrier Syst., 9, 249–304 (1992).
149. M. S. Hora, Parenteral delivery systems for proteins, Drug News Perspect., 4, 538–543 (1991).
150. R. Langer and N. A. Peppas, New drug delivery systems, Biomed. Eng. Soc. Bull., 16, 3–7 (1992).
151. D. Bodmer, T. Kissel, and E. Traechslin, Factors influencing the release of peptides and proteins from biodegradable parenteral depot systems, J. Controlled Release, 21, 129–138 (1992).
152. C. N. Tenhoor and J. B. Dressman, Oral absorption of peptides and proteins, STP Pharma Sci., 2, 301–312 (1992).
153. P. Wuthrich, S. Y. Ng, B. K. Fritzinger, K. V. Roskos, and J. Heller, Pulsatile and delayed release of lysozyme from ointment-like poly(*ortho*-esters), J. Controlled Release, 21:191–200 (1992).
154. T. G. Park, S. Cohen, and R. Langer, Controlled protein release from polyethyleneimine-coated poly(*l*-lactic acid)/pluronic blend matrices, Pharm. Res., 9, 37–39 (1992).
155. K. C. Lee, E. E. Soltis, P. S. Newman, K. W. Burton, R. C. Mehta, and P. P. DeLuca, In vivo assessment of salmon calcitonin sustained release from biodegradable microspheres, J. Controlled Release, 17, 199–205 (1991).
156. P. Edman and E. Bjork, Routes of delivery: Case studies (1) nasal delivery of peptide drugs, Adv. Drug Deliv. Rev. 8, 1165–1177 (1992).
157. F-D-C Reports—"The Pink Sheet," 55(2) (Jan. 11, 1993).
158. F-D-C Reports—"The Pink Sheet," 54(1) (Jan. 6, 1992).
159. *Physicians' Desk Reference*, 46th Ed., Medical Economics, Montvale, NJ, 1992.

23

Veterinary Pharmaceutical Dosage Forms: An Overview

J. Patrick McDonnell
Solvay Animal Health, Inc., Charles City, Iowa

I. INTRODUCTION

A. Veterinary Products: Economic Overview

Animal health issues and products are of interest to the pharmacist from the standpoint of sales, research (i.e., new products and new dosage forms of existing products), and zoonoses (diseases of animals that may secondarily be transferred to humans). The pharmacist, as the most readily and easily accessible community health representative, is in a position to provide information and products to customers and patients on issues of animal health. Approximately 58% of households in the United States now include at least one pet [1]. The American Veterinary Medical Association's 1992 survey showed the population of companion animals in the United States to be

Cats	60 million
Dogs	52 million
Birds	12 million
Horses	5 million

Food-producing (economic) animals by far outnumber companion animals. United States Department of Agriculture National Agricultural Statistics Service reported for 1993 the following numbers:

Hogs	56.8 million
Cattle and calves	100 million

Poultry
    Chicks-layer   281.5 million
    Chicks-broilers   6.4 billion
    Turkeys   285.4 million
    Sheep and lambs   10.2 million

All of these animals during their lifetime, will receive or be treated with at least one of the following drugs: feed additives, biologics (vaccines, bacterins), growth stimulants, pharmaceutics, or pesticides. These products are developed and manufactured by a multitude of pharmaceutical (human and animal), biological, and other companies, who employ pharmacists and pharmaceutical scientists. The pharmacy student should have an understanding and basic knowledge of veterinary products to enhance and complete his or her area as an expert in pharmaceutical dosage forms and drug products.

B. Veterinary Medicines, Drugs, and the Pharmacist

Students of pharmacy and pharmacists have had a long history of association with veterinary medicine and veterinary drugs. In the 1960s and 1970s, at least two colleges of pharmacy's offered courses in agricultural pharmacy and veterinary product development research [2–4]. This interest was expressed again in 1987, with the joining of a large veterinary pharmaceutical and biological company and one of the Midwest's largest colleges of pharmacy (the University of Iowa College of Pharmacy) into a collaborative venture in veterinary dosage form development [5]. The benefits of these types of programs was to expose pharmacy students to the world of animal health medications and dosage forms. They would then be better equipped to understand and provide information and counsel to persons who have questions and needs concerning animal drug issues, be it companion animals, such as dogs or cats in a metropolitan area, or chickens, hogs, or cattle in rural practices. Exposing students to veterinary dosage form development may also spark an interest in practicing as an industrial pharmacist in the animal health industry.

A little known facet of the alliance of pharmacy and the animal health industry is that the Animal Health Institute, the premier trade organization of animal health industries, had its beginnings in Des Moines, Iowa, sharing offices and staff with the Iowa Pharmacists Association. The executive secretary of the Iowa Pharmaceutical Association and Animal Health Institute (AHI) was a pharmacist who eventually became the full-time executive secretary for the Animal Health Institute when the AHI moved its headquarters to Washington, DC.

Several of the colleges of pharmacy in the United States are located on the campus of the agricultural land grant university in their state. Included in this group are Purdue University College of Pharmacy and Pharmacal Sciences, West Lafayette, Indiana; University of South Dakota, College of Pharmacy, in Brookings, South Dakota; University of Georgia, College of Pharmacy, Athens, Georgia; and the North Dakota State University, College of Pharmacy, Fargo, North Dakota. Thus, academic pharmacy has a physical presence within academic agricultural settings, and collaborative research and other programs between these colleges are common and productive.

In summary, familiarity with veterinary medicine and animal drugs is helpful to the pharmacist in practice as an advisor and reference source to his or her patients, the community, and other members of the health care team.

II. ANIMAL ANATOMY AND PHYSIOLOGY

There are many similarities, as well as substantial differences, in the anatomy and physiology of animals, avian species, and humans. It is outside the scope of this chapter to elaborate extensively on the individual differences. However, two systems that I feel of significance and interest will be discussed: the respiratory and digestive systems.

A. Respiratory System

The respiratory system of animals and humans have numerous similarities. However, the respiratory system of the fowl is different from animals in several aspects. Avian lungs are small in comparison with their body size. When comparing the lung capacity of a bird with a comparable-sized animal, the avian lung capacity is approximately one-half. The lung adheres to the rib cage, which would be abnormal in mammals. They are passive in action and expand and deflate with the thoracic cage, since they are relatively fixed and incapable of the elastic recoil found in mammals. Mammals have a respiration rate of 12 respirations per minute for the horse to 30 for the cow, whereas birds vary from 46 to 380 respirations per minute. Another important difference between avian and mammalian respiratory systems is the presence of air sacs in birds. These air sacs are connected to the lung. There are nine air sacs, which are grouped into the thoracic air sacs and the abdominal air sacs.

Birds breathe by a two-stage respiration, which is different from mammals. When birds inhale, incoming air is directed past the lung tissue and into the abdominal air sacs. As this occurs, they expand their abdominal cavity, and because of the lack of a diaphragm, fresh air does not go directly to the lungs, but to the abdominal air sacs. When they exhale, the abdominal air sacs act as a billows and push fresh air through the lungs from the abdominal end to the thoracic end. The air sacs expand and contract as the bird is breathing so the lungs do not have to. Air is forced passively through the lung tissue for gas exchange in a unidirectional flow (Fig. 1).

Birds have a high-respiratory efficiency; a fowl at rest does not need to breathe frequently. A healthy bird at rest will take a breath about one-third as often as a comparatively sized mammal. Because the air sacs connect and communicate with the long bones of the wings and legs, respiratory diseases may also cause musculoskeletal system problems. This is why respiratory infection drugs used for avian disease control must be considered separately from drugs used for mammalian respiratory infections.

The respiratory systems of mammals have many common features and are similar to humans. For detailed information on the anatomy and physiology of respiration in mammals, refer to Michel et al. *The Viscera of the Domestic Mammals* [6], and *Dukes' Physiology of Domestic Animals* [7].

B. Digestive System

As with humans, most animal drugs are administered orally, thus the formulator needs to pay particular attention to the gastrointestinal tract of animals. Animals are classified according to their eating habits. Carnivores (e.g., cats and dogs) in their native or wild state, obtain most of their food by eating other animals. Their digestion is enzymatic. The domesticated herbivores (e.g., cattle and sheep) have a rumen where extensive microbial fermentation of a plant diet takes place. The horse is also herbivore, but has a simple stomach. The microbial fermentation takes place in the posterior portion of the digestive tract of ruminant animals. Omnivorous animals have mainly enzymatic digestion. The pig is omnivorous, eating either plants or meat; however, under domestication is essentially herbivorous. Microbial degradation takes place in

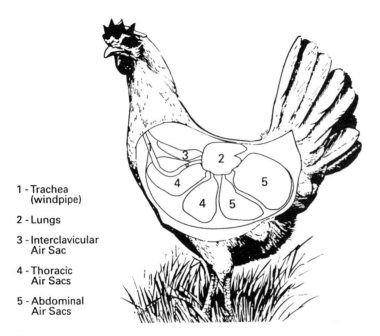

1 - Trachea
    (windpipe)

2 - Lungs

3 - Interclavicular
    Air Sac

4 - Thoracic
    Air Sacs

5 - Abdominal
    Air Sacs

Fig. 1   Diagram of respiratory system in birds. (*Courtesy of Solvay Animal Health, Inc.*)

the large intestine for plant material. Of all animals, the gastrointestinal tract of the pig (exclusive perhaps of some primates) is closest to humans.

Cattle, sheep, and goats have large, complex compartmentalized stomachs. These compartments are the rumen, reticulum, omasum, and abomasum. The rumen, reticulum, and omasum compose the forestomach (proventriculi) and are lined with a nonglandular mucous membrane. The abomasum is lined with a glandular mucous membrane (Fig. 2).

The intestines extend from the stomach to the anus or cloaca (birds) and are lined with a glandular mucous membrane arranged as villi. Carnivores, in general, have much shorter intestines than herbivores or the omnivorous pig. The horse is an exception and has a relatively short, small intestine, but has a large cecum and colon.

The anatomy of the avian alimentary canal is different from mammals in the mouth area, since the oral cavity of birds has a direct communication (choana) with the nasal passages. Birds also have a crop in the esophagus, and a muscular stomach called the gizzard (Fig. 3). In the mouth, there is no soft palate in most species (i.e., pigeons are an exception). Teeth are absent, but the horny beak is used to perform the functions of teeth in animals. The crop, which is an extension of the esophagus, acts as a storage place for food. Little or no digestion takes place in the crop. The gizzard, also known as the muscular stomach, remains quiet when empty. When food enters it, muscular contractions begin. The larger the particles of food, the more rapid the contractions. Since the gizzard usually contains some abrasive material, such as rocks and gravel, the food material is soon ground or reduced to small particles capable of being taken into the intestinal tract. When fine material enters the gizzard, it passes through relatively fast. Coarse food remains in the gizzard until broken down into fine particles, which may take several hours [8].

The role of food passage through the digestive tract from ingestion to elimination as feces is as follows: human, 15–25 hr; cow, 72–84 hr; goat, 14–17 hr; pig, 17–50 hr; chicken, 4–

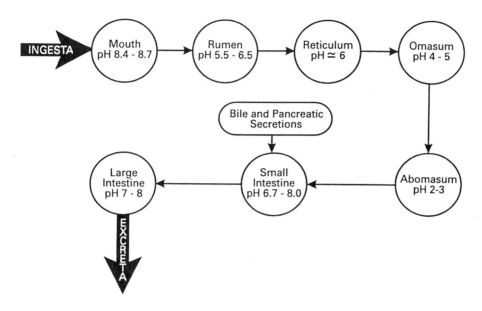

Fig. 2 Diagrammatical delineation of the ruminant gastrointestinal tract. (From Ref. 20.)

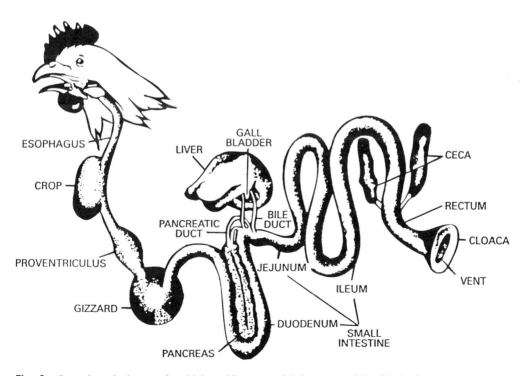

Fig. 3 Gastrointestinal tract of a chicken. (*Courtesy of Solvay Animal Health, Inc.*)

8 hr [9]. These transit times need to be weighed, as well as other anatomical and physiological factors previously mentioned, when designing an oral animal drug dosage form.

The digestive system must be considered not only from the standpoint of absorption, bioavailability, and site of drug action, but from the standpoint of disease processes. Digestive diseases of cattle include hypocalcemia (milk fever), enteritis (calf scours), and bovine coccidiosis. Species-specific gastrointestinal tract disease examples are: chickens, cecal coccidiosis; swine, swine dysentery, caused by *Treponema hyodysenteriae*; sheep, enterotoxemia; and horses, bots infestations. Diseases of the digestive tract are a very important cause of financial losses in animal production [10].

III. ANIMAL DOSAGE FORMS

A. Tablets and Boluses

Solid dosage forms, such as compressed tablets, are one of the most common means of administering medications to humans. These are less popular for animals because administration may be time-consuming, hazardous, and uncertain. Uncertain because one cannot be sure the tablet is swallowed, being spit out or dropped from the mouth after the administrator has left or moved on to another animal. For tablets that are accepted voluntarily by the animal, typically they are chewed, which exposes the disagreeable taste of some drugs. Thus, the advantage of the dosage form may be lost. This can occasionally be overcome by use of odors, flavors, or sweeteners. However, the formulator cannot fall into the trap of flavoring or masking based on human perceptions. Formulations must be tested on the species for which it is intended. An interesting thesis on animal flavors and flavoring was written by Talmadge B. Tribble of Flavor Corporation of America [11], who, himself, was a pharmacist and founded one of the largest animal feed flavor companies in the world.

All drugs are given on the basis of weight, be it for animal, avian species, or humans. The amount of drug needed for a large animal, such as a St. Bernard dog, cow, or horse, can be considerable. Therefore, one tends to find labeling for solid dosage forms to be stated in milligram or gram tablet per pounds (kilograms) of body weight. Drugs, such as sulfonamides, are dosed at relatively high rates; it is not unusual to prescribe as much as 15 g of drug for each 150 lb (68 kg) of body weight. A 750-lb (340-kg) cow or horse would receive 75 g of drug. A special tablet called a 'bolus' is commonly used to provide these large dosages. A bolus is nothing more than a very large tablet, which can range from 3 to 16 g or more. Although commonly called "horse pills," they are not used exclusively with horses. Indeed, because of the difficulty in handling horses, which may be less docile than cows, the ease with which the horse can spit them out, and the possibility of choking, the bolus form must be used with special care in these animals. Boluses are capsule or cylindrical shaped because a round bolus would be unwieldy and difficult to administer or swallow. Boluses are administered by an apparatus called a balling gun. These are devices consisting of a barrel with a plunger that can hold one or more boluses. The tube is inserted into the animal's mouth over the base of the tongue and, as the animal swallows, the plunger is depressed to push the bolus into the gullet. The bolus or boluses are thus expelled gently into the gullet, from where it is swallowed by reflex.

Bolus formulation poses challenges because of the high drug/excipient ratio. Less room is left for diluent and adjuvants needed to overcome objectionable features of the drug. In ruminant animals, such as cattle or sheep, it is possible to employ the concept of long-acting boluses that stay in the gastrointestinal track for periods much longer than 12 hr (for days or even weeks). This is because solid objects can remain in the ruminoreticular sac indefinitely.

The density of the bolus is the critical factor for retention in the sac. A range of densities from 1.5 to 8.0, is believed to be desirable. This is achieved by including excipients, such as iron, clay, sodium sulfate dihydrate, and dicalcium sulfate, in these formulations [12]. Weight and size influence retention, but not as significantly as density.

B. Feed Additives

An animal dosage form that the pharmacist would be least familiar with, since there is no analogy in human drug dosage forms, is the feed-additive premix.

Feed-additive premixes are formulated to contain bulk drug and excipient, in a form that may be readily combined with an animal feed. The feed route is used mainly for prophylactic treatment against diseases of parasites, or for growth promotion. The drugs may be given from day of age (e.g., hatching—in poultry rations) to market. However, the majority of drugs will be required to be withdrawn from the feed several days to a month before the animal is sent to market for human consumption.

The formulation of a premix consists of combining a drug with a carrier, diluent, or absorbent. The most common form premix employs a grain carrier (i.e., rice hulls, corn germ meal, corn meal, corn gluten meal, wheat middlings) in which the active drug is lightly bound or absorbed onto the surface of the carrier. The carrier functions by absorbing the small fine particles of drug onto the surface and into the pores of the carrier particles. The carrier will usually be two-thirds or more of the formulation. An oil may also be added to bind the drug to the carrier. If needed for flow conditioning and to prevent caking, an anticaking agent, such as silicon dioxide or magnesium aluminum silicate, can be incorporated in the product.

Another excipient used in a feed-additive premix may be a diluent used to dilute or standardize activity. They are similar in composition to grain carriers, except the particle size of diluents is generally smaller than the particle size of the carrier. No attempt is made to absorb the active drug to the individual particles of diluents. If a liquid is used it is mainly for dust control. A diluent is considered for use when the level of the active ingredient components in the premix approaches or exceeds 50% of the product, or when two or more active components vary greatly from one another in density [13]. Examples of diluent materials are ground limestone, sodium sulfate, kaolin, corn cob flour, and ground oyster shells.

Absorbents are another class of excipient material used in feed additive premixes. They are used when the drug substance is a liquid, or is readily soluble in water, oil, or other solvent. The liquid is sprayed onto the absorbent in a mixer as the mixer is running. Examples of absorbents are vermiculite, fullers earth, corn cob fractions, and clay.

Factors that need to be considered in formulating premixes and the choice of carrier are

1. Drug concentration in the premix.
2. Drug concentration in the final feed: If a drug premix will be added to a feed so that the drug level is less than 150 ppm, a carrier is needed to ensure adequate dilution.
3. Moisture content of drug and carrier: If the drug is moisture-sensitive or the carrier is subject to breakdown or spoilage from moisture levels in the drug or carrier itself, appropriate drying or other steps may be required.
4. Electrostatic charges: Fine drug powders will often develop static charges during particle size reduction and flow-through materials-handling systems. These charges need to be minimized to prevent unmixing or loss of even distribution throughout the premix and subsequent feed.
5. pH extremes: These can frequently be compensated by use of sodium carbonate to neutralize acid mixtures, or of calcium phosphate monobasic or fumaric acid to neutralize basic mixtures.

6. Flow is important when automatic premix addition equipment is used in modern feed mills. Bridging (an organized structure of product that impedes flow) which inhibits addition of the premix to the feed batch will cause the mill to shut down until the correct amount of premix is added. This shutdown of the mill can cause considerable consternation to the operators of the mill, who are producing multiple batches of feed per day.

The normal standard of premix usage in feed is 1 part medicated premix to 1999 parts of feed. A properly formulated premix can be used directly in preparing a medicated feed without further dilution. It can be further diluted in the feed mill by the use of in-plant premixes, but this would be at the discretion of the feed mill operator.

Although the practicing pharmacist may not have frequent contact with this particular dosage form, there has been a movement to place some feed additive drugs into veterinary prescription status. This may have future implications for those pharmacists practicing in a rural environment.

C. Drinking Water Medications

A common form of medicating animals for herd or flock health is through the drinking water. The medications are formulated as (a) dry powders for reconstitution into liquid concentrates to be added to the drinking water, or to be added directly to the drinking water; or (b) concentrated solutions that are dispensed directly in drinking water or injected into the drinking water through medication proportioners incorporated into watering lines. The advantage of medicating through the drinking water versus the feed is that sick or unhealthy animals will continue to drink water, whereas they may not eat. The use of water as the drug medium is limited, however, by the solubility of the drug moiety. Since animals drink twice as much water as they consume feed, the concentration in the water needs to be only half that of feed. This factor may overcome the problem of limited solubility.

Automatic-metering devices or medication proportioners are used for treating large numbers of animals. The powder medication is dissolved at the time of administration into water to make a stock solution, which is proportioned into the drinking water system as the water is being consumed by the animals. The common dilution in the United States is 1 fl oz (29.6 ml) of stock solution (or liquid drug concentrate) to 127 oz (3.76 liters) of water, producing a 1 fl oz/gal (29.6 ml/2.59 liter) dilution.

Whether a product is formulated as a dry powder or liquid concentrate, the product development pharmacist must be concerned with effects of the properties of the diluting water media. Tablet or granule hardness, buffer capacity, pH, and total dissolved solids, all play a role in the solubility rate and availability of the drug substance, as well as stability. In addition, dry products are usually formulated with a sugar diluent, such as lactose or dextrose. The use of these may cause a buildup of bacteria and fungi in water lines when the sugar level is high for an extended period [14]. In the product development laboratory, medicated drinking water samples must be prepared from these formulations, using a range of hard and soft waters, and stored at 25°C and 37–40°C in metal containers or troughs (galvanized iron or rusty metal), to simulate the worst possible conditions of use [15]. The drug stability in the drinking water should be adequate for the storage length of time listed on the label [16]. Consideration also has to be given when formulating a liquid concentrate, in using solvents other than water, of the possibility of precipitation or recrystallization of the drug when diluted with water. All of the foregoing factors make the formulation of animal drinking water products an interesting and challenging experience.

D.   Oral Pastes and Gels

Pastes and gels are semifluid masses that can be administered from a flexible tube, syringe, package, or other specialized dosing device. The advantage of a paste or gel dosage form is that it cannot be expelled from the animals mouth as readily as a tablet or liquid. Also, mass-medicating of animals can be achieved rapidly and easily with the paste medication using a multiple-dose dispenser such as a syringe.

A paste of the proper consistency adheres to the tongue or buccal cavity and is not readily dislodged. The animal will eventually end up swallowing it. Characteristics of a suitable paste formulation are [17]:

1.   When placed in the palm of the hand and the hand is inverted (palm down), it should remain there without falling.
2.   When the paste is ejected from the applicator, it should break free cleanly when rubbed against a flat surface.
3.   No paste should continue to ooze from the applicator after the dose has been ejected.
4.   The paste or gel should be free from air bubbles or voids.
5.   Only a minimum of force should be needed to expel the paste from the dispensing device.

The three types of vehicles used in formulating a paste or gel are (a) aqueous bases, (b) oil or oleaginous bases, or (c) organic solvents.

An aqueous base is the least expensive and poses no toxicity problems. A solution of the drug in water or water and cosolvent is made. Glycerin, glycols, natural and synthetic gums, or polymers are used to increase viscosity, cohesiveness, and plasticity. To overcome syneresis, a common problem with aqueous bases, one can use absorbing materials, such as microcrystalline cellulose, kaolin, colloidal silicon dioxide, starch, and so on.

Oleaginous bases consist of vegetable oil thickened with agents such as aluminum monostearate, colloidal silica, and xanthan gums. The lubricant properties of the oil make these formulations less adhesive than water bases.

Glycerin, propylene glycol, and polyethylene glycol thickened with carboxyvinyl polymers (Carboxamer NF) provide organic solvent bases. Consistencies ranging from soft jelly to peanut butter can be achieved.

A paste is administered to an animal volumetrically. The drug level and density of paste must be known to determine the amount of drug delivered per given volume. This takes trial and error in the formulation process to arrive at the volume of paste necessary to provide the required dose.

E.   Drenches and Tubing Products

Horses are administered certain medications by running a lubricated tube up through their nostrils and down into the stomach. A funnel attached to the tube is held above the horses head and the liquid medication is poured down the tube. This is known as tubing.

The normal dose for a horse by this method is approximately 10 fl oz (296 ml). It needs to be formulated so that the amount will flow through the tubing (i.e., $6^{1}/_{2}$ ft $\times$ $^{3}/_{8}$ in) in 60 sec. Wetting agents are used to increase flow rate. Thickening or suspending agents are contraindicated, since the formulations will thicken and resist flow when shear is removed.

The administration of a drug to animals by pouring a liquid medication down an animals throat is called drenching. Drenches are dispensed by syringe or drenching guns. The viscosity should be such as to prevent drip from the syringe during movement from drug containers to the animal. Drenching guns require a formulation that is not as viscose, since leakage from the gun is not a major problem. Too viscose a product may cause administrator fatigue when large numbers of animals are dosed.

These drenches and tubing products are often given in extreme range of temperatures. The formulator must take this into consideration when developing the dosage formula and testing it in the administration equipment.

F. Topical Dosage Forms

There are several unique topical dosage forms for animals. Four types that the pharmacist should have a basic understanding or awareness of are (a) pour-on or spot-on applications, (b) dust bags, (c) dips, and (d) flea and tick collars. These are used for treatment and prevention of internal and external parasites.

Pour On or Spot On Applications

These liquid products effect systemic activity after being poured onto an animals backline or applied as a one-spot concentrate on the animals back or rump. They are mainly used for the control of cattle grubs and lice. However, there is one pour-on–spot, on product (levamisole) that has broad-spectrum anthelmintic activity. These formulations contain organophosphorus insecticides or the anthelmintic dissolved in organic solvents, such as dimethyl sulfoxide or aromatic hydrocarbons. Advantages for using these formulations are [18]:

1. Risk of trauma and inhalation pneumonia associated with drenching or damage at injection site (for parenteral products) are eliminated.
2. No special skills are required for application, since they are administered topically by use of sprays or spotter bottle (a bottle with a squeeze-on applicator).
3. Sterile precautions are not necessary.
4. Troublesome animals are dosed relatively easily with safety to the person performing the application.
5. Speed of treatment is quick.

Dust Bags

Cattle are treated with insecticide powders through use of a device called a dust bag. This is accomplished by the animal brushing against the bag as they walk beside or under it. The bag has an inner porous storage bag containing the insecticide dust formulation. This is protected from the elements by an outer protective waterproof skirt open to the porous dust bag at the bottom.

The cattle can have free-choice application, or are forced to use them, depending on where they are hung. Forced-use bags are hung in doorways, lanes, gateways, and such. Free-choice applications can be achieved by placing the bags suspended from overhead structures, such as a tree or pole. One would be surprised at the acceptance achieved by the majority of range cattle to come to a free-choice application site.

Dips

For control of ectoparasites in economic animals, dipping is a method used extensively. A dip formulation containing the drug is diluted in a large dipping bath through which the animal is driven. This bath must be long, wide, and deep enough to cause immersion of the animal. The formulation of the ectoparasiticide is challenging. It must not be inactivated by matter that accumulates in the dipping bath and should maintain stability throughout a range of concentrations and temperatures. In addition, it must be nontoxic to the animal, but toxic to the ectoparasites [19].

Flea and Tick Collars

This dosage form will be most familiar to the pharmacist, since it is used for companion animals (dogs and cats) and is sold in most drugstores, supermarkets, and animal health products centers. There are two types of flea and tick collars, also known as slow-release pesticide generators: vaporous and powder-producing collars. Both contain the insecticide and a plasticized solid thermoplastic resin.

1. The vaporous collar contains a relatively high vapor pressure liquid pesticide mixed throughout the collar. The pesticide is slowly released and fills the surrounding atmosphere adjacent to the animals surface with a vapor of pesticide that kills the pest, but is innocuous to the animal.
2. The powder-producing collar contains a solid solution of the drug in the resin. Shortly after the collar is processed, the particles or molecules of the pesticide migrate from within the body of the resin and form a coating of particles, known as 'bloom,' resembling a dust or powder on the collar surface.

Ticks and fleas tend to concentrate in or migrate through the neck area of the animal. As they do this, they contact the active pesticide on or released by the collar and are killed. Powder-producing collars have an advantage over vapor ones in that by the movement of the dog or cat, the powder crystals (bloom) are rubbed or wiped onto the fur, which expands the contact area for it to continue to control the ticks and fleas [20].

IV. UNITED STATES REGULATORY REQUIREMENTS FOR APPROVAL OF ANIMAL DRUGS

In the United States, the Center for Veterinary Medicine of the Food and Drug Administration has responsibility for the review and approval of animal drugs. Toxicology, pharmacology, and pharmacokinetic data for drug products used in food-producing animals is the responsibility of the Human Food Safety Division. The mechanism for review and approval of animal drugs is through the submission of Form FD356, New Animal Drug Application (NADA). Animal feeds containing drugs must also be approved by submission of Form FD 1900, Animal Feed Application. The requirements for which are detailed in the Food, Drug, and Cosmetic Act and the Code of Federal Regulations [21].

Before final approval of a new animal drug application, the Center for Veterinary Medicine may require a manufacturing facilities visit called a Preapproval Inspection. The Good Manufacturing Practices regulations [22] are used as a guide for this audit by inspectors from the compliance division of the Food and Drug agency. The requirements for laboratory data, manufacturing equipment, and facilities are the same whether the drugs are made for animals or humans. The quality of animal drugs is expected to be the same as that of their human counterparts. They are not relegated to second-class status.

V. CONCLUSION

Animal drug dosage forms can be as complex and sophisticated as drugs that are used in humans. They have their own requirements and characteristics, based on the unique aspects of animal and avian physiology. It is hoped that this overview will stimulate pharmacists and students of pharmacy to further study issues of veterinary medicine dosage form development and the proper use of veterinary drugs. In doing so, they will have the knowledge to be a source of information to their clientele and the community they serve.

ACKNOWLEDGMENTS

The author wishes to acknowledge the assistance of Stephen K. Muir, D.V.M. in reviewing the section, *Animal Anatomy and Physiology*, and Laura J. Gogg and T. Jean Peterson for typing the manuscript.

REFERENCES

1. Harvard Health Lett., 19(2), December (1993).
2. The Drake Post-Scrip, 18(1; 2–4; Spring) (1969).
3. J. P. McDonnell, Drake College of Pharmacy, Salsbury Laboratories Research Program, Iowa Pharm. 24, 10 (1969).
4. University of Iowa Bull., Catalog No. 1970–72.
5. Univ. Iowa Coll. Pharm. News, 4(2; Spring) (1988).
6. R. Mickel, A. Schummer, and W. O. Saik, *The Viscera of the Domestic Mammals*, 2nd Ed., Verlog Paul Parey, Berlin, 1979.
7. M. J. Swenson, ed., *Dukes' Physiology of Domestic Animals*, Cornell University Press, Ithaca, 1984.
8. M. O. North, *Commercial Chicken Production Manual*, The Avi Publishing Company, Westport, CO, 1972.
9. K. J. Hill, Developmental and comparative aspects of digestion, in *Duke's Physiology of Domestic Animals*, 8th Ed., Cornell University Press, Ithaca, 1970, pp. 409–422.
10. W. G. Huber and V. K. Reddy, Physiology and anatomic features of monogastric and ruminant animals, in *Animal Health Products Design and Evaluation* (D. C. Monkhouse, ed.), American Pharmaceutical Association, Washington, DC, 1978.
11. T. B. Tribble, *Feed Flavor and Animal Nutrition*, AgriAids, Inc., Chicago, IL, 1962.
12. J. Blodinger, Formulation of drug dosage forms for animals, in *Formulation of Veterinary Dosage Forms* (J. Blodinger, ed.), Marcel Dekker, New York, 1983, pp. 139–142.
13. W. L. Larrabee, *A Guide to Mixing Microingredients in Feed*, Marek Services Bulletin, Rahway, NJ, 1976, p. 12.
14. W. L. Larrabee, Formulation of drugs given in feed or water, in *Formulation of Veterinary Dosage Forms* (J. Blodinger, ed.), Marcel Dekker, New York, 1983, p. 197.
15. *Drug Stability Guidelines*, Center for Veterinary Medicine, Food and Drug Administration, 4th Rev., 1990, p. 2–23.
16. H. L. Newmark and E. DeRitter, Animal health dosage forms: Stability requirements, in *Animal Health Products Design and Evaluation* (D. C. Monkhouse, ed.), American Pharmaceutical Association, Washington, DC, 1978.
17. J. Blodinger, Formulation of drug dosage forms for animals, in *Formulation of Veterinary Dosage Forms* (J. Blodinger, ed.), Marcel Dekker, New York, 1983, p. 158.
18. D. G. Pope, Specialized dose dispensing equipment, in *Formulation of Veterinary Dosage Forms* (J. Blodinger, ed.), Marcel Dekker, New York, 1983, p. 99.
19. D. G. Pope, Specialized dose dispensing equipment, in *Formulation of Veterinary Dosage Forms* (J. Blodinger, ed.), Marcel Dekker, New York, 1983, pp. 101–102.
20. D. G. Pope, Animal health specialized delivery systems, in *Animal Health Products Design and Evaluation* (D. C. Monkhouse, ed.), American Pharmaceutical Association, Washington, DC, 1978, pp. 87–90.
21. *Code of Federal Regulations*, 21 Part 514, New Animal Drug Applications, April 1, 1993.
22. *Code of Federal Regulations*, 21 Part 211, Current Good Manufacturing Practice for Finished Pharmaceuticals, April 1, 1993.

24

A View to the Future

Gilbert S. Banker
University of Iowa, Iowa City, Iowa

Christopher T. Rhodes
University of Rhode Island, Kingston, Rhode Island

I. INTRODUCTION

Various factors—economic, legislative, political, sociological, regulatory, scientific, and technical—are impinging on the practice of pharmacy worldwide at a rate and to an extent that are unprecedented. People in many countries today view medical care of the highest quality as a right, rather than a privilege. Indeed, if the American Bill of Rights were written today, a fourth right would be added to our three inalienable rights of life, liberty, and the pursuit of happiness: namely, health. The individual is highly conscious of the value of his or her personal good health. Politicians recognize the appeal to the electorate for programs that will assure high-quality health care to the elderly, to low-income people and their families, and eventually to the public as a whole. Whenever programs have evolved, wherein the recipient of medical services does not directly pay for those services, but instead, they are paid by a second or third party, the demand for the services has grown at an astronomic rate. As noted in Chapter 1, in the early 1950s the majority of people in the United States paid directly for their own hospitalization expenses, and the total health care cost in the country was about 13 billion dollars. Today, fewer than 10% of the population directly pay for their hospitalization costs, and the total health care costs of the country have risen to 1000 billion, or 1 trillion dollars. This represents an increase in costs of more than 7500% over this 45-year span. The vast majority of the cost increases over this time has been in hospitalization and physicians services. Drugs and pharmacy services account for only about 6% of the nations total health care bill.

The American public, corporate employers, state and federal governments, all have become increasingly concerned about the cost of health care. For many employers, the cost of providing health care to their employees adds between 20 and 25% to their total personnel cost. For many industries, providing health care to workers is now one of the primary costs of doing business. The fastest-growing cost in state and federal budgets over recent decades has been that related to health care benefits. Health care reform is the number 1 priority of state and federal legislatures in the mid 1990s. The demand for reform is as much about gaining control over these escalating costs as it is about providing universal access to care for all Americans.

One method of reducing health care costs, as well as to increase access to care in rural and underserved areas, is to make fuller use of all health care professionals, including physicians assistants, nurses and nurse practitioners, and pharmacists, in a fully integrated medical delivery model. Pharmacy, as a profession, is currently heavily engaged in documenting the value to the patient, and the cost savings to the payor, of delivering pharmaceutical care. The pharmacy profession is also rapidly coming to the realization that its future does not lie in drug distribution; that function can be largely handled by technicians and even by robotics. The future of pharmacy is in the optimization of drug therapy in the individual patient, in the primary care community setting, in addition to the institutional setting. Pharmaceutical services *will* greatly expand in primary care settings, and as pharmacists document the value of their services, they will increasingly be remunerated for their contributions to health care delivery, quite apart from the payment for the drug product(s).

Drug delivery will continue to develop as a science and technology to optimize drug products, making them safer, more effective, and more reliable. Drug delivery advances will play a growing role in effectively delivering proteins, peptides, and other biotechnology products. Chemotherapy will continue to rapidly advance, being heavily influenced by the biotechnology scientific revolution. Major breakthroughs in disease control and prevention will follow. Life expectancy will increase by at least another 10 years over the next 30–40 years. The consciousness of the public to health and health-related programs will continue to grow, as will the participation of government in all health-related programs.

II. LEGISLATIVE IMPACTS, THE NEW HEALTH CARE ENVIRONMENT AND THE FUTURE OF PHARMACY

Two legislative health care initiatives coming less than 4 years apart are having, and have had, a very major effect on all segments of pharmacy and, certainly, on community pharmacy. The first of these was the Omnibus Budget Reconciliation Act of 1990 (OBRA '90). This act required pharmacists, for the first time in history, to counsel patients and conduct prospective drug utilization reviews of the appropriateness and safety of prescribed medications. The second legislative impact has been the administration's Health Security Act, with its emphasis on universal health coverage. Although this comprehensive federal health reform legislation has yet to be enacted in any form, at the date of this writing, it is influencing the health care system at the state level, as states move to anticipate the consequences of national health care reform. In addition, pharmacy and all other health professionals are working to address and anticipate the revolutionary changes that are seen for the future of health care. The way in which federal legislation can and does dramatically influence practice in a profession is exemplified by OBRA '90. Since OBRA '90, every state board of pharmacy in the United States has adopted rules and guidelines, including enforcement procedures, by which patient counseling and other activities of the pharmacist must be conducted in that state. Most states have extended the OBRA '90 legislation, which was limited to Medicaid patients, to include all patients. In any future federal legislation design to increase access to health care, it is widely anticipated that prescription drugs will almost certainly become a benefit under Medicare.

In addition to the foregoing changes, the pharmaceutical industry, the hospital industry both public and private, and managed care organizations, all are also moving to implement change in an effort to reduce negative consequences of federal health care reform, and to best position themselves for the inevitability of the change and the resulting effects that are seen to be coming with change. Among these repercussions are threatened price controls.

The goal of OBRA '90 was to control the rising cost of Medicaid. A leading goal of health care reform is to contain rising health care costs in general, which first reached 1 trillion dollars in 1995. Other objectives of health care reform are to improve access to care while maintaining quality in the delivery of health care services.

Vogenberg and Pisano [1] have recently examined how the new health care environment is affecting quality control issues, pharmacists' interventions and payment for cognitive services, outcomes research and pharmaceutical care, and the use of pharmacy technicians as well as the likely effect of these changes on the near and longer-term future of pharmacy practice.

In the quality control of professional services area, the Agency for Health Care Policy and Research (AHCPR), the health care policy research arm of the federal government, is studying the delivery of pharmaceutical care and its financing implications and patient benefits in the future. Similarly the *United States Pharmacopeia* (*USP*), the Joint Commission on Accreditation of Health care Organizations (JCAHO), and other accreditation bodies, are directing their attention to standards development on pharmacy products or services. The profession of pharmacy, especially pharmacists who practice in long-term care institutions, home care organizations, and retail pharmacy, will all be affected by the actions of these groups in the years ahead.

Also mandated by OBRA '90 was that state Medicaid programs implement a prospective drug use review (DUR) program by January 1993. A required component of prospective DUR was the provision of counseling. OBRA '90 also required each state to establish, by law or regulation, standards for the counseling of its Medicaid patients. These state and federal rules and regulations have clearly set a new standard for the quality of pharmacy practice. For many community pharmacists, counseling is already a part of their daily practice.

Although federal guidelines call for the pharmacist to offer to counsel, HCFA recently clarified this to allow ancillary personnel to make the counseling offer. Similarly, during the summer of 1993, HCFA said it would allow ancillary personnel to collect records and maintain profile information, but mandates that a pharmacist must review and interpret the information. In other words, it is the pharmacist's responsibility to be sure that the necessary steps are completed. It is the state, however, that is establishing the standards for counseling by pharmacists.

The new counseling requirements are providing pharmacy with an unprecedented opportunity. The goal of patient counseling is to encourage interaction and exchanges of information between the pharmacist and the patient. This process has the potential to improve the quality of patient care outcomes and reduce overall costs. As noted by Vogenberg and Pisano in their article [1], pharmacists have long clained that they add value to the health care dollar. Now federal and state government and health care providers, who collectively pay for the majority of health care costs in the United States are asking pharmacy to prove it. Pharmacy must prove it, and how and when it occurs, as the first step in defining pharmacy's role in managing patient care, and indeed in defining the very future of the profession. Where the proof exists and is documented, reimbursement for nondispensing functions will follow. If pharmacists cannot or do not practice at a level that documents their value in the health care delivery system, not only will they not be paid for cognitive services, many of their traditional roles can be assumed by others, or even by machines. Nurse practitioners and physician assistants are aggressively positioning themselves to assume more responsibilities for patient care in a reformed health care environment. As noted by Helper and Strand [2], if pharmacists are to attain the goal of receiving payment for professional (cognitive) services, they must concentrate on what is in the best interest of the patient when it comes to providing pharmaceutical care.

III. THE PHARMACEUTICAL INDUSTRY

The pharmaceutical industry has been responsible for the development of the vast majority of the modern drugs that have had the greatest bearing on human longevity, health, and well-being as noted in Chapter 1. In addition, by formulating drugs into drug products that are not only stable, effective, and reliable, but also convenient to use and readily acceptable by the patient, the drug industry has altered the role of the pharmacist from that of a compounder to that of a dispenser and councilor. Since the drug industry will continue to play the major role in future new drug discovery, and will provide the profession with its primary commodities, pharmacists are and should be concerned with the health and vigor of the industry worldwide and in the country in which they practice.

During the past 50 years, the pharmaceutical industry in Japan, North America, and Western Europe has experienced an unparalleled period of expansion and change. These developments have had a direct bearing on many of the professional activities of both the community and hospital pharmacist. Whereas, at the beginning of the present century, most dosage forms were prepared in the community pharmacy or hospital, the situation in North America and Europe is now quite different, with by far the greater proportion of all dosage forms being prepared by the pharmaceutical industry. In recent years, there has been some reversal in this trend, in that a number of hospital pharmacists are now producing some of their own parenteral products and are heavily involved in IV admixture programs, and some community pharmacists are increasingly engaged in the compounding of specialty products.

The period during and immediately following World War II saw an explosive development of new drugs. Several classes of new therapeutic agents were introduced during this period, including antibiotics, steroids, tranquilizers, and the first anticancer agents. However, from about 1960, until recently, there was a significant decline in the rate at which new drug substances were introduced. The number of new drugs introduced onto the American market during the triennium 1957–1959 was 168, whereas from 1969–1971 the number was only 39, and has in general continued at this lower introduction rate. There were several reasons for this decline, despite the fact that the U.S. pharmaceutical industry spends a higher percentage of its revenue on research and development than any other industry (18.8% of nearly 14 billion dollars in 1994) [3]. A major cause for the decline in new drug product introductions has been the incresing rigor with which regulatory agencies, such as the Food and Drug Administration (FDA), consider applications to market new drugs. Much more data are now required than in the 1950s. The time, labor, and money required for the development of a new drug has increased over 1000% during the past 35 years or so. It must also be remembered that the percentage of drugs that successfully pass the various stages of development is very small. It now takes an average of 12 years to develop a new drug. Thus, there is much more caution on the part of the industry to make the investment of over 359 million dollars, which is the average cost of bringing a drug from the research laboratory to the marketplace today [3].

One result of the high cost of bringing new drug substances to the market has been the increased attention that many pharmaceutical manufacturers are now giving to dosage form design. There is an increasing realization that the rate and intensity of the pharmacological response and the duration of drug action can often be radically modified by the dosage form. Some companies have been established whose major research thrust is the development of novel dosage forms, rather than the discovery of new drug substances. In a later section of this chapter, more detailed attention will be given to some of the contemporary trends in such formulations.

The previous comments should not be construed to indicate that the potential for development of new drug substances is ever-decreasing. There are still many areas of medicine for which the possibility of developing new drugs is the major hope for effectively combatting various disease states or affecting a substantial improvement in therapy. For example, successful antiviral drugs could revolutionize the prognosis of many diseases, the treatment of which is currently little more than symptomatic. It is true that the search for drugs that are effective against viruses in humans and have low toxicity is an expensive and scientifically demanding task, but it is our opinion that chemotherapeutic agents in this area will become increasingly available in the next decade. Drugs such as interferon and interferon inducers are presently (1995) being tested as chemotherapeutic agents for the treatment of a variety of diseases, including the much publicized acquired immunodeficiency syndrome (AIDS). Future developments in pharmacotherapeutics are likely to be characterized by an intelligent use of both new drugs and new delivery systems.

The vast increase in the cost of developing new drugs, which now exceeds one-third of a trillion dollars for the successful new compound, has caused many small- and medium-sized pharmaceutical companies to curtail or eliminate their research for new drugs. Even the largest companies now specialize in the types of research that they undertake. Thus, one company may work mainly on cardiovascular drugs and anti-infectives, whereas another may concentrate on central nervous system drugs. Companies also move vigorously to pursue "breakthrough" or "blockbuster" drugs, as opposed to working on drugs that would represent only modest enhancements. The high cost of new drug development, combined with marketing advantages, have driven a major consolidation within the pharmaceutical industry in the last 5 years, as larger or more cash-rich companies buy smaller or less successful companies. Glaxo has recently purchased Burroughs Wellcome. The new merged company has become the world's largest pharmaceutical manufacturer. Some predict that there will only be six or fewer major pharmaceutical companies in the United States in the foreseeable future as a result of these consolidations.

Competition in the drug industry, as well as marketing practices, are changing rapidly and dramatically. Whereas prescription pharmaceuticals have been traditionally marketed directly to individual physicians by pharmaceutical sales representatives, this practice is changing as a result of forces in the new health care environment. Health care provider groups are increasingly developing formularies. Not only do these formularies dictate the maximum cost that will be paid for a particular generically available drug, they also often specify which drug or drug combination they will allow to be covered for a particular therapeutic indication. Thus drug companies must convince the pharmacy benefits managers of these managed care companies (which may cover tens of millions of lives in a region), that their products are not only therapeutically very effective, but also the most cost-effective therapy among the various options. On the one hand, this will result in fewer job opportunities for pharmacists as pharmaceutical sales representatives; on the other hand, it has created new and perhaps higher level and more challenging job opportunities for pharmacists in industry who are engaged in marketing to the managed care companies, or who are serving as benefits managers for those companies.

For drugs no longer covered by patents, there is vigorous competition. Policies such as the "paper NDA" introduced by the FDA have assisted smaller pharmaceutical companies to market generic products once a drug substance is no longer subject to patent protection. However, many major drug companies have also created generic divisions to share in the opportunities of this market.

Pharmacy benefits companies have been created to assist managed health care companies with their drug benefits programs. Some of these pharmacy care benefits companies design drug programs for from 30 to 50 million people. The largest three pharmacy benefits companies in the United States were purchased by three major drug companies in 1993–1994 (Medco by

Merck; P.C.S. by Lilly; and United Health Care by Smith, Kline Beckman). The benefits of the ability to a drug company to potentially be able to influence the formularies of such pharmacy benefits companies is obvious. At this writing, the Federal Trade Commission is reviewing these acquisitions.

Another influence of the growing role being played by managed care in drug selection is seen in business agreements being struck by drug companies to combine or "bundle" their products together, to enhance their marketing attractiveness or clout to and with the managed care industry. As health care reform continues to advance, the pharmaceutical industry and its members will undoubtedly develop additional strategies to be competitive, let alone viable.

IV. COMMUNITY AND HOSPITAL PRACTICE

The present and future economic and professional trends in both community and hospital pharmacy are in a more dramatic state of flux and change than at any time in American pharmacy's history. The results of these forces are not easy to predict. The economic pressure caused by state and federal governments to contain health care costs might tend to indicate a bleak economic future for pharmacy, especially if pharmacy is seen only as a product distribution system. In that scenario, technicians, automated systems, and even robotics can handle dispensing functions, and distribution can be handled by Federal Express or the U.S. Postal Service. On the other hand, if pharmaceutical care becomes a reality, and pharmacists provide meaningful, cost-effective benefits through documented pharmaceutical services, that optimize pharmacotherapy and minimize drug misadventuring, and such care becomes an integral part of the nation's health care, the future of pharmacy will be the brightest in our history. In other words, the profession of pharmacy faces what could become the worst of times, or the best of times. There may not be a whole lot in-between. How this scenario plays out in the years immediately ahead will determine what pharmacists do in the majority of pharmacy's settings, how they are paid and for which activities, their levels of practice, the skills they need, and how many pharmacists are needed to service this country's health care needs. It is believed that this story will unfold on a state and regional level, rather than nationally. States that will lead the way have strong pharmacy leaders and state associations; visionary and courageous pharmacy practitioners who are willing to take the risks to move practice ahead, while knowing that payment for pharmacy services will lag; and colleges of pharmacy who are not only preparing students to meet tomorrow's challenges, but are also partnering with practitioners, organized pharmacy, and even health care providers to move the profession to an entirely new level. One hopes that these states will carry the day, moving the profession forward nationally, rather than having states or regions that resist change and lag, holding the process of practice advancement and conversion back. National pharmacy associations, working collaboratively perhaps for the first time, offer another hope of moving the practice of pharmacy ahead, beyond a survival mode, to a partnership role in health care at the national level.

V. FUTURE PHARMACY WORKFORCE NEEDS

In examining the future of pharmacy, it may be helpful to look at the demographics of the profession and examine how the workforce has changed over about the last 20 years. Manasse [4] recently examined the pharmacy manpower issue, utilizing the 1990 and 1991 pharmacy licensure revewal cycle in the 50 states and the District of Columbia, as it had been compiled through the Pharmacy Manpower Database Project. In summary, using this most recent national tabulation, and recognizing that the 1990–1991 figures would now be somewhat low, the following data was presented. In 1990–1991 there were nearly 195,000 licensed pharmacists in the United States. Of this number, slightly over 88% considered themselves to still be in active practice, meaning that the nation had approximately 172,000 actively practicing pharmacists. Of these, 63% were 49 years of age or younger: 30% were female, and approximately 15% represented minorities. Just over 10% of licensed pharmacists hold degrees other than the B.S. in pharmacy (e.g., Pharm.D., M.S., Ph.D., and M.B.A.). This number will be increasing fairly rapidly in the years ahead, as more graduates and a higher percentage of graduates each year will hold the Pharm.D. degree. Seventy-five percent of the active pharmacists were employees, and 10% were listed as self-employed (primarily community pharmacy owners). An equal percentage of pharmacists, 33% in each case, worked in independent community pharmacies and in chain pharmacies. About 24% worked in hospital pharmacies, and the remaining 10–11% worked in other pharmacy-related practice settings.

Comparing this data to 1970s data, when a similar nationwide personnel survey was conducted, there were 140,000 active practicing pharmacists in 1974, compared with 172,000 in 1990–1991. This represents a 23% increase in a 16- to 17-year period. Given the change in population over that same period, the number of pharmacists increased from 56 to 66 per 100,000 population; about an 18% increase. It is estimated that the number of pharmacists in practice per 100,000 population today is approaching 70. Over this same time span, the number of women in pharmacy has increased from fewer than 10% to over 30%. Whereas women constituted only about a third or fewer of the students in colleges of pharmacy, only three decades ago, today two-thirds or more of the students are women. Thus the number of women in the pharmacy workforce will continue to dramatically increase in the years ahead. Several surveys have shown that many women work only part-time in pharmacy during their childbearing and child-rearing years. This may account in part for the continuing shortage of pharmacists in many regions of the country, even though the number of pharmacists per 100,000 population has increased. Not all of the increase in numbers of pharmacists per unit of population can be attributed to pharmacists assuming new professional roles.

Probably at no time in the history of pharmacy has it been more difficult to make estimates for the future needs of the profession's future workforce. If pharmacists are unable or uninterested in assuming more responsibility for the delivery and outcome of pharmacy care, some estimates indicate that we would need only 50,000 or fewer pharmacists in the future. Expanded use of technicians, automation, and robotics will be a reality. If on the other hand, pharmacists truly become credible providers of primary health care, engaged in meaningful and cost-effective monitoring and optimization of pharmacotherapy, the current number of practicing pharmacists will be very inadequate in the future.

Another major challenge faced by the profession will be enhancing the skills of the nation's pharmacists to permit them to more fully engage in pharmaceutical care, encompassing the optimization of pharmacotherapy, including objective assessments of therapeutic outcomes in all manner of primary care, community-based settings. As noted in the previous section, this is a challenge that must be addressed on a state-by-state level, involving partnership efforts between colleges of pharmacy, state pharmacy associations, and practitioners. Among reasons that leadership must be exerted at the state level to bring about the necessary changes in pharmacy are issues of reimbursement, the practice environment, pharmacy organizational structure and strengths, professional leadership, the goals and directions of the college(s), and the political environment, all of which vary state by state. Under any future health care reform initiatives, federal legislation will probably provide only enabling regulations under which the states will operate. Consequently, as previously stated, implementation of future health care reform is expected to be largely at the state level, and it will be at that level that pharmacy's future will largely be determined.

VI. ADVANCES IN THE MONITORING OF PHARMACOTHERAPEUTICS AND IN DRUG DELIVERY SYSTEM DESIGN

The basic objective of dosage for design is to supply the drug, at the appropriate concentration, to the target organ. There are several factors that control the concentration of drug at a target organ: (a) patient compliance, (b) intersubject variations, and (c) control of drug release from the dosage form.

A. Patient Compliance

There is good reason to believe that in many cases when pharmacotherapy fails, the reason is not because of any deficiency in the drug or dosage form, and not because of any biochemical abnormality in the patient. Rather, the cause of failure is simply that the patient has failed to follow the instructions for the use of the medicine. In some cases, noncompliance is due to plain forgetfulness; in others, a patient may have incompletely understood the instructions or have become confused. Many elderly patients are often taking five or more medications at the same time. This factor, in combination with a reduced mental alertness frequently seen in the aged, often results in errors occurring. One of the most effective way to improve compliance is to provide supportive counseling when the medicine is dispensed. Studies in which hospital pharmacists have developed a structured system for the education of patients in the use of their prescribed or over-the-counter (OTC) drugs have shown that this approach can clearly have considerable success. With the mandate that patient counseling be provided, that now exists in most states, as well as through the increased provision of pharmaceutical care to ambulatory, primary care patients, it is anticipated that drug misadventuring resulting from noncompliance or poor compliance will be greatly reduced.

Pharmaceutical technology can also offer some assistance in improving compliance. First, the use of unit-dose systems in hospitals makes it much easier for the nursing staff to see that the patient takes the product as instructed. Also, a number of special devices, such as a set of plastic boxes labeled with dates and supplied to the patient with each box filled with appropriate drugs, can be used for ambulatory patients. One factor sometimes overlooked in endeavors to improve patient compliance is that the dosage regimen can have a significant effect on compliance success. Thus, if a patient is required to take a product three times daily, failure is more likely that if the regimen is one or twice a day. It is well documented that sustained-release products that reduce dosing frequency may improve patient compliance.

B. Intersubject Variations

The blood levels of drug produced by the same dose of the same manufacturer's lot of a given drug product can often show considerable intersubject variation. Such variations can be the result of physiological variables within and between individuals, disease state effects, or other differences. Fortunately, for many drugs, a high therapeutic index (the ratio of lethal dose for 50% of the population to the effective dose for 50% of the population) and a relatively shallow slope of the dose–response curve means that precise control over blood levels is not always essential. However, there are drugs for which quite precise control over blood levels is essential to ensure efficient pharmacotherapeutics and to avoid toxicity. The classic example here is digoxin, a most useful drug, widely used to treat cardiac patients, but one with a very low therapeutic index. There are many other drugs for which blood-level control is particularly important, with lidocaine, phenytoin, lithium, theophylline, and gentamicin being a few notable examples. As a result of an awareness of this problem, major hospitals have well-established clinical pharmacokinetic laboratories and services. These services, which should be directed by an experienced pharmacokineticist, provide a service that allows blood or other fluid or tissue levels of drugs in individual patients to be adjusted using appropriate pharmacokinetic principles. The benefit of such services has been well documented for many years [5].

The ultimate drug delivery system for critical drugs would continuously sense the individual patient's response to a drug and, through some feedback system, continuously adjust ongoing drug administration. To "close the loop" for such a drug delivery system, we need to incorporate into it a sensor, the output of which will be a linear or simple nonlinear function of the levels of drug in the blood, or of a pharmacological response of the drug. One possible device that could be used for such a purpose is an enzyme-immobilized electrode. These relatively new devices represent a fascinating development in electrochemistry. The reader is aware of the use of the conventional hydronium ion-sensitive glass electrode to measure pH. During the past 15 years, many remarkable advances have been developing electrodes containing enzymes. These electrodes respond to the activity of an enzyme substrate in the vicinity of the electrode [6]. Progress is being made in developing such electrodes that contain drug-metabolizing enzymes. Thus, it would seem quite feasible to develop an intramuscular drug delivery system that contains a miniature enzyme-immobilized electrode responsive to the appropriate drug. The potential of the electrode could in turn serve as the feedback to control the subsequent rate of release of drug into the body. A great advantage of this device would be that the electrode signal would be a function of free (non–protein-bound) drug concentration, rather than a total drug concentration. It is generally accepted that, for many drugs, the intensity of pharmacological response is a function of free drug concentration, rather than total drug concentration.

An accurately quantifiable pharmacological response, such as blood pressure, heart rate, electroencephalogram (EEG) or electrocardiogram (EKG) pattern, respiration rate, and skin or other body temperature, may be used to monitor or reflect drug responses and provide the basis for subsequent automatic administration according to need. Such systems are well along in development to provide for safer and more reliable monitoring of general anesthetics. A system known as a beta cell has been developed to continuously monitor the need for insulin by the diabetic and permit adequate subsequent drug administration automatically and according to need.

C. Control of Drug Release from Dosage Forms

A major contribution that the pharmaceutical formulator can currently make in designing drug products that provide well-controlled drug concentrations in a patient's bloodstream is to provide controlled absorption (drug delivery systems). Several chapters in this book have given comprehensive treatment to the topics of drug absorption and biological availability, as well as sustained and controlled drug delivery. The goal of an optimized drug product that was simultaneously as effective, safe, and reliable as possible was described in Chapter 1. Only two or three decades ago many pharmaceutical formulators would have considered their duties as complete if the product designed was potent (contained the labeled amount of drug), pure, and chemically and physically stable. Today, however, the formulator's professional horizons must be extended to include responsibility for the in vivo performance of the dosage form, including optimizing the absorption rate and reliability, and blood level versus time profiles. This change has been brought about very largely by the effects of biopharmaceutics and pharmacokinetics and by advances in pharmaceutical technology.

D. Contemporary Trends in Pharmaceutical Product Design and Production

In Table 1, various trends and modern approaches to contemporary pharmaceutical product design, formulation, and production are listed. The complexity of the modern dosage form and the multiplicity of factors that must be taken into account in dosage form design and evaluation necessitates a systematic approach to modern formulation. These approaches, as well as the many advancements noted in Table 1 are described throughout the various chapters in this book.

VII. ADVANCES IN PHARMACOTHERAPY IN THE NEW MILLENNIUM

As we enter the next thousand-year period in the history of humankind, a new millennium starting with the 21st century, it is interesting to consider advances in pharmacotherapy with a view to the future.

There have been four scientific and technological revolutions that have occurred in the 200-year-plus history of the United States, and three of these revolutions have occurred in this century. The first revolution was the mechanical and engineering revolution. It began inventions like the cotton gin and the steam engine. The steam engine changed the shipping industry from sailing to steam ships, and made possible a railroad system that opened a new country to settlement and industrialization. This was followed by the internal combustion engine, which

Table 1 Contemporary Trends in Pharmaceutical Product Design and Production

More reliance on scientific, rather than intuitive, techniques (including preformulation approaches to
   drug product design)
Use of physicochemical methods to improve control over rate and extent of drug release
Wider use of unit-dose products
Development of target organ-oriented dosage forms
Use of routes of administration other than oral for self-administration of drugs (such as transdermal
   delivery)
Improvements in the reliability of the oral route of drug delivery by minimizing physiological variables
   and other effects
Automation of production operations and use of in-process quality control for total production
   monitoring
Larger batch sizes and development of some continuous production processes
Development of higher-speed pharmaceutical production equipment with closed-loop production control
Application of optimization procedures to drug product design and manufacture
Use of validation procedures to analyze production operations
Development of manufacturing methods that are totally environmentally compatible
Increased emphasis on development of new improved drug delivery systems
Development of totally new methods of drug production (e.g., the use of recombinant DNA
   technology)

led not only to the automobile, but also the farm tractor, which revolutionized farming and largely made modern agriculture possible, and then to the airplane and today's aircraft and airline industries; and to jet and rocket engines, resulting in our early exploration of space to this date. This mechanical, engineering, and agricultural revolution was greatly facilitated in this country by Abraham Lincoln's Morrel Land Grant Act, which created land grant universities across the country. The charge for these universities was to "enhance agricultural and the mechanic arts."

The next three revolutions began or occurred primarily or exclusively over the last 60 years, and were *scientific* revolutions. The first of these, clearly beginning about 60 years ago, was the atomic revolution (Table 2). This revolution not only developed atomic and molecular theory, and ultimately led to the first nuclear weapons and atomic energy, but also to many other advances in the various fields of chemistry shown, not the least of which was medicinal chemistry. Without these advances in chemistry, the "golden age of drug discovery" could never have occurred (see Chapter 1). The second scientific revolution of the 20th century, that began in about the 1950s, and continues yet today, was the electronic revolution (see Table 2). Some describe the age in which we live today as the *electronic age*. Clearly, computers and electronics have reshaped our lives, from modern kitchens, to our offices and manufacturing plants, to the control systems on our automobiles, our televisions, radios and high-fi's, our word processors, and pocket computers. Indeed, the electronic revolution and what it has brought is heavily impinging on other fields of technology as well, through computer-aided drug design. This approach is being used in many industries, including the pharmaceutical industry, to design new drugs through molecular modeling, for advanced information systems, data management, and data networking systems.

The third scientific revolution, which began during the last two decades of the 20th century, is the biotechnology revolution (see Table 2). Many scientists and forecasters believe that biotechnology will have a far greater influence on Americans and the way we live, and indeed on humankind in the 21st century, than did either of the previous two scientific revolutions that have so greatly affected our lives in the 20th century. Biotechnology is revolutionizing medicine and the pharmaceutical industry, including methods of diagnosing, treating, preventing, and even now predicting disease. It will dramatically enhance the quality and duration of life, in ways that are virtually beyond our imagining.

As was true with all of the previous scientific revolutions, the biotechnology revolution is founded on a science base, including the new science of molecular biology. Molecular biology and biotechnology are helping us understand human biology and other life processes, both animal and plant, at a systems level, at a cellular level, and even at a molecular level. We are identifying complex proteins and peptides, which often are exemplified as enzyme systems, that function as the body's natural bioregulators. With use of a powerful tool of biotechnology, genetic engineering, it has become possible to produce some of these complex bioregulators in quantities that are useful in medicine.

Biotechnology is also helping us develop an understanding of the body's immune system. This will not only lead to much more effective treatment of allergies, but may also lead to prevention or total cures of immunologically related diseases, such as type II, or adult-onset diabetes, along with numerous other diseases thought to have an immunological basis.

A major project of national and international scope that will also drive the future of biotechnology, is the human genome project. This multibillion dollar project has as its goal, the complete deciphering of the approximately 100,000 genes that each of us possess. Each gene in turn is comprised of a complex DNA structure with 3 billion base pairs of double-stranded DNA chains that define out makeup and biological future. We are developing an ability to look at this genetic material and identify exactly where the genetic defect exists that results in such

Table 2 Three Scientific Revolutions of the 21st Century

Atomic

Electronic

Biotechnology

The first scientific revolution of the 20th century
The atomic revolution

Organic chemistry   Physical chemistry
  Quantum chemistry   Medicinal chemistry
  Polymer chemistry   Atomic energy
        Atomic and nuclear theory

The second scientific revolution of the 20th century
1950s, 1960s, 1970s, 1980s
The electronic revolution

The transistor replaces the vacuum tube

The computer
      The supercomputer
        The microcomputer

Automation, electronic control, computer aid design, high-speed computing, information systems,
modern telecommunication, data management (storage and control)

The third scientific revolution
of the 20th century
(1980s into the 21st century)
The biotechnology revolution

Molecular biology
Recombinant DNA technology
Monoclonal antibody technology   New drugs
Nucleotide blockage
  (antisense therapy)   New treatments
Genetically engineered vaccines
Gene therapy
Interferons
Interleukins
Vaccines
Growth factors

inherited diseases as cystic fibrosis, Huntington's chorea, diabetes, and many other diseases. This will soon allow us to look at the gene map of an individual to determine if they carry genetic defects that may predispose them for the many diseases that we now know have a hereditary basis. Work is already underway to replace defective genetic material with normal material, using gene therapy as a method of treating previously untreatable diseases. The unraveling of the human genome and developing the gene therapies that will follow, offer enormous potential to the future treatment and eradication of many major disease states.

Other research that is underway has the potential to affect life on this planet in ways that are hard to imagine. These are studies on aging. Fundamental work is underway to analyze why cells age and how the aging process can be altered and controlled. Once we understand the aging process at a molecular level, who is to say that it will not become possible to retard or even reverse the aging process? It is also becoming clear that genetics play a role in longevity. Indeed, in some simpler life forms, such as nematodes, researchers have discovered a gene that causes aging. When such a gene or combination of genes is found in humans, it could become a routine matter to alter such genes to extend life. Gene therapy will also clearly have the ability to influence the onset of the degenerative diseases that accompany aging. As noted previously, genes are made up of DNA chains, and these chains can be damaged. Damaged DNA results in mutated cells. When enough cells mutate, the whole organism can fall victim to degenerative diseases. As gene therapy advances and as we identify which genes and which DNA alterations produce the mutated cells leading to degenerative diseases, it will become a routine procedure to block and drastically forestall such aging effects.

Other breakthroughs in biotechnology include those in neuro- and psychopharmacology. By developing and understanding brain chemistry we have been able to develop dramatically more effective antidepressant drugs such as fluoxetine (Prozac). We are also coming closer to understanding the biochemical and molecular cellular causes of paranoia and schizophrenia. In the process we are also closing in on the chemistry of normal personality. Some leading neuropsychiatrists expect that a major future focus of drug development will not be at "patients," but at normal people, who are already functioning at a high level, but who wish to enhance their memory, their intelligence, heighten their concentration, or improve their day-to-day moods. Brain mapping, to provide an understanding of which regions of the brain and which biochemical factors influence both normal and abnormal human responses, is becoming well advanced. This brain mapping is leading to an understanding of such human traits as shyness and hypersensitivity, impulsiveness and obsession, anxiety and concentration, and other characteristics. Although this may sound like science fiction, alteration and control of such traits and characteristics is probably only a few decades away, at most.

A technique to recombine genes was patented in 1961, which set the stage for the biotechnology industry. In 1975, Genentech was founded as the first biotechnology company, and the first cancer-causing gene was identified. In 1983 the gene for human insulin was cloned, which led to the marketing of the first biotechnology-produced drug product, Humulin, in the mid-1980s. In the same year, the genetic marker for cystic fibrosis was found [7]. Currently, there are approximately 25 biotechnology therapeutic agents and vaccines approved for sale in this country by the FDA. As shown in Table 3, in a 1993 survey there were 143 biotechnology medicines in testing, representing 170 different research projects in a wide range of product classes [8]. Table 4 illustrates the leadership role of the U.S. pharmaceutical biotechnology industry, wherein it is shown that in the most recent available period to survey patents, 140 of the 178 patents in this field were from the United States, nearly 70% of the total [8]. The origins of these various patents are shown in Table 5 [8]. Nearly half of all biotechnology patents issued are in the pharmaceutical and health care areas. The biotechnology medicines now available are treating a wide range of indications: AIDS, diabetes, dwarfism, hepatitis,

Table 3 Biotechnology Products in Development (1993)

Clotting factors	1
Colony-stimulating factors	6
Dismutases	1
Erythropoietins	1
Gene therapy	1
Growth factors	9
Human growth hormones	4
Interferons	11
Interleukins	10
Monoclonal antibodies	50
Recombinant soluble CD4s	2
Tissue plasminogen activators	1
Tumor necrosis factors	3
Vaccines	20
Others	23

Table 4 Genetic Engineering Biotechnology/ Pharmaceutical Patents by Country of Origin in a Recent Year

Country	No. of patents
United States	140
Japan	16
European	14
Other	8
Total	178

Table 5 Genetic Engineering Biotechnology/ Pharmaceutical Patents by Location of Inventor and Patent Assignment

Location	No. of patents
Large U.S. drug company	43
Biotechnology company	37
Foreign	38
U.S. university	25
U.S. nonprofit organization	20
U.S. individual	8
U.S. government	7
Total	178

Table 6 Some Diseases and Patient Populations Targeted for Biotechnology Therapies in the United States

Disease	Population
Heart attack	1.5 million people every year
Cancer	1 million new patients a year
AIDS	1.5 million HIV-infected men, women, and children
Diabetes	600,000–1.1 million people
Anemia	500,000 people
Hemophilia	20,000 people
Transplant patients	4,500 kidney patients a year
Dwarfism	1,700 children

heart attacks, anemia, leukemia, renal cancer, organ transplant rejection, cystic fibrosis, multiple sclerosis, and sarcoma. Drugs and vaccines under development are targeted to some of our most intractable diseases, such as cancer, arthritis, Alzheimer's, osteoporosis, and various genetic disorders. Table 6 lists some of the diseases and disease populations targeted by the biotechnology industry for new therapies [9].

Health care costs in the United States reached 1 trillion dollars for the first time in 1995. Drug costs represent somewhat less than 7% of the total health care dollar, or about 70 billion dollars.

Heart disease (cardiovascular and cerebrovascular diseases) and cancer are the number 1 and 2 causes of death in the United States, having the prevalence and economic consequences shown in Table 7 [10, 11]. Other diseases of the elderly, with their prevalence and economic effects are shown in Table 8 [12]. The economic effect per year of the diseases listed in Tables 7 and 8 alone is nearly 500 billion dollars. As the life expectancy of Americans continues to increase, the prevalence and cost of these diseases will also continue to increase. The primary method of combatting all these diseases will be using drugs, and drugs with an origin in biotechnology will become increasingly more important and dominant.

VIII. INTERNATIONAL DEVELOPMENTS

One of the most notable developments that has occurred between the publication of the first edition of this book in 1980 and the publication of the present third edition has been a dramatic increase in the level of international coordination in the development, production, and regulation of drug products. In some instances, the improvement in international cooperation can reasonably be attributed to regional or global political developments, such as the maturation of the European Community; in others, the driving forces have come predominantly from within our industry and profession (e.g., the desire to reduce costs for standardizing the formulation of product X that is produced by a pharmaceutical company at various locations throughout the world).

The three most important areas of the world in terms of the production of pharmaceutical products are (in order of monetary value) (a) The European Community, (b) North America (United States, Canada, and Mexico), and (c) Japan [12]. If present trends continue, it is possible that within the early years of the 21st century Japan will move into second place. In 1960 the Japanese pharmaceutical industry was ranked fifth in the world and was responsible for somewhat less than 7% of total world production of pharmaceuticals. By 1978 Japan had

Table 7 The Number 1 and 2 Causes of Death in the United States: Their Prevalence and Economic Impact

Cardiovascular and cerebrovascular diseases
    Prevalence of all types of disease: 70 million plus Americans
    Deaths: about 1.1 million/yr
    Economic impact: 120 billion/yr
    Drugs in development: 86
    More than one in four Americans has some form of these diseases and two in five will die from this cause.
    Heart attack is the number 1 killer of both men and women.
    Stroke is the number 3 killer and the number 1 cause of adult disability.
    The stroke death rate has been cut by one-third since 1979.

Cancer
    1993 incidence: all types of cancer—1.2 million
    Deaths: 526,000/yr
    Economic impact: 104 billion
    Drugs in development: 124
    The number 2 cause of death in the United States.
    People older than 65 years of age are ten times more likely to develop cancer than those younger than 65.
    The incidence of cancer in the elderly is increasing.
      By age 85 one of nine women will develop breast cancer.
      By age 85 one of ten men will develop prostate cancer.
    One of every five deaths in the United States is from cancer.

moved to third position and was responsible for 20% of world production. Current estimates place Japanese production in excess of 27% of the world output.

    Several Japanese companies (e.g., Takeda) are now aggressively entering the U.S. market, and if the success of Japanese automobile sales is indicative of what may happen in the pharmaceutical field, we may expect to see Japanese pharmaceutical companies rapidly establishing themselves in the United States as entities of importance.

Table 8 Other Diseases of the Elderly in the United States

Disease	Prevalence	Deaths	Economic impact	Medicines in development
Arthritis (all types)	55 million		$35 billion/yr	42
Neurological disorders				
Alzheimer's disease	4 million	100,000+/yr	$90 billion/yr	19
Parkinson's disease	1 million		6 billion/yr	6
Respiratory Diseases				
Chronic bronchitis and emphysema	14.6 million	85,000/yr	8 billion/yr	9
Pneumonia	4 million/yr	77,000/yr	6 billion/yr	10
Other diseases				
Prostate diseases	10 million	2,500	3 billion/yr	5
Depression	11.6 million	31,000 (suicides)	30.5 billion	12

It may now be a misnomer to refer to large pharmaceutical companies as being "German," "Swiss," or "American." Their research, production, and marketing activities occur in many countries, employing personnel of many nationalities. Such companies are best described as international or multinational.

It is well established that there are some pharmagenetic differences between different ethnic groups. For example, isoniazid shows quite substantial differences in the rate of metabolism between Europeans and Eskimos. Similarly, there are cultural differences that can affect the acceptability of different drug delivery systems. Thus, although rectal suppositories are widely used in Francophone countries as a method of supplying drug to the general circulation, in Anglo-Saxon countries, the use of rectal suppositories is usually restricted for treatment of pathological conditions specifically involving the rectal vault. However, these differences between ethnic groups are generally of relatively minor importance, and thus for most pharmaceutical products there is good reason to believe the product will be equally as effective in Mongolia as in Patagonia. The only caveat that must be placed on this general statement is that the label and package or container may require modification for different countries. The label must obviously be printed in the language, or languages, that prevail in the country of use and must comply with all relevant legal requirements that that country places on label content. Also, for some countries in which climatic conditions are especially onerous, such as those where there are extensive periods of hot, wet weather, a special heavy-duty tropical container or foil pack may be required. However, for many areas of the world there is no compelling clinical, scientific, or technical reason why pharmaceutical products (with the exception of the label) should not be entirely standardized in terms of components, production, and evaluation. There is obvious economic advantage to such standardization, particularly for drug products designed for use by those suffering from relatively rare pathological conditions. If a pharmaceutical manufacturer is forced to substantially modify a product, the sales of which in even the largest single markets, such as the European Community or United States are relatively small, the chances of the product being available on a global basis are dramatically reduced.

The problem of developing drug products for rare conditions has been recognized by a number of countries. For example, in the United States the FDA provides Orphan Drug status for new drugs being developed for conditions experienced by only small numbers of patients. The time and cost of obtaining regulatory approval for such products is greatly reduced.

It will be readily appreciated that the profit for a pharmaceutical product increases as annual sales increase. Thus, the greater the potential market for a product the more likely it is that a free enterprise company will focus its attention on the product area. Consider the case of persons suffering from Wilson's disease, a relatively rare condition affecting a small proportion of persons of European descent who have an enzyme deficiency that results in an inability to control serum copper levels. If we were presented with a project for the design and production of a specialized drug delivery system of value to these suffering from this condition, the economic viability of the project would be greatly increased if a single formulation produced using common equipment could be manufactured and approved throughout the world. Global standardization of the development, production, and regulatory approval of pharmaceutical products has obvious value as a method of improving our ability to cure or alleviate the ills that afflict mankind.

A world-important aspect of the pharmaceutical industry is the increasing frequency of ad hoc joint development, or license agreements among smaller companies based in North America, Western Europe, and Japan. Such arrangements, by spreading the cost of development among a number of companies, allow relatively small concerns that would otherwise be totally excluded from research, to participate in a meaningful fashion in the development and production of new drug substances or novel drug delivery systems.

What are the factors that inhibit globalization of the development and production of drugs and drug products? Some of them originate within the industry; other are external. Even in multinational companies, different production groups in various parts of the world may have their own preferences, likes and dislikes, relative to formulation approaches and manufacturing methods. Thus, those in charge of formulation and production in France may perhaps tend to prefer wet granulation as the method of choice in the production of compressed tablets, whereas their colleagues in the United States may prefer, whenever possible, to use direct compression. It is not unknown for such differences of opinion, within a company, to block standardization. Also, the availability of equipment and the training and experience necessary to use it, can often vary quite significantly in different parts of the world. Many of the large multinational pharmaceutical companies, which play a pivotal role in the development of pharmaceutical products, are now making efforts to reduce this variability. They are working to ensure that components selected for their formulations are available for use in all of their production locations throughout the world. Also, efforts are being made to standardize the types of manufacturing equipment utilized by different units of the same company. Additionally, many companies are now establishing SOPs (standard operating procedures: detailed, specific instructions on how an analytical or manufacturing process is to be performed), such that they will be uniform in both content and format throughout the world. These attempts from within the industry to globalize drug delivery systems are to be commended and are indeed already proving to be fruitful; however, it is apparent that the most important element in the globalization of pharmaceutical products lies within the purview of those responsible for setting regulatory standards.

Internationally accepted standards for both drugs and excipients used in the manufacture of pharmaceutical products are clearly important prerequisites for the process of globalization. Pharmacopeias, for many years, have provided quite comprehensive standards for drugs and some pharmacopeias have attained international prominence. Thus, the *USP* is not only recognized by the Congress of the United States as providing official standards for drugs, drug products, and devices, but is also official in about 30 other countries. Similarly, the *British Pharmacopoeia* (*BP*) is used not only in the United Kingdom, but also in numerous other countries in Africa, Asia, and Australia. Also, *Pharmacopoeia Françiase* (*PF*) is recognized by a number of countries in which there are links to France. The growing importance of Japan as a pharmaceutical giant has given increased importance to the *Japanese Pharmacopeia* (*JP*) as a standard of international value. Finally, the *European Pharmacopeia* (*EP*) is becoming established as an important regional standard for drug substances.

In 1990, the USP Convention resolved that USP should "explore and implement ways and means for working with the European, British, and Japanese pharmacopeias to harmonize standards for drug substances and excipients." By 1993, substantial progress had been made in this, reaching goals specified in this initiative. The USP convention has played an important role in assisting the catalysis of the harmonization of standards. There have been international meetings and, for some excipients, notably magnesium stearate, lactose, powdered cellulose, and microcrystalline cellulose, substantial progress has already been made.

Internationally, there is now a growing recognition that, whereas pharmacopeias have often done a fine job in developing standards for drugs, excipients, in the past, have all too often been neglected. A joint initiative by the Royal Pharmaceutical Society of Great Britain and the American Pharmaceutical Association resulted in the publication of the *Handbook of Pharmaceutical Excipients*. Both the authors of this chapter have been privileged to serve on the committees that produced the first edition and are now (1993) completing work on the second edition. This book has answered a real need, and in many areas of the world it has achieved a quasi-official status for pharmaceutical excipients. It is to be expected that the second edition, which has been radically revised and expanded, will be even more useful.

One factor that inhibits globalization of pharmaceutical products and also the introduction of novel technologies and new dosage forms is the difficulty, especially in the United States, of persuading regulatory agencies to approve the use of novel materials as excipients. It is well recognized by pharmaceutical scientists that the safety of such materials must be rigorously evaluated. However, it is not acceptable to establish such a bureaucratic and rigid blockade as to effectively prevent the introduction of new substances for pharmaceutical uses. Countries that do not allow reasonable procedures for the introduction of new excipients of proven safety are doomed to suffer economic penalties in an increasingly competitive world. The establishment of the International Pharmaceutical Excipients Council (IPEC) may well be expected to assist in stimulating both the availability of approved new pharmaceutical excipients and the uniformity of their standards throughout the world [13]. Some of these new excipients (e.g., the unsubstituted and substituted cyclodextrins) have outstanding potential for a variety of functions in pharmaceutical formulation [14].

Another international development of considerable importance to industrial pharmacy is the much greater attention now being given to the chiral nature of drugs. Many drugs of both natural and synthetic origin have one or more chiral centers [15]. Some most important therapeutic agents (e.g., ibuprofen, propranolol, verapamil, oxazepam, and methadone) are in this category. Traditionally, it has been common to market such drugs in delivery systems that contain a racemic mixture of the drug. However, in many instances we know that this is not an optimal procedure, since therapeutic activity may reside largely or entirely in one isomer, with the other isomer (or isomers) being implicated in the possible production of toxic side effects. For example, with ibuprofen there is good reason to believe that the S (+) isomer is the therapeutically important species with the R (−) isomer not contributing directly to therapeutic activity, although in vivo conversion of some R (−) to S (+) does occur [16]. Since a number of manufacturers of bulk drug substance have, within the past decade, developed economically viable methods of producing single isomers in a high state of purity, regulatory agencies in both Europe and America are now requiring additional data on the chiral nature of drugs. For some drugs, at least, the formulation of a single isomer product may have to be significantly different from that used for the racemic mixture [17, 18].

One of the results of the dramatic increase in our knowledge of pharmacokinetics and biopharmaceutics that occurred in the 1980s and early 1990s has been that formulators throughout the world have been forced to give increasing attention to biological, rather than simply chemical, attributes of drugs. Pharmacokinetic and pharmacodynamic studies are an increasingly important element in the design of drug products, sometimes utilizing new drug delivery systems to produce products that are simultaneously maximally safe, effective, and reliable in their performance. However, because of ethical concerns, our use of animals, which play a large role in both drug development and routine testing of pharmaceutical products, is likely to diminish significantly [19].

Another international development that is of substantial importance is the process of classifying the world into four climate zones for the process of assigning shelf lives to pharmaceutical products. These and other developments are nurturing the globalization of pharmaceutical products. Those now starting careers in pharmaceutical formulation may expect to see accelerating internationalization of the development, production, and evaluation of drug products and drug delivery systems.

REFERENCES

1. F. R. Vogenberg and D. J. Pisano, Pharm. Times Suppl., pp. 1–12, Sept. 1994.
2. C. D. Hepler and L. M. Strand, Am. J. Hosp. Pharm., 47, 533–540 (1990).

3. G. J. Mossinghoff, Pharm. News, 1, 32 (1994).
4. H. R. Manasse, Jr., Pharm. Manage. Advisor, pp. 4–6 (1994).
5. L. Z. Benet (ed.), *The Effect of Disease State on Drug Pharmacokinetics*, American Pharmaceutical Association, Washington, DC, 1976.
6. K. Mosback (ed.), *Methods in Enzmology*, Vol. 44, *Immobilized Enzymes*, Academic Press, New York, 1976.
7. K. B. Lee and G. S. Burrill, *Biotechnology 95: Reform, Restructure, Renewal*, The Industry Annual Report, Ernst & Young, Palo Alto, CA, 1994.
8. *Biotechnology Medicines in Development*, Pharmaceutical Manufacturers Association, Washington, DC, 1993.
9. *From Biotechnology to Biotherapy: Definitions, Applications and Clinical Issues*, The Amgen Co., Thousand Oaks, CA, 1994.
10. *New Medicines for Older Americans in Development: Heart Disease and Stroke*, Pharmaceutical Manufacturers Association, Washington, DC, 1993.
11. *New Medicines for Older Americans in Development: Cancer*, Pharmaceutical Manufacturers Association, Washington, DC, 1993.
12. *New Medicines for Older Americans in Development: Alzheimer's, Arthritis, and Other Debilitating Diseases*, Pharmaceutical Manufacturers Association, Washington, DC, 1993.
13. L. Blecher, *Pharm. Technol.*, p. 55 (1991).
14. Proceedings of the Pharmaceutical Uses of Cyclodextrins Conference, Center for Professional Advancement, East Brunswick, NJ, 28 October, 1993.
15. M. R. Wright and F. Jamali, Clin. Res. Regul. Aff., 10, 1 (1993).
16. A. J. Romero and C. T. Rhodes, Chirality, 3, 1 (1991).
17. A. J. Romero, L. Savastano, and C. T. Rhodes, 99, 125 (1993).
18. A. J. Romero and C. T. Rhodes, Pharm. Acta Belg., 48, 27 (1993).
19. Strategic Initiatives 1990–1995 Revision, Cyde Update, February 15, 1993.

Index

Abbreviated New Drug Application (ANDA), 195, 763–764
  clinical studies, 763
  manufacturing procedures, 763
  specifications, 763–764
Abrasion, 388
Absolute bioavailability, 105, 107
Absorbents, 356
Absorption, 17–19, 21, 22, 24, 54, 75, 155, 288, 342, 577, 578, 811, 896
Absorption base, 273
Absorption of drugs in the eye, 502–511
Absorption enhance, 168
Absorption enhancers, 867
Absorption within the oral cavity, 822
Absorption process, 103
Absorption rate, 37, 38, 57
Absorption rate constant, 748
Absorption site, 121
Absorption window, 287
Acacia, 321, 364, 376
Accela-Cota, 380
Accelerated stability testing, 194
Accelerator, 710
Accogel process, 433
Accommodative exotropia, 494
Accordion-pleated, 563
Accumulation, 108
Accuracy, 214
Acetaminophen, 45, 49, 347, 733, 812, 813
Acetaminophen elixir, 815
Acetazolamide, 318, 494

Acetone, 563
Acetone sodium bisulfite, 203
Acetonitrile, 854
Acetylation, 55
Acetylcholine chloride, 491, 528
Acetyl cysteine, 518
Acetyl metabolite, 166
Acetylsalicylic acid, pH-rate profile, 200
Acetylsalicylic anhydride, 339, 341
Achlorhydria, 61, 64, 819
Achlorhydric, 811
Acid-degradation, 229
Acid hydrolysis, 398
Acne, 254
Acquired immunodeficiency syndrome (AIDS), 891
Acrylonitrile, 709
Act on Drug Product Distribution, 753–756
Activated carbon (charcoal), 648
Activated charcoal, 144
Active air sampling, 483
Active drug, 450
Active ingredients, 343
Active processes, 614
Active-targeted delivery, 299
Active targeting, 631
Active transport, 811
Active-transport process, 173
Activity coefficient, 192, 202, 270, 271
Actuator, 720
Acute arthritis, 653
Acyclovir, 600
Added substances in parenteral formulations, 452

Additives, 202
Adenosine deaminase (ADA), 866
Adhesion strength, 388
Adhesive dispersion, 601
Adjunct tonicity, 516
Adolescence, 812
Adrenocorticoid steroid, 180
Adsorption, 57, 187, 456, 469, 595
Advantages of the aerosols dosage form, 548
Advantages and disadvantages of suspensions as a
    dosage form, 310–311
Adverse drug reaction (ADR), 813, 823
Aerochamber, 571
Aerosol, 169, 271, 272, 276, 548, 551, 632, 719
Aerosol containers, 561
Aerosol dosage form, 547
Aerosols packaging components, 551
Aerosol valve, 556
Agar, 276, 350
Age, 62, 63, 823
Age-adjusted death rate, 7
Agency for Health Care Policy and Research
    (AHCPR), 889
Agglomerate, 337, 344, 345
Agglomeration, 302
Aggregate, 195, 336, 855
Aggregation, 312, 858
Aggregation kinetics, 313, 314
Aggregation and precipitation, 853
Aging, 187
Aging process, 899
Air classification measurement, 483
Albumin, 172, 597, 660
Albuterol, 547
Alcohol-free emulsion, 318
Algenic acid, 276, 364, 524
Alimentary canal, 667
Aliphatic hydrocarbons, 321
Alkaline hydrolysis, 398
Alkaloids, 32
Alkanolamides, 567
Alkylating agents, 658
Alkylation, 476
Alkyl phenoxy ethanols, 567
Allergenic extracts, 460
Allergic incident, 296
Allergic rhinitis, 548
Allergies, 239, 296, 897
Allergy test, 447
Alopecia, 254
Aloprenolol, 579
Alpha 1-acid glycoprotein, 174
Alpha-keratin, 243
Alpha-lactose monohydrate crystals, 344
Alteration of the dose, 118
Alteration of the dosing interval, 117
Alternative delivery systems, 828
Aluminum aspirin, 42

Aluminum containers, 562
Aluminum foil, 520, 698
Aluminum monostearate, 432
Alveoli, 569
Alzheimer's, 901
Amber, 692
Amber glass, 686
Ambulatory patients, 894
American Academy of Pediatrics (AAP), 816
American Drug Manufacturers Association, 754
American Pharmaceutical Manufacturers
    Association, 754
Amide formation, 186
Amino acids, 40, 45, 248, 441, 470, 845, 847
Aminoglycosides, 95, 813, 825
Aminopenicillin, 57
Aminophylline, 187, 355
Aminophylline suppositories, 187
Ammonium sulfate, 706
Amorphous, 203, 221, 464
Amphetamine, 56
Amphiphiles, 312, 325
Amphiphilic, 646
Amphiphilic molecule, 324
Amphiphilic substance, 286
    glyceryl monolaurate, 286
    methyl oleate, 286
    propylene glycol monolaurate, 286
Amphiphilic surfactant, 320
Amphoteric, 307
Amphotericin B, 324, 652
Ampicillin, 162, 337
Ampicillin tetrahydrate, 184
Ampicillin trihydrate, 460
Ampoules, 441, 455, 456, 704, 709
Amylase, 539
Amylopectin, 353
Amylose, 353
Anaerobic organisms, 60
Analgesic/antipyretic combinations, 763
Analgesics, 341
Analytical standards, 10
Anastomoses, 246
Anatomy of the eye and adnexa, 492–493
Anesthetic, 817
Anesthetics, 495
Angina pectoris, 341
Angiotensin, 653
Angiotension-converting enzyme (ACE), 596
Angle of repose, 364
Anglo-Saxon countries, 903
Anhydrous lactose, 336, 417
Anhydrous theophylline, 336
Aniline, 39
Animal data, 761
Animal stearates, 432
Annealing, 466
Annealing oven, 704

Annual product review, 208
Antiadherent, 356
Antibiotics, 4, 95, 249, 294, 341, 657, 755, 762, 890
Antibody, 657, 663, 845
Antibody-targeted systems, 598
Anticancer agents, 41, 890
Anticancer drugs, 599
Anticancer products, 442
Anticholinergic, 45, 824
Antidepressant, 824
Antidiarrheal, 57
Antidoting, 444
Antifrictional agents, 354
Antigen, 598, 657, 658, 669, 845, 857
Antigenicity, 655
Antigen-specific antibodies, 631
Antiglaucoma drugs, 494, 503
Antihistamine, 18, 359, 761, 762,
Anti-HIV agent, 309
Antihypertensive, 762
Antihypertensive medication, 821
Anti-infective drugs, 4
Anti-infectives, 251, 494
Anti-inflammatory, 494
Anti-inflammatory drugs, 158
Anti-inflammatory preparations, 563
Anti-invectives, 891
Antimetabolites, 36, 40, 658
Antimicrobial, 453
Antimicrobial activity, 309
Antimicrobial Agents, 452, 454, 539
Antimicrobial effectiveness, 498
Antimicrobial preparations, 818
Antimicrobial Preservative Effectiveness Test, 458
Antineoplastic, 309
Antineoplastic agents, 4
Antioxidants, 183, 203, 322, 342, 452, 453, 518, 694
    ascorbic acid, 322
    ascorbyl palmitate, 322
    butylated hydroxytoluene, 322
    and chelating agents, 203
    cysteine hydrochloride, 322
    lecithin, 322
    propyl gallate, 322
    thioglycerol, 322
    vitamin E, 322
Antipyresis, 813
Antipyrine absorption, 49
Antiseptics, 249
Antistatic agents, 694
Antitumor, 663
Antitumor drugs, 658
Antiviral drugs, 891
Apocrine gland, 247, 248
Apocrine sweat glands, 241
Apoproteins, 644

Apparent volume of distribution, 85, 99, 174
Application formats, 787
Applied optimization methods, 731–733
Aqueous degradation, 229
Aqueous film coating, 378
Aqueous (humor) flare, 497
Aqueous humor, 518
Aqueous solution capability, 231
Aqueous suspension, 142
Aqueous vehicles, 457
Area under the curve (AUC), 59, 106, 163
Area under the plasma concentration, 105
Aromatic amino acid, 628
Aromatic ester, 203
Aromatic hydrocarbons, 184
Aromatic substance myrrh, 492
Arrhenius equation, 193, 688
Arrhenius equation and accelerated stability testing, 194–196
Arrhenius plot, 194, 231, 232
Arrhenius relationship, 849
Arsphenamine, 179
Arterial blood, 169
Arterial supply, 631
Arteries, 246
Arthritis, 823, 827
Artificial tears, 599
Ascitic fluid, 660
Ascorbic acid, 203, 345, 518, 833
Ascorbic acid injection, 458
Ascorbyl palmitate, 203
Aseptic crystallization and dry powder filling, 468
Aseptic filling, 482
Aseptic manufacturing, 512
Aseptic processing, 496
Aseptic techniques, 478
Asparagine, 849
Asparagine residue, 848
Aspartic acid, 850
Aspergillus niger, 498
Aspirin, 10, 54, 108, 339, 827, 832
Aspirin sensitive, 815
Asthmatics, 814
Astringents, 256
Atenolol, 126
Atherosclerosis, 653
Athy-Heckel plots, 370
Atmospheric pressure, 551
Atomic energy, 897
Atomic revolution, 897
Atomization, 345, 563
Atomizer, 271
Atropic gastritis, 819
Atropine, 34, 51, 494
Atropine sulfate, 526
Attapulgite suspension, 144
Auger fill principle, 407
Autoclave, 475

Autoimmune hemolytic anemia, 653
Automatic filling machines, 407
Autoxidation, 183, 203
Avidin, 659
Avogadro's number, 314
Axial force, 372
Axial pressure, 362
Azo dye tartrazine, 815
Azone, 286
Azopolymers, 669

2-benzoylcholestan-3-one, 184
B-D-glucoside dexamethasone, 628
Bacillus subtilis, 474
Bacitracin, 255
Bactericidal, 454
Bacteriostatic, 249, 457, 490, 522
Bacteriostatic action, 454
Bacteriostatic water, 457
Barbiturates, 19, 57, 170, 185, 457
Barium sulfate, 398
Baroreflexive stimuli, 821
Barrier-competent skin, 258
Barrier layer, 267
Barrier Packs, 551, 562
Basal lamina, 620
Base-catalyzed reaction, 202
Base degradation, 229
BASIC program, 114, 118
Basophils, 653
Batch-to-batch variability, 468
Beclomethasone dipropionate, 547
Beer-Lambert law, 855, 855
Beeswax, 321, 376, 432
Bentonite, 304, 357
Benzalkonium chloride, 499
Benzene, 709
Benzocaine, 203, 257
Benzocaine hydrochloride, 183
Benzodiazepine, 216
Benzoic acid, 39, 322
Benzothenium 15, 322
Benzphetamine, 42
Benzyl alcohol, 814
Benzyl benzoate, 458
Beta-adrenergic blockers, 510, 822
Beta-adrenergic blocking agents, 494
Beta-blocker, 126
Beta-cyclodextrin-nitroglycerin tablets, 187
Beta-elimination, 470
Beta-keratin, 243
Beta-lactam ring, 181, 201
Betamethasone, 283
Betaxolol hydrochloride, 494
Bezalkonium chloride, 322
Bicota tablet press, 384
Bicuspid, 624

Biexponential, 97
Bilayer, 594
Bile, 43
Bile salts, 811
Biliary excretion, 167
Bimolecular layers of lipids, 324
Binary solvent, 216
Binders and granulating fluids, 345
Binding agent, 42, 52
Bioadhesive polymers, 668
Bioadhesive tablets, 388
Bioadhesive, 19, 158, 579, 618
Bioavailability, 13, 15–17, 80, 104, 158, 168, 179,
    185, 187, 281–291, 337, 388, 424, 506,
    579, 580, 601, 727, 801, 817, 819, 827
Bioavailability/bioequivalence, 800
Bioavailability standards, 10, 134
Bioburden, 205, 474
Biochemical catalysts, 538
Biocompatibility, 516
Biocompatible, 603
Biodegradable, 613, 655, 656
Biodegradable materials, 603
Biodegradable matrix, 600
Biodegradable microspheres, 866
Biodistribution, 848, 848
Bioequivalence, 108, 397, 289, 795, 801
Bioerodible, 589, 603
Bioerodible and combination diffusion and
    dissolution system, 589
Biological activity, 473, 856
Biological availability, 13
Biological factors influencing oral sustained-release
    dosage from design, 578
Biological and functional age, 810
Biological half-life, 84, 89, 94, 180, 444, 578
Biological indicators, 484
Biological membrane, 155
Biological processes and events involved in drug
    targeted, 613
Biological reactivity tests, 709
Biological testing, 205
Biologics, 865
Biologics control, 754
Biomicroscope, 489
Biopharmaceutical aspects, 234
Biopharmaceuticals, 843
Biopharmaceutics, 896
Bioregulators, 897
Biotechnological products, 828
Biotechnology, 4, 19, 22, 865, 888
Biotechnology-based pharmaceuticals, 863
Biotechnology drugs, 442
Biotechnology products, 801
Biotechnology revolution, 897
Biotransformation, 626
Bisulfite, 452
Bitolterol mesylate, 547

Bitter taste, 180
Bladder cancer, 816
Bleeding, 159, 292
Bleomycin, 634
Blister, 686
Blister materials, 685
Blister package, 700, 831
Blister packs, 804
Block copolymer, 320
Blood, 18
Blood-brain barrier, 180, 628, 629
Blood concentration, 75
Blood flow, 53, 55, 148, 169, 820
Blood-perfumed tissues, 251
Blood-perfusion, 259
Blood purifiers, 12
Bloodstream, 21, 22, 28, 633, 827
Blood supplies, 170
Blood volume, 37, 86
Bloom Gelometer, 398
Bloom strength, 398
Blown glass, 704
Body compartments, 75
Body fluids, 155
Body surface area, 813
Boltzmann's constant, 192, 302
Bolus dose, 626
Bonding index (BI), 339
Bonding mechanism, 335
Bone gelatin, 398
Bottles, 470, 686
Bottles and jars, 713
Boundary layer, 146
Bowman's membrane, 497
Bradykinin, 623
Brain mapping, 899
Bread-through dose, 59
Breaking strength, 385, 386
Breast cancer, 1
Breath-activated actuators, 570
Brinell hardness, 388
British Pharmacopoeia, 441, 496
Brittle, 376
Brittle fracture, 361, 369, 370
Brittle fracture index (BFI), 339
Bronchi, 168
Bronchoconstriction, 815
Brookfiled viscometer, 292
Brownian motion, 302, 315, 569
Brownian motion and sedimentation, 314
Brush border, 28
Bubble method, 433
Bubonic plague, 2
Buccal, 334
Buccal mucosa, 240
Buffer capacity, 455, 515
Buffers, 452, 453, 454
Buffer salts, 202

Bulbar conjunctiva, 503
Bulk density, 415
Bulking agent, 453, 467
Bulking materials, 452
Bulk volume, 364
Bulla, 254
Bunolol hydrochloride, 494
Burn wounds, 260
Burst force, 716, 717
Butoconazole nitrate, 733
Butylated hydroxyanisole, BHA, 203
Butylated hydroxytoluene, BHT, 203

Cachets, 686
Caffeine, 203, 337
Caking, 227
Caking of suspensions, 180
Calamine lotion, 255, 296
Calcitonin, 52, 620
Calcium carbonate, 274, 376
Calcium channel blockers, 57
Calcium chloride, 464, 696
Calcium orthophosphate, 348
Calcium salts, 55
Calcium sulfate, 424
Calcium sulfate tablets, 355
Calculus, 730
Calorimeter, 233
Canada, 773
Canalicula, 503
Canals of Schlemm, 531
Cancer, 1, 172, 899
Candida albicans, 498
Canonical analysis, 746–747
Capillaries, 148
Capillary action, 350
Capillary air expansion, 352
Capillary forces, 348
Capsular, 632
Capsule capacities, 403
Capsule filling machines and their role in
    formulation development, 417
Capsule-filling simulator, 420
Capsules, 395, 686, 827
Carbachol, 494, 528
Carbamazepine, 811
Carbamazepine tablets, 337
Carbenicillin sodium, 184
Carbidopa, 628
Carbohydrate, 45, 857
Carbohydrates, amino acids, and vitamins, 448
Carbomers, 518
Carbon dioxide, 476, 551
Carbonic anhydrase inhibitors, 494
Carbopol 940, 521, 276, 280
Carbowax, 273
Carboxylic acids, 214

Carboxymethylating potato starch, 353
Carboxymethyl cellulose, 56, 276
Carboxypolymethylene, 276
Carbuterol, 230, 231
Carcinogen, 816
Carcinogenic-oncogenic potential studies, 797
Cardiac glycosides, 149
Cardiovascular agents, 4
Cardiovascular collapse, 814
Cardiovascular drugs, 891
Carnauba wax, 376, 435
Carrageen, 276
Carragheen, 396
Carrier, 758
Carrier material, 524
Carrier matrix, 629
Carrier-mediated transfer systems, 618
Cataract surgery, 526
Catheters, 158, 445
Cation-exchange resin, 350
Cationic surfactants, 478
Causes of death, 2
Caustic poison act, 754
Caveat emptor, 5
Ceftazidime pentahydrate, 450
Celiac, 24
Celiac disease, 61
Cells of the skin, 245
Cellular carriers, 650
Cellular targets, 672
Cellular uptake and processing, 614
Cellulose acetate butyrate (CAB), 534
Cellulose acetate phthalate (CAP), 375
Cellulose derivatives, 345
Cellulosic gums, 522
Center for Biologics Evaluation and Research
    (CBER), 865
Center for Drug Evaluation and Research (CDER),
    865
Center of Drugs and Biologics, 755
Central nervous system (CNS), 75, 813
Central nervous system drugs, 891
Centralized procedure, 787
Centrifugation, 317, 326
Cephalosporin, 442, 818
Cerebrospinal fluid, 445
Cetomacrogol, 183
Cetylpyridinium chloride, 202
Cetyltrimethylammonium bromide, 202
cGMP:
    active ingredient, 764
    batch, 764
    buildings, 765
    Component, 764
    Control of components and drug product
        containers and closures, 767
    drug substance, 764
    equipment, 766

[cGMP]
    inactive ingredient, 764
    laboratory controls, 768
    lot, 764
    materials approval unit, 764
    personnel, 765
    product containers and their components, 768
    production and process control, 766
    scope of the regulation, 765
    strength, 764
Chain pharmacies, 893
Chamber pressure, 462, 468
Characteristics of bioerodible matrix systems, 589
Characteristics of matrix diffusion systems, 588
Characteristics of osmotically controlled devices,
    592
Characteristics of reservoir diffusional systems,
    584
Charcoal, 57
Chelating agents, 452, 453, 455
Chemical decomposition, 181
Chemical degradation, 179, 187
Chemical degradation reactions, 184
    decarboxylation, 184
    hydration, 184
    pyrolysis, 184
Chemical degradative routes, 181–185
Chemical kinetics, 180
Chemical kinetics and drug stability, 179–211
Chemical modifications, 450
Chemical properties, 554
Chemical reaction, 76
Chemical reaction kinetics, 214
Chemical stability, 16
Chemical structure, 29, 847
Chemical structure, delivery and clinical response,
    281
Chemical, pharmaceutical and biological testing,
    791–795
Chemical testing, 205
Chemotherapy, 1, 3, 4, 10
Chemotherapy agents, 40
Chemotherapeutic agents, 891
Chewable tablets, 815, 818, 826
Child resistance (CR), 698, 701
Children, 811
Children's diseases, 3
Chilled-oil column, 433
Chipping, 376
Chiral configuration, 184
Chiral nature of drugs, 905
Chloasma, 254
Chloral hydrate, 431
Chlorambucil, 598
Chloramphenicol esters, 337
Chloramphenicol palmitate, 43
Chloramphenicol, 134, 424, 494, 495, 812, 813
Chloramphenicol sodium succinate, 451

Chlordiazepoxide, 179
Chlorhexidine, 499, 502, 598
Chloride, 717
Chlorobutanol, 478, 501, 522
Chlorofluorocarbon (CFC), 550, 720
Chlorofluorocarbon (CFC) propellants, 547
Chlorofluorocarbons, 552
Chloroquine, 179
Chlorpromazine, 652
Chlorthalidone, 184
Choice of preservative, 499
Cholest-2-en-ol benzoate, 184
Cholesterol, 273, 325, 644, 645
Cholestyramine, 57
Chondroitin sulfate, 527, 663
Chromophore, 666
Chronological age, 810
Chylomicra, 460
Chylomicrons, 29, 644, 645, 646
Chyme, 26
Cidal test, 499
Ciliary movement, 168
Cimetidine, 815
Cinemicroscopy, 352
Ciprofloxacin, 528
Circularly polarized light, 856
Citric acid, 203, 351, 464
Claimed investigational exemption for a new drug,
        756–760
Clarity testing and particulate analysis, 485
Classes of ophthalmic products, 514–533
Classic optimization, 729–730
Classification of age groups, 809–810
Cleaners, 537
Clean in place (CIP), 514
Cleanroom, 512
Clearance of the organ, 160
Clearance rate, 88
Climate zones, 905
Clindamycin HCl, 128
Clindamycin palmitate, 128
Clinical chemistry, 747
Clinical considerations in parenteral product
        design, 477
Clinical documentation, 798
Clinical effectiveness, 16
Clinical efficacy and safety, 800
Clinical evaluation, 479, 759
Clinical pharmacokinetic laboratories, 895
Clinical pharmacokinetics, 811
Clinical pharmacology, 799
Clinical protocol, 479
Clinical studies, 761
Clinical trial formulae, 791
Clinical trial (CT) program, 479
Clorazepate, 58
Clostridium sporogenes, 474
Closures and pumps, 715–716

Coacervation, 634
Coagulation, 312, 313
Coalescence, 323
Coated tablets, 373, 829
Coating, 334, 758
Coating equipment, 379
Coating in fluidized beds, 382
Coating pan, 375, 379
Coating pan air configuration, 381
Cocaine, 34, 167, 441, 495
Cocoa butter, 435
Code of Federal Regulations, (CFR), 865
Codeine, 183, 813
Cod liver oil emulsion, 318
Coefficient of friction, 362
Coefficient of lubrication, 362
Cognitive services, 889
Cohesiveness, 359
Cohesive powders, 420
Cohesive strength, 347
Cold-fill process, 572
Cold incorporation, 279, 280
Cold welding, 369
Colestipol, 57
Collagen, 245
Collapse temperature, 465
Collapsible plastic bag, 563
Collapsible tube, 716, 717
Colligative property, 307
Collision efficiency, 314
Colloidal, 320
Colloidal particles, 596
Colloidal silica, 435
Colloidal silicates, 424
Colloidal silicon dioxide, 350
Colloid mill, 279, 432
Collyrium, 492
Colon, 26, 51, 578
Colonization, 811
Color, 292
Color additive amendments, 755
Colorants, 399, 694
Columnar cells, 25, 26
Combined kinetic effect of Binding Alterations,
        174
Comedo, 254
Comminuting mill, 364
Committee for Proprietary Medicinal Products
        (CPMP), 785, 786
Common constituents of dermatological
        preparations, 276
Community and hospital practice, 892
Community pharmacist, 889
Community pharmacy, 890, 892
Community policy, 780
Compaction, 371
Compaction characteristics, 335
Compaction profile, 372

Compaction simulator, 373
Compartment, 75, 625
Compatibility, 336, 340, 415
Compatibility tests, 227–233
Compatibility tests for solid dosage forms, 227
Compendial standards, 5
Complexation, 144, 306
Complexing agents, 202
Complex reactions, 190–192
Compliance aids, 828
   Calendar-Paks, 828
   Dosett Trays, 828
   Patient Med Paks, 828
Compliance issues, 818
Complications of contact lens wear and the need
   for care products, 534
Composition of contact lenses, 534
Composition of the shell, 432
Compressed gases, 556
Compressed gas systems, 550
Compressed tablet, 142, 334, 348
Compressibility, 363, 422
Compressibility and flow, 833–834
Compression, 336, 361, 369, 417
Compression aids, 363
Compressional force, 385
Compressional pressure, 388
Compression coating, 375, 383
Compression force, 421
Compression ratio, 420
Computer-aided drug design, 897
Computer-aided grid search, 747
Computer-aided optimization, 748
Computers, 206, 748
Computers and systems, 748
Concentrated oral solutions, 830
Concentration, 77
Concentration of the dissolved drug in bulk
   solution, 145
Concentration gradient, 141
Concentration procedure, 785
Conduction, 467
Congealed, 280
Congestive heart failure, 55, 834
Conjunctiva, 157
Conjunctival, 239
Conjunctival congestion, 497
Conjunctival discharge, 497
Conjunctival sac, 515
Conjunctival swelling, 497
Conjunctivitis, 495
Consolidation, 369
Consumer products, 5
Contact angles, 348
Contact lens care products, 491, 533–541
Contact lenses, 539
Contact lens polymers, 535
Container-closure System, 207, 473, 474, 691

Containers, 704–709
Containers and Closures, 691–699
Containment, 248–249, 682
Contemporary trends in pharmaceutical product
   design and production, 896
Content uniformity, 343, 384, 719
Contour plots, 730, 743
Contraceptive, 601
Contraceptive metals, 602
Contraindications and side effects, 789
Control of components and drug product
   containers and closures, 684
Control of drug release from dosage forms, 895–
   896
Controlled release, 327, 341, 576, 747
Controlled-release excipients, 356, 357
Controlled-release product, 17
Controlled release system, 18
Control methods, 789
Control tests applied to the finished product, 795
Controlled aggregation formulations of
   suspensions, 317
Controlled drug delivery, 15, 575
Controls on advertising of medicinal products, 802–
   803
Controls applied to the starting materials, 794
Convection, 467
Cool-end treatment, 706
Coomassie stain, 856
Copper, 492
Copper sulfate, 495
Coring, 713
Corn oil, 457
Cornea, 239, 489, 505
Corneal cloudiness, 497
Corneal endothelium, 529–532
Corneal epithelial barrier, 517
Corneal limbus, 526
Corn starch, 350, 424, 734
Corpus striatum, 628
Corrosion, 704, 715
Corticosteroids, 183, 257, 268, 290, 494
Cosmetic products, 318
Cosmetics, 256, 295
Cosolvent, 181, 306, 550, 859
Cotton pledget, 295
Cottonseed oil, 457, 461
Coulter counter, 301, 302
Coulter method, 323
Coulter principle, 301
Council of ministers, 780
Covalent protein destabilization, 848–851
Creaming of emulsions, 180
Creams, 188, 274, 548
Creatinine clearance, 95
Cresol, 478
Critical micelle concentration (CMC), 56, 307, 308
Critical packing Parameter (CPP), 643

Crohn's disease, 61
Cromolyn sodium, 547, 638, 864
Croscarmellose sodium, 425
Crospovidone, 425
Cross-contamination, 481, 498
Crossed-linked carboxymethylcellulose, 350
Cross-linked polymers, 353
Cross-linked povodone, 357
Crushing strength, 385, 386, 388
Cryoprotectants, 456
Cryoprotection, 863
Crystal change, 13
Crystal characteristics, 450
Crystal growth, 315, 456
Crystal habit, 337
Crystalline, 203, 221
Crystalline form, 337–339
Crystallization, 185, 319, 464, 517, 863
Cube-root dissolution equation, 581
Cumulative amount of drug, 86, 87
Cumulative urinary excretion, 107
Current, 158
Current Good Manufacturing Practices (cGMP)
    guidelines, 204, 322, 683, 764–769
Cyanocobalamin, 184
Cyclic temperature stability tests, 318
Cyclodextrins, 618, 858
Cyclopentolate, 494
Cyclophosphamide, 185,203
Cycloplegics, 494
Cyclosporine, 598, 813
Cysteine, 654, 850
Cystic fibrosis, 61, 818, 897
Cystic hygroma, 634
Cystine, 850
Cytochemical assay, 857
Cytoplasma, 28
Cytotoxic drugs, 26, 595

Dactinomycin, 597
Dapiprazole hydrochloride, 494
Data requirements for an European community
    marketing authorization application, 790
Data requirements for marketing authorization
    approval, 787
Daunomycin, 598
Daunorubicin, 659
Deaggregate, 422, 423
Deaggregation, 13
Deamidation, 470, 848
Decarboxylase, 628
Decarboxylation, 181, 371
Deep concave tooling, 375
Definitions, 548
Deformation, 352, 361
Deformation process, 834

Degradation and metabolism in the gastrointestinal
    tract, 128
Degradation rates, 180
Dehumidification systems, 482
Dehydration, 184
Delayed-release, 576
Deliquescent, 405
Deliver, 682
Delivery system, 14, 155, 575, 579, 686, 722
Deltoid muscles, 160, 444
Demecarium bromide, 494
Denaturation, 195, 456, 848
Denaturation of proteins, 858
Density, 314, 364
Dental caries, 599
Dental systems, 598
Deodorants, 256
Deoxyribonucleic Acid, 659
Deponit, 601
Depot, 444
Depressant, 19
Depth filters, 477
Depth-type filters, 482
Derivation, 100
Derivative, 729
Dermatin sulfate, 663
Dermatitis, 254
Dermatological, 239, 240, 281
Dermatological disorders, 253
Dermatological formulations, 289
Dermatological liquids, 294
Dermis, 241, 245
Dermis, composition, 246
Desamino-8-D-arginine vasopressin, 52
Desialyated fetuin, 633
Desiccant, 696
Design and formulation of compressed tablets,
    333–359
Design of hard gelatin capsule powder
    formulations and choice of excipients, 422
Desirable properties of raw materials, 335–341
Desmopressin, 52
Desolvation of natural proteins, 597
Detergent, 308
Determinants, 730
Detoxification, 709
Developmental age, 810
Development of legal controls in the European
    community, 778–782
Developments in protein drug delivery, 866–867
Dexamethasone, 491, 494
Dexamethasone sodium phosphate, 180, 547
Dexterity, 823, 830
Dextran, 465, 598, 662, 864
Dextran and other polysaccharides, 662
Dextrose, 345, 357, 358, 864
Dextrose injection, 445
Diabetes, 575, 815, 843, 814

Diagnostics, 494
Diagnostic testing, 443
Dialysis, 326
Dialysis technique, 597
Diaper rash products, 251
Diarrhea, 22, 815
Diatomaceous earth, 477
Diazepam, 42, 187, 825
Dibasic calcium phosphate, 833
Dicalcium phosphate, 345, 353, 364
Dicalcium phosphate dihydrate, 423
Dichlorodifluoromethane (Propellant 12), 547, 551
Dichlorotetrafluoroethane (Propellant 114), 547, 551
Die cavities, 368
Die force, 369
Dielectric breakdown, 651
Dielectric constant, 187, 201, 216, 304, 308
Diels-Alder reaction, 180
Die roll, 433
Diet, 3
Diethyltoluamide (DEET), 285
Difference spectra, 855
Differential pressure measurement, 483
Differential scanning calorimetry (DSC), 222, 464, 695, 857, 864
Differential thermal analysis (DTA), 695
Diffuse double layer, 304
Diffusion, 39, 148, 284, 555, 579, 583, 625
Diffusional lag time, 267
Diffusional resistor, 262
Diffusional systems, 582
Diffusion coefficient, 37, 147, 263, 264, 268, 281, 302, 314, 508, 581, 585, 587
Diffusion-controlled, 580, 587, 588, 589
Diffusion of drug, 261
Diffusion layer, 141, 584
Diffusion layer thickness, 581
Diffusion-limiting, 580
Diffusion rate, 586
Diffusion systems, 585
Diffusion theory, 313
Digestion, 22
Digitoxin, 149
Digoxin, 125, 126, 149, 179, 578, 580, 811, 813, 825, 895
Digoxin tablets, 335, 346
Dihydropyridines, 628
Dihydroxyphenylalanine, 628
Diluent, 344, 363
Dimethylacetamide (DMA), 285, 287
Dimethyl sulfoxide (DMSO), 285, 286, 287
Diminution of physical function and its effects on drug disposition, 819–825
Diphtheria, 2
Dipivaloyl derivative of epinephrine, 628
Dipivalyl epinephrine, 600
Dipivefrin hydrochloride, 494, 517

Dipropylene glycol, 563
Dipyridamole, 42, 64
Direct compression, 335, 359, 363, 365, 904
Direct payments, 9
Discoloration, 227
Disease states, 60–62
Disinfectant, 512, 537
Disintegrant, 42, 347, 348, 354, 425, 734
Disintegrant efficiency, 351
Disintegrant's porosity, 354
Disintegrants that propagate capillary effects, 348
Disintegrants that swell, 350
Disintegrate, 126, 341, 422, 728
Disintegrating agents, 341
Disintegrating medium, 352
Disintegration, 13, 129, 187, 342, 353, 833
Disintegration and deaggregation, 133
Disintegration testing, 385
Disintegration time, 347, 350, 363
Dispenser pump, 715, 716
Dispersed phase, 300
Dispersed solids, 296
Dispersed system, 639
Disperse systems, 305, 322
    emulsion, 300
    foam, 300
    liquid aerosol, 300
    solid aerosol, 300
    solid emulsion, 300
    solid foam, 300
    solid suspension, 300
    suspension, 300
Dispersible or soluble tablets, 830
Dispersion, 300
    colloidal, 300
    lyophilic, 300
    lyophobic, 300
    micellar systems, 300
Displacement pump, 715
Displacement transducers, 421
Disposable syringes, 470
Dissolution, 75, 103, 126, 129, 155, 180, 187, 218, 303, 334, 336, 342, 356, 424, 579, 589, 728, 747, 832, 833
Dissolution-controlled systems, 581
Dissolution of drug, 261
Dissolution of drug substance and dosage form, 233
Dissolution modifiers, 342
Dissolution rate, 17, 36, 131, 133, 145, 187, 337, 385
Dissolution rate-limited, 144, 589
Distribution, 444, 811
Distribution coefficient, 508
Disulfide bond exchange, 851
Disulfide bonds, 847
Disulfide exchange, 470
Disulfide linkage, 657

Diverticulosis, 61
Divinyl ether-maleic anhydride (DIVEMA), 666
DL-α-Tocopherol, 461
DLVO theory, 312
DNA fragments, 845
Documentation, 206
Dopa-decarboxylase enzyme, 579
Dopamine, 629
Dosage form, 179, 196, 200
Dosage form preferences, 826
Dosage forms, 121, 157, 515, 548
  creams, 157
  inhalation, 157
  ointments, 157
  ophthalmic preparations, 157
  oral, 157
  rectal enemas, 157
  rectal solutions, 157
  solutions, 157
  suppositories, 157
  topical, 157
  vaginal, 157
Dosage regimen adjustment in renal failure, 116, 117
Dosage regimen, 76, 169, 894
Dosage from variables, 21
Dosator machines, 411, 417
Dose-dumping, 50
Dose effect scheme, 155–157
Dose-efficacy relationship, 156
Dose-ranging, 479
Dose-response, 894
Dose-response relationship, 799
Dose size, 579
Dosing considerations determined by alterations in physiology with age, 825–826
Dosing-disk machines, 410
Dosing of fill material, 406
Dosing interval, 115
Dosing regimen, 113
Dossier, 791
Double-blinds studies, 480
Double-rotary press, 369
Doxorubicin, 597, 644
Doxorubicin (Adriamycin), 324
Draize test, 497
Driacoater, 380
Droplet evaporation, 569
Drop-Tainer, 519
Drug absorption, 21, 28, 35, 47, 55, 121, 147
Drug actions on the skin's glands, 256
Drug administration, 626
Drug availability, 121
Drug in body, 77
Drug-carrier Delivery Systems, 629
Drug or chemical toxicity, 448
Drug cost, 9
Drug delivery systems, 14–16, 19, 121, 156, 334, 826, 895

Drug delivery systems and compliance issues, 814–818, 826–831
Drug-diluent mixture, 136
Drug dissolution, 284
Drug distribution, 169–176
Drug-drug incompatibility, 184
Drug and drug product quality, 5
Drug efficacy study, 762–763
Drug elimination, 84
Drug elimination rate, 88
Drug-excipient interactions, 340, 341
Drug excreted, 89
Drug-food and drug-drug interaction, 55–58
Drug immunology, 2
Drug input at or close to the site of action, 157–159
Drug input into the systemic circulation, 159–169
Drug interactions, 16
Drug loaded microspheres, 319, 318
Drug-loading, 595
Drug master file (DMF), 795
Drug-metabolism, 58
Drug performance, 291
Drug-plastic interaction, 187
Drug powder degradation, 229
Drug product, 1, 8, 10, 179
Drug product containers and closures, 683
Drug product quality, 11
Drug product selection, 8
Drug quality, 9
Drug release, 747
Drug-release profiles, 347
Drug-release rate, 444
Drug reservoir, 523
Drug selection, 9, 892
Drug solubility, 143
Drug stability, 343
Drub substance, 214
Drug therapy, 75
Drug utilization, 9
Drycoata tablet press, 383
Dry eye syndrome, 521
Dry-heat, 474
Dry-heat sterilization, 475, 496
Drying process, 339
Dry-powder-filling, 468
Dry Powder Inhalers (DPI), 721, 722
Dry powders, 460
Dry sprays, 548
DSC thermogram, 465
Dualchamber Plastic container, 470
Dump system, 17
Dunken walk, 314
Duodenum, 22, 23, 24, 26, 42, 578
Durham-Humphrey Amendment, 755
Dwell time, 366
Dyes, 815
Dymed, 539

Dymel Propellants, 558
Dynamic-immersion challenge, 485

Ear, 157
Ease redispersibility, 316
Eccrine glands, 247
Eccrine sweat, 247
Eccrine sweat glands, 241
Eccrine sweating, 251
Echothiophate iodide, 494
Economic and social committee, 781
Eczema, 254
ED_{50}, 14, 17,
Edetic acid (EDTA), 455
Education and experience, 204
Effect on availability, 174
Effect on clearance, 173
Effect on distribution, 172
Effective surface area of the drug, 131
Effect of manufacturing process, 135
Effect of route of administration and distribution
    on drug action, 155–178
Effects in deep tissue, 257
Effects of maturational changes on drug
    disposition, 810–812
Effervescent tablets, 829
Efficacy, 491
Efficiency of mixing, 361
Egg lecithin liposomes, 642
Egg phospholipids, 461
Eigenvalues, 746
Eigenvectors, 746
Ejection force, 354, 417, 419, 420
Ejection-induced stress, 369
Ejection station, 417
Elastic deformation, 361
Elasticity, 248
Elastic recovery, 371, 372
Elastin, 245
Elastomer, 710, 713
Elastomeric closures, 685
Elastomeric components, 703
Elastomeric seals, 720
Elastomers, 702
Elderly, 809
Elderly population, 819–831
Electrical barrier, 250
Electrocardiogram (EKG), 895
Electrochemistry, 895
Electroencephalogram (EEG), 895
Electrolyte balance, 445
Electrolyte displacement method, 301
Electrolytes, 26, 45, 317, 445, 591
Electromechanical compounding units, 442
Electronic age, 897
Electronic revolution, 897
Electrophoresis, 856

Electrostatic charge, 424
Electrostatic filters, 482
Electrostatic forces, 848
Electrostatic interactions, 195, 853, 859
Electrostatic theory, 201
Elimination, 75, 92
Elimination half-life, 117
Elimination rate constant, 235, 748
Emboli, 159, 460
Emcompress, 345
Emollients, 251, 274
Emptying half-time, 43
Emulsifying agents, 321
Emulsion, 567, 632, 703
Emulsion aerosol, 566
Emulsion polymerization, 597, 638
Emulsions, 188, 299, 307, 309, 319, 460, 470
Emulsion-stabilizing surfactant, 320
Enantiotropic system, 221
Encapsulated dissolution products, 582
Encapsulation, 422, 641, 652
Endocytosis, 595, 615, 633, 651, 652
Endoskeleton, 243
Endosomes, 616
Endosomotropic receptors, 617
Endothelial cells, 620
Endothelium, 620, 621, 622, 654
Energetics of reactions, 192
Energy of activation, 193, 194
Enteric-coated tablets, 45
Enteric-coating, 576, 820
Enteric polymers, 668
Enterohepatic recirculation, 811
Enterohepatic recycling, 576
Enthalpy, 222
Enthalpy of activation, 193
Entrapment, 651
Entrapped air, 351
Entrapped gas, 276
Entropy of activation, 193
Environment, 511
Environmental, 474
Environmental factors that affect reaction rates,
    196–204
Environmentally controlled area, 511
Environmental monitoring, 482
Environmental Protection Agency (EPA), 771
Enzymatic action, 348
Enzymatic cleaners, 538
Enzymatic degradation, 128
Enzyme immobilized electrode, 895
Enzyme-mediated reactions, 625
Enzyme reactions, 202
Enzymes, 19, 351, 662
Eosinophils, 653
Epianhydrotetracycline, 179
Epidermis, 241, 250, 252
Epinephrine, 184, 494, 547

Epinephrine bitartrate, 491
Epinephrine hydrochloride, 517
Epinephrine prodrugs, 494
Epithelial cells, 23, 26, 30, 175
Epithelial membrane, 577
Epithelium, 617
Epoxy resins, 562
Equation, 729
Equilibrium moisture content, 337
Equilibrium moisture transmissions (EMC), 689
Equilibrium solubility, 335
Equipment, 205, 514
Equivalent spherical diameter, 300
    projected diameter, 301
    Stokes' diameter, 301
    surface diameter, 300
    volume diameter, 300
Ergotamine, 184
Erodible ocular inserts, 525
Erythema, 250
Erythritol, 39
Erythrocyte, 652
Erythrocytes, 304, 597, 650, 659
Erythromycin, 42, 58, 123, 124, 339, 495
Erythromycin tablets, 147
Escherichia coli, 498
Esophageal irritability, 565
Esophageal mucosa, 669, 824
Esophageal obstruction, GI diseases, 448
Esophageal transit, 49, 397
Esophagus, 168, 397
Essential amino acids, 445
Essential nutrients, 40
Establishment license, 865
Ester, 450
Ethane, 551
Ethanol, 59
Ethical pharmaceuticals, 293
Ethidium bromide, 659
Ethinamate, 424
Ethyl acetate, 79, 83
Ethyl acetate, hydrolysis, 79, 83
Ethylacrylate, 525
Ethyl alcohol, 458, 563
Ethylcellulose, 334, 432
Ethylene chlorhydrin, 476
Ethylenediamine, 187
Ethylenediaminetetraacetic acid, EDTA, 203
Ethylene glycol, 754
Ethylene glycol by-products, 476
Ethylene vinyl acetate (EVA), 524
Ethyl salicylate, 182
Europe, 773
European Community (EC), 773, 901
European cooperation, 774–778
European Economic Area (EEA), 776
European Economic Community (EEC), 777
European Free Trade Association (EFTA), 775

European law, 782
European Pharmacopoeia, 777, 791
Eutectic mixture, 145
Eutectic temperature, 464
Evaluation of semisolids, 292
Evaporative loss, 293
Evaporative shrinkage, 290
Evolution of contact lenses, 533–534
Evolution of drugs, 9
Evolutionary operations (EVOP), 730, 731–733
Examination, 497
Examples of erythrocyte-targeted drug delivery
    systems, 651
Excipients, 13, 203, 335, 343, 363, 423, 450, 758,
    814, 833, 904
Excitation, 813
Excretion, 75, 444
Exhaustive grid search, 741
Exoskeleton, 243
Expenditure, 7
Expert reports, 790
Expiry date, 208, 804
Exponential rate, 91
Extemporaneously, 499
Extent of absorption, 61, 104
Extent of availability, 121
External ear, 240
External scintigraphy, 427
Extra-cellular fluids, 86
Extractables, 709
Extragranular, 348
Extravasation, 620, 623, 666–667
Extravascular administration, 98
Extravascular dose, 97
Extravascular route, 21, 112
Extrusion, 747
Eye, 157
Eyedrops, 515
Eye irritation, 294

Fab′ fragments, 658
Facilities, 205
Factorial design, 731
Factors affecting the formulation of emulsions,
    318
Factors affecting functioning of the skin barrier,
    258
Factors affecting the release of drugs from
    particulate carriers, 639
Factors affecting the rate of dissolution, 129–147
Factors influencing the absorption of drugs, 619
Fasting, 234
Fat emulsion for parenteral nutrition, 318
Fat receptors, 45
Fatty acid esters, 567
Fatty acids, 128
    butanoic acid, 249

[Fatty acids]
  heptanoic acid, 249
  hexanoic acid, 249
  propanoic acid, 249
FDA, 809
FDA guideline, 204
FDA legislation, 10
FDA recall system, 770
  class I recall, 770
  class II recall, 770
  class III recall, 770
FDA stability guidelines, 195, 207
Feathering the curve, 96, 100
Federal Hazardous Substances Act (FHSA), 497
Federal status, 753
Federal Trade Commission, 892
Federal Food, drug and cosmetic act, 753
Fenestrated capillaries, 620
Ferrous sulfate, 578
Fibroaerolar tissues, 620
Fibroblasts, 650
Fibronectin, 615
Fick equation, 583
Fick's first law of diffusion, 37
Fillers, 343, 423
Filling of hard gelatin capsules, 406
Film coating, 374, 376
Film-coating materials, 377
Film plasticizers, 378
Filter membranes, 854
  cellulose diacetate, 854
  nitrocellulose, 854
  nylon, 854
  polysulfone, 854
  polyvinylidene fluoride, 854
Filters, 442
Filtration, 474
First-order, 576
First-order degradation, 201
First-order kinetics, 37, 38, 76, 163, 202
First-order pharmacokinetics, 84
First-order process, 76, 78, 79, 80
First-order rate constant, 76
First-order reaction, 189–191
First-order release, 577
First-order targeting, 633
First-pass biliary secretion, 167
First-pass bioavailability, 169, 174
First-pass effect, 53, 161, 334, 600, 612, 811
First-pass hepatic metabolism, 57, 60
First-pass metabolism, 64, 166, 289, 442, 601
Fisher Sub-Sieve Sizer, 226
Fixed oil, 295
Flammability of propellants, 555
Flammable, 555
Flavoring agents, 357, 816
Flexible hydrophilic contact lens, 499
Flight-or-fight response, 247

Flocculated system, 321
Flocculating materials, 316
Flocculation, 312, 313
Flowability, 408
Flow properties, 833
Fluidity, 415, 422
Fluidized air, 382
Fluidized bed drying, 364
Fluid mosaic model, 29
Flunisolide, 547
Fluorescein staining, 497
Fluorometholone, 491, 494
Fluorouracil (5-FU), 257, 597
Fluphenazine esters, 457
Fluprednisolone, 185
Flurazepam, 227
Foams, 548
Foam system, 566
Foam-type product, 551
Follicle, 247
Follicular probe, 266
Follicular shunt, 268
Food and drink law, 753
Food and Drug Administration (FDA), 11, 186, 240, 755
Food and drug adulteration, 753
Food and drug law, 10
Food, Drug and Cosmetic Act, 11, 12
Forced ward air oven, 364
Foreign particles, 768
Form-fill-seal, 709
Formulation, 14, 728
Formulation approaches to protein stabilization, 857–865
Formulation of pharmaceutical aerosol, 563
Formulation of soft gelatin capsules, 432
Fourier transform IR spectroscopy, 863
Fractional area, 262
Fractional-order equations, 192
Fracture resistance, 385
Fragmentation, 319
Francophone countries, 903
Free-flowing, 335
Free-radical reactions, 183
Free samples, 803
Freeze denaturation, 863
Freeze-dry, 830
Freeze-drying, 317, 456, 460, 461, 520, 858
Freeze-thaw, 315, 856
Freeze-thaw cycle, 517, 640
Freezing, 462
Frequency of side effects, 16
Friability, 370, 385, 388, 734
Friability of tablets, 365
Frozen micelles, 274
Fuller's earth, 357
Full-scale production, 481
Functional age, 810

Functionally testing, 709
Functions and organization of the Federal Food and Drug Administration, 755–756
Fungal keratitis, 495
Fungistatic, 249
Furosemide, 578
Fusion, 595
Fusion method, 279
Future Developments, 573
Future pharmacy workplace needs, 892–894
Future (or new) system, 786–787

Gallbladder, 22, 43
Gamma scintigraphy, 44
Gantez AN-139, 521
Gas, 474
Gas-producing disintegrants, 350
Gas sterilization, 496
Gas transmission, 455, 456, 692
Gastric acid, 811
Gastric emptying, 19, 23, 43, 45, 122, 127, 811
Gastric emptying rate (GER), 48, 820
Gastric fluid, 42, 58
Gastric irritation, 333
Gastric motility, 19
Gastric retention, 579
Gastric secretion, 819
Gastrointestinal absorption, 57
Gastrointestinal irritation, 820
Gastrointestinal membrane transport, 148
Gastrointestinal microbleeding, 832
Gastrointestinal motility, 125
Gastrointestinal tract (GIT), 17, 21, 22, 23, 162, 166, 333, 341
Gastrointestinal transit, 811
Gelatin, 321, 347, 376, 395, 398, 465, 597
Gelatin capsule, 142, 397
Gelatin-coated pill, 396
Gelatin cross linking, 405
Gelatin walls, 432
Gel filtration, 326
Gender-related differences, 51
General behavior of semisolids, 272
General methods of preparation of topical systems, 279
General safety considerations, 495–502
Generic drug, 6
Generic effectiveness, 16
Generic equivalent, 108
Gene therapy, 10, 899
Genetic engineering, 845, 897
Gentamicin, 184, 494, 895
Geriatric, 62
Geriatric community, 828
Ghost matrix, 583, 589
Giant papillary conjunctivitis (GPC), 537
Gibbs energy, 221

GI ulceration, 334
Glass, 685, 704
Glass containers, 704
Glass flaking, 707
Glass transition temperature, 465, 468
Glatt-Coater, 380
Glaucoma, 494, 599
Glidant, 355, 356, 408, 424, 834
Globalization, 904
Globular protein, 848
Globulin, 172, 657
Glomerular filtration, 95, 655, 812, 866
Glomerular filtration rate (GFR), 89
Glucagon, 442
Glucocorticoids, 246
Glucose, 456, 461, 466
Glucose tolerance, 834
Glucuronidation, 811
Glutaminase, 651
Glutamine, 849
γ-Glutamyl-L-dopa, 628
Gluteal, 444
Gluteus maximus, 160
Glycerin, 458
Glycerol, 461, 859
Glyceryl monostearate, 568
Glycine, 850
Glycoconjugates, 660, 661
Glycol ethers, 563
Glycoproteins, 495, 617, 633, 660, 661
Glycosaminoglycan (GAGs), 495, 620, 663
Glycoside linkages, 848
Glycosome, 324
Golden age of drug discovery, 897
Gonadotropin, 660
Good Laboratory Practice (GLP), 795
Good Manufacturing Practice (GMP), 481, 755
Grafting, 260
Granulating fluid, 336, 347
Granulation, 334, 833
Granulators, 364
Granules, 686, 829
Granulomatous reactions, 521
Grapefruit juice, 57
Griseofulvin, 43, 46, 145, 580
Gross national product (GMP), 8
Growth hormone deficiency, 843
Growth hormone-releasing factor, 597
Guidance on standard terms for marketing authorization application, 787
Gut metabolism, 59

Hair follicles, 241, 247
Half-life, 80
Hallucinations, 813
Halogenated hydrocarbons, 551
Haloperidol, 149

Hard gelatin capsules, 397–428, 434
Hardness of tablet, 350, 364, 385
Harmonization, 778, 784
Harmonization of standards, 904
HDL, 648
Headspace, 455, 696
Health care, 891
Health care cost, 887
Health care delivery, 888
Health care reform, 887, 893
Health professionals, 803
Health Security Act, 888
Heart failure, 149
Heat conservation, 250
Heat-resistant spores, 484
Heat-sealed layers, 717
Heat-shear, 345
Heat of wetting, 351, 352
Heavy metals, 695
HeLa cells, 663
Hematocrit, 652
Hemoglobin, 650
Hemolysis, 456, 460, 713, 843
Hemophilia, 843
Henderson–Hasselbach equation, 34, 215, 832
Heparan sulfate, 663
Heparin, 442
Hepatic blood flow, 164
Hepatic clearance, 57, 164
Hepatic drug extraction, 45
Hepatic elimination, 174
Hepatic enzyme, 164
Hepatic extraction ratio, 162
Hepatic metabolism, 162, 164, 816, 825
Hepatic process, 176
Hepatitis-infected blood, 845
Hepatocytes, 650, 660
Hepatocytic cells, 613
Hepatoma, 667
Hepatotoxicity, 813
Heptanoate ester, 282, 283
Hermetic seal, 702
Herpes simplex, 494
Heterocyclic analogs, 184
Hexachlorophene, 269
Hexosaminidase, 651
Hexylene glycol, 563
Hi-Coater, 380
Hiestand Compaction Indices, 340
Hiestand Tableting Indices, 339
High-density lipoprotein (HDL), 644
High-density polyethylene (HDPE), 686, 694, 715
High-efficiency particulate air (HEPA) filters, 482, 512
High-frequency spark test, 484
High-output tablet presses, 368
High-performance liquid chromatography (HPLC), 205
High-resting-state viscosity fills, 434

High-shear mixer, 361
High-shear mixing, 388
High-shear-mixing machines, 364
High speed rotary press, 373
High-viscosity dispensers, 715
Higuchi model, 588
Hirsutism, 254
Histamine, 18, 623
Histoplamin, 447
History of drugs, 10
HLB value, 307, 308
Homeostatic mechanisms, 819
Homodisperse, 314
Homogeneity, 359, 397
Homogenization, 319
Homogenizer, 279
Homologous series, 31
Hopper design theory, 416
Hormones, 660
Horseradish peroxidase, 618
Hospital, 890
Hospitalization, 3
Hot-air sterilizers, 475
Hot-end treatment, 706
Housekeeper wave, 44
HPLC, 747
Human antimouse antibody (HAMA), 845
Human fibroblast interferon, 853
Human genome project, 897
Human growth hormone (hGH), 843, 855
Human interferon gamma, 853
Human serum albumin (HSA), 456, 859
Human skin, 247
Humoral fluid, 158
Humulin, 469
Huntington's Chorea, 899
Hyaluronic acid, 663
Hyaluronidase, 444
Hybrid, 800
Hybridoma, 843
Hybridoma cells, 658
Hybridoma technology, 672
Hydralazine, 815
Hydrated sodium silioaluminate, 356
Hydrate formulation, 450
Hydrates, 450
Hydration, 349
Hydration rate, 351
Hydroalcohol, 547
Hydroalcoholic, 445
Hydroalcoholic medium, 345
Hydrocarbon-miscible solvent, 273
Hydrocarbons, 554
Hydrochloride, 547
Hydrochlorofluorocarbons and hydrofluorocarbons, 555
Hydrochlorothiazide, 748
Hydrocortisone, 281, 282, 494, 653

Hydrodynamic condition, 233
Hydrodynamic radius, 622
Hydrofluoroalkane (HFA), 720
Hydrogel, 158, 603
Hydrogen bonding, 181, 848
Hydrogen ion, 122, 189, 196
Hydrogen peroxide, 540
Hydrolysis, 58, 76, 181, 182, 189, 194, 470, 833, 848
  amides, 182
  esters, 182
  imides, 182
  lactams, 182
  lactones, 181
  malonic ureas, 182
  nitrogen mustards, 182
  oximes, 182
Hydrolytic attack, 706
Hydrolytic reactions, 849
Hydrophilic, 172, 307, 308, 491, 584
Hydrophilic-lipophilic balance (HLB), 435, 307
Hydrophilic ointment, 277
Hydrophilic polyvinylidene difluoride, 458
Hydrophilization, 427
Hydrophobic, 307, 491, 584, 834, 848
Hydrophobic drug, 132
Hydrophobic interactions, 848, 854
Hydrophobicity, 282, 348, 569, 633, 854
Hydrophobic materials, 456
Hydrophyllic materials, 355
Hydroquinone, 257
Hydrostatic pressure, 623
Hydrostatic pressure difference, 591
Hydrotropic agents, 306
Hydroxide ion, 76, 189
Hydroxy acids, 249
Hydroxyethyl cellulose, 276, 427, 518
Hydroxyethyl methacrylic acid (HEMA), 528, 534
Hydroxyl group, 349
Hydroxyl-modified Hydrophilic polyamide, 458
Hydroxypropylmethylcellulose (HPMC), 347
Hygroscopic, 357
Hygroscopicity, 214, 227, 405
Hygroscopicity potential, 452
Hygroscopic material, 336
Hygroscopic substances, 275
Hyperosmotic, 449
Hyperosmotic laxative, 815
Hyperplastic, 259
Hypersensitivities, 296
Hypersensitivity, 296, 815
Hypertension, 184, 575, 834
Hyperthermia, 594
Hyperthyroidism, 61
Hypertonic, 445, 456
Hypertonic solutions, 516
Hypodermoclysis, 444
Hypoglycemic activity, 855

Hypotensive, 55
Hypothermia, 251
Hypothyroidism, 61
Hypotonic, 445, 456
Hypotonic hemolysis, 651
Hypotonic solution, 477
Hysteresis loop, 372

Ibuprofen, 813
Ichthyosis, 254
Identification of the product, 804
Identity, 11
Idoxuridine, 494
Idoxuridine, pH-rate profile, 199
IgA, 657
IgD, 657
IgE, 657
IgG, 657
IgM, 657
Ileocecal junction, 26
Ileocecal sphincter, 26
Ileum, 44
Imipramines, 653
Immobilization, 858
Immune system, 897
Immunizing agents, 1, 4
Immunoassays and biological activity assays, 856–857
Immunodeficiency syndrome (AIDS), 845
Immunogenicity, 662
Immunoglobins, 598, 615, 633, 656, 657
Immunological response, 843
Immunomodulating agents, 642
Impaired-vision, 823
Impeded conductance, 301
Impeller, 361
Implantation therapies, 592
In vivo absorption, 337
In vitro dissolution, 135, 337, 364
In vitro release rate, 734
In vivo testing, 234
Inactive ingredients, 814
Inactive ingredients in topical drops, 516
Incompatibilities, 184, 213, 321, 397, 442
Independent variables, 738
Index of mixing, 361
Industrialization, 896
Industrial scale, 319
Indwelling needles, 445
Inert Gases, 452, 455
Infants, 811, 814
Infections, 172
Infectious diseases, 2
Inferior vena cava, 159, 167
Infiltration, 254
Inflammation, 172, 282, 444
Information necessary before taking product, 804

Infusion pumps, 669
InhalAid, 571
Inhalation, 864
Inhalation aerosol, 563, 569, 568
Inhalation therapy, 556
Inhale, 720
Injections and implants, 603
Injection molding of tablets, 388
Inorganic fibers, 477
Inorganic ions, 249
Insect repellents, 251
Insensible perspiration, 244
Insoluble pigments, 358
InspirEase, 571
Instructions for proper use, 804
Instrumented tableting machines, 369
Insulin, 187, 442, 454, 469, 843, 854
Insulin, crystalline, 185
Insulin-dependent diabetics, 167
Interactions, 800
Interface, 288
Interfacial area, 303
Interfacial polymerization, 597, 638
Interfacial stabilization, 307
Interfacial tension, 320
Interferon, 668, 891
Intermolecular forces, 32
Internal partitioning, 293
International Conference on Harmonization, (ICH), 204
International developments, 901–905
International Pharmaceutical Excipients Council (IPEC), 905
Intersubject variables, 894
Intestinal membrane, 28
Intestinal motility, 50
Intestinal transit, 49
Intestinal transit rates, 55
Intestinal wall, 579
    intra-articular, 443
Intra-arterial, 445
Intra-arterial administration, 667
Intra-arterial injection, 817
    intra-arterial, 443
    intracardial, 443
Intracervical system, 601
    intracisternal, 443
Intracisternal, 447
    intradermal, 443
Intradermal, 447
    intraepidural, 443
Intramuscular (IM), 441, 443
Intramuscular dose, 167
Intramuscular injection, 311
Intramuscular route, 444
Intranasal aerosol, 568
Intraocular, 287
Intraocular dosage forms, 525–533

Intraocular pressure (IOP), 489, 494, 522
Intraperitoneal, 648
    intrapleural, 443
Intraspinal, 447
Intrathecal, 447
Intrauterine device, 601
Intravenous (IV), 441, 443
Intravenous administration, 162
Intravenous fat emulsions, 461
Intravenous lipids, 441
Intravenous route, 444
Intrinsic clearance, 160, 173, 174
Intrinsic compressibility, 337
Intrinsic dissolution rate, 233
Introduction, 575
Investigational New Drug (IND), 206, 213, 759
Iodoxuridine, 600
Ion-exchange products, 593
Ion-exchange resin, 57, 593
Ion-exchange systems, 592
Ionic strength, 849
Ionizable drugs, 32
Ionizable substances, 215
Ionization, 29, 141
Ionization constant, 452
Ionization pK_a and aqueous solubility, 579
Ionizing radiation, 474
Ionophores, 324
Ion-pairing agents, 854
Iontophoresis, 168, 495
Iopanic acid, 43
Ipratropium bromide, 547
Iridocyclitis, 494
Iris involvement, 497
Iritis, 494
Irrigating cannulae, 533
Irrigating solutions, 448, 449
Islets, 650
Isobutane, 563
Isoelectric focusing, 856
Isoelectric point, 195, 398, 849
Isoetharine mesylate, 547
Isoniazid, 903
Isopropanol, 423
Isoproterenol hydrochloride or sulfate, 547
Isosorbide dinitrate, 187
Isothermal system, 250
Isotonic, 445, 456, 529, 652
Isotonicity, 491
IV administration, 107
IV admixture, 445, 890
    therapeutic incompatibilities, 445
IV bolus dose, 170
IV valium, 458

Japan, 773, 890, 901
Jejunum, 22, 23, 24, 578

Joint Commission on Accreditation of Health Care Organization (JCAHO), 889
Justification for the omission of the pharmacology data and clinical data, 800–802

Kanamycin, 43, 95, 111, 116, 117, 118, 184
Kaolin, 357, 376
Karaya, 350
Karl Fisher titration, 473
Kefauver–Harris Drug Amendment, 755, 758
Keloid, 254
Keratin, 248, 270
Keratin-denaturing temperature, 260
Keratin layer, 168
Keratinized cells, 247
Keratinocytes, 245, 247
Keratin sulfate, 663
Keratoconjunctivitis sicca, 495
Keratolytic, 269
Keratosis, 254
Kerckring folds, 24
Ketaconazole, 42
Kidney, 173, 825
Kinetic equations, 188–192
Kinetic expressions, 198
Kinetic parameters, 176
Kinetic pH profiles, 229
Kinetics, 76
Kinetics of drug absorption, 96
Kneading, 280
Kozeny equation, 314
Kunstaliches Zellen, 324
Kupffer cells, 613, 618, 633, 641, 648, 660

Labeled potency, 473
Labeling, 486, 803
Labeling of medicinal products and requirements for leaflets, 803–805
Lacrimal puncta, 503
Lacrimal sac, 503
Lacrisert, 525
Lactated Ringer's solution, 529
Lactic-glycolic acid copolymer, 866
Lactobacillaceae, 662
Lactose, 344, 345, 466
Lagrange function, 736
Lagrangian method, 734–737
Lag time, 748
Lakes, 358
Lamellae, 567
Lamellarity
    giant unilamellar vesicles (GUV), 326
    large unilamellar vesicles (LUV), 326
    medium unilamellar vesicles (MLV), 325
    multilamellar vesicles (MLV), 325
    multivesicular vesicles (MVV), 326

[Lamellarity]
    oligolamellar vesicles (OLV), 325
    small unilamellar vesicles (SUV), 325
    unilamellar vesicles (ULV), 325
Lamina Propria, 24, 25, 28
Laminar flow, 511, 514
Laminar flow units, 442
Laminate, 339
Laminated bags, 563
Laminated tubes, 717
Lamination, 335, 336, 361
Langerhans cells, 245, 296
Langmuir adsorption isotherm, 320
Lanolin, 273
Laplace transforms, 97
Large intestine, 23, 26
Large unilamellar vesicles (LUVs), 593, 641
Large-scale manufacture, 279
Large-volume parenteral (LVP), 84, 442, 443, 445, 458, 703
Laser light-scattering technique, 326
Lateralis, 444
Latex, 378
Lattice substitution, 304
Laws governing evaluation of new drug products, 756–764
Laws governing preparation and distribution of existing products, 764–771
Layered tablets, 369, 384
Layerpress, 384
LD_{50}, 14, 17, 19
LDL, 645, 646, 648
L-Dopa, 55
Leaching, 704
Leaflets, 804
Leakage, 490, 595
Leaker testing and sealing verification, 484
Lecithin cholesterylacyltransferase (LCAT), 645
Legislative impacts, the new health care environment and the future of pharmacy, 888
Lehr, 704
Leprosy, 4
Leukocytes, 159, 650, 653, 654
Levigated, 280
Levodopa, 578
Lichen planus, 254
Lidocaine, 160, 164, 817, 895
Lidocaine hydrochloride, 442, 495
Light, 158
Light-absorbent, 702
Light degradation, 229
Light and humidity, 203
Light protection, 702
Light-resistance, 520
Light scattering, 855, 858
Light-scattering method, 323
Light-sensitive product, 685

Light stability, 452
Light transmission, 692, 695
Limbal route, 503
Limulus Amebocyte Lysate (LAL), 205, 484
Lincomycin, 57
Lipase, 539
Lipase activity, 811
Lipid, 29
Lipid bilayers, 327
Lipid emulsions, 445
Lipidic, 580
Lipid-sieve membrane, 29, 31
Lipid-soluble, 628
Lipid-soluble drugs, 811
Lipophilic, 159, 307
Lipophilic compounds, 168
Lipophilic materials, 355
Lipoprotein lipase, 645
Lipoproteins, 172, 644
Liposomal formulations, 456
Liposome-encapsulated, 327
Liposomes, 158, 299, 300, 309, 312, 324–327,
        327, 470, 593, 598, 639, 659
Liposomes in pharmacy, 324
Liquefaction, 227
Liquefied gas, 550, 555
Liquefied gas systems, 550
Liquid-borne particle, 479
Liquid chromatography, 854–855
    ion exchange, 854
    reverse-phase, 854
    size exclusion, 854
Liquid compatibilites, 231
Liquid-filled capsules, 434
Liquid paraffins, 355
Liquid penetration, 388
Liquids and semisolids, 713–717
Liquid and suspensions, 827
Liquids, 416, 703
Lithium, 825, 895
Lithium aluminum hydride, 180
Liver, 22, 162
Loading dose, 111, 119
Local toxicity (tolerance) studies, 798
Log-normal equation, 303
Long-term stability, 184
Loss on drying, 473
Low-density, 686
Low-density lipoprotein (LDL), 616, 644
Low-density pellets, 579
Low-density polyethylene (LDPE), 520, 686, 694,
        717
Low-molecular-mass proteins, 659
Low-shear, 345
Lozenges, 334, 763
L-tryptophan receptors, 45
Lubricant, 135, 354, 423, 424, 734
Lubricant efficiency, 362

Lubricating agent, 135
Lubricating gel, 278
Lubrication, 397, 710
Lubricity, 415, 422
Luminal cells, 26
Lung first-pass effect, 160
Lungs, 159, 160, 161, 721
Lupus erythematosus, 254
Luteinizing human-releasing hormone (LH-RH),
        867
Lycium, 492
Lymph, 28, 623, 641
Lymphatic capillaries, 246
Lymphatic drainage, 624
Lymphatics, 296
Lymphatic uptake, 623
Lymph node, 296
Lymphocytes, 179, 245, 246, 654, 657, 663, 845
Lyophilization, 858, 859
Lyophilization and protein formulation
        development, 863
Lyophilized cake, 203
Lysolecithin, 43
Lysosome, 324, 613, 653, 655, 659, 665
Lysosomotropic, 672
Lysosomotropic drug, 659
Lysosomotropic systems, 638
Lysozyme, 537, 659

Machine speed, 408
Machine variables on fill weight, 416
Macrocrystalline, 124
Macromolecule-drug conjugate, 655
Macromolecules, 186, 300, 321, 598, 616, 617,
        622, 626, 629, 646, 655, 843
Macrophages, 245, 246, 616, 631, 633, 642, 650,
        653, 667
Macroscopic precipitate, 856
Magnesium carbonate, 357
Magnesium lauryl sulfate, 419
Magnesium oxide, 357
Magnesium stearate, 136, 339, 354, 356, 408, 419,
        424, 834
Magnesium sulfate, 355
Magnetic fields, 158
Maintenance dose, 111, 119
Malabsorption syndrome, 61
Malaria, 753
Maleic anhydride copolymers, 666
Malonamide, 39
Maltose, 466
m-aminophenol, 179, 184
Mammalian skin, 34
Mannitol, 345, 357, 464, 859
Manpower, 892
Manufacture, 333
Manufacture of hard gelatin capsules, 398–406

Manufacturers, 398
Manufacture of soft gelatin capsules, 433
Manufacturing authorizations, 805
Manufacturing consideration, 511–514
Manufacturing facilities, 806
Manufacturing formula, 794
Manufacturing and packaging the aerosol dosage form, 572
Manufacturing-scale, 318
Manufacturing techniques, 512
Marcogol 400, 357
Mass movement, 51
Mass transport, 281
Mathematical analysis, 748
Mathematical model, 731, 734
Mathematical tools, 76
Matrices, 730
Matrix algebra, 746
Matrix devices, 586
Matrix diffusion, 601
Matrix diffusional products, 589
Matrix system, 589
Matrix tablets, 363
Maxair, 571
Maximum allowable cost (MAC), 5, 771
Maximum drug delivery, 16
Maximum drug safety, 16
Mean particle size, 301, 303
Measles, 2
Measurement of rheological properties of disperse systems, 305–306
Measuring roll, 433
Meat inspection, 754
Mechanical pumps, 669
Mechanical strength, 345, 350, 361
Mechanical strength of tablet, 336
Mechanisms of action, 351
Mechanisms and causes of protein destabilization, 848–854
Mechanisms of drug absorption, 36–41
Medicaid, 7, 8, 889
Medicaid patients, 888
Medical sales representatives, 803
Medicare, 7, 8
Medicinal chemistry, 757
Medicinal products, 802
Medicine, 893
Medroxyprogesterone, 601
Medrysone, 494
Melanin, 250
Melanocytes, 250
Melanocyte-stimulating hormone (MHS), 245
Melanotropin, 660
Melting endotherm for ice, 464
Melting point, 452
Membrane, 28, 458
Membrane and depth filters, 457
Membrane filters, 458

Membrane fusion, 615
Membrane permeability, 22, 33
Membrane transport, 147
Menstruum, 758
Mercaptobenzothiazole (MBT), 703
Mercury compounds, 478
Mesa model, 508
Mesenteric arteries, 24
Mesenteric blood flow, 130
Metabolic acidosis, 814
Metabolic clearance rate (MCR), 90
Metabolic reactions, 58
Metabolism, 28, 58, 75, 89, 90, 164, 165, 444, 579, 812, 813
Metal cans, 686
Metal closures, 685
Metal die, 334
Metal ion chelation, 851
Metal oxide, 704
Metal particles, 490
Metal screw-cap closures, 686
Metaproterenol sulfate, 547
Metastable, 219
Metastable amorphous phase, 466
Metasulfite, 452
Meter-dosed aerosols (oral, inhalation, nasal), 549
Meter-dosed inhalers (MDIs), 547, 686, 719, 720
Metered-dose, 569
Metered valves, 556, 558, 569
Metered-dose valves, 573, 719
Metering valve, 568
Methane, 551
Methaqualone tablets, 187
Methazolamide, 494
Methionine, 850
Method of inspection, 104
Method of manufacture, 789, 794
Methods to evaluate protein pharmaceuticals, 854–857
Methods of evaluating emulsions, 322
Methods of evaluating suspensions, 316
Methotrexate, 123, 597, 660
Methyl cellulose, 276, 427, 518
Methylparaben, 454, 499, 501, 522
Metoclopramide, 45, 50, 51
Metronidazole, 598–599
Metronidazole phosphate, 451
Micellar-binding equilibrium, 306
   equilibrium dialysis, 309
   gel filtration, 309
   potentiometry, 309
Micellar-bound drug, 308
Micellarization, 57
Micellar solubilization, 274, 306
Micelles, 202, 306, 308, 643
Michaelis–Menten kinetics, 40
Microbial barrier, 249
Microbial growth, 295, 309

Microbial preservatives, 295
Microbial specifications and sterility, 294
Microbiological contamination, 339, 350
Microbiological stability, 16
Microbiological standards, 14, 322
Microcapsules, 388, 659
Microchondrion, 653
Microcomputer, 113
Microcrystalline cellulose, 186, 337, 339, 350, 363, 417, 425
Microelectrophoresis, 304
Microelectrophoretic measurements, 323
Microemulsion, 306, 307, 322
Microflora, 60
Micromeretics, 300
Micronization, 145, 319, 343
Microorganism kill rate, 475
Microorganisms, 60, 447, 457, 490
Microparticles, 669
Microparticles and nanoparticles, 634
Microreservoir dissolution-controlled system, 601
Microscope, 302
Microscopic, 459
Microscopy, 224
Microspheres, 470, 598, 603, 638, 864, 866
Microtablets, 416
Microvilli, 24
Microwave drying, 388
Microwave irradiation, 323
Microwave vacuum drying, 365
Migraine attack, 61
Migrating motor complex (MMC), 125
Miliaria, 254
Mineral oil, 273, 295, 432
Minimum effective concentration (MEC), 109
Minimum toxic concentration (MTC), 109
Miotic drug, 494, 528
Miotics, 158
Mitomycin, 596, 639
Mitosis, 243
Mixed thermal–thixotropic systems, 435
Mixing, 280
Mixing movements, 50
Model of the skin, 262
Modified-release coating, 378
Modified starch, 344
Moist granulation, 359
Moisture barrier, 702
Moisture content, 336, 690
Moisture content of ingredients, 565
Moisture loss, 13, 715
Moisture transmission rate, 689
Mold lubricants, 694
Molecular biology, 22 , 897
Molecular modeling, 897
Molecular structure, 353
Molecular structure and weight, 452
Molecular volume, 271

Mongolia, 903
Monitoring Programs, 482
Monoclonal antibodies, 10, 470, 658, 845, 857
Monocytes, 633
Monograph, 794
Monolithic, 632
Monomeric components, 709
Mononuclear leukocytes, 653
Mononuclear phagocyte system (MPS), 615
Monopalmitate, 311
Monotropic system, 221
Montreal protocol, 720, 721
Morbidity obese, 62
Morphine, 10, 441
Morphology, 224
Motion sickness, 258
Mottling of tablets, 180
Mouth fresheners, 569
Moving-boundary diffusion, 589
Moving the drug away from the site of absorption, 148
Mucin, 43, 617
Mucopolysaccharides, 24, 26
Mucopolysaccharidic gel, 245
Mucosa, 822
Mucosal membrane, 57
Mucosal pH, 52
Mucous flow, 168
Mucous layer, 28
Multidimensional space, 746
Multidose, 691
Multilamellar vesicles (MLVs), 593, 641
Multinational company, 903
Multiphase systems, 299
Multiple Internal Reflectance Infrared Spectroscopy (MIR-IR), 695
Multiple-dose, 481, 498
Multiple-dosing regimens, 108
Multiple regression analysis, 731, 739
Multistate procedure, 785
Multistation presses, 366, 367
Multivitamin preparation, intravenous, 310
Murine antibody, 845
Muscle, 822, 444
Mutagenesis, 858
Mutagenic potential studies, 797
Mutual recognition procedure, 787
Mydriatics drugs, 158, 494
Myeloma cells, 658
Myoelectric complex, 44

NaCl equivalency, 516
Nadolol, 600
Nails, 241
Nanoparticles, 158, 596, 638
Nanospheres, 659
Naproxin crystals, 337

Narcotic analgesics, 45
Narrow therapeutic index, 111
Nasal, 867
Nasal administration, 167, 618
Nasal aerosol, 571
Nasal aerosol solutions, 564
Nasal allergies, 572
Nasal delivery, 720
Nasal epithelia, 168
Nasal mucosa, 240, 572
Nasal spray, 715
Nasolacrimal drainage, 502
Natamycin, 495
National 1551, 344
National applications, 786
National center for health statistics, 7
National Formulary (NF), 441, 489, 754
National systems, 782–784
National total health expenditure, 8
Natural fibers, 477
Natural gums, 276
Nature of Package Evaluations, 684–685
NDA, New drug application, 213
n-demethylation, 812
N-desmethyldiazepam, 58
Necrosis, 444
Nedocromil sodium, 547
Needle penetration force, 713
Needle shields, 709
Neomycin, 43, 255
Neomycin–Polymyxin, 494, 528
Neonates, 62, 809, 811, 814
Nernst-Brunner Equation, 129, 140, 147
Neuraminidase, 651
Neutrophils, 653
Newborn, 62
New drug, 204
New Drug Application (NDA), 760–762, 865
New drug substance, 758
Newtonian flow, 321
Newtonian fluids, 305
Nicotine, 34, 168, 246
Nictating membrane, 497
Nifedipine, 45, 51
Niosomes, 312, 643
Nitrodisc, 601
Nitrodur, 601
Nitrofurantoin, 46, 123
Nitrogen, 551
Nitrogen mustards, 445
Nitroglycerin, 168, 186, 257, 258, 270, 334, 357, 568, 717
Nitroglycerin transdermal systems, 288
Nitrosamines, 703
Nitrous oxide, 551
Nodule, 254
Nogami's method, 219
Nomenclature, 552

Nonalkalized cocoa, 358
Nonaqueous liquid, 232
Nonaqueous and mixed vehicles, 457
Nonaqueous solvent, 376
Noncompliance, 894
Noncovalent protein destabilization, 851–854
Nonelectrolyte, 31
Non enzymatic hydrolysis, 181
Nonepidemic, 2
Nonerodible inserts, 523
Nonflammable, 555
Nonimmunogenic, 597
Noninvasive routes, 576
Nonionic surfactant, 307
Nonionizable substances, 216
Nonlinear model, 76, 79
Nonpowder filling, 416
Nonpressurized containers: dry powder inhalers, 721
Nonpressurized containers: spray pumps, 720–721
Nonsterile oral products, 704
Nonsteroidal agents, 813
Non-steroidal anti-inflammatory drugs, 830
Normal distribution, 303
Normal-dosing regimen, 119
Norplant, 603
North America, 890, 901
North Atlantic Treaty Organization (NATO), 778
Nose, 157
No sedimentation diameter (NSD), 315
Notice to applicants, 782
Novel drug delivery systems, 577
Novel formulations, 470
Novosomes, 312
Noyes-Whitney equation, 218, 581, 832, 833
Nozzle, 721
Nuclear magnetic resonance (NMR), 326
Nucleic acids, 651
Nucleotide blockage, 10
Nucleotides, 128
Nutrients, 445
Nystatin, 495

Objective function, 737
Occupational Safety and Health Administration (OSIER), 771
Octanol, 223
Octogenarians, 3
Ocular absorption, 505
Ocular Pilo-20 system, 523
Ocular route, 514
Ocular systems, 599
Ocular toxicity and irritation, 496–498
Ocusert, 524
Ocusert-pilo, 599
Odor, 292
Office of Compliance, 757–758

Official journal of European communities, 780
Official standards, 384
Ofloxacin, 598
Oil-in-water emulsifier, 308
Oil-in-water emulsion, 272, 305
Ointment, 10, 188, 272, 296, 548
Oleic acid vapor, 706
Oligonucleotides, 611
Oligopeptidyl linkages, 665
Oligosaccharide, 617, 660
Omentum, 618, 648
Omnibus Budget Reconciliation Act (OBRA), 888
One-compartment model, 85
Opacity creams, 296
Opaquing agents, 399
Operation of the aerosols package, 549
Ophthalmic, 511, 703
Ophthalmic cul-de-sac, 491
Ophthalmic dosage form, 489
Ophthalmic drugs, 493, 525
Ophthalmic formulation, 497
Ophthalmic ointment, 180, 294, 295, 489, 513
Ophthalmic solution, 180, 489, 517
Ophthalmic sterility, 294
Ophthalmic suspensions, 489
Ophylline hydrate, 184
Opsonins, 615, 616
Opsonization, 615
Optical activity, 452
Optical methods, 302
Optical spectroscopy, 855–856
Optimal dose, 155
Optimization, 155, 491, 727–731, 738, 741, 888
Optimization parameters, 728–729
Optimization study, 738
Optimization technique, 747, 751
Optimize, 727
Optimized drug product, 15
Optimum drug effectiveness, 16
Oral adapter, 570
Oral administration, 21, 34, 162
Oral aerosols, 568, 569
Oral cavity, 824
Oral contraceptive, 343, 699
Oral dosage forms, 826, 829
Oral dosing without a first-pass effect, 167
Oral drug delivery, 592
Oral drug products, 814
Oral, inhalation, and nasal aerosols, 568
Oral sustained- and controlled-release products, 580
Oral system, 577
Ordered mixing, 336
Organ flow, 174
Organic molecule, 22
Organic nitrates, 167
Organization, 204
Organization for European Economic Cooperation (OEEC), 776

Organ size, 169
Organ transplantation, 1
Oropharynx, 570
Orosomucoid, 172
Orphan drug, 903
Ortho-esters, 283
Orthogonal design, 739
Orthostatic hypotension, 821
Osmolarity, 652
Osmotic, 29, 456
Osmotically Controlled Systems, 590
Osmotic core, 591
Osmotic effect, 529
Osmotic pressure, 590, 592
Osmotic system, 592
Osteoporosis, 899
Ostwald ripening, 315
Other parenteral routes, 443, 445
Out-of-date drug, 144
Outpatients, 3
Over-the-counter (OTC) drug, 6, 252, 813, 894
Over-the-counter human drugs, 769–770
Overmixing, 359
Oxidants, 452
Oxidation, 181, 183, 194, 470, 812, 833, 848, 850
    aldehydes, 183
    carboxylic acids, 183
    catechols, 183
    ethers, 183
    nitrites, 183
    phenols, 183
    thioethers, 183
    thiols, 183
Oxophenarsine, 179
Oxygen-free, 719
Oxygen permeable contact lenses, 491
Oxygen scavengers, 203
Oxygen tension, 849
Oxyhemoglobin, 865

Package, 681
Package definition and function, 682–683
Package and label design, 831
Package permeation rates, 689
Package specifications and quality assurance, 685–686
Packaging, 470, 519, 522
Packaging of aerosols, 718–719
Packaging and labeling control, 684
Packaging of parental and ophthalmic dosage forms, 702–713
Packaging requirements, 761
Packaging of semisolids and topicals, 713–718
Packaging of solid oral dosage forms, 686–702
Packaging, stability, and shelf life, 687–691
Packing density, 136
Palatability, 818

Palpable particulates, 294
Palpebral fissure, 502
p-aminobenzoate, 218
P-aminobenzoic acid, 166, 579
p-aminosalicylic acid (PAS), 138, 179, 184
Pan coating, 374
Pancreas, 22
Pancreatic enzymes, 43, 811
Pancreatin, 43
Pannus, 497
Paper NDA, 891
Papilla, 254
Papilose, 244
Parabens, 322
Paraffin wax, 432
Parasympathomimetic, 494
Parenchymal cells, 633
Parenteral, 441, 858, 890
Parenteral administration, 159, 160, 311, 817
 depot, 160
 intramuscular (IM), 160
 intravascular, 159
  intraarterial, 159
  intravenous, 159
 subcutaneous (SC), 160
Parenteral applications, 299
Parenteral dispersions, 317
Parenteral dosage forms and invasive devices, 827
Parenteral drug, 814
Parenteral formulations, 866
Parenteral nutrition, 811
Parenteral product, 441, 704, 707
Parenteral routes, 234, 443
Parkinsonism, 823
Partial derivatives, 730
Particle-free, 707
Particle repulsion theory, 352
Particle shape, 214, 335
Particle size, 131, 132, 214, 300–304, 318, 335,
 353, 450, 556, 566, 569, 573, 832, 834,
 864
Particle size determination, 301
Particle size distribution, 226, 335, 363
Particle size measurement, 316
Particle size and shape, 452
Particulate, 707
Particulate counters, 486
Particulate drug delivery systems, 631, 632, 635
Particulate-free environment, 483
Particulate matter, 159, 485, 511
Particulates, 294
Partition coefficient, 28, 29, 37, 169, 170, 172,
 187, 223, 263, 265, 267, 271, 283, 431,
 580, 584, 585
Partitioning, 261
Passive air sampling, 483
Passive diffusion, 36, 260, 615, 618
Passive processes, 614

Passive-targeted delivery, 299
Pastes, 273
Pasty materials, 416
Patagonia, 903
Pathogen-free, 295
Pathological condition, 158
Pathological terms, 254
Pathways of drug absorption, 28–29
Patient compliance, 13
Patient counseling, 888, 894
Patients with kidney and liver disease, 94
Peanut oil, 457
Pectin, 276
Pediatric, 62
Pediatric age group, 809
Pediatric population, 810–819, 831
Pediatric safety, 819
Peel-apart blister, 520
Pellets, 416
Penetration enhancer, 618
Penicillin, 43, 58, 201, 755
Penicillin G, 179, 199
Penicillin solution, 81, 82
Penicillin suspensions, 818
Penicillin V, 141
Pentazocine, 825
Pentobarbital, 40, 86
Peptidases, 627
Peptide bonds, 847
Peptide drug, 167
Peptides, 19, 22, 41, 888
Percutaneous absorption, 260, 261, 262, 285, 822
Percutaneously absorbed drugs, 246
Perfluorocarbon emulsion, 317
Performance of topical therapeutic systems, 280
Perfused tissue, 175
Peripheral vein, 155
Peritoneal dialysis solutions, 448
Peritoneum, 448
Permeability, 31, 267, 709
Permeability coefficient, 54, 264, 265, 269, 506
Permeation, 715
Permeation profile, 268
Personal health care expenditure, 9
Personnel, 204, 482
Petroleum based system, 273
pH, 29, 32, 196, 292, 450, 849
pH adjustment and buffers, 517
pH-dependent reactions, 128
pH-effect, 138
pH-independent, 196
pH-partition hypothesis, 35, 36, 42
pH and pK$_a$, 452
pH profiles, 197, 198, 229
pH-sensitive, 42
pH solubility profile, 452
Phagocyte, 653
Phagocytosis, 615, 654, 668

Phagolysosomes, 616
Phagosomes, 616
Pharmaceutical aerosol, 547, 548, 550
Pharmaceutical colorants, 360
Pharmaceutical colors, 358
Pharmaceutical development, 630
Pharmaceutical elegance, 13, 295
Pharmaceutical factors in oral dosage formulations
    for select populations, 831–834
Pharmaceutical forum, 789
Pharmaceutical industry, 890–892
Pharmaceutical laws of the European community,
    782–807
Pharmaceutical product licensing requirements for
    Europe, 786
Pharmaceutical quality assurance, 807
Pharmaceuticals, 720
Pharmaceutical technology, 896
Pharmaceutics, 3
Pharmacodisposition, 612
Pharmacodynamic differences observed in pediatric
    patients, 813
Pharmacodynamics, 819, 821, 905
Pharmacodynamic studies, 798
Pharmacokinetic analysis of urine data, 86
Pharmacokinetic data, 798
Pharmacokinetic model, 626
Pharmacokinetic parameters, 748
Pharmacokinetic and pharmacodynamic
    considerations, 625
Pharmacokinetic studies, 798
Pharmacokinetics, 13, 14, 18, 19, 38, 63, 75, 84,
    189, 504, 505, 819, 825, 829, 896, 905
Pharmacokinetics in humans, 799
Pharmacological activity, 176
Pharmacological effect, 626
Pharmacological response, 21, 281, 895
Pharmacological and toxicological tests, 795–796
Pharmacology, 493
Pharmacology and therapeutics of ophthalmic
    medication, 493
Pharmacotherapeutics, 894
Pharmacotherapeutics in drug delivery system
    design, 894–896
Pharmacotherapy, 1, 892, 893, 896
Pharmacotherapy in the new millennium, 896
Pharmacy, 893
Phase-boundary, 323
Phase change, 185
Phase diagram of water, 463
Phase inversion, 323
Phase separation, 322
Phenacetin, 131, 132, 134
Phenobarbital, 19, 56, 455, 813
Phenol, 478
Phenolate anion, 182
Phenomenological considerations in percutaneous
    delivery, 269

Phenothazines, 580
Phenylalanine, 54
Phenylbutazone, 46
Phenylephrine, 494
Phenylephrine hydrochloride, 526
Phenylepherine pivalate, 600
Phenylethanol, 499
Phenylethyl alcohol, 501
Phenylethylamine, 629
Phenylmercuric acetate, 501
Phenylmercuric nitrate, 322, 500
Phenylpropanolamine, 734
Phenytoin, 162, 578, 423, 424, 813, 895
Phosphatidylcholine, 650
Phosphatidylethanolamine, 650
Phosphatidylinositol, 650
Phosphatidylserine, 650
Phosphofructokinase, 863
Phospholipids, 324, 325, 460, 593, 618, 634, 644
Photochemical degradation, 179
Photo-Nephelometer, 485
Photolysis, 181, 183, 194
Photolytic reaction, 183
Photomicrographs, 224
Photon correlation spectroscopy, 302, 323
p-hydroxybenzoate, 182
Physical and chemical properties, 555
Physical and chemical stability, 291
Physical decomposition, 181
Physical degradative routes, 185–188
Physical instability in heterogeneous systems, 188
Physical limitations, 823
Physical stability, 16
Physical stability of suspensions, 311–314
Physical system, 280
Physical targeting, 631
Physical testing, 205
Physical therapy, 1
Physician, 6
Physicochemical factors and components, 449
Physicochemical factors governing drug
    absorption, 29–55
Physicochemical factors influencing oral sustained-
    release dosage from design, 579
Physicochemical parameters, 214–227
Physicochemical systems used topically, 271
Physicochemical variables, 21
Physicochemical, 13, 14, 491
Physics of tablet compression, 369
Physiochemical properties of hydrocarbons, 554
Physiological considerations for various routes and
    pathways of drug input, 157–169
Physiological perfusion model, 161
Physiological surfactants, 144
Physiological variables, 21
Physiological volume, 86
Pi value, 267
Picking, 356

Piezoelectric load cell, 420
Pigmentation, 296
Pills, 10
Pilocarpine, 184, 494, 517, 524, 599, 600
Pilopine HS gel, 522
Pilorum, 251
Pilosebaceous glands, 241, 247, 252
Pilot-scale, 318
Pinocytic, 616
Pinocytic transport, 620
Pinocytosis, 615, 616, 654, 659
Pinosome, 655
Pirbuterol acetate, 547
Piston can, 720
Piston systems, 562
Piston-tamp principle, 410
Pituitary gland, 156, 843
Pivotal processor, 364
pK_a, 29, 32, 200
pK_a and solubility, 215
pK_a and stability, 832–833
Plank's constant, 192
Plaque, 599
Plasma concentration, 84, 748
Plasma expander, 665
Plasma half-life, 866
Plasmalemma, 650
Plasma proteins, 172
Plastibase, 273
Plastibase-50W, 519
Plastic, 702, 709
Plastic bags, 709
Plastic bottle, 686
Plastic-coated glass bottles, 561
Plastic containers, 704
Plastic containers for IV fluids, 470
Plastic deformation, 335, 361, 363, 369, 370, 372,
    834
Plastic dispensing bottle, 511
Plastic-lined metal screw cap, 692
Plasticizer, 376
Plastic syringes, 709
Plastic vials, 709
Plate process, 433
Platelets, 650, 653
Platinum, 598
Pledget, 271
Platelets, 653
Plungers, 709
Pluronic F-68, 461
Pneumatically driven piston, 421
Poison ivy products, 251
Poliomyelitis, 2
Polishing pan, 376
Poloxmers, 634
Poly-l-glutamic acid, 598
Polyacrylamide, 522, 525
Polyalkylcyanoacrylate, 597, 618

Polyamino acids, 663
Polyaminopropyl biguanide, 499, 502
Polybutylcyanoacrylate, 597
Polydispersity, 302
Polyelectrolytes, 202
Polyester, 475
Polyethylene glycol (PEG), 56, 355, 400, 429, 435,
    456, 518, 563, 834, 866
Polyethylene glycol 300, 183, 186, 458
Polyethylene glycol ointment, 273
Polyethylene spray, 706
Polyethylene terephthalate (PET), 686, 694
Polyhydric alcohols, 815, 864
Polyhydroxyproplmethacrylamide copolymer, 665
Polymerization, 634
Polymers, 592, 857
Polymers for controlled release tablets, 356
Polymer system, 588
Polymethyl methacrylate (PMMA), 527, 533, 597
Polymethyl methacrylate lenses, 491
Polymorphic form, 293
Polymorphic transition, 337
Polymorphism, 143, 214, 221, 337, 450
Polymorphism potential, 452
Polymorphonuclear leukocytes, 653
Polymorphs, 185
Polymyxin, 255
Polynomial function, 739
Polynuclear aromatics (PNAs), 703
Polyoils, 859
Polyolefin, 475
Polyoxyethylated castor oil, 311
Polyoxyethylated fatty acid, 311
Polyoxyethylene (POE) fatty esters, 567
Polyoxyethylene lauryl sulfate, 356
Polyoxyethylene monostearate, 356
Polyoxyethylene polymer (carbowax), 273
Polyoxyethylated sorbitan, 311
Polyoxyethylene sorbitan esters, 563
Polyoxyl 40, 521
Polyoxyl 40 stearate, 518
Polypeptide, 128, 455, 848
Polypeptide drug, 867
Polypropylene (PP), 520, 686, 694, 715
Polyquad, 499, 502, 539
Polyquat, 501
Polysaccharides, 663
Polysorbate, 518
Polysorbate 20, 311
Polysorbate 40, 311
Polysorbate 80, 57, 311, 858
Polyvinyl, 717
Polyvinyl acetate phthalate (PVAP), 375
Polyvinyl alcohol, 518, 522
Polyvinyl chloride (PVC), 187, 475, 694
Polyvinylpyrrolidone, 186, 466
Population, 6
Porcelain, 477

Pore size distribution, 226, 363
Pore structure and size, 388
Pork skin gelatin, 398
Porosimetry, 388
Porosity, 361, 370, 388, 588
Porous tablet, 830
Portal blood, 165
Portal blood flow, 149
Portal drug delivery, 15
Portal vein, 53, 163
Posology, 789
Postmarketing experience, 800
Postnatal age, 812
Potassium chloride, 464
Potassium iodide, 57
Potassium penicillin G, 107
Potato starch, 350
Potency, 11, 687, 865
Poultices, 10
Pourability, 306
Powder blending, 397
Powder compaction, 361
Powdered drug, 301
Powder filling, 407
Powder fluidity, 397
Powder hopper, 412
Powder mixing, 359
Powders, 226, 333, 548, 686
Powders for reconstitution, 516
Practical size distribution, 459
Pralidoximeiodide, 629
Prazosin, 821
Precipitation, 186
Precision, 214
Preclinical studies, 759
   acute toxicity, 759
   pharmacological screening, 759
   preformulation, 759
Precompression, 412, 420
Precompression force, 417
Prediction of solubility, 218
Prednisolone acetate, 494
Preferential hydration, 859
Prefilled syringes, 702, 704
Preformulation, 213
Pregelatinized starch, 417
Preparation of emulsions, 323
   continental method, 323
   english method, 323
Preparation of suspensions, 318
Prescoter, 383
Presentation of particle size data, 302
Preservation and preservatives, 498
Preservative challenge, 205
Preservative challenge test, 295
Preservatives, 399, 450, 454, 491
Preservatives and antioxidants, 322
Preserve, 682

Pressure-fill equipment, 572
Pressure-fill process, 572
Pressurized aerosol systems, 547
Pressurized containers: metered-dose inhalers, 719
Preswell dilution technique, 597
Prevention of disease, 1
Prickle layer, 244
Primaquine, 652
Primary drying, 461
Primary package, 684
Primary package component, 702
Primary parental routes, 443
Primary structure, 469
Principal component analysis (PCA), 742
Principles of drug absorption, 21–73
Printing, 401
Private health insurance, 9
Probanthine, 580
Probe model, 266
Procaine hydrochloride, 203
Procaine penicillin, 451
Process facilities, 482
Process validation, 751
Prodrug, 451
Prodrug modification, 628
Prodrugs, 596, 599, 613, 627
Product actuation, 567
Product license, 865
Product-package, 703
Product-package compatibility, 709
Product temperature, 462
Product testing and evaluation, 483
Progestasert, 603
Progesterone, 168
Program, 206
Prolonged release, 576
Prolonged-release dosage, 341
Propantheline, 51, 126, 127, 580
Proparacaine hydrochloride, 495
Propellant, 563, 565, 551, 720
Propellant 11, 572
Properties of chlorofluorocarbon, 553
Properties of pharmaceutical solubilized systems, 308–310
Properties of propellants, 557
Prophylactic, 449
Propranolol, 64, 164, 822
Propulsive movements, 50
Propylene glycol, 284, 291, 296, 458, 754
Propylene glycol monolaurate, 288
Propyl gallate, 203
Propylparaben, 501, 522
Prostaglandin, 526, 600
Prostaglandin E2, 184
Protamine, 442
Protease, 539
Protect, 682
Protectant, 453, 456, 467

Protein, 845, 846, 849
Protein aggregation, 850
Protein binding, 169, 171, 174, 857
Protein conformation, 852, 856
Protein drug, 167
Protein formulations, 469
Proteins, 19, 22, 195, 888
  primary structure, 847
  quaternary structure, 847
  secondary structure, 847
  tertiary structure, 847
Proteins of the film, 537
Protein stabilization in the dried solid state, 859
Protein stabilization in solution using additives, 858
Protein structure, 847–848
Protein surface, 858
Proteoglycan, 617
Proteolytic effect, 431
Proteolytic enzymes, 495, 851
Protocol design, 751
Protocols, 206
Proximal jejunum, 42
Prozac, 899
Pseudo-first-order, 196
Pseudo-first-order process, 77, 78, 80
Pseudo-first-order rate constant, 76
Pseudo-steady state, 587
Pseudolatex, 378
Pseudomonas, 294
Pseudomonas aeruginosa, 494, 498, 818
Psoriasis, 252, 254, 257, 285
Psoriatic scale, 252
Psychopharmacology, 899
Psychotherapeutic drug, 4
Psychotherapy, 1
Public health, 3
Pulmonary, 867
Pulmonary activity, 568
Pulmonary circulation, 159
Pulmonary drug delivery, 818
Pulmonary fluids, 569
Pulmonary infections, 818
Pulmonary inhalation, 168
Pulmonary mucosa, 565
Pulsed delivery, 581
Pump, 715, 720
Punch force, 369
Purified water, 513
Purity, 11, 339, 450, 687, 709
Pustule, 254
Pyloric stenosis, 61
Pylorus, 42, 44
Pyrogen-free, 449
Pyrogenic reaction, 460
Pyrogenic silica, 356
Pyrogens, 205, 457, 481, 713
Pyrogen test, 484
Pyrogen testing, 484

Qualified person, 806
Quality, 8
Quality assurance, 480
Quality control, 369, 480
Quantitation of rate degradation, 188–193
Quantitative content of the dosage form, 791
Quasi-elastic light scattering (QELS), 301, 302
Quaternary ammonium compounds, 43
Quaternary salts, 629
Quaternary structure, 469
Quick-breaking foam, 276, 566
Quinine, 10, 39, 40

Racemization, 181, 184, 186, 470, 848, 851
Radial die wall, 362
Radial die wall force, 362
Radiation, 1
Radiation barrier, 250
Radioactive decay, 76
Radiodiagnostic agents, 631
Radio-opaque contrast media, 445
Radio-opaque emulsion, 318
Radio-opaque tablet, 44
Random mixing, 336
Random sampling, 496
Raoult's law, 554
Rapid dissolution, 17
Rapid intravenous injection, 84
Rate of absorption, 17
Rate of appearance, 188
Rate controller, 524
Rate-controlling coats, 581
Rate-controlling membrane, 288
Rate-controlling release membrane, 718
Rate-determining step, 138, 187
Rate of disappearance, 188
Rate of dissolution, 335, 355
Rate of excretion, 90
Rate expression, 188
Rate-limiting step, 202
Rate of shear, 305
Rationale for targeted delivery, 611
Reaction rates, 188
Real time studies, 688
Receding boundary, 461
Receptor sensitivity and response, 813
Receptor-specific ligands, 631
Recombinant compounds, 157
Recombinant DNA, 843, 845
Recombinant DNA products, 10
Recombinant DNA technology, 611
Recrystallization, 221
Rectal administration, 52, 167, 817
Rectification, 406
Rectum, 167
Red blood cells, 86
Reference standard, 865

Reflection coefficient, 31
Regression analysis, 743, 748, 751
Regression coefficient, 734
Regression techniques, 734
Regulatory agency, 890
Regulatory aspects of biotechnology-based
    pharmaceuticals, 865
Regulatory and compendial requirements, 481
Regulatory concerns, 207
Regulatory enforcement, 5
Regulatory requirements, 730–704
Regulatory requirements in pharmaceutical
    packaging, 683–684
Regulatory submissions, 208, 703
Relative bioavailability, 107
Relative humidity, 203, 227, 690
Renal clearance rate (RCR), 88
Renal excretion, 812
Renal failure, 95
Renal insufficiency, 449
Repeated-dose toxicity studies, 796
Repetitive dosing, 108
Repetitive extravascular dosing, 112
Repetitive intravenous dosing, 109
Replacement of caps, 406
Reproductive function, 796
Repulsive and attractive forces between particles,
    312
Resealability, 713
Resealed erythrocytes, 597
Reservoir devices, 582
Reservoir diffusional products, 586
Reservoir diffusion systems, 586
Reservoir effect, 285
Residence time, 49
Residual die wall force, 372
Residual moisture, 470
Residuals method, 96, 100
Resin, 709
Resources, 204
Respiratory system, 569
Response surface, 728, 741
Reticuloendothelial system (RES), 324, 616
Reticulum, 245
Retina, 489
Retinoids, 256
Retinopathies, 495
Retrobulbar, 491
Rheogram, 305
Rheological behavior, 292
Rheological parameters, 459
Rheological properties, 272, 305, 316, 421
    dilatant flow, 305
    plastic flow, 305
    pseudoplastic flow, 305
    thixotropy, 305
Rheological quality, 295
Rheological studies, 316

Rheological tests, 318
Rheometers, 292
Rheumatoid arthritis, 4, 612
Ribbon blender, 364
Ribitol, 149
Riboflavin, 46, 48, 51, 820
Ribonuclease, 851
Ribosome, 324
Rigid foams, 276
Rigid gas-permeable (RGP) lenses, 534, 536
Role of drugs, 1
Role of experts and expert reports, 805
Roll compactors, 365
Roller coat, 706
Roller-compaction, 347
Rose bengal, 495
Rose water ointment, 277
Rotary die process, 429, 433
Rotary press, 334
Rotating turret, 367
Route of administration, 321
Route and method of administration, 789
Routes of administration, 17, 22, 96
Routes by which pharmaceuticals degrade, 180–
    181
RS/Discover (RS1), 748
Rubber stopper, 710
Rubber-stoppered vials, 470

Saccharin, 357, 816
Safety, 491
Safflower oil, 461
Salicylamide, 59, 165, 580
Salicylic acid, 35, 39, 269
Salivary glands, 822
Salt, 450
Salt/complex formation, 186
Salt form of the drug, 138
Salt and sucrose content, 834
Salting-out, 852
Salves, 10
Sanitation, 3
Saturation solubility of the drug, 138
Scanning electron microscope, 302, 326
Scatchard–Hildebrand equation, 218
Scattered light, 302
Schiff's base formation, 186
Schizophrenia, 899
Scleral contact lenses, 533
Scleral route, 503
Scleroderma, 254
Scope and goals, 206
Scopolamine, 168, 258, 494
Scopolamine and epinephrine, 442
Screening, 364
Sealing coat, 375
Sealing roll, 433

Sealing and self-locking closures, 402
Sealing temperature, 432
Seal integrity, 704, 710
Search methods, 737–745
Sebaceous duct, 248
Sebaceous glands, 241, 247
Seborrhea, 254
Secondary drying, 462
Secondary package, 684
Secondary structure, 469
Second-order rate constant, 76
Second-order reaction, 189–191
Second-order regression analysis, 745
Sedimentation, 301, 310, 316
Sedimentation field-flow fractionation, 302, 323
Sedimentation volume, 316–318, 351
Segregation, 336
Seizures, 813
Selection of the oil phase, 320
Self-administration, 823
Self-locking closures, 402
    Coni-Snap, 402
    Loxit, 402
    Posilok, 402
Self-locking hard gelatin capsules, 434
Self temperature, 462
Semiautomatic filling machine, 408
Semi-in vivo testing, 234
Semilogarithmic plot, 82, 83, 84
Semimicrosieves, 301
Semipermeable membrane, 590
Semisolid aerosol, 568
Semisolid dosage forms: ophthalmic ointments and
        gels, 521–523
Semisolids, 272, 276
Semisolid systems, 291
Sensitivity analysis, 736
Separation of caps from bodies, 406
Serine, 850
Serum, 123, 246
Serum albumin, 470, 859
Serum concentration, 101, 813
Serum creatinine, 95, 117
Sesame oil, 457
Severity, 497
Shape and fractal dimension, 224
Shape-volume factor, 364
Shearing stress, 305
Shear mixing zone, 361
Shear-thickening, 321
Shelf life, 681, 688, 789
Shellac, 375
Shell composition, 398
Shell-filling machine, 434
Shell manufacture, 399
Short-term tests, 685
Sieves, 302
Silica gel, 696, 855

Siliceous materials, 356
Silicon dioxide, 357
Silicone elastomer reservoir, 586
Silicone oil, 461
Silicones, 603
Silver protein, 494
Silver stain, 856
Simple reactions, 189
Simplex method, 730, 733
Simultaneous metabolism and excretion, 90
Single-component systems, 467
Single-dose, 234, 713
Single-dose administration, 115
Single-dose studies, 166
Single-dose toxicity studies, 796
Single-drug entities, 762
Single punch machines, 366
Single-punch press, 361
Single-stage-filling systems, 572
Single-station presses, 366, 372
Sinusoidal capillaries, 620
Site of absorption, 37, 122
Site-specific, 576
Site-specific delivery, 613, 614, 627, 671
Site-specific drug delivery systems, 630
Size-reduction techniques, 335
Sizes and shapes, 401
Size and size distribution, 326
Skeletal muscle, 170
Sketch of skin, 242
Skin, 239, 274
    circulatory system of, 246
    microcirculation of, 260
Skin appendages, 247
Skin cross section, 262
Skin functions, 240, 241, 248
Skin penetration enhancers, 260, 288
Skin permeability model, 284
Skin problems, 252
Skin sensitivity, 296
Skin surface, 252
Slit-lamp, 497
Slit lamp fluorophotometer, 519
Smaller unilamellar vesicles (SUVs), 641
Small intestine, 22, 24, 35
Smallpox, 2
Small unilamellar vesicles (SUVs), 593
Small-volume parenterals (SVPs), 443, 458, 685,
        709
Soda-lime glass, 692
Sodium benzoate, 142
Sodium bisulfite, 203, 518
Sodium bisulfite equivalent, 478
Sodium carbonate, 464
Sodium carboxymethyl cellulose, 350
Sodium chloride, 345, 464
Sodium chloride injection , 445
Sodium desoxycholate, 311

Sodium fluorescein, 495, 500
Sodium formaldehyde sulfoxylate, 203
Sodium glycine carbonate, 350
Sodium hyaluronate, 527
Sodium lauryl sulfate, 56, 135, 286, 355, 359
Sodium nitroprusside, 184
Sodium oleate, 320
Sodium phenobarbital, 424
Sodium phenytoin, 142
Sodium phosphate, dibasic, 464
Sodium salicylate, 142
Sodium starch glycolate, 350, 353, 354, 425, 833
Sodium sulfacetamide, 494, 495, 500
Sodium sulfathiazole, 142
Sodium sulfite, 203
Sodium sulfite equivalent, 478
Sodium thiosulfate, 518
Soft gelatin capsules, 395, 428–434
Soft hydrophilic lenses, 536
Soft/liquid filled hard gelatin capsules, 434
Solid dosage forms, 686
   ocular insets, 523–525
Solid lubricants, 354
Solid-solid interactions, 145
Solid solution, 145
Solid state degradation, 192
Solubility, 201, 214, 216, 306, 588, 733, 858
Solubility of active ingredients, 565
Solubility of metastable polymorphs, 220–223
Solubilization, 308
Solubilization isotherm, 309
Solubilized systems, 306–310
Solubilizing agents, 308, 452, 453
Solubilizing agents and surfactants, 455
Soluble compressed gas, 568
Soluble dyes, 358
Soluble excipients, 345
Soluble lubricants, 354, 355
Soluble macromolecule drug-delivery systems, 654
Solute distribution, 326
Solution aerosols, 563
Solution filtration, 482
Solutions, 458, 515
Solvate formation, 452
Solvent evaporation, 597
Solvent evaporation technology, 327
Solvent formation, 143
Solvents, 200
Solvent system, 378
Solvolysis, 181
Sorbic acid, 539
Sorbitan trioleate, 547
Sorbitol, 337, 461, 466, 815, 859, 864
Sorting, 400
Soybean lecithin, 461
Soybean oil, 461
Spansule, 581
Specialized large-volume parenteral and sterile
      solutions, 447

Specific surface area, 224, 301, 335
Specific volume, 408
Spectinomycin hydrochloride, 460
Spectrophotometer, 692
Spectrophotometric response, 733
Spectrophotometry, 215, 713
Spectroscopy
   circular dichrosim, 855
   fluorescence, 855
   infrared, 855
   optical rotatory dispersion, 855
   Raman, 855
   UV and visible, 855
Spermaceti, 321
Spheronization, 747
Spheronizing granules, 365
Sphingomylein, 650
Sphingosomes, 312
Splanchnic blood flow, 149
Splanchnic circulation, 53
Spore-air interface, 476
Sporogenes, 474
Spray-dried lactose, 336, 344, 363
Spray-drying, 347, 468, 460, 859
Spray-drying of protein pharmaceuticals, 864–865
Spray-on-protective films, 563
Spray pump, 686, 715, 721
Spray system, 566
Spray technology, 720
Spray valves, 556
Spreadable semisolids, 276
Stability, 214, 336, 442, 473, 491, 580, 709, 863
Stability-indicating assay, 473
Stability considerations, 16
Stability data, 180
Stability studies, 795
Stability testing, 709
Stability testing in the pharmaceutical industry, 204
Stability-testing protocols, 180
Stabilization, 859
Stabilizers, 518
Stable foam, 566
Stainless steel, 514
Standard-operating procedure, 207
Staphylococcus aureus, 498
Starch, 136, 274, 345, 349, 376
StaRx 1500, 344, 358
State board of pharmacy, 888
Static-aerosol challenge, 485
Static-immersion challenge, 485
Station of tooling, 365
Statistical analysis, 207
Statistical analysis systems (SAS), 748
Statistical designs, 730–731, 739
Statistical sampling, 208
Steady-state flux, 265
Steady-state release, 585
Stealth liposomes, 595

Steam, 474
Steam sterilization, 496
Stearate spray, 706
Stearic acid, 274, 275, 408, 734
Steric exclusion, 859
Sterile dosage form, 703
Sterile filtered, 322
Sterile filtration, 458
Sterile ophthalmic ointments, 523
Sterile powders, 460
Sterile suspension, 460
Sterility, 205, 495–496
Sterility and preservation, 522
Sterility testing, 483
Sterilization, 317, 457, 496, 512, 520, 703
    aseptic crystallization, 496
Sterilization cycle, 475, 496
Sterilization by dry heat, 475, 513
Sterilization by ethylene oxide, 476, 513
Sterilization by filtration, 477, 496
Sterilization by ionizing radiation, 496, 513
Sterilization methods, 474
Sterilization by steam, 474
Steroid hormones, 602
Steroids, 4, 184, 185, 547, 890
Steroids and fat-soluble vitamins, 456
Sticky film, 295
Stoichiometric coefficients, 188
Stokes' law, 314, 320
Stokes–Einstein equation, 302
Stomach, 17, 22, 23
Stomach-emptying rate, 123
Storage, packaging, and stability considerations, 405
Strain gauge, 417, 420
Strain index (SI), 339
Stratum corneum effects, 241, 242, 243, 249, 252, 255, 259, 265, 266, 267, 284, 285, 287
Stratum cornium composition, 244
Streptococcal organisms, 494
Streptomycin surface injection, 458
Stress cracking, 709
Striated border, 26
Strip packages, 699
Strip packaging, 686, 700
Structural fibers, 245
Structure and function of skin, 240–248
Study in humans, 480
Styrene, 709
Styrenemaleic anhydride (SMA), 666
Subcoat, 375
Subcutaneous (SC), 411, 443
Subcutaneous connective tissue, 246
Subcutaneous dose, 167
Sublimation, 461, 468
Sublingual, 167, 334
Sublingual and buccal tablets, 827
Sublingual tablets, 187

Submission of application in special circumstances, 800
Subpapillary network, 246
Sucrose, 357
Sufactants, 453
Sugar coating, 374, 375–376
Suggested soluble macromolecule drug-delivery systems and their uses, 657
Sulfaethidole, 149
Sulfamethoxazole, 127
Sulfamethoxypyridazine, 54
Sulfanilamide, 754
Sulfasalazine, 60
Sulfathiazole, 423
Sulfonamides, 185, 335, 494
Sulfoxides, 850
Sulfur dioxide, 452
Sulindac, 596
Summary of product characteristics (SPC), 788, 789
Sunscreens, 250, 251
Supercooled liquid, 270
Superdisintegrants, 353, 425
Superficial fascia, 442
Superior hemorrhoidal vein, 167
Superior vena cava, 159, 448
Supersaturated solutions, 319
Suppositories, 716–717
Supracolloidal, 320
Supranational procedures, 784
Surface active agents, 56, 57, 132, 316, 567, 569
Surface adsorption, 853
Surface area of the drug particle, 150
Surface area, 214, 234, 335, 581
Surface characteristics, 224
Surface effects, 255
Surface epithelium, 50
Surface monitoring, 483
Surface particle, 304
Surface tension, 132, 200, 857, 859
Surface treatments, 709
Surfactant classification and properties, 307–308
Surfactant or dispersing agents, 565
Surfactants, 202, 307, 320, 426, 450, 456, 470, 518
Surgery, 1
Surgical adjuncts, 494
Surlyn spray, 706
Suspending agents, 450
Suspensions, 188, 299, 307, 310–319, 516
Suspensions, flocculated, 305
Suspension system, 565
Sustained- and controlled-release drug delivery systems, 575, 577
Sustained release, 334, 388, 576, 699
Sustained-release dosage form, 104
Sustained-release formulation, 812
Sustained-release product, 15, 50, 768

Sustained-release system, 580
Sustained-release tablet, 104
Swallowing and chewing, 824
Sweat, 576
Sweeteners, 316, 815
Sweetening agents, 357
Swelling, 351
Swelling capacity, 351
Swelling-controlled matrix, 590
Swelling rate, 352
Sympathomimetic, 494
Symptomatic response, 502
Synthetic polymers, 663
Syringe, 707, 709
Syringeability or injectability, 459
Syringe components, 685
System of compartments, 75
Systemic activity, 568
Systemic administration, 167
Systemic availability, 811
Systemic circulation, 28, 53, 59, 299
Systemic clearance, 289, 626
Systemic delivery, 257
Systemic drug effect, 19
Systemic toxicity, 629
Systems for control of pharmaceuticals in the
    European community, 782–787

Tablet, 333, 335
Tablet-coating, 747
Tablet components, 341
Tablet compression, 352
Tablet crushing strength, 364
Tablet disintegration, 351, 352
Tablet ejection, 362
Tablet evaluation, 384
Tablet geometry, 362
Tablet hardness, 13, 205, 354, 734
Tableting, 334
Tableting equipment, 364–374
Tablet machine instrumentation, 369
Tablet manufacture, 359–364
Tablet matrix, 352
Tablet press, 334, 365
Tablet punch, 361
Tablets, 686
Tablet strength, 362, 372
Tackiness, 295
Tacky, 376
Talc, 274, 356, 358, 376, 424, 834
Talc or carbon, 458
Tamper evidence (TE), 698, 701
Tamper-evident, 405
Tamper-proof, 548
Tamper-resistant packaging, 771
Targeted-delivery, 576
Targeted delivery systems, 593, 627

Targeted organ, 573
Targeted-oriented drug delivery systems, 611
Target site, 627
Target-specific delivery, 158
Target tissue, 156, 175
Targeting in the Gastrointestinal Tract, 668
Tartaric acid, 203, 351
Taste preference and palatability, 818
Taste preferences in oral dosage forms, 830–831
Teledyne Wirz tube, 523
Temazepam, 429
Temperature-cycling, 315
Temperature-dependent transitions, 688
Temperature of fusion, 270
Temperature and gravitational stress tests, 316
Temperature stress, 323
Temporal inhabitancy, 290
Tensile strength, 335, 355, 387
Terbutaline, 600
Terbutaline sulfate, 547
Terminally autoclave-sterilized, 458
Terminal sterilization, 511
Terminology, 576
Ternary solvents and optimization, 217
Tertiary structure, 469
Testing of aerosols, 573
Testosterone, 334, 823
Tetracaine, 203
Tetracaine hydrochloride, 495
Tetracycline, 17, 55, 56, 184, 495, 827
Theophylline elixir, 179, 813, 815, 895
Therapeutic agent, 179
Therapeutic availability, 13
Therapeutic concentration, 21
Therapeutic effect, 9
Therapeutic effectiveness, 18
Therapeutic efficacy, 177, 180, 641
Therapeutic index, 17, 18, 168, 179, 503, 612, 895
Therapeutic indications, 789, 804
Therapeutic level, 341
Therapeutic response, 21
Therapeutics committee, 8
Therapeutic stratification of the skin, 252
Therapeutic system, 15
Therapeutic windows, 577
Thermal analysis, 464
Thermal barrier and body temperature regulation,
    250
Thermal denaturation, 857
Thermal profile, 452
Thermal radiation, 467
Thermal-setting formulations, 435
Thermal treatment, 466
Thermal welding process, 404
Thermodynamic activity, 269, 271
Thermodynamic control, 141
Thermodynamic equilibrium, 192
Thermodynamic factors, 284

Thermodynamic parameters, 193
Thermoforming plastic, 716
Thermoforming process, 700
Thermogravimetric analysis (TGA), 695
Thermoplastic resin, 720
Thiamine-tetrahydrofuryldisulfide, 628
Thiamin hydrochloride, 345
Thickness of the stationary layer, 147
Thimerosal, 499, 501, 539
Thin-layer chromatography, 227
Thioglycerol, 203
Thioglycolic acid, 203
Thiol group, 853
Thiopental, 170, 171
Third-countries, 806
Thixotropic formulations, 434
Thrombocytopenic purpura, 653
Thrombophlebitis, 445, 478
Thrombosis, 159
Thymic cortex, 618
Thymidine kinase, 651
Thyroid, 2, 169
Tiltabs, 830
Timed release, 576
Time of the peak, 104
Timing and goals of preformulation, 214
Timolol, 600
Timolol maleate, 494
Tinnitus, 257
Tin-plate containers, 562
Tissue, 18
Tissue binding, 172, 176
Tissue compatibility, 529
Tissue dehydration, 445
Tissue distribution, 176
Tissue irritation, 457
Tissue plasminogen activator, 849
Titanium dioxide, 695
Tobramycin, 494, 528
Tocopherols, vitamin E, 203
Tolbutamide, 138
Tonicity, 516
Tonicity-adjusting agents, 453, 456
Tonicity and tonicity-adjusting agents, 516
Topical administration, 502
Topical aerosols (dermal, vaginal, and rectal), 549, 564
Topical cream, 180
Topical delivery, 239
Topical delivery systems, 289
Topical dosage forms, 251
Topical drug, 248
Topical eye drops, 514–521
Topical pharmaceutical aerosols, 562
Topicals, 239
Topical sprays, 563
Topogranulator, 364
Topsyn gel, 276

Tortuosity, 588
Total clearance rate (TCR), 89
Toxicity Studies, 479
Toxic proteins, 658
Trace metals, 441
Trachoma virus, 494
Tragacanth, 276, 321, 350
Tragacanth gel, 280
Trahalose, 456
Tranquilizers, 890
Transcorneal flux, 506
Transcorneal transport, 503
Transdermal, 232
Transdermal absorption, 168
Transdermal administration, 817
Transdermal delivery, 257, 281, 287
Transdermal delivery systems, 287, 290, 827
Transdermal devices, 717–718
Transdermal patch, 269, 287, 718
    nitroglycerin, 269
Transdermal route, 261, 263, 265, 267
Transdermal systems, 600
Transdermal therapy, 251, 252
Transderm-Scop, 600
Transepidermal, 266
Transfer devices, 442
Transfollicular, 266
Transfollicular resistance, 264
Transfollicular route, 261
Transition state, 192
Transit time, 578
Transmembrane transport, 504
Transport, 143, 625, 682
Transport across the epithelial barrier, 617
Transport system, 148
Transposition of a drug, 121
Transseccrine, 266
Trapezoidal rule, 105
Treatment, 1
Treatment of an inflammation, 157
Treminal sterilization, 702
Tremors, 827
Triamcinolone acetonide, 547
Trichloromonofluoromethane (Propellant 11), 551
Triglyceride oil, 432
Triglycerides, 29, 460, 465
Triple point of water, 461
Tris buffer, 854
Tris-hydroxymethylaminomethane, 467
Triturates, 345
Troches, 763
Tropicamide, 494
True density, 234
Trypsin, 43
Tryptophan, 855
Tuberculin, 447
Tuberculosis, 4
Tube spacers, 570

Tubing glass, 704
Tubular absorption, 812
Tubular reabsorption capacity, 812
Tubular reabsorption, 89
Tubular secretion, 89, 812
Tumor cells, 633
Turbidity, 713, 855
Tween 80, 359
Two-component systems, 228
Two-dimensional system, 746
Tyloxapol, 518
Types of disintegrants, 348–351
Types of lens care products, 537–541
Tyrosine, 855
Tyvek, 520

Ufasomes, 312
Ulceration, 495
Ulcerative colitis, 60, 448, 653
Ultrasonic mixer, 279
Ultraviolet lamp, 512
Ultraviolet radiation, 245
Unbound drug, 173
Undesirable effects, 805
Unfrozen, 468
Uniformity of dosage units, 384
Unit-dose, 498, 691, 716
Unit dose packaging, 699–702
Unit-dose and pouch packing, 716
Unit-dose preparation, 16
Unit-dose system, 894
Unites States, 773
United States Pharmacopeia (USP), 441, 489,
     754, 889
Unstable thermodynamic state, 145
Upper punch, 362
Urea, 39
Urease, 651
Uricase, 651
Urinary excretion, 88, 94, 124, 138, 166, 576
Urinary excretion rate, 734
Urine specimen, 88
Use of emulsions in pharmacy, 318
UV spectrophotometry, 714

Vaccines, 1, 634
Vaginal ring, 601
Vaginal and uterine systems, 601
Validation, 200, 751
Validation data, 794
Valproic acid syrup, 815
Value-added tax, 778
Valves and actuators, 556
Van der Waals, 848
Van der Waals forces, 369
Vaporization, 185, 572

Vaporized propellant, 558
Vapor pressure, 223, 550, 552, 555
Vapor tap, 558
Variability, 339
Variables, 728
Variance-covariance matrix, 742
Vascular system, 155
Vasoconstriction, 149
Vasopressin, 52
Vastus lateralis, 160
Vehicle selection, 816
Velocity, 305
Venous supply, 631
Verapamil, 825
Very low-density lipoprotein (VLDL), 644
Vesicles, 300, 324
Veterinary medicinal products, 787
Vials, 704
Vibratory fill principle, 408
Vibrio cholerea, 618
Vidarabine, 600
Villi, 24, 25
Villous atrophy, 61
Vinablastine, 653
Vinca alkaloids, 653
Vincristine, 653
Vindesine, 598
Vinyl chloride, 709
Virology, 747
Virtual membrane pH, 35
Virus, 2
Virus-specific enzyme, 596
Viscometer, 305
     capillary, 306
Viscosity, 200
Vision, 824
Visual impairment, 827
VLDLs, 645
Volatile medicaments, 357
Volatilization, 185
Volume, 734
Volume of distribution, 85, 101, 115, 169, 173,
     576
Volume number mean, 301
Volume surface mean, 301
Vomiting, 22
Vulcanization, 710
Vulcanizer, 710

Wagner–Nelson method, 100
Wall friction, 371
Warfarin, 48, 578, 825
Washburn equation, 226
Water-insoluble derivatives, 49
Water-miscible solvents, 200
     ethanol, 200
     glycerin, 200

[Water-miscible solvents]
  polyethylene glycols, 200
  polymeric alcohols, 200
  propylene glycol, 200
Water-in-oil emulsifier, 308
Water-soluble drugs, 50, 51
Water uptake, 353
Water-vapor permeation test, 700
Weak acid, 140
Weak base, 140
WEB system, 388
Weibull equation, 304
Weight loss, 715, 716
Weight variation, 384, 407, 415, 420
Western Europe, 890
Western European Union (WEU), 778
Wet granulation, 339, 335, 347, 348, 364, 904
Wet massing, 364
Wet sprays, 548, 556
Wettability, 347
Wetting, 314
Wetting agent, 308, 342
Whooping cough, 2

Wicking, 350, 351
Wilson's disease, 903
Wood's apparatus, 219, 233
Wurster coating chamber, 382

Xenobiotics, 809
Xerostoma, 824, 827
X-ray fluorescence, 714
X-ray treatment, 61
Xylose tolerance test, 63

Young adult, 810

Zero-order kinetics, 202
Zero-order reaction, 15, 189–191
Zero-order release, 576, 589
Zeta potential, 304, 312, 313, 317
Zeta potential determination, 317
Zwitterion, 138